ADVANCED MATERIALS AND STRUCTURAL ENGINEERING

PROCEEDINGS OF THE INTERNATIONAL CONFERENCE ON ADVANCED MATERIALS AND ENGINEERING STRUCTURAL TECHNOLOGY (ICAMEST 2015), 25–26 APRIL 2015, QINGDAO, CHINA

Advanced Materials and Structural Engineering

Editor

Jong Wan Hu
Incheon National University, South Korea

CRC Press
Taylor & Francis Group
Boca Raton London New York

CRC Press is an imprint of the
Taylor & Francis Group, an **informa** business

A BALKEMA BOOK

Published by:
CRC Press/Balkema
P.O. Box 447, 2300 AK Leiden, The Netherlands
e-mail: Pub.NL@taylorandfrancis.com
www.crcpress.com – www.taylorandfrancis.com

First issued in paperback 2020

Typeset by V Publishing Solutions Pvt Ltd., Chennai, India

ISBN 13: 978-0-367-73734-4 (pbk)
ISBN 13: 978-1-138-02786-2 (hbk)

Visit the Taylor & Francis Web site at
http://www.taylorandfrancis.com

and the CRC Press Web site at
http://www.crcpress.com

Table of contents

Structural and civil engineering

Preface

The 2015 International Conference on Advanced Materials and Engineering Structural Technology (ICAMEST 2015) took place in Qingdao, China, on April 25–26, 2015. This conference was sponsored by the Incheon Disaster Prevention Research Center (IDPRC) in INU.

The ICAMEST 2015 is an annual international conference aimed at presenting current research being carried out in the fields of materials, structures and mechanical engineering. The idea of the conference is for the scientists, scholars, engineers and students from universities, research institutes and industries all around the world to present on-going research activities. This allows for the free exchange of ideas and challenges among the conference participants and encourages future collaboration between members of these groups. The conference also fosters the cooperation among organizations and researchers involved in the merging fields and provides in-depth technical presentations with abundant opportunities for individual discussions with the presenters.

The book is a collection of accepted papers. All these accepted papers were subjected to strict peer-reviewing by 2–3 expert referees, including a preliminary review process conducted by the conference editors and committee members before their publication by CRC Press (Taylor & Francis Group). This book is separated into five sessions including 1. Advanced material and application, 2. Structural and civil engineering, 3. Mechanical and industrial engineering, 4. Computer aided for engineering application, 5. Civil material and hydrology science application. The committee of ICAMEST 2015 expresses their sincere thanks to all authors for their high-quality research papers and careful presentations. All reviewers are also thanked for their careful comments and advices.

Thanks are finally given to CRC Press (Taylor & Francis Group) as well for producing this volume.

Organization

Incheon Disaster Prevention Research Center (IDPRC)

Recently, various efforts to prevent and prepare are vitally needed for the prevention of disasters and calamities. Because we understand the necessity for technology of disasters, we built up the Incheon Disaster Prevention Research Center (IDPRC) in Incheon National University (INU) in 1997.

Accordingly, the Incheon Disaster Prevention Research Center (IDPRC) in Incheon National University has made progress with the research on the prevention of disasters and calamities through the various seminars, conferences and lectures. This research could be conducted in cooperation with Incheon National University (INU) in various fields, such as structure, soil, hydraulics and environment.

Incheon Disaster Prevention Research Center (IDPRC) will try to be a leader in the disaster of industry through various research activities and global conferences.

Advanced material and application

Advanced Materials and Structural Engineering – Hu (Ed.)
© 2016 Taylor & Francis Group, London, ISBN 978-1-138-02786-2

Production of external thread by means of enveloping with hob cutter

Elena V. Glushko, Nina T. Morozova & Natalia A. Glushko
Far Eastern Federal University, Vladivostok, Russia

ABSTRACT: Thread is the most common type of connections. The technology of thread milling by means of enveloping implies the existence of machine tool accessories where the simple machine is used. The production of thread surface by means of enveloping is performed with a special hob cutter.

1 INTRODUCTION

Thread is the most common type of connections. Virtually there are no industries where fittings are not applied.

There are two basic ways of manufacturing threads:

- By plastic deformation without removing material.
- By cutting with the metal removing.

Methods of making the threads depend on the presence of equipment, machining system, tools, and other factors. Despite many years of experience in producing threaded connections, improving quality and productivity of threading is an important task.

There are correspondent and well-defined methods for threaded connections to all the modes of production (large-lot, serial, unit, and repair). The material presented below is dedicated to the second method of forming threads and is focused on the use of the serial, unit, and repair industries.

The technology of thread milling by enveloping (Lotsmanenko 2003) implies the machine tool accessories and allows the following:

- Easy installation and removal of tool accessories on the selected metal-cutting equipment;
- Quick and easy replacement of the threaded detail with a new one;
- Elimination of the tracer templates usage in conical thread milling.

Choosing the equipment for carrying out the method of thread milling, we follow the fact that the machine should be universal, the most widespread and available. To meet the requirements of both serial and single production, we use the methods of group technology.

The authors use the universal chasing lathe.

2 THE MILLING OF CYLINDRICAL THREAD

The kinematic diagram of a thread milling process with the method of enveloping (Glushko 2005) is shown in the Figure 1. Production tools for the thread consist of two main components: index change gear train and milling head (number 3). These nodes are kinematically rigidly connected to each other by the driveshaft (number 9).

The index change gear train includes two gear wheels – numbers 7 and 8, of which, number 8 is removable when changing the entries of the cut thread. The toothed wheel 7 is rigidly connected to a machine spindle, and the wheel 8 is installed on a bracket 11 that is fixed to the machine frame.

In the support of milling head number 3, a single-start hob cutter number 2 is installed. The head itself is fastened to the tool holder number 5 of machine support stand. Milling head number 3 together with the tool holder number 5 have two linear displacements (along the detail rotation axis and perpendicular to it) and one angular displacement that goes around the vertical axis "O" of the rotation of the tool holder.

The raw-part number 1 is fixed in the machine jaw chuck number 10 for further threading. When adjusting the milling head number 3, the worker achieves relative alignment of horizontal axial planes of the work piece for the thread and hob cutter. In such case, the axis of rotation of the cutter should be parallel to the axis of rotation of the items.

If the items are cut with short threads equal to the width of a miller, in the process of thread milling the hob cutter number 2 is sent a radial infeed along the arrow S_1. When cutting the "long" thread, the milling head is additionally sent with an axial feed.

$$S_2 = Z_1 \cdot P_1, \text{ mm/rev} \tag{1}$$

where Z_1 is a number of thread entries, and P_1 –is the thread pitch in mm.

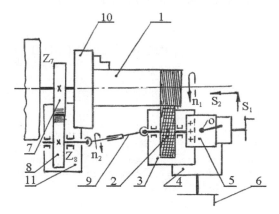

Figure 1. Scheme of an external thread milling (overhead view): 1— detail with a thread surface, 2—hob cutter, 3—milling head, 4—machine support stand 5—tool holder, 6—support cross-feed handle (tool holder), 7 and 8—thread entries index change gear train, 9—driveshaft, 10—jaw chuck (for details installation), 11—bracket of an index change gear train.

In the process of thread milling the angular velocities of the hob cutter and the detail, on the surface of which the thread is cut, are interconnected by an angular speed ratio that is equal to the number of thread entries. The number of hob cutter Z_2 entries=1, regardless of the number of cut thread Z_1 entries.

Depending on the number of the cut thread, entries interchangeable gear wheel number 8, situated in the index change gear train number 11, has teeth number

$$Z_8 = \frac{Z_7}{K} \qquad (2)$$

where K is a number of thread entries.

3 THE ANALYSIS OF CUTTING MODES FOR EXTERNAL THREAD MILLING

To provide the process of cutting the following conditions must be implemented at the cutting zone:

Arithmetical difference

$$V_2 > V_1$$
$$V_2 - V_1 = V_{21} > 0*$$

defines the speed of a relative slip of the thread surface and miller surface or, in other words, the speed of cutting, where

$$V_1 = \omega_1 \cdot R_1; \; V_2 = \omega_2 \cdot R_2 = \omega_1 \cdot \frac{R_2}{u_{12}},$$

where R_1 and R_2 are the radii of coinciding current points of a miller surface and thread surface, and $\omega_1(n_1)$ and $\omega_2(n_2)$ are the rotation speeds (rounds per minute) of a miller surface and thread as $\omega_2 = \frac{\omega_1}{u_{12}}$, where u_{12} is a progressive ratio of a machine-tool meshing.

* The relation considering linear speeds V_1 and V_2 is

$$V_{21} = \omega_1 \cdot \frac{R_2}{u_{12}} - \omega_1 \cdot R_1 = \omega_1 \cdot \left(\frac{R_2}{u_{12}} - R_1 \right)$$

Miller mid-radius is $R_2 = R_{a2} - H/2$, where H is the height of the thread (or router bit). Then the speed of cutting is

$$V_{21} = \omega_1 \left(\frac{R_{a2} - H/2}{u_{12}} - R_1 \right).$$

Considering that $\omega_1 = \frac{\pi n_1}{30}, u_{12} = \frac{Z_2}{Z_1} = \frac{1}{Z_1}$, as $Z_2 - 1$, the expression for calculating the speed of cutting becomes the following

$$Y_{cut} = Y_{21} = \frac{\pi \cdot n_1}{60} \cdot \left[(d_{a2} - 2H) \cdot Z_1 - d_{a1} \right] \qquad (3)$$

where d_{a1} and d_{a2} are the largest diameter of cut thread, and H is the height of turn of the thread.

The diameters of the hob cutter and thread are related by expression

$$d_{a2} \geq 2 \, d_{a1} \qquad (4)$$

Usually from the expression (3), it is possible to determine the rpm n_1 of thread surface rotation, setting the cutting speed V_{21}.

In the further study of cutting modes for external thread milling, assuming that the diameter of a hob cutter d_{a2} = constant, expression (3) in general can be presented as a function

$$V_{21} = V_{21} (n_1, Z_1, R_1) \qquad (5)$$

Because the cutting speed for external thread milling depends on rotation n_1 (rounds per minute), number of entries Z_1, and current radius of cutting surface $R_1 = d_1/2$.

The particular hob cutter and cut thread were picked for the analysis of thread milling modes (Yakuhin & Stavrov 1989):
d_{a2} =190 mm – external diameter of the cutter, P = 2 mm – helically cammed surface tool stepover.

4

Cut thread M64 ×2:

H = 0,86 · P = 1,72 mm – whole depth of cutter (thread)

$d_1 = d_{a1} - H = 62,3$ – mean diameter of a thread, mm.

Let us refer to expression (2), fixing there d_{a2}, d_1 and H. This way for $Z_1 = 1$, relation (2) becomes

$$V_{21} = (0,00659) \cdot n_1 \qquad (6)$$

For $Z_1 = 2$: $V_{21} = (0,01644) \cdot n_1 \qquad (7)$

For $Z_1 = 3$: $V_{21} = (0,02629) \cdot n_1 \qquad (8)$

The cutting speed during thread milling depends on many factors and requires an advanced study, especially for thread milling with profile-relieved milled tooth hob cutter by means of enveloping.

For a first approximation, it is recommended to define the cutting speed V21 from appropriate tables given in reference material (Kovan 1968), then refer to relations (6), (7), and (8) to determine the number of rotations n1 on a raw part prepared for thread.

Rotational movement of the raw part for thread (n1) and of hob cutter (n2) are kinematically connected by the progressive ratio

$$i_{12} = \frac{n_1}{n_2} = \frac{Z_2}{Z_1} = \frac{1}{Z_1}, \text{ because } Z_2 = 1$$

4 THE MILLING TAPERED THREADS

When cutting a tapered thread (Lotsmanenko 2000, Lotsmanenko & Lotsmanenko 1999) the same tooling is used for the milling of cylindrical thread. Instead of the tracer template a very simple-designed swivel head is additionally required and it is fixed on the machine frame. The cut item is fixed in the swivel head of jaw chuck. The swivel head gets rotation from the chuck number 10 that is connected with it by a separate driveshaft.

The cut detail, installed in the swivel head, turns horizontally with an angle equals to half the angle of the cone toward the axis of rotation of the chuck number 10 of the machine. The tuning up of the basic machine appliance is performed in the same way as when milling the cylindrical thread. The movement of the milling head remains unchanged that means parallel to the lathe-bed axis (the rotation axis of the chuck 10).

It should be noted that milling with this method almost does not limit the angle of cone thread, the thread profile is symmetric, and the quality of cutting is high.

We called this type of taper thread as "modified." There is more information about it provided and described in the recourse (Lotsmanenko 2000), which is listed in the references.

5 THE MILLER FOR THE THREAD MILLING

Thread milling of the cutting surface by means of enveloping is performed with a special hob cutter.

Cutting the thread surface on the workpiece can be performed only along with a special cylindrical single-thread hob cutter as shown in the Figure 2. While enveloping the two profiles (profile of milling tooth and profile of cut thread) with parallel and crossed axes of helical surfaces, the shape of these profiles (in any of the sections) is not the same.

For example, if the metric thread with a line profile in an axial section is milled on the workpiece, the hob cutter tooth profile in this same section will have some sort of curved shape.

Externally, the miller is presented as a thread surface of the limited length. On the thread surface parallel to the axis of rotation there are flutings forming gear chasing tools of a miller. The tooth of the hob cutter is sharp pointed, limited only by two cutting angles – front and rear. Front angle is $\gamma = 0$. Tooth pointing is made only at the front surface (Figure 2).

The material for the miller is instrument steel. Heat treatment of the miller is quenching HRC 55–60.

The quantity of chasing tools of a miller is limited. That's why the cut thread is not smooth, and it is the surface composed of individual facets. Qualitative characteristics of the cutting are the length of the threaded facets and the height of their joints.

Figure 2. Hob cutter for cutting by means of enveloping.

Figure 3.   Cylindrical cutting M64 × 2.

Figure 4.   Conical cutting MK64´2 j = 10°- cosines.

These qualitative parameters of the thread can be influenced by changing the amount of the miller chasing tools. These same characteristics are also influenced by a number of thread entries.

Since the thread milling method by means of enveloping simulates the machine-tool gearing of a screw thread surface with generating millers (Glushko & Lotsmanenko 2007), the directions of these surfaces are opposite.

6   CONCLUSION

Production of an external thread by means of enveloping is a new method of manufacturing thread surfaces with a hob cutter. The cutting is performed with cut-down milling, so the special production machine tool accessories are used.

Cylindrical thread surface is cut with parallel axes of the cutter and the workpiece rotation, and conical thread surface is cut with crossed axes.

The results of experimental thread milling with the method of enveloping:

While thread milling with hob cutter by enveloping the thread surface becomes faceted (composed of individual facets linked up with a break). The reason is that the hob cutter has a certain number of chasing tools. With increasing numbers of a miller chasing tools or their reduction the sizes of facets on the thread surface also change.

The experimental thread milling was performed with the cylindrical and conical external short threads (l = 30 mm); the miller feed during cutting is radial.

Thread is M64 × 12 taken from GOST (Russian State standard) 9150–81; miller da2 = 180 mm, number of chasing tools is "n" = 98. Screw-cutting lathe is 1 K62.

The parameters value, obtained by calculation, goes with an experimental one. Thread milling machine time was 15–20 seconds (on the details of bronze); practically it does not depend on the number of entries. The cooling mixture is Oil "Industrial 50".

The size Dl of facets on the surface of the thread with Z1 = 3 is not very visible.

The height of RH facets joint has no significant importance for the thread operation due to the smallness (when Z1 = 1, RH = 80 microns, with Z1 = 3, RH = 27 microns).

All the obtained threads are considered exploitable.

The parts made with cylindrical and conical threads are shown in Figures 3 and 4.

The technology of thread milling with this method allows cutting of multiple threads with a single-thread hob cutter, it increases the productivity of manufacturing thread surfaces on the workpieces.

REFERENCES

Glushko, E.V. & Lotsmanenko, V.V. 2007. Thread milling of – helically-cammed surfaces for a cutdown milling with the method of enveloping. Dimensional fettling, reliability, and effectiveness of machinery production processes, collection of articles. 3(3) FESTU, Vladivostok.

Glushko, E.V. 2005. Engineering support of thread milling by means of enveloping in the manufacturing environment. The research of increasing productivity of naval engineering and ship-repairing, collection of articles. FESTU, Vladivostok, 7.

Kovan, V.N. 1968. Reference book of production mechanic engineer. 2, Machinegiz, Moscow.

Lotsmanenko, V.V. & Lotsmanenko, M.V. 1999. Method of conical surfaces treatment, The Russian Federation patent 2131325.

Lotsmanenko, V.V. 2000. Modified method for manufacturing conical helical surface article, Machinostroitel 8, Virage, Moscow, 68: 28–31.

Lotsmanenko, V.V. 2003. The thread milling methods, The Russian Federation patent 2210470.

Yakuhin, B.G. & Stavrov B.A. 1989. Production of thread milling. Reference book. Mashinostroenie, Moscow.

Advanced Materials and Structural Engineering – Hu (Ed.)
© *2016 Taylor & Francis Group, London, ISBN 978-1-138-02786-2*

Comparing between OECD member countries based on S&T innovation capacity

S.R. Lee
Technology Foresight Division Office of Future Strategy, Korea Institute of S&T Evaluation and Planning, Korea

S.S. Chun
S&T Policy Planning Division, Korea Institute of S&T Evaluation and Planning, Korea

ABSTRACT: As Science & Technology (S&T) becomes a source of global competitiveness in knowledge-based economy, the level of S&T capacity determines a nation's competitive power. Therefor countries have been enhancing investment and political supports to strengthen S&T capacity. Most of all, accurate analysis and assessment of the level of nation's S&T ability is needed to make effective policy measures. On the basis of the framework of the NIS (National Innovation System), this paper suggests indexes to cover the entire cycle of S&T innovation. And it creates models to measure S&T capacity comprehensively, and tries to appraise 30 OECD members. And to conclude, in COSTII Score based on 2013 of Individual Nations, the United States took the first place by scoring 19.386 (out of 31) and was followed by Switzerland, Japan, and Sweden. Meanwhile, Korea ranked 8th with 11.866 points.

1 INTRODUCTION

S&T indicators are quantitative knowledge about the parameters of scientific, technological and innovation activity, at institutional, disciplinary, sectoral, regional, national or pluri-national levels (Barre 1997).

Once derived, S&T indicators can be used in various ways from decision making to research and analysis. Governments and corporations track their S&T resources and activities, assess how far these activities are meeting their goals and predict future trends and needs for finance and human resource development. S&T indicators can also inform public discussion on science resource allocation issues. If indicators are derived on a systematic basis and according to accepted definitions, S&T indicators can be used to compare investments and performances among countries.

Many OECD countries are already using national S&T indicators for their economic, industrial and human resource planning. Although concurrent efforts are being made to combine various indicators among different countries—some of them use OECD S&T indicators while others use Eurostat—statistics and indicators need to be reasserted according to the group members and purposes. This also applies to the OECD members. One reason for increasing the comparability of S&T indicators between

OECD member countries is to enhance their utility for monitoring joint research goals and strategies in S&T. These strategies specifically include supporting regional S&T programs of economic and social benefit, providing close coordination and management of S&T activities, developing S&T human resources and promoting networking and technology transfer between research institutions, and between the public research sector and industry.

The correlation with scientific discovery, technological innovation and economic development is of central policy concern to all countries. The level of R&D investment and skills has frequently been used as a proxy for the technological level of an industry or a country. However, this indicator alone cannot measure the outputs of the S&T system, nor the technological performance of industries or countries. Thus, many countries make their efforts to devise "innovation indicators", which are widely analyzed today.

In EU, the efforts to produce S&T statistics are materialized mostly on the "Statistics on Science, Technology and Innovation (STI Key Figures)" as one of the thematic studies of Eurostat and the "European Innovation Scoreboard (EIS)". STI Key Figures mainly cover R&D statistics, including statistics on Government Budget Appropriations and Outlays on R&D (GBAORD), innovation statistics (based on the Community innovation surveys/CIS), patent statistics, statistics on Human

Resources in Science and Technology (HRST statistics), statistics on the Career Development of Doctorate Holders (CDH statistics) and statistics on high-tech industries and knowledge-based services.

The OECD's Main S&T Indicators is a biannual publication that provides a set of indicators that reflect the level and structure of the efforts undertaken by OECD member countries and 9 non-member economies in the field of science and technology. These data include final and provisional results as well as forecasts established by government authorities. The indicators cover the resources devoted to research and development, patent families, technology balance of payments and international trade in highly R&D intensive industries. Also presented are the underlying economic series used to calculate these indicators. Series are presented for a reference year and the last six years for which data are available (paper publication) and beginning 1981 (electronic editions). It now categorizes S&T data into 18 dimensions and 149 indicators. Korea frequently uses this data to compare its S&T status with other countries.

2 CONCEPT

Today S&T is a main source of national competitive power in knowledge based economy. The necessity for an accurate diagnosis and evaluation of science and technology innovation capacity has been emphasized. For the improvement of a national S&T capability, it is needed to evaluate a present level of S&T accurately.

As we know, there are some surveys for inspect a national competitiveness, such as IMD, WEF and OECD STI. But they have a limitation on evaluation methods. In IMD report, S&T is regarded as infrastructure of internal enterprise's competitiveness. And it has No based model, No composite index. In case of OECD STI, it has difficulty in overall comparison of innovation capabilities levels among nations and R&D input & outcome is too centered on the private sector.

So, we tried to develop the COSTII (Composite Science and Technology Innovation Index) to overcome those limitations and to evaluate a nation's capability of S&T Innovation compositely by the medium of rational model, Based on National Innovation System model. COSTII is an indicator developed by Korea to look into the innovation capacity of 30 OECD members. It is created in order to obtain S&T information far beyond merely statistical numbers. Unlike simple statistical data that outlay all related S&T information, COSTII gathers innovation-related S&T statistics and reinterpret them in order to compare with

those from other countries. There are five dimensions for COSTII—resources, activities, network, environment, and performance—which are further categorized into human resources, organization, R&D investment, international cooperation, etc. 31 individual indicators comprise these dimensions, and the mean data for each dimension are rescaled to produce comparable international rankings.

3 METHOD

In, National Innovative Capacity is defined as the ability of a country to produce and commercialize a flow of innovative technology over the long term (Porter. & Stern 2001).

Science and technology capacity is defined for the purpose of this exercise as the ability of a country to absorb and retain specialized knowledge and to exploit it to conduct research, meet needs and develop efficient products and processes (Wagner et al. 2004). The ability to use specialized knowledge emerges from interactions of institutions and people, responds to public missions, and relies upon infrastructure. These bases can be represented by indicators, and it is possible to measure S&T capacity from a broad perspective of overlapping indicators representing direct and indirect measures. While it is possible to list countries merely by the percentage of investment in research and development (GERD), or by scientific papers or patents, which are direct measures of the outcomes of S&T, many countries would not be represented in such a list. These direct measures would provide little insight into the potential development of one country if it conducts various S&T activities, collaborates with other nations, or even uses existing resources to build additional capacity.

In this paper, we defines Science and Technology Innovation Capability as a nation's capability to produce outcomes that are of economic and social value at the final stage through innovation and improvement in the field of S&T, like OECD definition.

And our goal is to evaluate science and technology innovation capacity by developing a model and indicators that can give comprehensive diagnosis and later, identifies strengths and weaknesses to propose policy to improve science and technology innovation capacity.

Evaluated Nations are featured 30 member countries of the OECD (Organization for Economic Cooperation and Development). Although OECD now has 34 member countries, new members were excluded in COSTII due to low data availability. Information of additional members is expected to be reflected when relevant data can be collected.

Table 1. Evaluated nations.

Australia	Austria	Belgium	Canada	Czech Republic
Denmark	Finland	France	Germany	Greece
Hungary	Iceland	Ireland	Italy	Japan
Korea	Luxembourg	Mexico	Netherlands	New Zealand
Norway	Poland	Portugal	Slovak Republic	Spain
Sweden	Switzerland	Turkey	United Kingdom	United States

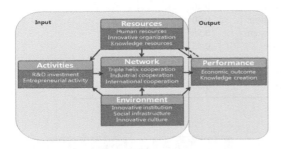

Figure 1. Evaluation model.

Evaluation Model is Based on the framework of the National Innovation System (NIS), the innovative process consists of five dimensions of innovation: resources, activities, network, environment, and performance. Innovation Resource, Innovation Activities, Innovation Network, Innovation Environment are in Input field, Innovation Performance is in Output. An arrow means that those 5 areas exchange an influence each other systematically.

The Concept of NIS is the elements and relationships which interact in the production, diffusion and use of new, and economically useful, knowledge ... And are either located within or rooted inside the borders of a nation State (Lundvall 1992).

On the basis of the framework of National Innovation System (NIS), the evaluation of science & technology innovation capacity consists of comprehensive review of the overall process of innovation, from input and activities to performance. The process assumes a systematic approach that regards the active interaction between the different elements as being a decisive factor of national science and technology innovation capacity.

Then put weights among 13 items, through expert surveys which based on fuzzy set theory. And Convert the ratio of weighting into integer numbers to allocate the number of indicators to each item. After Select 31 indicators out of the first selected 79 indicators pool, allocate the number of indicators by according to the importance of each items. And Select the indicators of each item, with conditions as follows.

For the selection of proper indicators, the possibility of acquiring statistical data is critical to compare OECD member country. Rationale for model and upper-level, like 5 areas and 13 items and distinction from other indicators is important, too. If the indicators possess high statistical relevance, the indicator expert committee selects most plausible and representative one. Then we draw 5 elements, 13 items, 31 indicators. It has 27 quantitative, 4 qualitative indicators.

Innovation Resource Indicator shows how much basic resources innovation entities can utilize for science and technology innovation. And it consists of human resources, innovation organization, and knowledge resources, such as researchers, top 100 universities and paper and patent stock.

Innovation Activities Indicator identifies innovation entities' activities of creating and utilizing new knowledge, and volition for innovation activities. It measures each entity's innovation activities according to the scale and distribution of material resources, such as R&D investment, the level of R&D activities, and start-up activities.

Innovation Network Indicator shows the network among innovation entities and cooperation through the network, such as flow of knowledge and technology diffusion, within the innovation system. So it identifies the status of cooperation among industry·academia·research institutes, major players of domestic research and development, and international cooperation.

Innovation Environment Indicator shows whether infrastructure is duly established for efficient innovation activities. Innovation environment is composed of various systems that support or facilitate innovation activities, innovation culture, and physical infrastructure, such as Tax advantage, protection of intellectual property right, broadband subscriber.

Innovation Outcome Indicator measures concrete outcomes of innovation activities. Innovation performance can be divided into knowledge creation and economic outcome. Knowledge creation is composed of indicators related with papers and patents. And economic outcome comprised of creation of added value, and improvement of trade balance.

In steps of collecting data, most data are from international statistical indicators for comparability with other countries. For the quantitative

Table 2. Structure of evaluation model.

Area	Items	Weights	# of indicators
Innovation resources (7)	Human resources	0.79	3
	Innovation organization	0.53	2
	Knowledge resources	0.59	2
Innovation activities (7)	R&D investment	0.99	5
	Start-up activities	0.60	2
Innovation networks (5)	Triple-helix cooperation	0.60	2
	Industrial cooperation	0.40	1
	International cooperation	0.50	2
Innovation environment (6)	Innovation support system	0.55	2
	Physical infrastructure	0.55	2
	Innovation culture	0.55	2
Innovation outcomes (6)	Knowledge creation	0.80	3
	Economic outcomes	0.80	3

indicators, data get from OECD MSTI, OECD scoreboard, USPTO, Thomson ISI, Global Entrepreneurship Monitor, and World Bank. And for qualitative data, use IMD competitiveness yearbook and WEF global competitiveness report.

The collected data are then "re-scaled" for standardization. The methodology is, for each country's indicator, the maximum data is designated "1", while the minimum data is "0".

Re-scaled standard value is,

$$\text{Standardized Value} = \frac{\text{Value}_{(object)} - \text{Value}_{(lowest)}}{\text{Value}_{(highest)} - \text{Value}_{(lowest)}}$$

(1)

※ To revise the missing value, if any, replaced it by the mean value of all indicators within the same dimension.

In order to produce the COSTII value, it is needed to calculate the standard value of 5 items. Items' value is draw through combining a standard value of indicators which are belonging to each item.

In this formula, weight of each indicator is equal

$$CI = \sum_{1}^{n} w_i X_i$$

(2)

CI = items index X_i: standard value of indicators
$w_i = 1$

(3)

Finally, COSTII is calculated by combining five items values from each dimension.

$$COSTII = \sum_{1}^{5} CI_i \quad CI = \text{items index}$$

(4)

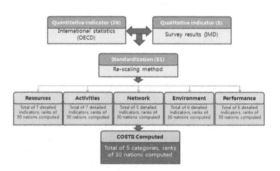

Figure 2. COSTII computation process.

This methodology is applied to OECD member countries, with values lying between 0 and 31.

4 RESULT AND CONCLUSION

In COSTII Score based on 2013 of Individual Nations, the United States took the first place by scoring 19.386 (out of 31) and was followed by Switzerland (14.476), Japan (13.661), and Sweden (13.236). Meanwhile, Korea ranked 8th with 11.866 points.

Putting United States, the best performer, at 100.0%, the relative level of Switzerland, the runner-up, is around 74.7% while South Korea stands at around 61.2%.

According to the analysis by Dimension, in resources, the United States ranked first with 5.853 points (out of 7 points), which was more than twice of the OECD average score. The US was followed by Japan (2.802), Germany (2.239), and United Kingdom (2.013).

As for activities, the leading group consists of the United States (5.338 points, out of 7 points),

10

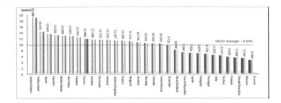

Figure 3.   COSTII score of 30 OECD member nations.

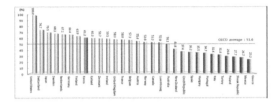

Figure 4.   Relative level of 30 OECD nations.

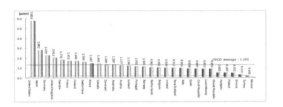

Figure 5.   Resources index of 30 OECD members.

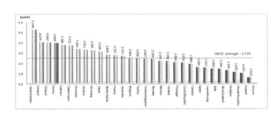

Figure 6.   Activities index of 30 OECD members.

Iceland (4.079), Finland (4.042), and South Korea (3.998). The OECD average was 2.539 points.

In Network, Belgium scored the highest in network with 2.757 points (out of 5 points) and the leading group includes Luxembourg (2.677), the Netherlands (2.587) and Switzerland (2.357).

In environment, the Netherlands ranked first with 4.175 points (out of 6 points), followed by Sweden (4.125), Canada (3.947), Finland (3.939), and the United States (3.929).

In case of performance, the leading group includes Switzerland (3.086, out of 6 points), Ireland (2.825), the United States (2.765 points), Japan (2.596), and the Netherlands (2.447). The OECD average was 1.496 points.

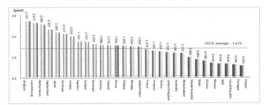

Figure 7.   Network index of 30 OECD members.

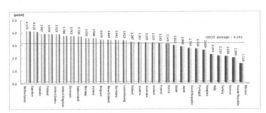

Figure 8.   Environment index of 30 OECD members.

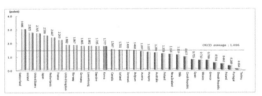

Figure 9.   Performance index of 30 OECD members.

Today S&T is a main source of national competitive power. The accurate evaluation of S&T innovation capacity has been emphasized, for the improvement of a national S&T capability. In this paper, we developed the COSTII to evaluate a nation's capability of S&T Innovation compositely by the medium of rational model, based on National Innovation System model. The COSTII can be a synthetic indicator for looking into the innovation capacity of 30 OECD members.

But, there are some challenges on the COSTII. First is the internationalization. The COSTII needs to be promoted an international recognition. It is necessary to seek ways to utilize not only Korean experts but also NESTI expert within OECD. And it need to develop an own survey indicators to overcome the limitation of quantitative indicators. And to enhance an application, revise a present S&T policy and establish a new one which is reflected the result of the COSTII Raising a rationality of methodology. Lastly to raise a rationality of this methodology, it needs to compare the outcomes by using such methods as AHP, Factor Analysis and fuzzy set theory.

REFERENCES

Barré, R. 1997. The European Perspective on S&T Indicators, Scientometrics, volume 38: 57~70.

EU, 2012. The 2012 EU Industrial R&D Investment Scoreboard.

Global Entrepreneurship Research Association, 2012. Global Entrepreneurship Monitor 2012 Global Report.

IMD, 2013. The World Competitiveness Yearbook.

KAIST, 2013. SCI research analysis.

KISTEP, 2004. A Study on National Innovation Assessment Indicator Development.

Korea Institute of Patent Information, 2013. USPTO Patent analysis.

Lundvall, B. 1992. *National System of Innovation-Toward a Theory of Innovation and Interactive Learning,* Chap1: 2~15. Printer Publisher, London.

MEST, 2004. A Plan for National Innovation System.

OECD, 2011. Science, Technology and Industry Scoreboard.

OECD, 2013. International Direct Investment Statistics.

OECD, 2013. Main Science and Technology Indicator.

Porter, M.E. & Stern, S. 2001. National Innovative Capacity, The global competitiveness report 2002: 102~118.

Quacquarelli Symonds, 2012. QS World University Rankings.

USPTO, 2012. General Patent Statistics.

Wagner, C.S. Edwin, H. & Arindam, D. 2004. *Can Science and Technology Capacity be Measured?* Input for Decision-making, RAND corporation.

WEF, 2011. The Global Competitiveness Report.

Advanced Materials and Structural Engineering – Hu (Ed.)
© 2016 Taylor & Francis Group, London, ISBN 978-1-138-02786-2

Dependence of the coercive force on the size of the core/shell nanoparticles magnetite/titanomagnetite

M. Shmykova, L. Afremov & I. Iliushin
Far Eastern Federal University, Vladivostok, Russia

ABSTRACT: This work is devoted to the modeling of the coercive field dependence on the core size of Magnetite/Titanomagnetite core/shell nanoparticles. Nanoparticles with sizes up to 100 nm and a different portion of magnetite have been studied. It has been shown that increasing the magnetite portion in particle does not affect coercive field up to some critical size of the core, after which it rapidly grows to a maximum value.

1 INTRODUCTION

Titanomagnetite is the main source of the remanent magnetization in most rocks and sometimes found in the oxidized state in the continental basalts and soils. Magnetic anomalies observed at the sea are due to the magnetization of the crust and provide fundamental information about the age of the rock in terms of the theory of plate tectonics. Titanomagnetite is a very interesting system to study the role of fine microstructure of Earth's magnetic field and magnetic properties of rocks. It is an important basalt mineral and well represented on Earth, Moon, and Mars. Remanent magnetization of these minerals can be a part of the planetary magnetic field and contains information about the evolution of the geomagnetic fields in the solar system. Portion of the titanium in solid solutions strongly depends on temperature and pressure that determine the equilibrium state of titanomagnetites. Magnetite, with magnetization determined as its volume fraction, is obtained as the result of titanomagnetite decay and can be used as an indicator for decay process. The first mention of existence of two-phase germination of magnetite oxides in titanomagnetite spinels was in the work of Mogensen (1946). Subsequently, Hjelm-Kwist and Ramdohr (1965) shown that magnetite exsolutions are a common feature of the titanomagnetite decay. Germination is highly interesting due to the possibility of changing of magnetic properties of rocks.

Data obtained during the study of titanomagnetite provide information about physical and chemical properties of oxides, environmental conditions during their formation and subsequent cooling. Magnetic properties of titanomagnetite can depend on low-temperature conversion of oxides, e.g., oxidizing or decay of solid solution leads to the formation of ulvospinel (Fe_2TiO_4) and magnetite (Fe_3O_4). It is known, that increasing of iron fraction in titanomagnetite leads to a number of variations in the magnetic, electronic and structural properties. For example, increasing of a fraction of divalent iron leads to the increase of magnetostriction and, thus, higher values of coercive field (Banerjee 1991, Pearce et al. 2012). Titanomagnetite plays a significant role in paleomagnetic researches and interesting for technological application (Kakol et al. 1991, Pearce et al. 2006). Among the existing methods of theoretical study of the dependence of the magnetic properties of the two-sublattice magnetic materials, which include titanomagnetite, it should be noted that the method of calculation for the exchange interaction of random fields is used for solving the problem of the concentration phase transitions of two-sublattice systems (Belokon et al. 2012), as well as numerical simulation within the Ising model (Nefedev & Kapitan 2013).

The aim of this work is to study the dependence of coercive field on magnetite core size in core/shell nanoparticles. Modeling of hysteresis loops behavior of titanomagnetite particles with sizes upto 100 nm (size of magnetite core is varied from 0 nm to almost 100 nm) has been carried out within our model of core/shell nanoparticle.

2 MODEL OF THE MAGNETITE/TITANOMAGNETITE NANOPARTICLE

Model, which is used for modeling of magnetization process of nanoparticles, can be described in the following ways (Nefedev & Kapitan 2013):

1. Uniformly magnetized titanomagnetite particle $Fe_{2,56}Ti_{0,44}O_4$ (phase 1) has an ellipsoidal shape with elongation Q and volume V, which

contains uniformly magnetized magnetite core (phase 2) of an ellipsoidal shape with elongation q and volume $v = \varepsilon V$;

2. It is assumed that the axes of crystallography anisotropy of both phases are parallel to the long axes of ellipsoids, and magnetization vectors of phases $I_S^{(1)}$ and $I_S^{(2)}$ are in the yOz plane (Fig. 1);

3. External magnetic field is applied along axis Oz;

4. The total energy of nanoparticle is composed of the anisotropy energy, magnetostatic interaction energy, exchange interaction energy, and energy of the external field (Afremov & Ilyushin 2013):

$$F = \left\{ -\frac{(I_s^{(1)})^2}{4} K^{(1)} \cos 2\vartheta^{(1)} \right.$$
$$-\frac{(I_s^{(2)})^2}{4} K^{(2)} \cos 2\vartheta^{(2)}$$
$$+ I_s^{(1)} I_s^{(2)} \left(-\mathcal{U}_1 \sin \vartheta^{(1)} \sin \vartheta^{(2)} \right.$$
$$\left. \left. + \mathcal{U}_2 \cos \vartheta^{(1)} \cos \vartheta^{(2)} \right) \right\} V, \qquad (1)$$

where the effective anisotropy constants $K^{(1,2)}$ of phases and the constants of interphase interaction $\mathcal{U}_1$ and $\mathcal{U}_2$ are defined by the following equations:

$$K^{(1)} = (1 - \varepsilon) k_A^{(1)} + (1 - 2\varepsilon) k_N^{(1)} + \varepsilon k_N^{(2)},$$
$$K^{(2)} = \varepsilon (k_A^{(2)} + k_N^{(2)}) \qquad (2)$$

$$\mathcal{U}_1 = \varepsilon \left(\frac{k_N^{(1)} - k_N^{(2)}}{3} + \frac{2s A_{in}}{\upsilon \delta I_s^{(1)} I_s^{(2)}} \right),$$
$$\mathcal{U}_2 = \varepsilon \left(\frac{2(k_N^{(1)} - k_N^{(2)})}{3} - \frac{2s A_{in}}{\upsilon \delta I_s^{(1)} I_s^{(2)}} \right) \qquad (3)$$

In these equations $k_A^{(1,2)} = K_1/(I_s^{(1,2)})^2, k_N^{(1,2)}$ — are dimensionless crystallography anisotropy

constants and shape anisotropy constants of phases, respectively. Where K_1—first anisotropy constant, V—particle volume, s—surface area separating the phases, ε—core volume to particle volume ratio, A_{in}—interphase exchange interaction constant, δ—the width of the transition area of the order of the lattice constant. Note that shape of anisotropy constant $k_N = 2\pi (1 - 3N_z)$ is calculated by using demagnetization factor N_z along the long axis, depending only on elongation of ellipsoid q:

$$N_z = \left[\sqrt{q^2 - 1} - q \, arc \, \cos(q) \right] / (q^2 - 1)^{3/2}.$$

According to the model (Nefedev & Kapitan 2013, Afremov & Ilyushin 2013), if thermal fluctuations can be neglected and in the absence of external field these nanoparticles can be in one of four states characterized by the different orientation of the magnetic moments:

– magnetic moments of both phases are parallel and directed along the axis Oz;
– magnetic moment of the first phase is directed along axis Oz, and second is directed in the opposite way;
– both the magnetic moments are antiparallel to the axis Oz;
– magnetic moment of the second phase is parallel to the axis Oz, and first phase is antiparallel.

If thermal fluctuations cannot be neglected, a system of two phase nanoparticles, remaining in the non-equilibrium state after some time t must come into a state with a population, defined by vector.

$N(t) = \{n_1(t), n_2(t), n_3(t), n_4(t)\}$, components $n_i(t)$, can be described as probabilities of finding the nanoparticle in one of the above-mentioned states. According to Afremov & Ilyushin, (2013), population vector is defined by the following equation:

$$\frac{dN(t)}{dt} = W \, N(t) + V,$$

$$N(t = 0) = N_0 = \begin{pmatrix} n_{01} \\ n_{02} \\ n_{03} \end{pmatrix},$$
$$\qquad (4)$$

$$N(t) = \begin{pmatrix} n_1(t) \\ n_2(t) \\ n_3(t) \end{pmatrix}, \widetilde{W}_{ik} = \begin{cases} -\sum\limits_{j \neq i}^{4} W_{ij} - W_{4i}, i = k, \\ W_{ki} - W_{4i}, \quad i \neq k, \end{cases}$$

$$V = \begin{pmatrix} W_{41} \\ W_{42} \\ W_{43} \end{pmatrix}, \qquad (5)$$

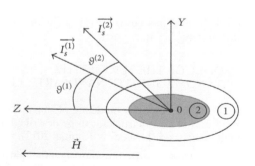

Figure 1. Illustration for the model of two-phase magnetite/titanomagnetite nanoparticles.

$n_4(t)$ can be defined from the normalization condition: $n_1(t) + n_2(t), n_3(t) + n_4(t) = 1$,

$W_{ik} = f_0 \exp(-E_{ik}/k_B T)$ —matrix elements of transition probability matrix from i equilibrium state to k, $f_0 = 10^{10}$ s^{-1} frequency factor,

$E_{ik} = E_{ik}^{(max)} - E_i^{(min)}$—potential barrier, where $E_{ik}^{(max)}$ is the smallest of maximal energy values, corresponding to the transition of magnetic moment from i equilibrium state with energy $E_i^{(min)}$ to k state. To calculate E_{ik}, we use the equation for the total energy of two-phase nanoparticle (Equation 1).

The solution of system 4 can be conveniently written by using matrix exponents:

$$N(t) = \exp(\widetilde{W}t) \cdot N_0 + \int_0^t \exp\left(\widetilde{W}(t-\tau)\right)d\tau \cdot V. \quad (6)$$

Equation 6 allows us to get an equation for the magnetization of the system of two-phase nanoparticles:

$$I(t) = \left[(n_1(t) - n_3(t))\left((1-\varepsilon) I_s^{(1)} + \varepsilon I_s^{(2)}\right) \right.$$
$$\left. + (n_2(t) - n_4(t))\left((1-\varepsilon) I_s^{(1)} - \varepsilon I_s^{(2)}\right) \right],$$
$$n_4(t) = 1 - n_1(t) - n_2(t) - n_3(t). \quad (7)$$

2.1 Calculation of the potential barriers

Thermal stability of magnetic carriers and magnetic storage elements becomes significant at low sizes of magnetic structures (Nefedev & Kapitan 2013). Calculation of thermal stability requires an assessment of the rate of transition between stable states of equilibrium of the nanoparticle. To accurately calculate the probability of the particle transition between states, one needs to own the most complete information about the value of the potential barrier between these states. The microstructure of magnetite can contain lots of local minimums that complicate the calculation of the optimal energy barrier (Fig. 2).

The solution to this problem is various numerical methods widely used in solid state physics, theoretical chemistry and material study. At the moment, there are many different algorithms used to calculate energy barriers. For example MEPs, NEB, etc. Many of these methods are focused on finding the special points—saddle points (Afremov & Ilyushin 2013, Afremov & Panov 2004). In our work, we used this idea of finding the saddle point to obtain the values of the energy barrier separating states by using simple calculations.

To determine the possible states of the nanoparticles is necessary to determine the minimums and maximums of this function as well as to determine whether there are the saddle points through which

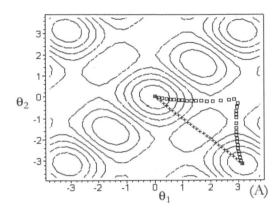

Figure 2. Illustration of the problem of finding the minimal barrier.

the particle can move from one state to another. It is necessary to investigate the points of extremum of this function.

After getting a set of solutions, we need to understand which points are minima and which are maxima. To do this, we use the standard mathematical condition for finding the points of extremum of functions of several variables.

$$f(\vartheta^{(1)}, \vartheta^{(2)}) = \frac{\partial^2 E(\vartheta^{(1)}, \vartheta^{(2)})}{\partial^2 \vartheta^{(1)}} \frac{\partial^2 E(\vartheta^{(1)}, \vartheta^{(2)})}{\partial^2 \vartheta^{(2)}}$$
$$- \left(\frac{\partial^2 E(\vartheta^{(1)}, \vartheta^{(2)})}{\partial \vartheta^{(1)} \partial \vartheta^{(2)}} \right)^2, \quad (8)$$

If $f(\vartheta^{(1)}, \vartheta^{(2)})$ and $\partial^2 E(\vartheta^{(1)}, \vartheta^{(2)})/\partial^2 \vartheta^{(1)} > 0$ (or $\partial^2 E(\vartheta^{(1)}, \vartheta^{(2)})/\partial^2 \vartheta^{(2)} > 0$) then these points are related to maximum energy $E = E(\vartheta^{(1)}, \vartheta^{(2)})$, if $f(\vartheta^{(1)}, \vartheta^{(2)}) > 0$ and $\partial^2 E(\vartheta^{(1)}, \vartheta^{(2)})/\partial^2 \vartheta^{(1)} < 0$ (or $\partial^2 E(\vartheta^{(1)}, \vartheta^{(2)})/\partial^2 \vartheta^{(2)} < 0$), then these points are minimum energy. All other points require further analysis. The solution of this problem is to calculate the value of the function in the neighborhood of a given point. If the given function is decreasing in all directions except for one point this can be attributed to a saddle value and use it in the calculation of the energy barrier, in the opposite case, the remaining points do not fulfill any of the above-mentioned conditions it is possible to calculate the energy barrier, moving the particle along the boundary of the function. Changing α or β and taking the maximum value of the barrier out of 2 presented.

$$\max(E[\alpha,0]) \text{ at } \{\alpha,0,\pi\} \quad \max(E[\pi,\beta]) \text{ at } \{\beta,0,\pi\}$$
$$(9)$$

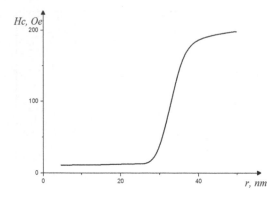

Figure 3. Dependence of coercive field H_c of magnetite core radius of magnetite/titanomagnetite core/shell nanoparticles.

The same in the opposite direction:

$$\max(E[0,\beta]) \text{ at } \{\beta,0,\pi\} \; \max(E[\alpha,\pi]) \text{ at } \{\alpha,0,\pi\}$$

(10)

Finding the minimal value from Equations 3 and 4 we obtain the barrier. Given operation allows us to accurately determine value of the minimal energy barrier and show if it is possible for the particle to move to this state.

3 COERCIVE FIELD OF THE MAGNETITE/ TITANOMAGNETITE NANOPARTICLES

Naturally coercive field H_c of the nanoparticles increasing with the increase of fraction of the magnetite in the core/shell nanoparticle (Fig. 3). It should be noted that there is an area where growth of magnetite core does not affect coercive field of the system. Only at $r > 30$ nm coercive field sharply increases up to the maximum value, corresponding to the H_c of magnetite. It is due to the blocking of the magnetic moments of nanoparticles at $r \approx 30$ nm, which results in sharp increase of coercive field. The dependence of coercive field on the size of magnetite core is in good agreement with experimental results (Pearce et al. 2012, Kakol et al. 1991).

4 CONCLUSIONS

In this work, the study of dependence of coercive field H_c on a size of magnetite core has been carried out within the model of magnetite/titanomagnetite core/shell nanoparticles. It is shown that H_c of core/shell nanoparticles sharply increases with an increase of magnetite fraction of the core sizes of the order of 60 nm due to the magnetic moment blocking.

It has been shown that increasing of magnetite portion in particle does not affect coercive field up to some critical size of the core, after which it rapidly grows to maximum value. These results can be used in the modeling process of disintegration titanomagnetites. Thus, our model of core/shell nanoparticles can be used to study the products of titanomagnetite decay.

ACKNOWLEDGMENT

This work supported by Ministry of Education and Science, project No 559.2014.

REFERENCES

Afremov L.L., A.V. Panov 2004. *Residual magnetization of ultrafine magnetic materials,* Vladivostok, Publishing House of the Far Eastern University, 192.

Afremov. L.L. & Ilyushin I.G. 2013. Effect of Mechanical Stress on Magnetic States and Hysteresis Characteristics of a Two-Phase Nanoparticles System, *Journal of Nanomaterials,* 2013: 687613.

Banerjee, S.K. 1991. Oxide Minerals: Petrologic and Magnetic significance, *Mineralogical Society of America,* 25: 107–128.

Belokon V.I. Nefedev K.V. & Dyachenko O.I. 2012. Concentration Phase Transitions in Two-Sublattice Magnets, *Advanced Materials Research,* 557–559: 731–734.

Bochardt-Ott, W. 1990. *Crystallography. An Introduction for Scientists.* Springer Verlag, Berlin.

Hjelniquist, S. // J. Geophys., 40:435–465. In S. Sweden. Sver. Geol. Unders. Arsb., Vol. 43:1, 1949, pp. 55.

Kakol, Z. J. Sabol, & J.M. Honig, Magnetic anisotropy of titanomagnetites $Fe_{3-x}Ti_xO_4$, $0 \le x \le 0.55$, *Physical Review B,* 43: 2198–2204.

Nefedev K.V. & Kapitan V.Y. 2013. Spin-Glass-Like Behavior and Concentration Phase Transitions in Model of Monolayer Two-Sublattice Magnetics, *Applied Mechanics and Materials,* 328: 841–844.

Pearce, C.I. et al. 2012. Synthesis and properties of titanomagnetite (Fe3-xTixO4) nanoparticles: A tunable solid-state Fe(II/III) redox system, *Journal of Colloid and Interface Science,* 387: 24–38.

Pearce, C.I. Henderson, C.M.B. Pattrick, R.A.D. van der Laan, G. & Vaughan, D.J. 2006. Direct determination of cation site occupancies in natural ferrite spinels by L(2,3) X-ray absorption spectroscopy and X-ray magnetic circular dichroism, American Mineralogist, 91(5–6): 880–893.

Advanced Materials and Structural Engineering – Hu (Ed.)
© 2016 Taylor & Francis Group, London, ISBN 978-1-138-02786-2

The interaction between contacting barrier materials for containment of radioactive wastes

H.C. Chang, C.Y. Wang & W.H. Huang
Department of Civil Engineering, National Central University, Taiwan

ABSTRACT: Zhisin clay is used as raw clay material in this study. This clay is mixed with Taitung area argillite to produce the backfill material for potential application such as barrier for the disposal of low-level radioactive wastes. The interactions between the concrete barrier and the backfill material are simulated by an accelerated migration test to investigate the effect of contacting concrete on the expected functions of backfill material. The results show that backfill material near the contact with the concrete barrier exhibited a significant change in the ratio of calcium/sodium exchange capacity, due to the release of calcium ions from the concrete material. Also, some decrease in swelling capacity of the backfill material near the concrete-backfill interface was noted.

1 INTRODUCTION

The disposal of low-level radioactive wastes requires multi-barrier facilities to contain the wastes and prevent contamination. Typically, the engineered barrier is composed of a concrete vault backfilled with sand/bentonite mixture. The backfill material is a mixture of bentonite and sand/gravel produced by crushing the rocks excavated at the site. With a high swelling potential, bentonite is expected to serve the sealing function, while the crushed sand/gravel improves the workability and stability of the mixture. Due to the nature of radioactive wastes, the disposal site is designed for a service life of 300 years or more, which is much longer than typical engineering or earth works. With such a long service life, the site is subject to groundwater intrusion and geochemical evolution, making the near-field environment evolution of the disposal site a complex problem. (Han, K. et al. 1997)

In the vicinity of the concrete vault in a disposal site, the high-alkali concrete environment can cause changes in the pore solution and alter the nature of backfill materials. Although the interaction between the concrete barrier and the backfill material does not affect the two barriers immediately, the interaction is reacting continually over a long period of time. The physical characteristics of the two barriers can be changed by this long-term interaction. Takafumi and Yukikazy (2008) used a migration technique to simulate the interaction between different types of concrete barrier and the backfill materials. It was found that the swelling capacity is reduced with the increase of accelerated migration test periods for the bentonites. The swelling capacity of the bentonite in contact with fly ash concrete is higher than that with OPC concrete. Therefore, the interaction between the concrete and the backfill material needs to be assessed, such that the barriers serve the expected functions for a pro-longed period of time. In this research, an accelerated migration test was devised to understand the effects of leaching from the concrete on the characteristics of backfill material. The two barrier materials (concrete and backfill) were placed in contact and then an electric gradient applied to accelerate the move of actions between the two barriers. A direct current was used for a composite specimen with a cylindrical section in which an electrical potential gradient was applied. The physical characteristics of bentonite are carefully examined so as to assure that the long-term contact with these two barriers does not cause severe degradation. The analysis includes swelling capacity and calcium and sodium exchange capacity (CEC) of the bentonite material.

2 MATERIALS AND METHODS

2.1 Materials

Locally available Zhisin clay originated from Taitung, Taiwan was used as raw clay material in this study. Zhisin clay is mixed with Taitung area argillite, which is originated from the rock around the disposal site to produce the backfill material. The chemical compositions of the clay are given in Table 1. The Taitung area argillite is crushed to a maximum size of 2.36 mm and a minimum size of 150 μm. (Sivapullaiah, P. V. et al. 1996) The mix proportion of concrete used in this study is given in Table 2. The compressive strength of the concrete

Table 1. Chemical compositions in Zhisin clay.

Compositions	(%)
SiO_2	55.4
Al_2O_3	20.1
Fe_2O_3	5.5
CaO	2.6
Na_2O	0.7
MgO	1.7
K_2O	1.4

Table 2. Mix proportions of concrete (kg/m^3).

Mineral	(kg/m^3)
Cement	243
Slag	88
Fly ash	88
Silica fume	28
Water	210
Coarse aggregate	1003
Fine aggregate	658
Water reducing admixture	4.4

at the age of 28 days was determined to be higher than 35 MPa. Hardened concrete specimens with a diameter of 70 mm were sliced at a thickness of 30 mm for the accelerated migration test.

2.2 Migration test

In this research, a migration technique was applied to accelerate the move of calcium ions from the pore solution of concrete so as to investigate the alteration of backfill material in contact with the concrete. A direct current voltage of 15 Volts was used for a composite specimen with a cylindrical section in which an electrical potential gradient was applied. Figure 1 shows a schematic diagram of the accelerated migration test. The cathode was embedded in the compacted bentonite gravel mixture, while the anode was immersed in a saturated calcium solution. Migration tests were conducted in a temperature controlled room (25 ± 2°C). The current was measured by multi-tester when the accelerated migration test was in progress.

2.3 Analysis

At the end of the accelerated migration period, the composite cell was dismantled so that the specimens were ready for analysis. The compacted bentonite was sliced into seventh layers with different

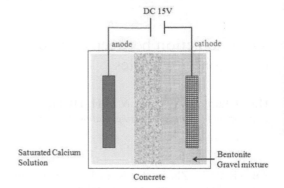

Figure 1. Schematic view of accelerated migration test.

thickness. The thickness of the first five layers was 8 mm, and that of the sixth and the seventh layers was 15 mm for the sliced compacted bentonites.

In order to remove the sand in composite specimen, each slice of the backfill material was sieved such that only powder material with particle size less than 150 μm was used for testing of swelling capacity and exchange capacity of calcium and sodium (CEC). According to ASTM D5890, the powder samples of the bentonite were mixed with 100 mL distilled water and then the volume increase was measured after 24 hours. The swelling capacity was measured by the increase in the free volume of the bentonite. The CEC of Zhisin clay was measured by Inductively Coupled Plasma (ICP) analysis.

3 RESULTS AND DISCUSSION

3.1 Cumulative electric charge on migration test

Figure 2 shows the change of current with time in an accelerated migration test. The measured current became stable when the test periods lasted over 216 hours. In order to confirm the interaction has been completed between the concrete barrier and the backfill materials, the accelerated migration test continued for 500 hours.

3.2 The ratio of calcium to sodium content

Figure 3 gives the exchangeable capacity of calcium and sodium in Zhisin clay along the depth of backfill materials. Over the test period lasting 500 hours, the exchangeable capacity of calcium increased, while that of sodium remained about the same.

Figure 4 shows the variation in the ratio of calcium to sodium exchangeable capacity in Zhisin clay. The interface between concrete and backfill

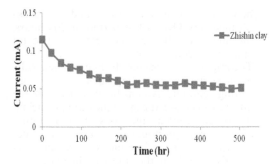

Figure 2. Measured current during the accelerated migration test.

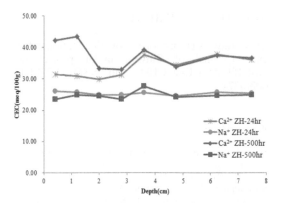

Figure 3. The exchange capacity of calcium and sodium in Zhisin clay at different test durations.

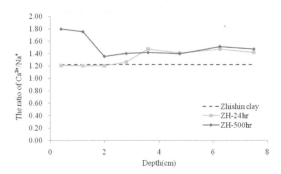

Figure 4. Variation of the ratio of calcium/sodium content with distance from the interface.

is at depth 0 cm, and the depth in the horizontal axis represents the distance from the interface at which the exchangeable capacity was measured. In Figure 4 the ratio of calcium to sodium content increases with the increase of accelerated migration test period. The ratio of calcium to sodium content for Zhisin clay shows an increase to 1.8

over 500 hours of the test. This indicates that there is an increase in the calcium content, especially in the layers close to the contacting interface. And the change in the ratio of calcium to sodium content in Zhisin clay occurs within 2.5 cm from the interface. As the distance from the interface increases, the change in the ratio of calcium to sodium content becomes less obvious.

3.3 Swelling capacity

The ions released from concrete tend to change the swelling capacity of bentonite. The change of the swelling capacity in the corresponding layer of the compacted backfill made with Zhisin clay is given in Figure 5, the swelling potential of the raw bentonite material is shown in these figures in dash line.

It is observed that the swelling capacity is reduced with the increase of accelerated migration test period. In Figure 5, the change of the swelling capacity in Zhisin clay occurs within a distance of 2.0 cm from the interface. This observation is most pronounced in the first layer after a test period of 500 hours. The swelling capacity decreases to 6 mL/2 g. This results from the accelerated migration of ions from the pore solution of concrete to backfill in an accelerated migration test. The ions react with montmorillonite, the main mineral component of bentonite, and converted it to non-swelling minerals such as zeolites, resulting in a decrease in swelling capacity. And the distance from the interface decreases, we observe more reduction of swelling capacity.

Figure 6 shows the relationship between the swelling capacity ratios (R_{EP}) and the ratio of cations using Equation (1).

$$R_{EP} = EP/EP_0. \tag{1}$$

where EP is the swelling capacity at each layer and EP_0 is the mean value of the swelling capacity

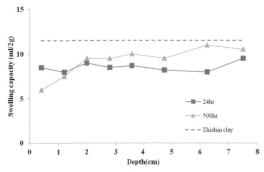

Figure 5. Change of swelling capacity of Zhisin bentonite with depth.

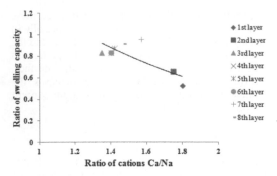

Figure 6. Reduction of swelling capacity with increase in calcium ions.

obtained from the pure Zhisin clay. The ratio of the swelling capacity reduces with the increase in the ratio of cations,—indicating that the swelling capacity becomes lower when the relative content of calcium ions gets higher. This tendency is more noticeable in the first and the second layers after 500 hours of testing for Zhisin clay. The swelling capacity ratio of the first and second layers was 0.52 and 0.65, respectively, while the corresponding ratio of cations is 1.80 and 1.75.

4 CONCLUSIONS

In this research, a migration technique was applied to accelerate the migration of calcium ions from concrete to investigate the alteration of compacted Zhisin clay-sand mixture in contact with the concrete. The followings conclusions were drawn from the experimental results:

1. The experimental results indicate that, the accelerated migration test could effectively simulate the long-term behavior of the interaction between the concrete barrier and the back-fill materials. Over a test period of 500 hours, the interaction between concrete and backfill approaches completion.
2. The migration of calcium from concrete results in the reduction of swelling capacity of the contacting bentonite. And as the distance from the interface decrease, the more the ratio of calcium to sodium content increases. This alteration effect is more pronounced for bentonite material near the contact interface with concrete. As the distance from the interface increases, the effect of the migration of calcium on the bentonite decreases.

ACKNOWLEDGEMENT

This study was supported by the National Science Council of Taiwan under project no. NSC100-2221-E-008-109- and NSC103-2221-E-008-077-.

REFERENCES

Han, K., Heinonen, W.J. & Bonne A. 1997. Radioactive waste disposal: global experience and challenges, *IAEA Bulletin*, 39: 41–99.
Sivapullaiah, P.V., Sridharan, A. & Stalin, V.K. 1996. Swelling behavior of soil-bentonite mixtures, *Canadian Geotechnical Journal*, 33: 808–814.
Takafumi, S. & Yukikazu, T. 2008. Use of a migration technique to study alteration of compacted sand-bentonite mixture in contact with concrete, *Physics and Chemistry of the Earth*, 33: S276–S284.

Structural and technological patterns of formation of surface nanostructured layers TiNiZr by high-speed flame spraying

P.O. Rusinov & Zh.M. Blednova
Kuban State Technological University, Krasnodar, Russian Federation

ABSTRACT: Based on the analysis of the phase composition, the average grain size was determined by using high-resolution electron microscopy that showed the correlation of the properties of the coatings with their structural phase state. The influence of structure and the mechanical properties of the steel composition were shown—in terms of wear of the coating. It is shown that the deposition of mechanically activated powder of optimal size provides increased durability.

1 INTRODUCTION

Materials with a Shape Memory Effect (SME) have been successfully implemented in modern units and structures. One of the areas for application of this development is the production of semi-finished products and technologies that use shape memory alloys for the creation of detachable joints and parts (Likhachev 1997, Blednova & Rusinov 2014, Rusinov & Blednova 2015, Rusinov et al. 2015, Blednova et al. 2014). Due to the effects of the power generation and the stress relaxation of surface-modified layer by alloy with thermo-mechanical memory, which is only a fraction of the total mass and may provide new features and details of the structural elements. It is widely known that the intermetallic nickel-aluminum-based NiAl possesses a high-temperature shape memory effect (the temperature of martensitic transformations in NiAl alloys with shape memory effect can reach 1000 K) with the formation of several different variants of martensite with different structures (Kositsyn et al. 2006). A less-known alloy with shape memory is TiNiZr. The substitution of the titanium with zirconium increases the temperature of martensitic transformations more than 400 K. Thus, the TiNiZr alloy has a high-temperature shape memory effect (Firstov et al. 2004), which is greater than the alloy strength properties of NiAl, and this is of interest to a number of different industries.

The aim of this work is to investigate the possibility of designing the structure of the surface layers of the alloys of TiNiZr with a high-velocity flame spraying of mechanically activated powder to ensure its functional and mechanical properties, and on this basis to create efficient functional materials and components.

(a)

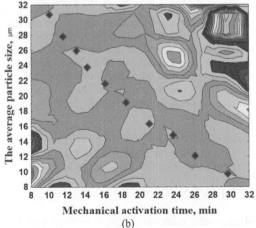

(b)

Figure 1. The PN47T26C27 powder mechanically pulverized in a GEFET-2 attritor, which was activated for 30 minutes: (a) ×300; (b) the effect of time of mechanical activation on the particle size of the powder PN47T26C27.

The formation of the surface layers was produced by high-speed flame spraying of mechanically activated powder materials with shape memory effect based on the modernized PN47T26C27 GLC. The material used for the mechanical activation of powder was PN47T26C27. The steel powder 1045 was used as a base. The size of the fractions of PN47T26C27 in the initial state is 50–70 microns. The structure of the newly formed PN47T26C27 powder consisted of the austenitic phase (~65%) and the martensitic phase (~35%). Mechanical activation and grinding of the PN47T26C27 powder was carried out by using ball mill Hephaestus-2 (AGO-2U) with the following parameters: frequency of rotation of the drum was 1200 min^{-1}, the rotational speed of the carrier was 900 min^{-1}, the diameter of the steel balls was 6 mm, while the running time was 10–30 min. After the mechanical activation, PN47T26C27 powder got the form of flat discs ranging from 10 to 30 μm (Fig. 1a and 1b).

2 THE TECHNOLOGY OF FORMING OF THE SURFACE LAYERS

Before the high-speed flame spraying of PN47T26C27 powder, the cleansing of the surfaces of steel samples from contaminants was performed, the blasting process, followed by an immersion in a 15–20% solution of HNO_3 was carried out. High-speed flame spraying was carried out in a vacuum chamber filled with argon.

The main process parameters of the high-speed flame spraying are the following: propane flow is 60–85 l/min, oxygen flow is 120–160 l/min, flow of powder in the carrier gas (argon), the distance and angle of the deposition, the feed speed of the torch, and the speed of rotation of the coated strip (Fig. 2). They define the characteristics of the coating such as its strength of adhesion to the substrate, its cohesive strength, the level of residual stress, porosity, structure, and thickness of the deposited layer.

3 FEATURES OF STRUCTURE FORMATION OF THE SURFACE-MODIFIED LAYERS OF SHAPE MEMORY TiNiZr

Macro- and micro-analysis of surface layers of the alloy of TiNiZr obtained by the established technology showed that the coating is a sufficiently dense structure. The interface between the coating and the substrate is without any visible defects (Fig. 3a). With the passage of the powder particles through the flame jet, they heat up

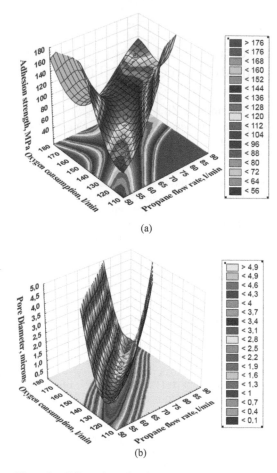

(a)

(b)

Figure 2.   Effect of combustible gases on the strength of adhesion of coating with the shape memory effect.

and strike the substrate in the form of solidified deformed discs with a diameter of 20–35 μm and a thickness of 5–20 μm. The grain size of the coating obtained by the flame spraying in the highly protective medium (argon) is from 80 to 100 nm (Fig. 3b and 3c).

The microhardness of the TiNiZr layer varies: Hμ = 9.5–12.7 GPa. Such increase of the microhardness is due to the high velocity of collision between particles and the substrate; the high speed of the cooling and the rapid quenching of the alloy, a high strength metastable nanostructure is formed.

The X-Ray Diffraction (XRD) analysis results showed that at room temperature the initial phase state of layer ($Ti_{33}Ni_{49}Zr_{18}$) after high-speed flame spraying of mechanically activated powder in a protective atmosphere (argon) is a B19' martensitic phase, with a monoclinic lattice,

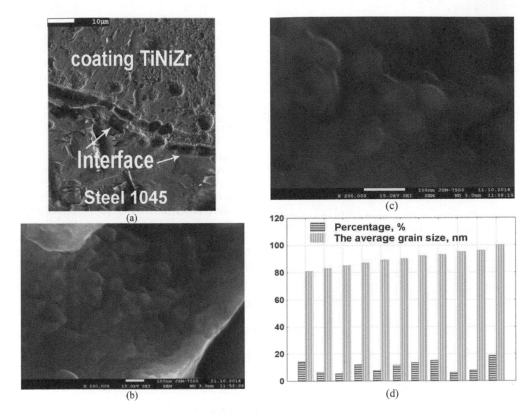

Figure 3.   Microstructure of $Ti_{33}Ni_{49}Zr_{18}$ coatings obtained with high-speed flame spraying: (a) ×1000; (b) ×100,000; (c) ×200,000; (d) the quantitative distribution of the grain size and the percentage of the TiNiZr coating.

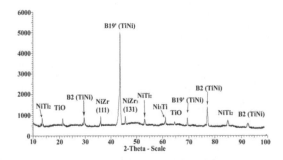

Figure 4.   X-ray analysis of the alloy $Ti_{33}Ni_{49}Zr_{18}$, after high-speed flame spraying with a local protection in argon.

austenitic phase with a B2 cubic lattice, and an intermetallic phase Ni_3Ti, Ti_2Ni, NiZr, and $NiZr_2$ with a cubic and hexagonal lattice, and there is a small amount of Titanium Oxide (TiO) of less than 2%.

When the Zr content is increased from 0% to 20% in the surface-modified layer with the shape memory effect, there is an increase in the TiNiZr

martensitic transformation temperature, which provides a high-temperature shape memory effect. On the basis of the analysis of experimental data with Statistica Program 10, the temperature A_s of the start of austenitic transformation in the surface-modified layer TiNiZr is described by the following equation:

$$A_s = 189.9842 - 29.085 \cdot C + 1.7836 \cdot M_s$$
$$- 0.2914 \cdot C^2 + 0.1116 \cdot C \cdot M_s$$
$$- 0.0037 \cdot M_s^2 \qquad (1)$$

where C is the percentage of Zr in the TiNiZr alloy, and M_s is the temperature of the martensitic transformation.

The test of the wear of surface layers $Ti_{33}Ni_{49}Zr_{18}$ was conducted under dry friction of the coated sample on the rotating hard disk on a 2070 CMT-1 test machine with a disk rotational speed of v = 0.5–2 m/s and with a pressure of P = 2–12 MPa, temperature registration nip. The evaluation of the wear rate was made on the basis of the experimental data using the application package Statistics v6.0 in SPSS (Fig. 6).

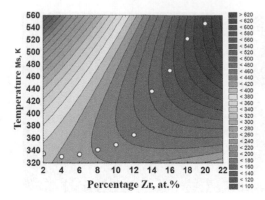

Figure 5. The effect of the content of Zr, on temperatures of martensitic transformations in surface-modified layers of shape memory effect TiNiZr.

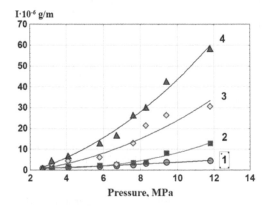

Figure 6. The dependence of the rate of wear on the pressure I drive F: drive sliding velocity 0.5 m/s^{-1}; 1 m/s^{-2}; 1.5 m/s^{-3}; 2 m/s^{-4}.

As a result of the tests, the wear resistance of the 1045 steel surface-modified by alloys with the shape memory effect $Ti_{33}Ni_{49}Zr_{18}$ is increased by 3.5–3.6 times.

4 CONCLUSIONS

The technology for the formation of the compound "Steel—a material with shape memory" by high-speed flame spraying of mechanically activated powder TiNiZr is worked out. The optimal processing parameters to ensure the formation of a nanostructured surface layer with a grain size of 80–100 nm with a Micro $H_\mu = 9.5 \div 12.7$ GPa and adhesive strength of 90–100 MPa are determined.

Experimental study of the effect of doping TiNi with the third component Zr showed that with increasing content of 20–22% Zr, an increase in the temperature of the beginning of the transformation of austenite takes place, which confirms the preservation of the shape memory properties of the surface-modified layer at high temperatures up to 450°C.

The simulation allows to predict the possibility of using surface modification of products by material SME TiNiZr in certain temperature conditions and to determine the appropriate coating composition that provides a positive effect.

Testing of steel 1045 with a surface-modified layer TiNiZr in conditions of dry friction, in which there is a significant increase in temperature, have confirmed the effect of increasing the wear resistance 3–3.5 times, due to the manifestation of SME under the test conditions and confirmed by the results of the structural and phase analysis.

ACKNOWLEDGMENT

This work was performed as a part of Russian Scientific Foundation (project № 15-19-00202), with the financial support of the Russian Scientific Foundation.

REFERENCES

Blednova, Zh.M. & Rusinov, P.O. 2014. Mechanical and Tribological Properties of the Composition "Steel—nanostructured Surface Layer of a Material with Shape Memory Effect Based TiNiCu", *Applied Mechanics and Materials*, 592–594: 1325–1330.

Blednova, Zh.M. Rusinov, P.O. & Stepanenko, M.A. 2014. Influence of Superficial Modification of Steels by Materials with Effect of Memory of the Form on Wear-fatigue Characteristics at Frictional-cyclic Loading, *Advanced Materials Research*, 915–916: 509–514.

Firstov, G.S. Van Humbeeck, J. & Koval, Y.N. 2004. High-temperature shape memory alloys some recent developments. *Materials Science and Engineering A*, 378: 2–10.

Kositsyn, S.V. Valiullin, A.I. Kataeva, N.V. & Kositsyna, I.I. 2006. Investigation of Microcrystalline NiAl-Based Alloys with High-Temperature Thermoelastic Martensitic Transformation: I. Resistometry of the Ni-Al and Ni-Al-X (X = Co, Si, or Cr) Alloys, *Physics of Metals and Metallography*. 102(4): 391–405.

Likhachev, V.A. 1997. Materials with shape memory effect. *NIIH SPSU*, 1: 424.

Rusinov, P.O. Blednova, Zh.M. & Chaevsky, M.I. 2015. Options for Forming of Nanostructured Surface Coatings, *Advanced Materials Research*, 1064: 154–159.

Rusinov, P.O. & Blednova, Zh.M. 2015. Technological Features of Obtaining of Nanostructured Coatings on TiNi Base by Magnetron Sputtering. *Advanced Materials Research*, 1064: 160–164.

Advanced Materials and Structural Engineering – Hu (Ed.)
© 2016 Taylor & Francis Group, London, ISBN 978-1-138-02786-2

Determination of a constitutive relation for damage of Al-Si-Cu alloy (ADC12)

Y.M. Hu
State Key Laboratory of Mechanical Transmission, Chongqing University, Chongqing, China
State Key Laboratory of Vehicle NVH and Safety Technology, Chongqing, China

H.R. Zheng, Z.F. Li & X.Q. Jin
State Key Laboratory of Mechanical Transmission, Chongqing University, Chongqing, China

Y.E. Zhou
State Key Laboratory of Vehicle NVH and Safety Technology, Chongqing, China

ABSTRACT: This paper presents an experimental study for obtaining the constitutive relation and damage model parameters of Al-Si-Cu alloy (ADC12). Several experiments were designed to produce different stress states at various strain rates. Digital Image Correlation (DIC) method was used to determine the strains with a high-speed camera and a high-speed testing machine which covered a wide range of strain rates 10^{-3}–10^3/s. The current Studies indicate that the effective plastic strains at fracture of aluminium die-cast alloy specimens will increase with increasing strain rates, and decrease with increasing stress triaxiality. This phenomenon is manifested in a constitutive relation and damage model used in the numerical simulations. A comparison of the results obtained from the Finite Element Method (FEM) simulations and experiments shows good agreement.

1 INTRODUCTION

The Al-Si-Cu die-cast aluminium alloys (ADC12) are recognized for their attractive mechanical properties, such as light weight, good surface finish and dimensional accuracy, excellent cast performance and wear ability. Therefore, they are widely used in aerospace and automobile industries as one of the most important structural materials (Stefanescu et al. 1988, Davis 1993). Particularly, the ADC12 alloys are desirable materials used for critical passive safety item (e.g., engine mounting brackets) on a vehicle. On the one hand, the ADC12 alloys have sufficient strength and stiffness during normal driving; on the other hand, they should be fractured promptly during the frontal vehicle crash, so the engine can sink instantaneously to provide sufficient space. It is, therefore, crucial to understanding the constitutive properties of the aluminium alloys and assess their fracture behavior.

Tensile properties and fracture behavior of ADC12 foams which exhibit brittle fracture have been investigated by image-based Finite Element (FE) analysis (Hangai et al. 2014). Mohr & Henn (2007) have studied the fracture behavior of cast aluminium alloys. In their model, the fracture strain as a function of stress triaxiality was established for the material. Fagerholt et al. (2010) proposed a user-defined material model to simulate fracture in cast aluminium alloys. In their model, the material behavior was described by the classical J2 flow theory while fracture was modeled by the Cock-croft–Latham criterion. In the complicated case of an impact load, fracture characteristics should be further investigated, because of the wide applications of ADC12 as safety critical automotive parts. The stress states and the strain rates of these parts generally differ from different positions and vary with time during a collision.

In this investigation, the Johnson-Cook constitutive relation for damage was determined experimentally, in conjunction with the explicit finite element code LS-DYNA. Experiments containing different stress states and different strain rates were conducted to obtain the parameters of the model. The constitutive relation was applied to the simulation at different strain rates. The results showed that the simulation data was consistent with the experimental measurements.

2 JOHNSON-COOK MODEL

2.1 *Johnson-Cook material model*

Based on visco-plastic mechanics, Johnson & Cook (1983) proposed a phenomenological model which defined the flow stress as a response to strain hardening, strain-rate effects, and thermal softening. Because of its simple identification, it is frequently used for impact problems. The Johnson-Cook relation reads:

$$\sigma_Y = (A + B\bar{\varepsilon}^{pn})(1 + C\ln\dot{\varepsilon}^*)(1 - T^{*m}) \qquad (1)$$

where, σ_Y is the flow stress, $\bar{\varepsilon}^p$ the effective plastic strain, and $\dot{\varepsilon}^* = \dot{\bar{\varepsilon}}^p / \dot{\varepsilon}_0$ the dimensionless plastic strain rate, with $\dot{\varepsilon}_0$ being a user-defined reference strain rate. Ideally, the magnitude of $\dot{\varepsilon}_0$ represents the highest strain rate for which no rate adjustment to the flow stress is needed (Schwer 2007). The homologous temperature T^* is defined as $T^* = (T - T_{room})/(T_{melt} - T_{room})$, where T_{melt} is the melting temperature and T_{room} is the room temperature. The adiabatic condition is assumed. Hence all the internal plastic work is converted into the temperature change, i.e., $\Delta T = \left(\int\sigma d\,\bar{\varepsilon}^p\right)/\rho C_p$ where ρ and C_p are mass density and specific heat, respectively.

It is noted above that there are five material constants: A, B, C, n and m; and two material characteristics: ρ, and C_p needed to define.

2.2 Johnson-Cook failure model

Based on cumulative-damage, Johnson and Cook further developed a fracture model on the basic material which defined the strain at fracture as a function of stress triaxiality, strain-rate, and homologous temperature. The damage fracture model reads:

$$\varepsilon_{failure} = (D_1 + D_2 e^{D_3\sigma^*})(1 + D_4\ln\dot{\varepsilon}^*)(1 + D_5 T^*) \qquad (2)$$

where, the stress triaxiality is defined as $\rho^* = P/\sigma_{eff}$, with P and σ_{eff} being mean stress and effective stress, respectively. Here, D_1, D_2, D_3, D_4, D_5 are additional five material constants.

The LS-DYNA material model (*MAT_015) provides another material constant EFMIN based fracture model. The value of EFMIN means the lower bound for calculated strain at fracture. The LS-DYNA implements the strain at fracture as:

$$\varepsilon^f = \max([D_1 + D_2 e^{D_3\sigma^*}][1 + D_4\ln\dot{\varepsilon}^*] \\ [1 + D_5 T^*], EFMIN) \qquad (3)$$

A fracture occurs when the damage parameter D reaches the value of 1, and D is given by the accumulated incremental effective plastic strains divided by the strain at fracture, i.e., $D = \sum\Delta\bar{\varepsilon}^p/\varepsilon^f$.

3 EXPERIMENTS

The raw materials for specimens were produced with Al-Si-Cu alloy using the die casting machine. The weight percentage of the alloy chemical composition is shown in Table 1.

Table 1. Mass fraction of chemical composition of aluminium alloy ADC12%.

Si	Cu	Mn	Mg	Fe
9.6~12.0	1.5~3.5	≤0.5	≤0.3	≤1.2
Ni	Zn	Pb	Sn	Al
≤0.5	≤1.0	≤0.1	≤0.1	The rest

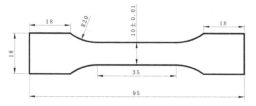

(a) Plate tension specimen (4mm thick)

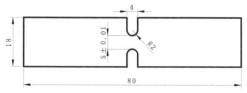

(b) 4mm notched tension specimen (2.6mm thick)

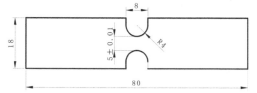

(c) 8mmnotched tension specimen (2.6mm thick)
(c) 8mmnotched tension specimen (2.6mm thick)

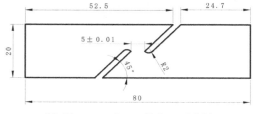

(d) Shear specimen (2.6mm thick)

(e) Compression specimen L = 8, 16, 20mm

Figure 1. Shapes and dimensions of the specimen for different stress states.

3.1 Different stress states experiments and results

The roughcast was machined into the following specimen by CNC wire-cut machine and the engine lathe. The geometry and dimensions of the specimen are shown in Figure 1.

An axial force is applied to the 15 mm length of the rectangular specimens held in the loading fixtures of the 50 KN microcomputer control electronic universal testing machine (Fig. 2). The loading speed of the tensile and compressive test machine was set as a constant 0.2 mm/min. Different shapes of the specimens were designed to determine the mechanical properties of Al-Si-Cu alloy under different stress states within the gage at the quasi-static test speed.

Before testing, the measured area of the sample surface was marked with a speckle pattern. During the test, a large number of images showing the micro deformation between the spots were traced by a CCD camera with a light source (Schreier et al. 2009). The images were post-processed based on the Digital Image Correlation (DIC) procedure to calculate the strain of the region of interest by using the VIC-2D software. Figure 3 illustrates the noncontact optical method for the measurement of deformation and the calculation of strain.

Conversion of the engineering stress-strain curve into the true stress-strain curve (Fig. 4) illustrates the result of plate tension. The Young's modulus, as interpreted by the slope of the initial linear portion, is 71 GPa. As a brittle material to a certain degree, e.g., with the ultimate tensile strength of 218 MPa and the elongation of 1.8%, the Al-Si-Cu die-cast alloy specimen exhibits no apparent necking after fracture; and there is only

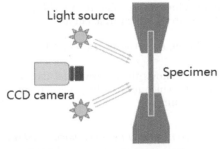

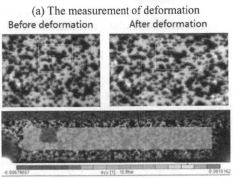

(a) The measurement of deformation

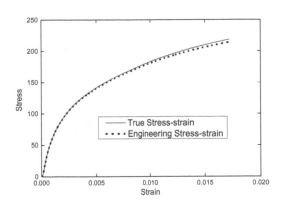

(b) The calculation of strain

Figure 3.   Noncontact optical method.

(a) Quasi-static testing machine   (b) Specimen holder

Figure 2.   Quasi-static test method.

Figure 4.   Stress-strain curve for plate tension.

negligible difference between the true strain and the engineering strain (Fig. 4).

A great deal of research (Johnson & Cook 1985, Jin et al. 2008) shows that the effective plastic strain at failure depends on the stress state. The stress triaxiality under different stress states is obtained by the finite element simulation of specimen tension. Table 2 provides the corresponding effective plastic strain at fracture. It indicates that the effective fracture plastic strain will decrease with the increase of the stress triaxiality. The parameters of the first item for Johnson-Cook failure model Equation (3) is fitted with the exponential curve:

$$D_1 = 0.00939, D_2 = 0.00001, D_3 = -29.4.$$

Figure 5 shows the experimental data and the fitted curve which reflects the general trend of the fracture strain.

3.2 *Different strain rates experiments and results*

The specimens were clamped on fixtures of the high-speed testing machine named Zwick HTM 5020 (Fig. 6) at room temperature. The geometry and dimensions of the specimen are shown in Figure 7. Similarly, the digital images were captured by a high-speed camera instead of CCD camera. The strain gages were plastered on both sides of the longer end of the test specimens which would be only used in the high strain ranges (Fig. 8). It is necessary to calibrate the gages before its application and to make curve fitting in calibration to obtain the voltage coefficient. This method can effectively eliminate oscillations in force signal under the impact load.

Table 2. Experimental data under different stress triaxiality.

Experimental type	Stress triaxiality σ^*	Effective fracture plastic strain ε_f
Compression	−0.33	0.217
Shear	0.02	0.0025
Plate tension	0.33	0.0153
8 mm notched	0.43	0.011
4 mm notched	0.51	0.0072

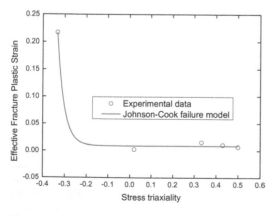

Figure 5. Experimental data and the fitted curves.

(a) Zwick HTM 5020　　　(b) Specimen holder

Figure 6. High-speed testing machine.

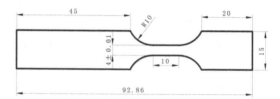

Figure 7. Different strain rates tension specimen (2 mm thick).

Figure 8. The way of sticking strain rages.

The speeds of the tension were set to 0.005 mm/s, 10 mm/s, 1 m/s, 2.5 m/s, 5 m/s and 8 m/s. Due to the deformation of the transition fillets and the brittleness characteristic of the cast aluminium alloy material, the specimen fractured when the strain rates did not reach the nominal value. We regard the slope of the strain-time history curve after yielding as the true strain rate (Table 3). Figure 9 illustrates the true stress-strain curves for the experiments at different strain rates.

It is seen that there is no obvious difference at the same true strain for the strain rate ranging from 0.0002/s to 650/s. It is hence concluded that

Table 3. Experimental results of tensile tests under different strain rates.

Tensile velocity	Nominal strain rate/s	True strain rate/s	Fracture stress MPa	Effective fracture plastic strain
0.005 mm/s	0.0005	0.0002	231	0.0176
10 mm/s	1	0.35	236	0.0143
1 m/s	100	60	232	0.0158
2.5 m/s	250	220	248	0.0207
5 m/s	500	400	285	0.0293
8 m/s	800	650	298	0.0363

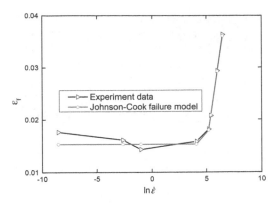

Figure 10. Experimental data and the damage model.

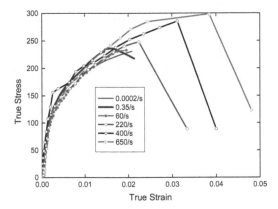

Figure 9. Stress-strain curve for different strain rates.

the strain rates do not have a significant effect on the amount of the true stress during the yielding stage. The similar phenomenon has been verified by Mukai et al. (1995) and Smerd et al. (2005). The parameters of the Johnson-Cook material model Equation (1) are obtained by using the lsqnonlin equation in MATLAB to fit all the curve data shown in Figure 9. Equation (1) is written as follows:

$$\sigma_e = 115.7 + 1644.6\bar{\varepsilon}^{p0.6232} \qquad (4)$$

The fracture stress and effective plastic strain remain stable at low-medium strain rates but increase obviously when the strain rate reaches a certain value. The rest parameters of Equation (3) are obtained by the nonlinear curve fit in the software Origin 9.0. As a result, $D_4 = 1.5$, $D_5 = 0$, $\dot{\varepsilon}_0 = 98$, EFMIN = 0.01537.

Figure 10 shows the effective plastic strain at fracture from experiments and the fitted curve. Table 3 and 4 show the fracture stress, effective plastic strain and the comparison between experiments

Table 4. Compare of fracture strain between experimental measure and equation calculation.

True strain rate/s	ε^f (experiments)	ε^f (Johnson-Cook)	$\Delta\varepsilon^f$
0.0002	0.0176	0.0154	0.0022
0.35	0.0143	0.0154	−0.0011
60	0.0158	0.0154	0.0004
220	0.0207	0.0208	−0.0001
400	0.0293	0.0292	0.0001
650	0.0363	0.0360	0.0003

Table 5. Johnson-Cook model constants of ADC12 cast aluminium alloy for LS-DYNA.

Johnson-Cook	Constants of models					
Constitutive relation	A/MPa	B/MPa	n	C	m	$\dot{\varepsilon}_0$
	115.7	1644.6	0.6232	0	0	98
Fracture model	D_1	D_2	D_3	D_4	D_5	EFMIN
	0.00939	0.00001	−29.4	1.5	0	0.01537

and Johnson-Cook failure model under different strain rates.

As seen from the Figure 10 and Table 4, good agreement is obtained between the measured and the damage model, which validates the present damage model.

Finally, the constitutive relation and damage model parameters of Al-Si-Cu alloy have been obtained without regard to the effect of temperature on the flow stress and fracture strain. Table 5 provides the material tables of Al-Si-Cu die-cast aluminium alloy for LS-DYNA.

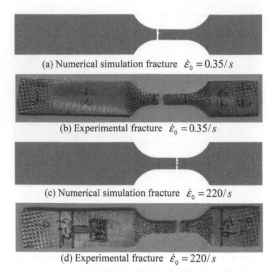

(a) Numerical simulation fracture $\dot{\varepsilon}_0 = 0.35/s$

(b) Experimental fracture $\dot{\varepsilon}_0 = 0.35/s$

(c) Numerical simulation fracture $\dot{\varepsilon}_0 = 220/s$

(d) Experimental fracture $\dot{\varepsilon}_0 = 220/s$

Figure 11. Comparison between the simulation and experimental fracture.

4 NUMERICAL SIMULATIONS AND EXPERIMENTAL VALIDATION

The uniaxial tensile tests were simulated in LS-DYNA using solid elements at different strain rates. Firstly, the mesh of the region of interest was created with characteristic element lengths of 0.2 mm. This gave 27489 hexahedral elements for the Al-Si-Cu die-cast aluminium alloy specimen. Secondly, the longer end of the finite element model was constrained to the same length in the experiments. Different speeds of tension were applied to the other end of the specimen, which turned out to be the same true strain rate as the experiments. The material constants shown in Table 5 were used in the calculation. Some simulations vs. the experimental observations of the crack patterns are shown in Figure 11. Figure 12 illustrates the comparison between experiments and simulation under different strain rates. Simulation results are consistent with the experimental data.

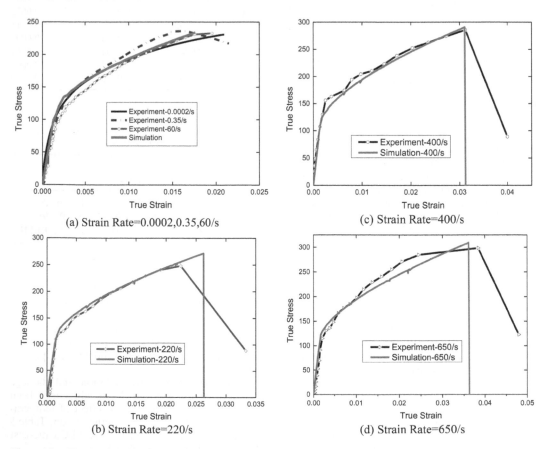

Figure 12. Simulation curves under different strain rates.

5 CONCLUSIONS

In this paper, a constitutive relation for damage of Al-Si-Cu alloy (ADC12) was studied based on experiments and computer simulations.

Experimental data showed that the strain at fracture decreases with the increase of the stress triaxiality. The strain rates do not have a significant effect on the magnitude of the true stress during the yielding stage. The fracture stress and effective plastic strain remain stable at low-medium strain rates but increase obviously when the strain rate reaches a certain value.

Nonlinear finite element simulations were carried out using solid elements. The simulation results were consistent with the experiments. Hence, the constitutive relation and damage model can be used to simulate the damage process of aluminium alloys components.

ACKNOWLEDGEMENT

The authors gratefully acknowledge the financial support from the State Key Laboratory of Vehicle NVH and Safety Technology. The supports from the Fundamental Research Funds of the Central Universities (No. CDJZR14285501), and the National Natural Science Foundation of China (Grant No. 51475055) are also acknowledged.

REFERENCES

Davis, J.R. 1993. *Aluminum and aluminum alloys*, ASM international.

Fagerholt, E. Drum, C. Brvik, T. Laukli, H. & HOPPERSTAD, O. 2010. Experimental and numerical investigation of fracture in a cast aluminium alloy. *International Journal of Solids and Structures*, 47: 3352–3365.

Hangai, Y. Kamada, H. Utsunomiya, T. Kitahara, S. Kuwazuru, O. & Yoshikawa, N. 2014. Tensile Properties and Fracture Behavior of Aluminum Alloy Foam Fabricated from Die Castings without Using Blowing Agent by Friction Stir Processing Route. *Materials*, 7: 2382–2394.

Jin, X. Chaiyat, S. Keer, L.M. & Kiattikomol, K. 2008. Refined Dugdale plastic zones of an external circular crack. *Journal of the Mechanics and Physics of Solids*, 56:1127–1146.

Johnson, G.R. & Cook, W.H. 1983. *A constitutive model and data for metals subjected to large strains, high strain rates and high temperatures*. Proceedings of the 7th International Symposium on Ballistics, The Netherlands, 541–547.

Johnson, G.R. & Cook, W.H. 1985. Fracture characteristics of three metals subjected to various strains, strain rates, temperatures and pressures. *Engineering fracture mechanics*, 21: 31–48.

Mohr, D. & Henn, S. 2007. Calibration of stress-triaxiality dependent crack formation criteria: a new hybrid experimental–numerical method. *Experimental Mechanics*, 47: 805–820.

Mukai, T. Ishikawa, K. & Higashi, K. 1995. Influence of strain rate on the mechanical properties in fine-grained aluminum alloys. *Materials Science and Engineering: A*, 204: 12–18.

Schreier, H. Orteu, J.J. & Sutton, M.A. 2009. *Image correlation for shape, motion and deformation measurements: basic concepts, theory and applications*. Springer-Verlag GmbH.

Schwer, L. 2007. *Optional Strain-rate forms for the Johnson Cook Constitutive Model and the Role of the parameter Epsilon_0. LS-DYNA Anwenderforum*, Frankenthal.

Smerd, R. Winkler, S. Salisbury, C. Worswick, M. Lloyd, D. & Finn, M. 2005. High strain rate tensile testing of automotive aluminum alloy sheet. *International Journal of Impact Engineering*, 32: 541–560.

Stefanescu, D. Davis, J. & Destefani, J. 1988. *Metals Handbook, Vol. 15—Casting*. ASM International, 937.

Advanced Materials and Structural Engineering – Hu (Ed.)
© 2016 Taylor & Francis Group, London, ISBN 978-1-138-02786-2

Study on phase transition of metallic materials by Specific Volume Difference Method

K.J. Liu, W.Y. Fu & S.L. Ning
Materials Science and Engineering, Shanghai Institute of Technology, Shanghai, China

ABSTRACT: In this paper, we propose a new method for measuring the phase transition of a metallic alloy by measuring the difference of the specific volume (or density) of metallic materials at different heat treatment conditions. This new measuring method gives a quantitative result with high accuracy, and small deviation. As for some aluminum alloys, we report the experimental measurements of this method, which are sensitive to the different age treatment temperatures. We checked this method with other ones, i.e., the volume change was verified by thermal expansion measurement; it was also discussed by Vegard's law. The results indicate that, the specific volume difference method is effective and practical for measuring the phase transition of metallic materials at different heat treatment conditions.

1 INTRODUCTION

It is well known that the Heat Treatment (HT) can change the mechanical properties of some metals and alloys (Xia 2008, Wang 2012). The nature of the HT is its Phase Transition (PT) (Lubarda 2003, Xu & Zhao 2004, Stolyarov et al. 1997). However, most of the PT also accompanied abrupt changes of other characteristics. The volume change is one of them. Therefore, one is able to represent the PT, and also represent the state of the HT. The change of the order of magnitude is in the range of $10^{-4} \sim 10^{-2}$, which can be measured by the electronic balance with the Archimedes Principle. While the order of magnitude is less than $10^{-4} \sim 10^{-2}$, the influence of materials performances would be poor; on the other hand, if the order of relative change is higher than 10^{-2}, the materials were unable to be used due to the crack and pulverization.

Thermal expansion measurement and X-Ray Diffraction (XRD) analysis are the common ways to test the phase transition (Wu et al. 2013, Huang et al. 2007, Zhou & Wu 1998, Blanc et al. 1998), however, in situ test is time-consuming for thermal expansion test, and the cost of the XRD analysis is expensive.

The difference of the Specific Volume (SV) ($v = 1/\rho$) of the materials under the different HT, Δv, was studied in this work. The PT and HT state of materials are able to be obtained by comparing the value of Δv. It is a new method for testing the phase transition which we named the Specific Volume Difference Method (SVDM).

The measure of the Specific Volume Difference (SVD) has some characters, firstly, the sample preparation is easy without specific preparation; secondly, the instrument is simple, a balance with sensitivity of 0.001 gram or higher is qualified; thirdly, the accuracy is high, the relative accuracy is 0.01%, if the sample has the mass 10 gram. Through the literature review, this new method is put forward by author (Liu), the new method has been applied patents, and there is no relevant methods were reported by others.

Past research has shown that during the process of heat treatment, there are significantly differences between some aluminum alloys and steel materials (i.e. carbon steel). The volume of steel is increased during the quenching process, while the aluminum-copper alloy is decreased. And the difference of the volume changes also exists between the tempering of steels and aging process of aluminum alloys. Therefore, the size of the volume changes of aluminum alloys were necessary to studied quantitatively after heat treatment.

In this work, the measurements of the new method for an aluminum alloy with the brand of 211Z was conducted, the practicality and reliability were verified by thermal expansion experiment and Vegard's law etc.. The new method were expounded in this paper, the alloy is used as an example of the testing method. The verification results indicate that this new method can be used to study the quantitative analysis research of materials phase transition. The results indicate that, during the process of heat treatment, the volume change of metallic materials, which caused by phase transition can be accurate quantitatively tested by the new testing method.

2 EXPERIMENTS

A type of Al-Cu alloy with a brand of 211Z was studied in this work. The samples of the alloy were be cut to the weight about 20 g, and then aged 5 hours at room temperature (25 °C), 100, 165, 300, 400, 500 °C after solid solution treated 8 h at 540 °C. Using Archimedes principle of water displacement, the specific volume of samples was measured by electronic balance of the sensitivity of 1 mg.

The tensile testing for two aged samples at 165 °C for 5 h, 12 h were done to determine 165 °C as aging treatment temperature is reasonable.

3 RESULTS AND DISCUSSIONS

The specific volume of the alloy under different aging temperature is indicated in Figure 1. It can be seen that, at the temperature range 25~300 °C, the SV is increase gradually with the temperature increasing; it can be considered that the decomposition of supersaturated solid solution and precipitates of second phase particles, which increase the volume of alloy during the aging treatment process (Lubarda 2003, John 1984). At the aging temperature of 165 °C, which was determined in previous studies, the slope of the line has an obvious change. Generally, the aging heat treatment of aluminum alloy is as far as possible to choose the low temperature to ensure that the precipitated phase dispersity (Gu 1985). On the other hand, it also needs to enhance the aging temperature to increase the amount of phase dispersity. At the temperature of 100 °C or bellow, precipitates of the second phase are so fewer that have little influence on improving the strength of materials. At the temperature of 300 °C, the amount of precipitates are more than before, however, those precipitates are almost gross secondary phase, and the strength of materials (Youn & Kang 2006, Huang et al. 2012, Zeng et al. 2008) will be decreased. Considering the two requirements, choose the aging temperature of 165 °C is reasonable, which was also proved by measurement of the phase transition by using SVDM.

The tensile strength of material under the aging process (165 °C 5 h) was reaching up to 521.7 MPa, tensile strength of material under the further aging treatment process (165 °C 12 h) was reaching up to 536.1 MPa, they were about 15.7% and 19.1% higher than this new kind of material that reported in patent (Lu et al. 2011). The tensile strength of a material can be improved steadily in our study, which made the material own a better mechanical performance.

In addition, the SV of the alloy decreases with the aging temperature increasing over the range of temperature from 300 to 500 °C, this is mainly due to that the solid solubility increases greatly, which causes the precipitates of second phase decreased, rather than increased (Hou et al. 2010, Feng et al. 2011, Liu et al. 2007, Gui 2012).

Figure 1 also shows that the specific volumetric change rate reaches the maximum value $\Delta v / v_0 = 0.477\%$ (v_0 is the SV of the sample aged at room temperature) at the aging temperature of 300 °C. Thus the dimensional change rate is $\Delta L / L_0 = 1/3(\Delta v / v_0) = 0.159\%$. It can be seen that the change of volume or length of samples after HT is measurable.

In order to validate the results measured by SVDM, they are compared with the results of previous thermal expansion experiments (Fig. 2). The value of thermal expansion tested by thermal dilatometer is superposed by linear expansion value and the dimensional change caused by the phase transition. At the temperature of 290 °C, the

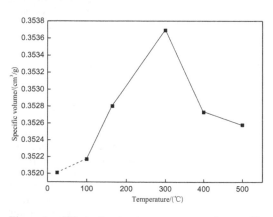

Figure 1. Effect of aging temperature on the specific volume of 211Z.

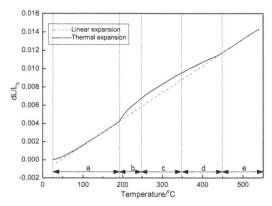

Figure 2. Thermal expansion curve of 211Z.

expansion caused by the phase transition reach the maximum value, $\Delta L' / L_0' = 0.109\%$. Compared with the maximum value $\Delta L / L_0 = 0.159\%$ measured by SVDM, they are the same order of magnitude though there exist some difference.

The main reason for the discrepancy is that the second phase can't exhale completely with non-isothermal, the temperature rise at the rate of 5 °C/min. However, the samples tested by SVDM were well precipitated after isothermal of 5 hours, which lead to a bigger precipitates. Therefore, the results of SVDM are confirmed by thermal expansion, the results also proved to be accurate.

On the other hand, the change in the size of solid solution alloy can be calculated by two theoretical calculations. One is the Vegard's law, which provides the estimate of the size change of Al-Cu alloy after solution treatment; the other provides a quantitative calculation of binary aluminum alloys solid solution (Wu et al. 2013). Toward the weight ratio of copper is 5.0% in Al-Cu alloy, the atomic ratio (x) of copper is 2.3%, the quantitative calculation of the Al-Cu alloy can be described as follows.

1. The quantitative calculation of Vegard's law. The law assumed that the variation of the lattice constant of the alloy is linear as $a_x = a_{Al} + x(a_{Cu} - a_{Al})$. The lattice constant of aluminum a_{Al} at room temperature is 4.049 Å, and of copper a_{Cu} is 3.615 Å, the crystal structure of those two materials are both FCC, which can form substitution solid solution. With the copper added in, the volume of aluminum decreases after the solid solution (Stolyarov et al. 1997), which means that the lattice constant of the aluminum alloy after solid solution is 4.039 Å. The lattice constant will return to a_{Al} after sufficient aging treatment, the size caused by phase transition will increase about 0.25% that the specific volume will increase about 0.75%, the calculative results are the same order of magnitude with the above conclusion 0.477% (in Fig. 1). The results (in Fig. 1) can be supported by the calculation of Vegard's law. The Vegard's law is an approximate calculation. Therefore, the calculations are quantitative different from the measurement of SVDM.
2. Reference (Wu et al. 2013) gives another calculation method. With 1 at.% of Cu added in, the lattice constant of aluminum alloys decreases 0.0051 Å, therefore, with the same amount of copper was precipitated, the lattice constant increases 0.0051 Å (Wu et al. 2013). The lattice constant of solid solution is 4.038 Å, however, after aging and precipitation of second phase completely, it will increase 0.27%, and namely the specific volume will increase 0.81%. The

result is the same order of magnitude with the experimental result of 0.477% above and similar with the result of Vegard's law.

The theoretical calculation results of those two methods are the same order of magnitude of the results of SVDM, which fully prove the new method of SVDM. The new method is an accurate and convenient way to measure the phase transition of materials. Through the experiment of thermal expansion and theoretical analysis, the results declare that SVDM is an effectively and accurately way to measure and analysis the phase transition of aluminum under the heat treatment process.

4 CONCLUSIONS

The phase transition of the 211Z aluminum alloy after heat treatment is studied by SVDM. Effect of aging temperature on the phase transition and the change of specific volume are main discussed in this paper, the results indicate that:

1. The method is accurate, simple, reliable and applicative and is a new method for measuring the size change of PT after HT quantificationally.
2. The SVDM is an effective and accurate method to measure and analysis the phase transition of aluminum alloys under the HT process, which is also a supplement to the traditional testing such as thermal expansion, etc.
3. As an application of SVDM, the new method proved that choose 165 °C as the aging temperature is reasonable. Using this aging process, the tensile strength of the 211Z aluminum alloy is about 15.7% higher than before.

REFERENCES

Blanc, C. Roques, Y. & Mankowski, G. 1998. Application of phase shifting interferometric microscopy to studies of the behavior of coarse intermetallic particles in 6056 aluminum alloy. *Corrosion science*, 40(6): 1019–1035.

Feng, Z.Q. Yang, Y.Q. & Huang, B. et al. 2011. Variant selection and the strengthening effect of S precipitates at dislocations in Al-Cu-Mg alloy. *Acta Materialia*, 59(6): 2412–2422.

Gu, J.C. 1985. Aging process of aluminum alloy. *Light alloy fabrication technology*, 3: 25–28.

Gui, Q.W. 2012. *Effect of heat treatment process on the properties of 2024 aluminum alloy precipitated phase.* Hunan university.

Hou, Y.H. Gu, Y.X. & Liu, Z.Y. et al. 2010. Modeling of the whole process of ageing precipitation and strengthening in Al-Cu-Mg-Ag alloys with high Cu-to-Mg mass ratio. *Transactions of Nonferrous Metals Society of China*, 20(5): 863–869.

Huang, J.W. Yin, Z.M. & Nie, B. et al. 2007. Investigation of phases and thermal expansivity of 7 A52 alloy in in-situ heating. *Ordnance material science and engineering*, 30(4): 9–12.

Huang, X.F. Jiang, M. Zhu & M.F. 2012. Analysis of microstructure of ZL107 alloy by thermodynamics calculation. *Special Casting & Nonferrous Alloys*, 32(2): 191–194.

John, E.H. 1984. A*luminum: properties and physical metallurgy*. Ohio: ASM International.

Liu, Z.Y. Li, Yun. T. & Liu, Y.B. et al. 2007. New research of precipitated phase of Al-Cu-Mg-Ag alloy. *Chinese journal of nonferrous metals*, 17(12): 1905–1915.

Lu, J.D. Zhang, D. & Che, Y. et al. *A kind of high strength casting aluminum alloy material*. China: 200810302668.6, 2011-01-12.

Lubarda, V.A. 2003. On the effective lattice constant of binary alloys. *Mechanics of Materials*, 35: 53–68.

Stolyarov, V.V. Latysh, V.V. & Shundalov, V.A. et al. 1997. Influence of severe plastic deformation on aging effect of Al-Zn-Mg-Cu-Zr alloy. *Materials Science and Engineering: A*, 234: 339–342.

Wang, X.L. 2012. *Heat treatment of metallic materials*. Beijing: Mechanical Industry Press.

Wu, Y.F. Liu, K.J. & Zhang, Z.K. et al. 2013. Research on phase transition of an Al-Cu alloy by thermal analysis. *Shanghai metals*, 35(3): 27–30.

Xia, L.F. 2008. *Heat treatment of metal*. Harbin: Harbin Institute of Technology Press.

Xu, Z. & Zhao, L.C. 2004. *Solid state phase transition principle of metal*. Beijing: Science Press.

Youn, S.W. & Kang, C.G. 2006. Characterization of age-hardening behavior of eutectic region in squeeze-cast A356-T5 alloy using nanoindenter and atomic force microscope. *Materials Science and Engineering: A*, 425(1): 28–35.

Zeng, Z.L. Ning, K.Q. & Peng, B.S. 2008. Second phase strengthening and its mechanism of high strength aluminum alloy. *Metallurgical collections*, 176(4): 5–7.

Zhou, Y. & Wu, G.H. 1998. *Materials analysis technology-Material X-ray diffraction and electron microscopic analysis*. Harbin: Institute of Technology Press.

Advanced Materials and Structural Engineering – Hu (Ed.)
© 2016 Taylor & Francis Group, London, ISBN 978-1-138-02786-2

A review on research and application development of Super Absorbent Polymer in cement-based materials

W.Q. Bai, J. Lv, Q. Du & H.H. Wu
School of Civil Engineering, Chang'an University, Xi'an, Shaanxi, China

ABSTRACT: Cement-based material has good mechanical properties and durable properties, which are widely used in construction project. In recent years, modern cement-based materials are gradually adopted in construction projects. However, due to the complicated composition in modern cement-based materials, volume deformation and premature cracking could be easily happened. Super Absorbent Polymer (SAP) is made from acrylic acid and various monomer copolymerization, its carboxyl (-COOH) and hydroxyl (-OH) mixed with water through its hydrogen bond to carry out hydration to improve water retention performance. This paper introduces the influence of SAP used in cement-based material in engineering construction, including the workability, mechanical property, durability and applications of SAP-Cement-based material. The mechanical properties containing compression and the bending were discussed in detail, the durability properties containing freezing and thawing resistant, chloride ion penetration resistant, carbonation resistant, shrinkage, creep, thermal expansion and high temperature resistant were also discussed. Finally, the construction applications and further research were summarized.

1 INTRODUCTION

Cement-based material utilizes cement as main gelled material in a man-made construction material.

It has good durable properties and mechanical property. Thus, it is widely used in construction projects. In recent years, with the deepening of research, super high-performance concrete and modern cement-based materials are gradually adopted in the construction industry. The complicated mixture composition in modern cement-based materials can easily cause deformation during the hydration process and lead to premature cracking affecting the durability. Many research indicate that the fineness of cement clinker mineral composition (Barcelo et al. 2005), mineral admixture (Burrows et al. 2004), curing temperature (Darquennes et al. 2011, Craeye et al. 2010, Gleize et al. 2007), humidity (Mounanga et al. 2006) and admixture (Hansen 2011, Meddah et al. 2011), can affect the volume change of cement-based material. Controlling curing temperature and humidity can ensure the hydration process continuing after cement-based material hardening, GB/T50081-2001 specified concrete has to be curried in 20±2°C, in relative humidity of 95% or more (GB/T50081-2002 2003). It is difficult to ensure cement-based material cured in above temperature and relative humidity in engineering practice, Philleo proposed the concept of inner curing

in 1991 (Skalny 1991), the media for inner curing mainly consists of pre-absorbent ultra-lightweight aggregate and super absorbent polymer, partly or totally replace coarse aggregate to pre-absorbent ultra-lightweight aggregate, the result shown that the concrete mix's self-drying and autogenous shrink is significantly reduced (Vaysburd 1996, Weber & Reinhardt 1997). However, pre-absorbent ultra-lightweight aggregate only can be used in the concrete mixture. This technique cannot be applied in cement-based materials such as mortar and fiber-reinforced cement-based composite. Jensen and Hansen (2001) proposed to add super absorbent polymer into cement-based materials to improve the shrinkage property of cement-based materials, this leads to a series of study and development in applying Super Absorbent Polymer (SAP) in cement-based materials.

2 PROPERTIES OF SUPER ABSORBENT POLYMER

Super Absorbent Polymer (SAP) is made from acrylic acid and various monomer copolymerization. Its carboxyl (-COOH) and hydroxyl (-OH) mixed with water to form hydrogen bonds to carry out hydration. The hydrogen bonds rapidly absorb dozens of and even thousands times of water content larger of its own weight. The gel mixture has a three-dimensional cross-linked network structure,

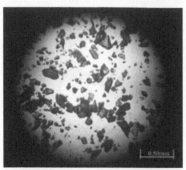

a. Before absorbing water

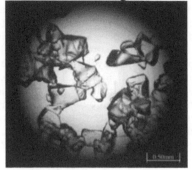

b. After absorbing water

Figure 1. Particle morphology before and after SAP absorbing water.

which can contain free water molecules inside the polymer network with a swelling effect. Thus, SAP has good water retention performance even under a pressure condition (Jensen & Hansen 2002, Liu & Meng 2013). The particle morphology before and after absorbing water is shown in Figure 1.

3 THE PROPERTY OF SUPER ABSORBENT POLYMER CEMENT-BASED MATERIALS

3.1 *The property of mixture*

The mixture property of cement-based materials can affect its construction property directly. Introducing SAP into a cement-based mixture can enhance the construction property of cement-based materials. Kong has described that adding SAP into cement mortar can reduce the mobility of mortar mixture (Kong & Li 2009). However, it is necessary to add additional water into cement mortar while using SAP, the amount of additional water depend on the category of SAP, when 0.3% of SAP was added into cement, the water–cement ratio increased by 0.02~0.06 (Christof & Viktor 2012). Although adding SAP can greatly reduce the workability of ultra-high performance concrete, but it can be resolved by selecting suitable additional water–cement ratio. Huang & Wang have supported that although adding SAP could reduce early workability of the ultra-high performance concrete, but its workability will regain after mixing for half an hour and the work-ability will regain in stages within a short time (Huang & Wang 2012). It was also experimented that after adding SAP and additional water into concrete mixture, the concrete slump increased by 20~30 mm, the work performance and water retention capability of concrete mixture has significantly improved (Ma et al. 2009). However, it was commented that, in order to ensure the similarity in the property of SAP-concrete mixture and concrete mixture, the increase dosage of SAP will lead to reduce the dosage of super plasticizer and constantly reduces the density of concrete mixture (Bart et al. 2011).

3.2 *Compression performance*

Argumentative comments have been reported on SAP influence of the compressive strength of cement-based material in the 28day. The category of SAP, practical size, dosage and water absorbed can affect the compressive strength of cement-based materials. Gao has compared the compressive strength of normal aluminate cement against aluminate cement with adding 0.2% and 0.6% of SAP without adding additional water. The compressive strength of aluminate cement with SAP increased from 36.1MPa to 40.5 MPa and 44.4 MPa (Gao et al. 1997). Another experiment shows that the compressive strength of cement mortar and concrete has improved after adding SAP (Ali et al. 2013, Agnieszka et al. 2013, Soliman & Nehdi 2013). Chen has also commented that in the ambient curing environment, the compressive strength of concrete significantly improved after adding SAP (Chen et al. 2007). However, it was noticed that the compressive strength will reduced first and then increase with the increase of age, and with the increase in water–cement ratio, the compressive strength will reduce.

In contrast, it has also reported that in both early age and ultimate strength, the compressive strength of mortar and concrete has significantly reduced after adding SAP (Ma et al. 2009, Pierard et al. 2006, Mechtcherine et al. 2009). Different curing environments have minor effect on the compressive strength of SAP added mortar, it was noted that with an increased amount of SAP dosage, the compressive strength gradually reduced (Lam & Hooton 2005, Liu & Meng 2013). Yan has reported that after adding SAP in ECC, with the increase of SAP dosage, the compressive strength of ECC constantly reduce; however, with the increase amount of coal ash in the mixture, the range of compressive strength reduction further increases (Yan et al. 2012).

3.3 *Flexural tensile property*

SAP can significantly improve the flexural tensile property of cement-based materials. The flexural strength of cement, which added 0.5% and 0.6% of SAP, has higher strength than the cement without adding SAP (Kong & Li 2009). However, Yan has commented that after adding SAP in ECC, the dosage of SAP increases; the flexural tensile strength of cement reduces but the bending capacity improves (Yan et al. 2012).

3.4 *Durability*

SAP shows a good improvement in the durability of cement-based materials. The relative dynamic elastic modulus of concrete containing SAP has greatly improved after freeze-thaw cycles, the chloride ion penetration resistant, carbonation resistant, shrinkage, creep, thermal expansion and high temperature resistant have also been improved. Compared with cement-based material without SAP, the electric flux of chloride ion diffusion reduced by 12.7% and carbonation depth reduced by 24.6% in cement-based materials containing SAP.

3.5 *Elasticity modulus*

With the increase of SAP dosage and inner curing water content, the compression elasticity modulus of concrete decreases. The growth rate of elasticity modulus of concrete is very fast, and the stiffness of concrete at 3 days age has reached 85% of concrete stiffness at 28 days age (Bart et al. 2011). The compressive elasticity modulus of mortar is decreased with the dosage of SAP increasing (Beushausen & Gillmer 2014).

3.6 *Shrinkage property*

The shrinkage properties of cement-based materials include plastic shrinkage, chemical shrinkage (autogenous shrinkage), drying shrinkage, and carbonated shrinkage (Kong & Li 2009). The chemical shrinkage and drying shrinkage problems of cement-based materials could be effectively eased by adding SAP into it.

The particle size of SAP can affect chemical shrinkage of concrete, compared with larger SAP particles, small particles absorb lesser amount of water in a faster time provided, but larger SAP particles have a better effect in reducing concrete autogenous shrinkage (Lura et al. 2006, Esteves

2010). In a cement-based material, SAP can shorten the water release duration and reduce the amount of water lease thus, leads to reduce the effect result from autogenous shrinkage.

The water retention ability of SAP is associated with the PH value of cement-based material; the water absorption capacity of SAP in distilled water is greater than potassium hydroxide solution (PH = 12.2), sulfate solution (PH = 12.2), and calcium salt solution (PH = 12.2) (Christ of & Viktor 2012). Compared with the concrete system mixed with coal ash, slag and mineral admixture, pure cement slurry system concrete alkalinity is higher. Where else, slag and mineral admixture concrete system has a better resistant to chemical shrinkage and drying shrinkage than pure cement slurry system concrete.

After adding pre-absorbent SAP in mortar, the autogenous shrinkage deformation significantly reduced. For different water–cement ratios, autogenous shrinkage can be reduced by 70%~90% (Ma et al. 2009). It shows an insignificant relationship to the total volume shrinkage when the water–cement ratio is between 0.5 and 0.7, even adding super plasticizer will greatly increase the volume shrinkage of mortar (Kong & Li 2009). SAP can extend the time of crack formation in mortar, with the increase in dosage of SAP, it significantly extends the cracking time of mortar (Zhu et al. 2013). Add SAP in ECC can significantly reduce the drying shrinkage of ECC, and drying shrinkage value decreases with the SAP dosage increases (Yan et al. 2012). Adding SAP which contains 0.2%~0.3% of gel material can significantly improve the rate of early expansion and limited expansion and decrease the difference between expansion and shrinkage deformation (Qin et al. 2011). SAP can significantly decrease autogenous shrinkage of ultra-high performance concrete, the reduction in shrinkage rate of 28 days can reach 32%~46%. It is also reported that the particle size of SAP have no significant or regular effect on the autogenous shrinkage value, only when the SAP dosage reaches a certain amount, the mix can achieve a better effect to reduce autogenous shrinkage, but it is known to be ineffective when SAP dosage is less than 0.4% (Huang & Wang 2012).

3.7 *Creep*

The creep of concrete increases along with the dosage of SAP and content of inner curing water, the creep of low water–cement ratio concrete is lower than high water–cement ratio (Bart et al. 2011).

3.8 *Hydration heat*

The particle size, dosage of SAP and additional water content have an effect on heat that produced during the hydration of cement-based materials.

The heat produces through hydration in a cement paste constantly increases with SAP particle size increasing. The hydration heat of cement paste is basically similar between the cement paste with SAP and without SAP before 2 curing days. The difference gradually expands with the growth of the age, the hydration heat of cement paste mixed with SAP is significantly higher.

It is shown that higher dosage of SAP will also increase the hydration heat of the cement paste (Zhu et al. 2013). With the increase of SAP dosage and inner curing water content, rising rate of internal temperature of concrete and internal temperature value both constantly increase.

3.9 Thermal expansion

Due to early hydration and self-drying process, the thermal expansion coefficient of low water–cement ratio cement-based materials will increase. The thermal expansion coefficient increment on the first day of hydration can reach above 70%. Adding SAP can effectively increase the humidity of a cement-based material and decrease the thermal expansion coefficient increment. The thermal expansion coefficient increment for the first day of hydration could be reduced about 60%~70% (Mateusz & Pietro 2013).

3.10 Other properties

Adding SAP can effectively reduce peel phenomenon of super performance concrete under high temperature.

4 APPLICATIONS OF SUPER ABSORBENT POLYMER IN CEMENT-BASED MATERIALS

4.1 The World Cup in Germany exhibition hall

To host the 2006 World Cup in Germany, Mechtcherine et al. (2006) have designed a shell structure exhibition hall with a slender column with no traditional rebars for the host city-Kaiserslautern. As reported by Mechtcherine et al. (2006), it was designed as a filigree, thin-walled structure with very slender columns (minimum wall thickness of 20 mm) and no conventional reinforcement. In order to meet the rigorous design requirements (including reduced autogenous shrinkage, high durability, enhanced ductility, self-compaction, and high-quality surface), self-compacting fiber-reinforced high-performance concrete with internal curing was developed by Dudziak and Mechtcherine. SAP made of covalently cross-linked acrylamide/acrylic acid copolymers was used for internal curing, and a polycarboxylate super plasticizer was used to

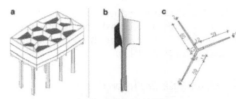

a. Geometric model figure

b. Exhibition hall actual effect figure

Figure 2. Exhibition hall model and actual effect figure.

ensure adequate self-compaction of the concrete (as shown in Fig. 2).

4.2 Lyngby, Denmark sprayed concrete wallboard

In Lyngby, Denmark, a project adopted a spray concrete to prepare wallboard that needs to overcome a series of technical issues. The test results of Jensen et al. (2008) indicate that SAP can be used as rheology modifiers of spray concrete. The concrete had an initial w/c of 0.4 and contained 0.4% SAP with a water absorption near 15 g of water per gram of dry SAP. It was observed that the uptake of water by SAP created a change in viscosity during placing and allowed the build-up of thick layers without the use of a set-accelerating admixture. In this case, SAP was added to shotcrete as a rheology modifier; however, other benefits may be found, such as internal water curing and mitigation of autogenous shrinkage.

5 CONCLUSIONS

SAP is a special function material with super high water absorption ability and water retention ability, it is widely used in sanitation, agriculture, forestry, gardening, soil and water conservation, medical, cosmetics etc.. However, there is very few existing research which investigated the properties of SAP-cement-based material, still requiring further studies and research on:

1. To establish a SAP index system that suited for the cement-based material and to determine the correlation between particle size, dosage, additional water of SAP and the properties of cement-based material.

2. The effects of SAP on mechanical properties and mechanism of cement-based material needs to be further studied.
3. The ability of SAP on releasing water inside cement-based material in different alkalinity environment needs to be systematic studied.
4. The durability of cement-based material with SAP under different engineering environmental conditions needs to be further researched.

ACKNOWLEDGMENT

The research work was jointly funded by the Central University Fund of Ministry of Education of China for a High-Tech Research Program (2014G228 0013), the Fundamental Research Funds for the Central Universities (CHD2012TD012), and Key International Cooperative Program of Shaanxi Province of China (2013KW13-01).

REFERENCES

Agnieszka, J. Klemm, K. & Sikora, S. 2013. The effect of Superabsorbent Polymers (SAP) on microstructure and mechanical properties of fly ash cementitious mortars, *Construction and Building Materials*, 49: 134–143.

Ali, P. Seyed, M.F. & Payam, H. et al. 2013. Interactions between superabsorbent polymers and cement-based composites incorporating colloidal silica nanoparticles, *Cement & Concrete Composites*, 37: 196–204.

Barcelo, L. Moranville, M. & Clavaud, B. 2005. Autogenous shrinkage of concrete: A balance between autogenous swelling and self-desiccation, *Cement and Concrete Research*, 35(1): 177–183.

Bart, C. Matthew, G. & Geert, De S. 2011. Super absorbing polymers as an internal curing agent for mitigation of early-age cracking of high-performance concrete bridge decks. *Construction and Building Materials*, 25: 1–13.

Bentz, D.P. Geiker, M. & Jensen, O.M. 2002. *On the mitigation of early age cracking*. In: Persson B, Fagerlund G, editors. Proceedings 3rd international seminar on self-desiccation and its importance in concrete technology. Sweden, 195–204.

Beushausen, H. & Gillmer, M. 2014. The use of superabsorbent polymers to reduce cracking of bonded mortar overlays, *Cement & Concrete Composites*, 52: 1–8.

Burrows, R.W. Kepler, W.F. & Hurcomb, D. et al. 2004. Three simple tests for selecting low-crack cement, *Cement and Concrete Composites*, 26(5): 509–519.

Chen, D.P. Qian, C.X. & Gao, G.B. et al. 2007. Mechanism and effect of SAP for reducing shrinkage and cracking of concrete, *Journal of Functional Materials*, 38(3): 475–478.

Christof, S. Viktor, M. & Michaela, G. 2012. Relation between the molecular structure and the efficiency of Superabsorbent Polymers (SAP) as concrete admixture to mitigate autogenous shrinkage, *Cement and Concrete Research*, (42): 865–873.

Craeye, B. de Schutter, B. & Desmet, B. et al. 2010. Effect of mineral filler type on Autogenous shrinkage of self-compacting concrete, *Cement and Concrete Research*, 40(6): 908–913.

Darquennes, A. Staquet, S. & Delpancke-Ogletree, M.P. et al. 2011. Effect of autogenous deformation on the cracking risk of slag cement concrete, *Cement and Concrete Composites*, 33(3): 368–379.

Dudziak, L. & Mechtcherine, V. 2008. *Mitigation of volume changes of Ultra-High Performance Concrete (UHPC) by using Super Absorbent Polymers*, 2nd Int Symp on Ultra High Performance Concrete, E Fehling et al. (eds) Kassel University Press GmbH, 425–432.

Esteves, L.P. 2010. *On the absorption kinetics of superabsorbent polymers*, in: O.M. Jensen, M.T. Hasholt, S. Laustsen (Eds.), International RILEM Conference on Use of Superabsorbent Polymers and Other New Additives in Concrete, RILEM Proceedings PRO 074, RILEM Publications S.A.R. L, Bagneux (France), 77–84.

Esteves, L.P. Cachim, P. & Ferreira, V.M. 2007. Mechanical properties of cement mortars with super-absorbent polymers. *advanced construction materials*, 451–462.

Gao, D. Heimann, R.B. & Alexander, S.D.B. 1997. Box-behnken design applied to study the strengthen- ing of aluminate concrete modified by a superabsorbent polymer/clay composite, *Advanced Chemistry Research*, 9: 93–97.

GB/T50081-2002. 2003. *Standard for test method of mechanical properties on ordinary concrete*, Beijing: China building industry press.

Gleize, P.J.P. Cyr, M. & Escadeillas, G. 2007. Effect of metakaolin on autogenous shrinkage of cement pastes, *Cement and Concrete Composites*, 29(2): 80–87.

Hansen, W. 2011. Report on early-age cracking: A summary of the latest document from ACI committee 231, *Concrete International*, 33(13): 48–51.

He, Z. Chen, Y. & Liang, W.Q. et al. 2008. Effects of internal curing on shrinkage and cracking of concrete, *New Building Materials*, 8: 6–10.

Huang, Z.Y. & Wang, J. 2012. Effects of SAP on the Performance of UHPC, *Bulletin of the Chinese Ceramic Society*, 31(2): 539–544.

Jensen, O.M. & Hansen, P.F. 2002. Water-entrained cement-based materials–II. Experimental observations, *Cement and Concrete Research*, 32(6): 973–978.

Jensen, O.M. & Hansen. P.E. 2001. Water-entrained cement-based materials. I. Principles and theoretical background, *Cement and Concrete Research*. 31(5): 647–654.

Jensen, O.M. 2008. *Use of superabsorbent polymers in construction materials*, 1st International Conference on Microstructure Related Durability of Cementitious Composites, Nanjing, China, 757–764.

Kong, X.M. & Li, Q.H. 2009. Influence of super absorbent polymer on dimension shrinkage and mechanical properties of cement mortar, *Journal of the Chinese Ceramic Society*, 37(5): 855–861.

Lam, H. & Hooton, R.D. 2005. *Effects of internal curing methods on restrained shrinkage and permeability*. In: Persson B, Bentz D, Nilsson L-O, editors. Proc. 4th int. sem. on self-desiccation and its importance in concrete technology. Lund, (Sweden): Lund University; 210–228.

Liu, F. & Meng, F.L. 2013. Effect of superabsorbent polymer on properties of concrete, *Concrete*, 10: 98–100.

Lura, P. Durand, F. & Jensen, O.M. 2006. *Autogenous strain of cement pastes with superabsorbent Polymers,* in: O.M. Jensen, P. Lura, K. Kovler (Eds.), International RILEM Conference on Volume Changes of Hardening Concrete: Testing and Mitigation, RILEM Proceedings PRO 052, RILEM Publications S.A.R.L, Bagneux (France), 57–66.

Ma, X.W. Li, X.Y. & Jiao, H.J. 2009. Experimental Research on Utilization of Super Absorbent Polymer Cement Mortar and Concrete, *Journal of Wu Han University of Technology,* 31(2): 33–36.

Mateusz, W. & Pietro, L. 2013. Controlling the coefficient of thermal expansion of cementitious materials– A new application for superabsorbent polymers, *Cement & Concrete Composites,* 35: 49–58.

Mechtcherine, V. Dudziak, L. & Hempel, S. 2009. *Mitigating early age shrinkage of ultrahigh performance concrete by using Super Absorbent Polymers (SAP).* In: Tanabe T, et al., editors. Creep, shrinkage and durability mechanics of concrete and concrete structures–CONCREEP-8. London, UK: Taylor & Francis Group; 847–853.

Mechtcherine, V. Dudziak, L. & Schulze, J. et al. 2006. Internal curing by Super Absorbent Polymers (SAP)— Effects on material properties of self-compacting fibre-reinforced high performance concrete, Int RILEM Conf on Volume Changes of Hardening Concrete: Testing and Mitigation, Lyngby, Denmark, 87–96.

Meddah, M.S. Suzuk, M. & Sato, R. 2011. Influence of a combination of expansive and shrinkage- reducing admixture on autogenous deformation and self-stress of silica fume high-performance concrete, *Construction and Building Materials,* 25(1): 239–250.

Mounanga, P. Baroghel-Bouny, V. & Loukili, A. et al. 2006. Autogenous deformation of cement pastes: Part I. Temperature effects at early age and micro-macro correlations, *Cement and Concrete Research,* 36(1): 110–122.

Piérard, J. Pollet, V. & Cauberg, N. 2006. *Mitigating autogenous shrinkage in HPC by internal curing using superabsorbent polymers.* In: Jensen OM, Lura P, Kovler K, editors. RILEM proc. PRO 52, volume changes of hardening concrete: testing and mitigation. Bagneux (France): RILEM Publications SARL; 97–106.

Pietro, L. & Giovanni, P.T. 2014. Reduction of fire spalling in high-performance concrete by means of super absorbent polymers and polypropylene fibers Small scale fire tests of carbon fiber reinforced plastic-prestressed self-compacting concrete, *Cement & Concrete Composites,* 49: 36–42.

Qin, H.G. Gao, M.R. & Pang, C.M. et al. 2011. Research on Performance Improvement of Expansive Concrete with Internal Curing Agent SAP and Its Action Mechanism, *Journal of Building Materials,* 14(3): 394–399.

Skalny, J.P. & Mindess, S. 1991. *Materials science of concrete (II),* American Ceramic Society, 1–8.

Soliman, A.M. & Nehdi, M.L. 2013. Effect of partially hydrated cementitious materials and superabsorbent polymer on early-age shrinkage of UHPC, *Construction and Building Materials,* 41: 270–275.

Vaysburd, A.M. 1996. Durability of Lightweight Concrete Bridges in Severe Environments, *Concrete Intenational,* 18: 33–38.

Weber, S. & Reinhardt, W. 1997. A New Generation of High Performance Concrete: Concrete With Autogenous Curing, *Advanced Cement Based Materials,* 6(2): 59–68.

Yan, Y. Yu, Z. & Yang. Y.Z. 2012. Incorporation Superabsorbent Polymer (SAP) particles as controlling pre-existing flaws to improve the performance of Engineered Cementitious Composites (ECC), *Construction and Building Materials,* 28: 139–145.

Zhu, C.H. Li, X.T. & Wang, B.J. et al. 2013. Influence of Internal Curing on Crack Resistance and Hydration of Concrete, *Journal of Building Materials,* 16(2): 221–225.

Advanced Materials and Structural Engineering – Hu (Ed.)
© 2016 Taylor & Francis Group, London, ISBN 978-1-138-02786-2

The effect of reducing agents on GO/manganese oxide composites for super-capacitors

F.F. Ding, N. Zhang & C. Zhang
Shanghai Institute of Technology, Shanghai, P.R. China

ABSTRACT: The effect of reducing agents on morphology, structure and performance of graphene oxide (GO)/manganese oxide for super-capacitors has been investigated. First, GO/needle-like MnO_2 composite is successfully synthesized in a water-isopropyl alcohol system. Second, sodium borohydride and Vc are utilized to reduce the as-obtained composite. Typically, the X-Ray Diffraction (XRD) and Scanning Electron Microscopy (SEM) are employed to investigate the change in the crystalline and morphology of the composite. Third, the electrochemical performance of the as-prepared electrodes is also detected via Cyclic Voltammetry (CV) and galvanostatic charge–discharge experiments. The results indicate that the phase transformation from needle-like MnO_2 to Mn_3O_4 nanoparticle has been induced in sodium borohydride system. And the GO/Mn_3O_4 reduction by sodium borohydride possession the higher specific values of 209.5 F/g. The above analysis suggests that reducing agents play an important role in the fabrication of high energy-density GO/ manganese oxide super-capacitors materials.

1 INTRODUCTION

Graphene, one-atom-thick two-dimensional layers of sp^2 hybridizes carbon atoms, is comprised of a single sheet of hexagonally packs carbon atoms. It has triggered a great interest due to its high electrical conductivity, large specific surface area, and good chemical stability (Chen et al. 2010). Manganese oxide is thought to be one of the most promising materials for super-capacitors because of its low cost, eco-friendliness, and abundant availability (Komaba et al. 2008). Combining the remarkable electrical conductivity of graphene with the high pseudo-capacitance of MnO_2 is expected to improve the electrochemical performance.

Thus, GO/MnO_2 composite has been widely investigated in recent years due to the excellent performance, and a lot of attempts have been made to improve the electrochemical performance. Chen et al. have prepared the needle-like MnO_2 with GO through a simple soft chemical route and had the improved specific capacitance of 216 F/g (Chen et al. 2010). Deng et al. have synthesized graphene/MnO_2 nanorods composite under a hydrothermal condition with the specific capacitance of 218 F/g (Deng et al. 2013). Yan et al. have obtained the rGO/MnO_2 composite of 310 F/g in hydrazine hydrate (Yan et al. 2010). Li et al. have synthesized the rGO/MnO_2 electrode materials of 211.5 F/g by hydrothermally (Li et al. 2011). They obtained the higher specific capacitance either by enhancing specific surface area or by exploiting the excellent electronic conductivity of graphene. However, the effect of different reducing agents on the phase transformation and the morphology change of GO/MnO_2 are rarely reported.

In this paper, sodium borohydride and vitamin c (Vc) are used to investigate the phase transformation of the GO doped with MnO_2. Herein, a series of measurements including XRD, SEM, CV and charge–discharge curves have been employed to characterize the corresponding resultant.

2 EXPERIMENTAL

2.1 Synthesis of GO/manganese oxide composite

Graphene oxide is synthesized using the modified Hummers method (Hummers & Offeman 1958). In a typical synthesis of GO/MnO_2 composite, GO (0.14 g) and $MnCl_2 \cdot 4H_2O$ (2.8 g) are dispersed in isopropyl alcohol (200 ml) with ultrasonication for 0.5 h, subsequently, the solution is transferred to 83°C in a water bath with vigorous magnetic stirring. Then, $KMnO_4$ (1.51 g) dissolved in 20 mL of deionized (DI) water is added to the above boiling solution. After stirring and refluxing for 0.5 h, 20 mmol sodium borohydride is dropped into the solution and kept overnight, and then the mixture is heated to 83°C in a water cooled condenser with vigorous magnetic stirring for 2 h. Finally, the GO/manganese oxide powder is obtained by filtering and washing with DI water for several times. The target product reduced by Vc (20 mmol) is obtained using a procedure similar to sodium borohydride.

2.2 Preparation of working electrode

Fabrication of working electrodes are carried out as follows: the active materials, acetylene black, and polytetrafluoroethylene (PTFE) are mixed at a mass ratio of 8:1:1, a small quantity of ethanol is added drop-wise into the mixture to form a slurry. Then, the resulting mixture is coated onto a nickel foam substrate and dried in a hot oven.

2.3 Characterization

The crystallographic structure of the composite is determined by a powder X-Ray Diffraction system (XRD) at room temperature. The microstructure of the composite is characterized by SEM. The electrochemical property is characterized with CVs and charge–discharge experiments using a CHI660D electrochemical workstation (Chenhua, Shanghai). CVs are recorded at scan rates ranging from 10 to 200 mV/s in the potential range of −0.12 to 0.88 V. Galvanostatic charge–discharge measurements are recorded at different current densities ranging from 0.5 to 1.5 A/g. The supercapacitors tests are conducted with a conventional three electrode systems in 0.5 M Na_2SO_4 electrolyte, where the composite electrode is used as the working electrode, a platinum wire as the auxiliary electrode, and a saturated silver chloride electrode as the reference electrode.

3 RESULTS AND DISCUSSION

Figure 1 shows the XRD patterns of GO, GO/MnO_2 precursor and GO/MnO_2 nanocomposite reduced by Vc (V-GO/MnO_2) and sodium borohydride (S-GO/Mn_3O_4), respectively. The intense peak of GO located at about $2\theta = 10.24°$ is ascribed to the (001) reflection, and the interlayer spacing is much larger than that of pristine graphite due to the introduction of oxygen-containing functional groups on the graphite sheets. The diffraction peaks of the as-synthesized GO/MnO_2 are similar to those of the nanotetragonal phase of α-MnO_2 (JCPDS 44-0141), where the (001) reflection peak of layered GO has almost disappeared. It is suggested that homogeneous composites are formed on the surface of GO, and are covered by MnO_2. After reduction of GO/MnO_2 to V-GO/MnO_2 using Vc, there is no obvious difference in XRD patterns between GO/MnO_2 and V-GO/MnO_2. However, in the case of S-GO/Mn_3O_4, the XRD patterns are ascribed to Mn_3O_4 (JCPDS 18-0803), which means that MnO_2 has transformed into Mn_3O_4, and indicates that there may be significant changes in morphology and crystallinity in the composite.

SEM images of GO, GO/MnO_2, V-GO/MnO_2, and S-GO/Mn_3O_4 are shown in Figure 2. Figure 2(a) and (b) shows GO and GO/MnO_2, respectively. In comparing Figure 2(a) and (b), the formation of GO sheets homogeneously decorates with nanoneedle MnO_2 structures are clearly observed, which is consistent with the XRD observations. There are no significant differences in the morphologies between GO/MnO_2 and V-GO/MnO_2 after reduction by Vc (Fig. 2(c)). Further, MnO_2 nanoneedles are well attached without obvious damage. While for the S-GO/Mn_3O_4 (Fig. 2(d)), there are significant changes in the morphology and crystallinity in the composite. It consists of nanoparticles with a small amount of irregular nanoneedle, and the MnO_2 component has transformed into Mn_3O_4, which coincides with the previous speculation. The differences in the results may be attributed to the different reducing agents, compared with Vc, sodium borohydride is a strong reducing candidate.

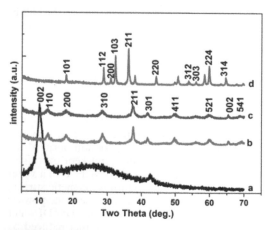

Figure 1. XRD pattern of (a) GO, (b) GO/MnO_2, (c) V-GO/MnO_2 and (d) S-GO/Mn_3O_4.

Figure 2. SEM images of (a) GO (b) GO/MnO_2, (c) V-GO/MnO_2 and (d) S-GO/Mn_3O_4.

It is not only reduced GO to graphene, but also reacted with MnO_2.

To explore potential applications for super-capacitors, samples are fabricated into super-capacitor electrodes and characterized with cyclic voltammograms and galvanostatic charge/discharge measurements. Figure 3a shows the CV curves of GO/MnO_2, $V-GO/MnO_2$, and $S-GO/Mn_3O_4$ electrodes at a scan rate of 10 mV/s. The CV curves of all the electrodes have the poor symmetry, which is attributed to combination of the double-layer and pseudo-capacitive contribution to the total capacitance (Chen et al. 2011). The $S-GO/Mn_3O_4$ electrode clearly exhibits a much larger integrated area than other electrodes, which indicates the higher specific capacitance of the composite. To our best knowledge, the specific capacitance can be determined by means of the charge–discharge cycles using the equation $C = (I \times \Delta t)/(\Delta V \times m)$, where ΔV is the voltage range of one scanning segment, and Δt is the time of a discharge cycle. The galvanostatic charge–discharge analytic results of various electrodes are shown in Figure 3b. All samples have triangular characteristics, revealing relatively ideal capacitor behavior in a neutral aqueous electrolyte. As we can see, $S-GO/Mn_3O_4$ has the longest charge–discharge period, indicating the highest capacitance, which is consistent with the CV data. The specific capacitance of the electrode is plotted in Figure 3c. The electrochemical performance of $S-GO/Mn_3O_4$ composite is more excellent than the others, this may be attributed to the following aspects: (1) The substrate of GO in sodium borohydride has a better electrochemical property than the others. (2) The doped MnO_2 has been reduced to the Mn_3O_4, and the diversity of the ionic state of Mn_3O_4 is favorable to the enhancement of specific capacitance.

To acquire more information on capacitive performance of the as-prepared composite, $S-GO/Mn_3O_4$ electrode is selected for the detailed measurements. Figure 4a shows the CV curves of the electrode at scan rates of 10, 20, 50, 100 and 200 mV/s. At a low scan rate, the CV curves exhibit an approximate rectangular shape, and no redox peaks are observed in all the CV curves, indicating an excellent capacitance behavior and low contact resistance in the capacitors. However, with the increasing of the scan rate, the CV curve deviation from rectangularity becomes obvious. This may be due to the internal resistance of the composites electrode. The galvanostatic charge–discharge curves of $S-GO/Mn_3O_4$ composite at different current densities are shown in Figure 4b, the charge curve is almost symmetric to its corresponding discharge counterpart. The specific capacitance at 0.5, 1.0, 1.25, and 1.5 A/g are 209.5 F/g, 187 F/g, 147.5 F/g, and 131 F/g, respectively.

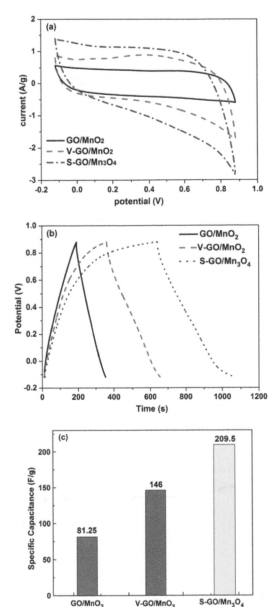

Figure 3. (a) CV curves at 10 mV/s; and (b) galvanostatic charge/discharge curves of at 0.5 A/g of GO/MnO_2, $V-GO/MnO_2$ and $S-GO/Mn_3O_4$ electrodes; (c) the value of specific capacitance calculated from (b).

Therefore, $S-GO/Mn_3O_4$ type has the wide application at the broad range of current density.

Cycle lifetime is one of the most critical factors in super-capacitor applications. The long-term cycle stability of $S-GO/Mn_3O_4$ composite is examined by repeating CV test at 10 mv/s for 1000 cycles

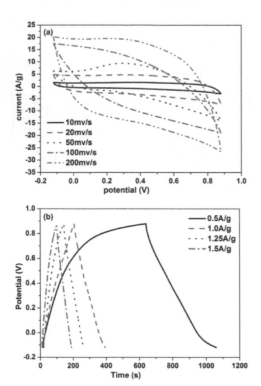

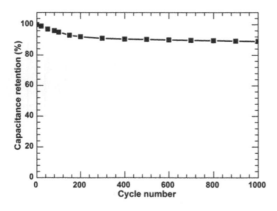

Figure 4. Electrochemical performance of S-GO/Mn_3O_4 composite electrode: (a) CV curves at different scan rates and (b) galvanostatic charge–discharge curves at different current density.

Figure 5. Cycle stability of S-GO/Mn_3O_4 electrode measured at 10 mV/s.

in 0.5M Na_2SO_4 solution. As shown in Figure 5, it is found that the specific capacitance decreases gradually with respect to number of cycles because of the loss of active material. And the capacitance retention rate is 89% of the initial after 1000 cycles,

which means S-GO/Mn_3O_4 as the electrode material has the good long-term cycle stability.

4 CONCLUSIONS

In summary, GO/manganese oxide composite are synthesized by a chemical reduction of GO/MnO_2 with sodium borohydride and vitamin c (Vc), it shows that the tuning microstructure and morphology of the composite are obtained in different reducing agents. Especially, it is found that the S-GO/Mn_3O_4 exhibits the best capacitive performance with the specific capacitance of 209.5F/g, and shows the good long-term cycle stability. These results demonstrate that for the GO/MnO_2 system, a reducing agent efficiently reduce GO or MnO_2 component is an essential factor to obtain an electrode material with the good electrochemical performance. This may be readily applicable on other transition metal oxides electrode materials for energy conversion and storage.

ACKNOWLEDGEMENT

The authors acknowledge the financial support by the Key discipline grant for composite materials from Shanghai Institute of technology (No. 10210Q140001), the Shanghai alliance project (LM201321).

REFERENCES

Chen, S. Zhu, J. & Wang, X. 2010. Graphene oxide–MnO_2 nanocomposites for supercapacitors, *ACS Nano*, 4: 2822–2830.
Chen, Y.L. Hu, Z.A. Chang, Y.Q. Wang, H.W. Zhang, Z.Y. Yang, Y.Y. & Wu, H.Y. 2011. Zinc oxide/reduced graphene oxide composites and electrochemical capacitance enhanced by homogeneous incorporation of reduced graphene oxide sheets in zinc oxide matrix. *The Journal of Physical Chemistry C*, 115:2563–2571.
Deng, S.X. Sun, D. & Wu, C.H. & Wang, H. 2013. Synthesis and electrochemical properties of MnO_2 nano-rods/graphene composites for supercapacitor applications. *Electrochimica Acta*, 111:707–712.
Hummers, W.S. & Offeman, R.E. 1958. Preparation of graphitic oxide. *Journal of the American Chemical Society*, 80:1339.
Komaba, S. Ogata, A. & Tsuchikawa, T. 2008. Enhanced supercapacitive behaviors of birnessite. *Electrochemistry Communications*. 10: 1435–1437.
Li, Z. Wang, J. Liu, S. Liu, X. & Yang, S. 2011. Synthesis of hydrothermally reduced graphene/MnO_2 composites and their electrochemical properties as supercapacitors. *Journal of Power Sources*, 196: 8160–8165.
Yan, J. Fan, Z. Wei, T. Qian, W. Zhang, M. & Wei, F. 2010. Fast and reversible surface redox reaction of graphene–MnO_2 composites as supercapacitor electrodes. *Carbon*, 48: 3825–3833.

Advanced Materials and Structural Engineering – Hu (Ed.)
© *2016 Taylor & Francis Group, London, ISBN 978-1-138-02786-2*

The properties of cementitious materials in prolonged curing

X.T. Yu, P. Gao, X. Wang & Y.D. Liao
College of Harbor, Coastal and Offshore Engineering, Hohai University, Nanjing, Jiangsu, China

ABSTRACT: The uniaxial compressive strength of the cementitious materials in prolonged curing ages was investigated by compression testing. The weight variations were recorded and the hydraulic conductivity was obtained from the permeability testing with pressure. The results showed that, except the hydraulic conductivity, the compressive strength and the weight of samples increased with the curing time at first and reached the approximate equilibrium when the cement particles hardly reacted. The diminution of hydraulic conductivity represents the smaller pores were filled with the hydration product gels and well explained the enhancement of the strength. The exponential relationship between the uniaxial compressive strength and the second curing time was also observed.

1 INTRODUCTION

In recent years, the application of the cementitious materials has got much development. The mechanical properties of the materials are seen as the foremost aspect for engineering application. Mengqiang Ma proposed the strength development principle and presented the concrete hardening strength formula (Ma 1957). However, Mingchu Yao thought that linearizing the strength development curve by the logarithmic function is inappropriate (Yao 1959). Also many reports on the mechanical properties are based on the short-term curing, such as 7 days, 28 days, etc. Few people have studied the long-term curing exceeding 90 days (Chen et al. 2013).

For the concrete structures in the inland river waters, the strength will higher than the anticipated compressive strength for the second curing in the water. Studying the variation in the mechanical properties of the cementitious materials cured for a long time will be helpful to predict the compressive strength of the concrete structures in the inland rivers. Also, it is beneficial to estimate the degree of the erosion and a certain time to begin the preventive maintenance.

2 EXPERIMENTS

2.1 Materials and samples preparation

Cylindrical samples with a size of φ50 mm × 100 mm, were prepared with ordinary Portland cement (P.O 42.5), the compositions are shown in Table 1. The ISO standard sand confirmed to GB1717671 and derived from Xiamen and tape water in the laboratory, was also used in this study.

Table 1. The major mineral composition in the ordinary Portland cement.

Mineral composition	Mass fraction
Tricalcium silicate	60.74
Dicalcium silicate	16.18
Tricalcium aluminate	6.66
Tetra calcium aluminoferrite	14.17

The sand–sand ratio and the water–cement ratio (W/C) of the mortars were separately chosen at 3 and 0.5. The samples were removed from their molds after 24h. The cement mortar samples were maintained for 28 days in a curing room, where the temperature and humidity were kept at 20±2°C and 95±3%, respectively. After this, the samples were placed in a tap water for a second curing, with the water surface being above the sample top surface. The water was renewed each two months.

2.2 Experiments

The Adaptive, Automatic and Multipurpose Rock Triaxial Rheological Apparatus of TOP INDUSTRY was used for the uniaxial compression testing. The displacement rate was configured at 0.06 mm/min. The experiment was conducted separately at 0, 15, 30, 60, 90, 120, 150, 180, 210, 240, 270 days in the water. And the change of compressive strength was calculated according to Equation (1)

$$\text{Change of compressive strength} = (q_{u(t)} - q_{u(0)})/q_{u(0)} \quad (1)$$

where $q_{u(0)}$ is the initial compressive strength of the samples cured for 28 days in the curing room.

$q_{u(t)}$ is the compressive strength after t days cured in the water.

The weights of the samples were recorded 56 times within 270 days with different time intervals in the balance of 0.01g. And the variation of relative weight could be calculated according to Equation (2).

Relative variation of weight $= (w_{(t)} - w_0)/w_0$ (2)

where w_0 is the initial weight of samples, and $w_{(t)}$ is the weight after t days.

The Adaptive, Automatic and Multipurpose Rock Triaxial Rheological Apparatus was also used for the permeability testing. The confining pressure was 5 MPa and the bottom hydraulic pressure was 2 MPa. This experiment was completed after the mortar samples were cured up to 0, 30, 60, 90, 120, 150, 180, 210, 240, 270 days in the tap water.

3 RESULTS

3.1 *Variation of mechanical properties of cementitious materials in prolonged curing*

As the results in Figure 1 shows, the evolution of the compressive strength can be divided into two stages. In the early stage, the compressive strength increased as the curing time increasing. However, after about 100 days of curing in water, the strength became stable. The Figure 2 indicates that increase of the strength was 30% or so at the early time, and then it became stable.

The clinker mineral reacted with water and generated hydration product such as crystalline C-S-H, Portlandite, and others as follows (Sun et al. 2011):

$$2(3CaO \cdot SiO_2) + 6H_2O \rightarrow 3CaO \cdot 2SiO_2 \cdot 3H_2O + 3Ca(OH)_2 \qquad ①$$

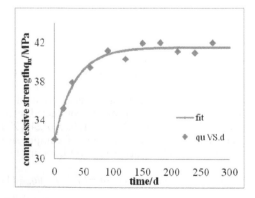

Figure 1. Compressive strength of samples cured in water at various ages.

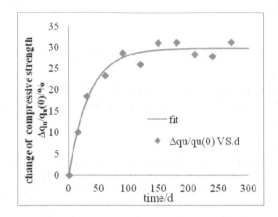

Figure 2. The change of compressive strength of samples immersed in water at various ages.

$$2(2CaO \cdot SiO_2) + 4H_2O \rightarrow 3CaO \cdot 2SiO_2 \cdot 3H_2O + Ca(OH)_2 \qquad ②$$

$$3CaO \cdot Al_2O_3 + 6H_2O \rightarrow 3CaO \cdot Al_2O_3 \cdot 6H_2O \qquad ③$$

$$4CaO \cdot Al_2O_3 \cdot Fe_2O_3 + 7H_2O \rightarrow 3CaO \cdot Al_2O_3 \cdot 6H_2O + CaO \cdot Fe_2O_3 \cdot H_2O \qquad ④$$

The hydration progress can be classified into four stages as follows: ① the dissolution stage, the cement began to dissolve and hydrated with water. ② The induction period, gel film layers grew round the Portland cement particles. ③ The setting stage, the gel film layers thickened, cement particles experienced further hydration. ④ The hardening stage, the gels filled the capillary pores.

Based on Paya's empirical correlation (Paya et al. 1997), Da Chen and Yingdi Liao et al. have presented a modified logarithm function between uniaxial compression strength and the curing time t, as Equation (3) (Chen et al. 2013):

$$q_u = a + b \, log_{10}(t + c) \qquad (3)$$

where q_u is the peak uniaxial compression strength, and a, b and c are the constants for a given cement and soil.

However, in our study, the samples had been cured for 28 days in the standard curing room before they were put into water for the second curing. The function was not suitable here. Figure 1 shows a stabilized state with the prolonged of the curing time. At that time, the hydration products had almost completed the transformation process. While the residual cement particles without hydration would be harder to react with water to reduce the porosity because of the hindrance of the harden cement paste. The reaction could be regarded to reach equilibrium. Presumed the exponential relationship between the compressive strength q_u and the curing time t, and present the following Equation (4):

$$q_u = a - be^{(ct)} \qquad (4)$$

where q_u is the peak uniaxial compression strength, and a, b and c are the constants for a given sample's size and cement.

According to the Equation (4), the calculated value and measured results of the uniaxial compressive strength were shown in Figure 1. The coefficient of determination R^2 is 0.975 and the fitting results had a good effect.

3.2 Variation of relative weight of cementitious materials in prolonged curing

As the Figure 3 shows, the relative weight increased with the curing time increasing. It was the cause of the formation of the hydration products such as crystalline C-S-H, Portlandite. Also, the water absorbed in the capillary pores would also result in gaining the weight of mortar samples. It was 100 days before the increasing tendency became stable. This was the evidence that proved the hydration process was reaching the equilibrium at a certain curing time to some extents.

3.3 Variation of hydraulic conductivity of the cementitious materials in prolonged curing

The change of hydraulic conductivity could represent the variation of the porosity of the mortar samples. It could be seen from the Figure 4 that the value of hydraulic conductivity decreased with the curing time increasing and it stabilized after around 100 days' curing in water. As the hydration

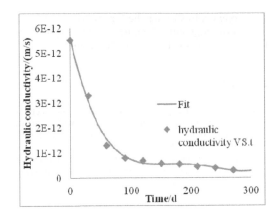

Figure 4. The hydraulic conductivity of samples cured in water at various age.

advanced, the larger pores were converted into smaller pores by the gradual filling of pores with cement hydrate materials. This also added the compactness of the mortars and enhanced the compressive strength of the samples. Also, the turning point of the hydraulic conductivity was around 100 days, which was corresponding with the turning time obtained from the above figures.

4 CONCLUSIONS

In the present study, the compressive strength, the relative weight and the hydraulic conductivity of cement mortars in prolonged curing in water have been investigated. The results showed the compressive strength and the weight of mortars increased with the curing ages in the early time. As the hydration process gained the equilibrium at a certain time, the compressive strength and the relative weight began to stabilize with the curing time going. On the contrary, the hydraulic conductivity corresponds with the mortar's permeability involved an initial decrease stage followed by the equilibrium. As the hydration advanced, the larger pores were converted into smaller pores by the gradual filling of pores with cement hydrate materials. This also added the compactness of the mortars and enhanced the compressive strength of the samples.

The exponential relationship between the uniaxial compressive strength and the second curing time was also observed, and the fitting curve is well consistent with the measured results.

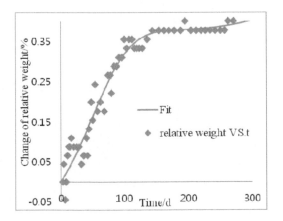

Figure 3. Variation of the relative weight of samples cured in water at different ages.

ACKNOWLEDGMENT

The financial support by the Ministry of Water Resources' Special Funds for Scientific Research on Public Causes under the Grant No. 201301052 is acknowledgment. Comments and suggestions from Prof. Chen Da are also appreciated.

REFERENCES

Chen, D. Liao, Y. Jiang, C. & Feng, X. 2013. The mechanical properties of coastal soil treated with cement, *Journal of Wuhan University Of Technology-Materials Science Edition,* 28(6): 1155–1160.

Ma, M. 1957. Cement, mortar and concrete hardening strength relationship with age, *China Civil Engineering Journal,* 02: 151–167.

Payá, J. Monzó, J. & Borrachero, M.V. et al. 1997. Mechanical Treatments of Fly Ashes, Part III: Studies on strength development of ground fly ashes (GFA)-cement mortars, *Cement and Concrete Research,* 27(9): 1365–1377.

Sun, G. Sun, W. Zhang, Y. & Liu, Z. 2011. Quantitative calculation on volume fraction of hydrated products in Portland cement, *Journal of Southeast University (Natural Science Edition),* 03: 606–610.

Yao, M. 1959. The discussion of 'Cement, mortar and concrete hardening strength relationship with age' (a), *China Civil Engineering Journal,* 02: 144–146.

Advanced Materials and Structural Engineering – Hu (Ed.)
© *2016 Taylor & Francis Group, London, ISBN 978-1-138-02786-2*

Dynamic tensile behavior of Flattened Brazilian Disc of TiB$_2$-B$_4$C composites

Y.B. Gao & W. Zhang
Hypervelocity Impact Research Center, Harbin Institute of Technology, Heilongjiang, Harbin, P.R. China

T.G. Tang & C.H. Yi
Laboratory for Shock Wave and Detonation Physics Research Institute of Fluid Physics, CAEP, Mianyang, Sichuan, P.R. China

ABSTRACT: Boron carbide has been widely used as engineering materials, especially the addition of Titanium Diboride. However, less study focuses on the dynamic tensile properties of TiB$_2$-B$_4$C composite. In the present study, the dynamic split tensile test of FBD of the composite was performed with the improved split Hopkinson pressure bar. The stress equilibrium of the specimen was measured by PVDF gauges, which were placed between inserts and compression bar. Two factors for the reason of non-stress equilibrium were analyzed: the micro cracks; mechanical parameters between the two flat ends and compression bar. Then the rupture mode of FBD was studied by using high-speed camera. And the dynamic tensile properties and strain-rate sensitivity of TiB$_2$-B$_4$C composite were obtained in the tests.

1 INTRODUCTION

Boron carbide (B$_4$C) has been widely used as wear-resistant components, lightweight armor products and neutron radiation shields, etc. (Wang, 2015). The addition of Titanium Diboride (TiB$_2$) in ceramic sintering will improve the fracture toughness and sintering characteristics of B$_4$C (Xu, 2012). However, most studies have focused only on the microstructure, hardness, fracture toughness, flexural strength, elastic modulus, etc. There is less study on dynamic tensile behavior of TiB$_2$-B$_4$C composite. Due to the high strength, high elasticity modulus, brittle and low failure strain of TiB$_2$-B$_4$C composite, the dynamic tensile tests are difficult to complete in directly. In 1978, the International Society for Rock Mechanics (ISRM) issued the suggested method for testing static rock tensile strength with the Brazilian disc specimen (ISRM, 1978). And Wang et al. have improved this traditional test, and proposed the Flattened Brazilian Disc (FBD) that machines two parallel flat ends on the disc circumference (Wang, 2004, 2006, 2009, 2011).

In the present study, the dynamic split tensile test of FBD of TiB$_2$-B$_4$C composite was performed with the improved Split Hopkinson Pressure Bar (SHPB) setup. The dynamic tensile properties of TiB$_2$-B$_4$C composite were obtained in the tests. And the rupture mode of FBD was studied by using high-speed camera.

2 EXPERIMENTAL PROGRAMME

2.1 *Specimen preparation*

The specimen of dynamic tensile tests adopted the form of Flattened Brazilian Disc. Due to the special mechanic properties of TiB$_2$-B$_4$C composite, the dimension of the specimen should be required in an absolute accuracy, which has a parallelism of 10^{-6} m and a flatness of 10^{-5} m in the two flat ends of FBD in the present study. The specimen has a diameter of 0.016 m, which was 2.42 times of the thickness.

Figure 1 shows the loading mode of the FBD specimen. The parallel flat ends can substantially replace the stress concentration as a distributed load. In order to prevent a premature breakage of

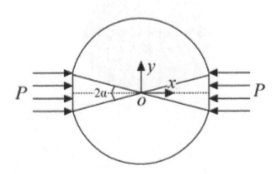

Figure 1. The loading mode of FBD specimen.

the specimen occurs at the contact point, a loading angle 2α was applied to ensure the crack initiated in the center of the disc. In the present study, $2\alpha = 20°$ was chosen for calculation given by Wang et al. (2004). P is the total value of the uniformly distributed load applied on the flat end.

In order to avoid the shortage of low failure strain, the strain gauges were glued on the same location at both lateral plane sides of the disc and perpendicular to the loading direction. The average recording of the two strain gauges at two opposite sides was adopted in the tests. The strain gauges have a length of 4.5×10^{-3} m, which is very small and thin compared with the size of FBD specimens, as shown in Figure 2.

2.2 Dynamic splitting tensile tests

The improved split Hopkinson pressure bar setup was used in the dynamic tensile tests to realize the one-dimensional stress wave loading of the FBD specimen. The material parameters for the steel compression bar are: diameter $d = 0.0127$ m, elasticity modulus $E = 210$ Gpa, density $\rho_0 = 7850$ kg/m^3, and velocity of sound $C_0 = 5200$ m/s. The length of the incident and transmitted bar is 1 m, and the striker bar is 0.2 m. In both sides of the incident and transmitted bar, the strain gauges are attached in order to record the corresponding strain waves.

The SHPB tests have the assumption of one-dimensional stress, stress uniformity of specimen, keeping elasticity of compression bar, etc. Due to the high strength, high elasticity modulus, brittle

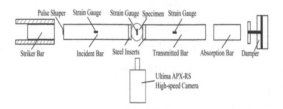

Figure 3. Schematic of modified SHPB apparatus and specimen.

and low failure strain of TiB$_2$-B$_4$C composite, the improved SHPB setup was used to satisfy these assumption, as shown in Figure 3. First, in order to avoid the effect of stress concentration and the damage of compression bar, the inserts of high strength steel were added in the tests. Second, the TiB$_2$-B$_4$C composite usually fails promptly at a small strain, and has a short time to destroy in the condition of high strain rate. So, a $\Phi3.5 \times 0.5$ mm pulse shaper was glued concentrically at the impacted end of the incident bar to ensure the rising front of the pulse not so steep. Then, an ideal waveform was gained to produce a nearly uniform distribution of the stress in the specimen after a certain time. Finally, two PVDF gauges were placed between inserts and compression bar in the part of tests to measure the stress equilibrium condition at both sides of the specimen.

3 EXPERIMENTAL RESULTS

3.1 Experimental recordings

Figure 4 was the recordings of strain waves for testing specimen 01. As shown in Figure 4(a), the reflected wave exists as an apparent turning point, which was the same time for the crush of specimen. When the specimen crushed completely, the residual signal of the wave totally turned back to the incident bar. Meanwhile, the response time of transmitted wave reached agreement with the step of reflected wave, as shown in Figure 4(b). Figure 4(c) was the record of tensile strain history at the center of the specimen. An obvious shock signal happened at 498 μs, which actually illustrated the specimen failed in this section. However, only the intermediate curve is useful as the shock line shows a cut-off for the measurement range of strain and hence of no use.

3.2 Stress equilibrium

In the present study, PVDF was made by three parts: sensitive piezoelectric thin film (8×8 mm), down-lead gauges (copper foil with a thickness

Figure 2. The specimen of dynamic tensile testing.

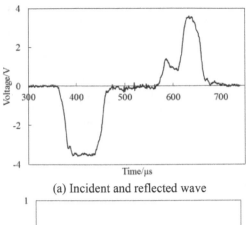

(a) Incident and reflected wave

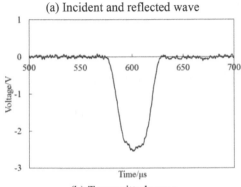

(b) Transmitted wave

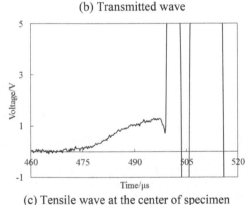

(c) Tensile wave at the center of specimen

Figure 4. The recordings of strain waves for testing specimen 01.

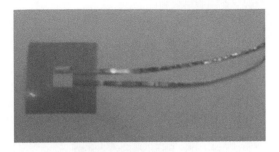

Figure 5. Packaged PVDF gauges.

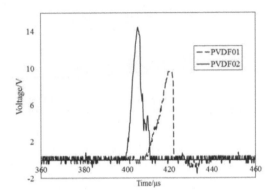

Figure 6. The typical signal of PVDF gauges.

of 0.02 mm), and insulating polyimide film. The package used the epoxy resin adhesive to solidify about 24 hours. The packaged PVDF is shown in Figure 5.

The stress equilibrium was measured by using the PVDF gauges that were placed between inserts and compression bar. It is important to note that PVDF gauges cannot be used in every test as it has an influence on the wave propagation, which the impedance matching of the gauges is different with

the compression bar. For TiB_2-B_4C composite, the whole response process just stays at the elastic stage. So it needs multi-reflect for the wave to reach the stress equilibrium as low failure strain of TiB_2-B_4C composite. Figure 6 is the typical signal of PVDF gauges. Only the loading stage of the signal exists as a damage of the compression bar to PVDF gauges. It can be seen that the forces acting on the left end was higher than the right end of the specimen. Two reasons exist to lead difference between them: 1) In the process of wave propagation, the micro cracks reduced the stress wave and resist its propagation as energy absorption consumed through refraction and reflection; 2) Many parameters between the two flat ends and compression bar were different, such as the density, the elastic modulus and wave speed, which would lead to the wave reflection and transmission. However, the influence of the difference can be neglected when the value of PVDF is smaller than 5% in fact. So it can be considered that the experimental data were satisfied by the stress equilibrium in the test.

3.3 The rupture modes of FBD specimen

Figure 7 is the crack propagation process of the FBD specimen caught by using high-speed camera.

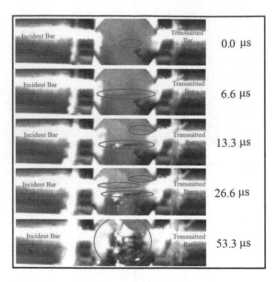

	0.0 µs
	6.6 µs
	13.3 µs
	26.6 µs
	53.3 µs

Figure 7. Crack propagation process of the FBD specimen (set the crack initiation stage as the time origin).

The rupture modes of the specimens can be provided at four stages: crack initiation; crack extension; crack generation in the other place of the disc; and broken extensively of disc. First, the crack initiated at the center of the disc, which is vital for FBD tests and was corresponding to 0 µs. Second, the crack was extended along the loading diameter, which was corresponding to 6.6 µs. Third, crack was generated in the other place of the disc, which was the location of initial micro cracks of disc (corresponding to 13.3 and 26.6 µs). Finally, the disc was broken completely, which was corresponding to 53.3 µs.

3.4 Dynamic tensile strength of TiB₂-B₄C composite

In the previous study, Wang et al. (2009) used the Griffith strength criterion to formulate the dynamic tensile strength of FBD specimen. The dynamic tensile strength was measured at the critical point when the tensile strain wave recorded at the disc center got peak value of the strain derivative with respect to time (Wang, 2009). The tensile strength σ_t of FBD specimen for the loading angle $2\alpha = 20°$ is given as follows:

$$\sigma_t = -0.95 \frac{2P_{max}}{\pi DB} \qquad (1)$$

where D and B is the diameter and thickness of specimen, respectively. P_{max} is the maximum load at the center of the disc:

Table 1. The dynamic tensile strength of the TiB₂-B₄C composite under different impact velocity V_0.

Specimens	V_0 (m/s)	$\dot{\varepsilon}$ (s⁻¹)	σ_t (Mpa)
01	34.4	33	340
02	34.4	34	348
03	30.4	36	340
04	32.8	47	341
05	35.2	81	363
06	35.2	88	368
07	34.2	97	371
08	37.6	162	380
09	36.8	183	408
10	36.8	200	418
11	38.4	254	421

$$P(t) = \frac{EA}{2} \left[\varepsilon_i(t) + \varepsilon_r(t) + \varepsilon_t(t) \right] \qquad (2)$$

where A is the cross-section area of compression bar; $\varepsilon_i(t)$, $\varepsilon_r(t)$ and $\varepsilon_t(t)$ are the strain of incident wave, reflected wave and transmitted wave, respectively.

The tensile strain rate at the center of the FBD specimen is calculated by the derivative of the tensile strain history $\varepsilon_s(t)$ with respect to the time:

$$\dot{\varepsilon}(t) = \frac{d\varepsilon_s(t)}{dt} \qquad (3)$$

Table 1 was the dynamic tensile strength of the FBD specimen under different impact velocities. As shown in Table 1, the impact velocity has a great influence on the strain rate of the specimen. And the dynamic tensile strength was increased with the strain rate of loading. That is to say that the TiB₂-B₄C composite is strain-rate sensitivity material in tensile tests. However, the increment of the dynamic tensile strength is limited as the value added is 81 Mpa corresponding to the stain rate from 33 to 254 s⁻¹.

4 CONCLUSIONS

In the present study, the dynamic split tensile test of FBD of the TiB₂-B₄C composite was performed with the improved split Hopkinson pressure bar. It can be concluded that:

1. The reason of non-stress equilibrium was analyzed for two reasons: the micro cracks; mechanical parameters between the two flat ends and compression bar.
2. The rupture modes of the specimens were provided into four stages: crack initiation; crack

extension; crack generation in the other place of the disc; broken extensively of disc.

3. The dynamic tensile strength of FBD specimen was formulated by using the Griffith strength criterion.

4. The dynamic tensile strength was increased with the strain rate of loading. And the TiB_2-B_4C composite is strain-rate sensitivity material in the tensile tests.

ACKNOWLEDGMENTS

The author gratefully acknowledges the support on financial and technical support of China Academy of Engineering Physics, and the support of specimen of Wuhan University of Technology.

REFERENCES

ISRM. 1978. Suggested methods for determining tensile strength of rock materials. *International Journal of Rock Mechanics and Mining Sciences & Geomechanics Abstract*, 15: 99–103.

Wang, D.W. Ran, S.L. Shen, L. Sun, H.F. & Huang, Q. 2015. Fast synthesis of B_4C-TiB_2 composite powers by pulsed electric current heating TiC-B mixture, *Journal of the European Ceramic Society*, 35: 1107–1112.

Wang, Q.Z. & Wu, L.Z. 2004. The flattened Brazilian disc specimen used for testing elastic modulus, tensile strength and fracture toughness of brittle rocks: experimental results. *International Journal of Rock Mechanics and Mining Sciences*, 41(3): 357–358.

Wang, Q.Z. Feng, F. Ni, M. & Gou, X.P. 2011. Measurement of mode I and mode II rock dynamic fracture toughness with cracked straight through flattened Brazilian disc impacted by split Hopkinson pressure bar. *Engineering Fracture Mechanics*, 78: 2455–2469.

Wang, Q.Z. Li, W. & Song, X.L. 2006. A method for testing dynamic tensile strength and elastic modulus of rock materials using SHPB. *Pure and Applied Geophysics*, 163: 1091–1100.

Wang, Q.Z. Li, W. & Xie, H.P. 2009. Dynamic split tensile test of Flattened Brazilian Disc of rock with SHPB setup. *Mechanics of Materials*, 41: 252–260.

Xu, C.M. Cai, Y.B. Flodström, K. Li, Z.S. Esmaeilzadeh, S. & Zhang, G.J. 2012. Spark plasma sintering of B_4C ceramics: The effects of milling medium and TiB_2 addition. *Int. Journal of Refractory Metals and Hard Materials*, 30: 139–144.

Advanced Materials and Structural Engineering – Hu (Ed.)
© 2016 Taylor & Francis Group, London, ISBN 978-1-138-02786-2

Dibenzothiophene adsorption on Activated Carbons and its water effect

Z.J. Li, S.L. Jin, S.M. Zhang, N. Jiang, X. Shao, M.L. Jin & R. Zhang
School of Materials Science and Engineering, Shanghai Institute of Technology, Shanghai, China

ABSTRACT: Four Activated Carbons (ACs) with different surface areas and pore structures were heat-treated at 600°C for 3 h to modify their surface chemical properties and the ACs before and after the treatment were characterized by nitrogen adsorption, XPS and dibenzothiophene adsorption. Results indicated that heat treatment causes a collapse of micropores less than 1 nm and an increase of volume of mesopores and micropores between 1.1 and 1.6 nm. The dibenzothiophene adsorption capacity increases with the percentage of oxygen-containing surface groups. The adsorption capacity has a linear relationship with the pore volume of pores in the range of 0.536–1.179 nm, indicating that micropore filling is the dominant mechanism for dibenzothiophene adsorption. Oxygen-containing functional groups and adsorbed water can also promote dibenzothiophene adsorption.

1 INTRODUCTION

Sulfur compounds in gasoline and diesel as environmental pollutants are the main source of SO_x during combustion, which not only causes acid rain but also poisons catalysts in catalytic converters for reducing CO and NO_x (Fallah et al. 2012) in vehicles. The increasingly stringent environmental regulations have been implemented to reduce the sulfur concentration in transportation fuels by government worldwide (Lorencon et al. 2014). Traditional removal method is hydrodesulfurization that has long been used at the refinery plants. It is effective to remove mercaptans and sulfides, but less effective for refractory sulfur compounds such as benzothiophene, dibenzothiophene and their derivatives because of their steric hindrances (Ania et al. 2005) during adsorption on active sites. In order to achieve deep desulfurization for fuel oils, many clean non-hydrogenated methods have been proposed, such as oxidative desulfurization, biodesulfurization, extractive desulfurization and adsorption desulfurization (Dan et al. 2014, Wei et al. 2012).

Recently, adsorption desulfurization was popular for its mild operation conditions and no need for hydrogen or oxygen. Activated carbons have been widely investigated in adsorption desulfurization, owing to their well-developed pores, large surface area, cheap and easy regeneration. The pore structure and surface functional groups of activated carbons play a major role in the adsorption (Yu et al. 2009). The influence of pore sizes and surface functional groups on adsorption capacity of dibenzothiophene was studied previously. Ania and co-workers (Ania et al. 2005) found that sulfur adsorption capacity was dominated by the volume of micropores with a width of <0.7 nm with

a linear correlation coefficient of 0.98. Zhang and coworkers (Zhang et al. 2014) revealed that dibenzothiophene adsorption capacity can be correlated linearly with the pore volume of pores between 0.6 and 1.2 nm with a correlation coefficient of 0.99, indicating that dibenzothiophene molecule can enter the pores with their sizes comparable to size of dibenzothiophene. Activated carbon adsorbs water when it is exposed to atmosphere, which might have impact on its adsorption properties. Fernandez found that it could improve the ability of adsorption for phenol (Femandez et al. 2003). However, the joint effect of oxygen functional groups and pore structure on thiophenic adsorption capacity has not been well understood, and the impact of adsorbed water on adsorption desulfurization is rarely reported.

In this study, a deep insight into the mechanism of adsorption desulfurization on activated carbons is explored and, the influence of high temperature treatment on pore structure and surface properties of the activated carbon is investigated. And the influence of adsorbed water on thiophenic adsorption was revealed.

2 EXPERIMENTAL METHODS

2.1 Materials

Four Activated Carbons (ACs) with different properties were used in the study. AC1 is a commercial activated carbon, AC2, AC3 and AC4 are high surface area activated carbons obtained by KOH activation from coke for AC2 and AC3, and from corn cob for AC4. ACs was heat treated at 600°C for 3 h to reduce surface oxygen containing functional groups. The resulting activated carbons were marked as AC1-600, AC2-600, AC3-600 and AC4-600. Before all experiments the activated carbons were dried at 80°C for 2 h.

2.2 Characterization

The surface areas and pore structures of samples were characterized by N_2 adsorption at –196°C using a Micromeritics ASAP 2020 instrument. The N_2 adsorption data at the relative pressure (P/P_0) between 0.05 and 0.2 was used to calculate surface area using the standard Brunauer, Emmet and Teller (BET) equation (Nguyen et al. 2014) and the pore size distributions were determined using the Density Functional Theory (DFT) method assuming a slit pore geometry (Chang et al. 2014). The volumes of pores in a certain range, such as between 0.536 and 1.090 nm were calculated based on the cumulative pore volume curves (Lorencon et al. 2014), resulting in six pore volumes between different pore size intervals ($V_{0.536-1.090\ nm} = V_{1.090\ nm} - V_{0.536\ nm}$, $V_{0.536-1.179\ nm} = V_{1.179\ nm} - V_{0.536\ nm}$, $V_{0.536-1.268\ nm} = V_{1.268\ nm} - V_{0.536\ nm}$, $V_{0.536-1.358\ nm} = V_{1.358\ nm} - V_{0.536\ nm}$, $V_{0.589-1.179\ nm} = V_{1.179\ nm} - V_{0.589\ nm}$, $V_{0.679-1.179\ nm} = V_{1.179\ nm} - V_{0.679\ nm}$).

The elemental compositions and surface functional groups on activated carbons were analyzed by a British Kratos Axis Ultra DLD X-ray Photoelectron Spectroscope (XPS). Spectra were acquired with a monochromatic Al Kα radiation (1486.6 eV). The peak positions were corrected on the basis of the most intense graphitic carbon peak, which is taken as 284.6 eV.

2.3 Quantification of dibenzothiophene content in model fuels

The concentration of dibenzothiophene in the model fuel were quantitatively analyzed by a GC-2014C gas chromatograph equipped with a capillary column (MXT-5, 60 in length, 0.25 mm in internal diameter and 0.25 μm film thickness, Restek, America) using a flame ionization detection as a detector and diphenyl sulfide as an internal standard. The column temperature was programmed to hold at 20°C for 3 min, heated at 5°C/min from 180°C to 280°C and held at 280°C for 10 min. The injector and detector temperatures were 250 and 310°C, respectively.

2.4 Effect of water on dibenzothiophene adsorption capacity

0.1 g activated carbon was dried at 80°C for 2 h in a flask that was added to 20 μL water using a micro-injector, which was kept at 35°C for 24 h to allow water adsorbed onto the activated carbons in the form of vapor. This corresponds to a water addition of 20 wt%. Model fuels with different contents of dibenzothiophene were added to the wet

activated carbons to evaluate the effect of water on the dibenzothiophene adsorption capacity.

2.5 Adsorption experiments

The model fuel (10 g) with a dibenzothiophene concentration of 10 μmol · g⁻¹ in n-octane was added to a volumetric flask containing 0.1 g above adsorbents, which was shaken at 30°C for 48 h in a water bath. The amount of dibenzothiophene adsorbed was calculated from the formula $q_e = (C_0 - C_e)$ m*/m, where q_e is the amount of sulfur adsorbed per gram of adsorbent (mgS/g-A) after 48 h, m* is the mass of the model fuel (g), m is the amount of the adsorbent (g), C_0 is the initial concentration of dibenzothiophene in the model fuel, C_e (ppmS) is the equilibrium concentration of the dibenzothiophene in bulk solution in the flask.

3 RESULTS AND DISCUSSION

3.1 Pore texture and surface properties of ACs

According to the N_2 adsorption/desorption isotherms and cumulative pore volumes of the samples shown in Figure 1, it is found that both

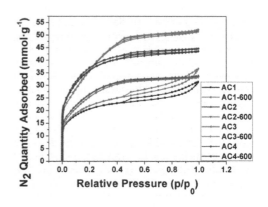

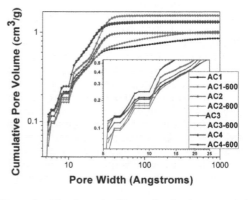

Figure 1. N_2 adsorption/desorption isotherms and the cumulative pore volumes versus pore size.

Table 1. Porous parameters of the ACs.

	S_{BET} (m²/g)	V_{micro} (cm³/g)	V_{meso} (cm³/g)	V_{macro} (cm³/g)	$V_{<1\,nm}$ (cm³/g)	$V_{1.1–1.6\,nm}$ (cm³/g)
AC1	1655	0.461	0.381	0.0182	0.2161	0.1408
AC1-600	1666	0.452	0.550	0.0230	0.2103	0.1478
AC2	2172	0.512	0.479	0.0004	0.2065	0.1508
AC2-600	2103	0.469	0.513	0.0007	0.1650	0.1539
AC3	3034	0.560	0.973	0.0005	0.1852	0.1911
AC3-600	3005	0.549	1.002	0	0.1761	0.1966
AC4	2935	0.777	0.517	0.0002	0.2493	0.2602
AC4-600	3023	0.800	0.457	0.0710	0.249	0.2739

micropores and mesopores in the four activated carbons are well developed and there are only a negligible amount of macropores. Micropore volumes of AC1, AC2, AC3 decreased and their mesopore volumes increased after the heat treatment while those of AC4 had an inverse changing trend after the heat treatment. The amount of oxygen-containing functional groups on the activated carbon surface decreased after the heat treatment, accompanied by a part collapse of micropores less than 1 nm (Tazibet et al. 2003) and an increase of volumes of the pores between 1.1 and 1.6 nm, indicating that a part of micropores less than 1 nm was converted to large micropores between 1.1 and 1.6 nm.

Figure 2 shows that oxygen contents all decreased significantly after the heat treatment. The surface oxygen contents of the activated carbons before and after the heat treatment can be calculated according to the binding energy of carbon and oxygen. The calculated results are listed in Table 2. The carbon precursor (cellulose) that has a high oxygen content might be the main cause that the AC4 had a high oxygen content even after the heat treatment.

3.2 Adsorption of dibenzothiophene

Table 3 lists the dibenzothiophene adsorption capacity of activated carbons before and after heat treatment. It can be found that the sulfur capacity was all reduced by the heat treatment. Figure 3 shows a linear relationship between the decreased amount of sulfur capacity and the decreased oxygen percentage by the heat treatment with a correlation coefficient of 0.94, indicating that surface oxygen functional groups favor the adsorption of dibenzothiophene.

The correlation between the dibenzothiophene adsorption capacity and micropore volumes of pores at different size ranges for the four activated carbons after the heat treatment is shown in Figure 4. It can be found that the capacity is best described by a linear equation for the micropore volumes in the pore range of 0.536–1.179 nm.

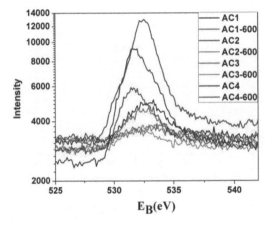

Figure 2. XPS patterns of activated carbons before and after heat treatment.

Table 2. Oxygen contents of activated carbons before and after heat treatment.

	AC1	AC2	AC3	AC4
Oxygen percentages before treatment (%)	3.67	16.15	4.20	17.03
Oxygen percentages after treatment (%)	1.01	1.43	1.78	7.10
Decreased oxygen percentages (%)	2.66	14.72	2.42	9.93

This optimum pore size range might be related to the size of dibenzothiophene, below which dibenzothiophene cannot enter inside and above which utilization rate of pores is not sufficient. The size of dibenzothiophene molecules is 0.65 nm reported by Zhang (2014), and 0.554 nm reported by Cychosz (2008). It can be seen that dibenzothiophene can enter the pores with diameter smaller than itself according to the dibenzothiophene molecular diameter and the fitting results, indicating that micropores can be inserted and expanded.

Table 3. Dibenzothiophene adsorption capacity of activated carbons before and after heat treatment.

	AC1	AC2	AC3	AC4
Sulfur capacity before treatment (mgS/g)	18.84	21.16	21.18	22.77
Sulfur capacity after treatment (mgS/g)	17.57	18.44	19.80	20.85
Reduction by the heat treatment (mgS/g)	1.27	2.71	1.38	1.92

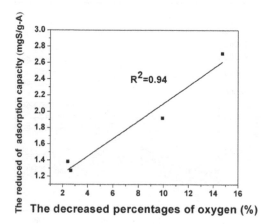

Figure 3. The decreased percentages of oxygen in relation to the decreased amounts of dibenzothiophene adsorption capacity.

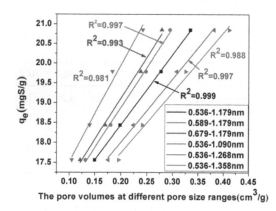

Figure 4. The relationship between pore volumes at different pore size ranges and the dibenzothiophene adsorption capacity.

3.3 The influence of surface moisture content on the desulfurization performance

Figure 5 shows the linear relationship between the percentage of oxygen of ACs and increased percentages of dibenzothiophene adsorption capacity

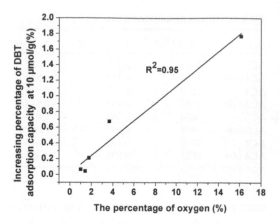

Figure 5. The relationship between the oxygen percentage and increased percentage of dibenzothiophene adsorption capacity with a water addition of 20 wt%.

via a water addition of 20 wt% in ACs at a dibenzothiophene initial concentration of 10 μmol/g with the correlation coefficient of 0.95. The higher is the oxygen content in ACs, the higher is the adsorbed water on surface of ACs. This indicated that surface adsorbed water can promote the dibenzothiophene adsorption on ACs. Fernandez (2003) considered that carbon surface oxygen functional groups can enhance the binding energy of carbon on the surface of the activated carbon. Nguyen (2014) believed that the water is potential oxygen functional groups of the activated carbon surface. The increase of dibenzothiophene adsorption could be enhanced by an increase of surface polarity of the activated carbon surface by water adsorption.

4 CONCLUSIONS

Heat treatment causes a collapse of micropores with their sizes less than 1 nm, leading to an increase of pore volumes of mesopores and large micropores between 1.1 and 1.6 nm.

The optimum pore size range of dibenzothiophene adsorption is 0.536–1.179 nm. Oxygen functional groups of surface are favorable for the adsorption of dibenzothiophene.

Adsorbed water can enhance surface polarity of activated carbons, which cause an increase of dibenzothiophene adsorption.

ACKNOWLEDGEMENTS

This work was supported by the Capacity Building Program of Shanghai local universities (No. 12160503600), the first-class discipline construction fund of Shanghai municipal education commission (No. J201212), Nature Science Foundation of China (U1332107) and Key discipline construction fund of composite materials of Shanghai Institute of technology (No. 10210Q140001).

REFERENCES

Ania, C.O. et al. 2005. Importance of Structural and Chemical Heterogeneity of Activated Carbon Surfaces for Adsorption of Dibenzothiophene. *Langmuir.* 21: 7752–7759.

Chang, M.Z. et al. 2014. Synthesis, Characterization, and Evaluation of Activated Carbon Spheres for Removal of Dibenzothiophene from Model Diesel Fuel. *Industrial & Engineering Chemistry Research.* 53: 4271–4276.

Cychosz, K.A. et al. 2008. Liquid Phase Adsorption by Microporous Coordination Polymers: Removal of Organosulfur Compounds. *Journal of the American Chemical Society.* 130: 6938–6939.

Dan, H.W. et al. 2014. Oxidative desulfurization using ordered mesoporoussilicas as catalysts. *Journal of Molecular Catalysis A: Chemical.* 393: 47–55.

Fallah, R.N. et al. 2012. Removal of thiophenic compounds from liquid fuel by different modified activated carbon cloths. *Fuel Processing Technology.* 93: 45–52.

Fernandez, E. et al. 2003. Adsorption of Phenol from Dilute and Concentrated Aqueous Solutions by Activated Carbons. *Langmuir.* 19: 9719–9723.

Lorencon, E. et al. 2014. Oxidative desulfurization of dibenzothiophene over titanate nanotubes. *Fuel.* 132: 53–61.

Nguyen, V.T. et al. 2014. Water as a potential molecular probe for functional groups on carbon surfaces, *Carbon.* 67: 72–78.

Tazibet, S. et al. 2003. Heat treatment effect on the textural, hydrophobic and adsorptive properties of activated carbons obtained from olive waste. *Microporous and Mesoporous Materials.* 170: 293–298.

Wei, Z. et al. 2012. Enhancement of dibenzothiophene adsorption on activated carbons by surface modification using low temperature oxygen plasma. *Chemical Engineering Journal.* 209: 597–600.

Yu, M.X. et al. 2009. Effect of thermal oxidation of activated carbon surface on its adsorption towards dibenzothiophene. *Chemical Engineering Journal.* 148: 242–247.

Advanced Materials and Structural Engineering – Hu (Ed.)
© 2016 Taylor & Francis Group, London, ISBN 978-1-138-02786-2

The preparation and desulfurization performance of carbon aerogels in-situ loaded with copper for fuel oils

S.M. Zhang, Z.J. Li, N. Jiang, H.F. Zhang & R. Zhang
School of Materials Sciences and Engineering, Shanghai Institute of Technology, Shanghai, China

ABSTRACT: Copper was in-situ loaded into carbon aerogels by sol-gel polycondensation of resorcinol with formaldehyde using sodium carbonate as a catalyst and copper acetate as copper source. The desulfurization performance of as-prepared carbon aerogels was evaluated by selective adsorption of dibenzothiophene (DBT) as a model sulfur compound and benzene as a competitive aromatic compound. It is shown that the introduction of copper leads to a generation of ultrafine micropores. The adsorption capacity for DBT is linearly related to the volume of micropores. And the selectivity is improved with the increase of copper content, at the same time the selectivity is reduced when copper is etched by dilute HNO_3 acid leaching, indicating that the presence of copper is favorable for DBT adsorption via its π-complexing effect.

1 INTRODUCTION

When a sulfur compound in transportation fuels is converted to SOx during combustion of fuel in an internal combustion engine, it not only makes a contribution to acid rain but also acts as a poison to the catalytic converter for exhaust emission treatment. Desulfurization of transportation fuels such as diesel has increasingly gained importance since most of the countries, particularly the developed ones, have implemented more stringent legislation to regulate the sulfur content of transportation fuels. More-over, Sulfur concentration in the fuel needs to be reduced to less than 1 ppm for Proton Exchange Membrane Fuel Cell (PEMFC) and less than 10 ppm for Solid Oxide Fuel Cell (SOFC) (Song 2003, Ma et al. 2002). Hydro-Desulfurization (HDS) is a traditional technique to produce low-sulfur liquid fuels at high temperature and pressure. However, the HDS approach is not suitable for removing refractory sulfur compounds such as DBT due to their low reactivity in the HDS. On the other hand, achieving the ultralow sulfur fuels will lead to increasing consumption of hydrogen and energy (Kabe et al. 1992, Nagai et al. 2005).

The preparation of copper-loaded carbon aerogels is paid more and more attention in the recent years. But most of them are using the impregnating or ion exchange method, and there is no report for fuel oil desulfurization. The copper-loaded carbon aerogels have high specific surface areas and high porosities, which are expected to have strong adsorption properties and catalytic ability.

In this paper, Cu-loaded carbon aerogels were pre-pared by the sol-gel polymerization of resorcinol with formaldehyde in aqueous solution to produce organic gels that are super-critically dried in n-hexane and subsequently pyrolyzed in an inert atmosphere, using sodium carbonate as catalyst and copper acetate as a copper source. The desulfurization performance of as-obtained carbon aerogels was also investigated by selective adsorption of DBT as a model sulfur compound and benzene as a competitive aromatic compound.

2 EXPERIMENTAL

2.1 *Preparation of Cu-loaded carbon aerogels*

Copper loaded organic aerogels were prepared by the sol-gel polymerization reaction of resorcinol with formaldehyde in water, using copper acetate as a copper source and sodium carbonate as a catalyst. Mixtures were stirred to obtain a homogeneous solution that was cast into glass molds. Glass molds were sealed and the mixtures were cured for 7 days at different temperatures up to 85°C. After the curing, the gel rods were placed in alcohol to displace water. The displaced alcohol gel rods were dried by supercritical drying method at 240°C and 6.0 MPa for 1 hour, using n-hexane as a supercritical fluid. The dried copper-loaded organic aerogels were carbonized under N_2 flow (300 $cm^3 \cdot min^{-1}$) at a heating rate of 1.5°C·min^{-1} up to 800°C with a soaking time of 3 hours. The obtained samples were designated as Cu-C-n, where then represents the addition amount of copper acetate in the precursor solutions. Sample powder of Cu-C-n and HNO$_3$ (98%) were put into the conical flask according to the 1:10 ratio and sealed at 25°C for 2 days to remove copper. The resulting slurry was then filtered, washed to neutral and dried. The samples obtained were marked as C-n.

2.2 *Characterization*

X-ray Diffraction (XRD) patterns of samples were recorded on a Rigaku-DMAX2200PC diffractometer using Cu Kα (λ = 0.1506 Å) radiation in the 2θ range from 10 to 80°. The samples were separately degassed at 200 °C under vacuum for at least 4 hours prior to the measurements. Nitrogen adsorption data in the relative pressure (P/P^0) range of 0.05–0.26 was used to calculate surface area using the Brunauer-Emmett-Teller (BET) equation. The microstructures of the samples were observed by a FEI Quanta200 FEG Scanning Electron Microscope (SEM). Trans-mission Electron Microscopic (TEM) observations were made with a JEOL 2000 EX electron micros-cope, equipped with a top entry stage.

2.3 *Adsorptive test*

The model fuel concentration of 320 ppm was prepared by mixing DBT and n-octane. For the competitive adsorption experiments, an extra 20 wt% of benzene was added to the model fuel. The model fuel with and without benzene were designated as MF-1 and MF-2, respectively. The desulfurization performance of adsorbents was evaluated by static saturation adsorption tests. About 0.1 g of tested adsorbent and 10 g of model fuel were added into a glass tube. The tube was capped and placed in a shaking bath at 30°C for 48 hours. Then the mixture was filtered, and the treated model fuel samples were analyzed by capillary chromatography to estimate the sulfur adsorption capacity of the adsorbents. The amount of sulfur adsorbed on per gram of adsorbent, q$_e$ (mg-S/g-A), was calculated with the following equation:

$$q_e = \frac{(C_0 - C_e)\, m^*}{m} \tag{1}$$

where, C_0 and C_e are the initial and equilibrium sulfur concentrations in the model fuel (mg/g), respectively, m* is the weight of model fuel (g) and m is the mass of the adsorbent (g).

In order to facilitate the quantitative discussion of the adsorptive selectivity, a selectivity factor K was used in the present study, which is defined as:

$$K = \frac{q_e^1 - q_e^2}{q_e^1} \tag{2}$$

where, q_e^1 is the sulfur adsorptive capacity for the model fuel MF-1 (mg-S/g-A) and of q_e^2 is the sulfur adsorptive capacity for the model fuel MF-2 (mg-S/g-A). Note that the larger is the K value, the lower is the selectivity.

3 RESULTS AND DISCUSSION

3.1 *Textural characteristics and chemical compositions of Cu-doped carbon aerogels*

SEM images of as-prepared carbon aerogels are shown in Figure 1. These aerogels were composed of interconnected particles, and the size of these particles increased with the increasing of the addition amount of copper acetate. The size of these particles was strongly influenced by the pH value that was adjusted by the amount of copper acetate in the initial solution, and the pH influenced the chemistry of the aerogel formation process. The polycondensation of resorcinol and formaldehyde involves two main reactions, addition and condensation. The former is catalyzed by a base and the latter by acid. Additionally, the pH value influences the molecular environment and the hydroxymethyl derivatives of resorcinol are negatively charged, whereas formaldehyde is a cation in acidic medium. The electrostatic attraction between these two species may be responsible for the aggregation of clusters and the formation of larger particles. Thus, the largest aerogel particles were obtained at the lowest pH with the largest addition amount of copper acetate. Moreover, it can be seen that there are some small particles anchored to the surface of the carbon aerogels Cu-C-0.25 and Cu-C-0.3, and the EDS pattern reveals the present of copper. During the carbonization, the metallic Cu particles may migrate to the external surface of the carbon aerogels. Figure 2a is

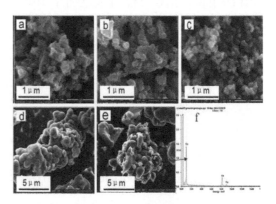

Figure 1. SEM images of Cu-loaded carbon aerogels: (a) Cu-C–0.1, (b) Cu-C-0.15, (c) Cu-C-0.2, (d) Cu-C-0.25, (e) Cu-C-0.3, and (f) the EDS pattern of Cu-C-0.3.

the TEM images of Cu-C-0.2, Figure 2b is the EDS pattern of Cu-C-0.2, and the EDS pattern reveals the presence of copper.

Figure 3 shows the powder XRD patterns of the resultant carbon aerogels. The peaks at 43.2°, 50.4° and 74.2° can be ascribed to the (111), (200) and (220) plane of metallic Cu (JCPDS No. 04-0836), respectively. This indicates that the copper ions added in the precursor solution have been reduced to metallic copper by carbon under 800°C. So the nanoparticles appearing on the surface of carbon particles from SEM observations are the copper particles. It is also found that a broad diffraction peak appears at about 26° for the Cu-loaded carbon aerogels, suggesting that some graphite-like carbon structure has been formed during the carbonization under catalysis of metallic copper. The crystalline copper content was determined by K value method with α-Al_2O_3 powder as a reference phase, and the unmarked diffraction peaks in the patterns are attributed to the α-Al_2O_3. The results are shown in Table 1.

Detailed characteristics of the pore structure of the obtained carbon aerogels are presented in Table 2. Analysis of the data indicates differences in the porosity of the samples. N_2 adsorption

Table 1. The copper content in the Cu-loaded carbon aerogels determined by different methods.

Samples	Copper contents by mass balance (wt%)	Crystalline copper contents by XRD (wt%)
Cu-C-0.1	0.5	0.14
Cu-C-0.15	0.8	0.16
Cu-C-0.2	1.0	0.19
Cu-C-0.25	1.3	0.21
Cu-C-0.3	1.6	0.24

Table 2. Porous parameters of samples.

Samples	S_{BET} (m^2/g)	$V_{<2nm}$ (cm^3/g)	V_{mes}(cm^3/g)
Cu-C-0	519.9	0.055	0.224
Cu-C-0.1	138.6	0.042	0.066
Cu-C-0.15	298.7	0.098	0.026
Cu-C-0.2	826.2	0.292	0.114
Cu-C-0.25	402.3	0.226	0
Cu-C-0.3	228.5	0.125	0.013

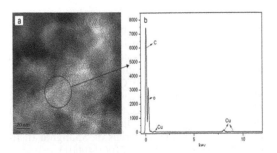

Figure 2. (a) TEM images of Cu-C-0.2 and (b) the EDS pattern of Cu-C-0.2.

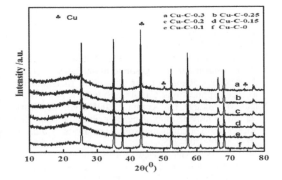

Figure 3. XRD patterns of Cu-loaded carbon aerogels.

desorption isotherms and the corresponding pore size distributions of samples are shown in Figure 4. As shown in Figure 4a, the samples Cu-C-0.25 and Cu-C-0.3 exhibit a type I isotherm according to the IUPAC classification. The major uptake at relative pressures less than 0.1 and an almost horizontal plateau at high relative pressures indicate that they are highly microporous materials. N_2 adsorption isotherms of the sample Cu-C-0, Cu-C-0.1, Cu-C-0.15 and Cu-C-0.2 belong to a Type II isotherm with reference to the IUPAC classification, which is typical of macro-porous solids. These isotherms show a steep uptake at high relative pressures with no sign of leveling off due to capillary condensation mainly in the macro-pores. There is also a small steep uptake of N_2 at low relative pressures in the initial part of these isotherms, which can be attributed to micropore filling of N_2. The hysteresis loop of these samples exhibits a type H3 hysteresis loop, characteristic of slit-shaped pores. The conclusions obtained based on the shape of the nitrogen adsorption isotherms at 77 K are in good agreement with the pore size distributions calculated using the DFT method (Fig. 4b). It is shown that the distribution of micropores of sample Cu-C-0.2, Cu-C-0.25 and Cu-C-0.3 is centered in the range of 0.5 to 0.7 nm, indicating the introduction of copper lead to a creation of ultrafine micropores.

3.2 Adsorptive desulfurization performance

The results of adsorptive desulfurization of Cu-loaded carbon aerogels are listed in Table 3.

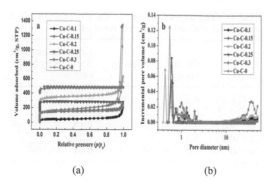

(a)   (b)

Figure 4. (a) N_2 sorption isotherms and (b) the corresponding DFT pore size distributions of carbon aerogels.

Table 3. The desulfurization capacity of Cu-C-n samples.

Samples	q_e (mg-S/g-A) (MF-1)	q_e (mg-S/g-A) (MF-2)	K (%)
Cu-C-0	10.83	4.67	56.9
Cu-C-0.1	5.31	3.08	42.0
Cu-C-0.15	6.76	4.25	37.1
Cu-C-0.2	13.96	9.77	30.0
Cu-C-0.25	9.72	7.01	27.9
Cu-C-0.3	7.16	5.3	26.0

Table 4. The desulfurization capacity of C-n samples.

Samples	q_e (mg-S/g-A) (MF-1)	q_e (mg-S/g-A) (MF-2)	K (%)
C-0.1	4.21	2.16	48.7
C-0.15	5.02	2.91	40.8
C-0.2	10.6	6.7	37.7
C-0.25	6.4	3.8	40.6

With the increasing of copper content, the desulfurization capacity of the Cu-loaded carbon aerogels first increased and then dropped, exhibited the largest value for the sample Cu-C-0.2. And the capacity decreased in the presence of benzene. This is because aromatics can strongly compete with thiophenic sulfur compounds, with their similar adsorption sites and size as sulfur compounds toward carbon surface via π-π dispersion interactions. It is found that the adsorption capacity is rather related to the pore size distribution than to the surface area. And the selectivity factor decreased with an increase of copper content, indicating the introduction of copper enhanced the selectivity of samples with its π-complexing effect.

The results of adsorptive desulfurization of Cu free carbon aerogels are listed in Table 4. The desulfurization capacity of the C-n samples first increased and then dropped, exhibited the largest value for the sample C-0.2. By comparing Table 3 and Table 4, it is found that the presence of copper increases both desulfurization capacity and selectivity.

4 CONCLUSIONS

In summary, Cu-loaded carbon aerogels were prepared by sol-gel method. The addition of copper leads to a generation of ultrafine micropores of the carbon aerogels, the distribution of micropores of sample Cu-C-0.2, Cu-C-0.25 and Cu-C-0.3 is concentrated in the range of 0.5 to 0.7 nm. The adsorption capacity for DBT is related to the volume of micropores, the selectivity and adsorption capacity are improved by presence of copper, indicating that copper is favorable to DBT adsorption on the surface via its π-complexing effect.

ACKNOWLEDGEMENT

This work was supported by the Capacity Building Program of Shanghai Local Universities (No. 12160503600), Youth Innovation Fund from Shanghai Institute of Technology (No. YJ2013-35), the First-class Discipline Construction Fund of Shanghai Municipal Education Commission (No. J201212), Nature Science Foundation of China (U1332107) and Key Discipline Construction Fund of Composite Materials of Shanghai Institute of Technology (No. 10210Q140001).

REFERENCES

Kabe, T. Ishihara, A. & Tajima, H. 1992. Hydrodesulfurization of sulfur-containing polyaromatic compounds in light oil, *Industrial & Engineering Chemistry Research*, 31: 1577–1580.

Ma, X.L. Sun, L. & Song, C.S. 2002. A new approach to deep desulfurization of gasoline, diesel fuel and jet fuel by selective adsorption for ultra-clean fuels and for fuel cell applications. *Catalysis Today*, 77: 107–116.

Nagai, M. Fukiage, T. & Kurata, S. 2005. Hydrodesulfurization of dibenzothiophene over alumina-supported nickel molybdenum phosphide catalysts. *Catalysis Today*, 106: 201–205.

Song, C.S. 2003. An overview of new approaches to deep desulfurization for ultra-clean gasoline, diesel fuel and jet fuel. *Catalysis Today*, 86: 211–263.

Advanced Materials and Structural Engineering – Hu (Ed.)
© *2016 Taylor & Francis Group, London, ISBN 978-1-138-02786-2*

Synthesis of Ni-loaded carbon aerogels by in-situ and incipient wetness methods and their adsorption performance for dibenzothiophene in model fuel oil

N. Jiang, S.L. Jin, X. Shao, H.F. Zhang, Z.J. Li, S.M. Zhang, M.L. Jin & R. Zhang
School of Materials Science and Engineering, Shanghai Institute of Technology, Shanghai, China

ABSTRACT: Nickel was loaded onto Carbon Aerogels (CAs) by an in-situ method in sol-gel process during the preparation of CAs and by an incipient wetness method after CA was prepared. In the in-situ method, resorcinol and furfural were used as carbon source, nickel nitrate as nickel source, absolute ethanol as a dispersion medium, 1, 2-epoxypropane as a gel initiator, polyacrylic acid as a chelating agent. In the incipient wetness method, EDTA-Ni was used as nickel source and the impregnated samples were heat treated at 400 °C for 3 h. The CAs were characterized by nitrogen adsorption and XRD. The effects of the molar ratio of 1, 2-epoxypropane to nickel on porosity of the Ni-loaded CAs were investigated. It was found that the Ni-loaded CAs had abundant mesopores and the loaded nickel was dispersed in samples in metallic state and oxide or pure metal depending on the preparation conditions in the in-situ method. The mesopore volume and average mesopore size all increased with the ratio of 1, 2-epoxypropane to nickel. The sulfur adsorption capacity for a model fuel oil containing dibenzothiophene was 2.56–6.40 mgS/g-adsorbent for the in-situ prepared Ni-loaded CAs. The nickel in the Ni-loaded CAs by the incipient wetness method was in an amorphous sate. The sulfur adsorption capacity decreased from 12.92 to 4.43 mgS/g-adsorbent with increasing Ni-loading level from 0 to 35.8 wt% for the Ni-loaded CAs prepared by the incipient wetness method. Nickel in amorphous state was more active for sulfur removal than that in crystal state.

1 INTRODUCTION

There are many reports about the property and preparation of Carbon Aerogels (CAs) at home and abroad. Meanwhile, there are some applied researches on electric absorption (Yan 2007). However, reports on transition metal-loaded CAs are rare. The loading level of metal in CAs prepared is also very low. In this paper, the CAs with high contents of metal loading was prepared by adding a complexing polymer during the sol-gel process (Fu 2005). And the characterization of the porosity and metallic state shows that the application of the complexing polymer can help to prepare the metal-loaded CAs with high contents of nickel. This method has characteristics of simple operation, one-pot and high content of metal loading. This is helpful in research on the study of the metal-loaded CA adsorption, catalysis and the electrochemical energy storage. As a comparison, nickel was loaded onto CAs after CA was prepared by an incipient wetness method. These two kinds of Ni-loaded CAs were used for desulfurization for a model fuel containing dibenzothiophene. Their desulfurization performance was compared.

2 EXPERIMENTAL

In the in-situ loading method, the furfural (F) and resorcinol (R) were used as carbon precursors, the absolute ethanol as a solvent, 1, 2-epoxypropane (PO) as a gel initiator, and the polyacrylic acid as a chelating agent to prepare Ni-loaded CAs. The molar ratios of PO to Ni were 2, 4, 5 and 7, corresponding to sample 1, 2, 3 and 4, respectively. The other formulations, including molar ratios of R/F, (R+F)/Ni, and Ni concentration in the sol, were fixed for all samples investigated. Briefly, resorcinol, 1, 2-epoxypropane and furfural were dissolved in absolute ethanol under stirring to form a clear solution, named as solution A. Polyacrylic acid and nickel nitrate were dissolved in absolute ethanol to form a clear solution, named as B. First, put the flask containing the solution A into an ice bath to reach thermal equilibrium. Then solution B was added into solution A drop by drop under an intense agitation. The resulting solution mixtures were put into glass ampoules, sealed, and placed in a water bath at room temperature for 3 days, then heated to 75 °C and held at this temperature for 5 days. The gels formed in the ampoules. The ampoules were broken, the gels inside were taken out and put into a closed container containing absolute ethanol for solvent displacement for 1 day. The absolute ethanol in the container was displaced once a day and this displacement was repeated 5 times. The solvent displaced gels were put into an autoclave, into which N-hexane was poured and heated up to 240 °C with a rate of 1 °C/min. During heating, the pressure inside the autoclave was increased by self-generated vapor pressure. The pressure inside the autoclave was kept no more than 6.0 MPa with an outlet valve until the temperature inside the auto-clave reached 240 °C. The temperature and pres-sure inside the autoclave remained unchanged for 1 hour. After that, the outlet valve of the autoclave was gradually opened to release pressure inside at a rate of 0.2 MPa/min to ambient pressure under 240 °C. The gels inside the autoclave after the tem-perature inside was lowered to room temperature were hybrid inorganic/organic aerogels. The hybrid aerogels were heated up to 800 °C with a rate of 5 °C/min under the protection of N_2 in a tubular reactor and held at this temperature for 3 h to obtain Ni-loaded CAs.

In the incipient wetness method, the process for CA preparation was similar to the above proce-dure, but the formaldehyde (F) and resorcinol (R) were used as precursors and sodium carbonate as a catalyst. When the CA had been made, EDTA-Ni was loaded with the CA by the incipient wetness method to a Ni content of 4, 11.9, 19.9, 27.2 and 35.8 wt%.

About 0.1 g of tested adsorbent and 10 g of model fuel were added into a glass tube and mixed. The tube was capped and placed in a shaking bath at 30 °C for 48 h. Then the mixture was filtered, and the treated model fuel samples containing dibenzothiophene were analyzed to estimate the sulfur adsorption capacity of the adsorbents. The amount of sulfur adsorbed per gram of adsorb-ent, qe (mg-S/g-A), was calculated by the following equation:

$$q_e = \frac{(C_0 - C_e)m^*}{m} \tag{1}$$

where, C_0 and C_e are the initial and equilibrium sul-fur concentrations in the model fuel (mg/g), respec-tively, m* is the weight of model fuel (g) and m is the mass of the adsorbent (g).

A Micromeritics ASAP 2020 instrument was used to characterize the surface areas and pore structures of the samples using N_2 adsorption/desorption at −196 °C. The samples were degassed under vacuum at 150 °C for 15 h prior to measurements. The Brunauer Emmett and Teller BET surface area (S_{BET}) were calculated from the N_2 adsorption data at the relative pressure (P/P_0)

range between 0.05 and 0.2 using the standard BET equation (Zhang et al. 2012). The pore size distri-bution was determined by BJH method from the desorption branch of isotherms. X-Ray Diffraction (XRD) patterns were recorded in the range from 10 to 80° on a Rigaku-DMAX2200PC diffractom-eter using Cu Kα (λ = 0.1506 Å) radiation.

3 RESULTS AND DISCUSSION

3.1 *Specific surface area, pore volume and pore size*

The absorption and desorption isotherms of all samples have obvious hysteresis loops. Figure 1a shows a typical absorption and desorption iso-therm of an in-situ loaded a sample. Figure 1b shows a typical absorption and desorption iso-therm of Ni-loaded CA with the incipient wetness method. These types of isotherms show that these samples have mesopore structure and a wide pore size distribution (Yan 2007).

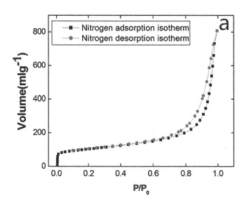

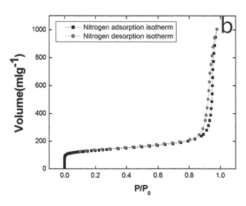

Figure 1. Nitrogen adsorption-desorption isotherm: (a) the in-situ loaded sample and (b) the incipient wetness loaded sample.

Figure 2a shows the BJH pore size distributions of all the in-situ loaded samples. The samples all have two sizes of pores around 4 nm and 20 nm. The pores of around 4 nm are all similar, but that of pore around 20 nm are quite different. With increasing PO/Ni molar ratio, the pore volume of pore around 20 nm increases, indicating that 1, 2-epoxypropane is favorable for the development of this kind of mesopores. Figure 2b shows the BJH pore size distributions of the incipient wetness loaded CA, the pore volume of pore around 20 nm is dominant in this sample.

Table 1 shows that specific surface area, pore volume, and average pore size of samples with increasing the ratio of PO/Ni. It is obvious that BET surface area, pore volume, and average pore size all increase with the PO/Ni ratio, indicating that 1, 2-epoxypropane is favorable for the development of mesopores. This is reasonable in that 1, 2-epoxypropane promotes the gel formation and possibly leads to the formation of stiff NiO network, which can endure intense heat

Table 1. BET surface area, pore volume and average pore size of in-situ loaded samples.

	S_{BET} (m^2/g)	V_{pore} (cm^3/g)	Average D_{pore} (nm)	Wt (Ni)%
1	250.8	0.688	15.8	9.3
2	307.1	0.7	15.4	7.9
3	308.5	0.816	16.8	7.3
4	364.2	1.208	18.8	6.4

during carbonization and reduce the collapse of pore network. The nickel content in the aerogels decreases with PO/Ni ratio, which might be caused by an incomplete hydrolysis of nickel nitrate that was leached out during ethanol displacement step.

Table 1 also shows that specific surface area, pore volume, and average pore size of the un-loaded CA that has a high BET surface area is appropriate for Ni loading. The lower surface areas of Ni-loaded CAs than that of unloaded CA might be caused by micropore blocking Ni species.

3.2 XRD

Figure 3a and 3b are the XRD patterns of the in-situ loaded samples. Figure 3a shows that there are obvious diffraction peaks in the 2θ of 37.28°, 43.28° and 62.859°, which conforms to PDF card (No. 44–1159) and belongs to nickel oxide. Moreover, sample 1 shows obvious diffraction peaks in the points of 44.46° and 51.801°, which conforms to PDF card (No. 04-0580) and belongs to metallic nickel. There is obvious diffraction peak in the 2θ of 44.58° and 51.919° in XRD pattern of sample 2, which shows that sample 2 contains only metallic nickel which is relatively pure and in a complete crystal form. The reason for the nickel oxide's appearance in sample 1 is probably because of air leakage during the carbonation, which causes an oxidation of metallic nickel to nickel oxide. Figure 3b shows the XRD patterns of the in-situ loaded sample 2, 3 and 4, all show the same diffraction peaks at 44° and 52°, which conforms to PDF card (No. 04-0580) and belongs to metallic nickel, which is relatively pure and in a complete crystal form.

Figure 3c shows the XRD pattern of the incipient wetness loaded sample, it belongs to an amorphous type after loading.

3.3 The desulfurization performance

Table 2 shows the desulfurization capacity of all in-situ loaded samples.

Table 3 shows the sulfur adsorption capacity of the incipient wetness loaded samples. The sulfur adsorption capacity decreases with increasing

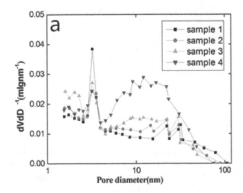

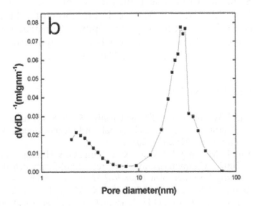

Figure 2. Pore size distributions of samples: (a) the in-situ loaded samples and (b) the incipient wetness loaded sample.

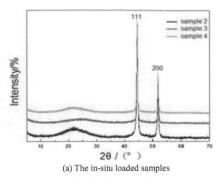

(a) The in-situ loaded samples

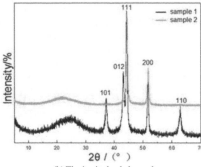

(b) The in-situ loaded samples

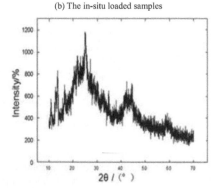

(c) The incipient wetness loaded sample

Figure 3.   XRD patterns of samples.

Table 2.   The sulfur adsorption capacity of the in-situ loaded CAs.

Sample	1	2	3	4
mgS/g-A	6.4	5.12	2.56	6.08

Table 3.   The sulfur adsorption capacity of the incipient wetness loaded CAs.

Wt% (Ni)	CA (0%)	4%	11.9%	19.9%	27.2%	35.8%
mgS/g-A	12.92	12.28	7.12	8.01	6.83	4.43

Ni content. This again confirms that Ni can block pores of CA and thus causes a decrease of sulfur adsorption capacity.

The sulfur adsorption capacity is higher for the incipient wetness loaded CAs than the in-situ loaded CAs under the similar Ni content. This could be caused by the high temperature treatment of the former at 800 °C, which is favorable for nickel growth and decrease accessible surface area. This is partly confirmed by the fact that the Ni in the latter is in an amorphous state while Ni in the former is in a crystal state with a larger size than the latter.

4   CONCLUSIONS

For the in-situ loaded CAs, the pore volume, specific surface, and pore diameter all increase with the in-creasing PO/Ni ratios. The nickel doping level in our samples is between 6.3 to 9.3 wt.% for the in-situ loading method, which decreases with increasing PO/Ni ratio. The loaded nickel is in a metallic state in most cases except for sample 1, in which there is nickel oxide that might be caused by air leakage during carbonization of the sample. The sulfur adsorption capacity for the Ni-loaded CA is low due to a pore blocking effect. For the incipient wetness loaded CAs, the nickel is in an amorphous sate and the sulfur adsorption capacity decreases with Ni-loading level, also indicating a pore blocking phenomenon.

ACKNOWLEDGEMENTS

This work was supported by the Capacity Building Program of Shanghai Local Universities (No. 12160503600), Youth Innovation Fund from Shanghai Institute of Technology (No. YJ2013-35), the First-class Discipline Construction Fund of Shanghai Municipal Education Commission (No. J201212), Nature Science Foundation of China (U1332107) and Key Discipline Construction Fund of Composite Materials of Shanghai Institute of Technology (No. 10210Q140001).

REFERENCES

Fu, R.W. Baumann, T.F. & Cronin, S. 2005. Formation of graphitic structures in cobalt and nickel-doped CA, *Langmuir*, 21: 2647–2651.

Yan, H.M. Tang, Y.J. Wang, C.Y. Jiang, G. & Huang, C.G. 2007. Characterization of resorcinol-formaldehyde aerogel powder, *New Chemical Materials*, 35(7): 29–30.

Zhang, H.X. Huang, H.L. Li, C.X. Meng, H. Lu, Y.Z. Zhong, C.L. Liu, D.H. & Yang, Q.Y. 2012. Adsorption behavior of metal–organic frameworks for thiophenic sulfur from diesel oil, *Industrial & Engineering Chemistry Research*, 51: 12449–12455.

Advanced Materials and Structural Engineering – Hu (Ed.)
© 2016 Taylor & Francis Group, London, ISBN 978-1-138-02786-2

An experimental technique to investigate gas-turbine blades dry-friction dampers efficiency

M. Nikhamkin, N. Sazhenkov & S. Semenov
PNRPU—Perm National Research Polytechnical University, Russian Federation

I. Semenova
VSB-TU Ostrava, University of Ostrava, Czech Republic

ABSTRACT: The experimental technique to investigate efficiency of gas turbine blades underplatform dampers is described. Dummy and real gas turbine blades with wedge, half-cylindrical and underplatform shaped dampers were used as the study objects. The experimental blade damping characteristics for dampers of various design under the simulated centrifugal and excitation forces were obtained. The presented experimental set-up enables to investigate blade dampers efficiency in two stages. At the first stage the most harmful blade vibration mode is identified using the 3-D scanning laser vibrometry technique. At the second stage the blade-damper interaction at the previously defined single high stress vibration mode is studied more detailed in a wide range of the modeled centrifugal and excitation forces. The forces correspond to operating points of real aircraft gas-turbine engine. The experimental technique is recommended to estimate underplatform dry-friction dampers efficiency and numerical models verifications.

1 INTRODUCTION

Underplatform friction dampers are commonly used in turbine blades to prevent high vibration stresses and high cycle fatigue failures. The main problem in the dampers design is the proper choice of damper mass and geometry to achieve the optimal working characteristics. The optimal characteristics suggest that the damping efficiency during the operational engine regimes is close to the maximum. The design optimization is a difficult task because dry friction presupposes nonlinear calculations of numerical models that need to be experimentally verified (Petrov & Ewins 2006, Marquina et al. 2008).

The main object of the article is to describe the experimental technique that enables to get experimental data for numerical models verification in laboratory conditions, make estimations of the damping efficiency using various real and dummy blades with various friction damper designs and to analyze basic regularities in vibrating turbine blade-damper systems.

2 EXPERIMENTAL SET-UP

2.1 *Dummy blades*

The block of two dummy blades with half-cylindrical and wedge-shaped dampers was used as

the investigation object. The massive blade cubic base simulated a part of the rotor disc (Fig. 1). The absence of blades locks eliminated additional friction pairs so the contact between the blade platform and the damper was the only friction element in the investigation process.

The special hole in the dummy blade cubic base was made to join two of them with the coupling bolt. The tightening torque of the joint was 180 N•m. The dummy blades were fixed in the high precision support on a massive steel base. The tightening torque when fixing blades in the support

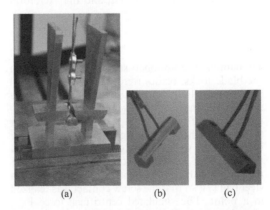

(a)   (b)   (c)

Figure 1. Investigation object a) dummy-blades block; b) cylindrical damper; c) wedged damper.

was 220 N•m. The support was joined to the metal beams with pulleys, cables and weight system for simulating centrifugal load F_{cent}, N to the damper.

The main idea of the experiment is to simulate the effect of the loaded by various centrifugal forces damper on the blades block vibration velocities.

The experimental set-up was based on the three dimensional scanning laser vibrometry (Zucca et al. 2008, Balakirev et al. 2008). Measurements of the vibration velocities in scan nodes were made with 3D scanning laser vibrometer Polytec PSV-400-3D.

Application of the excitation force was made through the elastic metal rod, the free end of it was set against the dummy blade foot and the other was connected to the magnet vibrator. The harmonic excitation force had frequency range of 20...3000 Hz with the maximum amplitude of 1600 N. The force amplitude control was made by piezosensor.

The designed experimental set-up makes it possible to simulate the centrifugal forces acting on the damper in range of 0...2000 N. It covers the whole spectrum of the real turbojet engine work regimes in case of damper mass lower than 0.005 kg. The experimental technique enables to determine vibration modes and frequencies so as damping decrements of the blade-damper system.

At the first stage of the experiment the modal analysis of the two dummy-blade block was made using three-dimensional scanning laser vibrometer. Frequency response was produced by three reference axes at 40 points (20 points on each blade). Scanning was made in frequency range of 250...350 Hz with frequency scan step of 0.488 Hz. The excitation force had sine-wave form with constant amplitude and increasing frequency.

At the second stage of the experiment the blade damping efficiency was studied. The experiment was conducted by the force vibration response analysis of the blade-damper system at various levels of centrifugal load and the excitation force.

2.2 Real turbine blades

To eliminate additional structural damping in the blade locks, blades were welded to a base (2) (Fig. 3) simulating the rotor disc. The base was fixed in a clamping device by side surfaces. Simulation of centrifugal force impacting on the damper (6) was provided by pressing a damper to the rotor blade shelves with clamping screw (3) installed in the bottom of the base (2) (Fig. 3,d). The force of the screw to the damper was transmitted through steel ball (7), so that the screw contact was realized in a point. The simulated centrifugal force F_{cent}, N acting on the damper depending on the engine operation regime was in range of 735...1520 N.

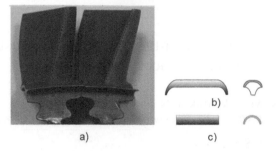

Figure 2. a) Real turbine blades and dampers: b) №1; c) №2.

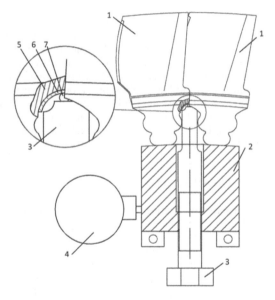

Figure 3. Force applying scheme.

The experimental technique was realized using 3D scanning laser vibrometer Polytec PSV-400-3D. Oscillations were excited by piezoshaker based on the blade block (see Fig. 2) in the frequency range of 20 to 3125 Hz. Scanning grid consisted of 29 nodes, which is sufficient to identify the lower natural modes of the blade. For each of the simulated centrifugal loads, the experiment was conducted three times to average random measurement errors.

3 EXPERIMENTAL RESULT

3.1 Dummy blades

During the experiment two resonance frequencies of the blades were detected, that are the synphase bend mode 306.3 Hz and the antiphase bend mode

308.1 Hz. Vibration response analysis was made for various levels of the simulated centrifugal force and the excitation force amplitudes for the case of the antiphase bend mode. The experimental relations between the relative damping decrement (normalized to the relative damping decrement of system without friction damper) and the centrifugal load F_{cent}, N on the cylindrical damper are shown in Figure 5. The relations were defined for three values of the excitation force amplitude 9.5, 15 and 40 N.

The plotted relations (Fig. 4) illustrate that the maximum damping efficiency can be achieved in centrifugal load range of 40...60 N for all kind of the excitation force amplitudes. With the excitation force amplitude increase from 9.5 N to 15 N, the maximum damping efficiency drops from 3.6% to 3%. Then with excitation amplitude increasing from 15 N to 40 N the maximum damping efficiency raises to 4%.

The next stage of the experiment was to compare damping efficiency of two types of dampers: half-cylindrical and wedge-shaped (Fig. 5).

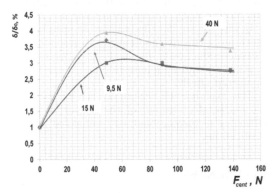

Figure 4. Relative damping decrement to centrifugal load dependence for various excitation force amplitudes.

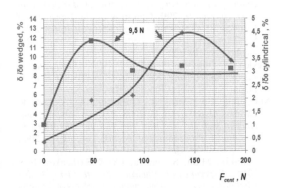

Figure 5. Relative damping decrement to centrifugal load dependence for wedged (blue) and cylindrical (red) dampers.

Maximum damping efficiencies of 12–13% for the dummy-blade block with wedge shaped damper were achieved in the centrifugal loads range of 130...160 N. The maximum damping efficiency with the cylindrical damper does not exceed 4% and the damper is suited to work at lower centrifugal loads (around 50 N) than wedged damper.

3.2 Real turbine blades

Frequency response and mode shapes of the studied real-blade block with two types of friction underplatform dampers (Fig. 2,b,c) were obtained. Figure 6 shows an example blade frequency response with the first damper type (Fig. 2,b) for several values of the simulated centrifugal load F_{cent}, N.

The reproducibility of the experimental results was estimated by the coefficients of variation of the natural frequencies and the logarithmic decrements. They were determined by three experiments conducted for each variant of modeling centrifugal load. The coefficient of variation of the natural frequencies did not exceed 0.03%, indicating good reproducibility of the natural frequencies. Coefficient of variation of the relative damping coefficient (normalized to the relative damping decrement of system without friction damper) was somewhat worse and was within 5.9%.

Without a damper two blade natural frequencies of 1480 Hz and 1891 Hz were revealed. The first frequency corresponded to synphase blade bend mode and antiphase blade mode as shown in Figure 7.

When installing the first type damper and increasing contact pressure the both natural frequencies increase due to system stiffness growing. The natural frequency of antiphase mode increases from 1891

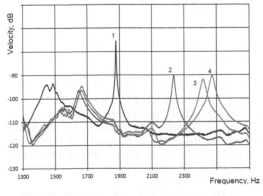

Figure 6. Frequency response of real blades with first type damper design for various centrifugal loads F_{cent}, N: 1—F_{cent}, N = 0; 2—F_{cent}, N = 162 N; 3—F_{cent}, N = 798 N; 4—F_{cent}, N = 1369 N.

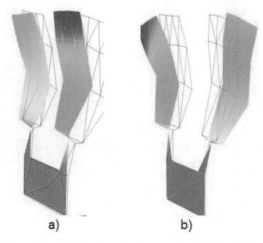

Figure 7. The first bending forms of real blades block a) synphase mode b) antiphase mode.

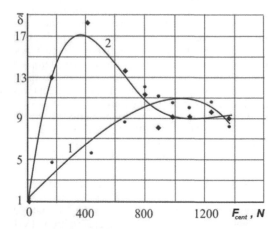

Figure 8. Relative damping decrement to centrifugal load dependence for dampers designs: (1), (2), antiphase vibration mode.

Figure 8 shows the dependence of the relative damping (normalized to the relative damping decrement of the system without a friction damper) on the simulated centrifugal load value F_{cent}, N for both dampers (1) and (2) for the case of blade antiphase vibration mode.

4 CONCLUSIONS

The experimental technique for damping efficiency estimation of gas turbine blades with friction dampers was developed. The designed experimental set-up enables to obtain damping characteristics in the range of 0...2 kN for centrifugal load value and 20...3000 Hz for excitation force frequency.

For the dummy-blade block the maximum damping efficiency of the wedged damper (12%) is 3 times greater than the damping efficiency of the cylindrical damper (4%) and it is located in the centrifugal load range of 130...160 N.

For the real gas-turbine blades the experiment performed for so called antiphase and synphase condition which simulates the two basically important motion types of neighboring blades.

It was shown that the second damper type is much more effective and enables to reach 28 and 18 times reduction in the blades dynamic stresses at synphase and antiphase resonant mode vibrations, respectively.

The experimental technique can be used for the independent studies of the friction damping efficiency in turbine blades or numerical models verifications.

Hz (without damper) to 2481 Hz (31%) for simulated centrifugal load on damper $F_{cent} = 1369$ N. The synphase natural frequency increases only by 18% from 1480 Hz (without damper) to 1750 Hz.

For damping effectiveness estimation the ratio of the logarithmic decrement of the blades with damper to logarithmic decrement of the blades without a damper can be used. Increasing the logarithmic decrement n times means a reduction in the blades dynamic stresses of resonant mode vibrations at n times.

For the synphase vibration mode the maximum ratio in case of the first damper type is 1.3, for the second type damper—28. For the antiphase blades bend mode the first damper type gives ratio of 11, the second damper type—18.

REFERENCES

Balakirev, A. Bolotov, B. Golovkin, A. Nikhamkin, M. Sazhenkov, N. Voronov, L. & Konev, I. 2014. *Experimental evaluation of the efficiency of gas turbine engine parts damping with dry friction dampers using laser vibrometer.* 29th Congress of the International Council of the Aeronautical Sciences, ICAS 2014, St. Petersburg; Russian Federation; 7–12 September 2014; Code 108502, 7.

Heylen, W. Lammens, S. & Sas, P. 1997. *Modal Analysis Theory and Testing*, Katholieke Universiteit Leuven, Departement Werktuigkunde, Leuven.

Marquina, F.J. Coro, A. Gutierrez, A. Alonso, R. Ewins, D.J. & Girini, G. 2008. *Friction damping modeling in high stress contact areas using microslip friction model.* Proceedings of ASME Turbo Expo 2008, GT2008-50359. 10.

Nikhamkin, M.Sh. Semenova, I.V. & Sazhenkov, N.A. 2014. *Experimental efficiency investigation of dry-friction dampers for turbine blades.* Problems of dynamic and strength in gas-turbine construction: Abstract of 5-th International Conference/Ed. A.P. Zhinkovsky. Kyiv: G.S. Pisarenko Institute for Problems of Strength of the National Ac. Sci. of Ukraine, 167–168.

Nikhamkin, M.Sh. Semenova, I.V. & Sazhenkov, N.A. 2014. *Mathematical simulation of underplatform dry-friction damper and turbine blades interaction.* Problems of dynamic and strength in gas-turbine construction: Abstract of 5-th International Conference/ Ed. A.P. Zhinkovsky. Kyiv: G.S. Pisarenko Institute for Problems of Strength of the National Ac. Sci. of Ukraine, 169–170.

Petrov, E.P. & Ewins, D.J. 2006. *Advanced modeling of underplatform friction dampers for analysis of bladed disc vibration.* Proceedings of ASME Turbo Expo 2006, GT-2006-90146 10.

Wu, J. Yuan, R. Zhao, P. & Xie, Y. 2013. Experimental study of dynamic characteristics of dry friction damping of turbine blade steel, *Applied Mechanics and Materials,* 316: 268–272.

Zucca, S. Botto, D. & Gola, M. 2008. *Range of variability in the dynamics of semi-cylindrical friction dampers for turbine blades,* ASME Turbo Expo 2008: Power for Land, Sea, and Air, 5: Structures and Dynamics, Paper № GT2008-51058, 519–529.

Advanced Materials and Structural Engineering – Hu (Ed.)
© 2016 Taylor & Francis Group, London, ISBN 978-1-138-02786-2

A nonlinear compressive response of polypropylene

H. Zheng, Y.G. Liao, K. Zhao & Z.P. Tang
Department of Modern Mechanics, University of Science and Technology of China, Hefei, China

ABSTRACT: The compressive mechanical behavior of polypropylene (PP) was invested in this paper. The quasi-static and dynamic tests were used, over a wide range of strain rates (10^{-4} s^{-1} to 10^{3} s^{-1}), using an MTS machine and a Split Hopkinson Pressure Bar (SHPB). The nonlinear viscoelastic characteristic was shown in the compression process. In addition, it was found that the compressive yield stress of polypropylene increased as strain rate increased.

1 INTRODUCTION

Polymer structures are usually used with multi-axial deformations in engineering. The linear viscoelastic theory of polypropylene as a common engineer polymer has been developed. But few studies about polymers' nonlinear viscoelastic response can be found.

It is well established that the yield behavior of polymeric materials subjected to external loads exhibits pronouncing hydrostatic pressure dependency, temperature and strain rate sensitivity (Anand et al. 2009, Rittel & Brill 2008, Rittel & Dorogoy 2008, Khan & Farrokh 2006, Khan & Zhang 2001, Caddell & Woodliff 1980, Pae 1977, Silano et al. 1974).

More efforts, however, have been made to investigate the effect of hydrostatic pressure on yield responses of these materials, both experimentally and theoretically.

The present work explores the strain rate dependence and temperature dependence of polypropylene in uniaxial compress process, to investigate how this important polymer fit into the Ree-Eyring framework. Quantitative comparison of experimental results and the theory will be reported elsewhere.

2 MATERIALS

The material used in this study is polypropylene obtained from Heifei Huasheng Rubber Co. Ltd.

All specimens were cylindrical, with a length of 7 mm and a diameter of 10 mm. They were cut using a precision diamond saw from stock materials and their ends were polished to ensure them parallel and flat. To cool the specimen, the cutting procedure was very slow to avoid the introduction of machining heat-induced anisotropy in the specimen.

Table 1. Material parameters of polypropylene.

Material	Density	Tg	Tf	Crystallinity ratio
Polypropylene	0.907	263 K	438 K	64.6%

3 EXPERIMENTAL DEVICES

3.1 Quasi-static tests

Quasi-static tests were performed by using an MTS 810 material testing system. During this test, grease was applied to both ends of the specimen to reduce friction and prevent barreling. Stress-strain curves were obtained from load-displacement measurements considering compensation for machine compliance taken into account. All quasi-static tests were performed at a specific temperature without attempting to measure the specimen's temperature change.

3.2 Dynamic tests

Dynamic loading was realized, using an SHPB (Split-Hopkinson pressure bar). The incident strain signal, reflected strain signal, and transmitted strain signal recorded by strain gauges on the SHPB were used to obtain the average stress and strain in a homogenously deforming specimen as follows

$$\sigma(t) = \frac{AE\varepsilon_t(t)}{A_s}, \tag{1}$$

$$\varepsilon(t) = \int_0^t -\frac{2c\varepsilon_r(\tau)}{L_s}d\tau, \tag{2}$$

where, A, E and c represent the cross-sectional area, Young's modulus and wave speed, of the

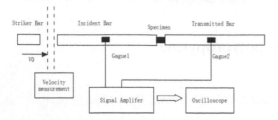

Figure 1. A schematic of conventional SHPB setup.

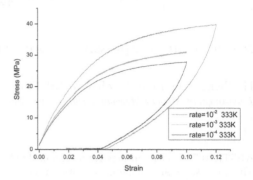

(a) Temperature=333K results

bar respectively. L_s and A_s denote the length and cross-sectional area of the specimen. Details of the SHPB can be found in Follansbee (Follansbee 1985) and will not be discussed in this paper. The data analysis procedure used for these experiments was discussed by Li and Lambros (Li and Lambros 2001). It includes both an experimental verification of homogeneous deformation, and a dispersion correction for all signals recorded by an incident and transmitted strain gauges.

4 RESULTS AND DISCUSS

4.1 Quasi-static results

All quasi-static uniaxial compression tests were performed, using an MTS 810 material testing system. These experiments were conducted at different strain rates of 10^{-2} S^{-1}, 10^{-3} S^{-1} and 10^{-4} S^{-1} and a different test temperature of 333 K, 363 K and 393 K. Stress-strain curves were obtained from load-displacement measurements considering compensation for machine compliance taken into account.

Stress-strain curves for polypropylene at all three strain rates and three temperatures are shown in Figure 2. Similarly with the tensile test's results which has been observed in the work of Arruda (Arruda et al. 2009), our results shows the response of polypropylene is consisted of three regimes: an initial linear portion, nonlinear viscoelastic transition portion, and finally a yielding and strain softening portion. The strain rate sensitivity and temperature sensitivity of polypropylene material are obvious in the quasi-static compression process.

4.2 Dynamic results

All results from 10 SHPB dynamic compression tests at room temperature were conducted in Table 2. An original signal from typical SHPB test is shown in Figure 3. Unlike the situation in the quasi-static tests, in the SHPB case, the value of maximum strain reached depends not only on the material

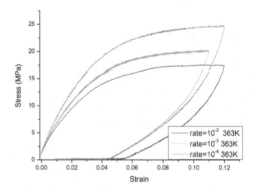

(b) Temperature=363K results

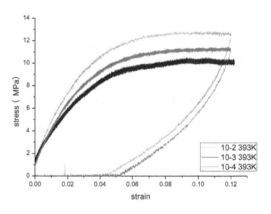

(c) Temperature=393K results

Figure 2. Stress-strain curve in quasi-static experimental.

properties, but also on the duration of the loading pulse. The longest loading pulse duration determined by the projectile length used in this study was 120 us in Figure 3. The pulse duration is too long to extract enough strain and stress information.

Table 2. Experimental parameters and results of high strain rate loading.

No	Bullet length	Bullet speed	Specimen size D*L	Average strain rate	Max stress	Max strain
#1	300 mm	3.98 m/s	10.01*7.01	281 S^{-1}	73.2 Mpa	3.4%
#2	300 mm	5.96 m/s	10.01*7.01	409 S^{-1}	103.3 Mpa	5.1%
#3	300 mm	9.1 m/s	10.01*6.99	672 S^{-1}	124.0 Mpa	8.3%
#4	300 mm	7.3 m/s	10.01*6.99	528 S^{-1}	116.0 Mpa	6.1%
#5	300 mm	11 m/s	10.02*7.02	914 S^{-1}	128.6 Mpa	11.1%
#6	300 mm	4.7 m/s	10.03*6.955	340 S^{-1}	96.9 Mpa	4.7%
#7	300 mm	8.2 m/s	10.03*6.95	626 S^{-1}	120.7 Mpa	7.5%
#8	300 mm	7.6 m/s	10.02*7.02	584 S^{-1}	117.9 Mpa	6.9%
#9	300 mm	9.2 m/s	10.005*7.00	734 S^{-1}	125.3 Mpa	8.4%
10	300 mm	10.6 m/s	10.005*6.965	800 S^{-1}	129.2 Mpa	9.1%

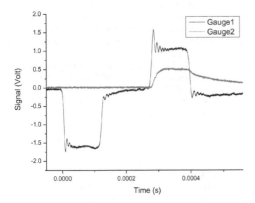

Figure 3. Experiment strain-time signal (9#).

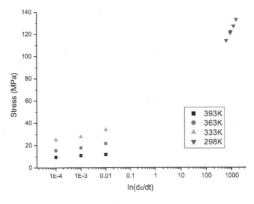

Figure 5. Stress-strain curve summary in dynamic experimental.

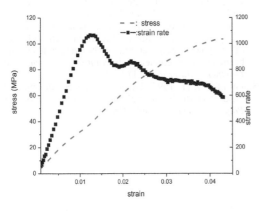

Figure 4. Stress-strain curve and strain rate-strain curve (3#).

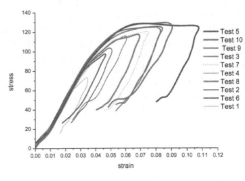

Figure 6. The relationship between yield stress σ_y and strain rate.

4.3 Rate and temperature dependence during compressive yield of polypropylene

Figure 5 summarizes all stress-strain curves of polypropylene over an average strain-rate from 281 S^{-1} to 914 S^{-1}. It should be noted that rate sensitivity obviously exists in shock compression, compared with the quasi-static compression.

For most polymers at room temperature, the dependence of yield stress on log (strain rate) is linear at modest rate ranges, hence Eyring's theory agrees with the observed response.

In constructing the Eyring plot for the wide range of strain rates studied in this work, the 'yield' stress, σ_y was identified as the maximum true stress in stress-strain curve. Figure 6 shows that the distribution of yield stress from the experiment in large strain rate and temperature range is approximately linear and consistent with the theory of Eyring.

5 CONCLUSIONS

In this study, the effect of strain rate sensitivity and temperature sensitivity in polypropylene, a semi-crystalline polymer, was investigated experimentally. The material was subjected to compressing loading at different strain rate and different temperature condition. The experiments show that the yield stress of polypropylene increases with increasing strain rate, at the same temperature, however, yield stress will decrease with the increase of temperature at the same strain rate loading condition. Moreover, yield stress in large strain rate (10^{-4} to 10^3 s^{-1}) and temperature range (298 K to 393 K), is consistent with the Eyring's theory.

REFERENCES

Anand, L. Ames, N.M. Srivastava, V. & Chester, S.A. 2009. A Thermo-mechanically coupled theory for large deformations of amorphous polymers. Part I: formulation. *International Journal of Plasticity*, 25(8): 1474–1494.

Bowden, P.B. & Jukes, J.A. 1972. The plastic flow of isotropic polymers. *Journal of Materials Science*, 7: 52–63.

Caddell, R.M. & Woodliff, A.R. 1980. Yield behavior of unoriented and oriented polycarbonate and polypropylene as influenced by temperature. *Materials Science and Engineering*, 43: 189–198.

Follansbee P.S. 1985. *The Hopkinson bar, in: Mechanical Testing, Metals Handbook*, vol. 8, 9th ed. American Society for Metals, Metals Park, OH, 198–217.

Ishai, O. & Bodner, S.R. 1970. Limits of linear viscoelasticity and yield of a filled and unfilled epoxy resin. *Transactions of the Society of Rheology*, 14(2): 253–273.

Khan, A.S. & Baig, M. The effect of strain rate and temperature on the formability of an aluminum alloy. *International Journal of Plasticity*, submitted for publication.

Khan, A.S. & Farrokh, B. 2006. Thermo-Mechanical response of nylon 101 under uniaxial and multi-axial loadings: part I, experimental results over wide ranges of temperatures and strain rates. *International Journal of Plasticity*, 22: 1506–1529.

Khan, A.S. & Zhang, H. 2001. Finite deformation of a polymer: experiments and modeling. *International Journal of Plasticity*, 17: 1167–1188.

Khan, A.S. Xiang, Y. & Huang, S. 1991. Behavior of Berea sandstone under confining pressure, Part I: Yield and failure surfaces, and nonlinear elastic response. *International Journal of Plasticity*, 7: 607–624.

Li, Z., & Lambros, J. 2001. Strain rate effects on the thermomechanical behavior of polymers. *International Journal of Solids and Structures*, 38(20): 3549–3562.

Mears, D.R. Pae, K.D. & Sauer, J.A. 1969. Effects of hydrostatic pressure on mechanical behavior of polyethylene and polypropylene. *Journal of Applied Physics*, 40: 4229–4237.

Pae, K.D. 1977. The macroscopic yielding behavior of polymers in multiaxial stressfield. *Journal of Materials Science*, 12: 1209–1214.

Quinson, R. Perez, J. Rink, M. & Pavan, A. 1997. Yield criteria for amorphous glassy polymers. *Journal of Materials Science*, 32: 1371–1379.

Raghava, R. Caddell, R.M. & Yeh, G.S.Y. 1973. The macroscopic yield behavior of polymers. *Journal of Materials Science*, 8: 225–232.

Rittel, D. & Brill, A. 2008. Dynamic flow and failure of confined polymethylmethacrylate. *Journal of the Mechanics and Physics of Solids*, 56/4: 1401–1416.

Rittel, D. & Dorogoy, A. 2008. A methodology to assess the rate and pressure sensitivity of polymers over a wide range of strain rates. *Journal of the Mechanics and Physics of Solids*, 56: 3191–3205.

Silano, A.A. Bhateja, S.K. & Pae, K.D. 1974. Effect of hydrostatic pressure on the mechanical behavior of polymers: polyurethane, polyoxymethylene, and branched polyethylene. *International Journal of Polymeric Materials*, 3: 117–131.

Stachurski, Z.H. 1997. Deformation mechanisms and yield strength in amorphous polymers. *Progress in Polymer Science*, 22: 407–474.

Advanced Materials and Structural Engineering – Hu (Ed.)
© *2016 Taylor & Francis Group, London, ISBN 978-1-138-02786-2*

Comparative investigation into the catalytic efficiency of Ni and Fe in the formation of carbon coils

H.J. Kim, G.H. Kang, J.K. Lee & S.H. Kim
Center for Green Fusion Technology and Department of Engineering in Energy & Applied Chemistry, Silla University, Busan, Republic of Korea

ABSTRACT: Carbon coils were synthesized using a thermal chemical vapor deposition system with C_2H_2 and H_2 as source gases and SF_6 as an additive gas. Thin-film Fe and Ni, prepared by the deposition of metal vapor onto a SiO_2-coated Si substrate, were employed to investigate the effects of metal catalysis on the formation of carbon coils. The formation densities, morphologies, and geometries of the carbon coils deposited on the substrate were investigated. In the case of Fe catalysts, wave-like carbon nano-coils were prevalent on the substrate surface. Conversely, under identical experimental conditions, Ni catalysts predominantly led to the formation of DNA-like carbon micro-coils. The surface morphologies of the samples were systematically investigated at each synthetic stage of the reaction process. The different geometries observed in carbon coils formed on different metal catalysts were rationalized with reference to different thermochemical properties of the metal carbides.

1 INTRODUCTION

Since the successful synthesis of Carbon Nanotubes (CNTs), the development of unique carbon morphologies has been proposed as a promising method to extend the application of carbon nanomaterials in diverse industrial fields (Dresselhaus et al. 1996, 2001, Dupuis 2005, Shaikjee & Coville 2012). Of the unique morphologies observed in carbon nanomaterials, helical carbon coils are particularly interesting because of their unique shape. They have been proposed as suitable materials for a diverse range of applications such as field emitters, electromagnetic absorbers, highly sensitive nano- and microsized detectors, reinforcing fillers for composites, and building blocks for the fabrication of nano-devices. (Coville, 2012, Pan et al. 2001, Hokushin et al. 2007, Hemadi et al. 2001).

Catalytic Chemical Vapor Deposition (CVD) using a metal catalyst is regarded as an effective method for the synthesis of carbon coils. The iron family elements (Fe, Co, and Ni) are known to be effective metal catalysts for the growth of carbon coils on solid substrates. Ni is usually considered as the most effective catalyst for the formation of carbon coils (Hou et al. 2003).

Co with Fe served as both the catalyst and substrate for the formation of carbon coils (Hemadi et al. 1996). Also, Fe, as $Fe(CO)_5$, has been used for the large-scale pyrolytic synthesis of carbon coils (Hwang et al. 2000). Furthermore, Chen et al. reported the formation of carbon micro-coils by the catalytic pyrolysis of acetylene using a Ni foam catalyst, which also functioned as the substrate (Chen et al. 1991).

The characteristics of the metal catalyst employed are regarded as a vital factor in determining the final growth morphology of carbon coils. Zhang et al. reported the effects of the metal catalyst size on the morphology of carbon fibers (Zhang et al. 2008). The dependence of the properties of helical carbon nanotubes on the metal catalyst size was also reported by Tang's work (Tang et al. 2010).

The formation of coiled fibers by the crossing or entwisting of two primary coils originating from diamond-shaped Ni seeds was studied by Motojima's group (Motojima et al. 1991). Also, Jang and Kim reported the optimal shape and size of Ni catalysts for the formation of carbon micro-coils (Jang & Kim 2013).

Thus, research on the influence of metal catalyst properties on the formation of carbon coils is considered to be of primary importance for the development and optimization of carbon coil synthesis.

In this work, our research focused on the structural influence of different metal catalysts for the formation of carbon coils. Specifically, we used Fe and Ni as metal catalysts. The formation of carbon coils was investigated step by step, and the morphologies of as-grown sample surfaces were investigated after every synthetic phase. Based on these results, the different geometries observed in carbon coils formed using different iron-family elements were rationalized.

Table 1. Experimental conditions of the deposition of carbon coils for the different samples.

Process Steps	Samples	C₂H₂ flow rate (sccm)	H₂ flow rate (sccm)	SF₆ flow rate (sccm)	Total pressure (Torr)	Total deposition time (min)	Source gases flow time (min)			Catalyst-Substrate &
							C₂H₂	H₂	SF₆	Temp.(°C)
(1)	A	250	35	35	100	1	1	1	1	Fe-SiO₂, 750
(1)	B	250	35	35	100	1	1	1	1	Ni-SiO₂, 750
(2)	C	250	35	35	100	5	5	5	5	Fe-SiO₂, 750
(2)	D	250	35	35	100	5	5	5	5	Ni-SiO₂, 750
(3)	E	250	35	35	100	60	60	60	5	Fe-SiO₂, 750
(3)	F	250	35	35	100	60	60	60	5	Ni-SiO₂, 750

2 EXPERIMENTAL DETAILS

SiO₂-layered Si substrates were prepared by the thermal oxidation of 2.0×2.0 cm² p-type Si(100) substrates. The thickness of the SiO₂ layer on the Si substrate was estimated to be ca. 300 nm.

Ni powder (0.1 mg, ca. 99.7%) was evaporated for 1 min to form a Ni catalyst layer on the substrate using a thermal evaporator. The estimated thickness of the Ni catalyst layer on the substrate was ca. 100 nm. The same method was used to form an Fe catalyst layer of ca. 100 nm thickness.

For the deposition of carbon coils, a thermal CVD system was employed. C₂H₂ and H₂ as source gases and SF₆ as an additive gas were injected into the reactor during the initial reaction stage. The flow rate of C₂H₂, H₂, and SF₆ were fixed at 250, 35, and 35 standard cm³/min, respectively. The reaction process was terminated after 1, 5, and 60 min reaction times. The reaction conditions of the different processes are shown in Table 1. The morphological details of the carbon coils deposited onto the different substrates were investigated using Field Emission Scanning Electron Microscopy (FESEM, Hitach 4500).

3 RESULTS AND DISCUSSION

Six samples (A–F) representing different reaction process steps (see Table 1) were prepared. FESEM images showing the surface morphologies of the samples were recorded following each step. The different metal (Fe or Ni) catalyst-layered substrates were mounted in the reaction chamber simultaneously. Thus, the carbon coils were formed on the different substrates under identical reaction conditions.

At step (1), i.e., carbon coil deposition at 1.0 min, the Fe catalyst layer appears to be converted into a number of nano-sized Fe-carbide grains. These grains are uniformly dispersed on the surface of sample A, as shown in Figure 1a. The diameters of these grains are around a few hundred nanometers in diameter (see the inset of Fig. 1a). In the case of the Ni-layered substrate, nano-sized carbon nano-filaments are prevalent on the surface of sample B

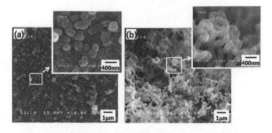

Figure 1. FESEM images for (a) sample A and (b) sample B after process step (1). Insets of Figure 1a and 1b show magnified images of the square areas indicated in Figure 1a and b.

Figure 2. FESEM images of sample C under magnification of (a) × 1,000, (b) × 5,000, and (c) × 50,000.

(see Fig. 1b). As shown in the inset of Figure 1b, these filaments are shown to be Wave-Like Carbon Nano-Coils (WCNCs) of less than 200 nm in diameter (Eum et al. 2012).

At step (2), i.e., carbon coil deposition after 5.0 min, a number of nano-sized Fe-carbide grains can still be observed on the surface of sample C (an Fe-layered substrate), as well as a small number of clusters bearing carbon nano-filaments (see Fig. 2a and 2b). The magnified image indicates that these clusters were made up of ball-type carbon nano-materials (see Fig. 2c).

For the Ni-layered substrate, however, both WCNCs and well-developed Carbon Micro-Coils (CMCs) are present on the entire surface of sample D (see Fig. 3a). As shown in Figure 3b and 3c, the CMCs appear to have formed above the WCNCs. This result indicates that CMCs might be formed after the formation of WCNCs. This conclusion is in agreement with a previous report (Jeon et al. 2013).

At step (3), i.e., the completion of carbon coil deposition after 60 min, the clusters formed by the carbon nano-filaments are frequently observed on the surface of sample E (an Fe-layered substrate) (see Fig. 4a). As shown in the magnified images of Figure 4a, the filaments that form the

Figure 3. FESEM images of sample D under magnification of (a) × 1,000, (b) × 5,000, and (c) × 50,000.

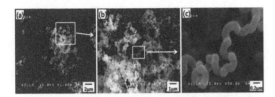

Figure 4. FESEM images of sample E under magnification of (a) × 1,000, (b) × 5,000, and (c) × 50,000.

Figure 5. FESEM images of sample F under magnification of (a) × 100, (b) × 1,000, and (c) × 5,000 and (d) × 10,000.

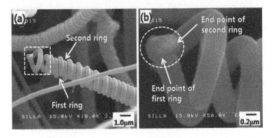

Figure 6. (a) Magnified FESEM image of the growth of a CMC and (b) the highly magnified FESEM image focused on the head area of a CMC.

clusters appear to be WCNCs (see Fig. 4b and 4c). Conversely, for the Ni-layered substrate, carbon nano-filaments are dominant on the surface of sample F (see Fig. 5a). The magnified images in Figure 5a show that CMCs are exclusively present on the entire surface of the substrate (see Fig. 5b and 5c). The length of the micro-sized carbon coils is more than 100 μm (see Fig. 5a), while the diameters of the carbon coils are within the range of a few micrometers, as shown in Figure 5d.

Based on the results presented in Figures 1–5, we suggest that the Ni catalyst is more efficient for the formation of CMCs than the Fe catalyst.

The details of the micro-sized carbon coils were also investigated using highly magnified FESEM images, as shown in Figure 6.

The individual carbon nano-filaments constituting the coil have sub-micro-meter diameters. The terminal and second rings can be clearly observed (see the rings in Fig. 6a). This indicates that the geometry of the micro-sized carbon coils in this work follows a typical double-helix structure. Circular morphology was observed in the rings constituting the coils. The head of the coil is clearly visible, as shown in the inside of the dotted circle in Figure 6b. This appears to be the initiation point for these two rings, as suggested in our previous report (Park et al. 2013). Two individual carbon nano-filaments appear to be independently formed on the Ni grains, and then grow in opposite directions.

To directly compare the efficiency of Ni and Fe catalysts for the formation of CMCs, we fabricated a half-and-half substrate surface bearing both Fe and Ni layers. To do this, we first deposited a Ni layer on half the surface of the substrate by masking the other half, and then deposited an Fe layer on the previously masked half of the surface area while masking the Ni-layered area of the substrate.

Figure 7 shows FESEM images of the carbon nanomaterials deposited on the half-and-half surface of the substrate. In the case of the Fe-layered substrate, WCNCs were frequently observed on the substrate surface, as shown in Figure 7b–7d. In addition, a large number of nano-sized Fe-carbide grains are also formed on the substrate (see Fig. 7e). In the case of the Ni-layered substrate, however, CMCs mainly developed on the substrate under the same experimental condition, as shown in Figure 7f–7g. These results confirm that the formation of CMCs is favored by Ni, rather than by Fe, as the metal catalyst.

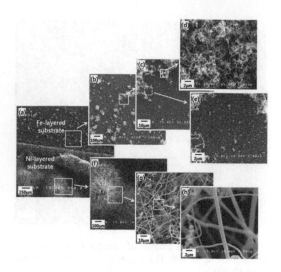

Figure 7. (a) FESEM image of carbon nanomaterials deposited on the half-and-half substrate surface, (b) formation of WCNCs on the Fe-layered substrate, (c) magnified image of the square area in Figure 7b, (d) high-magnification image of the square area in Figure 7c, (e) high-magnification image of the dotted square area in Figure 7c, (f) formation of CMCs on the Ni-layered substrate, (g) magnified image of the square area in Figure 7f, and (h) high-magnification image of the square area in Figure 7g.

The different thermochemical properties of Fe-carbide and Ni-carbide are proposed as the main cause for the geometry differences in carbon coils formed on different metal (Fe or Ni) catalyst-layered substrates. It is known that Fe_3C-type Fe-carbide is more stable than Ni_3C-type Ni-carbide (Shatynski, 1979). It is expected that the most unstable metal carbide plays a more active role in catalyzing the growth of the carbon nanomaterials. Consequently, Ni-layered substrates produce more abundant carbon nanomaterials than Fe-layered substrates. Ni-layered substrates can readily promote the growth of CMCs, whereas Fe-layered substrates promote the growth of WCNCs alone under the same experimental conditions.

4 CONCLUSIONS

By comparing the activity of Fe and Ni catalysts, it was shown that Ni catalysts are more efficient for the formation of CMCs on SiO_2 substrates. The different thermochemical properties of Fe-carbide and Ni-carbide were proposed as the main cause for the geometry differences in carbon coils formed on the different metal-layered substrates. We suggest that Ni_3C-type Ni-carbide, which is more unstable compared to Fe_3C-type Fe-carbide, may produce more abundant carbon nanomaterials, eventually leading to the growth of CMCs.

The geometry of CMCs by Ni catalyst follows a typical double-helix structure. The rings constituting CMCs have the circular morphology.

ACKNOWLEDGEMENT

This work was supported by the Korea Foundation for the Advancement of Science & Creativity (KOFAC) and funded by the Korean Government Ministry of Education (MOE).

REFERENCES

Chen, Y. Liu, C. Du, J.H. & Cheng, H.M. 1991. Preparation of carbon microcoils by catalytic decomposition of acetylene using nickel foam as both catalyst and substrate, Carbon, 29: 1874–1878.

Dresselhaus, M.S. Dressalhaus, G. & Eklund, P.C. 1996. Science of fullerenes and carbon nano-tubes, New York: Academic Press.

Dresselhaus, M.S. Dresselhaus, G. & Avouris, P. 2001. Carbon nanotubes: synthesis, structure, properties, and applications, Berlin: Springer-Verlag: 12–51.

Du, F. Liu, J. & Guo, J. 2009. Shape controlled synthesis of Cu2O and its catalytic application to synthesise amorphous carbon nanofibres, Materials Research Bulletin. 44: 25–29.

Dupuis, A.C. 2005. The catalyst in the CCVD of carbon nanotubes-a review, Progress in Materials Science, 50: 929–961.

Eum, J.H. Kim, S.H. Yi, S.S. & Jang, K. 2012. Large-scale synthesis of the controlled-geometry carbon coils by the manipulation of the SF6 gas flow injection time, Journal of Nanoscience and Nanotechnology, 12(5): 4397–4402.

Hernadi, K. Fonseca, A. Nagy, J.B. Bernaerts, D. & Lucas, A.A. 1996. Fe-catalyzed carbon nanotube formation, Carbon, 34: 1249–1257.

Hernadi, K. Thien-Nga, L. & Forro, L. 2001. Growth and Microstructure of Catalytically Produced Coiled Carbon Nanotubes, The Journal of Physical Chemistry B, 105: 12464–12468.

Hokushin, S. Pan, L.J. Konishi, Y. Tanaka, H. & Nakayama, Y. 2007. Field emission properties and structural changes of a stand-alone carbon nanocoil, Japanese Journal of Applied Physics. 46: L565.

Hou, H. Jun, Z. Weller, F. & Greiner, A. 2003. Large-Scale Synthesis and Characterization of Helically Coiled Carbon Nanotubes by Use of Fe(CO)5 as Floating Catalyst Precursor, Chemistry of Materials, 15: 3170–3175.

In-Hwang, W. Yanagida, H. & Motojima, S. 2000. Vapor growth of carbon micro-coils by the Ni catalyzed pyrolysis of acetylene using rotating substrate, Materials Letters, 43: 11–14.

Jang, C.Y. & Kim, S.H. 2013. Effect of the Ni Catalyst Size and Shape on the Variation of the Geometries for the As-grown Carbon Coils, *Journal of the Korean Institute of Surface Engineering,* 46(4): 175–180.

Jeon, Y.C. Ahn, S.I. & Kim, S.H. 2013. Developing Aspect of Carbon Coils Formation During the Beginning Stage of the Process, *Journal of Nanoscience and Nanotechnology,* 13(8): 5754–5758.

Motojima, S. Kawaguchi, M. Nozaki, M. & Iwanaga, H. 1991. Preparation of coiled carbonfibers by catalytic pyrolysis of acetylene, and its morphology and extension characteristics, *Carbon,* 29: 379–385.

Pan, L.J. Hayashida, T. Zhang, M. & Nakayama, Y. 2001. Field Emission Properties of Carbon Tubule Nanocoils, *Japanese Journal of Applied Physics,* 40: L235.

Park, S. Jeon, Y.C. & Kim, S.H. 2013. Effect of Injection Stage of SF6 flow on Carbon Micro Coils Formation, *ECS Journal of Solid State Science and Technology,* 2 (11): M56–M59.

Shaikjee, A. & Coville, N.J. 2012. The synthesis, properties and uses of carbon materials with helical morphology, *Journal of Advanced Research,* 3(3): 195–223.

Shatynski, S.R. 1979. The Thermo-chemistry of Transition Metal Carbides, *Oxidation of Metals,* 13(2): 105–118.

Tang, N. Wen, J. Zhang, Y. Liu, F. Lin, K. & Du, Y. 2010. Helical Carbon Nanotubes: Cata-lytic Particle Size-Dependent Growth and Magnetic Properties, *ACS Nano,* 4: 241–250.

Zhang, Q. Yu, L. & Cui, Z. 2008. Effects of the size of nano-copper catalysts and reaction temperature on the morphology of carbon fibers, *Materials Research Bulletin,* 43: 735–742.

Advanced Materials and Structural Engineering – Hu (Ed.)
© 2016 Taylor & Francis Group, London, ISBN 978-1-138-02786-2

The effect of orifice head loss coefficient on the discharge of throttled surge tank

S. Palikhe & J.X. Zhou
College of Water Conservancy and Hydropower, Hohai University, Nanjing, China

ABSTRACT: The throttle surge tank is the most widely used key components for checking and controlling transients in any hydraulic system. In this phenomenon flow occurs from the surge tank to main conduit and vice versa through orifice provided at the bottom of the throttled surge tank. This orifice obstructs the flow in and out of surge tank causing orifice head loss at this section and also makes the flow process complex. This paper investigates using three different methods for calculating orifice head loss coefficient, which is (i) Based on Specification Code Book, China, (ii) Based on Regimes Experimental Analysis and (iii) Based on T-Junction Gardel's Empirical Expression and its effect on the discharge of the surge tank. The Method of Characteristics is used for the solution of complex energy equation and continuity equation. The simple model consisting upstream reservoir, headrace tunnel, throttled surge tank, downstream penstock and the valve is simulated in FORTRAN.

1 INTRODUCTION

The hydraulic transient is variation in flow parameters due to change in the state of flow control components and their consequences in the hydraulic system. This is the physical process (Chaudhry 1987) which takes place between two steady states within a certain time period. Hydraulic transient and water hammer are synonymous terms given to explain unsteady flow with pressure fluctuation in the pipeline system. The transient phenomenon produced a high pressure surge, which can destroy the pipeline and damage hydraulic components. In any hydraulic system, transient control is one of the most important parts of its operation. A throttle surge tank is the most common and effective component to control and reduce the surge of incoming fluctuating pressure waves. The orifice provided at the bottom of throttled surge tank function is to offer head loss when the water is flowing in or out of the tank. This mechanism makes it more effective in surge control and damping phenomenon. This head loss at the bottom of the orifice is called orifice head loss, corresponds to orifice head loss coefficient. In many research works, it has been considered as minor loss, being small quantity compare to the frictional head loss and given less importance, but it has a significant effect on the discharge of the surge tank. This orifice head loss coefficient is obtained from three different methods and they are (i) Based on Specification Code, China, (ii) Based on Regimes Experimental

Analysis and (iii) Based on T-Junction Gardel's Empirical Expression. The first method is based on Specification Code followed in China to calculate head loss coefficient for design of surge tank and surge analysis of any hydraulic system like hydropower, water supply system, drainage pipeline system etc. This method considers head loss coefficient is constant and provides only one particular value for inflow and another for out flow from surge tank, so only two constant values. The second method is based on six different regimes. Each regime has their own equation for calculation of orifice head loss which is obtained from the experimental model of Huhehot Pump Storage Hydropower Station. The third method is based on empirical equations for T-junction obtained by Gardel and is applied to calculate orifice head loss at the T-junction formed at the intersection of surge tank and main conduit. The discharge calculation of different time intervals are calculated using The Method of Characteristics. The Method of Characteristic (Wylie 1993) is a numerical technique in which governing complex energy and continuity equations in partial differential forms is converted into simple differential equations and boundary conditions are applied to the solution. The simulation of a simple hydraulic system consists of upper reservoir, headrace tunnel, surge tank, downstream penstock and valve at the end is conducted and output results are compared with computation obtained from three different methods.

2 BASIC EQUATION FOR THROTTLE SURGE TANK

The basic differential equation for throttle surge tank water level fluctuation is given below:
Energy Equation,

$$\frac{L}{gA_1}\frac{d(A_1V)}{dt} = Z - \frac{fL}{2gA_1}*(A_1V)^2 - \frac{\xi}{2gA_S^2}Q_S^2 \quad (1)$$

Continuity Equation,

$$\frac{dZ}{dt} = \frac{A_1V - Q_2}{A_S} = -\frac{Q_S}{A_S} \quad (2)$$

Time period of Oscillation,

$$T = 2\pi\sqrt{\frac{LA_S}{gA_1}} \quad (3)$$

where, L = length of upstream headrace tunnel; Z = surge tank water level fluctuation; V = velocity in upstream headrace tunnel; A_1 = area of upstream headrace tunnel; A_S = area of surge tank; H_f = total head loss; Q_2 = discharge downstream tunnel; Q_S = discharge in surge tank; f = head loss coefficient of upstream headrace tunnel due to friction; ξ = orifice head loss coefficient; g = acceleration due to gravity.

3 METHOD OF CHARACTERISTIC AND BOUNDARY CONDITIONS

The momentum and continuity (Wylie 1993) equation for any hydraulic system is given as

$$E_1 = g\frac{\partial H}{\partial x} + \frac{\partial V}{\partial t} + \frac{f}{2D}V|V| = 0 \quad (4)$$

$$E_2 = \frac{\partial H}{\partial t} + \frac{a^2}{g}\frac{\partial V}{\partial x} = 0 \quad (5)$$

The Characteristics equations are

$$C^+ : \frac{g}{a}\frac{dH}{dt} + \frac{dV}{dt} + \frac{f}{2D}V|V| = 0; \frac{dx}{dt} = +a \quad (6)$$

$$C^+ : -\frac{g}{a}\frac{dH}{dt} + \frac{dV}{dt} + \frac{f}{2D}V|V| = 0; \frac{dx}{dt} = -a \quad (7)$$

The solution of the original system is given by Equation (6) and Equation (7) and can be represented on x-t plane as in Figure 1.
After applying Finite Difference technique above characteristic equations are reduced to

$$C^+ : H_i = C_P - B_P Q_i \quad (8)$$

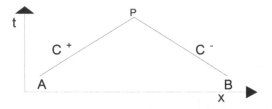

Figure 1.   Characteristic line in x and t plane.

$$C^- : H_i = C_M + B_M Q \quad (9)$$

where, C_P, B_P, C_M, B_M are known constants coefficient given by

$$C_P = H_{i-1} + BQ_{i-1}; B_P = B + R|Q_{i-1}|$$

$$C_M = H_{i+1} - BQ_{i+1}; B_P = B + R|Q_{i+1}|$$

$$B = \frac{a}{gH}; R = \frac{f\Delta x}{2gDH}$$

where, B and R = characteristics impedance and resistance coefficient, respectively. From Equation (8) and Equation (9)

$$H_{Pi} = \frac{C_P B_M + C_M B_P}{B_P + B_M}; Q_{Pi} = \frac{C_P - C_M}{B_P + B_M} \quad (10)$$

are the required equations for transient analysis in any hydraulic system. The subscript P on parameters Q and H indicates the latest required value.
Boundary Conditions are equations provided at the particular boundary in terms of Q_i and H_i which give complete response and behavior of hydraulic system in that transient period. C^- characteristic equation is available for upstream end and C^+ characteristics equation for downstream end. These boundary conditions are solved independently of interior section and other boundary. The boundary conditions of nodes in the hydraulic system which is simulated in this paper are given below.

3.1 Upstream reservoir

For the short duration of transient considered the hydraulic grade line elevation is constant. Also entrance losses and velocity head are small, so can be neglected.
Then, Boundary equations are
$H_P = H_R$ = Elevation of water in reservoir above datum;

$$Q_{P1} = \frac{H_R - C_M}{B_M} \quad (11)$$

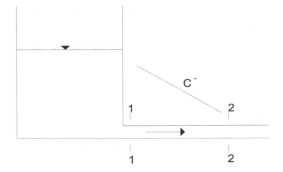

Figure 2.   Upper reservoir node.

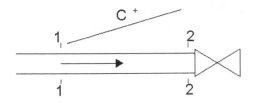

Figure 4.   Valve node.

3.3   *Downstream valve node*

Boundary equations for a valve node are

$$H_{P2} = C_P - B_P Q_{P2} \qquad (18)$$

$$Q_{P2} = -B_P C_V - \sqrt{(B_P C_V)^2 + 2C_V C_P}$$

$$C_V = \frac{(\tau Q_0)}{2H_0} \qquad (19)$$

where, Q_0 = initial maximum discharge when valve is completely open; H_0 = initial pressure head; τ = valve opening dimensionless parameter.

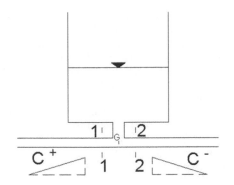

Figure 3.   Upper surge tank node.

3.2   *Surge tank node*

The boundary condition of throttled surge tank (Zhang 2007) deduce in this section can also be used for simple surge tank considering the orifice head loss equal to zero. The cross sectional area of surge tank is considered very large compared to the cross sectional area of the upstream pipe so the water hammer wave effect in the tank can be ignored. Boundary equations for a surge tank node are

$$C^+ : H_P = C_P - B_{P1} Q_{P1} \qquad (12)$$

$$C^- : H_P = C_M - B_{M1} Q_{P2} \qquad (13)$$

$$H_P = H_{PG} = H_{P1} = H_{P2} \qquad (14)$$

$$Q_{P1} = Q_{P2} + Q_{PS} \qquad (15)$$

$$H_P = H_{PS} + \zeta Q_S |Q_S| \qquad (16)$$

$$H_{PS} = H_{S0} + \frac{Q_S + Q_{S0}}{2A_S} * \Delta t \qquad (17)$$

where, H_{S0} = initial height of water level in the surge tank; Q_{S0} = initial discharge in the surge tank.

4   ORIFICE HEAD LOSS COEFFICIENT

It is a measure of head drop down between main conduit and surge tank. It affects the flow that takes place between throttled surge tank and main conduit through the orifice. It is similar to the frictional head loss coefficient, but smaller and also has an important role in oscillation and surge damping phenomenon in throttled surge tank. The three different methods of calculation are given below.

4.1   *Based on specification code*

Specification Code, China (1996) is employed to obtain the head loss coefficient of orifice in this method. In and out flow directions of surge tank are considered, providing only two possible values for a throttled surge tank establishing the head loss coefficient. This method only allows constant values for inflow and outflow of the surge tank. The head loss coefficient for orifice of the surge tank is expressed as:

$$\Delta h = \frac{1}{2g\phi^2 A_O^2} Q^2 = \xi |Q|Q; \xi = \frac{1}{2g\phi^2 A_O^2} \qquad (20)$$

where, Q = discharge flowing in or out of surge tank, A_O = area of orifice or throat; ξ = orifice head loss coefficient, ϕ = Flow coefficient = 0.6 for inflow and 0.8 for outflow.

4.2 Based on regimes experimental analysis

An experiment was conducted (Cai 2011) to analyze the surge tank of pump storage power plant situated in Inner Mongolia called Huhehot pump storage power station. This power plant is mainly consists of an upper reservoir, lower reservoir and an underground power house. For different operation conditions in the laboratory, the expressions for head loss coefficients considering different flow regimes, observed during experiments were obtained using Curve fitting method.

These equations obtained from experiments conducted in the laboratory were used considering six different flow regimes. Flow regimes (Arshenevskii 1984) are the patterns of flow that take place in the junction of surge tank and main conduit during transient. Each flow regime constitutes its own equation for calculation of head loss coefficient. Figure 5 shows the six different flow regimes. These different flow regimes are ignored by The Specification Code, only considering two flow directions, in and out flow from surge tank. This method includes the detail analysis and investigation of flow at the junction of surge tank and main conduit for computation of orifice head loss coefficients. These flow regimes (Cai 1996, 2001) take place in different modes and operating conditions of the pump and turbine in hydropower system or any hydraulic system.

Head loss coefficient for different flow regimes:

$$\xi_I = 5.564q_{21}^2 - 0.044q_{21} + 0.830; q_{21} = \frac{Q_2}{Q_1} \quad (21)$$

$$\xi_{II} = 3.262q_{23}^2 - 0.059q_{23} + 0.537; q_{23} = \frac{Q_2}{Q_3} \quad (22)$$

$$\xi_{III} = 2.971q_{12}^2 - 2.821q_{12} + 26.10; q_{12} = \frac{Q_1}{Q_2} \quad (23)$$

$$\xi_{IV} = 5.699q_{32}^2 - 2.821q_{32} + 0.368; q_{23} = \frac{Q_2}{Q_3} \quad (24)$$

$$\xi_V = 5.262q_{21}^2 + 2.008q_{21} - 0.522; q_{21} = \frac{Q_2}{Q_1} \quad (25)$$

$$\xi_{VI} = 13.74q_{12}^2 - 14.72q_{12} + 33.13; q_{12} = \frac{Q_1}{Q_2} \quad (26)$$

Equation (21), Equation (22), Equation (23), Equation (24), Equation (25) and Equation (26) give the orifice head loss coefficient for Regime(I), Regime(II), Regime(III), Regime(IV), Regime(V), Regime(VI), respectively.

4.3 T-Junction Gardel's Empirical Expression

The flow of fluid in the T-shape junction tube is widely observed in hydraulic engineering, so head loss coefficient of this type of section is important. For the computation of head loss coefficient, Vogel (1926) from Germany conducted different experiments. McNown (1954) and Gardel (1957) conducted a model test independently. Based on these tests, Gardel put forward an empirical expression for calculation of head loss coefficient of T-junction known as Gardel's Empirical Formula. Blaisdsel and Manson (1963) carried an experiment to study the characteristics of head loss in water conveyance system. Miller (1971) researched in comparison of the 30 years test results and concluded, except Vogel the rest of the research results were in good consistency with each other and also he suggested that Gardel's formula could be used for head loss calculation with satisfactory accuracy. Gardel's formula depends on the T-junction structural parameters size and angles of branched pipes. Based on flow directions, head loss coefficient is divided into two types: Diverse head loss coefficient and Confluent head loss coefficient (Hua Dong Shui Li Xue Yuan 1983).

4.3.1 Diverse head loss coefficient

According to Gardel's formula, a flow which is coming from one section is divided into two branched sections is called diverse flow and the diverse head loss coefficient is calculated as

$$\xi_D = -0.95(1-q_D)^2 - q_D^2\left(1.3\cot\frac{\theta}{2} - 0.3 + \frac{0.4-0.1\varphi}{\varphi^2}\right)$$

$$\times \left(1 - 0.9\sqrt{\frac{\rho}{\varphi}}\right) - 0.4\left(1+\frac{1}{\varphi}\right)\cot\frac{\theta}{2}(1-q_D)q_D \quad (27)$$

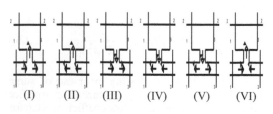

(I)   (II)   (III)   (IV)   (V)   (VI)

Figure 5. Six different flow regimes.

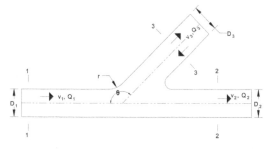

Figure 6. T-Junction with diversion and conversing flow.

where, ξ_D = orifice head loss coefficient for diverse flow; q_D = diverse flow ratio = Q_3/Q_1; θ = angle at the junction of branched pipes; φ = area ratio = A_3/A_1; ρ = ratio of radius of curvature at junction to diameter of main conduit = r/D_1.

4.3.2 Confluent head loss coefficient

The flow which is coming from different adjacent section combined and flow in one section is called confluent flow and the confluent head loss coefficient is calculated as

$$\xi_C = (1+q_C)\{0.92+q_C(2.92-\varphi)\}$$
$$+q_C^2\left[\left(1.2-\sqrt{\rho}\right)\left(\frac{\cos\theta}{\varphi}-1\right)\right.$$
$$\left.+0.8\left(1-\frac{1}{\varphi^2}\right)-(1-\varphi)\frac{\cos\theta}{\varphi}\right] \qquad (28)$$

where, ξ_C = orifice head loss coefficient for confluence flow, q_C = confluence flow ratio = Q_3/Q_2.

The inverted "T" shape is formed at the junction of surge tank and main conduit, so Gardel's empirical equations for T-shape are applied to calculate head loss coefficient at the orifice. These equations were established from basic principles of momentum applied to the main conduit, the continuity principle for fluid in whole T-junction and energy balance principle for flow coming from the branches. These empirical formulas are obtained by Gardel's through experiments and researches on T-shaped branched pipes for calculation of head loss coefficient. The head loss at the enlargement of the orifice and tank is ignored being very small in quantity compared to head loss at throat and simplifying the calculation. For the T-section formed at the junction of surge tank and main conduit in the simulation model, $\theta = 90°$ and $\rho = 0$ are considered.

5 SIMULATION AND ANALYSIS

The simulation model considered for this research consisted of upper reservoir, headrace tunnel,

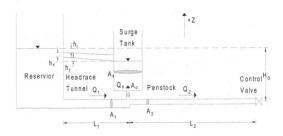

Figure 7. Simulation model.

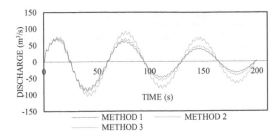

Figure 8. Discharge at surge tank VS time.

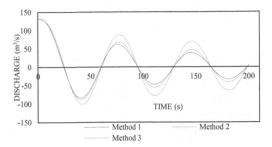

Figure 9. Discharge at section just downstream of surge tank VS time.

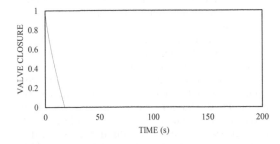

Figure 10. Valve closure VS time.

surge tank, tailrace penstock and downstream valve. The initial water level at the upper reservoir (H_0) = 1940 m, length of the head race tunnel (L_1) = 565.649 m, diameter of the head race tunnel (D_1) = 6.2 m, length of the tailrace penstock (L_2) = 1000 m, diameter of the tail race penstock (D_2) = 5.8 m, resistance coefficient of the head race tunnel (R_1) = 6.81687*10^{-06}, resistance coefficient of tail race penstock (R_2) = 9.3451*10^{-06}, diameter of the surge tank (D_s) = 9 m. In the beginning, the valve was opened full and closed gradually within (T_C) = 18 seconds, maximum discharge at the valve (Q_0) = 132.4 m^3/s and the total time of computation (T) = 200 seconds. The diameters of throttle orifice (A_O) = 4.3 m.

Table 1. Surge tank discharge output.

Surge tank	Method 1	Method 2	Method 3
Max. discharge in (m^3/s)	69.77	73.64	92.67
Max. discharge out (m^3/s)	84.80	92.92	105.21

The values of discharge for surge tank were obtained from three different methods as mentioned above. The numerical calculation was carried out using The Method of Characteristics and FORTRAN were used for simulation. The output results from the simulation were obtained at different time intervals. The graphs were plotted against time for three different methods and comparison table was also constructed. Method 1, Method 2 and Method 3 represent (i) Based on Specification Code, China, (ii) Based on Regimes Experimental Analysis and (iii) Based on T-Junction Gardel's Empirical Expression, respectively.

From the above graphs and table, we can conclude that the discharge fluctuation in surge tank is damped with time for all three methods. This signifies the damping property of surge tank which controls the transient in the system. Figure 9 represents the discharge of the section of the just after surge tank. Figure 8 and Figure 9 show that initially total flow is towards downstream, no discharge in the surge tank and with the closure of the valve there is fluctuation in discharge at the section close to surge tank and inside the surge tank. The maximum discharges obtained in and out of the surge tank from Method 1 are 69.77 and 84.80 m^3/s, respectively. The maximum discharges obtained in and out of the surge tank from Method 2 are 73.64 and 92.92 m^3/s, respectively. The maximum discharge obtained in and out of the surge tank from Method 3 are 92.67 and 105.21 m^3/s, respectively. The maximum in and out discharges from the surge tank obtained from Method 3 is greater than Method 2 and Method 1. The maximum in and out discharges from the surge tank obtained from Method 1 is least in comparison. The maximum discharges obtained from method one is in relative deviation of 5.25% and 8%, respectively with Method 2. The maximum discharges obtained from Method 3 is in relative deviation of 20.53% and 11.68%, respectively with Method 2. This indicates that Method 2 is in closer agreement with Method 1 than Method 3 comparatively though the deviation is not small even for Method 1.

6 CONCLUSION

The surge tank functions as a damping component in a hydraulic system controlling the transient measures. The output discharges from Specification Code Based Method is in close agreement with Regimes Based Experimental Analysis compared to Gardel's T-Junction results. This excludes Gardel's T-Junction which gives a greater value of discharge compared to the other two methods. The Specification based method is simpler, easier, efficient and effective method of calculation, so can be used for preliminary analysis and approximation. But, for precise, accurate and detailed analysis, it should be supported by regime based method. These maximum computed discharges can be used for design purpose, surge analysis and proper operation of the surge tank.

REFERENCES

Arshenevskii, N.N. Berlin, V.V. & Murav'ev, O.A. 1984. Mathematical Modelling of Hydraulic Regimes in the transition between the surge tank and conduits of a hydroelectric station, *Power Technology and Engineering*, 3: 105–110.

Blaidsdell, F.W. & Manson, P.W. 1963. Loss of Energy at Sharp edged pipe junction in Water Conveyance System, *US Department of Agriculture, Tech. Bulletin*: 1281–1292: 163.

Cai, F.L. & Zhou, J.X. 2011. *Hydraulic Model Test for Detail Analysis of Diversion Surge tank of HuheHot Pump Storage Power Station*, College of Water Conservancy And Hydropower, Hohai University.

Cai, F.L. Hu, M. & Cao, Q. 2001. Coefficient of Head Loss of Throttle Surge Tanks with Long Linking Pipe, *Water Resources and Power*: 19(4): 57–60.

Cai, F.L. Liu, Q.Z. Suo, L.S. & Liu, D.Y. 1996. Experimental Study on Hydraulic Characteristics of Rectangular Tailrace Surge Tank Shared by Two Units, *Journal of Hohai University*: 24(6): 66–71.

Chaudhry, M.H. 1987. *Applied Hydraulic Transients*, New York: Van Nostrand Reinhold Company.

Gardel, A. 1957. Pressure Drops in Flows through T-Shaped Fittings, *Bulletin Technique De La Suisse Romande*, 10: 123–130.

Hua, D. Shui, L. & Xue, Y. 1983. *Hand book of Hydraulic Structure Design Fundamentals 1*, Water Resources and Electric Power Press: 384–386.

McNown, J.S. 1954. Mechanics of Manifold Flow, *Transaction of American Society of Civil Engineers*, 119: 1119–1142.

Miller, D.S. 1971. *Internal Flow a Guide to Losses in Pipe and Duct System*, The British Hydromechanics Research Association, England.

Specification for Design of Surge Chamber of Hydropower Station, *Electric Power Industry Standard of the People's Republic of China*, DL/T 5058–1996: 18.

Vogel, G. 1926. Investigation of Loss in Right Angled Pipe Branches, Hydraulic Institute Tech, *Hoscheule-Munche*, 1: 75–90.

Wylie, E.B, Streeter, V.L. & Suo L.S. 1993. *Fluid Transient in Systems*, Prentice Hall, Englewood Cliffs, NJ 07632.

Zhang, Jian & Zhou, Jianxu.2007. Shui Li Ji Zu Guo Du Guo Cheng, Peking University Press: 33.

Advanced Materials and Structural Engineering – Hu (Ed.)

Effect of an enclosed cage structure on the chain characteristics of TSP-POSS/PU hybrid composites

R. Pan & L.L. Wang
Chemistry and Material Science College, Sichuan Normal University, Chengdu City, P.R. China

Y. Liu
Key Laboratory of Special Waste Water Treatment, Sichuan Province Higher Education System, Chengdu City, P.R. China

ABSTRACT: Molecular simulation was applied in elucidating the polymer chain characteristics of TSP-POSS/PU hybrid composites with Trisilanolphenyl Polyhedral Oligomeric Silsesquioxane (TSP-POSS) at various concentrations. The hybrid composite models were constructed and characterized by the Radial Distribution Function (RDF) and the Mean Square Displacement (MSD) at the molecular level. The results indicate that as TSP-POSS concentration increases up to 15 wt%, due to three phenyl groups in the humping cage structure, the distance between neighboring chains is increased and the motion of polymer chains is restricted apparently. It also demonstrates that the TSP-POSS cage linked to polymer backbones acts as a heavy core with less mobility.

1 INTRODUCTION

POSS can improve various properties of PU in application ranging from elastomers, coatings to adhesives (Liu et al. 2006). Though the improvement of incorporating POSS moieties into polymeric matrices is well known as stated above, the mechanism of interaction between the cage structure and polymer chain is seldom elucidated. Since polyurethane has a large range of applications and TSP-POSS with three phenyl groups is a good filler for modification, the effect of TSP-POSS incorporation on polymer chain characteristics is deserved to be investigated.

2 SIMULATION PROCEDURES

Accelrys Amorphous Cell module and COMPASS force field in Materials Studio software were applied in the simulation process of our research, which have been verified for polymer nanocomposites containing POSS simulation (Zhao et al. 2008). A total of 10 initial configurations for each sample were optimized by the molecular dynamics technique under the NPT (constant particle numbers, pressure and temperature) conditions at 4 Gpa with a minimization involving 30000 steps to relax and equilibration. After this minimization procedure, the density fluctuation of each system is less than 0.05 g/cm^3 under a given condition. Since these optimized configurations might not be in a local energy minimum state, an annealing procedure from 623 K to 273 K was applied to the above optimized configurations by conducting the velocity Verlet algorithm in NVT dynamics to reduce the possible potential energy. Finally, configurations with the highest energy were rejected, and 5 configurations of each sample were selected for further analysis of composites' characteristics. The mole ratio of TSP-POSS in each composite and sample code is listed in Table 1. The molecular structure of TSP-POSS/PU hybrid composites is shown in Figure 1 (Zheng et al. 2001).

Table 1. Characteristics of TSP-POSS/PU hybrid composites.

Sample code	TSP-POSS: MDI: Voranol301 (mole ratio)	Parameters of 3D boundary conditions	Initial density (g/cm^3)	Final density (g/cm^3)
3 PU	1:10:0.9	20.01	1.1672	1.1666
10 PU	2:0.6:0.5	18.99	1.1788	1.1789
15 PU	2:0.5:0.3	17.31	1.1904	1.1906
20 PU	1:0.3:0.1	14.80	1.2023	1.2044

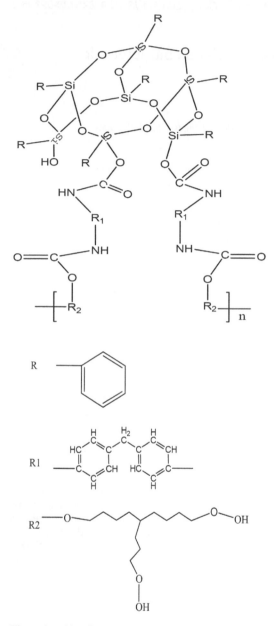

Figure 1. Chemical structure of TSP-POSS/PU hybrid composites.

3 3D BOUNDARY CONDITION STRUCTURE CONSTRUCTION

Recently, it has been reported that the structure and the energy of the POSS/PU hybrid composites were successfully simulated by using the COMPASS force field (Wang & Rui 2015). Model structure was generated through several

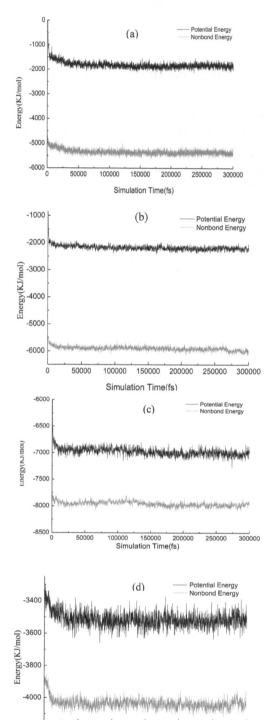

Figure 2. Plots of potential energy optimization and non-bonded energy optimization: (a) 3 PU; (b) 10 PU; (c) 15 PU; (d) 20 PU.

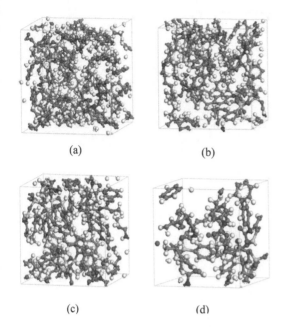

(a)   (b)

(c)   (d)

Figure 3. 3D periodic boundary conditions of TSP-POSS/PU hybrid composites: (a) 3 PU; (b) 10 PU; (c) 15 PU; (d) 20 PU.

cycles of molecular mechanics and molecular dynamics energy minimization. After the above energy optimization procedure, the density fluctuation of each system is less than 0.05 g/cm^3, which indicates that the generated structure is fully relaxed and in the equilibrium state.

4 RADIAL DISTRIBUTION FUNCTION ANALYSIS

In statistical mechanics, the radial distribution function (pair correlation function) g(r) is a measure of the probability of finding a pair of atoms (α, β) that is separated by a radial distance $r_{\alpha\beta}$, which is expected for a completely random distribution. Thus, details of polymer chain packing can be estimated by defining the atoms in the polymer main chain as given referenced particles in inter-molecular pair correlation function, respectively (Yin 2009). Figure 4 shows the inter-molecular pair correlation function based on all atoms in the main chain. With the increase in TSP-POSS concentration, the number of contacts between main chains is decreased. It can be concluded that at any given distance, the number of contacts between neighboring chains is decreased due to the presence of TSP-POSS with the humping enclosed cage structure, which makes the

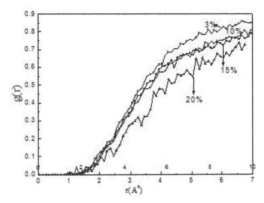

Figure 4. Radial distribution functions g(r) of 0 PU, 5 PU, 10 PU, 15 PU and 20 PU hybrid composites.

neighboring chains depart from each other and extends their distance.

5 MEAN SQUARE DISPLACEMENT ANALYSIS

The mobility of atoms (molecules) can be investigated by using the Mean Square Displacement (MSD) from Equation (1) (Bruce et al. 2000).

$$MSD = [|(r_i(t) - r_i(0)|^2).\qquad(1)$$

where $r(0)$ is the initial positional coordinate of atom i (or molecule) and $r_i(t)$ is the coordinate of time. Thus, the restrictions of polymer chains motion imposed by TSP-POSS can be evaluated by elucidating the mobility of atoms in polymer chains with TSP-POSS incorporation. In Figure 5(a)–(d), the MSD of all atoms in the TSP-POSS/PU samples is compared with that of the Si-O pairs in TSP-POSS cores. In all samples, the MSD of TSP-POSS core is lower than that of all atoms as a whole. This result confirms that the TSP-POSS cage linked to polymer backbones acts as a heavy core with less mobility. In Figure 5(e), it shows the mobility of all atoms in composites with various TSP-POSS concentrations. From the plot, as the concentration of TSP-POSS is increasing, the value of MSD decreases due to the restriction of the heavy cage structure effect. Interestingly, as TSP-POSS concentration reaches up to 10%, MSD of all atoms in 10 PU is higher than that in 3 PU. It can be deduced that due to the increased distance between neighboring chains, the mobility of polymer chains is released to some extent. However, as the concentration is still raising, the percentage of heavy cage structure is increasing in backbones, and finally, it leads to the apparent restriction on the whole chain mobility.

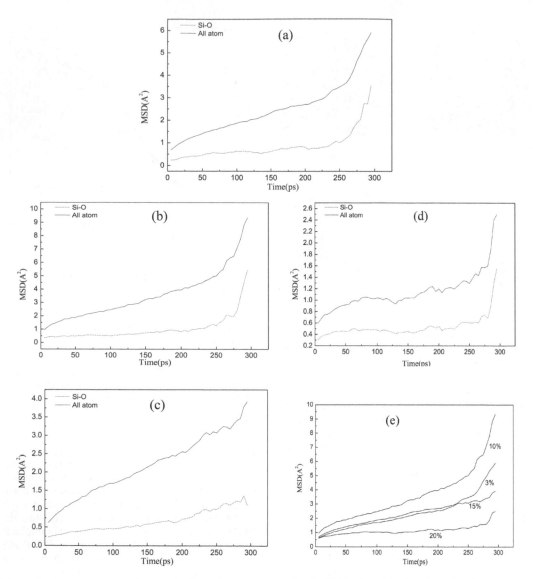

Figure 5. Mean square displacement of all atoms in composites (solid lines) and Si-O pairs in TSP-POSS cores (dotted lines): (a) 3 PU; (b) 10 PU; (c) 15 PU; (d) 20 PU. (e) Mean square displacement of all atoms in composites.

6 CONCLUSIONS

In this study, molecular simulation was applied to research the effect of the TSP-POSS cage structure on the chain characteristics of hybrid composites at the molecular level. As the TSP-POSS concentration reaches up to 15 wt%, the humping cage structure in the backbone apparently increases the distance between neighboring chains and restricts the mobility of the whole chain in hybrid composites. Furthermore, the MSD results also indicate that TSP-POSS is a rigid core with less mobility linked to the polymer backbones.

ACKNOWLEDGMENT

This work was supported by the Sichuan Education Office Foundation (Project No. 15ZA0040) in China and the Key Laboratory of Special Waste Water Treatment in Sichuan Province Higher Education System.

REFERENCES

Bruce, X.F. & Joseph, S. et al. 2000. Nanoscale reinforcement of Polyhedral Oligomeric Silsesquioxane (POSS) in polyurethane elastomer. *Polymer International*. 49: 437–440.

Liu, Y.H. & Ni, Y. et al. 2006. Polyurethane networks modified with octa (propylglycidyl ether) polyhedral oligoeric silsesquioxane, *Macromolecular Chemistry and Physics*. 207: 1842–1851.

Wang, L.L. & Rui, P. 2015. An Investigation on Structure and Thermal Properties of TSP-POSS/PU Hybrid Composites by Molecular Simulation Approach. *Acta Polymerica Sinica*. 0(3): 266–276.

Yin, Y. & Monica, H.L. 2009. Molecular dynamics simulation of mixed matrix nanocomposites containing polyimide and Polyhedral Oligomeric Silsesquioxane (POSS). *Polymer*. 50: 1324–1332.

Zhao, J.Q. et al. 2008. Polyhedral Oligomeric Silsesquioxane (POSS)-Modified thermoplastic and thermosetting nanocomposites: A review. *Polymers & Polymer Composites*. 16(8): 483–500.

Zheng, L. & Coughlin, E.B. et al. 2001. X-ray Characterizations of Polyethylene Polyhedral Oligomeric Silsesquioxane Copolymers. *Macromolecular*. 35: 2375–2379.

Advanced Materials and Structural Engineering – Hu (Ed.)
© 2016 Taylor & Francis Group, London, ISBN 978-1-138-02786-2

The preparation, characterization and properties of La$_2$O$_3$/TPU nano-composites

C.F. Wang & R.P. Jia
School of Materials Science and Engineering, Shanghai Institute of Technology, Shanghai, China

ABSTRACT: Rare earth due to its special chemical and physical properties has been tightly connected with polymers in recent year. In this work, nano-La$_2$O$_3$ was introduced into the TPU matrix via the one-step *in situ* bulk polymerization technique, thus La$_2$O$_3$/TPU nano-composites were obtained. The influences of nano-La$_2$O$_3$ on the mechanical and biological properties of the thermoplastic polyurethane matrix were investigated. It was found that both the tensile strength and elongation at the break of the corresponding La$_2$O$_3$/TPU nano-composites had great improvements. When the addition amount of nano-La$_2$O$_3$ was 0.3 wt%, the tensile strength and elongation at break reached up to 44 MPa and 756%, respectively, which were higher than those of the pure TPU elastomer by 46% and 28%, respectively. The results of the MTT test and cell proliferation experiments indicated that La$_2$O$_3$/TPU nano-composites had better cytocompatibility compared with pure TPU.

1 INTRODUCTION

Due to its excellent physical properties and chemical stabilities, Thermoplastic Polyurethane (TPU) elastomer is widely used in daily necessities, sports goods, decorative material, and biomedical material. Their desirable properties mainly come from the micro-phase separation between hard and soft segments (Frick & Rochman 2004). The soft segment consists of long flexible polyester or polyether that primarily influences the TPU's elastic properties (Buckley et al. 2010, Martin et al. 1996). The hard segment is composed of a short chain (diol or diamine) reacting with diisocyanate that acts as multifunctional tie points, working as both physical crosslinks and reinforcing fillers (Zdrahala et al. 1979). In order to improve the performance and exploit versatility of the TPU elastomer, more attention has been paid to the research on rare earth-modified polyurethane in recent years (Ciobanu et al. 2007).

Due to the fact that rare earth has a special structure of valence electron shell, doped polymers have attracted great interest. Lanthanum oxide (La$_2$O$_3$) is one kind of important, low-cost, non-poisonous and pollution-free rare earth, and has been widely studied with polymer materials. Li et al. studied the synergistic effects of La$_2$O$_3$ on the flame retardant of polypropylene composites (Li et al. 2008). Song et al. investigated the effect of La$_2$O$_3$ on the structure and crystallization of poly (vinylidene fluoride) (PVDF) (Song et al. 2010). Pazarlioglu et al. studied the microstructure and mechanical properties of Bovine-derived Hydroxyapatite (BHA) doped with nano-power of La$_2$O$_3$ (Pazarlio et al. 2011). Several attempts manifested that La$_2$O$_3$ has considerable influences on polymer's performances. However, to the best of our knowledge, there has so far been no report on La$_2$O$_3$/TPU nano-composites prepared via the *in situ* polymerization technique.

In this study, Polybutylene Adipate Glycol (PBA) was employed as the soft segment, and 4,4-diphenylmethane diisocyanate (MDI) and 1,4-Butanediol (BDO) were used as hard segments, respectively. La$_2$O$_3$/TPU nano-composites were synthesized via the one-step *in situ* bulk polymerization technique, and the influences of nano-La$_2$O$_3$ on the TPU's mechanical and biological properties were investigated.

2 EXPERIMENTAL

2.1 *Materials*

PBA (Mn = 1000 g/mol) was purchased from BASF Co. Ltd (Shanghai, China) and further dried at 120°C under vacuum for 2 h. MDI was provided by Wanhua Polyurethane Co. Ltd (Yantai, China). BDO, Phosphate Buffered Saline (PBS), glutaraldehyde, 4′,6-diamidino-2-phenylindole (DAPI) and Dimethyl Sulfoxide (DMSO) were purchased from Sinopharm Chemical Reagent Co. Ltd (Shanghai, China). Nanometer La$_2$O$_3$ was purchased from Hangzhou Wanjing New Material Co. Ltd.

2.2 Synthesis of La₂O₃/TPU nano-composites

The TPU nano-composites with different La_2O_3 contents were fabricated via the one-step *in situ* bulk polymerization technique. The NCO/OH ratio and hard segment content were set as 0.99 and 39%, respectively. The details of the synthesis were listed as follows: first, the desired amounts of La_2O_3 nanoparticles and PBA (20 g, 0.02 mol) were added to a glass beaker and further dispersed at an ultrasonic vibration generator with 200 W power for 30 min. Then, appropriate amounts of MDI (10.7 g) and BDO (2.1 g) were subsequently added, which were calculated according to the fixed NCO/OH ratio and hard segment content, and stirred quickly for several minutes. Then, the system was casted into a Teflon mold preheated at 140°C and kept at room temperature for 1 h, and then heated up to 120°C with a rate of 0.5°C/min and kept for 8 h. Figure 1 presents the manufacture procedure of La_2O_3/TPU nano-composites.

2.3 Cell viability and proliferation test

Sterilized TPU and La_2O_3/TPU films were first placed into a 24-well plate and put in a humidified incubator under standard culture conditions (5% CO_2 at 37°C, 1 h). Then, mouse Mesenchymal Stem Cells (MSCs) were cultured in the 24-well culture plates, and each well was injected with 1 ml of cell suspension containing 2×10^4 cells. After 2 days later, 1 ml of 4% glutaraldehyde was used to fix the cells on films for 10 min and washed with PBS solution for three times, and then 200μl DAPI was added to each hole to stain for 50 min. Finally, the films were taken out, rinsed with PBS solution for three times, and the cell proliferation was observed by using a confocal laser scanning microscope.

The viability of adherent cells on TPU nano-composites films was commonly determined by the MTT assay. Briefly, 400 μL of 3-(4,5-dimethylthiazol-2-yl)-2,5-diphenyltetrazoliumbromide (MTT) solution (0.5 mg/mL, 1 × PBS) were added into each plate, and the cells and films were incubated at 37°C in the dark for 4 h. Thereafter, the supernatant was removed and 600 μL of DMSO were added into each well to dissolve the formazan crystals for 10 min. A microplate reader was used to determine the Optical Density (OD)

values at the wavelength of 570 nm, which represented cell viability of each sample.

2.4 Physical measurement

The structure and crystallinity of La_2O_3 and its nano-composites with TPU were characterized by a Rigaku D/max 2500 X-Ray Diffractometer (XRD) with CuKα radiation (λ = 1.542 Å, Japan) in the 2θ range of 5°~80°. Fourier Transform Infrared (FTIR) spectra were scanned at room temperature with a Nicolet NEXUS870 FTIR spectrometer (USA) in the range of 500~4000 cm⁻¹. The Optical Density (OD) value was determined by a microplate reader (DV-990-BV, Italy). Their tensile properties were recorded with an Instron 4302 material testing machine (USA).

3 RESULTS AND DISCUSSION

3.1 XRD patterns

Figure 2 shows the XRD scattering of La_2O_3 (a), TPU (b) and La_2O_3/TPU nano-composites (c-e), and the influences of La_2O_3 on the TPU's structure and crystallization were investigated. The sharp reflection peaks of La_2O_3 nanoparticles at 2θ≈26.09°, 29.95°, 39.49°, and 46.04° correspond with the crystal surface (100), (101), (102), (110), which perfectly match with a six-party phase La_2O_3 JCPDS [No. 05-0602]. Due to the self-organization of hard segments, a strong and broad semicrystalline peak of TPU at 2θ≈20° can be observed. Additionally, it can be seen that as the La_2O_3 contents increase, the intensity of the TPU diffraction peak decreases gradually, while the peak width shows a broad trend and the position does not change. This

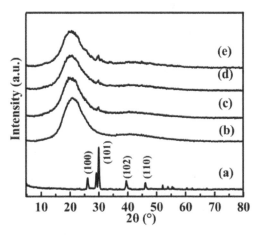

Figure 2. XRD patterns of La_2O_3 nanoparticles (a) and TPU nano-composites with different La_2O_3 contents: (b) 0 wt%; (c) 0.3 wt%; (d) 0.5 wt% and (e) 0.7 wt%.

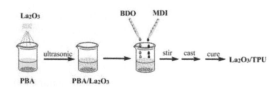

Figure 1. Synthesis procedure of La_2O_3/TPU nano-composites.

can be ascribed to the interference of the La_2O_3 with the ordering of the hard domain.

3.2 *FTIR analysis*

The chemical structures of La_2O_3 (a), TPU (b) and La_2O_3/TPU nano-composites (c-e) were confirmed by FTIR spectra (Fig. 3). The absorption bands at 3440 cm^{-1} and 1633 cm^{-1} were attributed to the O-H vibration in the absorbed water on the surface of La_2O_3 nanoparticles. For TPU and its composites, the characteristic absorption bands at around 3332 cm^{-1} and 2958 cm^{-1} were attributed to urethane N-H in hydrogen bonding and aliphatic (-CH-) asymmetric stretching vibrations, respectively. The peaks at 1726 cm^{-1} and 1702 cm^{-1} were attributed to the ester carbonyl group and the carbonyl stretching of the urethane group. It was found that with the introduction of La_2O_3 into TPU, the band at 3332 cm^{-1} shift to 3326 cm^{-1} and the band at 1702 cm^{-1} were strong compared with the band at 1726 cm^{-1}. Interestingly, these changes in the characteristic absorption band occurred in the hard segment of TPU, indicating that nano-La_2O_3 has impacted on the hard domain of TPU. The influences of nano-La_2O_3 on the TPU matrix can be attributed to the polar N-H and carbonyl groups easily generating chemical bonds with the rare earth element, which further impacts on the properties and applications of the TPU elastomer.

3.3 *Tensile properties*

The mechanical properties of TPU composites depend on many factors such as micro-phase separation of TPU, fillers dispersion, fillers concentration, orientation of fillers, interfacial interaction between TPU and the fillers.

We also investigated the mechanical properties of the TPU elastomer with different nano-La_2O_3 contents. Figure 4 shows the typical stress–strain curves of the pure TPU and La_2O_3/TPU nano-composites. With the introduction of nano-La_2O_3, mechanical properties of TPU are significantly enhanced compared with the pure TPU. When the La_2O_3 addition content was 0.3 wt%, the tensile strength and elongation at break reached up to 44 MPa and 756%, which were higher than those of the pure TPU elastomer by 46% and 28%, respectively. This can be attributed to the fact that La_2O_3 acts as both reinforcing fillers, and physical and chemical linkages in the TPU matrix, thus resulting in a positive effect on the TPU's mechanical properties.

3.4 *MTT cell viability assay*

MTT assay is commonly used to evaluate the cell viability of materials' surface after incubation. The number of living cells on the material surface can be reflected by the OD value at 570 nm. In this work, mouse MSCs were chosen as the living cell and cultured on the pure TPU and La_2O_3/TPU nano-composites films. Cell viability was quantified by measuring the MTT absorbance, and the results are shown in Figure 5. It can be found that La_2O_3/TPU nano-composites showed higher OD values than the pure TPU; therefore, La_2O_3/TPU nano-composites had better compatibility with cell viability in comparison with TPU. When the concentration of nano-La_2O_3 on the TPU matrix was 0.3 wt%, the OD value was significantly higher than that of the pure TPU, indicating that the appropriate addition of La_2O_3 has a positive effect on the TPU's cell viability.

3.5 *Cell proliferation test*

Figure 6 shows the Confocal Laser Scanning Microscope (CLSM) images of different La_2O_3 contents

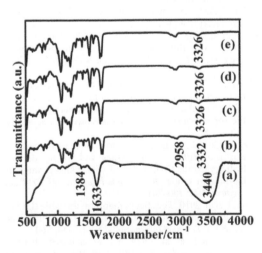

Figure 3. FTIR spectra of La_2O_3 nanoparticles (a) and TPU nano-composites with different La_2O_3 contents: (b) 0 wt%; (c) 0.3 wt%; (d) 0.5 wt% and (e) 0.7 wt%.

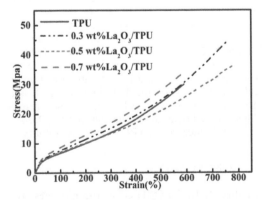

Figure 4. The strain-stress curves of TPU and La_2O_3/TPU nano-composites.

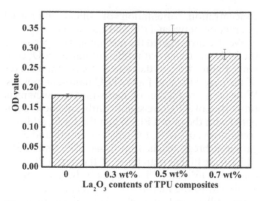

Figure 5. Viability test of MSCs on TPU and La$_2$O$_3$/TPU films.

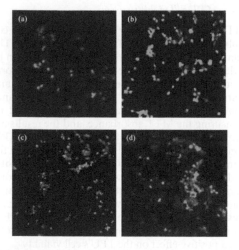

Figure 6. CLSM images (x 20) of TPU nano-composites with different La$_2$O$_3$ contents: (a) 0 wt%; (b) 0.3 wt%; (c) 0.5 wt% and (d) 0.7 wt%.

on the TPU matrix. It can be clearly seen that the amount of cells on the surface of the pure TPU is clearly less than that on the surface of La$_2$O$_3$/TPU nano-composites. The cells on the surfaces of La$_2$O$_3$/TPU nano-composites were still able to retain their typical cell morphology. The results revealed that La$_2$O$_3$ had a remarkable effect on the cell proliferation of the TPU matrix; which is in good agreement with cell viability assay results. This can be attributed to the fact that La$_2$O$_3$ has a good influence on the TPU's cytocompatibility.

4 CONCLUSIONS

In this study, La$_2$O$_3$/TPU nano-composites were successfully prepared via the one-step *in situ* bulk polymerization technique. With the introduction of nano-La$_2$O$_3$, the structure and crystallization of TPU has a slight shift; furthermore, it has a significant influence on the TPU's mechanical and biological properties. When the amount of nano-La$_2$O$_3$ was 0.3 wt%, the tensile strength and elongation at break reached up to 44 MPa and 756%, which were higher than those of the pure TPU elastomer by 46% and 28%, respectively. Meanwhile, MTT cell viability assay and cell proliferation test results indicated that the appropriate concentrations of La$_2$O$_3$ has positive effects on the biological properties of TPU, which has vital practical significance in the application of TPU in the biomaterial and surgical market.

ACKNOWLEDGMENT

This project was supported by the National Natural Science Foundation of China (21106083), the Shanghai Leading Academic Discipline Project (J51504), the Composite Materials Leading Academic Discipline Project from Shanghai Institute of Technology (10210Q140001), the Shanghai Affiliate Programs (LM201336, LM201450), and the Shanghai Teacher professional development project (201456).

REFERENCES

Buckley, C.P. et al. 2010. Elasticity and inelasticity of thermoplastic polyurethane elastomers: Sensitivity to chemical and physical structure. *Polymer*, 51(14): 3213–3224.

Ciobanu, C. et al. 2007. Polyurethane doped with low concentration erbium. *Journal of applied polymer science*, 103(2): 659–669.

Frick, A. & Rochman, A. 2004. Characterization of TPU-elastomers by thermal analysis (DSC). *Polymer testing* 23(4): 413–417.

Li, Y. et al. 2008. Synergistic effects of lanthanum oxide on a novel intumescent flame retardant polypropylene system. *Polymer Degradation and Stability*, 93(1): 9–16.

Martin, D.J. et al. 1996. The effect of average soft segment length on morphology and properties of a series of polyurethane elastomers. I. Characterization of the series. *Journal of applied polymer science*, 62(9): 1377–1386.

Pazarlioglu, S.S. et al. 2011. Microstructure and mechanical properties of composites of bovine derived hydroxyapatite (BHA) doped with nano-powder of lanthanum oxide. *International journal of artificial organs*, 34(8): 700–701.

Song, J. et al. 2010. The effect of lanthanum oxide (La$_2$O$_3$) on the structure and crystallization of poly (vinylidene fluoride). *Polymer International* 59(7): 954–960.

Zdrahala, R.J. et al. 1979. Polyether-based thermoplastic polyurethanes. I. Effect of the hard-segment content. *Journal of Applied Polymer Science*, 24(9): 2041–2050.

Advanced Materials and Structural Engineering – Hu (Ed.)
© 2016 Taylor & Francis Group, London, ISBN 978-1-138-02786-2

Analysis on the thermal decomposition process for the preparation of cobalt-doped zinc oxide from oxalate

J.W. Li & S.X. Guo
Jiaozuo Teachers College, Jiaozuo, Henan Province, China

ABSTRACT: A series of Co-doped oxalate powders of $Co_xZn_{(1-x)}C_2O_4 \cdot 2H_2O$ (x = 0, 0.01, 0.04) were prepared. The UV-VIS-NIR absorption spectrum and XRD patterns reveal that each of them is single-phase crystal, and that Co^{2+} doping does not change the crystal structure of zinc oxalate. The TG-DTA curves show that the thermal decomposition proceeds in two stages: dehydration of oxalate dihydrate and decomposition of anhydrous oxalate. For the Co-doped oxalate, decomposition processes are endothermic in the nitrogen atmosphere, while exothermic in the air atmosphere.

1 INTRODUCTION

Diluted magnetic semi-conductors (Dietl 2010, Chang & Xia 2004) are a kind of semiconductor material in which the minority cations are substituted by paramagnetic ions. Due to the interaction between local magnetic moments of paramagnetic ions and carriers, which has many new properties (Xu et al. 2012, Zhang & Pan 2009), this kind of material has become an important material in the field of spintronics (Wolf et al. 2001), and has very broad application prospects (Wang et al. 2007). Co-doped zinc oxide material $Co_xZn_{(1-x)}O$, due to the higher solubility of Co ions (Lee et al. 2002, Ueda et al. 2001) and high Curie temperature (Ueda et al. 2001, Yang et al. 2009), has become one of the key research objects of diluted magnetic materials, and has attracted increasing attention (Aljawfi & Mollah 2011, Köseôglu 2013, Thangeeswari et al. 2013, Saleem et al. 2012). $Co_xZn_{(1-x)}O$ belongs to a metastable system, and using a single source precursor, in which each ion is in close contact, is the most ideal preparation. Oxalate can not only meet this requirement, but also has a cleaner product for CO and CO_2 can be removed from the surface along with decomposing. The microstructure of $Co_xZn_{(1-x)}O$ is influenced by the preparation atmosphere, and then affects its performance (Schwartz & Gamelin 2004, Kittilstved et al. 2006). In this paper, we prepared cobalt-doped zinc oxide nanopowder by decomposing oxalate precipitation, and studied the effects of different atmospheric conditions on the thermal decomposition process.

2 EXPERIMENTS

2.1 Sample preparation

The reagents used in the experiments were of analytical grade, and the solvent is deionized water. The aqueous solutions of zinc acetate dihydrate $(Zn(CH_3COO)_2 \cdot 2H_2O)$ and cobalt acetate tetrahydrate $(Co(CH_3COO)_2 \cdot 4H_2O)$ were configured according to Zn and Co ratio (1-x): x (wherein x is, respectively, 0, 0.01, 0.04). Then, in the vigorous stirring conditions, respectively, 100 ml oxalic acid (0.07 mol) aqueous solution is added dropwise. The precipitate obtained was washed 3 times with deionized water to prepare oxalates A, B, C (corresponding to x = 0, 0.01, 0.04). In the preparation of the series of oxalates, the number of moles of cations 0.07 mol kept unchanged. The resulting oxalate was dried at 100 °C. Zinc oxalate that contains cobalt ions A is white, and that containing cobalt ions B and C is pale pink.

2.2 Physical measurements

All samples were characterized by powder X-Ray Diffraction (XRD) using an X'Pert Philips diffractometer, operating at power 40 kV × 40 mA and scanning wavelength 0.15406 nm. The ultraviolet-visible-Infrared(UV-VIS-NIR)absorption spectrum is measured by a Cary5000 spectrometer made in America. TG-DTA analysis was performed on a Japan DSC6200 integrated thermal gravimetric analyzer, with the heating rate of 10 °C/min, air or nitrogen atmosphere, the gas flow of 180 ml/min.

3 RESULTS AND DISCUSSION

3.1 UV-VIS-NIR absorption spectrum analysis of the prepared oxalates

The absorption spectra of Co-doped zinc oxalates B, C are similar. Figure 1 shows the ultraviolet-visible-infrared absorption spectrum of the oxalate C with the ratio of Zn to Co being 96:4. The absorptions at 500 nm, 1500 nm and 2000 nm are related to the Co^{2+} interatomic d-d transition. These d-d transitions are possible only in a Co^{2+} ion in an octahedral crystal field (Radovanovic et al. 2002).

3.2 XRD analysis of the prepared oxalates

As can be seen from Figure 2, the prepared oxalates A, B and C are all single phase, and the crystal structures of zinc oxalate are not altered by the Co^{2+} doped.

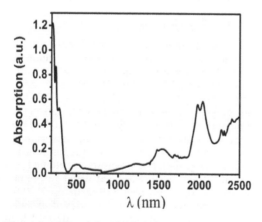

Figure 1. Ultraviolet-visible-infrared absorption spectrum of the oxalate c (ratio of Zn to Co 96:4).

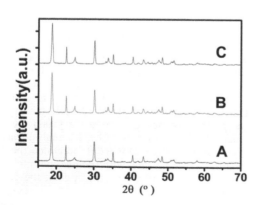

Figure 2. X-ray diffraction patterns of the oxalates A, B, C.

3.3 TG/DTA analysis of the prepared oxalates

3.3.1 Air atmosphere

Figure 3 show the thermal decomposition curves of the prepared oxalates A, B and C. According to the decomposition curves, the decomposition processes can be divided into two steps. The first step occurs in the vicinity of 150 °C, significant weight loss and an endothermic peaks correspond to the loss of crystal water into anhydrous oxalate. From weight loss ratio point of view, it should be $Co_xZn_{(1-x)}C_2O_4 \cdot 2H_2O$ (A, B, C, respectively, correspond to x = 0,0.01,0.04) lost two crystal water, i.e.

$$Co_xZn_{(1-x)}C_2O_4 \cdot 2H_2O \rightarrow Co_xZn_{(1-x)}C_2O_4 + 2H_2 \quad (1)$$

The second in the vicinity of 400 °C with significant weight loss, too, while corresponding to the decomposition of anhydrous oxalate. Pure oxalate decomposition is an endothermic process, while Co-doped zinc oxalate decomposition is an exothermic process, and the decomposition temperature decreased slightly with the increase in cobalt content.

The powder samples of $Zn_{0.99}Co_{0.01}O$ and $Zn_{0.96}Co_{0.04}O$ obtained by the decomposition of pale pink Co-doped oxalates B and C at 400 °C in air are dark green, and the color becomes darker with increasing cobalt content.

3.3.2 Nitrogen atmosphere

In order to discover the reason for the exothermic peaks near 400 °C for the Co-doped zinc oxalates B and C, thermal decomposition comparative experiments were done in the nitrogen atmosphere. The DTA curve of the sample B is similar to that of the sample C. Figure 4 shows the thermal

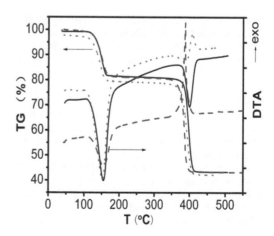

Figure 3. TG/DTA curves of the oxalates: solid lines, dotted lines and dashed lines correspond to the curves of $Zn_{(1-x)}Co_xC_2O_4 \cdot 2H_2O$ with x = 0, 0.01, 0.04, respectively.

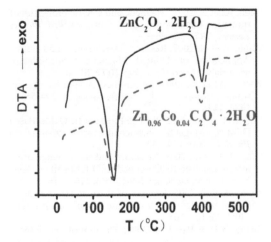

ZnC$_2$O$_4$ · 2H$_2$O

Zn$_{0.96}$Co$_{0.04}$C$_2$O$_4$ · 2H$_2$O

Figure 4. DTA curves of the oxalates in the nitrogen atmosphere.

decomposition DTA curves of samples A and C in the nitrogen atmosphere.

As shown in Figure 4, an endothermic peak appears near 400 °C for the pure zinc oxalate sample A, similar to its DTA curve in the air atmosphere shown in Figure 3, but for the Co-doped zinc oxalate C, the endothermic peak near 400 °C (Fig. 4) with the nitrogen atmosphere contrasts starkly with the exothermic peak in the air atmosphere (Fig. 3).

It has been reported (Mohamed et al. 2005, Zhan et al. 2005) that certain products (e.g. NiO, Fe$_3$O$_4$), platinum sample pool used in the decomposition process of zinc oxalate dihydrate, nickel oxalate, and iron oxalate can act as an catalyst to make CO released in the oxalate decomposition process be oxidized at a lower temperature and release heat. The material of the sample pool used in this experiment is alumina, which does not have the above catalytic action report. Our analysis, in the air atmosphere, the products of cobalt-doped oxalate thermal decomposition also have some catalytic oxidation of CO, reflected in the exothermic peak near 400 °C shown in Figure 4.

From the above results, it can be concluded that the thermal decomposition of anhydrous oxalate containing Co is as follows:

$$Co_xZn_{(1-x)}C_2O_4 \rightarrow Co_xZn_{(1-x)}O + CO + CO_2 \quad (2)$$

In Figure 3, in the air atmosphere, exothermic peaks near 400 °C reflects the exothermic oxidation of CO at $Co_xZn_{(1-x)}O$ catalysis, both generated in the oxalate decomposition process. The endothermic effect of thermal decomposition and the exothermic effect of catalytic oxidation are, respectively, reflected

by the endothermic and exothermic peaks corresponding to the dotted lines with x = 0.01 shown in Figure 3; when x = 0.04, with the increase of Co content, due to the enhanced catalytic oxidation, the exothermic peak near 400 °C reflects the overall thermal effects of oxalate endothermic decomposition and CO exothermic oxidation.

4 CONCLUSIONS

Each of the series of oxalate $Co_xZn_{(1-x)}C_2O_4 \cdot 2H_2O$ (x = 0, 0.01, 0.04) powders prepared is single-phase crystal. Co ions are in octahedral sites, and do not change the structure of zinc oxalate. In addition, these three kinds of oxalate decomposition proceed in two stages: the first stage, endothermic, occurring at around 150 °C, loses 2 molecules of water of crystallization; the second step is the decomposition of the anhydrous oxalate, occurring at around 400 °C: the decomposition of pure zinc oxalate is an endothermic process, and the decomposition of $Co_xZn_{(1-x)}C_2O_4$ (x = 0.01, 0.04) relates to the atmosphere: the decomposition processes are endothermic in the nitrogen atmosphere, while in the air atmosphere, the CO gas generated by decomposition will be oxidized by oxygen in the air and release heat under the catalysis of $Co_xZn_{(1-x)}O$ similarly generated in the decomposition process.

ACKNOWLEDGMENT

This work was supported by the Scientific Research Key Project in Colleges and Universities in Henan Province (No. 15B140005), the Science and Technology Program of Jiaozuo (No. 2014400035) and the Basic Research Projects and Advanced Technology of Henan Province (No. 132300410387 and 112300410326).

REFERENCES

Aljawfi, R.N. & Mollah, S. 2011. Properties of Co/Ni codoped ZnO based nanocrystalline DMS, *Journal of Magnetism and Magnetic Materials*, 323(23): 3126–3132.

Chang, K. & Xia, J.B. 2004. Diluted magnetic semiconductors—bridging spin and charge, *Physics*, 33(6): 414–418.

Dietl, T. 2010. A ten-year perspective on dilute magnetic semiconductors and oxides, *Nature Materials*, 9(12): 965–974.

Kittilstved, K.R. et al. 2006. Direct kinetic correlation of carriers and ferromagnetism in Co^{2+}: ZnO, *Physical Review Letters*, 97(3): 37203.

Köseoğlu, Y. 2013. Enhanced Ferromagnetic Properties of Co-doped ZnO DMS Nanoparticles, *Journal of Superconductivity and Novel Magnetism*, 26(2): 485–489.

Lee, H.J. et al. 2002. Study of diluted magnetic semiconductor: Co-doped ZnO. *Applied Physics Letters,* 81(21): 4020–4022.

Mohamed, M.A. et al. 2005. A comparative study of the thermal reactivities of some transition metal oxalates in selected atmospheres, *Thermochimica Acta,* 429(1): 57–72.

Radovanovic, P.V. et al. 2002. Colloidal Transition-Metal-Doped ZnO Quantum Dots, *Journal of the American Chemical Society,* 124(51): 15192–15193.

Saleem, M. et al. 2012. Origin of Ferromagnetism in Al and Ni Co-doped ZnO Based DMS Materials, *Chinese Physics Letters,* 29(10): 106103.

Schwartz, D.A. & Gamelin, D.R. 2004. Reversible 300 K Ferromagnetic Ordering in a Diluted Magnetic Semiconductor, *Advanced Materials,* 16(23–24): 2115–2119.

Thangeeswari, T. et al. 2013. Strong room temperature ferromagnetism in chemically precipitated ZnO:Co^{2+}:Bi^{3+} nanocrystals for DMS applications, *Journal of Materials Science: Materials in Electronics,* 24(12): 4817–4826.

Ueda, K. et al. 2001. Magnetic and electric properties of transition-metal-doped ZnO films, *Applied Physics Letters,* 79(7): 988–990.

Wang, Y. et al. 2007. Research Development and Future Application of Diluted Magnetic Semiconductor Materials, *Materials Review,* 21(7): 20–30.

Wolf, S.A. et al. 2001. Spintronics: A spin-based electronics vision for the future, *Science,* 294(5546): 1488–1495.

Xu, X.H. et al. 2012. Recent Progress In Oxide Based Diluted Magnetic Semiconductors, *Progress in Physics,* 32(4): 199–232.

Yang, J.H. et al. 2009. Structure and room-temperature ferromagnetism of Co-doped ZnO DMS films, *Solid State Communications,* 149(29–30): 1164–1167.

Zhan, D. et al. 2005. Kinetics of thermal decomposition of nickel oxalate dihydrate in air, *Thermochimica Acta,* 430(1–2): 101–105.

Zhang, Y.P. & Pan L.Q. 2009. Theoretical and Experimental Research Progress of Oxide Diluted Magnetic Semiconductors, *Materials Review,* 23(9): 1–8.

Advanced Materials and Structural Engineering – Hu (Ed.)
© 2016 Taylor & Francis Group, London, ISBN 978-1-138-02786-2

Compensating the shrinkage and expansive stress of CaO-based Expansive Additives treated by thermoplastic polyacrylic esters

R. Wang, W. Xu & Q. Tian
State Key Laboratory of High Performance Civil Engineering Materials, Jiangsu Research Institute of Building Science, Nanjing, China

J.Y. Jiang
College of Materials Science and Engineering, Southeast University, Nanjing, China

ABSTRACT: CaO-based Expansive Additive (CEA) was treated by thermoplastic polyacrylic esters (PAE) to form a protective film, and used to make paste samples for expansion tests. The compressive strength was not influenced by the encapsulation. The PAE-coated CEA paste expanded 30% than the CEA paste under a restrained condition and showed a remarkable stress increase by the double ring test, and the weathering rate was reduced by a maximum of 30%. This article offers a doable method to improve the efficiency of additives since such varieties of polymer can be utilized.

1 INTRODUCTION

Expansive additives are widely used for reducing cracking in concrete, for example by compensating for the drying shrinkage strain of concrete or providing chemical pre-stress (Collepardi et al. 2008, Gao et al. 2006, Collepardi et al. 2005, Yan & Qin 2001, Mo et al. 2014). The hydration of CaO-based expansive additives is sensitive (Serris et al. 2011), and the expansion is not useful during the plastic stage in the fresh state. For CaO-based expansive additives, it is useful to avoid the hydration of CaO-based expansive additive before the formation of the cement matrix, and a subsequent large expansion can be achieved.

Many methods have been taken to optimize the CaO-based expansive additives. Higuchi T. reported that high-temperature carbonation used to generate the calcite film surface of the expansive additive could increase expansion and improve the stability toward weathering (Higuchi et al. 2014). Lee Y.S. found that the granulated CSA-based expansive agent with PVA film coating could control the time of autonomic healing and can prevent water migration via crack closure (Lee & Ryou 2014).

If a protective membrane is applied to cover the expansive additive, the hydration of the CaO-based expansive additive may be decreased, and the invalid loss may be reduced in the plastic stage, which could also increase the expansion of the additive after the formation of the cement matrix and slow down the progress of weathering. This paper describes the expansive properties of a CaO-based expansive additive enclosed by thermoplastic polyacrylic esters (PAE).

2 EXPERIMENTS

2.1 Materials

Table 1 presents the chemical compositions of the materials. Ordinary Portland Cement (OPC), CaO-based expansive additives (CEA) sieved by the diameter of 160 um, Chinese standard sand and tap water were used. The thermoplastic polyacrylic ester (PAE) polymer was prepared by radical polymerization of acyclic acid, butyl acrylate and methyl methylacrylate according to the literature (Davis & Matyjaszewski 2000).

The CEA treated by PAE film (CEA@PAE) was prepared as follows: PAE polymer was dissolved in dichloromethane with the concentration of 20 mg/mL, a given mass of CEA was added and stirred for 5 min followed by removing the solvent by vacuum rotary evaporation. The solid obtained was milled and sieved by the diameter of 160 um.

Table 1. Chemical composition and physical properties of materials.

Material	Chemical composition							Density (cm³/g)	Blain (cm²/g)
	CaO	Al₂O₃	SiO₂	Fe₂O₃	MgO	SO₃	Loss		
OPC	64.02	4.57	22.13	3.05	2.00	2.44	1.79	3.16	3200
CEA	82.31	0.26	10.64	1.2	1.26	2.05	2.28	3.00	2800

2.2 Methods

2.2.1 Characterization of the expansive additives CEA and CEA@PAE

PAE content in the expansive additives was calculated by the weight of the polymer and the CEA added. The PAE content at CEA@PAE utilized was 2 wt% on subsequent tests, except the accelerated storing stability test. The scanning electron microscope (SEM) was employed to identify the morphological observation of CEA and encapsulated CEA@PAE under high vacuum (SEM, Quanta250, FEI Company, Czech Republic).

2.2.2 Compressive strength of the cement paste

According to the Chinese Standard GB 23439–2009, 6wt% of the OPC was replaced with expansive additives; the water power ratio was 0.35. The compressive strength of the cement paste specimens (40 mm * 40 mm * 160 mm) was evaluated in accordance with the Chinese Standard GB 17671. The tests were conducted at 0.5,1,2, 3, 5 and 7 day.

2.2.3 Expansion of restrained expansion pastes

Expansion deformation of restrained cement pastes including additives was measured on specimens (40 mm * 40 mm *140 mm) according to the Chinese Standard GB 23439–2009, where the cement paste is encapsulated by polyethylene films and tinfoil starting from the final setting. The deformation was monitored by the laser displacement sensor and acquired at 1 min intervals, and the ambient temperature for the measurement was 20°C.

2.2.4 Expansion stress of cement paste by the ring test

Expansion stress was evaluated on cement paste, in which 6 wt% of the OPC was replaced with expansive additives by the ring test. The annular gauge consists of two invar rings restraining the expansion. All of the pastes in the rings had an inner diameter of 180 mm, an outer diameter of 300 mm, and a height of 75 mm. The inner and outer wall thicknesses of the steel rings were 60 mm and 30 mm, respectively. Each steel ring used in this study had three strain gages attached at the mid-height of the inner surface of the inner ring and the outer surface of the outer ring. The strain gages were connected to a data acquisition system for continuous monitoring. Strain data were acquired at 1 min intervals, and the ambient temperature for the measurement was 20°C.

2.2.5 Accelerated storing stability test of expansive additive

All the three content CEA and CEA@PAE additives (0.5%, 2% and 5%) were tested. A 3.0 g sample was added into a 30 mL glass cup, uncovered and exposed to 25 °C, 60% humidity in the temperature and humidity chamber. The samples were analyzed and tested at 0.3, 1, 2, 3, 5, 7, 14, 21, and 35 days of exposure.

3 RESULTS AND DISCUSSION

3.1 Characterization of expansive additives

The thermoplastic polyacrylic esters were selected to encapsulate the CaO-based expansive additive as the PAE polymer was elastic and water vapor permeable. Figure 1 shows the schematic process of the preparation. PAE solution and CEA were mixed and stirred for 5 min, followed by removing the solvent by vacuum rotary evaporation.

The PAE content of the CEA@PAE expansive additives was calculated by the weight of the polymer and the CEA added. Figure 2 shows the SEM images of CEA and CEA@PAE. The edges of the CEA particle surface (A) were legible, but comparatively, the surface of the encapsulated CEA particle was smooth. The difference indicated that the polymer covered the surface of the CEA particles by the method.

3.2 Compressive strength of the cement paste

The compressive strength development is shown in Figure 3. Both the treated-CEA pastes and the CEA pastes showed similar strength increases with age.

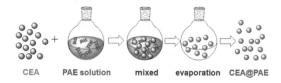

CEA   PAE solution   mixed   evaporation   CEA@PAE

Figure 1. The schematic preparation process of the expansive additive CEA@PAE.

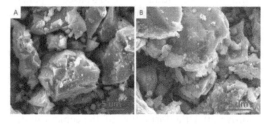

Figure 2. SEM images of expansive additives (A: CEA, B: CEA@PAE).

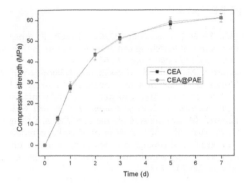

Figure 3. Compressive strength of pastes containing expansive additives.

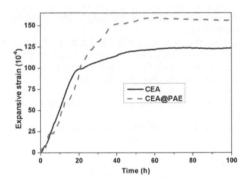

Figure 4. Expansion of restrained cement pastes containing expansive additives.

3.3 Expansion of restrained cement pastes

There are many methods to measure the expansion (Jensen & Hanse 1995, Appah & Reichetseder 2002, Darquennes et al. 2011, Subauste& Odler 2001), in which the method invented by Jensen can measure the whole process of the deformation and avoid the defect of the choice of time-zero. To acquire the whole process of the expansion of the restrained cement pastes containing additives, a similar method was chosen in which the laser displacement sensor was utilized to monitor the matrices from the final setting.

The expansion property of restrained cement pastes is shown in Figure 4. The expansion of both the CEA@PAE and CEA pastes showed rapid increases with age by 20 h, and thereafter, the speed was slowed down. The CEA@PAE paste expanded to 160 ε, while the CEA paste expanded to only 120 ε from the beginning to 40 h, which showed 30% more expansion of the CEA@PAE paste than that of the CEA paste. Then, the expansion of both the CEA@PAE and CEA pastes remained steady.

3.4 Expansion stress of cement paste by the ring test

Due to its simplicity and versatility, the ring test has become commonly used over the last two decades to assess the potential for shrinkage cracking (Grysbowski & Shah 1990, Kovler et al. 1993, Hossain & Weiss 2006, 2004). The ring test consists of a concrete annulus that is cast around a hollow steel cylinder. As the concrete curing, shrinkage or expansion is prevented by the steel ring, resulting in the development of tensile or expansive stresses in concrete. Figure 5 shows the expansive stresses of cement pastes containing additive samples calculated according to the literature (Hossain & Weiss 2004). The expansive stress of both the CEA@PAE and CEA pastes showed a similar tendency, indicating that the stress was about zero before setting and then increased drastically with age by the same 20 h, and the value of the treated additive paste was about 4.5 MPa higher than that of the non-treat additive paste, which was approximately 3 MPa. Then, the stress of CEA pastes remained steady, while the pressure of the CEA@PAE paste was slowed down gradually, which may be attributed to the relaxation.

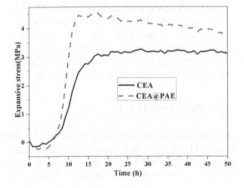

Figure 5. Expansion stress of cement pastes containing expansive additives by the ring test.

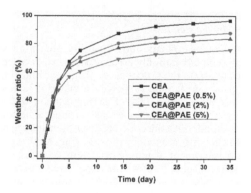

Figure 6. Accelerated storing stability test of expansive additives.

3.5 *Accelerated storing stability test of expansive additives*

Figure 6 shows the result of the accelerated storing stability test of expansive additives at 25 °C and 60% humidity. In the accelerated storage period of the expansive additives at day 35 of exposure, the weather ratio of the CEA paste increased rapidly to 97%, more than that of the CEA@PAE paste. The weather ratio of the CEA@PAE paste was related to the polymer content, and the more the polymer located on the surface of the expansive additive, the slower the weather was. The CEA@PAE paste with 5 wt% polymer was reduced by 30% compared with CEA at 35 days of exposure.

4 CONCLUSIONS

The CaO-based expansive additive was encapsulated by the thermoplastic polyacrylic ester (PAE) polymer to form a protective membrane, and its expansion was examined in restrained cement pastes. The PAE-coated CEA paste expanded 30% than the CEA paste, and the paste strength of both samples was not affected. The expansive stress also increased remarkably. The improvement was ascribed to the delayed hydration of the polymer-treated CEA, and the reasons will be explained by TAM and XRD methods in another study.

ACKNOWLEDGMENT

This work was financially supported by the Technological Development Program of China Railway (Grant No. 2014G001-C). The authors gratefully acknowledge the colleagues in the program team for their hard work. The authors also acknowledge some construction companies for their cooperation.

REFERENCES

Appah, D. & Reichetseder, P. 2002. Practical improvements in CaO-swelling cements, *Journal of Petroleum Science and Engineering*, 36: 61–70.

Collepardi, M. Troli, R. Bressan, M. Liberatore, F. & Sforza, G. 2008. Crack-free concrete for outside industrial floors in the absence of wet curing and contraction joints. *Cement & Concrete Composites*, 30: 887–891.

Collepardi, M. Borsoi, A. Collepardi, S. Jacob, J. Olagot, O. & Troli, R. 2005. Effects of shrinkage reducing admixture in shrinkage compensating concrete under non-wet curing conditions, *Cement & Concrete Composites*, 27: 704–708.

Darquennes, A. Staquet, S. Delplancke-Ogletree, M.P. & Espion, B. 2011. Effect of autogenous deformation on the cracking risk of slag cement concretes, *Cement & Concrete Composites*, 33: 368–379.

Davis, K.A. & Matyjaszewski, K. 2000. Atom Transfer Radical Polymerization of tert-Butyl Acrylate and Preparation of Block Copolymers, *Macromolecules*, 33: 4039–4047.

Gao, P.W. Wu, S.X. Lin, P.H. Wu, Z.R. & Tang, M.S. 2006. The characteristics of air void and frost resistance of RCC with fly ash and expansive agent, *Construction and Building Materials*, 20: 586–590.

Grysbowski, M, & Shah, S.P. 1990. Shrinkage cracking of fiber reinforced concrete. *ACI Materials Journal*, 87: 138–148.

Higuchi, T. Eguchi, M. Morioka, M. & Sakai, E. 2014. Hydration and properties of expansive additive treated high temperature carbonation, *Cement and Concrete Research*, 64: 11–16.

Hossain, A.B. & Weiss, J. 2006. The role of specimen geometry and boundary conditions on stress development and cracking in the restrained ring test, *Cement and Concrete Research*, 36: 189–199.

Hossain, A.B. & Weiss, J. 2004. Assessing residual stress development and stress relaxation in restrained concrete ring specimens, *Cement & Concrete Composites*, 26: 531–540.

Jensen, O.M. & Hanse, P.F. 1995. A dilatometer for measuring autogenous deformation in hardening Portland cement paste, *Materials and Structures*, 28: 406–409.

Kovler, K. Sikuler, J. & Bentur, A. 1993. Restrained shrinkage tests of fiber reinforced concrete ring specimens: effect of core thermal expansion. *Materials and Structures*, 26: 231–237.

Lee, Y.S. & Ryou, J.S. Self healing behavior for crack closing of expansive agent via granulation/film coating method, *Construction and Building Materials*, 71: 188–193.

Mo, L. Deng, D. Tang, M.S. & Al-Tabbaa, A. 2014. MgO expansive cement and concrete in China: Past, present and future, *Cement and Concrete Research*, 57: 1–12.

Serris, E. Favergeon, L. Pijolat, M. Soustelle, M. Nortier, P. Gärtner, R.S. Chopin, T. & Habib, Z. 2011. Study of the hydration of CaO powder by gas–solid reaction, *Cement and Concrete Research*, 41: 1078–1084.

Subauste, J.S. & Odler, I. 2002. Stresses generated in expansive reactions of cementitious systems, *Cement and Concrete Research*, 32: 117–122.

Yan, P.Y. & Qin, X. 2001. The effect of expansive agent and possibility of delayed ettringite formation in shrinkage-compensating massive concrete, *Cement and Concrete Research*, 31: 335–337.

Advanced Materials and Structural Engineering – Hu (Ed.)
© *2016 Taylor & Francis Group, London, ISBN 978-1-138-02786-2*

Effect of the pH level of the electrolyte on the photocatalytic performance of TiO$_2$ nanotubes

C.S. Chen & S.Z. Yi

College of Material Science and Chemical Engineering, Tianjin University of Science and Technology, Tianjin, China

C. Wang

College of Marine Science and Engineering, Tianjin University of Science and Technology, Tianjin, China

ABSTRACT: Highly ordered TiO$_2$ nanotubes can be prepared by the anodic oxidation method, and the pH levels of the electrolyte is one of the major influential factors in preparation. Changing the pH levels of the electrolyte and observing the aspect and photocatalytic performance of TiO$_2$ nanotubes, we determine the effect of the pH levels. The experimental results showed that at the pH level of 5.5, the degradation constant of an acid orange solution by TiO$_2$ nanotubes was the best. The degradation rate was 97.3% after 30 minutes. By analyzing the result, it was found that the sample's specific surface area and concentration of oxygen vacancy influenced the photocatalytic performance of TiO$_2$ nanotubes.

1 INTRODUCTION

Due to the chemical stability and the promoting effect, TiO$_2$ has been widely utilized as a support for metal-oxide catalysts or photocatalysts (Zhang et al. 2011). The anodization (Macak et al. 2005, Mor et al. 2005) is a simple way to synthesize TiO$_2$ nanotubes, due to its moderate reaction condition and simple preparation procedure, for a rapid preparation of highly structured and ordered TiO$_2$ nanotubes on the titanium substrate. There are many effects on the aspect of TiO$_2$ nanotubes including the composition of the electrolyte, anodization voltage and the pH of the electrolyte. All these parameters of TiO$_2$ nanotubes could influence the efficiency of their photocatalytic activity. Among those preparation conditions, the pH value is important to the size of the tubes, mainly referring to the length (Liu et al. 2008).

The aim of this study was to identify how the pH condition during the preparation of TiO$_2$ nanotubes affects their photocatalytic activities. TiO$_2$ nanotubes were fabricated at different pH levels and their photocatalytic activities were evaluated for the degradation of AO 7. The morphological, structural, optical and physico-chemical properties of TiO$_2$ nanotubes were characterized. Finally, the relationship between the pH, the concentration of oxygen vacancies, the relationship between concentrations of oxygen vacancies, and their photocatalytic activities were determined based on the findings.

2 MATERIALS AND METHODS

2.1 Preparation of TiO$_2$ Nanotubes

The thickness of titanium sheets (99.6% purity, China) was 0.3 mm. The sheets were rinsed with an ultrasonic bath with acetone, isopropanol, methanol and deionized water for 5 min in turn. Finally, the sample was rinsed with deionized water and then air-dried.

The anodization was performed in a two-electrode electrochemical cell with a direct current power supply (Keithley 2400, KEITHLEY, USA). The anodizing voltage rose from 0 to 60 V and was kept at 60 V for 2 hr, with a Ti sheet serving as the anode and a Ni sheet as the cathode. The electrolyte consisted of ethylene glycol, NH$_4$F (0.5 wt.%), and H$_2$O (8 vol.%). The pH of the electrolyte was adjusted by H$_3$PO$_4$. The samples were synthesized at pH levels of 6.0, 5.5, 5.0, 4.5, 4.0 and 3.5 were denoted as S$_{6.0}$, S$_{5.5}$, S$_{5.0}$, S$_{4.5}$, S$_{4.0}$, S$_{3.5}$, respectively. During the experiment, the electrolyte was agitated with a magnetic stirrer. After anodization, the sample was immediately rinsed with deionized water and dried under ambient conditions. Then, it was an annealed at 450 °C in the air for 2 hr with heating and cooling rates of 10 °C/min.

2.2 Characterization

The morphology of the TiO$_2$ nanotubes was characterized by a scanning electron microscope (SEM)

(SU-1015, HITACHI, Japan). The crystal structure of the sample was identified by XRD (XD-3, PER-ESS, China) using a diffractometer with Cu Kα radiation. The surface chemical composition of TiO_2 nanotubes was analyzed by X-ray photoelectron spectroscopy (XPS) with a monochromatic Mg Kα X-ray source (E = 1486.6 eV) at 250 W.

2.3 *Photocatalytic experiments*

The photoreaction system consists of a cylindrical glass reactor with a magnetic stirrer and a 500 W high-pressure mercury lamp with the main emission at 365 nm as an external UV light source. The distance between the light and the anode was 10 cm.

The photocatalytic activity of TiO_2 nanotubes was evaluated via the degradation of AO 7, respectively. The photocatalytic degradation experiment was performed by the following procedure: (1) TiO_2 nanotubes with an area of 1 cm^2 were placed in 80 ml of aqueous AO 7 solution with an initial concentration of 10 mg/L (denoted as C_0), (2) prior to degradation, the solution was magnetically stirred in the dark for 1 hr to establish an adsorption-desorption equilibrium, (3) the reaction with stirring was irradiated by the UV lamp vertically outside the reactor, and the whole process was conducted with leading air of 30 mL/min. The remaining dye concentration (denoted as C) was monitored by a Lambda 35 spectrophotometer. AO 7 was monitored at 486 nm.

3 RESULTS AND DISCUSSION

3.1 *Catalysis characterization*

3.1.1 *Structure of TiO_2 nanotubes*
Figure 1 shows the XRD patterns of TiO_2 nanotubes annealed at 450 °C for 2 hr. The phases of all TiO_2 nanotubes were anatase, and no rutile phase was detected. The peak with scattering angles of 25.28°, 37.80°, 48.05°, 53.89°, 55.06° and 62.69° corresponded to the crystal planes of anatase TiO_2. It also showed the peaks of titanium at scattering angles of 38.41°, 40.17° and 53.00° corresponding to the crystal planes. The titanium peaks related to the Ti substrate.

3.1.2 *Morphology of TiO_2 nanotubes*
Figure 2 shows the SEM images of TiO_2 nanotubes. At various pH levels, all of the synthesized TiO_2 nanotubes exhibited a straight shape and high order, and had similar pore diameters. To the best of our knowledge, a TiO_2 nanowire, which is deformed TiO_2 nanotube with a smaller diameter, would often be seen after the formation of TiO_2 nanotubes during anodization in the EG-based electrolyte and the morphologies of TiO_2 nanotubes would change clearly. As seen in Figure 2,

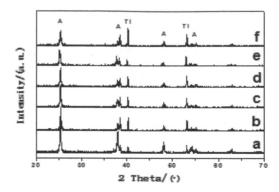

Figure 1. XRD patterns of TiO_2 nanotube arrays with different pH levels in the preparation after an annealing at 450 °C (A: anatase, Ti: titanium). (a) pH = 6.0; (b) pH = 5.0; (c) pH = 5.0; (d) pH = 4.5; (e) pH = 4.0; (f) pH = 3.5.

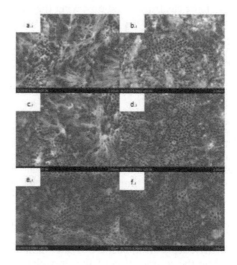

Figure 2. SEM images of anodized TiO_2 nanotubes annealed at 450 °C. (a) pH = 6.0; (b) pH = 5.5; (c) pH = 5.0; (d) pH = 4.5; (e) pH = 4.0; (f) pH = 3.5.

when the pH level increased above 5.0 (5.0--6.0), the surface morphologies became more and more rough and led to the formation of nanowires due to the split of nanotubes by etching. When the pH level was below 4.5 (3.5--4.5), the surface morphology became smooth and no nanowires were found. It is well known that nanotubes form through two competition reactions. They are the chemical dissolution of the oxide at the mouth of the tube and the inward movement of the metal/oxide boundary at the bottom of the tube. The dissolution rate of TiO_2 is higher at the mouth of tube than that at the bottom due to the high viscosity of EG + H_2O solution, leading to the dissolution of nanowires and a shorter length of TiO_2 nanotubes with the pH level decreasing. It is obvious that the pH of the electrolyte strongly influenced the morphology including the surface morphology and the length of TiO_2 nanotubes.

3.1.3 Surface compositions of TiO_2 nanotubes

XPS was used to determine the surface compositions of the TiO_2 nanotubes. Figure 3 shows the Ti 2p3/2 XPS spectra of the TiO_2 nanotubes after an annealing at 450°C. In this work, we detected $S_{6.0}$, $S_{5.5}$ and $S_{3.5}$ in order to find a difference among the samples. The typical peak of pure TiO_2 corresponding to Ti 2p3/2 is at 458.5 eV (Nefedov et al. 1974). As shown in Figure 3, $S_{6.0}$, $S_{5.5}$ and $S_{3.5}$ exhibited not only the Ti 2p3/2 peak at 458.5 eV, but also another peak at 457.6 eV. It is known that the peak at 457.6 eV matches the trivalent states of Ti. The formation of a low-binding energy of Ti 2p3/2 can be attributed to the decrease in the effective positive charge on Ti atoms. In order to keep charge neutrality, oxygen vacancies could be produced with the formation of Ti^{3+}(Li et al. 2012). This finding is inconsistent with the previous research suggesting that TiO_2 nanotubes synthesized by anodization would make a large number of vacancies.

As shown in Table 1, the increase in the Ti^{3+}/Ti^{4+} values indicates the increasing concentration of oxygen vacancies with the decreasing pH value. According to the mechanistic model of nanotube formation, field-assisted dissolutions will polarize and weaken the Ti-O bond that promotes the dissolution of Ti^{4+}cations, and the free O^{2-} anions migrate toward the metal/oxide interface. When the acid is added to an electrolyte, at which the pH level is below 5.5, more H^+ atoms could react with O^{2-} to hinder its movement and further make TiO_2 lattice to lack O atoms. However, the concentration of oxygen vacancies of $S_{6.0}$ is higher than that of $S_{5.5}$, when a small amount of phosphoric acid is added to the electrolyte. This may be ascribed to the PO_4^{3-} ions, which are strongly absorbed on the surfaces of TiO_2 and prevent the reaction between the H^+ and O^{2-} atoms. The reaction mechanism is still not so clear and needs to be further studied.

3.2 Photocatalytic activity studies

The photocatalytic efficiency was evaluated by the degradation of AO 7 in an aqueous solution under UV light irradiation. Figure 4 shows the degradation results of AO 7 by different samples of TiO_2 nanotubes. All the samples of TiO_2 nanotubes showed good photocatalytic activity. The highest level of photocatalytic activity is obtained with $S_{5.5}$, which has the AO 7 removal percentage at 97.3% after the irradiation for 30 min. For $S_{6.0}$, $S_{5.0}$, $S_{4.5}$, $S_{4.0}$ and $S_{3.5}$, the AO 7 removal percentages are equal to 93.5%, 91.1%, 84.0%, 82.7% and 81.7% after irradiation for 30 min. It was shown that the degradation rate reduced when the pH

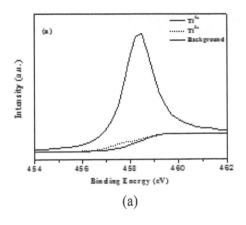

(a)

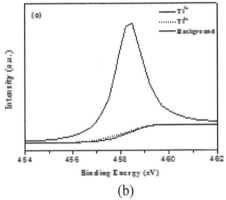

(b)

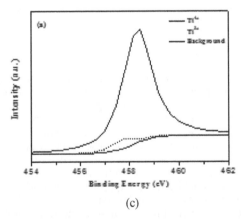

(c)

Figure 3. XPS spectra of Ti2p3/2 fitting peak of TiO_2 nanotubes after heating at 450°C. (a) S6.0; (b) S5.5; (c) S3.5.

level in the preparation process reduced. Because of the approximate surface area, the photodegradation rates of $S_{3.5}$, $S_{4.0}$, $S_{4.5}$ are very near. After 90 minutes, the AO 7 in the liquor was all degraded by the TiO_2 nanotubes of $S_{6.0}$, $S_{5.5}$, and $S_{5.0}$. The degradation rates of AO 7 by $S_{3.5}$, $S_{4.0}$, and $S_{4.5}$ are 93.7%, 95.2%, and 97.1%, respectively.

Table 1. The parameters of $Ti_2p_{3/2}$ fitting peak of TiO_2 nanotubes. $(S_{6.0})pH = 6.0$; $(S_{5.5})pH = 5.5$; $(S_{3.5})pH = 3.5$.

Samples	$Ti_2p_{3/2}$				
	$Ti^{4+} (TiO_2)$		$Ti^{3+}(TiO_{2-x})$		
	E_B	Peak area	E_B	Peak area	Ti^{3+}/Ti^{4+}
$S_{6.0}$	458.5	148322.8	457.6	3114.8	0.021
$S_{5.5}$	458.5	148324.0	457.6	2549.3	0.017
$S_{3.5}$	458.5	94480.6	457.7	6012.2	0.064

Experimental conditions: anodic voltage: 60 V; electrolyte: 0.5 wt%NH_4F + 2 vol%H_2O; anodic time: 2 hr; calcination temperature: 450 °C.

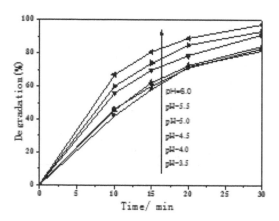

Figure 4. The degradation of AO 7 by the samples with different pH levels in the preparation process.

4 DISCUSSION

There are two factors that influence the efficiency of the photocatalytic activity of TiO_2 nanotubes affected by pH values through the anodic oxidation process. The one is the physical properties of TiO_2 nanotubes. The small crystallite size and large surface area could increase the photocatalytic activity. The large surface area is related to the size of the crystallite based on the present results. The small crystallite has a large surface area, which is consistent with the previous study. The vertical and highly ordered nanotubes have a large surface area, and are expected to have more catalytic active sites, absorb more reactants and make light penetration more efficiency.

Besides, the concentration of oxygen vacancies is another important factor that affects the photocatalytic activities of TiO_2 nanotubes. On the one hand, the concentration of oxygen vacancies could influence the crystal expansion that was further influenced the size of crystallites and the surface area of TiO_2 nanotubes. On the other hand, the concentration of oxygen vacancies acts as active

centers, causing the electron-hole recombination that could influence the photocatalytic activities of TiO_2 nanotubes. $S_{5.5}$ has the lowest concentration of oxygen vacancies, which could have the highest charge separation process and the highest photocatalytic activity among the samples. In addition, the low concentration of oxygen vacancies, small-sized titania crystallites, the large surface area of TiO_2 nanotubes have all led to a positive effect on the improvement of their photocatalytic activities. However, the concentration of oxygen vacancies shows a crucial effect on the photocatalytic activity, considering all the features of TiO_2 nanotubes.

5 CONCLUSIONS

Highly ordered TiO_2 nanotubes were fabricated by anodization at various pH values. In this paper, the relationship between the pH level, structure and the photocatalytic activity of TiO_2 nanotubes was studied. The sample $S_{5.5}$ (prepared with the pH of 5.5), which has a large surface area and a small-sized crystallite structure, exhibits an excellent catalytic activity for the degradation of Acid Orange 7(AO 7) under UV irradiation. Moreover, the low concentration of oxygen vacancies determines the low electron-hole pair recombination. Compared with other factors, the concentration of oxygen vacancies shows the crucial effects on the photocatalytic activity of TiO_2 nanotubes. In addition, $S_{5.5}$ showed high pseudo-first-order rate constants and a high photocatalytic activity based on the present results.

REFERENCES

Li, Z. Hu, R. Guo, J. Ru, L. & Wang, H. 2012.Effect on the formation of oxygen vacancy of TiO2, *Materials Science and Technology*, 20: 80–85.

Liu, Z. Zhang, X. Nishimoto, S. Jin, M. Tryk, D.A. Murakami, T. & Fujishima, A. 2008.Highly Ordered TiO2 nanotube arrays with controllable length for photoelectrocatalytic degradation of phenol, *The Journal of Physical Chemistry C*, 112(1): 253–259.

Macak, J.M. Tsuchiya, H. & Taveira, L. et al. 2005. Smooth anodic TiO2 nanotubes, *Angewandte Chemie International Edition*, 44(45): 7463–7465.

Mor, G.K. Varghese, O.K. & Paulose, M. et al. 2005. Transparent highly ordered TiO2 nanotube arrays via anodization of titanium thin films, *Advanced Functional Materials*. 15(8): 1291–1296.

Nefedov, V.I. Salyn, Y.V. Chertkov, A.A. & Padurets, L.N. 1974. X-ray electron study of the electron density distribution in hydrides of the transition elements, *Russian Journal of Inorganic Chemistry*, 16: 1443–1445.

Zhang, H. Pan, X. & Liu, J.J. et al. 2011. Enhanced Catalytic Activity of Sub nanometer Titania Clusters Confined inside Double Wall Carbon Nanotubes, *Chem. Sus. Chem.* 4(7): 975–980.

Advanced Materials and Structural Engineering – Hu (Ed.)
© 2016 Taylor & Francis Group, London, ISBN 978-1-138-02786-2

Forming limit prediction of 5182-O aluminum alloy sheet using finite element analysis

R. Kurihara, S. Nishida, H. Kamiyama & R. Okushima
Gunma University, Honcho, Ota City, Japan

ABSTRACT: Forming limit prediction of 5182-O aluminum alloy sheet was carried out by using finite element analysis. JSTAMP/NV was used in the finite element analysis to predict the forming limit of sheet metal. The length of the aluminum specimen was 120 mm, and the width was varied from 10 mm to 120 mm, and the thickness was 1.0 mm. Stretching test was operated by Erichsen test to measure the forming limit at both uniaxial and biaxial tensile areas. Forming limit prediction and necking limit prediction by using finite element analysis was proposed. Forming Limit Diagram (FLD) of analytical results indicated relatively good conformability with the FLD of experimental results. Thus, it would benefit sheet metal forming.

1 INTRODUCTION

Weight saving technology in automotive industries has played an important role especially from the viewpoint of improving the collision safety as well as consumption of fuel efficiency. Thus, the application of aluminum sheets in car bodies is increasing.

It has been reported that a 100 kg saving of a car weight improves a 1.0 km/l of fuel consumption. At the same time, the stiffness and the collision safety are required for a car body. Thus, aluminum alloy sheets are expected as the substitute of traditional steel sheets, because aluminum alloy sheets have a high specific strength compared with common steels. It has been considered that the application of complicated shapes of products of aluminum alloy sheets in car components is difficult, due to its poor formability. For increasing the press production of the aluminum alloy sheets, exact prediction methods of formability of aluminum alloy sheets are needed. Forming Limit Diagram (FLD) is one of the standards indicating the formability of sheet metal. Thus, a higher accuracy of the forming limit prediction is required. The comparative evaluation by using an experiment and finite element method has also been reported by many researchers (Ozturk et al. 2004).

In this paper, a forming limit prediction method as localized necking before the fracture for the forming limit was proposed. The advantage of this method is in its easy estimation compared with other methods when using the forming limit as a fracture criterion. An FLD of aluminum alloy sheets was obtained by stretching test with both experiment and FEM analysis. The FLD obtained was compared with an exact point of fracture found between the experimental values and the analysis values.

2 EXPERIMENTAL CONDITIONS

2.1 Erichsen test

Erichsen testing machine, ERICHSEN GmbH & Co. KG, model 142-20 is shown in Figure 1. Erichsen testing machine was used to investigate the height of the stretched specimen. Grid marking was stamped to the specimen, and the deformation of the grid was observed with four cameras. The strain was obtained by measuring the deformation of the grid with the three-dimensional measuring machine during the test. The strain ratio and the strain path were obtained by changing the width of the specimen. The three-dimensional measuring machine ViALUX, Auto Grid was used.

2.2 Test material

In this study, the aluminum alloy sheet 5182-O was used. A 5000 series aluminum alloy was used for automotive body panels. Table 1 lists the material properties of aluminum alloy 5182-O. The specimen length was 120 mm, and the thickness was 1.0 mm. Figure 2 shows the examples of the specimen. The specimen widths were 10 mm, 20 mm, 30 mm, 60 mm, 80 mm, and 120 mm.

2.3 Forming limit diagram

Forming Limit Diagram (FLD) of aluminum alloy sheet 5182-O obtained by the experiment is shown

Figure 1. Erichsen testing machine and measuring cameras.

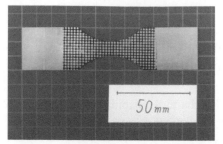

(a) 10mm width

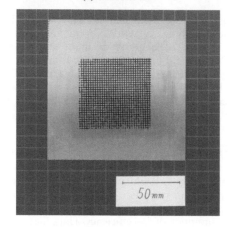

(b) 120mm width

Figure 2. Examples of the specimens for the Erichsen test.

Table 1. Material properties of the aluminum alloy 5182-O.

Young's modulus (GPa)	70.0
Yield stress (MPa)	132
Tensile strength (MPa)	272
Poisson's ratio	0.330
K-value (MPa)	534
r-value	0.753
n-value	0.295

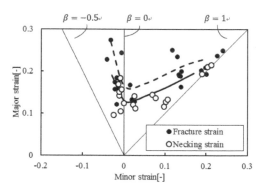

Figure 3. Forming limit diagram obtained by the experiment.

in Figure 3. In the figure, the black circle plots indicate the fracture strain distribution, and the white circle plots indicate the necking strain distribution. A straight line of $\beta = -0.5$ shows the uniaxial stress state, $\beta = 0$ shows the plane stress state, and $\beta = 1$ shows the biaxial stress state. When fracture and necking occurred at the uniaxial tensile side toward the plane tensile side, the strains were distributed in the linear downward-sloping. The strains were distributed in the upward-sloping at the plane tensile side toward the biaxial tensile side. Necking strains were distributed under the fracture strain distribution. In this study, it was regarded as the occurrence of localized necking when the difference in the thickness reduction rate of the adjacent grid strain was reached up to 1%.

3 FINITE ELEMENT ANALYSIS

Analytical Erichsen test was carried out by using the finite element method. The software used for the analysis was JSOL, JSTAMP/NV. The element size was 1 mm × 1 mm, and the element type was

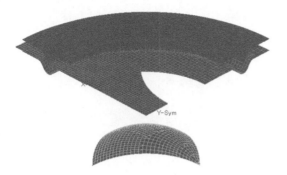

Figure 4.   Analytical model of the stretching test.

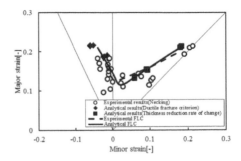

Figure 5.   Comparison between the analytical results and the experimental results.

shell type with 5 integration points in the thickness direction. The friction coefficient was 0.05 between the blank and the punch, and was 0.15 between the blank and the top-bottom die, respectively. The stress-strain curve was approximated with Swift's formula. An analytical model was applied quarter model, considering the symmetry. FEM analysis was conducted starting from the blank-holding process to the stretching process. The yield function was applied to Hill'48. Figure 4 shows the analytical model. In this paper, a ductile fracture criterion was applied for the prediction of localized necking at the uniaxial tensile side. In the biaxial tensile deformation, the occurrence of localized necking was evaluated by using the change in the thickness reduction rate, which was the same as that observed in the experiment.

4   DUCTILE FRACTURE CRITERION

Ductile fracture criteria express the stress strain state to fracture. In this study, the localized necking was predicted by applying Cockcroft's ductile fracture criterion, represented by Equation (1) as follows:

$$\int_0^{\bar{\varepsilon}_N} \sigma_{max} d\bar{\varepsilon} = C \qquad (1)$$

where $\bar{\varepsilon}_N$ is the equivalent strain of necking; σ_{max} is the maximum principal stress; $d\bar{\varepsilon}$ is the increment of the effective strain; and C is the material constant. In addition, σ_{max} is expressed in the relation (2) in a tensile state (Hayashi 2010):

$$\sigma_{max} = \sigma_m + \frac{2}{3}Y \qquad (2)$$

where σ_m is the mean normal stress and Y is the yield stress. In this paper, the C value of 71 was used, which was obtained from the result of the tensile test.

5   ANALYTICAL RESULTS AND DISCUSSION

FLD comparison between the experimental results and the analytical results is shown in Figure 5. The ductile fracture criterion was applied in the analytical FLD at the uniaxial tensile side. The change in the thickness reduction rate was used as a reference for the occurrence of localized necking, which was the same as the experiment in the biaxial tensile side. The analytical FLD agreed well with the experimental FLD. In the case of applying only a ductile fracture criterion in the biaxial tensile side, forming a limit was underestimated. It is possible to appropriate the evaluation of forming a limit to apply the change in the thickness reduction rate in the area that is from the plane strain deformation area to the biaxial tensile area.

6   CONCLUSIONS

Forming limit diagram of 5182-O aluminum alloy sheet was obtained by Erichsen stretching test and finite element method. JSTAMP/NV is used in the finite element analysis. The ductile fracture criterion is applied for the prediction of localized necking in the uniaxial tensile state, and the thickness reduction rate of change is used as a reference for the occurrence of localized necking in the biaxial tensile state. The FLD of the analytical results agrees well with that of the experimental results.

REFERENCES

Hayashi, H. 2010. *Stamping process of hard-to-form material High tensile strength steel*, Nikkan Kogyo Shimbun Ltd., Japan.
Ozturk, F. & Lee, D. 2004. Analysis of forming limits using ductile fracture criteria, *Journal of Materials Processing Technology*. 147(3): 397–404.

Advanced Materials and Structural Engineering – Hu (Ed.)
© 2016 Taylor & Francis Group, London, ISBN 978-1-138-02786-2

Fabricating polymer films on a super-hydrophobic surface by the blow film method

Y. Ma
State Key Laboratory for Modification of Chemical Fibers and Polymer Materials, Donghua University, Shanghai, P.R. China
College of Material Science and Engineering, Donghua University, Shanghai, P.R. China

J.L. Zhao, K. Li, J.J. Zhang, C.M. Kang, H.Z. Zhang & X.Y. Zhang
College of Material Science and Engineering, Donghua University, Shanghai, P.R. China

ABSTRACT: Facile fabrications of a super-hydrophobic surface with a water contact angle greater than 150° have attracted attention in both fundamental and engineering applications. In this paper, an effective and simple method was suggested to fabricate a transparent super-hydrophobic surface. The water contact angle of the super-hydrophobic surface reached up to 160° without coating any low-surface energy materials. The morphology of the surface exhibited a combination of micro- and nanoscale rough structures. Moreover, polymer films can be fabricated on the super-hydrophobic surface by a blow film method. This research will have a profound influence on the future preparation of polymer films.

1 INTRODUCTION

One of the most important properties of materials is their water repellency property or hydrophobicity. When the water contact angle is greater than 150°, the solid surface is called super-hydrophobic (Ahsan et al. 2013). Inspired by the super-hydrophobic property of many plants and animals in nature, great research efforts have been made to fabricate such a magic surface by controlling the hierarchical roughness and topography. Taking advantage of the micro- and nanoscale hierarchical roughness, the super-hydrophobic surface exhibits some attractive effects, such as anti-adhesion, anti-fouling, and anti-ice effects. However, it remains a challenge to develop a simple and broadly applicable approach to obtain a super-hydrophobic surface with a high transparency and strong adhesive force with a substrate (Zhang et al. 2013). Therefore, in this article, we first introduced an effective and simple preparation method to fabricate a transparent super-hydrophobic surface on glass, and then attempted to fabricate polymer films on this surface by the blow film method. So far, film formation on top of a super-hydrophobic surface has not yet been reported. However, such a surface could be an ideal substrate to assist the formation of a flat polymer film or membrane from the solution, due to its anti-contaminant properties and low adhesiveness to a polymer.

2 EXPERIMENTS

HCl and NH_4OH were used to adjust the pH value of deionized water. Tetraethoxysilane (TEOS) was hydrolyzed with deionized water at 60°C for 2h, with a fixed molar ratio of $1TEOS:48C_2H_5OH:4H_2O$ (pH = 2). Thereafter, 1,1,1,3,3,3-hexamethyldisilazane (HMDS) was divided into 20 parts, which were put into the sol every 9 minutes and super sonicated at 20°C for 3h (Wang & Luo 2012). The prepared sol-gel mixture was aged at 25°C for 48h, and then coated onto a glass slide by the dip-coating method. The slide was dried in the muffle with N_2 protection at 600°C for 4h. The contact angel of water reached up to 159° without coating any low-energy materials.

A certain amount of polyvinyl alcohol (PVA, M_w = 94000; Sigma-Aldrich) was dissolved in the distilled water at 100°C and then filtered. Finally, a series of PVA solution of different concentrations was obtained.

The size of the particles was characterized by a nanoparticle and potentiometric analyzer (Nano ZS). The contact angel of water on the surface was determined using a contact angle measurement (OCA 40 Micro). The surface structure of a super-hydrophobic and the thickness of the polymer film were characterized by SEM (HITACHI 8010, Japan). The high-speed camera (Motionpro Y4) was used to record the movement of water droplets on the substrate. The prepared super-hydrophobic substrates were installed into a home designed blow film device. A PVA liquid film was formed by driving the bubbles from one side to the other.

3 RESULTS AND DISCUSSIONS

Figure 1 shows the size of the sol-gel particle after aging for 48h. The particle size prepared was less than 10 nm, and the size distribution by number was rather narrow. Then, the sol-gel mixture was coated on a glass slide by a dip-coating method. After heat treatment, the surface with different water contact angles as well as the topography of the surface were obtained, as shown in Figure 2.

Figure 2a and 2b shows the relationship of contact angles and the thickness of the coating. The contact angle value was increased with the increase in coating thickness, and became a maximum after the coating thickness increased to a certain value (Debmita et al. 2011). The super-hydrophobic surface with high transparency was obtained by

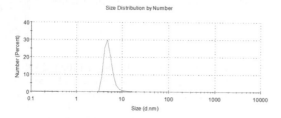

Figure 1. The size distribution of the sol-gel particles after aging for 48h.

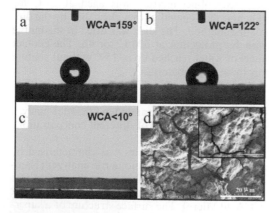

Figure 2. Water contact angle of glass surfaces: (a) the multi-coated glass surface was dried in the muffle with N2 protection; (b) the single-coated glass surface was dried in the muffle with N2 protection; (c) the multi-coated glass surface was dried in the muffle without N2 protection; (d) the SEM images of (a).

roasting. In addition, the contact angles had an apparent change from super-hydrophobic to super-hydrophilic if roasting without the protection of N_2, as shown in Figure 2c. It may be due to the fact the methyl group could react with O_2 in the air and change into hydroxyl (Satish et al. 2012). As a result, the super-hydrophobic surface would become a super-hydrophilic surface, although their surface topology is similar. The rough structure on the surface reflected the hydrophobic and hydrophilic effect of the surface. Therefore, in order to get a super-hydrophobic surface with high transparency and strong adhesive with glass slides, it is necessary to coat at least 3 layers of sol-gel dispersion and then dry in the muffle with N_2 protection.

Figure 2d shows the SEM image of super-hydrophilic surfaces. There were a large number of cracks of micrometer size on the coating surface, which provided the surface with micrometer-scale roughness. Besides, the nanoparticles that deposited on the surface of the microstructure would further increase the hydrophobicity of the sample. The contact angle of water on the surface was close to 160°, and the roll angle water on the surface was small so that the water droplets exhibited a sphere and could scroll freely on the surface, as shown in Figure 3.

The PVA polymer film was prepared by the spin-coating, dip-coating and bubble blown methods, respectively. When using the spin-coating and dip-coating methods, there were no polymer films left on the super-hydrophobic surface. This is due to the fast rotation during spin-coating, inhibiting the spread of the PVA solution into a film on the super-hydrophobic surface. This also reflected that the super-hydrophobic surface exhibited a self-cleaning effect, and the surface of the droplet could be easily removed. Similarly, when the dip-coating method was adopted, due to the fact that the contact angle of a substrate was very large, the PVA solution also could not be spread.

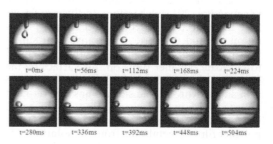

Figure 3. The movement of a water droplet on the substrate captured by a high-speed camera.

We successfully prepared the PVA polymer film on the super-hydrophobic surface via the blow film method, as shown in Figure 4, which showed the process of the fabricating polymer film by the blow film method. For the first time, we suggested that the polymer film could still be fabricated on the super-hydrophobic surface by the blow film method. Although the liquid film exhibited a strong retraction trends on top of a super-hydrophobic surface, the entanglements of molecules in the polymer solution provided a good ductility of a liquid film, and ensured the extension of a liquid film without fracture. Once a hole appeared in the liquid film, the force spread liquid film would re-wet the super-hydrophobic surface quickly and automatically, and finally retracted into droplets. In addition, in the course of a blow film, a super-hydrophobic surface was in its Cassie state with air pockets between the liquid film and the super-hydrophobic surface. The presence of these air layers could greatly reduce the frictional resistance of the polymer solution flows forward.

Interestingly, these polymer films could be peeled off from the super-hydrophobic substrates, as shown in Figure 4. The preparation of the polymer film on the super-hydrophobic surface depends extensively on the stability of the polymer fluid itself and the curing rate. That is, the faster the curing rate, the thicker the polymer film thickness. However, when a curing rate is too slow, the non-uniformity of the substrate surface can make the liquid film inclined to fracture in the process of blow thinning, especially on the super-hydrophobic surface. Meanwhile, there are many topological surface features on the super-hydrophobic surface, which make the liquid film free but still supported. So, the blow film is a facile method to prepare the polymer film with micrometer thickness from the solution. In order to prepare a nano-film, it is necessary to further improve the stability of the polymeric fluid to shear and the tensile

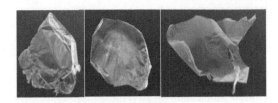

Figure 5. The polymer film was peeled off from the super-hydrophobic substrates. From left to right, the thickness was increased in turn (the thickness of the film was controlled by changing the concentration of the PVA water solution, which respectively was 20, 35, and 50 mg/mL).

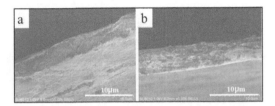

Figure 6. The SEM image of the cross section of the polymer films peeled off from the super-hydrophobic substrates.

process to avoid a rupture in case of small fluctuations. The thickness of the polymer film was also closely related to the concentration and the addition amount of the solution. Figure 6 shows the SEM image of the polymer films that were peeled off from the super-hydrophobic substrates. Obviously, the thickness of the polymer film was about 10 μm.

4 CONCLUSIONS

An effective and simple method was suggested to obtain a super-hydrophobic surface with high transparency. We demonstrate that the polymer film of 10 μm thickness can be fabricated from the solution on such a super-hydrophobic surface by the blow film method. The self-clean and low adhesive property of the super-hydrophobic surface will provide wide potential applications in the fabrication of the polymer film, especially from the solution.

ACKNOWLEDGMENT

We thank the financial support from "The Fundamental Research Funds for the Central Universities".

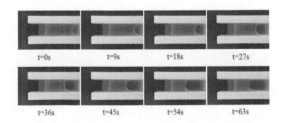

Figure 4. The figures of the blow film device and process. Both substrates were super-hydrophobic, the concentration of the solution of PVA is 50 mg/mL. The bubbles moved from right to left. A liquid polymer film was formed between the bubble and the substrate.

REFERENCES

Ahsan, M.S. Dewanda, F. & Lee, M.S. 2013. Formation of superhydrophobic soda-lime glass surface using femtosecond laser pulses, *Applied Surface Science*. 265: 784–789.

Debmita, G. Samar, K.M. & Goutam, D. 2011. Superhydrophobic films on glass surface derived from trimethylsilanized silica gel nanoparticles, *ACS Applied Materials & Interfaces*. 3: 3440–3447.

Satish, A. Mahadik, D.B. & Parale, V.G. 2012. Recoverable and thermally stable superhydrophobic silica coating, *Journal of Sol-Gel Science and Technology*. 62: 490–494.

Wang, S.D. & Luo, S.S. 2012. Fabrication of transparent superhydrophobic silica-based film on a glass substrate, *Applied Surface Science*. 258: 5443–5450.

Zhang, Y. Li, J.L. & Huang, F.Z. 2013. Controlled fabrication of transparent and superhydrophobic coating on a glass matrix via a green method, *Applied Physics A*. 110: 397–401.

Advanced Materials and Structural Engineering – Hu (Ed.)
© 2016 Taylor & Francis Group, London, ISBN 978-1-138-02786-2

Microelectrode-based sensor for detecting viability and activity of cells

X. Zhu

College of Mechanical and Electrical Engineering, Hohai University, Changzhou, Jiangsu, China
Department of Mechanical and Aerospace Engineering, University of California, Los Angeles, USA

ABSTRACT: A microelectrode-based Micro-electromechanical Systems (MEMS) sensor for detecting the cell activity is proposed. A spiral microelectrode array was designed, and the MEMS sensor was fabricated. The viability and activity of cultured human Jurkat cells were detected by observing the traveling velocity of the cell toward the center of spiral electrodes in this sensor. Under the traveling-wave dielectrophoresis (twDEP) effect, the human leukemia Jurkat cells with different activities moved with different velocities in the sensor chamber. Moreover, the DEP-induced velocities of cells decreased over time after these cells were taken out of the humidified CO_2 incubator, which indicated that the cell activities decreased over time in the out-of-incubator environment. Therefore, this microelectrode-based sensor could potentially serve as a sensor integrated in microfluidic systems for detecting the cell activity.

1 INTRODUCTION

In many biological microfluidic assays or applications, biological cells serve as the key components of microfluidic analytic systems (Lei 2014, Ma et al. 2009, Nie et al. 2007, Irimia et al. 2007). Furthermore, the viability and activity of cells play an important role in the studies of electroporation (Santra & Tseng 2013, Movahed & Li 2011), cell migration (Lamb et al. 2008), cell-cell interactions (Chung et al. 2010) and cellular responses in the micro/nano environment (Lei 2014) in microfluidic systems. In many microfluidic assays or applications, the activities of cultured cells are diverse from each other, and the cells with low activity may not be suitable for certain assays or specific operations. On the other hand, the humidified CO_2 atmosphere usually disappears after the cell sample is taken out of the incubator. Then, the cells in the microfluidic device (e.g. a chip) will experience the out-of-incubator environment during the washing or dilution of the sample, cell injection, and microfluidic dispensing, which could result in decreased cell activity. To ensure that the cells could satisfy the reaction or process requirements, cell viability and activity should be tested before the next operating step in the microfluidic system.

There are many methods to determine or test the cell viability and activity, including detecting the ability of cells to reproduce, determining membrane integrity, enzyme activities, respiration or membrane potential (Breeuwer & Abee 2000), staining with special dyes (Heo et al. 2003), and observing the degree of cell atrophy (Yeh et al. 2011). These methods have been demonstrated to be effective, and appear to be much conspicuous for the observation through a microscope. However, all these processes used for determining cell viability and activity involve chemical reactions that may change the physiological state of cells, and usually need fluorescence observation that requires configuration of an extra fluorescence module for the microscope and thus raises the test costs.

Here, a microelectrode sensor based on dielectrophoresis (DEP) for detecting cells with different activities is presented, which could be utilized to alleviate the above problems. The cellular activities could be characterized by the moving rates of cells on the electrode array involving no chemical reactions. The microelectrodes are designed to consist of four spiral strip-electrodes with the same widths and gaps. This electrode array in the spiral form is convenient for observing the particle motion in all directions in a compact small working area. It can be operated with only four electrical connections, which is easy for integrating with other electronic devices.

2 MICROSENSOR DESIGN AND FABRICATION

2.1 *Structure of the Microelectrode-based Sensor*

The sectional view and plan view of the developed microelectrode-based sensor for detecting cell viability and activity are shown in Figure 1a and Figure 1b, respectively. The planar electrode sensor system is comprised of a glass substrate with four spiral microelectrodes on its surface and an insulation spacer between the top coverslip and the

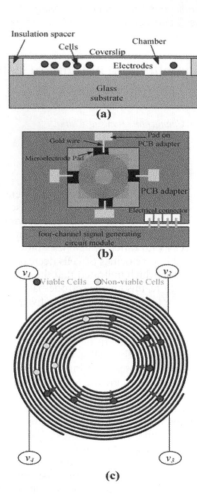

Figure 1. Structure and schematic of the spiral-electrodes based sensor for detecting the cell activity (schematic drawing is not to scale).

bottom substrate. To fabricate the microelectrodes, about 50 nm-thick Ti film was first sputtered onto the insulating glass substrate, and then a 200-nm thick Au film was sputtered on the Ti film. The Au/Ti (200 nm/50 nm thick) electrode pattern was fabricated on the glass substrate through the standard photolithographic process. The microelectrodes had a spiral configuration and each electrode strip had a width of 20 µm and a gap of 20 µm, as shown in Figure 1c. The microelectrode pads and the pads on the printed circuit board (PCB) adapter are connected via thin gold wires, as shown in Figure 1b. The electric potentials provided by a signal generating circuit module were applied through the pads on the adaptive PCB. When the electric potentials were on, the electrodes generated non-uniform electric fields spatially in the space above the electrode surface.

2.2 Basic principle

In this non-uniform electric field, the electric neutral particle (e.g. cells) can move toward the electrodes (high-field region) or away from the electrodes (to the low-field region) under the conventional DEP (cDEP) effect (Morgan & Green 2003) in the vertical directions. Furthermore, an electric field with a spatially varying phase gives rise to traveling wave dielectrophoresis (twDEP) (Lei et al. 2009). The directions for the twDEP motion of cells are theoretically towards (or away from) the center of the spiral electrode array, which are represented by the arrows in Figure 1(c). The twDEP forces act on cells that exhibit a close relationship with the electric fields, the dielectric attributes of the cells and the suspending medium. According to the electrokinetic theory (Morgan & Green 2003, Jones 2003), the time-averaged twDEP force in the AC field can be expressed as

$$\langle \mathbf{F}_{\mathrm{DEP}} \rangle = -(1/2)v\,\mathrm{Im}[\tilde{\alpha}]\nabla \times (\nabla \phi_{\mathrm{R}} \times \nabla \phi_{\mathrm{I}}) \qquad (1)$$

In Equation (1), Im indicates the imaginary part; v is the cell volume; and $\tilde{\alpha}$ is the induced effective polarizability, and

$$\tilde{\alpha} = 3\varepsilon_{\mathrm{m}}(\tilde{\varepsilon}_{\mathrm{p}} - \tilde{\varepsilon}_{\mathrm{m}})/(\tilde{\varepsilon}_{\mathrm{p}} + 2\tilde{\varepsilon}_{\mathrm{m}}) \qquad (2)$$

where $\tilde{\varepsilon}_{\mathrm{p}}$ and $\tilde{\varepsilon}_{\mathrm{m}}$ are the complex permittivity of the particle and the suspending medium, respectively. In Equation (2), $\tilde{\varepsilon}_{\mathrm{p}}$ is determined by the cell size and structure, membrane properties (permeability, capacitance) and internal conductivity of the cells, which could be utilized as the criterion for the DEP discrimination of cells (Gimsa et al. 1996, Becker et al. 1995, Wang et al. 2002). ϕ_{R} and ϕ_{I} are the real and imaginary parts of the complex phasor for the electrical potential (Morgan & Green 2003). ∇ is the del vector operator. In Equation (1), $\nabla \times (\nabla \phi_{\mathrm{R}} \times \nabla \phi_{\mathrm{I}})$ is referred to as twDEP vector. The cells in the electric field also undergo the opposing Stokes' drag (Morgan & Green 2003) as

$$\mathbf{F}_{\mathrm{drag}} = -6\pi\eta RV \qquad (3)$$

where η, R, V are the dynamic viscosity of the fluidic medium, particle radius, and particle velocity, respectively. At equilibrium, $\langle \mathbf{F}_{\mathrm{DEP}} \rangle + \mathbf{F}_{\mathrm{drag}} = \mathbf{0}$. Combining with Equation (3) and Equation (1) gives the traveling velocity of the cell as

$$V = -(R^2/9\eta)\,\mathrm{Im}[\tilde{\alpha}]\nabla \times (\nabla \phi_{\mathrm{R}} \times \nabla \phi_{\mathrm{I}}) \qquad (4)$$

where $\mathrm{Im}[\tilde{\alpha}]$ contains the information about the cell structure, composition, viability and activity. In another word, the traveling velocity of cells can be used to characterize the activity of the same species of cells that suspended with similar appearances.

3 OPERATION METHODS

3.1 Cell culture

Human leukemia Jurkat cells were grown in a RPMI-1640 medium supplemented with 100 U mL^{-1} of penicillin, 100 µg mL^{-1} of streptomycin and 10% Fetal Calf Serum (FCS). Cells were incubated at 37°C in a humidified incubator (5% CO_2 and 95% air).

3.2 Test operations

The culture solution containing Jurkat cells was diluted with the PBS solution and deionized water in a glass tube for the following viability tests. To generate a traveling-wave field in the microsensor, each alternating current (AC) signal applied on the spiral electrode had a successive phase shift of 90 degree, and the voltages of all the four AC signals had the same magnitude of 3V peak-peak. The phase-shifted signals were provided by a four-channel signal generating a circuit module. During the experiment, the frequency of 1MHz for the four-channel signals was chosen, because the low frequencies below 1MHz may give rise to the alternating current electro-osmosis (ACEO) effect that will induce the fluid flow (Ramos et al. 2005). The ACEO flow can also drag the particle to move. To avoid the disturbance of the ACEO flow pushing the cells, a higher frequency (1 MHz) was chosen. A CCD camera was used to capture the images for the motion of cells in this MEMS sensor placed on an upright microscope (Nikon). There is no need of fluorescence observation for the experiment.

4 RESULTS AND DISCUSSION

The sample Jurkat cells used in the first test was placed out of the incubator for about 1.5 hours. A droplet containing Jurkat cells of about 8–10 µL was perfused into the sensor chamber using a pipette. After the 1-MHz signals of 3V peak-peak with phase shifts were applied on this microelectrode sensor, the cells suspended in this sensor chamber simultaneously moved towards the center of the spiral electrode array (see Figure 2, where the arrows represented the twDEP force directions). As shown in Figure 2, the cells with high viability and activity moved with a traveling velocity of ~15 to 17.5 µm/s (e.g. the cells circled by the green and white rings). Other cells with a relatively low activity moved with a lower velocity of ~8 to 12µm/s (e.g. the cells circled by the black ring). After several minutes, the sedimentation of cells was obvious, which was caused by the combinational effect of the cDEP force and buoyancy acting on Jurkat cells in the vertical direction.

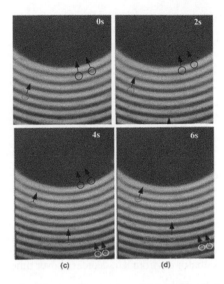

Figure 2. The traveling motion of the Jurkat cells in the first test.

The cells with different viabilities and activities had different membrane capacities, which could change the dielectric attributes of the cells. Moreover, the membrane of high viable cells had the ability of selective permeation and could maintain a difference in the concentration of ions between the intracellular and extracellular environments. However, cells with low viability and activity may lose part of the ability of selective permeation through the membrane, and thus had the dielectric properties different from those of the high viable cells. This resulted in a DEP-induced traveling velocity differential among the cells with a different activity. When the cellular viability and activity deceased, the polarizability differential between the cell and the suspending medium decreased correspondingly, and thus the twDEP force deceased (Morgan & Green 2003, Gascoyne et al. 2002). When a cell is dead, there is no polarizability differential between the cell and the suspending medium because the cell membrane loses its function of maintaining the ion concentration gradient across the membrane.

The second test was implemented about one hour after the first test. Another droplet of about 8–10 µL containing Jurkat cells was perfused into the sensor chamber. In the second test, several cells moved with a maximal velocity of ~14µm/s, which indicated that these cells had the highest viability and activity in the cell population. The other cells in the sensor chamber with various activities were indicated by the various traveling velocities of the cells. Similarly, the third test was implemented around one hour after the second test. The three test results are compared, which is shown in Figure 3. According to the results of maximum cell velocities, the

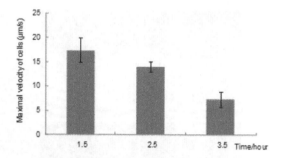

Figure 3. The maximum velocity of the cells with respect to the duration after the cell was out of the incubator.

3.5-hour-later viability of cells was obviously lower than the 1.5-hour-later viability of cells. It was followed that the viabilities of Jurkat cells in the out-of-incubator environment decreased over time.

5 CONCLUSIONS

From the above results, the following conclusions can be drawn:

1. The results for the three tests at different time instants with the same time interval (one hour) show that the maximum cell velocity of each test decreased at 1.5, 2.5 and 3.5 hours after the cells were out of the incubator successively, which indicates that the cellular activity decreased over time. Therefore, the viability and activity of cells could be detected and characterized by measuring the traveling velocity of each cell in this microelectrode-based MEMS sensor.

2. In each test, the cells exhibited various traveling velocities, which indicated that the cells had an individual difference in viability and activity. It lays a foundation for detecting very few cells that have intrinsically different physiological status involving no chemical reactions in a population of large number of cells.

ACKNOWLEDGMENTS

This work was supported by the Fundamental Research Funds for the Central Universities (2015B04414).

REFERENCES

Becker, F.F. Wang, X.B. Huang, Y. Pethig, R. Vykoukal, J. & Gascoyne, P.R.C. 1995. Separation Of Human Breast-Cancer Cells From Blood By Differential Dielectric Affinity, *Proceedings of the National Academy of Sciences of the United States of America*, 92: 860–864.

Breeuwer, P. & Abee, T. 2000. Assessment of viability of microorganisms employing fluorescence techniques, *International Journal of Food Microbiology*, 55: 193–200.

Chung, S. Sudo, R. Vickerman, V. Zervantonakis, I.K. & Kamm, R.D. 2010. Microfluidic Platforms for Studies of Angiogenesis, Cell Migration, and Cell-Cell Interactions, *Annals of Biomedical Engineering*, 38: 1164–1177.

Gascoyne, P. Mahidol, C. Ruchirawat, M. Satayavivad, J. Watcharasit, P. & Becker, F.F. 2002. Microsample preparation by dielectrophoresis: isolation of malaria, Lab on a Chip, 2: 70–75.

Gimsa, J. Muller, T. Schnelle, T. & Fuhr, G. 1996. Dielectric spectroscopy of single human erythrocytes at physiological ionic strength: Dispersion of the cytoplasm, *Biophysical Journal*, 71: 495–506.

Heo, J. Thomas, K.J. Seong, G.H. & Crooks, R.M. 2003. A microfluidic bioreactor based on hydrogel-entrapped E. coli: Cell viability, lysis, and intracellular enzyme reactions, *Analytical Chemistry*, 75: 22–26.

Irimia, D. Charras, G. Agrawal, N. Mitchison, T. & Toner, M. 2007. Polar stimulation and constrained cell migration in microfluidic channels, Lab on a Chip, 7: 1783–1790.

Jones, T.B. 2003. Basic theory of Dielectrophoresis and Electrorotation, *IEEE Engineering in Medicine and Biology Magazine*, 22: 33–42.

Lamb, B.M. Barrett, D.G. Westeott, N.P. & Yousaf, M.N. 2008. Microfluidic lithography of SAMs on gold to create dynamic surfaces for directed cell migration and contiguous cell cocultures, *Langmuir*, 24: 8885–8889.

Lei, K.F. 2014. Review on Impedance Detection of Cellular Responses in Micro/Nano Environment, *Micromachines*, 5: 1–12.

Lei, U. Huang, C.W. Chen, J. Yang, C.Y. Lo, Y.J. Wo, A. Chen, C.F. & Fung, T.W. 2009. A travelling wave dielectrophoretic pump for blood delivery, Lab on a Chip, 9: 1349–1356.

Ma, B. Zhang, G.H. Qin, J.H. & Lin, B.C. 2009. Characterization of drug metabolites and cytotoxicity assay simultaneously using an integrated microfluidic device, Lab on a Chip, 9: 232–238.

Morgan, H. & Green, N.G. 2003. *AC electrokinetics: colloids and nanoparticles*. Microtechnologies and microsystems series, 2. Research Studies Press, Philadelphia, PA.

Movahed, S. & Li, D.Q. 2011. Microfluidics cell electroporation, *Microfluidics and Nanofluidics*, 10: 703–734.

Nie, F.Q. Yamada, M. Kobayashi, J. Yamato, M. Kikuchi, A. & Okano, T. 2007. On-chip cell migration assay using microfluidic channels, *Biomaterials*, 28: 4017–4022.

Ramos, A. Morgan, H. Green, N.G. Gonzalez, A. & Castellanos, A. 2005. Pumping of liquids with traveling-wave electroosmosis, Journal of Applied Physics, 97: 084906.

Santra, T.S. & Tseng, F.G. 2013. Recent Trends on Micro/Nanofluidic Single Cell Electroporation, *Micromachines*, 4: 333–356.

Wang, X.J. Becker, F.F. & Gascoyne, P.R.C. 2002. Membrane dielectric changes indicate induced apoptosis in HL-60 cells more sensitively than surface phosphatidylserine expression or DNA fragmentation. *Biochimica Et Biophysica Acta-Biomembranes*, 1564: 412–420.

Yeh, C.H. Chen, C.H. & Lin, Y.C. 2011. Use of a gradient-generating microfluidic device to rapidly determine a suitable glucose concentration for cell viability test. *Microfluidics and Nanofluidics*, 10: 1011–1018.

Advanced Materials and Structural Engineering – Hu (Ed.)
© 2016 Taylor & Francis Group, London, ISBN 978-1-138-02786-2

Impact of the microstructure and texture on the elastic modulus in an electron beam-welded near-α titanium alloy joint

X.Z. Li
Department of Materials Engineering, Hubei University of Automotive Technology, Shiyan, China
State Key Laboratory of Material Processing and Die & Mould Technology,
Huazhong University of Science and Technology, Wuhan, China

S.B. Hu & J.Z. Xiao
State Key Laboratory of Material Processing and Die & Mould Technology,
Huazhong University of Science and Technology, Wuhan, China

ABSTRACT: The texture and elastic modulus variations in the different zones of the electron beam-welded near-α titanium alloy joint were measured by X-ray diffraction and the nano-indentation technology. The microstructure of the welded joint was observed by using the optical microscope. The results indicated that the {0001}<-2110> basal plane texture existed in the BM, and the orientation of the α' martensitic phase in the FZ was almost random. The elastic modulus values calculated at the hexagonal phase in the BM, HAZ and FZ by the nano-indentation were 150.2±7.1GPa, 144.1±8.4GPa, 135.4±4.2GPa, respectively. The higher elastic modulus values obtained were mainly related to the basal plane texture in the BM and HAZ.

1 INTRODUCTION

Owing to its excellent mechanical and chemical properties, titanium alloy is employed in the aerospace, chemistry, and medicine industries,. These properties are closely related to the microstructure of titanium alloy, and the microstructure always depends on the process technology. The preferred orientation of the crystal is often developed during the metal processing; especially, it has an important effect on the mechanical property. The hexagonal α phase has a lower crystal symmetry than the cubic β phase in the titanium alloy, so the texture can lead to a strong anisotropy of the mechanical property (Chen & Boehlert 2011, Yu et al. 2009). The yield and ultimate tensile strength, bend ductility and total fatigue life to failure along the transverse direction are significantly different from those along the rolling direction due to the stronger texture of the α phase in the unidirectional rolling Ti-6Al-4V plate (Bach & Evans 2001).

The elastic modulus is one of the important mechanical parameters, which plays a critical role in the design and application of the welded structure. However, research on the texture, elastic modulus and their relationship in the different zones of the titanium alloy welded joint by electron beam has been so far scarce. Thus, in this experiment, the texture and elastic modulus variations in the Base Material (BM), the Heat Affected Zone (HAZ) and the Fuse Zone (FZ) in the electron beam welded joint are measured by X-ray diffraction and the nano-indentation technology.

2 EXPERIMENTAL PROCEDURES

The near-α titanium alloy welded in this study had a veritable chemical composition of (by wt%) 6.65 Al, 2.15Zr, 2.10 V, 1.62Mo, 0.08 O, 0.004H, and balance Ti. The alloy plate was processed by unidirectional hot rolling in the (α + β) region, and supplied as a 45mm-thick plate, with no subsequent heat treatment. A single-pass, full-penetration electron beam butt welding was performed perpendicular to the hot Rolling Direction (RD). Hence, the Welded Direction (WD) was parallel to the Transverse Direction (TD). Vacuum annealing at 650 °C for 4 h was implemented for the relaxation of residual stress after the electron beam welding.

The textures from the close-packed hexagonal (HCP) phases were measured using a X-ray diffraction goniometer based on the Schulz reflection geometry. The crystallographic textures were generated, using two orthogonal pole figures, i.e. {0002} and {11–20}. Nano-indentation experiments were conducted to calculate the elastic modulus (E) values of the HCP phases. These experiments were

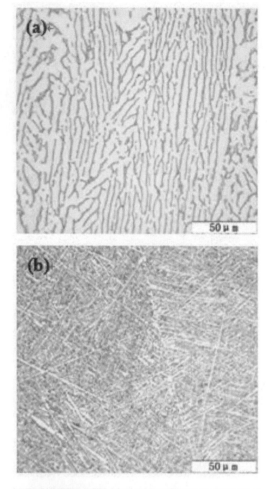

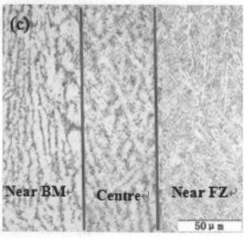

Figure 1. Optical micrographs of the TA15 EB-welded joint after the post-weld vacuum anneal treatment: (a) BM; (b) FZ; and (c) HAZ.

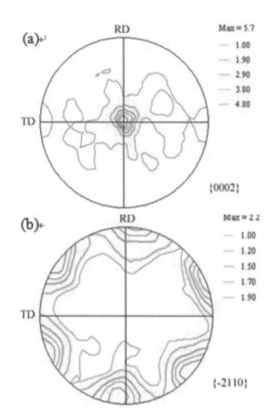

Figure 2. Texture of the α phase in the BM. (a) {0002} pole figure and (b) {-2110} pole figure.

performed on a CSM nano-indentation instrument with a Berkovich tip. The maximum load was 10 mN, and the loading/unloading rates were 20 mN/min. The E values were obtained from the analysis of the resulting load vs. depth of indentation (p vs. h) curves with the Oliver–Pharr method.

3 RESULTS

The optical micrographs of the BM, FZ, and HAZ are shown in Figure 1(a)–(c). The BM was composed of many lamellar α phases and several β phases, and the dark β phases were between the light lamellar α phases. The acicular α' martensitic phases in the FZ were transformed from the high-temperature column β grains, as shown in Figure 1(b). The narrow HAZ exhibited a continuously transitional microstructure from the BM to the FZ. The HAZ near the BM contained many lamellar α and β structures, whereas the HAZ near the FZ formed more acicular α' martensitic phases, as shown in Figure 1(c).

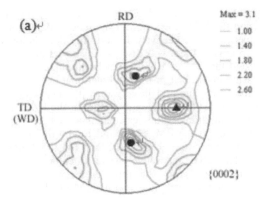

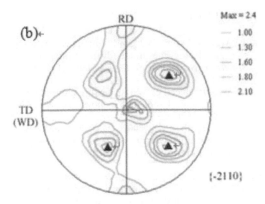

Figure 3. Texture of the α' phase in the FZ. (a) {0002} pole figure and (b) {-2110} pole figure ▲ {01–11} <-2110> ● {01–12} <1–212>.

The cubic β vol.% was only about 6.2% in the BM, so the HCP α phase was in the great majority. The microstructure was almost an acicular α' martensitic phase, with an HCP structure in the FZ. For this reason, the textures of the two HCP phases were considered and measured in the BM and FZ. The {0002} and {-2110} pole figures of the BM are shown in Figure 2, whose maximum pole intensities were 5.7 and 2.2, respectively. The pole figures suggested that the stronger {0001} basal plane texture was in the BM, and the <10–10> direction was parallel to the RD. The {0002} and {-2110} pole figures of the FZ are shown in Figure 3, with the maximum intensities of 3.1 and 2.4. The pole figure showed that two weaker textures existed in the FZ, namely {01–11} <-2110> and {01–12} <1–212>.

The sample was finally prepared by mechanical and chemical polishing to obtain an adequate surface for the measurements. Ten indentation points were measured for the HCP phases in the BM, HAZ and FZ. The results were averaged as the E value of the HCP phases in three different zones. The average E value of the α phase was 150.2±7.1Gpa in the BM, and that of the α' phase was 135.4±4.2Gpa in the FZ. Two HCP phases, α and α', gradually blended in the HAZ, and the average E value of the blending phase was 144.1±8.4Gpa.

4 DISCUSSION

The cooling rate in the electron beam welding was high, and the high cooling rate promoted the martensitic transformation (Karthikeyan et al. 2010), so the coarse column β grains in the FZ were transformed into the acicular α' martensitic phase below the transus temperature. However, the differences in lattice parameters were small between the HCP α' and α phases (Peng et al. 2010). The β and α phases were well known to be related by the Burgers orientation relations: {110} β // {0001} α, <111> β // <-2110>α. So, the strong α texture could inherit from the sharp high-temperature β texture during the β→α transformation along with the variant selection, when the material underwent a cold rolling prior to the α→β→α transformation sequence (Gey & Humbert 2002). However, the original α texture in the BM was removed during the fusion welding. For the cubic β phase, the <100> direction was the crystallographic direction favoring the fastest growth, and the other two orthogonal axes could be at any arbitrary rotation about the <100> axis. Hence, this maximum intensity distribution of the {h k l} <100> fiber texture could be illustrated, as shown in Figure 4.

According to the Burgers orientation relations, the dense concentration of the {0002} α pole figure should be similar to that of the {110} β (Figure 4). However, this case could not be observed in Figure 3(a), which suggested that the strong {h k l} <100> fiber texture had not formed during the solidification of the β phase. The nucleation rate of prior β grains was more rapid than the growth rate due to the high cooling rate in the electron beam welding, so the preferential growth along the <100> direction could not perform very well. Because of the crystal symmetry, twelve distinct α orientation variants could form in a single parent β grain, which were subjected to the Burgers relationship. The high cooling rate greatly promoted the homogeneous nucleation, so several orientation variants could occur with an equal statistical probability during the β→α' transformation in the FZ. Moreover, the higher the cooling rate, the more the number of variants. For these above reasons, the texture of the α' phase was very weak and nearly random orientations in the FZ.

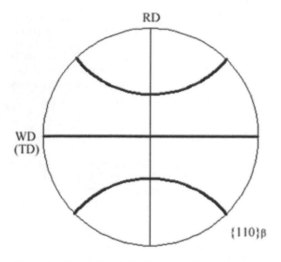

RD

WD
(TD)

{110}β

Figure 4. The deduced distribution of intensity points in the {110} β pole figure (indicated by black bold lines) if the strong {h k l} <100 > fiber texture could form in the FZ during the solidification.

A strong anisotropy existed in the HCP Ti because of its lower crystal symmetry. The nano-indentation hardness varied significantly with the crystal orientation. Near the [0001] stress axis, the hardness values were the highest, and the hardness values decreased as the stress axis deviated from the near [0001] orientations (Viswanathan & Lee 2005). Hence, the E of the α phase in the BM was higher than that of the α′ phase in the FZ because of the strong {0001} basal plane texture in the BM. The orientation of the α′ martensitic phase in the FZ was almost random, so the E value measured by the nano-indentation method was the average value, without the effect of the texture. The basal plane texture from the BM was partly remained in the HAZ, and it gradually changed the weak α′ texture from the FZ. As a result, the value of the nano-indentation E in the HAZ containing mixture microstructure ranged between the values found in the BM and FZ.

5 CONCLUSIONS

The microstructure of the electron beam welded joint was observed by using the optical microscope, the textures and the elastic modulus of the close-packed hexagonal (HCP) phases were measured using the XRD method and the nano-indentation experiments, respectively. In addition, the relationship among them was discussed. The results suggested that the {0001}<-2110> basal plane texture existed in the BM in the unidirectional hot rolling thick TA15 plate, and the very weak {01–11}<-2110> and {01–12}<1–212> textures were in the FZ. The elastic modulus values calculated at the hexagonal phase in the BM, HAZ and FZ by the nano-indentation were 150.2±7.1GPa, 144.1±8.4GPa, 135.4±4.2GPa, respectively. The difference in the elastic modulus in the three zones was closely related to the basal plane texture in the BM.

REFERENCES

Bach, M.R. & Evans, W.J. 2001. Impact of texture on mechanical properties in an advanced titanium alloy, Materials Science and Engineering: A, 319–321: 409–414.
Chen, W. & Boehlert, C.J. 2011. Texture induced anisotropy in extruded Ti-6Al-4V-xB alloys, Materials Characterization, 62(3): 333–339.
Gey, N. & Humbert, M. 2002. Characterization of the variant selection occurring during the α→β→α phase transformations of a cold rolled titanium sheet, Acta Materialia. 50: 277–287.
Karthikeyan, T. Dasgupta, A. & Khatirka, R. 2010. Effect of cooling rate on transformation texture and variant selection during β→α transformation in Ti-5Ta–1.8Nb alloy, Materials Science and Engineering: A. 528: 549–558.
Peng, P. Ou, K. & Chao, C. 2010. Research of microstructure and mechanical behavior of duplex (α+β) Ti-4.8Al-2.5Mo-1.4V alloy, Journal of Alloys and Compounds. 490: 661–666.
Viswanathan, G.B. & Lee, E. 2005. Direct observations and analyses of dislocation substructures in the α phase of an α/β Ti-alloy formed by nanoindentation, Acta Materialia. 53: 5101–5115.
Yu, L. Nakata, K. & Yamamoto, N. 2009. Texture and its effect on mechanical properties in fiber laser weld of a fine-grained Mg alloy, Materials Letters. 63(11): 870–872.

Advanced Materials and Structural Engineering – Hu (Ed.)
© 2016 Taylor & Francis Group, London, ISBN 978-1-138-02786-2

Considering air compressibility in analyzing gap pressure in the partially porous aerostatic journal bearing

T.Y. Huang
Department of Mechanical Engineering, Lunghwa University of Science and Technology, Taoyuan, Taiwan

S.Y. Hsu
Mechanical and Systems Research Laboratory, Industrial Technology Research Institute, Hsinchu, Taiwan

B.Z. Wang
Department of Mechanical Engineering, National Taiwan University of Science and Technology, Taipei, Taiwan

ABSTRACT: The influence of the rotational speed of the spindle on the compressibility of the air and the pressure in the gap between the spindle and a partially porous aerostatic journal bearing was studied. Based on the finite volume method and the pressure-velocity coupling scheme of the SIMPLE algorithm with the standard k-ε turbulent model, a software specialized in computational fluid dynamics was used to solve the Navier-Stokes equations to calculate pressure and velocity of the air flow. The results revealed there were positive pressure zones and vacuum pressure zones in the air gap between the bearing and the spindle. Under the same rotational speed, the pressure difference between the positive peak and the negative peak in the case of incompressible air was greater than that in the case of compressible air. However, the averaged pressure on the surface of the spindle with compressible air flow was higher than that with incompressible air flow. The average pressure and the net force on the inner ring of the bearing and the surface of the spindle increased when the spindle rotated faster.

1 INTRODUCTION

The aerostatic journal bearings are widely used in ultra-precision machine tool, semiconductor, panel display and precision instrumentation industries. Due to the remarkable ability of the porous material in flow restriction, the porous aerostatic journal bearings are better than other types of aerostatic bearings in characteristics such as load carrying capacity, stiffness, damping and dynamic stability. The aerostatic porous journal bearings are divided into two kinds, namely, the fully porous type and the partially porous type, depending on the area of the porous medium for restricting air flow. It is not easy to produce and assemble the fully porous aerostatic journal bearing since the porous material was adhered to the metallic inner ring of the bearing. On the contrary, the partially porous aerostatic journal bearing has the advantages of easy production, simple assemblage and low cost. For these reasons, the partially porous aerostatic journal bearing has attracted the attention of the industry.

Many research papers studying the characteristics of the fully porous aerostatic journal bearings have been published. However, the research reports regarding the partially porous aerostatic journal bearing are rarely seen. So far, Szwarcman and Gorez (1978) used a mathematical model to investigate the characteristics of the aerostatic journal bearing with porous inserts to find the relationships for the load capacity, the stiffness and the gas flow rate of the bearing as functions of the supply pressure, the relative eccentricity and the geometry of the bearing. Rao (2006) used FLUENT code to analyze the static characteristics of aerostatic journal bearings with an orifice restrictor, fully porous restrictor and partially porous restrictor. He found the load capacity and the stiffness of the fully porous journal bearing could be the highest among the three types of bearings if the permeability coefficient of porous material, thickness of air gap and structural parameter could be properly chosen. Wang (2007) applied the finite element method to study the effects of various parameters on static characteristics of the partially porous aerostatic journal bearing. He did experiments on the partially porous aerostatic thrust bearing to verify the numerical solutions and proved the partially porous bearing was better than the fully porous bearing in terms of stiffness and production. Huang et al. (2014) studied the effects of the size of porous medium, the bearing gap, the eccentricity and the spindle speed on the gap pressure, the load capacity and the stiffness of the fully porous, the annular porous and the partially porous aerostatic journal bearings. They found the partially porous aerostatic journal bearing had the highest operational efficiency.

As the Mach number of the air velocity is less than 0.3, the air can be considered as incompressible. When the Mach number of the air velocity is not less than 0.3, the air should be considered as compressible. It was found the porous aerostatic journal bearings pressurized with incompressible fluid had been studied. However, there was no report on the effect of air compressibility on the gap pressure in any type of the porous aerostatic journal bearing due to changes in rotating speed. Thus, it is the purpose of this study to investigate the effect of rotating speed on air compressibility and pressure in the gap between the partially porous journal bearing and the spindle. The gap pressure and the load capacity of the bearing calculated with compressible and incompressible air were compared.

2 THEORY AND APPROACH

Based on the finite volume method and the pressure-velocity coupling scheme of the SIMPLE algorithm for the standard k-ε turbulence model, this study utilized the CFD software, FLUENT, to solve the incompressible and the compressible three dimensional Navier-Stokes equations to analyze the velocity and pressure fields of a partially porous aerostatic journal bearing. In order to simplify the numerical simulation, the following assumptions were made for the air flow:

1. The air was assumed to be a Newtonian flow. The Mach number of the velocity of air flowed through the air gap was figured out based on the rotational speed of the spindle. When the Mach number was less than 0.3, the air, with an invariant density, was assumed to be incompressible. Otherwise, it was considered as compressible and its density was variable.
2. The effects of gravity, surface roughness and heat radiation were all neglected.
3. All physical properties were independent of the temperature. The ambient temperature was a constant. The properties of the porous material were isotropic and the porosity was a constant.
4. Inside the porous material, the air flow obeyed the Darcy's law and was considered laminar due to its low velocity. And, the inertia resistance was neglected for the same reason.

The momentum equation in a conservative form is expressed as follows.

$$\rho\frac{\partial u_i}{\partial t}+\rho\frac{\partial(u_iu_j)}{\partial x_j}=$$

$$-\frac{\partial p}{\partial x_i}+\frac{\partial}{\partial x_j}\left[\mu\left(\frac{\partial u_i}{\partial x_j}+\frac{\partial u_j}{\partial x_i}\right)\right]+\rho g_i+F_i \quad (1)$$

where, ρ is the fluid density, μ the dynamic viscosity, x_i and x_j the coordinate directions, u_i and u_j the velocity tensors, g_i the gravity and F_i the body force.

When the fluid inertia affects the flow field more significantly than the fluid viscosity, the flow develops into a turbulent flow. The k-ε equation for turbulence proposed by Launder and Spalding (1972) is employed to solve the Navier-Stokes equations.

$$\frac{\partial}{\partial t}(\rho u_i)+\frac{\partial}{\partial x_j}(\rho u_iu_j)=$$

$$-\frac{\partial p}{\partial x_i}+\frac{\partial}{\partial x_j}\left[\mu\left(\frac{\partial u_i}{\partial x_j}+\frac{\partial u_j}{\partial x_i}-\frac{2}{3}\delta_{ij}\frac{\partial u_i}{\partial x_i}\right)\right]$$

$$+\frac{\partial}{\partial x_j}(-\rho\overline{u_i'u_j'}) \quad (2)$$

where, the Reynolds stresses are modeled employing the Boussinesq hypothesis (Hinze 1975).

$$-\rho\overline{u_i'u_j'}=\mu_t\left(\frac{\partial u_i}{\partial x_j}+\frac{\partial u_j}{\partial x_i}\right)-\frac{2}{3}\left(\rho k+\mu_t\frac{\partial u_i}{\partial x_i}\right)\delta_{ij} \quad (3)$$

There are two important turbulent parameters associated with the k-ε turbulent model, namely the turbulent kinetic energy k and the turbulent dissipation rate ε. The turbulent kinetic energy equation is in the form as

$$\frac{\partial}{\partial t}(\rho k)+\frac{\partial}{\partial x_i}(\rho k u_i)=$$

$$\frac{\partial}{\partial x_j}\left[\left(\mu+\frac{\mu_t}{\sigma_k}\right)\frac{\partial k}{\partial x_j}\right]+G_k+G_b-\rho\varepsilon-Y_M+S_k \quad (4)$$

The turbulent dissipation rate equation is in the form as

$$\frac{\partial}{\partial t}(\rho\varepsilon)+\frac{\partial}{\partial x_i}(\rho\varepsilon u_i)=$$

$$\frac{\partial}{\partial x_j}\left[\left(\mu+\frac{\mu_t}{\sigma_\varepsilon}\right)\frac{\partial\varepsilon}{\partial x_j}\right]+C_{1\varepsilon}\frac{\varepsilon}{k}(G_k+C_{3\varepsilon}G_b)$$

$$-C_{2\varepsilon}\rho\frac{\varepsilon^2}{k}+S_\varepsilon. \quad (5)$$

where, σ_k and σ_ε are the turbulent Prandtl coefficients in the turbulent kinetic energy equation and the turbulent dissipation rate equation, respectively. G_k and G_b are the turbulent kinetic energies generated by the mean velocity gradients and the buoyancy, respectively. S_k and S_ε are the momentum source terms. Y_M represents the contribution of the fluctuating dilatation incompressible turbulence to the overall dissipation rate, and $C_{1\varepsilon}$, $C_{2\varepsilon}$, $C_{3\varepsilon}$ and C_μ are model constants. The turbulent viscosity μ_t is defined as a function of k and ε in the form of $\mu_t = \rho C_\mu (k^2/\varepsilon)$.

The following boundary conditions were considered in the analyses:

1. On the inlet boundary, the air was pumped into the bearing at a gage pressure of 400 kPa. On the outlet boundary, the air was discharged to the ambient atmosphere at a gage pressure of 0 kPa.
2. Wall function: The flow passing by the solid wall must meet the no-penetration condition and the no-slip condition. On selecting the standard $k - \varepsilon$ turbulent model, the standard wall function in the option of Near-Wall Treatment was used.
3. Slip wall boundary condition: It was assumed the surface of the porous material was under slip wall boundary condition, and the rest of solid surface was under no-slip condition. Based on the theory of Beavers and Joseph (1967), the permeability coefficient was defined in FLUENT software to deal with the slip wall boundary condition.
4. Moving wall boundary condition: The moving wall boundary condition was defined for the surface of the spindle.

3 SIMULATION TECHNIQUES

Figure 1(a) illustrated the schematic view of an aerostatic spindle module with multiple partially porous journal bearings. A spindle with an outer diameter of 60 mm and a section with an axial length of 41.5 mm were considered in the analysis. In Figure 1(b), the partially porous aerostatic journal bearing was composed of eight porous inserts equally distributed around the spindle. Each porous insert with a porosity of 0.1 was assumed a thickness of 5 mm and a diameter of 5 mm as shown in Figure 2(a). The computer model of the partially porous journal bearing with the air inlet and the outlet was shown in Figure 2(b), where the uniform bearing gap was assumed 10 μm in thickness for a spindle without eccentricity.

When the Mach number of the air velocity is less than 0.3, the air can be considered as

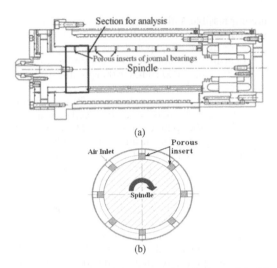

Figure 1. Sections of the aerostatic spindle module (a) and the partially porous journal bearing (b).

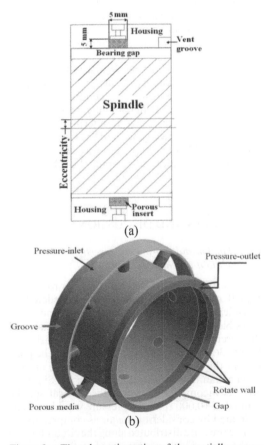

Figure 2. The schematic section of the partially porous journal bearing (a) and its CAD model (b).

incompressible. Otherwise, the air should be considered as compressible flow. In this study, the pressure in the air gap and the load capacity of the bearing were analyzed for a spindle rotating at 10000 rpm, 33000 rpm, 60000 rpm and 120,000 rpm, respectively.

As the ambient temperature is 20°C and the atmosphere pressure is 1 atm, the speed of sound is 343 m/s. As the spindle rotated at 10,000 rpm, the tangential velocity of the air on the periphery of a spindle with a diameter of 60 mm was 31.42 m/s. Compared with the speed of sound, the Mach number of the air velocity was 0.09. Thus, the air was considered as incompressible. As the spindle rotated at the critical speed of 33,000 rpm, the air flow could be considered as compressible since the velocity of the air on the periphery of the spindle was 103.67 m/s and the Mach number was 0.302. When the spindle rotated at 60,000 rpm, the air was considered as compressible for sure.

To analyze the steady state characteristics of the bearing and to simplify the calculation of the flow field, the velocity of air on the solid surface was assumed to be zero due to the non-slip condition. The air flow was assumed to be laminar and isotropic inside the porous material. Based on the Navier-Stokes equation, the laminar flow module in FLUENT software was selected for simulating the performance of the bearings. The k-ε equation for turbulence was used to analyze the physical phenomena under the steady state condition. The second order Upwind Differentiating Technique was adopted to solve the convection terms. In order to have optimum convergence and satisfactory accuracy, the SIMPLEC approach was used to solve the iterative equations to compute the velocity and the pressure.

4 RESULTS AND DISCUSSION

4.1 *The journal bearing and a vertical spindle without eccentricity*

As a vertical spindle (eccentricity = 0) rotated at 10,000 rpm, the gap pressure was calculated by considering the air as incompressible and compressible flow and shown along the axial and the circumferential directions of the bearing in Figure 3(a) and Figure 3(b), respectively. The pressure in the bearing gap with compressible air was higher than that with incompressible air. Since the air was incompressible when the vertical spindle rotated at 10,000 rpm, the pressure would be overestimated by considering the air as compressible.

The pressure distributed along the circumference of the circle, which was generated on the surface of the vertical spindle by connecting the projections

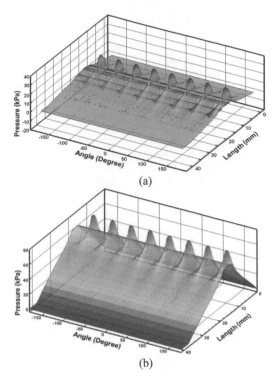

(a)

(b)

Figure 3. At a speed of 10,000 rpm, the gap pressure was computed by considering the air as incompressible and compressible flow and shown along the axial and the circumferential directions of the bearing in plot (a) and plot (b), respectively.

of the centers of the porous inserts on the spindle rotating at 33,000 rpm, was calculated by considering the air as incompressible and compressible respectively and shown in Figure 4 for comparison. It was seen there were positive pressure zones and vacuum pressure zones in the bearing gap. At the critical speed of 33,000 rpm, the air flow was considered as compressible, the high-peak pressure in the case of compressible air was higher than that in the case of incompressible air; while the low-peak pressure in the case of incompressible air was lower than that in the case of compressible air. It indicated, as the rotating speed exceeded its critical value, the air flow should be considered as compressible in an analysis to avoid underestimation of the average pressure in the bearing gap. Otherwise, the faster the spindle rotated, the larger the error due to underestimation would be.

When a vertical spindle rotated at 60,000 rpm, the velocity of the air on the periphery of the spindle was 188.5 m/s and the Mach number was 0.55. The air should be considered as compressible. The pressure was computed by considering the air as incompressible and compressible flow and shown

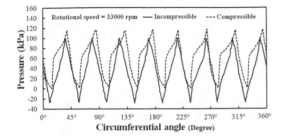

Figure 4. At a speed of 33,000 rpm, the pressure along the circle on the surface of the vertical spindle was computed by considering the air as incompressible and compressible, respectively.

along the axial and the circumferential directions of the bearing in Figure 5(a) and 5(b), respectively.

Comparing Figure 5 with Figure 4, it was noticed the faster the rotational speed, the larger the pressure difference between the front edge and the rear edge of the same porous insert would be. Though the peak to peak amplitude of pressure fluctuation in Figure 5(a) was larger than that in Figure 5(b), the net force and the average pressure applied on the inner surface of the bearing in the former were much less than those in the latter. It means the air should be considered as compressible when the rotating speed of the spindle is highly above its critical speed, otherwise, the average pressure, which is directly proportional to the load carrying capacity, would be significantly underestimated, as is shown in Table 1.

4.2 The journal bearing and an eccentric horizontal spindle

When a horizontal spindle rotated at high speed, the effects of self-weight and rotation causes the spindle to be eccentric. The stiffness of the bearing is defined as the force needed to cause one unit of transverse displacement of the journal bearing. In order to calculate the stiffness of the journal bearing, an eccentricity for the spindle under loading was assumed.

As the horizontal spindle with an eccentricity of 0.3 rotated at 33000 rpm, the gap pressure was computed by considering the air as incompressible and compressible and displayed along the axial and the circumferential directions of the bearing in Figure 6(a) and Figure 6(b), respectively. The pressure distributions in Figure 6(a) and Figure 6(b) were quite different. In Figure 6, with an eccentric horizontal spindle rotating at 33000 rpm, if the air was considered as incompressible, a half of the air gap filled with positive pressure, while the rest of the gap filled with vacuum pressure. Since the low-peak pressure canceled out the high-peak pressure, the

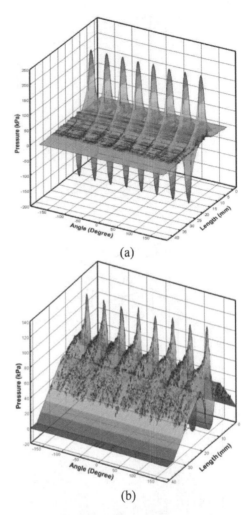

(a)

(b)

Figure 5. At a speed of 60,000 rpm, the gap pressure was computed by considering the air as incompressible and compressible flow and shown in the axial and the circumferential directions of the bearing in plot (a) and plot (b), respectively.

Table 1. Considering the air as incompressible and compressible flow respectively, the average pressure on the surface of the spindle was computed for different spindle speeds.

Rotating speed (rpm)	Average pressure (kPa) (Incompressible air)	Average pressure (kPa) (Compressible air)
10,000	4.34	22.84
33,000	16.89	37.94
60,000	22.77	60.94
120,000	73.65	156.62

135

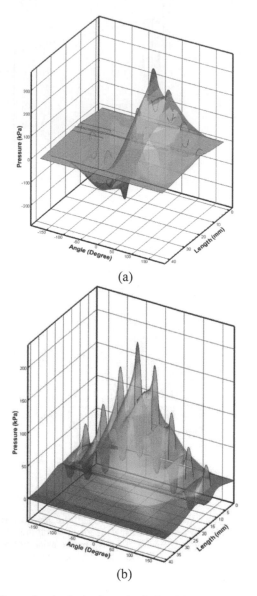

(a)

(b)

Figure 6.   As the horizontal spindle with an eccentricity of 0.3 rotated at 33000 rpm, the gap pressure in the axial and the circumferential directions of the bearing was computed by considering the air as incompressible and compressible flow and shown in plot (a) and plot (b), respectively.

average pressure became less. However, if the air was considered as compressible, the average pressure would be larger because the positive pressure was dominant through the air gap. Obviously, if the air was considered as incompressible, the net force subjected to the bearing would be underestimated. So was the load carrying capacity of the bearing.

To further explain the effect of eccentricity of spindle on gap pressure, the pressure distributed on the circumference of the fictitious circle generated by connecting the projections of the porous inserts on the surface of the horizontal spindle with a rotating speed of 33000 rpm and an eccentricity of either 0 or 0.3 was calculated, by considering the air as compressible flow, and shown in Figure 7. When the eccentricity was zero, the profile of pressure distribution was quite similar to that in Figure 5(b). The pressure was evenly distributed over each porous insert. However, when the eccentricity of the spindle was 0.3, the highest peak pressure occurred at the angular location of −7.5°. It could be seen that, around the upper half of the spindle, there were vacuum pressure zones happened to the porous inserts at the angular locations of ±135° and ±180°. Around the lower half of the spindle, the pressure distributed over the porous inserts at the angular locations of 0° and ±45° were positive for high peak and a low peak of pressure amplitude. With the assumed rotating speed and eccentricity, the average pressure on the surface of the spindle and the stiffness of the bearing were calculated and listed in Table 2 for the case of incompressible air and the case of compressible air respectively.

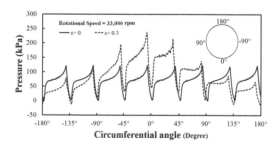

Figure 7.   With compressible air, the pressure distributed on the circumference of a circle on the surface of a horizontal spindle (eccentricity = 0 or 0.3) rotating at 33000 rpm was shown.

Table 2.   The average pressure on the surface of the spindle and the bearing stiffness were computed for different spindle eccentricity and air compressibility.

Spindle speed (rpm)	Eccentric ratio	Average pressure on spindle surface (kPa)	Bearing stiffness (N/μm)
0	0	10.39	0
0	0.3	12.36	11.15
33,000	0	16.89	0
33,000	0.3	18.61	11.61

From Figure 5 through Figure 7 and Table 2, it was seen the eccentricity not only changed the distribution of pressure, but also increased the magnitude of the average pressure. Since the spindle rotated at its critical speed of 33,000 rpm, the air should be considered as compressible. The stiffness of the bearing with compressible air, which flowed through the gap between the bearing and the eccentric spindle, was much larger than that with incompressible air.

5 CONCLUSIONS

The averaged pressure on the spindle and the bearing became higher as the spindle rotated faster. The pressure distribution on the surface of the spindle calculated by considering the air as compressible flow was quite different from that calculated by considering the air as an incompressible flow. When the Mach number of the air velocity was less than 0.3, the air should be considered as an incompressible flow in analysis to avoid overestimation of the pressure in the bearing gap. Otherwise, the air should be considered as compressible flow to avoid underestimation of the gap pressure.

The eccentricity not only changed the distribution of the pressure in bearing gap, but also increased the magnitude of the average pressure and the stiffness of the bearing. When the eccentricity of the spindle was as sizable as 0.3, the stiffness of the bearing with compressible air became much larger than that with incompressible air.

Without properly recognizing the compressibility of the air flowing through the journal bearing could end up with miscalculation of the characteristics of the bearing and wrong judgment on evaluation of the bearing. Therefore, care must be taken in deciding the air compressibility in analysis of pressure in the gap of partially porous aerostatic journal bearings.

REFERENCES

Beavers, G.S. & Joseph, D.D. 1967. Boundary conditions at a naturally permeable wall, *Journal of Fluid Mechanics*, 30(1): 197–207.

Hinze, J.O. 1975. *Turbulence*, McGraw-Hill, New York.

Huang, T.Y. Wang, B.Z. Lin, S.C. & Hsu, S.Y. 2014. Performance of the partially porous and the fully porous aerostatic *journal bearings with incompressible air flow. Key Engineering Materials*, 625: 384–391.

Launder, B.E. & Spalding, D.B. 1972. *Lectures in mathematical models of turbulence, 1st ed.*, Academic Press, London.

Rao, H. 2006. *Simulation based on FLUENT and experimental research of porous aerostatic bearing*, Master Thesis, School of Mechatronics Engineering, Harbin Institute of Technology, China.

Szwarcman, M. & Gorez, R. 1978. Design of aerostatic journal bearings with partially porous walls, *International Journal of Machine Tool Design Research*, 18(2): 49–58.

Wang, J.M. 2007. *A study on part of key techniques in ultra-precision aerostatic bearing axes system*, Ph.D. Thesis, National University of Defense Technology, China.

Advanced Materials and Structural Engineering – Hu (Ed.)
© 2016 Taylor & Francis Group, London, ISBN 978-1-138-02786-2

Structural and optical studies of thermally evaporated NiPc thin films

V. Mekla
Program of Physics, Faculty of Science, Ubon Ratchathani Rajabhat University, Ubon Ratchathani, Thailand

C. Saributr
College of Nanotechnology, King Mongkut's Institute of Technology Ladkrabang, Chalongkrung, Bangkok, Thailand

ABSTRACT: Thin films of Nickel Phthalocyanine (NiPc) were fabricated at a pressure of 7.5×10^{-6} mbar and a deposition rate of 0.3 Å/sec by thermal evaporation in an ultra-high vacuum system. The films were deposited on the glass ITO and Si substrates at various temperatures of 27, 50, 80, 100 and 120°C. The surface morphologies and roughness of NiPc grown by different techniques were investigated. The surface morphologies of thin films were also examined using an atomic force microscope (AFM). The thickness of NiPc thin films increases with an increase in temperatures. The optical properties of thin films were studied using the UV-Visible spectroscopy. Thin films absorb in the B-band peak at about 300–400 nm and absorb in the Q-band peak at about 525–750 nm. These data indicate that the energy band gap of the Q-band at the β phases is 1.68 eV, that at the α phases is in the range of 1.79–1.87 eV, and that of the B-band is in the range of 3.23–3.27 eV.

1 INTRODUCTION

The phthalocyanine (Pc) polymers have become one of the most studied materials of all organic functional materials (Bloom & Bruke 1980). Phthalocyanine exists in several crystalline polymorphs, including the α-, β- and γ-structures (Soliman et al. 2007). Metal-substituted and metal-free phthalocyanine polymers such as MnPc, CuPc, NiPc, FePc, and CoPc have also been found. Nickel phthalocyanine (NiPc) is an organic semiconductor with a relatively high mobility value ($\sim 1 \times 10^{-5} m^2 V^{-1} s^{-1}$) (Abdel-Malik et al. 1995) when compared with the equivalent for other MPcs ($7.6 \times 10^{-9} m^2 V^{-1} s^{-1}$ for ZnPc (Sharma et al. 1996), $3 \times 10^{-7} m^2 V^{-1} s^{-1}$ for PbPc (Gravano et al. 1991), and $10^{-8} m^2 V^{-1} s^{-1}$ for CuPc (Gould 1986)), which makes it a promising candidate for the development of future electronic devices. There are several studies of the electrical properties of NiPc thin films for transistors and photovoltaic devices (BenChaabane et al. 1997), but only a few investigations for the interfaces of NiPc with metals and Si, although the part of the interfaces is crucial in the device performance. Since the structure has a strong impact on the functional properties, it is very important to understand the growth process of the organic films, and find ways to optimize their surface morphology. To our knowledge, there are no enough efforts made with respect to the determination and characterization of optical constants, spectral and structural features of NiPc. In this work, we have studied the structural and optical studies of thermally evaporated NiPc thin films.

2 EXPERIMENTAL

The NiPc powder used in this study was obtained from Aldrich chemical company, and used as the source material for thermal evaporation. Before starting the deposition, the glass substrate was cleaned in an ultrasonic bath for 10 min using acetone, followed by rinsing in distilled water. The substrates were dried in open air in a cleaned room. A molybdenum boat was used as a heating source. Thin films of NiPc were deposited by the vacuum evaporation technique on a thoroughly cleaned glass substrate at different substrate temperatures (27, 50, 80, 100, 120°C) using Kurt J. Lesker company system. During deposition, the pressure in the vacuum chamber was kept constant at about 7.5×10^{-6} mbar, and the deposition rates of all the films were kept at about 0.3 Å/sec. Also, the substrates are placed at a distance of 15 cm from the source. The thickness of the films was 200 nm. The thickness of the films was measured using the Tolansky's multiple beam interference technique (Maissel & Glang 1985). The deposition rate of the films to the substrates seems to be extremely good.

The surface morphology of the NiPc films at different substrate temperatures (Ts) was investigated by means of atomic force microscopy (SEIKO Ltd. model. SPA400). The absorbance spectra of NiPc films were measured at normal incidence at room temperature in the range of 300–850 nm by using a double beam spectrometer (UV/Vis Spectrometer; PG Instruments Ltd).

2.1 Results and discussion

Surface Morphology: Figure 1 shows the AFM images of surface morphology and the roughness of as-deposited NiPc thin films with different substrate temperatures. The scan size of each AFM image was 800 nm × 800 nm. The AFM images showed that NiPc films were in the molecular cluster form. The lateral mode images showed that the surface roughness of NiPc thin films increased with the increase in the temperatures. The surface morphologies and roughness of NiPc grown by different techniques were investigated (Joseph & Menon 2007). The experiments showed that different growth techniques led to differences in morphology and roughness.

The absorption spectra in the UV–VIS region of spectra for NiPc thin films of different temperatures of the substrate of 27, 50, 80, 100 and 120°C showed that the absorbance of the films increased as the temperature increased and the band gap shifted slightly to smaller energy, as shown in Figure 2(a), (b) and Table 1.

As shown in Figure 2 (a), the absorption spectrum of the UV-visible of NiPc exhibited 2 phases, namely B-band and Q-band: the B-band absorption spectrum of the UV-visible in the wavelength ranged from 300 to 400 nm and the Q-band absorption spectrum of the UV-visible was in the wavelength range of 525–750 nm. It was found that the absorption spectrum of the UV-visible Q-band was divided into two phases, namely α phases and β phases. The spectral lines of α phases were pronounced in the wavelength range of 525–650 nm, and the spectral lines of β phases were pronounced in the wavelength range of 525–650 nm. Considering the spectral absorption spectra, the length of α phases showed that the absorption spectrum was slippery when the temperature of the substrate increased. As shown in Figure 2 (b) the absorption spectrum of normalized absorption was in the wavelength range of 675–725 nm, and confirmed that the absorption in the wavelength was more slippery on the temperature of the substrates. We found that the energy band gap of the Q-band at the β phases was 1.68 eV, that of α phases was in the range of 1.79–1.87 eV, and that of the B-band had a value in the range of 3.23–3.27 eV, as shown in Table 1. These data were used

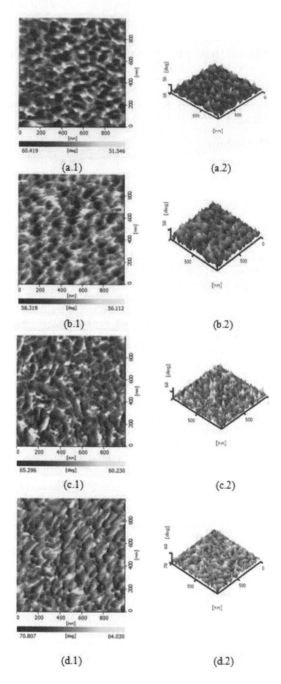

Figure 1. (*Continued*).

to calculate the absorption coefficient (α), band gap energy (Eg) and the optical constants (extinction coefficient, refractive index, and real and imaginary parts of dielectric constant). The relation between the intensity of the incident light (I_o)

140

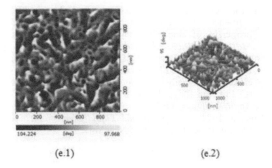

(e.1)   (e.2)

Figure 1. AFM images of the surface morphology of NiPc thin films with the substrate temperatures of 27°C of (a.1) 2D and (a.2) 3D; substrate temperatures of 50°C of (b.1) 2D and (b.2) 3D; substrate temperatures of 50°C of (c.1) 2D and (c.2) 3D; substrate temperatures of 80°C of (d.1) 2D and (d.2) 3D; substrate temperatures of 80°C of (e.1) 2D and (e.2) 3D.

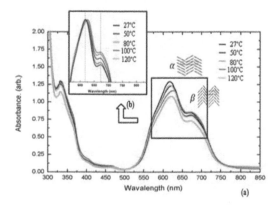

Figure 2. Nickel Phthalocyanine spectra (a) absorbance and (b) normalized absorption for NiPc thin films of different substrate temperatures of 27, 50, 80, 100 and 120°C.

Table 1. The value energy gap of the Q-band is divided into two phases, namely α phases and β phases and B-band.

Condition (°C)	Egβ(eV)	Egα(eV)	EgB(eV)
27	1.68	1.87	3.27
50	1.68	1.85	3.26
80	1.68	1.83	3.26
100	1.68	1.82	3.25
120	1.68	1.79	3.23

and the transmitted intensity (I_T) is represented by an exponential form:

$$I_T = I_0 \exp(-\alpha t) \tag{1}$$

where α is the absorption coefficient and t is the film thickness. According to this equation, the optical absorption coefficients of thin films were evaluated from the transmittance data using the relation:

$$\alpha = \ln(1/T)/t \tag{2}$$

where $T = I_T/I_o$ is defined as the transmittance (Hummel 2001). As a result of the absorption coefficient data, the nature of transition (direct or indirect) was determined according to the Tauc relation (Hummel 2001, Tauc 1974).

3 CONCLUSIONS

We succeeded in preparing NiPc thin films, and discovered that the surface roughness of NiPc thin films increase with the increase in temperatures. The absorption spectrum of the UV-visible of NiPc thin films exhibits 2 phases, namely B-band and Q-band, which can in turn be divided into two phases, i.e. α phases and β phases. The energy gap of β phases is a constant value of 1.68 eV for every condition, and α phases is a maximum value of 1.87 eV for the substrate temperature of 27°C, and a minimum value of 1.87 eV for the substrate temperature of 120°C. The maximum value of the energy gap of the B-band is 3.27 eV for the substrate temperature of 27°C and the minimum value of 3.23 eV for the substrate temperature of 120°C.

ACKNOWLEDGMENT

This work was supported by the Program of Physics, Faculty of Science, Ubon Ratchathani Rajabhat University and College of Nanotechnology, King Mongkut's Institute of technology Ladkrabang, chalongkrung. The authors gratefully thank them for their support.

REFERENCES

Abdel-Malik, T.G. Abdel-Latif, R.M. El-Samahy, A.E. & Khalil, S.M. 1995. Transport properties in nickel phthalocyanine thin films using gold electrodes, Thin Solid Films. 256: 139.
BenChaabane, R. Guillaud, G. & Gamoudi, M. 1997. Study of the electrical – properties of thin film transistors based on Nikel phthalocyanine, Thin Solid Films, 296: 145.
Bloom, A. & Bruke, W.J. 1980. U.S. Patent, 4,241,355.
Gould, R.D. 1986. Dependence of the mobility and trap concentration in evaporated copper phthalocyanine thin films on background pressure and evaporation rate, Journal of Physics D: Applied Physics, 9: 1785.

Gravano, S. Hassan, A.K. & Gould, R.D. 1991. Effects of Annealing on the Trap Distribution of Cobalt Phthalocyanine Thin Films, *International Journal of Electronic.* 70: 477.

Hummel, R.E. 2001. *Electronic Properties of Materials*, 3rd edition, Springer, New York.

Joseph, B. & Menon, C.S. 2007. Studies on the Optical Properties and Surface Morphology of Nickel Phthalocyanine Thin Films, *E-Journal of Chemistry.* 4: 255–264.

Maissel, L.I. & Glang, R. 1985. *Handbook of Thin Film Technology*, McGraw-Hill, New York.

Sharma, G.D. Sangodkar, S.G. & Roy, M.S. 1996. Influence of iodine on the electrical and photoelectrical properties of zinc phthalocyanine thin film devices, *Materials Science and Engineering: B.* 41: 222–227.

Soliman, H.S. El-Barry, A.M.A. Khosifan, N.M. & El Nahass, M.M. 2007. Structural and electrical properties of thermally evaporated cobalt phthalocyanine (CoPc) thin films, *The European Physical Journal— Applied Physics.* 37: 1–9.

Tauc, J. 1974. *Amorphous and Liquid Semiconductors*, Plenum Press, London and New York.

Advanced Materials and Structural Engineering – Hu (Ed.)
© 2016 Taylor & Francis Group, London, ISBN 978-1-138-02786-2

Effect of carbon content on the properties of LiFePO$_4$/C synthesized by hydrothermal stripping technique

G.Q. Wan, P.F. Bai, J. Liu, X.M. Zu, L. Chen & X.Y. Wang
Department of Chemistry, School of Science, Tianjin University, Tianjin, China

ABSTRACT: LiFePO$_4$ is a very important cathode material for lithium ion batteries. A novel and green method based on hydrothermal stripping technique were used for synthesizing LiFePO$_4$/C and the effect of carbon content on the properties of LiFePO$_4$/C was systematically investigated. The samples were characterized by X-Ray Diffraction (XRD), Scanning Electron Microscopy (SEM), High-Resolution TEM (HRTEM), initial discharge and cycling performance and Cyclic Voltammetry (CV). The results reveal that the carbon content does not affect the structure of LiFePO$_4$/C and has no significant effect on the product morphology, but the electrochemical performance of LiFePO$_4$/C was highly dependent on the carbon content. The 10% glucose-added LiFePO$_4$/C had an excellent rate capability and can deliver a discharge capacity of 155.1 mAh g^{-1} at 0.1 C and 140.1 mAh g^{-1} at 1.0 C. In addition, the CV shows the LiFePO$_4$/C with 10% glucose-added also had the smallest polarization.

1 INTRODUCTION

In recent years, olivine-structured LiFePO$_4$ has been widely accepted as a promising cathode candidate for lithium-ion batteries, because of its high theoretical capacity (170 mAh g^{-1}), operating voltage (3.4V vs. Li$^+$/Li), environmental compatibility, low-cost, and thermal stability (Doeff et al. 2006, Wong et al. 2010, Yu et al. 2010). However, its inherently low electronic and ionic conductivities seriously restrict the electrochemical performance, especially at a high rate. Considerable attempts have been made to solve these problems, such as reducing the particle size, optimizing the morphology, carbon coating and lattice doping (Choi & Kumta 2007, Hsu et al. 2004, Wang et al. 2005). Among these strategies, carbon coating is considered as one of the most effective methods to improve the electrochemical performance. Carbon coating not only increases the surface electrical conductivity but also prevents the iron dissolution, which both results in the improvement of the electrochemical performance of LiFePO$_4$ cathode material (Chen & Dahn 2002, Huang et al. 2001). So the suitable carbon layer thickness is an important factor on the improvement of electrochemical performance of LiFePO$_4$. In this paper, the influence of carbon content on LiFePO$_4$/C composites synthesized by hydrothermal stripping method was systematically investigated.

2 EXPERIMENTAL

The hydrothermal stripping process, in which metal-containing species are hydrolyzed and precipitated directly from the organic phase with an aqueous phase, is a promising method for synthesizing LiFePO$_4$. The organic phase is the iron (II)-loaded naphthenic acid prepared by mixing naphthenic acid and isooctyl alcohol, saponifying by 1:1 (v/v) ammonia, and adding a certain concentration of FeSO$_4$·7H$_2$O. Then the solution was separated into the organic phase and aqueous phase after stirring 0.5 h. The organic phase was washed with deionized water for each hydrothermal stripping experiment. The aqueous phase is composed of certain amount of H$_3$PO$_4$, LiOH·H$_2$O (the Li:P = 1.2) and ascorbic which is used as a reducing agent. In hydrothermal stripping process, transfer the aqueous phase and organic phase into a stainless steel autoclave and then heat to 250 °C for 3 h with vigorously stirring. The precipitate obtained was washed several times with deionized water and alcohol, dried at 80 °C for 12 h. The precipitate with a different amount of glucose was sintered at 650 °C for 4 h in a tubular furnace under an argon atmosphere.

The crystal structure of samples was identified by X-ray diffraction (XRD) (Rigaku D/max 2500v/pc made in Japan). The amount of residual carbon in the composite was determined by thermo-gravimetric analysis (TG, NETZSCH STA 449 F3

Jupiter) from room temperature to 800 °C with the heating rate of 10°C min⁻¹ in air atmosphere. The particle microstructure, morphology, and carbon-coating layer of the samples were investigated by a scanning electron microscopy (SEM, S-4800 made by Japan) and transmission electron microscopy (TEM, 100CX-IImade by JOEL, Tokyo, Japan). The electrochemical measurements were made in a two-electrode cell, using lithium as counter electrode. The electrolyte consists of $1.0 \, mol \cdot L^{-1} \, LiPF_6$ dissolved in ethylene carbonate (EC)/dimethyl carbonate (DMC)/methyl ethyl carbonate (EMC) in a 1:1:1 (v/v) ratio. Galvanostatic cycling tests of the assembled cells were carried out on an LAND system (made in China) in the voltage range of 2.0–4.0 V (vs Li⁺/Li). The Cyclic Voltammetry (CV) curves were performed on an electrochemical analyzer (Gamry, PC14-750 made by USA).

3 RESULTS AND DISCUSSION

3.1 Structure and carbon content analysis of the LiFePO₄ and LiFePO₄/C powders

Figure 1 gives the XRD patterns of the LiFePO₄ and LiFePO₄/C samples which were synthesized using a different mass ratio of glucose and LiFePO₄. All samples show well-crystallized diffraction peaks indexed to orthorhombic olivine LiFePO₄ (JCPDS No. 40-1499). There is no evidence of diffraction peaks for carbon due to their amorphous structure and low content in samples b, c, d and e. This shows that the carbon content cannot affect the structure of LiFePO₄/C composites. The final carbon content in LiFePO₄/C increases with the glucose adding and the content is 0.68 wt.%, 2.68 wt.%, 3.5 wt.%, 3.61 wt.% for adding 5 wt.%, 10 wt.%, 15 wt.%, 20 wt.% glucose, respectively.

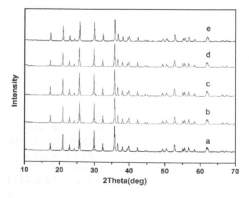

Figure 1. XRD patterns of samples prepared with different amount of glucose, a) 0, b) 5%, c) 10%, d) 15% and e) 20%.

3.2 The morphology of the LiFePO₄ and LiFePO₄/C powders

Figure 2 gives the SEM images of LiFePO₄ and LiFePO₄/C samples. Clearly, the rodlike samples can be produced on a large scale for LiFePO₄ and LiFePO₄/C. The particle size distribution of LiFePO₄ (a) uncoated carbon is very wide, which is not beneficial to the electrochemical performance of a material. Compared to pure LiFePO₄ (a), the LiFePO₄/C (b-e) are smaller because of the presence of carbon which can effectively inhibit the crystal growth to a great extent. However among the LiFePO₄/C (b-e), there is little difference in the matter of morphology and particle size, although which were synthesized respectively by adding different measured amount of glucose. Thus, this would suggest that the carbon content has no significant effect on the morphology and particle size of LiFePO₄/C.

3.3 HRTEM analysis of LiFePO₄/C

Figure 3 displays the TEM images of LiFePO₄/C composite. The High-Resolution TEM (HRTEM) observed that the surface of the LiFePO₄/C composite is uniformly coated with a layer of carbon derived from glucose. The appropriate carbon layer has been confirmed to be efficient in

Figure 2. SEM of samples prepared with different amount of glucose, a) 0, b) 5%, c) 10% d) 15% and 20%.

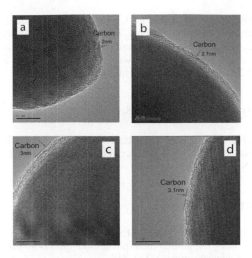

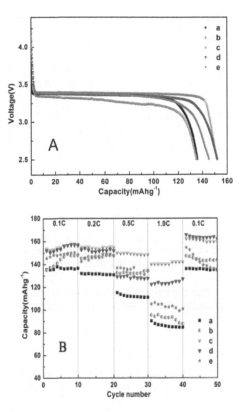

Figure 3. HR-TEM images of samples prepared with different amount of glucose, a) 0, b) 5%, c) 10%, d) 15% and e) 20%.

Figure 4. Initial discharge curves (A) and cycling performance (B) of LiFePO4 and LiFePO4/C samples prepared with different amount of glucose, a) 0, b) 5%, c) 10%, d) 15% and e) 20%.

achieving excellent rate capabilities for $LiFePO_4/C$ (Liang et al. 2008, Chen & Whittingham 2006). The thin carbon layer is not enough to form a continuous conductive layer for enhancing the electronic conductivity, while the thick carbon layer would decrease the tap density of the sample. When adding 5% glucose (a), the thickness of carbon is about 2 nm and adding 10% glucose (b), it increased to 2.7 nm. However, increasing to 15% glucose (c) and 20% glucose (d), it is found that the thickness of carbon changed a little. With adding 10% glucose may be in most favor of the improvement of electrochemical performance of $LiFePO_4/C$.

3.4 Effects of carbon content on the initial discharge and cycling performance of $LiFePO_4/C$

Figure 4(A) shows the initial discharge curves of the prepared $LiFePO_4$ and $LiFePO_4/C$ electrodes obtained in the voltage range of 2.5–4.2 V versus lithium at 0.1 C. The samples prepared by adding different amounts of glucose were used for comparison. When the adding glucose is 0% (a), 5% (b), 10% (c), 15% (d) and 20% (e), which is versus the mass of $LiFePO_4$, the initial discharge capacities are 135.5 mAh g^{-1}, 135 mAh g^{-1}, 151.7 mAh g^{-1}, 152 mAh g^{-1} and 145.3 mAh g^{-1}, respectively. Obviously, the 10% glucose-added $LiFePO_4/C$ (c) and 15% glucose-added $LiFePO_4/C$ (d) display higher initial discharge capacity, but the former presents the longest plateau around 3.4 V vs Li/Li+. It can be concluded that the suitable carbon content is an important factor in the improvement of electrochemical performance of $LiFePO_4/C$.

In order to illustrate above results, the cycling performances of the prepared $LiFePO_4$ and $LiFePO_4/C$ were investigated at various rates. Figure 4(B) shows the cycling performance of samples that were charged-discharged at different rates between 2.5–4.2V. Clearly, comparing to $LiFePO_4/C$ (b-e), bare $LiFePO_4$ (a) displays a low discharge capacity and bad rate performance, especially at a high rate. Among the $LiFePO_4/C$ (b-e), the highest discharge capacities are demonstrated by the 10%- $LiFePO_4/C$ (c) sample, which shows 151.7, 155.1, 149.8, 140.1 mAh g^{-1} at 0.1 C, 0.2 C, 0.5 C and 1.0 C after 10 cycles, respectively, better than the other samples. Here is an interesting phenomenon that when the lower rat of 0.1C was again applied, the 15%-$LiFePO_4/C$ (d) displayed the highest discharge capacity. Discharge capacity only displayed the122.4 Ah g-1 at 1.0 C and it is lower than 140.1 mAh g-1 with the amount 10% of glucose (c). So it can be considered that the added 10%-$LiFePO_4/C$ (c) possesses the best rate capability and good cycle life. This is also in correspondence with previous SEM results, which again

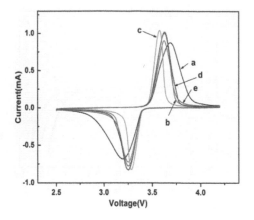

Figure 5. Cyclic voltammorgrams of LiFePO4 and LiFePO4/C of samples prepared with different amount of glucose, a) 0, b) 5%, c) 10%, d) 15%, e) and 20%.

proves that the suitable carbon content can greatly influence the electrochemical performance of the LiFePO$_4$/C sample.

3.5 CV analysis of LiFePO$_4$/C

To further determine the effect of carbon content on the electrochemical properties of LiFePO$_4$/C composites, the cyclic voltammetry (CV) was carried out in the voltage range of 2.5–4.2 V at a scan rate of 0.1 mV s^{-1}, as shown in Figure 5. The CV results show a single pair of oxidation and reduction peaks, which correspond to a Fe^{2+}/Fe^{3+} redox couple for all the samples. The five CV curves almost have the good symmetry, which demonstrate that all samples have electrochemical reaction reversibility during lithium ion insertion and extraction. But it should be noticed that the 10%-LiFePO$_4$/C (c) presents the best symmetrical, the sharpest shape of the anodic/cathodic peaks and the smallest potential separation between the anodic and the cathodic peaks, which indicate the high electrochemical reactivity of modified LiFePO$_4$ composites, a small polarization, high Li-ion diffusion and low inner resistance (Fey et al. 2009, Yin et al. 2011). These results further confirmed why 10%-LiFePO$_4$/C (c) presents the best rate performance among all samples.

4 CONCLUSIONS

The LiFePO$_4$/C composites with different carbon content were synthesized by hydrothermal stripping technique and characterized by XRD, SEM, TEM, CV and Galvanostatic cycling tests. It is found that the carbon content has no important influence on the structure and morphology of LiFePO$_4$/C composites, but the electrochemical

properties of LiFePO$_4$/C are highly dependent on the carbon content. The 10% glucose-added LiFePO$_4$/C has an excellent rate capability and can deliver a discharge capacity of 155.8 mA hg^{-1} at 0.2 C and 140.1 mAhg^{-1} at 1.0 C.

ACKNOWLEDGEMENT

This work was financially supported by the National Natural Science Foundation of China (No. 21276185).

REFERENCES

Chen, J. & Whittingham, M.S. 2006. Hydrothermal synthesis of lithium iron phosphate, *Electrochemistry Communications*. 8: 855–858.

Chen, Z. & Dahn, J.R. 2002. Reducing Carbon in LiFePO4/C Composite Electrodes to Maximize Specific Energy, Volumetric Energy, and Tap Density, *Journal of The Electrochemical Society*, 149: A1184–A1189.

Choi, D.W. & Kumta, P.N. 2007. Surfactant based sol-gel approach to nanostructured LiFePO4 for high rate Li-ion batteries, *Journal of Power Sources*. 163: 1064–1069.

Doeff, M.M. Wilcox, J.D. Kostecki, R. & Lau, G. Optimization of carbon coatings on LiFePO4, *Journal of Power Sources*. 163: 180–184.

Fey, G.T.K. Chen, Y.G. & Kao, H.M. 2009. Electrochemical properties of LiFePO4 prepared via ball—milling, *Journal of Power Sources*. 189: 169–178.

Hsu, K.F. Tsay, S.Y. & Hwang, B.J. 2004. Synthesis and characterization of nano-sized LiFePO4 cathode materials prepared by a citric acid-based sol-gel route, *Journal of Materials Chemistry*, 14: 2690–2695.

Huang, H. Yin, S.C. & Nazar, L.F. 2001. Approaching Theoretical Capacity of LiFePO4 at Room Temperature at High Rates, *Electrochemical and Solid-State Letters*, 4: A170–A172.

Liang, G.C. Wang, L. Ou, X.Q. Zhao, X. & Xu, S.Z. 2008. Lithium iron phosphate with high-rate capability synthesized through hydrothermal reaction in glucose solution, *Journal of Power Sources*. 184: 538–554.

Wang, D.Y. Li, H. Shi, S.Q. Huang, X.J. & Chen, L.Q. 2005. Improving the rate performance of LiFePO4 by Fe-site doping, *Electrochimica Acta*. 50: 2955–2958.

Wong, H.C. Carey, J.R. & Chen, J.S. 2010. Physical and electrochemical properties of LiFePO4/C composite cathode prepared from aromatic dike tone-containing precursors, *International Journal of Electrochemical Science*, 5: 1090–1102.

Yin, Y.H. Gao, M.X. Ding, J.L. Liu, Y.F. Shen, L.K. & Pan, H.G. 2011. A carbon-free LiFePO4 cathode material of high-rate capability prepared by a mechanical activation method, *Journal of Alloys and Compounds*, 509: 10161–10166.

Yu, F. Zhang, J.J. Yang, Y.F. & Song, G.Z. 2010. Porous micro-spherical aggregates of LiFePO4/C nanocomposites: A novel and simple template-free concept and synthesis via sol-gel-spray drying method, *Journal of Power Sources*. 195: 6873–6878.

Advanced Materials and Structural Engineering – Hu (Ed.)
© *2016 Taylor & Francis Group, London, ISBN 978-1-138-02786-2*

Optimization of ultrasound-assisted extraction conditions for active substances with anti-tyrosinase activity from tomatoes

J. Han & G.M. Gong
School of Perfume and Aroma Technology, Shanghai Institute of Technology, Shanghai, China

ABSTRACT: The high concentration of active substances in tomato extraction is correlated with its anti-tyrosinase and antioxidant effect activity. We established the optimum ultrasound extraction conditions for maximizing the yield of active substances. The 4 important factors (ethanol concentration, extraction time, extraction temperature, and solid-liquid ratio) that had great effects on ultrasonic extraction were selected to set up a single-factor test and investigate the effects of inhibitory activity on tyrosinase. According to the result of the single-factor test, the orthogonal experiment was conducted to select the optimum process parameter of extracting a total active substance from tomatoes by the alcohol extraction method. The factors affecting the extraction of active substances from tomatoes by an ultrasonic assistant method in order were ethanol concentration > solid-liquid ratio > extraction temperature > extraction time. The optimum process condition from the orthogonal experiment was as follows: ethanol concentration of 80%, solid-liquid ratio of 1:6, extraction temperature of 50°C and extraction time of 50 min, under which the average anti-tyrosinase activity of active substances was 32.77% through the tyrosinase inhibition assay.

1 INTRODUCTION

Tomato, as a fresh or processed product, has important roles such as anti-tyrosinase activity and antioxidants in the body, thus whitening skin and slowing down aging, and preventing tissue damage, heart disease and certain cancers (Harrison et al. 2003, Palozza et al. 2011, Tan et al. 2010). It possesses an array of important micronutrients such as carotenoids, phenolic compounds, vitamin C and folic acid (Pedro & Ferreira 2005).

Nowadays, the interest in skin whitening has grown quickly, so the development of preparations for bleaching hyperpigmented lesions or to safely achieve overall whitening is one of the challenges for the cosmetic industry. Therefore, tomato alcohol extraction is considered to be used for whitening active substances. In the whitening effect evaluation, tyrosinase has been found to be a key enzyme in melanin biosynthesis, which plays a crucial role in determining the color of the mammalian skin and hair.

Several environmentally friendly extraction technologies have been developed for the extraction of bioactive components of plants (Han et al. 2005, Jacquemin et al. 2012, Chen et al. 2012). Ultrasound-assisted extraction is one of the most popular methods because it is an inexpensive, simple, and efficient extraction technique. It offers high reproducibility, saves time, has low solvent consumption, and requires low energy input (Jiao & Zuo 2009, Wang & Zuo 2011).

The purpose of this study was to develop an efficient and environmentally friendly ultrasound-assisted extraction procedure for active substances with anti-tyrosinase activity in tomatoes.

2 MATERIALS AND METHODS

2.1 Reagents

All the chemicals and reagents were purchased from Sinopharm Chemical Reagent Co., Ltd. Tyrosinase was obtained from Worthington Biochemical Corporation, and dopamine from Shanghai Treasure Biological Technology Co., Ltd.

2.2 Extraction and sample preparation

Each extraction experiment used 20 g of fresh tomato pulp, and was performed in triplicate. The sample was mixed with the water–ethanol mixture in a 250 ml Erlenmeyer flask. The extraction process was placed in the ultrasonic water bath for 10 min with a rated power of 60 W. Then, extraction conditions were set according to an experimental design plan explained below. The independent variables were extraction time (X1) varying within a range of 30–110 min, the ethanol percentage X2 of the water–ethanol solvent (35–95%, v/v), temperature X3 (30–70°C), and the liquid-to-solid (L/S) ratio X4 (1:2–

1:10 m/m). Once the extraction was completed, the supernatant was separated from the insoluble solid by centrifugation at 8,000 rpm for 20 min at 4°C. The remaining solid phase was then filtered with a vacuum machine. The filtrate was concentrated under reduced pressure using a rotary evaporator. The concentrated liquor was constant volume to 10 ml with distilled water (the liquor was labeled as CL). After the single-factor experiment, in the orthogonally designed experiment (Table 1) of four factors, three levels were adopted to optimize the ethanol extraction technology. This experiment was repeated 3 times, the data were averaged. The inhibitory activity of CL was determined with the tyrosinase inhibition assay.

2.3 Tyrosinase inhibition assay

Tyrosinase inhibition was assayed according to the method of Masamoto (Masamoto et al. 2003). Briefly, aliquots (10μL) of CL (27–109 mg/mL) were mixed with 100μL of L-DOPA solution (0.5 mM), 60uL of sodium phosphate buffer (50 mM, pH 6.8), and preincubated at 25°C for 10 min. Finally, 30 μL of an aqueous solution of mushroom tyrosinase (333 U/mL) were added to the mixture. The optical density (OD) of the samples at 475 nm was measured and compared with the control without an inhibitor, demonstrating a linear color change with time during the 20 min of the experiment. Control incubations represented 100% enzyme activity and were conducted in a similar way by replacing extracts by buffer. Inhibitory activity was determined by comparing the enzyme activity in the absence and presence of the evaluated inhibitor. The extent of inhibition by the addition of samples was expressed as the inhibition percentage, and calculated as follows:

$$\text{Inhibition } (\%) = \frac{(A - B)}{A} \times 100 \qquad (1)$$

where A and B represent the absorbances at 475 nm for the blank and CL, respectively.

Table 1. The design of the orthogonal experiment.

Levels	(A) Ethanol concentration (%)	(B) Extraction temperature (°C)	(C) Extraction time (min)	(D) Solid-to-liquid ratio
1	50	40	50	1:4
2	65	50	70	1:6
3	80	60	90	1:8

3 RESULTS AND DISCUSSION

3.1 Single-factor experiments

The single-factor experiment was performed by varying one factor at different levels, while other factors were fixed. There are many factors affecting the extraction yields of target compounds, including ethanol concentration, extraction time, extraction temperature and solid-liquid ratio. All the results of the single-factor experiment are shown in Figure 1.

The results shown in Figure 1a indicate that the inhibition of tyrosinase increased when the ethanol concentration was increased from 35% to 65%. However, further increases in the ethanol concentration resulted in the decrease of the inhibition of tyrosinase. As the ethanol concentration increased, its polarity declined, which would make it difficult for the solvent to extract some components and decrease the inhibition of tyrosinase. The inhibition of tyrosinase was the highest when ethanol concentration was 65%. Finally, the ethanol concentration range of 50%–80% was selected for the subsequent experiments.

The extraction time is another crucial factor that should be studied to increase the inhibition of tyrosinase. As shown in Figure 1b, when the extraction time increased from 30 to 70 min, the inhibition of tyrosinase of the CL increased slightly. When the time variable was changed from 70 min to 110 min, the inhibition of tyrosinase of the CL reduced dramatically. Therefore, the extraction time of 50–90 min was selected for further experiments.

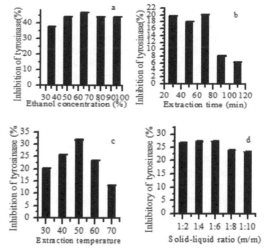

Figure 1. Effects of the concentration of ethanol concentration (a), extraction time (b), extraction temperature (c) and solid-liquid ratio (d) on the extraction efficiencies of the target analytic. The inhibition of tyrosinase was expressed as the observed values of the target analytic.

As shown in Figure 1c, the inhibition of tyrosinase of the CL increased with the increase in the extracting temperature from 30 to 50°C, which might be because the increasing extracting temperature contributes to enhancing the spread ability and solubility of the extract, which was beneficial in dissolving and extracting the target compounds. However, when the temperature was higher than 50°C, the inhibition of tyrosinase of the CL decreased slightly. The main reason may be the decomposition of some extract at the high temperature. Thus, a temperature of 50–90°C was selected for further experiments.

The solid–liquid ratio is an important factor in determining the extraction efficiency. Generally, in conventional extraction techniques, a high solvent volume will increase the inhibition of tyrosinase, and a low solid–liquid ratio will result in the incomplete extraction of the analytic and low inhibition of tyrosinase. However, larger solvent volumes could complicate the procedure and result in unnecessary waste. As shown in Figure 1d the inhibition of tyrosinase increased as the solvent volume increased up to a solid–liquid ratio of 1:6. Further increases in the solvent volume did not greatly increase its inhibition. Thus, a solid–liquid ratio range of 1:4–1:8 was selected for the subsequent experiments.

3.2 Optimization of parameters by the orthogonal experiments

The first step in the extraction procedure of the CL is to optimize the operating conditions to obtain an efficient extraction of the target compounds and avoid the co-extraction of the undesired compounds. Since various parameters potentially affect the extraction process, the optimization of the experimental conditions is a critical step in the development of a solvent extraction method. In fact, ethanol concentration, extraction time, extraction temperature and the solid–liquid ratio are generally considered to be the most important factors. Optimization of the suitable extraction conditions in the extraction of CL can be carried out by using an experimental design. In the present study, all the selected factors were examined using an orthogonal L 9(34) test design. The total evaluation index was analyzed by the statistical method. The results of the orthogonal test are summarized in Table 2.

As shown in Table 2, the effect of factors on the inhibition of tyrosinase of CL decreased in the order A> D> C> B, according to the R values. The ethanol concentration was found to be the most important determinant of the inhibition of tyrosinase, while the extraction time was least. According to the outcomes and analyses, the optimized extraction method was A3B2C1D2, when ethanol concentration, extraction time, extraction temper-

Table 2. The analysis results of the orthogonal experiments.

No.	(A) Ethanol concentration (%)	(B) Extraction temperature (°C)	(C) Extraction time (min)	(D) Solid-liquid ratio	Inhibition of tyrosinase (%)
1	1	1	1	1	21.44
2	1	2	2	2	17.39
3	1	3	3	3	7.28
4	2	1	2	3	5.42
5	2	2	3	1	26.21
6	2	3	1	2	32.77
7	3	1	3	2	30.77
8	3	2	1	3	30.95
9	3	3	2	1	26.72
K1	15.37	19.21	28.387	24.79	
K2	21.467	24.85	16.51	26.977	
K3	29.48	22.257	21.42	14.55	
R	14.11	5.64	11.877	12.427	

ature and the solid–liquid ratio were 80%, 50°C, 50 min, and 1:6, respectively, and the inhibition of tyrosinase was the highest.

4 CONCLUSIONS

The inhibition of tyrosinase of CL was selected as the index for optimizing the ultrasonic extraction method in this study. By a comprehensive assessment of ethanol concentration, extraction time, extraction temperature, and the solid–liquid ratio, the effects of these factors on the extraction yields decreased in the order: A > D > C > B. It was shown that ethanol concentration affected the extraction most and extraction time least. According to the outcomes and analyses, the optimized extraction method was $A_3B_2C_1D_2$, when ethanol concentration, extraction time, extraction temperature and the solid–liquid ratio were 80%, 50°C, 50 min, and 1:6, respectively, and the inhibition of tyrosinase was the highest.

ACKNOWLEDGMENT

The authors sincerely thank Professor Zhang Wangping for his helpful comments and advice. This work was financially supported by the Shanghai Institute of Technology.

REFERENCES

Chen, R.Z. Li, S.Z. Lium C.M. Yangm S.M. & Li, X.L. 2012. Ultrasound complex enzymes assisted extraction and biochemical activities of polysaccharides from Epimedium leaves, *Process Biochemistry*, 47(12): 2040–2050.

Han, H.W. Cao, W.L. & Zhang, J.C. 2005. Preparation of biodiesel from soybean oil using supercritical methanol and CO_2 as co-solvent, *Process Biochemistry*, 40(9): 3148–3151.

Harrison, D. Griendling, K.K., Landmesser, U. Hornig, B. & Drexler, H. 2003. Role of oxidative stress in atherosclerosis, *The American Journal of Cardiology*, 91(3 A): 7 A–11 A.

Jacquemin, L. Zeitoun, R. Sablayrolles, C. Pontalier, P.V. & Rigal, L. 2012. Evaluation of the technical and environmental performances of extraction and purification processes of arabinoxylans from wheat straw and bran, *Process Biochemistry*, 47(3): 373–380.

Jiao, Y. & Zuo, Y. 2009. Ultrasonic extraction and HPLC determination of anthraquinones, aloe-emodine, emodine, rheine, chrysophanol, and physcione, in Radix Poly-goni multiflori, *Phytochemical Analysis*, 20(4): 272–278.

Masamoto, Y. Ando, H. Murata, Y. Shimoishi, Y. Tada, M. & Takahata, K. 2003. Mushroom tyrosinase inhibitory activity of esculetin isolated from seeds of Euphorbia lathyris L, *Bioscience, Biotechnology, and Biochemistry*, 67(3): 631–634.

Palozza, P. Simone, R.E. Catalano, A. & Mele, M.C. 2011. Tomato lycopene and lung cancer prevention: From experimental to human studies, *Cancers*, 3(2): 2333–2357.

Pedro, A.M.K. & Ferreira, M.M.C. 2005. Nondestructive determination of solids and carotenoids in tomato products by near-infrared spectroscopy and multivariate calibration, *Analytical Chemistry*, 77(8): 2505–2511.

Tan, H.L. Thomas-Ahner, J.M. Grainger, E.M. Wan, L. Francis, D.M. Schwartz, S.J. Erdman, J.W. Jr. & Clinton, S.K. 2010. Tomato-based food products for prostate cancer prevention: What have we learned?, *Cancer and Metastasis Reviews*, 29(3): 553–568.

Wang, C. & Zuo, Y. 2011. Ultrasound-assisted hydrolysis and gas chromatography–mass spectrometric determination of phenolic compounds in cranberry products, *Food Chemistry*, 128(2): 562–568.

Advanced Materials and Structural Engineering – Hu (Ed.)
© *2016 Taylor & Francis Group, London, ISBN 978-1-138-02786-2*

Purification and properties of nitrite reductase from lactobacillus plantarum

Y. Zhou & G.M. Gong

School of Perfume and Aroma Technology, Shanghai Institute of Technology, Shanghai, China

ABSTRACT: Nitrite reductase is defined as a class of enzymes catalyzing reduction of nitrite. The nitrite reductase from Lactobacillus plantarum was an intracellular enzyme. The enzyme was purified to electrophoretic homogeneity with 12.7-fold purification by ammonium sulfate precipitation, DEAE-52 cellulose column chromatography and Sephadex G-150 gel filtration chromatographic techniques. The properties of the purified enzyme were investigated. The results showed that the subunit molecular mass of nitrite reductase was about 67.6 kDa. The enzyme showed an optimum temperature of 35°C and an optimum pH of 6.0. It was stable within the scope of pH 5.6–7.2 below 40°C.

1 INTRODUCTION

Lactobacillus plantarum is regarded as an important microorganism in industrial production (Sabo et al. 2014). It is a safe organism and has been widely used in food-related technologies (Brinques et al. 2010, Bove et al. 2012, Sauvageau et al. 2012). Nitrite is regarded as an important material in the nitrogen cycle (Kmět 2006, Jetten 2008, Galloway et al. 2008). Processed food such as meat products and pickled food also have the problem of excessive nitrite content. People have made a higher intake of nitrite through vegetables, water, and processed food (Johnson & Kross 1990, Ward 2009). Nitrite is a precursor of carcinogenic nitrosamines. Nitrite will react with secondary amines, the decomposition intermediate of protein in food, to form nitrosamines. Nitrosamines are able to induce many types of cancer such as liver cancer, stomach cancer, esophagus cancer, and so on (Moghaddam 2012). People are trying to find out the effective method to control or degrade nitrite. Enzymatic elimination of nitrite in food is considered a feasible way. Nitrite reductase is defined as a class of enzymes catalyzing the reduction of nitrite. It was reported that Lactobacillus plantarum would generate nitrite reductase under anaerobic or aerobic conditions (Wolf & Hammes 1988). In this paper, the purification and properties of nitrite reductase from Lactobacillus plantarum DS31 were investigated. It provides a foundation for further study of nitrite reductase and its application.

2 MATERIALS AND METHODS

2.1 Organism and preparation of cell-extracts

Strain Lactobacillus plantarum H2 was originally isolated in our laboratory from Chinese pickle bought in the market in Shanghai, China. It would generate nitrite reductase when it was induced by nitrite under anaerobic condition. To improve the yield of nitrite reductase, the original strain was treated with UV and Diethyl Sulphate (DES), and a mutant Lactobacillus Plantarum DS31 with an increased enzymatic activity was screened. Cells were collected by centrifugation at $12,000 \times g$ for 15 min at 4°C. They were suspended in a 1/10 volume of Tris-HCl buffer (50 mM, pH 7.0). Then 5 mg/mL lysozyme was added into the suspension, and incubation for 30 min at 37°C. The mixture was subjected to the ultrasonic cell disruption device (SONICS, USA) for 25 min, and any intact cells were sedimented by centrifugation at $12,000 \times g$ for 30 min at 4°C. The high speed supernatant was decanted and labeled the crude extract.

2.2 Purification of nitrite reductase

Unless otherwise noted, all operations were performed at 4°C. The crude extract was first brought to 20% saturation with the gradual addition of powdered ammonium sulfate with continuous stirring for 4 h. The precipitate was removed by centrifugation at $12,000 \times g$ for 15 min. The supernatant was made 40%, 60%, and 80% saturated with respect to ammonium sulfate in the same way. The precipitate of each step was dissolved in a small volume of Tris-HCl buffer (50 mM, pH 7.0) and was desalted by dialysis, using a dialysis tube with a MWCO 14,000 Da, against the same buffer for 24 h with 6 changes. The dialyzed solution was applied to a DEAE-52 cellulose (Whatman, U.K.) column (1.6×40 cm) equilibrated with Tris-HCl buffer (50 mM, pH 7.0) for 1 h. The proteins were eluted with a continuous NaCl gradient (from 0 to 0.5 M). The flow rate was 1 mL/min and the eluted fractions (3 mL each) were collected by automated fraction collector. Those fractions containing nitrite reductase were combined and concentrated. The concentrated fraction was applied to a Sephadex G-150 (Pharmacia, Sweden) column (1 × 40 cm) which was equilibrated with Tris-HCl buffer (50 mM+150 mM NaCl, pH 7.0). The enzyme was eluted from the gel with the same buffer. A flow rate of 0.2 mL/min was maintained, and 115 fractions of 2.0 mL were collected. The protein content was determined at 280 nm.

2.3 Assays

Nitrite reductase activity was determined by measuring the disappearance of nitrite using N-(1-naphthyl) ethylenediamine dihydrochloride method at 35°C. The reaction was carried out at pH 6.0 in a mixture that contained the following additions to a total volume of 200 μL: sodium citrate (4 μmol), sodium phosphate (8 μmol); $NaNO_2$ (2 μmol); NaCl (8 μmol); sodium dithionite (0.8 μmol); methyl viologen (0.8 μmol); and the enzyme. The reaction was started by adding the enzyme to the assay mixture. After incubation at 35°C for 15 min, the reaction was terminated by shaking vigorously to oxidize the remaining hydrosulfite and MV. Nitrite was determined by a diazotization reaction (Showe & DeMoss 1968, Mancinelli et al. 1986). All values were corrected for any nitrite disappearance that occurred when the enzyme was omitted from the reaction mixture. One unit of nitrite reductase activity was equivalent to the amount of enzyme that caused the net disappearance of 1 nmol of nitrite/min. Protein concentration was measured as described by Bradford (Bradford 1976) with UV absorption spectrometry and refer to bovine serum albumin as a protein standard.

2.4 SDS-polyacrylamide gel electrophoresis

SDS-polyacrylamide gel electrophoresis of the purified protein sample was performed for the determination of purity and molecular mass of the enzyme according to Laemmli (1970), using 12% and 5% acrylamide concentration for separating and stacking gel, respectively. Before loaded onto the gel, the protein samples were denatured by heating them with sample buffer at 100°C for 5 min. The electrophoresis was performed at 20 mA in Tris-glycine buffer (pH 8.3) until tracking dye reached the bottom of the gel. The relative subunit molecular mass of the enzyme was calculated with low molecular weight markers. Following electrophoresis, protein bands were visualized by staining with 0.25% (w/v) Coomassie Brilliant Blue R–250.

2.5 Effects of temperature on enzyme activity and stability

The optimum temperature was determined with nitrite (100 μg/mL) as substrate at different temperatures (from 25 to 65°C) in citrate (0.1 M)-phosphate (0.05 M) buffer (pH 6.0). Temperature stability was determined by measuring the residual activities after incubation of the purified enzyme at the same temperatures. The control was designed without incubation. These studies were run in the 1.5 mL microcentrifuge tubes, and the tubes with the reaction mixture were completed sealed and immersed in a water bath heating to the required temperatures for 6 h. Samples were removed from the water bath after incubation and then assayed for enzyme activity.

2.6 Effects of pH on enzyme activity and stability

The optimum pH for the enzyme activity was determined in citrate (0.1 M)-phosphate (0.2 M) buffer with the pH ranging from 5.2 to 7.6 at the optimal temperature. The pH stability of nitrite reductase was studied by pre-incubating the mixture of the purified enzyme and different prepared buffer at the same pH values for 2 h at 4°C. The residual activities of the purified enzyme were measured at pH 6.0 by the method described previously. The relative activity was determined as the percentage of the residual activity referred to the enzyme activity measured at pH 6.0 before incubation.

3   RESULTS AND DISCUSSION

The bacterial Strain Lactobacillus plantarum H2 was originally isolated from the Chinese pickle. By combination treatment of the wild strain with UV and DES, an excellent mutant strain DS31 was screened, and its enzyme activity increased by 31.2%. In this study, a novel nitrite reductase from Lactobacillus plantarum DS31 was used for further purification and characterization.

3.1 Purification of nitrite reductase

Nitrite reductase was purified to homogeneity by four steps, i.e., 30%–60% ammonium sulfate precipitation, dialysis, ion exchange chromatography

Table 1. Summary of purification of nitrite reductase from Lactobacillus plantarum DS31.

Procedure	Total activity [U]	Total protein [mg]	Specific activity [U/mg]	Recovery [%]	Purification
Crude extract	234776	1976.3	118.8	100	1
Ammonium sulfate saturation (0–60%)	201480	1092.5	184.4	85.8	1.6
DEAE-52	80520	162.1	496.7	34.3	4.2
Sephadex G-150	12507	8.3	1506.9	5.3	12.7

and gel filtration chromatography. A summary of the purity and yield of the enzyme after each step was shown in Table 1. The crude extract was precipitated with ammonium sulfate (0–20% and 20–60% saturation). The results indicated that 20–60% saturation achieved a yield of 85.8%. Next, the resulting fractions were purified by DEAE-52 anion exchange chromatography. It was observed that the enzymatic activity was eluted in one peak which met with the protein peak (Figure 1a). The enzyme was purified 4.2-fold with a specific activity of 496.7 U/mg of protein (Table 1). The active fractions were subsequently separated using sephadex G-150 gel filtration chromatography (Figure 1b). The enzyme eluted out with 12.7-fold purification and a specific activity of 1334.06 U/mg of protein. The enzyme had been purified to homogeneity as confirmed by the presence of one single band by SDS-PAGE, and the subunit molecular mass was about 67.6 kDa (Figure 2). The result was very similar to the enzymes from fungus Cylindrocarpon tonkinense (Kubota et al. 1999) and plant corn (Dalling et al. 1973) while larger than that from other organisms (Yamazaki et al. 1995, Abraham et al. 1993, Kobayashi & Shoun 1995, Han et al. 2012, Suzuki et al. 2006, Denariaz et al. 1991).

3.2 Effect of temperature on enzyme activity and stability

It was revealed that the enzyme exhibited maximum enzyme activity at 35°C (Figure 3a). It was a little higher than the nitrite reductase from Lactobacillus lactic TS4 with an optimum temperature of 30°C (Dodds & Thompson 1985) and was lower than the nitrite reductase from eukaryotic microalga Monoraphidium braunii at 40°C (Vigara et al. 2002), Hydrogenobacter thermophilus TK-6 at 70–75°C (Suzuki et al. 2006). More than 50% of the initial activity was retained at 55°C. It was indicated that the enzyme was much more stable than the nitrite reductase purified from Lactobacillus lactic TS4 which retained only 14% of the maximum activity at 43°C (Dodds & Thompson 1985). The analysis of the thermal stability of the purified enzyme indicated that the enzyme retained more than 60% of initial activity in the temperature range of 25–55°C (Figure 3b).

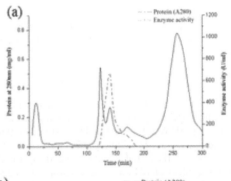

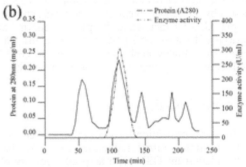

Figure 1. Elution profiles of nitrite reductase from DEAE-52 column (a) and sephadex G-150 column (b).

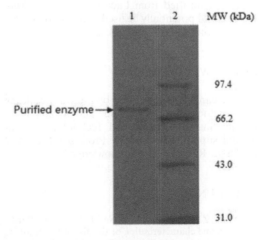

Figure 2. SDS-PAGE of nitrite reductase isolated from Lactobacillus plantarum DS31.

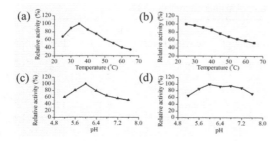

Figure 3. The optimum temperature (a) optimum pH (c) thermal stable (b) and pH stable of purified nitrite reductase.

3.3 Effect of pH on enzyme activity and stability

The nitrite reductase from Lactobacillus plantarum exhibited optimal pH for 100% of the initial enzyme activity at pH 6.0 (Figure 3c). The residual activities of the enzyme at pH 5.2 and pH 7.6 were only 60.3% and 51.6% of the initial activity, respectively. In addition, the effect of pH on enzyme stability was determined by the measurement of the residual activities at pH 6.0 after incubating the purified enzyme at various pH values for 2 h at 4°C (Figure 3d). The enzyme was highly stable in a pH range of 5.6–7.2 and retained more than 85% of the initial activity.

4 CONCLUSION

With the purification procedure described, we have separated nitrite reductase from Lactobacillus plantarum DS31 with a specific activity of 1506.9 U/mg of protein. Our present work provides the first kinetics of enzyme-catalyzed reactions of a novel dissimilatory nitrite reductase. The nitrite reductase purified from Lactobacillus plantarum DS31 could potentially be used in the food, fodder, and drinking water industries.

ACKNOWLEDGMENTS

This research was supported by the team-building fund of the School of Perfume and Aroma Technology, Shanghai Institute of Technology. Instrumental support provided by Professor Xiao and Professor Rong are also acknowledged.

REFERENCES

Abraham, Z.H.L. Lowe, D.J. & Smith, B.E. 1993. Purification and characterization of the dissimilatory nitrite reductase from Alcaligenes xylosoxidans subsp. Xylosoxidans (N.C.I.M.B. 11015): evidence for the presence of both type 1 and type 2 copper centres, Biochemical Journal, 295: 587–593.

Bove, P. Gallone, A. Russo, P. Capozzi, V. Albenzio, M. Spano, G. & Fiocco, D. 2012. Probiotic features of Lactobacillus plantarum mutant strains, Applied Microbiology and Biotechnology, 96: 431–441.

Bradford, M.M. 1976. A rapid and sensitive method for the quantitation of microgram quantities of protein utilizing the principle of protein-dye binding, Analytical Biochemistry, 72: 248–254.

Brinques, G.B. Peralba, M.C. & Ayub, M.A.Z. 2010. Optimization of probiotic and lactic acid production by Lactobacillus plantarum in submerged bioreactor systems, Journal of Industrial Microbiology & Biotechnology, 37: 205–212.

Dalling, M.J. Hucklesby, D.P. & Hageman, R.H. 1973. A comparison of nitrite reductase enzymes from green leaves, scutella, and roots of corn (zea mays L.), Plant Physiology, 51: 481–484.

Denariaz, G. Payne, W.J. & LeGall, J. 1991. The denitrifying nitrite reductase of Bacillus halodenitrificans, Biochimica et Biophysica Acta, 1056: 225–232.

Dodds, K.L. & Thompson, D.L.C. 1985. Characteristics of nitrite reductase activity in Lactobacillus lactic TS4, Canadian Journal of Microbiology, 31: 558–562.

Galloway, J.N. Townsend, A.R. Erisman, J.W. Bekunda, M. Cai, Z.C. Freney, J.R. Martinelli, L.A. Seitzinger, S.P. & Sutton, M.A. 2008. Transformation of the nitrogen cycle: recent trends, questions, and potential solutions, Science, 320: 889–892.

Han, C. Wright, G.S.A. Fisher, K. Rigby, S.E.J. Eady, R.R. & Hasnain, S.S. 2012. Characterization of a novel copper-haem c dissimilatory nitrite reductase from Ralstonia pickettii, Biochemical Journal, 444: 219–226.

Jetten, M.S.M. 2008. The microbial nitrogen cycle, Environment Microbiology, 10(11): 2903–2909.

Johnson, C.J. & Kross, B.C. 1990. Continuing importance of nitrate contamination of groundwater and wells in rural areas, American Journal of Industrial Medicine, 18: 449–456.

Kmět, T. 2006. Model of the nitrogen transformation cycle, Mathematical and Computer Modelling, 44: 124–137.

Kobayashi, M. & Shoun, H. 1995. The copper-containing dissimilatory nitrite reductase involved in the denitrifying system of the fungus Fusarium oxysporum, Journal of Biological Chemistry, 270: 4146–4151.

Kubota, Y. Takaya, N. & Shoun, H. 1999. Membrane-associated, dissimilatory nitrite reductase of the denitrifying fungus Cylindrocarpon tonkinense, Archives of Microbiology, 171: 210–213.

Laemmli, U.K. 1970. Cleavage of structural proteins during the assembly of the head of bacteriophage T4, Nature, 227: 680–685.

Mancinelli, R.L. Cronin, S. & Hochstein, L.I. 1986. The purification and properties of a cd-cytochrome nitrite reductase from Paracoccus halodenitrificans, Archives of Microbiology, 145: 202–208.

Moghaddam, S.N. Zaeem, H.N. Saniee, P. Pedramnia, S. Sotoudeh, M. & Malekzadeh, R. 2012. Oral nitrate reductase activity and erosive gastro-esophageal reflux disease: a nitrate hypothesis for GERD pathogenesis, Digestive Diseases and Sciences, 57: 413–418.

Sabo, S.D.S.S. Vitolo, M. González, J.M.D. & Oliveira, R.P.D.S. 2014. Overview of Lactobacillus plantarum as a promising bacteriocin producer among lactic acid bacteria, *Food Research International*, 64: 527–536.

Sauvageau, J. Ryan, J. Lagutin, K. Sims, I.M. Bridget, L.S. & Timmer, M.S.M. 2012. Isolation and structural characterization of the major glycolipids from Lactobacillus plantarum, *Carbohydrate Research*, 325: 151–156.

Showe, M.K. & DeMoss, J.A. 1968. Localization and regulation of synthesis of nitrate reductase in Escherichia coli, *Journal of Bacteriology*, 95: 1305–1313.

Suzuki, M. Hirai, T. Arai, H. Ishii, M. & Igarashi, Y. 2006. Purification, characterization, and gene cloning of thermophilic cytochrome cd1 nitrite reductase from Hydrogenobacter thermophilus TK-6, *Journal of Bioscience and Bioengineering*, 101: 391–397.

Vigara, J. Sánchez, M.I.G. Garbayo, I. Vílchez, C. & Vega, J.M. 2002. Purification and characterization of ferredoxin—nitrite reductase from the eukaryotic microalga Monoraphidium braunii, *Plant Physiology and Biochemistry*, 40: 401–405.

Ward, W.H. 2009. Too much of a good thing? Nitrate from nitrogen fertilizers and cancer, *Reviews on Environmental Health*, 24: 357–363.

Wolf, G. & Hammes, W.P. 1988. Effect of hematin on the activities of nitrite reductase and catalase in lactobacilli, *Archives of Microbiology*, 149: 220–224.

Yamazaki, T. Oyanagi, H. Fujiwara, T. & Fukumori, Y. 1995. Nitrite reductase from the magnetotactic bacterium Magnetospirillum magnetotacticum, *European Journal of Biochemistry*, 233: 665–671.

Advanced Materials and Structural Engineering – Hu (Ed.)
© 2016 Taylor & Francis Group, London, ISBN 978-1-138-02786-2

The influence of on semi-flexible airport pavement material working performance by porosity

B. Yang, X. Weng, J. Liu, L. Jiang, J. Zhang, P. Liu & X. Wen
Department of Airfield and Building Engineering, Air Force Engineering University, Xi'an, Shanxi, China

ABSTRACT: The stress distribution of semi-flexible material airport pavement panel which bears the cycling wheel load has been analyzed by the finite element software ANSYS, and the location of the maximum tensile stress on the sub face of the pavement panel has also been determined. The mechanic test has been done and the mechanical performance of semi-flexible pavement material has been recorded. The wheel cyclic load test has been conducted on the runway test bench by using 4 groups of semi-flexible material pavement specimens and their dimension is 300 mm × 300 mm × 50 mm. Meanwhile, the maximum deflection and the maximum tensile stress on the sub face of the semi-flexible material pavement in different porosity (20%, 23%, 26% and 30%) have been collected. The results demonstrate that: the test data and finite element analysis results matched well with each other. When the porosity is 26%, the mechanical perforce of the material will be superior. Based on the same base layer subgrade condition, the rigidity of the semi-flexible material will increase with the increase of porosity. On the contrary, the maximum deflection of the pavement will decrease. The number of cyclic load acting on the semi-flexible material which the porosity is 26% and 30% is much larger than the material which the porosity is 20% and 36%. Considered all the aspects, the reasonable value for the porosity of the semi-flexible material for airport pavement engineering should be 26%.

1 INTRODUCTION

In China, most of the airport pavements are made of cement concrete. Due to the presence of pavement joints, the pavement life and aircraft traveling comfort are affected. The semi-flexible pavement material has become a new kind of pave material. In the application of the pavement project, fewer joints or none joint will appear on the pavement and it could exhibit superior performance (Ahlrich & acderto 1991, Ai-Qadi et al. 1994). This surface material applied in airport pavement engineering has great potential (Setyawan 2003, Mayer & Thau 2001). France is the first country applying semi-flexible pavement material in airport pavement engineering (Ling et al. 2010). United States and Britain have done a lot of research on the semi-flexible pavement materials (Hassan & Setyawan 2003, Larsen 2004, Beer et al. 2012). In China, the research on semi-flexible pavement material has also been its infancy (Hao et al. 2003). The porosity of the semi-flexible pavement material is the important parameter and foreign counties are mainly based on experience to design the semi-flexible pavement material porosity, generally controlled at 20% to 25% (Dong 2009). However, the determination of the porosity is a lack of scientific experimental study to support.

For our research, the wheel cyclic load test which can simulate the aircraft loading on the pavement has been conducted to investigate the influence of porosity on semi-flexible pavement material working performance. And the recommended value of semi-flexible airport pavement material porosity has been determined. The test data can provide support for the design of semi-flexible airport pavement, and the test results have great practical value for applying the semi-flexible pavement material in airport pavement engineering.

2 ANALYSIS MODEL

2.1 *Structure of pavement*

Figure 1 shows the structure of the pavement, and it is composed of semi-flexible material pavement, cement stabilized gravel base and soil base. The parameters of each structure layer were listed in Table 1. Compared with concrete pavement, the semi-flexible material pavement has more flexibility. So the temperature joint was hardly seen on the semi-flexible material pavement. Generally, the whole semi-flexible material airport pavement can be seen as a boundless panel which is composed of homogeneous material. In order to calculate the mechanical response of the semi-flexible material pavement when the wheel load is acting, we supposed that the wheel load must act on the interior of the pavement panel.

2.2 Type of loading

Taking the wheel load of an aircraft for analysis, we supposed that the plane was gliding forward along the center-line of semi-flexible pavement panel and the contact surface of the tire and pavement surface was a rectangle which size was 0.47 m × 0.32 m. The tire pressure is 1.23 MPa and the load of a nose wheel was 185.0 kN. The position of maximum stress on the pavement panel can be gathered from the mechanical responses of semi-flexible material pavement when the nose wheel was passing at the center of the pavement panel.

2.3 Calculator model

According to the pavement structure and the structural parameters shown in Figure 1 and Table 1, the 3D model of the semi-flexible material pavement structure has been built. The size of pavement panel was 5 m × 5 m × 0.3 m, the size of lime-fly ash concrete was 10 m × 10 m × 0.3 m, and the size of solid base was 10 m × 10 m × 10 m. Figure 2 shows the 3D meshed model below. 3D physical structure model was used to simulate the structure of semi-flexible pavement, and the solid base was simulated by the model of elastic layered

system. The inter-layer contact elements which named conta170 and conta173 were set between sub-face of the semi-flexible panel and surface of lime-fly ash concrete base. Finite element model of semi-flexible pavement structural was all analyzed using eight-node solid elements named solid45. It was supposed that the X-axis positive direction was the direction of aircraft, Y-axis direction is a direction perpendicular to the wheel and Z-axis direction of the depth direction of the pavement.

2.4 Simulation results analysis

Stress distribution of pavement panel was simulated using finite element software ANSYS. Figure 3 shows the stress distribution of the panel when the aircraft was passing through the center of the pavement panel. Figure 4 shows the stress distribution of panel on center cross section. In order to find out the position of maximum stress on the pavement panel, the relationship curves between stress and position were drawn.

Figure 5 shows the relationship between position and stress of nodes which was along the X-axis center-line direction on the panel surface as well as panel subsurface. Figure 6 shows the relationship between position and stress of nodes which was along the Y-axis center-line direction on the panel surface as well as the panel subsurface. Seen from figures above, the position of maximum stress was the center of the panel.

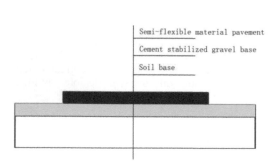

Figure 1. Structure of semi-flexible pavement.

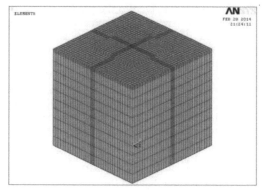

Figure 2. 3D model of semi-flexible pavement.

Table 1. Material parameters of each structural layer.

Structure name	Flexural-tensile strength/MPa	Flexural-tensile modulus/MPa	Poisson ratio	Thickness/m
Semi-flexible Material Pavement	2.5	3600	0.15	0.3
Lime-fly Ash Concrete	–	300	0.25	0.3
Soil Base	–	80	0.30	10

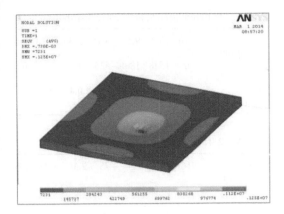

Figure 3. Pavement stress distribution.

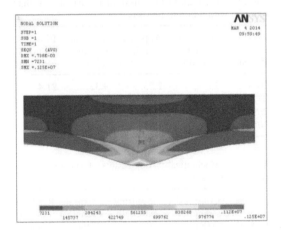

Figure 4. Pavement section stress distribution.

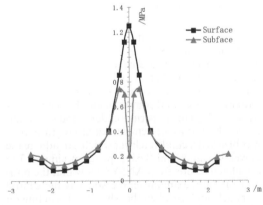

Figure 5. Stress curves of nodes along the central line of wheel.

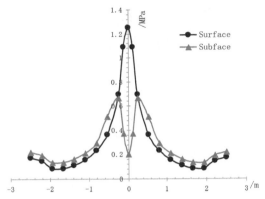

Figure 6. Stress curves of nodes perpendicular to the center-line of wheel.

3 WORKING PERFORMANCE RESEARCH

3.1 Test equipment

The test equipment was KPD-01 runway test bench. Figure 7 shows the sketch of runway test bench. It is composed of vertical wheel loading mechanism and horizontal moving mechanism. The vertical wheel loading mechanism has reaction frame system, walking beam and vertical static servo loading mechanism. The maximum vertical travel is 200 mm. The radius of the wheel is 200 mm and the width of the rubber tire is 100 mm. The horizontal moving mechanism has a loading cylinder which horizontal travel is ±500 mm. the size of runway box is 500(width) mm × 1200(length) mm × 1000(height)mm, which can contain the simulated structure of airport pavement. The electric strain gauge was BX120-20 AA and the electric resistance of the strain gauge is $120 \pm 0.1\Omega$, its sensitivity ratio is $2.08 \pm 1\%$ and the size is 20 mm (length) × 3 mm (width). We used HZ1008 A-60CH high speed programmable static resistance strain data collection to collect the stress at the sub-face of the specimen.

3.2 Technical performance and mechanical property

3.2.1 Cement slurry technical performance

The 42.5R ordinary Portland cement was used in order to make the cement slurry good fluidity, according to the Chinese standard (JTG E42 2005), fine sand has been selected and the fineness modulus is 2.1. The cement admixture is polycarboxylate water reducing agent. The tests have been conducted according to the standard method (JTG E30 2005) and (Cheng & Hao 2003). Table 2 shows the proportioning test results of cement slurry.

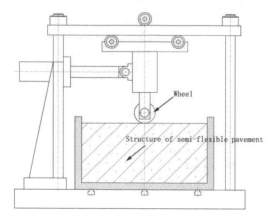

Figure 7. Sketch of runway test bench.

Table 2. Proportion of cement slurry.

Proportion	7d flexural strength /MPa	7d compressive strength /MPa	1h fluidity /mm
Water-binder Ratio: 0.40 Sand-binder Ratio: 0.25 Water Reducer: 0.3%	12.6	82.27	248

3.2.2 *Material asphalt material technical performance*

AH90 asphalt has been used to make the material asphalt material and it is made by Panjin Northern Asphalt Co., Ltd. Crushed limestone was selected as coarse aggregate which was composed of 5–10 mm and 2–15 mm graded gravel. Its apparent density is 2.72 g/cm^3 and clay content less than 1.0%. The particle size distribution is shown in Table 3 and Table 4. Volume method has been used to design material asphalt material with different porosity. The specific parameters are shown in Table 5.

3.2.3 *Semi-flexible pavement material mechanical properties*

According to the standard (JTG E20 2011), mechanical tests were conducted on the prepared specimens and the results were shown in Table 6.

3.3 *Test methods*

Semi-flexible material pavement panel specimens have been prepared with different porosity (20%, 23%, 26% and 30%) and their dimension is 300 mm × 300 mm × 50 mm; the thickness of cement stabilized gravel base is 200 mm; Subgrade compaction degree is 95%. The 20 mm~30 mm,

Table 3. 5–10 mm gravel particle size distribution.

Sieve Size/mm	16	13.2	9.5	4.75	2.36	1.18	0.6
Screen Margin/g	0	1538	1806	525	15	21	0
Sieve Percentage/%	0	39.4	46.2	13.4	0.4	0.5	0

Table 4. 2–15 mm gravel particle size distribution.

Sieve Size/mm	16	13.2	9.5	4.75	2.36	1.18	0.6
Screen Margin/g	0	153	300	4599	31	45	0
Sieve Percentage/%	0	3.0	5.9	89.7	0.6	0.9	0

Table 5. Marshall Test results of material matrix asphalt material with different porosity.

Porosity	Optimum asphalt content/%	Apparent Density /(g · cm^{-3})	Marshall stability /kN	Flow Value /(0.1 mm)
20%	3.6	2.23	9.32	21.4
23%	3.3	2.12	8.61	26.3
26%	2.9	2.02	7.17	27.3
30%	2.7	1.93	7.02	29.7

Table 6. Mechanical testing results of specimens with different porosity.

Porosity	Compressive strength /MPa	Splitting strength /MPa	Flexural strength /MPa	Resilient modulus /MPa
20%	6.06	1.84	7.1	2441.5
23%	6.23	2.31	7.3	2478.1
26%	8.64	2.49	8.5	3097.7
30%	7.34	2.52	8.6	2753.0

10 mm~20 mm, 5 mm~10 mm three kinds of aggregates were used, and the mixed ratio was 30:40:30. The 32.5R ordinary Portland cement was selected as the stable material which content is 4%. The soil base and the cement stabilized gravel base have been paved and compacted in the runway box with a 7-day curing period under standard conditions. The surface of cement stabilized gravel base was brushed a emulsion layer which can adhere the gravel base and semi-flexible panel specimens. The calculated results from ANSYS show that the position of the maximum tensile stress is the central of panel sub face when the wheel load is acting on the semi-flexible material pavement panel. In order to verify the reliability of the results, we set three strain gauges every 5 cm along the X-axis positive

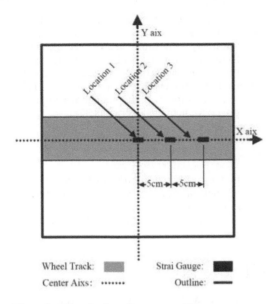

Figure 8. Sketch of strain gauge position.

direction the center-line of the specimen on the sub face. Figure 8 shows the specific positions of the three strain gauges.

4 RESULTS AND ANALYSIS

When the test begins, the speed of the moving wheel is set 200 mm/s and the move distance is set ±250 mm, and the vertical load is set 185.0 kN. The high speed programmable static resistance strain data collection will collect the stress from the strain gauges every a thousand times of cyclic load and the runway test bench will automatically record the maximum deflection value of specimen. Stop the test when the length of cracks appearing on surface is more than 20 mm and the date should be taken the previous record data. The test results are shown in Table 7 and Figure 9 showed the stress that collected from the sub face of each specimen.

Our test has been conducted in the same base condition, the test results can be used to analysis the influence of porosity on semi-flexible pavement material working performance. Seen from Table 6, the mechanical properties of semi-flexible material almost have no change when the porosity varies from 20% to 23% and 26% to 30%. But when the porosity varies from 23% to 26%, the test results have changed a lot. The compressive strength is increased by 38.7%, splitting strength is increased by 7.8%, flexural strength is increased by 16.4%, and resilient modulus has reached the maximum 3097.7 MPa which is increased by 25%. So when the porosity is 26%, the mechanical properties could be much superior.

Table 7 showed that the stiffness of the composite material is improved with the increase of porosity, but the maximum deflections of specimens decrease because of the increase in the cement slurry content. When the porosity varies from 20% to 23% and 26% to 30%, the changes of cyclic loading times and maximum deflection are not very obvious. On the contrary, when the porosity varies from 23% to 26%, the number of cyclic loads is increased dramatically by 153.5% and the maximum deflection is almost decreased by 8.9%. Additionally, the stress of Location 1 is greater than the others.

It can be seen from Figure 9 that when the porosity is 20% and 23%, the curves have the obvious segmentation. With the increase of the cyclic load number, the increment ratio of the tensile stress changes from large to small in the beginning and

Table 7. Wheel cyclic load test results.

Poro-sity	Numbers of cyclic load	Deflec-tion /μm	Loca-tion 1 /MPa	Loca-tion 2 /MPa	Loca-tion 3 /MPa
20%	37000	225	2.74	2.64	2.58
23%	43000	214	2.76	2.70	2.59
26%	66000	195	2.85	2.72	2.58
30%	71000	191	2.80	2.72	2.61

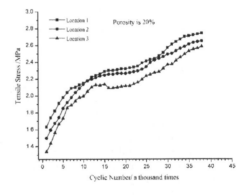

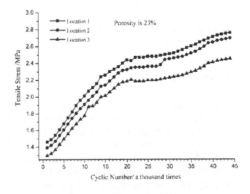

Figure 9. (Continued).

161

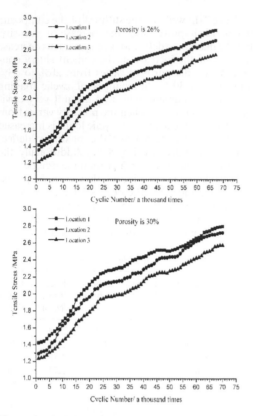

Figure 9. Relationship between the tensile stress and the times of cyclic load.

then increases again in at last. When the porosity is 26% and 30%, the curves are smoother. With the number of cyclic loads increasing, the increment of the tensile stress is becoming smaller. So when the porosity is small, the stiffness and flexibility of material are very sensitive to external force and the stiffness. The mechanical properties of the material asphalt material and cement slurry can be easily exhibited at the different stage of the cyclic load damage. When the porosity is high, the material asphalt material and cement slurry can be fully mixed each other. So the rigid material and flexible material can be fully mixed, the composite material can bear larger numbers of cyclic load damage.

5 CONCLUSIONS

The stress of Location 1 is larger than the others and the position of the maximum tensile stress is the center of the pavement panel on the sub face. The test data and finite element analysis results match well with each other. According to the mechanical performance test, the mechanical properties of the semi-flexible material could be superior if the porosity is 26%. When the porosity is 26% and

30%, the working performance of the semi-flexible pavement material is more superior to the material which porosity is 20% and 23%. But there is almost no increment of working performance when porosity varies from 26% to 30%. In view of working performance and energy saving, the reasonable value for the porosity of the semi-flexible material in airport pavement engineering is 26%.

ACKNOWLEDGEMENT

This material is based upon work funded by Zhejiang Provincial Natural Science Foundation of China under Grant No. LQ12E09002; Project (51308497) supported by National Natural Science Foundation of China.

REFERENCES

Ahlrich, R.C. & Anderto, G.L. 1991. *Construction and Evaluation of Resin Modified Pavement*, Technical Report G L-91-13, U.S Army Engineering Waterways Experiment Station, Vicksburg, MS.

Ai-Qadi, I.L. Gouru, H. & Weyers, R.E. 1994. Asphalt Portland Cement Concrete Composite: Laboratory Evaluation, *ASCE*, 120(1): 94–108.

Beer, M. de. Maina, J.W. & Netterberg, F. 2012. Mechanistic Modelling of Weak Interlayers in Flexible and Semi-flexible Road Pavements, *Journal of the South African Institution of Civil Engineering.* 54(2): 43–54.

Cheng, L. & Hao. P.W. 2003. Mixture of Cement Slurry With Semi-flexible Pavement, *J. Chang'an Univ. (Natur. Sci. Edit.)* 23(2): 1–4.

Dong, Y.Y. 2009. *Study on High-performance Semi-flexible Pavement Parametric Design and Construction Process*, Master Thesis. Chongqing Jiaotong University.

Hao, P.W. Cheng, L. & Lin, L. 2003. Pavement Performance of Semi-flexible Pavement in Laboratory, *J. Chang'an Univ. (Natur. Sci. Edit.)*, 23(2): 1–6.

Hassan, K.E. & Setyawan, A. 2003. Effect of Cementitious Grouts on the Properties of Semi-flexible Bituminous Pavement, *Performance of Bituminous and Hydraulic Materials in Pavements*, 113–120.

JTG E20. 2011. Highway Engineering Asphalt and Asphalt Mixture Test Procedures.

JTG E30. 2005. Highway Engineering Cement Concrete Test Procedures.

JTG E42. 2005. Test Methods of Aggregate for Highway Engineering.

Larsen, P. 2004. *Reinforced Semi-flexible Pavement*, US, US 20040101365 A1.

Ling, T.Q. et al, 2010. Research on Performance of Water-retention and Temperature-fall Semi-flexible Pavement Material, *China Journal of Highway Transport*, 23(2): 7–11.

Mayer, J. & Thau, M. 2001. *Jointless Pavements for Heavy-duty Airport Application: the Semi-flexible Approach*, Proceedings of the 27th International Air Transportation Conference, Chicago, Illinois, USA, 87–100.

Setyawan, A. 2003. *Development of Semi-Flexible Heavy-Duty Pavements*, PhD Thesis. University of Leeds.

Advanced Materials and Structural Engineering – Hu (Ed.)
© *2016 Taylor & Francis Group, London, ISBN 978-1-138-02786-2*

Electrochemical detection of baicalin at a carbon nanosphere-modified electrode

Z.F. Wang, L. Shi, G.Z. Gou, A.P. Fan, C. Xu & L. Zhang
Department of Chemistry, Honghe University, Mengzi, China

ABSTRACT: The carbon nanosphere-modified Glassy Carbon Electrode (GCE) for the determination of baicalin is described in this paper. The colloidal carbon nanospheres were successfully synthesized by the hydrothermal method. The resident porosity of porous carbon nanospheres will promote diffusion of baicalin molecules through interconnected micropores, and will be beneficial for the increase in detection sensitivity. The effect of the kind and pH of the supporting electrolyte were investigated by cyclic voltammograms. The results showed that the acceptable electrochemical redox reversibility was obtained in HAc-NaAc buffer solutions at the pH level of 4.6.

1 INTRODUCTION

Flavonoids are a class of polyphenolic compounds distributed throughout the plant kingdom, which has been widely studied in drug exploitation due to their specific effect on human health. Baicalein, a member of the flavonoid family, has been found in the root of *Scutellaria baicalensis* (Liu et al. 2014). Baicalin is an important anti-inflammatory and anticancer drug, which has been widely used in medicine. It has been demonstrated to possess a series of biological effects, including anti-inflammation (Rogerio et al. 2007), antitumor (Ikemoto et al. 2000) and inhibition of the proliferation of cancer cells or inducement of apoptosis in breast and prostatic cell lines (Chan et al. 2000). However, overdose of baicalein will result in severe side effects. Therefore, the establishment of highly sensitive analytical techniques for the determination of baicalin is of great significance in clinics and pharmaceutics.

Several methods have been developed for the analysis of baicalin, such as gas chromatography (Lin et al. 1999), high-performance liquid chromatography (Feng et al. 2010; Kotani et al. 2006), and thin layer chromatography (Okamoto et al. 1993). Nevertheless, these methods either required complicated sample preparation or suffered from low sensitivities and specificities. Therefore, developing a highly sensitive and simple detection method for baicalein is one of the most important analytical challenges. The electrochemical methods, with the merits of accuracy, simplicity and ease of on-site determination, have attracted considerable attention for the analysis of flavonoid drugs.

Carbon nanomaterials have attracted considerable attention in electrochemical sensors because of their extraordinary physical properties and remarkable conductivities (McCreery 2008). Recently, porous carbon nanospheres have also displayed unique advantages owing to the tunability of particle size and shape, as well as the resident porosity that promotes the diffusion of guest molecules through interconnected micropores. A "green" synthetic approach has been developed, which involves the transformation of sugars into homogeneous and stable colloidal carbon nanospheres, which are hydrophilic (Sun & Li 2004).

We report on the electrochemical determination of baicalin by using a glass carbon electrode that was doped with carbon nanospheres. Carbon nanospheres can act as an enhanced electrochemical material for the determination of baicalin. The effect of the kind and pH of the supporting electrolyte were investigated by cyclic voltammograms.

2 EXPERIMENTAL

2.1 *Reagents*

Baicalin was purchased from Certification Institute of Chinese Pharmaceutical and Biological Products. Other chemicals were of analytical grade and used without further purification. All aqueous solutions were prepared with doubly distilled water.

2.2 *Synthesis of colloidal carbon spheres*

The colloidal carbon spheres were prepared followed by our previously reported method (Sun & Li 2004): Glucose (3.6 g, analytical purity; Sinopharm Chemical Reagent Co., Ltd) was dispersed in double-distilled water (40 mL) by sonicating for 30 min, and formed a clear solution. Then, the resulting solutions were placed in a 40 mL Teflon-sealed autoclave and maintained at 180 °C for 4 h. The black or puce products were isolated by centrifugation (12000 rpm, 8 min), cleaned four times with double-distilled water, and finally redissolved in double-distilled water.

2.3 Electrochemical measurement

Electrochemical measurements were made using a CHI660D electrochemical workstation (ChenHua Instruments Co., Shanghai, China). The glassy carbon electrode (GCE, 3 mm in diameter) was polished carefully with alumina slurry (1.0, 0.3, and 0.05 μm) and rinsed with distilled water followed by sonication in nitric acid (1:1), ethanol, and distilled water, and then dried in a stream of nitrogen gas. A conventional three-electrode system was used, including an Ag/AgCl reference electrode, a Pt wire counter-electrode, and a modified electrode as the working electrode. Then, 6.0 μL of 0.5 mg mL^{-1} carbon nanospheres were carefully cast on the surface of the well-polished glassy carbon electrode and dried in air. Finally, the modified electrode was used as the working electrode for all electrochemical studies.

3 RESULTS AND DISCUSSION

The types of supporting electrolytes played a key role in the electrochemical detection of baicalein. The effect of electrolytes on the electrochemical response of the carbon nanosphere-modified glassy carbon electrode for the detection of baicalin was investigated by cyclic voltammetry in 0.56 μM baicalin solution. Figure 1 shows the differential pulse voltammograms studies of 0.56 μM baicalin in KCl solution at pH = 6.9, HAc-NaAc buffer solutions at pH = 4.6 and HCl-sodium citrate at pH = 3.0. A weak peak current was observed at +0.21 V in KCl solution at pH = 6.9 (Figure 1a) with the carbon nanosphere-modified glassy carbon electrode. Meantime, the carbon nanosphere-modified glassy carbon electrode had a weak current at +0.32 V in HCl-sodium citrate at pH = 3.0 (Figure 1c). However, a very strong peak current was observed at +0.26 V in the HAc-NaAc buffer solutions at pH = 4.6 (Figure 1b). The results show that a higher peak current could be obtained in HAc-NaAc buffer solutions at pH = 4.6. These results indicate that the type of supporting electrolytes has a significant effect on the electrochemical detection of baicalin.

Moreover, the types of supporting electrolytes with the same pH also have been investigated by cyclic voltammograms. A very weak oxidation peak was observed at +0.23 V but no reduction peak was observed between −0.2 V and 0.5 V in HCl-sodium

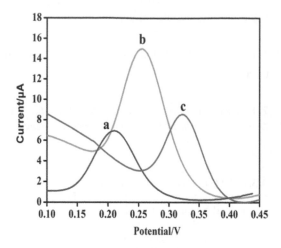

Figure 1. Differential pulse voltammograms of the carbon nanosphere-modified glassy carbon electrode in different solutions: (a) 0.5 M KCl solution (pH = 6.9), (b) 0.2 M HAc-NaAc buffer solutions (pH = 4.6) and (c) 0.1 M HCl-sodium citrate (pH = 3.0) with 0.56 μM baicalin.

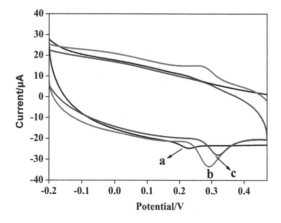

Figure 2. Cyclic voltammograms of 0.56 μM baicalin at the carbon nanophere-modified glassy carbon electrode with the same pH in different solutions: (a) 0.1 M HCl-sodium citrate (pH = 4.6), (b) 0.2 M HAc-NaAc buffer solutions (pH = 4.6), (c) 0.1 M citric acid-sodium citrate (pH = 4.6), and the scan rate is 50 mV s^{-1}.

citrate at pH = 4.6 (Figure 2a). An oxidation peak was observed at +0.32 V but no reduction peak was observed between −0.2 V and 0.5 V in citric acid-sodium citrate at pH = 4.6 (Figure 2c). However, a pair of redox peaks was observed in the potential window between −0.2 V and 0.5 V with the oxidation peak at 0.29 V and the reduction peak at 0.28 V in HAc-NaAc buffer solution at pH = 4.6 (Figure 2b). Therefore, the HAc-NaAc buffer solution with pH = 4.6 was used as the supporting electrolyte for the determination of baicalin.

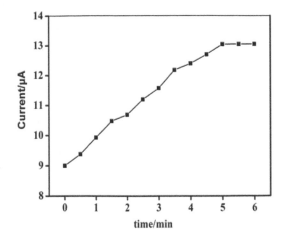

Figure 3. Effects of the accumulation time on the electrochemical response of 0.56 μM baicalin in 0.2 M HAc-NaAc buffer solutions (pH 4.6) at the carbon nanosphere-modified glassy carbon electrode.

The accumulation time is another important factor that influences the detection of baicalin. As can be seen from Figure 3, the peak currents of 0.56 μM baicalin on the carbon nanosphere-modified glass carbon electrode increased rapidly with the accumulation time in the range of 0–5 min. Then, the peak current remained almost constant. This indicates that the adsorption/extraction of baicalin on the film reaches an equilibrium. As a result, 5 min was chosen as the optimal accumulation time for the determination of baicalin.

4 CONCLUSIONS

In the current research, carbon nanospheres were obtained by a "green" hydrothermal method. The resident porosity of carbon nanospheres promoted the diffusion of guest molecules through interconnected micropores. The carbon nanospheres played an important role in accelerating the electron transfer. In addition, a simple and enhanced electrochemical sensing platform for the determination of baicalin was successfully constructed. The effect of the kind and pH of supporting electrolyte were investigated by cyclic voltammograms and differential pulse voltammograms. The results showed that the type of supporting electrolytes had a significant effect on the peak current of baicalin, and the acceptable electrochemical redox reversibility was obtained in HAc-NaAc buffer solutions at pH = 4.6.

ACKNOWLEDGMENT

This work was supported by the Scientific Research Fund Project of Honghe University (XJ14Z02), the Key Project of Open Fund for Master Construction Disciplines of Yunnan Province (No. HXZ1304), and the Youth project of Yunnan Province (No 2014FD054).

REFERENCES

Chan, F.L. Choi, H.L. Chen, Z.Y. Chan, P.S.F. & Huang, Y. 2000. Induction of apoptosis in prostate cancer cell lines by a flavonoid, baicalin, Cancer Letters. 160: 219–228.

Feng, J. Xu, W. Tao, X. Wei, X. Cai, F. Jiang, B. & Chen, W. 2010. Simultaneous determination of baicalin, baicalein, wogonin, berberine, palmatine and jatrorrhizine in rat plasma by liquid chromatography-tandem mass spectrometry and application in pharmacokinetic studies after oral administration of traditional Chinese medicinal preparations containing scutellaria-coptis herb couple, Journal of Pharmaceutical and Biomedical Analysis. 53: 591–598.

Ikemoto, S. Sugimura, K. Yoshida, N. Yasumoto, R. Wada, S. Yamamoto, K. & Kishimoto, T. 2000. Urology, Antitumor effects of Scutellariae radix and its components baicalein, baicalin, and wogonin on bladder cancer cell lines, Urology. 55: 951–955.

Kotani, A. Kojima, S. Hakamata, H. & Kusu, F. 2006. HPLC with electrochemical detection to examine the pharmacokinetics of baicalin and baicalein in rat plasma after oral administration of a Kampo medicine, Analytical Biochemistry. 350: 99–104.

Lin, M.C. Tsai, M.J. & Wen, K.C. 1999. Supercritical fluid extraction of flavonoids from Scutellariae Radix, Journal of Chromatography A, 830: 387–395.

Liu, Z. Zhang, A. Guo, Y. & Dong, C. 2014. Electrochemical sensor for ultrasensitive determination of isoquercitrin and baicalin based on DM-β-cyclodextrin functionalized graphene nanosheets, Biosensors and Bioelectronics. 58: 242–248.

McCreery, R.L. 2008. Advanced carbon electrode materials for molecular electrochemistry, Chemical Reviews, 108: 2646–2687.

Okamoto, M. Ohta, M. Kakamu, H. & Omori, T. 1993. Evaluation of phenyldimethylethoxysilane treated high-performance thin-layer chromatographic plates application to analysis of flavonoids in scutellariae-radix, Chromatographia. 35: 281–284.

Rogerio, A.P. Kanashiro, A. Fontanari, C. Silva, E.V.G. Lucisano-Valim, Y.M. Soares, E.G. & Faccioli, L.H. 2007. Anti-inflammatory activity of quercetin and isoquercitrin in experimental murine allergic asthma, Inflammation Research. 56: 402–408.

Sun, X.M. & Li, Y.D. 2004. Colloidal carbon spheres and their core/shell structures with noble-metal nanoparticles, Angewandte Chemie International Edition. 43: 597–601.

Advanced Materials and Structural Engineering – Hu (Ed.)
© *2016 Taylor & Francis Group, London, ISBN 978-1-138-02786-2*

Pt/C catalyst for methanol electro-oxidation and oxygen electro-reduction in DMFC

Z.F. Wang, L. Shi, G.Z. Gou, A.P. Fan, C. Xu & L. Zhang
Department of Chemistry, Honghe University, Mengzi, China

ABSTRACT: In this work, the electrocatalytic oxidation of methanol and electrochemical reduction of oxygen on the Pt/C catalyst are investigated. The electrochemically active surface area was measured using a hydrogen adsorption-desorption method in conjunction with Cyclic Voltammetry (CV). The electrocatalytic activity of the Pt/C catalyst toward methanol oxidation and oxygen reduction was investigated by cyclic voltammetry. The long-term stability of the Pt/C catalyst was also investigated by the chronoamperometry test. The Pt/C catalyst shows electrocatalytic activity toward methanol oxidation and oxygen reduction.

1 INTRODUCTION

A Direct Methanol Fuel Cell (DMFC) is an electrochemical device that converts the chemical energy into electricity using methanol and oxygen as the anode and cathode reactants, respectively. The usage of a liquid fuel is a superior characteristic, in that it allows easy handling of fuel with a very high specific energy density (Jeon et al. 2008). High energy density requires the utilization of high methanol concentration (Kamarudin et al. 2007). However, the usage of concentrated methanol in fuel cell applications causes several problems such as low catalytic activity at the anode electrode and methanol crossover to the cathode electrode (Jeon et al. 2008, Arico et al. 2001). Meanwhile, the high anodic overpotential of methanol oxidation is due to the poor reaction kinetics (Loffler et al. 2001), which prevents the widespread commercial application of DMFC. Platinum (Pt) has a high activity for methanol oxidation, and has been used as an anode electrocatalyst for many years. However, the Pt electrocatalyst becomes poisoned by the intermediate products of methanol oxidation, in particular CO (Hussein et al. 2010). Additionally, both Nafion and Pt are expensive materials, and the cost of DMFCs containing Pt and Nafion is extensively high. Much effort has been made toward overcoming these problems with the goal of developing less expensive anode materials with significantly better catalytic activity, proton conductivity, and reduced susceptibility to CO poisoning. To overcome these problems under DMFCs conditions, it is desirable to develop catalyst supports. It is well known that the specific activity of catalysts is strongly related to their size, distribution, and the support. Among the possible supports, carbon black has been widely used as an electrode, which disperses Pt nanoparticles (Liu et al. 2004).

Herein, the methanol electro-oxidation on the commercial Pt/C catalyst was investigated. The electrochemically active surface area was measured using a hydrogen adsorption-desorption method in conjunction with Cyclic Voltammetry (CV). The electrocatalytic activity of the Pt/C catalyst toward methanol oxidation and oxygen reduction was investigated by cyclic voltammetry. The long-term stability of the Pt/C catalyst was also investigated by the chronoamperometry test.

2 EXPERIMENTAL

2.1 Reagents

Pt/C catalyst with an average particle size less than 3.0 nm was bought from Johnson Matthery Inc. H_2SO_4, methanol and other reagents were of analytical grade and used without further purification. All aqueous solutions were prepared with doubly distilled water.

2.2 Preparation of the working electrode

A Glassy Carbon Electrode (GCE), 3 mm in diameter, was polished with alumina slurry (1.0, 0.3, and 0.05 μm) and rinsed with distilled water followed by sonication in nitric acid (1:1), ethanol, and distilled water, before each experiment.

2.3 Electrochemical measurements

Electrochemical measurements were performed using a CHI660D electrochemical workstation (ChenHua Instruments Co., Shanghai, China). A conventional three-electrode system was used including an Ag/

AgCl reference electrode, a Pt wire counter electrode, and the modified electrode as the working electrode. The commercial Pt/C catalyst was mixed in water to form a homogeneous ink (1 mg mL⁻¹), and 7 μL of the ink were deposited on the GCE. Subsequently, 3 μL of the Nafion (0.2%) solution were added to fix the catalyst on the GCE surface. The electro-chemical active surface area (ECSA) of Pt NPs was calculated from the hydrogen electrosorption curve, which was recorded between −0.2 and +1.2 V in a 0.5 M H_2SO_4 solution at a scan rate of 50 mV s⁻¹. The electrocatalytic activity for methanol oxidation was characterized by cyclic voltammograms (CVs) in a 0.5 M H_2SO_4 solution containing 1.0 M CH_3OH at a scan rate of 20 mV s⁻¹. The activity for the oxygen reduction reaction (ORR) was measured by CVs in an O_2-saturated 1.0 M H_2SO_4 solution at a scan rate of 20 mV s⁻¹.

3 RESULTS AND DISCUSSION

Pt/C catalysts have been widely used in direct methanol fuel cells. Cyclic Voltammetry (CV) is a convenient and efficient tool used to estimate the ECSA of the Pt catalyst on an electrode. The ECSA of an electrocatalyst not only provides important information regarding the number of electro-chemically active sites per gram of the catalyst, but also is a crucial parameter to compare different electrocatalytic supports. Hydrogen adsorption/desorption peaks are usually used to evaluate the ECSA of the catalyst. The CV curves for the Pt/C catalyst, in the 0.5 M H_2SO_4 solution at a scan rate of 50 mV s⁻¹, are shown in Figure 1A.

By using the hydrogen adsorption-desorption method in conjunction with cyclic voltammetry (CV), the electrochemically active surface area (ECSA) was estimated by measuring the charge associated with H_{upd} adsorption (Q_H) between −0.2 and +0.2 V by assuming 210 μC cm⁻² for the adsorption of a monolayer of hydrogen on a Pt surface (q_H) (Schmidta et al. 1998). The H_{upd} adsorption charge (Q_H) can be determined using $Q_H = 0.5 \times Q$, in which Q is the charge in the H_{upd} adsorption-desorption area obtained after double-layer correction. The specific ECSA was then calculated based on the following relationship (Lee et al. 2007):

$$Q = \int I dt = \int \frac{I[A] \times dE[V]}{v[V/s]} \tag{1}$$

$$\text{Specific ECSA} = \frac{Q_H}{m \times q_H} \tag{2}$$

where Q_H is the charge for H_{upd} adsorption; m is the loading amount of the metal; and q_H is the charge

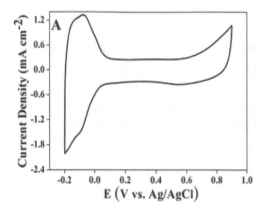

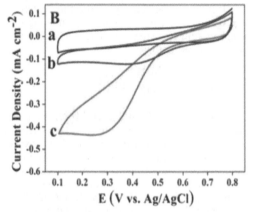

Figure 1. (A) CV of the Pt/C/GCE catalyst-modified glassy carbon electrode in a N_2-saturated 0.5 M H_2SO_4 solution at a scan rate of 50 mV s⁻¹. (B) CVs of oxygen reduction at the Pt/C/GCE in a 1.0 M H_2SO_4 solution at 20 mV s⁻¹ in the N_2-saturated (a), air-saturated (b) and O_2-saturated (c) solutions.

required for the monolayer adsorption of hydrogen on a Pt surface. Thus, the obtained ECSA of the Pt/C catalyst was 41 m² g⁻¹.

The electrocatalytic activity of the Pt/C catalyst was also investigated for oxygen reduction. Figure 1B shows the typical cyclic voltammograms (CVs) of oxygen reduction at the Pt/C catalyst-modified GCE in the 1.0 M H_2SO_4 solution in the presence of air, saturated N_2 and O_2. No catalytic reduction current can be observed in the N_2-saturated solution (Figure 1B, a). In the presence of air, a remarkable catalytic reduction current occurs at 0.41V (Figure 1B, b) at a scan rate of 20 mV s⁻¹. A higher catalytic current for dioxygen reduction is observed at 0.30 V in the presence of saturated oxygen (Figure 1B, c). It should be noted that the Pt/C catalyst-modified GCE exhibits a higher electrocatalytic current for dioxygen reduction. In addition, the oxygen reduction potential (0.30 V) observed at the Pt/C-modified electrode is more positive.

The electrocatalytic activity of the Pt/C catalyst for the methanol oxidation reaction was studied in an acid medium by CV. The resulting voltammograms are shown in Figure 2A. The bare GCE does not show any electrocatalytic activity toward the oxidation of methanol in the studied potential range (Figure 2A, a). The Pt/C shows catalytic behavior for the electro-oxidation of methanol in the presence of an oxidation current in the positive potential region. The onset potentials are around 0.3 V (versus Ag/AgCl). The current peak at about 0.62 V (versus Ag/AgCl) in the forward scan is attributed to methanol electro-oxidation on the catalyst. In the reverse scan, an oxidation peak is observed around 0.49 V (Figure 2A, b), which is probably associated with the removal of the residual carbon species formed in the forward scan (Lin et al. 2005). As is known, the ratio of the forward oxidation current peak (I_f) to the reverse current peak (I_b), I_f/I_b, is an index of the catalyst tolerance to the poisoning species, Pt = C = O (Zheng et al. 2007). A higher ratio indicates more effective removal of the poisoning species on the catalyst surface. The I_f/I_b ratio of the Pt/C catalyst is 1.13, which is higher than that of the E-TEK catalyst (0.74), with 13e showing a better catalyst tolerance of the Pt/C catalyst. The long time stability of the Pt/C catalyst is also investigated by the chronoamperometry test (Figure 2B). The polarization current for the methanol oxidation reaction shows a slow decay during the initial period because the catalyst has a large number of active sites initially available for methanol activation, and then reaches an apparent steady state within 200s. The result shows that the catalyst favors a long-term application as the anode material in DMFC.

4 CONCLUSIONS

The electrochemically active surface area of the Pt/C catalyst is calculated to be 41 m^2 g^{-1}. The results indicate that the Pt/C catalyst has electrochemical catalytic activity toward methanol oxidation and oxygen reduction.

ACKNOWLEDGMENT

This work was supported by the Scientific Research Fund Project of Honghe University (XJ14Z02), the Key Project of Open Fund for Master Construction Disciplines of Yunnan Province (No. HXZ1304), and the Youth project of Yunnan Province (No. 2014FD054).

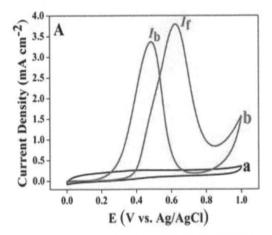

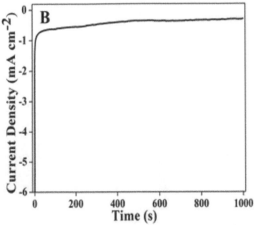

Figure 2.   (A) CVs of 1 M methanol in a 0.5 M H₂SO₄ solution on bare GCE (a) and Pt/C catalyst-modified GCE (b) at 20 mV s⁻¹. (B) Chronoamperometric curves of Pt/C/GCE in 1.0 M CH₃OH + 0.5 M H₂SO₄.

REFERENCES

Arico, A.S. Srinivasan, S. & Antonucci, V. 2001. DMFCs: From fundamental aspects to technology development, *fuel cells*, 1: 133–161.
Hussein, G. Karim, K. & Mohammad, K. 2010. Platinum nanoparticles supported by a Vulcan XC-72 and PANI doped with trifluoromethane sulfonic acid substrate as a new electrocatalyst for direct methanol fuel cells, *The Journal of Physical Chemistry C*, 114: 5233–5240.
Jeon, M.K. Lee, K.R. Lee, W.S. Daimon, H. Nakahara, A. & Woo, S.I. 2008. Investigation of Pt/WC/C catalyst for methanol electro-oxidation and oxygen electroreduction, *Journal of Power Sources*, 185: 927–931.
Kamarudin, S.K. Daud, W.R.W. Ho, S.L. & Hasran, U.A. 2007. Overview on the challenges and developments of micro-direct methanol fuel cells (DMFC), *Journal of Power Sources*, 163: 743–754.
Lee, E.P. Peng, Z. Cate, D.M. Yang, H. Campbell, C.T. & Xia, Y. 2007. Growing Pt nanowires as a densely packed array on metal gauze, *Journal of the American Chemical Society*. 129: 10634–10635.

Lin, Y.H. Cui, X.L. Yen, C.H. & Wai, C.M. 2005. PtRu/carbon nanotube nanocomposite synthesized in supercritical fluid: A novel electrocatalyst for direct methanol fuel cells, *Langmuir.* 21: 11474–11479.

Liu, Z.L. Ling, X.Y. Su, X.D. & Lee, J.Y. 2004. Carbon-supported Pt and PtRu nanoparticles as catalysts for a direct methanol fuel cell, *The Journal of Physical Chemistry B.* 108: 8234–8240.

Loffler, M.S. Gross, B. Natter, H. Hempelmann, R. Krajewski T. & Divisek, J. 2001. Synthesis and characterization of catalyst layers for direct methanol fuel cell applications, *Physical Chemistry Chemical Physics.* 3: 333–336.

Schmidta, T.J. Gasteigera, H.A. Stabc, G.D. Urbanc, P.M. Kolbb, D.M. & Behma, R.J. 1998. Characterization of high-surface-area electrocatalysts using a rotating disk electrode configuration, *Journal of The Electrochemical Society.* 145: 2354–2358.

Zheng, S.F. Hu, J.S. Zhong, L.S. Wan, L.J. & Song, W.G. 2007. In situ one-step method for preparing carbon nanotubes and Pt composite catalysts and their performance for methanol oxidation, *The Journal of Physical Chemistry C.* 111: 11174–11179.

Advanced Materials and Structural Engineering – Hu (Ed.)
© 2016 Taylor & Francis Group, London, ISBN 978-1-138-02786-2

Pt/C-modified Glass Carbon Electrode for the determination of Dopamine, Uric Acid and Ascorbic Acid

Z.F. Wang, L. Shi, G.Z. Gou, A.P. Fan, C. Xu & L. Zhang
Department of Chemistry, Honghe University, Mengzi, China

ABSTRACT: In this study, a Pt/C-modified Glassy Carbon Electrode (GCE) was created to characterize Ascorbic Acid (AA), Dopamine (DA), and Uric Acid (UA) via Cyclic Voltammetry (CV). The Pt/C catalyst with an average particle size less than 3.0 nm was obtained from Johnson Matthery Inc. Cyclic voltammetry was used to evaluate the electrocatalytic activity toward the oxidation of AA, DA, UA, respectively. Compared with a bare GC electrode, a pair of well-defined redox peaks was observed at the Pt/C-modified glassy carbon electrode for the detection of AA, DA, and UA, respectively. The results indicated that the proposed sensor laid the foundation for the further research in electrochemical analysis.

1 INTRODUCTION

Dopamine is naturally produced and widely distributed in the central nervous system of mammals. Abnormal levels of DA in body fluids are the indications of many serious diseases such as Schizophrenia, Huntington's disease and Parkinson's disease (Wu et al. 2012). DA coexists with ascorbic acid and uric acid in the extracellular fluids of the central nervous system and serum in mammals. The concentrations of AA and UA are much higher (100–1000 times) than that of DA in body fluids, and offer greater interference during the determination of one in the presence of the other two (Tian et al. 2012). Uric acid is the major final product of purine catabolism in the human body. In a healthy human, the normal level of UA in the urine is in the millimolar range, whereas in serum, it is in micromolar range (Manjunatha et al. 2009). Abnormal levels of UA in body fluids are the symptoms of many diseases such as gout and Lesch-Nyhan syndrome. Ascorbic acid is present in many vegetables, citrus fruits and biological fluids, where it acts as an antioxidant and free-radical scavenger. AA concentration in body fluids can be used to assess the level of oxidative stress, and excessive stress is related to diseases such as cancer, diabetes mellitus and hepatic disorders (Magdalena et al. 2014).

Electrochemical methods have been proved to be a very promising approach for the determination of AA, DA and UA by virtue of the electroactive nature of these biomolecules. However, the major problem with this approach is that AA, DA and UA usually require a very high over potential to undergo electrochemical oxidation at bare electrodes, and, furthermore, the electrode surface suffers from the fouling effect due to the accumulation of oxidation products (Ramesh et al. 2004). To overcome these difficulties, many types of materials have been employed to modify electrodes for the electrochemical detection of AA, DA and UA.

Recently, carbon nanomaterials have been widely used in electroanalytical investigations and have enormous potentials for constructing electrochemical sensing platforms with high sensitivity to detect different target molecules, because of their chemical inertness, relatively wide potential window, low background current, and suitability for different types of analysis. Additionally, metal nanoparticles (NPs) have continued to receive considerable interest due to their particular optical, electronic, and catalytic properties and their important applications in many fields such as nanosensors, catalysis (Suo et al. 2008). Therefore, functionalizing 1D supporting nanomaterials with metal NPs that combine the properties of two functional nanomaterials, such as high conductivity and surface area of 1D nanomaterials and unique catalytic properties of metal NPs, to achieve a wider range of applications, will probably play an important role in the development of nanoscience and nanotechnology.

In this study, an electrochemical sensor was fabricated using the Pt/C-modified glassy carbon electrode (Pt/C/GCE). The electrochemical behavior of AA, DA and UA at the surface of the Pt/C/GCE was investigated. It was found that DA shows a pair of redox peaks appearing at 129 mV and 181 mV at the Pt/C/GCE, corresponding to the two-electron oxidation of *o*-dopaminoquinone and subsequent reduction of *o*-dopaminoquinone to DA. AA show a weak oxidation peak and a well-defined reduction peak appearing at −74 mV and −6 mV at the Pt/C/GCE. In addition, UA shows a quasi-reversible process at the Pt/C/GCE. This sensor shows an excellent electrochemical activity for the determination of DA, AA and UA.

2 EXPERIMENTAL

2.1 *Reagents*

Dopamine and uric acid were purchased from Sigma-Aldrich. The Pt/C catalyst with an average particle size less than 3.0 nm was bought from Johnson Matthery Inc. All other reagents used in this study were of AR grade, which were purchased from the Sinopharm Chemical Reagent Co., Ltd. (Shanghai China). The phosphate buffer solution (PBS) was prepared from $Na_2HPO_4 \cdot 2H_2O$-KH_2PO_4. All aqueous solutions were prepared using double-distilled water.

2.2 *Preparation of the working electrode*

A Glassy Carbon Electrode (GCE), 3 mm in diameter, was polished with alumina slurry (1.0, 0.3, and 0.05 μm) and rinsed with distilled water followed by sonication in nitric acid (1:1), ethanol, and distilled water, before each experiment.

2.3 *Electrochemical measurements*

Electrochemical measurements were performed using a CHI660D electrochemical workstation (ChenHua Instruments Co., Shanghai, China). A conventional three-electrode system was used including an Ag/AgCl reference electrode, a Pt wire counter electrode, and the modified electrode as the working electrode. The commercial Pt/C catalyst was mixed in water to form a homogeneous ink (1 mg mL⁻¹), and 7 μL of the ink were deposited on the GCE. Subsequently, 3 μL of the Nafion (0.2%) solution were added to fix the catalyst on the GCE surface.

3 RESULTS AND DISCUSSION

Cyclic voltammetry was used to evaluate the ability and electrochemical behavior of the modified electrode for the detection of AA in a neutral solution. Cyclic voltammograms of AA at the bare electrode and the Pt/C/GCE are shown in Figure 1. At the bare GCE, AA show broad oxidation peaks with the peak potentials at 64 mV, but no reduction peaks can be observed (Figure 1a). However, a weak oxidation peak and a well-defined reduction peak

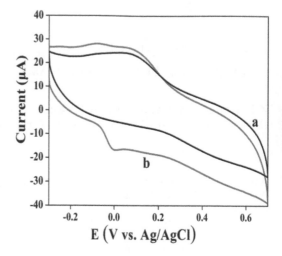

Figure 1. Cyclic voltammograms of (a) bare GCE and (b) Pt/C/GCE in PBS containing 1.26 mM AA at pH = 7.0.

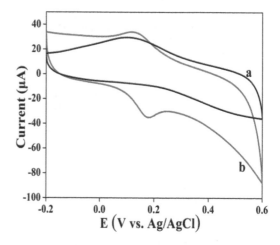

Figure 2. Cyclic voltammograms of (a) bare GCE and (b) Pt/C/GCE in PBS containing 0.86 mM DA at pH = 7.0.

appear at −74 mV and −6 mV at the Pt/C/GCE, respectively (Figure 1b). The ΔEp value is 68 mV and the oxidation peak current is 1.16 times higher than that obtained at the bare GCE, suggesting the electrocatalytic activities of Pt/C toward AA.

At the bare GCE, DA shows broad oxidation peaks with the peak potentials at 103 mV, but no reduction peaks can be observed (Figure 2a). However, the Pt/C/GCE show a pair of redox peaks appearing at 129 mV and 181 mV, corresponding to the two-electron oxidation of DA to *o*-dopaminoquinone and subsequent reduction of *o*-dopaminoquinone to DA (Figure 2b), respectively (Luczak

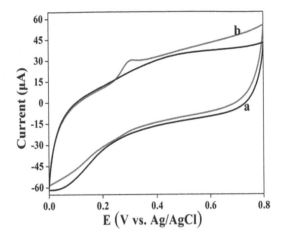

Figure 3. Cyclic voltammograms of (a) bare GCE and (b) Pt/C/GCE in PBS containing 2.92 mM UA at pH 7.0.

2008). The ΔEp value is 52 mV and the oxidation peak current is 1.15 times higher than the obtained at the bare GCE, suggesting the excellent electrocatalytic activities of Pt/C toward DA.

In the case of UA, the bare GCE shows no oxidation peak and reduction peak (Figure 3a). However, a well-defined oxidation peak and a weak reduction peak appear at 304 mV and 247 mV at the Pt/C/GCE (Figure 3b), respectively. Meanwhile, the oxidation peak current increased about 1.41 times, indicating the electrocatalytic activities of Pt/C toward UA when compared with the results obtained at the bare GCE. The result indicates that UA shows a quasi-reversible process at the Pt/C/GCE, which can be interpreted as that UA is first oxidized to its diimine intermediate and then reduced back to uric acid (Yang et al. 2014).

4 CONCLUSIONS

In this work, the Pt/C nanocomposites were used to investigate the electrochemical behaviors of AA, DA and UA by cyclic voltammograms, respectively. It can be clearly concluded that the Pt/C nanocomposite-modified glass carbon electrode had electrocatalytic activities toward the oxidation of AA, DA and UA. DA shows a pair of redox peaks appearing at 129 mV and 181 mV at the Pt/C/GCE, corresponding to the two-electron oxidation of o-dopaminoquinone and subsequent reduction

of o-dopaminoquinone to DA. AA show a weak oxidation peak and a well-defined reduction peak appearing at -74 mV and -6 mV at the Pt/C/GCE. In addition, UA shows a quasi-reversible process at the Pt/C/GCE. Thus, it is confirmed that the proposed sensor lays a foundation for the further research in electrochemical analysis.

ACKNOWLEDGMENT

This work was supported by the Scientific Research Fund Project of Honghe University (XJ14Z02), the Key Project of Open Fund for Master Construction Disciplines of Yunnan Province (No. HXZ1304), and the Youth project of Yunnan Province (No. 2014FD054).

REFERENCES

Guo, S.J. Dong, S.J. & Wang, E.K. 2008. Polyaniline/Pt hybrid nanofibers: high-efficiency nanoelectrocatalysts for electrochemical devices, Small. 5: 1869–1876.

Luczak, T. 2008. Preparation and characterization of the dopamine film electrochemically deposited on a gold template and its applications for dopamine sensing in aqueous solution, Electrochimica Acta. 53: 5725–5731.

Magdalena, P.A. Aneta, P. Iren, S.A. & Cornelia, S.A. 2014. Electrochemical methods for ascorbic acid determination, Electrochimica Acta. 121: 443–460.

Manjunatha, H. Nagaraju, D.H. Suresh, G.S. & Venkatesha, T.V. 2009. Detection of uric acid in presence of dopamine and high concentration of ascorbic acid using PDDA modified graphite electrode, Electroanal. 21: 2198–2206.

Ramesh, P. Suresh, G.S. & Sampath, G.S. 2004. Selective determination of dopamine using unmodified, exfoliated graphite electrodes, Journal of Electroanalytical Chemistry. 561. 173–180.

Tian, X.Q. Cheng, C.M. Yuan, H.Y. Du, J. Xiao, D. Xie, S.P. Choi, M.M.F. & Li, C.H. 2012. Simultaneous determination of L-ascorbic acid, dopamine and uric acid with gold nanoparticles- β- cyclodextrin-graphene-modified electrode by square wave voltammetry, Talanta. 93: 79–85.

Wu, L. Feng, L.Y. Ren, J.S. & Qu, X.G. 2012. Electrochemical detection of dopamine using porphyrin-functionalized graphene, Biosensors and Bioelectronics. 34: 57–62.

Yang, L. Liu, D. Huang, D. & You, T.Y. 2014. Simultaneous determination of dopamine, ascorbic acid and uric acid at electrochemically reduced graphene oxide modified electrode, Sensors Actuators. B. 193: 166–172.

Advanced Materials and Structural Engineering – Hu (Ed.)
© *2016 Taylor & Francis Group, London, ISBN 978-1-138-02786-2*

Electrochemical sensor based on Polyaniline-modified Graphene Nano-composites for dopamine determination

L. Shi, Z.F. Wang, G.Z. Gou, Q.S. Pan, X.L. Chen & W. Liu
Department of Chemistry, Honghe University, Mengzi, China

ABSTRACT: High-quality Polyaniline-modified Graphene Nano-sheets (PANI/GN) were successfully synthesized using liquid-liquid interface polymerization method. The interfacial polymerization at a liquid-liquid interface allows PANI to grow uniformly on the surface of the GN. The obtained nano-composites were characterized by scanning electron microscopy and UV/Vis absorbance spectra. The electrochemical sensor based on PANI/GN nano-composites was constructed to determine dopamine (DA). The obtained PANI/GN-modified GCE showed high catalytic activity for the oxidation of dopamine.

1 INTRODUCTION

Graphene is a two-dimensional sheet of carbon atoms bonded through sp² hybridization. GN has attracted intensive interests in recent years since its discovery by Geim and coworkers in 2004, owing to its large specific surface area, high thermal and electrical conductivities, great mechanical strength, and potential low manufacturing cost (Geim & Novoselov 2007, Novoselov et al. 2004). Furthermore, GN-based nano-composites as enhanced sensing material for fabricating electrochemical sensors have received increased attention, because these kinds of nano-composites film may generate synergy on electrocatalytic activity and thus enhance the sensitivity of the sensors (Zhang et al. 2011).

On the other hand, conducting polymers also have been extensively studied and widely applied. Among the known conducting polymers, polyaniline (PANI) is one of the mostly studied conducting polymers owing to its good environmental stability, tunable conductivity switching between insulating and semiconducting materials, facile synthesis, and potential application in many areas (Zhao et al. 2009).

The polymer has the similar electronic, magnetic, and optical properties of metals, whereas it can retain the flexibility and processibility of conventional polymers. The doping level of polyaniline can also be readily controlled through an acid-doping/base dedoping process. It has been extensively studied for many potential applications including secondary battery electrodes, supercapacitors, electromagnetic shielding devices, conducting molecular wires, sensors, and so forth (Wan 2008). In recent years, polyaniline-modified graphene

nano-sheets have been successfully prepared by in situ chemical or electrochemical polymerization and covalent or noncovalent functionalization (Qiu et al. 2012). Although these PANI/GN composites have been documented, the fabrication of this kind of composite material with multiple functions by using simple and effective methods remains scientifically challenging.

Dopamine (DA), an important neurotransmitter, plays a significant role in the function of the central nervous, renal and hormonal systems. Extreme abnormalities in DA levels are symptoms of several disease states such as schizophrenia, Parkinson's and Alzheimer's diseases (Thomas et al. 2011, Cao et al. 2010). The determination of DA has been performed by using versatile technology, for example high performance liquid chromatography (Carrera et al. 2007), electrophoresis (Huang & Lin 2005), chromatography (Uutela et al. 2009), and so on. However, these detection methods generally require time-consuming sample preparation and expensive instrumental equipment, and have poor specificity. Recently, electrochemical method has been applied to detect DA for a long time. However, the fouling of electrode surface by the oxidation product can result in poor performance at the conventional electrodes (Liu et al. 2012). Therefore, it is essential to develop a sensitive and selective method for the determination of DA in biological fluids such as serum for clinical diagnosis.

In this work, we reported a simple electrochemical sensor for dopamine determination based on PANI/GN nano-composites. The PANI/GN nano-composites were prepared by liquid-liquid interface polymerization method. The obtained PANI/GN nano-composites simultaneously possessing the unique properties of GN (large surface area) and PANI (high conductivity) through combining their individual characteristics, which will provide good opportunities for applications in the fields of sensors. The SEM and UV-vis spectroscopy were used to characterize the prepared nano-composites. The electrochemistry performance of the resulted sensor was discussed. The nano-composites show excellent electrochemical oxidation activity toward DA. That maybe due to the combination of graphene with PANI could improve the conductivity, stability and the performance of electrochemical sensors for determination of DA. The nano-composites could be used as a platform for biosensor and biocatalyst applications.

2 EXPERIMENTAL

2.1 *Reagents*

Graphite flake (99.8%, 325 mesh) was provided by Alfa Aesar. Other reagents were of analytical grade and used without further purification. Aniline was fresh distillation prior to use. All aqueous solutions were prepared with doubly distilled water.

2.2 *Preparation of PANI/GN nano-composites*

The PANI/GN nano-composites were prepared followed our previously reported method in the literature (Qiu et al. 2012).

2.3 *Electrochemical measurement*

Electrochemical measurements were carried out on CHI660D electrochemical workstation (ChenHua Instruments Co., Shanghai, China). The glassy carbon electrode (GCE, 3 mm in diameter) was polished carefully with alumina slurry (1.0, 0.3, and 0.05 μm) and rinsed with distilled water followed by sonication in nitric acid (1:1), ethanol, and distilled water, then dried in a stream of nitrogen gas. A conventional three-electrode system was used, including an Ag/AgCl reference electrode, a Pt wire counter-electrode, and the modified electrode as the working electrode. The 6.0 μL of 1.0 mg/mL PANI/GN nano-composites were carefully cast on the surface of the well-polished glassy carbon electrode and dried in air. And then, the modified electrode was used as the working electrode for all electrochemical studies.

3 RESULTS AND DISCUSSION

The morphology of as-prepared PANI/GN nano-composites was investigated by SEM (Figure 1A). It displayed that graphene nano-sheets are homogeneously surrounded by PANI film and a large wrinkled paper-like structure appears. The morphology of PANI on the surface of GN is completely different from the pure PANI. The reason for this is that the liquid/liquid interface provides a good soft template for the polymerization of aniline on the surface of GNs. That will be a benefit for the improvement the electrochemistry performance of PANI/GN. The PANI/GN nanocomposites were further revealed by UV/Vis spectra. The PANI/GN nano-composites exhibit three absorbance peaks at $\lambda = 260$, 332, and 632 nm (Figure 1B), which can be attributed to the resonance-absorption peak of GN, the π-π^* transition of the benzenoid rings, and the π-polaron transition, respectively (Qiu et al. 2012).

The charge transfer property of PANI/GN nanocomposites modified electrode was characterized using an electrochemical method. Figure 2 showed the Cyclic voltammetry curves obtained at GN modified GCE and PANI/GN nano-composites modified GCE in 1.0 mM $Fe(CN)_6^{3-/4-}$ a solution containing 0.1 M KCl. A pair of redox peaks was

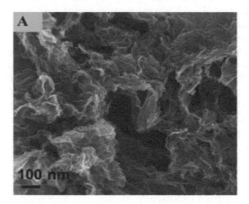

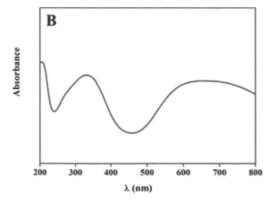

Figure 1.   SEM images (A) and UV/Vis spectra (B) of PANI/GN nanocomposites.

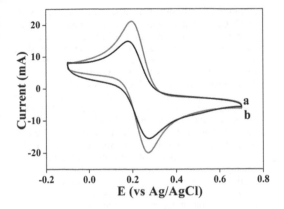

Figure 2. Cyclic voltammetry curves for 1.0 mM $Fe(CN)_6^{3-/4-}$ in 0.1 M KCl solution recorded at GN modified GCE (curve a), PANI/GN modified GCE (curve b), scan rate 50 m V s^{-1}.

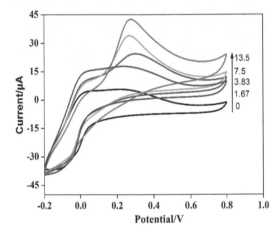

Figure 3. Cyclic voltammetry curves of PANI/GN modified GCE in 0.1 M NaH_2PO_4-HCl buffer solution (pH 2.0) with 0, 1.67, 3.83, 7.5, and 13.5 mM DA at a scan rate of 20 mV/s.

observed at the electrode with the potential difference of 95 mV appearing at the GN modified GCE (curve a). While a pair of well-defined quasi-reversible redox peaks was observed at the PANI/GN nano-composites modified GCE (curve b), the ΔEp was decreased to 75 mV. Moreover, the peak current was much higher than GN modified GCE. These results indicated that the PANI/GN nano-composites facilitated electron transfer rate, it might be attributed to the high electric conductivity of GN and PANI, the large surface area and the plenty of various edge defects presented on the surface of graphene.

The electrocatalytic activity of PANI/GN modified GCE towards DA was investigated and results are shown in Figure 3. In the absence of DA, no

oxidation peak current change was observed over the potential range employed. After addition of different concentrations of DA from 1.67 mM to 13.5 mM, the oxidation peak current increases with increasing DA concentration. The oxidation peak corresponds to the oxidation of DA to dopamine-quinone (Kalimuthu & John 2009). The π-π interaction between phenyl structure of DA and two-dimensional planar hexagonal carbon structure of graphene makes the electron transfer feasible. Additionally the PANI/GN nano-composites simultaneously possess the unique properties of GN (large surface area) and PANI (redox properties). That will be enhanced the conductivity of the electrode to accelerate the probe to reach the electrode surface.

4 CONCLUSIONS

In summary, a green and fast approach to the synthesis of PANI modified graphene nanosheets is reported. This method is green and will not result in contamination of the product PANI/GN nano-composites. The PANI/GN nano-composites were successfully used in sensing of dopamine. It was attributed to that the PANI/GN nano-composites could enhance the conductivity of the electrode to accelerate the probe to reach the electrode surface. The result shows that graphene was a good advanced electrode materials and could be combined with other functional materials to fabricate the sensing interface for electroanalysis. Moreover, valid response to dopamine obtained in present work also indicates the prospective performances of graphene to other biological molecules, such as nucleic acids, proteins and enzymes.

ACKNOWLEDGEMENT

This work was supported by the Key Project of Open Fund for Master Construction Disciplines of Yunnan Province (No. HXZ1304), the Youth project of Yunnan Province (No. 2014FD054), the National Natural Science Foundation of China (No. 61361002) and the scientific research fund project of Honghe University (XJ14Z02).

REFERENCES

Cao, X.H. Zhang, L.X. Cai, W.P. & Li, Y.Q. 2010. Amperometric sensing of dopamine using a single-walled carbon nanotube covalently attached to a conical glass micropore electrode, *Electrochemistry Communications.* 12: 540–543.

Carrera, V. Sabater, E. Vilanova, E. & Sogorb, M.A. 2007. A simple and rapid HPLC–MS method for the simultaneous determination of epinephrine, norepinephrine, dopamine and 5- hydroxytryptamine: Application to the secretion of bovine chromaffin cell cultures, *Journal of Chromatography B.* 847: 88–94.

Geim, A.K. & Novoselov, K.S. 2007. The rise of graphene, *Nature Materials.* 6: 183–191.

Huang, H.M. & Lin, C.H. 2005. Methanol plug assisted sweeping-micellar electrokinetic chromatography for the determination of dopamine in urine by violet light emitting diode-induced fluorescence detection, *Journal of Chromatography B.* 816: 113–119.

Kalimuthu, P. & John, S.A. 2009. Modification of electrodes with nanostructured functionalized thiadiazole polymer film and its application to the determination of ascorbic acid, *Electrochimica Acta.* 55: 183–189.

Liu, S. Xing, X.R. Yu, J.H. Lian, W.J. Li, J. Cui, M. & Huang, J.D. 2012. A novel label-free electrochemical aptasensor based on graphene-polyaniline composite film for dopamine determination, *Biosensors and Bioelectronics.* 36: 186–191.

Novoselov, K.S. Geim, A.K. Morozov, S.V. Jiang, D. Zhang, Y. Dubonos, S.V. Grigorieva, I.V. & Firsov, A.A. 2004. Electric field effect in atomically thin carbon films, *Science.* 306: 666–669.

Qiu, J.D. Shi, L. Liang, R.P. Wang, G.C. & Xia, X.H. 2012. Controllable deposition of a platinum nanoparticle ensemble on a polyaniline/graphene hybrid as a novel electrode material for electrochemical sensing, *Chemistry - A European Journal.* 18: 7950–7959.

Thomas, T. Mascarenhas, R.J. Nethravathi, C. Rajamathi, M. & Swamy, B.E.K. 2011. Graphite oxide bulk modified carbon paste electrode for the selective detection of dopamine: A voltammetric study, *Journal of Electroanalytical Chemistry.* 659: 113–119.

Uutela, P. Reinila, R. Harju, K. Piepponen, P. Ketola, R.A. & Kostiainen, R. 2009. Analysis of intact glucuronides and sulfates of serotonin, dopamine, and their phase I metabolites in rat brain microdialysates by liquid chromatography-tandem mass spectrometry, *Analytical Chemistry.* 81: 8417–8425.

Wan, M.X. 2008. A template-free method towards conducting polymer nanostructures, *Advanced Materials.* 20: 2926–2932.

Zhang, Y. Sun, X. Zhu, L. Shen, H. & Jia, N. 2011. Electrochemical sensing based on graphene oxide/Prussian blue hybrid film modified electrode, *Electrochimica Acta.* 56: 1239–1245.

Zhao, M. Wu, X.M. & Cai, C.X. 2009. Polyaniline nanofibers: synthesis, characterization, and application to direct electron transfer of glucose oxidase, *The Journal of Physical Chemistry C.* 113: 4987–4996.

Advanced Materials and Structural Engineering – Hu (Ed.)
© 2016 Taylor & Francis Group, London, ISBN 978-1-138-02786-2

Electrochemical sensor for ascorbic acid based on graphene-polyaniline nano-composites

L. Shi, Z.F. Wang, G.Z. Gou, Q.S. Pan, X.L. Chen & W. Liu
Department of Chemistry, Honghe University, Mengzi, China

ABSTRACT: In this study, the polyaniline-functionalized graphene nano-composite (GN/PANI) was synthesized by a facile interfacial polymerization method. An electrochemical sensor was constructed based on a glassy carbon electrode modified with graphene/polyaniline (GN/PANI) nano-composites. Ascorbic acid was used as representative analysis to demonstrate the sensing performance of GN/PANI-modified electrode. Cyclic voltammetry and amperometric current responses were used to evaluate the electrocatalytic activity towards the oxidation of ascorbic acid. The results show that the GN/PANI nano-composites show electrocatalytic activity toward ascorbic acid oxidation.

1 INTRODUCTION

Ascorbic acid, which is a water soluble vitamin, widely present in both the animal and plant kingdoms. Ascorbic acid is commonly used to supplement inadequate dietary intake, as anti-oxidant and plays an important role in the human metabolism as a free-radical scavenger, which may help to prevent radical induced diseases such as cancer and Parkinson's disease (Pakapongpan et al. 2014). Furthermore, deficiency of ascorbic acid can cause scurvy disease. It is administered in the treatment of many disorders, including Alzheimer's disease, atherosclerosis, cancer, infertility and in some clinical manifestations of HIV infections (Arrigoni & Tullio 2002). Thus, the development of a simple and rapid method for the determination of ascorbic acid is desirable for diagnostic and food safety applications. Diverse analytical methods have been developed and used for the detection of ascorbic acid: for example, fluorescence, chromatography, electrophoresis, and electrochemical methods (Wu et al. 2012). Among these techniques, electrochemical methods have attracted considerable attention due to the advantages of simplicity, low cost, high sensitivity, rapid analysis, easy operation, etc.

In order to enhance the sensitivity and selectivity of the electrochemical sensors, a variety of materials have been employed to modify electrode. Graphene (GN) is a two-dimensional layer of sp² bonded carbon atoms closely packed into a honeycomb lattice (Geim & Novoselov 2007). Due to its unique properties such as high surface area, high electrical conductivity, and strong mechanical strength, graphene has been widely employed for various applications such as supercapacitors (Zhang et al. 2010), nano-electronics (Ho et al. 2014), sensor (Yuan et al. 2014), batteries (Cheng et al. 2014). However, graphene generally tends to form irreversible agglomerates or even restack to form graphite through strong π-π stacking interaction.

Polyaniline (PANI) is one of the most widely studied conducting polymers. It is highly conductive, exhibits good environmental stability and can easily be prepared. PANI can act as a suitable matrix for immobilization of biomolecules and mediator for redox and enzymatic reactions and it exhibits impressive signal amplification and antifouling properties (Zhao et al. 2009). And also, PANI is environmentally friendly, water-soluble, and can improve the solubility and stability of functional materials. If graphene are modified with PANI, it is possible to obtain new materials simultaneously possessing the unique properties of graphene (large surface area) and PANI (high conductivity) through combining their individual characteristics, which will provide good opportunities for applications in the fields of sensors, electrocatalysis, luminescence, and electronics, etc (Qiu et al. 2012).

In this work, GN/PANI nano-composites were prepared by liquid-liquid interface polymerization method. The prepared nano-composites were characterized by UV-vis spectroscopy. Cyclic voltammetry was employed to investigate the electrochemical behaviors of ascorbic acid. The nano-composites show excellent electrochemical oxidation activity toward ascorbic acid. That maybe due to the combination of graphene with PANI could improve the conductivity, stability and the performance of electrochemical sensors for determination of ascorbic acid. The nano-composites could be used as a platform for biosensor and biocatalyst applications.

2 EXPERIMENTAL

2.1 Reagents

Graphite flake (99.8%, 325 mesh) was provided by Alfa Aesar. Other reagents were of analytical grade and used without further purification. Aniline was fresh distillation prior to use. All aqueous solutions were prepared with doubly distilled water.

2.2 Preparation of GN/PANI nano-composites

The GN/PANI nano-composites were prepared followed our previously reported method in the literature (Qiu et al. 2012).

2.3 Electrochemical measurement

Electrochemical measurements were carried out on CHI660D electrochemical workstation (ChenHua Instruments Co., Shanghai, China). The glassy carbon electrode (GCE, 3 mm in diameter) was polished carefully with alumina slurry (1.0, 0.3, and 0.05 μm) and rinsed with distilled water followed by sonication in nitric acid (1:1), ethanol, and distilled water, then dried in a stream of nitrogen gas. A conventional three-electrode system was used, including an Ag/AgCl reference electrode, a Pt wire counter-electrode, and the modified electrode as the working electrode. The 6.0 μL of 1.0 mg/mL GN/PANI nano-composites were carefully cast on the surface of the well-polished glassy carbon electrode and dried in air. And then, the modified electrode was used as the working electrode for all electrochemical studies.

3 RESULTS AND DISCUSSION

The typical synthetic process was carried out at the interface of dichloromethane (containing 1 g/mL of aniline) and 1.0 M of hydrochloric acid (containing H_2O_2 and $FeCl_3$). The aqueous phase was carefully spread onto the organic phase forming an aqueous/organic interface. In this reaction, the GNs act as an efficient template for aniline nucleation and polymerization. The electron-accepting GN and the electron-donating aniline combine to form a kind of weak charge-transfer complex. PANI can be covalently attached to the surface of GN through an in situ interfacial polymerization in H_2O/CH_2Cl_2 with the help of H_2O_2 and $FeCl_3$. The polymerization occurs at the interface and terminates as the GN/PANI nano-composites diffuse into the top aqueous phase. Interfacial

polymerization represents one effective method to suppress secondary growth. The obtained GN/PANI nano-composites show excellent water solubility and stability. Figure 1 show the illustration of GN and GN/PANI nano-composites aqueous solutions were left undisturbed for 0 h and 96 h. The results show that graphene sheets are hydrophobic and readily agglomerate in hydrophilic solvents. The graphene nano-sheets lose their water dispersibility, aggregate, and eventually precipitate after standing for 48 hours. However, the GN/PANI nano-composites are well-dispersed in solvents. The GN/PANI nano-composites show excellent water solubility and only a few GN/PANI are aggregated after left 96 h. The introduction of PANI is greatly improving the water solubility of graphene.

The kinetics of electrode reaction was investigated by evaluating the effect of scan rate on the oxidation peak current and peak potential. As shown in Figure 2, the scan rate of cyclic voltammetry

Figure 1. Illustration of (a) GN and (b) GN/PANI nano-composites aqueous solutions were left undisturbed at ambient temperature for 0 h. Illustration of (c) GN and (d) GN/PANI nano-composites aqueous solutions were left un-disturbed at ambient temperature for 96 h.

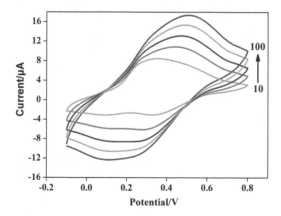

Figure 2. Effect of the scan rate on the cyclic voltammetric response of 0.5 mM ascorbic acid at the GN/PANI/GCE with 10, 20, 40, 60, 80, 100 mV/s in 0.1 M NaH_2PO_4-HCl buffer solution (pH 2.0).

exhibits a profound effect on the oxidation peak current of ascorbic acid. For the scan rates in the range of 10–100 mV/s, a linear relationship is established between the peak current and the scan rate, indicating the surface controlled mechanism is significant at the low scan rate. Additionally, the oxidation peak potentials of the molecules shift positively as increasing the scan rate.

The electrocatalytic activity of GN/PANI modified GCE towards ascorbic acid was investigated and results are shown in Figure 3. In the absence of ascorbic acid (curve a), no oxidation peak current change was observed over the potential range employed. After addition of different concentrations of ascorbic acid from 0.74 mM to 12.7 mM, as shown in curve b to f, the oxidation peak current increases with increasing ascorbic acid concentration. The oxidation peak corresponds to the oxidation of hydroxyl groups to carbonyl groups in furan ring of ascorbic acid on GN/PANI/GCE. The results show that the GN/PANI nano-composites can catalyst the oxidation of ascorbic acid and the GN/PANI modified GCE has good electrocatalytic activity toward ascorbic acid oxidation. That maybe due to the combination of graphene with PANI could improve the conductivity, stability and the performance of electrochemical sensors for determination of ascorbic acid.

Figure 4 shows the typical amperometric responses of the GN/PANI/GCE for the successive additions of ascorbic acid at an applied potential of 0.4 V. The sensor exhibited rapid step increases and reached a steady-state current within 5 s. The amperometric signal showed a good linear correlation with ascorbic acid concentration in the range from 0.15 M to 3.34 mM. The ascorbic

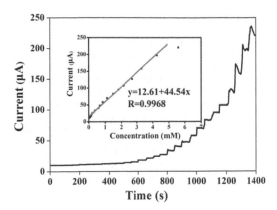

Figure 4. Typical amperometric current responses of a GCE modified with GN/PANI nano-composites on successive injection of H_2O_2 into stirred 0.1 M NaH_2PO_4-HCl buffer solution (pH 2.0). Detection potential: 0.4 V. The inset shows calibration curve of ascorbic acid detected by the GN/PANI/GCE.

acid detection limits are 0.042 mM (S/N = 3). The linear regression equation was expressed as y = 12.61 + 44.54x with a correlation coefficient of R = 0.9968.

4 CONCLUSIONS

We have successfully prepared GN/PANI nano-composites by liquid-liquid interface polymerization method and applied it for constructing an ascorbic acid sensor. GN/PANI provides a large surface area and a high conductivity. The nano-composite shows an excellent electrocatalytic activity toward ascorbic acid. The results indicate that the graphene and PANI can promote the electron transfer of ascorbic acid at the electrode and also improve the conductivity and stability. This sensor platform combines easy fabrication and excellent electrocatalytic activity toward ascorbic acid, which has great potential for sensor applications of several analyses in clinical diagnosis, pharmaceutical analysis and in the field of bioelectrochemistry.

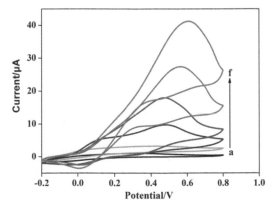

Figure 3. Cyclic voltammetry curves of GN/PANI/GCE in 0.1 M NaH_2PO_4-HCl buffer solution (pH 2.0) without ascorbic acid (a) and with 0.74 mM (b), 1.94 mM (c), 4.5 mM (d), 7.4 mM (e) and 12.7 mM (f) AA at a scan rate of 20 mV/s.

ACKNOWLEDGEMENT

This work was supported by the Key Project of Open Fund for Master Construction Disciplines of Yunnan Province (No. HXZ1304), the Youth project of Yunnan Province (No. 2014FD054), the National Natural Science Foundation of China (No. 61361002) and the scientific research fund project of Honghe University (XJ14Z02).

REFERENCES

Arrigoni, O. & De Tullio, M.C. 2002. Ascorbic acid: much more than just an antioxidant, *Biochimica et Biophysica Acta-General Subjects*. 1569: 1–9.

Cheng, B. Zhang, X.D. Ma, X.H. Wen, J.W. Yu, Y. & Chen, C.H. 2014. Nano-Li$_3$V$_2$(PO$_4$)$_3$ enwrapped into reduced graphene oxide sheets for lithium ion batteries, *Journal of Power Sources*, 265: 104–109.

Geim, A.K. & Novoselov, K.S. 2007. The rise of graphene, *Nature Materials*. 6: 183–191.

Ho, K.I. Huang, C.H. Liao, J.H. Zhang, W.J. Li, L.J. Lai, C.S. & Su, C.Y. 2014. Fluorinated graphene as high performance dielectric materials and the applications for graphene nanoelectronics, *Scientific Reports*, 4: 5893.

Pakapongpan, S. Mensing, S. Phokharatkul, D. Lomas, T. & Tuantranont, A. 2014. Highly selective electrochemical sensor for ascorbic acid based on a novel hybrid graphene-copper phthalocyanine-polyaniline nanocomposites, *Electrochimica Acta*. 133 (2014): 294–301.

Qiu, J.D. Shi, L. Liang, R.P. Wang, G.C. & Xia, X.H. 2012. Controllable deposition of a platinum nanoparticle ensemble on a polyaniline/graphene hybrid as a novel electrode material for electrochemical sensing, *Chemistry—A European Journal*. 18: 7950–7959.

Wu, G.H. Wu, Y.F. Liu, X.W. Rong, M.C. Chen, X.M. & Chen, X. 2012. An electrochemical ascorbic acid sensor based on palladium nanoparticles supported on graphene oxide, *Analytica Chimica Acta*. 745: 33–37.

Yuan, C.X. Fan, Y.R. Tao, Z. Guo, H.X. Zhang, J.X. Wang, Y.L. Shan, D.L. & Lu, X.Q. 2014. A new electrochemical sensor of a nitro aromatic compound based on three-dimensional porous Pt-Pd nanoparticles supported by graphene-multiwalled carbon nanotube composite, *Biosensors and Bioelectronics*. 58: 85–91.

Zhang, K. Zhang, L.L. Zhao, X.S. & Wu, J.S. 2010. Graphene/polyaniline nanoriber composites as supercapacitor electrodes, *Chemistry of Materials*. 22: 1392–1401.

Zhao, M. Wu, X.M. & Cai, C.X. 2009. Polyaniline nanofibers: synthesis, characterization, and application to direct electron transfer of glucose oxidase, *The Journal of Physical Chemistry C*. 113: 4987–4996.

Advanced Materials and Structural Engineering – Hu (Ed.)
© 2016 Taylor & Francis Group, London, ISBN 978-1-138-02786-2

Influence of low firing temperature on the characteristics of Positive Temperature Coefficient of Resistance and the Ni internal electrode of multilayer $Ba_{1.005}(Ti_{1-x}Nb_x)O_3$ ceramics

X.X. Cheng, B.W. Li, Z.X. Zhao & X.X. Li
College of Electronic Information and Mechatronic Engineering, Zhaoqing University, Zhaoqing, Guangdong, P.R. China

ABSTRACT: The influences of sintering conditions and Nb-dopant concentration on the characteristics of Positive Temperature Coefficient of Resistance (PTCR) and electrical properties of $Ba_{1.005}(Ti_{1-x}Nb_x)O_3$ (BTN) ceramics were investigated. The ceramics were sintered at 1100 °C for 2 h under a reducing atmosphere and then reoxidised at 750 °C for 1 h. The low sintering temperature affected the electrical properties, PTCR effect and Ni internal electrodes of the BTN specimens. The room-temperature resistance of the BTN specimens initially decreased (x ≤ 0.5 mol%) and then increased (0.5 mol% < x < 0.8 mol%) with increased doping concentration (0.2 mol%–0.8 mol% Nb^{5+}). In addition, the influence of Ni internal electrodes in the interface layer of the BTN samples obtained at a low sintering temperature on the PTCR characteristics was investigated using EDS analysis. The BTN ceramics showed a remarkable PTCR effect, with a resistance jump greater by 3.1 orders of magnitude and a low RT resistance of 7.5 Ω at a low reoxidation temperature of 750 °C after sintering under a reducing atmosphere.

1 INTRODUCTION

The semiconducting $BaTiO_3$-based ceramics exhibit Positive Temperature Coefficient of Resistance (PTCR) characteristics (Mancini & Paulin 2006, Heywang 1964, Illingsworth et al. 1990). Heywang (1961) explained this phenomenon, and an improved discussion was later provided by Jonker (1964).

Decreasing the resistance of a $BaTiO_3$-based PTCR thermistor material has recently emerged as a research hotspot. PTCR ceramics can be applied to low-voltage integrated circuits as overcurrent-protection elements. At present, the demand for elements with low room-temperature resistance and high resistance jump has rapidly increased. However, reducing resistance and enhancing the resistance jumping ratios of samples involves contrasting concepts. Moreover, reducing the room-temperature resistance of samples is challenging according to conventional processes. Multilayer specimens (Kanda et al. 1994, Niimi et al. 2007, Niimi et al. 2008) had been prepared by a reduction-reoxidation method to reduce further room-temperature resistance.

The reoxidation temperature has been reported to be above 900 °C (Jo & Han 2007, Pu et al. 2010, Xiang et al. 2008). However, a high reoxidation temperature is harmful in the preparation of multilayer $Ba_{1.005}(Ti_{1-x}Nb_x)O_3$ (BTN) ceramics because of the oxidation of the Ni electrodes. Ho & Fu (1992) recently suggested that acceptor-state density is related to annealing time. Al-Allak et al. (1987) found a resistance layer during an annealing treatment. The PTCR effect is influenced by the interfacial segregation of cation vacancies (Kolodiazhnyi & Petric 2003). Langhammer et al. (2006) claimed that the height of the potential barrier can be enhanced during cooling. The precipitation of a Ti-rich phase is due to a vacancy-compensation mechanism of V_{Ti}'''' during reoxidation (Makovec et al. 2001).

2 EXPERIMENTAL PROCEDURES

The starting materials were high-purity $BaCO_3$ (>99.8%), TiO_2 (>99.8%), Nb_2O_5 (>99.99%) and SiO_2 (>99.99%). They were weighed according to the ratio of the following formula: $Ba_{1.005}(Ti_{1-x}Nb_x)O_3 + 0.05$ mol% SiO_2. Then the compositions were mixed by high energy ball milling for 90 minutes (2400 r/min) in deionized water using zirconia balls and then calcined at 1150 °C for 2 h in air. After drying and sieving, the calcined powder was ground again by wet ball milling for 5 h in a polyurethane jar. Next, the dried powder was blended with dispersant, solvent, binder, and defoamer by ball milling for 18 h in a nylon pot and cast into green sheets of 55-μm thickness by the doctor-blade method.

These sheets were printed Ni internal electrodes and stacked with 15 MPa pressure at 50 °C to form a ceramic block. Then they were cut into rectangular blocks (3.8 mm × 1.6 mm × 1.4 mm). Subsequently, the binder was removed by heating at 330 °C in air. Sintering was conducted in an aluminum tube at 1100 °C for 2 h in a reducing atmosphere (3% H_2/N_2), with the heating and cooling rates being 3.33 °C/min. The bulk densities of the sintered samples were measured using the Archimedes method. The sintered BTN ceramics was re-oxidized at 750 °C in air, and the surfaces were rubbed with In–Ga alloy (60:40) to form an electrode. Resistance was measured by a digital multimeter, and the temperature dependence of resistance was measured in a temperature-programmable furnace (ZWX-B, Huazhong University of Science and Technology, China) at a heating rate of 1.6 °C/min in the range of 25–250 °C. The surface microstructure of the as-sintered ceramics was observed using scanning electron microscopy (SEM; TESCAN VEGA 3 EasyProbe, Czech). The Energy-Dispersive X-ray Spectrometry (EDS) was used to obtain the morphology and chemical composition of the samples. The mean grain size of the ceramics was estimated by the line-intersection method.

3 RESULTS AND DISCUSSION

3.1 Microstructure of the ceramics

The microstructure of the samples is significantly affected by sintering temperature, and a noticeable difference in the grain size is observed (Fig. 1). The average grain sizes of the Ni internal electrode and samples shown in Figure 1 are 0.34 and 0.79 μm, respectively. Furthermore, the grain size becomes uniform and increases with increased firing temperature, probably because of the decrease in porosity and increase in the grains with increased sintering temperature. Therefore, grain size may have been due to the differences in the degrees of grain growth of the samples. The mass-transfer process becomes faster because of the increasing sintering temperature.

4 INFLUENCE OF THE DOPANT CONCENTRATION ON ELECTRICAL PROPERTIES AND PTCR EFFECT

The dependence of the room-temperature resistance and the resistance-jumping ratio [Lg(R_{max}/R_{min})] of the BTN-based ceramics on the dopant concentration is shown in Figure 2. Note that the BTN-based ceramics were reoxidised at 750 °C for 1 h in air after firing at 1100 °C for 30 min in a reducing atmosphere. The influence of the donor dopant on the PTCR effect is shown in Figure 3. The room-temperature resistance of the samples first decreases and then increases with an increase in the dopant content from 0.2 mol% to 0.8 mol% Nb^{5+} (Fig. 2). This phenomenon signifies that the conductive mechanism can be transformed from the electronic compensation of the replaced Ti ions by Nb ions and the formation of oxygen vacancies (Chan et al. 1986, Pu et al. 2011) into ionic compensation depending on the formation of the cation vacancies (Brzozowski & Castro 2004, Brzozowski et al. 2002, Nowotny & Rekas 1994). In addition, the critical dopant concentration is 0.5 mol% Nb^{5+}, and its corresponding room-temperature resistance is 7.5 Ω. The high value of the donor-doped content indicates that the electrical properties of

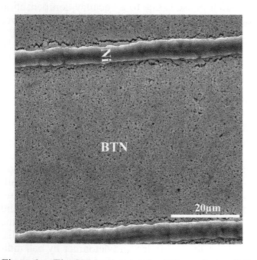

Figure 1. The SEM micrograph of the surfaces of the as-sintered specimen sintered at 1100 °C for 2 h.

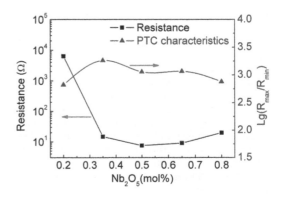

Figure 2. The room-temperature resistance and the jumping ratio are as a function of the dopant concentration.

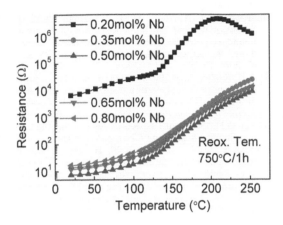

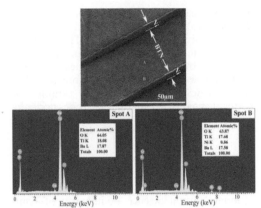

Figure 3. The temperature dependence of resistance for different donor doped content.

Figure 4. SEM–EDS images of the ceramics sintered at 1100 °C for 2 h.

the samples may be affected by the low sintering temperature and oxygen vacancy formation. Oxygen vacancy forms according to the following reaction:

$$O_O \rightarrow V_O^{\bullet\bullet} + 1/2\,O_2 + 2e' \qquad (1)$$

By contrast, the resistance jumping ratio increases with the doped-Nb content for specimens doped with a low donor-doped concentration (<0.35 mol%). The ratio is initially invariant and then decreases at high-dopant concentration (≥0.35 mol%). This result demonstrates that the higher the donor-doped content, the lower the resistance jump. Additionally, the corresponding resistance jumping ratio of the minimum resistance is 3.1 orders of magnitude. The corresponding dopant content of the ceramics with a maximum resistance jump of 3.3-orders of magnitude is 0.35 mol% Nb. This result shows that the low- and high-resistance jumps of the same samples cannot be matched well because of the low sintering temperature and numerous oxygen vacancies.

5 SEM–EDS ANALYSES

SEM–EDS analyses are used to determine detailed information based on the surface of the as-sintered multilayer BTN ceramic that was fired at 1100 °C for 2 h under a reducing atmosphere. Only Ba, Ti, O, or Ni were found in the matrix (Fig. 4), which indicates that the number of Ba and Ti ions present at the centre of the spot between the adjacent Ni electrodes (A) are slightly more than those near the Ni electrode area (B). A larger amount of oxygen is also found around the adjacent Ni electrodes than at the midgrain region. This condition indicates

that low firing temperature leads to a larger number of Ni losses near the Ni electrode region, because the low sintering temperature can improve the diffusion of the Ni electrodes and change the mechanism of the defect reaction near the B spot area (Brzozowski et al. 2002). However, the Ba:Ti:O ratio at spot A (at the middle of the BTN layer) is approximately 1:1.01:3.58, which is also nearly consistent with the composition of BaTiO$_3$ and Nb$_2$O$_5$. Ti^{4+} has been substituted by the Nb^{5+} in the matrix. The Ba:Ti:O:Ni ratio at spot B was approximately 1:1.01:3.63:0.05, which is approximately consistent with the composition of BaTiO$_3$ and NiO. These results confirm the oxidation of Ni (near Ni internal electrodes). Thus, the room-temperature resistance of the ceramics sintered at a low temperature (1100 °C) will increase. Moreover, the consent of Ti for spot B is slightly less than that of spot A. This condition suggests that a small quantity of Ti beside the Ni internal electrode may be substituted by a spot of Ni, which can react with acceptor states. Thus, the PTCR effect of the samples will be improved to a certain extent.

6 CONCLUSIONS

The ceramics were sintered at 1100 °C for 2 h under a reducing atmosphere and then reoxidised at 750 °C for 1 h. The room-temperature resistance of the BTN specimens initially decreased (x ≤ 0.5 mol%) and then increased (0.5 mol% < x < 0.8 mol%) with increased doping concentration. The concentration of the critical donor dopant of the specimens is quite high under low sintering temperature. However, the resistance jump of the samples firstly increased and then reduced as a function of the dopant content. The room-temperature resistance and resistance jump of BTN samples doped with 0.5 mol% Nb^{5+} are 7.5 Ω and 3.1 orders of magnitude, respectively. In addition, Ba, Ti, and Ni elements exist in the interface between BTN matrix and Ni electrodes. Moreover, Ni near the BTN matrix can be easily oxidised. Therefore, the PTCR characteristics of multilayer BTN ceramics are greatly affected by Ni internal electrodes under low sintering temperature.

ACKNOWLEDGEMENT

This work was financially supported by the National Natural Science Foundation of China (no. 51402258), the Creative Talents Planning Project of the Outstanding Young in University of Guangdong Province (no. 2013 LYM0099), the Characteristic Creative Project in University of Guangdong Province (no. 2014 KTSCX189), the Grant of Science and Technology Planning Project of Guangdong Province (no. 2012B040303007), the Youth Fund of Zhaoqing University (no. 201320), and the Scientific Research Initial Foundation for the Doctor of Zhaoqing University.

REFERENCES

Al-Allak, H.M. Russell, G.J. & Woods, J. 1987. The effect of annealing on the characteristics of a semiconducting $BaTiO_3$ positive temperature coefficient of resistance devices, *Journal of Physics D: Applied Physics.* 20: 1645–1651.

Brzozowski, E. & Castro, M.S. 2004. Influence of Nb^{5+} and Sb^{3+} dopants on the defect profile, PTCR effect and GBBL characteristics of $BaTiO_3$ ceramics, *Journal of the European Ceramic Society. Soc.* 24: 2499–2507.

Brzozowski, E. Castro, M.S. Foschini, C.R. & Stojanovic, B. 2002. Secondary phases in Nb-doped $BaTiO_3$ ceramics, *Ceramics International,* 28: 773–777.

Chan, H.M. Harmer, M.P. & Smyth, D.M. 1986. Compensating Defects in Highly Donor-Doped $BaTiO_3$, *Journal of the American Ceramic Society.* 69(6): 507–510.

Heywang, W. 1961. Barium titanate as a semiconductor with blocking layers, *Solid-State Electronics.* 3: 51–58.

Heywang, W. 1964. Resistance anomaly in doped barium titanate, *Journal of the American Ceramic Society.* 47: 484.

Ho, I.C. & Fu, S.L. 1992. Effect of reoxidation on the grain-boundary acceptor-state density of reduced $BaTiO_3$ ceramics, *Journal of the American Ceramic Society.* 15: 728–730.

Illingsworth, J. Ai-Allak, H.M. Brinkman, A.W. & Woods, A.W. 1990. The influence of Mn on the grain-boundary potential barrier characteristics of donor-doped $BaTiO_3$ ceramics, *Journal of Applied Physics,* 67: 2088–2092.

Jo, S.K. & Han, Y.H. 2007. Effects of reoxidation process on positive temperature coefficient of resistance properties of Sm-doped Ba0.85Ca0.15TiO3, *Japanese Journal of Applied Physics.* 46(3 A): 1076–1080.

Jonker, G.H. 1964. Some aspects of semiconducting barium titanate, *Solid-State Electronics.* 7: 895–903.

Kanda, A. Tashiro, S. & Igarashi, H. 1994. Effect of firing atmosphere on electrical properties of multilayer semiconducting ceramics having positive temperature coefficient of resistance and Ni-Pd internal electrodes, *Japanese Journal of Applied Physics,* 33: 5431–5434.

Kolodiazhnyi, T. & Petric, A. 2003. Effect of PO2 on bulk and grain boundary resistance of n-type $BaTiO_3$ at cryogenic temperatures, *Journal of the American Ceramic Society.* 86(9): 1554.

Langhammer, H.T. Makovec, D. Pu, Y. Abicht, H.P. & Drofenik, M. 2006. Grain boundary reoxidation of donor-doped barium titanate ceramics, *Journal of the European Ceramic Society.* 26: 2907.

Makovec, D. Ule, N. & Drofenik, M. 2001. Positive temperature coefficient of resistivity effect in highly donor–doped barium titanate, *Journal of the American Ceramic Society.* 84(6): 1273–1280.

Mancini, M.W. & Paulin Filho, P.I. 2006. Direct observation of potential barriers in semiconducting barium titanate by electric force microscopy, *Journal of Applied Physics.* 100: 104501.

Niimi, H. Hikone, A. & Omihachiman, A. 2008. U.S. Patent 7,348,873 B2.

Niimi, H. Mihara, K. Sakabe, Y. & Kuwabara, M. 2007. Preparation of multilayer semiconducting $BaTiO_3$ ceramics Co-fired with Ni inner electrodes, *Japanese Journal of Applied Physics,* 46(10 A): 6715–6718.

Nowotny, J. & Rekas, M. 1994. Defect structure, electrical properties and transport in Barium Titanate. VII. Chemical diffusion in Nb-doped $BaTiO_3$, *Ceramics International.* 20: 265–275.

Pu, Y. Chen, X. Xu, N. Wu, H. & Zhao, X. 2010. Influence of Ti/Ba ratio on the positive temperature coefficient resistance characteristics of Mn-doped semiconducting $BaTiO_3$ ceramics, *Ferroelectrics.* 403: 181–186.

Pu, Y.P. Wu, H.D. Wei, J.F. & Wang, B. 2011. Influence of doping Nb^{5+} and Mn^{2+} on the PTCR effects of Ba0.92Ca0.05 (Bi0.5 Na0.5)0.03TiO3 ceramics, *Journal of Materials Science: Materials in Electronics.* 22: 1480.

Xiang, P.H. Harinaka, H. Takeda, H. Nishida, T. Uchiyama, K. & Shiosaki, T. 2008. Annealing effects on the characteristics of high Tc lead-free barium titanate-based positive temperature coefficient of resistance ceramics, *Journal of Applied Physics.* 104: 094108.

Advanced Materials and Structural Engineering – Hu (Ed.)
© 2016 Taylor & Francis Group, London, ISBN 978-1-138-02786-2

Effects of interstitial impurity content on the plastic deformation behavior in austenitic steel monocrystals

S. Barannikova, A. Malinovsky & D. Pestsov
Institute of Strength Physics and Materials Science, SB RAS, Tomsk, Russia
Tomsk State University of Architecture and Building, Tomsk, Russia
Tomsk State University, Tomsk, Russia

ABSTRACT: The effects of interstitial impurity content (nitrogen or carbon) and of extension axis orientation were studied in single austenitic steel (γ-Fe) crystals. The shape of plastic flow curves plotted for the test samples of γ-Fe and the mechanisms involved in the deformation are discussed. Using the technique of speckle photography, the regular features of plastic deformation macro-localization were established for the tensile single γ-Fe crystals containing nitrogen or carbon as alloying additions. Thus, the distribution of local elongations over the sample extension axis was obtained for the deforming sample. The result suggests that at the macro-scale level, the plastic deformation tends to localize in the deforming sample from the yield point to the failure. It is also found that localized plasticity waves are generated in the deforming sample; the wave rate and wavelength are defined experimentally for the waves of localized plasticity. The investigation provides convincing evidence that due to the occurrence of interstitial impurity, the strength properties of steels are enhanced significantly, thereby precluding the possible adverse effect of plastic deformation localization on the material.

1 INTRODUCTION

During recent decades, a series of studies have been carried out on the plastic deformation in solids. A wealth of experimental data have been collected, which deepens the understanding of plasticity problem. It is pertinent to mention herein a few experimental and theoretical studies related to dislocation physics and deformation and flow mechanics (Lazar 2013, Kuhlmann-Wilsdorf 2002 and Rizzi & Hahner 2004). Recent findings have suggested that the deforming medium is a self-organizing system, which has not reached a thermodynamic equilibrium (Mudrock et al. 2011). The plastic flow occurring in the deforming solid is characterized by space-time periodicity (see Barannikova 2000, Danilov et al. 2003). At the macro-scale level, the plastic deformation will exhibit an inhomogeneous localization behavior from the yield point to the failure of material. Hence, the localization behavior is thought to be a unique and defining characteristic of the plastic flow process. On the basis of abundant experimental evidence, a new approach to plastic flow kinetics is developed (Zuev & Barannikova 2014).

Of particular interest are the investigations of plastic deformation localization, which were carried out on the single crystals of austenitic stainless steels. An enhancement in the strength properties

of these materials was attributed to the presence of interstitial impurity, i.e. nitrogen or carbon (Saussan & Degallaix 1991).

2 MATERIALS AND EXPERIMENTAL METHODS

The regular features of plastic deformation behavior were studied for the single FCC crystals of stainless chromium-nickel steel containing nitrogen or carbon as alloying additions. The test samples of γ-Fe$_I$ contained 0.35% or 0.5% N; the test samples of γ-Fe$_{II}$ contained 0.93% or 1.03% C (Table 1). The experiments were conducted for the test samples of stainless steel Fe-18%Cr-12%Ni (by mass percentage). The deformation occurred in these samples via the mechanism of dislocation glide. Thus, the solid γ-Fe solution would form no phase of its own

Table 1. Chemical composition of the studied steels.

Material	Content [wt%]					
	Cr	Ni	Mo	Mn	C	Si
γ-Fe$_I$	18.0	12.4	2.3	1.2	0.013	0.06
γ-Fe$_{II}$	–	–	–	13.0	> 0.900	–

since nitrogen impurity was placed in interstitial sites in the lattice. The test samples had a longitudinal axis oriented along the [$\bar{1}$11] direction; the sample work face had an index (110). In view of the above, the slip systems (111) [10$\bar{1}$], ($\bar{1}$11) [011] and [$\bar{1}$0$\bar{1}$] (111) had the same Schmidt factor value, m = 0.27. The hardening of the material was due to the high content of nitrogen. With the increasing stress level, the fraction of extended dislocations would grow in the same material. Hence, in the early stage of deformation, the preferential glide plane is the system ($\bar{1}$11) [011]. The test sample of Fe-13%Mn alloy contained ~1% C; it had a gauge having the index (011) and extension axis orientations [$\bar{3}$77], [$\bar{3}$55], [$\bar{1}$23] or [$\bar{1}$11]. The test sample of single Fe-13%Mn crystal contained ~1% C; it had the extension axis orientation [377]. The latter sample was tested in tension at room temperature. As soon as the yield point was attained, the deformation occurred in the sample via a twinning mechanism in the system (111)[$\bar{2}$11] (Karaman et al. 1998).

The investigation was carried out on the single crystals of stainless Cr-Ni steel (γ-Fe) containing nitrogen or carbon as alloying additions. The test samples had a gauge of dimension 30 × 5 × 1 mm. These were tested in tension in an 'Instron-1185' testing machine; the movable grip had a constant rate, V_m = 0.2 mm/min, and the temperature was 300 K. The investigations of plastic flow macro-localization were performed by the method of speckle photography (Jones & Wykes 1983), which enabled us to record the displacement vector field, $R(x, y)$, for the sample surface. Numerical differentiation of data was performed in the x- and y co-ordinates. As a result, the space-time distributions of plastic distortion tensor components, $\varepsilon_{xx}(x, y, t)$, $\varepsilon_{xy}(x, y, t)$ and $\omega_z(x, y, t)$, were obtained for the plane stressed state. These were matched against the flow stages, which were revealed on the curve σ(ε) plotted for the test sample (Barannikova 2000, Danilov et al. 2003, Zuev & Barannikova 2014). An analysis was performed for the spatial patterns $\varepsilon_{xx}(x, y)$, which enabled us to define the distributions of local strain zones (Fig. 1). The kinetics of macro-localization patterns was examined using the local nuclei's positions, X, as a function of time, t (Fig. 2, a).

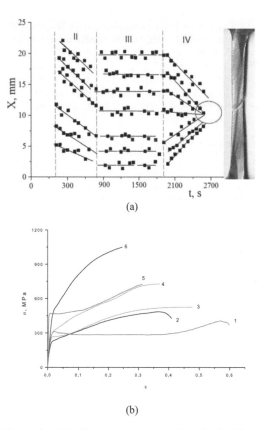

(a)

(b)

Figure 2. The diagrams X-t plotted for the localized plasticity zones, which were observed at plastic flow stages II–IV for the test sample of γ-Fe$_I$ (0.35% N); [111] (a); Designations: ■—nucleus position, X vs t. The flow curves plotted for the single crystals having different orientations (b): 1—γ-Fe$_{II}$ (0.93%C); [$\bar{3}$77]; 2—γ-Fe$_I$ (0.35%N); [001]; 3—γ-Fe$_I$ (0.35% N); [111]; 4—γ-Fe$_I$ (0.5%N); [111]; 5—γ-Fe$_{II}$ (1.03% C); [$\bar{3}$55] 6—γ-Fe$_I$ (0.5% N); [001].

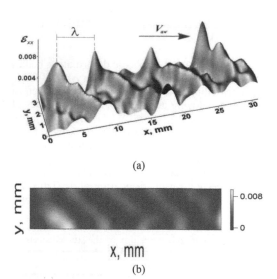

(a)

(b)

Figure 1. (a) A set of active localization zones, ε_{xx}, moving in a concerted manner along the extension axis. (b) A halftone photographic image of strain distribution, ε_{xx}, observed for the sample of high manganese austenite steel containing 1.03% C [$\bar{1}$11].

The test samples were cut out from single crystals grown by the Bridgman method in the atmosphere of inert gas. The samples had a dog-bone shape; these were 1.3–1.5 mm thick; their gauge had dimensions of 28 × 5 mm. The samples were tested in tension on a universal test machine, 'Instron-1185', at a constant rate at room temperature. The loading curves plotted for the test samples are illustrated in Figure 2, b. The test samples had different nitrogen and carbon contents; these also differed in the extension axis orientation. The above factors determined the type of shearing, i.e. single or multiple slip, and the deformation mechanisms involved, i.e. dislocation glide or twinning (Barannikova 2000, Danilov et al. 2003).

3 RESULTS AND DISCUSSION

Steels containing alloying additions, e.g. nitrogen or carbon, are known to have enhanced strength characteristics. It is found that the general behavior of plastic flow curves plotted for such steels is different (Fig. 2, b). The effect of the extension axis orientation on the mechanical properties of steels was also studied. It is found that the shape of the flow curve, the amount of straining at the respective flow stage and the work hardening coefficient, θ, also depend on the extension axis orientation in steels (Barannikova 2000).

The single γ-Fe$_I$ crystals oriented in the [001] or [$\bar{1}$11] direction were tested in tension. These steels are found to deform by dislocation shear; hence, the likelihood of multiple slips in these materials is also high. In the case of single γ-Fe$_{II}$ crystals oriented in the [$\bar{3}$77], [355] or [$\bar{1}$11] direction, the condition for the Schmidt factor ratio $m_{tw}/m_{gl} > 1$ is satisfied (Karaman et al. 1998). The plastic deformation generally occurs in these materials at room temperature via a twinning mechanism.

Using the technique of speckle photography, the local strains were measured for the macro-localization zones observed for the single austenitic steel crystals (Fig. 1). The analysis of speckle photography data indicates that the plastic deformation tends to localize in equidistant zones arranged on the sample surface at intervals $\lambda \approx 3...10$ mm. At the stage of linear strain hardening (stage II), the zones travel at the velocity $V_{aw} \approx 10^{-5}...10^{-4}$ m/s; at the stage of parabolic strain hardening (stage III), the equidistant localization zones would become stationary (Fig. 2, a). At stage IV, the localization zones start moving in a concerted manner to merge finally together at a point where a failure of material will occur (Fig. 2, a).

The plastic flow process is known to progress in a stepwise fashion. The deformation curve of the material will appear at several stages. Thus, all the deformation curves plotted for the test samples of single γ-Fe crystals have a linear straining stage, regardless of the composition of the material or the mechanism involved in the deformation. At the linear straining stage (stage II), the plastic flow tends to localize at the macro-scale level. As a result, an arrangement of localized plasticity zones would emerge along the sample extension axis (Fig. 1). These would travel in a concerted manner along the sample extension axis at a constant rate of $V_{aw} \approx 10^{-5}...10^{-4}$ m/s. The mobile zones are spaced at regular intervals, $\lambda \approx 3...10$ mm. Hence, an arrangement of mobile localization traveling along the deforming sample at the linear straining stage may be regarded as a specific wave process involved in the plastic deformation.

On the basis of experimental evidence, obtained for the test samples of single γ-Fe$_I$ and γ-Fe$_{II}$ crystals, the following dependence is established; the velocity of active zones, V_{aw}, is a function of the work hardening coefficient, θ. This dependence holds for stage II (Fig. 3, a); it has the form

$$V_{aw}(\theta) = V_0 + \frac{\Xi}{\theta_*}, \qquad (1)$$

where V_0 and Ξ are constants and $\theta_* = \theta/G$ (here, G is the shear modulus); hence, $V_{aw} \sim \theta^{-1}$.

The physical sense of dependence (1) has been discussed in detail elsewhere (Zuev & Barannikova 2014). The form of dependence $V_{aw}(\theta)$ was generalized for a wide range of materials, both single crystals and polycrystals, which differ in the crystal lattice type (FCC, BCC or HCP) and in the mechanisms involved in the deformation, i.e. dislocation glide, twinning or martensitic transformation-induced plasticity (Barannikova 2000, Danilov et al. 2003, Zuev & Barannikova 2014).

Next, the characteristics of the waves in question are matched against those of Kolsky's waves, i.e. plasticity waves (Kolsky 1963). Let us consider the propagation rate of plasticity waves, i.e. $V_{pw} \approx \sqrt{\theta/\rho_0} \sim \sqrt{\theta}$ (here, ρ_0 is the material density). Apparently, the latter dependence differs fundamentally from that established for wave processes involved in the plastic deformation, i.e. $V_{aw} \sim \theta^{-1}$. It is thus contended that a new type of wave processes has been discovered, which are called 'localized plasticity waves'.

It was found previously (Zuev & Barannikova 2014) that the wavelength of waves in question depends on the grain size, D, the sample length, L, and the characteristic size of the dislocation substructure, $\bar{d}$. The power dependence of the wavelength, $\lambda(\sigma)$, is discussed below.

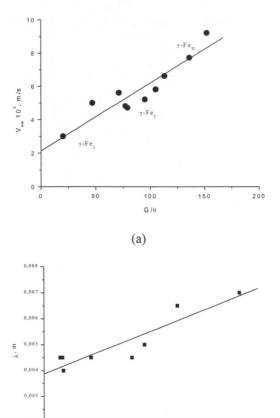

(a)

(b)

Figure 3. Dependence of the localization zones' motion rate on the dimensionless work hardening coefficient θ/G (a) and dependence of the wavelength on the averaged stress level (b) at stage II in the γ-Fe$_I$ and γ-Fe$_{II}$ monocrystals.

On the basis of the experimental data, it was established that the wavelength, λ, depends on the averaged stress level as

$$\lambda = \lambda_0 + \frac{\Omega}{\sigma_*}, \qquad (2)$$

where λ_0 and Ω are constants and $\sigma_* = \langle\sigma\rangle/G$ (here, G is the shear modulus), i.e. $\lambda \sim 1/\langle\sigma\rangle$ (Fig. 3, b).

An enhancement in the strength properties of single γ-Fe$_I$ and γ-Fe$_{II}$ crystals is due to the presence of an interstitial impurity in these materials. It is found that the strength properties of austenitic steels containing nitrogen as alloying additions are

determined by the orientation of single crystals [Karaman et al. 1998]. Thus, the 'soft' single crystals oriented in the direction [$\bar{1}$11] deform via the mechanism of sliding; hence, the level of critical cleaving stresses in these single crystals is significantly lower relative to the 'rigid' single crystals oriented in the direction [001]. From the onset of yielding, the single Hadfield steel crystals oriented in the directions [$\bar{1}$11] and [$\bar{3}$77] will deform by twinning; the level of critical cleaving stresses in the former is lower relative to the single Hadfield steel crystals, which deform by sliding in the direction [012].

Due to the effect of external stresses, splitting of pure dislocations occurs in the single crystals oriented in the [$\bar{1}$11] direction to yield Shockley dislocations, which interact with interstitial impurity atoms. In the single crystals oriented in the [001] direction, perfect dislocations $a/2 < 110 >$ interact with interstitial impurity atoms (Karaman et al. 1998). Dependence (2) suggests that in the case of the 'rigid' single γ-Fe$_I$ crystals oriented along the [001] direction, the localized plasticity wave has the length $\lambda \approx 4$ mm; in the case of the 'soft' single γ-Fe$_I$ crystals oriented in the [$\bar{1}$11] direction, the localized plasticity wave has the length $\lambda \approx 7$ mm. Hence, $\lambda \sim 1/\langle\sigma\rangle$.

4 CONCLUSIONS

Using the single crystals of high-nitrogen and high-manganese steels, the regular main features are established for the evolution of the deformation localization wave process.

i. On the basis of experimental evidence, the characteristics of localized plasticity waves have been defined.
ii. It is also found that the strength characteristics of steels are determined by the interstitial impurity content.
iii. The propagation rate of localized plasticity waves is found to be inversely proportional to the coefficient of work hardening; it also depends on the interstitial impurity content.
iv. The localized plasticity wave is found to have the length $5 \leq \lambda \leq 10$ mm, which is inversely proportional to the averaged stress level obtained for the linear strain hardening stage, i.e. $\lambda \sim 1/\langle\sigma\rangle$. The value λ is found to depend on the geometry of a single crystal.

On the basis of new experimental evidence, specific wave processes involved in the plastic deformation have been discovered. The studies were carried out for high-nitrogen and high-manganese steels. An enhancement in the strength properties of steels is found to be due to the impurity content.

The regular features of strain material hardening and the micro-mechanisms involved in the deformation are also discussed.

ACKNOWLEDGMENT

This work was performed in the framework of the Tomsk State University Academic D.I. Mendeleev Fund Program and the Program for Fundamental Research of State Academies of Sciences for the period 2013–2020.

REFERENCES

Barannikova, S.A. 2000. Localization of stretching strain in doped carbon gamma-Fe single crystals, *Technical Physics,* 45: 1368–1370.

Danilov, V.I. Barannikova, S.A. & Zuev, L.B. 2003. Localized strain autowaves at the initial stages of plastic flow in single crystals, *Technical Physics,* 48: 1429–1435.

Jones, R. & Wykes, R. 1983. *Holographic and Speckle Interferometry,* Cambridge University Press, Cambridge.

Karaman, I. Sehitoglu, I. & Chumlyakov, Y.I. 1998. On the deformation mechanisms in single crystal Hadfield manganese steel, *Scripta Materialia.* 38: 1009–1015.

Kolsky, H. 1963. *Stress waves in solids,* Dover, New York.

Kuhlmann-Wilsdorf, D. 2002. *The low energetic structures theory of solid plasticity*, in Dislocations in Solids, Elsevier, Amsterdam, 213–338.

Lazar, M. 2013. On the non-uniform motion of dislocations: the retarded elastic fields, the retarded dislocation tensor potentials and the Lienard-Wiechert tensor potentials, *Philosophical Magazine A,* 93: 749–776.

Mudrock, R.N. Lebyodkin, M.A. Kurath, P. Beaudoin, A. & Lebedkina, T.A. 2011. Strain-rate fluctuations during macroscopically uniform deformation of a solid strengthened alloy, *Scripta Materialia.* 65: 1093–1095.

Rizzi, E. & Hähner, P. 2004. On the Portevin-Le Chatelier effect: theoretical modeling and numerical results, *International Journal of Plasticity,* 20: 121–165.

Saussan, A. & Degallaix, S. 1991. Work-hardening behavior of nitrogen-alloyed austenitic stainless steels, *Materials Science and Engineering: A,* 142: 169–176.

Zuev, L.B. & Barannikova, S.A. 2014. Experimental study of plastic flow macro-scale localization process: pattern, propagation rate, dispersion, *International Journal of Mechanical Sciences,* 88: 1–8.

Advanced Materials and Structural Engineering – Hu (Ed.)
© *2016 Taylor & Francis Group, London, ISBN 978-1-138-02786-2*

An experimental method of the cutting force coefficient estimation of grey cast iron FC25

N.T. Nguyen, M.S. Chen & S.C. Huang
Department of Mechanical Engineering, National Kaohsiung University of Applied Sciences, Taiwan, R.O.C.

Y.C. Kao
Advanced Institute of Manufacturing with Hi-Tech Innovations, National Chung Cheng University, Taiwan, R.O.C.

ABSTRACT: In this paper, the linear force model was applied to investigate the cutting force coefficients for milling processes. This study was performed in HSS flat-end mill cutter and grey cast iron FC25 work-piece by three cutting types such as half-up, half-down, and slotting. By performing the milling tests at stable cutting condition, all components of cutting force coefficients were determined from experimental data. The developed cutting force calculation model has been modeled and successfully verified by both simulation and experiment with very promising results. The best value of each cutting force coefficient was used to predict the cutting forces and can be applied in the development of machine tool in industrial manufacturing. This method can be applied to determine the cutting force coefficients of the flat-end mill for every pair of tool and work-piece material.

1 INTRODUCTION

In theories of metal-cutting mechanics, the cutting mechanics can be analyzed by orthogonal and oblique models that have been researched in many studies. The procedure of cutting force modeling is generally realized by developing the experiential chip-force relationship through the cutting force coefficients. In the traditional mechanistic approach, edge force and shear force coefficients are calibrated for different pairs of the cutter and work-piece through the cutting tests. There were two methods for the calibration of cutting force coefficients.

In the first method, the shear angle, friction angle and shear yield stress resulted from orthogonal cutting test were used to estimate the cutting force coefficients, (Altintas 2012, Budak et al. 1996). The cutting force coefficients were calculated from the oblique cutting model, (Li et al. 2004). In the second method, the cutting force coefficients were determined directly from cutting tests. The instantaneous cutting force coefficients were determined depending on the instantaneous uncut chip thickness, (Cheng et al. 1997). The measurement of cutter deflection was used to calculate the cutting force coefficients, (Larue & Anselmetti 2003). The cutting force coefficients were estimated by considering the instantaneous cutting force, (Azeem et al. 2004). Besides, by using the measured average cutting forces, the cutting force coefficients were also determined, (Wang et al. 2014).

Two models are often used to calculate the cutting force coefficients in direct calibration method. First one, the cutting force coefficients are relatively dependent on the average chip thickness. By using this model, some researchers mentioned in the calculation method of cutting force coefficients, (Wan et al. 2012, Wan et al. 2010). In the second model, the cutting force coefficients are relatively dependent of the average cutting force (linear-force model). This model is quite suitable to be applied to many types of milling tool such as the flat-end mill (Wang et al. 2014), ball-end mill, (Narita 2013), and the general-end mill, (Gradišek et al. 2004).

By analysis of the effect of cutter's helix angle on the cutting force coefficient, the linear force model was used to determine the cutting force coefficients for the flat-end mill, (Kao et al. 2015). The authors performed the cutting test in only one milling type that is half-down milling; so, the cutting tests in other cutting types as half-up and slotting were not carried out, and the cutting force coefficient values were not compared and evaluated in all cutting types.

In this study, the cutting force coefficients were determined through the linear force model of the flat-end mill. The main contributions of this study lie in two aspects: (1) Completing the method to determine and evaluate the cutting force coefficients and (2) modeling and verifying the cutting forces in flat-end mill processes.

2 IDENTIFICATION OF CUTTING FORCE COEFFICIENTS AND MODELING CUTTING FORCE

2.1 Identification of cutting force coefficients

In order to determine cutting force coefficients in flat-end mill processes, the linear force is applied. The cutting force coefficients were calculated from the experimental data. The detail of this method is described clearly in (Kao et al. 2015). By this method, in flat-end mill processes, the cutting force coefficients were determined by Equation (1).

$$\begin{cases} K_{tc} = \dfrac{C_1\bar{F}_{fc} - C_2\bar{F}_{nc}}{C_1^2 + C_2^2} \quad K_{te} = \dfrac{C_3\bar{F}_{fe} - C_4\bar{F}_{ne}}{C_3^2 + C_4^2} \\[2mm] K_{rc} = \dfrac{C_2\bar{F}_{fc} + C_1\bar{F}_{nc}}{C_1^2 + C_2^2} \quad K_{re} = \dfrac{C_4\bar{F}_{fe} + C_3\bar{F}_{ne}}{C_3^2 + C_4^2} \\[2mm] K_{ac} = -\dfrac{\bar{F}_{ac}}{C_4} \quad K_{ae} = \dfrac{\bar{F}_{ae}}{C_5} \end{cases} \quad (1)$$

where, $\bar{F}_{fc}, \bar{F}_{fe}, \bar{F}_{nc}, \bar{F}_{ne}, \bar{F}_{ac}$, and $\bar{F}_{ae}$ are the designate constants of the linear force model that can be calculated by a linear regression of the average of cutting forces as expressed by Equation (2).

$$\begin{cases} \bar{F}_f = \bar{F}_{fc}f_t + \bar{F}_{fe} \\ \bar{F}_n = \bar{F}_{nc}f_t + \bar{F}_{ne} \\ \bar{F}_a = \bar{F}_{ac}f_t + \bar{F}_{ae} \end{cases} \quad (2)$$

with, $\bar{F}_f$, $\bar{F}_n$, and $\bar{F}_a$ are the average cutting forces in feed, normal, and axial directions, respectively. And, C_1, C_2, C_3, C_4, C_5 are the setting constants that can be calculated by Equation (3).

2.2 Modeling of cutting forces

In the flat-end mill, the immersion is measured clockwise from the normal axis. Assuming that the bottom end of flute number 1 is designated as the reference immersion angle ϕ_1 and the bottom end point of the remaining flute number j is at an angle ϕ_j as shown in Figure 1. Then ϕ_j can be expressed as in Equation (4).

$$\phi_j = \phi_1 - (j-1)\phi_P, \quad j = 1 \sim N_f \quad (4)$$

The lag angle $\Psi_j(z)$ at each axial depth of cut z, can be expressed in Equation (5), (Kao et al. 2015).

$$\Psi_j(z) = \frac{2\tan\beta}{D}z \quad (5)$$

For, the immersion angle of flute number j, at an axial depth of cut z can be expressed by Equation (6).

$$\phi_j(z) = \phi_j - \Psi_j(z) = \phi_1 - (j-1)\phi_P - \frac{2\tan\beta}{D}z \quad (6)$$

Assuming that the nose radius of the cutter is zero, for flute number j, at the rotation angle ϕ_j, the tangential, radial, and axial forces acting on a differential flute element can be expressed as in Equation (7), (Kao et al. 2015).

$$\begin{cases} dF_{t,j}(\phi_j, z) = K_{te} * dz + K_{tc} * h_j(\phi_j(z)) * dz \\ dF_{r,j}(\phi_j, z) = K_{re} * dz + K_{rc} * h_j(\phi_j(z)) * dz \\ dF_{a,j}(\phi_j, z) = K_{ae} * dz + K_{ac} * h_j(\phi_j(z)) * dz \end{cases} \quad (7)$$

$$\begin{cases} C_1 = \dfrac{N_f}{4\pi}\left(-\int_{\phi_{st}}^{\phi_{st}+\Psi_a}\left[\int_0^{\frac{D}{2\tan\beta}(\phi-\phi_{st})}(\sin 2\phi_j(z))dz\right]d\phi - \int_{\phi_{st}+\Psi_a}^{\phi_{ex}}\left[\int_0^a(\sin 2\phi_j(z))dz\right]d\phi - \int_{\phi_{ex}}^{\phi_{ex}+\Psi_a}\left[\int_{\frac{D}{2\tan\beta}(\phi-\phi_{ex})}^a(\sin 2\phi_j(z))dz\right]d\phi\right) \\[4mm]
C_2 = \dfrac{N_f}{4\pi}\left(-\int_{\phi_{st}}^{\phi_{st}+\Psi_a}\left[\int_0^{\frac{D}{2\tan\beta}(\phi-\phi_{st})}(1-\cos 2\phi_j(z))dz\right]d\phi - \int_{\phi_{st}+\Psi_a}^{\phi_{ex}}\left[\int_0^a(1-\cos 2\phi_j(z))dz\right]d\phi - \int_{\phi_{ex}}^{\phi_{ex}+\Psi_a}\left[\int_{\frac{D}{2\tan\beta}(\phi-\phi_{ex})}^a(1-\cos 2\phi_j(z))dz\right]d\phi\right) \\[4mm]
C_3 = \dfrac{N_f}{2\pi}\left(-\int_{\phi_{st}}^{\phi_{st}+\Psi_a}\left[\int_0^{\frac{D}{2\tan\beta}(\phi-\phi_{st})}(\cos\phi_j(z))dz\right]d\phi - \int_{\phi_{st}+\Psi_a}^{\phi_{ex}}\left[\int_0^a(\cos\phi_j(z))dz\right]d\phi - \int_{\phi_{ex}}^{\phi_{ex}+\Psi_a}\left[\int_{\frac{D}{2\tan\beta}(\phi-\phi_{ex})}^a(\cos\phi_j(z))dz\right]d\phi\right) \\[4mm]
C_4 = \dfrac{N_f}{2\pi}\left(-\int_{\phi_{st}}^{\phi_{st}+\Psi_a}\left[\int_0^{\frac{D}{2\tan\beta}(\phi-\phi_{st})}(\sin\phi_j(z))dz\right]d\phi - \int_{\phi_{st}+\Psi_a}^{\phi_{ex}}\left[\int_0^a(\sin\phi_j(z))dz\right]d\phi - \int_{\phi_{ex}}^{\phi_{ex}+\Psi_a}\left[\int_{\frac{D}{2\tan\beta}(\phi-\phi_{ex})}^a(\sin\phi_j(z))dz\right]d\phi\right) \\[4mm]
C_5 = \dfrac{N_f}{2\pi}\left(\int_{\phi_{st}}^{\phi_{st}+\Psi_a}\left[\int_0^{\frac{D}{2\tan\beta}(\phi-\phi_{st})}dz\right]d\phi + \int_{\phi_{st}+\Psi_a}^{\phi_{ex}}\left[\int_0^a dz\right]d\phi + \int_{\phi_{ex}}^{\phi_{ex}+\Psi_a}\left[\int_{\frac{D}{2\tan\beta}(\phi-\phi_{ex})}^a dz\right]d\phi\right) \end{cases}$$

$$(3)$$

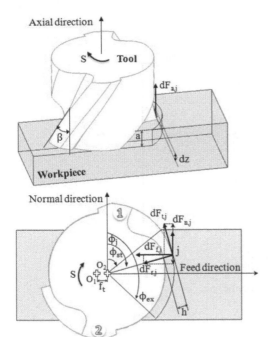

Figure 1. Flat-end mill processing.

where, the instantaneous chip thickness h_j (ϕ_j (z)) that was expressed as in Figure 1 and can be calculated by Equation (8).

$$
\begin{cases}
dF_{t,j}(\phi_j,z) = K_{te} * dz + K_{tc} * h_j(\phi_j(z)) * dz \\
dF_{r,j}(\phi_j,z) = K_{re} * dz + K_{rc} * h_j(\phi_j(z)) * dz \\
dF_{a,j}(\phi_j,z) = K_{ae} * dz + K_{ac} * h_j(\phi_j(z)) * dz
\end{cases} \quad (8)
$$

At each point in cutting edge, the cutting forces consist of three components, including radial force, tangential force than can be calculated by using the transformation as in Equation (9).

$$
\begin{Bmatrix}
dF_{f,j}(\phi_j,z) \\
dF_{n,j}(\phi_j,z) \\
dF_{a,j}(\phi_j,z)
\end{Bmatrix}
$$
$$
=
\begin{bmatrix}
-\cos(\phi_j(z)) & -\sin(\phi_j(z)) & 0 \\
\sin(\phi_j(z)) & -\cos(\phi_j(z)) & 0 \\
0 & 0 & 1
\end{bmatrix}
\begin{Bmatrix}
dF_{t,j}(\phi_j,z) \\
dF_{r,j}(\phi_j,z) \\
dF_{a,j}(\phi_j,z)
\end{Bmatrix} \quad (9)
$$

The differential cutting forces are integrated analytically along the in-cut portion of the flute number j. So, the total cutting forces in flute number j can be calculated by Equation (10).

$$
F_{q,j}(\phi_j) = \int_{z_1(\phi_j)}^{z_2(\phi_j)} dF_{q,j}(\phi_j,z), \quad q = f, n, a \quad (10)
$$

Considering the case that having more than one flute executing the cutting processes simultaneously, the total cutting forces on the feed, normal, and axial direction can be determined by Equation (11).

$$
\begin{cases}
F_f(\phi_j) = \sum_{j=1}^{N_f} F_{f,j}(\phi_j) \\
F_n(\phi_j) = \sum_{j=1}^{N_f} F_{n,j}(\phi_j) \\
F_a(\phi_j) = \sum_{j=1}^{N_f} F_{a,j}(\phi_j)
\end{cases} \quad (11)
$$

3 EXPERIMENTAL METHOD

3.1 Experimental procedures

The research procedure was performed by a scheme as shown in Figure 2. Firstly, the cutting depth and spindle speed were selected as the stable cutting condition through the cutting tests and the evaluation of the cutting force's stable. This section was detailed in the study (Kao et al. 2015).

By this method, the effect of vibration, chatter, and other factors on milling process was reduced. Secondly, the cutting tests were performed in three types of milling (half-up, half-down, and slotting). For each cutting type, the experiment was carried out with variations of feed per flute to determine the cutting force coefficients. Thirdly, the cutting force coefficients were calculated and evaluated from the experimental data. Finally, the predicted cutting forces were verified by the measured cutting forces.

3.2 Setup of the experiment

A series of flat-end mill experiments were performed to determine the cutting force coefficients. The cutter was chosen as follows. Cutter: a HSS flat-end mill with number of flute $N_f = 2$, helix angle $\beta = 30°$, rake angle $\alpha_r = 5°$, and the diameter was 10 mm.

The work-piece material was grey cast iron FC25 (Ferrum Casting). The compositions of FC25 are listed in Table 1 and the properties of the FC25 were the following: hardness 197-269 HB, Young's modulus = 109.5 GPa, Poisson's ratio = 0.29, tensile strength = 250 MPa.

A Dynamometer (Type: XYZ FORCE SENSOR, Model: 624-120-5 KN), signal filter and processing system, and a PC were used to measure cutting forces. The experiments were performed at a Three-axis vertical milling center (Tongtai TMV-720A). The detail is illustrated in Figure 3.

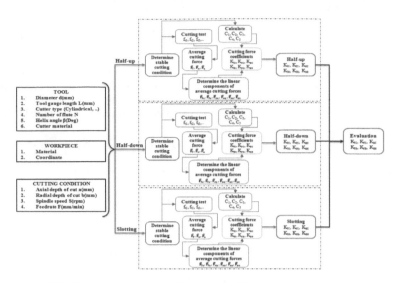

Figure 2. Approach to identify and evaluate the cutting force coefficients.

Table 1. Chemical compositions of grey cast iron FC25.

	Composite (%)					
	C	Mn	Si	P	S	Fe
Min	2.44	0.39	1.83	0.15	–	
Max	3.02	0.52	2.03	0.30	0.15	Balance

a. CNC machine b. Dynamometer
c. Signal filter and processing system
d. PC and Display System

Figure 3. Setup of cutting force measurement.

Table 2. Cutting conditions in experimental plan.

Test no	Cutting type	a [mm]	Spindle speed [rpm]	Feed per flute [mm/flute]
1	Half-up	0.5	1000	0.05
2	Half-up	0.5	1000	0.10
3	Half-up	0.5	1000	0.15
4	Half-down	0.5	1000	0.05
5	Half-down	0.5	1000	0.10
6	Half-down	0.5	1000	0.15
7	Slotting	1.0	1000	0.05
8	Slotting	1.0	1000	0.10
9	Slotting	1.0	1000	0.15

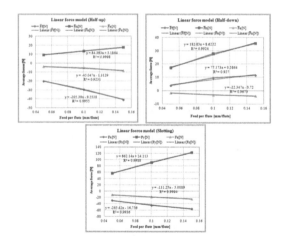

Figure 4. The average cutting forces versus feed per flutes.

The experiments were carried out in three cutting types: half-up, half-down, and slotting. At each depth of cut, the experiments were repeated with variations of feed per flute. The spindle speed was held constant at each experiment. The experiments were performed with the different cutting conditions as shown in Table 2.

4 EXPERIMENTAL RESULTS AND DISCUSSIONS

4.1 Verification of linear force model

By the measured cutting force data, the average cutting force in feed, normal, and axial directions

196

Table 3. Calculation values of cutting force coefficients.

Group	Note	Shearing force coefficient [N/mm²]			Edge force coefficient [N/mm]		
		K_{tc}	K_{rc}	K_{ac}	K_{te}	K_{re}	K_{ae}
1	Half-up	1224.813	865.182	−275.248	45.899	15.054	−5.452
2	Half-down	1384.610	282.901	−144.666	28.193	26.073	−2.990
3	Slotting	1324.712	531.013	−206.626	36.315	28.478	−5.200
4	Average values	1311.378	559.699	−208.847	36.802	23.202	−4.547
5	With maximum absolute values	1384.610	865.182	−275.248	45.899	28.478	−5.452

were determined for each value of feed per flute. The relationship of the average cutting force and the feed per flute was estimated and illustrated in Figure 4. This figure showed that all the absolute values of average cutting forces increase with the increasing of feed per flute for all milling types (half-up, half-down, and slotting). In all milling types, the relationship between average cutting forces and feed per flute are very close to the linear function. Therefore, in all milling types of flat end-mill tool, the measured average cutting forces can be expressed by the linear function of feed per flute. The measured data of average cutting forces can be used to estimate the cutting force coefficients for each pair of tool and work-piece.

4.2 Determination of cutting force coefficients

Using Fitting Toolbox of MATLAB software and using linear regression, the best lines passing through the values of average cutting forces were determined and the designate constants as $\bar{F}_{fc}$, $\bar{F}_{fe}$, $\bar{F}_{nc}$, $\bar{F}_{ne}$, $\bar{F}_{ac}$, and $\bar{F}_{ae}$ of the best lines were estimated. Besides, in this study, the constants C1 to C5 were calculated by Equation 3; so, all cutting force coefficient components can be determined by Equation 1. Five cutting force coefficient groups were calculated and listed as in Table 3.

4.3 Verification of the cutting model and evaluation of the cutting force coefficient values

Using five groups of cutting force coefficient, the cutting forces were predicted and compared with the measured results. The predicted cutting forces and measured cutting forces were shown in Figure 5. For all groups of cutting force coefficient, the predicted cutting forces were quite close to the measured cutting forces.

In all directions, the predicted cutting forces with cutting force coefficients of group 5 were the most reasonable predicted results of cutting forces because the amplitudes of those predicted cutting forces are very close to the amplitudes of meas-

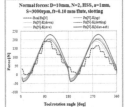
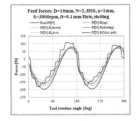
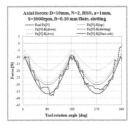

Figure 5. Comparison of real cutting force and simulation cutting forces.

ured cutting forces. Besides, the borders of those predicted cutting forces cover almost the measured cutting forces. The predicted results from research model agree satisfactorily with experimental results. Therefore, the cutting force models and cutting force coefficient values in this study can be used to predict the cutting forces and other machining characteristics in milling processes.

By verification of the cutting force model, the calculated cutting force coefficients were compared and evaluated. The verified results showed that the cutting force coefficients in group 5 are the best values to predict the cutting forces when milling the grey cast iron FC25 work-piece by HSS flat-end mill cutter.

Although exiting some different points between predicted cutting forces and measured forces, the differences are quite small. The reasons for the above differences were mostly originated from the noise, the temperature, the friction, the deflection, the inconstancy of cutting depth, the inhomogeneous distribution of tool and work-piece hardness, and so on.

5 CONCLUSIONS

In this study, through experimental method, the linear force model was used to determine the cutting force coefficients for HSS flat-end mill tool and the grey cast iron FC25 work-piece.

The cutting force coefficients that were formed from the maximum absolute values of cutting force coefficients are the best values to predict the cutting forces. This method can be applied to determine cutting force coefficients for every pair of tool and work-piece material in the flat-end mill processes.

The predicted force model has been successfully verified by both the amplitude and the shape of cutting forces. Therefore, this research results can be used to analyze the machine tool development, milling simulation, milling operation optimization in industrial manufacturing.

This research model can be extended to more complex type of milling tool such as the ball-end mill, the bull-end mill, etc. and will be the futuristic study of the extended research.

NOMENCLATURE

D = tool diameter [mm]
Nf = number of flutes on the cutter.
β = helix angle on the cutter [deg].
ϕ_{st} = cutter entry angle [deg].
ϕ_{ex} = cutter exit angle [deg].
ϕ_{j} = instantaneous immersion angle [deg].
a = maximum axial depth of cut [mm]
dz = differential axial depth of cut [mm]
$z_1 (\phi_j)$ = lower axial engagement limit of the in-cut portion of the flute number j [mm]
$z_2 (\phi_j)$ = upper axial engagement limit of the in-cut portion of the flute number j [mm]
Ψ_a = the maximum lag angle [deg]
f_t = feed per flute [mm/flute]
K_{tc} = tangential shearing force coefficient [N/mm^2]
K_{rc} = radial shearing force coefficient [N/mm^2]
K_{ac} = axial shearing force coefficient [N/mm^2]
K_{te} = tangential edge force coefficient [N/mm]
K_{re} = radial edge force coefficient [N/mm]
K_{ae} = axial edge force coefficient [N/mm]
$h_j (\phi_j)$ = instantaneous chip thickness [mm]
$dF_{t,j} (\phi_j,z)$ = differential tangential cutting force [N]
$dF_{r,j} (\phi_j,z)$ = differential radial cutting force [N]
$dF_{a,j} (\phi_j,z)$ = differential axial cutting force [N]
$dF_{f,j} (\phi_j,z)$ = differential feed cutting force [N]
$dF_{n,j} (\phi_j,z)$ = differential normal cutting force [N]
$F_{t,j} (\phi_j)$ = total tangential cutting force in flute number j [N]
$F_{r,j} (\phi_j)$ = total radial cutting force in flute number j [N]

$F_{a,j} (\phi_j)$ = total axial cutting force in flute number j [N]
$F_{f,j} (\phi_j)$ = total feed cutting force in flute number j [N]
$F_{n,j} (\phi_j)$ = total normal cutting force in flute number j [N]
$F_f (\phi_j)$ = total feed cutting force of N_f flutes [N]
$F_n (\phi_j)$ = total normal cutting force of N_f flutes [N]
$F_a (\phi_j)$ = total axial cutting force of N_f flutes [N].

REFERENCES

Altintas, Y. 2012. *Manufacturing Automation: Metal cutting mechanics, machine tool vibrations, and CNC design,* Cambridge University Press, 2nd ed, ISBN 978-1-00148-0.

Azeem, A. Feng, H.Y. & Wang, L. 2004. Simplified and efficient calibration of a mechanistic cutting force model for ball-end milling, *International Journal of Machine Tools & Manufacture,* 44(2): 291–298.

Budak, E. Altintas, Y. & Armarego, E.J.A. 1996. Prediction of milling force coefficients from orthogonal cutting data, *Journal of Manufacturing Science and Engineering,* 118(2): 216–224.

Cheng, P.J. Tsay, J.T. & Lin, S.C. 1997. A study on instantaneous cutting force coefficients in face milling, International *Journal of Machine Tools & Manufacture,* 37(10): 1393–1408.

Gradišek, J. Kalveram, M. & Weinert, K. 2004. Mechanistic identification of specific force coefficients for a general end mill, *International Journal of Machine Tools & Manufacture,* 44(4): 401–414.

Kao, Y.C. Nguyen, N.T. Chen, M.S. & Su, S.T. 2015. A prediction method of cutting force coefficients with helix angle of flat-end cutter and its application in a virtual three-axis milling simulation system, *The International Journal of Advanced Manufacturing Technology,* 77(9–12): 1793–1809.

Larue, A. & Anselmetti, B. 2003. Deviation of a machined surface in flank milling, *International Journal of Machine Tools & Manufacture,* 43(2): 129–138.

Li, X.P. & Li, H.Z. 2004. Theoretical modeling of cutting forces in helical end milling with cutter run-out, *International Journal of Mechanical Sciences,* 46(9): 1399–1414.

Narita, H. 2013. A Determination Method of Cutting Coefficients in Ball End Milling Forces Model, *International Journal of Automation Technology,* 7(1) 39–44.

Wan, M. Lu, M.S. Zhang, W.H. & Yang, Y. 2012. A new ternary-mechanism model for the prediction of cutting forces in flat end milling, *International Journal of Machine Tools & Manufacture,* 57: 34–45.

Wan, M. Zhang, W.H. Dang, J.W. & Yang, Y. 2010. A novel cutting force modeling method for cylindrical end mill, *Applied Mathematical Modelling,* 34(3): 823–836.

Wang, M. Gao, L. & Zheng, Y. 2014. An examination of the fundamental mechanics of cutting force coefficients, *International Journal of Machine Tools & Manufacture,* 78: 1–7.

Advanced Materials and Structural Engineering – Hu (Ed.)
© 2016 Taylor & Francis Group, London, ISBN 978-1-138-02786-2

Changes in the state of stress in enclosing panels after additional thermal insulation

C. Radim, B. Kamil, M. Petr & L. Jana
Department of Structures, Faculty of Civil Engineering, VŠB—Technical University Ostrava,
Ostrava—Poruba, Czech Republic

ABSTRACT: Changes in the indoor and, particularly, outdoor temperatures during the year result in thermal gradient loads in the external cladding. Thus, the differences between the interior and exterior surfaces need to be considered. This situation is even more pressing in old panel block buildings without any external thermal insulation. Varying temperatures changes the volume of the panels, and cracks start appearing in the panel joints. Within the reconstruction of the panel block buildings, cracks are repaired and additional thermal insulation is applied by means overcladding. The thermal overcladding reduces the thermal gradient and eliminates the cracks. Thus, the thermal overcladding can save funds for repairs as fewer cracks spread in the panel.

1 INTRODUCTION

Changes in the indoor and, in particular, outdoor temperatures during the year load unevenly the external components in the structure with a temperature gradient. This results in volume changes in the panel structure. So, this situation is highly undesirable, for instance, in original buildings without any thermal insulation. Because of the volume changes in the panel structure, cracks appear in the gaps between the panels. Since the external panel cannot deform freely, internal forces co-occur there. If the strength of a material is exceeded, cracks will appear on the panel surfaces as well (Fig. 1).

The panel buildings are being reconstructed now. Within the reconstruction, cracks are often repaired and additional thermal insulation (abbreviated overcladding) is applied. Studies (Low et al. 2011, Gastmeyer 2004, Cajka & Mateckova 2013, Cajka 2014, Fabian et al. 2013, Perina et al. 2014, Perina et al. 2014) have dealt with the optimization of building materials for reconstruction and its effect on the quality of the indoor microclimate. The additional overcladding reduces the thermal load and, in turn, volume changes.

Renovation and repairs of cracks are rather expensive. Therefore, it is recommended to take into account positive effects of the overcladding and to save funds. The designing of additional overcladding is necessary to consider as well as fire safety (Bradacova & Kucera 2013, Cajka & Zidkova 2005, Cajka & Mateckova 2013, Cajka 2014). Overcladding of a building can be regarded as a kind of renovation work.

Figure 1. Distribution of real cracks in the external panel.

The influence of the temperature on the additional overcladding can be evaluated by using the infrared thermography. The utilization of this method has been described elsewhere (Perina et al. 2014, Perina et al. 2014).

Deformations and state of stress of the structure exposed to thermal loading were calculated using the Finite Element Method for an enclosing element of a panel building, G 57 type, built in Ostrava-Zábřeh in 1962. Forty years after the construction of the building, extruded polystyrene was applied there as the additional overcladding; the purpose was to meet the thermal performance specified in the Czech Standard ČSN 73 0540: Thermal Protection of Buildings (CSN73 0540 2011).

2 CALCULATION MODEL

The temperature field, deformation and state of stress of the external panel were calculated using an FEM model, similar to that described previously (Cajka & Zidkova 2005, Cajka & Mateckova 2013, Cajka 2014). Numerical analysis of the slabs by means of the FEM has been described generally in Sucharda & Kubosek (2014). The deformation and state of stress of the external panel can be calculated using an FEM model only if the parameters of structural materials are known. The external wall panel is made from slag pumice concrete. The bulk density is $\rho = 1{,}400 \text{ kgm}^{-3}$ and the average ultimate compressive strength is 5.0 MPa. Another important parameter that needs to be known is the modulus of elasticity of the material, which is E = 4.685 MPa. The modulus of transversal extension is assumed to be $\nu = 0.2$. The coefficient of thermal expansion can be assumed to be lower than that of plain concrete. To be on the safe side, $\alpha = 10 \cdot 10^{-6}$ is used in calculations.

Because normal forces and bending moments will occur in the external panel structure that is subject to temperature changes, a plated structure is used for modeling. As it is rather difficult to describe the real fastening of the panel in the structure, two limit states of the fastening are dealt with herein. The first case assumes that the panel moves freely in the gaps. The second case describes a fixed support where shifts in the gaps are not possible.

The external panel structure is loaded with temperature changes. Such load needs to be calculated separately for summer and winter. In Ostrava, where the panel block building is located, the maximum and minimum temperatures of air are obtained from the map given in CSN EN1991

(2010). Once the additional overcladding is fitted, the temperatures at the outdoor side of the external panel will go down. Such a decrease in the temperature depends, in particular, on the overcladding system and material used there. In this case, the contact system consists of extruded polystyrene, with a thickness of 160 mm.

The temperature in the x distance from the indoor surface can be obtained as follows (CSN EN1991 2010):

$$T(x) = T_{in} - \frac{R(x)}{R_{tot}} \cdot \left(T_{in} - T_{out} \right) \quad (1)$$

The final temperatures that load the external panel from the outdoor side need to be specified separately for summer and winter:

$$T_{out,s} = T_{in,s} - \frac{R(x)}{R_{tot}} \cdot \left(T_{in,s} - T_{out,s} \right) \quad (2)$$

$$T_{out,w} = T_{in,w} - \frac{R(x)}{R_{tot}} \cdot \left(T_{in,w} - T_{out,w} \right) \quad (3)$$

where
R_{tot} total resistance of the element including heat transfer resistance for both surfaces;
$R(x)$ heat resistance in the overcladding/panel contact point;
$T_{in,s}$, $T_{in,w}$ temperature of the inside environment in summer/winter; and
$T_{out,s}$, $T_{out,w}$ temperature of the outdoor environment in summer/winter.

The structure is calculated for 4 ultimate states of thermal parameters: the structure with/without the overcladding in summer/winter. Figure 2 shows clearly the changes caused by the temperature at the outdoor side of the external panel.

If the temperature at the upper face and lower face of the structure is decreased by T_h and T_d, respectively, from the basic temperature, T_0, and if the material is assumed to be homogeneous and isotropic and stress is distributed linearly along the thickness of the element, the change in the temperature can be divided into two states (Fig. 3).

Uneven increases in temperatures result in extension and bending of the external panel structure. The extension, ε_0, and bending, k, caused by the heat load can be calculated from (4, 5):

$$\varepsilon_0 = \alpha \cdot \Delta T_u \quad (4)$$

$$k = \frac{\alpha \cdot \Delta T_M}{h} \quad (5)$$

Summer

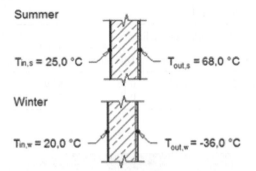

$T_{in,s} = 25,0\ °C$ — — $T_{out,s} = 68,0\ °C$

Winter

$T_{in,w} = 20,0\ °C$ — — $T_{out,w} = -36,0\ °C$

Original structure without overcladding

Summer

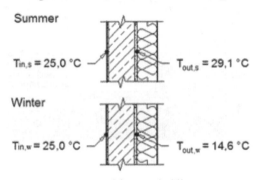

$T_{in,s} = 25,0\ °C$ — — $T_{out,s} = 29,1\ °C$

Winter

$T_{in,w} = 25,0\ °C$ — — $T_{out,w} = 14,6\ °C$

Structure with overcladding

Figure 2. Thermal parameters of the structure.

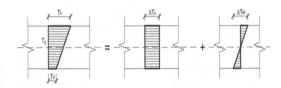

Figure 3. Components of the additional increase in the temperature of the structure.

Table 1. Deformation load of a panel.

Deformation load	Original structure without overcladding		Structure with overcladding	
	Summer	Winter	Summer	Winter
Strain ε_0 [mm/m]	0.365	−0.180	0.171	0.073
Curvature [mrad/m]	1.521	−2.333	0.171	−0.225

Deformation load caused by the heat changes is given in Table 1.

As shown in Table 1, there is a considerable decrease in the heat load intensity for the external panel after the overcladding is fitted. The decreasing load results in less deformation and lower internal forces.

3 FEM AND FINAL DEFORMATION AND INTERNAL FORCES

The final deformation plays a major role in the structures exposed to a thermal gradient. If the deformation is too high, cracks may appear in places where the panels contact each other. The heat load of such structures results in the development of internal forces, which may cause cracks to propagate along the panel surface, once the specified strength of the material is exceeded.

3.1 Deformation

Two limit states were considered for the fastening of the structure (Fig. 4). In the first case, shifting along the panel perimeter was possible (Fig. 4a). Table 2 gives the maximum deformation for each thermal state in the left bottom corner of the panel. For the distribution of the shift in different directions, see Figures 5 and 6: the structure is the original structure without any overcladding in summer. In the second case (Fig. 4b), the distribution will be similar. There is a zero shift along the perimeter of the panel (Fig. 4b). The deformation is the same as shown in Figure 7 and Figure 8. The real shifts in the structure are somewhere between the limit states.

As shown in Table 2, once the overcladding is fitted, the deformation caused by temperature changes decreases by as much as 80 per cent. The decrease in deformation is directly related to the reduced internal forces.

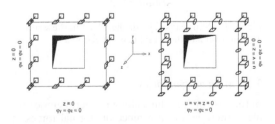

Figure 4. Computational model of panel: a) shift along the panel perimeter is possible, b) shift along the perimeter of the panel is impossible.

Table 2.   Comparison of the shift of the external panel for different loads (FEM).

Quantity	Original structure without overcladding			Structure with overcladding			Total decrease [%]
	Summer	Winter	Difference	Summer	Winter	Difference	
Shift u_x [mm]	+0,71	−0,34	1,05	+0,32	+0,14	0,18	82,9
Shift u_y [mm]	+0,37	−0,18	0,55	+0,20	+0,08	0,12	78,2

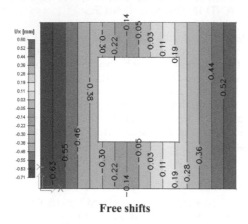

Free shifts

Figure 5.   Shift caused by the heat load in the structure without overcladding in summer: direction u_x.

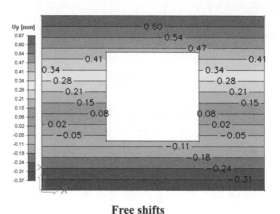

Free shifts

Figure 6.   Shift caused by the heat load in the structure without overcladding in summer: direction u_y.

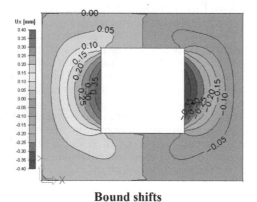

Bound shifts

Figure 7.   Shift caused by the heat load in the structure without overcladding in summer: direction u_x.

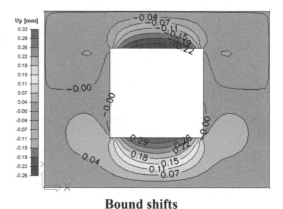

Bound shifts

Figure 8.   Shift caused by the heat load in the structure without overcladding in summer: direction u_y.

3.2   Main normal forces

In the case of free shifts along the perimeter of the plate, there is no occurrence of normal forces. If the structure is fastened along the perimeter, the normal forces in winter will be as shown in Figure 9 to Figure 12. The influence of the overcladding is clear from the distribution of main normal forces. The reason is that the overcladded panel is loaded with a positive deformation, while before the overcladding, it was exposed to a negative deformation (see Table 1).

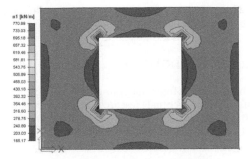

Structure without overcladding

Figure 9.   Main normal forces (n_1) caused by the heat load—the structure without overcladding in winter.

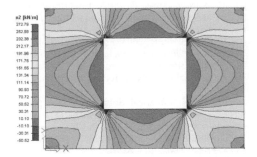

Structure without overcladding

Figure 10.   Main normal forces (n_2) caused by the heat load—the structure without overcladding in winter.

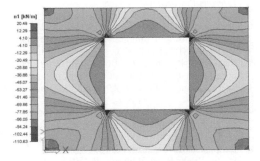

Structure with overcladding

Figure 11.   Main normal forces (n_1) caused by the heat load—the structure without overcladding in winter.

3.3   *Main bending moments*

Main bending moments are identical for both cases. The distribution of bending moments before and after the overcladding is very similar. After the overcladding is fitted, the main

Structure with overcladding

Figure 12.   Main normal forces (n_2) caused by the heat load—the structure without overcladding in winter.

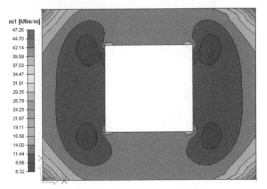

Structure without overcladding

Figure 13.   Main moments (m_1) caused by the heat load—the structure without overcladding in winter.

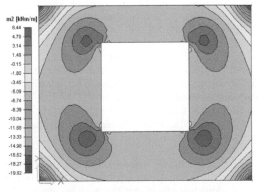

Structure without overcladding

Figure 14.   Main moments (m_2) caused by the heat load—the structure without overcladding in winter.

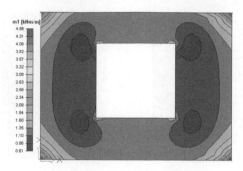

Structure with overcladding

Figure 15. Main moments (m_1) caused by the heat load—the structure without overcladding in winter.

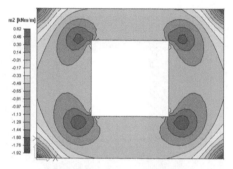

Structure with overcladding

Figure 16. Main moments (m_2) caused by the heat load—the structure without overcladding in winter.

moments will decrease considerably (see Fig. 13 to Fig. 16).

4 CONCLUSIONS

The extensive thermal loading of the external panel results in the development of considerable deformations and internal forces. The deformation can cause cracks to develop in the contact joints of the panels. Cracks can also appear on the panel surfaces after the ultimate strength of the material is exceeded in consequence of the excessive internal thermal forces.

The additional overcladding considerably reduces the thermal loading of the structures as well as the deformation; if cracks are to be repaired in buildings during reconstructions, such repairs will be rather expensive. Therefore, it is recommended to take into account the positive effects of the over-cladding and save funds. Such repairs and renovations can include the additional overcladding.

ACKNOWLEDGMENT

This paper was financially supported by the project of "The conceptual research and development 2015" in the Faculty of Civil Engineering, VŠB TU—Technical University of Ostrava.

REFERENCES

Bradacova, I. & Kucera, P. 2013. Concrete structures restoration from the fire safety point of view. *Advanced Materials Research*, 688: 113–119.

Cajka, R. & Mateckova, R. 2013. Fire Resistance of Ceiling Slab Concreted in Trapezoidal Sheet. *Procedia Engineering*, 65: 393–396.

Cajka, R. & Zidkova, R. 2005. *Fire resistance of garage plate-wall prefabricated structure*. Final Conference of COST Action C12: Improvement of Buildings' Structural Quality by New Technologies, Innsbruck, Austria.

Cajka, R. 2014. Numerical Solution of Temperature Field for Stress Analysis of Plate Structures, *Applied Mechanics and Materials*, 470: 177–187.

ČSN 73 0540, 2011. *Thermal protection of buildings, Part 1 to 4*, Prague 2005, (in Czech).

ČSN EN 1991, 2010. *Eurocode 1: Actions on structures—Part 1-5: General actions—Thermal Actions*. ČNI 05/2005, (in Czech).

Fabian, R. Kozakova, R. Cmiel, R. & Rykalova, E. 2013. Effect of the geometric solution of cladding on the quality of the indoor microclimate. *Advanced Materials Research*, 649: 65–68.

Gastmeyer, R. 2004. Load carrying behaviour of partially precast concrete panels with additional site-cast concrete and integrated thermal insulation. *Bautechnik*, 81(11): 869–873.

Halirova, M. & Rykalova, E. 2014. Use of multi-criteria optimization for selection of building materials for reconstruction. *Advanced Materials Research*, 899: 474–478.

Ng, S.C. Low, K.S. & Tioh, N.H. 2011. Thermal insulation property of newspaper membrane encased soil-based aerated lightweight concrete panels. *Advanced Materials Research*, 261–263: 783–787.

Perina, R. Plsek, R. & Wolfova, R. 2014. Verification of range of chemical grouting of masonry by non-destructive method using infrared thermography. *Advanced Materials Research*, 923: 191–194.

Perina, Z. Solar, J. Cmiel, F. & Fabian, R. 2014. The elimination of reflected radiation in an infrared thermographic measurement in the exterior. *Advanced Materials Research*, 923: 187–190.

Sucharda, O. & Kubosek, J. 2014. Numerical analysing the slabs by means of the finite difference method and the finite element method. *International Journal of Mechanics*, 8(1): 167–175.

Advanced Materials and Structural Engineering – Hu (Ed.)
© 2016 Taylor & Francis Group, London, ISBN 978-1-138-02786-2

Slag from Biomass Combustion (BCS)—chemical properties in accordance with BCS utilization in cement concrete

F. Khestl, P. Mec, M. Turicová & V. Šulková
Faculty of Civil Engineering, VŠB—TU Ostrava, Ostrava, Czech Republic

ABSTRACT: During the combustion, the newly emerged ash partially melts and causes slagging and fouling of boiler parts. In this case, the slag is very porous, glossy and contains large amount of ash and unburned part of straw. The use of Biomass Combustion Slag (BCS) in concrete by replacing natural aggregates and/or partially binders is a promising concept. However, to successfully use the available materials, they must be suitable for the planned purpose, i.e. to be placed in concrete. This paper deals with the characterization of slag from smaller biomass heating plant in the town Velký Karlov, and describes the behavior of BCS in the cement matrix by using of isothermal calorimetry.

1 INTRODUCTION

To meet the global demand of concrete in the future, it is becoming a more challenging task to find suitable alternatives to standard bonding agents and natural aggregates used for making of concrete. Therefore, the use of alternative sources is becoming increasingly important. Slag, as a co-product of the iron making process, is, nowadays, commonly used in the building industry due to its beneficial properties and also its sufficient quantity. The slag and fly ash are produced during the biomass combustion, which is very interesting alternative way in the energy generation. Herbaceous fuels contain silicon and potassium as their principal ash-forming constituents. They are also commonly high in chlorine content relative to other biomass fuels. The chemical composition of this ash is in large measure from SiO_2, CaO, MgO, Na_2O, K_2O, Al_2O_3, Fe_2O_3, P_2O_5 and other oxides. (Jenkins et al. 1998) The slag properties are influenced by plenty of factors, mostly on the heat of combustion, type of fuel, moisture and type of used fertilizers, the amount of soil (contamination) or time period in the boiler. That is, all these properties significantly influence porosity, coloring, the type and volume of mineral composition, and amount or mechanical properties. These constituents cause ash and slag formation and deposition problems at high or moderate combustion temperatures. The slag could be, on the one hand, very solid, compacted, and glossy material and, on the other side, very porous material that is often polluted by ash and unburned pieces of straw. Properties of the slag from biomass combustion are very hard to predict.

In this paper, a slag from a smaller biomass heating plant in the town Velký Karlov was used. The main type of fuel used in this heating plant is rape straw. Chemical analysis of the whole rape straw ash—the main fuel in the Velký Karlov heating plant—has been described elsewhere (Volakova 2010). During the combustion, the newly emerged ash partially melts and causes slagging and fouling of boiler parts. In this case, the slag is very porous and glossy, and contains large amount of ash and unburned part of straw. The biomass combustion slag is gained directly from the heating plant in the form of very porous brittle stones of maximum size approximately 30 cm, polluted by ash and unburned straw residues. For the use in the concrete, especially reinforced cement concrete, standard methods of analysis leading to an accurate and consistent evaluation were examined. The specific weight according to the ČSN EN 1097-7 is 2.41 g.cm⁻³. The amount of chlorides in the sample was examined according to ČSN EN 1744-1: the analysis suggested that the value of 0.010% is very low and could be used in the reinforced cement concrete without any problem. Positive influence of fine milled BCS on the strength characteristics in the cement mortars prepared and tested according to ČSN EN 196-1 showed results of bending strength and compression strength. The results of specimens with partial substitution of the Portland cement by CBS were more than 20 percent better than the reference cement specimens.

The chemical analysis was carried out on a very fine milled specimen.

2 ANALYSIS OF BIOMASS COMBUSTION SLAG (BCS)

Ground Granulated Blast Furnace Slag (GGBFS) is primarily made up of silica, alumina, calcium oxide, and magnesia (95%). Other elements such as manganese, iron, sulfur, and trace amounts of other elements make up about other 5% of the slag. The exact concentrations of elements vary depending on where and how the slag is produced. BCS is slightly different due to the chemical composition of biomass used during the combustion. The states of CBS, crystalline or amorphous, are examined with X-ray diffraction. For the chemical constitution of CBS two specimens were prepared: first, one specimen was washed with water; second, the untreated specimen was not washed. Both specimens were finely milled. The chemical constitution of those specimens was examined by X-Ray Fluorescence (XRF). In order to observe how the materials affect the cement hydration when mixed with cement paste, isothermal calorimetric measurements were made. Finally, for the building materials, the leachate analysis is important. The leachate of BCS was prepared according to the ČSN 72 0102, ČSN ISO 10523, ČSN EN 459-2 and ČSN EN 196-2. The finely milled BCF samples were mixed with water at a 1:10 ratio. After 24 hours of leaching, the specimen was filtered and immediately tested. Except the pH value and sulfide amount, the CBS water leachate analysis was examined by Inductively Coupled Plasma-Atomic Emission Spectrometry.

2.1 Inductively Coupled Plasma-Atomic Emission Spectrometry (ICP-AES)

Chemical composition of water leachate was measured on ICP-AES Spectro vision EOP (SPECTRO Analytical Instruments GmbH). This equipment works in wavelength range 120–800 nm and allow to measure 72 elements.

2.2 X-ray powder diffraction

The crystalline phases were identified and quantified by powder X-ray diffraction D8 ADVANCE XRD. Radiation source is CoK(α).

2.3 X-Ray Fluorescence (XRF) analysis

For the chemical composition of elements of solid sample the handheld X-Ray fluorescence spectrometer Niton XL3t (Thermo Scientific) was used. This equipment can measure elements with atomic number 12–92. Due to low energy of handheld machines the elements with lower atomic number than 12 is not possible to measure.

2.4 Isothermal calorimeter TAM Air

Calorimetry test was done on TAM Air device. TAM Air is an eight-channel, isothermal, heat conduction calorimeter operating in the milliwatt range.

Process of cement hydration is characterized by strong exothermic reactions. The shape of the heat flow curve resulted from measurement reflect the cement hydration process. The partial substitution of cement by admixtures changes the shape of the heat flow curve, so the admixture effect can be quantified.

Testing was done at temperature 25°C. All samples were tested in 20 mL ampoules. Every ampoule was filled with 7.5 g of tested sample made from 5 g of binder (Portland cement CEM I 42.5R and proportional part of BCS) and 2.5 g of demineralized water. All tested samples were compared to the reference sample made from 5 g Portland cement CEM I 42.5R and 2.5 g of demineralized water.

3 RESULTS AND DISCUSSION

3.1 X-ray diffraction analysis

The XRD analysis shows that the CBS consists of nearly 97% of the amorphous phase and only 3% of the crystalline phase. The results are given in Table 1.

From X-Ray diffraction analysis, it was found thah except for α-Cristobalite and Magnesite, there is Merrihueite (K, Na)$_2$ (Mg, Fe^{2+})$_5$ Si$_{12}$O$_{30}$ of which X-RAY data are nearly identical to those for osumilite, (K, Na, Ca) (Mg, Fe)$_2$ (Al, Fe)$_3$ (Si, Al)$_{12}$O$_{30}$-H$_2$O, and very similar to those for synthetic K$_2$ Mg$_5$Si$_{12}$O$_{30}$. Merrihueite is interpreted as an alkali-ferromagnesian silicate mineral of the osumilite type (Dodd et al. 1965).

3.2 X-Ray Fluorescence (XRF) analysis

The chemical constitution of CBS was examined by X-Ray Fluorescence (XRF). The results are given in Table 2.

3.3 CBS water leachate analysis

A thorough analysis of trace elements present in the leachate samples involves the use of atomic emission spectroscopy. The detectable examined

Table 1. Mineralogy phases of BCS.

Phase	Content (% mass)
α-Cristobalite	39.7 ± 1.8
Magnesite	22.3 ± 1.7
Merrihueite	38.0 ± 2.3

Table 2. X-Ray Fluorescence (XRF) analysis results.

Main chemical constituents	Untreated CBS [%]	Water treated CBS [%]
Si	37.85	38.50
K	15.92	16.27
Ca	3.54	3.77
P	1.33	1.46
Mg	0.84	0.91
Al	0.39	0.48
S	0.19	0.09
Cl	0.14	0.04
Fe	0.10	0.17
Mn	0.10	0.11
Ba	0.07	0.05
Others	<0.05	<0.05
Bal*	39,49	38,12

*Ballast, due to low energy, the elements with lower atomic number than 12 is not possible to measure—Na, C, O_2, and others.

Table 3. X-Ray Fluorescence (XRF) analysis results.

Parameter	Units	Results	Uncertainty
Free CaO	[%]	0.12	± 0.02
Sulfides	[%]	0.67	± 0.07
pH		10.60	± 0.5
Conductivity	mS/m	206	± 11
C	mg/l	211	± 10
Al	mg/l	< 0.1	
Ca	mg/l	2.38	± 0.08
Fe	mg/l	0.13	± 0.01
K	mg/l	559	± 28
Mg	mg/l	2.42	± 0.15
Na	mg/l	4.49	± 0.36
P	mg/l	68	± 7
S	mg/l	30	± 4
Si	mg/l	219	± 25

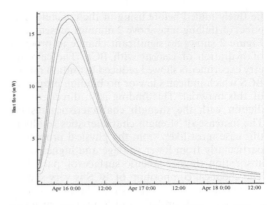

Figure 1. Heat flow curves of 10%, 20% and 40% of CBS substitution compared with the reference cement curve. From the highest peak to the lowest: CEM reference (5 g), 10% BCS, 20% BCS, and 40% BCS.

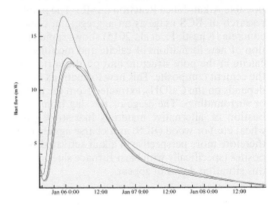

Figure 2. Heat flow curves of 20% CBS substitution compared with the reference cement curve. From the highest peak to the lowest: CEM ref, 15 min. milled BCS, 2 min. milled BCS, 9 min. milled BCS and 20 min. milled BCS.

elements with the highest level in the leachate samples included, in descending order, K, Si, C, P, S, Na, Mg, Ca, Al.

3.4 Calorimetric measurements

Isothermal calorimetric measurements were performed in order to observe how the CSB affects the cement hydration when mixed with the cement paste. Figure 1 shows decreasing heat flow depending on the substitution of cement amount by CBS. Portland cement CEM I 42.5 R was gradually substituted by 10, 20 and 40 weight percent of CBS. Pure Portland cement was used as reference material. As shown in Figure 2 the CBS retards the

heat development with no significant influence of the milling time duration. The hydration process in all samples reached a steady state temperature within 3 days. It was observed that with the higher amount of BCS the heat of hydration decreased.

4 CONCLUSION

The results from chemical analyses of the BCS and its water leachate do not indicate harmful influence of BCS to the cement hydration. The pH value of 10.6 is similar or even higher than fly ashes or granulated blast furnace slags that are often used in cement composites. Just like the Portland cement, also granulated blast furnace slags or BCS must be finely milled before using in the concrete. The effect of milling time above 2 minutes presented in Figure 2 shows no significant change in progress of hydration of cement with BCS. The calorimetry experiments showed reduced hydration rate of BCS which indicates low or no binding properties of this material. This finding is in direct contradiction with the strength characteristics results. The increase of strength characteristics results in this case most likely from the physical properties, particularly from lower voidage and higher water absorption (higher specific surface of particles). The chemical composition of BCS is very similar to the composition of fly ashes. The BCS is composed mainly from SiO_2 and K_2O, then in lower amount CaO, MgO, Al_2O_3, and other oxides. But the chemical composition of BCS is highly influenced by the high amount of amorphous SiO_2. This high amount of amorphous phase makes more difficult to use BCS in cement composites due to the possible Alkali-Silica Reaction (ASR). The parallel research of BCS usage as an aggregate in cement concrete (Khestl, F., et al., 2015) shows rapid creation of new formations of calcite and monohydrocalcite in the pore structure and on the surface of the cement composite. This new formations highly depends on the $Ca(OH)_2$ extracted from the matrix or surroundings. The usage of the slag from combustion of alternative materials like straw (rape wheat etc.) or wood (BCS) as a coarse aggregate is therefore more perspective in alkali activated composites (specifically with blast furnace slag), where this situation does not appear.

ACKNOWLEDGMENT

This work was financially supported by the Ministry of Education, support of specific academic research—Student grant competition VŠB—TU Ostrava, the identification number SP2014/161—Utilization of glassy waste material from biomass combustion.

REFERENCES

ČSN 72 0102, *Basic analysis of silicates—Determination of loss by drying*, (2009).

ČSN ISO 10523, *Water quality—Determination of pH*, (2010).

ČSN EN 459-2, *Building lime—Part 2: Test methods*, (2011).

ČSN EN 196-1, *Methods of testing cement—Part 1: Determination of strength*, (2005).

ČSN EN 196-2 *Method of testing cement—Part 2: Chemical analysis of cement*, (2013).

ČSN EN 1097-7, *Tests for mechanical and physical properties of aggregates—Part 7: Determination of the particle density of filer—Pyknometer method*, (2008).

ČSN EN 1744-1, *Tests for chemical properties of aggregates—Part 1: Chemical analysis*, (2013).

Dodd, R.T., W.R. van Schmus a U.B. Marvin. Merrihueite, A New Alkali-Ferromagnesian Silicate from the Mezo-Madaras Chondrite. Science. 1965, vol. 149, issue 3687, s. 972–974. DOI: 10.1126/science.149.3687.972.

Jenkins, B.M., L.L. Baxter, T.R. Miles a T.R. Miles. Combustion properties of biomass. In: Fuel Processing Technology. 1998, pp. 17–46. DOI: 10.1016/s0378-3820(97)00059-3.

Khestl, F., Šulková, V. and Mec, P. Use of Glassy Slag from Biomass Combustion in the Building Industry—Coarse Aggregate. *Advanced Materials Research*. Vol. 1122, 2015. pp 219–224. Trans Tech Publications. doi:10.4028/www.scientific.net/AMR.1122.219.

Merrihueite [online]. Available at: http://www.webmineral.com/data/Merrihueite.shtml.

Voláková, P.: Biomass fly ash—chemical composition and possibilities of usage. (In Czech). Biom.cz [online]. ISSN: 1801-2655.

Advanced Materials and Structural Engineering – Hu (Ed.)
© 2016 Taylor & Francis Group, London, ISBN 978-1-138-02786-2

Fundamental thermo-elastic solutions for 2D hexagonal Quasicrystal

T. Wang

School of Mechanics and Engineering, Southwest Jiaotong University, Chengdu, China
Dongfang Electric Wind Power Co. Ltd., Tianjin, China

ABSTRACT: This paper presents the fundamental solutions for an infinite space of 2D hexagonal Quasicrystals (QCs) in the framework of thermo-elasticity. Based on general solutions of quasi-harmonic, by assuming appropriate potential functions, corresponding fundamental solutions of an infinite space subjected to a thermal load are explicitly derived in terms of elementary functions. The obtained solutions will benefit the study of QC for crack, indentation and dislocation problems. In addition, these solutions can serve as guidelines of relevant numerical simulations.

1 INTRODUCTION

Quasicrystal (QC) is a kind of aperiodic and orderly structure between glass and crystal, which was firstly discovered by Shechtman et al. (1984), who was awarded the 2011 Nobel Prize in chemistry for this great discovery. Unlike traditional crystalline materials, QC owns the distinct characteristics of low porosity, low adhesion, capacity of reducing friction coefficient and high wear resistance (Dubois et al. 1991), this may lead to the widely promising applications in engineering and industry. In the past three decades, mechanical properties of QC are in-depth studied in a variety of aspects, such as contact (Wu et al. 2013), crack (Li 2014) and inclusions (Wang 2004) problems.

Fundamental solutions, in the framework of QC elasticity, can be obtained on the basis of corresponding general solutions. Li (2012) studied thermal effects of 1D hexagonal QC through the method of Green's function. However, the reports on the fundamental solutions of 2D hexagonal QCs in the thermo-elastic field are very limited, according to the literature survey.

In the present study, the thermal solutions of infinite space subjected to a thermal load for 2D hexagonal QCs are obtained explicitly in closed forms by utilizing the general solutions. By assuming appropriate potential functions, corresponding fundamental solutions are expressed in terms of elementary functions. The analytical solutions of 2D hexagonal QC in thermo-elastic field will provide guidelines of 2D hexagonal QC in thermo-elastic field in further researches.

2 BASIC EQUATIONS OF 2D QC AND GENERAL SOLUTIONS

In the Cartesian coordinate (x,y,z), atomic sequence of 2D QCs is arranged by the means that coincident with the quasi-periodic plane and the z-axis identical to the periodic direction. According to Yang et al. 2013), a set of general solutions of 3D thermo-elasticity for 2D QCs in closed form is derived as

$$U = u_x + \mathrm{i} u_y = \Lambda\left(-\mathrm{i}\sum_{j=1}^{2}\phi_j + \sum_{i=1}^{4}\varphi_i \right),$$

$$W_1 = u_z = \sum_{i=1}^{4} K_{2i} s_i \frac{\partial \varphi_i}{\partial z_i},$$

$$W = w_x + \mathrm{i} w_y = \Lambda\left(-\mathrm{i}\sum_{j=1}^{2} K_{4j}\phi_j + \sum_{i=1}^{3} K_{1i}\varphi_i \right),$$

$$T = \sum_{i=1}^{4} K_{3i} s_i^2 \frac{\partial^2 \varphi_i}{\partial z_i^2}, \tag{1}$$

$$\sigma_{zm} = \sum_{i=1}^{4} \alpha_{im} \frac{\partial^2 \varphi_i}{\partial z^2}, \; (m = 1\text{--}3)$$

$$\tau_{zk} = -\mathrm{i}\sum_{\alpha=1}^{2} \mu_{jk} \frac{\partial \phi_j}{\partial z} + \Lambda\sum_{i=1}^{4} \upsilon_{ik} \frac{\partial \varphi_i}{\partial z}, \; (k = 1, 2)$$

$$\sigma_2 = -\mathrm{i}\Lambda^2 \sum_{j=1}^{2} \beta_{j1}\phi_j + \Lambda^2 \sum_{i=1}^{4} \gamma_{i1}\varphi_i,$$

$$S_2 = -\mathrm{i}\Lambda^2 \sum_{j=1}^{2} \beta_{j2}\phi_j + \Lambda^2 \sum_{i=1}^{4} \gamma_{i2}\varphi_i, \tag{2}$$

where, $i=\sqrt{-1}$, $U=u_x+iu_y$, $W=w_x+iw_y$, $W_1=u_z$, $\Lambda=\partial/\partial x+i\partial/\partial y$, $\sigma_{z1}=\sigma_{zz}$, $\sigma_{z2}=\sigma_1=\sigma_{xx}+\sigma_{yy}$, $\sigma_{z3}=S_1=H_{xx}+H_{yy}$, $\tau_{z1}=\sigma_{xz}+i\sigma_{yz}$, $\tau_{z2}=S_{z1}=H_{xz}+iH_{yz}$, $\sigma_2=\sigma_{xx}-\sigma_{yy}+2i\sigma_{xy}$, $S_2=H_{xx}-H_{yy}+i(H_{xy}+H_{yx})$.

The constants K_{mi}($m=1-3$), α_{im}, μ_{jk}, υ_{ik}, β_{jk} and γ_{ik} are defined in Appendix A.

3 FUNDAMENTAL SOLUTIONS

In this section, we consider an infinite/half-infinite space of the above-mentioned QC, which are subjected to a point heat source at the origin of the Cartesian coordinate system (x,y,z), as shown in Figure 1 and 2. The fundamental solutions are

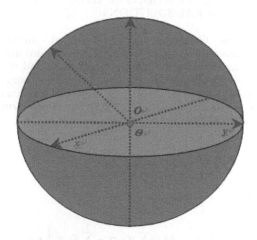

Figure 1. A schematic figure for an infinite space of 2D hexagonal QC, subjected to heat loadings applied at the origin of the Cartesian coordinate system. (x,y,z).

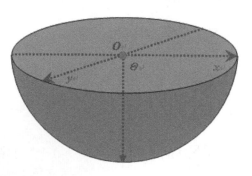

Figure 2. A schematic figure for a half-infinite spaces of 2D hexagonal QC, subjected to heat loadings applied at the origin of the Cartesian coordinate system. (x,y,z).

derived owing to the general solutions and the specific boundary conditions.

3.1 Infinite space of 2D QC

To solve the axisymmetric problem, we assume the potential functions referring to Hou et al. (2008) for thermo-elastic materials as

$$\phi_j=0,\ \varphi_i=C_i[sign(z)z_i\ln R_i^*-R_i]\ (i=1-4;j=1,2),$$
(3)

where, $r^2=x^2+y^2$, $R_i=\sqrt{r^2+z_i^2}$ and $R_i^*=R_i+z_i$, sign is the signum function, and C_i are constants to be determined.

Inserting (3) into (1) and (2), the expressions of the generalized displacements and stress can be derived

$$U=-(x+iy)\sum_{i=1}^{4}C_i\frac{1}{R_i^*},\ W_1=sign(z)\sum_{i=1}^{4}K_{2i}C_is_i\ln R_i^*,$$

$$T=\sum_{i=1}^{4}K_{3i}C_is_i^2\frac{1}{R_i},\ W=-(x+iy)\sum_{i=1}^{4}K_{1i}C_i\frac{1}{R_i^*},$$
(4)

$$\sigma_{zm}=\sum_{i=1}^{4}\alpha_{im}C_is_i^2\frac{1}{R_i},$$

$$\tau_{zk}=(x+iy)sign(z)\sum_{i=1}^{4}\upsilon_{ik}C_is_i\frac{1}{R_iR_i^*},$$

$$\sigma_2=(x+iy)^2\sum_{i=1}^{4}\gamma_{i1}C_i\frac{1}{R_iR_i^{*2}},$$

$$S_2=(x+iy)^2\sum_{i=1}^{4}\gamma_{i2}C_i\frac{1}{R_iR_i^{*2}}.$$
(5)

Considering the continuity conditions at $z=0$ of W_1, τ_{zk} with a signum function, we have

$$\sum_{i=1}^{4}K_{2i}C_is_i=0,\ \sum_{i=1}^{4}\upsilon_{1i}C_is_i=0,\ \sum_{i=1}^{4}\upsilon_{1i}C_is_i=0.$$
(6)

By considering the equilibrium of the layer $\varepsilon_1\leq z\leq\varepsilon_2$ (with $\varepsilon_1\leq 0\leq\varepsilon_2$), we can obtain that

$$\int_{-\infty}^{+\infty}\int_{-\infty}^{+\infty}\left[\left(-K_{33}\frac{\partial T}{\partial z}\right)_{z=\varepsilon_1}-\left(-K_{33}\frac{\partial T}{\partial z}\right)_{z=\varepsilon2}\right]dxdy=\Theta.$$
(7)

which, with the help of (4), gives rise to

$$\sum_{i=1}^{4}K_{3i}C_is_i^3=\frac{\Theta}{4\pi K_{33}}.$$
(8)

From (5) and (7), it is seen that

$$
\begin{Bmatrix} C_1 \\ C_2 \\ C_3 \\ C_4 \end{Bmatrix} = \frac{\Theta}{4\pi K_{33}} \begin{bmatrix} K_{21}s_1 & K_{22}s_2 & K_{23}s_3 & K_{24}s_4 \\ \upsilon_{11}s_1 & \upsilon_{12}s_2 & \upsilon_{13}s_3 & \upsilon_{14}s_4 \\ \upsilon_{21}s_1 & \upsilon_{22}s_2 & \upsilon_{23}s_3 & \upsilon_{24}s_4 \\ 0 & 0 & 0 & K_{34}s_1^3 \end{bmatrix}^{-1} \begin{Bmatrix} 0 \\ 0 \\ 0 \\ 1 \end{Bmatrix}
$$
(9)

Substituting (9) back into (4) and (5), the resultant fundamental filed is obtained.

3.2 Half-infinite space of 2D QC

To derive the corresponding fundamental solutions, we assume the functions to be in the form

$$
\overline{\varphi}_j = 0 , \quad \varphi_i = D_i \left[z_i \ln R_i^* - R_i \right],
$$
(10)

where, $R_i = \sqrt{x^2 + y^2 + z_i^2}$, $R_i^* = R_i + z_i$ and D_i are constants to be determined.

Inserting (10) into (1) and (2), we got the displacement components

$$
U = -(x+iy) \sum_{i=1}^{4} D_i \frac{1}{R_i^*},
$$

$$
w_1 = u_z = \sum_{i=1}^{4} K_{2i} D_i s_i \ln R_i^*,
$$

$$
T = \sum_{i=1}^{4} K_{3i} D_i s_i^2 \frac{1}{R_i},
$$

$$
W = -(x+iy) \sum_{i=1}^{4} K_{1i} D_i \frac{1}{R_i^*}
$$
(11)

and the stress components

$$
\sigma_{zm} = \sum_{i=1}^{4} \alpha_{mi} D_i s_i^2 \frac{1}{R_i}, \quad \tau_{zk} = (x+iy) \sum_{i=1}^{4} \upsilon_{ki} D_i s_i \frac{1}{R_i R_i^*},
$$

$$
\sigma_2 = (x+iy)^2 \sum_{i=1}^{4} \gamma_{1i} D_i \frac{1}{R_i R_i^{*2}},
$$

$$
S_2 = (x+iy)^2 \sum_{i=1}^{4} \gamma_{2i} D_i \frac{1}{R_i R_i^{*2}}.
$$
(12)

In the current case, the boundary conditions at the plane $z = 0$ can specify as

$$
\sigma_{z1}|_{z=0} = 0, \quad \tau_{z1}|_{z=0} = 0, \quad \tau_{z2}\big|_{z=0} = 0, \quad \frac{\partial T}{\partial z}\bigg|_{z=0} = 0.
$$
(13)

where, the last condition has been satisfied automatically.

Form (20) and (21), it is readily arrived at

$$
\sum_{i=1}^{4} \alpha_{1i} D_i s_i^2 = 0, \quad \sum_{i=1}^{4} \upsilon_{1i} D_i s_i = 0, \quad \sum_{i=1}^{4} \upsilon_{1i} D_i s_i = 0. \quad (14)
$$

Considering the thermal equilibrium of the layer $0 \le z \le \xi_3$, we have

$$
\iint \left(-K_{33} \frac{\partial T}{\partial z} \right)\bigg|_{z=\xi_3} dxdy = \Theta.
$$
(15)

which, in light of (11), give rise to

$$
\sum_{i=1}^{4} K_{3i} D_i s_i^3 = \frac{\Theta}{2\pi K_{33}}.
$$
(16)

By (14) and (16), all constants can be determined as

$$
\begin{bmatrix} D_1 \\ D_2 \\ D_3 \\ D_4 \end{bmatrix} = \frac{\Theta}{2\pi K_{33}} \begin{bmatrix} \upsilon_{11}s_1 & \upsilon_{12}s_2 & \upsilon_{13}s_3 & \upsilon_{14}s_4 \\ \upsilon_{21}s_1 & \upsilon_{22}s_2 & \upsilon_{23}s_3 & \upsilon_{24}s_4 \\ \alpha_{11}s_1^2 & \alpha_{12}s_2^2 & \alpha_{13}s_3^2 & \alpha_{14}s_4^2 \\ 0 & 0 & 0 & K_{34}s_4^3 \end{bmatrix}^{-1} \begin{bmatrix} 0 \\ 0 \\ 0 \\ 1 \end{bmatrix}.
$$
(17)

Substituting (17) back into (11) and (12), we can obtain the corresponding fundamental solutions in the phonon-phason coupling field in terms of elementary functions.

4 CONCLUSIONS

On the basis of the general solutions of Yang et al. (2014) and combining appropriate potential functions, the fundamental solutions of infinite space of 2D hexagonal QCs which subjected to a point heat source are presented in terms of elementary functions. With the help of the boundary conditions and the thermal equilibrium equation, the constants involved in the potential functions are determined. The integral boundary equations for phonon-phason coupling field are established in terms of elementary functions. It should be pointed out that the present solutions can be served as fundamental solutions to crack or contact problems. Furthermore, the present solutions also play an important role in numerical simulations. The follow-on work of the present solutions can be expected.

ACKNOWLEDGEMENT

This work is supported by the National Natural Science Foundation of China (Nos: 11102171) and Program for New Century Excellent Talents in University of Ministry of Education of China (NCET-13-0973). The support from Sichuan Provincial Youth Science and Technology Innovation Team (2013-TD-0004) and Scientific Research Foundation for Returned Scholars (Ministry of Education of China) are acknowledged as well.

APPENDIX A

This section details mentioned constants in the derivation processes in the article.

$$P_j = \frac{R_6}{s_j'^2} - R_4, \quad m_j = \frac{-K_3}{s_j'^2} + K_4, \quad n_i = -\left(\frac{c_1}{s_i^4} - \frac{b_1}{s_i^2} + a_1\right),$$

$$q_i = -\left(\frac{c_2}{s_i^4} - \frac{b_2}{s_i^2} + a_2\right), \quad l_i = -\left(\frac{c_3}{s_i^4} - \frac{b_3}{s_i^2} + a_3\right),$$

$$\gamma_i = -\left(a - \frac{b}{s_i^2} + \frac{c}{s_i^4} - \frac{d}{s_i^6}\right),$$

$$K_{1i} = \frac{q_i}{n_i}, \quad K_{2i} = \frac{l_i}{n_i}, \quad K_{3i} = \frac{r_i}{n_i}, \quad K_{4j} = \frac{p_j}{m_j};$$

$$\alpha_{1i} = (C_{33}K_{2i} - \beta_3 K_{3i})s_i^2 - (C_{13} + R_3 K_{1i}),$$

$$\alpha_{2i} = (2C_{13}K_{2i} - 2\beta_1 K_{3i})s_i^2 - 2\left[C_{12} + C_{66} + (R_2 + R_6)K_{1i}\right]$$

$$\alpha_{3i} = 2R_3 K_{2i} s_i^2 - 2\left[R_2 + R_6 + \left(K_2 + \frac{K_3 + K_6}{2}\right)K_{1i}\right]$$

$$\mu_{j1} = C_{44} + R_4 K_{4j}, \mu_{j2} = R_4 + K_4 K_{4j},$$

$$\upsilon_{i1} = C_{44}\left(1 + K_{2i}\right) + R_4 K_{1i}, \quad \upsilon_{i2} = R_4\left(1 + K_{2i}\right) + K_4 K_{1i},$$

$$\beta_{j1} = 2(C_{66} + R_6 K_{4j}), \quad \beta_{j2} = 2R_6 + (K_3 + K_6)K_{4j},$$

$$\gamma_{1i} = 2(C_{66} + R_6 K_{1i}), \quad \gamma_{2i} = 2R_6 + (K_3 + K_6)K_{1i},$$

where, $a_i, b_i, c_i (i = 1 - 3)$, s_i and s_i' are defined by Yang et al. (2013) and C_{ij}, K_n and $R_n (n = 1, 2, 3, 4, 6)$ are respectively phonon, phason and phonon-phason elastic constants and β_i are thermal constants.

REFERENCES

Dubois, J.M. Kang, S.S. & Stebut J.V. 1991. Quasicrystalline low-friction coatings. *Journal of Materials Science Letters* 10: 537–541.

Hou, P.F. Leung, A.Y.T. & Chen, C.P. 2008. Fundamental solution for transversely isotropic thermoelastic materials. *International Journal of Solids and Structures*, 45: 392–408.

Li, X.Y. & Li, P.D. 2012. Three-dimensional thermalelastic general solutions of one-dimensional hexagonal Quasi-crystal and fundamental solutions. *Physics Letters A*, 376: 2004–2009.

Li, X.Y. 2014. The elastic field in an infinite medium of one-dimensional hexagonal quasi-crystal with a planar crack. *International Journal of Solids and Structures*, 51: 1442–1455.

Schechtman, D. Blech, I. Gratias, D. & Cahn, J.W. 1984. Metallic phase with long-range orientational order and no translational symmetry. *Physical Review Letters*, 53: 1951–1953.

Wang, X. 2004. Eshelbys problem of an inclusion of arbitrary shape in a decagonal quasicrystalline plane or half-plane. *International Journal of Engineering Science*, 550(42): 1911–1930.

Wu, Y.F. Chen, W.Q. & Li, X.Y. 2013. Indentation on one-dimensional hexagonal quasicrystals: general theory and exact complete solutions. *Philosophical Magazine*, 93(8): 858–882.

Yang, L.Z. Zhang, L.Z. Song, F. & Gao, Y. 2014. General solutions for three dimensional thermo-elasticity of two-dimensional hexagonal quasicrystals and an application, *Journal of Thermal Stresses*, 37: 363–379.

Advanced Materials and Structural Engineering – Hu (Ed.)
© 2016 Taylor & Francis Group, London, ISBN 978-1-138-02786-2

Effect of induction heating power on zinc-coating quality

F. Fang, Y.M. Chen & L.X. Wang
Research and Development Center, Wuhan Iron and Steel Co., Wuhan, China

ABSTRACT: An important parameter—induction heating power—conducted on Ultra Low Carbon Interstitial Free Steel (ULC-IF steel) by the HDPS (Hot Dipping Process Simulator) in the laboratory was studied, to analyze its effect on Zn-coating quality. The results show that, in conducting four parameters of induction heating power, the Zn coating distributed uniformly on the sample surface with no bare spots (defects commonly exit on Zn-coating steel products). The as-received coatings attached to the matrix well with some microcracks, but none of them extended to the matrix. It was not obvious to specify different zinc-iron intermetallics from viewing the surface morphology. The weight and thickness of coating displayed a maximum value and a minimum value of 10% and 15%, respectively, when conducting the power. However, the iron amount in the coating had a little difference (all within the range of 3.7 to 4%), which was lower than 10% for steel plate produced in the industry.

1 INTRODUCTION

Recently, the automotive industry started to require lightweight, enhanced crash safety and durability in vehicles in addition to fundamentally good performance in relation to regulations and fuel efficiency (Santos et al. 2008). Therefore, car makers are increasingly using various Advanced High-Strength Steels (AHSS) to reduce the Body-in-White (BIW) weight of vehicles. As Zn coating has traditionally been used for decelerating or impeding the corrosion rate of underlying ferrous alloy substrates (Bindiya et al. 2012), galvanized (GI) and galvannealed (GA) steels are being used in vehicles to prevent the external corrosion of the automotive body. GA sheets are mainly used in exposed and non-exposed car panels, as well as in appliances, due to their superior corrosion resistance. GA coating offers several advantages compared with GI coating, such as better weld ability, paint ability and, in some cases, better reported corrosion resistance (Almeida & Morcillo 2000, Bandyopadhyay et al. 2006, Marder 2000).

Because testing in large-scale production plants would require large amounts of sheet metal and cost-valuable production capacity, a highly integrated experiment—Hot Dipping Process Simulator—has been created, which is a miniature pilot plant that serves the purpose of developing processes and products to be used in the preparation of production processes for industrial-scale hot-dip galvanizing facilities. Using miniature-scale test samples, we thus examined to determine optimum process parameters for industrial-scale production with a view to developing and testing new products at moderate expense,

which could simulate multiple coatings based on zinc as the main alloy element. There are numerous parameters for the equipment to simulate galvanizing and galvannealing processes, and the galvannealing process is executed in the induction heating furnace, so that a key parameter in the controlling suitable heating procedure—the heating power of the inductor—is studied in this paper.

2 EXPERIMENTAL PROCEDURE

Based on the above-mentioned simulator, all the analyses were executed on the standard samples of the laboratory—Ultra Low Carbon Interstitial Free Steel (ULC-IF steel), cut by 220 mm × 110 mm × 0.7 mm. The samples were partially immersed in the zinc bath for coating analysis. The applied heating treatment and dipping cycles were as follows: heating rate as 5 °C/s from room temperature to 800 °C for 60 s, cooling rate as 25 °C/s to 470 °C, holding for 30 s to stabilize sample temperature, immersing in a Zn-0.12% Al bath for 3.5 s, with no saturated iron. The atmosphere during heating and cooling was N_2 5% H_2, and the dew point was −10 °C. After immersing in the zinc bath, the samples were lifted upside and heated quickly in the induction heating furnace at a temperature of 510 °C for 10 s. To ensure that the coating thicknesses were almost within 20 μm, the air knife flow rate was set to a maximum value of 500 l/min. The induction heating power was conducted in four steps from 5% to 20% to study its effect on the coating quality. The details of the parameters are provided in Table 1.

Table 1. Galvannealing process parameters for experimental steel.

Code	Final cooling temperature, °C	Dipping time, s	Air knife flow rate, l/min
1	470	3.5	500
2	470	3.5	500
3	470	3.5	500
4	470	3.5	500

Code	Galvannealing temperature, °C	Galvannealing time, s	Induction heating power, %
1	510	10	5
2	510	10	10
3	510	10	15
4	510	10	20

After simulation in the induction furnace for the galvannealing process, the samples were cut by 15 mm×15 mm to analyze the surface state completed by using the Scanning Electron Microscope (SEM). Coating quality was assessed by visual appearance. The element depth profiling was obtained by Glow Discharge Optical Emissivity Spectroscopy (GDOES) using a Horiba Jobin-Yvon Profiler, with a polychromater of 1-meter focal length (38 channels) and a monochromator of 1-meter focal length.

3 RESULTS AND DISCUSSIONS

3.1 Coating morphology

The morphologies of the samples galvannealed in the laboratory were analyzed by SEM for both cross section and surface state, as shown in Figures 1 and 2. When the samples were heated in

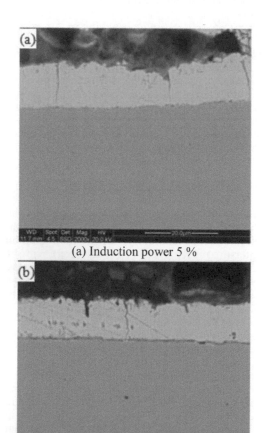

(a) Induction power 5 %

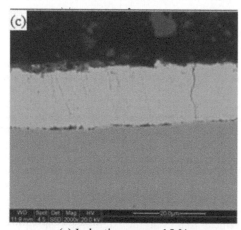

(c) Induction power 15 %

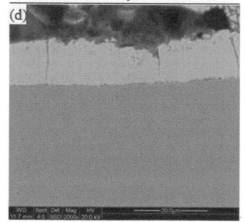

(b) Induction power 10 %

(d) Induction power 20 %

Figure 1. Cross-section morphology of experimental steel.

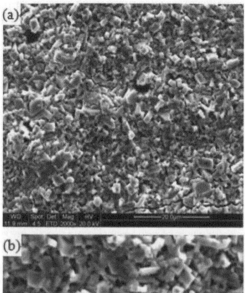

Figure 2. Surface morphology of experimental steel.

the induction heating furnace for a short time (10 s), there were no significant differences for samples 1 to 4 with four induction heating powers. The coating thickness difference will be discussed later. All the coatings attached to the matrix well. Some micro-cracks were generated in the coatings, but none of them extended to the matrix. The cracks may be caused by cutting or inlaying damage in the sample machining procedure. As coatings are too thin and having different hardnesses with the matrix, they can be very easily destroyed. The coatings distributed uniformly on the surface with no bare spots in the middle area of the sample, whereas the observation area was effective beneath the thermo-couple within a 90 × 90 mm area. From Figure 1, it can be seen that there is little effect of induction heating power on the cross-section state of coating.

Moreover, the surface quality is analyzed in detail, as shown in Figure 2: the size of the zinc granule demonstrated homogenously with no major dimensional discrepancy. For different morphologies of zinc-iron intermetallics as the ζ phase, δ phase, η phase and Γ phase, it was not obvious to specify the different intermetallics, as shown in Figure 2, but the columnar ζ phase and granular δ phase were relatively obvious in the surface, with most area of the surface exhibiting as granular with little size differences. The experimental steel galvannealed in the laboratory had a different coating state with industrial produced steel plate, whose zinc-iron intermetallics could be observed and identified (Fang et al. 2014).

3.2 Coating thickness

As shown in Figure 1, there was a little difference in the coating thickness for samples 1 to 4. Then, the effect of induction heating power on the coating thickness was analyzed, as shown in Figure 3a.

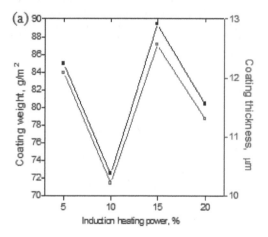

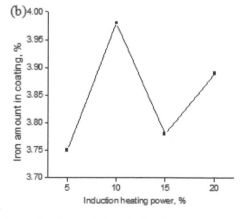

Figure 3. Effect of the induction heating power on the coating weight and thickness.

215

When the induction heating power was increased from 5% to 20%, as shown in Figures 1 and 2, the coating quality had no obvious difference but distributed homogenously in the surface with no obvious bare spots. However, the coating weight and thickness had a maximum value when the induction power was 10% and a minimum value when the power was 15%. With increasing power, the induction furnace will receive more energy from the controlling system and release correspondingly, so that the elements in the coating and the matrix will diffuse more sufficiently. If the power is too high so that the diffusion is over developed, the coating quality will be destroyed with more bare spots, and the sample will be thrown away.

The iron amount in the coating is displayed in Figure 3b. Interestingly, as the induction heating power increased from 5 to 20%, the iron amount had a little difference, so that all were within the range of 3.7 to 4% (wt%), lower than ~ 10% that is produced in the industry. The reason may be that there was no saturated iron in the zinc pot, and the induction furnace energy was less than the galvannealing furnace in the real producing plant, and the iron in the matrix could not diffuse to the coating sufficiently.

4 CONCLUSIONS

1. While conducting four parameters of induction heating power, the Zn-coating distributed uniformly on the sample surface with no bare spots. All the coatings attached to the matrix well, and some microcracks generated in the coating but none of them extended to the matrix.

2. Columnar ζ phase and granular δ phase in the coating were relatively obvious to specify in the surface for experimental steel, having a different coating state with industrial produced steel plate, whose zinc-iron intermetallics could be observed and identified.

3. The weight and thickness of the coating displayed a maximum value and a minimum value of 10% and 15%, respectively, when conducting the power. The iron amount had a little difference, so that all were within the range of 3.7 to 4% (wt%).

REFERENCES

Almeida, E. & Morcillo, M. 2000. Lap-Joint Corrosion of Automotive Coated Materials in Chloride Media. Part 2: Galvannealed Steel, *Surface and Coatings Technology,* 124: 180–189.

Bandyopadhyay, N. Jha, G. Singh, A.K. Rout, T.K. & Rani, N. 2006. Corrosion Behaviour of Galvannealed Steel Sheet, *Surface and Coatings Technology,* 20: 4312–4319.

Bindiya, S. Basavanna, S. & ArthobaNaik, Y. 2012. Electrodeposition and Corrosion Properties of Zn-V2O5 Composite Coatings, *Journal of Materials Engineering and Performance,* 21: 1879–1884.

Fang, F. Wang, L.X. & Wang, L.H. et al. 2014. Study on the Galvanealing Process for High Strength Low Alloying Auto Steel, *Applied Mechanics and Materials,* 692: 450–453.

Marder, A.R. 2000. The Metallurgy of Zinc-Coated Steel, *Progress in Materials Science,* 45: 191–291.

Santos, D. Raminhos, H. Costa, M.R. Diamantino, T. & Goodwin, F. 2008. Performance of finish coated galvanized steel sheets for automotive bodies, *Progress in Organic Coatings,* 62(3): 265–273.

Advanced Materials and Structural Engineering – Hu (Ed.)
© 2016 Taylor & Francis Group, London, ISBN 978-1-138-02786-2

High-temperature deformation and strain measurement for aircraft materials using digital image correlation

A.P. Feng
Foshan Polytechnic, Foshan, Guangdong, P.R. China

J. Liang, H. Hu & X. Guo
Xi'an Jiaotong University, Xi'an, Shaanxi, P.R. China

ABSTRACT: To solve the problem of thermal deformation measurement of the aircraft materials and\or components at high-temperature, a non-contact measurement method based on the digital image correlation is proposed and implemented. A high-performance experimental system, with an electric current heating device and an active imaging setup, is designed to reproduce the transient thermal environments experienced by hypersonic vehicles and to capture the digital images of the test specimen surface at various temperatures. A band pass filter, a linear polarizing filter, and some neutral density filters are used to reduce the intensity and noise in the images. The captured images are processed by using a seed-point diffusion algorithm and a fractionized facet tracking algorithm is proposed recently to extract the full-field thermal deformation. The experimental results show that the proposed technique is effective and robust for thermal deformation and strain measurement at the temperatures up to 2600 °C, and has a great potential in characterizing the thermo-mechanical behavior of the aircraft materials and components.

1 INTRODUCTION

High temperatures are often generated in the hypersonic vehicles (e.g., rockets and space crafts) due to transient aerodynamic heating. In order to ensure the safety and reliability of the hypersonic vehicles, it is vital to obtain the deformation and strain of the materials and structures at high temperature. Although the strain of a structure under high temperature can be measured using appropriate strain gauges, it is noted that the strain gauge technique only provides a point-wise measurement which is average over the gauge length, and the preparation and attachment of a high-temperature strain gauge is time-consuming (Heckmann et al. 2009, Huang et al. 2010).

The Digital Image Correlation (DIC) is an important non-contact strain measurement method that can be employed at high temperatures. Turner and Russell (1990) were the first to measure the full-field strains using the DIC at elevated temperatures. Subsequently, M.A. Sutton et al. (1998) used the DIC to measure the creep deformation up to 650 °C. Considering the CCD camera is very sensitive to the infrared radiation emitted by the heated specimen, Grant et al. (2009) proposed a full-field strain measurement up to 900 °C, with some band-pass filters. And Pan et al.

(2011) proposed a method by using a band-pass optical filter to eliminate the influence of black-body radiation of high-temperature objects at 1200 °C. In 2012, Chen et al. (2012, 2013) developed a monochromatic light illuminated stereo DIC system for high-temperature strain measurement of the titanium alloy up to 800 °C. Leplay et al. (2012) conducted a four-point bending test for ceramic materials by using DIC, the results have a strong consistency with the constitutive equations. In 2013, Garrett et al. (2013) used DIC to investigate the high-temperature fatigue crack growth of Haynes 230 at room temperature and 900 °C.

Although DIC has great advantages, such as non-contact, 3-D measurement and increasing measured points significantly over the conventional strain gauges, the weak-correlation phenomenon occurs and DIC computation fails when the image was captured at a higher temperature (>1300 °C). In this paper, a reliable high-temperature deformation measurement method was developed. The principle challenges in performing DIC at high temperature, including speckle preparation, imaging and reliable DIC method, were investigated and improved. The experimental results show that the method is capable of providing accurate DIC measurements up to 2600 °C.

2 EXPERIMENTAL DETAILS

2.1 *Experimental setup*

Figure 1 illustrates the schematic of the established experimental setup for a high-temperature strain measurement, which consists of a electric current heating device, a high temperature test chamber, a specimen stretching device and a self-developed 2D-DIC imaging system. Figure 2 shows the actual scenarios of the experiment, the heating lamps can heat the test specimen from room temperature to a maximum transient temperature of 2800 °C, and the imaging system composed of a CCD camera and a band pass filter is used to acquire the images of the test specimen during the deformation processes. In our experiment, the CCD camera is placed outside the chamber and the filters are fixed on the lens, air at the high-temperature chamber is first pumped into vacuum and filled with protective gas to prevent the reaction of tungsten at high temperatures.

2.2 *Speckle preparation*

Because the surface of the carbon fiber specimen is porous, the traditional speckle preparation method that involves painting may be out of place, especially under high-temperature. Speckle preparation is the key challenge in performing DIC for high-temperature deformation measurement. In this paper, Tungsten powder is used to make a speckle pattern by using the plasma spraying method (as shown in Fig. 3). Tungsten is an excellent candidate armor material for plasma components in nuclear fusion reactors or other equipment operating in high-temperature environments, because of its high physical sputtering threshold energy, high melting point (3410 °C), good thermal conductivity (180 W/mK) and suitable thermal expansion coefficient that is quite close to the thermal expansion coefficient of the measuring carbon material. During the spraying process, the tungsten ions will penetrate into the gaps of the carbon fibers. Thus the bonding strength between tungsten and carbon will be very high. Since the spraying angle of each specimen is different, and the size and shape of each mesh are different, the sprayed pattern is formed randomly. By using the plasma spraying method, a good speckle pattern can be achieved (see Fig. 4).

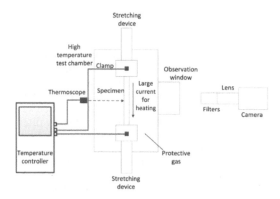

Figure 1. The schematic of experimental setup.

(a) Temperature controller

(b) Image acquisition device

Figure 2. High temperature controller and image acquisition device.

Figure 3. Nonlinear imaging model.

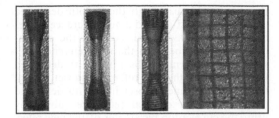

Figure 4. The test specimen with speckle.

In addition, to verify how the spraying speckle influences the properties of the carbon fiber specimen, a tensile test is conducted to investigate the mechanical properties of the specimens before and after speckle spraying, at the room temperature. The experimental results show that the stress-strain curves of the specimens before and after speckle spraying are in a good agreement, the influence of the plasma spraying on the mechanical properties of specimens can be ignored.

2.3 High-fidelity imaging method

In principle, DIC can be directly extended to reliable and accurate high-temperature displacement/strain measurement, provided that high-fidelity speckle images of the specimen with sufficient contrast can be recorded at elevated temperatures. But it is non-trivial to capture the high-fidelity speckle images at very high temperature, due to the thermal radiation of the specimen and the 'heat haze effect' (optical distortion of the sample's image) induced by the transient variations of the heated air's refractive index.

To solve this problem, researchers (Grant et al. 2009, Bing et al. 2011) have used a band-pass filter for high-temperature image acquisition. However, the light of carbon radiation suppressed by the band-pass filter is still too high, and the tungsten is difficult to distinguish from the captured image at 2600 °C, due to the thermal radiation. Therefore, a Linear Polarizing (LP) filter and three Neutral Density (ND) filters are used to suppress the black-body radiation in conjunction with a 450-nm band-pass filter. From Figure 5, we can see that the radiated light within the band-pass range of the optical band-pass filter is greatly increased with the increase of temperature, resulting in a sudden increase in overall brightness and a decrease in image contrast.

Figure 5(a) shows the image recorded at room temperature, the black spots are carbon fibers and the white spots are tungsten powder. Figure 5(b)~(f) shows the images recorded with different filter conditions at 2600 °C, the black spots are tungsten

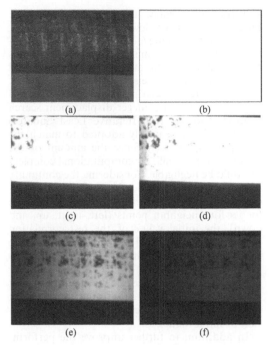

Figure 5. Images of the test specimen. (a) At room temperature. (b) Without filters, at 2600 °C. (c) With one PL filter, at 2600 °C. (d) With filter one PL filter and one 450-nm band-pass, at 2600 °C. (e) With one PL filter, one 450-nm band-pass filter, and two ND8 filters, at 2600 °C. (f) With one PL filter, one 450-nm band-pass filter, two ND8 filters, one ND2 filter and one ND4 filter, at 2600 °C.

powder and the white spots are carbon fibers. ND2, ND4, and ND8 are three different ND filters that can be used to achieve suppression of black-body radiation. Figure 5(d) shows the band-pass filter can reduce the intensity of the captured image to a certain extent, but the reduction is still not enough for DIC calculation. To sum up, more filters should be used to reduce the radiated light intensity sufficiently and to provide a high-fidelity speckle image.

2.4 Reliable DIC method

Generally, DIC can effectively obtain the full-field strains of a heated specimen at a temperature up to 1200 °C. However, when a temperature exceeds 1200 °C, the quality of digital images recorded inevitably decrease, direct use of DIC will fail. In this paper, to further improve the applicability of the DIC method, a seed-point diffusion algorithm and a fractionized facet tracking algorithm we proposed recently (Tang et al. 2010, Hu et al. 2014) were used for "weak-correlation image" matching and tracking reliably. The two algorithms are briefly described as follows.

As shown in Figure 6, before the image correlation analysis, one or more subsets are chosen as seed point(s), when the measuring area is specified and divided into subsets in the reference image. Then the selected seed point(s) can be searched out by an automatic integer displacement search method, and the ILS (Iterative Least Squares) algorithm is subsequently adopted to match the seed point(s) precisely. Since the amount of the seed point(s) is small, the computational complexity could be negligible. Considering the continuity of deformation, the seed point is then used to calculate the initial value of correlation parameters for its four neighbor points (left, right, up, and down), the initial values of the rest correlation parameters are set equal to the seed point. Once the four neighbor points are matched successfully, they can act as seed points for their neighbor points, until all points are matched. It is clear that the seed-point diffusion algorithm improves not only the computing efficiency but also the precision of initial values.

In addition, to further improve the performance of DIC, a fractionized facet tracking strategy is applied for DIC calculation. Taking the correlation analysis of a point (image facet) with the ith deformed images, for example, the measurement procedures are as follows. When the search in the ith deformed images for the seed point fails, the reference facet is first reshaped using the correlation parameter with the $(i-1)$ th deformed images (provided that the correlation calculation between the reference subset and

the deformed subset with the $(i-1)$th deformed images is successful), and a new reference subset can be obtained. Second, conduct the correlation calculation between the new reference subset and the deformed subset in the ith deformed images through the new reference subset. Third, match the deformed subset in the ith deformed images with the original reference subset, by using the correlation parameter of the new reference subset. Finally, repeat the above processes until all the deformed subsets during the whole deformation have been processed.

3 EXPERIMENTAL RESULTS AND DISCUSSION

In this experiment, the specimen was mounted to a fixture (as shown in Fig. 7) and heated to 2600 °C by electric current at first, next the specimen was stretched until fracturing at 2600 °C. A CCD camera equipped with 25 mm focal-length lens and the filters were used to acquire images of the specimen during the high-temperature deformation process.

The specimen after heating at high-temperature is shown in Figure 8, a color of the heating region and that of the clamping region is different. Figure 9(a) shows the measured strain fields of the specimen without stretching, no deformation happened and the strain value is 0. Figure 9(b) shows the time at which the average strain is 0.16% and uneven deformation of the specimen occurs. Figure 9(c) shows the time at which the specimen was broken, the temperature fell sharply and the image turns dark. It is clear that the greatest deformation appears at the lower section of the upper part of the specimen, and the overall deformation there is about 3.81%.

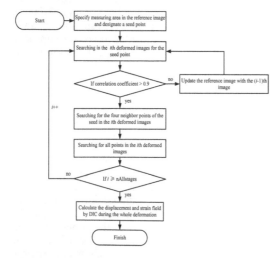

Figure 6.   Flow diagram of the improved DIC.

Figure 7.   Installation of the test specimen.

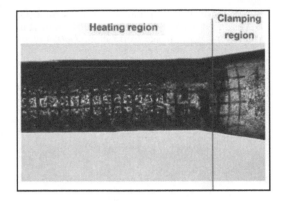

Figure 8. The test specimen after heating.

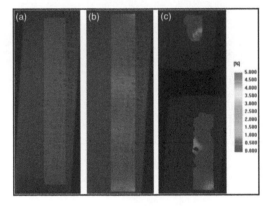

Figure 9. The surface strain fields of the specimen.

Table 1. The resulting thermal expansion coefficient.

Temperature	Measurement value coefficient $(10^{-6}C^{-1})$	Temperature	Nominal value $(10^{-6}C^{-1})$
10 °C~115 °C	16.7	20 °C~115 °C	16.6
10 °C~200 °C	17.2	20 °C~200 °C	17.0
10 °C~300 °C	16.8	20 °C~300 °C	17.2
10 °C~400 °C	17.5	20 °C~400 °C	17.5
10 °C~500 °C	17.7	20 °C~500 °C	17.9
10 °C~600 °C	18.4	20 °C~600 °C	18.6
10 °C~700 °C	18.9	20 °C~700 °C	
10 °C~800 °C	19.9	20 °C~800 °C	

Based on the above results, the material's thermal expansion coefficient can be calculated. A thermoscope was used to measure the temperature of the specimen center position. As shown in Table 1, the thermal expansion coefficient measured by our method is consistent with the nominal value found in a material manual. The relative deviation is about 2.3%. In addition, the thermal expansion coefficient at higher temperature (>600 °C) can be obtained by using our method.

4 CONCLUSIONS

A non-contact measurement method is used in this study to evaluate the deformation and strain measurements of the aircraft materials at high-temperature effects. The experimental results demonstrated that the proposed method is capable of measuring the full-field deformation of the carbon fiber materials at the temperatures ranging from room temperature to 2600 °C. Furthermore, the material's thermal expansion coefficient can also be calculated. In addition, it can be seen clearly that the greatest deformation appears at the upper part of the specimen, which provides a good solution for investigating the mechanical properties of the high-temperature materials.

ACKNOWLEDGMENT

The authors acknowledge the support of the National Natural Science Foundation of China (Grant No. 51275378) and the National Natural Science Foundation of China (Grant No. 51275389).

REFERENCES

Bing, P. et al. 2011. High-temperature digital image correlation method for full-field deformation measurement at 1200 °C, *Measurement Science and Technology*. 22(1): 015701.

Chen, X. Xu, N. Yang, L.X. & Xiang, D. 2012. High temperature displacement and strain measurement using a monochromatic light illuminated stereo digital image correlation system, *Measurement Science and Technology*. 23(12): 125603.

Chen, X. Yang, L.X. & Xu, N. et al. 2014. Cluster approach based multi-camera digital image correlation: Methodology and its application in large area high temperature measurement, *Optics & Laser Technology*. 57: 18–326.

Garrett, J.P. Huseyin, S. & Hans J.M. 2013. High temperature fatigue crack growth of Haynes 230, *Materials Characterization*, 75: 69–78.

Grant, B.M.B. Stone, H.J. Withers, P.J. & Preuss, M. 2009. High-temperature strain field measurement using digital image correlation, *The Journal of Strain Analysis for Engineering Design*. 44(4): 263–271.

Hao, H. Jin, L. & Tang, Z.Z. et al. 2014. Digital speckle based strain measurement system for forming limit diagram prediction, *Optics and Lasers in Engineering*. 55: 12–21.

Heckmann, U. Bandorf, R. & Gerdes, H. 2009. New materials for sputtered strain gauges, *Procedia Chemistry*. 1(1): 64–67.

Huang, Y.H. Liu, L. & Sham, F.C. 2010. Optical strain gauge vs. traditional strain gauges for concrete elasticity modulus determination, *International Journal for Light and Electron Optics*. 121(18): 1635–1641.

Liu, J. Sutton, M.A. & Lyons, J.S. 1998. Experimental investigation of near crack tip creep deformation in alloy 800 at 650^C, *International Journal of Fracture*. 91(3): 233–268.

Paul, L. Julien, R. & Sylvain, M. 2012. Identification of asymmetric constitutive laws at high temperature based on Digital Image Correlation, *Journal of the European Ceramic Society*. 32(15): 3949–3958.

Tang, Z.Z. Liang, J. & Xiao, Z.Z. et al. Three-dimensional digital image correlation system for deformation measurement in experimental mechanics, *Optical Engineering*, 49(10): 103601.

Turner, J.L. & Russell, S.S. 1990. Application of digital image analysis to strain measurement at elevated temperature, *Strain*. 26(2): 55–59.

Advanced Materials and Structural Engineering – Hu (Ed.)
© 2016 Taylor & Francis Group, London, ISBN 978-1-138-02786-2

Material selection of auto-body plates based on simplified models

Y.W. Luo, A. Cui, Q. Xu & S.Z. Zhang
State Key Laboratory of Automotive Simulation and Control, Jilin University, Changchun, China

ABSTRACT: In this paper, material selection method is studied for auto-body plates by using the simplified models instead of the actual plate models in the concept design stage. A car engine hood is applied in the paper to analyze and compare the performance change trends of simplified plate models and detailed actual models with different materials and thickness. The performances considered include static bending stiffness, static torsional stiffness, and modal frequency. The results show that the performance trends between simplified models and detailed actual models are consistent, which proves that it is feasible to apply simplified models in the material selection of auto-body plate structures. Finally, the framework of the material selection system using the simplified models of auto-body parts is designed.

1 INTRODUCTION

As weight reduction will make a contribution to the fuel consumption and carbon emissions reduction, the lightweight automobile has become a new trend in the automobile manufacturing industry. The application of lightweight materials plays a key role in the field. The body is one of the only three automotive assemblies (Huang & Huang 1992). In general, the BIW weight comprises around 20–25% of the total vehicle curb weight. Nowadays, the focus of the major automobile manufacturers and research institutions has been transferred to the research and application of lightweight materials.

To obtain the best performances and reduce weight, suitable materials are selected and applied in appropriate positions (Ermolaeva & Kaveline 2002, Lam & Behdinan 2003). However, the materials are selected based on a traditional experience, which cannot form a scientific guidance method. At present, the urgent problem is how to select materials of main auto-body parts scientifically and accurately.

If some of is not important details of automotive parts are focused greatly in the research of material selection, the difficulty of modeling and the number of cells of finite element models must be increased, and the sizes of finite element models are changed heavily, which will affect the calculation accuracy. There is a great advantage to select simplified structures with simple calculation in the early development stage. The aim of this paper is to select materials preliminarily rapidly and scientifically by using simplified models during early design stage.

2 MATERIAL SELECTION BASED ON SIMPLIFIED MODELS

Material selection is a process of making decision essentially based on the application functions and the design constraints (Huang & Dong 2006). The optimization process of selecting suitable materials is determined by analyzing the performances of materials.

Sohnshetty et al. (2011) proposed a method to select materials for auto body beams primarily and rapidly based on simplified models in the concept design stage. The method is aimed at some beam structure and the specific performances. The performance trends of simplified beam models with different materials are analyzed to sequence alternative materials.

This paper attempts to study the material selection for auto-body plates by using simplified models based on Sohnshetty et al. (2011). In this work, the main performance indexes are bending stiffness, torsional stiffness, and modal frequency. The five different materials are Bake Hardening

Table 1. Properties of materials.

Material	Young's modulus [E/MPa]	Poisson's ratio [μ]	Density [ρ/(kg/mm³)]
BH	2.10E+05	0.28	7.87E-06
AL	0.70E+05	0.30	2.70E-06
Mg	0.45E+05	0.35	1.84E-06
Ti	1.14E+05	0.34	4.41E-06
CFRP	0.80E+05	0.30	1.90E-06

steel (BH), Aluminum alloy (AL), magnesium alloy (Mg), titanium alloy (Ti), and Carbon Fiber Reinforced Plastic (CFRP). The main properties of these materials are listed in Table 1.

3 SIMPLIFIED MODELS OF AUTOBODY PLATES

In this paper, the auto-body plate is simplified as a rectangular thin plate with regular and equal cross-sections. The structure of the simplified plate model and the shape of the cross-sections are shown in Figure 1.

When the plate is cantilever structure, the formula of maximum bending deformation is:

$$\delta_{max} = 0.37239 * \frac{Fa^2}{D} \qquad (1)$$

where D is the bending stiffness of the thin plate, $D = Et^3/12(1-\mu^2)$.

The calculation formula of equivalent static bending stiffness at the stress point is:

$$k = \frac{F}{\delta_{max}} = \frac{D}{0.37239 * a^2} \qquad (2)$$

The formula of the largest torsion angle is:

$$\vartheta = \frac{3M_t a}{Gt^3 b} \qquad (3)$$

where M_t is the torque; G is the shear modulus, $G = E/2(1+\mu)$.

The formula of equivalent torsional stiffness is:

$$k = \frac{M_t}{\theta} = \frac{Gh^3 b}{3a} \qquad (4)$$

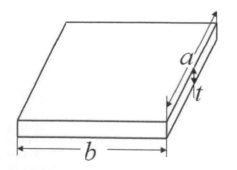

Figure 1. The simplified model of plate and the shape of the cross-section.

The formula of modal frequency in the free vibration condition is:

$$\varpi_{mn} = \frac{k_n^2}{a^2}\sqrt{\frac{D}{\overline{m}}} \qquad (5)$$

where k_n^2 is the coefficient of modal frequency; a and b are the length and width of the simplified plate; m and n are the vibration mode along direction a and direction b of the plate edge; $\overline{m}$ is the quality of the plate.

4 APPLICATION AND DISCUSSION

4.1 Application

Engine hood is applied to verify the feasible of material selection method for auto-body plates based on simplified models. Engine hood includes the outer plate, the inner plate, and the local reinforcing plate. All the thicknesses are 0.85 mm.

Bending on the analysis of engine hood, the model shown in Figure 2a (left) was used to analyze the bending stiffness using FEA. All freedom degrees of the two rear support points were constrained and the concentrated load applied at nodes arranged the lock latch (along the Z-axis negative direction). The size of concentrated load is 100 N.

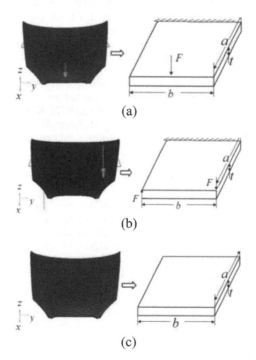

Figure 2. FE Model (left) and simplified model (right).

Table 2. Performance indexes of between FE and simplified models.

Performance		BH	AL	Mg	Ti	CFRP
EI	Detailed	1.23E-01	4.17E-02	2.71E-02	6.83E-02	4.71E-02
	Simplified	8.46E-01	2.91E-01	1.90E-01	4.78E-01	3.26E-01
GI	Detailed	8.90E+08	3.00E+08	1.99E+08	5.01E+08	3.43E+08
	Simplified	4.53E+05	1.45E+05	9.21E+04	2.35E+05	1.70E+05
Quantity	Detailed	2.18E-02	7.46E-03	5.08E-03	1.22E-02	15.25E-03
	Simplified	2.01E-02	6.89E-03	4.69E-03	1.12E-02	4.85E-03
Frequency	Detailed	3.11E+01	3.08E+01	3.01E+01	3.09E+01	3.93E+01
	Simplified	1.94E-03	1.95E-03	1.91E-03	1.95E-03	2.46E-03

The model was then simplified to calculate the bending stiffness (Equation 2) under similar loading conditions as in the FE model. The dimensions and loading conditions of the simplified model are shown in Figure 2a (right).

Torsion analysis of engine hood. The model shown in Figure 2b (left) was used to analyze the torsional stiffness using FEA. All the freedom degrees of the two rear support points are constrained (Cao 1983) and a moment of 100 N is applied at one end of the model. The model was then simplified to calculate the torsional stiffness (Equation 4) under similar loading conditions as in FE model. The dimensions and loading conditions of the simplified model are shown in Figure 2b (right).

Modal analysis of engine hood. The model shown in Figure 2c (left) was used to analyze the free vibration using FEA. There are no any constraint and load in the model. The model was then simplified to calculate the modal frequency (Equation 5) under similar loading conditions as in FE model. The dimensions and loading conditions of the simplified model are shown in Figure 2c (right).

4.2 Discussion

The performance indexes of the simplified model and detailed finite element model are listed in Table 2. The unit of bending stiffness is N/mm; the unit of torsional stiffness is $N \cdot mm/rad$; the unit of quantity is t; the unit of modal frequency is Hz.

According to the performance indexes comparison of the simplified model and detailed finite element model, materials can be sequenced preliminarily. According to bending stiffness, materials can be sequenced preliminarily as BH > Ti > CFRP > AL > Mg; according to torsional stiffness, materials can be sequenced preliminarily as BH > Ti > CFRP > AL > Mg; according to quality, materials can be sequenced preliminarily as Mg > CFRP > AL > Ti > BH; according to modal frequency, materials can be sequenced preliminarily as CFRP > BH > AL > Ti > Mg.

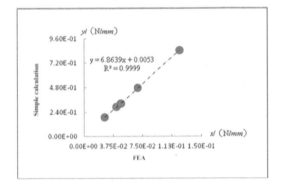

Figure 3. Bending stiffness trend.

From the performance indexes, it can be seen that, when the simplified models and the detailed finite element models use the same thickness and the same material, the absolute value of the performance indexes is greatly different, when the simplified models and the detailed finite element models use the different materials, the change trends of performance need to be studied. If the change trends are consistent, the material selection method of auto-body plates by using the simplified models is feasible. In this paper, regression analysis is used to analyze the consistent issues and the coefficient R^2 is used to determine the fitting degree of the independent variable and the dependent variable. If the fitting degree is greater than 0.5, the relationship is linear.

The scatter diagram of all the performance indexes can be made by regression analysis. The performance trends are shown in Figures 3–6.

As we can seen from Figures 3~6, the R^2 value of the bending condition is 0.9999; the R^2 value of the torsional condition is 0.9976; the R^2 value of the quantity trend is 1; the R^2 value of the free vibration condition is 0.9976. Therefore, the performance trends of the simplified model and detailed finite element model are consistent. The material selection method for auto-body plates by using simplified plate model is feasible in the concept design stage.

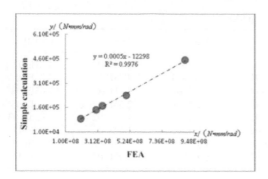

Figure 4. Torsional stiffness trend.

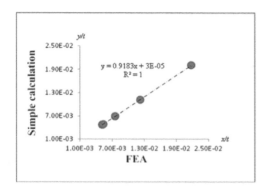

Figure 5. Quantity trend.

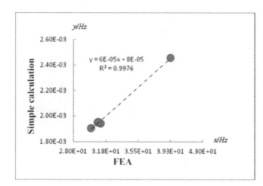

Figure 6. Modal frequency trend.

5 FRAMEWORK DESIGN OF MATERIAL SELECTION SYSTEM

The system includes the following modules as shown in Figure 7:

1. Size parameters module: according to the reference automobile, the basic dimensions of simplified models for auto-body plates are determined to construct the simplified plate models.

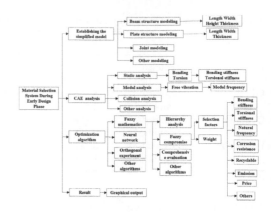

Figure 7. Frame of function modules.

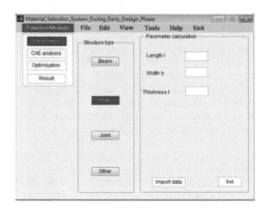

Figure 8. Size input interface of plate structure.

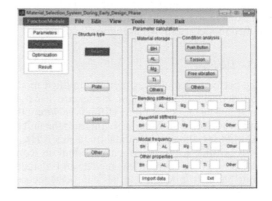

Figure 9. CAE analysis interface.

2. CAE analysis module: the boundary conditions of simplified models are added and analyzed; the performance indexes of bending stiffness, torsion stiffness, and modal frequency are computed in the module.

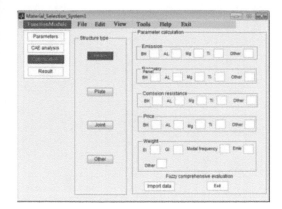

Figure 10. Optimization algorithm interface.

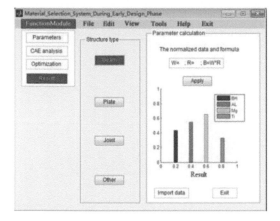

Figure 11. Result interface.

3. Selection factors and optimization algorithms module: in order to select the appropriate materials for the auto-body plate, there are other selection factors should be considered besides bending stiffness, torsional stiffness, and modal frequency. The materials can be selected synthetically by using optimization algorithms (Huang 1997).
4. Result module: the selection result is output in the graphic form.

The main module interfaces of material selection system are designed as shown in Figures 8–11.

6 CONCLUSIONS

This paper proposed a material selection method for auto-body plates based on the simplified models in the concept design stage. The simplified models were constructed, and the calculation formulas of the performance indexes are deduced. The feasibility of the material selection method is verified by the application of the engine hood. The results showed that: the values of the performances of the simplified models and detailed actual models with same material are greatly different. In addition, the performance trends of the simplified models and detailed actual models with different materials are the same. The application demonstrated the feasibility of the material selection method. A material selection system for auto-body beams and plates in the concept design stage based on the simplified models has been researched preliminarily.

ACKNOWLEDGMENT

The work is supported by the project of National Science and Technology, P.R. China (2011BAG03B01).

REFERENCES

Cao, G.X. 1983. *Elastic vibration of rectangular thin plate.* Chinese Architectural Publications, China.
Ermolaeva, N.S. & Kaveline, K.G. 2002. Materials selection combined with optimal structural design concept and some results. *Material & Design,* 23(5):459–470.
Huang, H.Z. 1997. *Fuzzy optimization theory and application of mechanical design.* Science Publications, China.
Huang, J.R. & Dong, H.H. 2006. *Pair analysis and application of material selection method,* Mechanical Design Publications, China.
Huang, T.Z. & Huang, J.L. 1992. *The structure and design of autobody,* Machinery Industry Publications, China.
Lam, K.P. & Behdinan, K. 2003. A Material and gauge thickness sensitivity analysis on the NVH and crashworthiness of automotive instrument plate support. *Thin-walled Structures,* 41(11):1005–1018.
Sohmshetty, R. Ramachandra, R. Mariappasamy, R & Karuppaswamy, S. 2011. Material Selection During Early Design Phase Using Simplified Models, *SAE International,* 4(1): 759–770.
Yan, Q.W. & Zhang, X.K. 2008. *The finite element modeling and lens characteristics analysis of autobody in white.* TIABJIN AUTO, China.

Advanced Materials and Structural Engineering – Hu (Ed.)
© *2016 Taylor & Francis Group, London, ISBN 978-1-138-02786-2*

MFC/NFC aerogel-Ag composite and its application

J.H. Yan & G.B. Kang
Guangdong Industry Technical College, Guangdong, China

R.M. Xu
Ball State University, IN, USA

ABSTRACT: MFC/NFC aerogel was prepared, silver nanoparticles (AgNPs) were loaded in the aerogel. Ag-NPS were obtained successfully in the matrices without adding any other binders. The aerogel AgNPS size was defined by TEM. The thermal-stability was marked by TGA. This aerogel showed superior properties for Ag-NPS loading and excellent antibacterial activity. MFC/NFC aerogel will have potential application in antibacterial substrate.

1 INTRODUCTION

Metal particles in the nanometer size range have attracted considerable interest in recent years, as they have many attractive applications in various fields (Shervani et al. 2008, Zhong & Maye 2001, Kim et al. 2001, Zhu et al. 2000, Li et al. 1999, Chen et al. 2007) due to their unique size-dependent optical, electrical, magnetic, and antimicrobial properties. Silver nanoparticles (AgNPs) are believed to be size-dependent to their antimicrobial properties, where smaller AgNPs provide stronger antimicrobial effect than larger AgNPs (Chen et al. 2007, Mahndra et al. 2009). The size of the nanoparticle implies that it has a large surface area to come in contact with bacterial cells and hence, it will have a higher percentage of interaction than bigger particles (Morones et al. 2005, Raimondi et al. 2005, Pal et al. 2007). In this case, aggregation of AgNPs must be avoided because it will drastically decrease the accessibility of nanoparticles' surfaces, resulting in insufficient functionality. It is basically difficult to disperse metallic nanoparticles in a solvent, as nanoparticles tend to aggregate due to their high surface energy. Facile synthesis of morphologically controlled nanoparticles is a significant challenge in the field of nanotechnology.

In this work, we provide a different method to load nanoparticles using MFC/NFC aerogel as templates. MFC/NFC aerogel was prepared, then nanoparticles were generated on the aerogel. Silver was an example. This biocompatible AgNPs/ matrix composite was found to exhibit excellent anti-microbial activities for Escherichia coli. Generally, MFC-based AgNPs aerogel showed the best properties for nanoparticles loading than NFC aerogel.

2 EXPERIMENTAL PART

2.1 Silver NPs (AgNPs) loading

MFC/NFC aerogel was prepared in our lab. In situ loading of silver nanoparticles onto the aerogel was carried out through the reduction of 10 mM $AgNO_3$ solution. Under ambient temperature (~25°C), several drops of $AgNO_3$ solution were spread on the porous matrix, keeping totally wet for enough long time till the substrates were not an absorbing solution. It is similar to that the substrates were soaked in $AgNO_3$ solution for saturation and uniformity, but much less solution was used in this experiment.

After a certain time of air drying, a partially dehydrated matrix was obtained. It was then immersed in aqueous solution of $NaBH_4$ (50 mM) for 20 min. Meanwhile, the color of the resultant samples turned to yellow or dark brown due to the reduction of Ag^+ into silver nanoparticles. The composite was rinsed with Milli-Q water three times to remove water-soluble substances and free silver particles. Finally, the composite was freeze dried.

2.2 TEM preparation of MFC/NFC aerogels with ANPs

A JEOL 2000FX Transmission Electron Microscope (TEM) operating at 20.0 kV was utilized to define the AgNPs in the MFC/NFC aerogels. A specimen can be prepared by cutting the sample into thin slices using a diamond saw, then cutting 3-mm-diameter disks from the slice, thinning the disk on a grinding wheel, dimpling the thinned disk, then ion milling it to electron transparency.

2.3 Antibacterial activity test

The antibacterial activity of the aerogels was tested against Escherichia coli (E. coli) Gram negative bacteria, using the viable cell counting method. Briefly, about 100 μL E. coli was cultivated in 100 mL of a broth nutrient solution, to give a bacterial concentration of about 7×1011 CFU/mL. Then, 1 mL of the bacteria/nutrient solution was added to 9 mL of sterilized broth nutrient solution (0.8%). Several decimal dilutions were performed until the bacterial concentration increased from 7×103 to 7×107 CFU/mL. NFC, NFC/chitosan, MFC/chitosan, and their silver loaded counterparts, NFC/chitosan/Ag-NPs, MFC/chitosan/Ag-NPs, were used in the antibacterial tests. The accurate weight of the aerogel samples (100 mg) was assigned to the experiment. To perform the antibacterial testing, aerogels of the respective samples were put into 10 mL of the bacteria/nutrient solution incubated in a shaker at 37 °C for 12 h. After the exposure of the bacteria to films, 100 μL of the bacterial solution was taken out and quickly spread on a plate containing nutrient agar. Plates containing bacteria were incubated at 37 °C for 24 h, and then the numbers of the surviving colonies were counted. These results were compared to the number of bacteria colonies of the untreated control.

3 RESULTS AND DISCUSSIONS

3.1 Formation of AgNPs on MFC/NFC aerogel

The MFC aerogel and NFC aerogel microstructures were shown below in Figure 1. In Figure 1, we can see the difference of MFC aerogel and NFC aerogel. MFC aerogel has open structure and has more pores, the texture is more uniform. NFC aerogel seems tighter and has fewer pores. NFC holds more carboxyl groups and more easily to form hydrogen bonds. We can image that MFC aerogel will have a good performance to nanoparticles loading and filling.

After AgNPs loading, the aerogels structure and nanoparticle distribution were shown in Figure 2. In Figure 2, MFC aerogel has better particles distribution than NFC aerogel, which the nanoparticles scattered more evenly in MFC aerogel.

In order further to know AgNPs size and distribution for MFC aerogel, TEM was conducted (see Fig. 3). The particles are mono-dispersed and have very small size with a mean diameter of 7.08 nm, most particles in the range of 4–9 nm. This is because the aerogel was treated by very dilute $AgNO_3$ solution (10 mM), there were less Ag+ surrounding nanofibrils, typically at negatively charged groups which are the anchors to be

Figure 1. MFC aerogel (left) and NFC aerogel (right) SEM images.

a nucleus of particle growth. Nanofibrils placed an important role to stop physically the Ag+ free move, than mono-dispersed small particles were formed.

3.2 TGA test

In view of the importance of thermal stability in many applications of AgNPs loaded MFC/NFC aerogels, we examined thermal decomposition of composites loaded AgNPs by Thermogravimetry (TGA) in a nitrogen atmosphere under 500°C, as shown in Figure 4. In all TGA curves, the small weight losses below 150°C apparently resulted from evaporation of adsorbed moisture. Under nitrogen, the most weight loss of the MFC-Ag aerogel taking place at 280°C, meanwhile, NFC-Ag aerogels have a lower decomposition

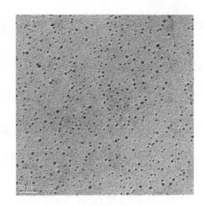

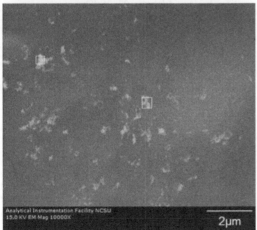

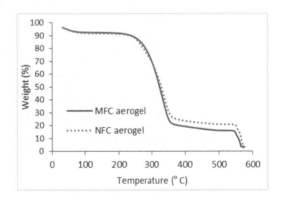

Figure 3. TEM images of MFC aerogels AgNPs distribution.

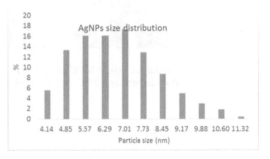

Figure 2. MFC aerogel (left) and NFC aerogel (right) SEM images after AgNPs loading.

Figure 4. MFC/NFC aerogels TGA.

temperature of 250°C around. In order to obtain the AgNPs amount in aerogel, TGA under oxygen in the range of 500–575°C was continued. The isothermal time for 25 min at 575°C was set up. The AgNPs amount in MFC aerogel has more than that in NFC aerogel.

3.3 *Antibacterial activity test*

The antibacterial activity of MFC/NFC aerogels was tested against E. coli. by using the viable cell-counting method. The effects of aerogels on the growth of the recombinant bacteria E. coli are shown in Figure 5. As shown in the plates, no bacterial colonies were observed at concentrations of 7×10^7 CFU/mL for both MFC aerogel-AgNPs and NFC aerogel-AgNPs which represent

the highest antibacterial activity. A lot of bacterial colonies were observed for MFC and NFC aerogels which implied the poor antibacterial activity. AgNPs in these aerogels imported great inhibition of bacterial. MFC aerogel-AgNPs may have better antibacterial property than NFC aerogel-AgNPs, since the former has excellent nanoparticles distribution. This should be further examined.

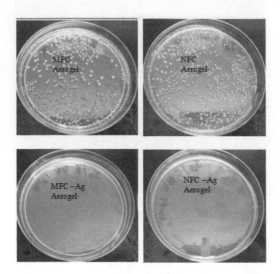

Figure 5. Antibacterial activity of MFC aerogel, NFC aerogel, MFC-AgNPs aerogel, and NFC-AgNPs aerogel against *E. coli*.

4 CONCLUSIONS

In this work, we have shown that MFC/NFC aerogel can be facilely used as substrate for preparation and stabilization of nanoparticles of AgNPs. AgNPs are well dispersed on the MFC aerogel. The MFC/NFC aerogel-AgNPs hybrid performed strong antibacterial properties. We, therefore, propose that the procedure described here, the metal ions are anchoring on the aerogel due to their negative charged groups, to first perform chemical reactions as the core, and then grow particles, will provide a technically more feasible procedure. The work is also a demonstration that highly fibrillated MFC/NFC aerogel shows desirable properties that lead to new methods in technology. MFC aerogel is better for AgNPs loading than NFC aerogel here.

REFERENCES

Chen, C. Wang, L. Yu, H. Wang, J. Zhou, J. Tan, Q. & Deng, L. 2007. Morphology-controlled synthesis of silver nanostructures via a seed catalysis process, *Nanotechnology,* 18: 115612–115620.

Chen, C. Wang, L. Yu, H. Wang, J. Zhou, J. Tan, Q. Deng, L. 2007. Morphology-controlled synthesis of silver nanostructures via a seed catalysis process, *Nanotechnology,* 18: 115612.

Kim, Y. Johnson, R.C. & Hupp, J.T. 2001. Gold Nanoparticle-Based Sensing of "Spectroscopically Silent" Heavy Metal Ions, *Nano Letters.* 1(4): 165–167.

Li, M. Schnablegger, H. & Mann, S. Coupled synthesis and self-assembly of nanoparticles to give structures with controlled organization, *Nature,* 402: 393–395.

Mahndra, R. Alka, Y. & Aniket, G. 2009. Silver nanoparticles as a new generation of antimicrogials, *Biotechnology Advances*, 27: 76–83.

Morones, J.R. Elechiguerra, J.L. Camacho, A. & Ramirez, J.T. 2005. The bactericidal effect of silver nanoparticles. *Nanotechnology,* 16: 2346–2353.

Pal, S. Tak, Y.K. & Song, J.M. 2007. Does the antibacterial activity of silver nanoparticles depend on the shape of the nanoparticle? A study of the gram-negative bacterium Escherichia coli. *Applied and Environmental Microbiology,* 27(6): 1712–1720.

Raimondi, F. Scherer, G.G. Kotz, R. & Wokaun, A. 2005. Nanoparticles in energy technology: examples from electochemistry and catalysis. *Angewandte Chemie International Edition,* 44: 2190–2209.

Shervani, Z. Ikushima, Y. Sato, M. Kawanami, H. Hakuta, Y. Yokoyama, T. Nagase, T. Kuneida, H. & Aramaki, K. 2008. Morphology and size-controlled synthesis of silver nanoparticles in aqueous surfactant polymer solutions, *Colloid and Polymer Science,* 286(4): 403–410.

Zhong, C. & Maye, M.M. 2001. Core-Shell Assembled Nanoparticles as Catalysts, *Advanced Materials.* 13(19): 1507–1511.

Zhu, J. Liu, S. Palchik, O. Koltypin, Y. & Gedanken, A. Shape-controlled synthesis of silver nanoparticles by pulse sonoelectrochemical methods, *Langmuir,* 16(16): 6396–6399.

Advanced Materials and Structural Engineering – Hu (Ed.)
© *2016 Taylor & Francis Group, London, ISBN 978-1-138-02786-2*

Nanopaper with Ag composite and its application

J.H. Yan
Guangdong Industry Technical College, Guangdong, China

R.M. Xu
Ball State University, IN, USA

ABSTRACT: Nanopaper with silver nanoparticles loading was explored in this paper. Ag ions were spread into nanopaper, followed by an immersion in $NaBH_4$ solution. Silver nanoparticles (AgNPs) were obtained successfully in the matrices without adding any other binders. The size of AgNPs was defined by TEM. Its antibacterial activity was assessed. This nanopaper showed superior properties for AgNPs loading and antibacterial activity. Nanopaper will have potential applications in nanoparticles loading in the functional paper field.

1 INTRODUCTION

Micro/Nanofibrillar Cellulose (MFC/NFC) is isolated from natural cellulose fibers by using basic mechanical action after enzyme or chemical pretreatment. MFC/NFC has been widely investigated recently due to its sustainable, renewable, and biocompatible nontoxic properties for low cost and relatively high physical-chemical stability materials (Eichhorn et al. 2010, Moon et al. 2011, Dong et al. 2013). Moreover, MFC/NFC typically has a higher aspect ratio. MFC/NFC has an attractive application of providing a more beneficial template to accommodate nanoparticles. It has been used as nanoparticles templates successfully, for example, NFC with silver nanoparticles using UV reduction (Dong et al. 2013, Schlabach et al. 2011), NFC with magnetic particles (Nypeloet al. 2012), and NFC with drug nanoparticles (Valo et al. 2013, Kolakovic et al. 2012). All NFC templates here are in slurry, that is, huge amount of wash water is necessary to remove unattached reactants and resultants. Since reaction happens in the whole system, a lot of reactants are consumed in the water and a big post-treatment load is used.

In this work, we provide different methods to load nanoparticles using MFC/NFC as templates. Nanopaper was prepared using MFC/NFC, and then nanoparticles were generated on the nanopaper. Silver was an example. This biocompatible AgNPs/matrix composite was found to exhibit high antimicrobial activities for *Escherichia coli*. Generally, MFC-based AgNPs nanopaper showed the best properties for nanoparticles loading between NFC and regular paper-based membranes.

2 EXPERIMENTAL PART

2.1 Silver NPs (AgNPs) loading

Nanopaper was prepared in our lab. *In situ* loading of silver nanoparticles into the nanopaper was carried out through the reduction of $AgNO_3$ solution with varied concentrations (10 mM, 50 mM, 100 mM, and 500 mM). Under ambient temperature (~25°C), several drops of $AgNO_3$ solution were spread on the porous matrix, being kept totally wet for long enough time till the substrates were not absorbing solution any more. It was similar to that the substrates were soaked in $AgNO_3$ solution for saturation and uniformity, but much less solution was used in this experiment.

After a certain time of air drying, a partially dehydrated matrix was obtained. It was then immersed in ethanol for 5 min and then in an aqueous solution of $NaBH_4$ (50 mM) for 20 min. Meanwhile, the color of the resultant samples turned to yellow or dark brown due to the reduction of Ag+ into silver nanoparticles. The composite was rinsed with Milli-Q water for three times to remove water-soluble substances and free silver particles. Finally, the composite was air-dried. The nanopaper microstructures are shown in Figure 1.

2.2 TEM preparation of nanopaper with AgNPs

A JEOL 2000FX Transmission Electron Microscope (TEM) operating at 20.0 kV was utilized to define the AgNPs in the nanopaper.

Figure 1. SEM of nanopaper (top for nanopaper from NFC and bottom for nanopaper from MFC).

2.3 Antibacterial activity

Antibacterial testing was performed against Gram negative bacteria, *E. coli*, using the disk diffusion method. The assessment was conducted based on the disk diffusion method of the US Clinical and Laboratory Standards Institute (CLSI). Both the neat (pure cellulose from Whitman filter paper) and the AgNPs-containing samples (filter paper, NFC nanopaper, and MFC nanopaper) were cut into circular disks (1 inch in diameter). Each of the specimens and the control samples were placed on nutrient agar in a Petri dish, and then incubated at 37°C for 24 h. If inhibitory concentrations were reached, there would be no growth of the microbes, which could be seen as a clear zone around the disk specimens. These were photographed for further evaluation.

In order to avoid the disk diffusion method limitation, a second test of the surviving colonies numbers was conducted. The antibacterial activity of the fiber mats against *Escherichia coli* (*E. coli*) Gram negative bacteria, which is commonly found on burned wounds, was measured by using the viable cell counting method. Briefly, about 100 μL of *E. coli* was cultivated in 100 mL of a nutrient broth solution, to give a bacterial concentration of about 7×1011 CFU/mL. Then, 1 mL of the bacteria/nutrient solution was added to 9 mL of sterilized nutrient broth solution (0.8%). Several decimal dilutions were performed until the bacterial concentration increased from 7×10^3 to 7×10^7 CFU/mL. Paper/AgNPs, NFC/AgNPs, and MFC/AgNPs were used in the antibacterial tests. The weight and size of the films were 100 mg as disks with a 2.8-cm diameter. To perform the antibacterial tests, the films of the respective samples were put into 10 mL of the bacteria/nutrient solution incubated in a shaker at 37°C for 12 h. After the exposure of the bacteria to the films, 100 μL of the bacterial solution was taken out and quickly spread on a plate containing nutrient agar. The plates containing bacteria were incubated at 37°C for 24 h, and then the numbers of the surviving colonies were counted. These results were compared with the number of bacteria colonies of the untreated control that had not been exposed to the films.

3 RESULTS AND DISCUSSIONS

3.1 Nanoparticles formation on nanopaper

The formation of AgNPs into MFC/NFC nanopaper was confirmed by TEM observation (Fig. 2). The existence of AgNPs can be seen clearly in the films. For the paper, we also found a lot of particles dispersed well while we hardly found them in the SEM images, although the paper composite has changed color to such as yellow, brown, etc., containing impurities for sure. It implies that AgNPs are not located on the surface but inside the paper.

In order to define the sizes of the nanoparticles and their distribution, the Image J software was used. The analyzed results of TEM images are obtained and also showed in Figure 2. The particle size on the MFC is obviously larger than that on the regular paper. The sizes of most of the AgNPs on the MFC are in the range of 10–30 nm, which accounts for more than 60% particles. The sizes of AgNPs on the regular paper are much smaller of which 70% or more are smaller than 15 nm, and 45% of the AgNPs particles are smaller than 10 nm. Both MFC and paper can be good supports for mono-dispersed AgNPs composites.

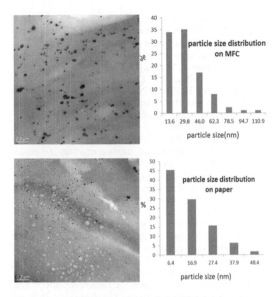

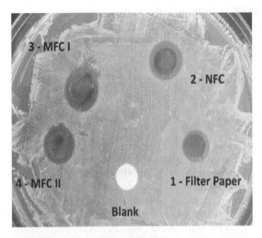

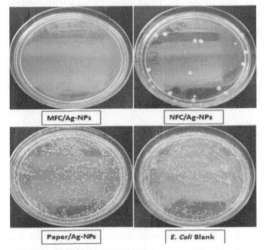

Figure 2. TEM of AgNPs-loaded MFC (top) and regular paper (bottom) and their particle size distributions.

However, there exists an optimum AgNPs size to reach the best antibacterial performance (Panacek et al. 2006), which will be explained in the following section.

3.2 Antibacterial activities

The potential of using AgNPs-containing substrates as functional materials was assessed by observing their antibacterial activities (based on the disk diffusion method) against common bacteria, *E. coli*. All tested samples were loaded with the same concentration of silver nitrate (100 mM). The activity of the neat cellulose sample against these bacteria was used as a control. Both the diameters of the inhibition zone of the blank and those of the specimens were analyzed accordingly, as seen in Figure 3. According to the results obtained, the untreated blank showed no activity against the tested bacteria. For the AgNPs-containing specimens, inhibitory zones were evident, and MFC/NFC obviously had a bigger clear zone than the filter paper.

It is accepted that the antibacterial mechanism of Ag is from the Ag+. Feng et al. (2000) demonstrated that, upon interacting with Ag^+ ions, both *E. coli* and *S. aureus* underwent a series of events that led to their demises. In addition, they postulated that the bactericidal mechanism was based on the ability of Ag^+ ions to bind with certain chemical functionalities of the cell wall, the cytoplasm and the nucleus. This causes DNA molecules to condense, the cytoplasmic membrane to detach from

Figure 3. Antibacterial activities of AgNPs-containing MFC/NFC and paper samples by disk diffusion (top) and colony number (bottom).

the cell wall, and the cell wall to become severely damaged (Morones et al. 2005). A similar observation of the bactericidal effect of AgNPs and/or Ag^+ ions has been reported by Shrivastava and his co-workers (2007).

Here, AgNPs are located inside the filter paper samples, while on the surfaces of the MFC/NFC samples, according to the TEM images in Figure 2. AgNPs of MFC/NFC are mostly exposed to air, and can more easily be oxidized into Ag+, which is the element to produce the antibacterial effect, leading to higher antibacterial activities than paper samples.

Since MFC and NFC expressed close antibacterial activities, the second test method was conducted. The test result was shown in Figure 3 (right). From Figure 3, we obviously see that the

MFC-AgNPs has the highest antibacterial activity since there is no colony in the plate, while the NFC-AgNPs plate has several grown colonies. That is, MFC nanopaper is a better substrate for nanoparticles loading than NFC nanopaper, typically for silver nanoparticles.

4 CONCLUSIONS

In this work, we have shown that nanopaper can be facilely used as a substrate for preparation and stabilization of AgNPs. AgNPs are well dispersed on the film with a range diameter of 10–30 nm. The hybrid nanopaper films performed strong antibacterial properties, and MFC nanopaper is better than NFC nanopaper. We, therefore, propose that the procedure described here, in which the metal ions are anchoring on the nanopaper due to their negatively charged groups, to first perform chemical reactions as the core and then grow particles, will be a technically more feasible procedure. The work is also a demonstration that highly fibrillated nanopaper shows desirable properties that can lead to new methods in technology.

REFERENCES

Dong, H. Snyder, J.F. Tran, D.T. & Leadore, J.L. 2013. Hydrogel, aerogel and film of cellulose nanofibrils functionalized with silver nanoparticles, *Carbohydrate Polymers*, 95: 760–767.

Eichhorn, S.J. Dufresne, A. Aranguren, M. Marcovich, N.E. Capadona, J.R. & Rowan, S.J. 2010. Current international research into cellulose nanofibres and nanocomposites. *Journal of Materials Science*, 45(1): 1–33.

Feng, Q.L. Wu, J. Chen, G.Q. Cui, F.Z. Kim, T.N. & Kim, J.O. 2000. A mechanistic study of the antibacterial effect of silver ions on Escherichia coli and Staphylococcus aurous, *Journal of Biomedical Materials Research Part A*, 52: 662–668.

Kolakovic, R. Peltonen, L. Laukkanen, A. Hirvonen, J. & Laaksonen, T. 2012. Nanofibrillar cellulose films for controlled drug delivery, *European Journal of Pharmaceutics and Biopharmaceutics*, 82: 308–315.

Moon, R.J. Martini, A. Nairn, J. Simonsen, J. & Youngblood, J. 2011. Cellulose nanomaterials review: Structure, properties and nanocomposites. *Chemical Society Reviews*, 40(7): 3941–3994.

Morones, J.R. Elechiguerra, J.L. Camacho, A. Holt, K. Kouri, J.B. & Ramírez, J.T. 2005. The bactericidal effect of silver nanoparticles. *Nanotechnology*, 16: 2346–2353.

Nypelo, T. Pynnönen, H. Sterberg, M.O. Paltakari, J. & Laine, J. 2012. Interactions between inorganic nanoparticles and cellulose nanofibrils, *Cellulose*, 19: 779–792.

Panacek, A. Kvitek, L. & Prucek, R. 2006. Silver Colloid Nanoparticles: Synthesis, Characterization, and Their Antibacterial Activity, *The Journal of Physical Chemistry B*, 110: 16248–16253.

Schlabach, S. Ochs, R. Hanemann, T. & Szabo, D.V. 2011. Nanoparticles in polymer-matrix composites, *Microsystem Technology*, 17: 183–193.

Shrivastava, S. Bera, T. Roy, A. Singh, G. Ramachandrarao, P. & Dash, D. 2007. Characterization of enhanced antibacterial effects of novel silver nanoparticles, *Nanotechnology*, 18: 225103.

Valo, H. Arola, S. Laaksonen, P. Torkkeli, M. Peltonen, L. Linder, M.B. Serimaa, R. Kuga, S. Hirvonen, J. & Laaksonen, T. 2013. Drug release from nanoparticles embedded in four different nanofibrillar cellulose aerogels, *European Journal of Pharmaceutical Sciences*, 50: 69–77.

Advanced Materials and Structural Engineering – Hu (Ed.)
© *2016 Taylor & Francis Group, London, ISBN 978-1-138-02786-2*

Effective thermal conductivity of multiple-phase transversely isotropic material having coupled thermal system

S.A. Hassan, A. Israr & H.M. Ali
Department of Mechanical Engineering, Institute of Space Technology, Islamabad, Pakistan

W. Aslam
Department of Materials Science and Engineering, Institute of Space Technology, Islamabad, Pakistan

ABSTRACT: In this study, effective thermal conductivity is obtained for two-phase transversely isotropic bulk material. Generalized self-consistent model and simple energy balance principle are used to derive the system of equations. Spherical inclusions are introduced into the model as a void phase to investigate the effects of imperfections on the thermal conduction of bulk materials. The spherical voids are imagined to be carrying process generated gasses or containing entrapped air within the material for observing the thermal conduction properties of the materials. Instead of conduction mode, convection conduction coupled mode is considered to implement a more realistic thermal conduction system. The effects of volume fraction of bulk material V_s and thermal convection coefficient h, on the effective thermal conductivity of the material are discussed in detail. The results show an extremely good agreement with the existing available models.

1 INTRODUCTION

Thermal conduction plays a vital role in the thermal characterization of the material and its use as a thermal sink, insulators (Xu & Yagi 2004) and in the electronics industry (Graebner 1995). A lot of work has been done in the past to find empirical and mathematical relations to predict the effective transverse thermal conductivity of the multi-phase materials and composites. Analytical models based on micro mechanics of the constituent phases such as, Differential model, Mori Tanaka's model, Budiansky theoretical model, Halpin Tsai's theoretical model, self-consistent model, and Generalized Self-Consistent Model (GSCM) successfully predicted thermal properties of composite materials over the years (Progelhof et al. 1976). Hashin (1962) developed GSCM to determine the effective thermal conductivity. In addition, a group of researchers (Christensen & Lo 1979, Lee & Kim 2010) considered the effects of the interaction between reinforcement materials and matrix to achieve more realistic results. Researchers (Lee et al. 2006) also reported the use of GSCM to find the effective thermal conductivities based on the spherical and cylindrical models for particulate and transversely isotropic fiber composites with thermal contact conductance. The method of solving one-step differential equations in the GSCM model to derive respective equations for

thermal conduction in longitudinal and transverse direction is also reported (Lee & Kim 2013). Recent (Hassan et al. 2013) study based on the possible effects of the spherical imperfections in the form of micro-voids on the thermal conduction of the transversely isotropic fiber composites using GSCM was reported. Olanrewaju et al. (Olanrewaju et al. 2011) employed the convective surface boundary conditions to study heat generation effects on thermal boundaries in the plates. Jiang et al. presented the effect of radiations in the overall thermal conduction of the porous materials using GSCM (Jiang et al. 2012).

The present study is the extension of the work presented earlier by Hassan et al. (2013) to determine the effects of the voids in the thermal conductivity materials, such as in silicon chips in the electronics industry (Graebner 1995). Instead of composite, an imperfect two-phase matrix-void material is considered. This time an analytical approach is considered using convection-conduction coupled thermal system and hence, associated boundary conditions are generated for measuring the effects of convection in the region of voids. Effects of the volume fraction of solid material V_s and thermal convection coefficient h between solid material and void phase on the effective thermal conductivity of the material K_{eff} are discussed in the subsequent sections followed by important results and conclusions.

2 COUPLED CONVECTION, CONDUCTION THERMAL MODEL

The three-phase system including void, solid bulk material containing void, and effective medium are shown in Figure 1.

The concentric sphere is shown in Figure 1 with the void imagined to be at the center of the Representative Volume Element (RVE) has a radius a. The void is imagined to be carrying some process generated gas or possibly air, entrapped in the matrix (in case of composites) or in any solid bulk material, for example, such as in Silicon Wafer electronic board. The gray-shaded region is the solid material phase with radius b carrying the previously mentioned void. The outermost region is the infinite effective medium with homogenized properties but unknown required thermal conductivity k_{eff}. The radii ratio $(a/b)^3$ similar to Lee's work (Lee et al. 2006) between void and solid material determines the volume fraction of void phase V_v and similarly for V_s it can be calculated by subtracting V_v from 1.

Steady state conduction equations for void, solid and effective media ignoring any heat generation or radiation effects are as follows (Cengel 2003):

$$\nabla^2 T_v = 0 \quad \text{where } r \text{ is, } 0 \leq r \leq a \tag{1}$$

$$\nabla^2 T_s = 0 \quad \text{where } r \text{ is, } a \leq r \leq b \tag{2}$$

$$\nabla^2 T_{eff} = 0 \quad \text{where } r \text{ is, } b \leq r \leq \infty \tag{3}$$

Following equations satisfy the heat flux and temperature continuity Boundary Conditions (BCs) at the two interfaces, void solid and solid effective medium are:

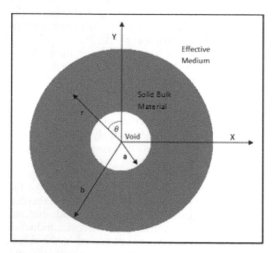

Figure 1. GSCM derived model showing void-solid-effective medium.

$$T_v = T_s \tag{4}$$

$$h(T_s - T_v) = k_s \left(\frac{\partial T_s}{\partial r} \right)_{r=a} \tag{5}$$

$$T_s = T_{eff} \tag{6}$$

$$k_s \left(\frac{\partial T_s}{\partial r} \right)_{r=b} = k_{eff} \left(\frac{\partial T_{eff}}{\partial r} \right)_{r=b} \tag{7}$$

where T, h and k represent the temperature, thermal convection coefficient and thermal conductivities, respectively with subscripts v, s, eff, a and b representing the void, solid material, effective medium, radii a and radii b, respectively.

The temperature solutions are rewritten as in Lee et al. (2006):

$$T_v = A_v r \cos\theta \tag{8}$$

$$T_s = \left(A_s r + \frac{B_s}{r^2} \right) \cos\theta \tag{9}$$

$$T_{eff} = \left(\beta r + \frac{B_{eff}}{r^2} \right) \cos\theta \tag{10}$$

Substituting the values of temperature are from Equations (8) to (10) into the Equations (4) to (7), following a set of equations are obtained.

$$A_v - A_s - \frac{B_s}{a^3} = 0 \tag{11}$$

$$A_s(h - k_s) - A_v h a + B_s \left(\frac{h}{a^2} + \frac{2k_s}{a^3} \right) = 0 \tag{12}$$

$$A_s - \beta + \frac{B_s}{b^3} - \frac{B_{eff}}{b^3} = 0 \tag{13}$$

$$A_s k_s - \frac{2k_s B_s}{b^3} + \frac{2k_{eff} B_{eff}}{b^3} - \beta k_{eff} = 0 \tag{14}$$

The thermal conductivity of the material k_{eff} can be only determined if the unknown coefficients A_v, A_s, B_s, β and B_{eff} are known. The unknown coefficients are eliminated as discussed in (Beck et al. 1992), keeping $B_{eff} = 0$, as derived in (7). They derived the value of $B_{eff} = 0$ using the energy balance and equated the thermal energy in the heterogeneous and equivalent homogeneous medium to find the effective thermal conductivity k_{eff} for composites having multiple phases. After subsequent substitution of BCs into the equation of thermal energy and solving, the term B_{eff} gets equal to zero, making it possible to determine k_{eff} using this method.

From Equation (11), A_v can be written as:

$$A_v = A_s - \frac{B_s}{a^3} \tag{15}$$

Now substituting the value of A_v from Equation (15) into Equation (12) to have an equation containing only two unknown coefficients A_s and B_s.

$$A_s(h - k_s - ha) + B_s\left(\frac{2k_s}{a^3}\right) = 0 \qquad (16)$$

Similarly from Equation (13), β can be written as:

$$\beta = A_s + \frac{B_s}{b^3} \qquad (17)$$

Now substituting the value of β from Equation (17) into Equation (14) to have an equation containing only two unknown coefficients A_s and B_s.

$$A_s k_s - \frac{2k_s B_s}{b^3} - \left(A_s + \frac{B_s}{b^3}\right)k_{eff} = 0 \qquad (18)$$

Finally, solving Equations (16) and (18) for the effective thermal conductivity k_{eff}, eliminating the remaining two unknown coefficients A_s and B_s, yields:

$$k_{eff} = 2k_s\left(\frac{3k_s}{k_s(V_s + 2) - h(a-1)(V_v - 1)}\right) \qquad (19)$$

Equation (19) contains the effect of convection on the effective thermal conductivity of the material along with the effects of volume fraction and size of the voids.

3 RESULTS AND DISCUSSION

Equation (19) can be used to determine the effective thermal conductivity of the solid materials containing imperfections in the form of voids carrying air or any fluids, which conducts heat through convection principle. The test value used for k_s is 149 W/mK that is approximately equivalent to that of silicon and for h it is 2 W/m² K, the void size of a is taken as 2 microns. The results are computed at different volume fractions of solid material V_S for effective thermal conductivity. In addition to the effects of void size, thermal convection coefficient is also studied. The Lee et al. model assumed that at the center of the model, a solid particle is present, whereas, in the present study a hollow spherical void is considered, which is carrying some process generated fluid such as entrapped air. Hence the thermal conduction mode at the core of both models differs significantly, as one is conduction-conduction and other is convection-conduction. To compare the results of the present model with Lee's results thermal conductivity of air is used instead of the solid particle in the Lee's model. Furthermore, interfacial contact conductance C is assumed as infinity for perfect contact between the two phases.

Figure 2 shows the effect of volume fraction of solid material V_S on the effective thermal conductivity of the material k_{eff}. The increasing trend matches closely with Lee's model. The present study shows a slight increase in effective thermal conductivity as compared with Lee's model and the difference becomes almost equal as V_S approaches 1. The change is appreciable at the beginning of the dominant phase at low values of V_S, which is because of void containing air. The mode of thermal conduction between the void-solid phases in Lee's model is in the form of conduction only, as compared with the convection in the present model.

The difference is maximum at higher values of V_V as there are more voids, hence there are more surfaces and boundaries available for the entrapped gas (fluid) molecules to interact with the higher temperature phase (solid material) hence increases effective thermal conduction, however the case is different in comparison with Lee et al. (2006). It assumes the central sphere having a fixed thermal conductivity with conduction as the thermal conductivity mode stays steady throughout the thermal cycle and provides little help in thermal conduction because of its low thermal conductivity (for air it is reported as approximately equivalent to 0.024 W/mK), even at higher void volume fractions.

The effective thermal conductivity with varying thermal convection coefficient h is studied at two different volume fractions of void phase (i.e., V_V is 0.5 and 0.6) and is shown in Figure 3. It can be seen that an increase in the value of h, increases the heat flowing through the constant cross-sectional area for each °K, hence the increase in effective thermal conductivity. Furthermore, it is observed that h has a more pronounced effect

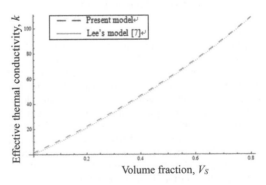

Figure 2. Effect of volume fraction of a solid phase on effective thermal conductivity.

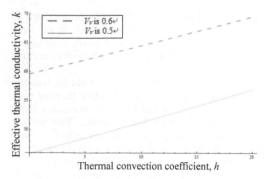

y-axis: Effective thermal conductivity, k

legend: V_V is 0.6; V_V is 0.5

x-axis: Thermal convection coefficient, h

Figure 3. Effect of thermal convection coefficient on effective thermal conductivity.

on the effective thermal conductivity when the volume fraction of voids V_V is high. This can be visualized as by increasing the number of imperfect voids, there is a high probability of void-solid material interaction. Hence, overall thermal conductivity is highly dependent on the parameter controlling the convection-conduction coupled system, and the effect of h becomes significant at high values of V_V.

4 CONCLUSIONS

A new form of an analytical model for coupled convection conduction is obtained with more realistic consideration. The model can be used to determine the effective thermal conductivity of two-phase material containing imperfection in the form of voids as the second phase by the use of GSCM and can further be extended for more than two phases of materials. The generalized self-consistent scheme has not been used before with convective boundary conditions to determine the effective thermal conductivity of the porous materials. This study assumes the effect of convection instead of conduction, due to a presence of entrapped gas. The convection phenomena at a submicron level were ignored before which is important to be considered for better results where conduction-based assumptions were taken. Thus, this study can be utilized in the industries dealing with the thermal systems to fine tune their materials and can provide with the data regarding level of heat fluxes each phase of material can bear for optimum applications. The percentage level of voids can be manipulated within the material that leads to Microstructure Sensitive Designs (MSD) where overall thermal conductivity can be controlled with the change in microstructures and phases, hence, insulating industries can make great use of this study.

The trends, for the effect of V_s on the values of k_{eff}, match well with that in the literature. In addition, the voids tend to decrease the overall thermal conductivity of the material. However, in comparison, the present model shows a slight increase in thermal conduction than that of the existing literature, since the present study contains convection coefficient h that is dependent on the interaction between two phases having temperature gradient. As this convection dependent phase V_s increases, greater is the chance for the successful collision between gas molecules with the walls/boundary of the solid phase having higher kinetic energy molecules and higher temperatures.

Furthermore, the increase in the value of h tends to increase the overall effective thermal conductivity of the material; it represents the quantity of heat that could flow between the void and solid material at the interface, for a given area per °K. However, the increase is more pronounced at higher values of V_v.

This study improves the results and further can be employed to measure the temperature gradients in the various phases of the model with the help of the unknown temperature gradients. Further study is planned to compare the results of the present model with the aid of finite element method. In addition, another study is also being conducted to include the transient conditions instead of steady state to have a more realistic picture in the analysis, as thermal conductivities change with temperature as time progresses.

REFERENCES

Beck, J.V. Cole, K.D. Sheikh, A.H. & Litkouhi, B. 1992. *Heat Conduction Using Green's Function*, Hemisphere Publishing Corporation, London, 346–351.

Cengel, Y.A. 2003. *Heat Transfer, A Practical Approach, 2nd Ed.*, McGRAW HILL, 74–76.

Christensen, R.M. & Lo, K.H. 1979. Solutions for Effective Shear Properties in Three Phase Sphere and Cylinder Models. *Journal of the Mechanics and Physics of Solids*, 27: 315–330.

Graebner, J.E. 1995. Thermal-conductivity-of-printed-wiring-boards, *Electronics Cooling*, October 1st, 1995.

Hashin, Z. 1962. The Elastic Moduli of Heterogeneous Materials. *Journal of Applied Mechanics*. 29: 143–150.

Hassan, S.A. Ahmed, H. & Israr, A. 2013. An Analytical Modeling for Effective Thermal Conductivity of Multi phase Transversely Isotropic Fiberous Composites using Generalized Self Consistent Method, *Applied Mechanics and Materials*, 249–250: 904–909.

Jiang, C.P. Chen, C.P. Yan, P. & Song, F. 2012. Prediction of effective stagnant thermal conductivities of porous materials at high temperature by the generalized self-consistent method, *Philosophical Magazine*, 92(16): 2032–2047.

Lee, J.K. & Kim, J.G. 2010. Generalized Self Consistent Model for Predicting Thermal Conductivity of Composites with Aligned Short Fibers, *Material Transactions*, 51(11): 2039–2044.

Lee, J.K. & Kim, J.G. 2013. Model for predicting effective thermal conductivity of composites with aligned continuous fibers of graded conductivity, *Archive of Applied Mechanics*, 83(11): 1569–1575.

Lee, Y.M. Yang, R.B. & Gau, S.S. 2006. A Generalized Self consistent Method for Calculation of Effective Thermal Conductivity of Composites with Interfacial Contact Conductance, *International Communications in Heat and Mass Transfer,* 33(2): 142–150.

Olanrewaju, P.O. Gbadeyan, J.A. Hayat, T. & Hendi, A.A. 2011. Effects of internal heat generation, Thermal radiation and buoyancy force on a boundary layer over a vertical plate with a convective surface boundary condition, *South African Journal of Science*, 107(9–10): 1–6.

Progelhof, R.C. Throne, J.L. & Ruetsch, R.R. 1976. Methods for Predicting the Thermal Conductivity of Composite Systems: A Review, *Polymer Engineering and Science*, 76(9): 615–619.

Xu, Y. & Yagi, K. 2004. Automatic FEM Model Generation for Evaluating Thermal Conductivity of Composite with Random Materials Arrangement, *Computational Materials Science*, 30: 242–252.

Advanced Materials and Structural Engineering – Hu (Ed.)
© 2016 Taylor & Francis Group, London, ISBN 978-1-138-02786-2

Using optical methods to determine high resolution Coefficient of Thermal Expansion of materials

S.A. Hassan & H.M. Ali
Department of Mechanical Engineering, Institute of Space Technology, Islamabad, Pakistan

M.A.A. Khan, W. Aslam & M.H. Ajaib
Department of Materials Science and Engineering, Institute of Space Technology, Islamabad, Pakistan

ABSTRACT: High technology markets demand more reliable products. This has propelled the need for materials to be designed and tuned with more precision and control. Thermal effects play a major role in deciding the life cycle of the composite structures that made up of dissimilar materials. The coefficient of thermal expansion of the material is one such dominant property of the material, which dictates its behavior under the action of thermal gradients and cycles. This paper discusses the utility of a new and improved optical method for finding coefficient of thermal expansion of materials. Using diode laser beam as a profiling tool, the strain produced in the material resulting due to temperature change is projected on the photo quadrant after passing through the glass lever. The relative positions of the two laser beam spots are calculated, which providing very high resolution measurements, typically in the range of sub-micron levels for the strain produced.

1 INTRODUCTION

The increased use of advanced materials has proliferated across the industrialized world and is increasingly finding its way in most of the industries from Military, Aerospace, Medical, Electronics and/or Civil and Manufacturing. However, prior to acceptance; material properties need to be thoroughly analyzed and verified that they meet the requirements of the user. Considering, for example, the case of BOEING's newest commercial aircraft the B787 that is composed of Aluminum, Carbon Composite/Silicon Carbide mated or sandwiched with Titanium alloy (Graebner 1995). To make jigsaw work reliably and it meets structural integrity requirements for the products projected life, specialized hardware is built to verify claims. For this purpose, Highly Accelerated Stress Screening (HASS) and Highly Accelerated Lifetime Testing (HALT) methodologies are used which brings out in a matter of hours or weeks the failures that otherwise will exhibit and/or manifests itself as much as decades later (Jiang et al. 2012). The HALT and HASS studies have revealed that one of the dominant reasons for products failure is due to degradation caused by thermal effects or mismatch in the Coefficient of Thermal Expansion (CTE) values of two or more dissimilar materials, used in a product when subjected to temperature fluctuation (Lee & Kim 2013).

Similarly, a standard circuit board comprising of FR4 base with Cu lining and Pb solder joint: all have different CTE values. The standard failure on the circuit board—assuming no manufacturing defects—is due to the dislodging of soldering contacts. The dislodging occurs in the assembly either because of frequently induced stresses due to CTE mismatch during temperature variation or the vibration that physically breaks contact between FR4/ Solder/Cu Lining (Lee & Kim 2010). Temperature fluctuations, when considered from the context of materials CTE mismatch, cause the mated composites panels/structure to lose gradually its integrity, eventually resulting in product failure. Today, manufacturer develops hardware that has a minimum CTE mismatch between different materials or the CTE of the material itself is reduced during development such as in the selection of proper resin and epoxy. However, the highly precise and accurate tuning of the components and materials are the major requirements of the aerospace industries.

Several methods are documented and discussed by the authors in the past such as Push rod quartz dilatometer, Michelson Interferometer Probing on 2 sides of sample, Michelson Interferometer Probing on 1 side of sample, Absolute Michelson Interferometer Probing on 1 side of sample, Diverging beam interferometer probing on flat mirror on sample and the Capacitance cell (Beck et al. 1992,

Olanrewaju et al. 2011, Progelhof et al. 1976). These are high resolution measurement methods for measurement of the material CTE. One such approach being manifested by authors in the past is using optical lever phenomena to aid in taking measurements for the changes in the length of the columns, materials and blocks in consideration, to calculate the linear CTE of the samples (Christensen & Lo 1979, Hassan et al. 2013).

An idea and study is presented in this paper, which can be implemented to calculate the CTE of the components at high resolutions to accurately measure the CTE of the modern engineering materials. This will leverage the proposed CTE measurement technology to be employed as an economical tool for improving the reliability of the overall structures by selecting components of the composites with lesser mismatch in their CTE values and hence, greater product end life.

2 PROPOSED CTE MEASURING METHOD

The schematic shown in Figure 1 presents the proposed design of the CTE measuring tool. The physics of this experimental procedure is based upon the measurement of the change in the linear dimensions of the specimen under test (S_T) with the change in the temperature, using the principal of an optical lever. The length change of the two specimens for, the reference material and the S_T, at specific temperature change is detected with the help of the separation distance between the laser beam spots.

The dimensional changes; expansion or contraction is associated with the coefficient of thermal expansion, which is an intrinsic property of the material specimen, where CTE is mathematically

defined as (Cengel 2003) "the change in the value of strain at a certain temperature change".

$$\alpha = d\varepsilon/dT \tag{1}$$

where α is CTE of the material, $d\varepsilon$ is the strain produced, and dT represents the change in temperature.

The equipment comprises of, pigtailed diode laser of a fixed wavelength, preferably of higher wavelengths in the visible range, powered by a suitable direct current source. The laser beam that emerges from the source is collimated and is incident on the flat parallel surfaced, i.e., tilt-mirror. The tilt-mirror is one of the critical pieces of the equipment of set-up and acts as a pivoting element. This tilt mirror comprises of a flat Fresnel plate that has three legs to stand, where two legs lie on the S_T and the third leg rests on the Zerodur, a reference material. Zerodur has negligible or well-documented value of CTE (Lee et al. 2006) in the temperature range of −2000 C to +2000 C. Both of these materials (S_T and the reference) are housed and mounted inside a flask with its base. As the temperature in the flask is changed, the S_T changes its length considerably as compared with the reference material Zerodur; and expansion or contraction is observed depending on the nature of the temperature change. The difference in the CTE value of the Zerodur and S_T causes a dimensional change of different magnitude in each material and hence causes the tilt mirror which is stated earlier, rests on both the material columns, to tilt and causes the reflected beam to shift from its initial path. The continuous changes in the length of the material due to changing temperature will cause the tilt mirror to tilt further from its initial rest position. The laser beam emerging from its source is incident on the top surface of the tilt mirror which is 100% anti-reflection coated to achieve full transmission. After entering into the mirror, it undergoes the phenomena of refraction and bend towards the normal because it is going from lighter (air) to denser (tilt mirror) medium. Now this beam strikes the lower surface of the tilt mirror. It is 50% anti-reflection coated.

2.1 *This beam divides into two parts*

As the beam strikes the lower surface, a part of it reflects back and termed as laser reflected from internal surface of tilt mirror. After passing through the thickness of the mirror, it comes out and afterward hits the screen. This is the first spot and is intense as compared with the second spot.

The other part of beam refracts out of the tilt mirror's bottom moves away from the normal, as it goes from denser (glass) to lighter (air) medium.

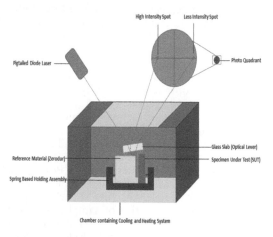

Figure 1. Schematic of proposed design.

It hits the 100% reflective aluminum coated surface of the zerodur. After getting reflected from the surface, the beam now again strikes the bottom of the tilt mirror but now going into the glass where it refracts, comes out of the top after traveling through glass and hits the screen.

These laser beams are projected over relatively long distance; this amplifies even a little change in the path difference of the two laser patterns produces due to the tilt. After getting guided by pair of reflecting mirrors, the beams are projected onto the photo quadrants or camera lens attached to the computer to record the positions of the spots. The camera is programmed with the computer software to interpret and calculate the movement of the laser dots on the pixel scale. A laser source, Moveable tilt mirror, Zerodur as a reference, test specimen and combination of reflecting mirrors all are set to deflect the beam onto the desired plane that is the camera's lens. The slightest tilt in the mirror due to change in dimension of test sample (i.e., S_T) results in the beam pattern to deviate significantly. This deviation is mapped and amplified through proper optical arrangements as laser dots or centroid and hence in this way strain produced at a certain temperature change can be calculated.

3 LASER BEAM PATH ANALYSIS

Considering two cases to analyze the path taken by the laser beam, while in zero tilt position see Figure 2 (no expansion or contraction) and when in tilt position see Figure 3 (some expansion or contraction) and at different temperature than the first case.

The beam spot in the first case that is the glass slab at total horizontal level after going through a series of refraction and reflection steps come out at the top of the screen. The both emerging beams, i.e., one coming out after striking the lower surface

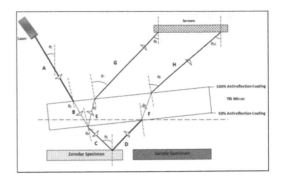

Figure 3. Schematic of path followed by laser beam when slab is at an elevated angle.

of the glass slab and the one after striking the polished surface of the Zerodur reference material, make an identical angle with the glass slab's normal plane. This is the same angle that incident laser beam makes with the top of the glass slab.

However, the situation differs in the expansion or contraction, i.e., case 2, where after simple calculation, it can be determined that the first emerging beam coming out of the slab after striking the lower surface makes the same angle as that of the incident beam. But, for the second beam emerging out of the glass slab this angle changes its magnitude with the same fraction with which the glass slab/optical lever tilts. So in general, in the tilted case the second laser beam spot moves a distance away from the first beam spot depending upon the fraction of the tilt produced. Thus, this relative movement of the spots can be projected on the screen depending on the resolution required. As greater the distance of the screen/photo quadrant from the top surface of the optical lever the greater is the relative movement of the beam spots achieved, hence, higher resolutions. However, this has practical implications for increasing the distance will also result in lowering of the laser beam spot intensity. Hence, the solution is chosen accordingly.

4 CONCLUSIONS

The proposed system can measure and determine the CTE of the material blocks manufactured with the flat surfaces quite impressively. The limitation on the size of the sample, which can be put into the chamber, can be controlled by the size of the chamber that is used. The only condition is that at the rest position both the reference material and the S_T must have their upper surfaces at the same horizontal level, so that the glass slab lies at the top in perfectly horizontal level. This is the

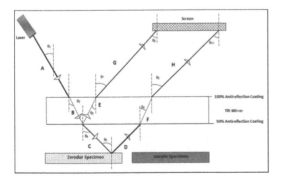

Figure 2. Schematic of path followed by laser beam when slab is at 0 degree.

fundamental piece of engineering design in this CTE tool. The use of the zerodur as a reference material can be substituted for the use of any ultra-low thermal expansion/contraction material such as quartz crystal or some defect free single crystals having known well-documented CTE values for the specified range.

The sensitivity and resolution greatly depends on the factors such as; centroid movement of the beam spot, the laser beam spot size, in general shorter is the spot size greater will be the measuring resolution. Apart from it, the laser beam electronics play a major role in the stability of the laser beam, where beam deviation can deteriorate the experimental accuracies. The quality of the glass used as the lever also plays an important role. It needs to manifest low thermal expansion itself so that its effect on the overall function of the tool is not affected. Its legs need to be made from the same material from which it is made and joined using ultra-low CTE epoxy or resin.

The real time experimental readings that we took showed that the majority of the limitation is being introduced by the unstable laser source resulting in the beam spot centroid to behave erratically at the photo quadrant or camera lens, which if can be controlled can result in the very stable measurements and higher resolutions. The maximum change in the value of Δl (i.e., change in the length of the S_T) due to X-movement of centroid was calculated to be equivalent to 1.1 Å ($1.1 \times 10-10$ m), which is well below the limits and confirms that accurate measurements can be made by the proposed setup. The maximum change in the value of Δl due to Y-movement of centroid was calculated to be equivalent to 8.709 Å ($8.709 \times 10-10$ m). These values were taken from the large amount of data captured by our team for the X- and Y-direction centroid deviations from its mean position at the CMOS area for a constant exposure time (1.4 ms), gain (1) and frame average (10 s-1). Hence this shows that highly accurate measurements can be taken without the introduction of errors from the proposed pieces of equipment.

REFERENCES

Beck, J.V. Cole, K.D. Sheikh, A.H. & Litkouhi, B. 1992. *Heat Conduction Using Green's Function,* Hemisphere Publishing Corporation, London, 346–351.

Cengel, Y.A. 2003. *Heat Transfer, a Practical Approach, 2nd Ed.*, McGRAW HILL, 74–76.

Christensen, R.M. & Lo, K.H. 1979. Solutions for Effective Shear Properties in Three Phase Sphere and Cylinder Models. *Journal of the Mechanics and Physics of Solids*, 27: 315–330.

Graebner, J.E. 1995. Thermal-conductivity-of-printed-wiring-boards, *Electronics Cooling*, October 1st, 1995.

Hassan, S.A. Ahmed, H. & Israr, A. 2013. An Analytical Modeling for Effective Thermal Conductivity of Multi phase Transversely Isotropic Fiberous Composites using Generalized Self Consistent Method, *Applied Mechanics and Materials*, 249–250: 904–909.

Jiang, C.P. Chen, C.P. Yan, P. & Song, F. 2012. Prediction of effective stagnant thermal conductivities of porous materials at high temperature by the generalized self-consistent method, *Philosophical Magazine*, 92(16): 2032–2047.

Lee, J.K. & Kim, J.G. 2010. Generalized Self Consistent Model for Predicting Thermal Conductivity of Composites with Aligned Short Fibers, *Material Transactions*, 51(11): 2039–2044.

Lee, J.K. & Kim, J.G. 2013. Model for predicting effective thermal conductivity of composites with aligned continuous fibers of graded conductivity, *Archive of Applied Mechanics*, 83(11): 1569–1575.

Lee, Y.M. Yang, R.B. & Gau, S.S. 2006. A Generalized Self consistent Method for Calculation of Effective Thermal Conductivity of Composites with Interfacial Contact Conductance, *International Communications in Heat and Mass Transfer*, 33(2): 142–150.

Olanrewaju, P.O. Gbadeyan, J.A. Hayat, T. & Hendi, A.A. 2011. Effects of internal heat generation, Thermal radiation and buoyancy force on a boundary layer over a vertical plate with a convective surface boundary condition, *South African Journal of Science*, 107(9–10): 1–6.

Progelhof, R.C. Throne, J.L. & Ruetsch, R.R. 1976. Methods for Predicting the Thermal Conductivity of Composite Systems: A Review, *Polymer Engineering and Science*, 76(9): 615–619.

Advanced Materials and Structural Engineering – Hu (Ed.)
© 2016 Taylor & Francis Group, London, ISBN 978-1-138-02786-2

Study on the properties of W-C infiltrated strengthening layer on H13 steel surface formed by plasma alloying

H.P. Zou
School of Electromechanical and Architectural Engineering, Jianghan University, Wuhan, Hubei, China

M.W. Chen
Dongfeng Peugeot Citroen Automobile Company Ltd., Wuhan, China

ABSTRACT: W-C infiltrated strengthening layer was formed on the surface of H13 alloy steel by double glow plasma surface metallurgy to improve the performance of the substrate. The microstructure, thickness and ingredient distribution of the strengthening layer were observed using SEM and XRD. The micro-hardness, friction coefficient, abrasion and corrosion resistance properties of the strengthening layer were also systematically investigated. The results showed that W-C infiltrated strengthening layer and the surface of H13 steel substrate were metallurgy union firmly; the thickness of the strengthening layer was about 25 μm; the average micro-hardness of strengthening layer was up to 1285.4 HV, which enhanced more than 4 times compared with H13 steel substrate; the element ingredient of the strengthening layer was a gradient distribution. The corrosion and experimental abrasion results showed that W-C infiltrated strengthening layer significantly modified the corrosion resistance of H13 steel and the average friction coefficient was 0.25, the average wear resistance was 6 times in comparison with the H13 steel substrate.

1 INTRODUCTION

H13 alloy steel is a high-strength steel widely used in industry as a raw material for both cold-working dies and hot-working dies domain such as aluminum dies and casting dies, and dummy blocks used in aluminum extrusion because of its good temperature resistance, high hot hardness and strength, good ductility and impact strength, resistance to thermal fatigue, erosion and wear (Yang et al. 2008, Wang et al. 2010, Cui et al. 2011). In majority situations, wear and failure of H13 steel dies are significant issues on the material surface of the dies. Major causes of premature failure were distortion, thermal fatigue, erosion and corrosion, and so on (Jia et al. 2015, Telasang et al. 2015, Cong et al. 2014, Taktak 2007, Papageorgiou et al. 2013).

The plasma surface alloying technique can penetrate and deposit desired source alloy elements that come from the source electrode to an active surface of the work piece through ion stimulation. Stimulated ion of source alloy elements can infiltrate and proliferate on the surface of the target work-piece and obtain alloying strengthening layer with special physical and chemical properties (Ding et al. 2012).

This paper discussed about the properties of the W-C infiltrated strengthening layer on H13 steel surface obtained by penetrating and depositing tungsten and carbon on H13 alloy steel substrate using the plasma surface alloying technique.

2 EXPERIMENTAL

2.1 Materials

The W-C infiltrated strengthening experiment on H13 steel surface employed the DLZ-20 multipurpose plasma immersion ion implantation furnace made by Wuhan Anders heat treatment equipment limited company. The experimental samples were made from an anneal condition of H13 steel with 25 mm in diameter and 9 mm thickness. The source electrode materials used in plasma double glow penetration metal process are commercially pure tungsten stick and graphite stick. All the test samples were polished to a mirror finish and cleaned in an ultrasonic bath with acetone and ethanol.

2.2 Experimental method

The metallographic microstructures and thickness profile of the samples were observed using the Quanta 200 scanning Electron Microscope (SEM), the ingredient distribution of the strengthening layer was observed by using the Energy-Dispersive X-ray Spectroscopy (EDS). The micro-hardness of the samples was obtained by using the HVS-1000S micro-hardness tester. Meanwhile, the friction coefficient and abrasion test were carried out using the MMW-1A friction abrasion testing machine.

3 RESULTS AND DISCUSSION

3.1 *Morphological aspects*

Figure 1 shows cross-section morphology of the W-C infiltrated strengthening layer. The layer thickness from the surface to H13 steel substrate is approximately 25 μm. There is no obvious boundary and flaw such as micro-hole or micro-crack between the layer and substrate, the layer and substrate are metallurgy union firmly. The W-C infiltrated strengthening layer was relatively very thin and the reason is that tungsten is a strong carbide forming tendency element and the carbon atom prevents the carbide penetration inward and the carbon atomic radius was very small, so it can penetrate inward as an interstitial atom, finally meet the Fe atom and few W atom in the substrate then formed the carbide or the intermetallic compound. Eventually, the W-C infiltrated strengthening layer's effective thickness is approximately about 25 μm.

3.2 *Ingredient distribution*

Figure 2 shows the W, Fe, and C element intensity distribution curve along W-C infiltrated strengthening layer cross-section obtained through EDS line scanning. As showed in Figure 2, the W-C infiltrated strengthening layer is principal consisted of W, Fe, and C three kinds of elements. The W and C element contents are relatively high and the Fe element content is low, the element intensity distribution changes continuously from the substrate surface to the interior. There are no ingredient points of discontinuity.

As shown in Figure 2, the carbon ingredient intensity is lower in the region of 25 μm~30 μm because the carbon atom in the substrate occurred overflowing away from the substrate under 1150°C high temperature experimental condition. At the distance of

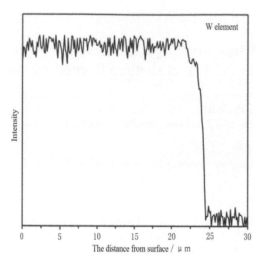
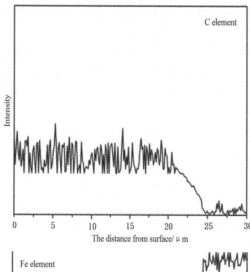
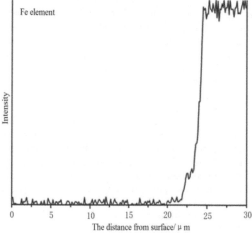

Figure 2. W, C, and Fe element intensity distribution curve.

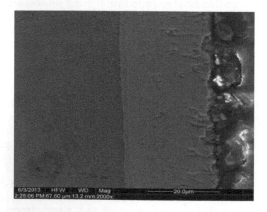

Figure 1. SEM micrograph of W-C infiltrated strengthening layer.

20 μm, the carbon and the tungsten element ingredient intensity starts to present the descent the tendency of descent, the Fe ingredient intensity appears increases sharply, at the distance of 25 μm, the carbon and the tungsten ingredient intensity tends to be stable, formed a thickness approximately 4 μm of carbon tungsten diffusion layer, thus completed the ingredient transition from the H13 steel substrate to the W-C infiltrated strengthening layer.

3.3 Micro-hardness characterization

Micro-hardness tests were carried out by using the HVS-1000S micro-hardness tester. The average skin micro-hardness of H13 steel substrate is 262.1 and the average micro-hardness of the W-C infiltrated strengthening layer is 1285.4. So the micro-hardness of W-C infiltrated strengthening layer was enhanced more than 4 times compared with H13 substrate.

3.4 Fiction coefficient

The fiction coefficient experiment is carried out in the WWM-1 A miniature friction attrition tester. Figure 3 shows the friction coefficient relational testing curve of H13 steel substrates and W-C infiltrated strengthening layer sample under room temperature dry friction condition along with the time. The H13 steel substrate's average friction coefficient is very big, is approximately 0.51, after W-C infiltrated strengthening, the sample's average fiction coefficient is approximately 0.25.

3.5 Abrasion resistance

The abrasion resistance property was tested using a weight method. Table 1 shows the weightlessness contrast of H13 alloy steel substrate and W-C infiltrated strengthening specimen. The H13 alloy steel substrate weightlessness reaches 1.2 mg while the W-C infiltrated strengthening specimen's weightlessness is only 0.2 mg under the same test conditions. So it can be seen after W-C infiltrated strengthening, the abrasion resistance performance enhanced more than 6 times.

3.6 Corrosion analysis

The corrosion behavior of H13 alloy steel specimen and W-C infiltrated strengthening specimen was investigated through solution electrochemistry experiment using 5% HCl. Figures 4 and 5 show the SEM observations respectively of H13 specimen original and H13 specimen with W-C strengthening layer after the 5% HCl corrosion tests.

Figure 4 shows the surface morphology of the H13 specimen original has a large corrosion area

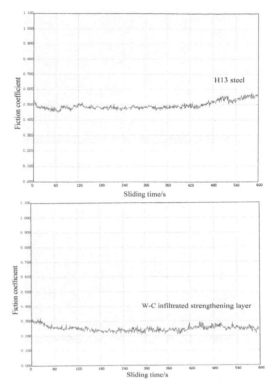

Figure 3. Friction coefficient relational test curve.

Table 1. Wearing capacity and relative wear rate under room temperature.

Sample	H13 Steel	W-C infiltrated specimen
Quality before attrition (mg)	5743.0	5919.3
Quality after attrition (mg)	5741.8	5919.1
Wearing capacity (mg)	1.2	0.2
Relative wear rate	1	0.167

and the specimen substrate surface was corroded seriously, the surface corrosion products are very much which means that its surface has not formed the deactivated membrane. In general, a uniform corrosion is the large area of destructive morphology, and it will quickly decrease the tool life. So the anticorrosion performance of H13 alloy steel is bad.

Figure 5 shows the surface morphology of specimen with W-C strengthening layer after 5% HCl corrosion test. Obviously, the corrosion damage and corrosion area significantly decrease, the test specimen surface is still smooth and compact, with only scattered particles of several bulges. There have almost no corrosion products and does not have obvious etch pit. Therefore means the W-C

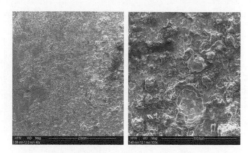

Figure 4. SEM observation of H13 specimen original after 5% HCl corrosion tests.

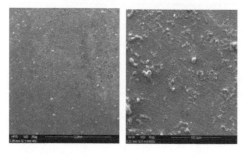

Figure 5. SEM observation of H13 specimen with W-C strengthening layer after 5% HCl corrosion tests.

strengthening layer has a good anticorrosion performance in this corrosion solution.

The results demonstrated that the W-C infiltrated strengthening layer significantly modified the corrosion resistance of H13 steel, which was due to the passive film formed during the corrosion test. When the specimens were tested in chloride solution, the oxide and hydroxide of W and C were formed on a surface of alloyed layer to retard the anodic dissolution process. Once the protective film was formed, it would protect the surface against the transport of aggressive Cl- to the surface and the film could prevent metallic ions away from the surface simultaneously. Thus, the W-C strengthening layer, formed on the surface of H13 steel by double glow plasma surface metallurgy, significantly improved the corrosion resistance of the substrate.

4 CONCLUSIONS

1. The obtained W-C infiltrated strengthening layer by penetrating and depositing tungsten and carbon on H13 alloy steel substrate using the plasma surface alloying technique was metallurgy union firmly with the H13 steel substrate.
2. The thickness of the W-C infiltrated strengthening layer was about 25 μm; the average microhardness of strengthening layer skin was up to 1285.4 HV, which enhanced more than 4 times compared with H13 steel substrate.
3. The element ingredient of the strengthening layer was a gradient distribution. The abrasion experimental results showed that the strengthening layer average fiction coefficient was 0.25 and the average wear resistance was 6 times in comparison with the H13 steel substrate.
4. The W-C infiltrated strengthening layer on H13 alloy steel significantly modified the corrosion resistance of H13 steel and effectively improve the corrosion resistance in the 5% HCl corrosion solutions.

ACKNOWLEDGMENT

The research described in this paper was financially supported by the Wuhan Science and Technology Plan Project (201250499145-18) and Academy level project (JD023).

REFERENCES

Cong, D. Zhou, H. Ren, Z. Zhang, Z. Zhang, H. Meng, C. & Wang, C. 2014. The thermal fatigue resistance of H13 steel repaired by a biomimetic laser remelting process, *Materials and Design*. 55: 597–604.

Cui, X.H. Wang, S.Q. Wei, M.X. & Yang, Z.R. 2011. Wear characteristics and mechanisms of h13 steel with various tempered structures, *Journal of Materials Engineering and Performance*. 20: 1055–1062.

Ding, L. Xiang, D.P. Li, Y.Y. Zhao, Y.W. & Li, J.B. 2012. Phase, microstructure and properties evolution of fine-grained W-Mo-Ni-Fe alloy during spark plasma sintering, *Materials and Design*. 37: 8–12.

Jia, Z. Liu, Y. Li, J. Liu, L. & Li, H. 2015. Crack growth behavior at thermal fatigue of H13 tool steel processed by laser surface melting, *International Journal of Fatigue*. 78: 61–71.

Papageorgiou, D. Medrea, C. & Kyriakou, N. 2013. Failure analysis of H13 working die used in plastic injection moulding, *Engineering Failure Analysis*. 35: 355–359.

Taktak, S. 2007. Some mechanical properties of borided AISI H13 and 304 steels, *Materials and Design*. 28: 1836–1843.

Telasang, G. Dutta Majumdar, J. Padmanabham, G. & Manna, I. 2015. Wear and corrosion behavior of laser surface engineered AISI H13 hot working tool steel, *Surface and Coatings Technology*. 261: 69–78.

Wang, C. Zhou, H. Lin, P.Y. Sun, N. Guo, Q. Zhang, P. Yu, J. Liu, Y. Wang, M. & Ren, L. 2010. The thermal fatigue resistance of vermicular cast iron coupling with H13 steel units by cast-in process, *Materials and Design*. 31: 3442–3448.

Yang, J.H. Li, S. Cheng, M.F. & Luo, X.D. 2008. Characterization of surface composition and microstructure of h13 steel implanted by ti ions using masking implantation procedure, *Surface Review and Letters*. 15: 481–485.

Advanced Materials and Structural Engineering – Hu (Ed.)
© 2016 Taylor & Francis Group, London, ISBN 978-1-138-02786-2

Petrographic and chemical assessment of siliceous limestone for cement production suitability

N. Bouazza, A. El Mrihi & A. Maâte
Department of Earth Sciences, Faculty of Sciences of Tetouan, Tetouan, Morocco

ABSTRACT: The cement plant in Tetouan, North East of Morocco, shows two varieties of limestone, high and low grade siliceous limestone. Knowing that the high grade is homogeneous, only low-grade samples were collected at El Mashar quarry. They have been analyzed in order to assess their suitability for cement production. Petrography as well as X-Ray Diffraction (XRD) analysis indicates that samples are dominantly composed of calcite, quartz and ankerite. The identified petro-facies is bio-micritic limestone with flinty beds, and bio-mictitic limestone with cryptocrystalline irregular flint nodules. The geochemical analysis displays a high variability of the major and minor oxides in siliceous limestone. The high average of CaO and SiO_2 respectively (39.88%), and (19.29) indicating that calcite is the principal carbonate mineral followed by silica mineral. High value of MgO (2.73%) content in siliceous limestone suggests a lack of dolomitization process. The content of Fe_2O_3 with an average of 0.74% illustrate reducing conditions unfavorable for Fe^{3+} formation, those conditions are linked to the pH of the water. Low alumina (Al_2O_3 2.13%) content probably reflects a low energy environment and the presence of impurities (clay). According to Moroccan standard specifications, both studied siliceous limestone need to be mixed with a proportion of high-grade limestone, clay, and iron oxide to be fertile for cement production.

1 INTRODUCTION

Morocco possesses large deposits of limestone along the Dorsale Calcaire Complex (El Kadiri 2002b, Zaghoul et al. 2005). However, the suitable limestone deposits for cement production are not uniformly distributed in the entire Dorsale Calcaire Complex. Nowadays, the high-grade limestone deposits are depleting because of continuous exploitation over the years. Hence, the cement plant of Tetouan is exploring the use of existing low-grade siliceous limestone to conserve mineral reserves as well as a sustaining environment. The plant has two varieties of limestone; high and low grade siliceous limestone. Firstly, the proportion of siliceous limestone that will be exploited must be compatible with the quarry-operating plan in order to perpetuate the life of the plant. Secondly, access to high-grade limestone (used as a raw material of the cement clinker and for adding limestone to the cement) which is geologically below the siliceous limestone is conditioned by the rational exploitation of the latter. According to the specifications of the Moroccan standard NM 10.1.004, Portland cement clinker is a hydraulic material which must be composed of at least two-thirds by mass of calcium silicate [$(CaO)_3.SiO_2$], [$(CaO)_2.SiO_2$], containing the remaining part of iron oxide (Fe_2O_3), aluminum oxide (Al_2O_3) and other oxides.

The mass ratio $(CaO)/(SiO_2)$ should not be less than two. The magnesium oxide (MgO) must not exceed 5% by mass.

In order to attain these specifications, this study focuses only on the petrographic and chemical characterization of siliceous limestone.

2 GEOLOGICAL SETTING

The Internal Domain of the Rif Chain (Fig. 1a) is organized into three superimposed structural complexes, from the bottom to the top: Sebtides

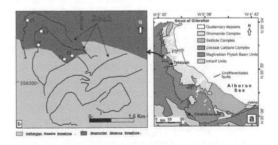

Figure 1. (a) schematic geological map of northern Rif showing the study area (from Vitale et al. 2014). (b) simplified geological map showing a location of studied limestone samples.

(Negro et al. 2006); Ghomaride (Chalouan & Michard 1990) and Dorsale Calcaire Complex (El Kadiri 2002b). The Dorsale Calcaire Complex includes several stacked tectonic slices forming the highest units of the Rif Internal Domain, defining the boundary between Internal and Flysch Basin Domains in the Beltic-Rif orocline (Bouillin et al. 1986). The Dorsale Calcaire is geographically subdivided into three different sectors through the Rif Chain, aligned from the Gibraltar Strait to the Al Hoceima town (Fig. 1a): the Haouz Dorsale (Durand Delga 2006, Durand Delga & Olivier 1988, Lallam et al. 1997, Leikine 1969), between Sebta and Tetouan, the Dorsale Calcaire s.s. (El Kadiri 2002b, Vitale et al. 2014, Zaghloul et al. 2005), between Tetouan and Jebha and finally the Bokoya Dorsale (Galindo-Zaldívar et al. 2009) outcropping east of Al-Hoceima town.

Classically the Haouz Dorsale is subdivided, according to the original position with respect to the paleogeographic setting, into (i) Internal ii) Intermediate and (ii) External Dorsale Calcaire, both formed by Triassic-Jurassic and Paleogene-Neogene successions (Durand Delga 2006, Durand Delga & Olivier 1988, El Hatimi et al. 1991, Leikine 1969). The External Dorsale Calcaire usually overthrusts a transitionalsub-Domain located in the inner part of the Flysch Basin Domain, and known in the literature as Pre-Dorsalian Flysch (Durand Delga & Olivier 1988), but locally both are overturned by latter backthrusts (Hlila 2005).

The study area is a limestone quarry of the cement plant in Tetouan located in the Saddina village, and belongs to El Mashar Unit of the External Haouz Dorsale Calcaire (The Internal Domain) about 10 km Northwest of Tetouan (Fig. 1b). Triassic early-Jurassic Limestone Formation is a carbonate unit in the Haouz Chain. Earlier researches (Durand Delga 2006, Durand Delga & Olivier 1988, Leikine 1969), have provided a first geological background on these rocks and established the first geological maps. They have defined all units in the Dorsale Calcaire. Later, research work was mainly focused on sedimentology, biostratigraphy, depositional environment(s) of the dolomitic, calcareous successions and structural geology (El Hatimi et al. 1991, Hlila 2005). The overlying Hettagian massive limestone show a lateral variation in thickness and are made of about 100 m, followed by Sinemurian-Pleinsbachian thin-bedded cherty limestone with 60 m of thickness.

3 SAMPLING AND ANALYTICAL PROCEDURES

A reconnaissance survey was conducted at outcrop locations, taking into account lithological descriptions of color, texture, bedding characteristics, prominent sedimentary structures. A total of five limestone samples (each weighting 20 kg) were collected at three benches where the siliceous limestone is exposed (Fig. 1b). The position of each sample was noted on the Geographic Positioning System (GPS). The limestone samples have been subjected to petrographic and chemical investigations. The classifications of Folk (1959) is used here. After, thin sections of preparation rock samples were processed as follow:

1. Elementary major XRF: Samples were fused by the RSS LT66/LM34 to prepare the beads and analyzed by XRF Magix 2400.
2. Mineralogical characterization by XRD of the sample was performed using powder diffractometer PHILLIPS (PANALYTICAL) XRD XPERT MPD.

4 RESULTS AND DISCUSSION

4.1 *Petrographic and mineralogical characterization*

Petrographic examination of the limestone samples illustrates two main petro-facies: siliceous limestone with flinty beds (Fig. 2A) and siliceous limestone with flinty nodules (Fig. 2B). The first is in the form of limestone benches gray-green in

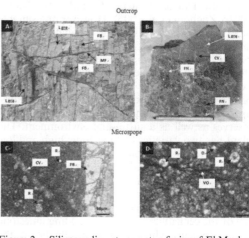

Figure 2. Siliceous limestone petro-facies of El Mashar quarry (outcrop and thin section). (A) limestone with flinty beds. (B) limestone with irregular flint nodules and calcitic veins. (C) biomicritic limestone with radiolarians ghosts, silicified bands and calcitic recrystallized vein. (D) biomictric limestone with radiolarians ghosts, rhombohedra of dolomite and opaque minerals vein. Lgre: green limestone, Lgr: gray limestone, FB: flinty bed, FN: flinty nodule, MF: micro-fold, R: radiolarian ghosts, CV: calcite vein, VO: opaque vein minerals, D: dolomite.

color, grained fine and smooth to the touch, with dark black intercalation flint beds of up to 10 cm. It is affected by a tectonic fabric demonstrated by dislocated lamination, with strange dipping and micro-folds paces to disharmony. Sometimes these micro-folds are faulted. Flint benches were imputed by Kornpobst (1966) to the formation of cherty type. This micro-facies (Fig. 2C) is a biomicrite with high percentage of micrite dominated by ghost radiolarians and silicified bands. The second one, with flinty nodules, is characterized by thin greenish limestone benches also intercepted by cracks of millimetric size filled with calcite. These facies are thicker compared to siliceous limestone with flinty beds. These benches are dolomitized and speckled with flinty centimeter nodules. In thin sections, these facies (Fig. 2D) shows a biomicritic matrix with radiolarians ghosts and some skeletal debris. The dark micritic matrix is dolomitized corresponding to hemipelagites deposited by decantation basins. Clay and iron oxide are present in trace amounts forming a fine-grained disseminated material in many samples.

Mineralogical characterization of siliceous limestone samples by X-Ray diffraction are presented in Figures 3a, b. This characterization revealed that this limestone is dominantly composed of three main minerals. Calcite ranging between 54% and 74%, Quartz between 13% and 29%, and Ankerite with value ranges from 10% to 33%. The average wt% Calcite, Quartz and Ankerite of the studied limestone are respectively 64%, 18.20% and 33%.

4.2 Chemistry characterization

The Geochemical analysis of high-grade limestone is presented in Table 1, based on the high-grade limestone, which is homogeneous, only the siliceous

Table 1. Geochemical analyses of El Mashar high-grade limestone.

SiO_2	Al_2O_3	Fe_2O_3	CaO	K_2O	MnO
1,34	0,39	0,18	54,4	0,03	0,00
TiO_2	P_2O_5	Na_2O	SO_3	MgO	LOI
0,00	0,00	0,04	0,04	0,69	42,6

limestone is presented in this paper. Table 2 shows major and minor elements analysis of the siliceous limestone samples from the studied area. The compositional chemistry of cement depends largely on geochemistry of its raw materials, i.e., limestone. Approximately 75% of the cement's raw material consists of lime (CaO)-bearing material (Lea 1970). Geochemical analyses of the major elements revealed that limestone samples contain lime (CaO) as the major constituent, followed by silica (SiO_2) and magnesia (MgO). Alumina and iron oxide form as minor constituents. Alkalis (K_2O and Na_2O) are present in traces.

The geochemical assay of siliceous limestone shows Calcium Oxide (CaO) contents ranging from 37.66% to 42.36%, which is because the limestone is primarily calcite (Pettijohn 1975). Silica oxide (SiO_2) ranges from 15.38% to 26.27% with an average of 19.29%. Magnesium Oxide (MgO) has values ranging from 1.8% to 4.04% (averaging 2.73%). Iron oxide (Fe_2O_3) values are generally low from 0.57% to 0.88% with an average of 0.74%. Alumina (Al_2O_3) in these samples varies from 1.54% to 2.39%.

The CaO content in this siliceous limestone shows a negative correlation with silica (Fig. 4a) and a positive correlation with LOI (Fig. 4b). The positive correlation between CaO and LOI may be due to the fact that LOI is generated mainly by the carbonate content of calcite. The negative correlation between CaO and SiO_2 is based on the fact that the CaO (from calcite) and SiO_2 (from quartz) are forming two different mineral phases that are not related. Alumina (Al_2O_3) in these samples with an average of 2.13% shows a strong positive correlation with iron, potash and titanium oxide (Table 3). These could be related to clay material presence in the limestone samples. Among other constituents that are commonly important is MgO (1.80% to 4.04%), it might have been derived either from the magnesium contained in skeletal debris or linked to post depositional additions or formed during digenesis. In the last case, the magnesium could be added by dolomitization process. Furthermore, the appreciable mean values of SiO_2, Al_2O_3 and K_2O from the samples suggest the presence of detrital materials including silt size (grains) as impurities

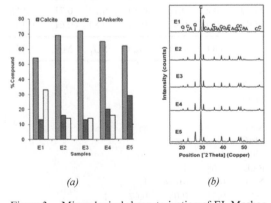

(a)   (b)

Figure 3. Mineralogical characterization of EL Mashar limestone samples: (a) mineralogical composition (b) XRD patterns: C = Calcite, Q = Quartz, A = Ankerite, (E1, E2, E3, E4 and E5 = sample from Table 2).

Table 2. Geochemical analyses of El Mashar siliceous limestone samples.

Sample no.	SiO$_2$	Al$_2$O$_3$	Fe$_2$O$_3$	CaO	K$_2$O	MnO	TiO$_2$
E1	15,38	2,38	0,88	40,36	0,91	0,03	0,12
E2	16,73	2,39	0,84	41,25	0,91	0,25	0,11
E3	15,70	2,00	0,64	42,36	0,77	0,02	0,09
E4	22,38	2,33	0,76	37,77	0,91	0,02	0,11
E5	26,27	1,54	0,57	37,66	0,57	0,03	0,07
Average	19,29	2,13	0,74	39,88	0,81	0,07	0,10
Min	15,38	1,54	0,57	37,66	0,57	0,02	0,07
Max	26,27	2,39	0,88	42,36	0,91	0,25	0,12

Sample no.	P$_2$O$_5$	Na$_2$O	SO$_3$	MgO	LOI	Tot
E1	0,05	0,02	0,20	4,04	35,10	99,47
E2	0,05	0,09	0,24	2,49	33,80	99,15
E3	0,04	0,07	0,30	2,48	34,50	98,97
E4	0,04	0,09	0,25	2,83	32,31	99,80
E5	0,04	0,03	0,09	1,80	31,20	99,87
Average	0,04	0,06	0,22	2,73	33,38	99,45
Min	0,04	0,02	0,09	1,80	31,20	98,97
Max	0,05	0,09	0,30	4,04	35,10	99,87

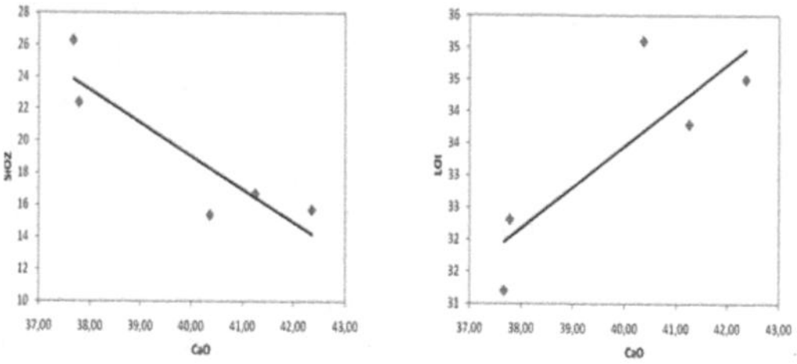

(a)   (b)

Figure 4. (a) CaO vs. SiO$_2$ plotting showing negative correlations. (b) CaO vs. LOI plotting showing a positive correlation.

Table 3. Correlation coefficients of major and minor oxides of El Mashar limestone samples.

	SiO$_2$	Al$_2$O$_3$	Fe$_2$O$_3$	CaO	K$_2$O	MnO
SiO$_2$	1					
Al$_2$O$_3$	−0,66	1				
Fe$_2$O$_3$	−0,62	0,93	1			
CaO	−0,90	0,34	0,26	1		
K$_2$O	−0,65	0,99	0,90	0,33	1	
MnO	−0,29	0,39	0,44	0,35	0,34	1
TiO$_2$	−0,64	0,98	0,96	0,26	0,97	0,27
P$_2$O$_5$	−0,61	0,64	0,85	0,40	0,59	0,63
Na$_2$O	−0,11	0,42	0,12	0,18	0,44	0,46
SO$_3$	−0,71	0,63	0,33	0,67	0,66	0,13
MgO	−0,62	0,72	0,80	0,22	0,71	−0,15
LOI	−0,98	0,62	0,62	0,84	0,61	0,14

	TiO$_2$	P$_2$O$_5$	Na$_2$O	SO$_3$	MgO	LOI
SiO$_2$						
Al$_2$O$_3$						
Fe$_2$O$_3$						
CaO						
K$_2$O						
MnO						
TiO$_2$	1					
P$_2$O$_5$	0,68	1				
Na$_2$O	0,23	−0,14	1			
SO$_3$	0,51	0,05	0,67	1		
MgO	0,84	0,60	−0,29	0,28	1	
LOI	0,64	0,61	−0,05	0,62	0,72	1

and indicate serious contamination of quartz and shaly materials (Greensmith 1978). The geochemical composition of the limestone reflects its mineralogical composition. In comparison, dolomite is more important in siliceous limestone with flinty nodules than siliceous limestone with flinty beds.

The X-ray diffraction pattern analysis indicated that the limestone sample was dominantly composed by calcite, quartz and ankerite. Petrographic study also demonstrates that the limestone is essentially biomicritic. The calcite, in form of matrix cement constitutes the carbonate principal component thought to be precipitated from a solution of organisms and their skeletal remains. The relatively high level of silica, calcite and relatively low values of MgO, show a highly siliceous limestone with magnesia. Magnesium could be added by dolomitization process. The medium concentration of alumina greater than 2% in this siliceous limestone is also an indication of clay presence. The increase of MgO, prove that this limestone cannot be used directly by the cement manufacture because of its high magnesium contents (more than 1.5%). Sodium oxide (Na_2O) and potassium oxide (K_2O) are considered as traces with respect to the normal limestone chemistry indicated in Table 2. However, the low values illustrate that the depositional environment is a reducing type and suggest that the water pH as well as the redox potential of the environment do not support the precipitation of (Fe^{3+}) to Fe^{2+} and that oxides is thus leached away (Ingram & Daugherty 1991).

The geochemical analysis showed that the levels of SiO_2 and MgO reach important values. The presence of flint in the form of kidneys or narrow parallel stratification bands indicates a heterogeneous distribution of SiO_2 in the siliceous limestone. The distribution of MgO in the siliceous limestone also seems to be quite heterogeneous and high levels exist at the contact point with the high-grade limestone. These results condition the deposit operating method and the establishment of raw materials. The contents of the deposit in SiO_2 and MgO should be strictly monitored during processing to control variations. In particular, in areas with high concentrations simultaneously mixing is the siliceous limestone and high-grade limestone by exploiting several benches at the same time. The siliceous limestone which was considered as a waste rock in various limestone mines, can be effectively utilized not only for conservation of mineral resources, but also for sustaining the environment (Rao et al. 2011).

From the results, it is clear that siliceous limestone, as indicated by its name, has a high amount of silica oxide, medium content of lime, relatively high amount of magnesium oxide and low content of iron oxide. This limestone can easily be used in cement manufacturing due to its high silica content. As the alumina and iron levels in this limestone are low, this makes it more suitable for manufacturing sulfate-resisting cement. But its relatively high content of magnesium is not advisable as raw material for cement, which suggests that this siliceous limestone will need to be mixed both with a proportion of high-grade limestone from the study area, clay and iron oxide to be added for cement manufacturing.

5 CONCLUDING REMARKS

The geochemical composition of carbonate rocks is an important factor of its use in the cement industry.

Generally, limestone with 35%–65% carbonate is suitable for the manufacture of Portland cement. However, Petrographic as well as X-ray diffraction pattern studies indicated that the limestone samples were biomicritic and dominantly composed of calcite, quartz and ankerite. Two main petro-facies, siliceous limestone with flinty nodules and siliceous limestone with flinty beds, have been characterized. From geochemical analysis, it was concluded that calcium oxide (CaO) is the dominant constituent of the siliceous limestone. The fact that the limestone is primarily calcite supports its suitability for cement production. It is apparent that this siliceous limestone sample can be used directly for the cement industry if this limestone is mixed with high-grade limestone to conserve mineral resources as well as to make it sustainable for the environment.

REFERENCES

Bouillin, J.P. Durand Delga, M. Olivier, Ph. 1986. Betic-Rifain and Tyrrhenian arcs: distinctive features, genesis and development stages. In: Wezel, F. (Ed.), *The Origin of Arcs*: 281–304. Amsterdam: Elsevier Science Publishers.

Chalouan, A. & Michard, A. 1990. The Ghomarides nappes, Rif coastal Range, Morocco: a Variscan chip in the Alpine belt, *Tectonics* 9: 1565–1583.

Durand Delga, M. & Olivier, P. 1988. Evolution of the Alboran block margin from Early Mesozoic to Early Miocene time, in Jacobshagen, V.H. (Ed.), The Atlas system of Morocco. *Lecture Notes in Earth Sciences* 15: 465–480.

Durand Delga, M. 2006. Geological adventures and misadventures of the Gibraltar Arc. *German Journal Geology* 157: 687–716.

El Hatimi, N. Duée, G. & Hervouet, Y. 1991. The limestone Dorsal of Haouz: old Tethyan passive continental margin (Rif, Morocco). *Bulletin of the Geological Society of France* 162: 79–90.

El Kadiri, K. 2002b. Tectono–eustatic sequences of the Jurassic successions from the Dorsale Calcaire (Internal Rif Morocco) evidence from a eustatic and tectonic scenario. *Geologica Romana* 36: 71–104.

Folk, R.L. 1959. Practical petrographic classification of limestones. AAPG Bulletin 43(1): 1–38.

Galindo-Zaldívar, J. Chalouan, A. Azzouz, O. Galdeano, C.S.D. Anahnah, F. Ameza, L. Ruano, P. Pedrera, A. Ruiz-Constán, A. Marín-Lechado, C. Benmakhlouf, M. LópezGarrido, A.C. Ahmamou, M. Saji, R. Roldán-García, F.J. Akil, M. & Chabli, A. 2009. Are the seismological and geological observations of the Al Hoceima (Morocco, Rif) 2004 earthquake (M = 6.3) contradictory?. Tectonophysics 475(1): 59–67.

Greensmith, J.T. 1978. *Petrology of the Sedimentary Rocks* (6th Edition): 241. London: Goerge Allen & Unwin Limited.

Hlila, R. 2005. *Tertiary tectono-sedimentary evolution of the western front of the Alboran domain (Ghomarides and limestone Ridge)*: 351. Tetouan: PhD thesis University at Tetouan.

Ingram, K. & Daugherty, K. 1991. A review of limestone additions to Portland cement and concrete. *Cement and Concrete Composites* 13: 165–170.

Kornprobst, J. 1966. The Haouz Chain, of the Hafa Queddana to the col of Azlu of Arabia. *Notes and Memoirs of the Geological Service of Morocco* 194: 9–60.

Lallam, S. Sahnoun, E. El Hatimi N. Hervouet, Y. & De Leon, J.T. 1997. Evidence of Tethyian margin dynamic from Hettangian to Aalenian age in the Dorsale calcaire (Tetouan, Rif, Morocco). *Comptes Rendus de l'Académie des Sciences*, Paris, 324(IIa): 923–930.

Lea, F.M. 1970. *The Chemistry of Cement and Concrete* (3rd Edition): 727. London U.K: Edward Arnold Publishers Limited.

Leikine, M. 1969. The Haouz Chain in the North of Tetouan (Jbel Dersa). *Notes and Memoirs of the Geological Service of Morocco* 194: 7–43.

Negro, F. Beyssac, O. Goffé, B. Saddiqi, O. & Bouybaouene, M.L. 2006. Thermal structure of the Alboran Domain in the Rif (northern Morocco) and the Western Betics (southern Spain). Constraints from Raman spectroscopy of carbonaceous material, *Journal of Metamorphic Geology* 24(4): 309–327.

Pettijohn, F.J. 1975. *Sedimentary Rocks* (3rd Edition): 628. New York: Harper and Row.

Rao, D.S. Vijayakumar, T.V. Prabhakar, S. & Bhaskar Raju, G. 2011. Geochemical assessment of a siliceous limestone sample for cement making. Chinese Journal of Geochemistry 30: 33–39.

Vitale, S. Zaghloul, M.N. Tramparulo, F.D.A. Ouaragli, B.E. & Ciarcia, S. 2014. From Jurassic extension to Miocene shortening: An example of polyphasic deformation in the External Dorsale Calcaire Unit (Chefchaouen, Morocco). *Tectonophysics* 633: 63–76.

Zaghloul, M.N. DiStaso, A. El Moutchou, B. Gigliuto, L.G. Puglisi, D. 2005. Sedimentology, provenance and biostratigraphy of the Upper Oligocene-Lower Miocene terrigenous deposits of the internal "Dorsale Calcaire" (Rif, Morocco): palaeogeographic and geodynamic implications. *Bollettino della Società Geologica Italiana* 124: 437–454.

Structural and civil engineering

Advanced Materials and Structural Engineering – Hu (Ed.)
© *2016 Taylor & Francis Group, London, ISBN 978-1-138-02786-2*

A study on the fatigue strength of a truck in railway applications

S.C. Yoon
Korea Railroad Research Institute, Uiwang, Gyeonggi, South Korea

ABSTRACT: This paper describes an experimental study to evaluate the fatigue strength of the truck frame for urban railway vehicles. The truck of the railway vehicle is situated between the vehicle body and the rail, consists of the truck frame and a wheelset composed of an axle and a wheel, and has a function of supporting the weight of the body and running the vehicle directly. In relation to the operation of the vehicle, it requires the securing of durability and running safety as one of major components directly related to the safe operation of railway vehicles. In the static load test, various loading conditions were set based on the JIS E 4207 standard. The experimental study was carried out by means of two methods such as static load test and fatigue load test.

1 INTRODUCTION

The truck for railway vehicles is an important unit that transports passengers safely by supporting the load of the vehicle body with excellent stability, good riding comfort, and smooth running performance on both curved and straight tracks. The truck should be designed and manufactured considering continuous vertical and horizontal vibrations, i.e., pitching and rolling, while the vehicle is running, in relation to its specifications, weight and load condition, and speed, besides the track conditions. The structure of the vehicle is divided into bolster and bolsterless trucks depending on the body support system. Looking at the changes in the truck structure, the bolster truck was widely used at an early stage, but the lightweight bolsterless truck with the bolster part removed has recently been developed and used. The truck frame is considered one of the most important parts of the truck. It is subjected to static and dynamic loads. That is, it is under composite fatigue loads in which the amplitude and frequency are changed. Therefore, for the strength design of the truck frame, an analysis should be performed in consideration of the repeated fatigue road. In this regard, this paper seeks to evaluate the safety by analyzing the fatigue strength of the truck frame. To this end, a static load test and a structural analysis of the truck frame for multiple electric units were performed based on the JIS E 4207 standards, which present the fatigue strength evaluation criteria for the truck frame. Based on the results, the static strength and the fatigue strength were evaluated. In the JIS E 4207 or UIC 615-4 standards that present the fatigue strength evaluation criteria for the truck frame, an infinite life design concept that considers the fatigue limit is used, and the design specifications require control of the stress amplitude on the conditions presented by each standard within the allowable stress, considering the fatigue limit of the materials (Park et al. 2000).

The truck in this study was that for the air-spring-powered bolsterless vehicle of a medium-sized multiple electric unit, which is under operation in Bundang, Korea.

2 STRUCTURAL ANALYSIS OF THE TRUCK

For the analysis, the fatigue strength of the truck frame was evaluated using a finite element method with respect to the truck frame with a bolsterless-type welding structure. The analysis model is shown in Figure 1.

Table 1 shows the load conditions applied to the design of a truck for multiple electric units. Each test load, including the vertical load, was calculated using the values in Table 1. The load that acted on the truck frame was calculated according to the truck test method of the performance test. The values are as follows. The structural analysis and the static load test were carried out using the load conditions in Table 3 (MLTMA 2008, JIS 1992, JIS 1988).

The strength of the truck had to be less than the allowable stress of the material of each part of the truck frame shown in Table 2, and the safety was evaluated in accordance with the location on the fatigue endurance diagram in terms of the combined stress.

The stress distribution for each condition is shown in Table 4 and Figure 2. In Table 4, the

Figure 1. Analysis model of the truck frame.

Table 1. Weight of the truck frame.

No.	Type	Weight	Remarks (unit: kg)
1	Empty weight	40,525	
2	Max. passenger weight	30,000	
3	Truck weight	13,954	2 sets

Table 2. Mechanical property of the truck frame.

| Material | Yield stress | Tensile stress | Fatigue limit | | | Remarks (unit: kgf/mm²) |
			Base metal	Welding and grinding	Welding	
SM490A	33	50	16	11	7	Side frame, transom support bracket
STKM18B	32	50	14			Transom pipe
SS400	25	41	14			Stiffener, seat

Table 3. Load conditions of the truck frame.

No.	Load condition	Load	Direction of the load
1	Vertical load	28,314 kg	Down
		39,640 kg	Down
2	Twist load	28,314 kg	Down
		28,314 kg	Down
3	Longitudinal load	8,494 kg	Forward
		8,494 kg	Backward
4	Lateral load	8,494 kg	Left
		8,494 kg	Right
5	Driving gear load	3,493 kg	Up and down
		3,493 kg	Up and down
6	Traction motor load	3,866 kg	Down (5.0 g)
		2,515 kg	Up (3.0 g)
7	Brake load	4,118+	Forward
		1,029 kg	Backward
		4,118+	Forward
		1,029 kg	Backward

Table 4. Maximum stress of the load condition (unit: kgf/mm²).

No.	Load condition	Max. stress	Yield stress	Material	Remarks
1	Vertical load	16.43	33.0	SM490A	Side frame
2	Twist load	11.84	33.0	SM490A	Side frame
3	Longitudinal load	7.84	33.0	SM490A	T/M mounting bracket
4	Lateral load	6.30	33.0	SM490A	Transom support bracket
5	Driving gear load	4.90	33.0	SM490A	Gear hanger bracket
6	Traction motor load	9.12	32.0	SM490A	T/M mounting bracket
7	Brake load	2.28	33.0	SM490A	Brake bracket

Figure 2.   Vertical load.

maximum stress is 16.43 kgf/mm², which occurs at the point from the upper side frame when the vertical load is applied. It is within the allowable stress of SM490A (33.0 kgf/mm²), which is the material of such part.

3   LOAD TESTS OF THE TRUCK

Strain gauges were attached to the areas where the stress concentration caused by the change in the shape was expected and where a high concentration was expected under each load condition by referring to the results of the structural analysis of the truck frame. Then the vertical load, torsional load, longitudinal load, lateral load, driving gear load, main motor load, and brake load were measured. Strain gauges were intensively attached to the half part in consideration of the symmetry of the truck frame, and more than 60 strain gauges were attached to the area where high stress was expected. The weight of the truck frame is shown in Table 1 (Yoon et al. 2011, MLTMA 2008).

3.1   Vertical load test

The vertical load of 39.64 tons was added to the air spring location by considering the dynamic effect of 0.4 g in the vertical static load for each truck. The maximum stress under the vertical load condition occurs at the area where the air spring seat comes in contact with strain gauge no. 6, and is 18.55 kgf/mm², which is within the allowable stress (33 kgf/mm²) of the material used (SM490A).

3.2   Twist load test

The twist load is caused by the imbalance of the rail. Its size corresponds to the height difference at the position of the primary spring. First, forced displacement is given toward the upward direction in the same state as that of the vertical load test, and then the static vertical load is applied to the air spring position to measure the stress. The

maximum stress under the torsional load condition occurs at the area where the air spring seat comes in contact with strain gauge no. 6, and is 13.57 kgf/mm², which is within the allowable stress (33 kgf/mm²) of the material used (SM490A).

3.3   Longitudinal load test

The longitudinal load is caused by longitudinal vibration while driving, and is 8.49 kg in size, which is ±30% of the vertical static load. The longitudinal load is applied to the T/M mounting bracket. The maximum stress under the longitudinal load condition occurs at the T/M mounting bracket (strain gauge no. 39) in the reverse direction, and is 7.20 kgf/mm², which is within the allowable stress (33 kgf/mm²) of the material used (SM490A).

3.4   Lateral load test

The lateral load is caused by lateral vibration while driving, and is 8.49 kg in size, which is ±30% of the vertical static load. The lateral load is applied to the transom support bracket. The maximum stress under the lateral load condition occurs at the transom support bracket (strain gauge no. 24) at the time of the load test in the left direction, and is 5.34 kgf/mm², which is within the allowable stress (33 kgf/mm²) of the material used (SM490A).

3.5   Driving gear load test

The driving gear load is a reaction force that acts on the T/M mounting bracket towards the maximum torque of the driving motor during the operation of a vehicle. The maximum stress under the driving gear load condition occurs at the gear hanger bracket (strain gauge no. 18) in the forward direction, and is 4.63 kgf/mm², which is within the allowable stress (33 kgf/mm²) of the material used (SM490A).

3.6   Traction motor load test

The load caused by longitudinal vibrations of the motor is 3.0 g in the upward direction and 5.0 g in the downward direction, considering the empty weight. The maximum stress under the main motor load condition occurs at the transom pipe (strain gauge nos. 77, 78, and 79) in the downward direction, and is 6.87 kgf/mm², which is within the allowable stress (32 kgf/mm²) of the material used (SKTM18B).

3.7   Brake load test

The load is generated by braking includes the braking reaction force caused by the buoyancy of the

Table 5. Combined stress result (mean stress).

S/G no.	Combined stress (driving)		Combined stress (braking)		Remarks
	Mean stress	Stress amplitude	Mean stress	Stress amplitude	
6	−14.41	5.85	−13.13	5.79	Welding
70	−13.59	5.94	−13.83	5.85	Welding
5	−13.42	5.56	−12.22	5.52	Welding

Table 6. Combined stress result (stress amplitude).

S/G no.	Combined stress (driving)		Combined stress (braking)		Remarks
	Mean stress	Stress amplitude	Mean stress	Stress amplitude	
77,78,79	−9.21	8.83	−8.02	8.30	Grinding
39	−2.40	7.20	−2.21	7.17	Base metal
51	−0.03	6.40	0.05	6.42	Base metal

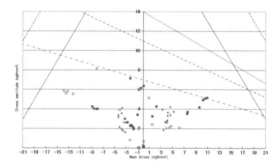

Figure 3. Fatigue endurance diagram for braking mode.

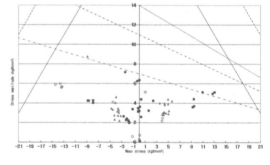

Figure 4. Fatigue endurance diagram for driving mode.

piston and a couple of forces caused by the friction between the brake shoe and the wheel. The maximum stress under the brake load condition occurs at the brake bracket (strain gauge no. 10) in the forward direction, and is 2.52 kgf/mm², which is within the allowable stress (33 kgf/mm²) of the material used (SM490A).

3.8 *Combined stress*

Among the test results, that which showed a greater stress value was used to obtain the combined stress and the fatigue endurance diagram. According to the calculation results, the measurement points at which the mean stress of the combined stress was more than 12.2 kgf/mm² and those at which the stress amplitude was more than 6.4 kg/fmm² are shown in Tables 5 and 6, and the fatigue endurance diagram, in Figure 3 and 4. The results reveal that in combined stress, the mean stress appears high (−14.41 kgf/mm²) at the area where the upper side frame comes in contact with air spring seat (strain gauge no. 6, the welding part), the stress amplitude appears high (8.83 kgf/mm²) at the gear hanger bracket (strain gauge nos. 77, 78, and 79, the grinding part), and they are all located in the safety margin for each part, as shown in Figure 3 and 4.

4 CONCLUSIONS

The results of the structural analysis and the load tests on the truck of an urban rail vehicle are as follows.

1. According to the results of the structural analysis, the maximum stress (16.43 kgf/mm²) occurs at the air spring seat of the side frame when vertical load is applied, and is 11.84 kgf/mm² when

twist load is applied. These values are all within the allowable stress of the material used.

2. According to the results of the load tests, the maximum stress (18.55 kgf/mm²) occurs at the air spring seat of the side frame when vertical load is applied, and is 13.57 kgf/mm² when twist load is applied. These values are all within the allowable stress of the material used.

3. The results of the combined stress calculation show that the maximum mean stress occurs at the area (welding part) where the upper side frame comes in contact with the air spring seat while driving, and is −14.41 kgf/mm², and the maximum stress amplitude (8.83 kgf/mm²) occurs at the gear hanger bracket (grinding part) while driving.

4. The test results show that all combined stresses are located in the safety margin on the fatigue endurance diagram.

The above-stated results confirm that the truck frame of the train has sufficient static strength and fatigue strength.

ACKNOWLEDGEMENT

This work was supported by Korea Research Council for Industrial Science and Technology to Korea Railroad Research Institute (KRRI) with contract No. PK14004C.

REFERENCES

Japanese Industrial Standards, 1988. *Test Methods of Static Load for Truck Frames and Truck Bolsters of Railway Rolling Stock*, E 4208.
Japanese Industrial Standards, 1992. *Truck Frames for Railway Rolling Stock-General Rules for Design*, E 4207.
MLTMA, 2008. *Standard for performance tests for urban railway rolling stocks*, load tests for the truck.
Park, K.J. & Lee, H.Y. et al. 2000. Evaluation of fatigue strength of the bogie frame of the standard electrical multiple unit, *Journal of the Korean Society for Railways*, 3(3): 170–176.
Yoon, S.C. & Kim, J. et al. 2011. A Study on the Fracture Test in Running System of Railway, *Key Engineering Materials*, 452–453: 49–52.

Advanced Materials and Structural Engineering – Hu (Ed.)
© *2016 Taylor & Francis Group, London, ISBN 978-1-138-02786-2*

A method for constructing a simplified model of wire strand cable

Y.M. Hu, X. Tan & B.W. Zhou
State Key Laboratory of Mechanical Transmission, Chongqing University, Chongqing, China

J.S. Li
State Key Laboratory of Heavy Mining Equipment, Henan University of Science and Technology, Luoyang, Henan Province, China

ABSTRACT: This paper presents a simplified model of the simple '6+1' straight spiral wire strand cable, it consists of two types of elements including solid element and beam element in LS-DYNA. This simplified model can be used for studying the static behaviors of the wire strand cable. A full 3D FE (finite element) model is developed first, and then constructs the simplified model based on the results of the full 3D FE model. By comparing the results of between the simplified model, full 3D FE model and the experiments of Utting & Jones (1987a, b), it shows that the simplified model has high accuracy for studying the static behaviors of the wire strand cable. Considering that this simplified model has less quantity of elements and nodes much, it provides a feasible method to study the dynamic global behaviors of the wire strand cable used in deep mining.

1 INTRODUCTION

Wire strand cable is an integral part of mine hoisting system. It is necessary to accurately assess its mechanical behaviors, including structural strength and dynamic response properties accurately, for the reason that these behaviors have a direct bearing on the operating stability, reliability, safety and hoisting capacity of the mine hoisting system.

Experimental work on large diameter and extraordinarily long cables requires specific, large and expensive testing devices (Ghoreishi et al. 2007). With the development of FE (Finite Element) methods during the last decades, together with the development of computer capacity, many methods for constructing full 3D model of the stranded wire ropes have developed very fast in the past few years, e.g., Ghoreishi et al. (2007), Judge et al. (2012), Yu et al. (2014). Most of these models showed good agreements with the experimental results reported by Utting & Jones (1987a, b). But it should be noted that most of these models were used to study the full mechanical response of spiral strand cables when subjected to quasi-static axial loading. As we know, cables often suffer from vibration (Gu et al. 2009) or dynamic loads in certain circumstances, it is very hard to investigate the response by such a full 3D FE model due to the large quantity of elements and nodes in the FE model.

For the purpose of studying the dynamic behaviors, the model of wire strand cable should be simplified. In the early years, cable was usually modeled as a bar element or beam element with its end freedom released, which often had a uniform cross-section, no compressive strength, and lateral stiffness for simplification, e.g., Carlson et al. (1973), Cutchins et al. (1987), Nawrocki (1997) and Nawrocki & Labrosse (2000).

In this paper, a simplified method for constructing a wire strand cable consisting of six helical wires wrapped around a straight core is developed. To study the global response of the wire strand cable, a full 3D FE model of the simple '6+1' straight spiral wire strand cable is developed first, and then compared it with the results reported by Utting and Jones (1987a, b). Based on the results, solid elements and beam elements are used to construct an equivalent simplified model that has the same static behaviors, including bending behavior and coupling behaviors between tension and torsion. Because there is much less quantity of elements and nodes in this simplified model, it provides a feasible manner for studying the dynamic global response in a short time. By the same method proposed in this paper, any type of wire strand cables can be studied without the expensive experiments.

2 OVERALL STATIC BEHAVIORS

2.1 *Axial behavior*

Considering a single '6+1' straight spiral wire strand cable consisting of six helical wires with a circular cross-section wrapped around a straight core. Because of the helical design of the wire strand, the axial behavior of such a structure shows coupling between tension and torsion. Thus, the elastic axial behavior can be expressed in the form (Ghoreishi et al. 2007):

$$\begin{Bmatrix} F_z \\ M_z \end{Bmatrix} = \begin{bmatrix} k_{\varepsilon\varepsilon} & k_{\varepsilon\theta} \\ k_{\theta\varepsilon} & k_{\theta\theta} \end{bmatrix} \begin{Bmatrix} u_z \\ \theta_z \end{Bmatrix} \quad (1)$$

where u_z = overall axial strain; θ_z = twist angle per unit length; F_z = axial force and M_z = torque. The four components $k_{\varepsilon\varepsilon}, k_{\varepsilon\theta}, k_{\theta\varepsilon}, k_{\theta\theta}$ in the stiffness matrix are pure tensile, torsion and coupling terms, respectively.

2.2 Lateral behavior

As shown in Figure 1, the wire strand specimen is simply supported at its two ends. Meanwhile, applying lateral load P at the middle location and applying an axial load T at the right end, which could not rotate about its axial axis. L is the length of the specimen, $y(x)$ is the deflection at x which denotes the distance from the left end. F_1, F_2 denote the reaction force in y the direction at each end, respectively, and $F_1 = F_2$. Under the small deformation case, the differential equation for the deflection at x is:

$$\frac{d^2 y}{dx^2} = \frac{M(x)}{EI} \quad (2)$$

where $M(x)$ denotes the moment at x, and can be expressed as follows:

$$M(x) = Px/2 + Ty \quad (3)$$

Applying Equation (3) into Equation (2), the differential equation can be rewritten as:

$$\frac{d^2 y}{dx^2} = \frac{Px/2 + Ty}{EI} \quad (4)$$

The boundary conditions of this differential equation are:

$$y(0) = 0, y'(L/2) = 0 \quad (5)$$

Finally, the deflection function at x is:

$$y_c = \frac{P\sqrt{EI}}{2T^{3/2}} \frac{e^{\frac{L}{2}\sqrt{T/EI}} - e^{-\frac{L}{2}\sqrt{T/EI}}}{e^{\frac{L}{2}\sqrt{T/EI}} + e^{-\frac{L}{2}\sqrt{T/EI}}} - \frac{PL}{4T} \quad (6)$$

3 DESCRIPTION OF FE MODEL

3.1 Method of modeling the 3D FE model

In this research, use the same method as the model of Yu et al. (2014) to build the full 3D FE model of the simple '6+1' straight spiral wire strand cable. Material parameters are primarily settled according to the data of Utting and Jones (1987a, b), and the *MAT_PLASTIC_KINEMATIC model is used. The geometrical and material parameters used for all models are listed in Table 1. Figure 2 shows the completely meshed model and Table 2 shows the information of each model, including the total nodes and total elements.

Table 1. Geometrical and material parameters.

Strand diameter (mm)	11.4
Central wire diameter (mm)	3.94
Helical wire diameter (mm)	3.73
Lay angle (°)	17.03
Pitch length (mm)	78.67
Pitch number	5
Model length (mm)	393.34
Young's modulus (GPa)	197.9
Poisson's ratio	0.3
Density (Kg/m³)	7850

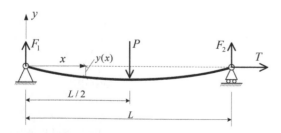

Figure 1. Wire strand specimen under simple supported.

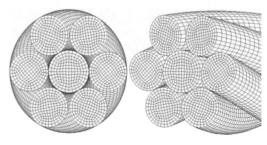

Figure 2. The 3D FE model of the simple '6+1' straight spiral wire strand cable.

Table 2. Model information.

Specimen	Cable length (mm)	Nodes number	Elements number
$\alpha = 17.03°$	157.33	353,293	319,479
	393.34	880,753	796,279
$\alpha = 12.2°$	222.90	405,331	366,519
	557.24	1,012,441	915,319

3.2 Contact conditions

There are two physical types of contacts existed in the '6+1' straight spiral wire strand cable, namely, wire/wire contact between helical wires and wire/core contact between helical wires and the core. The automatic single surface contact algorithm, *AUTO-MATIC_SINGLE_SURFACE in LS-DYN A, was used in the simulations to prevent surface penetration. The research of Yu et al. (2014) shows that, with different frictional coefficients, namely, 0.115, 0.2 and 0.4, the axial stretch force coincided with each other, which means friction coefficient had a little influence on longitudinal stiffness. In this research, the friction coefficient is set to 0.115.

3.3 Boundary conditions

3.3.1 Tension

In order to compare the results reported by Utting and Jones (1987a, b), apply the same BCs to the wire strand specimens. All of the strands have one end fixed in all displacement degrees to simulate the clamped end. At the loading end, two boundary conditions, namely, fixed-end and free-end, were modeled for the '6+1' wire strand specimens. The fixed-end condition constrains all degrees of freedom except for the translation along the longitudinal axis of the specimen. The free-end condition permits all nodal translations and rotations, allowing the wires to rotate and straighten out.

3.3.2 Bending

The boundary condition in the bending case is the same as shown in Figure 1, and the axial force T = 1000 N, lateral force P = 100 N.

The quasi-static solution procedure in LS-DYNA Explicit was used in all simulations.

4 RESULTS AND DISCUSSION

4.1 Axial results

The results of the 3D FE model are compared with the experimental data reported by Utting and Jones (1987a, b) and the numerical results studied by Ghoreishi et al. (2007). Both the rotations (θ_z)

in free-end tests and the torque (M_z) generated in fixed-end tests are recorded when the reacting axial force at the fixed end is equal to $40kN$. As shown in Tables 3 and 4 for the lay angle $\alpha = 17.03°$, the results of the 3D FE model is very close (the difference is within 4.4% comparing with the experimental data, and 2.2% versus the numerical results), but the difference of the twist angle per unit length θ_z between the experimental data and the results of the 3D FE model increases to 10% for the lay angle $\alpha = 12.2°$. For this reason, the results and the basis for building the simplified model that will be discussed later are all based on the results of the wire strand cable for the lay angle $\alpha = 17.03°$.

4.2 Lateral results

The lateral displacement at the loading place is −0.531 mm. Substituting the concerned parameters into Equation (6), and then the equivalent bending stiffness (k_{eq}) of the wire strand cable for the lay angle $\alpha = 17.03°$ could be obtained, which is $k_{eq} = EI = 12.8 \ Nm^2$.

As shown in Table 5, all the stiffness components are computed based on the results as illustrated before.

Table 3. Comparison between 3D FE model, experimental data (reported by Utting and Jones (1987a, b)) and numerical results (reported by Ghoreishi et al. (2007)), for fixed-end boundary conditions.

Specimen	M_z (Nm) (Fixed end)		
	Experimental	Numerical	3D FE model
$\alpha = 17.03°$	34.4	36.7	35.9
$\alpha = 12.2°$	26.0	27.2	25.4

Table 4. Comparison between 3D FE model, experimental data (reported by Utting and Jones (1987a, b)) and numerical results (reported by Ghoreishi et al. (2007)), for free-end boundary conditions.

Specimen	θ_z (N/m) (Free end)		
	Experimental	Numerical	3D FE model
$\alpha = 17.03°$	2.49	2.51	2.50
$\alpha = 12.2°$	2.20	2.12	1.98

Table 5. Static stiffness components of the simple '6+1' straight spiral wire strand cable.

$k_{\varepsilon\varepsilon}$ (kN)	$k_{\varepsilon\theta}$ (Nm)	$k_{\theta\varepsilon}$ (Nm)	$k_{\theta\theta}$ (Nm²)	k_{eq} (Nm²)
11,200.0	9,464.2	9,860.0	22.70	12.80

5 CONSTRUCTION OF THE SIMPLIFIED MODEL

In deep mining, the length of wire strand cable exceeds 1500 m, and the full 3D FE model cannot be used for computing the dynamic global response of the wire strand cable, because of its large quantity of the elements and time consuming. In order to get the dynamic global response of the wire strand cable used in deep mining accurately in a short time, it is necessary to build an equivalent simplified cable model. At the same time, this simplified model should have the same properties (geometry, mass, mechanical properties, etc.) as the full 3D FE model.

5.1 Description of the simplified model

As shown in Figure 3, on account of the coupling behavior between the tension and the torsion of the wire strand cable, as well as the helical wire numbers (it is 6 here), an hexagonal shape solid bulk is used to mainly represent the bending characteristic of the cable, and then six spiral beam elements are distributed on the bulk surface to represent the coupling characteristic between tension and torsion. Finally, put beams along the axial direction in the middle of the hexagonal shape solid bulk to mainly represent the tensile characteristic. This simplified model is named as an SBS (Spiral Beam Solid) model. It should be noted that all the added beams share the same nodes with solid elements at the corresponding place. By assigning the elasticity modulus of the solid elements and the axial stiffness of each beam, different static properties of this simplified model can be obtained.

5.2 Results and discussion

Up to now, a group of parameters has been found to build the SBS model, and its static stiffness components are all listed here and compared with the results of the full 3D FE model in Table 6 respectively.

As shown in Table 6, the results of the SBS model is very close to the full 3D FE model, and the differences between them are less than 5.66%, which confirm the validity of the SBS model.

As illustrated in Table 7, it shows the information about the full 3D FE model and the SBS

Table 6. Comparisons of the static stiffness components between the full 3D FE model and the SBS model.

	$k_{\varepsilon\varepsilon}$ (kN)	$k_{\varepsilon\theta}$ (Nm)	$k_{\theta e}$ (Nm)	$k_{\theta\theta}$ (Nm²)	k_{eq} (Nm²)
3D FEM	11200	9,464.2	9,860.0	22.70	12.80
SBS model	11600	9,377.4	9,302.3	22.81	12.78
Differences	3.56%	0.92%	5.66%	0.45%	0.17%

Table 7. Comparison of the computing time between 3D FE model and SBS model.

	Model length	Nodes	Elements	Computing time
3D FEM	157 mm	353,293	319,479	18 hours
SBS model	157 mm	638	624	10 minutes

model. When both of them share the same length, the SBS model has less quantity of nodes and elements much. Consequently, it will cost much less time to solve completely.

6 CONCLUSION

A new method for constructing the simplified model of wire strand cable has been proposed for the analysis of the overall static behaviors. By contrast with the experimental data and numerical results, the simplified model shows high accuracy and greatly improves the calculating efficiency as well due to its much less quantity of elements and nodes. What's more, it is of great significance when the method is applied to the analysis of the dynamic global performance of the wire strand cable.

ACKNOWLEDGMENTS

The authors gratefully acknowledge the support from the National Key Basic Research Program of China (No. 2014CB049401).

REFERENCES

Carlson, A.D. Kasper, R.G. & Tucchio, M.A. 1973. *A structural analysis of a multiconductor cable.* Technical report, Naval Underwater System, New London Laboratory, AD-767 963.
Cutchins, M.A. Cochran Jr, J.E. Guest, S. Fitz-Coy, N.G. & Tinker, M.L. 1987. An investigation of the damping phenomena in wire rope isolators. *American Society of Mechanical Engineers, Design Engineering Division DE, 5,* 197–204.

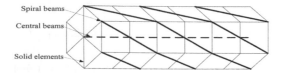

Figure 3. The structure of the SBS model: including solid elements, central beams and spiral beams.

Ghoreishi, S.R. Messager, T. Cartraud, P. & Davies, P. 2007. Validity and limitations of linear analytical models for steel wire strands under axial loading, using a 3D FE model. *International Journal of Mechanical Sciences,* 49: 1251–1261.

Gu, M. Du, X.Q. & Li, S.Y. 2009. Experimental and theoretical simulations of wind–rain-induced vibration of 3-D rigid stay cables. *Journal of Sound and Vibration,* 320: 184–200.

Judge, R. Yang, Z. Jones, S.W. & Beattie, G. 2012. Full 3D finite element modelling of spiral strand cables. *Construction and Building Materials,* 35: 452–459.

Nawrocki, A. 1997. *Contribution à la modélisation des câbles monotorons par éléments finis.* PhD thesis, Ecole Centrale de Nantes, France.

Nawrocki, A. & Labrosse, M. 2000. A finite element model for simple straight wire rope strands. *Computers & Structures,* 77: 345–359.

Utting, W.S. & Jones, N. 1987a. The response of wire rope strands to axial tensile loads—Part I. Experimental results and theoretical predictions. *International Journal of Mechanical Sciences,* 29: 605–619.

Utting, W.S. & Jones, N. 1987b. The response of wire rope strands to axial tensile loads—Part II. Comparison of experimental results and theoretical predictions. *International Journal of Mechanical Sciences,* 29: 621–636.

Yu, Y. Chen, Z. Liu, H. & Wang, X. 2014. Finite element study of behavior and interface force conditions of seven-wire strand under axial and lateral loading. *Construction and Building Materials,* 66: 10–18.

Advanced Materials and Structural Engineering – Hu (Ed.)
© 2016 Taylor & Francis Group, London, ISBN 978-1-138-02786-2

Synergistic interaction of scaling and abrasion on the fly ash concretes

A. Nowak-Michta
Institute of Building Materials and Structures, Cracow, Poland

ABSTRACT: In this paper, the influence of freeze-thaw with de-icing salt (scaling) on the abrasion resistance of hardened concretes with siliceous fly ash addition is analyzed. Abrasion resistance was measured in the reference Wide Wheel test according to EN 1338: 2003/AC: 2006. Before abrasion tests, air-entrained fly ash concretes were subjected to 56 cycles of freeze-thaw with de-icing salt in Borås test. Cement was replaced with 20, 35, and 50% of Class F siliceous fly ash in three categories of losses on ignition A, B and C by mass. The water-binder ratio, the air-entraining and the workability of mixtures were maintained constant at 0.38, 4.5% and 150 mm, respectively. The test results indicated that all the tested concretes not subjected to scaling according to EN 1338: 2005 could be classified as the fourth—the highest class of abrasion resistance. Synergistic interaction of scaling and abrasion reduces abrasion resistance of concretes.

1 INTRODUCTION

Fly ash is one of the most commonly used additions to the concrete, however, due to the effect of dust. It is not widely used in road construction and industrial floors (Wesche 1991). Two of the most intensive factors affecting this type of constructions are freeze/thaw attack with de-icing salt (scaling) and abrasion.

Loss on ignition is indicated as the main quality parameter of fly ash determining frost resistance of concrete with its addition (Luehr 1972; Wesche 1991).

EN 450-1: 2012 'Fly ash for concrete. Part 1: Definitions, specifications and conformity criteria.' defines three categories of loss on ignition: A: LOI (LOI—loss on ignition) ≤5%, B: LOI ≤7% and C: LOI ≤9%. In addition, in the standard for the categories of loss on ignition there is a note that the amount of loss on ignition may influence the effectiveness of air-entraining admixtures used for the manufacture of concretes resistant to freezing and thawing. Defined three categories of loss on ignition of fly ash allow the user to take this into account by selecting the appropriate category for application and exposure class.

According to EN 206: 2013 'Concrete—Specification, performance, production and conformity.' and its Polish supplement PN-B-06265: 2004 durability of these concretes is provided thanks to the requirements included in exposure classes, for concretes exposed to freeze/thaw attack XF with (XF2&XF4) or without (XF1&XF3) de-icing agents and for concretes exposed to abrasion XM1÷XM3.

The requirements for composition and properties of concretes exposed to freeze/thaw attack according to EN 206: 2013 are presented in the Table 1 and to abrasion according to PN-B-06265: 2004 are shown in Table 2.

Studies by Liu et al. (2002) and Sidigue'a (2004) confirm that commonly used in mass construction fly ashes to 15% cement replacement in order to reduce the heat of hydration do not decrease the abrasion resistance of concrete with their addition. Thus, fly ash in limited amounts could be used in concretes subjected to erosion type of abrasion.

The main parameters responsible for ensuring the concrete resistance to scaling are air-entraining and correspondingly low water-binder ratio (Fagerlund 1997, Kurdowski 2010, Pigeon & Pleau 1995, Rusin 2002).

Table 1. Requirements for concretes exposed to freeze/thaw attack according to EN 206: 2013.

Properties	Exposure classes			
	XF1	XF2	XF3	XF4
Maximum water-cement ratio	0.55	0.55	0.50	0.45
Minimum strength class	C30/37	C25/30	C30/37	C30/37
Minimum cement content (kg/m³)	300	300	320	340
Minimum air content (%)	–	4.0	4.0	4.0
Other requirements	Aggregate in accordance with EN 12620 with sufficient freeze/thaw resistance			

Table 2. Requirements for concretes exposed to abrasion according to PN-B-06265: 2004.

	Exposure classes		
Properties	XM1	XM2	XM3
Maximum water-cement ratio	0.55	0.55	0.45
Minimum strength class	C30/37	C30/37	C35/45
Minimum cement content (kg/m^3)	300	300	320

The realized research programs have shown that the air-entrained fly ash concretes with water-binder ratio of 0.38 are independently resistant to both scaling (Nowak-Michta 2013) and abrasion (Nowak-Michta 2014).

According to standard requirements EN 1338: 2003/AC: 2006 'Concrete paving blocks—Requirements and test methods.' both these effects are taken into account in the tests separately.

Therefore, the objective of this paper was to check whether also these concretes meet the requirements for a synergistic attack of scaling and abrasion, which would allow their use in road construction or industrial floors.

2 EXPERIMENTAL SECTION

2.1 Purpose and scope of research

The research program aims to assess the influence of fly ash quality (loss on ignition) and quantity of synergistic interaction of scaling and abrasion of concrete resistance with their addition. Recipes for mixes were designed with the following assumptions:

– four quantities of added fly ash φ: 0, 20, 35 and 50% of the weight of cement (using a simple method to replace the cement with fly ash),
– water-binder ratio w/b = 0.38 (binder = cement + fly ash),
– three types of siliceous fly ashes compatible with EN 450-1: 2012, in terms of the loss on ignition A (LOI = 1.9%), B (LOI = 5.1%) & C (LOI = 9.0%),
– air-entraining with admixture based on modified wood resins (concrete mixes without addition of fly ash: non-air-entraining and air-entraining on the level of 4.5%, air-entraining concrete mixes with the addition of fly ash on the level 4.5%),
– fixed consistency of concrete mixes S3 (100÷150 mm slump) adjusted by the superplasticizer based on polycarboxylates.

The mixtures are made of cement CEM I 32.5R and a natural aggregate of sand point 35% and the maximum grain D = 16 mm. The amount of fly ashes addition is in the range of 0–50% due to the widespread use in practice (Bouzoba et al. 2001, Liu et al. 2002, Neville 2000, Siddique 2004). The main criterion used in choosing the fly ashes was to generate a representative domestic production with the loss on ignition in three categories A, B and C, while having a similar chemical composition and fineness. The physical and chemical properties of the selected three ashes are shown in Nowak-Michta (2013).

Abrasion resistance tests were carried out on air-entraining fly ash concretes with water-binder ratio w/b = 0.38, due to their scaling resistance (Nowak-Michta 2013) and fulfilment of the composition requirements for the abrasion classes XM1÷XM3 according to PN-B 06265 (Table 2).

Compositions and the compressive strength of concretes after 90 days of maturation are given in Table 3.

2.2 Test samples

The concretes were performed in the laboratory at 20°C and relative humidity above 60%. The quantities of additives, both air-entraining and superplasticizer were being added in order to achieve consistency in the class S3 (100÷150 mm slump) and the required level of air-entraining. The amounts of admixtures were used within the ranges recommended by the manufacturer.

The test samples were removed from moulds after 24 hours and stored in a chamber at 20 ± 2 °C and humidity of 95% in accordance with EN 12390–2: 2009 'Testing hardened concrete—Part 2: Marking and curing specimens for strength tests.'

2.3 Scaling test

Freeze/thaw test with de-icing salt (scaling) was carried out with an automatic chamber for freezing and thawing of samples, using the Slab test, according to PKN-CEN/TS 12390-9: 2007 'Testing hardened concrete—Part 9: Freeze-thaw resistance—Scaling.' and started after 90 days. For each of a series of concretes, the test was conducted on four samples of 140 × 72 × 50 mm, which were subjected to an impact of 56 freeze-thaw cycles in the presence of 3% NaCl (Nowak-Michta 2013). To assess the resistance of concretes to scaling, the criteria of EN 1338: 2003/AC: 2006 were used (Table 4).

2.4 Abrasion test

The test for abrasion resistance was carried out with the reference Wide Wheel test according to EN 1338: 2003/AC: 2006. For each of a series of concretes, the test was conducted on eight standard samples—four samples previously subjected

Table 3. Compositions and properties of concretes.

Type of ash	Without	Fly ash A		
Series	U0N	U2AN	U3AN	U5AN
Share of ash in binder φ [%]	0	20	35	50
Water-binder ratio w/b	0.38			
Cement [kg/m³]	450	360	293	225
Fly ash [kg/m³]	0	90	158	225
Water [kg/m³]	171	171	171	171
Aggregate [kg/m³]				
Sand 1	170	115	87	115
Sand 2	431	431	431	431
Gravel 2/8	712			
Gravel 8/16	518			
Admixture [% m.s.]				
Superplasticizer	1.2	1.3	1.4	1.1
Air-entraining	1.0	1.0	1.1	1.0
Compressive strength at 90 days [MPa]	64.1	63.2	44.6	37.0

Type of ash	Fly ash B			Fly ash C	
Series	U2BN	U3BN	U5BN	U2CN	U3CN
Share of ash in the binder φ [%]	20	35	50	20	35
Water-binder ratio w/b	0.38				
Cement [kg/m³]	360	293	225	360	293
Fly ash [kg/m³]	90	158	225	90	158
Water [kg/m³]	171	171	171	171	171
Aggregate [kg/m³]					
Sand 1	87	96	72	96	112
Sand 2	431	431	431	431	431
Gravel 2/8	712				
Gravel 8/16	518				
Admixture [% m.s.]					
Superplasticizer	1.1	1.3	1.2	1.3	1.6
Air-entraining	0.95	1.0	1.0	1.0	1.2
Compressive strength at 90 days [MPa]	61.8	49.4	42.8	66.9	59.7

Table 4. Resistance to freeze-thaw with de-icing salt according to EN 1338: 2003/AC: 2006.

Class	Marking	Mass loss after freeze/thaw test kg/m²
3	D	≤1.0 as a mean with no individual value > 1.5

Table 5. Abrasion resistance classes according to EN 1338: 2003/AC: 2006.

Class	Marking	Requirements for the Wide wheel abrasion test Maximum groove length [mm]
1	F	No performance measured
3	H	≤23
4	I	≤20

and four samples not subjected to an impact of 56 freeze-thaw cycles with 3% de-icing salt. Cast surfaces of samples were tested.

To assess the abrasion resistance of concretes, the criteria of EN 1338: 2003/AC: 2006 were used (Table 5). According to EN 1338 every single result should be higher than the permissible value.

3 RESULTS AND DISCUSSION

3.1 *Resistance to scaling*

All the tested air-entrained concretes with the water-binder ratio of 0.38 without and with fly ash addition, in light of the evaluation according to the classification EN 1338: 2003/AC: 2006 (Table 4) are class '3' of freeze-thaw resistance with de-icing salts marked 'D'. For all the tested concretes mass losses after 56 freeze/thaw cycles were not greater than 1.0 kg/m² (Nowak-Michta 2013).

3.2 *Abrasion resistance*

Figure 1 shows the results of abrasion (groove length) for concretes not subjected to scaling dependently on fly ash (A, B&C) content in binder. According to the classification EN 1338: 2003/AC: 2006 (Table 5), all the tested concretes are of the highest class '4' of abrasion resistance marked 'I'. The groove length measured by the Wide Wheel test reached levels in the range of 13 to 18 mm. The greatest lengths of grooves are observed for concretes with 35% addition of fly ash A and B.

3.3 *Influence of scaling on the abrasion resistance of fly ash concretes*

Figures 2–4 show the dependence of groove length and the content of fly ash in a binder for concretes

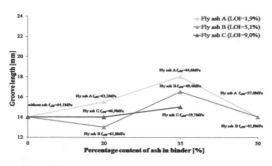

Figure 1.   Dependence of groove length and the content of fly ash (A, B & C) in a binder for concretes with their addition no subjected to scaling.

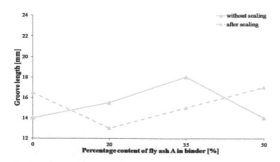

Figure 2.   Dependence of groove length and the content of fly ash A in a binder for concretes with their addition no subjected and subjected to scaling.

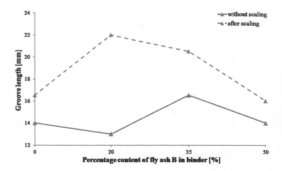

Figure 3.   Dependence of groove length and the content of fly ash B in a binder for concretes with their addition no subjected and subjected to scaling.

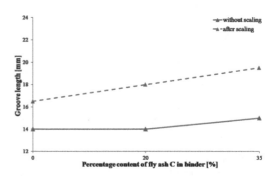

Figure 4.   Dependence of groove length and the content of fly ash C in a binder for concretes with their addition no subjected and subjected to scaling.

with their addition not subjected and subjected to scaling for all three types of fly ashes. In case of concrete with fly ash B and C, all the concretes previously subjected to scaling showed lower resistance to abrasion.

Grooves length increases by 2 mm in case of concrete with the addition of 50% fly ash B to 9 mm for concrete with 20% addition of also fly ash B. In concretes with 20 and 35% addition of fly ash A, the abrasion resistance after scaling increased compared with the concrete not subjected to freeze/thaw test. Groove lengths decreased by 2 and 3 mm. In contrast, the concretes with 50% of the fly ash A, as well as concretes with fly ash B and C indicated a decrease in abrasion resistance.

4   CONCLUSIONS

The carried out research program and analysis of the obtained results were the basis for the following conclusions:

– All the tested air-entrained fly ash concretes not subjected to scaling with the water-binder ratio of 0.38 and 20, 35 and 50% addition of fly ashes A, B & C according to EN 1338: 2003/AC: 2006 are resistant to abrasion and could be classified to the highest class '4' of abrasion resistance.

– Freezing and thawing attack with de-icing salt reduces the abrasion resistance of concretes. In most tested concretes groove lengths caused by abrasion of the surface previously subjected to scaling increased by 2.5 to 9 mm in comparison with concretes not subjected to scaling.

Both the quantity and quality (loss on ignition) of fly ashes, which significantly affected the compressive strength and frost resistance of concretes do not indicate the impact in case of abrasion resistance.

REFERENCES

Bouzoubaa, N. Zhang, M.H. & Malhotra, V.M. 2001. Mechanical properties and durability of concrete made with high-volume fly ash blended cements using a coarse fly ash, Cement and Concrete Research, 31: 1393–1402.

Fagerlund, G. 1997. Durability of concrete structures. Arkady, Warsaw (in Polish).

Kurdowski, W. 2014. *Cement and Concrete Chemistry*, Springer Netherlands.

Liu, Y.W. Tsong, Y. Tsao-Hua, H. Jeng-Chuan, L. 2002. *On the abrasion resistance of low cement high performance concrete*. 6th Int. Symp. on HS/HPC, Leipzig, 1139–1147.

Luehr, H.P. 1972. Use of fly ash (electrostatic precipitator) as a concrete additive. Betonwerk und Fertigteil-Technik, 7: 511–517 (in Deutch).

Neville, A.M. 2000. Properties of Concrete, Polish Cement Association (in Polish).

Nowak-Michta, A. 2013. Water-binder Ratio Influence on De-icing Salt Scaling of Fly Ash Concretes. Procedia Engineer-ing pp. 823–829, DOI: 10.1016/j. proeng.2013.04.104.

Nowak-Michta, A. 2014. Influence of fly ash on abrasion resistance of concretes with their additions. *Key Engineering Materials*, 592–593: 651–654.

Pigeon, M. Pleau, R. 1995. *Durability of concrete in cold climates*, E&FN SPON.

Rusin, Z. 2002. Technology of frost-resistant concretes, Polish Cement Association (in Polish).

Siddique, R. 2004. Performance characteristics of high-volume Class F fly ash concrete, *Cement and Concrete Research*, 34: 487–493.

Wesche, K. 1991. *Fly Ash in concrete, Properties and Performance*, E&FN SPON, RILEM.

Advanced Materials and Structural Engineering – Hu (Ed.)
© *2016 Taylor & Francis Group, London, ISBN 978-1-138-02786-2*

Performance comparison analysis of SUP-13 Lucobit asphalt mixture

Y.M. Zhang, M.L. Zheng & K. Wang
*Key Laboratory for Special Area Highway Engineering of Ministry of Education, Chang'an University,
Xi'an, Shanxi, China*
Jiang Su Province Communications Planning and Design Institute Co. Ltd., Nanjing, P.R. China

ABSTRACT: To research the influence of Lucobit on the performance of asphalt mixture, the method
of experimental comparison was conducted. The indexes of immersion Marshall, freeze thaw split, bending of the small beam, rutting and uniaxial static load creep experiments were tested with different dosages
of Lucobit, and the results were compared with SBS asphalt mixture. The results prove that, the indexes
of immersion Marshall, freeze thaw split, bending of the small beam have a little influence and the performance of anti-rutting increases with the increase of Lucobit. The anti-rutting performance of asphalt
mixture with Lucobit is higher than with SBS at the same grading. Moreover, the dosages of Lucobit can
increase the rutting performance of asphalt mixture by increasing the rebound modulus. Temperature has
a remarkable effect on high temperature deformation resistance of SUP-13 asphalt mixture.

1   INTRODUCTION

Rutting is a common phenomenon in the damage
of asphalt pavement. Compared with the crack and
flushing, the damage of rutting is bigger that will
influence road smoothness, cut down the whole
strength of the pavement structure and surface
layer, reduce the sliding resistance of pavement and
even influence the safety of the high-speed driving
for water accumulation in the rutting. Also, it may
affect the operation stability of vehicles while they
overtake or change lanes (Han & Yao 2001, Kong
2008). Thus, the problem of rutting is urgent necessary to be solved.

In order to improve the anti-rutting ability
of high-way asphalt pavement, several kinds of
modifiers are widely used in all levels of the road.
For example, Xiao Qiyi, Rui Shaoquan, Wang
Hang et al. (2006) researched the effect of the PR
PLASTS contents on asphalt mixture performance by tests. N.Y. Yin et al. (2013) researched the
rutting depth prediction for asphalt Mixture with
PCF anti-rutting. Wang Wanping, Wu Heping and

Zhao Sheng et al. (2009) researched the effect of
the asphalt mixture intermingled anti-rut agent
CheZhe Wang. The text selects Lucobit as a modifier and researches the performance of modified
asphalt mixture. The results will provide a certain technical support for the anti-rut agent of
Lucobit.

2   MATERIALS

The aggregate of the above layer is basalt produced
by Liyang Company, the aggregate size of 1# is
10–15 mm; 2# is 5–10 mm; 3# is 3–5 mm; 4# is
0–3 mm. The mineral powder is made up of limestone aggregate produced by Liyang Shuangyou
Company. According to the Test Methods of
Aggregate for Highway Engineering (JTG E42-
2005), relevant tests of aggregate and mineral
powder were conducted. Test results of basalt and
mineral powder are shown in Tables 1 and 2.

Asphalt is 70# produced by Nanjing Refinery
Limited Liability Company. According to the

Table 1.   Test results of limestone mineral aggregates.

Name of mineral aggregates	Apparent relative density	Bulk relative density	Water absorption (%)	Sand equivalent	Particle content of needles and flakes (%)	Clay content of aggregates (%)
1#	2.961	2.882	0.93	/	5.8	0.9
2#	2.980	2.893	1.02	/	4.8	1.0
3#	2.977	2.851	1.49	/	/	1.0
4#	2.925	2.765	1.99	74	/	/
Mineral powder	2.695	/	/	/	/	/

Table 2. Screening results of limestone mineral aggregates.

Diameter of the sieve aggregates	The pass rate (%)									
	16	13.2	9.5	4.75	2.36	1.18	0.6	0.3	0.15	0.075
1#	100	86.1	29.7	1.3	1.0	1.0	1.0	1.0	1.0	1.0
2#	100	100	100	22.7	3.4	2.0	2.0	2.0	2.0	1.0
3#	100	100	100	99.4	5.4	3.0	1.8	1.4	1.3	1.1
4#	100	100	100	100	85.4	56.9	34.6	16.9	12.2	10.0
Mineral powder	100	100	100	100	100	100	100	97.5	93.1	72.3

Table 3. Technical indicators of A grade 70# asphalt.

Test items	Technical requirements	Actually measured values	Conclusions
Softening point (TR&B)(°C)	46	48.5	Qualified
Ductility (5 cm/min, 15°C)(cm)	100	>100	Qualified
Penetration (25°C, 100 g, 5 s) (0.1 mm)	60~80	64	Qualified
Penetration index PI	−1.5~+1.0	−0.99	Qualified
Dynamic viscosity (60°C)(Pa·s)	180	362	Qualified
The relative density (25°C/25°C)	/	1.032	/
The solubility (%)	99.5	99.62	Qualified
Flash point (COC)(°C)	260	298	Qualified
Wax content	2.2	2.0	Qualified
RTFOT			
Quality changes (%)	±0.8	−0.08	Qualified
Penetration ratio (%)	61	65.6	Qualified
Ductility (5 cm/min, 15°C)(cm)	15	20	Qualified

Table 4. Technical performance of the modifier.

Types	Colors of appearance	Size of diameter (mm)	Melting point (°C)	Density (g·cm⁻³)
Lucobit	Granular, black	4	80–100	0.97

Standard Test Methods of Bitumen and Bituminous mixtures for Highway Engineering (JTG E20-2011), tests of the asphalt are conducted. The technical indicators are shown in Table 3.

Table 3 indicates that all of the technical indicators of A grade 70# asphalt can meet the requirements of the Specifications for Design of Highway Asphalt Pavement (JTG D50-2006).

The technical performance of Lucobit modifier is shown in Table 4.

3 RESULTS AND DISCUSSIONS

The gradation of mixture adopts SUP-13. The asphalt respectively adopts A grade 70# asphalt with 5% Lucobit, A grade 70# asphalt with 7% Lucobit, A grade 70# with 9% Lucobit, and SBS asphalt. Pavement performance test results are shown in Table 5.

Table 5 indicates that the rutting performance of asphalt mixture with Lucobit is better than the SBS modified at the same gradation and gradually increases with the increase of Lucobit content.

Through the indoor experimental research and analysis of a large number of literature at home and abroad, the text selects dynamic stability, relative deformation parameters and comprehensive stability index as evaluation indexes of anti-rutting performance. Rutting test results are shown in Table 6.

Table 6 shows that dynamic stability and dynamic stability ratio increase with the increase of Lucobit content. The dynamic stability ratio of Lucobit asphalt mixture is higher than SBS asphalt mixture. It indicates that the rutting performance and durability of Lucobit asphalt mixture is higher than SBS asphalt mixture. The comprehensive stability index of Lucobit asphalt mixture is higher than SBS asphalt mixture. The relative deformation rate of Lucobit asphalt mixture is lower than SBS asphalt mixture.

Table 5. Pavement performance test results of Lucobit asphalt mixture.

Mixture type	Dynamic stability (time/mm)	Residual stability (%)	Splitting tensile strength ratio (%)	Flexure tensile strain ($\mu\varepsilon$)
70# with 5% Lucobit	5568	88	87	3122
70# with 7% Lucobit	7733	90	88	2849
70# with 9% Lucobit	10430	93	88	2747
SBS modified	4344	91	90	3226

Table 6. Rutting test results.

Mixture type	60°C dynamic stability (times/mm)	70°C dynamic stability (times/mm)	Dynamic stability ratio/%	Deformation at 45 min/mm	Deformation at 60 min/mm	Comprehensive stability index/ mm^2	Relative deformation parameters/%
70# with 5% Lucobit	5568	3296	59.2	2.985	3.098	1865	6.196
70# with 7% Lucobit	7733	4972	64.3	1.563	1.644	4948	3.289
70# with 9% Lucobit	10430	6665	63.9	1.364	1.424	7647	2.849
SBS modified	4344	1659	38.2	4.652	4.797	934	9.594

Modulus is one of the main factors influencing the pavement deformation performance of asphalt. The main modulus indexes include rebound modulus, dynamic modulus, and static resilient modulus. In abroad, designing of highway asphalt pavement adopts dynamic modulus for the most part. However, in our country, the specifications still adopt static resilient modulus, which is using to reflect the elastic deformation and elastic hysteresis deformation.

The paper uses the uniaxial compression test of asphalt mixture to a compressive modulus of resilience. The test results are used to represent the resistance capacity to deformation of Lucobit asphalt mixture.

Test results of a uniaxial compression test are shown in the following Table 7.

Table 7 shows as follows: The rebound modulus of SBS modified asphalt mixture is higher than 70# asphalt mixture. From this, it can be elicited that the types of asphalt have a great influence on rebound modulus. Moreover, the rebound modulus gradually increases with the increase of the asphalt viscosity.

It can be drawn the following conclusions from the test in which the asphalt mixture has the same grading and different dosages of Lucobit. First, the rebound modulus gradually increases with the increase of Lucobit content. Then, it indicates that the dosages of Lucobit can increase the rutting performance of asphalt mixture by increasing the rebound modulus.

Test results of uniaxial creep are shown in Table 8.

Table 8 shows as follows: With the increase of Lucobit content, the creep rate of the same gradation and temperature gradually decreases and viscous stiffness modulus gradually increases. It indicates that the rutting performance of mixture gradually increases with the increase of Lucobit content.

The creep rate of the same gradation and temperature with Lucobit asphalt mixture is lower than SBS modified and viscous stiffness modulus of the former is higher than the latter. It indicates that the rutting performance of Lucobit modified asphalt mixture is higher than SBS modified.

With the increase of temperature, the creep rate of the same gradation and Lucobit content gradually increases and viscous stiffness modulus gradually decreases. It indicates that the temperature has a great influence on high temperature performance.

Table 7. Test results of a static modulus.

Mixture type	15°C compressive strength (MPa)	15°C modulus of resilience (MPa)	20°C compressive strength (MPa)	20°C modulus of resilience (MPa)
70#	2.826	2064	2.332	1259
70# with 5% Lucobit	3.271	2348	2.975	1777
70# with 7% Lucobit	3.846	2504	3.413	1898
70# with 9% Lucobit	4.266	2617	3.935	1978
SBS modified	3.001	2117	2.611	1332

Table 8. Test results of uniaxial creep.

Mixture type	60°C creep rate $(10^{-6}$ MPa/s)	60°C viscous stiffness modulus (MPa)	70°C creep rate $(10^{-6}$ MPa/s)	70°C viscous stiffness modulus (MPa)
70# with 5% Lucobit	7.149	46.26	20.685	15.42
70# with 7% Lucobit	4.526	57.58	13.474	19.20
70# with 9% Lucobit	3.248	66.48	9.664	22.01
SBS modified	11.359	36.44	26.174	12.15

4 CONCLUSIONS

To analyze the performance of SUP-13 Lucobit and SBS asphalt mixture, Immersion Marshall Test, freeze thaw split test, bending test of small beam, rutting test, and uniaxial static creep test were respectively carried out with different dosages of Lucobit. The conclusions are as follows.

1. The rutting performance of SUP-13 asphalt mixture with Lucobit is better than with SBS modified at the same gradation. It gradually increases with the increase of Lucobit content.
2. Temperature has an outstanding effect on high temperature deformation resistance of SUP-13 asphalt mixture.
3. The types of asphalt have a great influence on rebound modulus of asphalt mixture.
4. The dosages of Lucobit can increase the rutting performance of asphalt mixture by increasing the rebound modulus.
5. Dynamic stability, relative deformation parameters, and the comprehensive stability index were used to evaluate the performance of Lucobit and SBS modified asphalt mixture of SUP-13. The test results show that the performance of SUP-13 Lucobit asphalt mixture is higher than SUP-13 SBS asphalt mixture.

REFERENCES

Han, P. & Yao, L.Y. 2001. Necessity of asphalt pavement using modified asphalt, *Shanxi Science & Technology of Communications*, 2(144): 28–31.
Kong, Z.G. 2008. New Type of Modified Asphalt Additives-Lucobit 1210A, *Transpoworld*, 2008(17): 104–105.
Ministry of Transport of the People's Republic of China. JTG D50-2006, *Specifications for Design of Highway Asphalt Pavement*. Beijing: China Communications Press.
Ministry of Transport of the People's Republic of China. JTG E20-2011, *Standard Test Methods of Bitumen and Bituminous mixtures for Highway Engineering*. Beijing: China Communications Press.
Ministry of Transport of the People's Republic of China. JTG E42-2005, *Test Methods of Aggregate for Highway Engineering*. Beijing: China Communications Press.
Wang, W.P. Wu, H.P. & Zhao, S. 2009. Research on the Effect of the Asphalt Mixture Intermingled Anti-Rut Agent CheZhe Wang, *Highway Engineering*, 34(2): 127–130.
Xiao, Q.Y. Rui, S.Q. & Wang, H. et al. 2006. Research on the Effect of the PR PLASTS Contents on Asphalt Mixture Performance, *Journal of wuhan university of technology*, 28(7): 36–39.
Yin. N.Y. 2013. *Study of Evaluation Index of High-temperature Stability and Rutting Depth Prediction for Asphalt Mixture with PCF Anti-rutting*, Ji Lin University.

Advanced Materials and Structural Engineering – Hu (Ed.)
© *2016 Taylor & Francis Group, London, ISBN 978-1-138-02786-2*

Quantitative inspection of ferroconcrete damage by using ground-penetrating radar

C.W. Chang, C.H. Lin, C.C. Jen, C.A. Tsai, H.Y. Chien & P.S. Huang
Department of Civil Engineering, Chung Hua University, Hsinchu, Taiwan, R.O.C.

ABSTRACT: The main goal of this research is focused on the applicability, reliability, and practically of detecting the reinforcement steel bar (rebar) corrosion in concrete components by using the nondestructive Ground Penetrating Radar (GPR) method and comparing with the electrochemical half-cell potential conventional method. The accelerated corrosion tests are conducted to different concrete corrosion degrees in concrete samples. The frequency of 1 GHz GPR is used to detect the reflected electromagnetic voltage to induce significant corrosion levels. The obtained interface characteristic parameters specimens (Reflected voltage) for various thickness of the concrete specimens are compared with the half-cell potential measurement data. The nondestructive technique (GPR) and the half-cell potential method show a good agreement in evaluating the degree of ferroconcrete. The results also indicate that GPR method is capable of quantitative inspection of the ferroconcrete in the practical field.

1   INTRODUCTION

The environment of Taiwan belongs to that of a coastal region or maritime climate. In such an environment of high humidity and high corrosion, the corrosion rate of ferroconcrete components is extremely high. The most relevant ferroconcrete corrosion detection methods employ the half-cell corrosion potential or corrosion current method to measure the occurrence rate of rebar corrosion. The corrosion potential method mainly uses electrochemical pathways (a reference electrode, such as copper or silver) and material impedance characteristics to evaluate the degree of corrosion of steel (rebar) within concrete. The above corrosion detection methods can only qualitatively forecast the probability of rebar corrosion. The half-cell potential detection processes must partially drill destroyed the surface of the concrete component to form the electrochemical pathway to complete the measurement of the rebar corrosion potential and corrosion current. The electromagnetic wave reflection of ground penetrating radar can effectively determine the ferroconcrete's defective and deteriorated areas. Without drill, the concrete material comparing these two features based on the different characteristics of reflected waves from intact and defective ferroconcretes can identify the defect and determine the degree of deterioration. By comparing the ground penetrating radar to detect the rebar corrosion problems of different ferroconcrete components with corrosion potential measurement methods is also relevant.

2   ALGORITHM

The electromagnetic wave emits from ground penetrating radar transmitting in the medium to capture the reflection voltage of the concrete surface and rebar interface, and its value is mainly affected by the value of the characteristic impedance/current of the wave of the material medium interface. The ratio relationship between reflection voltage is defined as follows:

$$R_I = \frac{r_I(t)}{s(t)} \tag{1}$$

where $r_I(t)$ is the reflection voltage from the air to the concrete interface, $s(t)$ is the incident voltage of interface I, and R_I is the reflection coefficient of interface I. In the second layer medium, the incident wave of concrete medium to interface II (the rebar or corrosion rebar interface) will generate another reflection wave, as shown in Figure 1. R_{II} is the reflection coefficient of interface II. The equation is as follows:

$$R_{II} = \frac{r_{II}(t)}{s(t) \cdot w_i} \tag{2}$$

For the reflection voltage of rebar or corrosion rebar interface II, the equation is as follows:

$$r_{II}(t) = R_{II} \cdot s(t) \cdot w_i \tag{3}$$

where $r_{II}(t)$ is the reflection voltage of the concrete and rebar (corrosion rebar) interface, $s(t)$ is

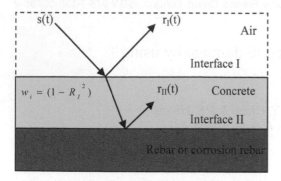

Figure 1. Incidence and reflected of electromagnetic wave between interfaces I and II.

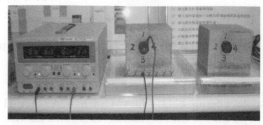

Figure 2. Rebar accelerated corrosion test.

the incident voltage of interface II, w_i is the incident power ($w_i = (1 - R_I^2)$), and R_I^2 is the reflection power of interface I.

3 EXPERIMENTAL INVESTIGATION

In this experiment, through the incidence of the electromagnetic wave of ground penetrating radar into the rebar concrete specimen containing rebar (diameter = 2.2 cm) and with concrete coverage thicknesses of 4 cm, 6 cm, 7 cm and 9 cm. The electromagnetic wave's reflection signal characteristic and rebar (corroded/non-corroded rebar interface) characteristic in the rebar concrete material measure both the half-cell potential and GPR acquire at the same time, in order to the frequency 1 GHz of the ground penetrating radar and capture the electromagnetic parameters such as the reflection voltage of the electromagnetic wave. The corresponding reflected voltage can judge the degrees of light corrosion, medium corrosion, and severe corrosion at different stages.

3.1 Rebar accelerated corrosion test

Utilize the impressed current of DC power supply to accelerate the corrosion rate of the rebar. The test method is to submerge the specimen in water. Without, the rebar contacting the water, connect the positive pole of the power supply to the rebar and the negative pole to the titanium mesh, and then adjust the impressed current of power supply, as shown in Figure 2.

3.2 Ground-penetrating radar rebar corrosion test

This places the specimens of different accelerated corrosion stages (0~408 hrs) and different concrete coverage thicknesses (4 cm, 6 cm, 7 cm, and 9 cm).

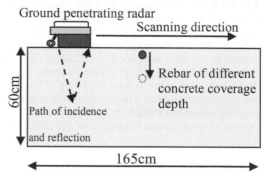

Figure 3. Ground-penetrating radar scanning ferroconcrete.

Figure 4. Rebar corrosion potential measurement.

The ferroconcrete specimens are scanned by GPR as shown in Figure 3.

3.3 Half-cell potential rebar corrosion test

This test uses a half-cell potentiometer, uses (copper/copper sulfate) as the reference electrode, and referring to ASTM C876 Code, and conducts measurement of the rebar corrosion potential, as shown in Figure 4.

4 RESULTS AND CONCLUSIONS

4.1 Ground-penetrating radar for ferroconcrete of different corrosion degrees

The sectional diagram of ground penetrating radar for conducting the accelerated corrosion test on concrete containing a single rebar, and for conducting ground-penetrating radar scanning at different concrete coverage thickness (4 cm, 6 cm, 7 cm, and 9 cm) and at different accelerated corrosion times with light, medium, and severe degrees of corrosion is shown in Figure 5. The results indicate that, according to the captured sectional diagram of the ground-penetrating radar, using corrosion potential to measure different accelerated corrosion times, of rebar with light, medium, and severe degrees of corrosion cannot directly determine the corrosion status of the rebar inside the concrete with different concrete coverage thicknesses (4 cm, 6 cm, 7 cm, and 9 cm) and different accelerated corrosion times (0~408 hrs).

According to the results of the reflection voltage of the rebar interface, under different thicknesses of concrete coverage, the reflection voltage of the rebar interface increases with the increment

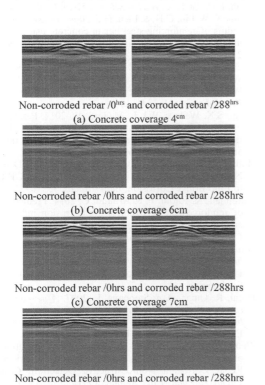

Non-corroded rebar /0^{hrs} and corroded rebar /288^{hrs}
(a) Concrete coverage 4^{cm}

Non-corroded rebar /0hrs and corroded rebar /288hrs
(b) Concrete coverage 6cm

Non-corroded rebar /0hrs and corroded rebar /288hrs
(c) Concrete coverage 7cm

Non-corroded rebar /0hrs and corroded rebar /288hrs
(d) Concrete coverage 9cm

Figure 5. Ground-penetrating radar profile on non-corroded/corroded rebar with different coverage thickness.

of corrosion time. The concrete coverage thickness reflected electromagnetic of the reflection voltage of the rebar interface from non-corroded to severe corrosion is 160 mV, 201 mV, 215 mV, and 174 mV respectively, according to the sequencing of the concrete coverage (4 cm, 6 cm, 7 cm, and 9 cm), the concrete coverage thickness of 7 cm is the water soaking surface of corrosion test specimen, thus the rebar corrosion interface is the most severe, as shown in Figure 6.

4.2 Electrochemical half-cell potential method in corrosion potential

The test uses corrosion potential/current density to measure the rebar concrete specimen containing rebar with concrete coverage thicknesses of 4 cm, 6 cm, 7 cm, and 9 cm. The test process and results are described as follows. For the different concrete coverage thicknesses, the accelerated corrosion time reached 288 hours, the rebar corrosion degrees all reached the severe corrosion status, as shown in Figure 7.

4.3 Comparison GPR/half-cell potential and code results

The characteristics parameter of electromagnetic waves on the rebar corrosion interface has

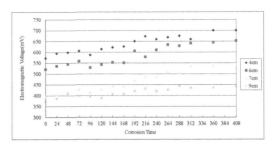

Figure 6. Reflection voltage of different concrete coverage thicknesses/rebar corrosion interface (GPR).

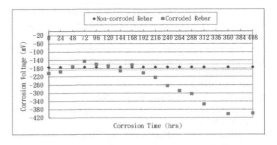

Figure 7. Various concrete coverage thicknesses/rebar corrosion potential voltage (half-cell potential).

conducted normalized study and analysis on the characteristic parameter obtained from the experiment through the incidence of the electromagnetic wave into rebar corrosion interfaces with different concrete coverage thicknesses.

After relatively normalizing the electromagnetic reflection voltage and corrosion potential degree, the results have compared the corrosion degree of overall qualitative rebar, and determined the scope of the GPR electromagnetic reflection signal of the rebar corrosion interface characteristic parameter captured at different points in time during the stages of light corrosion, medium corrosion, and severe corrosion, as shown in Figure 8. It may have defined the reflected voltage scope of the overall rebar corrosion phenomenon and compared it with the electrochemical measurement results, as shown in Table 1.

The normalized characteristics parameter reflected electromagnetic voltage with various concrete coverage thickness is shown in Figure 8. The result shows that the ferroconcrete reflected voltage is independence of concrete coverage thickness. By comparison with ASTM C876 Code, the results can be summarized in Table 1. The various corrosion degree of the GPR reflection electromagnetic

voltage is shown in Figure 8. In this study, using GPR scan the ferroconcrete component, the signal to evaluate the interface different reflection electromagnetic waves in the non-corroded/corroded situation. The algorithm of development for digital GPR can measure the status of the rebar interface with/without corrosion status.

ACKNOWLEDGMENT

The authors gratefully acknowledge the financial support given the grant by National Science Council in Taiwan.

REFERENCES

AL-Qadi, I.L. & Lahouar, S. 2005. Measuring layer thicknesses with GPR—Theory to practice, *Construction and Building Materials*, 19(10): 763–772.

Arndt, R. & Jalinoos, F. 2009. *NDE for corrosion detection in reinforced concrete structures—a benchmark approach.*, NDTCE.

Chang, C.W. Lien, H.S. & Lin, C.H. 2010. Determination of the stress intensity factors due to corrosion cracking in ferroconcrete by digital image processing reflection photoelasticity, *Corrosion Science*, 52(5): 1570–1575.

Chang, C.W. Lin, C.H. & Lien, H.S. 2009. Measurement radius of reinforcing steel bar in concrete using digital image GPR. *Construction and Building Materials*, 23(2): 1057–1063.

Eisenmann, D. Margetan, F. Chiou, C.P.T. Roberts, R. & Wendt, S. 2013. GPR-Based Water Leak Models in Water Distribution Systems. *Source: AIP Conference Proceedings*, 1511: 1341–1348.

Hong, S.X. Wallace, W.L.L. & Gerd, W. 2014. Periodic mapping of reinforcement corrosion in intrusive chloride contaminated concrete with GPR, Construction and Building Materials, 66: 671–684.

Hudson, S.G. Kumke, C.J. Beacham, M.W. Hall. D.D. & Narayanan, R.M. 2003. Nebraska DOR Tests GPR to Find Bridge Corrosion, *Better Roads*, 73: 70–73.

Inamullah, K. Raoul, F. & Arnaud, C. 2014. Prediction of reinforcement corrosion using corrosion induced cracks width in corroded reinforced concrete beams. *Cement and Concrete Research*, 56: 84–96.

Jamal, R. 2007. *Condition assessment of existing concrete bridge decks: the GPR versus the-half-cell potential.* University de Sherbrooke, Department of Civil Engineering.

Jamal, R. Omar, D. & Stéphane, L. 2007. *A new application of the gpr technique to reinforced concrete bridge decks*, 4th Middle East NDT Conference and Exhibition, Kingdom of Bahrain.

Zhang, Q.W. Chen, B.H. & Lin, Z.H. 2008. Application of Ground Penetrating Radar Method to Bridge Rebar Corrosion Non-destructive Detection, *Civil Engineering and Water Resources Periodical*, 35(5).

Zhang, Q.W. Lin, Z.H. & Chen, Y.D. 2007. *Study on the Behavior of Ground Penetrating Radar Electromagnetic Waves on Corroded Ferroconcrete*, Taiwan Concrete Institute 2007 Concrete Engineering Seminar.

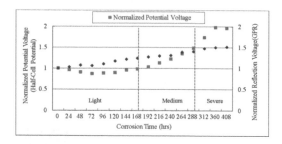

Figure 8. Level of rebar corrosion degree.

Table 1. Assessment of rebar corrosion probability. (Comparison with ASTM C876 Code).

Corrosion status	Concrete coverage thickness	ASTM C876 Code (half-cell potential)	Reflation voltage (GPR)
Corrosion probability less than 10%	4 cm~9 cm	>−200 mV	79 mV
Corrosion probability at 10%~90%	4 cm~9 cm	−200~ −350 mV	79~ 148 mV
Corrosion probability higher than 90%	4 cm~9 cm	<−350 mV	>148 mV

Advanced Materials and Structural Engineering – Hu (Ed.)
© 2016 Taylor & Francis Group, London, ISBN 978-1-138-02786-2

Properties of the advanced sulfur-bituminous pavements

E.V. Korolev, V.A. Gladkikh & V.A. Smirnov
Moscow State University of Civil Engineering, Moscow, Russian Federation

ABSTRACT: Currently, there is a strong need for advanced materials in road construction. The most used building material for pavements is bituminous concrete. Thus, the existing production facilities are mostly for this type of road concrete. Obviously, it is preferable for newly developed advanced materials to utilize the existing production cycle. Because of this, modification of traditional binder (bitumen) is the most promising direction during the design of advanced materials for pavements. The sulfur-bituminous concrete (sometimes called sulfur-advanced asphalt concrete)—constructional composite with matrix formed from a sulfur and bitumen mix—is one of such materials. To eliminate some negative technological aspects of the sulfur-bituminous mix (emission of the toxic gasses), earlier we had proposed the special admixture. There was a lack of essential knowledge concerning mechanical properties of the sulfur-bituminous concrete with such an admixture; therefore, we had carried out the necessary examination. It is revealed that a new material satisfies local regulations in terms of compressive and tensile strength, shear resistance, and internal friction.

1 INTRODUCTION

The continuous increase of the population all over the world leads to the problem of effective transportation. So far, the automotive transport is one of the most used. Increasing the number of vehicles causes the necessity both for construction of additional roads and for the improvement of properties of the pavements.

The observed trend in the area of the modern building materials for road construction and pavement are the R&D works, which are directed to design the material with complex of mutually exclusive properties. In one group there are properties related to the thermal stability of the pavement: upper limit of exploitation temperature, durability, low wearing, high resistance to rutting, etc. In another group there are properties related to plasticity of the bituminous matrix and asphalt concrete: low bound of the operational temperatures, low temperature of concrete mix compaction, etc.; such properties are of importance in cold regions of Russian Federation.

Of cause, the eradicative decision is to exchange the traditional bitumen by completely different matrix materials. Unfortunately, such decision can not be considered as a practical one for the entire road network just because of relatively high cost of the pavements made of other construction materials, including cement concrete. Also, newly developed materials have to utilize the existing production cycle and equipment. Thus, modification of traditional asphalt concrete is the promising direction during design of the advanced materials for pavements.

There exist numerous ways to improve the operational properties of concretes based on thermoplastic binder. By means of application of polymers distributed on the surface of fine filler it is possible to outcome several drawbacks of the traditional material with thermoplastic matrix (Korolev et al. 2012, Korolev et al. 2013). The disadvantage of such approach is the consequence of extra processing stage during preparation of concrete mix.

The bitumen matrix of the asphalt concrete can be enhanced by polyolefins. In particular, addition of the grinded polyethylene decreases penetration increases the softening and melting points (Shedame et al. 2014, Ali et al. 2014). As a consequence, the resulting material is characterized by increased rutting resistance and durability. However, admixture of polyolefins makes the production more complex and costly.

One of the most promising admixtures for the asphalt concrete is technical sulfur. Complex of operational properties of sulfur-based and sulfur-extended building materials may lead both to economical efficiency of construction and to reduce load on the environment. The feasibility of the sulfur-extended materials is primarily caused by properties, availability, and low cost of sulfur (Korolev et al. 2014).

Today, technology of sulfur-bituminous materials develops intensively in different countries. Asphalt modified with sulfur, due to several factors (changes in the chemical composition of bitumen and varying the intensity of the interaction at the interface boundaries) is characterized by increased values of operational properties at elevated temperatures, especially—by high resistance to rutting. The long-term water resistance is also at high level for concretes with binder that includes no more than 45% of sulfur (Gladkikh et al. 2013).

The sulfur-extended asphalt concretes often have to be characterized by an improved technical and economical efficiency; their usage leads to several benefits: increasing the material's application area, lowering the pollution of the environment (technical sulfur can be considered a waste because in Russia there are numerous sulfur dumps near the oil industry enterprises), decreasing the costs of construction and exploitation of roads (due to long periods between maintenance works).

The main factor limiting the widespread practical use of the mentioned technology was the lack of solutions to neutralize toxic gasses—hydrogen sulfide and sulfur dioxide—formed during the manufacture and installation of sulfur-bituminous concrete. It was shown (Gladkikh et al. 2014) that oxides of amphoteric metals are effective suppressors for hydrogen sulfide and sulfur dioxide. The obtained experimental data allowed the selection of most prospective suppressors for toxic gasses emitted during the production and laying of sulfur-bituminous concrete.

But till now, there was limited knowledge about physical and mechanical properties of the concrete with sulfur-bituminous matrix and aforementioned suppressors.

2 EXPERIMENTAL SETUP

The standard samples (Fig. 1) of asphalt concrete were prepared according to RU GOST 12801 from the concrete mixes with different amount of sulfurs (20%–45%). Quality of the concrete mixes was determined according to RU GOST 31015. The prepared samples were examined for physical-mechanical and operational properties according to RU GOST 12801 (crack resistance), AASHTO

Figure 2. AASHTO TP 63 testing during examination of samples.

TP 63-06 (resistance to rutting), and RU ODM 218.3.018-2011 (fatigue life).

The UNIFRAME device was used during determination of compressive strength. Resistance to rutting is determined with Asphalt Pavement Analyzer (Fig. 2). Examination of fatigue life is performed with Dynapave 130 equipment.

3 RESULTS AND DISCUSSION

One of the most important mechanical properties of any building material is compressive strength. Experimentally obtained dependencies between the amount of sulfur-based admixture and compressive strength are shown in Figure 3.

As it follows from Figure 3, addition of the small amount of sulfur does not change the compressive strength. Starting from concentration of 20%, there is a linear dependence between compressive strength and amount of sulfur. Adding the sulfur notably alters the properties of the asphalt binder (the concentrations above 40% can not be used in practice because several other operational properties of the concrete are out of allowed limits). Such behavior is observed for both temperatures of experiment (20 and 50 °C).

This can be described as follows. Part of the sulfur is dispersed in the bitumen and leads to only a small improvement of the properties. Another part of the sulfur (when concentration is about 20% and higher) is undissolved and coagulates. Formation of large sulfur droplets takes place; these droplets crystallize upon cooling. The additional links of the crystallization are formed and because of this there is a hardening of asphalt material. At the operating temperatures (both 20 and 50 °C), sulfur is solid and plays the role of disperse filler. This filler organizes the structure of the bitumen and increases thermal stability, rigidity, and resistance to rutting of asphalt concrete.

At the technological temperatures, sulfur is a highly mobile melt and increases the efficiency of

Figure 1. Samples of asphalt concrete after AASHTO TP 63 test.

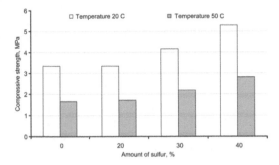

Figure 3. Compressive strength of sulfur-bituminous pavement.

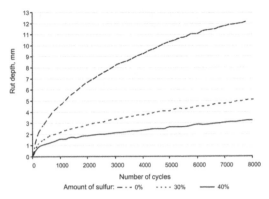

Figure 4. Dependencies between number of cycles and depth of the rut.

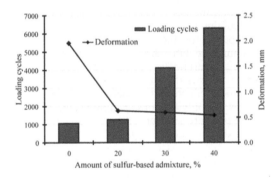

Figure 5. The dependencies between fatigue limit, horizontal deformation and number of load cycles for different amounts of sulfur-based modifier.

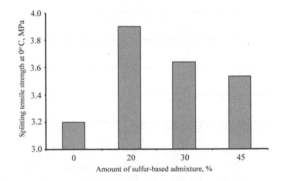

Figure 6. Splitting tensile strength.

mixing and laying of asphalt concrete mixes; thus, the goals of two different aforementioned groups are achieved simultaneously.

There is one operational property that is tightly correlated with mechanical properties—resistance to rutting. Low values of rutting resistance significantly reduce durability of the pavement under load of heavy vehicles.

The experimental dependencies between number of processing cycles (according to AASHTO TP 63 test procedure) and depth of the rut are presented in Figure 4.

As it follows from Figure 4, even the small amount of sulfur significantly increases rutting resistance. Therefore, the area of primary application of sulfur-bituminous concrete is the construction of pavements with heavy load.

Material's fatigue life, together with rutting resistance, is the primary indicator of suitability for durable and long-life pavement. It is found during the study of fatigue life by means of multiple cyclic loading with indirect tensile scheme that designed sulfur-bituminous materials are of high resistance (Fig. 5).

It is clear from the experimental data in Figure 5 that fatigue life increases together with amount of sulfur in sulfur-bituminous mix. As before, starting from 20% concentration, we can observe linear dependence between operational property (in this case—fatigue life) and concentration. The number of cycles preceding the start of irreversible deformation induced by main crack grow, if compared with such number for reference sample, increases significantly for compositions with 30% and 40% of sulfur-based modifier (about 4 and 6 times, respectively). At the same time, enlargement of the content of sulfur modifier leads to decrease of deformability.

So far, all the obtained data clearly demonstrate that there is the ultimate way for production of advanced pavement material—namely, complete elimination of bitumen. Obviously, such a conclusion would be wrong. Some consequences of such decision were already mentioned, and necessity of massive modification in current production facilities is not the only reason for that. Several important operational properties (e.g., water and frost resistance) are negatively affected by excess of the sulfur. Moreover, not every mechanical characteristic are in a positive correlation with amount of

Table 1. Properties of the developed sulfur-bituminous concrete

Property	RU GOST 31015	Reference sample	Modified concrete, sulfur content		
			20%	30%	40%
Compressive strength at 20 °C, MPa	> 2.2	3.3	3.3	4.2	5.3
Compressive strength at 50 °C, MPa	> 0.65	1.67	1.72	2.19	2.82
Average density, kg/m^3	–	2620	2640	2650	2650
Clutch shear (50 °C), MPa	> 0.18	0.29	0.29	0.38	0.54
Water resistance	> 0.85	0.90	0.89	0.95	0.90

sulfur in bitumen. One of such properties is crack resistance. This parameter can be characterized by experimentally obtained values for splitting tensile strength at 0 °C (Fig. 6).

It is evident from Figure 6 that the replacement of most bitumen by sulfur may have a negative impact on the low-temperature properties of sulfur-bituminous mixtures.

Properties of the sulfur-bituminous materials are summarized in Table 1.

The experimental results summarized in Table 1 show that compressive strength of sulfur-bituminous concrete at temperatures 20 and 50 °C exceeds not only the corresponding values for reference concrete without sulfur, but also the values required by RU GOST 31015.

It was also revealed during additional examination that sulfur-bituminous concrete after storage in water shows very low absorption. The absorption is 1.5–1.8 times lower if compared with reference composition. Therefore, freezing and thawing causes only a small negative effect if compared with conventional asphalt concrete. It is established that under identical operating temperatures the modulus of elasticity of sulfur-bituminous concrete is higher than modulus of conventional asphalt concrete. Moreover, with increasing temperature this difference also increases.

4 CONCLUSIONS

In the present work, we have discussed the results of several laboratory tests. The tests were carried out to answer the question concerning dependence between the amount of sulfur-based modifier, mechanical and important operational properties of sulfur-bituminous concrete for pavements.

It is shown that low amounts of sulfur are dissolved in bitumen, the compressive strength of concrete with such matrix is close to the strength of reference sample. Extra amounts of sulfur may lead to the formation of additional spatial cross-linked network. The formation of such a network, in turn, causes the increase of both strength and modulus of the concrete; rutting resistance

also increases significantly. Thus, almost all the mechanical properties, including compressive strength and resistance to rutting, are positively correlated with concentration of designed multifunction sulfur-based admixture. Taking into consideration values of such properties as crack resistance (especially at low temperatures), water and frost resistance, and also such factors as necessity for preservation existing production infrastructure, the recommended amount of sulfur-based admixture with gas-suppression additives is about 30–40%.

ACKNOWLEDGMENTS

This work is supported by the Ministry of Science and Education of Russian Federation, job No 2014/107, project title "Structure formation of sulfur composites: phenomenological and ab initio models".

REFERENCES

Ali, T. Iqbal, N. Ali, M. & Shahzada, K. 2014. Sustainability assessment of bitumen with polyethylene as polymer. *IOSR-JMCE,* 5: 1–6.

Gladkikh, V.A. & Korolev, E.V. 2013. Technical and economical efficiency of sulfur-bituminous concretes. *Vestnik MGSU,* (4): 76–83.

Gladkikh, V.A. & Korolev, E.V. 2014. Suppressing the hydrogen sulfide and sulfur dioxide emission from sulfur-bituminous concrete. *Advanced Materials Research,* 1040: 387–392.

Korolev, E.V. & Inozemcev, S.S. 2013. Design and investigation of the nanomodifiers for the bituminous concretes. *Vestnik MGSU,* (10): 131–139.

Korolev, E.V. Smirnov, V.A. & Albakasov, A.I. 2012. Nanomodified composites with thermoplastic matrix. *Nanotechnologies in Construction,* (5) 4: 81–87.

Korolev, E.V. Smirnov, V.A. & Evstigneev, A.V. 2014. Nanostructure of matrices for sulfur constructional composites: methodolody, methods and research tools. *Nanotechnologies in Construction,* 6 (6): 106–148.

Shedame, P. & Pitale, N. 2014. Experimental study of bituminous concrete containing plastic waste material. *IOSR-JMCE,* 3: 37–45.

Advanced Materials and Structural Engineering – Hu (Ed.)
© 2016 Taylor & Francis Group, London, ISBN 978-1-138-02786-2

Application research on concrete recycled aggregate of construction waste in the road base

T.Z. Ming, K. Wang & G.W. Hu
Jiang Su Province Communications Planning and Design Institute Co. Ltd., Nanjing, P.R. China

ABSTRACT: In order to study the application of concrete recycled aggregate of construction waste in the road base, related tests were studied on the basis of the existing highway industry standards. The test results show that the density, crushing value, Los Angeles abrasion value of recycled aggregate have a certain gap compared with natural aggregate. But all of them can meet the requirements of standards. Water absorption of recycled aggregate is three times more than the natural aggregates. So, it is recommended to add a certain proportion of natural aggregate when used in road base.

1 INTRODUCTION

With the development of urban construction, the amount of construction waste increased with years. Now, the processing mode of construction waste is mainly landfilling and piling up, or is used as a filling material for the building foundation in China. The method not only takes up a lot of lands and causes environmental pollution, but also causes a huge waste of resources. On the other hand, natural aggregate resources will be increasingly dried up with the rapid development of road and construction. But the construction waste of concrete can be used for road construction as a replacement of natural aggregate resources by a certain processing (Arulrajah et al. 2012).

In recent years, many researchers have done a lot of study on the recycling of construction waste, but the achievements of application research in road base are still less and it is also lack of relevant technical standards (Tam & Tam 2007, Qi et al. 2014). To guide a large number of applications of recycled aggregate of concrete construction waste in the road base, in this paper a relatively comprehensive test study is conducted on the performance of recycled aggregate according to the related technical standards of highway industry.

2 MATERIALS AND TESTS

The concrete recycled aggregate of construction waste used in the tests is continuously graded aggregate produced by Jiangsu Wujin Lvhe environmental protection building materials technology CO.LTD. Single-stage gradations are respectively 9.5–26.5 mm, 4.75–9.5 mm, and 0–4.75 mm.

Because the recycled aggregate considered in this paper is mainly considered for using in road base, the comparison of test methods and evaluation standards of concrete recycled aggregate and natural aggregate was done. The studies on the three kinds of recycled aggregate, included chemical composition analysis, screening test, density and water absorption, crushing value, Los Angeles abrasion, etc. were conducted. Test methods and test equipment were on the basis of Industry standard of the People's Republic of China—Test Methods of Aggregate for Highway Engineering (JTG E42-2005).

3 TEST RESULTS AND DISCUSSIONS

3.1 The composition of recycled aggregate

Compared with natural aggregate, recycled aggregate have a complex composition containing partially hardened cement mortar (porosity and high water absorption). Also, the concrete block produces a large number of micro-cracks when it is damaged in disintegration and fragmentation process. So the above characteristics show that the recycled aggregate are quite different with the natural aggregate. The compositions of coarse aggregate of 4.75 mm particle above were detected. The composition and proportion of coarse aggregate are shown in Table 1 and Figure 1.

Most of the coarse aggregate of construction waste can be recycled. Waste concrete, waste brick, waste ceramics, waste glass, and other materials can be used as building materials after a certain treatment. But wood, plastics, and other materials must be removed.

Table 1. Composition and proportion of concrete recycled aggregate.

Composition	Proportion (%)
Waste concrete	95.2
Waste brick	1.0
Waste ceramics	2.9
Waste glass	0.7
Waste wood	0.1
Other	0.1

Figure 1. Composition of concrete recycled aggregate.

Table 2. Gradation of recycled aggregate.

Aggregate type	Passing rate through the sieve pore (square hole sieve, mm) (%)							
	31.5	26.5	19	9.5	4.75	2.36	0.6	0.075
1#	100	100	65.0	2.4	1.9	1.9	1.6	0.4
2#	100	100	100	88.2	40.8	13.6	10.7	5.0
3#	100	100	100	100	100	93.7	57.8	16.0

3.2 Gradation of recycled aggregate

The gradations of recycled aggregate used in the text are shown in Table 2.

3.3 Chemical analysis of recycled aggregate

In this paper, ARL9800XP type X-ray fluorescence (in Fig. 2) was used to detect the chemical composition of recycled aggregate. The samples were placed in an oven at 105 °C for 4 hours. X-ray fluorescence was used to detect the chemical

Figure 2. ARL9800XP type X-ray fluorescence.

Table 3. Chemical composition of recycled aggregate of construction waste.

Test items	The proportion (%)
SiO_2	63.9
Al_2O_3	8.24
Fe_2O_3	3.39
CaO	11.17
MgO	1.10
K_2O	1.64
Na_2O	0.57
TiO_2	0.47
SO_3	1.20
P_2O_5	0.12
MnO	0.08
SrO	0.045
ZrO_2	0.022
BaO	0.062
Cl	0.029
Cr_2O_3	0.023
ZnO	0.014
V_2O_5	0.012
Loss on ignition	7.89

composition of recycled aggregate. The results are shown in Table 3.

According to the chemical composition, the main components of recycled aggregate of construction waste include SiO_2, Al_2O_3, Fe_2O_3 and $CaCO_3$ (using the amount of CaO and loss on ignition to express in the test results). According to the test results of SiO_2 content, it can be preliminary judged that the recycled aggregates are from neutral aggregate.

3.4 Density of recycled aggregate

The recycled aggregate calculation formula of apparent relative density is shown in Formula 1, th test results are shown in Table 4.

$$\gamma_a = \frac{m_a}{m_a - m_w} \qquad (1)$$

Formula:

γ_a—Apparent relative density of recycled aggregate, no dimensional;
m_a—Drying quality of aggregate (g);
m_w—Water quality of aggregate (g).

As is shown in Table 4, the apparent relative density of recycled coarse aggregate was lower than natural coarse aggregate because of the effect of mortar that attached to the surface of aggregate. The apparent relative density of recycled coarse aggregate was higher than 2.50 and meets the requirements of the apparent relative density of low-grade road base material in "Specifications for Design of highway asphalt pavement" (JTG F40-2004, JTG D50-2006).

The lowest limit (>2.450) of apparent relative density of the class I recycled coarse aggregate in "Recycled aggregate application of technical regulation (JGJ/T240-2011)" meets the requirements of apparent relative density of low-grade highway coarse aggregate in "Technical Specifications for Construction Design of Highway Asphalt Pavements (JTG F40-2004)".

3.5 Water absorption of recycled aggregate

The water absorption of recycled coarse aggregate was calculated by Formula 2. The results are shown in Figure 3.

$$\omega_x = \frac{m_f - m_a}{m_a} \qquad (2)$$

Formula:

ω_x—The water absorption of recycled aggregate (%);
m_f—The skin drying quality of recycled aggregate (g);
m_a—The drying quality of recycled aggregate (g).

Table 4. Results of apparent relative density of recycled aggregate.

Aggregate type	Apparent relative density γ_a		
	1#	2#	3#
Recycled aggregate	2.670	2.682	2.667
Natural aggregate	2.912	2.921	2.917

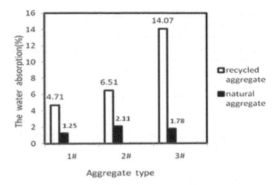

Figure 3. Water absorption of recycled aggregate.

Table 5. Test results of crushing value of aggregate.

Aggregate type	Crushing value (%)	Technical requirements (%)
Recycled aggregate	22.9	/
Natural aggregate	12.5	≤30

As shown in Figure 3, the water absorption of recycled aggregate was far greater than natural aggregate for the influence by cement mortar and micro-cracks. The water absorption was three times more than the natural aggregate.

3.6 Crushing value of recycled aggregate

The crushing values of recycled coarse aggregate and natural aggregate were calculated by Formula 3. The results were shown in Table 5.

$$Q_a = \frac{m_1}{m_0} \times 100 \qquad (3)$$

Formula:

Q_a—The crushing values of recycled coarse aggregate (%);
m_0—the quality of the sample before the test (g);
m_1—The quality of fine material by 2.36 mm sieve after the test (g).

The test results show that the crushing value of recycled aggregate is greater than the crushing value of natural aggregate, which can be attributed to the crushing value of mortar which wrapped the surface of recycled coarse aggregate, was lower than the crushing value of natural gravel. Test method for crushing value of recycled coarse aggregate is different from the coarse aggregate of asphalt concrete in "Technical Specifications for Application of Recycled Aggregate (JGJ/T240-2011)".

Table 6. Test results of Los Angeles abrasion value of recycled aggregate.

Aggregate type	Angeles abrasion index (%)
Recycled aggregate	14.3
Natural aggregate	12.5

Therefore, the crushed value is not compared in the section.

3.7 Los Angeles abrasion value of recycled aggregate

The results are shown in Table 6.

Because Los Angeles abrasion value of recycled coarse aggregate is not specified in the "Technical Specifications for Application of Recycled Aggregate (JGJ/T240-2011)". Thus the recycled aggregate cannot be compared with the natural aggregate.

4 CONCLUSIONS

1. More than 99% of the concrete recycled aggregate were waste concrete, brick and ceramic materials.
2. The gradation composition of recycled aggregate can meet the requirements for cement-stabilized soil in the road-base standard. But the gradation is not excellent. So, it is recommended that sub-file processing was needed before the designs of mixing ratio.
3. The recycled aggregate belongs to a siliceous material. The chemical composition of recycled aggregate contains SiO_2 of 63.9%, a small amount of SO_3, Cl^-, Fe, Al, Ca, K, etc.

4. Although the density, crushing value, and Los Angeles abrasion value of recycled aggregate have a certain gap with the natural aggregate, they can meet the requirements of a standard. The water absorption of third gear recycled aggregate is three times more than the natural aggregate. It is recommended to add a certain proportion of natural aggregate when recycled aggregate used in road base.

REFERENCES

Arulrajah, A. Piratheepan, J. & Disfani, M.M. et al. 2012. Resilient moduli response of recycled construction and demolition materials in pavement subbase applications. *Journal of Materials in Civil Engineering*, 25(12): 1920–1928.

Ministry of Housing and Urban-Rural Development of the People's Republic of China. JGJ/T240-2011, *Technical Specifications for Application of Recycled Aggregate*, Beijing: China Architecture and Building Press.

Ministry of Transport of the People's Republic of China. JTG D50-2006. *Specifications for Design of highway asphalt pavement*, Beijing: China Communications Press.

Ministry of Transport of the People's Republic of China. JTG E42-2005. *Test Methods of Aggregate for Highway Engineering*, Beijing: China Communications Press.

Ministry of Transport of the People's Republic of China. JTG F40-2004, *Technical Specifications for Construction Design of Highway Asphalt Pavements*. Beijing: China Communications Press.

Qi, F. Han, R.M. & Zhang, M.C. 2014. Construction Waste Inorganic Regenerated Aggregate Performance Tests, *Transportation Standardization*, 42(5): 96–100.

Tam, V.W.Y & Tam, C.M. 2007. Crushed aggregate production from centralized combined and individual waste sources in Hong Kong. *Construction and Building Materials*, 21: 879–886.

Advanced Materials and Structural Engineering – Hu (Ed.)
© 2016 Taylor & Francis Group, London, ISBN 978-1-138-02786-2

Design an angle-inserting joint for steel sandwich panel

X.J. Wu & F.L. Meng
*School of Materials Science and Engineering, Shenyang Ligong University, Hunnan New District,
Shenyang City, P.R. China*

X.G. Meng
*China Nuclear Industry 22nd Construction Co. Ltd., Steel Structure Branch, Jiaxing City,
Zhejiang Province, P.R. China*

ABSTRACT: Designing a steel sandwich panel joint is one of the most difficult problems associated with the structures in question. This paper presents the searching for process of optimum geometry of a panel-to-panel angle-inserting joint along longitudinal arrangement, which performed by means of the as-low-as possible values of maximum von mises stress at acceptable mass and deformations of the structure.

1 INTRODUCTION

The steel sandwich panels can be used as structural elements of the ship hull smooth, flat surfaces like decks, walls and, in certain cases, side shells due to their superior performance in comparison to other structural materials in terms of improved stability, high stiffness and strength to weight ratios, excellent thermal insulation and acoustic damping (Metschkow 2006). From the ship building point of view, I-core sandwich panels seem to be the most suitable because they have an optimal relationship to its mass to stiffness both in the longitudinal and transverse directions, and are relatively easy for manufacturing. Generally, the I-core panels can be made of relatively small size in the industry, the panel width is 500 ~ 3000 mm and a length is 1000 ~ 10000 mm (Wang et al. 2012). In order to form bigger sections of the ship structure, connection of some panels is necessary. Design of panel-to-panel joints of sandwich structure is one of the key issues for applying sandwich in ship building, which affect not only the hull structure assembly but also the strength and fatigue resistance of sandwich structure.

In order to design the optimal sandwich panel joint, Pyszko (2006) established a typical two-dimensional finite element model for selected cover-plate joint, and analyzed the variation of ultimate bearing capacity with main design parameters under tensile, compressive and bending load. Niklas (2008) searched for optimum geometry of cover plate and rectangular profile joints by analyzing the geometrical stress concentration coefficients at acceptable mass and deformations of the structure. The results show that rectangular profile joint has an excellent performance.

In our research, one typical angle-inserting joint was selected for I-core steel sandwich panel in longitudinal arrangement. By comparing the maximum von mises stresses at the joint, the optimum geometry of angle-inserting joint was obtained at different design parameters under out-plane loads.

Q345 steel sandwich structure unit width 2800 mm and length 2200 mm was adopted in our research. The geometry of the selected angle-inserting joint was described using parameters shown in Figure 1, in which the adopted coordinate system is marked as well. Thickness "d" and width "L" of angle steel, overlap distance "L_1" between angle steel and a cover plate of sandwich panel were variable parameters. d varied in the range of 2, 2.5 and 3 mm, the values of L were 20, 30, 40, 50 and 60 mm, respectively. L_1 depended on the values of L, L_1 always was less than L. For example, when L was equal to 60 mm, the value of L_1 could be any one among 10, 20, 30, 40 and 50 mm.

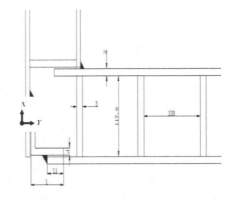

Figure 1. Geometry parameters of angle-inserting joint.

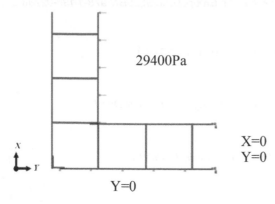

29400Pa

x

r

X=0
Y=0

Y=0

Figure 2. 2-D geometric model of angle-inserting joint sandwich panel.

2 MODELING

2.1 Modeling geometry, boundary conditions and load

The calculations were performed in an ABAQUS environment. It is possible to analyze 2-D model due to plane strain condition of sandwich panel, which considerably speed up the calculations. Figure 2 shows the 2-D model for angle-inserting joint. In order to improve the accuracy of the simulation results, the contact interaction property was employed (Kozak 2006). 29400 Pa compressive stress was applied in the direction perpendicular to the surface of steel sandwich structure. Boundary conditions were X = 0 and Y = 0 at the right of the plate and X = 0 at the bottom of the plate.

2.2 Meshing and element selection

The model contains approximately 45000 CPE4R elements, the global element size is about 1 mm. Considering the stress concentration is usually located in fillet weld of joint, the element sizes around fillet weld are refined to 0.1 mm.

3 RESULTS AND DISCUSSION

Fixed d = 2.0 mm, then L and L_1 were changed in a range. We assumed that all of the deformation and stress were elastic for comparing the stress concentration. In all simulation results, the position of maximum von mises stress always located at the fillet weld leg, as shown in Figure 3. The reason is this location is the maximum tensile stress state according to the load and boundary condition.

The variation of maximum von mises with L and L_1 is shown in Figure 4a. The diagrams reveal that the maximum von mises stress is insensitive

to L or L_1 values, it is a single-valued function of $(L - L_1)$ when the angle-steel width is fixed. That is, the maximum von mises stress reaches the minimum value about 500 MPa when $(L - L_1)$ is 10 mm, and then it reaches the maximum value about 1310 MPa when $(L - L_1)$ is 50 mm. It illustrated in Figure 3 that angle-inserting joint has high stiffness at smaller width of angle steel. As a consequence, smaller deformation and stress occurs at joint when L is relatively smaller.

The thickness of angle steel chosen as 2.5 and 3 mm, the effect of parameters L and L_1 on the maximum stress is similar to d = 2 mm (As shown in Fig. 4b and 4c). However, the maximum von mises stress increases with increasing of thickness at the same $(L - L_1)$ value. It can be seen in Figure 4d that, fixed L = 20 mm and L_1 = 10 mm, the maximum von mises stress and thickness are an approximately linear relationship. The von mises stresses were used as the measure of joint's ability to carry a compressive load at an acceptable mass of the joint (up to 10% of panel mass) (Kozak 2009, Boronski & Kozak 2004). If we only consider the stress, then the optimal parameters are t = 2 mm and $(L - L_1)$ = 10 mm that include

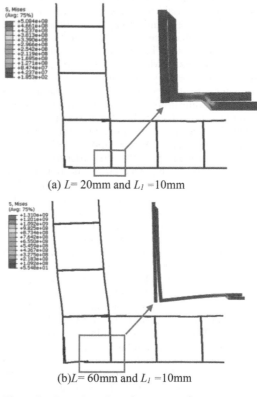

(a) L= 20mm and L_1 =10mm

(b)L= 60mm and L_1 =10mm

Figure 3. Location of maximum von mises.

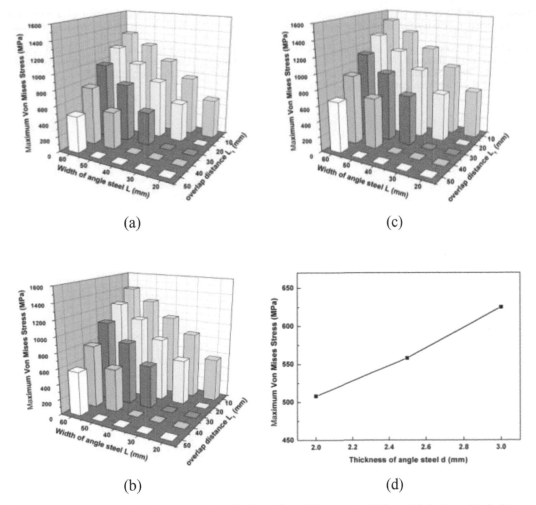

Figure 4. The variation of maximum von mises with L and L_1 at different angle thickness (a) $d = 2$ mm (b) $d = 2.5$ mm (c) $d = 3$ mm and (d) fixed $L = 20$ mm and $L_1 = 10$ mm, maximum von mises vs. thickness of angle steel.

five sets of parameters. The joint with smaller mass was selected in order to reduce the weight of joints. The optimal parameter $d = 2$ mm $L = 20$ mm $L_1 = 10$ mm is selected finally.

4 CONCLUSIONS

The steel sandwich panel angle-inserting joints were designed for optimization using finite element method. Thickness "d" and width "L" of angle steel, overlap distance "L_1" between angle steel and cover plate of sandwich panel were investigated. The optimal geometric parameters of the joint were obtained by comparing maximum stress of sandwich panel under out-plane compressive load. Conclusions are as follows:

1. The maximum von mises stress is insensitive to L or L_1 values, it is a single-valued function of $(L - L_1)$ when the angle-steel width is fixed.
2. The maximum von mises stress increases with increasing of thickness at the same $(L - L_1)$ value.
3. The optimal parameter $d = 2$ mm $L = 20$ mm $L_1 = 10$ mm is selected finally.

REFERENCES

Boronski, D. & Kozak, J. 2004. Research on deformations of laser-welded joint of a steel sandwich structure model, *Polish Maritime Research*, 11: 3–8.

Kozak, J. 2006. Problems of strength modeling of steel sandwich panels under in-plane load, *Polish Maritime Research*, 13: 9–12.

Kozak, J. 2009. Selected problems on application of steel sandwich panels to marine structures, *Polish Maritime Research*, 16: 9–15.

Metschkow, B. 2006. Sandwich panels in shipbuilding. *Polish Maritime Research*, 13: 5–8.

Niklas, K. 2008. Search for optimum geometry of selected steel sandwich panel joints, *Polish Maritime Research*, 15: 26–31.

Pyszko, R. 2006. Strength assessment of a version of joint of sandwich panels, *Polish Maritime Research*, 13: 17–20.

Wang, H. Cheng, Y.S. & Liu, J. 2012. Strength Analysis on Model I-Core Steel Sandwich Panel Joints. *Chinese Journal of Ship Research*, 7(3): 51–56.

Advanced Materials and Structural Engineering – Hu (Ed.)
© 2016 Taylor & Francis Group, London, ISBN 978-1-138-02786-2

Green Building Development system modeling and simulation

J.Y. Teng, L.M. Zhang, J.B. Zhong & C.Y. Du
School of Civil Engineering and Mechanics, Huazhong University of Science and Technology, Wuhan, China

H.Y. Chen
Institute of International Education, Wuhan University of Technology, Wuhan, China

ABSTRACT: Green buildings have many problems may emerge in the complex and dynamic process of green building development. In order to evaluate the current state and future trend of variation of green building for sustainable building development, a "Green Building Development (GBD)" model is first constructed with the System Dynamics (SD) method and implemented using the Vensim software. After the GBD-SD model has been verified with historic data and by Vensim, it is then used to simulate and evaluate the state and trends of variation of green building development in Wuhan during the years 2008–2050. This study provides a basic model to facilitate better understanding of the internal relationships in green building development, and may be used to coordinate the relationship between participants and provide a reference for green building policy-makers.

1 INTRODUCTION

Green building as a new type of building, green buildings have many problems may emerge in the complex and dynamic process of green building development. In-depth understanding of the complex behavior and mechanism of such a process may be achieved by developing a system dynamics model.

As many scientific studies have concluded, System Dynamics offers an effective approach to simulate and analyze the complex and dynamic relationships among multiple factors in macro-perspective (Forrester 1985). Nowadays, the SD method has also been successfully applied to various settings, such as the sustainable development of energy (Blumberga et al. 2014), urban planning of carbon dioxide emissions (Fong et al. 2009), analyzing the carrying capacity of water resources (Feng et al. 2008), and construction performance analysis (Wan et al. 2013, Han et al. 2013).

However, the regional green building development has neither been found to be treated as a system, nor has a system dynamics model been developed for regional green building development. In order to evaluate the current state and future trend of variation of regional green building and optimize the strategy for sustainable building development, in this study, a "Green Building Development (GBD)" model is constructed using the System Dynamics (SD) method implemented in the Vensim software; this model can facilitate our understanding of the underlying dynamic structure and inner behaviors of GBD systems over time,

and provide guidelines for regional green building development policy-making. This model is then used to study the green building development in Wuhan, demonstrating the features of the regional GBD system and identifying the emerged problems during the process of green building development in Wuhan.

2 METHODOLOGY

2.1 System dynamics method

System Dynamics (SD) provide a system analysis and simulation approach that quantitatively represents complex dynamic behaviors that are inherent in real-world systems, such as nonlinearity, hierarchy, and time-lag, by creating feedback models (Forrester 1985, Liang 2008). This method has a low-level requirement to data accuracy, but can describe complex, dynamic, high-order, and highly non-liner relationships among the factors in a huge system (Yuan et al. 2011, Zhang et al. 2014).

Vensim is an effective platform for implementing SD simulations. This software provides user-friendly interfaces for causal loop diagram design, stock-flow diagram construction, model quantification, result output visual display, and policy test (Zhang et al. 2012).

2.2 Scope of study

Green building development involves developers, consumers, contractors, technology suppliers and the government; these participants can influence each other and form a complex dynamic system. A regional system of green building development in Wuhan is the case studied in this paper, and the cause–effect relationships between multiple factors in such a system are identified. More specifically, "building" represents "civil building", and the area of Wuhan, the capital city of Hubei, China, is 8,494 square kilometers.

2.3 Study process

The process of constructing the GBD-SD model and policy optimization has three main steps: 1) Model construction and parameter quantification. 2) Model validation. 3) Analyzes of simulation results.

3 MODEL CONSTRUCTION AND VALIDATION

Constructing a Green Building Development (GBD)—System Dynamics (SD) model involves SD model construction, model quantification, and verification. Once the SD model has been verified to be reliable, it can be then applied to the further analysis.

3.1 System Dynamics (SD) model construction

Stock-flow model of the GBE is the final form of the GBD-SD model constructed in this paper (Fig. 1). The stock-flow model is in equations (formulae) and computer code, which allows us to simulate the model and conducting the quantitative analysis (Yuan et al. 2011). The GBD-SD model built in using the Vensim platform is used to quantitatively describe the relationships of major factors in the system.

3.2 Model quantification

The real data collected from Wuhan are used in SD model quantification to establish the mathematical

equations relating different factors in the system. Before running the GBD-SD model on the Vensim platform, the simulation time is set to be 2008–2050. The simulation results in 2008–2013 are used to analyze the current state of green building eco-environment in Wuhan, and those in 2015–2050 are used to study the trend of variation.

3.3 Model validation

3.3.1 Comparison with historical data

A quantitative factor, regional total areas of the green building, in the GBD-SD model is taken as the object of model validation, and the errors between historical data and a preliminary simulation result over time are shown in Figure 2. It is generally considered (Liang 2008; Zhang et al. 2014) that relative errors within +10% are acceptable because building development is a complex activity and accurate simulation is difficult; moreover, the SD method is good at deriving consistent trend of variation with lower data requests. Figure 2 shows that the average error of the GBD-SD model is within 4%, suggesting that the model is reasonable.

3.3.2 Model validation by Vensim

Vensim has a model validation function that can be used to check the formulas, units and structures of an SD model. This function shows the following results: ① the formulas in the GBD-SD model are valid without warnings; ② "Units Check" show no errors; ③ "Check Model" is okay.

Both methods have verified that the GBD-SD model (Fig. 1) can reflect the real system behaviors and conditions in Wuhan, and it can be used to analyze the state and trend of variation of the GBD system.

Curve 1 in both figures show that the development cycle of green buildings is about 37 years based on current policies in Wuhan, and this period can be divided into four stages: 1) the slow

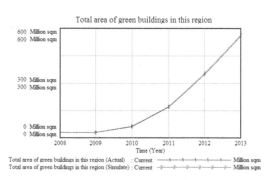

Figure 2. The errors between historical data and a preliminary simulation result.

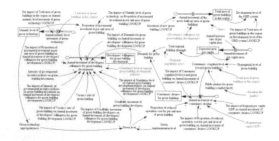

Figure 1. GBD-SD model.

growth stage (2008–2015) (the development is slow during this time because green buildings were first advocated in China in 2008); 2) the rapid development stage (2016–2030); 3) the mature stage with lower rising speed (2031–2044); and 4) the stable stage (2045–2050) with very slow rate of increase.

Curve 2 in both figures show that the maturity level of green technology and the total area of green buildings in this region have similar trends of growth, except that the former gradually increases with the continual increase of the latter, i.e., green technology gets mature with the continual increase of green buildings. Compared with the total area of green buildings, green technology would delay about 10 years entering into the mature stage.

Curve 3 in both figures show that the rapid growth of developers' willingness for green building development will start by 2020, and the rate of growth will tend to be stable by 2042. The general level of development willingness is expected to reach 0.6 by 2050.

Curve 4 in both figures show that compared with development willingness, consumers' desires for green buildings has a lower rate of increase before 2030, and it is expected to reach a relatively high level of 0.8 by 2050. The simulation results show that the level of consumption desires will be 0.2 higher than that of development willingness by 2050, which suggesting that although consumption desires would increase slowly before 2030, it is more rapid increase later with the increase of green buildings indicates a larger potential of growth.

4 SIMULATION RESULTS AND DISCUSSION

The current status and future trend of variation of green building development in Wuhan are simulated with the verified GBD-SD model (Fig. 1) as above mentioned above, and the problems existing in the current GBD system are analyzed, for which optimal strategies and advice are proposed.

The simulation results of main state variables and rate variables are shown in Figure 3a and b, and the current status (2008–2013) and future trend of variation (2015–2050) of these variables are analyzed as follows:

The simulation results of main auxiliary variables are shown in Figure 4 and the current status (2008–2014) and future trend of variation (2015–2050) of these variables are analyzed as follows:

Curve 1 show that a rapid growth of green level would start in 2024, and reach Two-Star level by 2050. Curve 2 shows that the proportion of reduced operation cost for per unit area of green building would rise gradually, and is expected to reach 12% by 2050. The result shows that operation cost

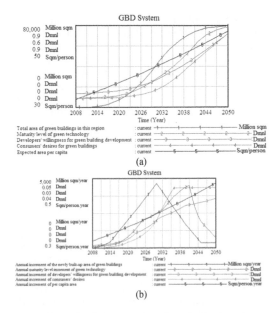

Figure 3. Simulation results of main level and rate variables.

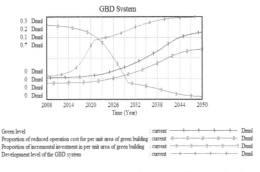

Figure 4. Simulation results of main auxiliary variables.

reduction is already significant at the Two-Start green level, suggesting that green building can bring higher economic efficiency. Curve 3 shows that the proportion of incremental investment in per unit area of green building generally keeps decreasing: it falls slowly from 2008 to 2017, plunges after 2018, and tends to decrease steadily after 2031; with the maturity of green technology, it is expected to be dropped to 0.3% by 2050. Curve 4 shows that the development level of the GBD system is low, 0.2, in 2008. It would rise rapidly in 2015, tend towards a steady state by 2045, and reach a relatively high level of 0.71 in 2050.

The simulation results and analysis reveal three major problems in the development of green building in Wuhan up to now: 1) compared with other cities such as Chongqing (Liang 2008), the development of green buildings in Wuhan would step into the stable stage late, not until 2045, owing to relatively slow rate of growth; 2) the ability of green construction is low, and the overall green level in Wuhan needs to be enhanced; 3) the gap between development willingness and consumption desires is too big, causing supply-demand imbalance, which hinders the development of green buildings in Wuhan.

5 CONCLUSIONS

In this paper, a Green Building Development (GBD system) System model is constructed with the System Dynamics (SD) method and implemented using the Vensim software. After the GBD-SD model has been verified with historic data and by the internal function of the Vensim software, it is used to simulate the GBD system in Wuhan. Analysis of the simulation results shows that the development cycle of green building is about 37 years based on current policies in Wuhan, and it can be divided into four stages: the slow growth stage (2008–2015), the Rapid development stage (2016–2030), the mature stage with lower rate of increase (2031–2044), and the stable stage (2045–2050). Three major problems in the development of green building in Wuhan are also revealed and in this study. This should be solved in the near future.

REFERENCES

Blumberga, A. Blumberga, D. & Bazbauers, G. et al. 2014. Sustainable development modelling for the energy sector. *Journal of Cleaner Production.* 63: 134–142.

Forrester, J.W. 1958. Industrial dynamics: a major breakthrough for decision makers. *Harvard Business Review.* 36(4): 37–66.

Fong, W.K. Matsumoto, H. & Lun, Y.F. 2009. Application of System Dynamics model as decision making tool in urban planning process toward stabilizing carbon dioxide emissions from cities. *Building and Environment.* 44(7): 1528–1537.

Feng, L.H. Zhang, X.C. & Luo, G.Y. 2008. Application of system dynamics in analyzing the carrying capacity of water resources in Yiwu City, China. *Mathematics and Computers in Simulation.* 79(3): 269–278.

Han, S. Love, P. & Pena-Mora, F. 2013. A system dynamics model for assessing the impacts of design errors in construction projects. *Mathematical and Computer Modelling.* 57(9–10): 2044–2053.

Liang, H.Q. 2008. *Study for system evaluation and simulation of energy and land saving housing development.* Chongqing University.

Ling, Y. Cheng, Z.J. Wang, Q.Q. Lin, W.S. & Ren, F.F. 2013. Overview on green building label in China. *Renewable Energy.* 53: 220–229.

Wan, S. Kumaraswamy, M. & Liu, D. 2013. Dynamic modelling of building services projects: A simulation model for real-life projects in the Hong Kong construction industry. *Mathematical and Computer Modelling.* 57(9–10): 2054–2066.

Yuan, H.P. Shen, L.Y. & Jane, J.L. et al. 2011. A model for cost-benefit analysis of construction and demolition waste management throughout the waste chain. *Resources, Conservation and Recycling.* 55: 604–612.

Zhang, Z. Lu, W.X. & Zhao, Y. et al. 2014. Development tendency analysis and evaluation of the water ecological carrying capacity in the Siping area of Jilin Province in China based on system dynamics and analytic hierarchy process. *Ecological Modelling.* 275: 9–21.

Zhang, B. Yu, Z.H. & Sun, Q. etc. 2012. The review of System Dynamics and the related software. *Environment and Sustainable Development.* 02: 1–4.

Advanced Materials and Structural Engineering – Hu (Ed.)
© 2016 Taylor & Francis Group, London, ISBN 978-1-138-02786-2

Corrosion measurement of galvanic cell-type concrete sensor

J.A. Jeong
Department of Ship Operation, Korea Maritime and Ocean University, Busan, South Korea

J.M. Ha
Conclinic. Ltd., Kangdongku, Seoul, South Korea

ABSTRACT: Reinforced concrete structure is damaged by several factors. Corrosion among the factors of the reinforced concrete structures is a major one in the deterioration. Steel corrosion in concrete, such as salt damage must be checked for investigating the condition of reinforced concrete structure. There are several ways how to measuring corrosion condition of reinforced concrete structure. The potential measurement among the several methods is a simple, rapid, and non-destructive method of checking the corrosion to evaluate the severity of corrosion in reinforced concrete structure. However, it produces limit information not only about corrosion rate but also corrosion probability because of environment factors as the oxygen concentration, the chloride content, and the concrete resistance in consideration of the exposure conditions submerged zone, splash zone, tidal zone, and atmospheric zone. Therefore, this study represents the results of corrosion measurement of galvanic cell-type corrosion monitoring sensor in mortar specimens. The corrosion rate, potential, and galvanic current measurements in accelerated corrosion conditions were carried out. It was found that the function of the sensor had well reacted with environmental severity.

1 INTRODUCTION

Nowadays, the reinforced concrete has been used as a constructive material, and it becomes one of the most frequently utilized materials due to its cost-effectiveness and abundance, respectively (Ahmad 2003, Baessler & Ralph 2007). Normally, the pH of concrete is 12–14 in a high alkaline condition. In this environment, the steel is not corroded because compact passive layers are formed in the steel surface. However, the reinforced concrete structures can suffer from the severe corrosion problems (Bjegovic et al. 2003). Various detrimental factors in harsh environments have influences on the deterioration of concrete structures. The major factors in corrosion of the reinforced concrete are the oxygen, water, chloride, and carbon dioxide. Especially, chloride is the very critical element on the corrosion of concrete structures. Chloride acts as a catalyst, and destroys passive layers locally (Broomfield 2007). A number of cases have been reported about premature failures before the design life (Presuel-Moreno et al. 2005). However, chloride or carbon dioxide can reduce the pH of the reinforced concrete either locally or over the entire structure, which leads to corrosion (Qian et al. 2003, Sagues & Powers 1996). When the corrosion of reinforced concrete has already occurred, it is difficult to solve the corrosion problem. The condition survey becomes very important for the appropriate management of those structures before losing the right time for maintenance of corrosion damage. This is the primary reason that the condition monitoring of concrete structures is needed for conservation of their lifetime. It is desirable to check the initial corrosion and the application of corrosion monitoring system. By inspecting inner state regularly, it is easy to detect the initial corrosion time and it reduces cost to spend conservations.

The study related to the corrosion and cathodic protection monitoring has been mainly focused on the qualitative methods by means of electrochemical measurements such as corrosion potential measurement. The quantitative methods measuring the corrosion, such as polarization resistance method, corrosion current measurement, and AC impedance test have been suggested in laboratory experiments so far; however, it was difficult to monitor corrosion and protection state in reinforced concrete structure field (Jeong et al. 2013, Jeong et al. 2013). In addition, an accurate measurement and a detailed analysis are necessary to confirm the corrosion and monitor the deterioration process quantitatively and qualitatively in inhomogeneous conditions such as reinforced concrete. In order to solve the problems, various monitoring techniques have been developed. However, most of the monitoring techniques have many disadvantages such as high expense of measuring equipment, false decision, and difficulties in analysis (Jeong et al. 2013).

Therefore, based in part on conventional studies concerning the corrosion monitoring, this study has been focused on the corrosion properties in reinforced concrete in salt water. And galvanic cell-type corrosion monitoring sensors were applied to detect corrosion initiation and propagation. In addition, many corrosive conditions were simulated to verify the effectiveness of monitoring sensors in the mortar specimen.

Since there are limitations to obtaining accurate corrosion rate from outside of the real concrete structures, the embedded galvanic cell-type corrosion monitoring sensors installed inside of the specimens may be able to monitor the precise corrosion behaviors.

2 EXPERIMENTAL PROCEDURE

Figure 1 shows the schematic drawing of the proposed galvanic cell-type corrosion monitoring sensors in this study for corrosion monitoring in concrete. The galvanic cell-type corrosion monitoring sensors were embedded in cement mortar specimens, illustrated the sensor embedded in the cube slab mortar specimen. In an overall view, four working electrodes of flat round rod sensor were ordinary deformed rebar (D10) acting as a working electrode, and installed at every 0.5 cm distance of depth from the surface, and the working electrodes of flat round rod sensor with 1 cm diameter and 4 cm long were machined.

One counter electrode made of Stainless Steel (STS) 316 L was a type of flat thin plate.

The working electrodes and counter electrode have a galvanic couple each other. The round plastic spacer was used to fix the distance between working electrodes and counter electrode, and to make corrosion factors such as chlorides and water penetrate easily from the surface of the mortar specimen reach to the working electrode.

The dimension of the mortar specimen was 25 cm × 20 cm × 10 cm with rectangular as shown in Figure 2. To mix mortar, washed sand for a general building construction, ordinary Portland cement, and fresh tap water have been used, and the mixing ratio of sand:cement:water was 2:1:0.5, and the water/cement (w/c) ratio of mortar was 0.5 in order to give the less density of concrete and an easy arrival of corrosion elements to rebar to improve sensitivity. The mortar was cured for 7 days in the air. In case of cube slab specimen, working electrodes were located every different cover thickness at intervals of 0.5 cm from the surface, and the corrosion tests were carried out in several kinds of electrolytes. Test environments were four kinds, i.e., Air (Air), Tap Water (TW), natural Seawater (SW), 5% Salt Water (5%SW), and 15% Salt Water (15%SW) that was added more salt to natural seawater up to 5% or 15%. The test temperature conditions were −5, 10, 25, and 40°C.

The test parameters were corrosion potential, galvanic cell current and corrosion rate measurements, those parameters have been investigated both qualitatively and quantitatively. Measurements have been carried out by the potentiostat with Gamry Instruments (ASTM Standard G59-91). The reference electrode for potential measurement was silver/silver chloride electrode (SSCE, Ag/AgCl).

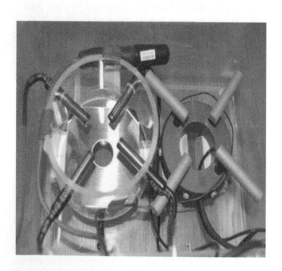

Figure 1.   Arrangement of galvanic cell type sensor.

Figure 2.   Entire view of mortar specimen.

3 EXPERIMENTAL RESULTS

Figure 3 shows the variation of corrosion rate in salt water conditions at 25°C with time. The initial data from the all conditions were nearly the zero mpy, but the value of the corrosion rate from the galvanic cell-type corrosion monitoring sensors were increased with time and salt contents of salt water, especially corrosion rate of the 15% salt water condition indicated more than 1.0 mpy.

Figure 4 is the corrosion potential variations for working electrodes of the galvanic cell-type corrosion sensor in different cover thickness and same temperature with time. The potentials for working electrodes of the sensor were decreased with time and it have maintained the most active potential at cover thickness 2.0 cm rebar. The potential of

cover thickness 2.0 cm was –500 mV/SSCE in early stages, and finally decreased to –650 mV/SSCE. It was thought that water and chloride have been penetrated at the early stage of experiment to compare with other cover thickness of working electrode.

Figure 5 is the corrosion potential variation for working electrodes of the galvanic cell-type corrosion sensor in same cover thickness and different corrosion conditions. The initial potentials were around –50 mV/SSCE, but the potentials were decreased with the different slopes depending on the salt contents of salt water. It was thought that water and chloride have been reached and break the

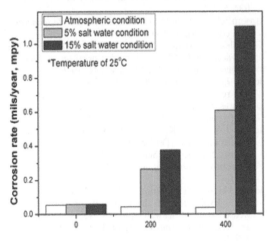

Figure 3. Corrosion rates of the sensors with time.

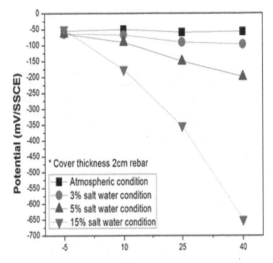

Figure 5. Potential variations of the sensors with corrosion condition.

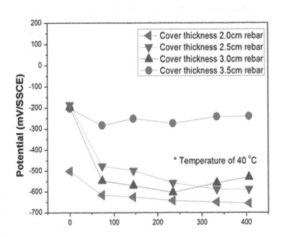

Figure 4. Potential variations of the sensors with cover thickness.

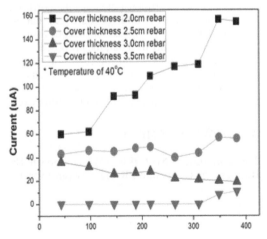

Figure 6. Galvanic current variations of the sensors with time.

passive film formed on the surface of the working electrode depending on the salt content of water and temperature, and thus potentials became lower with severity of corrosion (Gonzalez et al. 2007).

Figure 6 showed the galvanic current variations of working electrodes in different cover thickness with time.

The galvanic current for working electrode of cover thickness 2.0 cm was increased from 60 to 180 μA, respectively, that in cover thickness 3.5 cm was maintained until 300 hours, and finally increased to about 10 μA in 400 hours in Figure 6. Consequently, it was assumed that the galvanic cell-type corrosion monitoring sensor respond to the penetrated corrosive facts of different depths from the surface of the specimen.

4 CONCLUSIONS

From the results regarding the electrochemical performance of the proposed galvanic cell-type corrosion monitoring sensors to measure the corrosion of reinforced concrete structures related to the design technology of bridge longevity, the following results have been obtained.

1. The corrosion rates from the galvanic cell-type corrosion monitoring sensor were increased with time and chloride contents of salt water, especially corrosion rate of the 15% salt water condition indicated more than 1.0 mpy.
2. The potentials were decreased with the different slopes depending on the salt contents of salt water, and the potential of cover thickness 2.0 cm was −500 mV/SSCE in early stages, and finally decreased to −650 mV/SSCE because water and chloride have been penetrated at early stage of experiment.
3. Through the corrosion potential, the corrosion rate measurements and galvanic cell current measurements in various corrosion environments, it was confirmed that the proposed sensor was reasonable to obtain the corrosion state of structures with environmental severity depends on the salt contents, cover thickness, and temperature conditions qualitatively and quantitatively.

ACKNOWLEDGEMENT

This work (Grant No. C0186315) was supported by Business for Cooperative R&D between Industry, Academy, and Research Institute funded Korea Small and Medium Business Administration in 2014.

REFERENCES

Ahmad, S. 2003. Reinforcement corrosion in concrete structures, its monitoring and service life prediction—a review, *Cement and Concrete Composites*, 25: 459–471.

ASTM Standard G 59-91, 1994. *Annual Book of ASTM Standards*, 3(2): 230–233.

Baessler, R. et al. 2007. 07377 Electrochemical Devices for Determination of Corrosion Related Values for Reinforced Concrete Structures, *NACE, Corrosion Conference*, 07377: 1–13.

Bjegovic, D. Jackson, J. Mikulic, D. & Sekulic, D. 2003. Corrosion measurement in concrete utilizing different sensor technologies, *NACE, Corrosion Conference*, 03435: 1–9.

Broomfield, J.P. 2007. *Corrosion of steel in concrete*, 2nd ed. Taylor & Francis.

Gonzalez, J.A. Miranda, J.M. & Feliu, S. 2007. *Corrosion Science, 2004. J.P. Broomfield, Corrosion of Steel in Concrete*, 2nd ed. edited by T&F, 46: 10–64.

Jeong, J.A. & Jin, C.K. 2013. The effect of temperature and relative humidity on concrete slab specimens with impressed current cathodic protection system, *Korean Society of Marine Engineers*, 37: 260–265.

Jeong, J.A. Jin, C.K. Kim, Y.H. & Chung, W.S. 2013. Electrochemical Performance Evaluation of Corrosion Monitoring Sensor for Reinforced Concrete Structures, *Journal of Advanced Concrete Technology*, 11: 1–6.

Jeong, J.H. Kim, Y.H. Moon, K.M. Lee, M.H. & Kim, J.K. 2013. Evaluation of the corrosion property on the welded zone of seawater pipe by A.C shielded metal arc welding, *Korean Society of Marine Engineers*, 37: 877–885.

Presuel-Moreno, F.J. Kranc, S.C. & Sagues, A.A. 2005. Cathodic prevention distribution in partially submerged reinforced concrete, *Corrosion*, 61(6): 548–558.

Qian, S.Y. Cusson, D. & Chagnon, N. 2003. Evaluation of reinforcement corrosion in repaired concrete bridge slabs—A case study, *Corrosion*, 59(5): 457–468.

Sagues, A.A. & Powers, R.G. 1996. Sprayed-Zinc Sacrificial Anodes for Reinforced Concrete in Marine Service, *Corrosion*, 52(7): 508–522.

Advanced Materials and Structural Engineering – Hu (Ed.)
© 2016 Taylor & Francis Group, London, ISBN 978-1-138-02786-2

The analysis of air flows near the windbreak constructions on offshore structures

I.V. Dunichkin

Educational and Research-Production Laboratory on Aerodynamic and Aeroacoustic Tests of Building Construction, Department of Building Design and Urban Planning, National Research Moscow State University of Civil Engineering (MGSU), Moscow, Russian Federation

ABSTRACT: The research of air flows near the windbreak constructions of oil platforms presents an analysis of wind conditions close to offshore structures. The numerical simulation of offshore structures wind protection made it possible to identify promising areas for special deck's safety shield. The results of velocity field calculations include the data for various constructions and their location.

1 INTRODUCTION

Different types of offshore structures are used for strong environmental conditions, such as water depth, soil, ice, atmospheric ice, wind conditions, and so on. The wind velocity wave affects not only load but also splashing and atmospheric icing of constructions. The wind velocity distribution on the deck is not only a question of the comfort, but is also a very important safety factor.

2 THE RELEVANCE OF RESEARCH

The most important place for wind conditions examination seems to be a safety shield.

Windbreak constructions on offshore structures require collaborative research of loads and impacts on structure, accommodation spaces design, and platforms' topsides composition. Wind flow analysis on the deck is essential for the design of new offshore structures and reconstruction of existing offshore oil platforms. In accordance with the Tutorial "Marine engineering structures, Part I. Sea Rigs" (Borisov et al. 2003), it refers to the three categories of objects: 1) stationary—permanent bases, ramps, artificial islands; 2) semi-floating (self-raising) platform; 3) mobile platform and semi submersible duplicate objects. In the latter case, windbreak constructions may be used in the complex reconstruction, which has good prospects: «For the purpose of offshore infrastructure maintenance and further development hypothesis of oil platforms' functionality change is made. In prospect, it allows oil and gas companies of excluding abandonment costs from the project of the offshore field development. More than that,

it allows coastal states of improving the marine environment» (Dunichkin & Kalashnikov 2015). The complex reconstruction of offshore structure can be divided into 3 phases (equipment dismantling, topside upgrade, and new spaces creation). The last phase «New spaces creation» implies a safety shield installation, as well as monitoring and security systems of the windbreak constructions installation.

3 THE CLASSIFICATION OF LOADS AND IMPACTS

Wind loads and effects are classified according to:

A. The offshore structure height levels:
   A.1. Above sea level 15 meters—Supporting units;
   A.2. Above sea level 15–35 meters—Mooring with folding safety shield;
   A.3. Above sea level part of more than 35 meters from the water surface—Decks with the windbreak constructions are an object of research.
B. The main impacts of climate processes related to the wind:
   B.1. Insolation and solar radiation level. Solar radiation forms a pressure difference in the atmosphere, which creates the wind flows at the macro level and meso-level.
   B.2. Wave loads and impacts. The wind determines the formation of wind waves and wind wave damping sea.
   B.3. Ice loads and impacts. Interaction of rough seas and wind generates splash icing of offshore structures in subzero temperatures.

The atmosphere near the water surface is divided into two layers by icing conditions. The bottom layer has splash icing conditions. Icing structural elements comes from splashes of waves, which are carried by the wind, after a blow on the edge of the construction or directly from the crest of the wave. These are typical processes in wave interaction with structures (Wellen et al. 2010). Waves splashing icing on a mooring system, cranes, the safety shield prevails up to 26 meters above sea level.

The overlying layer of the atmosphere has only conditions of the atmospheric icing. The important factor for the atmospheric ice is the wind velocity above 5–7 m/s with negative temperatures. Atmospheric icing of decks and windbreak constructions occurs in large amounts over 26 meters above sea level.

A sample scenario changing parameters of windbreak constructions due to the influence of wind is as follows.

The wind creates waves and affects the nature and volume of waves splashing icing (6).

Icing changes weight, shape of windbreak constructions, and other aerodynamic characteristics (the natural frequencies of the structure, force coefficients).

Shape of windbreak constructions becomes a reflex silhouette.

Windbreak constructions with a new shape change zone of increased wind velocities and resonance parameters ripple of wind flow.

It should be mentioned that the design of topside modules in general and drilling derrick especially are affected by wind and ice. Notably, that's why offshore platform derrick (for platforms working in arctic conditions) is usually covered with special protecting shields (Kartamishev et al. 1990). They protect derrick equipment not only from strong winds that can reach the value of 40–50 m/s with wind gusts up to 75 m/s but also from salt droplets that are carried by wind and that induce equipment corrosion and icing. Shapes of protecting shields and windbreak constructions define their effective application.

Constructions parameters vary due to icing and different other climatic conditions and raise the question of accounting for the wind loads, climate comfort and safety of people on offshore structures. The aerodynamic research of topsides offshore structures is possible by analogy with bridges. In turn, the aerodynamics of bridges as well as offshore structures cover static aerodynamics and aeroelasticity (Churin & Poddaeya 2014). Wind effects of aeroelasticity should be considered in order to provide safety of the platform's staff. It is important to take into account the frequency characteristics of the air flow and frequencies of

the structure, as well as the effect of the form of the topside. Active communication zone of wind flow over topsides usually consists of two layers with different turbulences. The parameters of these layers are determined by the size and duration of the vortex shedding. The increase in the number of interacting layers of wind flow over topsides leads to a more rapid completion of the process of the vortex shedding (Churin et al. 2015).

4 WIND CONDITIONS MODELLING

As noted above, a significant element that effects the wind regime seems to be windbreak constructions (deck's safety shield). Height of windbreak constructions is 35–45 m above sea level. Wind velocities are 10–35 m/s at deck level. 35 m/s is a characteristic value 1 in 50 years and should be multiplied by an appropriate safety factor depending upon local construction standards, site supervision and any allowance for critical infrastructure. Typical safety factors internationally are between 1.35 and 1.5. To ensure a robust design, an appropriate safety margin is required for catastrophic failure. Turbulence levels are between 7% and 15%.

5 THE CORNICE DESIGN OF WINDBREAK CONSTRUCTIONS

The consistent turbulence at an altitude of fewer than 2 meters from the deck suggests that the air flow is not separating equally above and below the deck. This type of flow instability problem can be better dealt with through changes in the leading edge of the deck, Figure 1.

Cornice design of windbreak constructions has a height of 2 meters. Computer modeling of the wind impact on the deck of the offshore structure demonstrates velocity fields in the sections along the upper deck platform.

Case without windbreaks has a height speed flow on the deck, Figure 2.

The base case with vertical solid fencing has a big zone with the pulsation of air flow on the deck and zone with comfort conditions, Figure 3.

In this research work was carried out numerical simulations of other similar windbreaks constructions on the offshore structure:

1. Case with slanted (45 °) solid fencing, installed on the edge, has not so big zone with comfort conditions and pulsation of air flow on the deck, Figure 4.
2. Case with slanted (45 °) solid fencing, installed with an offset into the platform, has a stability zone with comfort conditions and without frequent pulsation of air flow on the deck, Figure 5.

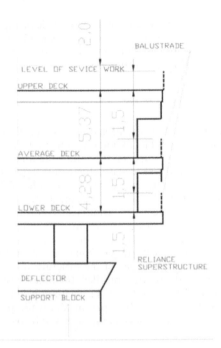

Figure 1.   Scheme of the offshore structure section.

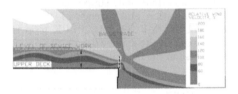

Figure 2.   Velocity fields in the sections along the platform with balustrade.

Figure 3.   Velocity fields in the sections along the platform with vertical solid fencing.

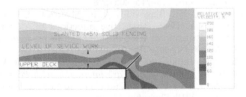

Figure 4.   Velocity fields in the sections along the platform with slanted (45 °) solid fencing.

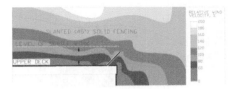

Figure 5.   Velocity fields in the sections along the platform with slanted (45 °) solid fencing (offset into the platform).

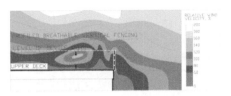

Figure 6.   Velocity fields in the sections along the platform with profiled breathable vertical fencing.

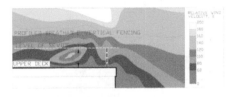

Figure 7.   Velocity fields in the sections along the platform with profiled breathable vertical fencing (offset into the platform).

3. Case with profiled breathable vertical fencing, installed on the edge, has a big zone of turbulence, Figure 6.
4. Case with profiled breathable vertical fencing, installed with an offset into the platform (on 0,5 h of fencing) has a big zone (h = 1,5 m) of turbulence with an offset into space of platform, Figure 7.
5. Case with profiled breathable vertical fencing, installed with the removal of the outside of the platform (on 0,5 h of fencing), has a small zone of turbulence (h = 0,5 m), Figure 8.
6. Case with profiled breathable vertical fencing, installed with the removal of outside of the platform (on 0,5 h of fencing) and omitted below the deck level, does not have a zone of turbulence. Best windbreak, Figure 9.
7. Case with reflector is located at the level of the platform and has a big zone of turbulence, Figure 10.
8. Case with reflector offset to outside of the platform (on 0,5 h of fencing) has a small zone of turbulence, Figure 11.

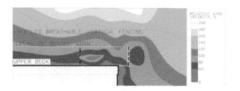

Figure 8. Velocity fields in the sections along the platform with profiled breathable vertical fencing (outside of the platform at half height of fencing).

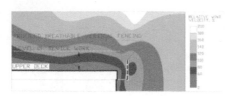

Figure 9. Velocity fields in the sections along the platform with profiled breathable vertical fencing (outside of the platform on half height of fencing and below deck level).

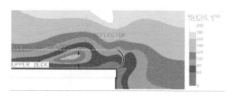

Figure 10. Velocity fields in the sections along the platform with reflector.

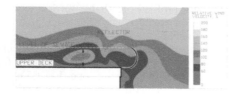

Figure 11. Velocity fields in the sections along the platform with reflector (outside of the platform on half height of fencing).

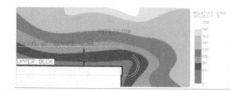

Figure 12. Velocity fields in the sections along the platform with reflector (outside of the platform on half height of fencing and below deck level).

9. Case with reflector offset to outside of the platform and omitted below the deck level (on 0,5 h of fencing) does not have a zone of turbulence. Best windbreak, Figure 12.

6 CONCLUSIONS

Analysis of the conditions on offshore platforms makes it possible to identify the most extreme pressures and impacts. It justifies the importance of the windbreak constructions and the scenario changing parameters constructions due to wind action study. Analysis of numerical modeling of wind flows near the windbreaks constructions demonstrated design solutions with an optimal performance. Assessment of cases windbreaks constructions with reflex shape was made taking into account the shape of the structure after icing.

Case № 6 with profiled breathable vertical fencing (Fig. 9) and case № 9 with reflector offset (Fig. 12) have the best parameters while placing the construction outside the deck and below it.

ACKNOWLEDGMENT

This work was financially supported by the Ministry of Education and Science of the Russian Federation, the task № 7.11.2014/K to perform research work in the framework of the project of the state task in the field of scientific activity.

REFERENCES

Borisov, R.V. Makarov, V.G. Makram, V.V. Nikitin, V.S. Portnoy, A.S. Simonenko, A.S. Sokolov, V.F. Stepanov, I.V. & Timofeev, O.J. 2003. *Marine engineering structures. Part I. Sea Rigs: Tutorial*, Under Society. Ed. V.F. Sokolov.—St. Petersburg.: Shipbuilding, 535.

Churin, P. & Poddaeva, O.I. 2014. Aerodynamic testing of bridge structures, *Applied Mechanics and Materials*. 477–478: 817–821.

Churin, P. Kapustin, S. Orechov, G. & Poddaeva, O. 2015. Experimental Studies of Counter Vortex Flows Modeling, *Applied Mechanics and Materials*, 756: 331–335.

Dunichkin, I.V. & Kalashnikov, P.K. 2015. Accounting for climate and typology of reuse of offshore structures with a change of function, *Applied Mechanics and Materials*. 713–715: 205–208.

Kartamishev, P.I. Blagovidov, L.B. Morozov, E.P. & Perez, N.Y. 1990. Perspective directions of design and construction of fixed offshore platforms. *Journal Shipbuilding Technology*, 9: 1990.

Wellen P.R. Borsboom, M.J.A. & van Gent, M.R.A. 2010. *3D Simulation of Wave Interaction with Permeable Structures*. Proc. of 32nd Conf. on Coastal Engineering—2010, Shanghai, China.

Advanced Materials and Structural Engineering – Hu (Ed.)
© 2016 Taylor & Francis Group, London, ISBN 978-1-138-02786-2

Research on the construction technique for integrated control of embankment settlement due to extending of pile platform into subgrade

X.Q. Wang, Y.L. Cui & S.M. Zhang
Department of Civil Engineering, Zhejiang University City College, Hangzhou, China

B.C. Qian & F.L. Li
Shanghai JASO Group Limited Company, Shanghai, China

ABSTRACT: Under the demand for urban transport and as more and more viaducts and overpasses appear, we often face the pavement distress caused by differential settlement because the bridge platform extends into subgrade. To address the problem, in this paper, a construction technique for integrated control of embankment settlement due to extending of pile platform into subgrade was studied. In the early stage road is put into use, integrated measures can be taken to the control subgrade settlement. Coordinated deformation ability of subgrade will be enhanced by using weak soil compaction pile, vertical and horizontal bent cap and backfill of light soil. Its features and principle were also introduced, the emphatically operation points during construction.

1 INTRODUCTION

With the expanding demand for urban transport and rapid growth of road traffic volume, municipal road construction has been paid more and more attention to it (Li 2008). At the same time, there are much more viaducts, overpasses and other road side independent foundation buildings. Since the bridge platform often extends into subgrade, uneven settlement of pile foundation and subgrade will result in uneven settlement deformation of subgrade and then longitudinal cracks at the edge of platform, fluctuation of pavement and other engineering defects.

Lu Wei-dong (Lu 2010) studied the application of three kinds of new piles in the municipal road embankment, which were monitored and verified to be effective to reduce embankment settlement; Based on the three-dimensional simulation result, Yi Yao-lin et al. (2009) adopted an additional stress calculation method in which the stress concentration effect of composite foundation substratum was included; Liu Ji-fu (2003) considered embankment differential settlement and proposed the formula of pile soil stress ratio, which was a differential settlement function of surface pile soil; Huang Zhe et al. (2007) studied benefits and drawbacks of the calculation method for bridge single pile and put forward their own suggestions; Han Mei (2007) systematically studied the settlement of bridge pile foundation on soft soil foundation.

An overall view of research status at home and abroad tells us that, in spite of many researches on municipal road and bridge settlement, there are few related studies about construction measures to avoid poor pavement quality caused by differential settlement because bridge platform often extends into subgrade. For the situation, a construction technique for integrated control of embankment settlement due to extending of pile platform into subgrade was studied in this paper. In the proposed new embankment structure, vertical and horizontal connection bent cap is installed on weak soil compaction pile as well as horizontal reinforced layer of solidified soil and light soil is back filled to improve the performance of filler and enhance coordinated deformation capacity of subgrade. In this paper, its technical principle and features, construction technology as well as operation points were introduced in detail.

2 TECHNOLOGY PRINCIPLES AND APPLICATION SCOPE

2.1 Technology principles

In the construction technique for integrated control of embankment settlement due to extending of pile platform into subgrade, weak soil compaction pile is installed on subgrade, with prefabricated connecting steel bars at its top. Vertical and horizontal connection bent cap is poured at the top

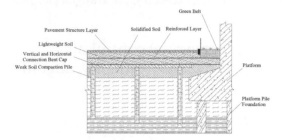

Figure 1. Structure schematic for integrated control of embankment settlement due to extending of pile platform into subgrade.

of weak soil compaction pile and connected with pile body through steel bars reserved at the top; solidified soil and light soil is back filled into subgrade in layers, with horizontal reinforced layer, as shown in Figure 1.

Vertical and horizontal cast-in-site bent cap on weak soil compaction pile can support plate solidified soil and light soil layer above to form a cover structure to improve embankment integrity and anti-settling performance; weak soil compaction pile can be successively shortened according to the distance from pile body to platform to guarantee upper load and smooth transfer of foundation settlement as well as reduce or even eliminate violent fluctuation or fracture of pavement near pile platform; the soil at the lower part of pavement structure layer is replaced by bubble light soil and solidified soil with a high bearing performance. There is an additional reinforced layer to improve further shear and crack resistance of pavement structure substratum. And the bubble light soil, for its small weight, can reduce subgrade settlement. Embankment settlement due to extending of pile platform into subgrade can be controlled by the above measures.

2.2 *Application scope*

The construction technique for integrated control of embankment settlement due to extending of pile platform into subgrade is applicable to subgrade treatment engineering for construction of highways and municipal roads in the areas of soft soil, in which, it may cause pavement cracks and fluctuation that adjacent pile platform extends into subgrade.

3 TECHNICAL FEATURES

The construction technique for integrated control of embankment settlement due to extending of

pile platform into subgrade has the following technical features:

1. Three measures for settlement control to strengthen subgrade, reduce its overlying load and enhance a collaborative stress performance of subgrade structure can effectively prevent uneven settlement caused by extending of pile platform into subgrade.
2. Weak soil compaction pile is used to transmit the load to subgrade bearing stratum. Vertical and horizontal connection bent cap is set to enhance embankment integrity and anti-settling performance. Solidified soil and light soil are back filled as well as horizontal reinforced layer are added to improve the performance of fillers.
3. Weak soil compaction pile and non-displacement pile are used for subgrade treatment to reduce the influence of pile body construction on adjacent viaduct pile platform; the length of weak soil compaction pile can be reduced or be constant according to smooth transition principle of settlement to coordinate better deformation of subgrade soil.
4. Vertical and horizontal connection bent cap is cast in place as a whole and connected to weak soil compaction pile through steel bars reserved at the top to effectively improve embankment integrity and reduce the load of soft soil.
5. Light soil and solidified soil are adopted to fill embankment with light weight and high strength to reduce the volume of work to move earth for embankment. Light soil has high strength, water resistance and durability, so construction for water stable layer is not required.

4 CONSTRUCTION TECHNOLOGICAL PROCESS AND OPERATING POINTS

4.1 *Materials*

The main materials used for construction technique for integrated control of embankment settlement due to extending of pile platform into subgrade are shown in Table 1.

4.2 *Construction technological process*

Main steps of construction technique for integrated control of embankment settlement due to extending of pile platform into subgrade are as follows: preparations for construction → installation of weak soil compaction pile → solidified soil backfill → construction of bent cap → filling with solidified soil → paving of reinforced layer → paving of light soil → subsequent construction.

Table 1. Main materials.

No.	Name of equipment	Unit	Note
1	Concrete	m³	Construction of bored cast-in-place pile and bent cap
2	Soil stabilizer	Kg	
3	Cement	Bag	Soil solidification and mixed lightweight soil
4	Lime	Kg	Soil solidification
5	Bubble light soil	m³	
6	Geogrid	m²	Steel plastic geogrid and glass fiber geogrid

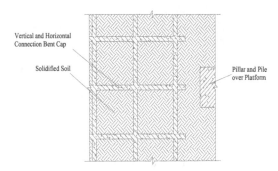

Figure 2. Layout schematic for vertical and horizontal connection bent cap at the top of pile.

4.3 Construction operating points

4.3.1 Installation of weak soil compaction pile

Generally, weak soil compaction pile adopts bored cast-in-place pile with a diameter of 400–600 mm and the nearest distance from pile body and pile platform shall not be less than 500 mm. The depth that weak soil compaction pile is driven into bearing stratum can be gradually decreased with increase in the distance. The construction of bored cast-in-place pile should conform to the requirements of current provisions and should not affect pile platform and pile foundation.

4.3.2 Solidified soil backfill

After the strength of bored cast-in-place pile exceeds 75% of design strength, embankment filling is dug with the way to combine artificial excavation with mechanical excavation to about 1000 mm below the top of bored cast-in-place pile.

After leveling, compaction and excavation of a site, waste soil produced during subgrade excavation is mixed with cement, paved evenly and compacted with basically the same height of the pile top after cutting to form a solidified soil layer. Compressive strength of solidified soil at the seventh day should be greater than 0.8 MPa. Cement content ratio is 5%. 42.5# ordinary portland cement can be adopted, spraying stabilizer or mixing with lime if necessary.

4.3.3 Construction of bent cap

Vertical and horizontal connection bent cap at the top of pile is cast in place as a whole using reinforced concrete with strength grade of C30. Its width is 300–500 mm and height is 200–400 mm. With the vertical and horizontal connection, it is well-shaped seen from above. Bent cap at the top of edge pile near pile platform extends outward from 500 mm short distance suspension arm to the position above pile platform, as shown in Figure 2.

The structural steel bars in vertical and horizontal connection bent cap are bound and connected to longitudinal steel bars in reinforcement cage of bored cast-in-place pile.

Construction sequence for vertical and horizontal connection bent cap: positioning and lining → binding of steel bars in bent cap → setting up and reinforcing of lateral molding → pouring of concrete → maintenance.

4.3.4 Filling of solidified soil

After bent cap construction, upper surface of solidified soil layer is chiseled properly and a new mix of solidified soil is filled in the bent cap to enclose the space formed. Solidified soil is, produced under the same requirements of original solidified soil, paved, compacted and smoothed, with the same height of the upper surface of a bent cap.

4.3.5 Paving of reinforced layer

Steel plastic geogrid or glass fiber geogrid can be used as a reinforced material. Reinforced layer is fixed with U-type nail and paved. First, fix its one end with the U-type nail to solidified soil layer. Then, tense geogrid longitudinally and fix it segment by segment, with the length of each segment of 2–5 m. The segment can also be divided according to the space of contraction joint. Steel nail is arranged on the joint. Geogrid should be tensioned state vertically and horizontally while it is tightened.

Vertical and horizontal lap width of the geogrid should be not less than 15 cm. For the geogrid at the lap, farther geogrid should be placed above closer geogrid according to the direction away from pile platform. After fixation, U-shaped nail should be fixed again if it is found to be broken or loose. Glass fiber geogrid should be moderately rolled using rubber roller to be stable after it is paved, so that the geogrid bonds firmly with the surface of solidified soil.

311

4.3.6 *Paving of light soil*

Lightweight materials are filled into embankment above solidified soil layer. Before paving, the template should be set up to control molding quality and avoid environment pollution due to outflow. Light soil is paved layer by layer and 1–2 reinforced layers are paved with the same method. Sparse hole lattice material can be used for geogrid.

4.3.7 *Subsequent construction*

Subsequent construction will be carried out after lightweight soil is paved, including isolation fence and green belt. Root separation under filling soil of green belt can be completed according to the planting of vegetation. During the construction of upper pavement structure, due to the feature of light filler, water stable layer can be selectively skipped to save costs.

5 CONCLUSIONS

1. The construction technique for integrated control of embankment settlement due to extending of pile platform into subgrade was applied to actual engineering. There is no engineering in which longitudinal cracks near the edge of a platform, fluctuation of pavement, pavement rutting or other engineering defects caused by uneven settlement of subgrade occur. Therefore, a pavement maintenance cycle is shortened and late repair costs are saved.
2. Weak soil compaction pile is used for a subgrade treatment to reduce the influence of pile body construction on adjacent pile platform; its length can be reduced or be constant according to smooth transition principle of settlement to better coordinate deformation of subgrade soil.
3. Vertical and horizontal connection bent cap is cast in place as a whole and connected to weak soil compaction pile through steel bars reserved at the top to effectively improve embankment integrity and reduce the load of soft soil.
4. Lightweight filler used to back fill subgrade can reduce the weight of subgrade. Horizontal reinforced layer between solidified soil and bubble light soil can improve shear capacity, enhance filler integrity, adjust the upper load distribution, and coordinate embankment deformation.

ACKNOWLEDGMENT

This material is based upon work funded by Zhejiang Provincial Natural Science Foundation of China under Grant No. LQ12E09002; Project (51308497) supported by National Natural Science Foundation of China.

REFERENCES

Huang, Z. Zhang, F.Q. & Wang, X. 2007. Research on the calculation methods of the settlement of pile foundation of bridges, *Shanxi Architect.* 33(19): 106–108.

Li, J.L. 2008. *Theories and Experiment Research to the Upgrading and Revising Projects of Municipal Roads.* Nanjing University of Science and Technology, China Nanjing.

Liu, J.F. 2003. Analysis on Pile-Soil Stress Ratio for Composite Ground under Embankment, *Chinese Journal of Rock Mechanics and Engineering.* 22(4): 674–677.

Lu, W.D. 2010. Applied Study on New Technology of Subgrade Treatment for Municipal Roads in Coastal City Area, *Building Construction China.* 32(6): 574–577.

Mei, H. 2007. *Settlement Analysis of Pile Group Foundation of Bridge under Vertical Loading*, Chengdu: Southwest Jiaotong University.

Yi, Y.L. & Liu, S.Y. 2009. Settlement Calculation Method of Composite Foundation under Embankment Load, *Engineering Mechanics*, 26(10): 147–153.

Advanced Materials and Structural Engineering – Hu (Ed.)
© 2016 Taylor & Francis Group, London, ISBN 978-1-138-02786-2

Research on the construction technology of self-water stopping pit support structure of pre-stressed pipe piles

X.Q. Wang, Y.L. Cui & S.M. Zhang
Department of Civil Engineering, Zhejiang University City College, Hangzhou, China

J. Song & M.G. Zhang
RiseSun Construction Engineering Co. Ltd., Langfang, China

R.G. Lin
College of Civil and Transportation Engineering, Hohai University, Nanjing, China

ABSTRACT: In order to overcome the defects of cast-in-situ bored piles as the deep foundation pit support structure, at present the pre-stressed pipe piles are usually adopted to replace the cast-in-situ bored piles. Yet, the pre-stressed pipe piles have the defects of limited bending resistance and poor water-stopping effect. In this paper, the foundation pit support structure of pre-stressed pipe piles is innovated to overcome the defects of the pre-stressed pipe piles as the foundation pit support structure. The construction technology of the self-water stopping foundation pit support structure of pre-stressed pipe piles is developed, and the technological principles, the application scope, the technical characteristics, the construction technology, and the operating points thereof are introduced.

1 INTRODUCTION

In recent years, with the development of the urban construction, the scale of the building is continuously expanded, and the deep foundation pit in the building design becomes much popularized, yet for a long time, cast-in-situ bored pits are usually adopted as the main support piles of the deep foundation pit. However, the construction of the bored piles tends to pollute the environment, and the slurry disposal is always a difficult problem in the urban pile foundation construction and the disposal charge of every cubic meter of slurry costs from 60 to 70 Yuan (Chen et al.). Therefore, in some areas, the pre-stressed pipe piles are adopted to replace the cast-in-situ bored piles as the support structure of the deep foundation pit.

The research carried out by Tian et al. (2012) proved that the combination body of piled anchor and water-stopping curtain has achieved excellent effects in application, and in the research carried out by Yu et al. (2008) the pre-stressed pipe piles are adopted to replace the cast-in-situ bored piles as the main piles of the foundation pit support, consequently, the construction period is shortened, the cost of the support charge is lowered by 50% compared with the cast-in-situ bored piles, and the economic advantage is obvious. At the same time, the pre-stressed pipe pile can effectively reduce the compaction effect, so as to avoid the deformation,

breakage and dislocation (Poulos & Davis 1980) frequently occurred to the solid piles like the prefabricated pile and the driven pipe pile. However, the usage (Feng & Xie 2005, Zhang 2004) of the pre-stressed pipe pile is limited to the pile foundation of bridges, and is rarely seen in the foundation pit support project.

However, the pre-stressed pipe pile has the defects as limited bending resistance and poor water-stopping effect. Therefore, we have researched and developed the construction technology of the self-water stopping foundation pit support structure of pre-stressed pipe piles, so as to overcome the shortcomings of the pre-stressed pipe piles as the foundation pit structure. Concrete water-stop is adopted to replace the deep-mixed piles, so as to accelerate the construction, and pre-stressed anchor cables are used for dynamically controlling the bearing capacity and deformation of the foundation pit support structure. This paper also introduces the characteristics and principles of the technology, and focuses on the operating points of the construction.

2 TECHNOLOGY PRINCIPLES AND APPLICATION SCOPE

2.1 Technology principles

The main process principles of the construction technology of the self-water stopping foundation pit support structure of pre-stressed pipe piles are as follows: components like pile toes, pre-stressed pipe piles, and concavo-convex connecting bands are prefabricated in the factory and welded into a whole; and the first pre-stressed pipe pile is pressed into the foundation pit through static pressing on the construction site, the follow-up pre-stressed pipe piles are pressed into through special pile toes; and a division plate and a pre-stressed anchor cable are pressed, at the same time the pre-stressed pipe pile is pressed in. The division plate forms a cavity between two pre-stressed pipe piles; after all the pre-stressed pipe piles are pressed in, rebar is inserted in and concreted, so as to form concrete water-stop, thereby reducing the construction steps as water-stopping curtain drilling in the conventional method and playing the role of water stopping.

The pre-stressed anchor cable presets in the pre-stressed pipe pile, the bottom end of the pre-stressed anchor cable is connected with the anchor cable hook on the prefabricated pile toe, and after all the pre-stressed pipe piles are pressed in, the bearing capacity detection and deformation observation of the support structure are carried out, so as to selectively pour concrete into the pipe piles and dynamically adjust the pre-stress value of the support structure based on the deformation and stress of the foundation pit, and then realize the dynamic control of the bearing capacity and deformation of the foundation pit support structure. The foundation pit support structure is shown in Figure 1.

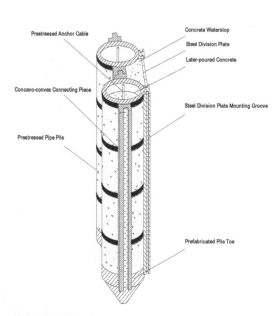

Figure 1.   Foundation pit support structure.

2.2   *Application scope*

The construction technology of the self-water stopping foundation pit support structure of pre-stressed pipe piles is applicable to the construction of foundation pit support structure, particularly to the foundation pit support structure with deep foundation pit and high groundwater level in high-rise projects, and can also be used for the water-facing support construction.

3   CONSTRUCTION TECHNOLOGICAL PROCESS AND OPERATING POINTS

3.1   *Materials and equipment*

The main materials needed for the construction technology of the self-water stopping foundation pit support structure of pre-stressed pipe piles are as follows: 1) cement, 425# ordinary Portland cement should be adopted; 2) pre-stressed pipe pile, both the concavo-convex connecting piece and the steel division plate mounting groove need to be welded in the factory; 3) Steel division plate. The steel plate serves as the steel division plate, and two holes for lifting anchor cable are formed in the top of the plate.

The main mechanical equipment adopted in the construction technology of the self-water stopping foundation pit support structure of pre-stressed pipe piles is as follows: static pressure piling machine (YZJ-600), CO_2 shielded welding machine (BX3-300-4), and grouting pump (UBJ3-R).

3.2   *Construction technological process*

The construction steps of the self-water stopping foundation pit support structure of pre-stressed pipe piles are as follows: a first pre-stressed pipe pile is pressed in through the static pressure piling machine, the follow-up pre-stressed pipe piles and steel division plates are pressed in, the water-stop is concreted, the division plate is pulled out, the deformation of the foundation pit is detected, the later-poured concrete is poured, and the pre-stress is applied. The construction steps are shown in Figure 2.

3.3   *Construction operating points*

3.3.1   *Factory prefabrication of pre-stressed pipe piles and pipe toes*

Concavo-convex connecting pieces are symmetrically welded on the pre-stressed pipe pile, a groove is formed between tenons, and a toothed steel plate structure is formed between the tenons and the groove. PN water-stop is attached to the interior of the groove. A vertical I-shaped mounting groove

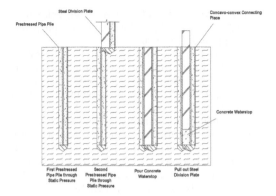

Figure 2. Construction Steps of self-water stopping foundation pit support structure of pre-stressed pipe piles.

is designed and welded on the pre-stressed pipe pile for mounting the steel division plate, and the PN water-stop is also attached to the interior of the mounting groove. The preservative is painted on the concavo-convex connecting pieces and the mounting groove.

The pre-stressed pile toe is prefabricated into a special-shaped pile toe, a pre-stressed anchor cable connector presets into the center of the prefabricated pile toe corresponding to the pre-stressed pipe pile, and a steel division plate mounting groove is formed in the position corresponding to the steel division plate. The prefabricated pile toe structure is shown in Figure 3.

The pre-stressed pipe pile is connected with another pre-stressed pipe pile. One end of the pre-stressed anchor cable is bound on the anchor cable connector and further fixed by welding, and the pre-stressed anchor cable is little longer than the pre-stressed pipe pile. Then the pile toe is welded to the pre-stressed pipe pile through electric welding, and the upper end of the pre-stressed anchor cable is temporarily fixed at the upper end of the pre-stressed anchor cable.

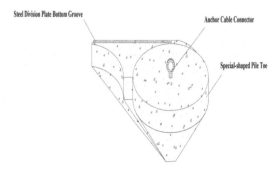

Figure 3. Prefabricated pile toe structure.

3.3.2 Pre-stressed pipe pile piling

An ordinary round pipe toe is used for piling the first pre-stressed pipe pile of the support structure; the special-shaped pile toes shown in the Figure 3 are adopted for piling the remaining of the pre-stressed pipe piles, before pressing pile, one end of the pre-stressed anchor cable is fixed at the anchor cable connector on the pile toe and the other end thereof is fixed at the pile top through a temporary fixing device, and the steel division plate needs to be hammered into the foundation while pressing pile.

The length of every section of the pile and the pile pressing order are chosen according to the designed pile length in every hole, and the piles need to be numbered. The piles are lifted to the position according to the numbered order with the self-owned crane, the lower end of the concavo-convex connecting piece of the pre-stressed pipe pile is closely connected with the top end of the concavo-convex connecting piece of the previous pre-stressed pipe pile, the pre-stressed pipe piles apart from the first one should lift together with the steel division plates, and one side of the steel division plate is closely connected with the mounting groove of the previous pre-stressed pipe pile.

3.3.3 Pouring concrete water-stop

Concrete is poured into the cavity formed by the pipe pile, the concavo-convex connecting piece and the steel division plate to form the concrete water-stop, and tremie pouring is adopted for pouring concrete: the tremie is kept 30 to 50 cm from the hole bottom, the water stopper is 0.2 to 0.3 m from the water surface, and the operation should be finished within 0.5 h. First pouring amount should be controlled during concrete pouring and does not exceed 1 m above the hole bottom as strictly required by the design document and standard, concrete pouring and steel division plate pulling-up should be carried out at the same time until the concrete water-stop is poured to the pile top.

3.3.4 Foundation pit monitoring

The foundation pit settlement observation points are arranged and the set elevation is observed one week before foundation excavation, and one day before the foundation excavation, the observation is carried out for once and the elevation is compared with the previous one as the initial value for observation, and the observation is carried out for once every meter is excavated or based on the deformation condition during the foundation excavation.

3.3.5 Pile core concrete pouring and pre-stress

According to the foundation deformation and stress, the pile core concrete is poured and the pre-stress is applied for enforcing the seriously deformed section, and tremie pouring is adopted for concrete pouring. The specific operating points are similar to the operating points of the concrete water-stop.

Pre-stress is applied to the pile core concrete after the completion of the concrete pouring. Stepwise tension is adopted for anchor cable tension, every step of load is added, 5–8 min is needed to last stably before unloading to the load of previous-step, and the load is updated to the designed load, 10 min is needed to last before unloading to the pre-stress value to be locked; thus, the whole tension process is completed.

After the anchor cable tension is completed, the extra steel strands are cut off, the anchor heads are sealed with concrete, and the protective layer is 10 cm in thickness.

4 CONCLUSIONS

1. The construction technology of self-water stopping foundation pit support structure of pre-stressed pipe is used and verified in the specific projects that the defects of the ordinary foundation pit support structure of the pre-stressed pipe piles are effectively overcome, and an excellent effect is achieved.
2. In the self-water stopping foundation pit support structure of pre-stressed pipe the concavo-convex connecting pieces are adopted and mutually locked between the pre-stressed pipe piles, and the water-stop is attached to the interior of the concavo-convex connecting pieces, so as to enhance the water stopping effect of the foundation pit support structure.
3. Concrete is poured into the cavity formed between the steel division plate, the pre-stressed pipe piles and the special-shaped pile toe of the self-water stopping foundation pit support structure of pre-stressed pipe piles, so as to form the concrete water-stop, and further enhance the waterproofing capability of the foundation pit.
4. After the completion of the foundation pit support structure, the deformation observation and the stress detection are carried out in the foundation pit, and the pile core concrete pouring and stress application can be carried out

to control the foundation pit from deforming further, so as to have the foundation pit deformation under control.
5. By adopting the construction technology of the self-water stopping foundation pit support structure of pre-stressed pipe piles, the water-stop curtain piles need not to be piled between the pre-stressed pipe piles, so as to reduce the construction equipment and the construction staff number and then reduce the construction cost; and the construction period of the foundation pit support can be shortened, so as to make the building come into use as soon as possible and have a better economic benefit.
6. The construction technology of the self-water stopping foundation pit support structure of pre-stressed pipe piles adopts the pre-stressed pipe piles as the support structure, so as to avoid the slurry pollution caused by the bored pile construction compared with the conventional drilled piles.

ACKNOWLEDGMENT

This material is based upon work funded by Zhejiang Provincial Natural Science Foundation of China under Grant No. LQ12E09002; Project (51308497) supported by National Natural Science Foundation of China.

REFERENCES

Chen, W.H. Guo, T. & Tong, N. 2009. Application Research of Large Diameter pre-stressed pipe pile in deep foundation pit engineering. *Technology of Highway and Transport Technology,* 2009(1): 125–128.

Feng, Z.J. & Xie, Y.L. 2005. *Technique of Bridge Pile Foundation.* Beijing: China Communication Press.

Poulos, H.G. & Davis, E.H. 1980. *Pile Foundation Analysis and Design.* New York: John Wiley & Sons.

Tian, X.M. 2012. Renwang Liang: Application of Combination system of pileanchor adding water-stop curtain in deep foundation pit. *Shanxi Architecture,* 38(5): 83–84.

Yu, J.M. Feng, C.H. & Yan, Y.G. 2008. Application of pressed-in piped piles in deep foundation pit engineering. *Railway Engineering,* 6: 80–82.

Zhang, M.Y. 2004. *Application and Research of Pressed-in Piped Pile.* Beijing: China Building Materials Press.

Advanced Materials and Structural Engineering – Hu (Ed.)
© 2016 Taylor & Francis Group, London, ISBN 978-1-138-02786-2

Incentive contract design of construction engineering time based on the principal-agent model

L. Yan & X.J. Tang

Hebei Finance University, Baoding, Hebei, China

ABSTRACT: Because of the asymmetric information of the two main objects of construction engineering: the owner and the contractor, the contractor pursues his own interests for a certain purpose, and forms a tense non-cooperative state with the owner. The owner's (principal) effective incentive for the contractor (agent) plays a significant role in completing the project. The author uses the principal-agent theory in this paper, mainly studies the factors affecting the construction time and the linear correlation founded between the compressed time and increased cost, gives its incentive coefficients, and establishes an incentive model this is how the owner encourage the contractor in the case of asymmetric information. Then the model has been solved, and incentive coefficients have been calculated. Finally, the author discusses the incentive issues of the construction time.

1 INTRODUCTION

There are two objects in construction project, one is the owners, and the other is the contractor. Because there is the principal—an agent relationship between them, the author uses a principal-agent theory to solve the existing problems. In the principal-agent theory, because principal and agent seek to maximize their own interests, and there also exist factors such as asymmetric information and uncertainty between them, the incentive mechanism is needed to maximize the owner's interests. At home, many scholars have studied the contractor's excitation on the problems of construction schedule. Lu Gongshu and Yi Tao have used principal-agent theory to analyze the excitation of construction contractor (Bubshait 2003). Wang Jiahao, Sun Yongguang, and Wu Zongxin have mentioned the problem of contractor's compressed time in the optimization of revenue incentive and the selection of optimal time (Ann 1993). Chai Guorong has studied the optimization of flexible resource-constrained project schedules and incentive problems (Ell-Rayes & Kandil 2005). In abroad, there are also a lot of researches on the theory of incentives for contractor. Shtub A vraham has studied the incentives with a single factor of project schedule (Wei & Kong 2012).

The author mainly uses the principal-agent theory in this paper to study the excitation that the owner impels the contractor in the part of compressed time, and establishes a performance model that contains specific compressed time after studying the excitation of the construction compressed time conducted by previous scholars, and considers the cost factors of the construction time in contract wages. Finally, the author analyzes the incentive's pros and cons of the compressed time. The author considers separately about the incentive coefficients of compressed time, and studies their influences more deeply, which is also a simplified methodology of the trust and agency.

2 THEORETICAL BASIS

In terms of the industry of engineering project, the owner commissions the contractor to build the project, and it can reflect intuitively that the contractor controls the engineering quality, schedule, risk, cost, safety, and other aspects in the process of construction; as for the contractor's work, the owner cannot determine whether the contractor works hard to achieve the fixed objectives, especially in terms of cost and schedule control, he can measure the efforts paid by the contractor only by virtue of the final results and completion of construction costs, as well as construction schedule. Therefore, if the information of engineering construction is asymmetrical, the owner will act as the "principal" who cannot fully get the information, and the contractor will be as the "agents" who has private information.

The relevant theories of principal-agent are used to study the schedule issues that the owner encourages the contractor in the phase of construction. When the incentive intensity is too small, the funds that the contractor obtained in advance of completion of the project are not enough to compensate for the extra cost of compression, then the contractor will not compress the schedule; on the contrary, due to technical limitations, the compression of the project has certain restrictions, if the incentive intensity is too large, although the owner pays a little more, the desired results may also not be achieved.

3 MODELING AND SOLUTION

3.1 *Modeling*

Assumption 1. There is a certain relationship between construction time and construction cost. Assuming that the output performance of the contractor is mainly determined by compressed time, its function shall be:

$$\pi = \Delta T + \theta \tag{1}$$

In the above formula, π stands for the output of the principal's task, ΔT stands for the changes of construction time, and θ represents a normal distribution with mean 0, variance is σ^2's dynamic random variables.

$$\Delta T = ge_t + \theta_1 \tag{2}$$

In the formula, e_t represents the agent's effort level of compressed time, g represents the marginal contribution rate of construction time, and θ_1 stands for the dynamic random variables that affect the compressed time.

By formula (1) and (2), we can get the production function of the contracted performance:

$$\pi = ge_t + \theta_t \tag{3}$$

In the case of the compressed time, the calculation formula of the difference between the actual cost and the estimated cost is as below:

$$c_a - c_b = \Delta C \tag{4}$$

$$\Delta C = e_t + \theta_2 \tag{5}$$

Assumption 2. Contractor's contract of remuneration:

$$S = x + y\Delta T + z\Delta C$$
$$= x + y(ge_t + \theta_1) + z(e_t + \theta_2) \tag{6}$$

In the formula, x stands for the fixed part in contractor's remuneration, y represents for the incentive coefficient of the compressed time, and z is the incentive coefficient of the project cost reduction, $0 \le y, z \le 1$.

Assumption 3. The cost function of the contractor:

$$C = \frac{1}{2}me_t^2 - \frac{1}{2}\rho y^2 \sigma^2 - \frac{1}{2}\rho z^2 \sigma^2 \tag{7}$$

In the formula, $m/2$ (>0) represents the agent's effort cost coefficient, $1/2\,\rho y^2 \sigma^2$ stands for the agent's cost of risk, ρ is the avoidance and measure of absolute risk. Namely, the contractor's benefits are the cost of its own effort deducted from the contract amount and the actual cost of the project.

Assumption 4. For the risk neutral owner, its expected utility EV is equal to the expected revenue, it is expressed as below:

$$EV = E(\pi) - S(\pi) = ge_t - x - yge_t - ze_t \tag{8}$$

Assumption 5. Contractor of risk avoidance, assuming that its utility function has the feature of constant absolute risk avoidance.

Based on the above assumptions, the contractor's certainty equivalent wealth $C(E)$ can be obtained:

$$C(E) = x + y(ge_t + \theta_1) + z(e_t + \theta_2)$$
$$- \frac{1}{2}me_t^2 - \frac{1}{2}\rho y^2 \sigma^2 - \frac{1}{2}\rho z^2 \sigma^2 \tag{9}$$

Contractor's maximized expected utility function is equivalent to the certainty equivalent wealth in the above formula. In the principal-agent theory, the principal chooses the incentive contract, which aims to maximize his own expected utility. Combined with the above description of the expected utility of the owner, the owner's objective function is to maximize his expected utility. Meanwhile, the participation constraint (IR) requires the expected utility of the contractor who acts as an agent in this incentive contract should be not less than his reservation utility; combined with the actual situation of the project that the contractor involved in, his reservation utility is equivalent to the fixed remuneration $S(\pi)$ that he obtained from the incentive contract; Incentive Constraint (IC) makes the contractor select the effort level that can maximize his own expected utility.

$$(IR): x + y(ge_t + \theta_1) + z(e_t + \theta_2) - \frac{1}{2}me_t^2$$
$$- \frac{1}{2}\rho y^2 \sigma^2 - \frac{1}{2}\rho z^2 \sigma^2 \ge \varpi \tag{10}$$

$$(IC): E[x + y(ge_t + \theta_1) + z(e_t + \theta_2)$$
$$- \frac{1}{2}me_t^2 - \frac{1}{2}\rho y^2 \sigma^2 - \frac{1}{2}\rho z^2 \sigma^2] \tag{11}$$

So the principal-agent model of the owner and the contractor on the project cost and construction time can be expressed as follows:

$$\max Ev = E(\pi) - C(\pi)$$

$$= ge_t - \frac{1}{2}me_t^2 - \frac{1}{2}\rho y^2\sigma^2 - \frac{1}{2}\rho z^2\sigma^2 \quad (12)$$

3.2 Model solution

According to the contractor's equivalent income, calculate the effort level of the compressed time e_t:

$$\frac{\partial C(E)}{\partial e_t} = gy + z - met = 0 \quad (13)$$

$$et = \frac{gy + z}{m}.$$

Then according to the effort level, the incentive coefficient that the owner encourages the contractor in compressed time and the incentive coefficient of cost of reduction can be calculated:

$$E(v) = ge_t - \frac{1}{2}me_t^2 - \frac{1}{2}\rho y^2\sigma^2 - \frac{1}{2}\rho z^2\sigma^2 - \varpi$$

$$\frac{\partial E(v)}{\partial y} = g\frac{\partial e_t}{\partial y} - me_t\frac{\partial e_t}{\partial y} - \rho y\sigma^2 = 0$$

$$\frac{\partial E(v)}{\partial z} = g\frac{\partial e_t}{\partial z} - me_t\frac{\partial e_t}{\partial z} - \rho z\sigma^2 = 0 \quad (14)$$

$$y = \frac{g^2}{1 + g^2 + \rho m\sigma^2}$$

$$z = \frac{g}{1 + g^2 + \rho m\sigma^2}$$

4 MODEL ANALYSES

1. Relationship between the optimal incentive level y^* and the marginal contribution rate of contractor's efforts g

$$\frac{\partial y}{\partial g} = \frac{2g + 2\rho mg\sigma^2}{(1 + g^2 + \rho m\sigma^2)^2} > 0 \quad (15)$$

Because g, $\rho mg\sigma^2$ and $(1 + g^2 + \rho m\sigma^2)^2$ are greater than zero, $\partial y / \partial g$ is always greater than zero, namely, there is a positive correlation between the optimal incentive level y^* and the marginal contribution rate of contractor's efforts g, indicating that the compressed time will increase largely when the marginal contribution rate of contractor's efforts increases. Therefore,

the optimal incentive level between the owner and the contractor should also increase so as to shorten the construction time.

2. Relationship between the optimal incentive level y^* and the cost coefficient of the contractor's effort m

$$\frac{\partial y}{\partial m} = -\frac{g^2\rho\sigma^2}{(1 + g^2 + \rho m\sigma^2)^2} < 0 \quad (16)$$

From the partial derivative with respect to y^* to m, we can see the relationship between the optimal incentive level y^* and the cost coefficient of contractor's effort is in an inverse proportion, which means that when the cost coefficient of contractor's effort is higher, its engineering output will be affected. So, the owner will reduce the contractor's optimal incentive level, otherwise he will give a higher incentive level to the contractor.

3. Relationship between the optimal incentive level y^* and contractor's risk aversion coefficient as well as the volatility of performance.

$$\frac{\partial y}{\partial \rho} = -\frac{g^2 m\sigma^2}{(1 + g^2 + \rho m\sigma^2)^2} < 0 \quad (17)$$

$$\frac{\partial y}{\partial \sigma} = -\frac{2\sigma g^2\rho m}{(1 + g^2 + \rho m\sigma^2)^2} < 0 \quad (18)$$

From the above partial derivatives, we can see $\partial y / \partial \rho$ and $\partial y / \partial \sigma$ are less than zero, i.e., y^* will decrease with the increasing of ρ and σ^2. If contractor's risk aversion or the volatility of construction increases, the contractor's risk of cost will increase, which will affect the contractor's job activities; eventually the owner will reduce the contractor's optimal incentive level.

5 CONCLUSIONS

We can see from the result of derivation of relevant factors of incentive coefficient, that the owner pays more attention to contractor's cost cuts, but gives insufficient emphasis to the compression of the time. Most owners think if the time is compressed, and the cost of the project will certainly increase, so they ignore the importance of compressed time. In fact, as long as the contractors do as well as planned, they can reduce costs when the time is compressed. Of course, this also requires owners to improve their incentive policies and increase their incentives of the contractor's compressed time.

The author assumes that the contractor's performance output is only related to the compressed time in the paper, and gives the compressed time and the linear representation, and separately analyzes that each coefficient affects the compressed time, and reflects the changes of cost indirectly. The complex relationship between compression cost and compressed time cannot reflect how to motivate the contractor without coefficients of determination. So, the author mainly researches the issue of compressed time from the perspective of cost cuts in the paper.

REFERENCES

Ann, V.A. 1993. The principal/agent paradigm: its relevance to various functional fields, *European Journal of Operational Research*, 70(1): 83–103.

Bubshait, A.A. 2003. Incentive/disincentive contracts and its effects on industrial projects, *International Journal of Project Management.* 21(1): 63–70.

El-Rayes, K. & Kandil, A. 2005. Time–cost–quality trade-off analysis for highway construction, *Journal of Construction Engineering and Management.* 131(4): 477–486.

Wei, D. & Kong, Z. 2012. School of Management Chongqing Jiaotong University Chongqing, 400074, China. Research on the Bid of Evaluating Method of the Lowest Price Based on the Game Theory, Proceedings of 2012 IEEE 3rd International Conference on Software Engineering and Service Science.

Advanced Materials and Structural Engineering – Hu (Ed.)
© 2016 Taylor & Francis Group, London, ISBN 978-1-138-02786-2

Studying of the influence to property of road base with cement stabilized gravel by the iron tailings sand

L. Hongbin & Z. Jiannan

The Communications Research Institute of Liaoning Province, Shenyang, China

ABSTRACT: This paper analyzes the mechanical properties and pavement performances of the road base on cement stabilized gravel with different dosages of iron tailings sand based on the experimental research. By comparing test, after cement stabilized gravel mixture is mixed with the iron tailings sand, the dynamic modulus and static modulus reduce, the compressive strength first increase then decrease, the splitting strength decrease, the freezing-thawing resisting ability and fatigue resistance performance enhance, and the change of shrinkage resistance performance is not obvious.

1 INTRODUCTION

Iron tailings sand is the rest of the solid waste after selecting iron concentrate. A large number of iron tailings sand not only occupied the land but also caused the waste of resources. It brought serious environment pollution and harm to the human living. It damages the balance of nature and has received the extensive attention of the whole society (Luis & Ivars 2006). Applying the iron tailings sand in the semi-rigid base will recycle waste material and reduce the cost of the road engineering construction, which has an important social significance (Yang et al. 2009).

By comparing test, we studied the influence of iron tailing sand on the mechanical properties and pavement performance to cement stabilized gravel mixture.

2 MIX DESIGN ON THE BASE WITH CEMENT STABILIZED IRON TAILINGS SAND

2.1 Raw material properties

2.1.1 Gravel
Gravel used in the test came from Hune River in Shenyang. The quality test results could satisfy the requirements of the current specification.

2.1.2 Iron tailings sand
Iron tailings sand for testing came from Qidashan iron tailings bank in Anshan Liaoning Province. The screening results are shown in Table 1.

2.1.3 Cement
The ordinary P·O32.5 cement with "GongYuan" brand produced by the cement plant in Benxi

Table 1.   Screening results of iron tailings sand.

Sieve size (mm)	2.36	1.18	0.6	0.3	0.15	0.075
Mass percentage passing (%)	100	99.4	95.1	77.8	56.7	25.2

Table 2.   Grading of aggregates.

Sieve size (mm)	Mass percentage passing (%)				Grading limitation
	1	2	3	4	
37.5	100	100	100	100	100
31.5	93.6	92.7	92.3	92.5	90~100
19.0	65.7	61.0	59.4	60.5	54~79
9.5	49.0	43.6	41.8	43.8	39~64
4.75	44.6	40.1	39.2	41.9	28~54
2.36	36.5	34.3	35.2	39.1	20~46
0.6	12.8	18.0	24.3	31.7	8~32
0.075	1.3	3.3	5.5	7.9	0~15

Liaoning Province was used. Various technical indicators met the requirements of the current specification.

2.2 Mix design

Four kinds of grading of aggregates were designed according to requirements of distribution range on cement stabilized gravel base of the ordinary highway. Iron tailings sand contents were 0, 10%, 20%, and 30%, respectively. The gradations of aggregates are shown in Table 2.

Table 3. Design results.

Iron tailings sand content (%)	0	10	20	30
Optimum water content (%)	6.2	5.6	6.1	6.4
Maximum dry density (g/cm³)	2.290	2.326	2.312	2.291

The optimum water contents and the maximum dry densities of the mixture with different blending ratios of iron tailings sand were determined by adopting a heavy compaction test method for inorganic binders stabilized materials, the cement content was 4.5% (China Communications Press 2000). The design results are shown in Table 3.

3 MECHANICAL PROPERTIES RESEARCH ON THE BASE WITH CEMENT STABILIZED IRON TAILINGS SAND

3.1 Dynamic modulus test

The formed cylinder specimens were 150 mm high and the diameters of them were 100 mm. They were cured for 90 days under the standard curing condition. On the last day, the specimens were soaked in water for 24h. The tests were carried out with continuous half versed sine load waveform without intermittent. The test temperature was 20°C. Dynamic modules tests were carried out by using the Simple Performance Tester. The results are shown in Table 4.

3.2 Static modulus test

The top surface method was adopted to test the static modulus of semi-rigid base materials. The results are shown in Table 5.

From Figures 1 and 2, it can be seen that dynamic modulus and static modulus gradually reduce with the increase of the dosage of iron tailings sand.

3.3 Compressive strength and splitting strength test

The compressive strength and splitting strength tests were carried out for specimens with different iron tailings sand contents. The results are shown in Table 6.

Figure 3 indicates that the compressive strength first increase then decrease and Figure 4 indicates that the splitting strength decrease with the increase of the dosage of iron tailings sand.

Table 4. Test results on dynamic modulus (MPa).

Frequency rate (Hz)	Iron tailings sand content (%)			
	0	10	20	30
25	17865	17623	15467	14812
20	17643	17432	15034	14645
10	17447	17256	14622	14483
5	17233	16831	14376	14332
2	16878	16678	13925	14109
1	16755	16321	13733	13956
0.5	16532	16124	13445	13837
0.2	16521	15953	13222	13722
0.1	16411	15862	13017	13644

Table 5. Test results on static modulus.

Iron tailings sand content (%)	0	10	20	30
Static modulus (MPa)	1220.7	1171.9	938.1	675.4

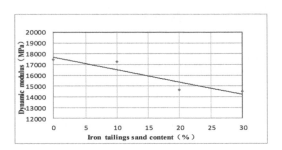

Figure 1. Relation curve on dynamic modulus and iron tailings sand content (10Hz).

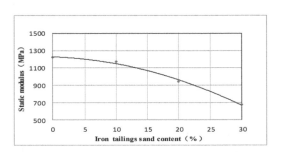

Figure 2. Relation curve on static modulus and iron tailings sand content.

Table 6. Test results on strength.

Iron tailings sand content (%)	0	10	20	30
Compressive strength (MPa)	4.67	5.25	4.96	3.47
Splitting strength (MPa)	0.612	0.565	0.549	0.376

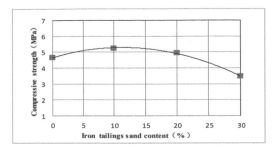

Figure 3. Relation curve on compressive strength and iron tailings sand content.

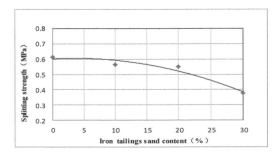

Figure 4. Relation curve on splitting strength and iron tailings sand content.

4 PAVEMENT PERFORMANCES RESEARCH ON THE BASE WITH CEMENT STABILIZED IRON TAILINGS SAND

4.1 *Freezing-thawing resisting ability*

The formed cylinder specimens were 150 mm high and the diameters of them were 150 mm. They were cured for 90 days under the standard curing condition. Then, the specimens were soaked in 3% salt water for 24h. Moisture on the surfaces of the specimens was wiped with a wet cloth before they were weighted. A set of specimens for freezing and thawing were placed in −18°C low temperature box for 16h, after that they were soaked in 20°C water tank for 8h. It was a freezing and thawing cycle. After undergoing 8 cycles, the specimens were weighed. The cleavage strength was tested and compared with the specimens curing in the standard condition for 90 days. The loss rate of strength and weight were calculated. The specimen pictures are presented in Figure 5. The results are shown in Table 7.

From Figure 5 we can see that picture is the specimen of cement stabilized gravel mixture without adding iron tailings sand, its surface loss is the worst. Pictures b, c and d are the specimens of cement stabilized gravel mixture adding 10% or

Figure 5. Specimen pictures.

20% or 30% iron tailings sand, respectively. Surface loss on specimens gradually reduced with the increase of the dosage of iron tailings sand.

From Figures 6 and 7, it can be seen that splitting strength and weight loss decrease with the increase of iron tailings sand content after undergoing freezing and thawing cycles. The specimens with adding iron tailings sand have better freezing-thawing resisting ability.

4.2 *Test on fatigue resistance performance*

The formed cylinder specimens were 150 mm high and the diameters of them were 150 mm. They were cured for 90 days under the standard curing condition. On the last day, the specimens were soaked in water for 24h. Splitting fatigue tests were carried out in the stress control mode with sine wave to load and load frequency for 10 Hz by the material testing machine named UTM100. The intermittent time of test load to load was 0 second. The results are shown in Table 8.

According to the experimental data in Table 8, the fatigue life curves on semi-rigid base with different dosages of iron tailings sand were drawn and shown in Figure 8. From the figure, we can see that the fatigue property of adding iron tailings sand is better than not adding iron tailings sand in the semi-rigid base. The fatigue resistance performance with 30% of iron tailings sand in the semi-rigid base is best.

4.3 *Test on shrinkage resistance performance*

Forming 400 mm long and 100 mm wide and 100 mm high specimens, curing for 7 days under the environment with temperature is 20°C±1°C and Relative Humidity (RH) is 60%±5%. On the last day, the specimens were measured length and their

Table 7. Test results on splitting strength after freezing and thawing.

Iron tailings sand content (%)	Splitting strength (MPa)	Freeze-thaw splitting strength (MPa)	Strength loss (%)	Weight loss (%)
0	0.612	0.310	−49.37	−13.63
10	0.565	0.302	−46.59	−10.45
20	0.549	0.324	−40.98	−7.72
30	0.376	0.246	−34.61	−5.56

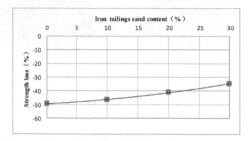

Figure 6. Relation curve on strength loss and iron tailings sand content after freezing and thawing.

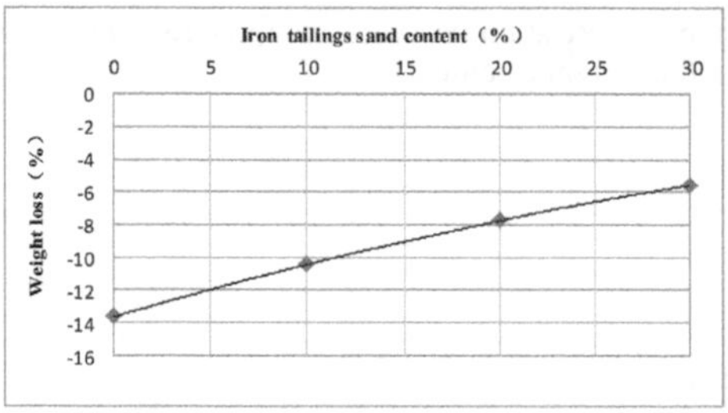

Figure 7. Relation curve on weight loss and iron tailings sand content after freezing and thawing.

Table 8. Test results on fatigue.

Stress ratio		0.85	0.75	0.65	0.55
Iron tailings sand content (%)	0	170	31221	85623	482325
	10	250	43245	114632	643562
	20	190	33346	102543	553467
	30	420	47564	123455	734562

Table 9. Test results on dry shrinkage.

Iron tailings sand content (%)	Total shrinkage ($\times 10^{-3}$ mm)	Total water loss ratio (%)	Total dry shrinkage strain ($\times 10^{-6}$)	Total dry shrinkage index ($\times 10^{-6}$)
0	342.6	5.72	856.5	149.7
10	288.9	5.03	722.3	143.6
20	340.8	5.67	852.0	150.3
30	394.5	6.15	986.3	160.4

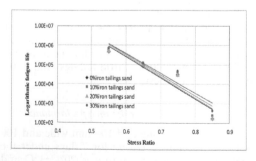

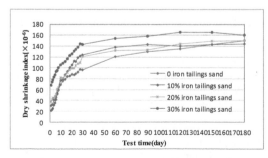

Figure 8. Fatigue life curve of semi-rigid base with different dosage iron tailings sand.

Figure 9. Relation curve on dry shrinkage index with test time.

weights were examined. Then, they were put into a drying shrinkage chamber and measured weight and deformation after different times. The water loss ratio and dry shrinkage strain and dry shrinkage index were calculated (China Communications Press, 2009). The results are shown in Table 9 and Figure 9.

It can be seen from Figure 9 that the cement stabilized gravel mixture mixed with iron tailings sand have similar air shrinkage resistance with cement stabilized gravel mixture.

5 CONCLUSIONS

After cement stabilized gravel mixture is mixed with iron tailings sand, its dynamic modulus and static modulus reduce. Its compressive strength first increase then decrease and its splitting strength decrease. Its freezing-thawing resisting ability and fatigue resistance performance enhance. The change of shrinkage resistance performance is not obvious. The cement stabilized gravel mixture with iron tailings sand has good mechanical properties and pavement performances, which can be used in the semi-rigid base of asphalt pavement.

REFERENCES

China Communications Press, 2000. Technical Specifications for Construction of Highway Road bases.
China Communications Press, 2009. Test Methods of Materials Stabilized with Inorganic Binders for Highway Engineering.
Luis, M. & Ivars, N. 2006. Long-term Environmental Impact of Tailings Deposits, *Hydrometallurgy*, 83: 176–183.
Yang, Q. Pan, B. & He, Y. 2009. *Research of Iron Tailings Sand Applied to Base of Highway Engineering*, Transportation Science & Technology.

Advanced Materials and Structural Engineering – Hu (Ed.)
© *2016 Taylor & Francis Group, London, ISBN 978-1-138-02786-2*

Durability of concrete to sulfate attack under different environments

B.H. Osman, E. Wu & B. Ji
College of Civil and Transportation Engineering, Hohai University, Nanjing, China

A. Ishag
College of Civil Engineering, Southeast University, Nanjing, China

ABSTRACT: The spread of damages in concrete structures due to the sulfate attack represents a topic of increasing significance in both urban and industrial areas. This paper examines the mass change and carbonation depth under different environmental conditions, to clarify the effects of a sulfate acid environment on a hardened concrete body. The concrete cub specimens were immersed in different sulfate acid solutions and experimentally examined. Both concentrations of sulfate acid solutions and Wet-dry cycles versus continuous immersion are compared. The measurement items were the variations of the concrete cubs dimensions, strength, and weight under different acid concentration. Experimental results showed that the sulfate acid has significant effects on the durability and strength of hardened concrete.

1 INTRODUCTION

One of the important engineering properties of concrete is its durability. It significantly determines the service life of concrete structures (Zivica & Bajza 2001, Zivica & Bajza 2002). Inadequate durability manifests itself by deterioration, which can be caused by external factors or internal forces within the concrete (Adesanya & Raheem 2010). The rate of acidic attack of cement based materials depends on numerous factors conditioning the aggressiveness of the acting medium and resistance of the attacked material (Zivica 2004).

Concrete is susceptible to attack by sulfate acid produced from either sewage or sulfur dioxide present in the atmosphere of industrial cities. This attack is due to the high alkalinity of Portland cement concrete, which can be attacked by other acids as well. Resistance to chemical attack is improved with the quality of the concrete. By observing fundamental rules for producing concrete of high quality, contractors can be assured of having the best natural resistance to aggressive chemicals. Sulfate acid is particularly corrosive due to the sulfate ion participating in sulfate attack (Meyer & Ledbetter 1970). The concrete strength loss is more significant at high acid concentrations (Chang et al. 2005, Santhanam et al. 2001, Dehwah 2007). The sulfate resistance of concrete structures can be improved by controlling sulfate permeation into concrete and the sulfate attack can be prevented either by changing cement from ASTM Type I to Type II or Type V or by introducing

Pozzolans such as fly ash, blast furnace slag, Volcanic Ash (VA) and finely ground Volcanic Pumice (VP) in concrete (Kalousek et al. 1972, Al-Amoudi et al. 1994, Naik 1996).

Sulfate acid ion in each environment decreases according to the immersion period of the test specimen. Especially, the rate of decreasing in the early stages was remarkable compared to that for other periods (JSCE 2001, Kawahigashi 2007). The moisture along with chlorides and dissolved oxygen will be absorbed by the concrete cover by capillary forces depending on the degree of the concrete saturation, which initiates the chloride-induced corrosion of the reinforcing steel. Hence, an assessment of the rate of ingress of chlorides has become very important for evaluating the long-term performance of concrete structures (Wee et al. 1999, Rahmani & Ramazanianpour 2008). The attacks of deleterious agents cause rapid deterioration of concrete structures leading to premature failure (Razak et al. 2004, Siddique & Kadri 2011, Vinod et al. 2013). The objective of this work is to investigate and evaluate the response of concrete specimens to sulfate acid attack under different environments. The degrees of concrete deterioration of both the physical and chemical indicators are experimentally examined using the changes in weight and thickness of the specimens. Comprehensive laboratory tests were conducted. The experiment procedures involved alternate acid immersion and drying of test specimens, as well as continuous acid immersion of other test specimens.

2 MATERIALS AND METHODS

Crushed aggregates of angular shape and rough texture with a maximum size of 20 mm were used. The specific gravity of the aggregate fractions was determined according to ASTM C127 (2012). The specific gravity of coarse aggregate was 2.67. The aggregates were separated into different size fractions and varying the percentage of individual fractions, to achieve the most effective particle size distribution. River sand passing through a 4.75 mm sieve with specific gravity (2.50) and fineness modulus (2.67) was used. The chemical composition and physical properties of Ordinary Portland Cement (OPC) used in this study is given in Table 1 and Table 2. Ordinary drinkable tap water with pH value of 6.2 ± 0.1 was used. Solutions of sulfate acid with the concentrations of 0.0%, 2%, 5% and 7% were used. The temperature of the solutions was $20 \pm 2°C$ during a period of 12 weeks.

The materials used were of commercially available ordinary Portland cement (specific gravity: 3.15). In addition, the best available quality reagent of sulfate acid was used as the environment solution.

2.1 Mix proportion and test procedure

The procedure of the experiment utilized three different acid concentrations. The concrete mix was made by weight with proportion of 1: 2: 4 for cement, sand and course aggregate, respectively. Standard $150 \times 150 \times 150$ mm concrete cubs were made from the concrete mix. All specimens were cured in water for 28 days to reach its hardness point. At the end of the curing period, three cubs were used to determine the compressive strength of each group (ASTM 2012) by using Compression Testing Machine (CTM) under pure uniaxial compression loading (Fig. 1a). Compressive strength test is a major strength test conducted on concrete. Strength usually expresses the quality of the concrete because it is directly related to the structure of cement paste.

The specimens were dried in an oven at 100°C and the initial weights were measured. Then the specimens were stored in solutions of sulfate acids with three different concentrations. For each acid and concentration, nine specimens were subjected to each solution. Group A, B, and C were alternately immersed in the acid solutions and dried. Group D was continuously immersed until the end of the test period. To represent the real actions in sites, Group F was continuously immersed in soil with different acid solution until the end of the test period same as in group D. Soil used in the test was collected from contaminated sites and filled in selected tank. Then, the soil was mixed with sulfate acid solution till it reached its saturation point. Group F cubs were covered with the saturated soil. Soil tank was kept completely covered to prevent evaporation and soil drying. The Compressive strength of the control (non-exposed) specimens after 28 days of their hardening in water and the test specimens subjected to solutions of acids for 12 weeks were determined.

After exposure to acids, all specimens were washed in order to remove the porous layer of the corrosion products such as soft and crystallized acidic materials or calcium salts. Then specimens were dried in an oven at 100°C. Losses in weight of the specimens were measured every two weeks during the test period. The compressive strength values of specimens were calculated by using the original cross-sectional area of the original cubes. The loss of strength values for all types of acids also depends on the concentration of the solutions. The test specimens were oven-dried to a constant weight for 24 hours and their weights and thicknesses were measured prior to the start of the acid attack test. The weight was measured using a balance with 0.013 g sensitivity. The thickness was measured with a pair of calipers for two locations along the sides for each specimen.

Table 1. Chemical composition of the cement.

Compound	SiO_2	Al_2O_3	Fe_2O_3	MgO	CaO	Na_2O	K_2O	SO_3
Value	20.4	5.4	3.5	2.4	64.5	0.2	0.5	2.1

Table 2. Physical properties of the cement.

Test name	Consistency	Setting times [min]	f_{cu} 28 days [N/mm²]	Color	Specific gravity
Value	29.75%	Initial 1:53 Final 3:25	41	Gray	3.15

(a)

(b)

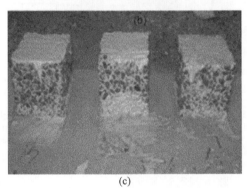

(c)

Figure 1. Test setup (a) Compression Testing Machine (CTM) (b) specimens in the glass mold (c) disintegrated specimens in the acid.

3 ENVIRONMENTAL CONDITION

The test specimens of each group were kept in glass mold filled with equal quantities of same percentages of sulfate acid solutions. The test results were obtained after 12 weeks after some of the test specimens were completely disintegrated in the acid solutions. Each environmental tank was kept completely covered to prevent evaporation from the solution surface in the indoor air environment as showed in Figure 1b.

4 RESULTS AND DISCUSSION

Each specimen was removed from the mold on the day after casting. It was cured in water for 28 days. The specimens were classified into groups according to acid concentration. Specimens were immersed in the sulfate acid solution tank. Items such as compressive strength, mass weight and carbonation depth were examined.

4.1 Mass change and carbonation depth

The mass weight of each specimen was measured. To assess the carbonation depth, specimens were measured using the following procedure. The test specimen was taken out at 2, 4 and 6 week from the start of exposure in solution and the loose part was cleaned using the dry cutter. These cut surfaces were sprayed with solution and were measured from the corroded surface to the boundary position of carbonation/non-carbonation (Fig. 1c). Each carbonation depth was used as an average measured value of three cubs. The value was subtracted from the dimensions of non-carbonation from the initial dimension before exposure. Measurements of the specimens were performed after selected periods of immersion in the acid. In order to measure the weight and thickness, the specimens were removed from the acid and immersed in fresh water several times and oven-dried for 24 hours.

4.2 Thickness change of the specimens

The change in thickness of the specimens versus immersion time in the acid solutions is shown in Figure 2. Each data point is an average of three specimens of the same environment. The data for the individual specimens varied not more than 5 percent from the average values.

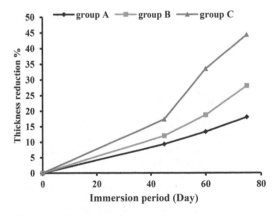

Figure 2. Percentage change in thickness.

The increases in thickness indicate that the specimens are susceptible to volume expansion or swelling as a consequence of the sulfate acid attack. The thickness loss (T_L) of the specimens is determined as: (Mansour et al. 2011 & K.M.A. Hossain, M. Lachemi. 2006).

$$(T_L)\% = \frac{T_1 - T_2}{T_1} * 100 \qquad (1)$$

where, T_1 is the thickness of the specimen before immersion and T_2 is the thickness of the cleaned specimen after immersion. The thickness loss of the specimens is measured every 15 days.

4.3 Wet-dry cycles versus continuous immersion

Figure 3 compares the expansion of specimens, which were alternately immersed in the acid solutions and dried with that continuously immersed (Group D) until the end of the test period.

The former specimens showed a greater expansion than those that were continuously immersed.

Figure 4 shows that when the acid solution is mixed with soil, it gives less effect on the concrete deteriorations than that directly immersed in acid solutions. This concludes that wet-dry cycles of exposure to sulfate acid solutions increase the degree of concrete deterioration. The increased permeability of concrete due to drying cracks would lead to a greater volume of material being attacked by the sulfate acid.

4.4 Loss of compressive strength

The elementary compressive strength (N/mm²) of the control specimens after 28 days is 41.54 N/mm².

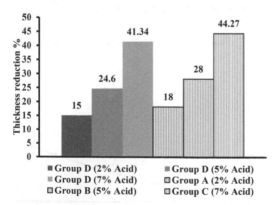

Figure 3. Expansion of specimens that were subjected to wet-dry cycles of acid exposure compared to an expansion of specimens that were continuously immersed in the acid solutions.

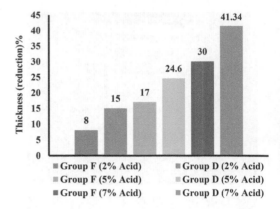

Figure 4. Change in the thickness of specimens that were continuously immersed in the acid solution compared with that covered by soil.

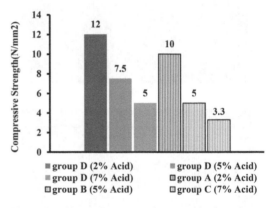

Figure 5. Change in compressive strength of specimens that were subjected to wet-dry cycles of acid exposure compared to specimens that were continuously immersed in the acid solution.

We observed that the concrete strength loss is more significant at high acid concentrations. For example, the loss of strength was higher than 40%, at a concentration of 7%. It worth mentioning that these values would be smaller for specimens immersed continuously in acid as shown in Figure 5 and Figure 6.

The results showed that there is a meaningful relationship between the loss of weight and loss of compressive strength for all specimens. The results in Figure 7 indicates that the specimens covered by soil save the concrete strength more than 8% than that directly immersed in acid solution because the distribution of acid through the soil reduce the acid concentration near concrete surface and decrease the rate of deterioration of concrete subjected to acid attack.

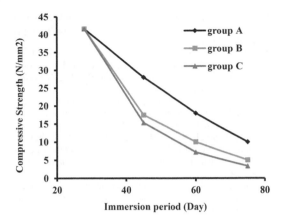

Figure 6. Change in compressive strength of specimens that were subjected to wet-dry cycles of acid exposure.

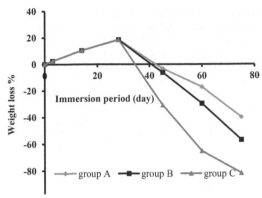

Figure 8. Variation of weight loss with immersion period of specimens immersed in different acid concentrations.

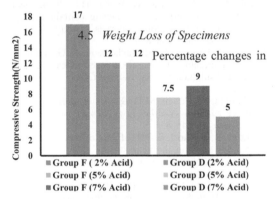

Figure 7. Change in compressive strength of specimens that were continuously immersed in the acid solution compared to that were covered by soil.

4.5 Weight loss of specimens

Percentage changes in weight of the specimens are shown in Figure 8. Results show that weight loss is greater for the specimens with high acid concentration than other ones. However, the initially increase of specimens weight is followed by weight loss with respect to time. The attacked portions of the concrete specimens were cleaned with de-ionized water. The acid attack was evaluated through measurement of the weight loss (W_L) of the specimens and determined as: (Mansour et al. 2011)

$$Weight\ loss\ (W_L)\% = \frac{W_1 - W_2}{W_1} * 100 \qquad (2)$$

where W_2 is the weight of the specimen before immersion and W_1 are the weight of the cleaned specimen after immersion. The weight loss of the specimens is measured every 15 days. The reaction is in between sulfuric acid and the concrete constituent results in the conversion of calcium hydroxide to calcium sulfate (gypsum) (Neville 2004). Each of these reactions involves an increase in volume of the reacting solids. The formation of calcium sulfate leads to softening (decrease in density) of the concrete.

Since weight depends on both volume and density, the initial weight gain of the specimens is probably due to the relative increase in volume being greater than the relative decrease in density. Both the increase in volume and the decrease in density of the concrete due to the sulfate acid-cement paste reaction would be higher acidity (lower pH) of the acid solution (Meyer & Ledbetter 1970). This indicates that a stronger acid solution can produce a greater weight loss in a concrete specimen than that produced by a weaker solution due to a significant increase in the volume of the concrete in comparison to the reduction in the density.

5 CONCLUSIONS

In this paper, we investigated the effects of sulfate acid in concrete durability by immersing concrete samples in different concentration of sulfate acid solution. The experimental results showed that the weight and strength losses of the specimens accordingly increase with the values of acid concentration. The increase in sulfate concentration gives a proper indicator of the extent of damage in concrete due to exposure to sulfate acid. The relationship between the degree of concrete deterioration and depth of penetration of sulfate acid is represented by the variation in sulfate concentration with the depth of acid penetration. Moreover, the wet-dry cycles of exposure to sulfate acid increase the degree of concrete deterioration. From the results, there are clear indicator that when the acid concentration exceed 5% total damage of specimens may occur, that means we should not subject normal concrete to that ratio before any treatment. Generally, the concrete resistant to sulfate attack should meet a criterion of low expansion and strength reduction, and little or no deterioration.

Future works can concentrate on the usage of other testing techniques, such as ultrasound to determine the rate of deterioration, penetration depth and strength values. Furthermore, other works can focus in investigation of concrete additives to reduce the sulfate effects on the concrete structure.

REFERENCES

Adesanya, D.A. & Raheem, A.A. 2010. A study of the permeability and acid attack of corn cob ash blended cements, *Construction and Building Material.* 24: 403–409.

Al-Akhras, N.M. 2006. Durability of metakaolin concrete to sulfate attack, *Cement and Concrete Research.* 36: 1727–1734.

Al-Amoudi, O.S.B. Maslehuddin, R.M. & Abduljauwad, S.N. 1994. Influence of chloride ions on sulphate deterioration in plain and blended cements, *Magazine of Concrete Research.* 46(167): 113–123.

ASTM C127,2012. Standard Test Method for Density, Relative Density (Specific Gravity), and Absorption of Coarse Aggregate.

ASTM C39/C39M, 2012. Standard Test Method for Compressive Strength of Cylindrical Concrete Specimens.

Attiogbe, E.K. & Rizkalla, S.H. 1989. Response of Concrete to Sulfuric Acid Attack, *Material Journal.* 1: 481–488.

Chang, Z.T. Song, X.J. Munna, R. & Marosszeky, M. 2005. Using limestone aggregates and different cements for enhancing resistance of concrete to sulphuric acid attack, *Cement Concrete Res.* 35: 1486–1494.

Dehwah, H.A.F. 2007. Effect of sulphate concentration and associated cation type on concrete deterioration and morphological changes in cement hydrates, *Construction and Building Material.* 21: 29–39.

Hossain, K.M.A. & Lachemi, M. 2006. Performance of volcanic ash and pumice based blended cement concrete in mixed sulfate environment, *Cement and Concrete Research 36 (2006) 1123–1133.*

JSCE Concrete Committee, 2001. Japan Concrete Standards, their Maintenance and Management.

Kalousek, G.L. Porter, L.C. & Benton, E.J. 1972. Concrete for long-time service in sulfate environment, *Cement and Concrete Research.* 2(1): 79–89.

Kawahigashi, T. 2007. *Effects of Sulfuric Acid Solution on Cement Mortar*, Institute for Science and Technology, Kinki University Kowakae.

Mansour, M.S. Kadri, El-H. Kenai, S. Ghrici, M. & Bennaceur, R. 2011. Influence of calcined kaolin on mortar properties, *Construction and Building Material.* 25: 2275–2282.

Meyer, A.H. & Ledbetter, W.B. 1970. Sulfuric Acid Attack on Concrete Sewer Pipe, *J. Sanitary Engineering Division*, 96(5): 1167–1182.

Naik, T.R. Singh, S.S. & Hossain, M.M. 1996. Enhancement in mechanical properties of concrete due to blended ash, *Cement and Concrete Research.* 26(1): 49–54.

Neville, A. 2004. The confused world of sulfate attack on concrete, *Cement and Concrete Research.* 34: 1275–1296.

Rahmani, H. & Ramazanianpour, A.A. 2008. Effect of binary cement replacement materials on sulphuric acid resistance of dense concretes, *Magazine of Concrete Research.* 60(2): 145–155.

Razak, H.A. Chai, H.K. & Wong, H.S. 2004. Near surface characteristics of concrete containing supplementary cementitious materials, *Cement Concrete Composite.* 26(7): 883–889.

Rozière, E. Loukili, A. El Hachem, R. & Grondin, F. 2009. Durability of concrete exposed to leaching and external sulphate attacks, *Cement Concrete Research.* 39: 1188–1198.

Santhanam, M. Cohen, M.D. & Olek, J. 2001. Sulphate attack research—whither now? *Cement and Concrete Research.* 31: 845–851.

Siddique, R. & Kadri, E. 2011. Effect of metakaolin and foundry sand on the near surface characteristics of concrete, *Construction and Building Materials.* 25(8): 3257–3266.

Vinod, P.A. Lalu, M. & Jeenu, G. 2013. Durability Studies on High Strength High Performance Concrete, *International Journal of Civil Engineering and Technology. (IJCIET)*, 4(1): 16–25.

Wee, T.H. Kand, S.A. & Tin, S.S. 1999. Influence of aggregate fraction in the mix on the reliability of the rapid chloride permeability test, *Cement Concrete Composite.* 21(1): 59–72.

Zivica, V. & Bajza, A. 2001. Acidic attack of cement based materials—a review. Part 1. Principle of acidic attack, *Construction and Building Material.* 15: 331–340.

Zivica, V. & Bajza, A. 2002. Acidic attack of cement based materials—a review. Part 2. Factors of rate of acidic attack and protective measures, *Construction and Building Material.* 16: 215–222.

Zivica, V. 2004. Acidic attack of cement based materials—a review. Part 3. Research and test methods, *Construction and Building Material.* 18: 683–688.

Experimental study on particle drift velocity in single-stage double-vortex collecting plate ESP

C.W. Yi, Y.N. Huang, J. Zhang, H.J. Wang & C.X. Lu
School of Environment and Safety Engineering, Jiangsu University, Zhenjiang, China

ABSTRACT: Drift velocity is an important indicator of the ESP performance. This paper studied the effective drift velocity of the particle in a novel laboratory ESP with double-vortex collecting plates. Single factor experiments were conducted to investigate the effect of applied voltage U, average airflow velocity v, space between adjacent collecting plate b_1, collecting plate distance b_2, effective collecting area S_e and initial dust concentration C_0 on effective drift velocity ω_e. The experimental results demonstrate that every factor has a different effect on particle drift velocity. Based on single factor experimental results, an orthogonal experiment was designed to evaluate the interaction effect of these factors on ω_e and then obtain the optimal parameter combination. The orthogonal experimental results showed that: when $U = 18$ kV, $v = 4.5$ m·s^{-1}, $S_e = 0.48$ m^2, $b_1 = 60$ mm, ω_e can reach 158.9 cm·s^{-1}. The drift velocity in double-vortex collecting plate ESP is clearly higher than it in conventional ESP.

1 INTRODUCTION

The maximum soot emission concentration in coal-fired boiler was reduced from 50 mg·m^{-3} to 30 mg·m^{-3}, according to the new standard promulgated by China in 2011. To meet the stricter environmental requirement, scholars have proposed many new methods and done a lot of researches (Byeon et al. 2006, Jaworek et al. 2007, Nakajima & Sato 2003, Parker et al. 2009, Dumitran et al. 2002, Biskos et al. 2005, Kim et al. 2010). Deutsch formula (Deutsch 1922) indicates that drift velocity is a key factor to the collection of particles. By increasing particle drift velocity, the problem of low collection of fine particles can be fundamentally solved (Zhang et al. 2002, Yi et al. 2010, Jedrusik et al. 2003, Jedrusik et al. 2001). Zou et al. (1990) stud-ied on a transverse plate ESP and demonstrated the linear relationship between drift velocity and average corona density, plate spacing; and the variable relationship between drift velocity and airflow velocity, collecting area. The single-zone double-vortex collecting plate ESP, which is characterized by the use of double-vortex collecting plates arranged perpendicularly to the airflow direction and discharge electrodes placed in the gap of adjacent collecting plates in the same row (Fig. 1), was studied to reveal the influence rules of the selected factors on drift velocity of charged particles and provide a theoretical basis for the modification of this ESP.

2 EXPERIMENTAL SETUPS AND PROGRAMS

The experimental system mainly consists of powder feeding device, power supply device, the detected equipment and the ESP (Fig. 1). The ESP body size is 1200 mm × 345 mm × 360 mm; each collecting plate area is 140 × 250 mm^2. Power supply device is GYDY-50 High-voltage Power Source with output voltage of 0~50 kV. The talcum powder is injected by spray gun at the pipe entrance and be pumped into the ESP by induced draft fan. Ankersmid CIS-50 Particle Size Analyzer is used to measure particle diameter (Fig. 2). Gas flow velocity and flow rate are measured by Testo350M/XL Multi gas Analyzer. The inlet and outlet dust concentration is measured by CEM50C Online Measuring

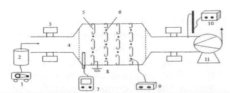

1.Air compressor 2.Dust spraying apparatus 3. Online measuring instrument of dust concentration 4.Velocity distribution plate 5.Collecting plate 6.Discharge electrode 7.Multi-gas analyzer 8.Grounding 9.High voltage power 10.Pitot tube automatic dust sampler 11.Induced draft fan

Figure 1. Schematic diagram of experimental setups.

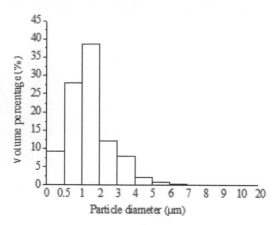

Figure 2. Particle diameter distribution at the entrance of ESP.

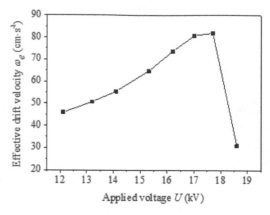

Figure 3. The effective drift velocity as a function of applied voltage.

Instrument of Dust Concentration to calculate the total collection efficiency η (Dumitran et al. 2002).

Single-factor experiments were conducted to identify the effect of each factor on effective drift velocity ω_e by changing factors of applied voltage U, average airflow velocity in ESP v, space between adjacent collecting plate b_1, collecting plate distance b_2, effective collecting area S_e and initial dust concentration C_0. ω_e can be derived from Deutsch formula (Zhuang et al. 2000).

$$\eta = 1 - \exp\left(-\frac{S_e}{Q}\omega_e\right) \qquad (1)$$

where Q is gas flow rate. According to single-factor experimental results, factors which have the relatively significant effect on ω_e were selected to design an orthogonal experiment with three levels to find out the optimum parameter combination for the maximum ω_e in this novel ESP.

3 EXPERIMENTAL RESULTS AND DISCUSSION

3.1 Effect of applied voltage v on the effective drift velocity ω_e

Operational parameters: $b_1 = 60$ mm, $b_2 = 90$ mm, $v = 2.8$ m·s⁻¹, $S_e = 1.12$ m², $C_0 = 235$ mg·m⁻³. As shown Figure 3, ω_e increases gradually with U, then plummet at 18.6 kV for an electric field breakdown. When the applied voltage is lower, the ion concentration is too low to ensure the particle saturated charged, leading to a lower ω_e. With the increase of U, the ion generation ability is enhanced, causing the improvement of particle charging probability and capacity, so ω_e can be increased.

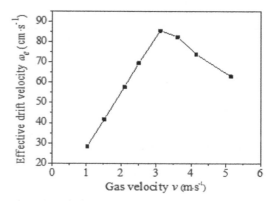

Figure 4. The effective drift velocity as a function of gas velocity.

3.2 Effect of gas velocity v on the drift velocity ω_e

Operational parameters: $U = 17.3$ kV, $b_1 = 60$ mm, $b_2 = 90$ mm, $S_e = 1.12$ m², $C_0 = 195$ mg·m⁻³. Figure 4 shows that, ω_e linearly increases with v of 1.0~3.1 m·s⁻¹. Because v can enhance particle inertia force, leading to the increase of particle drift velocity toward collecting plates. And higher v can reduce ion recombination rate which will be conducive to particle charging. When v is above 3.1 m·s⁻¹, ω_e begin to decrease. The higher airflow velocity is likely to cause re-entrainment, which goes against the collection of particle and the increase of ω_e.

3.3 Effect of adjacent collecting plate space b_1 on the drift velocity ω_e

Operational parameters: $b_2 = 90$ mm, $v = 2.9$ m·s⁻¹, $S_e = 1.12$ m², $C_0 = 235$ mg·m⁻³, U is the maximum applied voltage varying with b_1. Figure 5 shows that, ω_e increases first and then decreases with the increase of b_1. The maximum ω_e is 83.6 cm·s^{-1} with b_1 of 60 mm. Since the breakdown voltage is determined by b_1, when b_1 is small, spark discharge is likely to generate in the ESP at low voltage, which is not conducive to the generation of ions. The ion concentration is relatively reduced while b_1 is too wide, resulting in the decline of particle charging capacity and ω_e.

3.4 Effect of collecting plate distance b_2 on the drift velocity ω_e

Operational parameters: $U = 17.5$ kV, $b_1 = 60$ mm, $v = 1.61$ m·s^{-1}, $S_e = 1.12$ m^2, $C_0 = 195$ mg·m^{-3}. Figure 6 shows that, ω_e increases to 45.5 cm·s^{-1} with b_2 of 90 mm. When b_2 is not wide enough for

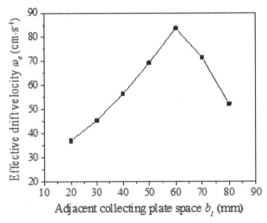

Figure 5. The effective drift velocity as a function of adjacent collecting plate space.

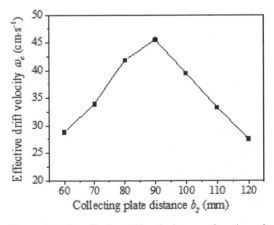

Figure 6. The effective drift velocity as a function of collecting plate distance.

the airflow to form vortexes on the collecting plate, particle residence time in the ESP will be reduced. Thus particle agglomeration and collecting efficiency are both low. And the gas stream with high velocity flows through the collecting plate surface, which may cause re-entrainment in the ESP. Then ω_e decreases to 27.5 cm·s^{-1} with the increase of b_2 up to 120 mm. Maintaining applied voltage constant, the electric field strength decreases with the increase of b_2, which will result in the reduction of ω_e (Zou et al. 1990).

3.5 Effect of effective collecting area S_e on the drift velocity ω_e

Operational parameters: $U = 17.2$ kV, $b_1 = 60$ mm, $b_2 = 90$ mm, $v = 3.03$ m·s^{-1}, $C_0 = 202$ mg·m^{-3}. Figure 7 presents the curve how ω_e changes upon S_e. The general trend is ω_e decreases with the increase of S_e and eventually levels off. The reason is that, if S_e is changed, η changes accordingly, so the value of ω_e calculated from Deutsch formula is determined by S_e and η jointly (Deutsch 1922).

3.6 Effect of initial dust concentration C_0 on the drift velocity ω_e

Operational parameters: $U = 17.1$ kV, $v = 3.14$ m·s^{-1}, $b_1 = 60$ mm, $b_2 = 90$ mm, $S_e = 1.12$ m^2. Figure 8 reveals that, ω_e increases first and then decrease with the increase of C_0. Because space charge carried by charged particles will inhibit the generation of corona currents, and excessive charged particles may even result in corona occlusion inside the ESP (Jaworek et al. 2007), leading to the decline of ion concentration and particle charged capacity, so that ω_e has a significant reduction.

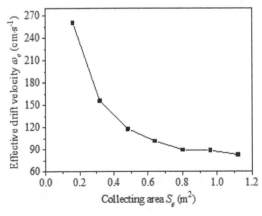

Figure 7. The effective drift velocity as a function of effective collecting area.

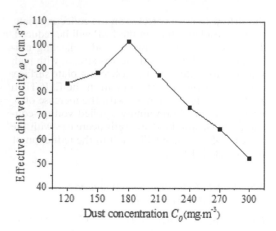

Figure 8. The effective drift velocity as a function of initial dust concentration.

Table 1. Factors and levels.

Level	A: v [m·s^{-1}]	B: S_e [m^2]	C: U [kV]	D: b_1 [mm]
1	2.5	0.48	14	40
2	3.5	0.8	16	60
3	4.5	1.12	18	80

Table 2. Results of the orthogonal experiment.

Num	A	B	C	D	ω_e [cm·s^{-1}]
1	1	1	1	1	99.6
2	1	2	2	2	72.0
3	1	3	3	3	42.0
4	2	1	2	3	126.5
5	2	2	3	1	89.8
6	2	3	1	2	69.3
7	3	1	3	2	158.9
8	3	2	1	3	85.0
9	3	3	2	1	64.2
K_1	71.2	128.3	84.6	84.5	
K_2	95.2	82.3	87.6	100.1	
K_3	102.7	58.5	96.9	84.5	
R	31.5	69.8	12.3	15.6	

4 ORTHOGONAL EXPERIMENT RESULTS

According to the single-factor experimental results, v, S_e, U and b_1 were selected as the factors of orthogonal experiment with three levels, as shown in Table 1. Each experiment was repeated three times to get the average ω_e. The results of the orthogonal experiments are shown in Table 2.

The R values in Table 2 represent the effect degree of each factor. The impact of each factor on ω_e from primary to secondary is v, S_e, U and b_1. The analysis of K values show that, the optimum values for factor A is level 3, for factor B is level 1, for factor C is level 3 and for factor D is level 2. So the optimum parameter combination is A_3, B_1, C_3, D_2. The ω_e can reach up to 158.9 cm·s^{-1} in this novel ESP, but it's usually within 30 cm·s^{-1} in the conventional ESP (Nakajima & Sato 2003).

5 CONCLUSIONS

1. The results of single factor experiments in single-stage double-vortex collecting plate ESP indicate that, ω_e increases and then decreases with the increase of U, v, b_1, b_2 and C_0, while with the increase of S_e, ω_e reduces gradually and finally tends to be stable.
2. The orthogonal experiment results reveal that, the impact of each factor on ω_e, from primary to secondary, is v, S_e, U and b_1. For the maximum ω_e of 158.9 cm·s^{-1}, the parameter combination should be as following: $U = 18$ kV, $v = 4.5$ m·s^{-1}, $S_e = 0.48$ m^2, $b_1 = 60$ mm. Compared with conventional ESP, particle drifts velocity in single-stage double-vortex collecting plate ESP can be effectively improved.

ACKNOWLEDGEMENT

This work was supported financially by National Natural Science Foundation of China (Project No. 51278229) and Six Talent Peaks Funded Project of Jiangsu Province, China (Project No. JNHB-018).

REFERENCES

Biskos, G. Reavell, K. & Collings, N. 2005. Electrostatic characterisation of corona-wire aerosol chargers, *Journal of Electrostatics*. 63: 69–82.

Byeon, J.H. Hwang, J.H. & Park, J.H. et al, 2006. Collection of submicron particles by an electrostatic precipitator using a dielectric barrier discharge, *Journal of Aerosol Science*. 37: 1618–1628.

Deutsch, W. 1922. Movement and charge of the electric exchanger in cylinder capacitor, *Annals of Physics*. 68: 335–344.

Dumitran, L.M. Atten, P. & Blanchard, D. et al, 2002. Drift velocity of fine particles estimated from fractional efficiency measurements in a laboratory-scaled electrostatic precipitator. *IEEE Transactions on Industry Applications*. 38: 852–857.

Jaworek, A. Krupa, A. & Czech, T. 2007. Modern electrostatic devices and methods for exhaust gas cleaning: A brief review, *Journal of Electrostatics*. 65: 133–155.

Jedrusik, M. Gajewski, J.B. & Swierczok, A.J. 2001. Effect of the particle diameter and corona electrode geometry on the particle migration velocity in electrostatic precipitators, *Journal of Electrostatics*. 51: 245–251.

Jedrusik, M. Swierczok, A. & Teisseyre, R. 2003. Experimental study of fly ash precipitation in a model electrostatic precipitator with discharge electrodes of different design, *Powder Technology*. 135: 295–301.

Kim, J.H. Lee, H.S. & Kim, H.H. et al, 2010. Electrospray with electrostatic precipitator enhances fine particles collection efficiency, *Journal of Electrostatics*. 68: 305–310.

Nakajima, Y. & Sato, T. 2003. Electrostatic collection of submicron particles with the aid of electrostatic agglomeration promoted by particle vibration, *Powder Technology*. 135: 266–284.

Parker, K. Haaland, A.T. & Vik, F. 2009. Enhanced fine particle collection by the application of SMPS energisation, *Journal of Electrostatics*. 67: 110–116.

Yi, C.W. Dou, P. & Wu, C.D. et al, 2010. Experimental Study on a Laboratory-Scale Transverse-Plate Electrostatic Precipitator, *Fresenius Environmental Bulletin*. 19: 2472–2479.

Zhang, X.R. Wang, L.Z. & Zhu, K.Q. 2002. An analysis of a wire-plate electrostatic precipitator. *Journal of Aerosol Science*, 33: 1595–1600.

Zhuang, Y. Kim, Y.J. & Lee, T.G. et al, 2000. Experimental and theoretical studies of ultra-fine particle behavior in electrostatic precipitators, *Journal of Electrostatics*. 48: 245–260.

Zou, Y.P. Zhou, Y.A. & Zhang, G.Y. 1990. Study on particle drift velocity in the transverse plate ESP, *Journal of Environmental Engineering*. 01: 26–28.

Advanced Materials and Structural Engineering – Hu (Ed.)
© 2016 Taylor & Francis Group, London, ISBN 978-1-138-02786-2

Synthesis and modification of polyurethane type track

X.H. Zeng
Institute of Packaging and Materials Engineering, Hunan University of Technology, Zhuzhou, Hunan, China

J.P. Luo
College of Science, Hunan University of Technology, Zhuzhou, Hunan, China

H.Q. Hu
Institute of Machinery, Hunan University of Technology, Zhuzhou, Hunan, China

ABSTRACT: The polyurethane track was prepared by a reaction between polyether polyol and diphenyl methane diisocyanate, different inorganic micro-particles $CaCO_3$ and SiO_2 were filled into the system as filler. The effect of different ratios of raw materials, the kind and mass of microparticles on the characters of polyurethane track was investigated. The results showed that the Shao type hardness, tensile strength, elongation at break all show a first rise and then fall trend with the increase of N_{NCO}/N_{OH}, the compression recovery rate and rebound value decreases. The optimal ratio of N_{NCO}/N_{OH} is 1.5~1.8. Filling microparticles into the polyurethane could improve the mechanical characters of the materials. Micro-sized SiO_2 is better filler than $CaCO_3$, and the appropriate mass fraction of SiO_2 is 35% of polyurethane raw materials.

1 INTRODUCTION

Polyurethane track shows a good elasticity, wear-resistant, skid resistance, and brightness. The track is easy to maintain and manage, which is hard to be affected by the climate. It is suitable to be used to prepare a superior running track. The construction of polyurethane track is always a double-component process. The first component is the isophorone pre-polymer, and the second component is the composition with hydroxyl (Tang & Bao 2009, Lu 2004).

So far, the domestic plastic track is prepared by the toluene diisocyanate (TDI). While, the toluene diisocyanate owns a high vapor pressure about 1.33 pa at 20 °C, which shows a high volatility of noxious gas, which is harmful to the health of constructors and environment. International sports runway technology association (IST) have advised that "whether could use no-TDT monomer to prepare polyurethane track." The use of low volatile diphenyl methane diisocyanate in synthesis is an effective method to prepare the environmental polyurethane track. In MDI, the 2,4-MDI and 4,4-MDI were both exploited and used. The structural formula was shown below.

Due to that the molecular structure of 4,4-MDI is symmetrical, the good crystallinity causes obvious difficult in producing the track. Inversely, the molecular structure of 2,4-MDI is asymmetry and spatial interaction, which exhibits poor crystallinity. Therefore, the mixture of 4,4-MDI and 2,4-MDI could combine the characteristic of both monomers and endow the product better properties including flexibility and elongation. However, the ratio of raw materials in the reaction also affects the properties of the track, especially the N_{NCO}/N_{OH} of diphenyl methane diisocyanate and polyether polyol (Liu 2007).

Besides, adding nano-sized materials into the system could endow the materials better mechanical strength and other properties such as resistance to ultraviolet ray, heat insulation due to its small size effect. In this paper, microparticles SiO_2 and $CaCO_3$ modified by a coupling agent were added to track, and the characters of micro-composites materials were researched.

2 EXPERIMENT

2.1 Experimental materials

2-4 Diphenyl methane diisocyanate (2, 4-MDI), 4-4 diphenyl methane diisocyanate (4, 4-MDI), catalyst zinc 2-ethylhexanoate, 3, 3'-dichloro-4, 4'-diamino diphenyl methane (MOCA) were purchased from TCL Co., Ltd, Japanese; polyether polyol GE210 with average molecular 2000 and hydroxyl value 54.5~57.5 was purchased from Aladdin Co., Ltd, America; Micro-sized SiO_2, $CaCO_3$, particle size: 20~60 um were purchased High Technology Nano Co., Ltd, China.

2.2 Sample preparation of microparticles filled polyurethane track

Putting polyether polyol GE210 into the three-necked flask with stirrer, the temperature was fixed at 100–115 °C, evacuating to the vacuum and dehydrating for 1.5h, when the content of water decreases below 0.05wt%, decreasing the temperature to 50 °C, adding MDI (the ratio of 2,4-MDI and 4,4-MDI is 1) and catalyst with the mass of 0.5wt% of MDI into the system and then rising the temperature to 80 °C, after reacting for 2 hours, decreasing the temperature to 40 °C, and the MDI type-PU pre-reaction glue stock was obtained. After that, the pre-reaction glue stock, micro-sized particles, and other additions including antioxygen, plasticizer, and pigment were mixed and added into the mold with mold release. The mixture was formed at room temperature and cured for 1 week to measure the characteristic.

2.3 Property representation

The Shao type hardness was measured by using the Shao type hardness tester (Misker Co., Ltd, China), according to standard GB/T 531-1999, the tensile strength, tearing strength and elongation at break were measured by an universal mechanical tester (Sans Co., Ltd, China), according to standard GB/T 528-1998 and standard GB/T 14833-1993.

3 RESULTS AND ANALYSIS

3.1 The effect of N_{NCO}/N_{OH} on the mechanical properties of PU track

For a clear research with respect to the effect N_{NCO}/N_{OH} ratio on the mechanical properties of materials, the content of microparticles SiO_2 was fixed at 35%, the properties of track with different N_{NCO}/N_{OH} ratio in prepolymer are shown in Figure 1. As shown in Figure 1a, with the increase of N_{NCO}/N_{OH}, the tensile strength increases slightly when N_{NCO}/N_{OH} below 1.8 and then increases rapidly, it can be explained that when the value of N_{NCO}/N_{OH} is low, there are a great number of soft segments in the molecular chain, which shows obvious soft character, bigger N_{NCO}/N_{OH} corresponds to the bigger content of hard segments in molecule chain, which should result in a high strength. According to the same reason, the elongation at break increases significantly. Meanwhile, we can see the Figure 1b, the hardness of materials increases slightly, and the compression recovery rate increases first and decreases after the N_{NCO}/N_{OH} higher than 1.5. This is attributed to the harder molecular structure of MDI with benzene ring. Moreover, from the Figure 1c, it was found that the rebound value of

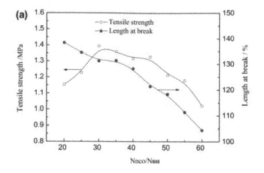

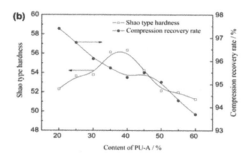

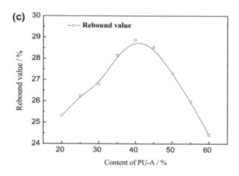

Figure 1. The effect of NNCO/NOH on mechanical characters of PU track. (a): tensile strength; (b): Shao type hardness; (c): rebound value.

materials also increases first and then decreases. Comprehensively, incorporating the mechanical properties of materials and the need of athletic running track, the optimal N_{NCO}/N_{OH} ratio of raw materials is 1.5~1.8.

3.2 The effect of kind and content of microparticles on the mechanical properties of PU track

The effects of adding different kinds of microparticles on the mechanical properties of PU track

340

are shown in Figure 2. In this research, the value of N_{NCO}/N_{OH} is fixed at 1.8. As shown in Figure 2, the trend related to the effect of both micro-sized $CaCO_3$ and SiO_2 particles on the properties of PU track is similar. With the increase of inorganic particles, the tensile strength, compression recovery rate and rebound value of materials first increase and then decrease.

Obviously, the content of microparticles in materials exist an optimal value. The existence of inorganic particles is in favor of the improvement of the elasticity due to the better dispersion of stress. However, the excessive content of inorganic particles should make the materials become

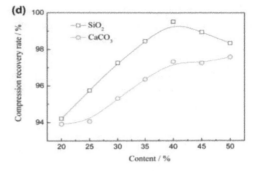

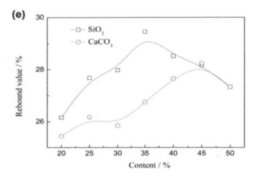

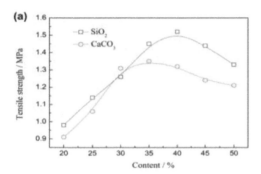

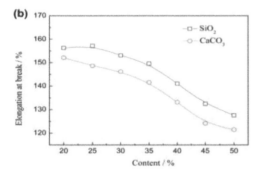

Figure 2. The effect of content of micro SiO_2 and $CaCO_3$ particles on mechanical characters of PU track. (a): tensile strength; (b): elongation at break; (c): Shao type hardness. (d): compression recovery rate; (e): rebound value.

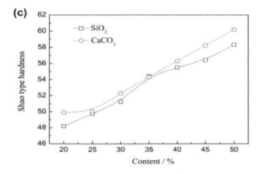

Figure 2. (*Continued*)

harder, which should affect the elasticity of track significantly. For the same reason, the hardness of materials increases drastically and the elongation at break decreases. Comparing the modification effect of $CaCO_3$ and SiO_2, it was found that the materials with SiO_2 show a better property for every index. The best content of SiO_2 particles in PU track is 35%.

4 CONCLUSION

In this paper, the polyurethane track was prepared by a reaction between polyether polyol and diphenyl methane diisocyanate, inorganic micro-particles $CaCO_3$ and SiO_2 were filled into the system as filler. The effect of different ratios of materials, the kinds and mass of microparticles on the characters of polyurethane track were investigated. The results showed that the tensile strength, Shao type hardness, elongation at the break all show a first rise then fall trend with the increase of N_{NCO}/N_{OH}, the compression recovery rate and rebound value decreases. The optimal value of

N_{NCO}/N_{OH} is 1.5~1.8. Filling microparticles into the polyurethane could improve the mechanical properties of the materials. Micro-sized SiO_2 is better filler than $CaCO_3$, and the appropriate mass fraction of SiO_2 was 35% of polyurethane raw materials. Comprehensively, comparing to the ideal track, the prepared materials with this formula is suitable to prepare the athletic running track.

ACKNOWLEDGMENT

The authors acknowledge the financial supports of the National Natural Science Foundation (21104017), Hunan Province Natural Science Foundation (2015 JJ4021).

REFERENCES

Liu, H. 2007. Analysis of the Replacement of MDI and TDI, Polyurethane, 11: 16–18.
Lu, H.F. 2004. Technical Development in the Runway, The Industry of Polyurethane, 4: 34–36.
Tang, W.W. & Bao, J.J. 2009. Technological Development in Synthetic Rubber Sports Track. Polyurethane, 9: 74–79.

Advanced Materials and Structural Engineering – Hu (Ed.)
© *2016 Taylor & Francis Group, London, ISBN 978-1-138-02786-2*

Recognition of traffic signs using local feature multi-layer classification

Y. Ye & X. Hao
School of Electronic and Information Engineering, Beijing Jiaotong University, Beijing, China

ABSTRACT: At present, the traffic sign recognition algorithms classify the signs mainly by extracting the features of the whole traffic signs. And there are many types of traffic signs and they are vulnerable to an external environment, which causing problems with low recognition accuracy and classification rate. Under this circumstance, the recognition of traffic signs using a local feature multi-layer classification method is proposed in this work. The method consists of multi-layer classification. First, the first-layer classifies the traffic signs into four categories including circle, triangle, diamond, and inverted triangle. Second, the algorithm classifies the circular ones into prohibition signs, mandatory signs, derestriction signs and others. Finally, the internal components of circular and triangular traffic sign are extracted and also subdivided in the use of a similar template matching method. This method makes each classification no longer require a lot of features and samples. The whole recognition algorithm is implemented for classification of 2601 traffic signs images, which are selected from the GTSRB database and its recognition accuracy reaches 98.3% higher than other three ways. The experimental results indicate that the proposed algorithm is robust, effective and accurate.

1   INTRODUCTION

How to detect and identify traffic signs correctly from the traffic video images have become an unresolved problem and also a hot and difficult research direction at home and abroad in recent years. Traffic sign recognition technology can be mainly divided into two methods: classification method based on template matching algorithm and classification method based on machine learning algorithms. The classification method based on machine learning algorithms takes advantage of neural network (Ciresan et al. 2011, Sermanet & LeCun 2011) and gets corresponding weights through an iterative algorithm to finish the classification. This method also takes the advantages of extracting features like HOG features, LBP features, Gabor features and haar feature and training classifier (Zaklouta et al. 2011, Maldonado et al. 2010) like SVM, random forest, Adaboost, and so on to complete the classification. The classification method based on template matching algorithms (Kus et al. 2008, Qin & Zhang 2014) have SURF, SIFT, and other technologies. Corner points or some certain feature points to establish descriptors for feature points matching, there are some matching of two images using the square difference, correlation coefficient, etc. Most of the above methods begin to classify right after extracting features of the entire traffic signs. However, the existing of similar traffic signs is bound to increase the number of required feature dimensions and, therefore require more training samples to guarantee the classifiers' performance. However, the degree of difference between classes can get significantly improved while the number of required feature dimensions can also be reduced if traffic signs can be classified into several sorts of extracting features.

2   THE DESCRIPTION OF CLASSIFICATION

Traffic signs have distinctive colors, standard geometrical shapes and different internal components. Taking these factors into account, a multi-layer classification method based on local features is presented in this work. The flow chart of the classification algorithm is showed in Figure 1. This algorithm mainly has two steps.

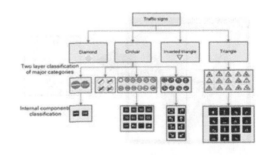

Figure 1.   The flow charts of classification algorithm.

2.1 Two layer classification of major categories

In the first-layer category, traffic signs are classified into four categories including circle, triangle, diamond, and inverted triangle based on local shape features and expert committee determination laws. And the circular signs can be classified into prohibition signs, derestriction signs, blue mandatory signs and other types of circular signs (STOP (octagon) sign and no access sign); in the second-layer category, the algorithm classifies the circular signs into prohibition signs, mandatory signs, derestriction signs and others based on colors and textures combined with expert committees determination laws.

2.2 Internal components classification

After two-layer classification, we only need to classify the circular and triangle traffic signs. Among these signs, the derestriction signs are identified by local texture features. While MSER algorithm is utilized to extract inner components of signs including triangular prohibition signs, prohibition signs, mandatory signs and other signs, and use of similar template matching algorithm to find the best matching component in the class, then the sub-categories will be finished.

3 TWO-LAYER CLASSIFICATION OF MAJOR CATEGORIES

3.1 The judgment of expert committee

Due to the blocking, blur and uneven illumination of traffic signs, traffic sign's local features are affected to some extent. Also, it may lead to a misjudgment on the types of traffic signs when we extract entire features of traffic signs. In this case, we adopt judgment from an expert committee. First, the panel of experts determines the type of traffic signs based on local features. Then according to the conclusion of the panel experts, the expert committee will take a final decision. And apparently the classification accuracy will be increased in this way. As shown in Figure 2, the traffic signs

are divided into several parts. Features of each part are extracted respectively and then specific classifiers are trained based on those features. Each classifier is considered as one expert of the expert panel. And the judgment of each figure part needs a panel of experts. Assuming that each expert can just judge two types and the expert committee will need the votes and scores after expert panels make their decisions. The flow chart of the judgment of an expert committee is shown in Figure 3.

The criteria of expert committee:

1. If the highest number of votes in category reaches certain credibility (credibility can be obtained by dividing the actual vote numbers by the maximum number of available votes), the expert committee will classify the sign into this type. The judgment process is over.
2. If the votes for each category were lower credibility, the expert committee functions would be used to judge after removing types of which credibility is lower than the threshold.

In this work, through extracting local features of traffic signs, we adopt one versus one way to train SVM classifiers in the use of LIBSVM. Every two-class classifier can be viewed as an expert. The Equation 1 is SVM discriminant.

$$f(x) = w^T x + b = \sum_{k=1} a_k(x, x_k) + b \qquad (1)$$

where $w = \sum_{k=1}^{l} a_k x_k$, b is the bias, $x_k : k \in \{1, 2 ..., l\}$ is the support vector, l is the number of support vectors, x is the eigenvectors, and a_k is the coefficient.

Calculate each class frequency of expert panels' votes $w = (w_1, w_2, ..., w_c)$. According to the vote of expert panels, the expert committee will output types with high credibility directly, remove types with low credibility and judge the rest again. And expert committee Equation 2 is used to obtain the largest value of uncertain type as the output.

$$F_{score} = \alpha F_A + \beta F_B + \gamma F_c \qquad (2)$$

where F_{score} is the score of uncertain type, F_A is the credibility of this type, F_B is the weight of results

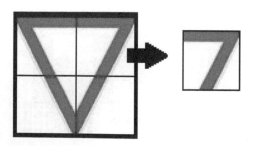

Figure 2.   Local feature.

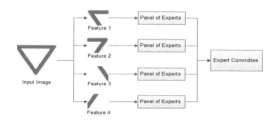

Figure 3.   The judgment of an expert committee.

of each expert panel voting for every type, F_C is the expert judgment score (Equation 1 obtained score) as feature, and train the classifier and use it to obtain the number of votes, $\alpha + \beta + \gamma = 1$.

3.2 Two-layer classification

First, according to the shape features of 43 kinds of traffic signs in the GTSRB database, as shown in Figure 3 obtained each image block of the traffic sign as the local features. In the first-layer classification, divided the traffic signs into circular, diamond, triangle and inverted triangle by the committee of experts and the local shape features (HOG). And the circular signs include prohibition signs, derestriction signs, blue mandatory signs and other types of circular signs (STOP (octagon) sign and no access sign). In the second-layer classification, using local texture (LBP) of traffic signs and the expert committee to divide the derestriction signs into four types. Then, using the color characteristics of traffic signs and computing the proportion of blue or red in the comparing area to determine whether blue or red traffic signs. Finally, separation of other type traffic signs by judging the scope of the internal non-red region of the red traffic signs comparison region. Now we have four types of circular traffic signs and triangle, inverted triangle and diamond traffic signs.

4 INTERNAL COMPONENTS CLASSIFICATION

4.1 Stable region extraction

MSER extraction algorithm includes extremal region detection and the maximally stable extremal region determination. The specific extraction process includes three steps that are image pixels sorting, extremal region detection and the maximally stable extremal region determination. It takes threshold of gray-scale images from 0 to 255, sets the point larger than the threshold value as 1 and the one smaller as 0 and gets 255 binary images after a threshold. A 255-dimensional sequence is produced which each unit stores the pixels with the same gray level and coordinates in the image. The detection algorithm calculates the extremal region on every threshold image in order according to the threshold value's gray level from small to large order. This process finally generates a tree data structure called a component tree. Each layer of the component tree corresponds to a threshold image and each node of the layer represents an extremal region of the corresponding threshold image. $Q_1, \ldots Q_{i-1}, Q_i, \ldots$ the nested extremal region, such as. $Q_{i-1} \subset Q_i$.

If $q_i = |Q_i + \Delta \backslash Q_{i-\Delta}|/|Q_i|$ exists local minimum in i^*, extremal region Q_{i^*} the maximally stable region

(where |.| represents pixel image area Δ the method parameters). The traffic sign internal components have good stability. Internal components are obtained by using MSER algorithm.

4.2 Internal components registration

Internal components of triangle prohibition sign are small and it needs a registration of traffic sign due to lean or the influence in classification task that the shooting angle causes the rotating in the pictures. First, choose the bottom area of the triangle sign by enhancing the color of an image. Then, obtain contour information by binarization and use a weighted least squares algorithm for fitting a straight line by iteration and obtaining the $f_{fit_{line}}(x)$ of Equation 3. Finally, use Equation 4 to calculate the deflection angle based on the straight line fitted, extract the stable region of traffic sign by MSER algorithm and use the bilinear interpolation to rotate the image. Thus it meets the goal of registration. The registration process is as shown in Figure 4.

$$f_{fit_{line}}(x) = \min \sum_i p(r_i), \; p(r) = 2 * \left(\sqrt{1 + \frac{r^2}{2}} - 1 \right) \quad (3)$$

where r_i represents the distance between the line and the i-th point.

$$\theta = \arctan\left(\frac{dy}{dx}\right) \quad (4)$$

where dx, dy are a unit vector in the x-direction and y-direction, and θ is the deflection angle.

4.3 The similar template matching method

The similar template matching method composed of two parts. In the first part, an internal standard components image is used as training samples. Extract its features training classifier to classify internal components. In the second part, the template matching for internal components is done. The template matching uses the window with the same template and size to slide on the internal component region get. According to Equations 5, 6, and 7, calculate the similarity value of each window.

$$I_{img} = I_{win} - I_{win_{mean}} \quad (5)$$

Figure 4. The registration process.

where I_{win} is the current sliding window. $I_{win_{mean}}$ is the mean of the current sliding window.

$$C = \sum_{i=1,j=1}^{i=M,j=N} I_{img(i,j)} {}^* T_{(i,j)} \qquad (6)$$

where C is the correlation coefficient between window image and component. T is the component image. M and N are the width and height of the window.

$$w = \frac{\alpha C}{\sum_{i=1,j=1}^{i=M,j=N} I_{img}(i,j) {}^* I_{img}(i,j)} \\ - \beta \frac{(R - R_T)^2}{R_T} - \gamma (P - P_T)^2 \qquad (7)$$

where w is the similarity of the current window. The larger the value is, the higher the matching degree is. M and N are the width and height of the window. R is the aspect ratio of the internal components. R_T is the aspect ratio of a template. P is the center of the current window position. P_T is the center position of the template image in the Figure. $\alpha + \beta + \gamma = 1$ ($\beta = 0$ If it is circular sign).

4.4 Internal components classification

The internal components classification not only can reduce background and traffic signs edges interference effectively but also can increase the difference between the categories. Also, traffic signs internal components are more standard and do not require a lot of training samples. After the classification of main categories, according to internal icon features, the circular traffic sign is divided into blue mandatory signs, red prohibition sign and other signs. The traffic signs internal components are extracted by combining S channels in HSV color space with the gray-scale image after brightness adjusted and MSER algorithm. For circular traffic sign internal components, first use the first part of the similar template matching method to judge the category of the circular internal components. Because the speed limit sign exists similar internal components, it needs the template matching part of the similar template matching method to do the second judge of the similar number (such as speed limit signs 30 and 80 need template matching to judge 3 and 8). To derestriction sign, it needs to classify based on the local texture features in the top left and bottom right corner. Because of triangular traffic sign, the sign internal components are small and in the extraction process it may lose some information, the second part of the similar template matching method is used to match for smaller internal components. Extract its internal components after internal components registration. Reduce the matching region by cutting it out and use the template matching to judge the category of a traffic sign.

4.5 Performance comparison

As we can see in Table 2, under conditions in same total number of dimensions, the correct number in using SVM one versus one method to do the shape classification (divided into circle, triangle, inverted triangle, and diamond) is less than the number of expert committee judged act, in which 128-dimensional feature of each team of experts is totally 512 dimension.

As it is shown in Table 3, the classification results obtained by four methods are close but the correct rate of the algorithm proposed is slightly higher than the other three comparison methods. In the CNN method, we normalize the image to 40*40 and calculate the gray scale of the images which are set as the input image. Convolution neural network

Table 1. The correct rate of the proposed method for various types of traffic signs.

Sign type	Number of sign	Number of correct	Correct rate
Circular prohibition signs	801	772	96.37%
Blue mandatory signs	539	532	98.7%
Triangle prohibition signs	761	753	98.54%
Derestriction signs	30	30	100%
Other circular signs	87	87	100%
Diamond signs	168	168	100%
Inverted triangle signs	225	225	100%
All traffic signs	2601	2557	98.3%

Figure 5. Some of the test data.

Table 2. Shape classification method comparison.

Methods	Number of signs	Number of correct	Feature dimension
One versus one	12630	12579	512
Expert committee	12630	12618	4*128

Table 3. Comparison of different methods (43 categories classification of 2601 tested images).

Methods	Correct rate
CNN	96.71%
HOG+SVM	96.96%
HOG+Random Forest	97.88%
Algorithm Proposed	98.3%

uses two convolutions, two down sampling layers and a link layer and obtains the weights for classification by iterative training the gray value of the input grayscale image. It can have a 96.29% classification correct rate in the 12630 test images of GTSRB database when using this CNN structure. And for the chosen data gets a 96.71% correct rate. In the HOG+SVM method, HOG feature uses HOG-2 feature (1568 dimensions) provided by the GTSRB database and uses LIBSVM to do multi classification training of one versus one method, with a 96.96% correct rate. In the HOG+Random Forest method, it uses the same HOG feature as SVM method training 500 trees to determine the type of traffic signs and gets a 97.88% correct rate. The algorithm proposed getting a 98.3% correct rate in all classification.

5 CONCLUSIONS

A traffic sign recognition using local feature multi-layer classification method is proposed in this paper. With traffic signs features and expert committee judged action to classify the traffic signs into several categories including circle, triangle, diamond and inverted triangle and classify circle traffic sign again by the local feature of shape, color and texture. Extract internal components of traffic signs with the same category and do subdivisions in the method of similar template matching for the feature of internal components. The method proposed having a higher recognition accuracy at the same time narrowing the characteristics of classification desired. In addition, it reduces the training time. The results demonstrate the effectiveness of the algorithm.

REFERENCES

Ciresan, D. Meier, U. & Masci, J. et al. 2011. *A committee of neural networks for traffic sign classification, C. Neural Networks (IJCNN)*, The 2011 International Joint Conference on. IEEE, 1918–1921.

http://benchmark.ini.rub.de/?section = gtsrb&subsection = news.

Kus, M.C. Gokmen, M. & Etaner-Uyar, S. 2008. *Traffic sign recognition using Scale Invariant Feature Transform and color classification*, Comput. Inform. Sci. ISCIS. 23rd Int. Symp. IEEE, 1–6.

Maldonado, B.S. Acevedo, R.J. & Lafuente, A.S. et al. 2010. An optimization on pictogram identification for the road-sign recognition task using SVMs, *Computer Vision and Image Understanding*, 114 (3): 373–383.

Qin, J. & Zhang, X.F. 2014. Hierarchical traffic sign recognition system based on improved shape context, *J. Comput. Eng. Des.* 1: 183–187.

Sermanet, P. & LeCun, Y. 2011. *Traffic sign recognition with multi-scale convolutional networks, C. Neural Networks (IJCNN)*, The 2011 International Joint Conference on. IEEE, 2809–2813.

Zaklouta, F. Stanciulescu, B. & Hamdoun, O. 2011. *Traffic sign classification using kd trees and random forests, C. Neural Networks (IJCNN)*, The 2011 International Joint Conference on. IEEE, 2151–2155.

Detection of traffic signs based on multi-feature high credibility regions

Y. Ye & X. Hao
School of Electronic and Information Engineering, Beijing Jiaotong University, Beijing, China

ABSTRACT: At present, for the traffic sign area detected the presence of an offset, missing and so on, the traffic sign detection algorithm will reduce the effect of detecting and recognizing. To let the circumstances above not to affect the credibility of the traffic sign area, the traffic sign detection algorithm based on multi-feature high confidence regions is proposed. This algorithm utilizes the characteristic of the traffic sign in shape, grain and color to find the traffic sign area, and uses the credibility function through an adaptive iterative adjustment to get the optimal traffic sign area in the image so that it can eliminate the effect in the presence of an offset and a missing, and improve the getting traffic sign area extremely. The result shows that the recall of the algorithm proposed got in GTSDB and STSDB is 97.1% and 98.5%, which exceeds most traffic sign detection algorithms at present. The average coverage of the algorithm exceeds 94%, which face up to robust demand.

1 INTRODUCTION

The traffic sign detection and recognition in traffic scene are one of the key research areas of intelligent transportation systems, which have significant theoretical value and broad market prospects. The traffic sign detection at home and abroad mainly contains methods based on color and shape. At abroad, it chooses the maximum between R and B channels from the normalized RGB space, combines with the RGB image to extract MSER region and uses SVM to the traffic sign judgment in reference (Greenhalgh & Mirmehdi 2012), which has a better real-time but a not very high detection rate because it is limited to the color space and the choose of threshold. It is recognized in reference (Kim 2013) that color and shape are easy to be affected by the circumstance around. Moreover, increasing the visual saliency model to detect traffic sign will have a high real-time performance, but the detection rate will be impacted easily in complex circumstance. In reference (Malik et al. 2007), it exchanges RGB image to HSV color space and carries on the fuzzy shape detection after filtering the color to determine whether or not a traffic sign. In reference (Loy & Barnes 2004), it strikes gradient on gray image and edge-based approaches to detect traffic signs, which can overcome the effects of illumination to some extent, but the presence blocking traffic sign will impact its detection rate. Interiorly, the reference (Wang et al. 2014) determines whether it is a traffic sign against detecting the red area of traffic signs by using a color filter and extracting an area that is without red internal

components. It gets a better detection result. In reference (Chen et al. 2007), it detects triangles and circles mark by using a vector filter to divide the region of interest and the inflection point, geometric features and symmetry. However, the methods above miss the coverage of the areas detected. A detection area with high coverage will get a good test result. The usual traffic sign detection algorithm detects the traffic sign area correctly as long as it gets on subsequent classification task, which will reduce the rate of detection and recognition of traffic signs. If these detected areas are selected in a high credibility, it will improve the detection rate and the recognition rate of traffic signs to some extent.

2 MULTI-FEATURE HIGH CREDIBILITY REGIONS DETECTION

Traffic sign has distinctive color information, specific shape, and similar texture. The block diagram of this algorithm, the traffic sign detection algorithm based on multi-feature high credibility regions proposed by the paper, is shown in Figure 1. It is divided into the following two steps:

2.1 *Pretreatment*

The Equations 1 and 2 are used to enhance the input color image respectively. Then it uses MSER (Maximally Stable Extremal Regions) algorithm to take thresholds from 0 to 255, setting the point greater than the threshold to 1 and the point less

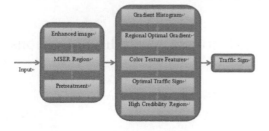

Figure 1. The block diagrams of the algorithm.

than the threshold to 0, to get the binary image after 255 thresholds and gets the relationship between area and the threshold changes by comparing the image area adjacent to the threshold. The chosen area when the changes in the area with the changes of the threshold less than the area detected above a certain threshold are the maximally stable extremal regions, in which the traffic signs are. It will produce two different extreme areas because the threshold can change in the two opposite directions. The traffic signs have good aspect ratio and size. The stable region covers a variety of shapes of objects such as vehicles, billboards, trees, and windows. It can significantly reduce the number of regions by the single-dimensional characteristics of traffic signs.

$$\Omega_{RB} = \max\left(\frac{R}{R+G+B}, \frac{B}{R+G+B}\right) \qquad (1)$$

$$S_i = (X_{avg} - X_i)^2 + (Y_{avg} - Y_i)^2 + (Z_{avg} - Z_i)^2 \qquad (2)$$

where S_i is the significance of its i-th pixel. X_i, Y_i, Z_i are the three-channel image in XYZ color space. $X_{avg}, Y_{avg}, Z_{avg}$ are the mean of the three-channel image.

2.2 High confidence region extraction

First, the low credibility region is filtered out based on the simple features such as length and width of traffic signs. Second, use the gradient histograms and linear SVM classifier on the shape detection of traffic signs and get the optimal gradient direction histogram of the same region. Third, determine whether it is the traffic signs with high credibility by extracting color and texture features from the image. Finally, do the search based on credibility function for the traffic sign area and get the optimal traffic sign area in the image.

3 GET HIGH CREDIBILITY REGIONS

3.1 The detection of the shape of the traffic sign

The traffic sign has good shape characteristics. The shape of the traffic sign can be well-depicted

by HOG. Calculated manner similar to the HOG features is adopted to calculate the local histogram of the image. The image is normalized to the size of 40*40 that divided into 16 sub-images with the size of 10*10. Meanwhile, it will calculate HOG of each sub-image with 64-dimensional feature (the gradient directions equally divided to 0°, 45°, 90°, 135° four directions here in order to do histogram statistics). Linear SVM (Support Vector Machines) classifier will take the shape judge. The reasons for that are: according to the VC dimension theory, the larger the given numbers of samples, the more likely correct the study result, at this moment the smaller the confidence risk. For the VC dimension of classification function, the larger the VC dimension, the worse the generalization obviously and meanwhile the confidence risk will get greater. Increasing the number of samples and reduce the VC dimension can decrease the confidence risk. It can speed up the computation of its characteristic if the feature dimension of the image is reduced to 64 dimensions. The shape of the traffic signs is fairly standard, which does not require too much characteristic to describe. The excellent performance of SVM is from maximal margin classification theory based on VC dimension and structural risk minimization principle. The Equation 3 is SVM discriminant.

$$f(x) = w^T x + b = \sum_{k=1}^{l} a_k(x, x_k) + b \qquad (3)$$

where $w = \sum_{k=1}^{l} a_k x_k$, b is the bias, $x_k : k \in \{1, 2 ..., l\}$ is the support vector, l is the number of support vectors, x is the eigenvectors, and a_k is the coefficient.

3.2 The fuzzy symmetry of the traffic signs shape

Fuzzy entropy is usually used to measure the fuzzy degree of fuzzy sets, which have a wide range of applications in pattern recognition, language analysis and many other areas. The fuzzy entropy index is used to indicate the fuzzy degree of the traffic signs so that the image distribution of the region is depicted. Therefore, the fuzzy symmetry can describe the symmetry within a target area around two sub-regions well. The equations as follows:

$$\begin{cases} F = -\frac{1}{MN} \sum_{i=1}^{M} \sum_{j=1}^{N} \left[\begin{array}{c} u(x_{ij}) \, lgu(x_{ij}) \\ +(1-u(x_{ij}))lg(1-u(x_j)) \end{array} \right] \\ S = \frac{|F_L - F_R|}{\max(F_L, F_R)} \end{cases} \qquad (4)$$

where F is the fuzzy entropy. $u(x_{ij})$ is the membership degree of x_{ij}. F_L and F_R are the fuzzy entropy of left and right sub-region. S is the Symmetry of the fuzzy entropy. When an image is completely symmetrical, its distribution strictly obeys symmetry. So, the symmetry value of the fuzzy entropy is small that tends to 0. On the contrary, the value tends to 1.

3.3 The texture symmetry of the traffic sign

LBP is an operator used to describe the image local texture features with significant advantages of rotational invariance and brightness invariance. The basic idea of LBP is to add the result of the comparison between the pixels of the image and its local surrounding pixels. Set this pixel as a center and compare the threshold value between the adjacent pixels. If the luminance of the center pixel is no less than its adjacent pixel, set it as 1 otherwise 0. In 1962, moment invariant theory and moment invariants based on algebraic invariants are proposed by Hu et al. A group of linearly independent geometric moment group builds up a nonlinear moment group, by which export a group of moments with translation, rotation and scale invariance called moment invariants. However, among 7 moment invariants, the first two are more stable. So take the first two into consideration. The texture of the left and right sub-region of the traffic sign is very close. Use LBP to extract textural properties of the traffic signs and use Hu moments to determine the symmetry of traffic signs around the sub-region.

$$TZ = (M1_L - M1_R)^2 + (M2_L - M2_R)^2 \qquad (5)$$
$$(IF\ TZ>1=1)$$

where Mi_L, Mi_R represent the i-th moment invariants of the left and right sub-region of Hu moments, where subscript L represents the left sub-region and R represents the right. Set the TZ value larger than 1 as 1. When image is completely symmetrical, its distribution strictly obeys symmetry. So, the texture similarity value of sub-region is small that tends to 0. On the contrary, the value tends to 1.

3.4 The proportion of the traffic signs color

The characteristic color can be distinguished between traffic signs and other objects. The traffic sign is basically divided into red and blue. According to different shapes, P is used to represent the possession of image proportion of the characteristic color. The possession of proportion of the traffic signs color has a better symmetry, which can keep a good proportion though it is rotated. According to the passage, it turns left on the image 45° and turns right 45°, calculate the share of the left and right sub-region characteristic color on the image

after rotating and non-rotating and the left and right sub-region proportion of symmetry S_c.

$$S_c = \frac{|P_L - P_R|}{\max(P_L, P_R)} \qquad (6)$$

3.5 High credibility area

An area with high credibility means the traffic sign of the traffic sign area got has a good sign and profile, in which there is no shelter, uneven illumination, position offset, blur, and so on. Credibility is used to indicate the condition of traffic sign. The higher the credibility, the easier will traffic signs identify. However, tilt, occlusion, fuzzy, positional deviation, etc., will reduce the credibility of traffic signs. Because MSER is binarization processing on multi-level threshold, it will cause the phenomenon of overlapping in the stable area and traffic sign area that can get pieces which conform to the rules of shape, color and texture determination near the same traffic sign. So it needs to choose the area with higher credibility in the same traffic sign area, as the initial position to find the optimal traffic sign.

In order to simplify the calculation, the shape feature is used as a determination of basis. A best area is chosen from MSER overlapping areas. In the selection process, the area when SVM value is maximum near the same traffic sign area is chosen based on the value calculated by SVM discriminant (based on Equation 3, is SVM judging value). Non-traffic sign area with the similar shape will be removed via the nature of color and texture so that it will get a high relative credibility traffic sign area. It is recognized that the high relative credibility traffic sign area is the convinced traffic sign area. The high relative credibility traffic sign area is not usually the optimal area in the image, which often exist the situation of position offset and deletions. In order to eliminate this situation, the credibility function is used to search the region with the highest credibility in the image iteration based on the location Information of the high relative credibility traffic sign area got. The credibility function:

$$f(x) = \alpha V_{svm} - \beta V_{Fentropy} + \gamma S_c \qquad (7)$$

where V_{svm} the value after SVM judging (the training of SVM classifiers use the high credible image as the training samples). $V_{Fentropy}$ is the fuzzy symmetry. S_c is the symmetry degree of traffic signs characteristic color. α, β, γ are the weights, $\alpha + \beta + \gamma = 1$. Because the location information of the shape is dominant which give α higher weight value and $V_{Fentropy}$ and S_c, the adjustable parameters, is the reference of judging the credibility of its location,

β and γ are given a the lower weight value. In experiment, it chooses $\alpha = 0.6, \beta = 0.2, \gamma = 0.2$.

Set the $f(x)$ of the high relative credibility traffic sign area as the initial value. Gradually increase and decrease this area and find the relative maximum of $f(x)$ in this process. After getting the relative maximum, find the best possible position by midpoint trimming. Get the optimal area in the image through multiple adaptive. The algorithm flowchart of getting the high credibility area is shown in Figure 2. The schematic algorithm of searching the highest credibility area is shown in Figure 3, in which θ^t is the initial value of moment t and θ^{t+1} is the initial value of moment $t+1$. When it is at the same time as moment t, the maximum is achievable after optimization by multiple iterations.

4 THE EXPERIMENTAL RESULTS AND ANALYSIS

4.1 *Method validation*

In order to verify the effectiveness and feasibility of the method, in this paper, the experiment takes utilization of MATLAB and OPENCV2.0 in Windows7 system for algorithm simulation. The experimental data are from the GTSDB Database, in which 600 pieces are used for training and 300 pieces are for image testing and 83 pieces for testing in Validation SET. Figure 4 is the process of using the algorithm in the paper to do the traffic sign detection for a traffic scene image. The Figure 4(a) is the original image. Figure 4(b) is the stable area got by MSER in the enhanced image. Figure 4(c) is the high relative credibility traffic sign area.

After Figure 5(a) gets convinced traffic sign area, it does the iterative calculation of the optimal traffic sign area in the image based on reliability function.

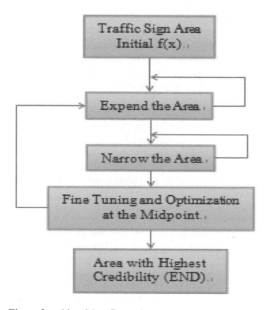

Figure 2. Algorithm flows charts.

Figure 4. Detection process.

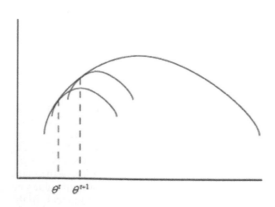

Figure 3. Schematic.

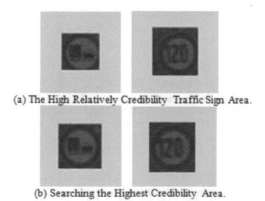

(a) The High Relatively Credibility Traffic Sign Area.

(b) Searching the Highest Credibility Area.

Figure 5. Comparisons between before and after search.

A traffic sign image with a better location seen in Figure 5(b) compared with (a) which is closer to area to be recognized calculated by traditional traffic sign detection algorithm and has a lower credibility than (b). Figure 6(a–d) are the results of different scenes.

4.2 *Performance comparison*

Figure 7 is Precise-Recall image of four kinds of an algorithm. Figure 7(a) is the detection result of using MSER algorithm and 1764 dimension HOG feature training linear SVM to judge. Figure 7(b) is the detection result of the method in the paper.

Figure 6.   Part of the image after processing.

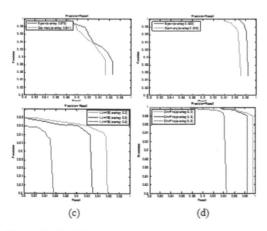

Figure 7.   Precision-Recalls.

In Figure 7(c), the method is that extract HAAR-LIKE feature in HSI color space by Viola-Jones' cascade classifier. What Figure 7(d) is using an integral channel feature with the method of cascade classifier. It can be seen from Figure 7(a) and (b) that the method in the paper can reduce the error number of the sample more effectively compared with the method that only use HOG feature and SVM judgment.

The traffic sign area coverage got by finding high credibility traffic sign area will exceed (a). The average coverage of Figure 7(a) is 85.3% and Figure 7(b) is 94.6%. It can be seen from the method (c) and method (d) that single HAAR-LIKE feature cannot get very good effect while (d) use the method of multi-feature fusion to do cascade detection, which gets a good effect. However, due to the extraction calculation of multi-feature, it is required to sacrifice the running speed and the better the area got (higher coverage rate), the lower the detection rate is. Figure 7(c) and (d) are the detection rate got when the coverage is 20%, 30%, and 50%. Though the detection rate of the algorithm in this paper slightly lower than that of (d), the average coverage got by the method in this paper is much higher.

5   CONCLUSIONS

A traffic sign detection algorithm based on multi-feature high credibility regions is proposed in this paper. It obtains relatively optimal area according to shape, color and texture features, and uses the credibility function to get globally optimal area, outperform the case that the area to be recognized by the usual traffic sign detection algorithm, which has positioned offset or deletion. It obtains the traffic sign area with a high detection rate and a low false detection rate. The detection rate is not highest when compare with other methods, but, its average coverage rate (over 94%) is highest and the detection rate also not bad. The experimental results demonstrate effectiveness of the algorithm.

REFERENCES

Chen, W. Li, C. & Wang, Z. 2007. Road Traffic Sign Detection Using Color and Shape, *Journal of Xiamen University (Natural Science),* 46(5): 640.

Greenhalgh, J. & Mirmehdi, M. 2012. Real-time detection and recognition of road traffic signs, *Intelligent Transportation Systems, IEEE Transactions on.* 13(4): 1498–1506.

http://benchmark.ini.rub.de/?section = gtsdb&subsection = news.

http://agamenon.tsc.uah.es/Segmentation/.

Kim, J.B. 2013. Detection of traffic signs based on eigen-color model and saliency model in driver assistance systems, *International Journal of Automotive Technology*. 14(3): 429–439.

Loy, G. & Barnes, N. 2004. Fast shape-based road sign detection for a driver assistance system, *Journal of Intelligent and Robotic Systems. RSJ Int. Conf. on. IEEE*. 1: 70–75.

Malik, R. Khurshid, J. & Ahmad, S.N. 2007. Road sign detection and recognition using colour segmentation, shape analysis and template matching, *Machine Learning and Cybernetics*, 6: 3556–3560.

Wang, G. Ren, G. & Jiang L. et al, 2014. Hole-based traffic sign detection method for traffic signs with red rim, *The Visual Computer*. 30(5): 539–551.

The role of cementitious materials in building materials

F.Q. Zhang
College of Material Science and Engineering, Shandong Jianzhu University, Jinan, China

ABSTRACT: As an important branch of materials science, the cementitious material is a theoretically scientific research in its composition, structure and properties, which is also known as the binder material. Under the physical and chemical effects, it can become into a sturdy stone-like buck, and bind with other materials, finally form composite solid substances with a certain mechanical strength. The cementitious materials science acts as a connecting link in building materials. Also, it is the foundation for further professional courses. Besides, time arrangement is the key factor to take full advantage of this course.

1 INTRODUCTION

The cementitious material is a branch of materials science. It refers to the theoretically scientific research in its composition, structure and properties. The cementitious material is also known as the binder material. Under the physical and chemical effects, it can become into a sturdy stone-like buck, and bind with other materials, finally form composite solid substances with a certain mechanical strength. The development of cementitious material has a long history. The first cementitious material is clay, which was used to build simple buildings. Then, the cement and other building materials have a lot to do with cementitious materials. The cementitious material is widely used in daily life due to its some excellent performances. With the development of cementitious materials science, cementitious materials and its relevant industry will have a new leap forward. Development of cementitious material is along with the development of humans' civilization, especially with the gradual deepening of China's socialist modernization. Various novel cementitious materials are increasingly being applied (Yue & Zhang 2013, Zhang & Wang 2014, Yuan 1996).

2 RESULTS AND DISCUSSION

The cementitious materials science was approval for undergraduate teaching in inorganic non-metallic materials, construction materials and products by colleges' inorganic non-metallic materials specialty teaching guidance committee. On basis of the Portland cement, cementitious materials science is based on Portland cement, describes the relationship between the composition of the material, structure and properties, which introduces the hydration of cementitious materials and its hardening mechanism, and the way to improve the structure and properties of the hardened body. Above all, it is a very important program for students who are major in materials science (Ma & Liu 2010).

But there are few universities with construction materials professional or inorganic nonmetallic materials professional have set up this course. It is apparently that the cementitious materials science plays a different role in universities or curriculum systems in colleges.

For author's better understanding of Universities in Shandong, it greatly varies in curriculum systems and study in the University of Jinan and Shandong Jianzhu University. Although these two universities have opened specialty inorganic nonmetallic materials. For example, the Inorganic Nonmetallic Material major is the feature in University of Jinan. It is established in its early years and has come a long way. The main career prospects of the graduates are the cement, ceramic and glass factories. Therefore, the professional courses are orientated very much towards the requirement of these companies. The main professional courses include Inorganic Materials Science Base, hydromechanics, Portland Industrial Thermal Engineering, Inorganic Non-metallic Materials Technology, etc. However, the school of material science and engineering is newly established in Shandong Jianzhu University. Considering the cultural accumulation and the hardware infrastructure of the School of Material Science in University of Jinan, there will be no competitive edge if Shandong Jianzhu University sets up the same specialty or training. Moreover, Shandong Jianzhu University is an architectural university that features Civil Engineering and Architecture. So, its students are mainly oriented towards architectural companies in its early years, rather than cement, ceramic and glass factories. Therefore, its main courses are Cementitious Materials Science, Housing Architecture, Inorganic Non-metallic Materials Technology, Building and Decoration Materials, Structural Building Materials, etc. Figure 1 is the distribution of student employment (Wang et al. 2011).

The Cementitious Materials Science mainly introduces Portland cement, explaining the relationship of composition, structure and properties between various types of cementitious material, introducing the hydration mechanism of cementitious material and the method to improve the structure and properties of hardened. As the professional foundation course of the building materials direction in Materials Science and Engineering in Shandong Jianzhu University, Cementitious Materials Science acts as an extension of the basics and groundwork for specialized knowledge. It lays a significant culture foundation for the following professional courses such as Introduction to Building Architecture, Structural Building Materials and Building Construction, etc. The Cementitious Materials Science in Shandong Jianzhu University is arranged to the first semester for third-year students. It is arranged simultaneously with Inorganic Materials Science Base and earlier than Materials Testing Methods settled in the next semester. With the author's experience in teaching, it would be best for Cementitious Materials Science to be arranged at the second semester of the third year, and Materials Testing Methods arranged at the first semester of the third year. This is because there are a lot of contents related to material performance testing such as SEM, XRD, etc. It would be difficult to understand the contents if the relevant courses are not taken. Meanwhile, Materials Testing Methods conflict with Inorganic Non-metallic Materials Technology in the second semester of the third year. Similarly, in the inorganic non-metallic materials technology

course which is also involved in a lot of aspects of the content of the test. Although Cementitious Materials Science can lay the foundation for the study of Inorganic Non-metallic Materials Technology, the curriculum can give rise to knowledge barriers for students. Therefore, it is strongly recommended that Cementitious Materials Science should come after Materials Testing Methods. All the theory courses are introduced after Materials Testing Methods in University of Jinan, which is pretty adoptable.

It is also found some problems to overcome or avoid for timely in the learning process of cementitious materials science as follows:

1. Adding teaching hours and experimental teaching. As an important basic course, there are fewer hours of cementitious materials science in the syllabus, only the most basic 32 hours, which is obviously not enough. And courses related experiments and experimental weeks are not arranged in students' training program, which makes students to learn about production processes hydration and hardening process of cementitious materials only in the paper. And in this way, it will not meet expectations for teaching effect only by abstract learning instead of hands-on understanding.

2. Outstanding students' protagonist position. In the past teaching, teachers were dominant, which means that everything was based on teaching, and students' enthusiasm and creativity was bounded. It also made it possible for students to lose interest in the course and the right attitude. After class, content in class would be analyzed and discussed by few students. Therefore, the ability to analyze and solve problems independently for students was very hard to be achieved effectively. The dominant position of students should be prominent in the process of teaching. Teachers only play inspiring, guiding and answering role in encouraging the students to find and solve problems. The student should have a basic grasp of course by a preview and the teacher should explain the key points for the students in the classroom. In the mean while students should raise their own concerns and difficulties for teachers to give proper guidance and some answers.

3. Increasing students' interests in learning by combining production. Teachers should try to create the conditions for students to be the actual construction site and to explain some major construction materials such as cement, steel and concrete, including the production, construction and maintenance procedures of them. In this way, it will make a deep impression for students by combing with the physical building materials, especially new building materials.

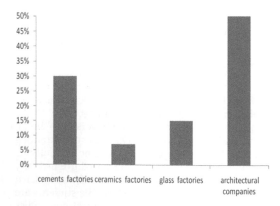

Figure 1.   The distribution of students employment.

This can avoid the simply boring caused by classroom and laboratory teaching and improve students' enthusiasm and creativity, which is beneficial for students to think independently. There would be an excessive number in a single group and could not provide hands-on opportunities for everyone because of the limited experimental apparatus and venues. There is often a serious gap between experimental and theoretical sectors and on-site teaching can overcome these shortcomings.

3 SUMMARY

As an important branch of materials science, the cementitious material is a theoretically scientific research in its composition, structure and properties, which is also known as the binder material. Under the physical and chemical effects, it can become into a sturdy stone-like buck, and bind with other materials, finally form composite solid substances with a certain mechanical strength. Cementitious materials science acts as a connecting link in building materials. Also, it is the foundation for further professional courses.

Besides, time arrangement is the key factor to take full advantage of this course.

ACKNOWLEDGMENTS

This work was supported by funding from National Natural Science Foundation of China (Grant Nos. 51272142).

REFERENCES

Ma, Q.S. & Liu, G.J. 2010. Experimental Technology and Management. *The New Campus*, 27(6): 150–152.
Wang, Y.L. Qi, X.W. Luo, S.H. & Liu, X.W. 2011. Experimental Technology and Management. *Education Forum*, 28(6): 300–302.
Yuan, R.Z. 1996. *Cementitious Materials Science*. Wu Han: Wuhan University of Technology Press.
Yue, X.T. & Zhang, F.Q. 2013. Discussion the importance of Cementitious materials science in building material students. *Education Forum*, 27: 100–101.
Zhang, F.Q. & Wang, J.W. 2014. Experimental teaching in physical properties of inorganic materials. *The New Campus*, 324(8): 57–58.

Advanced Materials and Structural Engineering – Hu (Ed.)
© *2016 Taylor & Francis Group, London, ISBN 978-1-138-02786-2*

Time-domain buffeting analysis of suspension bridge under non-stationary typhoon excitation

Y.L. Qiu, W.H. Guo & J.Q. Wang
School of Civil Engineering, Central South University, Changsha, China

ABSTRACT: To simulate normal smooth space wind spectrum as stationary incentive, the Harmonic synthesis method is used to the Fortran90 programming software, while the "dujuan" typhoon measured data are fitted as a non-stationary excitation. The large general finite element software is used to establish a 3-D model of Shidengmen suspension Bridge, and the time-domain analysis method based on numerical solution is chosen to analyze the suspension bridge buffeting under stationary or non-stationary excitations. The analysis results show that the main girder lateral and torsional displacement peak as well as RMS values will reach the maximum at the middle span under both stationary and non-stationary excitations, and its values will reduce extending to the bearing place. The vertical displacement values are bigger at the 1/8, 3/8 span, and the RMS value reaches the maximum at 1/4 span of the bridge. The whole bridge (vertical, lateral, and torsional) RMS of displacement values under the typhoon excitation is significantly higher than that under the normal wind excitation. The highest RMS ratio comes up to 1.47; while the displacement peak ratio reaches 2, which is detrimental to a bridge. The calculation results are consistent with the actual situation. And the feasibility and necessity of non-stationary calculation method for buffeting response of long-span bridges are verified.

1 INTRODUCTION

Research on buffeting of long-span bridge under a non-stationary excitation is still in its infancy. Nowadays, people mainly applied the time-varying dynamic control equations on modal decomposition, after which the virtual excitation was constructed by using the non-stationary fluctuating wind spectrum. Therefore, the modal equation in numerical solution can be obtained. (Ding 2001, Boonyapinyo et al. 1999). In addition, there are also scholars using simulated non-stationary fluctuating wind field to form the non-stationary buffeting force, then applied it to the bridge to analyze the effect of bridge structure buffeting (Liang et al. 2007, Wen & Gu 2004, Li & Li 2009). These existed analysis of non-stationary chattering mostly concentrated in the time-varying system solution which is based on the pseudo excitation method. However, studies on non-stationary chattering analysis through simulating non-stationary time history remain insufficient. In this paper, fitted a spectrum of the "dujuan" typhoon measured data is utilized to simulate non-stationary spatial fluctuating wind considering the non-steady buffeting force caused by time history. Suspension bridge time-domain buffeting is analyzed using the large universal finite element software ANSYS.

2 THE TIME-DOMAIN ANALYSIS OF BUFFETING

Most of the large flexible structures are proved to have various non-linearity caused by various factors, such as spatial correlation of fluctuating wind speed, influences of structural deformation on aerodynamics, etc. Hence, frequency-domain analysis method would be difficult for handling non-linear factors; however, time-domain analysis method based on the numerical integration would be easy and straightforward in dealing with these issues.

By simulating the statistical properties of random load, the time-domain analysis method transforms the incentive into time series to confirm the response of the structure through the dynamic finite element method. In order to conduct a time-domain analysis, it is critical to propose a model of the wind load in a time domain. The wind load applied to the bridge is divided into the following categories: static wind load caused by fluctuating wind, buffeting load caused by average wind and self-excited load derived from pneumatic coupling. Among these forces, the average wind force can be calculated by three coefficient parameters, and the fluctuating wind force can be referred to Scan land quasi-steady aerodynamic formula, while the self-excited aerodynamic force can resort to time-domain simulation processing.

2.1 Establishment of non-stationary wind buffeting model

2.1.1 The typhoon load processing

The wind loads applied to traditional bridges are practically decomposed into time-invariant average wind load, buffeting force with smooth Gaussian property and self-excited forces generated uniform flow. In order to improve the accuracy of the calculation results of buffeting force, the aerodynamic admittance is introduced to fix the buffeting force. However, when the loads, for instance, the typhoon load with apparent time-varying mean properties and characteristics of non-stationary fluctuating wind loads, are applied to the structure, it is required to draw a more sophisticated classification for wind loads to consider such impact of the non-stationary wind loads have imposed on. Therefore, the non-stationary wind loads can be disposed as time-varying mean force non-stationary buffeting wind loads with Gaussian properties and loads of self-excited force (Li & Li 2009, Clough & Penzien 2006, Hu et al. 2009, Xu & Chen 2004).

2.1.2 Non-stationary buffeting force model

Based on the existing researches (Clough & Penzien 2006, Hu et al. 2009, Xu & Chen 2004, Jain et al. 1996, Huang 2012), the buffeting force caused by typhoon non-stationary fluctuating wind can be expressed as follows:

$$L_b(t) = \frac{1}{2}\rho\bar{U}(t)B\{2C_L(\alpha)\mu(t) \\ + [C_L'(\alpha) + C_D(\alpha)]w(t)\} \quad (1)$$

$$D_b(t) = \frac{1}{2}\rho\bar{U}(t)B\{2C_D(\alpha)\mu(t) \\ + C_D'(\alpha)w(t)\} \quad (2)$$

$$M_b(t) = \frac{1}{2}\rho\bar{U}(t)B^2\{2C_M(\alpha)\mu(t) + C_M'(\alpha)w(t)\} \quad (3)$$

where B—width of bridge section; $L_b(t)$, $D_b(t)$ $M_b(t)$—the non-stationary lift, drag and torque respectively; $U(t)$—time-varying mean wind speed along the wind direction; $\mu(t)$, $w(t)$—the fluctuating wind speed of typhoon along the wind direction and vertical non-stationary time history; C_L, C_D, C_M—the three component coefficients of three component forces; C_L', C_D', C_M' derivative of three component coefficients about angle.

Considering the overall non-stationary characteristics of the typhoon, the buffeting analysis of dynamic control equation can be expressed as follows:

$$M\ddot{X}(t) + C\dot{X}(t) + KX(t) = F_m(t) + F_{se}(t) + F_b(t) \quad (4)$$

In which: M, C and K respectively represent the quality, damping and stiffness matrix of the structure itself with the corresponding dimension N × N; $\ddot{X}, \dot{X}, X$ stand for the structure dynamic response under load—acceleration, velocity and displacement response vector respectively; $F_m(t)$, $F_{se}(t)$, $F_b(t)$ are time-varying mean wind load, time-varying self-excited force load and non-stationary buffeting force load.

2.2 Solution of the dynamic equation

In this paper, the author applied non-stationary buffeting force generated by non-stationary wind speed time-history to the bridge nodes, then used the method of time-domain integral to solve the dynamic control equation. The time-domain step-by-step integral can be applied to solve the equation by assuming that the structure constitutive characteristics fit in with linear constitutive relation in a certain period. The implicit NEWMARK-β method is used for a dynamic solution (Li & Li 2009, Huang 2012). The bridge buffeting response analysis is calculated in two steps: first, under the effect of weight, the structure of the initial state is acquired; second, the non-linear buffeting response of the bridge will be analyzed based on the previous step.

3 THE BUFFETING ANALYSIS OF SHIDENGMEN SUSPENSION BRIDGE

To analyze the suspension bridge buffeting, the space finite element model of the Shidengmen suspension bridge is established through ANSYS. Shidengmen suspension bridge consists of two concrete towers of door type, with an inter-node length of 6 meters simply supported steel truss girder and concrete bridge deck, and its main-span is 276 meters. The center distance of two main cables is 7.5 meters, and the lane width is 6 meters, with 1/10 vertical span ratio. The left bridge tower is 43.042 meters tall, while the right bridge tower is 40.142 meters tall. The sketch of the suspension bridge is shown in Figure 1.

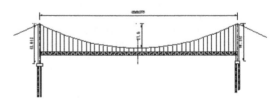

Figure 1. Shidengmen suspension bridge elevation.

3.1 Buffeting analysis and comparison under normal and non-stationary excitation

Based on the above analysis and theoretical elaboration, the buffeting force generated by stationary or non-stationary wind speed time-history series is applied to each node of the bridge for non-linear buffeting analysis.

3.1.1 The calculation results of buffeting displacement

Figures 2–7 show that the time-history displacement curve of the suspension bridge mid-span node under pulsating the wind and the non-stationary typhoon excitation. According to Figures 2 and 3 we can know that: the maximum transverse vibration displacement amplitude of

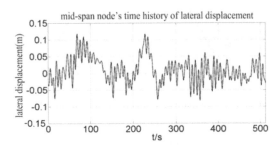

Figure 2.   Under stationary excitation.

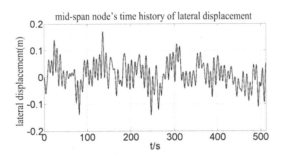

Figure 3.   Under non-stationary excitation.

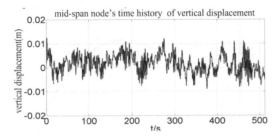

Figure 4.   Under stationary excitation.

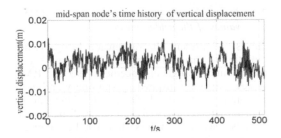

Figure 5.   Under non-stationary excitation.

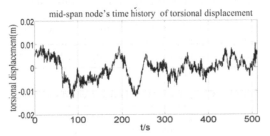

Figure 6.   Under stationary excitation.

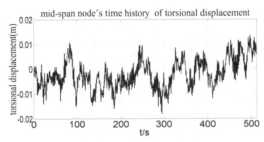

Figure 7.   Under non-stationary excitation.

the suspension bridge reaches 0.12 m under stationary excitation while under non-stationary excitation its value reaches 0.16 m and both frequency of transverse vibration are low. Figures 4 and 5 show that the vertical vibration maximum displacement amplitude of suspension bridge is 0.015 m under non-stationary or stationary excitations, but the frequency of vertical vibration is high. Figure 6 shows the time-history of angle displacement around the axis of the bridge under stationary excitation; and the maximum value is 0.12 rad. Figure 7 depicts the time-history of angle displacement under non-stationary excitation; while the maximum value is 0.16 rad. The vibration frequency is higher than the lateral vibration frequency; but smaller than the vertical vibration frequency, which is consistent to the structure natural vibration properties.

3.2 Comparison of displacement peak and RMS values of the bridge key nodes

Tables 1–3 present the suspension bridge girder buffeting horizontal displacement, vertical displacement and angle displacement peak and RMS values, which are inspired by stationary and non-stationary excitations. Tables 1 and 3 show that the main girder lateral as well as torsional displacement peak as well as RMS values will reach the maximum at the middle span under both stationary and non-stationary excitations, and its values will reduce extending to the bearing place. The maximum displacement peak is 0.1712 m under non-stationary excitation; however, the maximum displacement peak is 0.1177 m under stationary excitation. While in Table 2 the vertical displacement values are bigger at the 1/8, 3/8 span, and the RMS value

reaches the maximum at 1/4 span of the bridge. Whole bridge (vertical, lateral, and torsional) displacement RMS values under the typhoon excitation are significantly higher than the normal wind excitation, and the highest ratio reaches 1.47; while the displacement peak ratio reaches 2.

In order to gain an insight into the hazards that typhoon loads applied to the suspension bridge, Figures 8 to 10 show the buffeting displacement

Table 1. Comparison of the bridge lateral displacement.

| Position | Peak of displacement (m) | | RMS of displacement | |
	Stationary	Non-stationary	Stationary	Non-stationary
Mid-span	0.1177	0.1712	0.0383	0.0485
3/8 span	0.1120	0.1721	0.0362	0.0459
1/4 span	0.0878	0.1412	0.0285	0.0362
1/8 span	0.0484	0.0806	0.0158	0.0201

Table 2. Comparison of the bridge vertical displacement.

| Position | Peak of displacement (m) | | RMS of displacement | |
	Stationary	Non-stationary	Stationary	Non-stationary
Mid-span	0.0157	0.0125	0.0038	0.0041
3/8 span	0.0126	0.0168	0.0045	0.0049
1/4 span	0.0152	0.0200	0.0054	0.0059
1/8 span	0.0126	0.0151	0.004	0.0044

Table 3. Comparison of the bridge torsional displacement.

| Position | Peak of displacement (rad) | | RMS of displacement | |
	Stationary	Non-stationary	Stationary	Non-stationary
Mid-span	0.0013	0.0018	0.00050	0.00062
3/8 span	0.0012	0.0017	0.00044	0.00060
1/4 span	0.0010	0.0016	0.00035	0.00051
1/8 span	0.0006	0.0012	0.00021	0.00031

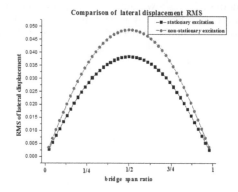

Figure 8. Comparison of lateral displacement RMS.

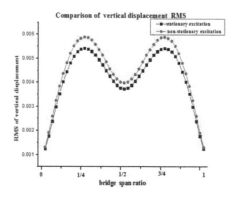

Figure 9. Comparison of vertical displacement RMS.

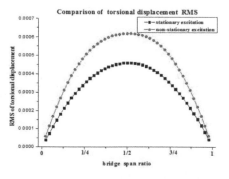

Figure 10. Comparison of torsional displacement RMS.

(horizontal, vertical, and torsional) RMS value of the whole bridge nodes under stationary and non-stationary excitations.

From the three figures, we can see the root mean square value of main girder node displacement under the excitation of non-stationary is obviously higher than that under the incentive of stationary, which suggests that the analysis of the large span bridge response under typhoon loads is necessary and the measured non-stationary characteristics of typhoon cannot be ignored.

4 CONCLUSIONS

In this paper, the author using the numerical solution method analyzed the suspension bridge time-domain buffeting response under the typhoon non-stationary and stationary incentives with the large general finite element software of ANSYS. Get the following conclusion: the whole bridge (vertical, lateral, and torsional) displacement RMS values under the typhoon excitation are significantly higher than the normal wind excitation. The highest RMS ratio comes up to 1.47; while the displacement peak ratio reaches 2, which is detrimental to bridge especially when the design safety coefficient of our country for a long-span bridge is about 2.5. The calculation results are consistent with the actual situation. And the feasibility and necessity of non-stationary calculation method for buffeting response of long-span bridges are verified.

REFERENCES

Boonyapinyo, V. Miyata, T. & Yamada, H. 1999. Advanced aerodynamic analysis of suspension bridges by state-space approach, *Journal of Structural Engineering*. 125(12): 1357–1366.

Clough, R.W. & Penzien, J. 2006. *Structure dynamics*, Higher Education Press.

Ding, S.Q. 2001. *Refinement of coupled flutter and buffeting analysis for long-span bridges*, Tongji University.

Hu, L. Xu, Y.L. & Huang, W.F. 2009. Typhoon-induced non-stationary buffeting response of long-span bridges in complex terrain, *Journal of Structural Engineering*, 57: 406–415.

Huang, W.F. 2012. *Typhoon wind field simulation and typhoon induced non-stationary response analysis of long-span bridge*, Harbin Institute of Technology.

Jain, A. Jones, N.P. & Scanlan, R.H. 1996. Coupled flutter and buffeting analysis of long-span bridges, *Journal of Structural Engineering*. 122(7): 716–725.

Li, J.H. & Li, C.X. 2009. Numerical simulation of non-stationary fluctuating wind, *Journal of Vibration and Control*, 28(1): 18–23.

Liang, J. Chaudhuri, S.R. & Shinozuka, M. 2007. Simulation of non-stationary stochastic processes by spectral representation, *Journal of Engineering Mechanics*, 133(6): 616–627.

Wen, Y.K. & Gu, P. 2004. Description and simulation of non-stationary processes based on Hilbert spectra, *Journal of Engineering Mechanics*, 130(8): 942–951.

Xu, Y.L. & Chen, J. 2004. Characterizing non-stationary wind speed using empirical mode decomposition, *Journal of Structural Engineering*. 130(6): 912–920.

Advanced Materials and Structural Engineering – Hu (Ed.)
© 2016 Taylor & Francis Group, London, ISBN 978-1-138-02786-2

Mechanism analysis for punching shear strength of hollow flat slabs

K.H. Yang
Department of Plant Architectural Engineering, Kyonggi University, Suwon, Kyonggi-do, South Korea

S.H. Yun
Kwangjang VDS Co. Ltd., Seoul, South Korea

S.T. Park
Winhitech Co. Ltd., Kolon Digital Tower Aston, Geumcheon-gu, Seoul, South Korea

ABSTRACT: A mechanism analysis based on the theory of plasticity is developed to predict an optimum failure surface generatrix and concrete punching shear capacity of a hollow flat slab system. To validate the proposed mechanism analysis, five two-way hollow flat slabs were tested under punching failure. From the mechanism analysis and experimental observations, the starting position of hollow materials in such slabs is recommended to be more than $2d$ away from all columns to avoid the reduction of concrete punching shear capacity, where d is the effective depth of the slab section.

1 INTRODUCTION

The practical application of hollow flat slabs to apartments or similar buildings has gradually grown owing to their structural and economic advantages (MacGregor & Wight 2005). The principle of this slab system is to connect lightweight materials (hollow plastic balls or expanded polystyrene (EPS)) with reinforcing elements in an industrial prefabrication phase to produce an internal void near the centroid axis of the concrete slab. In addition, the flat slab should retain its ability to transmit loads from the slab to the columns resting on the slab. Hence, the structural performance of the flat slab is governed by the punching shear as well as by flexure. Punching shear involves a truncated cone or pyramid-shaped area around the column. The punching shear capacity of a slab is generally less than the one-way shear capacity. It implies that the position of the hollow materials from the column surface is a very important consideration in controlling two-way shear cracks and preventing punching failure. Specifically, the punching shear capacity of a slab can be considerably deteriorated owing to hollow materials near the column. However, very few, if any, investigations (Schnellenbach-Held & Pfeffer 2002) are available on the punching shear capacity and failure surface generatrix.

This study presents a numerical technique using the upper-bound theorem of the theory of plasticity to predict the optimum geometry of the failure surface and, hence, to obtain an upper bound on the punching shear capacity of hollow flat slabs

under vertical loads. To examine the validity of the proposed numerical analysis, five biaxial hollow flat slabs were tested under punching failure. Based on the numerical analysis and the test results, the starting position of hollow materials in relation to the column surface is recommended to avoid the reduction of the punching shear capacity of the concrete because of the hollow materials.

2 MECHANISM ANALYSIS

2.1 *Failure mechanism*

The maximum moments in uniformly loaded flat slabs occur around the columns and lead to a circular crack around each column (MacGregor & Wight 2005). After additional loading, the cracks necessary to form a fan-shaped yield line mechanism develop, and at about the same time, inclined or shear cracks form on the truncated conical surface. Figure 1 shows an axisymmetric surface

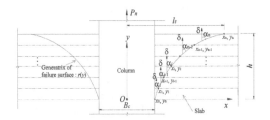

Figure 1. General failure area under punching shears.

of concrete punching failure observed in a widespread number of concrete flat slab specimens. Prior to the formation of the inclined cracks, the shear is transferred by the shear stresses in the concrete. Once the cracks have formed at the interface between column and slab, the cracks propagate nonlinearly toward the bottom surface of slab owing to the non-linear distribution of tensile stresses over the slab depth. At failure, a concrete member can be idealized as two rigid blocks separated by the failure surface, one of which moves vertically relative to the longitudinal axis of the other fixed rigid block, by an amount δ. The separated block containing the column has a truncated cone shape having a nonlinear generatrix of failure surfaces. Therefore, the horizontal extent l_F of the failure plane at the concrete surface and the angle α between the relative displacement δ and the failure surface are variable. In the idealized failure mechanism, the strain ε_{zz} in the circumferential direction is zero; hence, the generatrix can be considered to be a yield line, representing the zones of intense concrete separation, in a state of plane strain (Yang & Ashour 2008, Yang 2014).

2.2 Upper bound solution

To evaluate the geometry of the failure surface generatrix, the failure depth is divided into n segments (Fig. 1). For each layer, the vertical coordinate y_i is fixed, depending on the number of segments, and the horizontal coordinate x_i is variable and is obtained after minimizing the collapse load in accordance with the plasticity theory. The normal condition of a modified Coulomb material in the state of plane strain requires that the angle α_i between the relative displacement and yield line should be larger than a friction angle φ at the failure plane (Yang 2014);

$$\alpha_i = \tan^{-1}\frac{x_i - x_{i-1}}{y_i - y_{i-1}} \ge \varphi \tag{1}$$

The upper bound analysis uses the energy principle to calculate load capacity for the kinematically admissible failure mechanism. Taking into account the plasticity of concrete, the external work W_E at failure and the internal energy $(W_I)_i$ dissipated in each concrete layer i is estimated as follows:

$$W_E = P_n\delta \tag{2}$$

$$(W_I)_i = (W_A)_i A_i = \frac{1}{2}v_s v_c f_c'\delta[l - m\sin\alpha_i] \tag{3}$$

where, P_n is the failure load applied to column, v_s (= 0.5) is the factor explaining the reduced friction resistance caused by sliding failure, v_c is the effectiveness factor for concrete compressive strength.

$f_c', l = 1 - 2\mu\,(\sin\varphi/1 - \sin\varphi)$, $m = 1 - 2\mu\,(1/1 - \sin\varphi)$, and $\mu = f_t^*/f_c^*$, which is the ratio between the effective tensile and compressive strengths of concrete. Equating the total internal energy dissipated in concrete to the external work done, the punching shear capacity of concrete without hollow materials can be derived in the following form:

$$P_n = \frac{\pi}{2}v_s f_c^*\sum_{i=1}^{n}[l - m\sin\alpha_i]\frac{(y_i - y_{i-1})[(y_i - y_{i-1})\tan\alpha_i + 2x_{i-1}]}{\cos\alpha_i} \tag{4}$$

2.3 Concrete modeling

Concrete is regarded as a rigid perfectly plastic material obeying a modified Coulomb failure criteria with effective compressive and tensile strengths for the plane strain case. It is generally known that reasonable rupture conditions are attained by combining Coulomb's frictional hypothesis with a bound for the maximum tensile stress. Yang (2014) carried out numerical analysis to determine the values of v_c and μ using his stress-strain models of concrete in compression and tension; as a result, the following equations for normal-weight concrete have been obtained:

$$v_c = 0.8\left[f_c'/f_0\right]^{-0.12} \tag{5}$$

$$\mu = 0.06\left[f_c'\left(25/d_a^{0.5}\right)/f_0\right]^{-0.4} \tag{6}$$

where, f_0 (= 10 MPa) is the reference value for concrete compressive strength and d_a is the maximum aggregate size. For sliding failure of concrete, Yang derived the relationship of μ and φ as below:

$$\varphi = 22.89(\mu)^{-0.185} \tag{7}$$

3 FAILURE SURFACE GENERATRICES

The punching shear capacity of concrete is implicitly expressed as a function of the geometry of the failure-surface generatrix. According to the upperbound theorem, the collapse occurs at the least strength. The minimum value for concrete punching capacity can be obtained by varying the horizontal coordinate x_i of each layer since the vertical coordinate y_i of each layer is known. It is noted that the angle α_i of each layer should be larger than the friction angle φ, as explained above. The process of adjusting the horizontal coordinate x_i of each layer to evaluate the optimum geometry of failure—surface generatrix is achieved by reliable numerical optimization procedures programmed in the Matlab software. Examples of optimum failure-surface generatrices for different μ values are plotted in Figure 2. In the same figure, the failure-surface generatrix specified in ACI 318-11 (2011) is also shown. The optimum failure-surface generatrices are significantly dependent on the μ values, showing that l_F obtained from the mechanism analysis is commonly larger than that specified in the ACI 318-11 provision. When μ values are greater than 0.02, l_F is evaluated to be above $2.0d$, where d is the effective depth of the slab.

As plotted in Figure 2, the optimum failure-surface generatrices can be simplified as two straight lines (Fig. 3). From Equation (4), P_n for a cone-shaped failure surface having two straight generatrices can be produced as follows:

$$P_n = \frac{\pi}{2} v_s v_c f_c' \left[\begin{array}{l} \dfrac{h_0 (B_c + h_0 \tan \varphi)}{\cos \varphi}(1 - \sin \varphi) \\[2mm] + \dfrac{(h_{ef} - h_0)\left\{ \begin{array}{l} (h_{ef} - h_0)\tan \alpha \\ + B_c + 2h_0 \tan \varphi \end{array} \right\}}{\cos \alpha}(1 - m \sin \alpha) \end{array} \right] \tag{8}$$

where, h_0 is the depth of the bottom failure zone and B_c is the width of the column section. Therefore, the first and second terms on the right-hand side of Equation (8) give the dissipated energies in zone AB, having angle φ, and zone BC having angle α, respectively. Thus the lowest upper-bound solution can be obtained by letting $\partial P_n / \partial h_0 = 0$ and

$\partial P_n / \partial \alpha = 0$. When $\partial P_n / \partial h_0 = 0$, h_0 can be obtained as follows:

$$h_0 = -\cfrac{\begin{array}{l} B_c \{\cos \alpha (\sin \varphi - 1) + \cos \varphi (l - m \sin \alpha)\} \\ + 2h \cos \varphi \{l(\tan \alpha - \tan \varphi) + m \sin \alpha (\tan \varphi - \tan \alpha)\} \end{array}}{2\left[\begin{array}{l} \tan \varphi \cos \alpha (\sin \varphi - 1) - \cos \varphi \tan \alpha (l - m \sin \alpha) \\ + 2 \cos \varphi \tan \varphi (l - m \sin \alpha) \end{array} \right]} \tag{9}$$

As $\partial N / \partial \alpha = 0$, it is difficult to straightforwardly determine the value of α because of its non-linear equation. Yang (2014) employed numerical techniques for solving non-linear equations for α; as a result, the following expression is obtained:

$$\alpha = 429(\mu)^{-0.117} \tag{10}$$

On the other hand, h_0/h is also dependent on μ and B_c/h, indicating that h_0/h increases with an increase in μ and with the decrease in B_c/h, where h is the overall depth of the slab. Although h_0 could be calculated from Equation (9), for most practical applications, h_0 is simply expressed by nonlinear regression analysis as follows:

$$h_0/h = 0.96\left[\mu (B_c/h)^{-3} \right]^{0.11} \leq 1.0 \tag{11}$$

3.1 Effect of hollow materials on failure surface generatrices and punching shear capacity

The most suitable arrangement of internal hollow materials begins from location l_F. P_n of a hollow flat slab can be determined using Equation (8). However, the arrangement of internal hollow materials in a slab can change the failure-surface generatrix and

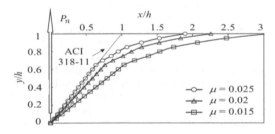

Figure 2. Optimized failure-surface generatrices (Prediction).

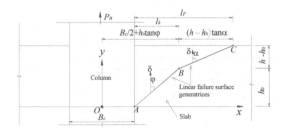

Figure 3. Idealized failure-surface generatrix.

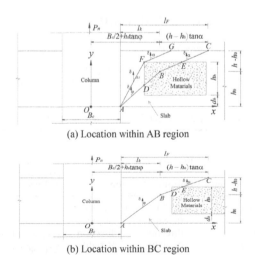

(a) Location within AB region

(b) Location within BC region

Figure 4. Transition of the failure surface generatrix owing to the arrangement of hollow materials.

deteriorate the punching shear capacity owing to the reduced concrete section area. When the hollow materials are arranged within a distance l_F from the column surface, failure-surface generatrices can be idealized by the following two classifications: the hollow materials are arranged within l_k (Fig. 4a); or they are arranged between l_k and l_F (Fig. 4b). From the upper-bound theorem, for a hollow flat slab with an internal void within l_k, the failure—surface generatrix is transited from surface AB-BC to surface AF-FG and the corresponding P_n of Equation (8) can be written as follows:

$$P_n = \frac{\pi}{2} v_s v_c f_c' \left[\frac{\left[x_h \tan\left(\frac{\pi}{2}-\varphi\right)\left(x_h \tan\left(\frac{\pi}{2}-\varphi\right)\tan\varphi\right) + B_c\right]}{\cos\varphi}(1-\sin\varphi) + \frac{(h-h_h-d_h)[(h-h_h-d_h)\tan\alpha + B_c + 2\{l_k+(h-h_h-d_h)\tan\alpha\}]}{\cos\alpha}(1-m\sin\alpha) \right]$$

(12)

On the other hand, for a hollow flat slab with an internal void between l_k and l_F, the failure-surface generatrix is transited from surface AB-BC to surface AB-BD-EC, and the corresponding P_n of Equation (8) can be written as follows:

$$P_n = \frac{\pi}{2} v_s v_c f_c' \left[\frac{(d_h+h_h)[(d_h+h_h)\tan\alpha_1 + B_c]}{\cos\alpha_1}(1-m\sin\alpha_1) + \frac{(h-h_h-d_h)[(h-h_h-d_h)\tan\alpha + B_c + 2\{x_h+(h-h_h-d_h)\tan\alpha\}]}{\cos\alpha}(1-m\sin\alpha) \right]$$

(13)

4 EXPERIMENTAL VERIFICATION

4.1 Test specimens

Five hollow flat slabs were prepared to examine the effect of the starting position of hollow materials on the failure-surface generatrix and the punching shear capacity of concrete. A hollow deck-plate slab system, which was developed for a simplified construction procedure controlling the buoyancy of the hollow materials, was used for test specimens (Yang 2014). To shorten the construction time, steel-deck plate was used for a concrete form, instead of wood panels, and lightweight EPS was used for the internal void of the slab. The main parameter investigated was the starting position (which varied from 1.0d to 3.0d) of the EPS from the column surface, as given in Table 1. For comparison, a solid slab without EPS was also tested.

All two-way flat slab specimens had the same dimensions. The nominal size of the slab specimens was 3000 × 3000 mm, and the h and d were 280 mm and 230 mm, respectively, as shown in Figure 5. Value specified by the ACI 318-11 provision was arranged.

For temperature reinforcement in the top region of the slabs, reinforcing bar with a 13 mm diameter was arranged at the spacing of 300 mm. The designed compressive strength of the ready-mixed concrete mixture was 30 MPa. The yield strength of the steel bars was measured to be 518 MPa and 525 MPa for bars with diameters of 13 mm and 19 mm, respectively.

4.2 Comparisons with mechanism analysis

The typical crack propagation and the horizontal extent l_F of the failure plane at the concrete surface, as measured from the slab specimens, are plotted in Figure 6. On the same figures, l_F predicted from the present mechanism analysis is also provided, represented by bold dotted lines. At failure, cracks spread out like the spokes of a wheel, with the column as the hub. The l_F of the radial cracks was significantly affected by the starting position of the EPS. The l_F tended to decrease with the decrease in the starting position of the EPS, whereas specimens S-2

Table 1. Specimen details and summary of test results.

Specimens	Reinforcement		Starting position of EPS	L_F (mm)	P_n (kN)	Predicted P_n(kN)		$(P_n)_{EXP}/(P_n)_{PRE}$	
	Bottom	Top				ACI 318-11	This study	ACI 318-11	This study
S	$\phi19@70$	$\phi13@300$	–	1.9d	1067.2	890	898.9	1.19	1.19
S-1	(at each	(at each	d (= 230 mm)	1.8d	766.4	890	754.8	0.86	1.12
S-1.5	direction	direction	1.5d (= 345 mm)	2.4d	768.9	890	792.7	0.86	0.97
S-2	of slab)	of slab)	2d (= 460 mm)	1.9d	1019.5	890	898.9	1.15	1.13
S-3			3d (= 690 mm)	1.9d	1038.9	890	898.9	1.17	1.16

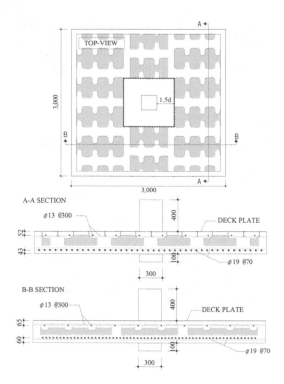

Figure 5. Details of slab geometry and arrangement of component materials (Specimen S-1.5).

and S-3 gave similar crack propagation and l_F location to those measured in the solid specimen S. The l_F location for specimens S, S-2, and S-3, as predicted from the mechanism analysis, was 2.7d.

The measured and predicted P_n values of the slab specimens are given in Table 1. The punching shear capacity of specimens S-2 and S-3 was very close to that of the solid specimen, indicating that the arrangement of the EPS does not affect the shear transfer capacity of concrete. Meanwhile, the P_n measured in specimens S-1 and S-1.5 was less than that of the solid specimen by approximately 28%. The ACI 318-11 equation gives a constant P_n for all specimens, because it does not consider the reduced shear transfer capacity of concrete due to internal voids. As a result, the ACI 318-11 equation is not conservative for specimens S-1 and S-1.5. The mechanism analysis predicts P_n more accurately than ACI 318-11 does, regardless of the starting position of the EPS. Overall, it can be recommended that the starting position of the EPS in hollow flat slabs to be more than 2d away from all columns to avoid the reduction of punching shear capacity owing to the EPS.

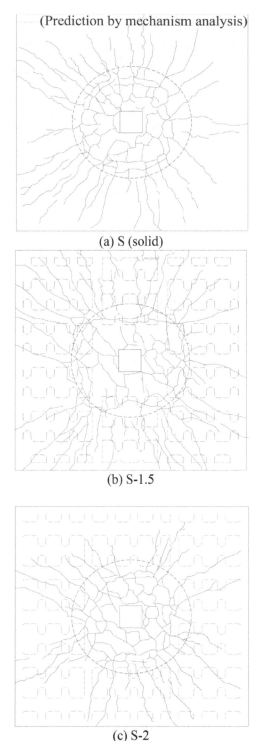

(a) S (solid)

(b) S-1.5

(c) S-2

Figure 6. Typical crack propagation and comparisons of predicted and measured l_F locations.

5 CONCLUSIONS

1. The optimum failure-surface generatrices of hollow flat slabs can be simplified as two straight lines considering the starting position of hollow materials.
2. The horizontal extent l_F of the failure plane at the concrete surface, as measured in tested slab specimens, tended to decrease with the decrease in the starting position of the EPS. However, when the starting position of the hollow materials exceeded $2.0d$, the l_F of the hollows slabs was not affected by the hollow materials; as a result, the punching shear capacity of those hollow slabs was equivalent to that of the solid slab.
3. The starting position of hollow materials in hollow flat slabs is recommended to be more than $2d$ away from all columns to avoid the reduction of concrete punching shear capacity.

ACKNOWLEDGEMENT

This research was supported by the Public Welfare & Safety Research Program through the National Research Foundation of Korea (NRF) funded by the Ministry of Science, ICT & Future Planning (2013067519). The authors also greatly appreciate the efforts and support provided by Winhitech CO. Ltd.

REFERENCES

ACI Committee 318. 2011. Building code requirements for structural concrete (ACI 318-11) and commentary (ACI 318R-11). American Concrete Institute.

MacGregor, J.G. & Wight, J.K. 2005. *Reinforced concrete: mechanics and design,* Prentice-Hall, Inc.

Schnellenbach-Held, M. & Pfeffer, K. 2002. Punching behavior of biaxial hollow slabs, *Cement and Concrete Composite.* 24: 551–556.

Yang, K.H. & Ashour, A.F. 2008. Mechanism analysis for concrete breakout capacity of single anchor in tensile loads, *ACI Structural Journal.* 150: 609–616.

Yang, K.H. 2014. *Evaluation of serviceability and structural performance of hollow deck slab system.* Technical Report, Kyonggi University.

Advanced Materials and Structural Engineering – Hu (Ed.)
© 2016 Taylor & Francis Group, London, ISBN 978-1-138-02786-2

Establishing a Decision Support Module for bridge maintenance in Taiwan

N.J. Yau & H.K. Liao

Graduate Institute of Construction Engineering and Management, National Central University, Taiwan, R.O.C.

ABSTRACT: Possessing an inventory of 24,300 bridges, the Taiwan Bridge Management System (TBMS) is a web-based bridge management system developed in the year 2000 for all the bridge management agencies in Taiwan. This research establishes a Decision Support Module (DSM) for the TBMS with four important functions: (1) budgeting for bridge inspection, (2) budgeting for repairing bridges, (3) prioritizing repair work, and (4) distributing fund for repairing. A new index DI (Danger Index), calculated based on the deterioration condition of the bridge, a structural safety factor, and a traveler's safety factor, is introduced in this research as a major criterion for prioritizing bridge repair work.

1 INTRODUCTION

Taiwan has more than 24,300 bridges, including freeway, highway, and railway bridges but excluding culverts and pedestrian bridges. These bridges deteriorate with time and are threatened by multiple natural disasters such as typhoons and earthquakes that occur frequently. In the early 1990s, Taiwan began its efforts on developing Bridge Management Systems (BMS). Early versions of BMS did not have the same bridge inventory formats, standards for bridge inspections, nor evaluation criteria; they were designed for one specific agency only. Development of the Taiwan Bridge Management System (TBMS) was motivated by the collapse of two dozen bridges during the Ji-Ji earthquake in 1999 (Yau & Liao 2007). Funded by the MOTC and developed by the Center for Bridge Engineering Research at National Central University (NCU), the TBMS was on line since 2000 and is now maintained by the Graduate Institute of Construction Engineering and Management at NCU.

2 DETERIORATION AND EVALUATION

In the TBMS, regular inspection of a bridge is performed by inspecting each component of the bridge. The number of components may vary depending on the type of bridge. The regular inspection is based on a DER&U methodology and the inspection results are then input into the inspection data module of the TBMS. The DER&U methodology was firstly invented by a joined effort of two consulting companies, CSIR (1994) and Join Engineering and has become a national inspection standard in Taiwan. According to the inspection results, a Condition Index (CI) that represents the overall condition of the bridge is calculated, as described in Table 1.

2.1 *Inspection of bridge components*

As shown in Table 1, 21 components are identified for a concrete bridge when performing a regular inspection. Notably, for the concrete bridge, components No. 1, No. 2 and No. 4 to No. 7 are divided into two inspection items, i.e., the two ends of the bridge. Components No. 12 to No. 20 are inspected and rated section by section; the rest of the components are inspected as a single inspection item. Each of these components has a corresponding weight of importance, also shown in Table 1, for calculating the CI of the bridge. These weights were determined jointly by a group of bridge experts without incorporating special methodologies. A higher weight denotes higher importance of the component. The summation of weights for all the 21 components is 100.

2.2 *DER&U methodology*

The TBMS incorporates the DER&U methodology for regular bridge inspection. In the methodology, four indices are used to evaluate the condition of a bridge component: "D" represents the degree of deterioration; "E" represents the extent of the deterioration; "R" represents the deterioration's relevancy to safety; and "U" represents the urgency for repairing the deterioration. All of these indices are numerically rated on an integer scale from 1 to 4 to describe the status of the deterioration, as shown in Table 2.

For a deteriorated component, in addition to its rating by the DER&U values, the inspector should take photos of the deterioration, measure applicable dimensions of the deterioration, and suggest an appropriate repair method. The overall bridge Condition Index (CI) can also be obtained accordingly based on the evaluated DER&U ratings of bridge components.

2.3 Condition index

After all the 21 components of a concrete bridge are inspected and rated, a component condition index Ic_{ij} is calculated based on the evaluated integers of D, E, and R for each component. The calculation

Table 1. Twenty-one components and their weights of a concrete bridge.

No*	Bridge component	Weight**
1	Approaching embankment	3
2	Approaching guardrail	2
3	Waterway	5
4	Protection works for the embankment	3
5	Abutment foundations	6
6	Abutments	5
7	Retaining walls	5
8	Pavements	3
9	Superstructure drainages	4
10	Sidewalks	2
11	Guardrails	3
12	Scouring protection of pier	6
13	Pier foundation	8
14	Piers & column	7
15	Bearings	5
16	Earthquake brakes	5
17	Expansion joint	6
18	Longitudinal girder	8
19	Transversal beam	6
20	Decks & slab	7
21	Other	1

* Component's serial number, denoted as i in all equations in this article.
** Component's weight, denoted as w in all equations in this article.

is based on a point-deduction mechanism; i.e. deficiencies of a component will deduct points from a perfect score of 100. Equation (1) shows the formula for calculating the Ic_{ij} value for the "jth" item of component "i".

$$Ic_{ij} = 100 - 100 \times \frac{D \times E \times R}{4 \times 4 \times 4} \qquad (1)$$

Herein, the condition index of a component "i" (Ic_i) will be an average value of all items; i.e. Ic_i equals to the sum of the condition indices of all items divided by the number of items of such component (n), as calculated by Equation (2).

$$Ic_i = \frac{\sum_{j=1}^{n} Ic_{ij}}{n} \qquad (2)$$

The overall bridge Condition Index (CI) can be obtained once condition indices of these twenty-one components are calculated. The CI is a weighted average of all twenty-one components Ic_i, as shown in Equation (3). The weights of all the 21 components are illustrated in Table 1.

$$CI = \frac{\sum_{i=1}^{21} Ic_i \times w_i}{\sum_{i=1}^{21} w_i} \qquad (3)$$

3 BRIDGE MAINTENANCE DECISION SUPPORT MODULE

The DER&U methodology was incorporated into TBMS and has become the national standard for regular bridge inspection in Taiwan. However, due to its average weighting nature, the Condition Index (CI) may not precisely reflect the actual bridge conditions in some scenarios, and thus can only serve as a reference to the bridge management agencies when performing maintenance tasks. Therefore, a Decision Support Module (DSM) for bridge maintenance is needed in the TBMS.

Table 2. The DER&U evaluation criteria.

		0	1	2	3	4
D		Component not existing	Good	Fair	Bad	Serious
E		Unable to inspect	Less than 10%	10~30%	30~60%	Over 60%
R		Relevancy uncertain	Minor	Limited	Major	Large
U		Urgency uncertain	Routine	In 3 years	In 1 year	Immediately

Yehia et al. (2008) depicted that fully developed decision support systems should be capable of accepting facts from the users, processing these facts, and delivering solutions that are close to the solutions that are offered by human experts. In this research, before the DSM was developed, nine bridge experts from various bridge management agencies were interviewed twice for collecting opinions and suggestions. Thus, by employing the existing bridge inventory, regular bridge inspection data and repairing methods in TBMS, the bridge maintenance DSM are then determined to include four functions as shown below.

3.1 Budgeting for bridge inspection

Many bridge management agencies respond that there is a need for an automated budgeting for bridge inspections due to the fact that their personnel turnover rates are very high and that rookie engineers are always lacking of experience to plan the bridge inspection budgets. Thus, such function is built in the DSM. A list of unit costs of bridge inspection was created by averaging previous 30 inspection contracts, as shown in Table 3.

The unit costs, varied by bridge lengths, are calculated based on unit slab area (in square meters) of the bridge and have been converted into US dollars. In the module, a click on this function will automatically generate a budget for bridge inspection for a management agency. The budget is a summation of every bridge's slab area multiplied by its corresponding unit cost according to the bridge's length.

3.2 Budgeting for repairing bridges

This function in the DSM is very straight forward. Firstly, a unit cost for each of the repairing methods is pre-determined according to the market prices; secondly, the inspector inputs the amount of repairing work required for a deteriorated bridge component; finally, the repairing cost is calculated by a simple multiplying operation of the amount and the pre-determined unit cost. The components that need to be repaired can be grouped by the U (urgency) index and also summarize their repairing costs for each group. Notably, the U index, ranging

Table 3. Unit price of bridge inspection.

Bridge length (m)	Price (USD/m²)
Less than 25	1.125
26~100	0.938
101~150	0.875
Over 150	0.75

from 1 to 4, is showing urgency of repairing for bridge components; thus, the repairing costs are categorized by components not by bridges. Indirect costs such as overhead, tax, and profit, will be a percentage of the repairing costs that can be determined by the user before obtaining the budget of repairing.

3.3 Prioritizing repair work

In the DER&U methodology, condition index (CI) represents the overall deterioration condition of a bridge. However, it's a weighted average index that can be misrepresented in some cases while a few components of a bridge are severely deteriorated but the others are in perfect condition. For example, if a bridge's longitudinal girders were severely damaged, then the conditions of this component Ic_{18} (Table 2, weight of longitudinal girder is 8) will equal 0 (Equation (2)); if the rest components have no deterioration, then CI of this bridge can still get a very high score of 92 according to Equation (3). Therefore, the CI only serves as a reference in the DER&U methodology; i.e., a high CI value does not guarantee the safety of the bridge.

In this research, a new index called Danger Index (DI) is established, as shown in Equation (4); where ds_i is a deterioration score of a component, w_i is the weight of component shown in Table 1, and f_i is a traveler's safety factor. DI is intended to represent how dangerous a bridge might become from the safety prospective. The DI of a bridge is represented by the highest DI score among its 21 components.

$$DI = Max\left[(ds_i \times w_i \times f_i / 24)\right] \quad \forall i \in \{1..21\} \qquad (4)$$

Instead of using condition of a component (Ic_{ij}) in the DER&U methodology, an index JI, representing a component's condition considering only the three DER indices was artificially determined by the bridge experts of a consultant company. The deterioration score (ds_i) of the bridge is then obtained by subtracting the JI from 100. Thus, a 100 score of ds_i represents the worst deterioration condition of a bridge component while a 0 score means no deterioration. In order to differentiate and emphasize the deterioration, ds_i is calculate for different sets of DER values using only when E equals to 4, are shown in Table 4.

In Equation (4), w_i is deemed as a factor of structural safety, same as the weight defined in DER&U methodology (Table 1), and has a maximum value of 8. The traveler's safety factor, f_i, is designed to magnify the seriousness of injuring travelers when component i failed. Determined by a group of bridge experts, the f_i is determined to have a range from 1 to 3, as shown in Table 5. Thus,

Table 4. List of deterioration score.

No	D	R	ds_i *
1	4	4	100
2	3	4	96
3	2	4	90
4	4	3	75
5	3	3	71
6	2	3	65
7	4	2	50
8	3	2	46
9	2	2	40
10	4	1	25
11	3	1	21
12	2	1	15
13	1	N/A	0

* Higher score represents poor bridge condition.

Table 5. Traveler's safety factor of bridge component.

No	Bridge component	Condition	f_i
1	Approaching embankment	None	2
2	Approaching guardrail	None	2
3	Waterway	None	2
4	Protection works for the embankment	None	2
5	Abutment foundations	D = 4, R = 4	3
		Others	2
6	Abutments	None	2
7	Retaining walls	None	1
8	Pavements	D = 4, R = 4	3
		Others	2
9	Superstructure drainages	None	2
10	Sidewalks	None	2
11	Guardrails	D = 4, R = 4	3
		Others	2
12	Scouring protection of pier	None	1
13	Pier foundation	D = 4, R = 4	3
		Others	2
14	Piers & column	D = 4, R = 4	3
		Others	2
15	Bearings	None	2
16	Earthquake brakes	None	2
17	Expansion joint	D = 4, R = 4	3
		Others	2
18	Longitudinal girder	D = 4, R = 4	3
		Others	2
19	Transversal beam	None	1
20	Decks & slab	D = 4, R = 4	3
		Others	2
21	Other	None	1

in Equation. (4), since the maximum value of w_i is 8, and the maximum value of f_i is 3, a divisor 24 is then put into the equation to normalize DI into a range between 100 and 0. The DI is calculated for each component of the bridge and the safety condition of a bridge is determined by the worst deteriorated components. Therefore, the worst score among all components is chosen to represent the bridge's DI. A higher value of DI depicts more dangerous of the bridge.

3.4 Distributing fund for repairing work

In this research, once all bridges' DI values in an agency are obtained, it is a relatively easy task to decide which bridges have the highest priority to repair. Apparently, bridges having higher value of DI have higher priority when repairing funds are available. However, if two bridges having the same DI values, then, two simple factors, road class and slab area of the bridge, both can be easily accessed from bridge inventory, have become the criteria for determining the priority. Road class is the first criteria to be considered. The list of road classes in Taiwan is shown in Table 6.

It is believed that bridges on higher class of roads, such as national freeway, should have higher priority. Again, if two bridges are in the same class of road, then, the size of slab area becomes the second criteria; i.e., the bridge with larger slab area gets higher priority.

As an example, there are ten bridge need to repaired, listed in Table 7, and the total amount of fund available is $100,000. DI values of the ten bridges are calculated first based on their inspection results, also shown in Table 7. Then, these bridges are ranked by DI values, from more dangerous to less dangerous, the fund for repairing bridges can be distributed accordingly. Herein, number 3, 4, and 5 have the same DI value of 32.5, but the number 3 bridge is on higher class of road and the number 4 bridge has larger slab area than the number 5 bridge; these three bridges are thus ranked accordingly. The given fund is distributed by the ranked priority. As a result, the available

Table 6. Road classes and priority.

Class	Priority
National freeway	1st
Highway	2nd
County/City road	3rd
Country road	4th
Rural road	5th
Others	6th

Table 7. Bridges with different condition.

No	DI	Road class	Slab	Fund required	Priority
1	41.67	Highway	1500 m^2	$26,000	1st
2	36.30	Highway	2680 m^2	$25,000	2nd
3	32.50	Highway	1000 m^2	$11,500	3rd
4	32.50	City road	1880 m^2	$34,020	4th
5	32.50	City road	980 m^2	$12,000	5th
6	20.85	County road	1100 m^2	$10,340	6th
7	16.00	Country road	1620 m^2	$28,800	7th
8	12.12	City road	650 m^2	$19,600	8th
9	11.05	Rural road	400 m^2	$8,500	9th
10	8.95	Rural road	300 m^2	$9,800	10th

fund only can repair the bridges with priority 1st to 4th, with a sum of $96,520 and with $3,480 left.

4 CONCLUSIONS

The TBMS has been online for 14 years with 24,300 bridges in its inventory. TBMS is a successful system provides functions for inventory, inspection, and maintenance information for the bridges. The decision support module, bridge maintenance DSM, developed in this research facilitates bridge management agencies in budgeting for bridge inspections and repair tasks, as well as the choice of repairing methods.

The newly introduced index DI (Danger Index) quantifies the degree of danger of bridge deterioration and utilized as a critical criterion for prioritizing repairing tasks and distributing maintenance fund. This module serves as a decision tool not only for personnel who are responsible for bridge inspection and repair tasks, but also for managers in developing budgeting and maintenance strategy. Verification and modification of the bridge maintenance DSM is a continuous task for improving the TBMS in the future.

REFERENCES

CSIR, 1994. *TANFB Bridge Management System Operation Manual*, Taiwan Area National Freeway Bureau, Taipei.

Flaig, K.D. & Lark, R.J. 2000. The development of UK bridge management systems, *Proceedings of the ICE-Transport*. 141: 99–106.

MOTC, (2012). *Highway Maintenance Codes, Ministry of Transportation and Communication.* Taipei.

Yau, N.J. & Liao, H.K. 2007. *Development of Bridge Management System in Taiwan,* in: Proc. of the 5th International Conference on Construction Project Management & 2nd International Conference on Construction Engineering and Management, CMA International Consultants Pte. Ltd., Singapore, 92–101.

Yehia, S. Abudayyeh, O. Fazal, I. & Randolph, D. 2008. A decision support system for concrete bridge deck maintenance, *Advances in Engineering Software,* 39: 202–10.

Advanced Materials and Structural Engineering – Hu (Ed.)
© 2016 Taylor & Francis Group, London, ISBN 978-1-138-02786-2

Strut-and-Tie Model for shear capacity of squat shear walls

J.H. Mun & W.W. Kim
Department of Architectural Engineering, Kyonggi University Graduate School, Seoul, South Korea

K.H. Yang
Department of Plant Architectural Engineering, Kyonggi University, Suwon, Kyonggi-do, South Korea

ABSTRACT: The objective of the present study was to establish a simplified Strut-and-Tie Model (STM) based on the crack band theory of fracture mechanics of concrete for use in evaluating the shear capacity of squat Reinforced Concrete (RC) shear walls. The shear transfer capacity of concrete is determined by the strut and tie action contributed by the web concrete and longitudinal reinforcement of the boundary elements of the walls, whereas that of shear reinforcement is determined by solving the equations of a statically indeterminate truss system defined by the vertical and horizontal shear reinforcing bars. The basic equations for strut-and-tie action are established using the energy equilibrium condition in the stress relief strip and the crack band zone. To validate the proposed model, three squat RC shear walls were tested under a constant axial load and cyclic lateral load. The proposed STM was found to predict the shear capacity of shear walls more accurately than ACI 318-11.

1 INTRODUCTION

Squat shear walls with aspect ratios (α_s) of less than 2.0 are required to resist a lateral load mainly by the in-plane strut-and-tie forces. The lateral and axial loads applied to squat shear walls are commonly considered to be directly transferred to the foundation through the concrete strut (Hwang et al. 2001). As a result, the Strut-and-Tie Model (STM) becomes a good and rational design tool for squat shear walls (Hwang et al. 2001). However, in the seismic design provision of ACI 318-11 (American Concrete Institute 2011), the nominal shear strength of shear walls is defined as the sum of the shear capacities of the web concrete and the horizontal shear reinforcement, regardless of α_s. This indicates that the ACI 318-11 shear provision does not consider the effectiveness of vertical shear reinforcement on diagonal crack control and that shear capacity increases with decreasing α_s. The purpose of the present study was to establish a reasonable and simplified strut-and-tie model for use in evaluating the shear capacity of shear walls.

2 SIMPLE STRUT-AND-TIE MODEL

2.1 *Shear transfer mechanism*

As Reinforced Concrete (RC) shear walls with aspect ratios (α_s) of less than 2.0 are commonly classified as D-region members, the shear transfer mechanism of such walls can be identified using

an STM, as shown in Figure 1 (Hwang et al. 2001). The lateral force applied to the free end of a shear wall is primarily transferred to a support through a concrete strut that represents the compression stress field and the tie action. The tie action is generated along the longitudinal reinforcement at the boundary elements of the shear wall and the shear reinforcement at the web. Axial forces in the struts and ties intersect at nodes. The following assumptions were made in developing the STM described in this paper: 1) fracture of concrete is caused by a crack band formed from a number of axial splitting microcracks, as shown in Figure 2; 2) stress relief strips in the concrete struts are concentrated along diagonal cracks in the web; 3) the compatibility condition between concrete and the reinforcement

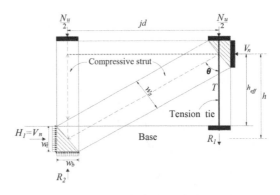

Figure 1. Schematic strut-and-tie model for shear walls.

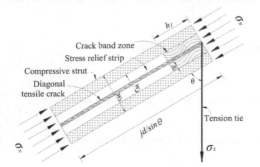

Figure 2. Stress relief strip and crack band zone at wall web without shear reinforcement.

is maintained until the failure of the strut; and 4) the stresses in reinforcing bars at wall failure are less than the bars' yield strength.

2.2 Shear capacity by strut-and-tie action

According to the fracture mechanics of concrete (Bažant and Planas 1998), the loss of strain energy (ΔU_c) in a shear wall without shear reinforcement owing to stress relief during the formation of a crack band is obtained from the following equation:

$$\Delta U_c = -\frac{\left\lfloor \sigma_N^2 b_w w_f jd \right\rfloor}{2E_c}\sin\theta - \left\lfloor \sigma_s^2 A_s w_f \sin\theta \right\rfloor 2E_s \tag{1}$$

where b_w is the width of the wall web; E_c and E_s are the moduli of elasticity of the concrete and reinforcing bars, respectively; A_s is the area of the longitudinal reinforcement at the boundary element; jd is the distance between the top and bottom nodes; w_f is the width of the stress relief strip; θ is the angle between the concrete strut and the longitudinal tie; $\sigma_N = \left((V_c/\sin\theta)/(b_w w_s)\right)$ and $\sigma_s = \left((V_c/\tan\theta - 0.5N_u)/(A_s)\right)$ are the axial stresses in the concrete strut and longitudinal reinforcement, respectively; V_c is the shear transfer capacity of concrete; N_u is the applied axial load; and w_s is the width of the concrete strut. The energy release rate (I_c) per unit wall thickness owing to the growth of the crack band at the stress relief strip is expressed in the following form:

$$I_c = -\frac{1}{b_w}\left[\frac{\partial(\Delta U_c)}{\partial w_f}\right] = \frac{\left(\dfrac{V_c^2}{sin^3\,\theta}\right)jd}{2E_c b_w^2 w_s^2}$$
$$+ \frac{\left(\dfrac{V_c^2 cos\theta}{tan\,\theta} - V_C N cos\,\theta + \dfrac{N^2 sin\theta}{4}\right)}{2nE_c b_w A_s} \tag{2}$$

where $n(= E_s/E_c)$ is the modular ratio between the reinforcement and the concrete. If the number of axial splitting microcracks in a crack band is calculated as w_f/s_c, where s_c is the average spacing between the microcracks, the total energy dissipated in the crack band can be obtained from the following equation (Bažant and Planas 1998):

$$W_{cc} = \left[w_f/s_c\right]b_w h_f G_f \tag{3}$$

where h_f is the length of the crack band and G_f is the fracture energy of concrete. The energy (R_c) dissipated in the crack band per unit length of the band and unit depth of the wall can be written as follows (Bažant and Planas 1998):

$$R_c = [1/b_w]\left[\partial W_{cc}/\partial w_f\right] = \left[h_f/s_c\right]G_f \tag{4}$$

From energy equilibrium ($I_c = R_c$), the shear transfer capacity of concrete can be obtained from the following equations:

$$V_c = \left[-B_1 \pm \sqrt{B_1^2 - 4A_1 C_1}\right]/2A_1 \tag{5.a}$$

$$A_1 = \left[nA_s jd\right]/\sin^3\theta + b_w w_s^2 \left[\cos\theta/\tan\theta\right] \tag{5.b}$$

$$B_1 = -N_u b_w w_s^2 \cos\theta \tag{5.c}$$

$$C_1 = N_u^2 b_w w_s^2 \left[\sin\theta/4\right] - 2nA_s E_c b_w^2 w_s^2 \left[h_f/s_c\right]G_f \tag{5.d}$$

2.3 Shear transfer capacity of shear reinforcement in web

To reflect the effect of redistribution of cracks at a closer spacing and over a wider area in concrete struts due to shear reinforcement in the web, Yang and Ashour (Yang and Ashour 2011) idealized the crack band extension zone as shown in Figure 3.

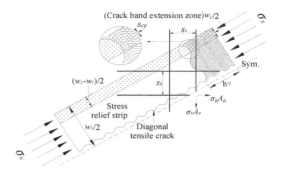

Figure 3. Idealization of crack band extension zone by shear reinforcement.

In this zone, the loss of strain energy (ΔU_s) in the shear reinforcing bars is expressed as follows:

$$\Delta U_S = -\frac{\left[\sigma_v^2 A_v jd(w_f + w_i)\sin\theta\right]}{2E_s s_v}$$
$$-\frac{\left[\sigma_h^2 A_h jd(w_f + w_i)\cos\theta\right]}{2E_s s_h \tan\theta} \tag{6}$$

where A_v and s_v are the area and spacing, respectively, of the vertical shear reinforcement; A_h and s_h are the area and spacing, respectively, of the horizontal shear reinforcement; and w_i is the width of the crack band extension zone. Because the proportion of shear transfer by the horizontal and vertical shear reinforcing bars resulting from tie action depends on the ratio of lateral and axial loads applied to the walls, the average stresses of the vertical (σ_v) and horizontal (σ_h) shear reinforcing bars can be calculated using the following equations:

$$\sigma_v = \left[(\alpha_{v1}V_s - \alpha_{v2}N_u)s_v\right]/A_v jd \tag{7}$$

$$\sigma_h = \left[(\alpha_{h1}V_s - \alpha_{h2}N)s_h \tan\theta\right]A_h jd \tag{8}$$

where V_s is the shear transfer capacity of the shear reinforcement; α_{v1} and α_{v2} are the proportions of lateral and axial loads, respectively, transferred to the vertical shear reinforcement; and α_{h1} and α_{h2} are the proportions of lateral and axial loads, respectively, transferred to the horizontal shear reinforcement. Using the above procedure for deriving Equation (5), the shear transfer capacity (V_s) of the shear reinforcement due to the tie action can be expressed as follows:

$$V_s = \left[-B_1 \pm \sqrt{B_1^2 - 4A_1C_1}\right]/2A_1 \tag{9.a}$$

$$A_1 = \alpha_{v1}^2 \rho_h + \alpha_{h1}^2 \rho_v \tag{9.b}$$

$$B_1 = -2N(\alpha_{v1}\alpha_{v2}\rho_h + \alpha_{h1}\alpha_{h2}\rho_v) \tag{9.c}$$

$$C_1 = N^2\left(\alpha_{v2}^2\rho_h + \alpha_{h2}^2\rho_v\right)$$
$$-\left[2nE_c jd\frac{b_w^2 \rho_h \rho_v}{\sin\theta}\right]\left[\frac{h_0}{S_{ce}}\right]G_f \tag{9.d}$$

where s_{ce} is the spacing of the axial splitting microcracks in the crack band extension zone; h_0 is a certain characteristic value representing the final length of the crack band; and ρ_v and ρ_h are the vertical and horizontal shear reinforcement ratios, respectively. Reasonable values of the factors α_{v1} and α_{v2} in Equation (9) can be obtained by solving the following equations of a statically indeterminate truss system defined by the vertical and horizontal shear reinforcing bars.

$$\alpha_{v1} = 0.2(\alpha_s)^3 - 0.87(\alpha_s)^2 + 1.2(\alpha_s) \tag{10}$$

$$\alpha_{h1} = -0.12(\alpha_s)^2 + 0.61(\alpha_s) \tag{11}$$

The proportion of axial load transfer is calculated to be zero for the horizontal shear reinforcement and less than 0.03 for the vertical shear reinforcement.

2.4 Determination of dimensions of concrete strut

The effective strut width depends on the width (w_b) and depth (w_t) of the bottom node at the reaction point of the interface between the wall and the foundation, as shown in Figure 4. As the bottom node can be classified as a CCC-type node that having equal stresses on all in-plane sides, the ratio of each face width of the hydrostatic node has to be the same as the ratio of forces meeting at the node to make the state of stress in the entire node region constant (Marti 1985). From the equivalence of the linear distribution of compressive stresses to the uniform state, w_b can be obtained from the following equation (see Fig. 4):

$$w_b = c/(2\nu_e) \tag{12}$$

where $\nu_e(= 0.85)$ is the effectiveness factor of concrete. The depth of the neutral axis (c) in this equation is determined from the equation proposed by Mun and Yang (2014). The modulus of elasticity and fracture energy of concrete are determined from the equations proposed by Yang et al. (2014) and Sim et al. (2014), respectively.

2.5 Determination of constants for shear capacities

As test data for squat shear wall specimens without shear reinforcement are very limited, the

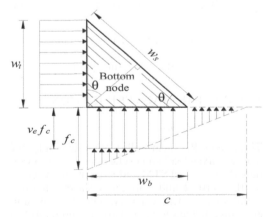

Figure 4. Equivalent effective width of the bottom node.

Table 1. Specimen details and test results.

Specimen	f_c' (MPa)	$N_u/A_g f_c'$	ρ_v	ρ_h	Measured V_n (kN)	Predicted V_n		$(V_n)_{EXP}/(V_n)_{PRE}$	
						ACI 318-11	This study	ACI 318-11	This study
C	40.8	0.070	0.0042	0.0042	1117.6	672.1	1093.5	1.36	1.07
V2	40.0	0.072	0.0084	0.0042	1203.0	668.9	1190.9	1.48	0.96
H2	40.1	0.072	0.0042	0.0084	1129.1	861.2	1107.7	1.39	1.05

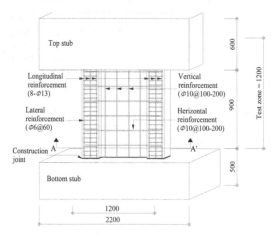

Figure 5. Details of specimens tested (all dimensions in mm).

constants h_f/S_c of Equation (5) were determined from the equation $0.4(l_w/d_a)^{1.5}$ proposed by Yang and Ashour (2011), where as that of Equation (9) was determined from the test data compiled by Mun (2014). The following expression for (h_o/s_{ce}) was obtained from regression analysis of the data:

$$h_o/s_{ce} = \sqrt{36\left(\rho_v^{0.85}/\rho_{vo} + \rho_h^{0.5}/\rho_{ho}^2\right)\left(1 + R_N^{0.1}\right)} \quad (13)$$

where ρ_{ho} and ρ_{vo} are the factors to minimize over fitting in the regression analysis by proposed by Bažant and Sun (Bažant & Sun 1987).

3 EXPERIMENTAL VERIFICATION

3.1 Test specimens

Three squat RC shear walls were tested under the constant axial load and cyclic lateral load to validate the proposed STM and to examine the effect of the vertical and horizontal shear reinforcement on the shear capacity. The main parameters investigated were ρ_v (varied from 0.0042 to

0.0084) and ρ_h (varied from 0.0042 to 0.0084), as given in Table 1 and Figure 5. The boundary elements of all specimens had barbell-shaped cross-sections with a 250-mm width (b_{eff}) and 250-mm length (l_c).

3.2 Comparison of prediction with test results

The measured and predicted V_n values of the shear walls are given in Table 1. The shear capacity of specimen V2 was 8% higher than that of the companion specimen C. The measured value of V_n for specimen H2 was similar to that for specimen C.

This indicates that vertical shear reinforcement is more effective than horizontal shear reinforcement in enhancing the shear capacity of a shear wall with $\alpha_s = 0.76$. The V_n value for specimen V2 predicted by the ACI 318-11 shear provision was similar to that for specimen C, because the provision does not consider the shear transfer capacity contributed by vertical shear reinforcement. The proposed STM predicts the shear capacity of shear walls quite accurately, with ratios of measured to predicted values of 0.96–1.07.

4 CONCLUSIONS

A simplified Strut-and-Tie Model (STM) was proposed by identifying the shear transfer capacities of concrete and shear reinforcement based on the crack band theory of fracture mechanic of concrete. To validate the proposed model, three squat RC shear walls were tested under the constant axial load and cyclic lateral load. The following conclusions were drawn from the results obtained:

1. Vertical shear reinforcement is more effective than horizontal shear reinforcement in enhancing the shear capacity of a shear wall with an aspect ratio of 0.76.
2. The proposed STM yields a better agreement with test results than the ACI 318-11 provision, with ratios of measured to predicted shear capacity of 0.96–1.07.

ACKNOWLEDGMENT

This work was supported by Kyonggi University's Graduate Research Assistantship 2014 and the National Research Foundation of Korea (NRF) grant funded by the Korea government (MSIP) (No. NRF-2014R1A2A2A09054557).

REFERENCES

ACI Committee 318, 2011. *Building code requirements for structural concrete (ACI 318-11) and commentary*, Farmington Hills, MI, American Concrete Institute.

Bažant, Z.P. & Planas, J. 1998. *Fracture and size effect in concrete and other quasibrittle materials*, New York, CRC.

Bažant, Z.P. & Sun, H.H. 1987. Size effect in diagonal shear failure: influence of aggregate size and stirrups, *ACI Materials Journal*, 84: 259–272.

Hwang, S.J. Fang, W.H. Lee, H.J. & Yu, H.W. 2001. Analytical model for predicting shear strength of squat walls, *Journal of Structural Engineering ASCE*, 127: 43–50.

Marti, P. 1985. Basic tools of reinforced concrete beam design, *ACI Journal*, 82: 46–56.

Mun, J.H. & Yang, K.H. 2014. Plastic hinge lengths model for reinforced concrete slender shear walls, *Magazine of Concrete Research,* 67: 414–429.

Mun, J.H. 2014. *Flexure and shear design approach of heavy-weight concrete shear walls, Kyonggi University, Ph. D thesis.*

Sim, J.I. Yang, K.H. Lee, E.T. & Yi, S.T. 2014. Effect of aggregate and specimen sizes on lightweight concrete fracture energy, *Journal of Materials in Civil Engineering ASCE*, 26: 845–854.

Yang, K.H. & Ashour, A.F. 2011. Strut-and-tie model based on crack band theory for deep beams, *Journal of Structural Engineering ASCE*, 137: 1030–1038.

Yang, K.H. Mun, J.H. Cho, M.S. & Kang, T.H.-K. 2014. Stress-strain model for various unconfined concretes in compression, *ACI Structural Journal*, 111: 819–826.

Advanced Materials and Structural Engineering – Hu (Ed.)
© *2016 Taylor & Francis Group, London, ISBN 978-1-138-02786-2*

Abnormal vibration analysis of a metro vehicle

H.Q. Liu & H.Y. Dai
State Key Laboratory of Traction Power, Southwest Jiaotong University, Chengdu, China

ABSTRACT: Abnormal lateral vibration appeared during the operation of a metro vehicle at constant speed, which had a negative effect on ride quality and the strength of vehicle. By analyzing the field test data with FFT and STFT, possible causes of lateral vibration were found. The test results showed that there was a lateral vibration frequency of car-body. Assuming it is a hunting frequency, the equivalent conicity can be deduced. Then, a dynamic simulation model was built to reproduce the phenomenon of car-body's hunting instability. The simulation result was in consistent with the analyzing result of field test data, which proved that the excessive lateral vibration of car-body was caused by hunting.

1 INTRODUCTION

Since the official opening and the operation of Line6, the whole equipment has been running in a good condition. But in some section of the track, the transverse vibration of the vehicle is large, which affected the passenger ride comfort.

In this paper, the vehicle-track coupling dynamic model is established according to the given parameters, with the use of dynamic analysis software SIMPACK. The phenomenon of car-body's lateral vibration is reproduced. And the comparison is made between the simulation result and the field test result. Also, conclusion can be made that it is the hunting instability, which causes the large lateral vibration.

2 FIELD TEST

The test section was selected between Dalianpo station and Qingnianlu station. Under the load case, the lateral vibration acceleration at the location of 1 m distance away from the bogie was measured by the using the acceleration sensor. Testing data at the stage of the boot process, constant speed process and braking process are chosen to do the transformation of FFT. At the same time, the entire time-domain signal is analyzed with the transformation of STFT.

Figures 1–3 show that, in the processes of startup and braking, the lateral vibration energy of the car-body is evenly distributed, and there is no main frequency. But in the process of constant and high speed, a main lateral vibration frequency of 4.9 Hz

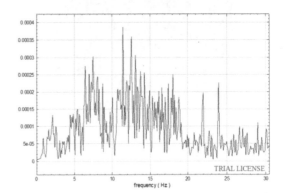

Figure 1. FFT of lateral vibration acceleration in the boot process.

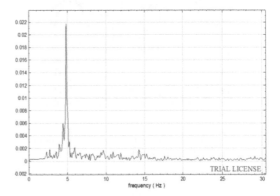

Figure 2. FFT of lateral vibration acceleration in the constant speed process.

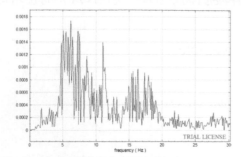

Figure 3. FFT of lateral vibration acceleration in braking process.

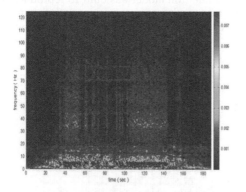

Figure 4. STFT of lateral vibration acceleration in entire process figure.

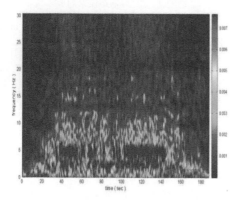

Figure 5. STFT of lateral vibration acceleration in entire process (0–30 Hz filtering).

exists, and most part of the energy is contained under this frequency.

Figures 4 and 5 show that the energy of the carbody's lateral vibration is mainly concentrated in constant speed process, and the frequency is about

4.9 Hz, equaling to the result of the FFT analysis. So, it can be primarily judged that it was a hunting instability frequency of the vehicle.

3 CALCULATION OF THE EQUIVALENT CONICITY

According to the tooth mesh frequency and the wheel diameter, the constant operation speed of the vehicle is calculated, and it is about 100 km/h. Combined with the vibration frequency of 4.8 Hz, the hunting wavelength is calculated, and it is 5.787 m.

The hunting wavelength of flexible bogie can be expressed as follows:

$$S = S_1 \cdot \sqrt{1 + \left(l_1/b\right)^2 \left(1 - \delta\right)} \qquad (1)$$

where the free wheel-set hunting wavelength can be expressed as:

$$S_1 = 2\pi \sqrt{\frac{br_0}{\lambda}} \qquad (2)$$

and δ can be expressed as,

$$\delta = \left(1 - \sigma Z\right)/\left(1 + Z^2\right) \qquad (3)$$

where l_1 is the half distance between the first and second wheel-set; b is the half lateral distance between the right and left wheel-rail contact point.

σ and Z can be expressed as,

$$\sigma = \left(2P - 3\right)\beta^2 \qquad (4)$$

$$Z = \beta K^P \qquad (5)$$

where,

$$\beta = b/l_1 \qquad (6)$$

$$p = 1 + \left[k_\psi / \left(k_\psi + k_y l_1^2\right)\right] \qquad (7)$$

$$K = \frac{1}{2fb^2} \cdot \frac{k_\psi \cdot k_y l_1^2}{k_\psi + k_y l_1^2} \cdot \frac{S_1}{2\pi} \qquad (8)$$

and $k_\psi = k_x b_1^2 \qquad (9)$

where b_1 is the half distance between the right and left axle box spring, k_x is the vertical suspension stiffness of axle box (per axis), and k_y is the lateral suspension stiffness of axle box (per axis).

Then f can be expressed as:

$$f = (f_{11} + f_{22})/2 \qquad (10)$$

And $f_{11} = Ea_0b_0C_{11} \qquad (11)$

$$f_{22} = Ea_0b_0C_{22} \qquad (12)$$

where f_{11} is the longitudinal creep factor; f_{22} is the lateral creep factor; E is the Young's modulus of elasticity; C_{ii} is the Kalker coefficient that can be determined in table of Kalker; a_0, b_0 are the major and minor axis of contact ellipse.

According to nonlinear Hertz elastic contact theory, the major and minor axis of contact ellipse can be expressed as:

$$\left(\frac{a_0}{m}\right)^3 = \left(\frac{b_0}{n}\right)^3 = \frac{6N(1-\sigma^2)}{4(A+B)E} \qquad (13)$$

where N is the normal load of contact ellipse (the pressure between one wheel and the rail), and is half axle load simply; σ is the Poisson's ratio; m and n are the coefficients that are relative to A,B and can be regarded as the function of A and B, and like Kalker coefficient, can be obtained in the table.

According to nonlinear Hertz elastic contact theory, A and B can be expressed as:

$$\begin{cases} (A+B) = \frac{1}{2}\left(\frac{1}{r_0} + \frac{1}{R'} \pm \frac{1}{R}\right) \\ (B-A) = \frac{1}{2}\left(\frac{1}{r_0} + \frac{1}{R'} \mp \frac{1}{R}\right) \end{cases} \qquad (14)$$

where R is the radius of the cross-sectional shape of the wheel tread; R' is the radius of the cross-sectional shape of the rail head; c is the radius of the wheel rolling circle.

According to the above formulas, the equivalent conicity of wheel tread could be calculated and it is 0.4447, which matches the hunting wavelength of 5.787 m.

4 DYNAMIC MODEL SIMULATION

4.1 Establishment of the vehicle dynamic model

The first wheel-set in the forward direction of vehicle is defined as the 1st wheel-set. Then, the coordinate system can be built as follows: the forward direction of the vehicle is x-axis; y-axis is parallel to the plane of the track and toward to the right; z-axis is perpendicular to the plane of the track and is downward.

Vehicle dynamic model which has 15 rigid bodies is consisted of one body, two frames, eight axle boxes, and four wheel-sets. Car-body has six degrees of freedom, which include vertical, horizontal, vertical, roll, pinch and yaw; frame has six degrees of freedom, vertical, horizontal, vertical, roll, pinch and yaw; axle box has one degree of freedom, which is yaw; the wheel-set has six degrees of freedom, vertical, horizontal, vertical, roll, pinch and yaw (the vertical and lateral movement are non-independent), a total of 42 degrees of freedom is included. The vehicle model is shown in Figure 6.

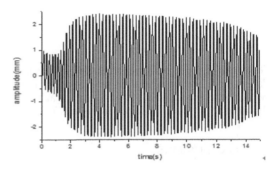

Figure 6. Vehicle dynamics model.

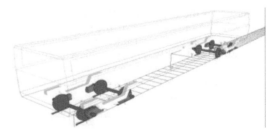

Figure 7. $v = 99$ km/h, $f = 4.73$ Hz.

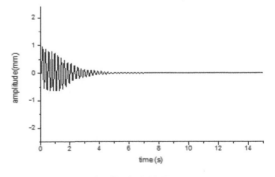

Figure 8. $v = 100$ km/h, $f = 4.93$ Hz.

385

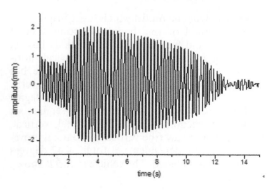

Figure 9. $v = 101$ km/h, $f = 4.97$ Hz.

4.2 Calculation of critical speed and hunting frequency

The taper of wheel tread is set to 0.4447 in the simulation model. The time history of car-body's lateral vibration at the speed of 100 km/h can be achieved by simulation, which is shown in Figures 7–9.

5 CONCLUSIONS

The following conclusions can be summarized from the field test in this paper and dynamic model formulation:

1. Based on the field test of Beijing Subway Line 6 vehicles and the data of the lateral vibration acceleration of the car-body, the frequency of hunting instability can be achieved with the FFT transformation analysis in the constant speed process. Most part of the energy is contained in this frequency.

2. The operation speed is calculated by the tooth mesh frequency and wheel diameter. Combined with the hunting frequency, the hunting wavelength is calculated. Then according to the flexible wavelength formula, the equivalent conicity of wheel-set is deduced.

3. A vehicle dynamic model is established and the critical speed and hunting frequency of the car-body is calculated by simulation. The result of simulation is in accordance with the field test results. The phenomenon of hunting instability is successfully reproduced, which proves that the hunting instability leads to the excessive lateral vibration of the metro vehicle's car-body.

ACKNOWLEDGMENT

This work has been supported by the State Key Program of National Natural Science of China (61134002), the National Key Basic Research Program of China (973 Program) (2011CB711100) and Innovation Group of Ministry of Education funded project (IRT1178).

REFERENCES

Dai, H.Y. 2008. *Locomotive Vehicle Bodywork Flutter Mechanism Analysis and Solution*, Diesel Locomotives.

Koyanagi, S. 1993. *The Approach to Design Operating Characteristics of Flexible Bogie*, Foreign Rolling Stock.

Wang, B.J. 2009. *Analysis and Solutions of Metro Vehicle Vibration*, Electric Locomotives Mass Transit Vehicles.

Wang, F.T. 1994. *Vehicle System Dynamics*, China Railway Publishing House, Beijing.

Advanced Materials and Structural Engineering – Hu (Ed.)
© 2016 Taylor & Francis Group, London, ISBN 978-1-138-02786-2

Structure health monitoring using the matched chaotic excitation

D.D. Yang, H.G. Ma & D.H. Xu
Xi'an Research Institute of High Technology, Xi'an, P.R. China

ABSTRACT: This work presents an efficient procedure for fault measurement of the Single-Input-Single-Output (SISO) system by driving the system directly with the output of a chaotic oscillator. First, we propose a method to adjust the Lyapunov Exponents (LEs) of the chaotic excitation by multiplying each equation with different acceleration factors. These acceleration factors constitute an Acceleration Factor Field (AFF). Second, in order to evaluate the degree of activation to the SISO system, which is excited by the chaotic signal with the different acceleration factors, we partition the AFF into fully matched, partially matched and non-matched areas by using the criteria derived from Kaplan–Yorke conjecture. Third, an attractor-based fault feature "Prediction Error (PE)" is used to find the optimal acceleration factor's location in the matched area, and then the fault feature is signified by using the chaotic excitation with optimized acceleration factors. Finally, the results are presented for an experimental cantilevered beam instrumented with fiber-optic strain sensors.

1 INTRODUCTION

Health monitoring (e.g., diagnosis and identification) has become an active domain since 1970's, and in the past decades we have seen the compelling progress in this domain. The process of health monitoring is to detect betimes if the abnormal phenomena exists during the operation of the System Under Test (SUT) (Hwang et al. 2010).

The SUT is excited by the assigned signal. The corresponding output is used to detect faults in the SUT (see Fig. 1). The process of fault measurement for such a system can be described in four steps: (1) excitation of the SUT; (2) observation of the system through periodic measurement of the structure's dynamic response; (3) extraction of damage sensitive features from these measurements; (4) investigation of these features to classify the system status appropriately (e.g., damaged or undamaged).

In recent years, methods that use a chaotic signal as the excitation come into fashion (Hwang et al. 2010). The chaotic signal has many unique features with respect to normally used signals (e.g.,

multi-tone signal and white-noise signal) in fault measurement. In some special cases, the chaotic signal can make the process of fault measurement much easier, e.g., chaotic signal has a narrow band and it can be easily generated compared with white noise, and its spectrum is dense due to the strange attractor in phase space (Kolumban et al. 1996). Moreover, the transformation from the time domain to a phase space geometric domain, which has long been of the most interest in nonlinear dynamics, has recently received close attentions in the field of damage assessment and brought about some unique advantages in developing useful features in a way similar to the transformation from time domain to frequency domain (Logan & Joseph 1996).

In this paper, we study which kind of chaotic excitations can sufficiently activate the specified SUT. If the specified feature reflects the change of a SUT's parameter, then the parameter is considered to be activated by the corresponding excitation. Here, we proposed the concept of activation degree to describe the degree of the matching between the excitation and the SUT.

The remainder of the paper is organized as follows. In Section 2, methodology used for fault measurement in this paper is introduced, including design of the excitation, feature selection. In Section 3, results are presented for an experimental cantilevered beam instrumented with the fiber-optic strain sensors. Finally, a conclusion is drawn in Section 4.

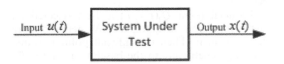

Figure 1. Conceptual model of a SISO system.

2 METHODOLOGY

As mentioned above, the SUT is excited by the output of a chaotic oscillator, which can be written as follows (Casdagli 1992).

$$\dot{\mathbf{x}} = \mathbf{F}(\mathbf{x}(t))$$
$$u(t) = h_1(\mathbf{x}(t)) \tag{1}$$

where $\mathbf{x}$ is the state vector of the chaotic oscillator. $\mathbf{x}$'s evolution is governed by $\mathbf{F}$. The excitation $u(t)$ is derived by projecting $\mathbf{x}$ to one-dimensional time-series with observation function $h_1(\cdot)$.

The SUT can be written as follows

$$\dot{\mathbf{z}} = \mathbf{G}(\mathbf{z}(t), u(t))$$
$$x(t) = h_2(\mathbf{z}(t)) \tag{2}$$

where excitation $u(t)$ is acted on the SUT using $\mathbf{G}$. $\mathbf{z}$ is the state vector of the SUT. The output $x(t)$ is derived by projecting $\mathbf{z}$ to one-dimensional time-series with observation function $h_2(\cdot)$.

2.1 Design of the excitation

Parameter changes in SUT will induce changes in eigen-structure of the SUT, and the eigen-structure is closely related with the LEs of the SUT. The Kaplan–Yorke conjecture (Kaplan & Yorke 1979) relates the LEs of a system to the dimension of that system via

$$D_L = K + \frac{\displaystyle\sum_{m=1}^{K} \lambda_m}{|\lambda_{K+1}|} \tag{3}$$

where λ_m is the LE arranged in a descending order, and D_L is the Lyapunov dimension. K satisfies

$$\sum_{m=1}^{K} \lambda_m \geq 0, \quad \sum_{m=1}^{K+1} \lambda_m < 0 \tag{4}$$

For a SISO system, the LEs of the output $x(t)$ consist of two parts: $\lambda_i^E : i = 1, ..., d_1$ and $\lambda_j^S : i = 1, ..., d_2$. Where λ_i^E is the exponent associated with the d_1-dimensional Excitation and λ_j^S is the exponent of the d_2-dimensional SUT. Rearrange all of the LEs as follows:

$$\lambda_1 > \lambda_2 > \cdots > \lambda_{(d_1 + d_2)} \tag{5}$$

Assume $K = k_1$ is the number of exponents of the SISO system that satisfy (4), chaotic excitations are classified into three types by examining the first $k_1 + 1$ exponents in (5).

TYPE 1: Fully matched chaotic excitation. All of the LEs of the SUT are contained in the first $k_1 + 1$ exponents. According to Equation (3), output $x(t)$'s phase space structure (e.g., dimension) is closely related to the whole eigen-structure (all of the LEs) of the SUT. Any parameter changes in the SUT could be detected by analyzing output $x(t)$'s phase space structure.

TYPE 2: Partially matched chaotic excitation. The part of the LEs of the SUT is contained in the first $k_1 + 1$ exponents. Changes of the SUT's parameters associated with these LEs could be detected by analyzing output $x(t)$'s phase space structure.

TYPE 3: Non-matched chaotic excitation. None of the LEs of the SUT is contained in the first $k_1 + 1$ exponents. Now Equation (3) cannot establish a relationship between SUT's parameters and output $x(t)$'s phase space structure.

In reality, chaotic excitations do not always have some proper LEs that match with the SUT. Here, we propose a method to adjust the LEs of the chaotic excitation without changing its basic structure.

Chaotic excitation is generated by a chaotic oscillator, which is modeled by a set of Differential Equations (DEs). The LEs of the chaotic excitation can be adjusted by multiplying each equation's right part with an acceleration factor, which accelerates the evolution of the dynamical system. Those acceleration factors $(\mu_1, \mu_2, ..., \mu_{d1})$ constitute an Acceleration Factor Field (AFF), which is denoted by μ-field. A point in AFF is called the μ-vector. We partition the AFF into three categories by checking if the μ-vector induced chaotic excitation is matched with the SUT: 1) AFF fully match with the SUT (fully matched μ-field), 2) AFF partially match with the SUT (partially matched μ-field), 3) AFF does not match with the SUT (non-matched μ-field).

2.2 Feature used to distinguish the fault

As mentioned above, if the chaotic excitation fully matched with the SUT, the dimension of the output will change if there is a fault in the SUT. And then, any geometric features can be used to measure the fault. In this paper, Prediction Error (PE) is used as a feature in qualifying the fault level (Schreiber 1997).

Fiducial point $Q_i(t = t_f)$ is first selected on the comparison attractor, and N_1 neighbors are selected on it. Then the fiducial point is transferred geometrically to the baseline attractor P_1, and a set of N_2 nearest neighbors are also accumulated. Both sets of nearest neighbors are then evolved by s steps to yield the time-evolved neighborhood $\phi_{t_r+s}^b$ and $\phi_{t_r+s}^b$. The average or centroid of these two neighborhoods now evolves to Q_2 and P_2:

$$P_2(t_f + s) = \frac{1}{N_2} \sum_{X_b \in \Phi^b_{t_f+s}} X_b(n) \qquad (6)$$

$$Q_2(t_f + s) = \frac{1}{N_1} \sum_{X_c \in \Phi^c_{t_f+s}} X_c(n) \qquad (7)$$

P_2 is used as the predicted position where the fiducial point on the comparison attractor should evolve to, and the Euclidean distance between this predicted coordinate and the time-evolved neighborhood centroid Q_2 in comparison attractor

$$\gamma_{t_f} = \left\| P_2(t_f + s) - Q_2(t_f + s) \right\| \qquad (8)$$

is the prediction error corresponding to that fiducial point.

For each pair of responses, we use the method produced in (Olson et al. 2009) to create a Gaussian distribution of PE values. It randomly sample 4000 prediction error values from the underlying distributions that describe the local dynamical differences between the two attractors. As this distribution is unknown, then it produce a new nominally Gaussian distribution by uniformly selecting 30% of the PE values, calculating their mean, and repeating the procedure 4000 times to get a distribution of resampled and averaged PE values, thus a nominally Gaussian distribution of baseline vs. comparison is formed.

Assume that we got two attractors reconstructed from the output time series of a normal system with different noise levels (denoted by normal attractor and comparison attractor respectively), and one attractor reconstructed from the output of a fault system (denoted by fault attractor). Then, the Gaussian distribution of the PE calculated between normal attractor and comparison attractor is denoted as the N-distribution, and the Gaussian distribution of the prediction error between fault attractor and comparison attractor is denoted as the F-distribution. Thus, fault level of the system can be estimated by comparing these two distributions.

3 EXPERIMENT

The system under study is a cantilevered aluminum beam, clamped at the fixed end by four clamping bolts threaded into an aluminum clamping stage. The beam itself had a length of 5.000×10^{-1} m, a width of 5.000×10^{-2} m, and a thickness of $t = 3.175 \times 10^{-3}$ m. Springs were placed between the clamping stage and the clamping bolts so that the clamping strength could be varied. This provided a controlled mechanism by which the structure may be "damaged". Seven fiber-optic strain sensors (based on fiber Bragg gratings) were evenly spaced along the central axis of the beam and an accelerometer was attached to the forcing mechanism in order to record the excitation. The entire setup is shown in Figure 2. The shaker is a M.B. Dynamics "Modal 50" attached to the base by a shaft fixed in place by set screws. It was determined experimentally that the first two LEs of the beam were $\lambda_1 = -2$ and $\lambda_2 = -11$.

The excitation used is an accelerated Lorenz oscillator. In the accelerated Lorenz oscillator, each equation of the traditional Lorenz oscillator (Lorenz 1963) is multiplied by different acceleration factors.

$$\dot{x} = (\sigma(y - x))\mu_1$$
$$\dot{y} = (\gamma x - y - xz)\mu_2 \qquad (9)$$
$$\dot{z} = -(\beta z + xy)\mu_3$$

where μ_1, μ_2, μ_3 are the acceleration factors, and $\sigma = 16$, $\gamma = 45.92$, $\beta = 4$. when $\mu_1 = \mu_2 = \mu_3 = 1$, the LEs of the system are: $\lambda_1^E = 1.497$, $\lambda_2^E = 0.00$, $\lambda_3^E = -22.46$. y variable of the accelerated Lorenz oscillator is taken as the output to excite the circuit.

We adjust (μ_1, μ_2, μ_3) to find all the fully matched areas in the μ-field. In order to illustrate the process more intuitively, we fixed $\mu_3 = 1$, and vary (μ_1, μ_2) to find the matched μ-field in $\mu_1 - \mu_2$ plane. According to Wolf's method (Wolf 1985), three LEs of the accelerated Lorenz oscillator calculated on the $\mu_1 - \mu_2$ plane are shown in Figure 3 (a–c).

We use the method proposed in Section 2.1 to find the fully matched μ-field. The LEs of each point in the $(\mu_1, \mu_2, 1)$ field are checked if it is fully matched with the SUT. The result is shown in Figure 3 (d). From the figure, we can see that the fully matched μ-field is a set of areas distributed in the AFF.

As mentioned in Section 2.1, chaotic excitations generated with acceleration factors chosen from

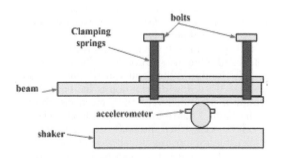

Figure 2.   Experiment setup.

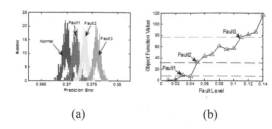

(a)　　　　　　　　　(b)

Figure 4. (a) Gaussian distributions of prediction errors with different fault levels, (b) Object function values calculated with different fault levels.

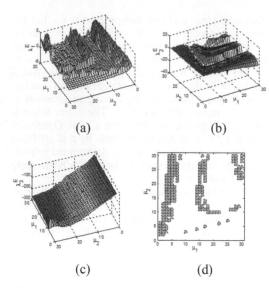

(a)　　　　　　　　　(b)

(c)　　　　　　　　　(d)

Figure 3. (a)–(c) The LEs of the accelerated Lorenz oscillator calculated on the $\mu_1 - \mu_2$ plane, (d) The fully matched μ-field.

the matched μ-field can sufficiently activate the SUT, that is to say, any changes of system's parameter can lead to an obvious change in the phase space structure and finally affect the trajectories reconstructed from the output time series.

As $(1, 1, 1)$ is in the fully matched μ-field, we first study the case of fault measurement using chaotic excitations with acceleration factors equal to $(1, 1, 1)$. The value of the capacitor C_2 is changed from $4/3F$ to $(4-\delta_f)/3F$ as the fault to be measured, and δ_f is the fault level. Use the routine mentioned above, we can calculate the Gaussian distribution of fault vs. comparison (denoted by F-distribution G_F) and normal vs. comparison (denoted by N-distribution G_N). When they satisfy the follow equation, we think the fault can be discerned.

$$\sum G_N G_F \leq \varepsilon_0 \tag{10}$$

where, ε_0 determines the maximal cross region between fault and N-distributions which should be set with a small number.

In Figure 4 (a), four Gaussian distributions are calculated: blue distribution is the N-distribution; red distribution is the Fault 1 distribution of fault level 0.03F; yellow distribution is the Fault 2 distribution of fault level 0.05F; green distribution is the Fault 3 distribution of fault level 0.11F. As it can be seen, when the fault become bigger, the F-distribution moves gradually to the right.

From Figure 4 (a), we can derive that Fault 1 cannot be discerned for there is a large cross area

between N-distribution and Fault 1 distribution. We say Faults with fault level greater than fault level 2 can be discerned for there is almost none cross region between the two distributions. Fault level 2 is the faults discern tolerance that makes the equality in (12) hold. Figure 4 (b) illustrates the object function values calculated with different fault levels. We can find a trend of increase with the increase of fault level. The object function value corresponding to the faults discern tolerance is the value of the yellow line. Faults with fault level larger than it can be discerned.

4 CONCLUSIONS

In the field of health monitoring, the main works are focused on design of the excitation, fault features and ways to improve the effectiveness of faults measurement. This paper is a new attempt to solve these problems with the usage of a chaotic excitation. The chaotic signals are thought to have many unique features fit to the use of health monitoring. There are two benefits when using the chaotic excitation matched or partially matched with the SUT. First, it guaranteed that changes in the system can be detected by analyzing the output time series; then it reflects that these matched excitations can strengthen the effect of fault diagnosis. The most significant aspect is that these matched excitations make a connection between system's parameter and the output's phase space structure.

REFERENCES

Casdagli, M. 1992. *A dynamical systems approach to modeling input-output systems.* Santa fe institute studies in the sciences of complexity-proceedings 12. Addison-wesley publishing co.

Hwang, I.S. et al. 2010. A survey of fault detection, isolation, and reconfiguration methods. *Control Systems Technology, IEEE Transactions on* 18.3: 636–653.

Kaplan, J.L. & Yorke, J.A. 1979. *Functional Difference Equations and Approximations and Fixed Points*, Springer-VerLag, Berlin, 730.

Kolumban, G. et al. 1996. *Chaotic systems: a challenge for measurement and analysis*. Instrumentation and Measurement Technology Conference, 1996. IMTC-96. *Conference Proceedings*. 'Quality Measurements: The Indispensable Bridge between Theory and Reality', IEEE. 2.

Logan, D. & Joseph M. 1996. Using the correlation dimension for vibration fault diagnosis of rolling element bearings—I. Basic concepts. *Mechanical Systems and Signal Processing* 10.3: 241–250.

Lorenz, E.N. 1963. Deterministic Non-periodic Flow, *Journal of the Atmospheric Sciences*. 20: 130–141.

Olson, C.C. Overbey, L.A. & Todd, M.D. 2009. An experimental demonstration of tailored excitations for improved damage detection in the presence of operational variability. *Mechanical Systems and Signal Processing*, 23(2): 344–357.

Schreiber, T. 1997. Detecting and analysing nonstationarity in a time series with nonlinear cross-predictions. *Physical review letters*, 78: 843–846.

Wolf, A. Swift, J.B. Swinney, H.L. & Vastano, J.A. 1985. Determining LEs from a Time Series, *Physica D*. 16: 285–317.

Advanced Materials and Structural Engineering – Hu (Ed.)
© *2016 Taylor & Francis Group, London, ISBN 978-1-138-02786-2*

Evaluation of consulting firms for Mass Rapid Transit projects

N.J. Yau & C.H. Sun
*Graduate Institute of Construction Engineering and Management, National Central University,
Jhongli, Taiwan, R.O.C.*

ABSTRACT: Engineering consultants are critical to a construction project because their project designs significantly affect each project phase. However, the quality of a consulting service is difficult to evaluate. This study aims to establish a performance evaluation mechanism for consulting firms that execute detailed designs in the design phase and provide consultancy in the construction phase for Mass Rapid Transit (MRT) projects in Taiwan. The evaluation is based on the concept of Balanced Scorecard (BSC) to establish a hierarchical structure for 24 (for the design phase) and 17 (for the construction phase) selected indicators which are consequently weighted by the Analytical Hierarchical Process (AHP) technique. When evaluating a consulting firm, the evaluator only needs to rate a score for each of the indicator in these two phases. The score of each indicator is then multiplied by its corresponding weight to come up with a total score for each of these two phases. Finally, an overall score for the consulting firm can be obtained by summing up 60% of the total weighted score in the design phase and 40% of that in the construction phase. Two real MRT projects in Taipei are tested against the mechanism and the results have demonstrated that the performance evaluation mechanism is effective and promising.

1 INTRODUCTION

In Taiwan, the Mass Rapid Transit (MRT) system in Taipei City offers millions of daily commuters who live in the outer Taipei metropolitan area a means of safe, economic, comfortable, and convenient way to commute to downtown. However, MRT projects are integration of civil, mechanical, electrical, signal, and control works which are procured via various contracts; they are much more sophisticate in construction interfaces than most public works such as highways and buildings, thus, the evaluation of the consulting firms are different from that for traditional projects.

This research can be described as follows. First, commonly used evaluation indicators for the design phase were selected by compiling 16 types of information, such as governmental agency regulations, legal requirements, contract provisions, research reports, and related international publications. Second, the concept of the Balanced Scorecard (BSC) was incorporated to reduce and structure these indicators by interviewing three experts of MRT projects. Third, these structured indicators were weighted using the Analytic Hierarchy Process (AHP) technique with input from experts from MRT projects. Last, scoring criteria for each indicator was established and the validity of the total evaluation mechanism was evaluated by comparing the performance and calculated scores of two engineering consulting firms that participated in real MRT projects.

2 PERFORMANCE EVALUATION

In this study, "performance evaluation" is defined as the quality of services provided by engineering consultants during the design and construction phases of MRT projects in Taipei, Taiwan. In this study, performance evaluation indicators from 16 types of literature are collected and organized. The collected vast amount of performance evaluation data serves as the basis of follow-up interviews, evaluation indicator selections and evaluation mechanisms. Finally, 40 (for design phase) and 35 (for construction phase) performance indicators of aforesaid perspectives were noted. Then one supervisor and two senior engineers were interviewed and asked to select the indicators through their years of professional knowledge. Evaluation indicators that were deemed as of "critical importance" by most professionals were then selected as the performance evaluation indicators in this research, as shown in Tables 1 and 2.

2.1 Balanced Scorecard (BSC)

BSC has been widely used by many enterprises as a method for performance evaluation management. It strongly suggested the performance evaluation be processed by the four perspectives of performance evaluation (finance, customers, internal processes, and learning and growth) so that the deficiency of evaluating performance executed only by financial

Table 1. Selected 24 indicators and their weights for the design phase.

Perspective	Strategic purpose	Evaluation indicator	Weight (%)
Customer (16.3%)	C_1 improving service performance	C_{1-1} progress of design	2.1
		C_{1-2} solutions to questions	3.0
	C_2 improving owner's satisfaction	C_{2-1} owner's satisfaction	8.3
		C_{2-2} communication & coordination	2.9
Financial (45.6%)	F_1 meeting project budget	F_{1-1} accuracy of quantity estimates	16.9
		F_{1-2} estimation of construction cost	11.8
	F_2 creating profits	F_{2-1} conformance of functions	13.0
		F_{2-2} value engineering analysis	3.9
Internal business processes (22.0%)	I_1 implementing project management strategy	I_{1-1} project management method	1.2
		I_{1-2} organization of human resources	1.0
	I_2 strengthening documentation control	I_{2-1} review and reply of submissions	1.0
		I_{2-2} management of document	0.8
		I_{2-3} accuracy of design documents	2.9
	I_3 guarantee of design quality	I_{3-1} quality assurance	1.5
		I_{3-2} responsibility of design change	2.6
		I_{3-3} environmental and ecological impacts	2.4
	I_4 completeness of design documents	I_{4-1} compiling of preliminary information	1.5
		I_{4-2} analysis of design	1.0
		I_{4-3} analysis of constructability	3.4
		I_{4-4} interface integration	2.7
Learning and growth (16.1%)	L_1 improving effectiveness of manpower	L_{1-1} certificates and academia degrees	1.4
		L_{1-2} professional experiences	5.2
	L_2 improving professional skills	L_{2-1} professional training	7.2
		L_{2-2} fees for research and development	2.3
Total			100.0

Table 2. Selected 17 indicators and their weights for the construction phase.

Perspective	Strategic purpose	Evaluation indicator	Weight (%)
Customer (22.5%)	CC_1 control of design progress	CC_{1-1} control of progress	6.8
		CC_{1-2} solutions to questions	4.8
	CC_2 coordination and compliance	CC_{2-1} interface coordination	6.1
		CC_{2-2} compliance of owner's intention	4.7
Financial (32.5%)	CF_1 meeting project budget	CF_{1-1} difference in quantities	9.3
		CF_{1-2} adequacy of unit price breakdowns	7.1
	CF_2 meeting contractual requirements	CF_{2-1} conformance of functions	9.8
		CF_{2-2} design omissions	6.3
Internal business processes (27.0%)	CI_1 control of documents	CI_{1-1} accuracy of design documents	6.5
		CI_{1-2} reply of review documents	3.1
		CI_{1-3} completeness of As-built documents	4.2
	CI_2 coordination of project manpower	CI_{2-1} human resources organization	6.6
		CI_{2-2} execution of site superintendence	6.6
Learning and growth (18.0%)	CL_1 improving effectiveness of manpower	CL_{1-1} certificates and academia degrees	5.8
		CL_{1-2} professional experiences	5.0
	CL_2 improving professional skills	CL_{2-1} professional training	3.8
		CL_{2-2} fees for research and development	3.4
Total			100.0

indicators can be avoided while the performance evaluation system possesses both strategic management and balance management (Karplan & Norton 1992). In Tables 1 and 2, the evaluation indicators are structured into the four perspectives of BSC.

2.2 Analytical Hierarchy Process (AHP)

AHP is a multi-objective decision making method developed by Professor Saaty when he conducted a problematic study on contingency plans for the United States Department of Defense. It is mainly applied in uncertainties and in the decision-making problems with several evaluation criteria, especially fit for the evaluation of qualitative information. After this method turned the complicated problems into a hierarchical chart, decision makers could make pairwise comparison according to the evaluation criteria of each level, create a comparison matrix so as to discover the relative weight for each criterion, and finally calculate the rating or weight of each selected alternatives with the relative weight to serve as a reference for decision makers (Saaty 1980, Saaty 1990) in Tables 1 and 2, weights of the indicators are obtained by the AHP technique contributed respectively by 13 and 15 experts of MRT projects.

2.3 Weight of indicator

Among the four perspectives, "Financial" perspective is weighted highest in both design (45.6%) and construction (32.5%) phases. Consequently, among all indicators in Table 1, "F_{1-1}, Accuracy of Quantity Calculation" (16.9%), "F_{2-1}, Conformance of Function" (13.0%), and "F_{1-2}, Construction Cost Estimate" (11.8%) are the highest ones in the design phase. Meanwhile, in Table 2, "CF_{2-1} Conformance of Functions" (9.8%), "CF_{1-1} Difference in Quantities" (9.3%), and "CF_{1-2} Adequacy of Unit Price Breakdowns" (7.1%) are among the highest in the construction phase.

2.4 Scoring criteria of indicator

In this research, the score of an indicator is rated in a scale between 55 and 100, where a higher score denotes a better performance of the consulting firm in the design phase. It was intended to differentiate the performance to five categories in which each category has two ratings with 5 points apart. The rationale for the score starting from 55 instead of 0 is that a consulting firm must have met certain qualification before entering into the project. Table 3 illustrates the scoring criteria of indicator F_{1-1} in the design phase.

The scoring criteria of indicator CF_{1-1} in the construction phase is illustrated in Table 4. The CF_{1-1} indicator means quantity differences between the quantities take-off in the design phase and the actual quantities that are constructed. Normally, the difference in these two quantities should be within 10%. If it is above 10% and the difference is attributable to the design consulting firm, the firm should be fined for such quantity errors.

2.5 Overall evaluation score

When evaluating a consulting firm, the evaluator rates a score for each of the indicators in these two phases. The score of each indicator is then multiplied by its corresponding weight to come up with a total score for each of these two phases. Finally, an overall score for the consulting firm can

Table 3. Scoring criteria of indicator F_{1-1} (Accuracy of quantity estimates).

Score rated (worse/better)		Descriptions of criteria
☐ below 55	☐ 60	Not in accordance with the design drawings for work item listing and quantity Estimates, missing items and quantity error over 10%.
☐ 65	☐ 70	In accordance with the design drawings for work item listing and quantity Estimates, quantity calculation error, missing items.
☐ 75	☐ 80	In accordance with the design drawings for work item listing and quantity Estimates, minor quantity calculation error, no missing items.
☐ 85	☐ 90	In accordance with the design drawings for work item listing and quantity Estimates, quantity calculated correct, no missing items.
☐ 95	☐ 100	In accordance with the design drawings for work item listing and quantity Estimates, the quantity calculation is complete, easy to control, distribution is correct after checking.

Table 4. Scoring criteria of indicator CF_{1-1} (Difference in quantities).

Score rated (worse/better)		Descriptions of criteria
☐ below 55	☐ 60	More than 20 items have a quantity difference more than 10%.
☐ 65	☐ 70	More than 10 items have a quantity difference more than 10%.
☐ 75	☐ 80	Less than 5 items have a quantity difference more than 10%.
☐ 85	☐ 90	No item has a quantity difference more than 10%.
☐ 95	☐ 100	No item has a quantity difference more than 5%.

be obtained by summing up 60% of the total score in the design phase and 40% of that in the construction phase. The percentiles for the two phases are determined by interviewed experts.

3 CASE STUDY

Two real MRT projects in Taipei, Taiwan were selected as the test cases to check the accuracy and effectiveness of the established performance evaluation mechanism. For each test case, five to seven project participants who were at management level in the government, such as site manager, section chief and project director, were invited to evaluate the consulting firm. i.e., those management individuals participating in the same project were invited to evaluate the same consulting firm of that project; the final score of each indicator and the overall weighed score of the consulting firm were then numerically averaged after evaluation.

3.1 *Case 1*

The performance evaluation results for the design and construction phases for Case 1 are shown in Figures 1 and 2. Notably, indicator F_{1-1} is relatively low with a score around 70. However, the overall score of the consulting firm for Case 1 is 86.96, which is calculated by summing up 60% of 85.91, the total weighted score of the design phase; and 40% of 88.53, the total weighted score of the construction phase. The high overall score represents a good performance of the consulting firm.

3.2 *Case 2*

The performance evaluation results for the design and construction phases for Case 2 are shown in

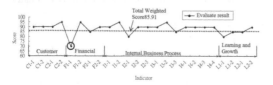

Figure 1. Scores of indicators in case 1 for the design phase.

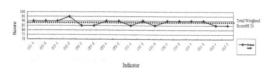

Figure 2. Scores of indicators in case 1 for the construction phase.

Figure 3. Scores of indicators in case 2 for the design phase.

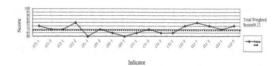

Figure 4. Scores of indicator in case 2 for the construction phase.

Figures 3 and 4. Notably, indicators F_{1-1} and I_{3-2} are very low with a score only 55, which reveal a very poor performance during the design phase. The overall score of the consulting firm for Case 1 is 69.08, which is calculated from summing up 60% of 69.00, the total weighted score of the design phase; and 40% of 69.22, the total weighted score of the construction phase. The low overall score depicts a poor performance of the consulting firm.

3.3 *Discussion*

In the evaluation mechanism, valuation indicators are different for the design and construction phase. However, the trend of scores is similar in both phases; i.e., if the consulting firm is professionally sound and competitive, it would perform well in both phases, even though there could be a single low score for certain indicator. As shown in Case 1, only indicator F_{1-1} scores low in Figure 1, but all the other indicators remain high scores, and there is no specific indicator in the construction phase scores low as shown in Figure 2. In Case 2, at least two indicators, F_{1-1} and I_{3-2}, are extremely low; and only two indicators (out of 17) score at 80.

4 CONCLUSIONS

This research incorporates expert interviews, BSC concepts, and AHP methodology to establish a mechanism for evaluating the performance of consulting firms of MRT construction projects. The evaluation processes are divided into design and construction phases where each phase has 24 and 17 evaluation indicators, respectively. These indicators were obtained from experts and categorized into BSC's four aspects. Each of the indicators was weighted by using the AHP approach and can be rated to a score using the rating criteria established in this research.

Test results of the two representative cases indicate that the score of each case indeed matches the actual contract-fulfillment by the consulting firms, whose individual performance was also manifested by their scores in different indicators. The overall score objectively represents the level of consulting service, thus it can be used as a reference when reviewing tenders in future MRT projects. In sum, test results are very promising and have demonstrate validity of the performance evaluation mechanism for consulting firms of MRT projects. The performance evaluation mechanism for other types of huge projects can be developed using similar approach developed in this research.

REFERENCES

Karplan, R.S. & Norton, D.P. 1992. The Balanced Scorecard-measures That Performance, *in: Harvard Business Review, Harvard Business Publishing, Boston*, 70(1): 71–79.

Saaty, T.L. 1980. *The Analytic Hierarchy Process*, McGraw Hill, Pittsburgh.

Saaty, T.L. 1990. How to Make a Decision: The Analytic Hierarchy Process, *European Journal of Operational Research*, 48: 9–26.

Advanced Materials and Structural Engineering – Hu (Ed.)
© *2016 Taylor & Francis Group, London, ISBN 978-1-138-02786-2*

Flexural behavior of hollow deck-plate slabs with simplified construction procedure

K.H. Yang
Department of Plant·Architectural Engineering, Kyonggi University, Suwon, Kyonggi-do, South Korea

M.K. Kwak
Department of Architectural Engineering, Kyonggi University Graduate School, Seoul, South Korea

S.H. Yun
Kwangjang VDS Co. Ltd., Seoul, South Korea

S.T. Park
Winhitech Co. Ltd., Kolon Digital Tower Aston, Geumcheon-gu, Seoul, South Korea

ABSTRACT: To overcome the limitations of the previous hollow flat slab system, the present study developed a new simplified system that is capable of shortening the construction time and controlling the buoyancy of the hollow materials. Six one-way slab specimens were prepared to evaluate the flexural behavior of the developed system, and we confirmed the experimental results with the predicted values of the ACI 318-11 design procedure. The test results clearly showed that the flexural behavior of the developed hollow deck-plate slab system is equivalent to that of the companion solid slabs with the same member dimension and material properties. Finally, the flexural design specification for the developed slab system was proposed based on test results.

1 INTRODUCTION

Since the introduction of "Bubble Deck," (Schnellenbach-Held & Pfeffer 2002, Teja et al. 2012) there has been an increased interest in the practical application of biaxial hollow flat slabs. Hollow flat slabs have various advantages including material-savings and reduction of dead load. Hollow flat slabs are constructed by distributing hollow plastic balls and reinforcing bars in concrete, and are designed to carry loads in any direction. A major disadvantage of current fabrication methods is that the buoyancy of the hollow plastic balls in concrete deteriorates workability. Decreased workability can result in construction delays. Furthermore, the buoyancy of the hollow plastic balls can worsen the structural performance of the slab by altering the designed layout of reinforcing bars.

The present study developed a hollow deck-plate slab system with a simplified construction procedure.

To shorten the construction time, steel-deck plate was used instead of wood panels for concrete forms and lightweight expanded polystyrene (EPS) was used to form internal voids. To control the buoyancy of the hollow materials, the EPS was

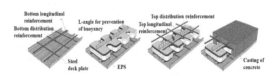

Figure 1. Hollow deck-plate slab construction procedure.

fixed in place by L-angle brackets connected into T-flange web of the deck-plate to avoid floating, as shown in Figure 1. The top and reinforcing bottom bars were arranged on the L-angle and T-flange of the deck-plate, respectively. This study examined the flexural behavior of the developed slab system.

2 EXPERIMENTAL DETAILS

2.1 Test specimen details

Six one-way slabs were prepared for flexural tests under uniform load, as shown in Figure 2. The parameters investigated were the longitudinal tensile reinforcement ratio (ρ_s) and the orientation of the expanded polystyrene (EPS) and steel deck-plate. The EPS and deck-plate were arranged either along the longitudinal or transverse axes of the slab specimens. The selected ρ_s for each group was the minimum and maximum values specified in ACI 318-11 (2011) flexure provisions (Table 1). In each group, solid slabs with minimum values of ρ_s were also prepared for comparisons.

All one-way slab specimens had the same geometry and dimensions. The nominal width and depth of section were 600 mm and 280 mm, respectively. The full length was 6400 mm and the length between the centers of both end supports was 6000 mm. For temperature reinforcement at the top region of the slabs, longitudinal reinforcing bars with 13 mm diameter and transverse reinforcing bars with 10 mm diameter were arranged at 300 mm spacing.

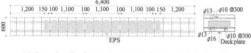

(a) Longitudinal arrangement of EPS and deck-plate

(b) Transverse arrangement of EPS and deck-plate

Figure 2. Details of slab geometry and arrangement of component materials (unit: mm).

2.2 Material properties

In the ready-mixed concrete mixture, the water-to-binder ratio by weight and fine-aggregate-to-total-aggregate ratio by volume were 47.8% and 47.5%, respectively. The maximum diameters of the coarse and fine aggregates were 25 mm and 5 mm, respectively. The measured concrete compressive strength (f_c') was 35 MPa, and the elastic modulus was 29709 MPa. A hot dip galvanized steel plate with 0.8 mm thickness and yield strength of 238 MPa was used for deck-plate production. A section of the deck-plate was composed of one bottom plate and two T-shaped webs, as shown in Figure 2. The flange width and web height of the T-shaped section were 30 mm and 58 mm, respectively. Mild deformed steel bars were used as reinforcing bars. The yield strengths of steel bars were measured to be 575 MPa, 518 MPa, 716 MPa, and 523 MPa for bars with diameters of 10 mm, 13 mm, 16 mm, and 32 mm, respectively. The moduli of elasticity of all metallic materials were approximately 200 GPa.

Lightweight EPS boxes with dimensions of $1100 \times 420 \times 120$ mm were used as a hollow material. The specific gravity and water absorption of the EPS were 0.016 and 0.008 g/cm^2, respectively. The compressive strength and flexural strength of the EPS were 0.09 MPa and 0.17 MPa, respectively.

2.3 Test procedure and instrumentation

All one-way slabs were tested to failure under symmetrical eight-point top loads designed to simulate

Table 1. Specimen details and summary of test results.

Name	Longitudinal reinforcement Top	Bottom	Direction of deck plate and EPS	ρ_s'	ρ_s	ρ_d	$\rho/(\rho_{max}-\rho_d)$	M_{cr} (kN·m)	M_n (kN·m)	μ
L-S1	2-φ13	2-φ13+2-φ16		0.0018	0.0047	0.0057	0.22	31.8	140.3	2.6
L-V1		2-φ13+2-φ16		0.0018	0.0047	0.0057	0.22	16.1	145.0	2.9
L-V2		2-φ32		0.0018	0.011	0.0057	0.78	22.1	247.1	1.6
T-S1	2-φ13	2-φ13+2-φ16		0.0020	0.0051	0.0029	0.18	10.1	88.5	1.8
T-V1		2-φ13+2-φ16		0.0020	0.0051	0.0029	0.18	14.9	88.3	1.7
T-V2		2-φ32		0.0020	0.012	0.0029	0.71	17.5	183.2	1.5

All of the specimens tested are identified by a three character alphanumeric code. For each sample, the first letter indicates the arrangement direction of deck-plate and EPS, the second letter indicates the type of slab (S = solid slab without EPS and V = hollow slab), and the third Figure refer to the longitudinal tensile reinforcement ratio (1 = minimum value and 2 = maximum value specified in ACI 318-11 provision.
[Note] ρ_s' = longitudinal compressive reinforcement ratio, ρ_s = longitudinal tensile reinforcement ratio, ρ_d = deck plate ratio, ρ_{max} = maximum longitudinal reinforcement ratio calculated in accordance with ACI 318-11 provision, Mcr = initial cracking moment, Mn = ultimate moment, and μ = displacement ductility ratio.

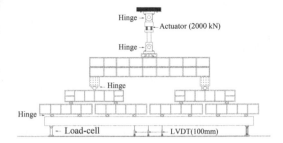

Figure 3. Test set-up under uniform load.

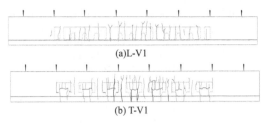

(a)L-V1

(b) T-V1

Figure 4. Typical crack propagation of slabs tested.

a uniform load (Fig. 3). The applied loads were controlled by a displacement rate of 1.25 mm/min, using a 2000 kN capacity actuator with a hydraulic shaft of 300 mm length. Each slab specimen was supported by a hinge at one end and on a roller at the other end. The reactions at both end supports were recorded using 1000 kN capacity load cells. Vertical deflections at mid-span were recorded using linear variable differential transducers. The propagation of flexural cracks in the slabs was also recorded using extensor meters within the flexural-critical region near mid-span. The strains of longitudinal tensile reinforcement were monitored using electrical resistance strain gages.

3 TEST RESULTS AND DISCUSSIONS

3.1 Crack propagation and failure mode

All one-way slabs tested showed crack propagation typically observed under flexure and no diagonal shear cracks (Fig. 4). In addition, the slabs failed by the crushing of compression concrete at the maximum moment zone after the peak load. These results indicate that the behavior of the one-way slabs is governed by flexure. The flexural cracks generally first occurred at the maximum moment zone and subsequently propagated upward as the load increased. At the peak load, the flexural crack in slabs with the minimum ρ_s propagated up to 0.95 h, where h is the overall height of slab. Increasing ρ_s was found to decrease the crack height. The propagation rate was higher in slabs with a transverse deck-plate than in slabs with a longitudinal deck-plate. With increasing load, the flexural cracks were evenly distributed with an average spacing between 200 and 90 mm. The distribution of flexural cracks was significantly affected by ρ_s but was marginally affected by the arrangement direction of the deck-plate. The distribution and propagation of flexural cracks were not significantly affected by the addition of EPS in the slabs.

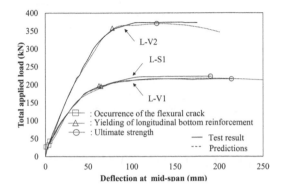

(a) Longitudinal arrangement of deck-plate

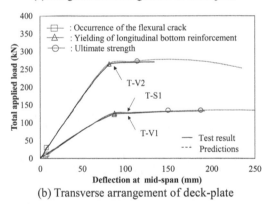

(b) Transverse arrangement of deck-plate

Figure 5. Mid-span deflection against the total applied loads.

3.2 Load–deflection relationship

The total applied load versus mid-span deflection curves of the different slab specimen are plotted in Figure 5. The self-weight (22.7 kN) of the steel loading beams was added to the total applied load. On the same figure, predicted curves determined by the section lamina method (Yang 2014) are also given for comparisons. In the section lamina method, the strain compatibility condition and equilibrium of forces were employed with the constitutive equation of each material. Before the occurrence of the first flexural cracks, all slabs exhibited a virtually linear response, showing that the initial slope of the load–deflection curve was only slightly affected by the addition of EPS. However, the initial slope was lower in slabs with a transverse deck-plate than in slabs with a longitudinal deck-plate. The rate of deflection sharply increased with the occurrence of the first flexural crack. This trend was more notable in slabs with the minimum ρ_s value than in slabs with the maximum ρ_s value. The yield strength of the steel deck-plate is only slightly affected the deflection rate. As the applied load increased, the longitudinal tensile reinforcement yielded, resulting in an abrupt change in the slope of the load–deflection curve. The response observed after the longitudinal tensile reinforcement yielded can be characterized by a plastic flow with a high increasing rate of deflection under a constant load. The plastic flow response continued up to the first visible crushing of the concrete in compression at the maximum moment zone.

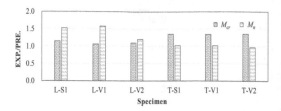

Figure 6. Comparisons of moment capacities with predictions obtained from ACI 318-11 procedure.

4 FLEXURAL CAPACITY AND DUCTILITY

The flexural capacity (M_n) of the slabs tested increased with the increase in ρ_s, regardless of the arrangement direction of the deck-plate (Table 1). The flexural capacity of the slabs was also significantly affected by the arrangement of the deck-plate. The M_n value of slabs with the longitudinal deck-plate was higher by an average of 53% than that of the companion slabs with the transverse deck-plate. The addition of EPS hardly affected the M_n value. The hollow slabs L-V1 and T-V1 gave very similar M_n values to the companion solid slabs L-S1 and T-S1.

The displacement ductility ratio $\mu\,(=\Delta_n/\Delta_y)$ was examined to evaluate the flexural ductility of the slab specimens, where Δ_n is the deflection at the peak load and Δ_y is the deflection at yielding of the longitudinal tensile reinforcement. The ductility ratio of the slab specimens decreased with the increase in ρ_s, as given in Table 1. This decreasing ratio was more notable in slabs with a longitudinal deck-plate than in slabs with a transverse deck-plate. When ρ_s increased from minimum to maximum, the μ value decreased by 40% and 17% for slabs with longitudinal and transverse deck-plates, respectively. Furthermore, the μ value was significantly affected by the arrangement of the deck-plate. Higher values were commonly obtained in slabs with a longitudinal deck-plate than in the companion slabs with a transverse deck-plate. The difference in the μ value according to the arrangement of the deck-plate decreased with the increase in ρ_s. The slab L-V1 had 1.7 times higher μ value

than the companion slab T-V1, whereas the slab L-V2 had only 1.06 times higher μ value than the companion T-V2. The μ value of the developed hollow slab system was independent of the existence of EPS. Very similar μ values were obtained in both slabs L-S1 and L-V1 and in both slabs T-S1 and T-V1.

5 COMPARISONS WITH PREDICTIONS OBTAINED FROM ACI 318-11 PROVISION

The initial cracking moment (M_{cr}) and ultimate moment (M_n) of the developed slab system were compared with the predictions obtained from the procedure specified in the ACI 318-11 provision (Fig. 6). The measured values include the moments due to the applied loads and self-weight of the steel loading beams. In the calculation of the moments, the contributions of the longitudinal and transverse deck-plates were ignored. The M_{cr} measured from the developed slab system was generally higher than the predicted values. The mean and standard deviation of the ratios between the experiments and calculations were 1.22 and 0.31, respectively. Furthermore, the measured M_n was usually higher than the predicted values. This indicates that the flexural capacity of the developed hollow deck-plate slab system can be conservatively evaluated using the equivalent stress block-based procedure specified in the ACI 318-11 provision.

6 FLEXURAL DESIGN CONSIDERATION OF THE DEVELOPED SLAB SYSTEM

Based on the comparisons of the load–deflection curves, moment capacities, and displacement ductility measured in both solid slabs and the developed hollow deck-plate slab, the following design considerations are recommended: 1) EPS can be neglected in predicting initial stiffness, failure mode, moment capacities, and ductility; 2) the contribution of the deck-plate to the moment capacities is disregarded for conservative design; and 3) the amount of longitudinal deck-plate is included in calculating the maximum ρ_s value but was excluded in calculating the minimum ρ_s value.

7 CONCLUSIONS

A new simplified system using a deck-plate and EPS was developed for shorten construction time. From the flexural tests of the developed one-way slabs, the following conclusions may be drawn:

1. The crack propagation rate was more notable in slabs with a transverse arrangement of the deck-plate than in slabs with a longitudinal deck-plate. The distribution and propagation of flexural cracks were not significantly affected by the addition of EPS.
2. Very similar load–deflection relationships were observed in both solid slabs and the companion hollow deck-plate slabs.
3. The flexural capacity (M_n) of slabs with the longitudinal deck-plate was higher by an average of 53% than that of the companion slabs with the transverse deck-plate. The addition of EPS hardly affected M_n.
4. The ductility ratio (μ) was higher in slabs with a longitudinal deck-plate than in the companion slabs with a transverse deck-plate. However, the difference of the μ value according to the arrangement of the deck-plate decreased with the increase in longitudinal tensile reinforcement

ratio. The μ value of the developed hollow slab system was independent of the addition of EPS.
5. The flexural capacity of the developed hollow deck-plate slab system can be conservatively evaluated using the equivalent stress block-based procedure specified in ACI 318-11 provision.

ACKNOWLEDGEMENT

This research was supported by the Public Welfare & Safety Research Program through the National Research Foundation of Korea (NRF) funded by the Ministry of Science, ICT & Future Planning (2013067519). The authors also greatly appreciate the efforts and support provided by Winhitech Co. Ltd.

REFERENCES

ACI Committee 318. 2011. *Building code requirements for structural concrete (ACI 318-11) and commentary (ACI 318R-11)*. American Concrete Institute.
Schnellenbach-Held, M. & Pfeffer, K. 2002. Punching behavior of biaxial hollow slabs, *Cement and Concrete Composites*, 24: 551–556.
Teja, P.P. Kumar, P.V. Anusha, S. Mounika, C.H. & Saha, P. 2012. *Structural behavior of bubble deck slab*, IEEE-Inter. Confer on ICAESM-2012, 383–388.
Yang, K.H. 2014. *Evaluation of serviceability and structural performance of hollow deck slab system*, Technical Report, Kyonggi University.

Advanced Materials and Structural Engineering – Hu (Ed.)
© 2016 Taylor & Francis Group, London, ISBN 978-1-138-02786-2

Diffusion model and damage development of concrete exposed to freeze-thaw cycles

L. Jiang, X.Z. Weng, B.H. Yang, R.Y. Zhang & J.Z. Liu
Department of Airfield and Building Engineering, Air Force Engineering University, Shaanxi, Xi'an, China

X.C. Yan
PLA Unit 95746, Qionglai, China

ABSTRACT: Based on the Fick's second law, a one-dimensional diffusion model is established, which subsequently infers the diffusion equation of semi-infinite media to the finite region. A relevant single-side freeze-thaw test is suggested to maintain the concrete soaking height at the same level so that the diffusion equation of semi-infinite media is available. Specifically, this paper examined the degree of saturation during the test, and got the relationship between the diffusion coefficient and the level of saturation as well as the freeze-thaw cycles by solving the diffusion equation. According to the trend of the gradient curve, the moisture diffusion is divided into three stages that are general diffusion, slow diffusion, and sharp diffusion. The magnitude range of the diffusion coefficient is 10^{-5}~10^{-4} cm²/s, which is decreased to the minimum at the time of 28 freeze-thaw cycles, and then increased until the end of the test.

1 INTRODUCTION

In cold regions, the concrete structures of airport pavement, bridge and other projects under wet conditions are always seriously damaged by the freeze-thaw cycles. To solve this problem, researchers focus on the mechanism of freeze-thawing. After a long-term research, the four typical theories, micro-ice-lens theory (Collins 1944), hydraulic pressure (Power 1945), osmotic pressure (Power & Helmuth 1949, Litvan 1980) and critical degree of saturation theory (Fagerlund 1977) are established to explain this degradation. Although these four typical theories have different views on the failure of concrete under freeze-thaw cycles, they all agree that the existence and diffusion of pore moisture are the main reason for the degradation. Meanwhile, the diffusion coefficient is an important parameter in the research of diffusion rule. Diffusion coefficient under freeze-thaw cycles needs more attention, some work have been carried out and regard diffusion coefficient as constant (Wang et al. 2002, Scherer & Valenza 2005, Cwirzen & Penttala 2005). However, Fagerlund (1993) holds that diffusion coefficient in concrete exposed to freeze-thaw cycles is related to water-binder ratio and pore structure. When concrete structure exposed to freeze-thaw cycles, the degradation of pore structure may change the diffusion coefficient. In this paper, based on the Fick's second law, a one-dimensional diffusion model is established, and the diffusion coefficient is regarded as a variable. Subsequently, a relevant single-side freeze-thaw test is suggested. According to the test, the degree of saturation is examined, and the relationship between diffusion coefficient, saturation and freeze-thaw cycles is displayed, in addition, different stages of diffusion are discussed by solving the diffusion equation.

2 DIFFUSION MODEL

The transmission of water in the porous material is a complex process including adsorption, diffusion, osmosis, and other modes. For compact pore structure, the main transmission mode is diffusion, and for sparse pore structure, the main mode is osmosis. In this paper, the concrete under research is used for pavement, which pore structure is compact, so that the main transmission mode in this paper is diffusion. Assume the pore water diffusion accord with the Fick's second law, and then the moisture diffusion differential equation is shown in Formula 1.

$$\frac{dC}{dt} = \frac{d}{dx}\left(D\frac{dC}{dx}\right) \tag{1}$$

Different initial condition and boundary condition result in different solutions of the diffusion differential equations. At the beginning of the diffusion, the total amount of substance M deposited at time $t = 0$ and in the plane $x = 0$, at time $t > 0$, the amount of substance diffusion to the both sides of the coordinate origin, as shown in Figure 1. Crank (1975) defined the one-dimensional diffusion in this condition as the plane source diffusion and solved the equation as shown in Formula 2.

$$C = \frac{M}{2(\pi Dt)^{1/2}}\exp(-x^2/4Dt) \tag{2}$$

In Formula 2, C is the concentration of the substance and D is the diffusion coefficient, if the initial conditions and boundary conditions are at time $t = 0$, the concentration at the region $x < 0$ is C_0 and the concentration at the region $x > 0$ is 0 at time $t > 0$, substance start to diffusion along the positive direction of axis X. In order to solve the equation in this condition, consider the diffusing substance in an element of width dx as a plane source, as shown in Figure 2(a).

Semi-infinite range of substance can be divided into infinite element section, and the distance from unit section to the diffusion point $P(x)$ is ξ, which range is $x\sim\infty$. According to Formula 2, the concentration at point P, at time t is shown in Formula 3.

$$dC = \frac{dM}{2(\pi Dt)^{1/2}}\exp\left(-\xi^2/4Dt\right) \tag{3}$$

In Formula 3, consider the diffusion substance in an element of width dx to be a line source of concentration $C_0 dx$, where C_0 is the initial concentration, which is substituted in Formula 3, and then Formula 4 is obtained.

$$dC = \frac{C_0 dx}{2(\pi Dt)^{1/2}}\exp\left(-\xi^2/4Dt\right) \tag{4}$$

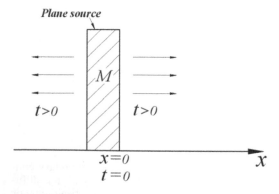

Figure 1.   One-dimensional diffusion of plane source.

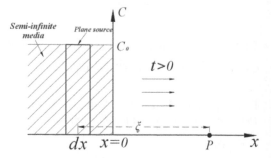

(a) Diffusion to the point

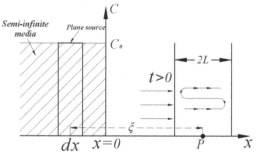

(b) Diffusion to the region

Figure 2.   One-dimensional diffusion of semi-infinite media.

Summing over the successive elements dx, and the concentration of semi-infinite substance at diffusion point P is shown in Formula 5.

$$C(x,t) = \frac{C_0}{2\sqrt{\pi Dt}}\int_x^\infty e^{-\xi^2/4Dt}dx = \frac{C_0}{\sqrt{\pi}}\int_{x/2\sqrt{Dt}}^\infty e^{-\eta^2}d\eta \tag{5}$$

$$\eta = \xi/2\sqrt{Dt} \tag{6}$$

A standard mathematical function, complementary error function is shown in Formula 7.

$$erfc(\beta) = \frac{2}{\sqrt{\pi}}\int_\beta^\infty e^{-\lambda^2}d\lambda \tag{7}$$

Then, Formula 5 is written in the form

$$C(x,t) = \frac{1}{2}C_0 erfc(x/2\sqrt{Dt}). \tag{8}$$

Formula 8 is available only for the diffusion in which diffusion region is a point but the diffusion region is normally a finite width region, therefore, as shown in Figure 2(b), the diffusion point is expanded into the 2 L width region. Because of the procedure of reflection and superposition

(Crank 1975), in this $2L$ width region, the diffusion distance of an element substance may be ξ, $\xi+2L$, $\xi+4L$, ..., $\xi+2nL$ ..., where n is the number of the propagation. Dividing the element section into n equal parts, and each part of the substance is M/n, the diffusion distance of the nth element is assumed $\xi+2(n-1)L$, therefore, at diffusion point P, the concentration of semi-infinite substance to $2L$ width region is shown in Formula 9.

$$
C = \int_{x}^{\infty} \frac{C_0}{2n(\pi Dt)^{1/2}} \exp(-x^2/4Dt)dx
$$

$$
+ \int_{x+2L}^{\infty} \frac{C_0}{2n(\pi Dt)^{1/2}} \exp(-x^2/4Dt)dx + ...
$$

$$
+ \int_{x+2nL}^{\infty} \frac{C_0}{2n(\pi Dt)^{1/2}} \exp(-x^2/4Dt)dx \qquad (9)
$$

The simplified equation is shown in Formula 10.

$$
C = \sum_{n=1}^{n} \frac{C_0}{2n} erfc\left(\frac{x+2nL-2L}{2\sqrt{Dt}}\right) \qquad (10)
$$

3 EXPERIMENTAL DESIGNS

3.1 Initial condition control

In order to satisfy the initial conditions of the one-dimensional diffusion model, a single-side freeze-thaw test is available. In the test, the specimen should be treated as follows:

1. As shown in Figure 3(a), all sides of the specimen, except bottom side, are sealed with a tape.
2. Maintain the water level 1 cm above the bottom side, especially during the freeze-thaw test. Adding water to maintain the water level unchanged, therefore, the sink seems to have an infinite number of water, and the distribution of diffusion source in the test is considered in the semi-infinite range.

3.2 Diffusion equation

In section 3.2, the concentration (C) is not easy to detect in freeze-thaw test. However, the degree of saturation (S) (Fagerlund 1993) is available in the test. The relationship between C and S is $S = C/\rho$, where ρ is the water density. Thus, we obtain the diffusion equation for freeze-thaw test, which is shown in Formula 11.

$$
S = \sum_{n=1}^{n} \frac{1}{2n} erfc\left(\frac{x+2nL-2L}{2\sqrt{Dt}}\right) \qquad (11)
$$

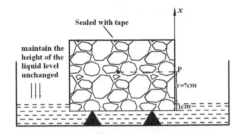

(a) Diagram of the test

(b) Specimens of the test

Figure 3. Single-side freeze-thaw test.

As shown in Formula 12, S can also be expressed by weight percent of pore water. In the Formula, M_w is the quality of pore water at test time t, M_p is the quality of pore water when the concrete in the test is fully saturated, M_d is the quality of specimen at test time t, M_s is the quality of a completely dry specimen, and M_s is the quality of a fully saturated specimen.

$$
S = \frac{V_w}{V_p} = \frac{V_w \times \rho}{V_p \times \rho} = \frac{M_w}{M_p} = \frac{M_t - M_d}{M_s - M_d} \qquad (12)
$$

The specimen is $15 \times 15 \times 15$ cm cube, and the water level is always 1 cm above the bottom side of the specimen, so that $2L = 14$, $L = 7$. This paper takes the geometric center pore of the specimen as the research object, so that $x = 7.5$. Parameter n is the water reciprocating propagation times, the value of complementary error function tends to 0 and the value of S tends to constant after finite propagation, in this paper, we set $n = 20$. The diffusion equation during freeze-thaw test is shown in Formula 13.

$$
S = \frac{M_t - M_d}{M_s - M_d} = \sum_{n=1}^{n} \frac{1}{2n} erfc\left(\frac{14n - 6.5}{2\sqrt{Dt}}\right) \qquad (13)
$$

3.3 Test time control

In the test, control of the freeze-thaw time per 12 hours a cycle, the change of the temperature is

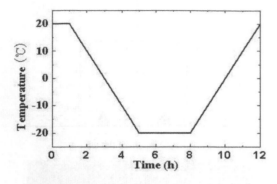

(a) Diagram of temperature cycle.

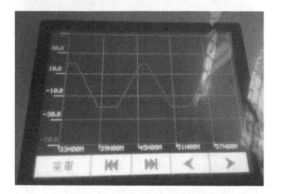

(b) Temperature cycle of the test.

Figure 4. Temperature cycle of the freeze-thaw test.

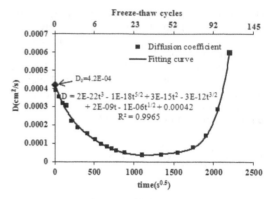

Figure 5. The diffusion coefficient during the freeze-thaw test.

shown in Figure 4. Weighing the exfoliation and the specimen quality at time 0.5d, 1d, 2d, 3d, 4d, 5d, 6d, 7d, 9d, 14d, 21d, 28d, 35d, 42d, 49d, and 56d, respectively.

4 RESULTS AND DISCUSSION

4.1 Diffusion coefficient

Through the detection of the specimen quality before and after freeze-thaw test, according to Formula 14, we can obtain the diffusion coefficient of the test, and the value of water content corresponding to the diffusion coefficient when time fixed. At the initial time, at time $t = 0$, however, the initial diffusion coefficient cannot be directly received through Formula 14. The value can be obtained graphically and analytically by the relation curve of the diffusion coefficient with time. As shown in Figure 5, the initial diffusion coefficient is 4.2×10^{-4} cm²/s by polynomial fitting. In order to verify the accuracy of the initial diffusion coefficient, substitute the estimated initial diffusion coefficient into Formula 13, calculate the water content and draw the calculation curve during the test, as shown in Figure 6, the calculated value is almost coincide with the test value, therefore, the initial diffusion coefficient in the test is 4.2×10^{-4} cm²/s.

As shown in Figure 5, the range of diffusion coefficient is $0.39{\sim}5.99 \times 10^{-4}$ cm²/s, which is decreased to the minimum at the time of 28 freeze-thaw cycles and then increased until the end of the test. At the early stage in the test, the degree of saturation of the pore is small, but the diffusion coefficient and the diffusion rate are lager because of the osmotic pressure. With the increase of pore water, the effect of the freeze-thaw cycle gradually increased (Mu et al. 2010), which hinders the diffusion of water from external to inner pore, therefore, the diffusion coefficient decreased in this stage. At the 28th freeze-thaw cycle, the diffusion coefficient reached the nick point in Figure 5. With the test continued, the increased osmotic pressure and hydraulic pressure are large enough to destroy the inner pore, cracking gradually occurs. At this stage, the resistance of moisture diffusion is reduced and the diffusion coefficient increased gradually until the obvious freeze-thaw damage occurs.

4.2 Moisture diffusion during the test

During the test, by observing the degree of saturation curve and the time gradient trend curve of S in Figure 6, the water content showed an increasing tendency, and the diffusion rate increased first, then decreased and finally increased again.

In the early period of the test, within the first 3 cycles, part of pore affected by differential concentration, moisture diffusion occurs, with the progresses of diffusion, water diffusion into the

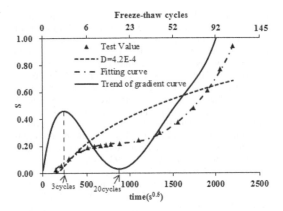

Figure 6. The degree of saturation during the freeze-thaw test.

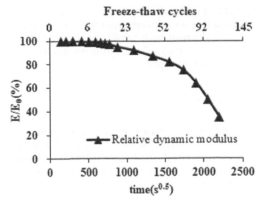

Figure 7. The relative dynamic modulus during the freeze-thaw test.

pore gradually, the diffusion rate increased gradually. In this stage, the calculated curve, which is based on the initial diffusion coefficient discussed in section 4.1, coincide with the measured curve. Therefore, in this stage, the freeze-thaw cycle has a little effect on moisture diffusion, and the diffusion of the concrete is similar to general diffusion, in which the diffusion coefficient is constant.

After the first 3 cycles, there is an obvious deviation between a calculated curve and a measured curve. The main reason is the effect of the freeze-thaw cycles, at a cooling stage, pore water tends to migrate to the surface, which hinders the diffusion of external moisture to inner pore. Therefore, the diffusion rate gradually decreased and moisture diffusion turned slowly. As the gradient trend curve and the degree of saturation curve are shown in Figure 6, at this stage (3 to 20 cycles), the concrete is in a slow diffusion stage.

With the test goes on, after 20 cycles, pore saturated gradually and the increased osmotic pressure and hydraulic pressure are large enough to destroy the inner pore, cracking gradually occurs, and cracks may connect with each other. Then, the inner diffusion channel increased; therefore, the diffusion coefficient gradually increased (as shown in Fig. 5). The deterioration of the inner structure accelerates the process of pore water saturation, and conversely, water saturation promotes the structural damage, such repetition result in the increased of diffusion rate obviously (as shown in Fig. 6). Therefore, the concrete is in a sharp diffusion stage, and the pore structure will be completely damaged within limited freeze-thaw cycles (Li et al. 2011).

4.3 Damage development

The relative dynamic modulus is frequently used as index to evaluate the extent of damage as shown

in Figure 7. The relative dynamic modulus is close to 100% in the initial stage of the test, it indicates that the pore structure is not damaged at that time. With the test goes on, the relative dynamic modulus decreased gradually, after 20 cycles, the decreased rate becomes rapid obviously, it indicates that the freeze-thaw cycle accelerates the structural damage, which is in agreement with the test results discussed in section 4.2. As the number of the test is more than 56 cycles, the relative dynamic is attenuated quickly, and the value is even decreased below 80%. It indicated that the structure of concrete was completely damaged at that time. The test result on diffusion coefficient (section 4.1) and moisture diffusion (section 4.1) is inconsistent with the decreased law of the relative dynamic modulus, which verity the accuracy of the diffusion model established in this paper.

5 CONCLUSIONS

A diffusion model is established, and a relevant single-side freeze-thaw test is suggested, diffusion coefficient, moisture diffusion law, and damage development law is analyzed, the main conclusions are as follows:

1. The initial diffusion coefficient is 4.2×10^{-4} cm^2/s, and the range of diffusion coefficient is $0.39 \sim 5.99 \times 10^{-4}$ cm^2/s, which is decreased to the minimum at the time of 28 freeze-thaw cycles, and then increased until the end of the test.
2. During the freeze-thaw cycle, the water content showed an increasing tendency, and the diffusion rate increased first, then decreased and finally increased again. According to the trend of the gradient curve, the moisture diffusion is divided into three stages that are general diffusion, slow diffusion, and sharp diffusion.

3. The test result of diffusion coefficient and moisture diffusion is inconsistent with the decreased law of the relative dynamic modulus, which verity the accuracy of the diffusion model established in this paper.

ACKNOWLEDGMENT

This material is based upon work funded by Zhejiang Provincial Natural Science Foundation of China under Grant No. LQ12E09002; Project (51308497) supported by National Natural Science Foundation of China.

REFERENCES

Collins, A.R. 1944. The destruction of concrete by frost. *Journal of Institution of Civil Engineers*, 23(1): 29–41.

Crank, J. 1975. *The Mathematics of Diffusion[M]. 2nd Edition,* Oxford: Oxford University Press.

Cwirzen, A. & Penttala, V. 2005. Aggregate-cement paste transition zone properties affecting the salt-frost damage of high-performance concrete. *Cement and Concrete Research*, 35(4): 671–679.

Fagerlund, G. 1977. The critical degree of saturation method of assessing the freeze/thaw durability of concrete. *Material and Structures*, 10(58): 217–229.

Fagerlund, G. 1993. The long time water absorption in the air-pore structure of concrete. Report TVBM.

Li, W. Pour-Ghaz, M. & Castro, J. et al. 2011. Water absorption and critical degree of saturation relating to freeze-thaw damage in concrete pavement joints. *Journal of Materials in Civil Engineering*, 24(3): 299–307.

Litvan, G.G. 1980. Freeze-thaw durability of porous building materials in durability of building materials and components. *American Society for Testing and Materials, Philadelphia*, 691: 293–298.

Mu, R. Tian, W.L. & Gao. X.D. 2010. Moisture Migration in Concrete Exposed to Freeze-thaw Cycles. *Journal of The Chinese Ceramic Society,* 38(9): 1713–1717.

Power, T.C. & Helmuth, R.A. 1949. Theory of volume change in hardened Portland cement pastes during freezing. *Proceedings of the Highway Research Board,* 32: 285–297.

Power, T.C. 1945. A working hypothesis for further studies of frost resistance. *Journal of the American Concrete Institute,* 16(4): 245–272.

Scherer, G.W. & Valenza, J.J. 2005. Mechanisms of frost damage. *Materials Science of Concrete,* 7: 209–246.

Wang, X.Y. Jiang Z.W. & Gao, X.D. et al. 2002. Review on the mechanism and model of moisture transfer in concrete. *Journal of Building Materials*, 5(1): 67–71.

Advanced Materials and Structural Engineering – Hu (Ed.)
© *2016 Taylor & Francis Group, London, ISBN 978-1-138-02786-2*

Fundamental properties of lightweight foam soil concrete using high volume Supplementary Cementitious Materials

K.H. Lee
Department of Architectural Engineering, Kyonggi University Graduate School, Seoul, Korea

K.H. Yang
Department of Plant Architectural Engineering, Kyonggi University, Suwon, Korea

ABSTRACT: The objective of this study was to determine the optimal mixture proportions of lightweight foam soil concrete. As the total weight of the mixture of dredged soil and binder varied with respect to the volume of the concrete, the flow, slurry, dried specific gravities, and compressive strength of the lightweight foam soil concrete were measured. In addition, the effect of the apparent slurry specific gravity on the compressive strength of the concrete specimens was established.

1 INTRODUCTION

Dredged soil from land development or harbor construction results in the environmental contamination of soil and water (Park et al. 2011). Attempts have been made to utilize such dredged soil in various fields, such as marine environment restoration, artificial habitat creation, and the development of eco-friendly waterfront spaces. Recently, dredged soil has been recycled to produce Lightweight Foam Soil (LWFS) concrete as back filler for underground utilities or structures (Kim et al. 2007, 2006). However, the production of LWFS concrete is based on field experience and does not include a systematic mixture proportioning approach, resulting in quality degradation including a decrease in volume owing to its antifoaming property and unreasonably low compressive strength.

The objective of the present study is to develop a reliable mixture proportioning method for LWFS concrete to obtain a target workability and compressive strength. To enhance the sustainability of concrete and minimize the use of cement, high-volume Supplementary Cementitious Materials (SCMs) were designed. Based on the test results, the effects of the apparent specific gravity of slurry on the development of the compressive strength of LWFS concrete were established.

2 EXPERIMENTAL DETAILS

2.1 *Materials*

Ordinary Portland Cement (OPC), Ground Granulated Blast-furnace Slag (GGBS), and Fly Ash (FA) were used to produce a high-volume SCM binder. The densities of the OPC, GGBS and FA were 3150, 2900 and 2210 kg/m^3, respectively, and their specific surface areas were 4000, 4400, and 3900 cm^2/g, respectively. The density of the dredged soil, which collected from a sandy soil series, was $2650 kg/m^3$. A hydrolyzable protein type animal foaming agent that has no chemical reactions was used to create the foam.

2.2 *Mixture proportions*

The mixture proportions of the LWFS concrete are given in Table 1. Five LWFS mixtures were prepared with different total weights of the dredged soil/binder mixture. To produce a high-volume SCM binder, 15% FA and 65% GGBS replaced some of the OPC content (Sim et al. 2014). For all LWFS concrete mixtures, other parameters were fixed as follows: a dredged soil/binder ratio of 5, a the water/binder ratio of 32.5%, a dredged soil moisture content of 40%, and a water-reducing agent content of 0.75%.

2.3 *Casting and testing*

To produce foam, the foaming agent was diluted with water in a 1:32.3 ratio by volume and subsequently aerated using a foam generator connected to a compressed air source. Preformed foam was added to the cementitious slurry produced by mixing the high-volume SCM binder, dredged soil, and water in a mixer pan equipped with rubber wiper blades. The water content of the preformed foam was considered as part of the water in the overall

Table 1. Mixing proportions.

No.	Soil-to-binder ratio	Unit binder contents (kg/m³)	Unit dredged soil (kg/m³)	Total amount of binder and dredged soil (kg/m³)	Foam (%)	W/B (%)	Moisture content of dredged soil (%)
A900	5	150	750	900	61	32.5	40
A1092		182	910	1092	53		
A1284		214	1070	1284	45		
A1476		246	1230	1476	36		
A1668		278	1390	1668	28		

mixture (Lee 2013). Fresh concrete produced using wet processing was cast in various steel molds. Most specimens were then sealed using a plastic bag to prevent evaporation and cured at room temperature until testing. The specimens intended for measuring dry density were cured under wet conditions after one day.

The flow and slurry specific gravity of the LWFS concrete were measured in fresh concrete, in accordance by the KS F4039 (2009) standard. For hardened concrete, the dried specific gravity and compressive strength development were measured.

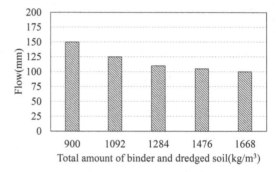

Figure 1. Flow of LWFS concrete.

3 RESULTS AND DISCUSSION

3.1 Flow and slurry specific gravity

The measured flow and slurry specific gravity of THE LWFS concrete are shown in Figures 1 and 2, respectively. Photographs of the fresh LWFS concrete are presented in Figure 3. The flow values tended to decrease as the dredged soil/binder composition increased. The slurry specific gravity increased in proportion to the dredged soil/binder composition.

3.2 Dried specific gravity and compressive strength

The dried specific gravity of the LWFS concrete tended to increase as the dredged soil/binder composition increased, as shown in Figure 4. The increase in the compressive strength of the LWFS concrete was found to be marginally dependent on the age; similar strengths were obtained for ages of 7 and 28 days, as shown in Figure 5. This may be attributed to the fact that soil disturbs the hydration of cementitious materials.

3.3 Relationship between slurry and dried specific gravities

The dried specific gravity of the LWFS concrete is indicating a linear relationship, as shown in Figure 6. Using regression analysis on the obtained

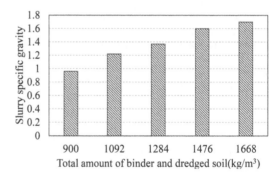

Figure 2. Slurry specific gravity of LWFS concrete.

test data, the relationship between the slurry specific gravity (S_g) and the dried specific gravity (ρ_c) was found to be

$$\rho_c = 0.32(S_g)^{1.12} \qquad (1)$$

3.4 Relationship between dried specific gravity and compressive strength

The compressive strength of the LWFS concrete tended to increase with increasing dried specific gravity, as shown in Figure 7. From the regression analysis of test data, the 28-day compressive

| (a) 900 | (b) 1092 | (c) 1284 | (d) 1476 | (e) 1668 |

Figure 3.   Flow shape of LWFS concrete.

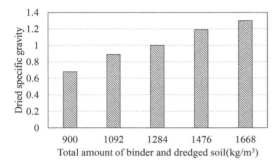

Figure 4.   Dried specific gravity of LWFS concrete.

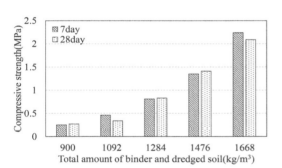

Figure 5.   Compressive strength of LWFS concrete.

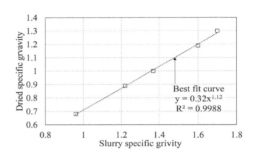

Figure 6.   Relationship between slurry and dried specific gravities.

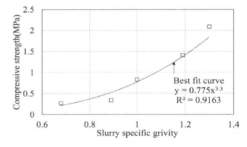

Figure 7.   Relationship between dried specific gravity and compressive strength of concrete.

strength (f_c') of the LWFS concrete using high-volume SCM was found to be

$$f_c' = 0.775(\rho_c)^{3.3} \qquad (2)$$

4   CONCLUSIONS

From testing of five LWFS concrete mixtures using high-volume SCM, the following conclusions may be drawn:

1. As the mass of the dredged soil/binder mixture per unit volume of the LWFS concrete increased, the flow and slurry specific gravity of the concrete tended to decrease.

2. As the slurry specific gravity increased, the dried specific gravity of the LWFS concrete increased linearly, which resulted in an increase in the compressive strength of the concrete.

3. The increase in the compressive strength of the LWFS concrete was marginally dependent on the age, as measurements indicated that similar strengths were obtained for ages of 7 and 28 days.

ACKNOWLEDGEMENT

This research was supported by a grant (14CTAP-C078666-01) from Technology Advancement Research Program funded by Ministry of

Land, Infrastructure and Transport of Korean government.

REFERENCES

Kim, Y.T. Han, W.J. & Jung, D.H. 2007. Development of Composite Geo-Material for Recycling Dredged Soil and Bottom Ash, *Korean Geotechnical Society.* 23: 77–85.

Kim, Y.T. Kim, H.J. & Kwon, Y.K. 2006. Compressive strength characteristics of cement mixing lightweight soil for recycling of dredged soil in Nakdong river estuary, *Korean Society of Ocean Engineers,* 20: 7–15.

KSF 4039, 2009. *Foamed concrete for cast in site*, Korean Industrial standard.

Lee, K.H. 2013. *Development of mixture proportioning model for low-density high-strength foamed concrete,* Kyonggi University.

Park, J.B. Lee, G.H. Woo, H.S. & Lee, J.W. 2011. Problems of disposal of dredged material and increasement of recycling, *KSCE Journal of Civil Engineering,* 59: 65–74.

Sim, S.W. Yang, K.H. Lee K.H. & Yoon, I.G. 2014. Evaluation of flow and engineering properties of high-volume supplementary cementitious materials lightweight foam-soil concrete, *Korean Recycled Construction Resources Institute,* 2: 247–254.

Advanced Materials and Structural Engineering – Hu (Ed.)
© 2016 Taylor & Francis Group, London, ISBN 978-1-138-02786-2

Fatigue stress–strain relationship of normal- and light-weight concrete mixtures under axial compression

J.S. Mun
Department of Architectural Engineering, Kyonggi University Graduate School, Suwon, Kyonggi-do, South Korea

K.H. Yang
Department of Plant Architectural Engineering, Kyonggi University, Suwon, Kyonggi-do, South Korea

S.J. Shin
HANMAC Engineering Co, Ltd., Geoyeo-dong, Songpa-gu, Seoul, South Korea

ABSTRACT: This study examined the fatigue stress–strain relationship of normal- and light-weight concrete mixtures according to the type of binder. For the fatigue test, the constant maximum stress level was varied to be 90%, 80% and 75% of the monotonic strength of concrete, whereas the minimum stress level was fixed at 10% of the monotonic strength. The test results showed that the fatigue life decreased in Light-Weight Concrete (LWC) than in Normal-Weight Concrete (NWC), which indicating a greater decrease in fatigue life with the increase in the maximum stress level. In the fatigue stress–strain curve, the slope of LWC was steeper than that of NWC because of a lower residual strain in the LWC.

1 INTRODUCTION

The fatigue performance of concrete generally depends on the properties of the concrete mixture, environmental conditions, and loading conditions (ACI 215 1997). It has been known that Normal-Weight Concrete (NWC) under 70% of its compressive strength has a fatigue life greater than 10×10^6. Because of this, the design of Reinforced Concrete (RC) structures does not consider the fatigue behavior of concrete. On the other hand, fatigue damage influences the structures permanently, resulting in deteriorating durability of concrete. Hence, for structures under cyclic loading, which include road pavements, girders supporting highway bridges with traffic, and offshore structures subjected to wind and sea waves, one should consider the fatigue damage and fatigue life of concrete (Xiao et al. 2013).

With the global efforts to reduce environmental loads including CO_2 emissions, energy consumption, and depletion of natural resources, many concrete industries have examined diverse approaches using Granulated, Ground, Blast-furnace Slag (GGBS) and Fly Ash (FA) as a partial replacement for Ordinary Portland Cement (OPC). In addition, lightweight aggregates produced from coal dust, incinerator bottom ash, and fly ash can contribute to the reduction of environmental loads because they are recycled by-products, and thus, reduce the consumption of the natural aggregates. In particular, lightweight aggregates have the advantage of reducing the density of concrete. However, the fatigue strength and fatigue life of Lightweight Concrete (LWC) and high-volume Supplementary Cementitious Material (SCM) concrete are still controversial owing to the lack of experimental data.

This study examined the fatigue performance of NWC and LWC in compression according to the type of binder including OPC or high-volume SCM. For the high-volume SCM binder, the weight ratios of FA and GGBS were selected to be 20% and 50%, respectively. The compressive strength of concrete was designed to be 40 MPa. For the fatigue tests, the minimum stress level was fixed at 10% of the static compressive strength of concrete, while the maximum stress level varied from 75 to 90% of the static strength.

2 EXPERIMENTAL DETAILS

2.1 Concrete mixtures

The main parameters selected for the concrete mixtures were the unit weight of concrete and the type of binder, as given in Table 1. Considering its practical application for offshore precast concrete, the targeted concrete compressive strength after aging for 91 days was 40 MPa. The target initial slump

Table 1. Details of concrete mixtures proportions (% by mass).

Specimens	Binder type	Concrete type	W/B (%)	S/a (%)	Unit weight (kg/m³)							
					B			W	F_N	C_N	F_L	C_L
					OPC	FA	GGBS					
OL	OPC	LWC	30	42	516	–	–	155	–	–	449	494
ON		NWC	35	45	442	–	–	155	824	1007	–	–
HL	High volume SCM	LWC	25	42	186	310	124	155	–	–	518	570
HN		NWC	30	45	155	258	103	155	772	943	–	–

Note: W/B = water-to-binder ratio by weight, S/a = fine aggregate-to-total aggregate ratio by volume, B = binder, OPC = ordinary Portland cement, FA = fly ash, GGBS = granulated ground blast-furnace slag, W = water, F_N = normal weight fine aggregate, C_N = normal weight coarse aggregate, F_L = lightweight fine aggregate, C_L = lightweight coarse aggregate.

Table 2. Physical properties of aggregates used.

Type	Maximum size (mm)	Unit volume weight (kg/m³)	Density	Water absorption (%)	Porosity (%)	Fineness
Coarse aggregate						
Expanded clay granule	19	729	1.21	18.96	68.17	6.56
Granite	19	1700	2.65	0.62	53.57	6.05
Fine aggregate						
Expanded clay granule	4	832	1.65	13.68	50.42	4.34
Natural sand	5	1750	2.60	1.85	56.21	2.51

and air contents were 200 ± 10 mm and 3 ± 0.5%. For high-volume SCM binder, the weight ratios of FA and GGBS were selected to be 20% and 50%, respectively, as a partial replacement for OPC. The targeted unit weight of all LWC mixtures was 1700 kg/m³. The water-to-binder ratios (W/B) were variably selected according to the types of aggregates and binder.

2.2 Materials

Two types of binders were prepared as follows: OPC (ASTM Type I) and high-volume SCM binder. GGBS, conforming to ASTM C989, had a high CaO content and a SiO_2-to-Al_2O_3 ratio by mass of 2.29. The FA had a low CaO and a SiO_2-to-Al_2O_3 ratio by mass of 1.91. The specific gravity and specific surface area were 3.15 and 3466 cm²/g, respectively, for OPC, 2.23 and 3720 cm²/g for FA, and 2.91 and 4497 cm²/g for GGBS.

For preparing NWC, locally available natural sand with a maximum particle size of 5 mm, and crushed granite with a maximum particle size of 19 mm were used as fine and coarse aggregates, respectively. For preparing LWC, the artificially expanded clay granules having maximum sizes of 19 mm and 4 mm were used as lightweight coarse

and fine aggregates, respectively. The density of the lightweight aggregates was approximately 1.6–2.2 times lower than that of natural aggregates, whereas the water absorption capacity of the lightweight aggregates was excessively high, as given in Table 2.

2.3 Testing

Fatigue testing in compression was conducted after the aging for 91 days. Monotonic uniaxial tests were applied with time at the frequency of 1 Hz. The minimum stress level was fixed at 10% of f'_c, and the maximum stress level was varied from 90% to 75% of f'_c. To measure the axial strain, a compressor meter installed with two 5 mm capacity Linear Variable Differential Transformers (LVDTs) was attached on the cylinder specimens. The test stopped after the specimen failed to maintain the specified maximum stress level.

3 TEST RESULTS

3.1 Crack propagation

In general, the initial crack occurred at the mid-height of the specimen and toward the top and bottom loading points. With the increase of the number of loading cycles, few vertical macroscopic cracks were developed, and then they were widened rapidly. The failure of all specimens under cyclic loading was governed by these continuous macroscopic cracks. The crack propagation was little affected by the binder type and stress level.

However, the crack propagation of NWC and LWC significantly differed. The crack in NWC propagated along the interface between paste and aggregate particles. Meanwhile, the cracks in LWC specimens penetrated through the lightweight coarse aggregate particles. Overall, the failure of LWC specimens was governed by the fracture of aggregates.

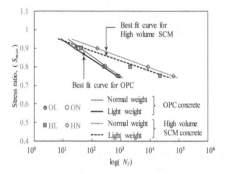

Figure 1.   Coordinate systems and loads distribution.

Table 3.   Summary of test results.

Specimens	Under monotonic loading			Fatigue life (N_f)		
				S_{max}		
	S_i	A_c	f_c'	0.9	0.8	0.75
	(mm)	(%)	(MPa)			
OL	180	3.5	38.61	29	257	775
ON	180	3.5	48.57	35	294	857
HL	206	2.6	35.29	43	2121	22522
HN	200	3.1	39.82	154	8500	63248

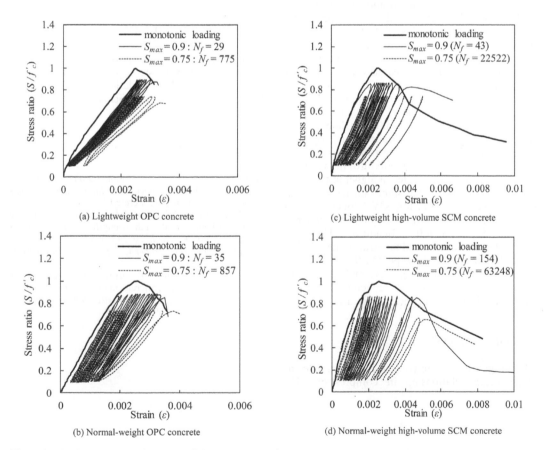

(a) Lightweight OPC concrete

(b) Normal-weight OPC concrete

(c) Lightweight high-volume SCM concrete

(d) Normal-weight high-volume SCM concrete

Figure 2.   Fatigue stress-strain curves of the concrete specimens.

417

3.2 S_{max}–N_f curves

Figure 1 shows the relationship of S_{max} and the logarithm of N_f according to the types of binder and concrete. In general, a linear negative relationship was observed, regardless of the types of binder and concrete, indicating that N_f decreased with an increase in S_{max}. Under the same stress levels, the N_f in LWC was slightly lower than that in the companion NWC, as given in Table 3. The weak cohesion and fracture of lightweight aggregate particles deteriorates the fatigue performance of concrete. This trend was more notable for high-volume SCM concrete mixtures than for OPC concrete mixtures.

3.3 Fatigue stress–strain curve

Figure 2 shows the fatigue stress–strain curves measured from the current specimens. For comparison, the monotonic stress–strain curve of each concrete mixture is plotted on the same figure. The shape and pattern of fatigue stress–strain curve are influenced by a binder type and concrete type. The slope of the ascending branch was commonly higher for NWC than for the companion LWC owing to a higher modulus of elasticity. The strain initially followed the monotonic stress–strain curve, but the fatigue strain at S_{max} and residual strain at S_{min} gradually increased with the increase in the number of loading cycles. The enclosed area of curves for each cycle was greater in the case of LWC than that in case of NWC. This may be attributed to the fact that the internal damage until fatigue failure was greatly localized for LWC compared with NWC. The fatigue strain was faster in high-volume SCM concrete mixtures than in OPC concrete. As may be expected, a higher initial strain and fatigue strain at the same normalized value of N/N_f were observed for NWC concrete under an S_{max} value of 0.9 than that for NWC concrete under an S_{max} value of 0.75. In addition, LWC generally showed larger residual strains than NWC, regardless of the binder type. On the other hand, the slope of the unloading curve and the residual strain were smaller for OPC concrete than those for high-volume SCM concrete mixtures.

4 CONCLUSIONS

The fatigue tests were conducted to evaluate the stress–strain relationship in compression. The selected parameters were the types of concrete (NWC and LWC), binder (OPC and high-volume SCM), and the applied maximum stress level. The conclusions are drawn may be summarized as follows:

1. The crack propagation was not affected much by the binder type and stress level. However, overall, the failure of LWC specimens was governed by the fracture of aggregates.
2. The fatigue life of LWC was slightly lower than that of the companion NWC, indicating that this trend was more notable for high-volume SCM concrete mixtures than for OPC concrete mixtures.
3. The enclosed area of curves for each cycle was greater in the case of LWC than that in the case of NWC.
4. The fatigue strain was faster in the high-volume SCM concrete mixtures than in the OPC concrete. On the other hand, the slope of the unloading curve and the residual strain were smaller for OPC concrete than those for high-volume SCM concrete mixtures.

ACKNOWLEDGMENT

This research was supported by the Public Welfare & Safety Research Program through the National Research Foundation of Korea (NRF) funded by the Ministry of Science, ICT & Future Planning (2013067519).

REFERENCES

ACI Committee 215, 1997. *Considerations for design of concrete structures subjected to fatigue loading (ACI 215R-92)* American Concrete Institute.
Lee, M.K. & Kim, Y.Y. 2004. Experimental study of the fatigue behavior of high strength concrete, *Cement and Concrete Research.* 26: 299–305.
Neville, A.M. 2011. *Properties of concrete*, Longman, England.
Xiao, J. Li, H. & Yang, Z. 2013. Fatigue behavior of recycled aggregate concrete under compression and bending cyclic loadings, *Construction and Building Materials.* 38: 681–688.

Advanced Materials and Structural Engineering – Hu (Ed.)
© 2016 Taylor & Francis Group, London, ISBN 978-1-138-02786-2

Performance parameters of concrete and asphalt-concrete surfaces

J. Rajczyk, M. Rajczyk & J. Kalinowski
Faculty of Civil Engineering, Czestochowa University of Technology, Czestochowa, Poland

ABSTRACT: This paper describes the measurement range and new diagnostic devices for determining the geometrical parameters of road surface: surface waviness and a device for measuring texture roughness. The devices have been constructed at Czestochowa University of Technology, Faculty of Civil Engineering. The paper also presents the measurement methods of surface geometrical parameters and the influence of these parameters on surface performance properties.

1 INTRODUCTION

Surface evenness or texture is one of the parameters for defining performance properties of concrete surfaces or asphalt-concrete roads.

Roughness parameter is directly connected with this surface geometrical feature; it defines the friction coefficient parameter appearing between processed surface and a means of transport or a machine type that directly influences the safety of use.

As reported by the Scopus Company from Gdynia, "The research done by a British company Pavestech that specializes in testing surfaces shows that increasing texture depth from 0.3 mm to 0.5 mm reduces the number of road accidents by 50% (Technical and Documentation Procedures Report/27). The Road Research Institute in Sweden demonstrated that the number of accidents is reduced as the friction coefficient increases, and the European Road Safety Commission demonstrated that the risk of a road accident resulting from skidding on a surface with the coefficient lower than 0.45 is 20 times as high as on a surface with the friction coefficient >0.60. Moreover, if the coefficient is lower than 0.30, the risk of an accident is 300 times as high!"

Surface geometrical parameters depend on properties of materials used to build it, and on technologies and tools used during shaping its micro geometry during which the parameters of inbuilt parts of working devices have a direct influence.

2 INFLUENCE OF SURFACE GEOMETRY ON PERFORMANCE PROPERTIES

Usable area irregularities as deviations from the designed flat surface geometry plane, which can be divided into components, have been presented in Figure 1.

The basic defect of geometry is convexity or concavity being a deviation of the highest wavelength.

A local deviation with lower wavelength takes place in the range of convexity or concavity and it is called as waviness. Surface is also described in a micro scale by an additional aberration of low regularity, the shortest distance between micro vertices called roughness, also described in literature as surface coarseness, e.g., concrete coarseness or asphalt-concrete abrasive layer.

Surface coarseness determines the friction coefficient, which is an important operational parameter that characterizes surfaces, especially road surfaces. It characterizes the amount of the force of friction that appears between two bodies sliding against each other. The friction coefficient value depends on the kind of materials and the micro geometry of adjacent surfaces. Regulations in this field of knowledge (Regulation of the Minister of Infrastructure in polish 2012) determine only the requirements for new roads as well as modernized and renovated ones. When it comes to roads that are in use, this feature is monitored and assessed according to various Systems (Regulation of the Minister of Transport and Maritime Economy in polish 1999).

A measurement of anti-skid properties is the so-called meaningful friction coefficient, which is the difference between the average value and standard deviation. Table 1 presents the values of

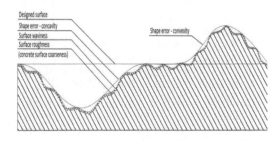

Figure 1. Components of surface irregularity geometry.

Table 1. Required friction coefficients of road surfaces according to (Regulation of the Minister of Transport and Maritime Economy in polish 1999).

Road class	Surface element	Meaningful friction coefficient at the speed of a blocked tyre in relation to the surface			
		30 km/h	60 km/h	90 km/h	120 km/h
A	Main traffic lanes, additional lanes, emergency lanes	0.52	0.46	0.42	0.37
	Slip roads, auxiliary road surfaces	0.52	0.48	0.44	–
S, GP, G	Traffic lanes, additional lanes, hard shoulders	0.48	0.39	0.32	0.30

Table 2. Classification of the condition of road surfaces with reference to anti-skid properties (for the tyre Barum Bravura) (Regulation of the Minister of Transport and Maritime Economy in polish 1999).

Class	Assessment of road surface condition	Meaningful friction coefficient μ_m
A	Good condition	≥ 0.52
B	Satisfactory condition	$0.37 \div 0.51$
C	Unsatisfactory condition (planning renovation works is recommended)	$0.30 \div 0.36$
D	Poor condition (instant intervention is recommended)	≤ 0.29

Figure 2. RK-4 prototype for measurement of the waviness and microprofile, constructed in the Department of Building and Material Processes Technology, Faculty of Civil Engineering at Czestochowa University of Technology.

the parameters that are required to be maintained after two months of road-use.

When it comes to anti-skid properties, the tested surfaces are categorized to one of four classes according to criteria defined for the meaningful friction coefficient μ_m, which is the difference between the average value of measurement results of friction coefficient $E(\mu_m)$ and standard deviation $D\mu$ (Table 2).

Testing anti-skid properties:

– Measuring the friction coefficient with model equipment in actual traffic conditions, e.g., a trailer-mounted Skid Resistance Tester (Polish conditions).
– Measuring factors that indicate anti-skid properties indirectly, e.g., measuring texture with the sand method and measuring coarseness with a British pendulum.
– Measurement with the photometric analysis method discussed in the author's monograph (Rajczyk 2004) or with the RK4 device (Fig. 2) described in the patent Project (Rajczyk).

Currently developed systems of road surface assessment in a given range require testing in order to determine objective ways of measuring surface evenness and coarseness before construction works acceptance and during road-use to ensure safe usage of different surface types.

3 SURFACE EVENNESS MEASUREMENT

The most popular and widely used method, which determines the evenness of flooring is the one using control straightedges (Rajczyk & Kowalska 2005). This is consistent with the directives of the standards that are in force in Poland PN-62/B-10144, DIN 18202, Germany and other European Union countries. In the United Kingdom, a device of profilograph is additionally used for evenness measurements, which is done according to the technical report of the British Concrete Society TR34.

The Polish Standard PN-62/B-10144 says that for the concrete flooring the maximum clearance between the flooring and the control straightedge of 2.0 m length, laid in various directions, in any place, should not exceed 5 mm. In order to test the deviations from a norm, the spirit level should be used. The permissible deviation from the level or from agreed drops is measured by the straight edge, and it should not exceed ±5 mm on the whole length or width of the floor (Gutkowski & Swietlicki 1986).

Table 3. Acceptable values of evenness parameters for the areas in fixed motion in that categories according to ACI-117-90 (Kaminski & Polorski 1983).

| | | Acceptable values in mm | | | |
| | | Feature 1 | | Feature 2 | |
Surface category	Location	A	B	A	B
Particular smoothness	Warehouses with very narrow passageways, maximum capacity, vehicles speed	0.75	1.0	1.0	1.5
Category 1	Warehouses as mentioned above, shelves height 8÷13 m	1.5	2.50	2.5	3.5
Category 2	Warehouses as mentioned above but with shelves height smaller than 8 m	2.5	4.0	3.25	5.0

Table 4. The classification of the flooring according to the standard ACI-117-90 with the assigned standard values of flatness and leveling parameters (Kaminski & Polorski 1983).

| | Required minimum values F_F and F_L | | | | Communicational paths on the flooring |
| | The whole flooring | | Minimal local values | | |
Classification of flooring	Flatness F_F	Leveling F_L	Flatness F_F	Leveling F_L	F_{min}
Conventional	20	15	15	10	20
Flat	30	20	15	10	30
Very flat	50	30	25	15	50
Super flat	100	66	50	33	100
Ultra flat	150	100	75	50	150

According to DIN 18202 standard, the tolerance of the flatness deviation in millimeters depends on the distance of measurement points, which is respectively:

a) 5 mm at a distance to 0.1 m,
b) 8 mm at a distance to 1 m,
c) 12 mm at a distance to 4 m,
d) 15 mm at a distance to 10 m,
e) 20 mm at a distance to 15 m.

Much better results of the measurement of flooring roughness can be obtained by applying the method proposed by the American Concrete Institute (ACI) and the Canadian Standards Association (CSA), which is defined by the standard included in ACI-117-90. These standards describe the evenness measurement method and floor leveling method, which is aimed at providing so-called F numbers. F numbers are the numbers of the floor surface profile which characterize its flatness and level. For this reason, straight lines are drawn on the floor surface and then the result of subtraction between the rises of the points lying on these lines at the distance of

12 inches (30.48 cm) are measured. After that, the results of the differences of the rises between all the neighboring points are calculated. In addition, the arithmetic differences between all neighboring discrepancies of the points measured in sections of 12 inches (30.48 cm) as well as elevation differences between all points, which are in the distance of 10 feet from each other, are measured. F numbers, flatness and level are determined by certain relations, which are described by statistics. According to this standard, the minimum of 36 measurements for each 100 m² should be carried out.

Depending on the size of F_F i F_L the floors have been properly classified, as shown in Table 4.

4 NEW DEVICE FOR MEASURING THE SURFACE GEOMETRY

A compact construct of a measuring apparatus was designed at the Faculty of Civil Engineering, Czestochowa University of Technology, which would also allow for automatic measurements of the surface geometry along the path covered by the measuring device.

In order to determine the above-mentioned geometry, the angle relations to the gravity vector and the angles between the elements of the machine moving on the flooring were measured, additionally aided by a precision laser rangefinder.

Designed, patented and constructed in the Department of Building and Material Processes Technology, Faculty of Civil Engineering at Czestochowa University of Technology, a new measuring device referred to as RK-4 (the 4th generation of the construction) consists of a self-propelled four-wheel drive train powered by electric motors, one for each wheel, which are fitted serially with two measurement carts. The measurement carts have a common axle with the measuring wheel. Measuring cart wheels are arranged in one plane, which assures that all of the wheels will follow the same tracks. The first measuring cart is fitted with a gauge, which is used to determine the angle of the frame, hanging firmly on the carts, and the gravity vector. Measurement of the angle between the measurement carts is carried out by means of an encoder variable transmission with timing belt, which increases the accuracy of the measurements. On the second measuring cart, a high precision laser range finder is mounted. One of the measuring carts coupled with another coder that calculates the rotation of the wheel to determine the traveled distance analogous to the measurement of traveled distance in a regular car. The process of measuring evenness with the RK-4 device is based on synchronous gathering of measurement data from all components and storing them for further processing in the computer memory. In order to render this possible, the device is equipped with a radio module that allows a two-way communication between separate components. Using wireless communication, the computer program can directly control the movement of the measuring device and each module and also receive measurement data from the measurement modules.

The device may work in three basic modes: corrugation measuring, profile measuring, and surface roughness micro profile measuring.

The corrugation measurement is based on synchronous reading of angle data from the angle encoder on a previously set interval of time triggered in equal intervals measured by a fixed angle of rotation of the cart wheel. The greater the amplitude of angle changes or frequency of these changes, the more corrugated is the flooring. Incorporating computer calculations using the statistical module allow high precision corrugation measurements.

Surface roughness profile measurement mode is based on continuous asynchronous distance measurement with a laser rangefinder on the baseline of length of 10 cm with a minimum movement

speed of the drive unit. Distance measurement is performed with a maximum frequency of the transmitter of the laser range finder that is synchronized locally by impulses from the distance measurement, and intermediate values are interpolated. In this way, the micro profile shape data are developed for further numerical designation of the selected parameter characterizing the micro profile.

The RK-4 construction presented in Figure 2 is a prototype of a measuring device that measures surface micro geometry profile.

5 CONCLUSIONS

Deviation from the assumed surface geometry and the arranged class of surface roughness is a big problem between the investor and the contractor connected with approving construction works.

The characteristics of the surface geometry of the flooring has a number of effects on dynamic loads on structures that are moving over the surface, generation of noise and safety of operation under varying atmospheric conditions. The features of new measuring devices should lead to introducing specific technical recommendations for technicians who responsible for the construction tasks and maintenance services and for whom the safety of operation is an essential purpose.

REFERENCES

British Concrete Society TR34, Technical Report.
DIN 18202 Tolerances in building construction, Buildings.
Guide for Concrete Floor and Slab Construction. Reported by ACI Committee 302.1R-96.
Gutkowski, R. & Swietlicki, W.A. 1986. *The dynamics and vibrations of mechanical systems*, PWN, Warsaw.
Kamiński, E. & Polorski, J. 1983. *The dynamics of suspensions and propulsion systems of motor vehicles*, Transport and Communications Publishing, Warsaw.
PN-62/B-10144, *Floors made of concrete and cement*, requirements and technical research upon receipt.
Rajczyk, J. & Kowalska, K. 2005. *Methods for evaluating the even-ness of the pavement concrete.* Highways Magazine, 12, Katowice, Poland.
Rajczyk, J. 2004. *The scientific basis for the selection of geometric structure and kinematics of machines for surface treatment of concrete working tools,* University of Częstochowa Publishing, Częstochowa, Poland.
Rajczyk, J. *An apparatus for measuring waviness and the surface roughness profile,* Patent Application No. UPRP P.397991, Poland.
Regulation of the Minister of Infrastructure dated 16.01.2012. On technical papers—for the construction of toll motorways *Journal of Laws,* 12: 116 (in polish).
Regulation of the Minister of Transport and Maritime Economy of 2 March 1999 on technical conditions to be met by public roads and their location, *Journal Law,* 43: 430 (in polish).

Advanced Materials and Structural Engineering – Hu (Ed.)
© 2016 Taylor & Francis Group, London, ISBN 978-1-138-02786-2

Pedestrian-induced vibrations in footbridges: A Fully Synchronized Force Model

M.A. Toso & H.M. Gomes
UFRGS, Sarmento Leite, Porto Alegre, RS, Brazil

ABSTRACT: In this paper, the pedestrian/structure interaction is investigated using an experimentally measured footbridge as the basis for comparisons. Kinetic parameters data from a design standard are used. Kinematic parameters of the human gait were obtained using a force platform. Two force models were used to represent the pedestrian loading: (a) a Simple Force Model (SFM) where it is assumed that the force from successive footfalls acts on a straight line along the direction of walking, which is represented by the Fourier series that comes from a pedestrian walking at a constant speed; (b) a Fully Synchronized Force Model (FSFM) wherein the load components are represented considering kinetic and kinematic parameters of a regular walking and are synchronized in time and space. Generally speaking, the footbridge responses evaluated by the two models presented similar behavior; however, longitudinal, vertical and lateral accelerations resulting from the dynamic response presented significant differences. The results show that there may be important differences in structural behavior when an FSFM is used, especially in footbridges with high flexibility.

1 INTRODUCTION

The interaction between people and structures occurs because humans are quite sensitive to vibration in a low frequency range of whole-body vibrations, in which natural frequencies of the human body can be observed. As such, they tend to change their behavior when structural vibrations are perceived in both vertical and horizontal directions. The latter can be classified as horizontal-lateral and horizontal-longitudinal. For the characterization of human-induced loads, most standards often consider the first two harmonics of the frequency spectrum of Ground Reaction Forces (GRF). Some of the design standards introduce load models for pedestrian loads applicable for simple structures, for instance, (SETRA 2006). On the other hand, the load modeling for more complex structures, are not well established, becoming threads for future studies. In this paper, the pedestrian/structure interaction is investigated using an experimentally measured footbridge with 34.08 m in length and 2.4 m in width as the basis for comparisons. It evaluated the mid-span acceleration (root-mean-square) of the structure to verify its serviceability. For the force model, kinetic (forces) and kinematic (speed, pacing rate, step length, stride length, and step width) parameters are used. The kinetic parameters are obtained from the design standard (SETRA 2006). The kinematic parameters of the human gait are obtained

using an especially designed force platform. The force signal from each foot is measured in separate allowing the evaluation of the step positioning. Two force models were used to represent the pedestrian loading: a) a Simple Force Model (SFM); b) a Fully Synchronized Force Model (FSFM), as will be described later on.

2 GROUND REACTION FORCE

The Ground Reaction Force (GRF) is equal in magnitude and opposite in direction to the force that the body exerts on the supporting surface through the foot. In general, the GRF can be represented by a three-dimensional vector that varies in time and in space due to the forward movement of the person. It should be mentioned that for engineering applications, the GRFs are usually quantified in the frequency domain through the body-weight normalized Fourier amplitudes, known as Dynamic Load Factors (DLFs). Some of these factors have been measured and compiled and can be found in the literature, such as: (SETRA 2006, Kerr 1998).

2.1 *Force platform*

Force platforms have been used to evaluate the pattern of applied human forces and to fit models for the interaction between pedestrians and structures.

These devices are designed to measure the forces exerted by a body in an external surface, the contact surface. For the experimental measurements, two independent instrumented plates were used, which was mounted side by side. In the design of the force platform, it was considered the acquisition of vertical forces only. When the pedestrian walks on the platform (left and right plates), the force applied to it is measured by load transducers, which generate electrical signals that are amplified and recorded by a data acquisition system. These signals allow the evaluation of the force position and intensity in each plate. Figure 1 presents the force platform and the collected vertical Ground Reaction Force (GRF) data during the experiments. Further design details, geometry and uncertainty quantification can be found in Toso et al. (2012).

The Total Force (Tf) represents the sum of the forces of the Left Foot (Lf) and Right Foot (Rf), and presents peaks when both feet are in contact with the ground that are greater than the individual foot peaks.

2.2 Kinematic parameters

The individual was asked to walk in a straight line along the force platform, starting and ending each crossing at rest. Tests were conducted with a pedestrian moving at his own natural pacing rate. So, kinetic and kinematic parameters could be measured simultaneously. The signals were recorded for repeated passages of the pedestrian (10 repetitions). For each crossing of the pedestrian, the following parameters were obtained: gait speed (V_s), pacing rate (f_p), step length (l_s), and step width (w_s). During the walking there is a separation of single and double stance phases. The walking and running can clearly be distinguished by the corresponding load patterns. In the running, a flight phase occurs, whereas for walking, the pedestrian

is in permanent contact with the ground and there are alternating double stance (ds) and single stance (ss) phases. Others parameters that are necessary to perform the pedestrian force synchronization were also evaluated: step time (st), single stance speed (sss), and double stance speed (dss). These parameters were measured for the lateral and longitudinal directions, according to Figure 2.

Table 1 presents these parameters (mean and standard deviation) considering 10 repetitions for the analysed pedestrian.

2.3 Simple Force Model (SFM)

In this force model, it is assumed that the force from successive footfalls acts on a straight line along the direction of walking, which is represented by the Fourier series that comes from a constant speed pedestrian. The SFM can be subdivided into: vertical, lateral and longitudinal force model. (SETRA 2006) assert that while walking, the vertical force induced by each foot has the same magnitude and the overall force is periodic, which is represented by Equation 1.

$$F(t) = G + \sum_{i=1}^{n} G\alpha_i \sin(2\pi i f_p t - \varphi_i). \qquad (1)$$

where $F(t)$ is the time varying vertical force, G it is the pedestrian weight, α_i is the Fourier's coefficient of the ith harmonic defined as Dynamic Load Factor (DLF), f_p is the pacing rate (Hz), t the time (s), φ_i is the phase shift of ith harmonic

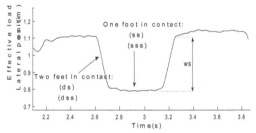

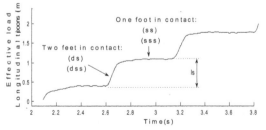

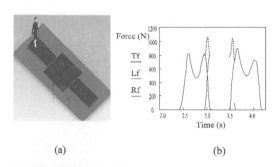

(a)   (b)

Figure 1. a) Force platform model for gait analysis (arrows indicate force a reaction from load cells); b) Acquired data of vertical Ground Reaction Force (GRF).

Figure 2. Change in walking speeds (lateral/longitudinal) during the different phases of walking.

Table 1. Summary for the kinematic parameters of analyzed pedestrian during normal walking.

	V_s (m/s)	f_p (Hz)	l_s (m)	w_s (m)	ds (s)
Mean	1.30	1.82	0.71	0.31	0.11
SD	0.06	0.09	0.03	0.03	0.01
		Longitudinal		Lateral	
	ss (s)	sss (m/s)	dss (m/s)	sss (m/s)	dss (m/s)
Mean	0.43	0.38	5.11	0.03	2.85
SD	0.02	0.04	0.32	0.02	0.24

with respect to the first harmonic, i is the the harmonic order number, and n is the total number of contributing harmonics. The guideline (SETRA 2006) suggests the following values for the vertical Fourier coefficients and phase shifts: $\alpha_1 = 0.4$; $\alpha_2 = \alpha_3 \approx 0.1$; $\varphi_2 = \varphi_3 \approx \pi/2$. In order to present Fourier's transform of the lateral component (at the frequencies $f_p/2$; f_p and $3f_p/2$) according to the basic pacing rate, this action also can be expressed in the frequency domain as a Fourier series, according to Equation 2.

$$F(t) = \sum_{i=1/2}^{n} G\alpha_i \sin(2\pi i f_p t). \qquad (2)$$

In this case, i have (non-integer) values of 1/2, 1, 3/2, 2, etc. The phase shifts for the horizontal (lateral and longitudinal) component are close to 0. According to (SETRA 2006), the main amplitudes for the lateral component are located at a frequency of about half the vertical component. The corresponding values of the Fourier coefficients are: $\alpha_{1/2} = \alpha_{3/2} \approx 0.05$; $\alpha_1 = \alpha_2 \approx 0.001$. For the longitudinal component, (SETRA 2006) affirm that the main frequency associated with this component is approximately the same as for the vertical direction. Its oscillations correspond, for each step, initially to the contact of the foot with the ground, then to the thrust exerted subsequently. For this component, the values of the Fourier coefficients are: $\alpha_{1/2} \approx 0.04$; $\alpha_1 \approx 0.2$; $\alpha_{3/2} \approx 0.03$; $\alpha_2 \approx 0.1$.

2.4 Fully Synchronized Force Model (FSFM)

The design guidelines present individual force models in each direction to represent the dynamic load of pedestrians over time. However, the reference time is not the same for the three force components (longitudinal, lateral, and vertical) these forces are not synchronized in time and space. In this proposed model, the three load components are represented considering kinetic and kinematic parameters of a regular walking and are synchronized in time and space. This force model considers the velocities

changes during the walking (single and double stance phases). The pedestrian speed in the double stance phase is larger than single stance phase. This speed is also larger than the average speed of the pedestrian. Another import point is that the human walking does not occur on a straight line along the direction of walking. There are parameters like step length and step width that cause influence in the application of the resultant force (see Fig. 4), these parameters are considered in this study (according to Table 1). Peak values from each force component should be placed accordingly on the contact surface and the model's reference time adjusted to the correct phase for each component.

3 NUMERICAL MODEL

The analyzed structure has 34.08 m in length. The structure was modeled as a space truss. The results of the numerical model (frequencies and mode shapes) were compared with the experimental measurements obtained by Brasiliano et al. (2008). The deck floor and handrails were modeled as distributed masses in the bottom chord. The roof was assumed as distributed masses along the top nodes. Additional information regarding to cross-sections, lengths, materials properties, etc. can be found in Brasiliano et al. (2008). The experimental values of the natural frequencies of the footbridge were taken as reference values in order to update the numerical model. The Young's modulus and the material density of the metallic components were modified accordingly. The baseline values were 210×109 N/m^2 for the Young's modulus and 7850 kg/m^3 for the material density (structural steel). The numerical model was updated to reproduce the dynamic features of the footbridge. The numerical results for frequencies were fairly close to the experimental ones and confirmed the adequacy of the model to represent dynamic characteristics of the analyzed structure. After the model updating using the sensitivity method (Mottershead et al. 2011), the Young's modulus was reduced to 193×109 N/m^2;

while the material density increased to 8080 kg/m³. The corresponding natural frequencies presented errors not greater that 21% for the first seven natural frequencies.

3.1 Footbridge response to single pedestrian

Figure 3 shows the three component pedestrian forces during walking: longitudinal, vertical and lateral direction. Figure 4 depicts the resultant force positions (in space) for normal walking. In the FSFM the three forces components were synchronized in time and space, according to Figures 3 and 4. The response of interest in this study is analyzing the mid-span acceleration (Root-Mean-Square, RMS) of the footbridge to verify the differences of acceleration between the models (SFM and FSFM). Table 2 shows the acceleration of the SFM and FSFM. The acceleration was measured in the longitudinal, vertical and lateral directions.

The results show the differences of about 90%, 30% and 2% corresponding to longitudinal, vertical and lateral accelerations, respectively. There are important differences in structural behavior when an FSFM is used.

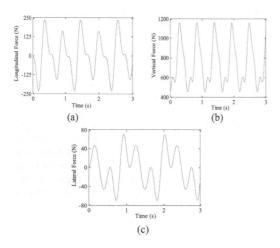

Figure 3. Pedestrian forces: a) longitudinal; b) vertical; c) lateral direction.

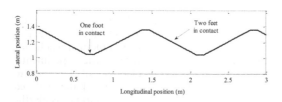

Figure 4. Resultant force positions for normal walking: lateral and longitudinal direction.

Table 2. Acceleration response of the footbridge to a 76.4 kg pedestrian with 1.82 Hz pacing rate.

| Model | RMS mid-span acceleration (m/s²) | | |
	Longitudinal	Vertical	Lateral
SFM	0.0011	0.0138	0.0325
FSFM	0.0021	0.0179	0.0332

4 DISCUSSION/CONCLUSIONS

In this study, a dynamic analysis of a footbridge is carried out considering the pedestrian–structure interaction. Two force models were used: a Simple Force Model (SFM) and a Fully Synchronized Force Model (FSFM). The dynamic analysis results using the SFM may underestimate the responses (accelerations), if compared with the FSFM, since more vibration modes are excited. It is possible to conclude that the use of the SFM may mask and underestimate important aspects of the vibrational behavior of the structure. For instance, for the longitudinal direction in the FSFM, the structure was excited due to the presence of the thrust and deceleration phases of the human step that are not modeled in the SFM. However, for the lateral direction, the FSFM did not presented substantial increase when compared with the SFM, perhaps due to the fact that there is no structural mode around the main frequency of the process (half the pacing rate). Regarding to mode shapes (spectrums were not presented), using the FSFM as a torsional mode was excited being detected by the vertical acceleration spectrum mainly due to the spatiality of application of the vertical load that produced torsion in the footbridge. As expected, this mode was not observed in the corresponding vertical acceleration spectrum for the SFM since this model assumes the application of the load along a straight line. Table 2 shows that the vertical acceleration presented a substantial increase when the FSFM is used. This is probably due to added effect of the spatiality of the force application in the FSFM and the different longitudinal and lateral velocities presented in a step length. The acceleration spectrum in this case showed the presence of sub-harmonics of the vertical force superimposed on the structural natural frequencies that is not present in the case of the SFM. The SFM seems to be an oversimplification of the process of human walking and their interaction with slender structures. Furthermore, it should be emphasized that the hypothesis of considering perfectly periodic pedestrian-induced forces (usually applied at a constant speed and along a straight line in the structure) may render a poor structural behavior representation. The FSFM may represent an alternative to enhance the walking force model that may render the structural behavior more realistic.

REFERENCES

Brasiliano, A. Doz, G. Brito, J.L.V. & Pimentel, R.L. 2008. *Role of non-metallic components on the dynamics behavior of composite footbridges.* 3rd International Conference Footbridge, Porto, 2–4 July 2008.

Kerr, S.C. 1998. *Human Induced Loading on Staircases.* PhD thesis, Mechanical Engineering Department, University College London, London, UK.

Mottershead, J.E. Link, M. & Friswell, M.I. 2011. The sensitivity method in finite element model updating: A tutorial, *Mechanical Systems and Signal Processing*, 25: 2275–2296.

SETRA. 2006. *Footbridges, assessment of vibrational behavior of footbridges under pedestrian loading, technical guide,* Service d'Etudes Techniques des Routes et Autoroutes, Paris.

Toso, M.A. Gomes, H.M. Silva, F.T. & Pimentel, R.L. 2012. *A biodynamic model fit for vibration serviceability in footbridges using experimental measurements in a designed force platform for vertical load gait analysis,* 15th International Conference on Experimental Mechanics, Porto, 22–27 July 2012.

Advanced Materials and Structural Engineering – Hu (Ed.)
© 2016 Taylor & Francis Group, London, ISBN 978-1-138-02786-2

Shrinkage of phosphoaluminate cement concrete

W.J. Long, X.W. Xu, J.G. Shi, S.F. Zhao & X.L. Fang
College of Civil Engineering, Guangdong Provincial Key Laboratory of Durability for Marine Civil Engineering, Shenzhen University, Shenzhen, China

ABSTRACT: Phosphoaluminate cement is a new kind of cementitious material. The shrinkage behavior of the phosphoaluminate cement concrete (PALC) was investigated in this study. The test results show that in general the shrinkage values of PALC are small, the shrinkage value of PALC at 360 days is 300 µstrain; the shrinkage occurs mainly in the early age, and reaches 50% of its ultimate values (1 year) at the age of 28 days; and 90% of its ultimate values at the age of 180 days. Moreover, shrinkage values of PALC increase with the increase of cement content, the decrease of water–cement ratio (w/c), and the decrease of the dosage of High Range Water Reducing Admixture (HRWRA), respectively.

1 INTRODUCTION

As the building material with the maximum use, concrete shows more and more durability issues during its service life. These problems incurred by shrinkage of normal Portland cement concrete, such as crack and deformation, etc. have become important factors influencing the durability of concrete, and they are hotspots in the current concrete research field (Qin 2001, You & He 2004, Yao 2011, Feng et al. 2012, Li et al. 2011, Gao et al. 2010). Cracks resulted from the shrinkage cover about 80% of the concrete cracks (Yao 2011, Feng et al. 2012, Li et al. 2011, Gao et al. 2010), these cracks not only influences the expression quality of concrete but also generates the flexibility increase and bearing capacity reduction of concrete components due to crack expansion.

Phosphoaluminate cement (PALC), a new kind of cementitious material, its advantages, such as early strength development, high compressive strength, adjustable setting time, outstanding compatibility of hydrated and hardened products with the environment, etc. have attracted the wide interests of researchers. Presently, the researches on the PALC concrete usually focus on the binding materials while those on the features and behaviors of concrete materials are still in the primary stage (Li et al. 2007, Li et al. 1999, Ma & Brown 1992, Wang et al. 2008, Wang 2008). The research on the shrinkage performance of the PALC concrete shows significance in accelerating the application of the PALC to the actual engineering.

2 EXPERIMENT DESIGN

2.1 Raw materials

Cement: Self-made in laboratory, with steps as follows: raw materials with aluminum (A), phosphorus (B) and calcium (C) are mixed uniformly according to a certain proportion; then fired under the temperature of 1250 °C–1280 °C in a high-temperature furnace; subsequently, the clinker is taken out for cooling; and then, the cement clinker is ground to reach the preset degree of fineness; Fine aggregate: River sand: fineness modulus of 3.1; medium sand: bulk density of 1540 kg/m³; close bulk density of 1720 kg/m³; Coarse aggregate: Granite: 5–31.5 mm continuous grading, bulk density of 1470 kg/m³, close bulk density of 1590 kg/m³; Retarder: borax; High-Range Water-Reducing Admixture (HRWRA): Sika® Visco-Crete 3301 high performance polycarboxylic water reducing agent.

2.2 Experimental method

At present, there is no standard testing method for the PALC concrete, so a method as that for normal Portland cement concrete is adopted for the testing according to GB/T50082-2009 Standard for Test Methods of Long-term Performance and Durability of Ordinary Concrete.

The concrete shrinkage for this test adopts a contact method, for which a prism specimen is used, with size of 100 mm × 100 mm × 515 mm. A horizontal-type concrete retractometer is adopted, and two ends of the specimen are embedded with gauge heads.

Table 1. Experimental program of PALC.

Mix. No	w/c	Cement kg/m³	Water kg/m³	Sand kg/m³	Coarse aggregate kg/m³
PALC1	0.33	400	132	793	1141
PALC2	0.33	400	132	639	1297
PALC4	0.38	400	152	770	1110
PALC5	0.33	500	165	582	1179
PALC6	0.33	500	165	720	1038
PALC7	0.36	450	162	671	1140

Mix. No	HRWRA, % of cement content	Retarder, % of cement content	S/A
PALC1	0.66	0.6	0.39
PALC2	0.33	0.6	0.33
PALC4	0.33	0.6	0.41
PALC5	0.66	0.6	0.33
PALC6	0.33	0.6	0.41
PALC7	0.49	0.6	0.37

The concrete shrinkage test result is calculated according to the Formula (1):

$$\varepsilon_{st} = \frac{L_0 - L_t}{L_b} \qquad (1)$$

where,

ε_{st}—Concrete shrinkage with t (d) as test age, t is calculated as of the measuring the initial length;

L_b—Measured gauge distance of the specimen, equaling to the inner distance between two gauge heads, namely measured gauge distance of the retractometer;

L_0—Initial reading of the specimen length (mm);

L_t—Length reading of the specimen measured in the test period t (d) (mm).

2.3 Mix proportions

The mix proportions of the PALC concrete is shown in Table 1.

2.4 Molding and curing

The molding and curing for this test are in accordance with GB/T50082-2009 Standard for Test Methods of Long-term Performance and Durability of Ordinary Concrete. Each mix proportion is for three prism test specimens (1000 mm × 100 mm × 515 mm), which are used to measure the dry shrinkage, specimens after molding shall be covered immediately by a waterproof film and placed for 24h in the environment with temperature of 20 ± 5 °C, then can de-mold and be numbered. And then, the specimens shall be placed in a curing room (temperature: 20 ± 2 °C; relative humidity of 95%) until the test age.

3 TEST RESULT AND ANALYSIS

The measured shrinkage values of the PALC concrete are shown in Table 2. In terms of the experimental result, the shrinkage value of PALC concrete at 1 year age can reach the stable state basically, totaling about 300 μstrain. The shrinkage value rises as time goes on, and early shrinkage value increases rapidly. After curing for half a year, its shrinkage increment is smaller. The shrinkage at the age of 28 days has reached 50% of 1 year μμshrinkage; basically 90% of the stable shrinkage value (1 year age value) as of 180 days.

3.1 Effect of cement content on shrinkage performance of PALC concrete

The cement content of PALC1 and PALC2 both are 400 kg/m³, PALC7 is 450 kg/m³, PALC5 and PALC6 both are 500 kg/m³. In terms of the experimental result, the less content of cement, its early shrinkage rate is smaller. At the age of 28 days, cement content has no significant effect on shrinkage. This is because the hydration of cement leads to the decrease of its volume; the less the cement content, the less the hydration volume reduction; on the contrary, the more the cement content, the more the hydration volume reduction. Especially in the early cement hydration is not completely, this kind of phenomenon is more obvious.

The PALC content has more obvious influences on the early shrinkage because its early high

Table 2. Measured shrinkage values of PALC (microstrain).

Mix. No	1d	3d	7d	14d	28d	90d
PALC1	12	47	63	77	148	179
PALC2	24	59	108	122	191	248
PALC4	53	110	134	165	191	246
PALC5	44	61	93	122	185	211
PALC6	43	51	61	118	181	225
PALC7	31	16	37	95	120	148

Mix. No	120d	150d	180d	270d	360d
PALC1	189	221	258	284	296
PALC2	307	319	333	345	376
PALC4	262	278	294	303	319
PALC5	234	254	296	307	313
PALC6	244	254	292	305	323
PALC7	171	201	211	221	240

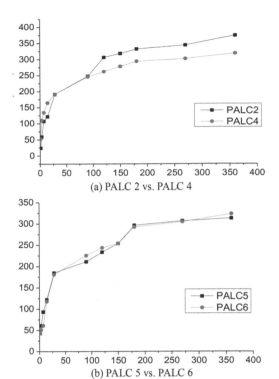

(a) PALC 2 vs. PALC 4

(b) PALC 5 vs. PALC 6

Figure 1. Effect of w/c on shrinkage of PALC.

strength makes it hydrate severely in the early stage, accordingly the shrinkage of concrete resulted from cement hydration is relatively obvious. At the completion of the late cement hydration, the volumetric reduction of the cement is competed without obvious further influences on the shrinkage of concrete.

3.2 Effect of w/c on shrinkage performance of PALC concrete

Under the same content of other raw materials, the shrinkages of PALC2 and PALC4 with different w/c vary accordingly, the final shrinkage of PALC2 with w/c of 0.33 is 57 μstrain higher than that of PALC4 with w/c of 0.38, rising by 15%, as shown in Figure 1(a); the cement content of PALC6 is 500 kg/m³, with w/c of 0.33; while the cement content of PALC7 is 450 kg/m³, with w/c of 0.36, namely, the cement content of the former is only 50 kg/m³ higher than the later, but its final shrinkage exceeds 83 μstrain, rising by 26%, as shown in Figure 1(b). It is obvious that the w/c shows evident effects on the shrinkage of PALC concrete, the lower the w/c is, the higher the shrinkage becomes in a certain range, vice versa.

Concrete shrinkage includes dry shrinkage and autogenous shrinkage, the dry shrinkage means an irreversible shrinkage due to loss of capillary bore water, gel pore water and adsorptive water from unsaturated air after concrete curing stops, and it increases with the rise of w/c; whereas the autogenous shrinkage means that shrinkage resulted from the spontaneous reduction of relative humidity of concrete which is caused by the insufficient water in the capillary bores inside the concrete under no outside dehydration, and it decreases with the rise of water–cement ratio. It cannot be neglected when the w/c is less than 0.4; besides, the lower the w/c is, the proportion of the autogenous shrinkage to concrete shrinkage becomes higher, so the result that shrinkage decreases with the rise of w/c is available.

3.3 Effect of HRWRA on the shrinkage performance of PALC concrete

The cement contents of PALC1 and PALC2 are identical, which totals 400 kg/m³, the HRWRA dosage of the former is 0.66 (% of cement mass), the later is 0.33, but the shrinkage value of PALC2 is 20% higher than that of PALC1, as shown in Figure 2(a); as for PALC5 and PALC6 with the same cement content of 500 kg/m³, the HRWRA dosage of the former is 0.67 (% of cement mass), and the later is 0.33, the shrinkage value of PALC6 is 3% higher than that of PALC5, so the shrinkages of the two are almost identical, as shown in Figure 2(b). It is obvious that polyocarboxy acid HRWRA is favorable for reducing the shrinkage of PALC concrete, because the incorporation of HRWRA makes that the water requirement of concrete reduces and the microstructure in the concrete is more uniform.

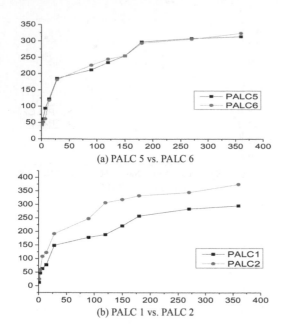

(a) PALC 5 vs. PALC 6

(b) PALC 1 vs. PALC 2

Figure 2.   Effect of HRWRA on shrinkage of PALC.

4   CONCLUSION

1. Integrally, the shrinkage of PALC concrete is unobvious, and its shrinkage value at the age of 360 days is about 300 μ strain.
2. The shrinkage of PALC concrete occurs in the early stage, and its shrinkage value is basically stable at the age of half a year; while its shrinkage value at the age of 28 days reaches 50% of its stable shrinkage value (one year value); 90% at the age of 180 days.
3. The shrinkage of PALC concrete increases with the rise of cement content, decrease of w/c and the HRWRA dosage in a certain range.

ACKNOWLEDGMENT

The authors gratefully acknowledge the financial support provided by the National Natural Science Foundation of China (No. 51278306), the Science Industry Trade and Information Technology Commission of Shenzhen Municipality (No. GJHZ20120614144906248).

REFERENCES

Feng, Z.W. Li, L.X. & Xie, Y.J. 2012. Experimental research on the shrinkage properties of concrete, *Concrete*, 4: 27–30.
Gao, X.J. Kan, X.F. & Yang, Y.Z. 2010. Shrinkage distribution of silica fume concrete prism with one-sided drying, *Journal of Jilin University: Engineering Science*, 40(3): 694–698.
Li, L.X. Xie, Y.J. & Feng, Z.W. et al. 2011. General comments of shrinkage of concrete and the methods of crack resistance, *Concrete*, 4: 113–117.
Li, S. Yi, Z. & Wang, W. et al. 2007. Fundamental study on chemical stability of phosphoaluminate cement hardened pastes, *Materials Research Innovations*, 11(2): 78–82.
Li, S.Q. Hu, J.S. & Liu, B. et al. 1999. Fundamental study on aluminophosphate, *Cement and Concrete Research*, 29(10): 1549–1554.
Ma, W.P. & Brown, P.W. 1992. Mechanical behavior and microstructure development in phosphate modified high alumina cement, *Cement and Concrete Research*, 22(1): 1192–1200.
Qin, W.Z. 2001. Shrinkage and cracking of concrete and its evaluation and prevention, *Concrete*, 7: 3–7.
Wang, W. Li, S.Q. Zhao, F.W. Liu, B. Hu, J.S. 2008. Study on the mechanical properties of carbon fiber and polyester fiber PALC concrete, *Journal of the Chinese Ceramic Society*, 27(4): 300–303.
Wang, W. Yi, Z.H. & Li, S.Q. et al. Sulfate resistance of novel phosphoaluminate cement, *Journal of the Chinese Ceramic Society*, 36(1): 82–85.
Yao, Y. 2011. *Volume stability and crack control of high performance concrete*, Beijing: China Architecture & Building Press.
You, Q.J. & He, M.H. 2004. Admixture affecting the concrete shrinkage and crack resistance, *Concrete*, 9: 3–7.

Advanced Materials and Structural Engineering – Hu (Ed.)
© *2016 Taylor & Francis Group, London, ISBN 978-1-138-02786-2*

Road surface micro- and macrotexture evolution in relation to asphalt mix composition

T. Iuele
Department of Civil Engineering, University of Calabria, Rende, Cosenza, Italy

ABSTRACT: Surface texture and friction greatly influence road functionality and have significant effects on user's safety and comfort, vehicle operational costs and environmental sustainability. Pavement surface characteristics include micro-texture, macrotexture, mega-texture and unevenness in relation to different wavelength classes. Friction depends on both macro and microtexture; in particular pavement initial friction is influenced by many factors, including aggregate properties (shape, texture and angularity), gradation, asphalt binder content, pavement construction, etc. Moreover, surface texture and, consequently, skid resistance change over time because of many deterioration phenomena. In the light of the above, this paper focuses on the evaluation of road surface micro and macrotexture evolution in relation to asphalt mix composition. Data were carried out from a two-year monitoring of an experimental road section where four different mixes were laid. Aggregates of different petrographic nature (limestone, basalt and expanded clay) were used for mixes production. Surface characteristics were monitored by several test methods and devices. Results highlighted significant differences in texture deterioration in relation to aggregate nature.

1 INTRODUCTION

Surface texture is defined as "the deviation of a pavement surface from a true planar surface" (ISO Standards 13473-1) and it can be seen as the superposition of many elementary harmonics, each one corresponding to a specific domain associated with a wavelength range: microtexture, macrotexture, megatexture and roughness (Boscaino & Pratico 2001, Pratico et al. 2012, Boscaino et al. 2005). In more detail, both micro and macrotexture influence road surface performance in terms of skid resistance, splash & spray and visibility, tyre-road noise, rolling resistance and type wear (Pratico et al. 2014, Pratico et al. 2014, Pratico & Vaiana 2012, Boscaino et al. 2005). All these phenomena are primarily related to road user safety but they also affect environmental sustainability increasing transportation costs). Pavement surface characteristics change over time. These variations are principally due to traffic actions (long-term variations), but they can also be identified as "short term variations" due to weather, rainfall and other environmental conditions, such as temperature variability (Masad & Rezaei 2009, Vaiana et al. 2012); the tendency follows a cyclical sinusoidal pattern throughout the year (Masad & Rezaei 2009). Generally, the evolution of skid resistance is characterized by an initial increase in friction coefficient that occurs in the months immediately after the laying of the road

surface because of the actions due to vehicular traffic: the bitumen film is gradually removed from the aggregates surface. This first phase is known as "early life skid resistance". Once the binder has been completely removed, the skid resistance evolution curve reaches a maximum. Afterwards (after 2 million of vehicle passes or 2 years of pavement service life) this higher exposure causes a decrease in skid resistance due to the smoothing and polishing actions under traffic loads (Do et al. 2007, Woodward 2008). Also macrotexture evolution shows a progressive decrease over time because of the voids occlusion due to binder migration and dust or oil accumulation. In addition, the aggregates are embedded in the asphalt matrix with a further reduction of the average texture depth (post-compaction actions). Texture deterioration rate is difficult to be estimated because it is affected by many factors, including aggregate properties, binder properties and aggregate-binder combination, road geometry and traffic, as discussed below. The ability of an aggregate to resist to traffic polishing depends on several parameters: hardness, mineralogical composition, crystalline structure, shape, angularity, resistance to abrasion, resistance to polishing, petrographic nature. It is a common practice to assume that aggregates with lower Los Angeles abrasion loss, lower sulfate soundness loss, lower freeze-thaw loss, lower absorption, and higher specific gravity have better resistance to polishing.

Table 1. Aggregate gradation for experimental asphalt mixes.

Sieve size [mm]	25	15	10	5	2	0.4	0.18	0.075
Percent passing [%]	100.0	100.0	99.4	46.9	32.3	16.8	12.5	8.3

Moreover, synthetic aggregates (slag or expanded clay), can also improve pavement frictional resistance. Aggregates composed of a hard mineral and a weak mineral behave differently: the softer mineral mass wears quickly, exposing the hard grains to the traffic loads, keeping good frictional properties for extended periods of time (Masad & Rezaei 2009, Vaiana et al. 2012, Do 2005). The quality of aggregate-bitumen should accelerate the removal of the bitumen film, especially for high levels of stressing under traffic loading. This phenomenon seems to be strictly associated with the binder type used in the asphalt mix. Several studies show that using polymer modified binders in asphalt mixtures results in a surface in which aggregate remain coated for a longer period of time when compared to an unmodified binder during the early life of surfacing (Woodward 2008, Woodward et al. 2011). Experimental studies on asphalt mixes produced with modified binders, confirm that aggregate with a lower Polished Stone Value (PSV) strips quickly; for this reason, in the phase of early life skid resistance, they may perform similarly, and in some cases better, than a higher PSV aggregate (Vaiana et al. 2012, Woodward et al. 2011). The variation of micro and macrotexture in the longitudinal direction is also due to road alignment (Hammoun 2008). In the light of the abovementioned facts, this paper investigates on asphalt pavements surface performance by experimental data carried out by a two-year monitoring of a test site. This work is part of a wider Research Project carried out by the Road Materials Research Laboratory of the University of Calabria in partnership with the Road Network Division of Provincial Administration of Cosenza (Italy).

2 MATERIALS AND METHODS

The test site is located in the district of Cosenza (SP 243, Rive Destra Crati). Four different dense graded friction courses were designed and laid on four road sections, characterized by a two-way single carriageway layout with both a straight stretch and a curve of radius less than 100 m.

The equivalent hourly flow rate ranged between 400 and 500 veh/h per lane with a percentage of heavy vehicles between 11% and 16%. Asphalt mixes variability is due to the use of aggregate of different petrographic nature: limestone, basalt

and expanded clay. The same aggregate gradation was used for all mixtures as reported in Table 1. Experimental mixes were designed and produced using the following percentages of aggregate on the total weight of the mix: i) M0 Mix: 100% limestone; ii) M1 Mix: 85% limestone, 15% basalt (aggregate size d > 5 mm UNI); iii) M2 Mix: 70% limestone, 30% basalt (aggregate size d > 5 mm UNI); iv) M3 Mix: 82% limestone, 8% expanded clay (aggregate size 3–11 mm UNI). A polymer-modified binder was used; the content was about 5.5% of the weight of the asphalt mix. Only for the M3 Mix the binder content reaches a value of 8%.

Micro and macrotexture measurements were carried out by means of the following devices and methods: British Pendulum Tester (BPT-CNR B.U. n°105/85), Sand Patch Test (SPT-CNR B.U. n°94/83) and Laser Profilometer (LP-ISO 13473-3). Surface profiles in terms of (x,z) coordinates were analysed in order to calculate both extrinsic and intrinsic indicators. Pavement sections were monitored every 6 months since they were laid, and overall 4 monitoring campaigns were done. Five measurement points for each wheel track were identified, for a total of 40 points for each section (20 in straight and 20 in the curve). Measurements were carried out on all trial sections in both directions along the wheel tracks.

3 EXPERIMENTS AND RESULTS

Experimental measurements were averaged along the wheel tracks for both North and South directions and for each trial section. Data were analysed considering straight and curve sections separately in order to better investigate the influence of road geometry on texture evolution over time. Microtexture results (in terms of BPN evolution trends) are summarized in Figure 1 where some differences between the mixes, especially in the phase of early life skid resistance, are highlighted; this variability is probably due to the composition parameters (aggregates and binder type) of the mixtures.

As it is possible to see, the M0 Mix is characterized by an initial increase in BPN, whereas for the other three mixes a decrease in microtexture was observed after six months. The different behaviour can be explained by considering the nature of aggregate: limestone has a lower PSV than basalt and expanded clay and, therefore, it has a lower stripping

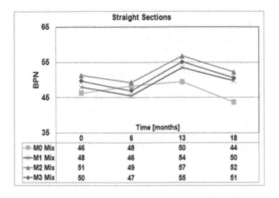

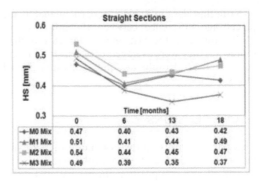

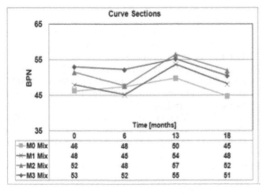

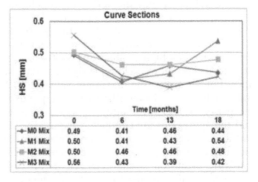

Figure 1. BPN evolution trends for straight and curve sections.

Figure 2. HS evolution trends for straight and curve sections.

resistance: the modified binder is easily removed and the aggregate becomes immediately more exposed (Woodward 2008, Woodward et al. 2011). For the other three mixes (M1, M2 and M3) the binder coating quickly becomes smoothed because the aggregate is more resistant to stripping (lower friction values after 6 months). For all mixes, the maximum value of BPN is reached after 13 months, that corresponds to a complete removal of the bitumen film so that aggregates are completely exposed to traffic actions; these same actions determine the next decrease (see Fig. 1). Note that for straight sections the behaviour of M1, M2 and M3 mixes is quite similar; as regards curve sections, instead, a lower increase is registered for the M3 mix after 13 months. Figure 2 plots the evolution trends of macrotexture, measured with the Sand Patch Test, for both straight and curve sections.

An early life decrease in macrotexture depth (after 6 months) was observed for all mixes in both straight and curve sections: binder smearing or migration together with a post-compaction action due to traffic loads has the effect of reducing the initial texture depth. This decrease is followed by a small but significant increase (after 13 months)

Table 2. Comparison between decrease rates for SH and other texture indicators.

Mix type	ΔHS [%]	ΔRt [%]	ΔRa [%]
M0	−11,5	−4,8	−5,9
M1	1,5	−5,2	−0,6
M2	−9,5	−1,4	−1,4
M3	−24,5	−12,0	−14,5

corresponding to the phase in which the migrated bitumen was removed by trafficking, with the exception of M3 mix for which this phase is shifted forward in time probably due to its higher binder content. 18 months after the pavements were laid an increase in texture depth for M1, M2 and M3 mixes can be observed in relation to a further removing of the migrated binder. For M0 mix, the differences are quite negligible.

Surface profiles were also analysed in order to calculate other macrotexture extrinsic indicators. In particular, the evolution trends observed for Sand Height measurements were compared to Rt and Ra values that represent the Peak-to-valley height of the surface profile and the Average Roughness, $Ra = \Sigma_i |z_i - z_{mean}| \cdot p(z_i)$, respectively (see Table 2).

Note that data in Table 2 were calculated as the percentage difference between the initial values (0 months) and the last ones (after 18 months); moreover, they were averaged between straight and curve sections. Results are quite consistent for all mixes.

4 CONCLUSIONS

This paper focuses on the evaluation of road surface micro and macrotexture evolution in relation to asphalt mix composition. In particular, 4 mixes were investigated and the following conclusions can be drawn: the mix produced with 100% of limestone aggregate had a lower resistance to binder stripping; on the contrary, mixes with aggregates characterized by higher values of PSV (basalt and expanded clay) showed a lower early life skid resistance. For all mixes, the maximum value of BPN was reached after 13 months that corresponds to the period in which the bitumen film is removed and the aggregates are completely exposed to the traffic actions; these same actions determine the next decrease. The general observed trend for macrotexture evolution is characterized by an early life decrease that is due to the binder migration inside the voids of the mix; after some months (around 6) the migrated bitumen is removed by trafficking and the texture depth begins to increase again. Only for the M3 mix texture depth begins to increase around 13 months after the pavement was laid, probably because of the higher percentage of binder content in the mix. Aggregate indicators were consistent with Sand Height results.

REFERENCES

Boscaino, G. & Praticò. F.G. 2001. A classification of surface texture indices of pavement surfaces. *Bulletin des Laboratoires des Ponts et Chaussees*, 234: 17–34.
Boscaino, G. Praticò, F.G. & Vaiana. R. 2005. *Tyre/road noise on different road pavements: synergetic influence of acoustical absorbing coefficient and surface texture*. In: International Federation of Automotive Engineering Societies, Proceedings of 10th EAEC European Congress. BELGRADE: JUMV (Yugoslav Society of Automotive Engineers), Beograd.
Do, M.T. Tang, Z. Kane, Z. & Larrard, F. 2007. Experimental simulation and modeling of pavement surface polishing by road traffic. *Bulletin de liaison des laboratoires des ponts et chaussées*, 267: 31–47.
Do, M.T. 2005. Relationship between microtexture and skid resistance. *Bulletin de liaison des laboratoires des ponts et chausses*, 255: 117–136.
Hammoun, F. Hamlat, S. & Terrier, J.P. 2008. *Laboratory evaluation of the resistance to tangential forces of bituminous surfacing*. Eurasphalt & Eurobitume Congress, Copenhagen, Denmark.
Masad, E. Rezaei, A. Chowdhury, A. & Harris, P. 2009. *Predicting asphalt mixture skid resistance based on aggregate characteristics*. Texas Department of Transportation and the Federal Highway Administration.
Praticò, F.G. & Vaiana, R. 2012. *Improving infrastructure sustainability in suburban and urban areas: is porous asphalt the right answer? And how?* Urban Transport 2012. Wit Transactions On The Built Environment, 128: 673–684, A Coruña, Spain.
Praticò, F.G. Vaiana, R. & Giunta, M. 2012. *Sustainable rehabilitation of Porous European Mixes*. ICSDC 2011: Integrating Sustainability Practices in the Construction Industry. 535–541.
Praticò, F.G. Vaiana, R. & Moro, A. 2014. The dependence of volumetric parameters of hot mix asphalts on testing methods. *Journal of Materials in Civil Engineering*, 26: 45–53.
Praticò, F.R. Vaiana, R. & Fedele, R. 2014. A study on the dependence of PEMs acoustic properties on incidence angle. *International Journal of Pavement Engineering*, ISSN: 1029-8436, doi: 10.1080/10298436.2014.943215.
Vaiana, R. Capiluppi, G.F. Gallelli, V. Iuele, V. & Minani. V. 2012. Pavement surface performance evolution: an experimental application. *Social & Behavioral Sciences*, 53: 1150–1161.
Woodward, D. 2008. *The effect of aggregate type and size on the performance of thin surfacing materials*. International Conference Managing Road and Runway Surfaces to improve safety, Cheltenham, England.
Woodward, D. Woodside, A.R. & Jellie, J.H. 2011. Predicting the early life skid resistance of Asphalt Surfacings. *University of Ulster, Northern Ireland, 200, Peat Journal*, 12(2): (2011).

Advanced Materials and Structural Engineering – Hu (Ed.)
© *2016 Taylor & Francis Group, London, ISBN 978-1-138-02786-2*

Effect of welding speed on the micro-hardness and corrosion resistance of similar laser welded (304/304) stainless steels and dissimilar (304/A36) stainless and carbon steels

M.M. Tash & K.M. Gadelmola

Department of Mechanical Engineering, Prince Sattam bin Abdulaziz University (Previously Salman bin Abdulaziz University), AlKharj, Saudi Arabia (on leave from Department of Mining, Petroleum and Metallurgical, Cairo University, Giza, Egypt)

ABSTRACT: An extensive study is carried out to investigate the effect of laser welding speed on the hardness and microstructure of the base metal, Heat Affect Zone (HAZ) and weld metal of plain-carbon and stainless steels. An understanding of the effect of welding speed on the corrosion resistance would help in selecting conditions required to achieve the best welding quality. Different laser welding speed i.e. 50, 80 and 100 mm/sec are applied for welding similar sheets and 65 and 100 mm/sec for welding dissimilar sheets. It is found that, the weld metal has the highest hardness levels in both similar and dissimilar laser welding. Also, increasing welding speed in the range (50–100 mm/sec) has increased the hardness levels in weld metal and HAZ. This can be attributed to the low heat input and high cooling rate produced when using laser welding. The finding results in this study showed that welding speed can affect the corrosion resistance in the range of 50–100 mm/sec. Increasing speed increases the corrosion resistance in the range of (50–80 mm/sec) following a decrease in the (80–100 mm/sec) range when laser welding of similar stainless steel (304) alloys.

1 INTRODUCTION

Welding of stainless steels, especially the austenitic grades, is important in energy related systems, for instance, power generation and petrochemical refining systems. In order to obtain the best performance of such joints, particular attention must be paid to the metallurgy of the weld metal (Medlock 1998). The ability to form chromium oxide in the weld region must be maintained to ensure stainless properties of the weld region after welding (Welding Handbook 1998). A major concern, when welding the austenitic stainless steels, is the susceptibility to solidification cracking. Materials that have a low tensile strength at temperatures near their melting point are said to exhibit "hot shortness," which often results in cracks appearing in the weld (Seaman 1977). In cases where fully austenitic welds are required, such as when the weld must be nonmagnetic or when it is placed in corrosive environments that selectively attack the ferrite phase, the welds will solidify as austenite and the propensity for weld cracking will increase (Rockstroh & Mazumder 1987).

Heat input is a relative measure of the energy transferred per unit length of weld. It is an important characteristic because, like preheat and interpass temperature, it influences the cooling rate, which may affect the mechanical properties and metallurgical structure of the weld and the HAZ (Mazumder 1983). Varying the heat input typically will affect the material properties in the weld. Cooling rate is a primary factor that determines the final metallurgical structure of the weld and Heat Affected Zone (HAZ), and is especially important with heat treated steels. Higher cooling rates for Laser Beam Welding process (LBW) are often needed to suppress precipitation of harmful intermetallic compounds during the solidification of the weld pool. High cooling rates also help to avoid sensitization during welding of stainless steels. Laser beam welding is a modern welding process; it is a high energy beam process that continues to expand into modern industries and new applications because of its many advantages like deep weld penetration and minimizing heat inputs. LBW is characterized by its low distortion and low specific energy input. It is an accurate method capable of high welding speeds for most materials, including many difficult-to-join materials. Some manufacturers of cigarette lighters are now using LBW as an alternative to resistance welding because of the lower porosity produced in laser welds (Harry 1974).

The scope of the present work is therefore to investigate the effect of welding speed during laser welding processes of similar and dissimilar materials on the hardness and corrosion resistance as well as the microstructure of base metal, Heat Affected Zone (HAZ) and weld metal positions, and particularly the alloy commercially known Stainless steel 304 and A36. The results of this study will provide a large input to existing data, in particular, by determining the effect of welding speed in both similar and dissimilar laser welding processes on the hardness, corrosion resistance and microstructure of base metal, HAZ and weld metal positions.

2 EXPERIMENTAL PROCEDURE

Sheets from stainless steels 304 of 1 mm thickness are laser welded. Also, sheets from stainless steels 304 and plain carbon steel A36 of 1 mm and 1.3 mm thickness, respectively are laser welded. Laser welding is carried out at different speed i.e. 50, 80 and 100 mm/sec for similar sheets and 65 and 100 mm/sec for dissimilar sheets. Micro-hardness measurements (HV) are performed on the base, HAZ and weld zones in all specimens with different conditions. Several reading are taken in the base metal, Heat Affected Zone (HAZ) and weld metal positions. Electrochemical measurements are carried out using Autolab PGSTAT 30. Linear polarization and Tafel plot tests are performed on welded specimens for similar and dissimilar sheets. All potentials are measured with respect to silver/silver chloride reference electrode (Ag/AgCl). Samples for metallographic examination are sectioned from the broken impact samples after hot rolling and hot forging (corresponding to each condition), mounted, polished and etched using Nital solution. The microstructure is analyzed using an optical and SEM microscope.

3 RESULTS AND DISCUSSIONS

Micro hardness measurements are performed on all specimens after laser welding with different conditions. Several reading are taken in the base metal, Heat Affected Zone (HAZ) and weld metal positions. The results are provided in Table 2. Results for micro hardness and corresponding standard deviation are listed in Table 2 and graphically presented in Figure 1. It is observed that the hardness is increased in both heat affected zone and weld metal zone as the welding speed is increased from 50 to 100 mm/sec., this can be attributed to the low heat input and high cooling rate produced as a result of increasing laser welding speed. Data fluctuation in some reading has been observed in micro-hardness results. This can be attributed to the microstructure variations effect where the measurement has done on the micro-hardness scale and measured values may influence by the presence of different phases in the microstructure of both HAZ and Weld metal zones. Figure 1 show the micro hardness results for different method i.e. similar and dissimilar welding of steels as a function of travelling speed. Again, it is observed that the hardness increases when welding speed increases. Also, the smaller HAZ produced and the high level of hardness in the weld metal at 100 mm/sec may be explained on the basis of the low heat input and high cooling rate produced with

Table 1. Chemical composition of low alloy steels grades used in the present work.

Alloy	C	Si	Mn	P	S	Ni	Cr	N
304	0.08	0.75	2.0	0.045	0.03	8.0–10.5	18.0–20.0	0.1
A36	≤0.2	≤0.35	0.6–1.0	0.02	0.008	–	–	–

Table 2. Micro-hardness (HV) results.

Welding type	Welding speed	Location	Hardness (HV)		
			Base	HAZ	Weld
Similar welding 304/304	50 mm/sec	304	209±4	220±4	232±0
	80 mm/sec	304	223±2.5	235±3.5	246±3
	100 mm/sec	304	250±5	269±3	280±4.5
Dissimilar welding 304/A36	65 mm/sec	A36	129±10	175±6	386±39
		304	216±5	242±13	395±12
	100 mm/sec	A36	122±1	182±3	382±19
		304	215±4	260±8	398±15

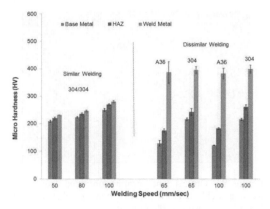

Figure 1. Micro-hardness profile for similar and dissimilar welding in the base metal, HAZ and weld metal when using laser welding speed of 50, 80 and 100 mm/sec for welding stainless steel 304 alloy sheets of 1 mm thickness and 65 and 100 mm/sec for welding stainless steel 304 alloy sheet of 1 mm thickness and plain carbon steel alloy A36 of 1.3 mm thickness, respectively.

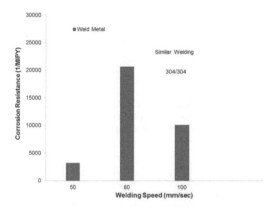

Figure 2. Corrosion resistance and corrosion rate of base and weld metal as a function of welding speed for similar welding using laser welding speed of 50, 80 and 100 mm/sec a) corrosion resistance.

increasing travelling speed. The maximum hardness of 398 HV has been obtained at laser speed of 100 mm/sec for dissimilar welding of stainless steel and plain carbon steel in the weld zone, however at lower travelling speed a significant decrease in hardness observed.

Variation in corrosion resistance as a function of travelling speed for similar welding is shown in Figure 2. The effect of laser speed (50–100 mm/sec) on the corrosion resistance is presented in Figure 2. Similar behavior is observed in dissimilar welding. In comparison for 50, 80 and 100 mm/sec travelling speed, the 80 mm/sec exhibit higher corrosion resistance levels than do the 100 mm/sec at similar

laser welding of stainless steel 304 alloys. However, the 100 mm/sec show higher micro hardness results than 50 and 80 mm/sec travelling speeds, Figure 1. It is observed from Figure 2 that increasing travelling speed (50–80 mm/sec) increases the corrosion resistance and following a decrease in corrosion resistance when using 100 mm/sec travelling speed. This can be attributed to the increase in hardness and the variation in microstructure produced at 100 mm/sec travelling speed in the weld metal zone.

3.1 Microstructure

Structure of the base metal, HAZ/weld metal and the weld metal for different welding speed for both similar and dissimilar welding of stainless and plain carbon steels are shown in Figures 3–8. Optical micrographs obtained from 304 stainless steel samples for the base metal, HAZ and weld metal when welding speed is 50 and 100 mm/sec are shown in Figures 3–6. Similarly, optical micrographs for the

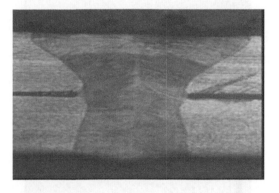

Figure 3. Optical micrographs obtained from laser welding: Weld metal at 100X. Laser welding speed is 50 mm/sec.

Figure 4. Optical micrographs obtained from laser welding: weld metal + base metal at 200X. Laser welding speed is 50 mm/sec.

Figure 5. Optical micrographs obtained from laser welding: Weld metal at 100X. Laser welding speed is 100 mm/sec.

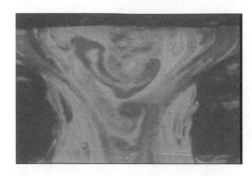

Figure 8. Optical micrographs obtained from laser welding: weld metal for plain carbon steel A36 at 100X. Laser travel speed is 100 mm/sec.

Figure 6. Optical micrographs obtained from laser welding: weld metal+ HAZ+ base metal at 200X. Laser welding speed is 100 mm/sec.

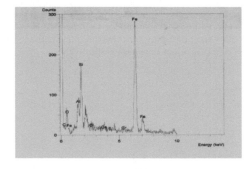

Figure 9. EDS Spectrum of a different zone for dissimilar welding (carbon steel A36/stainless steel 304) i.e. in the following position; In the half weld zone.

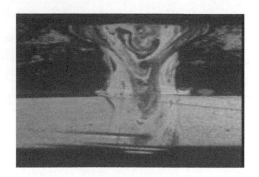

Figure 7. Optical micrographs obtained from laser welding: Weld metal for both stainless steel 304 and plain carbon steel A36 at 50X. Laser travel speed is 100 mm/sec.

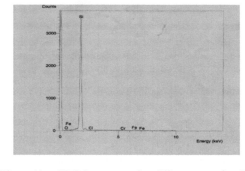

Figure 10. EDS Spectrum of a different zone for dissimilar welding (carbon steel A36/stainless steel 304) i.e. in the following position; In the weld zone close to stainless steel 304 interface.

base metal, HAZ and weld metal zones when dissimilar materials are laser welded at 100 mm/sec are shown in Figures 7–8. EDS Spectrum of different zone for dissimilar welding (carbon steel A36/ stainless steel 304) is presented in Figures 9–13

show EDS Spectrum in the weld zone/base A36 interface, in the weld zone close to A36 interface, in the HAZ/weld zone stainless steel 304 interfaces and in the base stainless steel 304, respectively. It is observed that the microstructure of the base metal

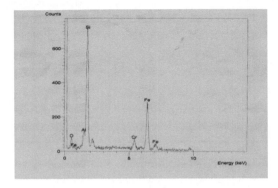

Figure 11. EDS Spectrum of different zone for dissimilar welding (carbon steel A36/stainless steel 304) i.e. in the following position; In the HAZ/weld zone stainless steel 304 interface.

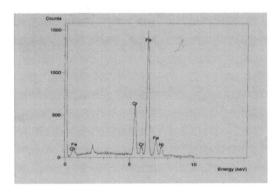

Figure 12. EDS Spectrum of different zone for dissimilar welding (carbon steel A36/stainless steel 304) i.e. in the following position; In the HAZ/weld zone stainless steel 304 interface.

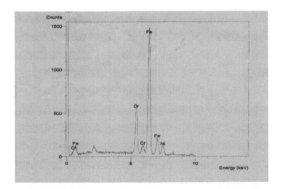

Figure 13. EDS Spectrum of a different zone for dissimilar welding (carbon steel A36/stainless steel 304) i.e. in the following position; In the base stainless steel 304.

of stainless steel 304 shows austenitic grains. The results in Figures 9–13 show that Cr and Ni peak start to appear when moving towards stainless steel side.

4 CONCLUSIONS

1. Increasing welding speed (50–100 mm/sec) increase the hardness of the weld metal and Heat Affected Zone (HAZ).
2. Low heat input and a high cooling rate are produced when using laser welding with increasing speed.
3. Increasing laser welding speed from 50 to 80 mm/sec increases the corrosion resistance.
4. The weld metal has the highest level of hardness in both similar and dissimilar welding by using laser welding.

ACKNOWLEDGEMENT

Financial and in-kind received support from the Deanship of Scientific Research, Vice Rectorate for Post Graduate and Scientific Research, Salman bin Abduaziz University (SAU) is gratefully acknowledged. This project was supported by the deanship of scientfic research at Sattam bin Abdulaziz University under the Research Program Project #.2300/01/2014".

REFERENCES

Harry, J.E. 1974. *Industrial Application of Lasers*, McGraw-Hill, United Kingdom.

Mazumder, J. 1983. *Laser Welding, Laser Materials Processing*, M. Bass, Ed., North Holland, Amsterdam, II 3–200.

Medlock, R.D. 1998. *Qualification of Welding Procedures for Bridges: An Evaluation of the Heat Input Method*. PhD Thesis, University of Texas.

Rockstroh, T. & Mazumder, J. 1987. Spectroscopic studies of plasma during CW laser materials interaction, *Journal of Applied Physics*, 61(3): 917–923.

Seaman, E. 1977. *Role of Shielding Gas in Laser Welding*, Technical Paper No. MR7, 982, Society of Manufacturing Engineer Dearborn, Michigan.

Welding Handbook, 1998. Vol. 4, 8th Edition. American Welding Society, 43–45.

Mechanical and industrial engineering

Advanced Materials and Structural Engineering – Hu (Ed.)
© 2016 Taylor & Francis Group, London, ISBN 978-1-138-02786-2

Reliability of detection of Glass Break in Intrusion and Hold-Up Alarm Systems

V. Nídlová & J. Hart
Czech University of Life Sciences Prague, Prague, Czech Republic

ABSTRACT: The problem of detecting glass break affects a large proportion of Intrusion and Hold-Up Alarm Systems (I&HAS). In a time of increasing property crime, it is highly important for Glass Break detector (GB) to be able to detect glass break within the guarded area reliably and free of error. In the case of installation of glass break detectors it is naturally important not only to ensure correct installation, to gauge the external influences impacting upon the detector and ensure proper maintenance, but also to guarantee their capability of detection under more arduous conditions. The tests that have been conducted to examine both the normal operation of the glass break detectors and the operation of these detectors under extreme conditions (different ways of breaking glass, foils on glass, etc.). These tests are important both from an informative perspective and due to the possibilities of development of potential counter-measures that could lead to their improvement and an enhancement of their level of security.

1 INTRODUCTION

Intrusion and hold-up alarm systems serve primarily for protecting buildings against unlawful conduct of third parties, and can be used as monitoring and control systems. They are, therefore, primarily a tool for ensuring a state of security. They operate in the material realm (physical protection of property, life and health) and in the emotional realm (providing a feeling of peace, safety and a certain security). As a result, it is important for them not to malfunction and for them to be sufficiently resistant to attack. The critical point of intrusion and hold-up alarm systems is predominant elements of the building envelope protection.

These elements are highly susceptible to poor installation, and as a result, it is very important to pay attention to this problem. One of the most widely used types of detector is the GB detector (glass break), which ranks among active detectors. On average, of all the types of the building envelope detectors used, the largest number of false alarms occurs on these detectors. This high error rate is primarily caused by incorrect installation.

2 MATERIALS AND METHODS

Several security risks may arise during the installation of intrusion and hold-up alarm systems, which impair the security of the entire building. The risks that occur due to poor installation or various sabotage techniques are always a serious danger for the guarded premises. They may jeopardise the guarded property or even the lives of the people who have the intrusion and hold-up alarm systems are intended to protect. Above all, however, they have an influence on determining the security risks of buildings.

Upon installation of GB detectors, it is necessary to take into account a number of fundamental prerequisites. The first prerequisite is for the detector must be installed on the opposite side than the guarded glass surface. The second prerequisite is for the cabling not to be visibly installed. In addition, the relevant norms must be adhered to upon implementation of the cable distribution mechanisms (Capel, 1999; Křeček, 2006; Uhlář, 2005). If the cable distribution mechanisms are installed in such a manner that enables access to them, it is possible to sabotage these systems and thus attack the entire installation of the intrusion and hold-up alarm systems.

If no End of Line (EOL) resistor is connected to the switchboard loop upon installation of the detector, the system is more vulnerable and can easily be bypassed. If a resistor is connected, bypassing is far more difficult than in the case of a simple loop (it is not possible to use simple short-circuiting). Upon sabotage, it is necessary to create a dual bypass and use it to replace the original loop at a single moment (Fig. 1).

Upon use of a bus bar (as wiring), sabotage is far more difficult than in the case of loop wiring. Successful sabotage would require, for example, the use of scanning communication (or decoding)

Zone   COM

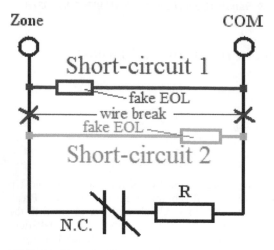

Figure 1. Short-circuit systems.

across the bus bar, with subsequent replacement of this communication with false reports which correspond to the communication of the existing system.

Wireless systems for communication most frequently use two unlicensed bands that comply with the Federal Commission for Communication (FCC) and the European Telecommunications Standards Institute (ETSI) (Powel & Shim, 2012). These are the bands 433 MHz and 868 MHz. These wireless transmissions should be protected by detecting disturbance of the frequency band, which monitors the load on the communication frequency. In the case of overloading of the frequency, the switchboard evaluates this fact and responds according to the setting (malfunction, alarm, etc.). The detectors are also mostly protected, namely by "wireless detector surveillance", which monitors the presence of the detector within the range of the switchboard (Petruzzellis, 1993; Cumming, 1994; Staff & Honey 1999).

The greatest risk upon use of wireless communication (between detectors and the switchboard) is a signal frequency jammer. This can overload the communication frequency by rendering the switchboard incapable of receiving the signal transmitted from the detector. This signal frequency jammer is dangerous above all because it can attack the system before the saboteur enters the guarded area, where he or she could be detected by one of the detectors.

Measurement of GB detectors should be focused primarily on tests that examine the capability of detection under more arduous conditions.

The GB detector detects pressure in the room and the characteristic sound of breaking glass.

The detectors GBS 210 and Glasstrek were used for measurement. These are frequently used

detectors, which are installed in both small buildings and large firms.

All the tested GB detectors are loop detectors with a simple type of sending of alarm information, which are cheap in comparison with other types of GB detectors (using a different type of data transmission).

During these tests an intrusion into the building was simulated, and a window was broken. To initiate the alarm, 60 × 60 cm glass plates were used, which were modified for various types of sabotage.

The GBS 210 detector (Fig. 2) uses the dual method for detection, wherein negligible changes to the air pressure in the room are evaluated (impact to the glass panel) and the subsequent sounds of breaking glass. The sensitivity of the pressure component of the detector can be easily configured according to the distance and dimensions of the protected windows.

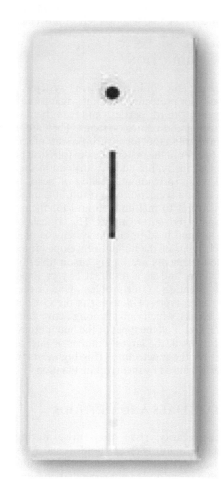

Figure 2. Detector GBS 210.

Like the GBS 210 detector, the Glasstrek Detector (Fig. 3) uses the dual method for detection, during which air pressure changes in the room are evaluated (impact to the glass panel) and the subsequent sounds of breaking glass. Although the sensitivity of the pressure component of the detector cannot be configured, the used installation distance (4 or 9 meters) can be configured. This configuration changes the evaluation characteristic of the breaking glass. The pressure compound of the detector is constant.

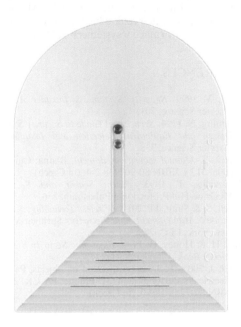

Figure 3.   Detector Glasstrek.

Table 1. Measured results for the detector GLASSTREK.

Method of measurement	Alarms
Breaking the glass	5/6
Breaking the glass with tape	4/6
Dent glass	5/6
Dent glass with tape	0/6

Table 2.   Measured results for detector 210 GBS.

Method of measurement	Alarms
Breaking the glass	6/6
Breaking the glass with tape	5/6
Dent glass	6/6
Dent glass with tape	2/6

Six detection ability methods of the detectors were tested with differently modified initialization materials—standard, with coating and a screen. Coating means that the initialization material is modified by being covered foil on one side. This modification changes the characteristic of breaking glass, and thus it also affects the functions of the detector. A screen is a barrier between the detector and initialization material that dampens the characteristic of the broken glass arising during an attack.

The testing was carried out for six times. During every detection method, both the classical breaking of the initialization material (using a metal rod) and the gradual denting of this material were tested. Through denting, the pressure component arises when the initialization material is punctured, is softened. The basic results of the measurements carried out are shown in Tables 1 and 2.

3   RESULTS AND DISCUSSION

The measured results and the overall comparison of GB detectors (Fig. 4) do not differ greatly, with the exception of the better elimination of false alarms. This is caused by the large demands of the building envelope detector, which leads a thorough checking during certification.

Until all the systems are tested, it is possible only to ask how many detectors and systems are at all secure. A further question is whether any system exists that could provide a reliable protection for a reasonable price.

The present state of development of security systems is at a point of stagnation. Although manufacturers are constantly attempting to develop systems, the majority copy old errors in the technical design of new products of a higher class, even despite the endeavours of customers to ensure manufacture is modified. Without innovative approaches and user feedback, this array will career into a blind alley.

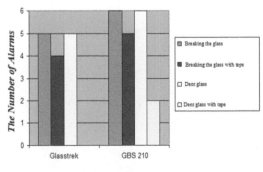

Figure 4.   Comparison of GB detectors.

4 CONCLUSIONS

Correct installation is one of the essential factors that can directly affect the security and functionality of the security system. It is necessary that the companies are providing the installation of security systems always follow the manuals for the given system and pay attention to the correct installation procedure according to the relevant standards. If the installation company does not follow the manuals and standards, then no equipment installed by them will meet the parameters of security systems and it is not suitable to use them.

The technical design of security systems is unique for the majority of manufacturers. In the case of every manufacturer, it is possible to find some poor technical designs that require modification. This deficiency can be resolved by technical development of the given product and adaptation to customer requirements.

The practical tests conducted on GB detectors brought an insight into their functionality and usability in practice. If a saboteur is instructed about the operation of these detectors, then they can be overcome. At the same time, the saboteur can also bypass the individual loops, and if skilled, can also bypass loops with an EOL resistor.

The only protection that would be usable in against current sabotage techniques is the development of new technologies. It is very important not to cast doubt on this development and to apply a constant endeavour to advance towards new technologies and greater security.

All of the measured data are also important for manufacturers of security systems as feedback on their products. In the future, there will be efforts to expand similar tests to other I&HAS manufacturers, as the reliability of these systems is very important, and it will be necessary to check them after deficiencies in the tested systems are ascertained.

ACKNOWLEDGMENT

It is a project supported by the IGA 2015 "The University Internal Grant Agency" (Minimalizace bezpečnostních rizik u bezdrátových přenosů poplachové informace v systémech PZTS).

REFERENCES

Capel, V. 1999. *Security Systems & Intruder Alarms.* Elsevier Science, 301.

Cumming, N. 1994. *Security: A Guide to Security System Design and Equipment Selection and Installation.* Elsevier Science, 338.

Křeček, S. *Manual security equipment.* Blatná: Circetus, 2006. 313 s. ISBN 80-902938-2-4. (in Czech).

Petruzzellis, T. 1993. *Alarm Sensor and Security.* McGraw-Hill Professional Publishing, 256.

Powell, S. & Shim, J.P. 2012. *Wireless Technology: Applications, Management, and Security.* Springer-Verlag New York, LLC, 276.

Staff, H. & Honey, G. 1999. *Electronic Security Systems Pocket Book.* Elsevier Science, 226.

Uhlář, J. 2005. Technical protection of objects, Part II, Electrical security systems II. Prague: PA ČR, 229 s. ISBN 80-7251-189-0. (in Czech).

Advanced Materials and Structural Engineering – Hu (Ed.)
© 2016 Taylor & Francis Group, London, ISBN 978-1-138-02786-2

A system utility analysis of Multiple Launcher Rocket System

H.F. Wang & C.W. Han
Northwest Institute of Mechanical and Electrical Engineer, Xi'an, Shaanxi, China

C. Zhao
School of Mechanical Engineering, Nanjing University of Science and Technology, Nanjing, Jiangsu, China

ABSTRACT: In order to establish the system efficiency evaluation model of weapon system, the multiple attribute utility theory and the complex multiple attribute utility model are introduced on the basis of WSEIAC model in this paper. Moreover, the system efficiency evaluation of the multiple launcher rocket system with equipped different reconnaissance subsystems is done, and a theoretical basis for the selection of the scheme is presented.

1 INTRODUCTION

The Multiple Launcher Rocket System (MLRS) has a strong firepower projection capability, and it is a useful method for fire suppression and annihilation between body tube artillery and tactical ballistic missile. Generally, MLRS consists of several subsystems. With the development of technology and constant update of equipments, each subsystem has a variety of options and it leads to forming weapon systems in different levels, which have great differences in system effectiveness. Such as unmanned reconnaissance aircraft instead of reconnaissance vehicle will improve the system effectiveness of MLRS. A multiple effectiveness models will be introduced in this paper based on the ADC effectiveness analysis model. The different schemes of subsystems where MLRS with unmanned reconnaissance aircraft or reconnaissance vehicle will be made an analysis of system effectiveness, and it will provide a theoretical basis and reference for the choice of subsystems of MLRS.

2 THE COMPOSITION OF MLRS

MLRS generally consists of the command vehicles, the reconnaissance vehicles, the meteorological vehicles, the rocket launchers, the ammunition loading vehicles, etc. This paper mainly studies different system effectiveness of MLRS that consists of different reconnaissance system. Since meteorological vehicle and ammunition loading vehicle use the same subsystem and have less influence on the weapon system effectiveness, they are not included in the system in our analysis. According to the difference in reconnaissance systems, MLRS

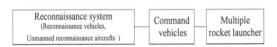

Figure 1. Architecture of system.

can be divided into scheme 1 which uses reconnaissance vehicles and scheme 2 which uses unmanned reconnaissance aircraft.

Thus, the whole system can be considered from the following three components (Fig. 1).

3 ANALYSIS MODEL OF WEAPON SYSTEM

ADC weapon system effectiveness analysis model proposed by US weapon system effectiveness advisory committee is a general model for assessing the effectiveness. In this model, the weapon system effectiveness E can be characterized by three parameters: validity vector A, reliability matrix D, and capability vector C. The model can be expressed as follows:

$$E = A \cdot D \cdot C \qquad (1)$$

In the model, the validity vector A represents the probability that the weapon system can be put into normal when it has been ready for the fight; the reliability matrix D is the conditional probability which indicates the probability that the system continues to work properly when the system status is known at the beginning of the task; the capability vector C is also the conditional probability which indicates the probability that the system finally successfully completed the desired task when the system status is known in the process of the task.

The validity vector A and reliability matrix D reflect the reliability and maintainability index of the weapon system. In this paper, the multiple effectiveness models are introduced for the capability vector C. In the model, the technical indicators of weapon system are the basis, the index system is set up for evaluation, and the capability vector C is characterized by the system total utility value.

4 MODELING FOR MLRS

4.1 System validity vector A

MLRS consists of reconnaissance systems (reconnaissance vehicle or unmanned reconnaissance aircraft), command vehicles, multiple rocket launchers (shown in Fig. 1). When the system is in the state of waiting for launching or launching in process, the system can be divided into two states:

1. Each subsystem can work normally, weapon systems are available;
2. Any subsystem is a failure, weapon systems are unavailable.

Let a_z, a_c and a_p denote the effectiveness of reconnaissance system, the effectiveness of command and control system (command vehicle) and the system effectiveness of MLRS. Thus:

$$a_1 = a_z \cdot a_c \cdot a_p \tag{2}$$

$$a_2 = 1 - a_1 = 1 - a_z a_c a_p \tag{3}$$

where,

$$a_z = \frac{MTBF_z}{MTBF_z + MTTR_z}$$

$$a_p = \frac{MTBF_p}{MTBF_p + MTTR_p}$$

$$a_c = \frac{MTBF_c}{MTBF_c + MTTR_c}$$

where,
$MTBF$—Mean Time Between Failure;
$MTTR$—Mean Time To Repair.

Thus, the system validity vector A can be expressed as follows:

$$A = [a_1, a_2] = [a_z a_c a_p, 1 - a_z a_c a_p] \tag{4}$$

4.2 System reliability matrix D

The reliability of each subsystem can be deemed to obey the exponential distribution. Since the series system consists of reconnaissance system, command vehicle and rocket launcher, the reliability can be expressed as R_z, R_c and R_p, where the expressions can be written as $R_z = \exp(-\lambda_z T)$, $R_z = \exp(-\lambda_c T)$ and $R_p = \exp(-\lambda_p T)$. In these equations, T is the system operation time, and λ is the fault rate of each subsystem. Thus, the probability that MLRS operates normally when enters the launching position and be normal throughout the firing process can be defined as follows:

$$d_{11} = R_z \cdot R_c \cdot R_p \tag{5}$$

Then, the fault probability of system can be expressed by:

$$d_{12} = 1 - d_{11} = 1 - R_z R_c R_p \tag{6}$$

It is impossible for the weapon system to maintain the unrepairable fault when launching in process, so $d_{21} = 0, d_{22} = 1$. Thus, the reliability matrix D of the whole weapon systems can be expressed by:

$$D = \begin{bmatrix} d_{11} & d_{12} \\ d_{21} & d_{22} \end{bmatrix} = \begin{bmatrix} R_z R_c R_p & 1 - R_z R_c R_p \\ 0 & 1 \end{bmatrix} \tag{7}$$

4.3 System capability vector C

Capability vector C is the comprehensive reflection of various kinds of the ability of weapon systems. It is the most important parameter in ADC effectiveness analysis model. This paper establish hierarchy target system to calculate the target effectiveness value based on multiple effectiveness models and use the total utility value of system as system ability vector.

4.3.1 Multiple utility model
The multiple attribute utility of model is a type of multiple criteria decision-making method. It can combine the utility that defined on each attribute into total utility. Multiple utility of model extends the concept of attribute utility to either level of the target. It also extends the regulation of inductive function (objective relation) to any level span. It permits the nodes which in the same span of each level to exist various kinds of relevant state (for instance full-correlation, uncorrelated, etc.). Moreover, different nodes can use different formats of utility function. Thus, the solving method of multiple attribute decision-making of model which is universal meaning can be established. It makes multiple attribute utility theory can be applied more convenient and quick to practical engineering project.

4.3.2 *Establishment of multiple utility model*

i. The Establishment of Hierarchy Relation of Evaluation Index.

The evaluation index of the weapon systems is various and complex and it needs to simplify according to specific problem generally. This paper mainly evaluates the influence of different reconnaissance subsystems to the weapon system. So the evaluation indexes about reconnaissance and command systems should be detailed and some other indexes can be simplified properly. This paper analyzes the main tactical and technical indexes that influence the operability of MLRS and establishes the hierarchy relation graph of the index (Fig. 2). The bottom of the figure is attribute indexes.

ii. Ascertaining the Relevant Relation of Index.

It is permitted to exist several types of relevant relation among the same level indexes in the multiple utility models. In this paper, there are two main relevant relations among the evaluation indexes, which are uncorrelated added style utility model and full-correlation multiplied style utility model. The uncorrelated added style utility model is expressed as follows:

$$U = \frac{\sum\limits_{i=1}^{I} U_i W_i}{W} \tag{8}$$

In this equation, i is the child node ordinal number, $i = 1, ..., I$; U is the utility value of parent node, W is the total weight of parent node, U_i is the utility value of the ith child node, and W_i is the weight value of the ith child node.

The full-correlation multiplied style utility model is expressed as follows:

$$U = K \prod_{i=1}^{I} U_i \tag{9}$$

$$K = \frac{U_{Ag}}{U_{Mg}} \tag{10}$$

In these equations, K is the utility demarcate constant of full-correlation multiplied structure; U_{Ag} and U_{Mg}, respectively, represent the utility value of the standard back-up scheme in added and multiplied structure. The standard back-up scheme chooses the back-up scheme that has the maximum utility value in multiplied structure.

In this paper, the indexes of the number of orientation locator and the time of salvo fire which are under the level of "fire performance" node are full-correlation multiplied relations, and it is uncorrelated with the other indexes. In addition, the other indexes are all uncorrelated with each other. So there are two relationships

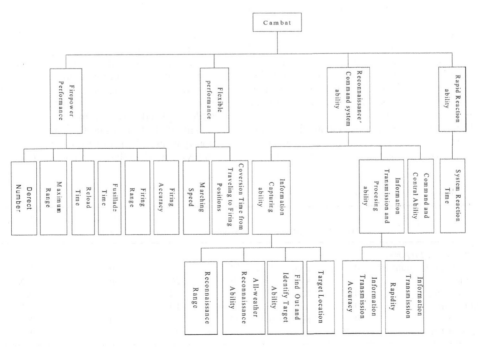

Figure 2.   The hierarchy relationship of the index of the system.

451

which are full-correlation and uncorrelated under the level of the "fire performance". It is an effectiveness model of composite mode. The calculation model of utility of the "fire performance" node is:

$$U_F = K_{14}U_1U_4W_{14} + U_2W_2 + U_3W_3 + U_5W_5 + U_6W_6 \quad (11)$$

In this equation, K_{14} is the utility calibration constant resulted of the full-correlation multiplied relation between the number of orientation locator and the time of salvo fire. It is calculated by (10); U_1, U_2, U_3, U_4, U_5, U_6, respectively, represent the index utility value of the number of orientation locator, the maximum range, the reloading time, the time of salvo fire, the field of fire and the accuracy of fire; W_{14} is the normalized relative weight since the indexes of the number of orientation locator and the time of salvo fire are synthesized; W_2, W_3, W_5, W_6, respectively, represent the normalized relative weight of the indexes which are the maximum range, the reloading time, the field of fire, and the accuracy of fire.

The other indexes of each level are uncorrelated added relation. The utility value can be calculated by (8).

iii. Ascertaining the Performance Value, the Utility Function and the Weight Value of Each Attribute Index.

When the hierarchical relation of indexes is established, the attribute index should be made clear and its performance value, utility function and the weight value of each index should be ascertained. Table 1 shows the performance value, the weight value and the utility function of each index which can affect the "fire performance" node. The weight value is obtained by experts surveying.

When the utility value of the attribute is calculated concretely, the performance value of the attribute of each scheme should be converted into dimensionless utility value. The utility value is the real number in the range of [0 1], and the transform is based on the utility function according to the utility theory.

iv. The Calculation of Target Utility Value

In the multiple utility of model, the calculation of the target utility value is a recurrent process which is from bottom to top. First, the attribute utility value should be calculated. Second, calculate the utility value of each back-up scheme which all the subgoals that defined on subordinate span are attribute target. Third, calculate the utility value of parent target by each level and span from the bottom of the structure. The concrete recurrent process is shown in Figure 3.

In the figure, I is the attribute number of parent node span, $f(x)$ is the utility function of the attribute index, and $F(U_1, U_2, \ldots U_I)$ is the function of relevant relation of index.

According to the figure of hierarchical relation of index and the recurrent process shown in Figure 3, the utility value of the fire performance U_h, the utility value of maneuvering performance U_{jd}, the utility value of reconnaissance and command system performance U_{zc} and the utility value of rapid responsive ability U_k can be calculated respectively from bottom to top. Thus, the total utility value U_b of the system operability can be expressed as follows:

$$U_b = W_hU_h + W_{jd}U_{jd} + W_{zc}U_{zc} + W_kU_k \quad (12)$$

In this equation, W_h, W_{jd}, W_{zc} and W_k represent the normalized relative weight of the index of the fire performance, maneuvering performance, reconnaissance and command system performance and the rapid responsive ability, the relevant relation of indexes is the uncorrelated added style effectiveness model.

The total utility value U_b of the system operability is just a factor of the capability vector. In the system, when the system is available, $C_1 = U_b$;

Table 1. The parameters of "fire performance" node.

	Weight	Unit	Scheme 1	Scheme 2	Worst value	Best value	Effectiveness function
Fire performance firepower							
Orientation locator number	0.25	Piece	12	12	0	12	X
Salvo time		s	20	20	30	15	1/X
Maximum range	0.25	Km	70	70	20	100	X
Reloading time	0.15	min	10	10	20	5	1/X
Field of fire	0.1	Scoring 0–1	0.8	0.8	0	1	X
Accuracy of fire	0.25	m	3	3	20	1	1/X

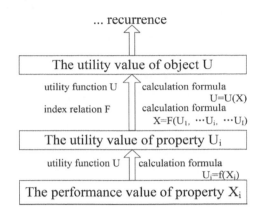

... recurrence

The utility value of object U

utility function U	calculation formula
	$U=U(X)$
index relation F	calculation formula
	$X=F(U_1, \cdots U_i, \cdots U_I)$

The utility value of property U_i

utility function U	calculation formula
	$U_i=f(X_i)$

The performance value of property X_i

Figure 3. Process of calculating the utility value of the object.

Table 2. Effectiveness evaluation result of multiple launcher rocket system.

Alternative scheme	Scheme 1	Scheme 2
System effectiveness	0.658	0.792
Scheme sorting	2	1

when the system is failure, $C_2 = 0$. Thus, the capability vector C can be described by:

$$C = [C_1, C_2] = [U_B, 0] \qquad (13)$$

5 ANALYSIS AND RESULTS

When the validity vector A, reliability matrix D and capability vector C are obtained and substituted into the (1), we can obtain the system utility value of MLRS that has different reconnaissance subsystems. The result is shown in Table 2.

Data can be drawn from Table 2, scheme 1 that unmanned reconnaissance aircraft is used in MLRS has a better weapon system effectiveness, and its utility value is 1.2 times the scheme 2 which reconnaissance vehicle is used in MLRS. Thus, it is recommended that MLRS with unmanned reconnaissance aircraft should be prior to adopting.

6 CONCLUSIONS

In this paper, a multiple effectiveness model is established for the system effectiveness analysis of MLRS. The paper gives the whole modeling process and an analysis result of MLRS with different reconnaissance subsystems. The model provides a quantitative angle to consider the combination of MLRS and it may conduct the product design of MLRS later.

REFERENCES

Hou, Y. 1999. *Weapons intelligent decision analysis support system based on knowledge*. Nanjing University of Science and Technology.
Li, M. & Liu, P. 2000. *Reasoning methods and application of weapon equipment development system*. Beijing: National Defence Industry Press.
Liu, M. 1999. Tactical missile weapon system effectiveness studies. *Journal of Projectiles, Rockets, Missiles and Guidance*, 3: 47–53.
Pei, Y. Ma, D.W. & Zhang, F.X. 2002. Effectiveness evaluation of the plan for multiple launcher rocket system on a warship. *Journal of Acta ArmamentarII*, 23: 212–214.
Wu, J.H. 2000. The reliability and maintainability in the application of anti-ship missile weapon system effectiveness evaluation. *Journal of Tactical Missile Technology*, 3: 18–23.

Advanced Materials and Structural Engineering – Hu (Ed.)
© 2016 Taylor & Francis Group, London, ISBN 978-1-138-02786-2

The influence of sintering temperature on microstructure and electrical performances of (1-x)BCZT-xCuO lead-free ceramics

J.J. Jiang & P. Xu
College of Materials and Metallurgy, Guizhou University, Guiyang, China

Q.B. Liu
College of Materials and Metallurgy, Guizhou University, Guiyang, China
School of Chemistry and Materials of Kaili College, Kaili, China

ABSTRACT: In order to obtain $Ba_{0.85}Ca_{0.15}Ti_{0.92}Zr_{0.08}O_3$ (BCZT) lead-free piezoelectric ceramics with good microstructure and electrical performances, BCZT lead-free piezoelectric ceramics doping CuO was synthesized to follow a conventional solid state reaction route. This text adopted the method of XRD, SEM, and quasi-static d_{33} mete, the effect of sintering temperature on the microstructure, piezoelectric and dielectric performances of $0.8(Ba_{0.85}Ca_{0.15})(Zr_{0.08}Ti_{0.92})O_3 \cdot 0.2CuO$(BCZT-CuO) lead-free piezoelectric ceramics were investigated. The results showed that the grain size and the dielectric loss of BCZT-CuO lead-free piezoelectric ceramics increase when the sintering temperature rises. When sintered at 1350 °C, BCZT-CuO lead-free piezoelectric ceramics exhibit the optimum electrical performances: $d_{33} = 286$ pC/N, $k_p = 12.3\%$, $\varepsilon_r = 3718$, and $\tan\delta = 1.36\%$, respectively.

1 INTRODUCTION

Lead-based piezoelectric ceramic has a dominant position in the area of piezoelectric materials because piezoelectric performances of lead-based piezoelectric ceramics are better than those of lead-free materials (Yin et al. 2004, Zhang et al. 2008, Jaffe 1971). As the main composition of PZT ceramics PbO has the property of high volatilization, it will cause an environmental pollution during firing and the disposal of PbO contaminated materials (Takanaka 2001). Therefore, it is necessary to find a new lead-free material system to exceed or compare the piezoelectric performances of lead-based materials.

In recent years, domestic and foreign researchers made great efforts to study and develop the lead-free piezoelectric ceramics. BaTiO₃-based ceramics are widely used in high dielectric constant of a ceramic capacitor, Positive Temperature Coefficient of Resistivity (PTCR) thermistor and lead-free piezoelectric ceramics (Wei et al. 2010, Xu et al. 2013). However, the pure BaTiO₃-based ceramics cannot achieve the desired performance, oxide and rare earth element usually should be doped into.

In this paper, CuO was doped into BCZT lead-free piezoelectric ceramics. The influence of sintering temperature on electrical performances and microstructure of the BCZT lead-free piezoelectric ceramics were researched. The dielectric properties of the samples were also measured as a function of temperature.

2 EXPERIMENTAL

BCZT-CuO lead-free piezoelectric ceramics was prepared by the conventional solid state reaction method. $BaCO_3$ (99%), $CaCO_3$ (99.9%), TiO_2 (99%), ZrO_2 (99%), and CuO (99%) powders were used as raw materials, weigh up these powders according to the chemical formula of (1-x)BCTZ-xCuO (x = 0.2 mol%). All the powders were mixed and ball (zirconias) milled with alcohol for 16h, which were then dried and calcined in an alumina crucible at 1280 °C for 3h. The calcined powders were milled and dried again. These weighed powders of BCZT-CuO were remixed with 4wt.% paraffin waxes and then pressed into 12 mm-diameter and 1.5 mm-thickness disks under 8 MPa pressure. After burning out paraffin waxes at 600 °C for 3h, the samples were finally sintered at 1320 °C to 1380 °C in steps of 15 for 4h.

The phases present in these ceramics were analyzed by using X-ray diffraction (XRD, PANalyticalXpert). The microstructures of ceramics were examined by using SEM (JSM-5900). For electrical measurements, the disk surfaces were coated with a silver paste on both sides and then poled under an electric field of 2 Kv at 50 °C in a silicone oil bath for 30 min. The piezoelectric constant d_{33} of the ceramics was measured by using a quasi-static d_{33} meter. Dielectric properties were carried out by the capacitance measuring tester.

3 RESULTS AND DISCUSSION

3.1 *Phase analysis*

The XRD patterns of BCZT-CuO lead-free piezo-electric ceramics samples within the range of 2θ from 20° to 70° at room temperature are shown in Figure 1. From Figure 1, all the samples have typical perovskite structure and no secondary phase can be found. This illustrates that Cu^{2+} diffuses into BCZT lattice and form a homogeneous solid solution.

3.2 *Microstructure analysis*

The SEM micrographs of BCZT-CuO lead-free piezoelectric ceramics sintered at different temperatures are shown in Figure 2. It is clearly shown that ceramic samples exhibit regular shaped grains with clear grain boundaries. With the increasing sintering temperature, the average grain size increases.

Thick ceramic grain size exerts good effects on piezoelectric properties, this is because that grain coarsening causes a decline in the grain boundary, the transmission of piezoelectric effect becomes prompter, space charge field effect weakened and domain wall motion are more freedom (Zhong 1998). The microstructure of ceramic materials is highly affected by the sintering temperature. However, the conductivity is closed related to it (Zhang et al. 2013).

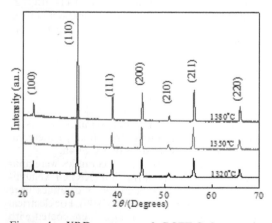

Figure 1. XRD patterns of BCZT-CuO ceramics sintered at different temperatures.

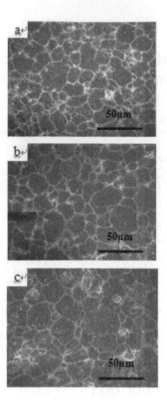

Figure 2. SEM micrographs of BCZT-CuO ceramics: (a) 1320 °C, (b) 1350 °C, (c) 1380 °C.

As is well known, clear boundary and uniform grain size are able to enhance the mechanical strength of the piezoelectric ceramic, and then improve the piezoelectric ceramics performance (Li et al. 2010).

3.3 *Analysis of piezoelectric properties*

The piezoelectric constant d_{33} and planar mode electromechanical coupling constant k_p of the BCZT-CuO lead-free piezoelectric ceramics at different sintering temperatures are shown in Figure 3(a) and (b). We can find that, with the increasing of sintering temperature from 1320 °C to 1350 °C, d_{33} and k_p increase to the maximum 286 pC/N and 12.3%. At the temperature of 1380 °C, with the increasing of sintering temperature, the d_{33} values decrease obviously. Grain growth inhomogeneous BCZT-CuO lead-free piezoelectric ceramics is over sintering when the sintering temperature is more than 1380 °C, which result in grain growth inhomogeneous. More difficult for small grain boundary domain wall movement (Xu et al. 2013), piezoelectric properties gradually deteriorated, representing as reduction of the piezoelectric coefficient.

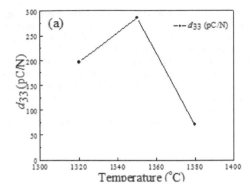

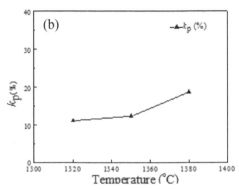

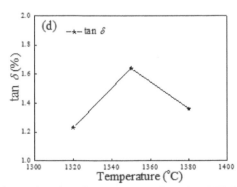

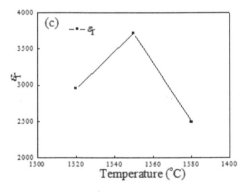

Figure 3. The electrical properties of BCZT-CuO ceramics sintered at different temperatures.

3.4 Analysis of dielectric constant and dielectric loss

The dielectric constant ε_r and dielectric loss $\tan\delta$ of the BCZT-CuO lead-free piezoelectric ceramics at different sintering temperatures measured at 1 kHz are shown in Figure 3(c) and (d). As can be seen, ε_r increases with the increasing of sintering temperature from 1320 °C to 1350 °C. ε_r reduced with the increasing of sintering temperature from 1350 °C to 1380 °C, and then, the variation tendency of the $\tan\delta$ is the same with ε_r. When the sintering temperature is higher than 1350 °C, the $\tan\delta$ decreases with the increasing of sintering temperature.

4 CONCLUSIONS

BCZT-CuO lead-free piezoelectric ceramics was prepared by the conventional solid state reaction method. Studying the influence of sintering temperature is on microstructure and electrical performances of BCZT-CuO lead-free piezoelectric ceramics. The author makes a conclusion: lead-free piezoelectric materials with excellent dielectric and piezoelectric belonging to pure perovskite structure. The electrical performance of BCZT-CuO lead-free piezoelectric ceramics increases when the sintering temperature rises and the sintering temperature is lower than 1350 °C. And then the electrical properties of BCZT-CuO ceramics decrease with the increasing of sintering temperature. When the sintering temperature reaches to 1350 °C, electrical properties perform the best: d_{33} = 286 pC/N, kp = 12.3%, ε_r = 3718, and $\tan\delta$ = 1.36%, respectively, BCZT-CuO ceramics show good electrical properties and hence they serve as excellent candidates for lead-free piezoelectric materials. In order to replace PZT systems, it is helpful to reduce environmental damages.

ACKNOWLEDGMENT

The authors gratefully acknowledge the supports of the Industrial Key Project of Guizhou Province (No. GY (2013)3027).

REFERENCES

Jaffe, B. 1971. *Piezoelectric Ceramics*. India: Academic Press.
Li, W. Xu, Z.J. & Chu, R.Q. et al. 2010. Dielectric and piezoe-lectric properties of $Ba(ZrxTi_1-x)O_3$ lead-free ceramics. *Brazilian Journal of Physics*, 40: 353–356.
Pu, Y.P. Wu, H.D. & Wei, J.F. 2012. Preparation and positive temperature coefficient of resistivity behavior of $Ba_{0.95}Ca_{0.05}TiO_3$-$BiYO_3$-$Na_{0.5}Bi_{0.5}TiO_3$ ceramics, *Powder Technology*. 219: 244–248.

Wei, C. Fu, C.L. & Gao, J.C. 2010. Effect of Mn doping on the dielectric properties of $BaZr_{0.2}Ti_{0.8}O_3$ ceramics. *Journal of Materials Science: Materials in Electronics,* 21: 317–325.

Xu, Q. Ding, S.H. Song & T.X. et al. 2013. Effect of Nd2O3 doping on microstructure and dielectric properties of BCZT ceramics. *Journal of the Chinese Ceramic Society*, 41: 292–297.

Yin, Q.Y. Lin, D.M. Xiao, D.Q. Zhu, J.G. & Yu, P. 2004. Researches on the Lead-free Piezoelectric Ceramics and Their Applications. *Metallic Functional Materials*, 11: 40–45.

Yoon, M.S. & Mahmud, I. 2013. Phase-formation, microstructure, and piezoelectric/dielectric properties of $BiYO_3$-doped $Pb(Zr_{0.53}Ti_{0.47})O_3$ for piezoelectric energy harvesting devices, *Ceramics International,* 39: 8581–8588.

Zhang, C. Liu, Q.B. & Huang, X.Q. 2013. Effect of sintering temperature on microstructure and electrical properties of (Mn, Nb)-doped BZT-BCT ceramics. *Advanced Materials Research,* 820: 59–62.

Zhang, T.J. Wang, J.Z. & Zhang, B.S. et al. 2008. Bottom electrodes dependence of microstructures and dielectric properties of compositionally graded (Ba1-xSrx)TiO_3 thin films. *Materials Research Bulletin,* 43: 700–706.

Zhong, W.L. 1998. *Ferroelectrica physics*. Beijing:Science Press.

Effect of Li on the microstructure and electrical properties of BCZT-xLi lead-free piezoelectric ceramics

L.L. Yao & X.Q. Huang
College of Materials and Metallurgy, Guizhou University, Guiyang, China

Q.B. Liu
College of Materials and Metallurgy, Guizhou University, Guiyang, China
School of Chemistry and Materials of Kaili College, Kaili, China

ABSTRACT: $0.5Ba(Zr_{0.2}Ti_{0.8})O_3$-$0.5(Ba_{0.7}Ca_{0.3})TiO_3$ $-x$Li (BCZT-xLi) lead-free ceramics with improved electrical performance were prepared by the conventional solid state reaction method. The effect of Li concentration on the microstructure, piezoelectric and dielectric properties of BCZT-Li ceramics were investigated by means of XRD, SEM, and quasi-static d_{33} meter, etc. The results show that single phase perovskite structure is obtained with Li^+ concentration higher than 2 wt.%. The emerge of a concentration additional peak was resulted by the formation of the secondary phase, and it is found that Li^+ substitutes the Ti-site as the acceptor dopants. The results showed that the BCZT-xLi ceramics with 0.5 wt.% Li^+ exhibit the optimum electrical properties: $d_{33} = 164$ pC/N, $k_p = 44.1\%$, $\varepsilon_r = 2365$, and tan $\delta = 1.78\%$, respectively.

1 INTRODUCTION

Since the Positive Temperature Coefficient of Resistance (PTCR) characteristics of semiconducting $BaTO_3$ materials was first observed in 1955, Barium titanate-based ceramics are widely applied in high dielectric constant ceramic capacitors, positive temperature coefficient of resistivity thermistors (Zhao et al. 2002, Tsur 2001). It is well known that the barium titanate-based ceramics is environmental friendly in comparison with the traditional lead zirconate titanate ceramics, which is highly toxic. With the recent growing demand of global environmental protection, many researchers have greatly focused on lead-free ceramics and their applications in various electric devices (Tian et al. 2013, Li et al. 2012).

Liu et al. (2009) first discovered the $Ba(Zr_{0.2}Ti_{0.8})$ O_3-$x(Ba_{0.7}Ca_{0.3})TiO_3$ lead-free piezoelectric ceramics, which exhibit an excellent piezoelectric property. It is known that the doping is an effective approach to improve the material performance in electroceramics. BCZT ceramics as a promising candidate for lead-free high piezoelectric ceramics, was studied with a special emphasis on the composition dependence of the piezoelectric properties and dielectric properties. Specifically, a recent report showed that the $0.5Ba(Zr_{0.2}Ti_{0.8})O_3$-$0.5(Ba_{0.7}Ca_{0.3})$ TiO_3 exhibits an excellent piezoelectric properties with a high piezoelectric coefficient up to 620 pC/N,

which appears to be a promising candidate for lead-free piezoelectric ceramics (Liu & Ren 2009). However, BCZT-based ceramics have higher sintering temperature and its microstructures lack excellent densification. So, it is necessary to study the microstructures and properties of BCZT-based lead-free piezoelectric ceramics.

Recently, BT-based ceramics with Li-doping have been investigated, which indicated that the doping is an effective way to decrease the sintering temperature and the compact texture (Kiang et al. 2013). However, the reports about the Li doped in BCZT-based ceramics are few. In this work, the effect of Li concentration on microstructure and electrical properties of $0.5Ba(Zr_{0.2}Ti_{0.8})$ O_3-$0.5(Ba_{0.7}Ca_{0.3})TiO_3$-$xLi_2CO_3$ (BCZT-xLi) lead-free ceramics were investigated.

2 EXPERIMENTAL PROCEDURE

$0.5Ba(Zr_{0.2}Ti_{0.8})O_3$-$0.5(Ba_{0.7}Ca_{0.3})TiO_3$-$x$ wt.%Li_2 CO_3 ($x = 0.5, 1, 2$ and 3) (BCZT-xLi) lead-free ceramics were synthesized by the conventional solid state reaction method. The $BaCO_3$, $CaCO_3$, ZrO_2, TiO_2 and Li_2CO_3 powders (purity 99.0%) were weighed in accordance with stoichiometric ratio. Then the powders were ball-milled for 12 h. After that, the mixed raw materials were dried at 90°C for 5 h, followed by being calcined at 1100°C for 4 h. Then, the calcined powders were again ball-milled for 12 h. The powers were finally dried and then pressed into 12 mm-diameter and 1.5 mm-thickness disks under 8 MPa pressure using 5 wt.% paraffin waxes as a binder. The disks were heated at 600°C for 4 h to remove the binder, followed by being sintered at 1350°C for 2 h in the air. The surface of the sintered ceramics disks was polished and coated with silver paste on both sides to act as the current collectors. To measure the relevant electric properties, the prepared ceramics samples were poled in silicone oil bath at 60°C under 2 Kv/mm for 30 min.

The phase structure and microstructure of the sintered disks were characterized by XRD and SEM, respectively. The piezoelectric constant d_{33} was measured by a quasi-static d_{33} meter. Dielectric properties and dielectric loss were carried out by the capacitance measuring tester. Electromechanical coupling factor k_p were calculated by using resonance–antiresonance method.

3 RESULTS AND ANALYSIS

3.1 Phase analysis

Figure 1 shows room temperature XRD patterns of the BCZT-xLi ceramics with different compositions in the range of 20°~70°. As it can be seen from these patterns, a single phase cubic perovskite structure is obtained for $0.5Ba(Zr_{0.2}Ti_{0.8})O_3$-$0.5(Ba_{0.7}Ca_{0.3})$ TiO_3-x wt.%Li_2CO_3 up to $x = 2\%$. This indicates that the Li has completely diffused into the BCZT lattice and thus maintained the perovskite structure of BCZT solid solution. The radius of Li^+ (0.076 nm) is much smaller than that of Ba^{2+} (0.135 nm) and Ca^{2+} (0.1 nm) but very close to those of Ti^{4+} (0.0605 nm)

and Zr^{4+} (0.072 nm). According to the principles of crystal chemistry, Li^+ ions generally enter the B-site for substituting the Zr^{4+} or Ti^{4+} ions of the ceramics, serving as acceptor-type dopants. When the weight percentage of Li ions surpasses 2 wt.%, a secondary phase appears in the XRD patterns, which can be clearly observed from Figure 1. From the XRD results, a limit solubility of Li was found in the $0.5Ba(Zr_{0.2}Ti_{0.8})O_3$-$0.5(Ba_{0.7}Ca_{0.3})TiO_3$. With the increase of Li ions, an additional phase emerges rather than perovskite phase.

According to the XRD results, it can be seen from Table 1 that the volume of the unit cell has no obvious changes with increasing Li_2CO_3 content. The lattice constants a and c are very smaller, which indicates that the ceramics have a trend to keep same phase. It is considered that the similar size of Li^+ and Ti^{4+} does not lead to the change of the BCZT lattice volume.

3.2 Microstructure analysis

Figure 2 shows the SEM images of polished cross-sections of BCZT-xLi piezoceramics sintered at 1350°C with different concentrations. It can be clearly seen that the grain size of the BCZT-Li ceramics exhibit relatively homogeneous. The grain sizes of the BCZT-Li ceramics are 5~10 μm.

Table 1. Lattice parameters for BCZT-xLi ceramics.

Compositions	a/Å	c/Å	c/a	a^2c/Å³
BCZT-0.5L	4.0310	4.0310	1	65.50
BCZT-1L	4.0310	4.0310	1	65.50
BCZT-2L	4.0310	4.0310	1	65.50
BCZT-3L	4.0310	4.0310	1	65.50

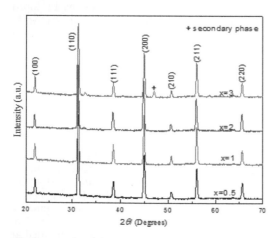

Figure 1. XRD patterns of BCZT-Li ceramics sintered at 1350 °C as a function of x wt.%.

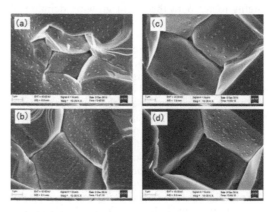

Figure 2. SEM micrographs of BCZT-Li piezoceramics with different Li+: (a) x = 0.5 wt.%, (b) x = 1 wt.%, (c) x = 2 wt.%, (d) x = 3 wt.%.

With the increase of Li content, the grains grow obviously and the grain size becomes larger. Moreover, there are obvious pores in the grain boundary with the increase of Li-doped. When the Li content is 0.5 wt.%, it was helpful for sintering of liquid phase, further promoting densification of the ceramics. When the Li content is up to 1 wt.%, 2 wt.% and 3 wt.%, the grain size continues to grow up, which results in lager grains and porosity. It can be confirmed that the excess addition of Li would foster the growth of the grain of the BCZTZ-Li ceramics.

In addition, it also can be seen from the Figure 2. That quite a number of distinct small particles spot exist in the grain for $0.5Ba(Zr_{0.2}Ti_{0.8})O_3$-$0.5(Ba_{0.7}Ca_{0.3})TiO_3$-$x$ wt.%Li_2CO_3. It was a spot that enrich Ti element in terms of EDS in Table 2. However, the concentration of Ti element in BCZT lattice was decreased. It is believed that, the Li^+ and Ti^{4+} ionic sizes are very close, not exactly but very close. When the Li^+ incorporated into BCZT-based lead-free ceramic, Li^+ would replace Ti^{4+} lattice. With the increasing the Li^+ content, Li^+ would squeeze out Ti^{4+} locations and stay there stably. It seems a reasonable deduction that this would also make Li^+ a more desirable in Zr^{4+} position. From the XRD patterns, we can see that no new phase was generated 2% of the Li content below.

3.3 Analysis of piezoelectric properties

The piezoelectric constant d_{33} and planar mode electromechanical coupling constant k_p of the BCZT-Li ceramics with different Li concentration, which sintered at 1350°C is shown in Figure 4. Planar electromechanical coupling factor (k_p) is calculated by the resonance frequency f_r and the antiresonance frequency f_a. It can be seen that the piezoelectric properties and planar mode electromechanical coupling constant exhibit obviously compositional dependence. The best piezoelectric properties of d_{33} ~164 pC/N and k_p ~44.1% appear when the Li content is 0.5 wt.%, which has dense and small grains. It can be clearly seen that with

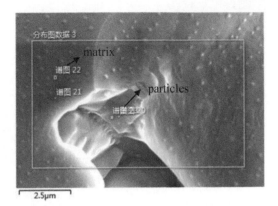

Figure 3. EDS micrographs of BCZT-Li piezoceramics with 3 wt.% of the Li content.

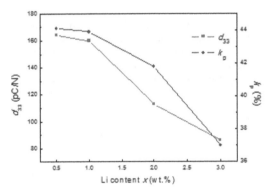

Figure 4. Piezoelectric constant d_{33} and planar electromechanical coefficient k_p of the $0.5Ba(Zr_{0.2}Ti_{0.8})O_3$-$0.5(Ba_{0.7}Ca_{0.3})TiO_3$-$x$ wt.%Li_2CO_3 as a function of x.

the increasing of Li concentration, the d_{33} and k_p decreased. Particularly, d_{33} and k_p values decrease dramatically when the concentration of Li ions surpasses 2 wt.%. Because the Li^+ could occupy the Ti-site, and the numerous Ti vacancies appeared, the domain wall move so easy that improve the piezoelectric properties. However, with excess Li content, the sintering character decreases and the grains grown larger, which reduce the density of the grains and decrease the electric properties of d_{33} and k_p. Nevertheless, the piezoelectric properties of this system are not high enough for industrial application.

3.4 Analysis of dielectric constant and dielectric loss

The dielectric constant ε_r and dielectric loss tan δ measured at 1 kHz of the BCZT-Li ceramics with different Li concentrations are shown in Figure 5. As it can be seen that ε_r increases with the addition

Table 2. Distribution of element for BCZT-3wt.% Li ceramics in terms of EDS.

Element	Matrix	Particles
C	1.50	0.58
O	12.27	9.28
Ca	2.09	1.79
Ti	19.46	38.95
Zr	6.22	1.91
Ba	58.46	49.50
Total	100.00	100.00

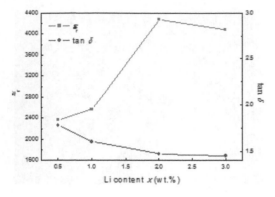

Figure 5. Dielectric constant and dielectric loss of the $0.5Ba(Zr_{0.2}Ti_{0.8})O_3-0.5(Ba_{0.7}Ca_{0.3})TiO_3-x$ wt.%Li_2CO_3 as a function of x.

of the content of Li (<2 wt.%). When the Li content is 2 wt.%, the ε_r reaches the maximum value $\varepsilon_r \sim 4275$. When the Li content exceeds 2 wt.%, the ε_r value decreases slowly. The decreased of dielectric constant may be attributed to the decrease in bulk density (shown in Fig. 2). However, the variation tendency of the tan δ is different from ε_r. The tan δ decreases with the increase of Li content. It can be observed that, the low dielectric losses tan δ 2% for all the samples below. The ε_r increases with the increase of doped Li^+ ions, which is due to the growth of grains. Larger grains and higher Li^+ concentration usually induce the increment of ε_r and also inhibit the deterioration of tan δ, so the dielectric properties are enhanced and tan δ decreased. From this work, it was also deducted that the addition of Li^+ causes the decrease of the dielectric loss. The mechanism for the effect of Li_2CO_3 is considered to be concerned with its valence state Li^+, with a radius of 0.076 nm. Li^+ occupies the B-site as Ti^{4+} in the BCZT lattice and functions as an acceptor, leading to some Ti-site vacancies, which suppresses the domain movement, resulting in a decrease in the dielectric loss.

4 CONCLUSIONS

The ceramics of $0.5Ba(Zr_{0.2}Ti_{0.8})O_3-0.5(Ba_{0.7}Ca_{0.3})$ TiO_3-x wt.%Li_2CO_3 were prepared via the conventional solid state reaction method. The effects of the Li_2CO_3 doping on the electrical properties were investigated. The XRD results show that the pure single phase cubic perovskite structure is obtained when the Li content below 2 wt.%. The dielectric constant increases with the addition of Li content. Furthermore, the dielectric loss decreases with the addition of Li content. The Li doping $0.5Ba(Zr_{0.2}Ti_{0.8})O_3-0.5$ $(Ba_{0.7}Ca_{0.3})TiO_3$ at the composition of 0.5 wt.% Li_2CO_3 shows the highest piezoelectric properties of $d_{33} = 164$ pC/N and $k_p = 44.1\%$.

ACKNOWLEDGMENT

This work was financially supported by a project of the Industrial Key Project of Guizhou Province (No. GY(2013)3027).

REFERENCES

Kiang, C. & Tan, I. et al. 2013. Effects of LiF on the Structure and Properties of $Ba_{0.85}Ca_{0.15}Zr_{0.1}Ti_{0.9}O_3$ Lead-Free Piezoelectric Ceramics, *International Journal of Applied Ceramic Technology*. 10(4): 701–706.

Li, W. & Hao, J.G. et al. 2012. Enhancement of the Temperature Stabilities in Yttrium Doped (Ba0.99Ca0.01) (Ti0.9Zr0.02)O3 ceramics, *Journal of Alloys and Compound*. 531: 46–49.

Liu, W. & Ren, X. 2009. Large Piezoelectric Effect in Pb-Free Ceramics, *Physical Review Letters*, 103: 257602–4.

Tian, Y. & Chao, X. et al. 2013. Phase Transition Behavior and Electrical Properties of lead-free (Ba1-xCax) (Zr0.1Ti0.9)O3 Piezoelectric ceramics, *Journal of Applied Physics*. 113: 184107.

Tsur, Y. & Dunbar, T.D. 2001. Crystal and Defect Chemistry of Rare Earth Cations in BaTiO3, *Journal of Electroceramics*. 7: 25–34.

Zhao, J. Li, L. & Gui, Z. 2002. A Study on the Properties of (Y, Mn) Co-doped Sr0.5Pb0.5TiO3, *Materials Science and Engineering. B*, 94: 202–206.

Advanced Materials and Structural Engineering – Hu (Ed.)
© *2016 Taylor & Francis Group, London, ISBN 978-1-138-02786-2*

The key controlling factors of tight sandstone reservoir of the lower Shihezi formation in Linxing area in eastern Ordos Basin

D. Zhao, H. Xu, D.Z. Tang, L. Li & T.X. Yu
School of Energy Resources, Key Laboratory of Marine Reservoir Evolution and Hydrocarbon Accumulation Mechanism, Ministry of Education, China University of Geosciences (Beijing), Beijing, China
School of Energy Resources, Key Laboratory of Shale Gas Exploration and Evaluation, Ministry of Land and Resources, China University of Geosciences (Beijing), Beijing, China

S.Z. Meng
School of Energy Resources, Key Laboratory of Marine Reservoir Evolution and Hydrocarbon Accumulation Mechanism, Ministry of Education, China University of Geosciences (Beijing), Beijing, China
School of Energy Resources, Key Laboratory of Shale Gas Exploration and Evaluation,
Ministry of Land and Resources, China University of Geosciences (Beijing), Beijing, China
China United Coalbed Methane Co. Ltd., Beijing, China

ABSTRACT: By employing the experimental methods of X-ray diffraction, scanning electron microscope, cathodoluminescence, casting slice, fluid inclusions and laboratory data in this study, the sandstone reservoirs of the lower Shihezi formation in Linxing area in eastern Ordos Basin are mainly litharenite and sub-litharenite. The pore types are mainly primary inter-granular pores and secondary dissolved pores. The reservoir physical properties are poor. The reservoir porosity is 5.9% on average and the average permeability is 0.073 md. The value of displacement pressure and median saturation pressure is very high. Moreover, the pore throat radius is small and the sorting is poor. The pore structure parameters are poor. Reservoir quality is mainly affected by the impact of deposition and diagenesis. The relatively high-quality formation mainly concentrated in the sand body of in the meandering river channel, and horizontally, the distribution of relatively high-quality formation mainly shaped in ribbon.

1 INTRODUCTION

Linxing area is located in the north-central part of Hedong Coalfield of the eastern margin of Ordos Basin, which is located in the transition zone between the Shanbei slope belt and the Jinxiflexure, and covering an area of about 2000 km² (Fig. 1). Since 2010, the Southwest block has 9 tight sandstone gas wells, and after fracturing, testing 6 wells obtained industrial

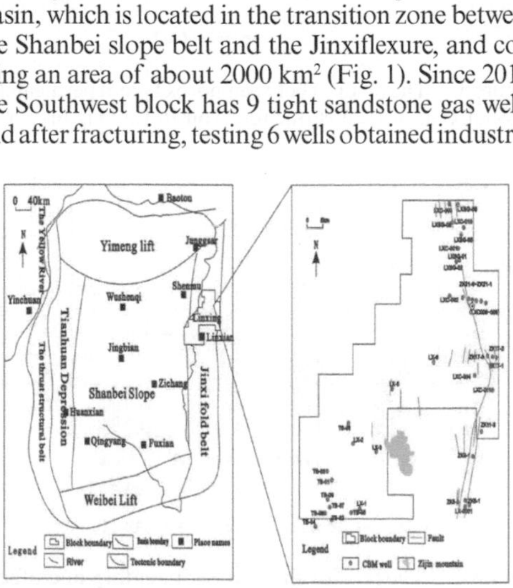

Figure 1. Location of Linxing area.

gas flow (He 2003, Guo 2012). Preliminary exploration and test show that the area has good coalbed methane and shale gas resource prospects (Gao et al. 2009, Zheng et al. 2014, Yang et al. 2014).

The lower Shihezi formation of lower Permian is a meandering river deposition. Sedimentation and sedimentary facies are controlled by a single source of NW. The lithology and lithofacies of tight sandstone change largely, which lead to the heterogeneity of the reservoir and the difficult exploration and development. Therefore, it is very important to study the key controlling factors of reservoir. There are applying X-ray diffraction, scanning electron microscope, cathodoluminescence, casting slice, fluid inclusions and laboratory data in this study. I made a detailed analysis on the characteristics of the reservoir characteristics and its evolution, which has a certain significance to guide the late exploration and development.

2 PETROLOGIC CHARACTERISTICS

Through the observation of 123 cast thin section of 8 wells and analysis of X-ray diffraction of whole rock, the tight sandstone reservoir lithology of Lower Shihezi Formation in Linxing area are mainly

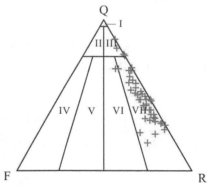

I: Quartz sandstone; II Secondary feldspar sandstone; III sub-litharenite; IV Feldspar sandstone; V Lithic arkose; VI Feldspar lithic sandstone; VII Lithic sandstone

Figure 2. Classification of the lower Shihezi group sandstone in Linxing region.

litharenite and sub-litharenite (Fig. 2). The content of quartz in Detrital composition of sandstones is the highest, which is 50.18% on average. The second content is debris, which is between 6 and 72% and is 42.73% on average. The debris is mainly quartz rock and acid volcanic rocks, which containing a small amount of mudstone and sandstone.

The content of feldspar is lowest, which is between 0 and 34% and is 6.97% on average. They are mainly plagioclase feldspar and potassium feldspar. Feldspar occurs argillation and alteration. The overall characteristics of the tight sandstone are high cutting-s, low quartz and feldspar.

Clay minerals are mainly mixed layer of illite and smectite, accounting for an average of 50%. The kaolinite and chlorite account for 41%. The calcite content of sandstone is extremely low, which is about 2% on average. Matrix is mainly some mud and is less than 15%. Cement is mainly quartz and clay minerals, and calcite cementation is relatively less. Cementation types are contact cementation and pore cementation. The sorting and roundness of rock particle are poor. The particles are mainly multi point line contact and line contact. In a word, the compositional maturity is low.

3 PHYSICAL CHARACTERISTICS

According to the porosity and permeability data statistical analysis of 43 samples of 21 wells, the porosity distribution of lower Shihezi reservoir is between 2.1% and 8.9%, and is 5.9% on average. The permeability distribution is between 0.001 md and 0.506 md, and is 0.073 md on average. It has a characteristic of low porosity and permeability. The porosity and permeability has a significant positive correlation, which

shows that the change of permeability is mainly controlled by the porosity development degree and that the contribution of micro fractures in the reservoir is very limited (Fig. 3).

4 INFLUENCE FACTORS OF RESERVOIR

4.1 *Influence of sedimentation*

The lower Shihezi formation in the study area of sedimentary period is mainly meandering river deposit, which mainly includes the riverbed deposition, beach sedimentary, naturallevee deposit and river swamp deposit. The sandstone reservoir is litharenite, which has a miscellaneous grain size and poor sorting, and resulting in a strong compaction (Liang et al. 2014, Yang et al. 2014, Li et al. 2012). And favorable reservoir is mainly located in the riverbed deposition and beach sedimentary (Figure 4).

In addition, many sets of coal seam developed of the lower Shihezi formation in the study area. The existence of coal makes an acidic or weakly acidic environment in the early stage of diagenesis, which led that the carbonate precipitation was less. Because of lack of carbonate sediments, the compression was very strong.

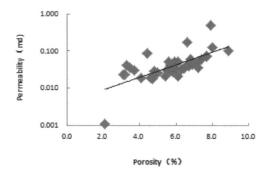

Figure 3. The relation between porosity and permeability of Shanxi group sandstone reservoir in Linxing region.

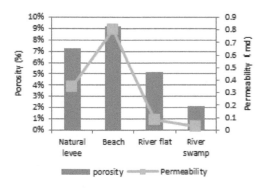

Figure 4. The relationships of Sedimentary microfacies with porosity and permeability.

4.2 The effect of diagenesis

4.2.1 Compaction

Through the casting slice observation, we can see the flexible composition of biotite with a strong compressional deformation-n, and part of rigid particles of quartz and feldspar with micro cracks. Compaction mainly occurs before the cement appears in large numbers (Zhou et al. 2010, Yao et al. 2013, Beard et al. 1973, Sun et al. 2011, Li et al. 2012).

4.2.2 Cementation

The content of clay minerals of sandstone reservoirs of the lower Shihezi formation in this area accounts for 13.1% which is the main fillings. Clay minerals mainly appear in the form of filling pore or in the form of rim by coating the clastic particles. X-ray diffraction analysis shows that the clay minerals are mainly Iraq-montmori-llonite mixed layer (58.1%), chlorite (20.3%), illite (11.6%), and kaolinite (10%).

Different clay minerals have different effects on porosity and permeability. The illite mixed layer and illite exist by filamentous filling the pores, which makes primary inter-granular pores into tiny inter-granular pores, the pore surface become rough, and circuitous degree increased, resulting in lower porosity and permeability (Figs. 5 and 6). Chlorites

exist in the film, which can reduce the early diagenetic compaction strength (Ji et al. 2008, Zhu et al. 2004, Wen et al. 2007, Ye et al. 2013, Zou et al. 2008). Moreover, chlorite film can block the pore water, and prevent the particles to increase itself, which is conducive to the preservation of primary porosity. There is book-like kaolinites filling pore, but their crystal is thick, which results in the intercrystalline pores. And the existence of pore-filling kaolinites can prevent other clay minerals, so the porosity and permeability increase (Figs. 7 and 8).

Authigenic Siliceous is very well developed in the study area, which is mainly in the form of secondary enlargement. Silicon content is generally 2.1%–6%, and 4.2% on average. Secondary quartz grows along the edge of the quartz particles, and there are visible mud ring edges between the different overgrowth periods.

4.3 Dissolution

Dissolution is the main diagenesis of the reservoir, which plays an important role in improving reservoir property. Dissolution includes feldspar dissolution, lithic dissolution, and minor authigenic quartz dissolution. The dissolution of feldspar

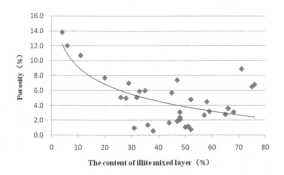

Figure 5. Relation between Illite mixed layer with porosity in the study area.

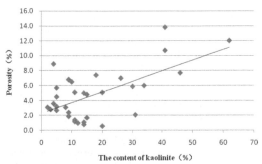

Figure 7. Relation between kaolinite with porosity in the study area.

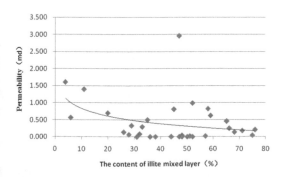

Figure 6. Relation between Illite mixed layer with permeability in the study area.

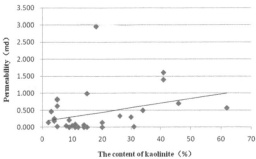

Figure 8. Relation between kaolinite with permeability in the study area.

along the cleavage generally forms intra-granular dissolved pores. In addition the kaolinite and illite aggregation have dissolution, which have a certain improving effect on reservoir.

5 CONCLUSIONS

1. The sandstone reservoir of the lower Shihezi formation in Linxing area is mainly litharenite and sub-litharenite. The pore types are mainly primary inter-granular pores and secondary dissolved pores.
2. The reservoir physical properties are poor, the reservoir porosity is 5.9% on average and the average permeability is 0.073 md. The value of displacement pressure and median saturation pressure is very high. Moreover, the pore throat radius is small and the sorting is poor.
3. Reservoir quality is mainly affected by the impact of deposition and diagenesis. The relatively high-quality formation mainly concentrated in the sand body of the meandering river channel, and horizontally, the distribution of relatively high-quality formation mainly shaped in ribbon.

REFERENCES

Beard, D.C. & Weyl, P.K. 1973. Influence of texture on porosity and permeability of unconsolidated sand, *AAPG Bulletin*, 57(2): 349–369.

Gao, X. Chen, H.D. & Zhu, P. et al. 2009. The diagenesis and evolution of He 8 reservoir of West- Surig gas field, *Natrual Gas Industry*, 29(3): 17–20.

Guo, B.G. Xu, H. & Meng, S.Z. et al. 2012.Geology condition analysis for unconventional gas co-exploration and concurrent production in Linxing area, *China Coalbed Methane*, 9(4): 3–6.

Ji, H.C. & Yang, X. 2008. Pore Types and Genetic Analysis of Shan-2 Member of Shanxi Formation in Eastern Ordos Basin, *Geological Journal of China Universities*, 14(2): 181–190.

Li, K. Zeng, T. & Pan, L. 2012. Reservoir characteristics of Xujiahe Formation in northeastern Sichuan Basin, *Lithologic Reservoirs*, 24(1): 46–51.

Li, W. Chen, G.J. & Lv, C.F. et al. 2012. Reservoir characteristics and controlling factors of lower member of Xiagou formation in jiudong depression, *Natural Gas Geoscience*, 23(2): 307–312.

Liang, Z.L. Zhang, S.C. & Jia, C.M. et al. 2012. Characteristics of the Triassic reservoirs in Cheguaiarea, northwestern margin of Junggar Basin, *Lithologic Reservoirs*, 24(3): 15–20.

Sun, J. Wang, B. & Xue, J.J. 2011. Reservoir characteristics and influencing factors of Permian Wutonggou Formation in eastern slope of Jimusar Sag, Junggar Basin, *Lithologic Reservoirs*, 23(3): 44–46.

Wen, H.G. Zheng, R.C. & Chen, H.D. et al. 2007. Characteristics of Chang 6 sandstone reservoir in Baibao-Huachi region of Ordos Basin, *Acta Petrolei Sinica*, 28(4): 46–51.

Yang, H. Xi, S.L. & Wei, X.S. et al. 2006. Exploration potential of natural gas in Sulige area, *Natural Gas Industry*, 26(12): 45–48.

Yang, Y.C. Li, F.J. & Dai, T.Y. et al. 2014. Reservoir characteristics of Chang 4+5 oil reservoir set in Hujianshan area, Ordos Basin, *Lithologic Reservoirs*, 26(2): 32–37.

Yao, J.L. Tang, J. & Pang, G.Y. et al. 2013. Quantitative simulation on porosity-evolution in member 8 of Yanchang formation of Baibao-Huachi Area, Ordos basin, *Natural Gas Geoscience*, 24(1): 38–46.

Ye, J.C. Sun, W. & Lu, D.G. et al. 2013. Diagenesis and Diagenetic Facies of Yanchang-81 Reservoir in Baibao Area, *Geological Science and Technology Information*, 32(4): 31–37.

Zheng, R.C. Wang, H.H. & Hou, C.B. et al. 2014. Diagenesis of sandstone reservoir of Chang 9oil reservoir set in Longdong area, Ordos Basin, *Lithologic Reservoirs*, 26(1): 1–9.

Zhou, X.F. Zhang, M, & Lv, Z.K. et al. 2010. The porosity evolution during the diagenesis of Chang 6 sandstone reservoir in Huaqing oil field, *Journal of Oil and Gas Technology*, 32(4): 12–17.

Zhu, H.Q. & Zhang, S.N. 2004. Reservoir diagenesis of Upper Paleozoic in northern Erdos basin, *Natural Gas Industry*, 24(2): 29–32.

Zixin, H.E. 2003. *Evolution of oil and gas in Erdos basin*, Beijing: Petroleum Industry Press.

Zou, C.N. Tao, S.Z. & Zhou, H. et al. 2008. Genesis, classification and evaluation method of diagenetic facies, Petroleum exploration and development, 35(5): 526–540.

Advanced Materials and Structural Engineering – Hu (Ed.)
© 2016 Taylor & Francis Group, London, ISBN 978-1-138-02786-2

Q&P process in manufacture of hollow products

B. Masek, K. Opatova & I. Vorel
Research Centre of Forming Technology—FORTECH, University of West Bohemia in Pilsen,
Plzeň, Czech Republic

ABSTRACT: All sectors of industry have a constant demand for complex-shaped products with the best possible mechanical properties delivered at low prices. The automotive industry has additional requirements for low weight and inertia mass. Parts that meet such demands can only be manufactured in new production chains, where new materials are combined with innovative elements of technology. One way of meeting the exacting requirements of engineering designers is the use of high-strength low-alloyed steels processed by means of advanced heat treatment techniques. Combining these with unconventional forming methods can enable producers to make complex-shaped parts with outstanding mechanical properties. One such new production chain comprises internal high pressure forming, press hardening and Q&P processing. In the present study, its capabilities were demonstrated by making thin-walled hollow products. The resulting ultimate strength levels were between 1950 MPa and 2300 MPa. Elongation was in the range of 10–18%.

1 INTRODUCTION

Advanced high-strength steels of AHSS and UHSS types find use in a wide range of industrial applications. Their excellent mechanical properties allow the mass of engineering parts made from them to be considerably reduced. New applications and ways of processing them are therefore continuously being sought. In this effort, various processes are combined in sophisticated production chains. In the present study, the Quenching and Partitioning Process (Q&P) was used to achieve excellent mechanical properties of the product.

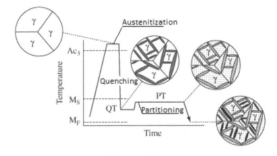

Figure 1. Schematic of Q&P process (Heilman 2011).

1.1 *Q&P process*

This process relies on incomplete quenching of steel and on subsequent partitioning of carbon. It is one of the modern heat treatment techniques for advanced high-strength steels (Kucerova et al. 2011) (Fig. 1). It consists of austenitizing and subsequent incomplete quenching to a temperature (QT), which is between the M_s and M_f temperatures. The resulting microstructure is a mixture of martensite and retained austenite (Speer 2003, 2005). The essential processes take place between the M_s and M_f temperatures and involve the diffusion of carbon from the super-saturated martensite into retained austenite. As a result, austenite becomes stable during isothermal holding at the Partitioning Temperature (PT) and remains stable even after cooling to ambient temperature (Tsuchiyma, 2012).

For the retained austenite to remain in the microstructure, it is essential that the carbon does not precipitate from martensite or form carbides during isothermal holding at the partitioning temperature. To prevent precipitation, the steels are alloyed with elements which prevent carbide formation. Silicon, aluminium and phosphorus are the most frequently used ones. The critical cooling rate can be reduced by adding chromium and manganese. These elements delay the pearlitic and bainitic transformations. However, the amount of chromium must be chosen so that it does not suppress the retarding effect that silicon has on carbide formation (Jirkova et al. 2013).

Manganese, silicon, chromium and aluminium have a favourable effect on solid solution strengthening. Manganese, an austenite stabilizer, effectively prevents formation of free ferrite in the microstructure and thus contributes to strengthening. Silicon, on the other hand, promotes free ferrite formation and causes a larger proportion of retained austenite to stabilize (Jirkova et al. 2012, Masek et al. 2010).

The Q&P process, an advanced heat treatment technique, can be combined with unconventional forming procedures. Complex-shaped parts with excellent mechanical properties can be manufactured in various ways, including a new technology chain comprising internal high pressure hot forming, press hardening and Q&P processing. The short term for this process is GFaP (Gas Forming and Partitioning) (Fig. 2), (Masek et al, 2011).

1.2 Internal high pressure forming

Forming by internal pressure of nitrogen gas is a process which enables complex-shaped hollow products to be made. The heated stock is enclosed in a die and shaped by internal pressure of nitrogen gas. The process can be altered by varying the die opening time: the period after which the die is opened and the product is removed at the corresponding temperature.

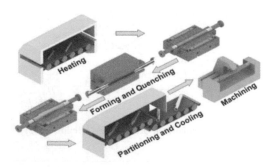

Figure 2. Technology chain which comprises internal high pressure hot forming, press hardening and Q&P processing.

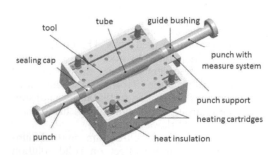

Figure 3. Forming of hollow parts by internal gas pressure.

2 MANUFACTURING ROUTE

In the present experimental programme, internal high pressure forming was trialled in conjunction with Q&P processing on low-alloyed high-strength 42SiCr steel (Table 1). The main alloying elements of the material were manganese, silicon and chromium.

The key phase transformation temperatures, M_s and M_f, were tentatively determined by calculation. The calculated M_s temperature was 289 °C and the M_f was 178 °C (Table 1). The microstructure of the tubular stock consisted of ferrite and pearlite, having a hardness of 295 HV10, ultimate strength of 981 MPa and elongation of 30%.

2.1 Process trials

The tubular stock diameter, length and wall thickness were 43 mm, 380 mm and 4 mm, respectively (Fig. 4). The sequence included 25 minute heating to 915 °C in a furnace. This heating step provided a fully austenitic structure in the stock. The stock was then placed in a die at ambient temperature. After the die had been closed, the internal high pressure forming step took place with the use of nitrogen gas. The pressure was 700 bar. The contact of the expanded stock with the die walls caused it to cool rapidly. In order to change the final temperatures in the work-piece hardening process, the die opening times were varied from 5 to 20 seconds. Once the products were removed from the die at

Table 1. Chemical composition of 42 SiCr steel (wt%).

C	Si	Mn	Cr	Mo	Nb	P	S	M_s [°C]	M_f [°C]
0.42	2.6	0.59	1.33	0.03	0.03	0.01	0.01	289	178

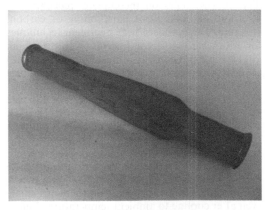

Figure 4. Final shape of the product.

temperatures between 180 and 250 °C, two process routes were used in order to map the effect of the Q&P process on the resulting microstructure and mechanical properties. The first group of products was cooled in still air. The second group was immediately placed in a furnace at 260 °C and held for 25 or 35 minutes to allow carbon partitioning and austenite stabilization.

3 RESULTS AND DISCUSSION

Products with various microstructures and properties were obtained by varying the process parameters. Two products representing two interesting process routes were selected for materials characterization and mechanical property measurement. The first process route was represented by a product with a die-opening time of 15 s ($Q_{210\,°C}$–C_{air}), which was quenched (press-hardened) to 210 °C and air cooled. The second route, in which the Q&P process was integrated, was represented by a product with the same die-opening time ($Q_{200\,°C}$–$P_{260\,°C}$) and with a quenching temperature of 200 °C. The partitioning process for stabilizing retained austenite took place in a furnace at 260 °C for 25 minutes (Table 2).

Samples for metallographic observation and for the measurement of mechanical properties were taken from both products after the process. The locations of these samples were selected in a way which permitted effective mapping of the effects of the process. Therefore, samples were taken from the gripped ends, from areas where the product cross-section, and thus the strain, was the largest and from transition zones between different diameters and cross-sections. A total of 14 test specimens were taken from each product. Statistical techniques were used to process the data from each location and the result was determined as the arithmetic mean.

3.1 Evaluation of microstructure and mechanical properties

Different microstructures were found in different locations. The product quenched to 200 °C and air-cooled ($Q_{210\,°C}$–C_{air}) contained martensite with a small amount of bainite in almost all the locations

Table 2. Parameters of experimental heat treatment.

	Austenitizing [°C]	QT [°C]	PT [°C]	Partitioning [min]
($Q_{210°C}$–C_{air})	915	210	–	–
($Q_{200°C}$–$P_{260°C}$)	915	200	260	25

examined (Fig. 5a). In locations where the product cooled more slowly due to incomplete contact with the die wall, a small proportion of free ferrite was found (Fig. 5b).

The product manufactured using the sequence with the integrated Q&P process contained primarily a mixture of martensite, retained austenite and bainite. Again, free ferrite was found in locations where cooling was slower. However, its amount was very small (Fig. 6).

Mechanical properties were measured in the locations of metallographic characterization. In order to map all parts of the product, the miniature tension test was chosen for the measurement.

The product which was removed from the die at 210 °C and cooled in air (Table 3) exhibited very high ultimate strength. Strengths at all measured locations exceeded 2260 MPa, combined with the A5 mm elongation of more than 10%. The lowest strength, 2263 MPa, and elongation of 12% were found in the transition zone of the largest cross-section. This area did not come into full contact with the die wall, and therefore did not cool as

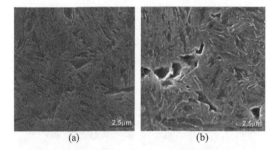

(a)   (b)

Figure 5. Product ($Q_{210°C}$–C_{air}) which cooled in still air: a) martensite-bainite microstructure (location T3/2); b) martensite-bainite microstructure with ferrite grains (location T3/12).

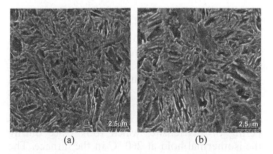

(a)   (b)

Figure 6. Product ($Q_{200°C}$–$P_{260°C}$) of the route with integrated Q&P process: a) martensite-austenite-bainite microstructure (location T7/2); b) martensite-austenite-bainite microstructure with a small amount of ferrite (location T7/12).

Table 3. Results of tension testing of the product ($Q_{210°C}$–C_{air}) cooled in still air.

Position	$R_{p0.2}$ [MPa]	R_m [MPa]	$A_{5\,mm}$ [%]	HV10 [–]
T3/2	1550	2324	14	709
T3/3	1534	2314	11	685
T3/5	1518	2263	12	667
T3/8	1641	2333	10	701
T3/12	1681	2345	10	738
T3/14	1652	2310	16	705

Table 4. Results of tension testing of the product ($Q_{200°C}$–$P_{260°C}$) which was Q&P-processed.

Position	$R_{p0.2}$ [MPa]	R_m [MPa]	$A_{5\,mm}$ [%]	HV10 [–]
T7/2	1576	1978	17	623
T7/3	1490	1956	21	615
T7/5	1379	1914	20	600
T7/8	1600	2006	18	596
T7/12	1584	1952	18	597
T7/14	1626	1978	17	608

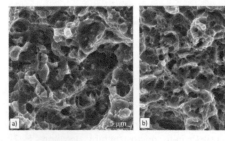

Figure 7. Analysis of fracture surfaces: a) air-cooled product ($Q_{210°C}$–C_{air}), location T3/6; b) Q&P-processed product ($Q_{200°C}$–$P_{260°C}$), location T7/3.

rapidly as the other locations. It also had the lowest hardness: 667 HV10. No other areas of the product showed an equally strong effect on the location of mechanical properties.

Products of the route with the integrated Q&P process exhibited ultimate strength levels 250–300 MPa lower than those of the other route (Table 4). This difference is due to partial tempering of martensite, which took place during the isothermal hold at 260 °C in the furnace. The lowest-strength location in the Q&P-processed product was in the same place as in the air-cooled product. In the first case, the strength and elongation were 1914 MPa and 20%, respectively. In other locations, the ultimate strength was close to

2000 MPa. On the other hand, the elongation levels after using the Q&P process were substantially higher: between 17 and 20%.

In order to obtain more complete information about the material's behaviour, the fracture surfaces were examined. Ductile fractures with dimples were found in all cases (Fig. 7). No signs of brittle fracture were detected even in the locations of the air-cooled product with the highest strength.

4 CONCLUSIONS

Hot forming by internal pressure of gas with subsequent press hardening and heat treatment was used for making complex-shaped hollow products without discontinuities and with high-quality surfaces. To demonstrate the process capabilities and obtain a wide range of mechanical properties, two sequences were selected. One of them included cooling of the product in still air after it was removed from the die. The other comprised isothermal holding, during which partitioning took place.

In the product ($Q_{210°C}$–C_{air}), which was air-cooled after removal from the die, the die opening time of 10 seconds led to press-hardening (quenching) with the quenching temperature of 210 °C. This was 79 °C higher than the M_s temperature. The resulting microstructure was a mixture of martensite, bainite and free ferrite. The values of mechanical properties were very high. The ultimate strength and elongation were 2300 MPa and 10%, respectively.

The product obtained from the route using Q&P processing ($Q_{200°C}$–$P_{260°C}$) was removed from the die at the temperature of 200 °C. Carbon partitioning due to diffusion took place over 25 minutes in a furnace at 260 °C. The resulting microstructure consisted of tempered martensite, retained austenite, bainite and a small amount of free ferrite. The resulting ultimate strength was lower than in the previous case: 2000 MPa. However, the elongation reached 18%.

Given the low complexity of the process and the excellent resulting mechanical properties, the produced parts promise an attractive potential for efficient production of a new-generation of high-strength products.

ACKNOWLEDGEMENTS

This paper includes results created under the projects CZ.1.05/3.1.00/14.0297 Technological Verification of R&D Results II, individual activity Hollow Shafts for Passenger Cars Produced by Heat Treatment with the Integration of Q-P

Process and CZ.1.05/2.1.00/03.0093 Regional Technological Institute. The projects are funded by the Ministry of Education, Youth and Sports from the European Regional Development Fund and using the resources of the state budget of the Czech Republic.

REFERENCES

Heilman, N. 2011. *Quenching and partitioning of high strength low alloyed steels containing Silicon.* TU Wien.

Jirkova, H. & Kucerova, L. 2013. *Q-P Process on Steels with Various Carbon and Chromium Contents.* (ed F. Marquis), In PRICM: 8 Pacific Rim International Congress on Advanced Materials and Processing: PRICM-8. Hoboken, NJ, USA.

Jirkova, H. Kucerova, L. & Masek, B. 2012. Effect of Quenching and Partitioning Temperatures in the Q-P Process on the Properties of AHSS with Various Amounts of Manganese and Silicon. *Materials Science Forum,* 706–709: 2734–2739.

Kucerova, L. Jirkova, H. Hauserova, D. & Masek, B. 2011. Comparison of Microstructures and Properties Obtained After Different Heat Treatment Strategies of High Strength Low Alloyed Steel. *Journal of Iron and Steel Research International,* 18: 427–431.

Masek, B. et al. 2011. Improvement of Mechanical Properties of automotive Components Using Hot Stamping with Integrated Q-P Process. *Journal of Iron and Steel Research International,* 18(1–2): 730–734.

Masek, B. Jirkova, H. Hauserova, D. Kucerova, L. & Klauberova, D. 2010. The Effect of Mn and Si on the Properties of Advanced High Strength Steels Processed by Quenching and Partitioning. *Materials Science Forum,* 654–656: 94–97.

Speer, J. et al. 2003. Carbon Partitioning into Austenite after Martensite Transformation. *Acta Materialia,* 51: 2611–2622.

Speer, J. et al. 2005. The "Quenching and Partitioning" Process: Background and Recent Progress. *Material Research,* 8(4): 417–423.

Tsuchiyma, T. et al. 2012. Quenching and partitioning treatment of a low-carbon martensitic stainless steel. *Materials Science and Engineering A,* 532: 585–592.

Advanced Materials and Structural Engineering – Hu (Ed.)
© *2016 Taylor & Francis Group, London, ISBN 978-1-138-02786-2*

Generalized-K (GK) distribution: An important general channel model for mobile fading channels

Y. Li

School of Automatic Control and Mechanical Engineering, Kunming University, China

ABSTRACT: The importance of channel models in radio mobile communication has long been recognized. Mobile radio channel models are essential for the development, evaluation and test of current mobile radio communication systems and also crucial for the realization of the future systems. Generalized-K distribution is very important for channel modeling in mobile communication. In this paper, we introduce a Generalized-K (GK) fading distribution. We simulate PDF of the received signal envelope characterized by the GK fading distribution, and simulation study the relays in Generalized-K fading under different scenarios.

1 INTRODUCTION

Radio wave propagation in wireless communication channels is a complicated phenomenon mainly characterized by the interaction of path loss fading, multipath fading and shadowing, which has presented a great challenge to researchers. A channel model is an abstract, simplified mathematical construct that describes a portion of reality (Tarokh 2009). A good channel model can make it possible to deep understanding the real radio wave propagation. Channel models are the basis for the software simulators, channel simulators, and RF planning tools that are used during the design, implementation, testing and deployment of wireless communication systems. They can also be used to precisely define the degree of impairment that a wireless system must be able to tolerate in order to meet the requirements for certification by standards groups and comply with contractual obligations (Tarokh 2009).

There are some well know distributions which are used to model fading channel models, such as Gaussian distribution, Rice distribution, Rayleigh distribution, and Nakagami distribution and so on. These fading models are typically used to fit the histogram of the experimental measurements of the envelope of the received random signals and are widely studied by scientists, researchers and engineers. Gaussian distribution is commonly used channel fading models under short-term and long term fading conditions; If there are strong LOS ray, the envelope of the signal at the receiver can be modeled by Rician distribution; The Nakagami distribution which has been first proposed in 1960 is a relatively new and advanced model, it can be used to model attenuation of wireless signals

traversing multiple paths. At the same time, lognormal distribution is used to model long term fading phenomena. However, these distributions can only investigate the effect, such as path loss, shadowing and multipath, of wireless channels on system performance separately. So, none of these models seems to be precisely enough to model mobile fading channel model. Generalized-K channel models (Abdi & Kaveh 1999) can simultaneously cope with all the above-mentioned three phenomena. It has better fitting ability to model radio channel, so there are particularly important for the design of future wireless communications systems.

In this paper, we focus on the simulation study of the Generalized-K distribution. The paper is organized as follows: Section II reviews Generalized-K distribution and its background. Section III provides simulation results of PDF for Generalized-K distribution using the different fading depth m, the average power Ω and coefficient K. Section IV simulation study of the relay in Generlized-K fading channel. Paper is finally concluded in section V.

2 THE GENERALIZED-K FADING CHANNEL MODEL

Generalized-K fading channel model is firstly formed by using the Gamma distribution to approximate the lognormal distribution (Simon & Alouini 2005), and then this model was improved by combining it with the Nakagami-M distribution.

When the channels at both the sender and the receiver are i.i.d. channels, due to the multipath propagation, the channel gains experience composite fading whose statistics follow a generalized-K distribution given by (Ansari et al. 2010):

$$f_x(x) = \frac{4m^{(k+m)/2}}{\Gamma(m)\Gamma(k)\Omega^{(k+m)/2}} x^{k+m-1} K_{k-m}\left(2\left(\frac{m}{\Omega}\right)^{1/2} x\right)$$

$$(1)$$

where, $\Gamma(\bullet)$ is the Gamma function, m and k are the Nakagami multipath fading and shadowing parameters, respectively. $K_n(\bullet)$ is the modified Bessel function of the second kind and order n. The parameters m and k are as measures of multipath fading and shadowing severity in radio channels, respectively, which means the smaller the values of m and k, the severer multipath fading and shadowing conditions will be, and vice versa. Furthermore, Ω is the mean power defined as:

$$\Omega \triangleq E\langle X^2\rangle/k \qquad (2)$$

where, $E[\bullet]$ denotes expectation.

Generalized-K channel model can describe different kind of fading and shadowing models by adjust the parameter(s) k and/or m. For example, When $k \to \infty$, it approximates Nakagami-m distribution; for $m=1$, it coincides with the K-distribution and approximate models R-L fading conditions; while when $m \to \infty$ and $k \to \infty$, it approximates the additive white Gaussian noise channel (Petros et al. 2006).

3 SIMULATION ANALYSIS OF PDF OF GENERALIZED-K DISTRIBUTION

In order to better understand the Generalized-K fading channel model, it is important to simulation study of the PDF of Generalized-K distribution with different m, k and Ω values.

The followings are PDFs of the Generalized-k Distribution with different m, k and Ω values.

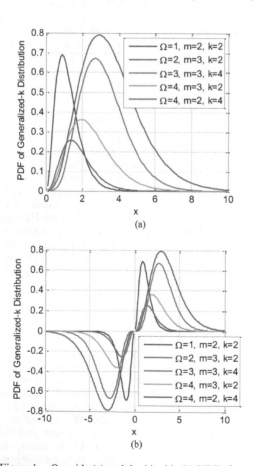

(a)

(b)

Figure 1. One side (a) and double side (b) PDF of generalized-K distribution.

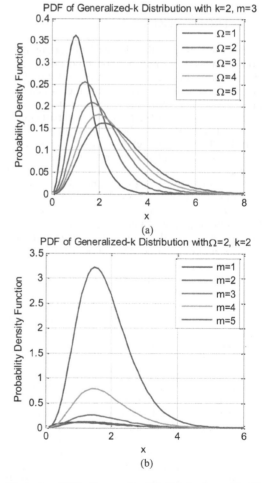

(a)

(b)

Figure 2. PDF of generalized-K distribution with different Ω (a) and different m (b).

From Figure 1 when can see, Generalized-K is an odd function symmetric to the origin.

Figure 2(a) is the PDF of Generalized-K distribution with $k = 2, m = 3$ and different Ω. From Figure 2(a) we can see, the bigger the value of Ω, the smoother the curve is, while, the longer the tail of the curve is. Figure 2(b) is the PDF of Generalized-K distribution with $\Omega = 2, k = 2$, and different m. From Figure 2(b) we can see, the bigger the value of m, the higher the peak of curve is.

Figure 3 is the PDF of Generalized-K distribution with $\Omega = 3, m = 3$ and different k. From Figure 3 we can see, the bigger the value of Ω, the higher the peak is, while, the longer the tail of the curve is. Figure 4 is the PDF of Gamma functions with different m when gamma is 2.

4 SIMULATION OF RELAYS IN GENERALIZED-K FADING CHANNELS

In this section, we simulation study the relays in Generalized-K fading under different scenarios.

Figure 5 is the SER of non-regenerative CSI relay in Generalized-K fading channel model; Figure 6 is the BER of non-regenerative CIS relay in generalized-K fading channel model; Figure 7 is the BER of fixed gain relay for QAM in generalized-K fading channel; Figure 8 is the SER of fixed gain relay for QAM in generalized-K fading channel.

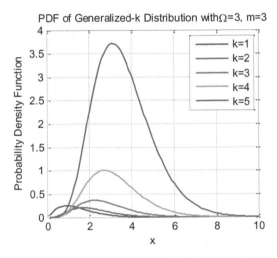

Figure 3. PDF of generalized-K distribution with different k.

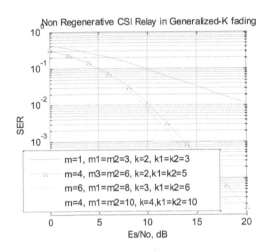

Figure 5. SER of non regenerative CSI relay.

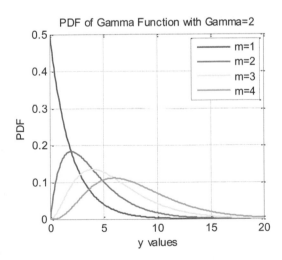

Figure 4. PDF of gamma function.

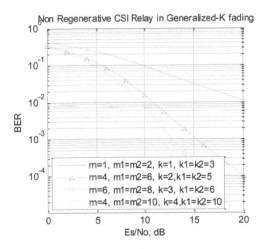

Figure 6. BER of non regenerative CSI relay.

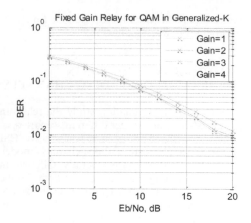

Figure 7.   BER of fixed gain relay for QAM.

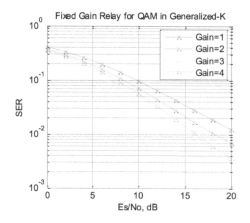

Figure 8.   SER of fixed gain relay for QAM.

5   CONCLUSIONS

Generalized-K channel model is an advanced model, which can simultaneously take into account of the propagation path-loss, shadowing and fast fading. It can usually cover more communications scenarios encountered in real mobile wireless systems, than other channel models, such as Gaussian, Rice, lognormal or the other composite channel models. In this contribution, Generalized-K fading channel is studied by both analysis and simulations.

REFERENCES

Abdi, A. & Kaveh, M. 1999. *On the utility of gamma pdf in modeling shadow fading (slow fading)*, Vehicular Technology Conference, 1999 IEEE 49th.

Ansari, I.S. Al-Ahmadi, S. Yilmaz, F. Alouini, M.S. & Yanikomeroglu, H. 2010. A New Formula for the BER of Binary Modulations with Dual-Branch Selection over Generalized-K Composite Fading Channels, *Communications, IEEE Transactions,* 59(10): 2654–2658.

Bithas, P.S. Mathiopoulos, P.T. Karagiannidis, G.K. & Rontogiannis, A.A. 2006. On the Performance Analysis of Digital Communications over Generalized-K Fading Channels, *IEEE Communications Letters,* 10(5): 353–355.

Simon, M.K. & Alouini, M.S. 2005. *Digital Communication over Fading Channels.* New York: Wiley.

Tarokh, V. 2009. *New Directions in Wireless Communications Research*, Springer US.

Yilmaz, F. & Alouini, M.S. 2010. *Extended Generalized-K (EGK): A New Simple and General Model for Composite Fading Channels*, IEEE Trans on Communications, Dec. 2010.

Advanced Materials and Structural Engineering – Hu (Ed.)
© *2016 Taylor & Francis Group, London, ISBN 978-1-138-02786-2*

Thin-layer element modeling method of aero-engine bolted joints

X.Y. Yao, J.J. Wang & X. Zhai
School of Energy and Power Engineering, Beihang University, Beijing, China

ABSTRACT: The bolted joints, which are very common in the structure of aero-engine, have efficient influences on the dynamic characteristics of the whole vibration. Based on thin-layer element method, this paper states general theory of modeling principles on aero-engine bolted joints structure. Then it gives application and compares the thin-layer element method with the detailed FE element method. Finally, the findings indicate that: the axial and bending stiffness are determined by using axial elastic modulus; Shear modulus has a significant influence on shear stiffness; thin-layer element method can simulate joint stiffness; and influence factors, element numbers, nodal numbers and calculation time are far less than detailed FE model.

1   INTRODUCTION

The bolted joints structure can connect two or more components together and has the characteristics of simplicity, practicability and good operability, so it is widely used in the mechanical structural system (Bickford 1990). In recent years, the influence of bolted joints on dynamic characteristic of whole model is greatly increasing. In the 1980s, NASA (Belvin 1987) has done systematic researches about the bolted joints modeling. In 2001, White Paper of Sandia Lab (Gregory & Martinez 2001) further stated the necessity of researching the model that contains connecting pieces. The modeling research of bolted joints structure has always been a hot issue in the field of structural dynamic research (Ibrahim & Pettit 2005).

At first, people set the bolted joints interface as rigid connection that ignores impact of all interfaces. After the development of FE method and Contact Theory, Jeong K et al. (Kim et al. 2007, Liu et al. 2010) established detailed FE model that retains the geometric characteristics of structure, and considers the contact and friction of interfaces fully so as to need smaller element sizes to mesh and more computing resources. In order to reduce the degrees of freedom, Ahmadiam H et al. (Ahmadian & Jalali 2007, Luan et al. 2012) used the spring, damping elements to simulate bolted joints and update the stiffness coefficients by experiment results. However, the surface–surface contact in bolted joints is simplified to the point–point contact, so the aforementioned methods have limitations in the process of modeling bolted joints in the complex mechanical structures.

In the complex structure, Ahmadian H (2006) used the thin-layer element method (Desai 1984) to

model the joints in the AWE-MACE structure and Ma Shuangchao (2013) applied it in welding parts of casing. Therefore, this method can simulate the joints in complex structures and maintain structure integrity. However, it has not been applied in aero-engine bolted joints, neither is the general theory of thin-layer element method of aero-engine bolted joints presented from basic mechanics principle.

Therefore, this paper discusses general theory of aero-engine bolted joints modeling with thin-layer elements method. The organization is as follows: the modeling principles of thin-layer element method are derived in Section 2; Section 3 gives the application of thin-layer element method and compares it with the detailed FE model; the conclusions are drawn in Section 4.

2   MODELING PRINCIPLES OF THIN-LAYER ELEMENT METHOD OF AERO-ENGINE BOLTED JOINTS

Figure 1(a) shows a typical aero-engine casing, there is something in common about the connections: (1) they are all with flange; (2) in many cases the components are thin cylinder; (3) the loads are mainly axial force, bending moment and shear force. Because of the complexity, the model should be simplified as a short and thick cylinder with bolted joints on the flange in the circumference direction, as shown in Figure 1(b). The aero-engine bolted joints play an important role in assembly and positioning, and they usually use the bolted joints with snap and precision bolted joints in order to decrease the stiffness loss caused by the bolted joints structure as much as possible. As a result, it could ignore the nonlinear properties of aero-engine bolted joints and make linearization when the bolt preload is large enough or the exciting force amplitude is relatively small (Ahmadian et al. 2006). Therefore, it models the bolted joints with thin-layer elements, the analytical model is shown in Figure 1(c). First, it introduces the FE equations formation of thin-layer element method; and then it gives the mathematical expressions of axial, bending and shear stiffness of bolted joints.

2.1 FE equations of thin-layer element method

The thin-layer elements can be generated by tetrahedron or hexahedron elements. It assumes that the thin-layer elements are 8 node hexahedron solid elements, and the nodal coordinate system is Cartesian coordinate system. The relationship between any node coordinate value and nodal coordinate value in an 8 node hexahedron solid element can be expressed as follows:

$$x = \sum_{i=1}^{8} N_i \cdot x_i \quad y = \sum_{i=1}^{8} N_i \cdot y_i \quad z = \sum_{i=1}^{8} N_i \cdot z_i \quad (1)$$

According to the basic equations of Theory of Elastic Mechanics, the relationship between element strain, stress and nodal coordinate is:

$$\{\sigma\} = [D]\{\varepsilon\} = [D][B]\{a\} \quad (2)$$

According to the principle of virtual work, the element stiffness matrix is:

$$[K^e] = \oint_{V_0} [B^t][D][B]dV_0 \quad (3)$$

Finally, all directions coupling stiffness of interface is neglected (Iranzad & Ahmadian 2012), and then the statics FE equation is assembled

$$[K] \cdot \{\delta\} = \{F\} \quad (4)$$

where,

$$[K_{ij}] = \begin{bmatrix} \sum_{m=1}^{k} \oint_{V_0} \begin{pmatrix} (A_{jx}A_{ix}c_{11} \\ +A_{jy}A_{iy}c_{44} \\ +A_{jz}A_{iz}c_{66})dV_0 \end{pmatrix} & \sum_{m=1}^{k} \oint_{V_0} (A_{jy}A_{ix}c_{44})dV_0 & \sum_{m=1}^{k} \oint_{V_0} (A_{jz}A_{ix}c_{66})dV_0 \\ & \sum_{m=1}^{k} \oint_{V_0} \begin{pmatrix} (A_{jy}A_{iy}c_{22} \\ +A_{jx}A_{ix}c_{44} \\ +A_{jz}A_{iz}c_{55})dV_0 \end{pmatrix} & \sum_{m=1}^{k} \oint_{V_0} (A_{jz}A_{iy}c_{55})dV_0 \\ & sym & \sum_{m=1}^{k} \oint_{V_0} \begin{pmatrix} (A_{jz}A_{iz}c_{33} \\ +A_{jy}A_{iy}c_{55} \\ +A_{jx}A_{ix}c_{66})dV_0 \end{pmatrix} \end{bmatrix},$$

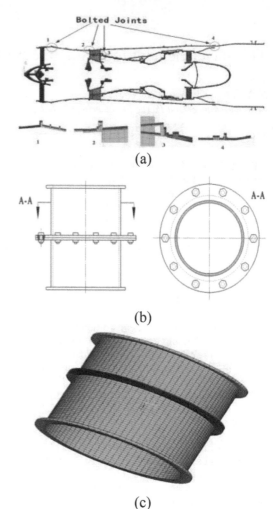

Bolted Joints

(a)

A-A   A-A

(b)

(c)

Figure 1. Bolted joints in aero-engine and the FE model.

$\frac{\partial N_i}{\partial x} = A_{ix}$, $\frac{\partial N_i}{\partial y} = A_{iy}$, $\frac{\partial N_i}{\partial z} = A_{iz}$, $i, j = 1, ..., n$, $k \leq 8$ is the node numbers in the element. The structure stiffness matrix $[K]$ is $3n \times 3n$ dimensions matrix that contains the thin-layer elements and the non-thin-layer elements.

For the choice of thin-layer elements thickness, Desai (1984) indicated that: if not considering the added mass and damping of interface, the influence of thickness is too small. Next the mathematical expressions of axial, bending and shear stiffness of bolted joints is given.

2.2 Axial stiffness

As shown in Figure 1(c), one end of analytical model is fixed and the other end is under axial force, so the axial stiffness of bolted joints can be defined as:

$$k_N = F_N/\Delta l_x \qquad (5)$$

where Δl_x is the axial displacement X_i of structure, F_N is the axial force F_{jx}. Solving the FE equation and get

$$\sum_{i=1}^{n} \begin{pmatrix} \sum_{m=1}^{k} (\oint_{V_0} (A_{ix}A_{jx}c_{11} + A_{iy}A_{jy}c_{44} + A_{iz}A_{jz}c_{66})dV_0 \cdot X_i \\ + \oint_{V_0} (A_{iy}A_{jx}c_{44})dV_0 \cdot Y_i + \oint_{V_0} (A_{iz}A_{jx}c_{66})dV_0 \cdot Z_i) \end{pmatrix} = F_N \qquad (6)$$

For the thin-layer elements, the thickness h is much smaller than the size of other directions, so it can be assumed approximately (Huang et al. 2008).

$$\varepsilon_y = \varepsilon_z = \gamma_{yz} \approx 0 \qquad (7)$$

Substitute the shape function N_i and get

$$\partial N_i/\partial y = \partial N_i/\partial z \approx 0 \qquad (8)$$

Substitute Equation (8) into Equation (6), the axial stiffness simplifies to

$$k_N = \sum_{i=1}^{n}\sum_{m=1}^{k} \oint_{V_0} (A_{ix}A_{jx}c_{11})dV_0 \cdot X_i/\Delta l_x \qquad (9)$$

The structure stiffness matrix $[K]$ contains the thin-layer elements and the non-thin-layer elements. From the Equation (9), the axial stiffness k_N is determined by the axial elastic modulus c_{11} if section size and type of elements are identified. Therefore, when the properties of non thin-layer elements are constant, the axial stiffness is determined by the axial elastic modulus of thin-layer elements.

2.3 Bending stiffness

According to the reference (Luan 2012), the equivalent bending stiffness is represented by axial stiffness:

$$k_M = M/\theta = b^2 k_N/2 \qquad (10)$$

where b is the outer diameter of cylinder, and k_N is the axial stiffness.

From Equation (10), it can be seen that the bending stiffness k_M is associated with the axial stiffness k_N and section size. So when the section sizes of structure are confirmed, the bending stiffness k_M can be represented by the axial elastic modulus of thin-layer elements.

2.4 Shear stiffness

As shown in Figure 1(c), one end of analytical model is fixed and the other end is under shear force, so the axial stiffness of bolted joints can be defined as:

$$k_{Sz} = F_{Sz}/\Delta l_z \qquad (11)$$

where Δl_z is the shear displacement of structure, and F_{SZ} is the axial force. Solving the FE equation and get

$$\sum_{i=1}^{n} \begin{pmatrix} \sum_{m=1}^{k} \oint_{V_0} (A_{jz}A_{ix}c_{66})dV_0 \cdot X_i + \sum_{m=1}^{k} \oint_{V_0} (A_{jz}A_{iy}c_{55})dV_0 \cdot Y_i \\ + \sum_{m=1}^{k} \oint_{V_0} (A_{jz}A_{iz}c_{33} + A_{jy}A_{iy}c_{55} + A_{jx}A_{ix}c_{66})dV_0 \cdot Z_i) \end{pmatrix} = F_{Sz} \qquad (12)$$

Similarly (Huang et al. 2008), according to

$$\partial N_i/\partial y = \partial N_i/\partial z \approx 0 \qquad (13)$$

Substitute Equation (13) into Equation (12), the shear stiffness of z direction simplifies to

$$k_{Sz} = \sum_{i=1}^{n}\sum_{m=1}^{k} \oint_{V_0} (A_{ix}A_{jx}c_{66})dV_0 \cdot Z_i/\Delta l_z \qquad (14)$$

in the same way, the y direction shear stiffness is

$$k_{Sy} = \sum_{i=1}^{n}\sum_{m=1}^{k} \oint_{V_0} (A_{ix}A_{jx}c_{44})dV_0 \cdot Y_i/\Delta l_y \qquad (15)$$

The Equations (14) and (15) show that the shear stiffness of bolted joints is related to the shear modulus. Therefore, when the properties of non thin-layer elements and section size are constant, the shear stiffness k_S is determined by the shear modulus c_{44} and c_{66} of thin-layer elements.

In conclusion, the axial, bending stiffness of aero-engine bolted joints are by the axial elastic modulus of thin-layer elements; and the shear stiffness is determined by the shear modulus of thin-layer elements.

3 APPLICATION

The modeling principles are described in details above. Now state the application and compare it with the detailed FE model in order to prove correctness and prospect of this method.

First, calculate the axial stiffness and bending stiffness by detailed FE model as the accuracy value; second, model the bolted joints using the thin-layer element method and adjust the axial elastic modulus to make the bending stiffness equal to the detailed FE model; third, under this condition, verify whether the axial stiffness is equal to the detailed FE model in order to verify the thin-layer element method and finally compare the results of the two models and draw the conclusions.

3.1 Detailed FE element method

As shown in Figure 2, it is a cylinder casing with 12 bolts on the flange in the circumference direction. Use detailed FE element method to model and consider preload of bolted joints (using PRETS179 element) and the contact (using TARGE170 and CONTA174 elements) between screw cap and flange, screw nut and flange, flange interfaces fully.

The elastic modulus is 210 GPa, the density is 7850 kg m^{-3} and the ratio is 0.3. The element type is SOLID185 with 88333 element numbers and 95917 node numbers.

Set one end of FE model fixed and apply axial force and bending moment on the other side, respectively. Finally, obtain change law of connection stiffness as shown in Figures 3 and 4, where x axes are the axial displacement and rotating angle respectively and y axes are the axial stiffness and bending stiffness, respectively. Because the shear stiffness of aero-engine bolted joints can be considered infinitely, it ignores the shear stiffness of this section. The definition of axial stiffness and bending stiffness is as Equations (5) and (10).

The conclusions are drawn that: (1) axial stiffness is invariable when axial force changes; (2) bending stiffness is constant when external load changes; (3) the axial stiffness is 1.95E9 N/m and the bending stiffness is 4.65E7 N·m/rad.

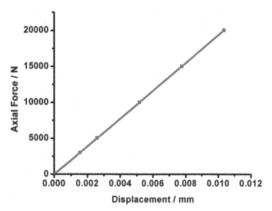

Figure 3. Change law of axial stiffness.

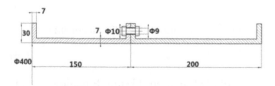

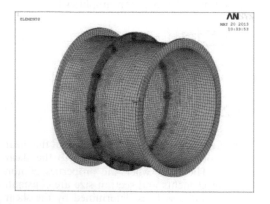

Figure 2. Detailed FE model.

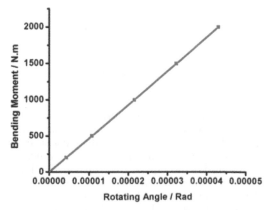

Figure 4. Change law of bending stiffness.

3.2 Thin-layer element method

Now use thin-layer element method to model bolted joints. The model is 19927 element numbers and 28161 node numbers and the size, element type, material properties are almost same as detailed FE model except that the bolted joints are modeled with thin-layer element. Set the thin-layer element as orthotropy and the thickness of thin-layer element is h ($h = 3$ mm). The material properties are shown in Table 1.

Because bending stiffness is determined by axial elastic modulus, change axial elastic modulus to adjust bending stiffness of thin-layer element model. There is an approximate linear relationship between bending stiffness and axial elastic modulus.

$$\frac{E_{Initial}}{K_{Initial}} \approx \frac{E_{equivalent}}{K_{equivalent}} \tag{16}$$

where, $E_{Initial}$ is initial axial elastic modulus, $K_{initial}$ is initial bending stiffness equivalent to $E_{Initial}$, $K_{equivalent}$ is equivalent bending stiffness and $K_{equivalent}$ is actual axial elastic modulus equivalent to $K_{equivalent}$.

After 9 times iterations shown in Table 2, $K_{equivalent}$ is 0.797 GPa, at the time bending stiffness is 4.65E7 N·m/rad equal to detailed FE model. Under this condition, axial stiffness is 1.93E9 N/m that has a difference of 1% with detailed FE model. Consequently, thin-layer element method is correct by verification.

3.3 Comparison

It compares detailed FE model with thin-layer element model as shown in Table 3.

The specific circumstances are that: (1) when E_x axial elastic modulus of thin-layer element is 0.797 GPa, bending stiffness of thin-layer element model is equal to detailed FE model, and at this moment axial stiffness is almost equivalent to detailed FE model; (2) axial stiffness and bending stiffness of detailed FE model are related to the diameter of bolts, bolt numbers, bolted hole position, flange thickness, preload of bolts and thin-layer element model is concerned with axial elastic modulus of thin-layer element. Therefore, thin-layer element model makes various stiffness

Table 1. Material properties of thin-layer element model.

	Element type	Density/ kg·m^{-3}	Property	Elastic modulus/ GPa			Shear modulus/ Gpa			Poisson ratio		
				E_x	E_y	E_z	G_{xy}	G_{yz}	G_{xz}	PR_{xy}	PR_{yz}	PR_{xz}
T-L elements	SOLID 185	7850	Orthotropy	100	100	100	300	300	300	0	0	0
Other elements	SOLID 185	7850	Isotropy	210			80.7			0.3		

Table 2. Process of iteration.

Number of iteration	1	2	3	4	5	6	7	8	9
E (GPa)	100	34.01	12.13	4.84	2.34	1.44	1.08	0.796	0.797
K (N·m/rad)	1.37E8	1.3E8	1.2E8	9.6E7	7.6E7	6.2E7	5.4E7	4.64E7	4.65E7

Table 3. Comparisons between detailed FE model and thin-layer element.

	Detailed FE model	Thin-layer element model	Percentage
Axial elastic modulus/GPa	210	0.797	
Bending stiffness/N·m/rad	4.65E7	4.65E7	0%
Axial stiffness/N/m	1.95E9	1.93E3	1%
Stiffness influence factors	Diameter of bolts, bolt numbers, preload of bolts et al	E_x	
Element numbers	88333	19927	78%
Nodal numbers	95917	28161	71%
Calculation time	49 min	1 min	98%

influence factors into one parameter; (3) the element numbers and nodal numbers of thin-layer element model are less than detailed FE model by 78% and 71%; (4) the calculation time of thin-layer element model is far less than detailed FE model by 98%. Above all, thin-layer element can simulate aero-engine bolted joints excellently and be applied in aero-engine modeling.

4 CONCLUSIONS

First, this paper studies general theory of structural modeling principles and puts forward a kind of parametric modeling method of aero-engine bolted joints based on thin-layer element method; second, it gives mathematical expressions of axial, bending stiffness of aero-engine bolted joints; and finally, it introduces application of thin-layer element method and compare with the detailed FE element method. It is clear that thin-layer element can simulate aero-engine bolted joints excellently and the physical significance of material parameters of thin-layer element is specific. The investigation results show that:

1. Shear modulus of thin-layer element has no influence on axial, bending stiffness;
2. Lateral elastic modulus has no influence on axial, bending stiffness;
3. Axial elastic modulus has a significant influence on axial, bending stiffness;
4. Shear modulus has a significant influence on shear stiffness.
5. Compared with the detailed FE model, the thin-layer element method can simulate joint stiffness, and influence factors, element numbers, nodal numbers and calculation time are far less than detailed FE model.

REFERENCES

Ahmadian, H. & Jalali, H. 2007. Identification of bolted lap joints parameters in assembled structures. *Mechanical Systems and Signal Processing,* 21(2): 1041–1050.

Ahmadian, H. Mottershead, J.E. & James, S. et al. 2006. Modelling and updating of large surface-to-surface joints in the AWE-MACE structure. *Mechanical Systems and Signal Processing,* 20(4): 868–880.

Belvin. W.K. 1987. *Modeling of joints for the dynamic analysis of truss structures.* NASA Technical Paper.

Bickford, J.H. 1990. An Introduction to the design and behavior of bolted joints, 2nd Edition. New York: Marcel Dekker, Inc.

Desai, C.S. Zaman, M.M. & Lightner, J.G. et al. 1984. Thin-layer element for interfaces and joints. *International Journal for Numerical and Analytical Methods in Geomechanics* 8(1): 19–43.

Gregory, D.L. & Martinez, D.R. 2001. *On the development of methodologies for constructing predictive models of structures with joints and interfaces. Sandia National Laboratories,* Technical Report No. SAND2001-0003P.

Huang, Y.Y. Wu, Z.R. & Wang, D.X. 2008. Discuss on fundamental assumption and simplification of thin-layer element. *Mechanics in Engineering,* 30(2): 49–52.

Ibrahim, R.A. & Pettit, C.L. 2005. Uncertainties and dynamic problems of bolted joints and other fasteners. *Journal of Sound and Vibration,* 279(3–5): 857–936.

Iranzad, M. & Ahmadian, H. 2012. Identification of nonlinear bolted lap joint models. *Computers and Structures,* 96: 1–8.

Kim, J. Yoon, J.C. & Kang, B.S. 2007. Finite element analysis and modeling of structure with bolted joints. *Applied Mathematical Modelling,* 31(5): 895–911.

Liu, S.G. Wang, J. & Hong, J. et al. 2010. *Dynamics design of the aero-engine rotor joint structures based on experimental and numerical study.* Proceedings of ASME Turbo Expo 2010, Glasgow, Scotland, GT2010-22199.

Luan, Y. 2012. *Study on dynamic modeling of bolted flange connections in aerospace structures.* Dalian: Dalian University of Technology.

Luan, Y. Guan, Z.Q. & Cheng, G.D. et al. 2012. A simplified nonlinear dynamic model for the analysis of pipe structures with bolted flange joints. *Journal of Sound and Vibration,* 331(2): 325–344.

Ma, S.C. Zang, C.P. & Lan, H.B. 2013. Dynamic model updating of an aero-engine casing. *Journal of Aerospace Power,* 28(4): 878–884.

Advanced Materials and Structural Engineering – Hu (Ed.)
© 2016 Taylor & Francis Group, London, ISBN 978-1-138-02786-2

The effect of controlled cooling after hot rolling on the microstructure and mechanical properties of LX72A

M. Jia & J. Liu

School of Materials Science and Engineering, University of Science and Technology Beijing, Beijing, China

ABSTRACT: The Continuous Cooling Transformation (CCT) curve of 72A tire cord steel was obtained by the Gleeble-3500 Thermal Mechanical Simulator. The effects of cooling rate and laying temperature on the materials microstructure and mechanical properties had been investigated through the comprehensive analysis of microstructure and micro-hardness via an Optical Microscope (OM), Scanning Electron Microscopy (SEM) and hardness gage. The results show that the beginning temperature and the finishing temperature of phase transformation both decrease with the increase of the cooling rate. Specially, when the cooling rate is 10 °C/s, the microstructure and mechanical properties are relatively optimal. Higher the laying temperature is, higher the sorbitzing rate is, finer the lamellar spacing is as well, finally leading to the increase in hardness. Though improve the laying temperature is beneficial to enhance the sorbitizing rate and decrease lamellar spacing, the troostite, a higher strength microstructure, will appear, when the laying temperature is over 900 °C.

1 INTRODUCTION

The wire rod of tire cord steel is the important raw materials for steel cord. Preparing for the follow-up drawing process, wire rod with high strength and good plasticity play a crucial role (Lee et al. 2009). In order to obtain uniform microstructure and high sorbite rate, it is of great importance to control the parameters while rolling and cooling process for high carbon steel wire rod. There will be a direct practical value to investigate the laying temperature and the relationship of transformation temperature, time and quantum at different cooling rate, because the laying temperature and cooling rate have a decisive influence on the final microstructure and mechanical properties of the wire rod of tire cord steel (Fan et al. 2010, Han et al. 2001, Meng et al. 2010). The Continuous Cooling Transformation curves (CCT) have been measured firstly in this paper, and then the effect of cooling rate and laying temperature on microstructure and mechanical properties were investigated by the Gleeble-3500 Thermal Mechanical Simulator.

2 EXPERIMENTAL PROCEDURE

2.1 *Preparing the new file with the correct template*

The tire cord steels were obtained from an iron and steel enterprise. The chemical composition is given in Table 1. The dimension of thermo-simulation specimens is shown in Figure 1.

The continuous cooling experiments were carried out in Thermal Mechanical Simulator. Several samples were heated to 1100 °C at a rate of 10 °C/s, then held 5 minutes for adequate austenitization. Subsequently, the samples were cooled to 1050 °C at a cooling rate of 5 °C/s for a deformation of 60% engineering strain at a strain rate of 20 s^{-1}. After cooling the samples to 880 °C at a cooling rate of 30 °C/s, the specimens were cooled continuously to the room temperature by the cooling rates of 0.8 °C/s, 3 °C/s, 5 °C/s, 10 °C/s, 25 °C/s, 35 °C/s, 50 °C/s respectively. The data of temperature and expansion were collected in real time during the whole process.

In order to investigate the influence of laying temperature on the microstructure of LX72A, the experiments were to heat several samples to 1100 °C at a rate of 10 °C/s and hold 5 minutes for adequate austenitization. Then the samples were cooled to 1050 °C at a rate of 5 °C/s for a deformation of 60% engineering strain at a strain rate of 20 s^{-1}. Afterward the specimens were respectively cooled to 840 °C, 860 °C, 900 °C, 930 °C at a cooling rate of 30 °C/s, followed by cooling the samples to room temperature at a cooling rate of 10 °C/s.

All samples were ground, polished and then etched using nitric acid and alcohol solution respectively. The Optical Microscope (OM) and Scanning Electron Microscopy (SEM) were used to examine the microstructure. The micro-hardness was tested

Table 1. Chemical composition of 72A tire cord steel.

Element	Mass fraction (%)	Element	Mass fraction (%)
C	0.72	S	0.006
Si	0.26	Cu	0.02
Mn	0.45	Cr	0.051
P	0.009	Ni	0.007

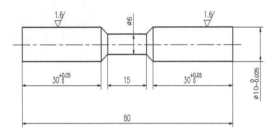

Figure 1. Dimension of continuous cooling transformation curves samples.

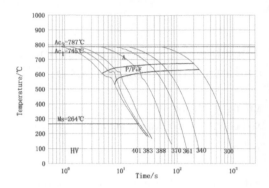

Figure 2. Continuous cooling transformation curves of 72A tire cord steel.

via HXD-1000TM hardness gage. The sorbite rate and lamellar spacing were analyzed by Image Tool software to investigate the effect of cooling rate and laying temperature on microstructure and mechanical properties.

3 RESULTS AND DISCUSSION

3.1 *The continuous cooling transformation*

3.1.1 *The continuous cooling transformation curves*
According to the thermal expansion, the tangents method was used to acquire the critical temperature of LX72A firstly. Then the CCT curves have been obtained, as shown in Figure 2.

Clearly, with the cooling rate enhancing, the beginning and finishing temperature of phase transformation decrease gradually. Furthermore, the phase transformation occurred mainly from 530 °C to 670 °C. The microstructure mainly consists of pearlite and a small amount of ferrite on the basis of investigation on Optical Microscope (OM) when the cooling rate is between 0.8 °C/s and 5 °C/s. The bright white matter existing in microstructure is martensite determined by microhardness, appears when the cooling rate is over 25 °C/s. Besides, the amount of martensite increases with the improvement of cooling rate further.

3.1.2 *Microstructure*
The microstructures of LX72A at Room Temperature (RT) under the different cooling rate

are shown in Figure 3. The ferrite almost exhibits netlike distribution when the cooling rate is 0.8 °C/s according to Figure 3(a). The volume fraction of ferrite is about 7%, and the pearlite lamella has been observed obviously under 500 times on OM. The content of ferrite increases with the cooling rate improving. Moreover, there appears some martensite (the bright white bulk), identified by microstructure tester when the cooling rate is more than 25 °C.

The typical pearlite lamellas under different cooling conditions are shown in Figure 4. According to the Figure 4(a)–(d), it can be seen that the pearlite lamellar spacing decreases with the increase of cooling rate. The lamellar spacing is about between 140 nm and 210 nm when the cooling rate is 0.8 °C/s, particularly, most of the lamellas are around 170 nm with the features of non-uniform lamellar spacing overall. This phenomenon is mainly attributed to small super-cooling degree at lower cooling rate, leading to the higher phase transformation temperature (Hao et al. 2011). Thus, the pearlite lamellas formed at high temperature is larger, but smaller during the subsequent process at low temperature. Microstructure with mean lamellar spacing about 108 nm is identified as sorbite when the cooling rate is 5 °C/s. When the cooling rate is up to 10 °C/s, the pearlite lamellas are finer and a small quantity of troostite appears with the lamellar spacing less than 100 nm, owing to the improvement of super-cooling degree. To increase the cooling rate further, there are few changes in the pearlite lamellar spacing which nearly is around 75 nm and the microstructure consist of troostite and martensite.

3.1.3 *Micro-hardness*
The micro-hardness of the black area and bright white area in Figure 3 were measured respectively

484

(Load 10 KG). The Vickers hardness values at different cooling rate are presented in Table 2.

It can be seen that when the cooling rate is more than 25 °C/s, the hardness of bright white area is over 600 HV, much higher than the values of the black area. Such high hardness is likely evident for martensite. The hardness of the black area shows an upward trend for the cooling rate improving, especially in the range of 0.8 °C/s to 10 °C/s, the pearlite lamellar spacing reduce significantly, leading to the hardness of a linear increase. When the cooling rate is more than 10 °C/s, the pearlite lamellar spacing have rare change and the hardness increase slow down as well.

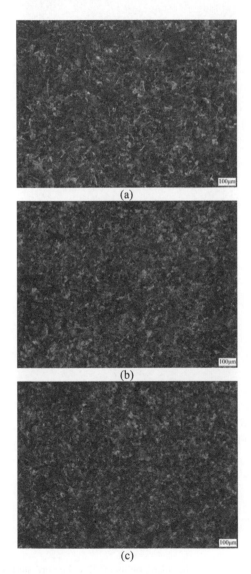

Figure 3. (*Continued*).

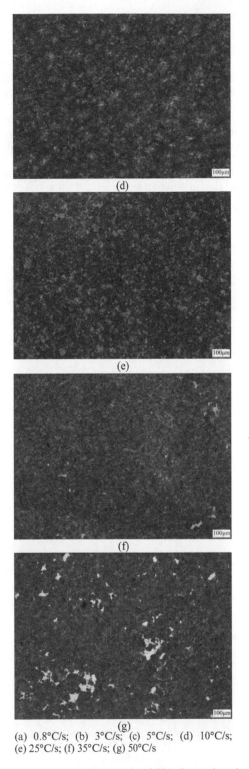

(a) 0.8°C/s; (b) 3°C/s; (c) 5°C/s; (d) 10°C/s;
(e) 25°C/s; (f) 35°C/s; (g) 50°C/s

Figure 3. Optical micrographs of 72A tire cord steel at different cooling rates.

3.2 *The effect of laying temperature on microstructure and mechanical properties*

The controlled cooling after hot rolling of wire rod is called on-line heat treatment, which is directly relevant to the final properties of materials (Liang et al. 2013). The top priority at this stage is to control laying temperature so as to make structural preparation for the subsequent phase transformation. Hence, the final microstructure and mechanical properties are highly influenced by the laying temperature.

Figure 5 presents the microstructure of 72A tire cord steel at different laying temperature. It is noted

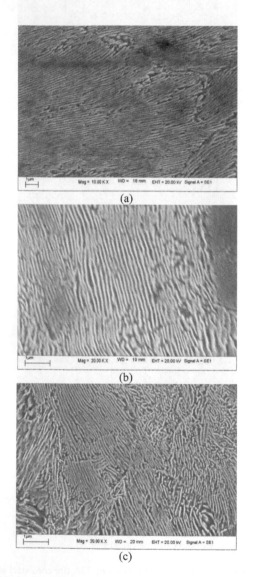

(a)

(b)

(c)

Figure 4. (*Continued*).

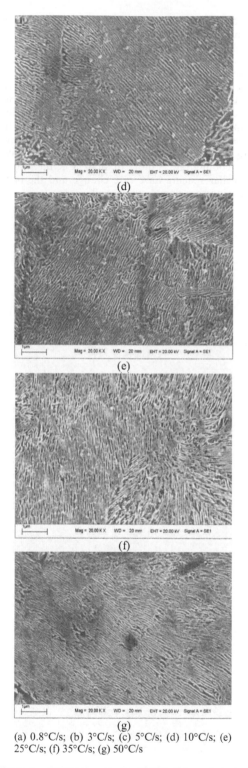

(d)

(e)

(f)

(g)

(a) 0.8°C/s; (b) 3°C/s; (c) 5°C/s; (d) 10°C/s; (e) 25°C/s; (f) 35°C/s; (g) 50°C/s

Figure 4. SEM micrographs of 72A tire cord steel at different cooling rates.

Table 2. Vickers hardness values at different cooling rates of 72A tire cord steel.

Cooling rate (°C/s)	Black area (HV)	White area (HV)
0.8	299.5	
3	339.8	
5	361.4	
10	370.1	
25	388.3	
35	383.4	620.3
50	400.9	609.0

that the pearlite fraction is incremental with the laying temperature increasing. As the temperature is up to 930 °C, the microstructures consist mainly of pearlite presented as black flocculent character. Because higher the laying temperature is, fewer precipitates appears in the austenite zone, resulting in coarser grains and a decrease of grain boundary area. In consequence, the proeutectoid ferrite fraction is relatively reductive, and the sorbite rate is correspondingly increased.

The Scanning Electron Microscope (SEM) was used to investigate the pearlite lamella of specimens. As shown in Figure 6, the lamellar spacing is apparently larger when the laying temperature is 840 °C. With the temperature increasing, the pearlite lamellar spacing decreases. Moreover, the pearlite lamellar spacing is less than 100 nm and the microstructure is mostly composed of troostite while the temperature is over 900 °C. This can be accounted for the compounds of austenite are homogenized with the temperature increasing, which promote the super-cooled austenite stability (Yan et al. 2014, Qian et al. 2013). As a result, the required super-cooling degree of phase transformation is enhancive. Namely, in same continuous cooling condition, the transformation temperature of pearlite decreases. With a lower transformation temperature, the rate of nucleation increases but the mobility of atoms is decreased (Kazeminezhad & Karimi 2003, Hwang et al. 2014). Therefore, it produces finer pearlite lamella. Unquestionably, the fine lamella is beneficial to the improvement of plasticity during the 72A productive process. However, excessive fine lamella will cause a substantial increase in strength with a slight decrease in plasticity correspondingly, which will result in fracture during the subsequent drawing process. Hence, it is very desirable to obtain the sorbite microstructure and avoid generating the troostite microstructure during the cooling process of wire rod (Zhao et al. 2014).

The results of Vickers hardness experiments at different laying temperature of LX72A are

(a) 840°C; (b) 860°C; (c) 900°C; (d) 930°C

Figure 5. Optical micrographs of 72A tire cord steel at different laying temperatures.

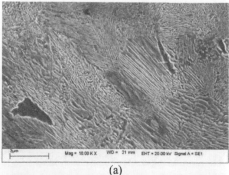

(a)

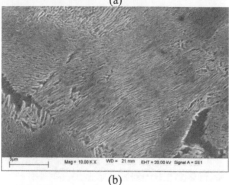

(b)

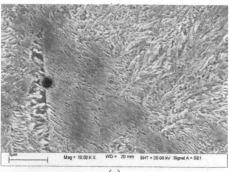

(c)

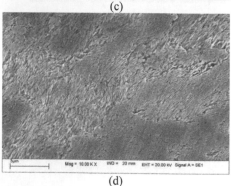

(d)

840°C; (b) 860°C; (c) 900°C; (d) 930°C

Figure 6. SEM micrographs of 72A tire cord steel at different laying temperatures.

Table 3. Vickers hardness values at different laying temperature of 72A tire cord steel.

Temperature (°C)	Hardness (HV)
840	363.2
860	373.4
900	382
930	399.9

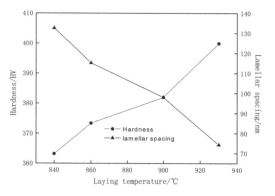

Figure 7. The hardness and lamellar spacing of 72A tire cord steel at different laying temperatures.

presented in the Table 3. It can be found that the hardness values are accessorial with the increase of laying temperature. The micro-hardness is largely dependent on the pearlite lamellar spacing. In fine pearlite microstructure, the ferrite and cementite plates are thinner. This would lead to the increase of phase interface (Liu et al. 2014). Consequently, it is attributed to the difficulty in dislocation motion as well as to the hardness enhancement in macroscopic. In accordance with the relationship between laying temperature and lamellar spacing discussed in the last section, a distinct relation can be observed between the micro hardness and pearlite lamellar spacing, as shown in Figure 7.

4 CONCLUSIONS

The CCT curves of 72A tire cord steel had been obtained by thermal mechanical simulation. The CCT curves reveal that the transformation temperature is decreased as the cooling rate increasing. When the cooling rates are in the range of 08 °C/s and 5 °C/s, a small amount of proeutectoid ferrite are produced and the fraction of ferrite is decreasing with the cooling rate accelerates. The microstructure mainly consists of pearlite, sorbite and ferrite. When the cooling rate is over 25 °C/s, the martensite appears and its fraction increases with the cooling rate improving. In this condition, the microstructure mostly contains troostite and martensite.

Raising laying temperature is beneficial to increase the sorbite rate. The higher the laying temperature is, the finer the pearlite lamella is, and thereby the micro-hardness is higher. However, when the temperature is over 900 °C, there appears a large amount of troostite which have a negative effect on the subsequent cold drawing process.

REFERENCES

Fan, G.B. et al. 2010. Pattern of continuous cooling transformation of 72A cord steel. *Journal of Wuhan University of Science and Technology*, 33(1): 43–46.

Han, K. et al. 2001. Optimization of mechanical properties of high-carbon pearlitic steels with Si and V additions. *Metallurgical and Materials Transactions A*, 32(6): 1313–1324.

Hao, F. et al. 2011. Effect of loop laying temperature on dynamic CCT curve of high carbon steel 82B. *Heat Treatment of Metals*, 36(12): 4–8.

Hwang, S.K. et al. 2014. The effect of grain refinement by multi-pass continuous hybrid process on mechanical properties of low-carbon steel wires. *Journal of Materials Processing Technology*, 214(7): 1398–1407.

Kazeminezhad, M. & Karimi Taheri, A. 2003. The effect of controlled cooling after hot rolling on the mechanical properties of a commercial high carbon steel wire rod. *Materials & Design*, 24(6): 415–421.

Lee, S.K. et al. 2009 Pass schedule of wire drawing process to prevent delamination for high strength steel cord wire. *Materials & Design*, 30(8): 2919–2927.

Liang, Y. et al. 2013. Effect of inclusion and sorbitizing rate on mechanical properties of 72a wire rod steel. *Hot Working Technology*, 42(16): 22–24.

Liu, L.Z. et al. 201 4. Transformation behavior and phase transformation kinetics model of a C82DA steel. *Transactions of Materials and Heat Treatment*, 35(2): 220–224.

Meng, X.Z. et al. 2010. Effect of isothermal transformation treatment on direct drawing capability of 72A wire rod for steel cord. *Transactions of Materials and Heat Treatment*, 31(4): 71–75.

Qian, Q.S. et al. 2013. Microstructure and mechanical properties of C92D2 steel wire rod for steel cord. *Heat Treatment of Metals*, 38(11): 21–26.

Yan, W. et al. 2014. Study on Inclusions in Wire Rod of Tire Cord Steel by Means of Electrolysis of Wire Rod. *Steel Research International*, 85(1): 53–59.

Zhao, T.Z. et al. 2014. Hardening and softening mechanisms of pearlitic steel wire under torsion. *Materials & Design*, 59: 397–405.

A finger-interaction-based LED control system

S.W. Jang, K.H. Seok & Y.S. Kim
Humane-Centered Interaction Laboratory (HILAB), Department of Computer Science and Engineering, Korea University of Technology and Education, Cheonan, Republic of Korea

ABSTRACT: This paper proposes a finger-interaction-based Light-Emitting Diode (LED) control system. The proposed system controls the illuminance level of an LED using the data sensed from finger actions via wireless communication. Based on finger-action sensing and recognition, the proposed LED control system allows users to easily control LEDs according to the number of fingers used and the events created with those fingers. To verify the validity of the proposed system, experiments were conducted using an industrial LED system. The functionality of the proposed system was verified by adjusting the dimming control of the LED to five levels.

1 INTRODUCTION

A Natural User Interface (NUI) does not use an input device such as a keyboard or mouse. Instead, the NUI employs a user's behavior as the input. The NUI is easier to manipulate than a Command Line Interface (CLI) or Graphic User Interface (GUI). Types of NUIs include a voice interface, sensor interface, touch interface, and gesture interface. Among these, the gesture interface provides more intuitive and easier manipulation than the other interfaces. In particular, the finger-based gesture includes more motions than other types of gestures.

Because the NUI has been introduced very quickly into many industries, the Light-Emitting Diode (LED) industry also requires a new type of control environment that allows users to manipulate various commands for LEDs. For general users, the control system for an LED must become easier to operate. This will expand the scope of applications for the smart LED in our everyday lives. Based on this requirement, a number of studies are being conducted regarding new types of user-friendly and intuitive methods that can render various motions (Nam et al. 2007, Seok & Kim 2008). In particular, gesture recognition using the Kinect™ line of motion-sensing input devices by Microsoft, as well as remote human–robot interaction using camera sensors, are actively being carried out (Qian et al. 2013, Pereira et al. 2013). In addition, there are studies that propose algorithms to improve hand gesture recognition (Licsar & Sziranyi 2005, Yu et al. 2011). However, user-intuitive or user-friendly LED control systems for the general user are still rare.

Therefore, this paper proposes an LED control system based on finger-action sensing and recognition. The proposed LED control system allows users to easily control LEDs according to the number of fingers and the events created with those fingers. The effectiveness of the proposed system was verified based on experiments using an actual industrial LED.

2 PROPOSED FINGER-INTERACTION-BASED LED CONTROL SYSTEM

2.1 Overview

Researchers of digital devices are actively studying new, convenient interfaces that provide features such as gesture recognition, face recognition, and voice recognition. There are a number of techniques for finger-action sensing, which is being used in various fields. With the exception of language, among our body parts (such as hands, eyes, mouth, arms, and legs), the hand is most frequently used for human communication. In this section, we describe an LED dimming control method using a finger count. The proposed system is composed of two parts: finger interaction and LED control. These are shown in Figure 1. The

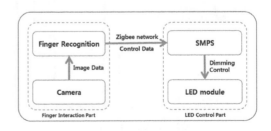

Figure 1. Block diagram of finger-interaction-based LED control system.

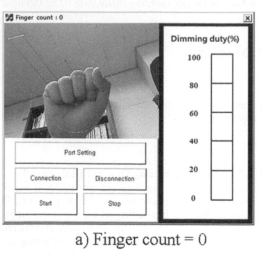

a) Finger count = 0

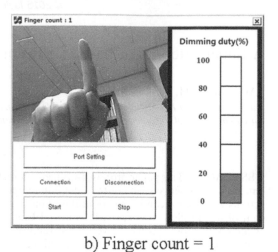

b) Finger count = 1

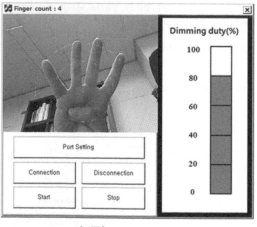

c) Finger count = 2

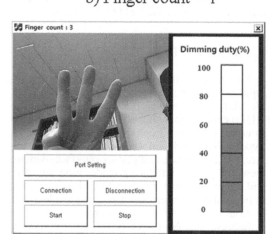

d) Finger count = 3

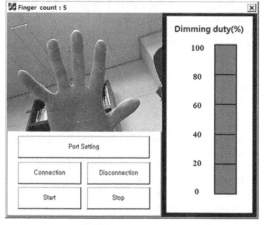

e) Finger count = 4

f) Finger count = 5

Figure 2.   Graphic user interface of finger interaction part.

finger-interaction part is composed of a camera and finger recognition, and it performs a real-time finger recognition using image data. The LED control part is composed of an LED module and an SMPS (unit module with 650 [W]), and it controls the SMPS using control data.

2.2 *Finger interaction part*

In this paper, a finger-interaction part capable of controlling the LED is implemented using finger-action sensing. The finger-interaction part, shown in Figure 2, consists of three panels: a finger interfacing panel, a communication panel, and an LED monitoring panel. The finger interfacing panel is a recognition unit that sees the finger actions from the finger images taken by the camera. The finger recognition unit is implemented using a color values scheme (Licsar & Sziranyi 2005, Kim et al. 2008). The finger recognition unit first converts RGB colors to grayscale and YCrCb for binary representation. Then the region inside the hand is filled by masking, and noise is removed. The binary image is examined in 25-pixel units. If the sum of 1's examined is greater than ten, every digit is set to 1. Using the image obtained from the masking performed by the finger recognition unit, the center of the hand can be calculated to identify the hand's location as well as the hand region farthest from the center (Seok & Kim 2014). The finger actions recognized by the finger interfacing panel allow the user to control the LED. The communication panel provides five functions (port setting, connection, disconnection, start, and stop) for SMPS communication. The LED monitoring panel allows the user to show the state of the LED dimming control in five levels (0%, 20%, 40%, 60%, 80%, and 100%).

2.3 *LED control part*

The LED control part is composed of an LED module, interface board, and an SMPS to control it, as shown in Figure 3. The output current control consists of a 650-[W] SMPS (Kim & Ko 2014), Zigbee module, and LED module, as shown in Figure 3. The LED control part provides the LED driver module and Zigbee module with DC 48 [V]

obtained via SMPS from the sup-plied 220 [V]. The Zigbee module is composed of one channel per module, and transmits acquired SMPS information to the server in the monitoring part.

3 EXPERIMENTAL RESULT AND DISCUSSION

In this section, to examine the feasibility of the LED dimming control in the proposed system, we conducted an experiment using an actual LED system. Figure 4 shows the experimental environment, which is composed of five components: a finger recognition tool, a PC camera, Zigbee communication module, an SMPS, an interface board, and an LED module, as shown in Figure 4. The control data were transmitted to the LED control system using wireless communication. We examined how the contents produced by the finger recognition tool controlled the LED according to the user's intentions.

Figure 4.   Experimental environment.

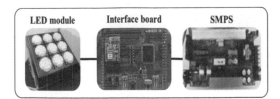

Figure 3.   Block diagram of LED control part.

Table 1. Experimental results.

Finger count	Dimming duty (%)	LED average current [A]
0	0	0.0
1	20	0.3
2	40	0.6
3	60	0.9
4	80	1.2
5	100	1.5

The dimming control results for the LED are shown in Table 1. The data in Table 1 show the switching operation, in which the LED current was off in the 0% section and on in 100% section (average current 1.5 [A]), and the LED dimming control results that correspond to various finger counts. The currents for finger counts 0, 1, 2, 3, 4, and 5 were 0.0 [A], 0.3 [A], 0.6 [A], 0.9 [A], 1.2 [A], and 1.5 [A], respectively. The LED operated well at an average current of 0.3 [A] in linear dimming of 20%. The average current was 0.6 [A] in linear dimming of 40%, and the average current was 0.9 [A] in linear dimming of 60%. The average current was 1.2 [A] in linear dimming of 80%, and the average current was 1.5 [A] in linear dimming of 100%. From the experimental results, we confirmed the effective operation of the LED dimming states in the finger-interaction-based LED control system.

4 CONCLUSIONS

This paper proposed a finger-interaction-based LED control system. The validity of the proposed LED control system was examined via experiments using an actual LED system. The output current control function was verified by an LED dimming control experiment that used a sensor network constructed with SMPS and LED dimming control results that corresponded to various finger counts. The experimental results for currents of finger counts 0, 1, 2, 3, 4, and 5 were 0.0 [A], 0.3 [A], 0.6 [A], 0.9 [A], 1.2 [A], and 1.5 [A], respectively. From the results, it was confirmed that the proposed system could easily control industrial LEDs by using a finger interaction. The proposed LED control method is expected to provide user-friendly and intuitive solutions for controlling industrial LED systems.

ACKNOWLEDGMENT

This work was supported by the Chungcheong leading industry promotion project of Korean Ministry of Knowledge Economy.

REFERENCES

Kim, K.W. Lee, W.J. & Jeon, C.H. 2008. *A Hand Gesture Recognition Scheme using WebCAM,* The Institute of Electronics Engineers of Korea, 619–620.

Kim, Y.S. & Ko, J. 2014. *Development of a Sensor Network-based SMPS System-a Smart LED Monitoring Application based on Wireless Sensor Network,* International Journal of Distributed Sensor Networks, Hindawi Publishing Corporation US.

Licsár, A. & Szirányi, T. 2005. User-adaptive hand gesture recognition system with interactive training. *Image and Vision Computing,* 23: 1102–1114.

Nam, S.Y. Lee, H.Y. Kim, S.J. Kang, Y.C. & Kim, K.E. 2007. A study on design and implementation of the intelligent robot simulator which is connected to an URC system, *Journal of Institute of Electronics Engineers of Korea,* 44(4): 157–164.

Pereira, F.G. Vassallo, R.F. & Salles, E.O.T. 2013. Human–Robot Interaction and Cooperation Through People Detection and Gesture Recognition. *Journal of Control, Automation and Electrical Systems,* 24(3): 187–198.

Qian, K. Niu, J. & Yang, H. 2013. Developing a Gesture Based Remote Human-Robot Interaction System Using Kinect. *Journal of Smart Home,* 7(4): 203–208.

Seok, K.H. & Kim, Y.S. 2008. *A new robot motion authoring method using HTM,* Proceedings of the International Conference on Control, Automation and Systems, 2058–2061.

Seok, K.H. & Kim, Y.S. 2014. Finger Actions Sensing-Based Robot Motion Authoring System, *International Journal of Multimedia and Ubiquitous Engineering,* 9(5): 203–214.

Yu, B. Park, S.Y. Kim, Y.S. Jeong, I.G. Ok, S.Y. & Lee, E.J. 2011. Hand Tracking and Hand Gesture Recognition for Human Computer Interaction. *Journal of Korea Multimedia Society,* 14(2): 182–193.

Advanced Materials and Structural Engineering – Hu (Ed.)
© *2016 Taylor & Francis Group, London, ISBN 978-1-138-02786-2*

Development of a SMPS system using LLC resonant converter for high efficiency LED driver

J.H. Ko & Y.S. Kim

*Humane-Centered Interaction Laboratory (HILAB), Department of Computer Science and Engineering,
Korea University of Technology and Education, Cheonan, Republic of Korea*

ABSTRACT: This paper proposes a SMPS System using LLC resonant converter for high efficiency LED driver. Due to the existence of the non-isolation DC/DC converter to control the LED current and the light intensity, the conventional three-stage LED driving system has the problem of low power conversion efficiency. To solve this problem, in this paper a system composed of PFC and LLC converter without any non-isolation DC/DC converter is proposed. By providing functions including isolation and LED current control, the proposed system can guarantee at least 90% of high efficiency. The effectiveness of the proposed system was verified by a performance measurement experiment on 200 W SMPS.

1 INTRODUCTION

Recently, cases replacing the conventional light to Light-Emitting Diode (LED) are increasing globally for energy saving. Method using AC voltage for LED driving is to use Switch-Mode Power Supply (SMPS), which supplies constant current or constant voltage to LED using active element such as a semiconductor switch (Hu & Zane 2010, Shin et al. 2010, Lee et al. 2013). The conventional three-stage LED driving systems consist of Power Factor Correction (PFC), isolation DC/DC converter, and non-isolation DC/DC converter. The biggest disadvantage of those conventional systems is that the efficiency is lower than 90% (Park et al. 2009).

Therefore, in this paper, we propose a SMPS system using Inductor-Inductor-Capacitor (LLC) resonant converter for high efficiency LED driver. The proposed system consists of PFC and LLC converter without non-isolation DC/DC converter. By providing the functions for isolation and LED current control, at least 90% of high efficiency can be achieved by the proposed system. An experiment on 200 W SMPS is conducted to examine the performance of the proposed system and the results are given and discussed.

2 SMPS SYSTEM USING LLC RESONANT CONVERTER FOR HIGH EFFICIENCY LED DRIVER

The proposed system uses a standard AC power required for light, and is composed of PFC and LLC converter as shown in Figure 1. PFC provides

a power factor correction function to support the harmonic regulations, and the LLC converter provides the functions for electrical insulation and LED current control.

The PFC operates continuous current-mode and its input voltage range is 90 V_{AC}~265 V_{AC}. The configuration of the PFC including inrush current limiting circuit and Electromagnetic Interference (EMI) circuit is shown in Figure 2. The operation procedure of PFC is as follows: If control Integrated Circuit (IC) turns on Metal-Oxide Semiconductor Field Effect Transistor (MOSFET), current is charged in inductor. The charged current in the inductor flows into diode, and it is charged in condenser of V_{DC}. If inrush current occurs during this input capacitor charging, varistor is operated (turned on) as a protection circuit.

LLC resonant converter is difficult to maintain a regulation function in a low load condition. To solve this problem, LLC converter composed of LLC control circuit, protection circuit, feedback circuit, and output current control circuit as shown in Figure 3 is used. LLC converter operates primary-side MOSFET at 50% duty cycle under a steady state condition. Regulation function is maintained by switching the frequency of converter changed by output voltage.

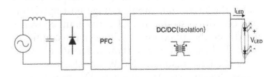

Figure 1. Block diagram of the proposed system.

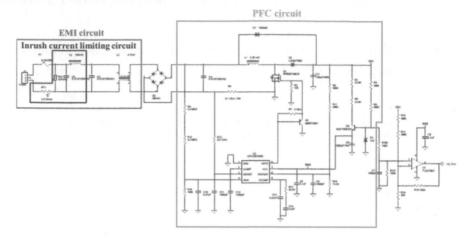

Figure 2.   Circuit diagram of the PFC.

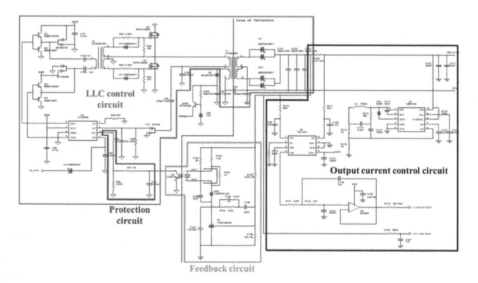

Figure 3.   Circuit diagram of the LLC resonant converter.

LLC resonant 200 W transformer was designed for Quality Factor (Q-Factor) selection guaranteeing region 2 under minimum input voltage and maximum load condition. Each ratio of magnetizing inductance (Lm) and resonant inductance (Lr) was designed to obtain the same output voltage under the minimum input voltage and maximum load condition so that the LLC resonant converter could get the maximum voltage gain condition. Table 1 shows the measurement results obtained using the LLC resonant 200 W transformer.

SMPS converts AC power to DC power using transformer. Power losses in the conversion

Table 1.   Measurement results obtained using LLC resonant 200 W transformer.

Factor	Unit	Design value	Measurement value
Input voltage	Vin	390 V	390 V
Output voltage	Vo	48 V	48.22 V
Output current	Io	4.17 A	4.17 A
Turn-ratio	n	4.0625	4.33
Magnetizing inductor	Lm	689 uH	673.6 uH
Resonant capacitor	Cr	10.9 nF	20 nF
Resonant inductance	Lr	138 uH	120.2 uH
Resonant frequency	Fo	130 kHz	130 kHz

process generate heat. The generated heat reduces both the life of SMPS and LED. To solve this problem, we installed aluminum heat radiation pads in the switching devices as shown in Figure 4.

(a) Aluminum heat radiation pads

(b) Heat radiation pad installed in PFC switching device

(c) Heat radiation pad installed in LLC hi-low switching device

(d) Heat radiation pad installed in LLC output switching device

Figure 4. Heat radiation pads installed in switching devices.

Table 2 shows the temperature measurement results of the 200 W SMPS. The measured results were satisfied with working temperature (0~70°C) of commercial SMPS products.

We developed a 200 W SMPS as shown in Figure 5. Figure 5(a) shows the inside of the developed SMPS, and Figure 5(b) shows the outside of the one integrated with LED.

We performed an experiment to examine the validity of the proposed SMPS system. The experiment was performed to measure the efficiency and power factor according to various input voltages (200–250 V_{AC}) under a full load (100%) condition. The experimental results are summarized in Table 3, and 90.88% of efficiency was obtained at 220 V_{AC}, the standard of South Korea which conventional three-stage system was not able to

Table 2. Temperature measurement results of the 200 W SMPS.

Measuring position	Temperature (°C)
PFC	64.9
LLC	66.2
Output diode	68.2
Inside of SMPS	64
Outside case of SMPS	55.3

(a) Inside of the 200W SMPS

(b) Outside of the 200W SMPS

Figure 5. The developed 200 W SMPS.

Table 3. Experimental results.

Input voltage (V$_{AC}$)	Power factor (PF)	Efficiency (%)
200	0.990	90.51
210	0.989	90.67
220	0.988	90.88
230	0.988	91.17
240	0.987	91.38
250	0.985	91.50

achieve. From the results, it is confirmed that the proposed system provided a better performance than the conventional system.

3 CONCLUSIONS

This paper proposed a SMPS system using the LLC resonant converter for high efficiency LED driver. The proposed system could provide the functions for isolation and LED current control without any non-isolation DC/DC converter. For such functions, we designed PFC, LLC and transformer. In addition, we experimented a heat radiation pad by considering LED application.

The performance of the proposed system was examined by the experiment on various input voltages. The experiment result obtained at 220 V$_{AC}$ is 90.88% of efficiency. Such experimental result confirmed that higher efficiency could be achieved by the proposed system than the one by the conventional system. In the future, we are expecting the proposed system to be applied to various LED system.

ACKNOWLEDGMENT

This work was supported by Chungcheong leading industry promotion project of the Korean Ministry of Knowledge Economy.

REFERENCES

Hu, Q.C. & Zane, R. 2010. LED Driver circuit with series-input-connected converter cells operating in continuous conduction mode, *IEEE Transaction on Power Electronics*, 25(3): 574–582.

Lee, E.S. Choi, B.H. Cheon, J.P. Kim, B.C. & Rim, C.T. 2013. The Design of Long-life and High-efficiency Passive LED Drivers using LC Parallel Resonance, *The Transactions of the Korean Institute of Power Electronics*, 18(4): 397–402.

Park, K.M. Lee, K.I. Hong, S.S. Han, S.K. & Roh, C.W. 2009. New LED Driver Circuit to Reduce Voltage Stress, *The Transactions of the Korean Institute of Power Electronics*, 14(3): 243–250.

Shin, D.S. Jung, Y.J. Hong, S.S. Han, S.K. Jang, B.J. Kim, J.H. Lee, I.O. & Roh, C.W. 2010. A High Efficiency LED Driver Circuit using LLC Resonant Converter, *The Transactions of the Korean Institute of Power Electronics*, 15(1): 35–42.

Advanced Materials and Structural Engineering – Hu (Ed.)
© 2016 Taylor & Francis Group, London, ISBN 978-1-138-02786-2

Numerical test derivation of the computational formula for the composite modulus of high-speed rail substructure composite foundation

H.R. Wei
School of Chemical Engineering, Lanzhou Institute of Arts and Science, Lanzhou, China

G.L. Zhang
School of Civil Engineering, Southwest Jiaotong University, Chengdu, China

ABSTRACT: The relationship between the composite modulus and the pile-soil modulus of the composite foundation for a high-speed rail substructure is far more than the simple relationship of the weighted mean. Using the numerical test, this paper consolidated the pile and the soil within the composite foundation area into the composite foundation, and made calculation, respectively, according to different replacement rates. The results of the calculation were obtained and comprehensively compared with the results obtained from the separate calculation on the pile and the soil, in order to study the difference in the influence of the pile and the soil on the settlement of the composite foundation. From this study, a computational formula for the composite modulus of the composite foundation was derived and some useful conclusions were suggested.

1 INTRODUCTION

At present, the soft foundation for a high-speed rail substructure in China is generally treated in two ways: the first is to improve the soil texture by ways such as drainage consolidation, dynamic compaction, *in situ* compaction and replacement; the second is to make a composite foundation with artificial reinforcement body and natural foundation, by ways such as addition of vertical reinforcement body or horizontal reinforcement body. The vertical reinforcement body is referred to as the pile composite ground, where the natural soil and the reinforcement body are consolidated into a composite foundation to bear the upper load (Wang et al. 2003).

Before calculating the load-bearing capacity and the settlement of the composite foundation for a high-speed rail substructure, the composite modulus of the composite foundation need to be made clear. The common methods used to calculate the composite modulus include the weighted mean, weighted mean modification, stress ratio, area ratio modification and the minimum potential energy (Deng et al. 2005). However, under most conditions, the calculation is made by using the weighted mean based on the replacement rate, assuming that the pile and the soil have the same strain and the underlying stratum is stiff, while the

pile body has no tip resistance and the pile length is finite. The results on these calculations are to a certain degree inconsistent with those reported previously (Deng et al. 2005, Chow 1996).

2 SETTING OF THE NUMERICAL TEST

To solve the relationship between the composite modulus and the pile-soil modulus of the composite foundation, this paper took the test section of the Yangling Station on the Xi'an-Baoji High-speed Railway as an example to actually design and build a three-dimensional full-section model according to the field test conditions. A series of numerical tests were confirmed, by using the boundary constraint, initial conditions and field test data on related parameters, in order to analyze the changes in the composite modulus of the composite foundation.

2.1 *Setting of the numerical test model*

The field test data of the Yangling Station were obtained, in which the embankment is found to be 3.2 m in thickness and divided into two layers as follows: broken stone hardcore and cement-improved soil, from the upper to the bottom layer. Between the underneath embankment and the composite foundation, there is the cemented soil bedding. DDC cement-soil piles are adopted to make the composite foundation, with the pile diameter of 0.6 m, the pile spacing of 1.1 m and the pile length of 8 m.

Related model parameters were obtained from the field test. A three-dimensional full-section model was set up, consisting of the underlying layer, DDC cement-soil pile reinforced area, reinforced bedding course, bottom layer of the sub-grade bed and surface layer of the sub-grade bed. The Mohr-Coulomb model was adopted as the stratum of the constitutive model, while the linear elastic model as the DDC cement-soil pile. To further fit the actual conditions of the project based on which this paper is composed, loads to be calculated in the model include the dead load of the embankment, the dead load of the ballast less track and the load of the train. The double-track loading center spacing was 5 m, while the loading width was 3.4 m. The dead load of the track was converted into a computational uniform load with an intensity of 52.07 KN/m and the load of the train was converted into a uniform load with an intensity of 15.31 KN/m. Model boundaries were generally supported, with the unit length of grids (mostly quadrangular) being 1 m.

The direction X showed the cross section of the embankment, with a length of 30 m, while the direction Y was the vertical section that was designed as a single-row 5 times of pile spacing, that is, 9.674 m; the direction Z of the composite foundation was designed as 3 times of the pile length, that is, 30 m, as shown in Figure 1.

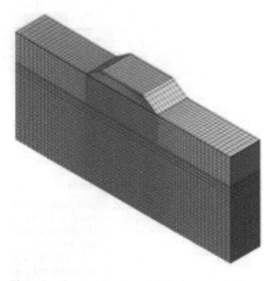

Figure 1.   Computational model for the numerical test.

2.2   *Setting of the test parameters*

The replacement rate was changed by changing the values of pile body diameter, represented by 0.45 m, 0.5 m, 0.55 m, 0.6 m and 0.65 m. Correspondingly, the replacement rates were 0.151796, 0.187403, 0.226757, 0.26986 and 0.316711.

3   RESULTS OF SEPARATE CALCULATIONS AND CONSOLIDATED CALCULATIONS

To further analyze the relationship between the pile, the pile-soil modulus and the composite modulus, the pile and pile-soil within the composite foundation area were consolidated into the composite foundation in the finite element model, and calculations were made separately based on different replacement rates. A comprehensive comparison was made between the results of the consolidated calculations and those of the separate calculations on piles and pile-soil in order to find the similarities and differences between the impacts of the two on the settlement of the composite foundation.

While choosing the finite element model, all parameters were the same as those used for the separate calculations on the original pile and pile-soil, and the constitutive model was the same. The only difference was that at the time of the consolidated calculations, the pile body and the network it contacts were removed and modification was made to some elastic modulus of the pile-soil to change its properties and treat the same as the composite foundation as a whole. The surrounding soil mass and upper load parameters were the same as those for the separate calculations. The results of these calculations are summarized in Table 1.

It can be seen from the test results that the conclusions derived from the separate calculations and consolidated calculations were basically the same, regardless of the settlement of the whole foundation, the settlement of the underlying layer or the settlement of the composite foundation range, and all conclusions were included in 5%. Especially for pile deformation or composite foundation deformation, differences in the results of the two calculations were all within 0.05%; thus, the results of the two calculations can be deemed as consistent.

Figures 2 and 3 show the vertical deformation of the foundation from the separate calculations and consolidated calculations on the pile and the pile-soil on the premises of pile diameter 0.6 m, pile spacing 1.1 m and replacement rate 0.18740.

It can be seen from the figures that the size, scope and form of deformation from the two calculations were basically the same. In other words, the composite foundation treatment area had a relatively

Table 1. Correlation table of the settlement of the composite foundation.

Replacement rate	Pile modulus (KPa)	Pile-soil modulus (KPa)	Composite modulus (KPa)	Settlement of composite foundation top (mm)		
				Separate	Consolidated	Difference (%)
0.02998	200000	15000	18500	178.84	177.48	0.76
		25000	28530	160.95	159.58	0.85
		35000	38542	152.20	150.81	0.92
		45000	48526	146.99	145.58	0.95
0.06747		15000	22252	172.15	168.93	1.87
		25000	32267	158.75	155.67	1.94
		35000	42258	151.66	148.57	2.04
		45000	52299	147.23	144.13	2.11
0.11994		15000	25844	167.83	163.07	2.83
		25000	35861	157.57	152.70	3.09
		35000	45813	151.83	146.77	3.33
		45000	55842	148.11	142.93	3.50
0.18740		15000	29250	166.08	159.29	4.09
		25000	39266	157.78	151.25	4.14
		35000	49247	152.99	146.21	4.43
		45000	59265	149.81	142.60	4.81

Replacement rate	Pile modulus (KPa)	Pile-soil modulus (KPa)	Composite modulus (KPa)	Settlement of composite foundation bottom (mm)		
				Separate	Consolidated	Difference (%)
0.02998	200000	15000	18500	126.28	124.82	1.16
		25000	28530	126.45	125.06	1.10
		35000	38542	126.45	125.08	1.08
		45000	48526	126.40	125.05	1.07
0.06747		15000	22252	128.18	124.95	2.52
		25000	32267	128.19	125.08	2.43
		35000	42258	128.14	125.08	2.39
		45000	52299	128.07	125.03	2.37
0.11994		15000	25844	130.70	125.02	4.34
		25000	35861	130.60	125.09	4.22
		35000	45813	130.50	125.06	4.16
		45000	55842	130.40	125.01	4.13
0.18740		15000	29250	131.12	125.06	4.62
		25000	39266	131.21	125.08	4.67
		35000	49247	131.22	125.04	4.71
		45000	59265	131.34	124.98	4.84

Replacement rate	Pile modulus (KPa)	Pile-soil modulus (KPa)	Composite modulus (KPa)	Settlement of composite foundation range (mm)		
				Separate	Consolidated	Difference (%)
0.02998	200000	15000	18500	52.56	52.66	0.00
		25000	28530	34.50	34.52	0.00
		35000	38542	25.75	25.72	0.00
		45000	48526	20.59	20.53	0.00
0.06747		15000	22252	43.96	43.98	0.00
		25000	32267	30.55	30.59	0.00
		35000	42258	23.52	23.49	0.00
		45000	52299	19.17	19.10	0.00
0.11994		15000	25844	37.13	38.05	−0.02
		25000	35861	26.97	27.62	−0.02
		35000	45813	21.34	21.71	−0.02
		45000	55842	17.72	17.91	−0.01
0.18740		15000	29250	34.96	34.23	0.02
		25000	39266	26.57	26.17	0.02
		35000	49247	21.77	21.17	0.03
		45000	59265	18.47	17.62	0.05

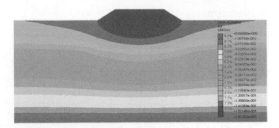

Figure 2. Vertical deformation from the consolidated calculations.

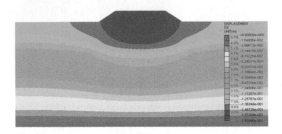

Figure 3. Vertical deformation from the separate calculations.

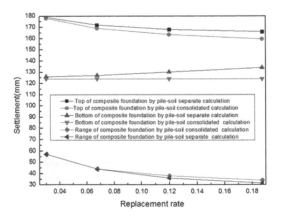

Figure 4. Comparison analysis of the modulus and settlement calculation results.

great settlement, and the central area of foundation had the biggest deformation, and this deformation weakened along with the distance from the central area. Furthermore, the settlement area from the consolidated calculations was slightly greater than that from the separate calculations.

Figures 4 and 5 show the comparison analysis on the settlement of composite foundations under different replacement rates.

It can be seen from the figure that the results of the separate calculations and consolidated

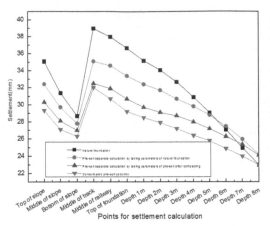

Figure 5. Comparison among the calculation methods.

calculations were basically the same. Along with the increase in the replacement rate, the two results showed slight differences. The main factor contributing to these differences is the mutual influences from friction between the soil and the pile (Randolph 2003). As for the composite foundation treatment area, the two results were basically equivalent to the pile body area.

4 RELATIONSHIP BETWEEN THE COMPOSITE MODULUS AND THE PILE-SOIL MODULUS

Data shown in the above table were obtained from the pile-soil separate calculations and the pile-soil consolidated calculations based on the finite element model built according to field test parameters. These data were nonlinearly fit with Matlab by using the formula beta = nlinfit (x,y, 'program name', beta0) [beta,r,J] = nlinfit (X,y,fun,beta0), in which the modulus was identified as KPa, thus the following conclusions can be obtained:

$$E = mE_p^{0.9431} + (1-m)E_s^{1.0091} - 780.4098$$

The correlation coefficient was 0.9988.
The relationship between the composite modulus and the pile-soil modulus can be expressed as:

$$E = mE_p^a + (1-m)E_s^b + c. \tag{1}$$

5 VALIDATION

Three methods were adopted for the validation according to the field test data: 1) separate

502

calculation on the pile and the pile-soil; 2) consolidated calculation on the pile and the soil; 3) separate calculation on the pile and the pile-soil. In these methods, the soil was calculated with the original data of foundation before compaction. Later, a comparison analysis was carried out between the results of the three calculations and the actually measured values, as shown in Figure 5.

The computational formula for the composite modulus of the foundation can be obtained after the nonlinear regression simulation:

$$E = mE_p^{0.9431} + (1-m)E_s^{1.0409} - 35.764.$$

6 CONCLUSIONS

It is known from the numerical test that the conclusions derived from the separate calculations and consolidated calculations were basically the same, and all conclusions were included in 5%. Especially for pile deformation or composite foundation deformation, differences in the results from the two calculations were all within 0.05%; thus, the results of the two calculations can be deemed as consistent. The relationship between the composite modulus and the pile-soil modulus can be expressed as Formula (1).

Then, the relationship between the composite modulus and the pile-soil modulus of the composite foundation of the high-speed rail is far more than the simple relationship of the weighted mean. When carrying out the calculation of the load-bearing capacity and the settlement of the composite foundation, the mutual influence of friction between the pile and the soil should be taken into consideration sufficiently to modify the computational formula of the composite modulus, and a reasonable derivation should be made according to Formula (1), to confirm related parameters before determining the composite modulus of such foundation.

REFERENCES

Chow, Y.K. 1996. Settlement Analysis of Sand ComPaetion Pile, Soils and Foundations, 36(1): 111–113.

Deng, Y.F. Liu, S.Y. & Hong, Z.S. 2005. A New Method for Calculating Deformation Modulus of DJM Composite Foundation. *Journal of Highway and Transportation Research and Development.* 22(3): 13–16.

Randolph, M.F. 2003. Science and empiricism in pile foundation design. *Geotechnique,* 53(10): 847–875.

Wang, F.C. Zhu, F.S. & Wang, X.C. 2003. Theoretical Analysis on the Modulus of Construction Composite Foundation. *Journal of Northeastern University (Natural Science),* (5): 491–494.

Advanced Materials and Structural Engineering – Hu (Ed.)
© *2016 Taylor & Francis Group, London, ISBN 978-1-138-02786-2*

Thermal diffusivity and electrical resistivity measurements of a 30CrMnSiA rolled steel sheet under different thermal conditions

W.Q. Khan & Q. Wang
College of Materials Science and Engineering, Beijing University of Technology, Beijing, China

ABSTRACT: The thermal diffusivity and electrical resistivity of as-rolled 30CrMnSiA aerospace grade high-strength low alloy steel were investigated under different thermal treatments. Thermal treatment parameters were selected as per the actual microstructural and mechanical properties of various components (fasteners, rivets, and pressure vessel) to determine the effect of the microstructure and the temperature on the thermal and electrical properties. The results were correlated with different conditions of annealing, normalizing and quenching from 500°C to 900°C. Electrical resistivity values ranged from 2.90451×10^{-6} Ω-m to 2.93046×10^{-6} Ω-m, on the lower side after annealing-normalizing treatments and increased markedly to 3.69695×10^{-6} Ω-m and 4.02203×10^{-6} Ω-m after quenching in oil and water, respectively. Thermal diffusivity increases from 0.0142 m²/s to 0.0828 m²/s when measured from 100°C to 200°C. From 200°C to 700°C, diffusivity decreases to 0.0356 m²/s due to the heat produced by the laser. It is almost equal to the sample temperature and decreasing onwards. The microstructural evolution from the ferrite and pearlite (annealing-normalizing) to martensitic (quenching) phase contributed to the increase in thermal diffusivity and electrical resistivity.

1 INTRODUCTION

Thermal diffusivity and electrical resistivity have been widely investigated in different engineering materials applications (Gnanasekaran & Balaji 2013, Dawson et al. 2015, Liu et al. 2010). Conventional medium carbon steels have been replaced by high-strength low alloy steels in the last two decades (Garcia et al. 1992, Hulka et al. 1988). Importance of steels in the development of industry has been enormous as they are related to engineering straightforwardly. The heat treatment can affect the microstructure and mechanical properties such as tensile strength, hardness, ductility, and toughness (Krauss 1989). It is important to compare the behavior of materials (e.g. iron, plain carbon steel, alloy steels, and high-strength low alloy steels) just as they are subjected to the same kinds of tests (e.g. tensile test, hardness test, and electrical resistivity measurement) and conditions (e.g. temperature and pressure) (Lee & Su 1999). The increase in annealing temperature and time during thermo-mechanical processing showed an increased grain size when compared with normalizing treatment due to the decreased cooling rate (Srivastava et al. 2006). Many factors affect texture evolution during the grain growth. These include a degree of texturing, initial volume fractions, grain sizes, size distributions of different texture components, anisotropy in grain boundary energy and mobility. Even though all the factors considered affect the texture evolution, it has been found that the key parameter that controls the texture evolution is the grain boundary energy density (Ray et al. 2003). Considerable knowledge exists, which suggests how the alloy composition and the heat treatment affect the mechanical properties of engineering materials (Barbacki & Mikolajski 1998, Lee & Su 1999, Srivastava et al. 2006, Ray et al. 2003, Ma et al. 2004, Dhua et al. 2001).

Thermal diffusivity (α) is the ratio of thermal conductivity and the product of the density and heat capacity per unit mass. Thermal conductivity (λ) is the time rate of steady heat flow through the unit thickness of an infinite slab of a homogeneous material in a direction perpendicular to the surface, induced by a unit temperature difference. Thermal diffusivity is an important consideration in many applications such as a design of components under transient heat flow conditions, determination of safe operating temperature, process control and quality assurance. Equation 1 shows the relationship between these thermal parameters (Parker et al. 1979, Watt 1966).

$$\lambda = \alpha\, C_p\, \rho. \tag{1}$$

2 EXPERIMENTAL PROCEDURE

The chemical composition of steel was determined experimentally by using an X-ray fluorescence machine (S2 ranger Bruker AXS A20 A2), as shown in Table 1. Different heat treatment techniques were performed at various temperatures and soaking times. Dimensions of all the heat-treated samples were 10 mm × 10 mm × 2 mm.

Heat treatments carried out included full annealing, oil quenching, water quenching and normalizing using a muffle furnace (Nabertherm 3000). Full annealing of the samples was done at 900°C for 60 min, and then cooled in the furnace to 500°C followed by air cooling. Normalizing was done at 900°C for 60 min followed by air cooling. In contrast, oil and water quenching was performed at 900°C for 45 min. After thermal treatments, all the samples were polished and etched for 5 seconds in Nital, and microstructures of the prepared samples were observed by using an optical microscope LX-31. After polishing, the samples were etched. The specimens were etched for 5 seconds in 2% Nital for general microscopic examination.

Tensile and hardness testing was carried out to observe the response of the thermal treatments to the mechanical properties. Rockwell hardness testing (HRC) of the samples was done by using the Qualitest (Qualirock) machine according to the ASTM E 18-08b standard. Three measures were carried out in each sample, and then the average was calculated.

Specimens for a tensile test were made by using bench type EDM (wire cutting) according to the ASTM standard E8 considering sub-size specimen dimensions, as shown in Figure 1. Tensile tests were conducted at room temperature using a universal testing machine (SCHIMADZU 20 kN) to compare the steel studied under different treatment conditions.

Table 1. Chemical composition analyzed by using the Ion Trap Mass Spectrometer (ITMS) Model 220.

Elements	C	Mn	Cr	Si	P	S	Fe
Wt%	0.30	1.05	0.95	1.1	<0.030	<0.030	Balance

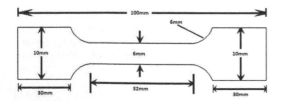

Figure 1.   Dimensions of the tensile test specimen.

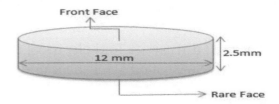

Figure 2.   Dimensions of the thermal diffusivity specimen.

The electrical resistivity of the heat-treated samples was measured by using the Source meter (KEITHLEY 2400). It was done by using the standard four-probe method. First, the voltage was regulated and then the electrical resistance was recorded at 10 mA current. It was then converted to resistivity by using the following formula:

$$\rho = RA/L, \tag{2}$$

where R is the electrical resistance; A is the area (thickness × width); L is the length of a specimen; and ρ is the resistivity (Ω-m).

The thermal diffusivity of the samples subjected to various heat treatment processes was measured at 200, 400, 500 and 700°C by using a Thermal Diffusivity Measuring System (Anter FL 3000). The sample was placed in a sample holder and then lowered inside the furnace where argon atmosphere was provided to obtain accurate results. After the sample was stabilized at the desired temperature, a nearly instantaneous pulse of energy, usually laser, was incident on its front face and the temperature increase on the rare face of the sample was recorded as a function of time. The dimensions of the specimen were 12 mm in diameter and 2.5 mm in thickness, as shown in Figure 2.

3   RESULTS AND DISCUSSION

3.1   *Effect of heat treatments on hardness*

The effect of heat treatments on Hardness of Steel was determined. The results showed that when the sample was annealed at 900°C and soaked for 60 minutes (average), the hardness value was 59.97 HRC owing to the fact that the microstructure consisted of slice pearlite and ferrite. When the sample was normalized at 900°C and homogenized for 60 minutes (average), the hardness value was 70.25 HRC due to the reason that the microstructure consisted of two phases, namely pearlite and ferrite, with different volume fractions. Oil quenching at 900°C for 60 minutes resulted in a high hardness value of 85.125 HRC, which was due to the martensitic structure with retained austenite. In contrast, water quenching at 900°C for 60 minutes resulted in a hardness value of 90.125 HRC, which was due to the complete martensitic structure that is relatively hard. Figure 6 shows the microstructures under various heat treatments.

3.2 Effect of heat treatments on tensile strength

The effect of heat treatments on the tensile strength of the samples was determined. The results showed that when the sample was annealed at 900°C, its tensile strength value was 4.2 KN, and when the sample was normalized at 900°C, its tensile strength was 7.5 KN. Oil quenching and water quenching at 900°C provided the tensile strength value of 15.1 KN and 16.075 KN, respectively, as shown in Figure 3. The water-quenched sample had a maximum tensile strength value because in water quenching, the cooling rate was so high that the grain size became very fine. It should be noted that as the finer the grain size, the greater the strength. Also, it is easier to produce deformation in a BCC structure (ferrite) than in a BCT structure (martensite).

3.3 Effect of heat treatments on electrical resistivity

The effect of heat treatments on the electrical resistivity of steel was determined. The results showed that when the sample was normalized to 900°C, its electrical resistivity value was 2.93046×10^{-6} Ω-m. When the sample was annealed at 900°C, its electrical resistivity value was 2.90451×10^{-6} Ω-m. When the sample was oil quenched at 900°C, its electrical resistivity value was 3.69695×10^{-6} Ω-m. When the sample was water quenched at 900°C, its electrical resistivity value was 4.02203×10^{-6} Ω-m, as shown in Figure 4. The water-quenched sample had a maximum electrical resistivity value since the cooling rate was very high, which created

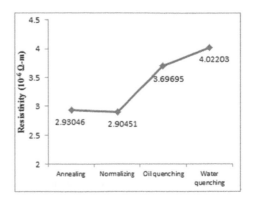

Figure 4. Effect of heat treatments on electrical resistivity.

imperfections in the crystal structure. The movement of electrons was retarded due to these imperfections, hence electrical resistivity increased. The above results confirmed that the annealing-normalizing treatments resulted in the lowest resistivity value, while in water quenching, the values were relatively higher. In fact, in annealing, the microstructure was comprised of ferrite and pearlite, which were relatively softer phases and also free from stress. While in the case of quenching, the microstructure had a BCT structure, which was more resistive in nature and provided more hurdles to the flow of electrons, hence more electrical resistivity comparatively.

3.4 Effect of the temperature on thermal diffusivity

The effect of the temperature on thermal diffusivity is very dramatic. Thermal diffusivity increased from 0.0142 m²/s (measured at 100°C) to 0.0828 m²/s (measured at 200°C), due to increasing thermal vibration of atoms. From 200°C to 700°C, diffusivity decreased to 0.0356 m²/s due to the fact that the heat produced by a laser was almost equal to the sample temperature and decreasing onwards. Also, the formation of the slight oxide layer after 500°C acted as an insulation barrier and consequently decreased the thermal transfer through the sample. Here, the results are in accordance with GuiBing Wang's work. Also, phonons were responsible for heat transportation. Conductivity was low at a high temperature and became nearly temperature independent when it reached the minimum possible value, as shown in Figure 5.

3.5 Microstructures

The microstructures of 30CrMnSiA under different heat treatments are shown in Figure 6.

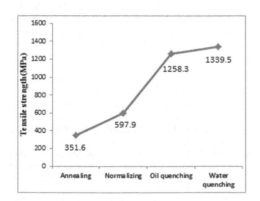

Figure 3. Effect of heat treatments on the tensile strength of 30CrMnSiA.

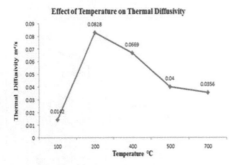

Figure 5. Effect of the temperature on thermal diffusivity.

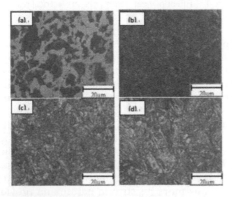

Figure 6. Microstructures of 30CrMnSiA (500x) (a) as received sample, fully annealed at 900°C, homogenized for 60 minutes, cooled up to 500°C in a furnace followed by air cooling, slice pearlite and ferrite, (b) normalized sample, heated at 900°C, homogenized for 60 minutes, then cooled in air, pearlite and Ferrite, (c) oil-quenched sample, heated at 900°C, homogenized for 60 minutes, martensite, (d) water-quenched sample, heated at 900°C; homogenized for 60 minutes, large grained martensite.

4 CONCLUSIONS

This investigative work showed very important structural modifications that could be made by using different heat treatment techniques, and new perspectives of the study and of applications are now being opened for 30CrMnSiA. The main conclusions drawn from this work are as follows:

a. Electrical resistivity values ranged from 2.90451×10^{-6} Ω-m to 2.93046×10^{-6} Ω-m on the lower side under the annealing-normalizing treatments.
b. Electrical resistivity increased markedly to 3.69695×10^{-6} Ω-m and 4.02203×10^{-6} Ω-m after quenching in oil and water, respectively.
c. Thermal diffusivity increased from 0.0142 m²/s to 0.0828 m²/s when measured at a temperature ranging from 100°C to 200°C.

d. From 200°C to 700°C, diffusivity decreased to 0.0356 m²/s and then remained constant.

REFERENCES

Barbacki A. & Mikolajski E. 1998. Optimization of heat treatment conditions for maximum toughness of high strength silicon steel. *Journal of Materials Processing Technology*, 78: 18–23.

Dawson, A. Rides, M. Maxwell, A.S. Cuenat, A. & Samano, A.R. 2015. Scanning thermal microscopy techniques for polymeric thin films using temperature contrast mode to measure thermal diffusivity and a novel approach in conductivity contrast mode to the mapping of thermally conductive particles, *Polymer Testing*, 41: 198–208.

Dhua, S.K. Ray, A. & Sarma, D.S. 2001. Effect of tempering temperatures on the mechanical properties and microstructures of HSLA -100 type copper bearing steels. *Materials Science and Engineering: A*, 318: 197–210.

Garcia, C.I. Lis, A.K. Pytel, S.M. & Deardo, A.J. 1992. Ultra-low carbon bainitic steel plate steels: Processing, microstructure and properties. *Transactions of the Iron & Steel Society of AIME*, 13: 103–112.

Gnanasekaran, N. & Balaji, C. 2013. Markov Chain Monte Carlo (MCMC) approach for the determination of thermal diffusivity using transient fin heat transfer experiments, *International Journal of Thermal Sciences*, 63: 46–54.

Hulka, K. Hesterkamp, F. & Nachtel, L. 1988. *Processing, Microstructure and Properties of HSLA Steels*, TMS, Warrendale, PA, 153.

Krauss, G. 1989. *Steels: Heat Treatment and Processing Principles*, ASM International, OH, USA.

Lee, W.S. & Su, T.T. 1999. Mechanical properties and microstructural features of AISI 4340 high strength alloy steel under quenched and tempered conditions. *Journal of Materials Processing Technology*, 87: 198–206.

Liu, X.B. Shiwa, M. Sawada, K. Yamawaki, H. & Watanabe, M. 2010. Thermal diffusivity measurement of 2.25Cr–1Mo steel with internal friction, *Materials Letters*, 64(11): 1247–1250.

Ma, N. Karzaryan, A. Dregia, S.A. & Wang, Y. 2004. Computer simulation of texture evolution during the grain growth effect of boundary properties and initial microstructure. *Acta Materialia*, 52(13): 3869–3879.

Parker, W.J. Jenkins, R.J. Butler, C.P. & Abbott, G.L. 1961. Flash Method of Determining Thermal Diffusivity Heat Capacity and Thermal Conductivity, *Journal of Applied Physics*, 32(9) 1679–1684.

Ray, P.K. Ganguly, R.I. & Panda, A.K. 2003. Optimization of mechanical properties of an HSLA-100 steel through control of heat treatment variables. *Materials Science and Engineering: A*, 346: 122–131.

Srivastava, A.K. Jha, G. Gope, N. & Singh, S.B. 2006. Effect of heat treatment on micro-structure and mechanical properties of cold rolled C–Mn–Si Trip aided steel. *Materials Characterization*, 57: 127–135.

Watt, D.A. 1966. Theory of Thermal Diffusivity of Pulse Technique, *British Journal of Applied Physics*. 17(2): 231–240.

Advanced Materials and Structural Engineering – Hu (Ed.)
© 2016 Taylor & Francis Group, London, ISBN 978-1-138-02786-2

Microstructure and mechanical property of TiC/Ti composite layer fabricated by laser surface alloying

J.J. Dai, J.Y. Zhu, S.Y. Li, L. Zhuang, A.M. Wang & X.X. Hu
Qingdao Binhai University, Qingdao, Shandong Province, P.R. China

ABSTRACT: TiC/Ti composite layers were fabricated on CP Ti substrate by laser surface alloying using a continuous wave CO_2 laser. The phase and microstructure of the alloyed layer were analyzed, and the micro-hardness and friction coefficient were measured. The result of X-ray diffraction analysis showed that the alloyed layer was consisted of TiC and Ti (martensite). Typical TiC morphologies including globular microstructure, well-developed dendrite, cross-petal microstructure, cellular dendrite and acicular microstructure were observed. Micro-hardness of the laser surface alloyed layer was improved to 420 Hv as compared to 200 Hv of the substrate. A lower friction coefficient of the alloyed layer was obtained due to the formation of hardening reinforcing phase TiC and the lubrication of residual carbon.

1 INTRODUCTION

Titanium and its alloys are extensively used in aero-nautical, marine and chemical industries owing to their specific properties such as high strength, excellent corrosion, and high temperature resistance. However, low hardness and poor tribological properties, such as higher friction coefficient and lower wear resistance of titanium and its alloys limit their practical applications (Kulka et al. 2014, Makuch et al. 2014, Tian et al. 2005). It is necessary to eliminate these issues for many engineering applications.

Many available techniques could increase the surface hardness of titanium and its alloys through modification of composition or microstructure. However, conventional surface modification processes, such as electro-plating, deposition, ion plating and thermal spraying, have some disadvantages such as a long processing time, the easy deformation of the work-piece being treated, low coating density and limited bond strength between the coating and the substrate. Laser surface modification processes can be free from the above-mentioned shortcomings. In addition, laser surface modification processes have several other advantages including precise control over the width and depth of processing, ability to process complex parts, and to selectively process specific areas of a component (Adebiyi & Popoola 2015, Filip 2006, Wu et al. 2014).

Due to excellent physical and mechanical properties of TiC, TiC is widely used as reinforcement in titanium matrix composites. In this paper, a continuous wave CO_2 laser was used to irradiate commercially pure titanium surface with pre-placed active carbon powders in argon atmosphere. A compact, well-adherent, crack-free and homogeneous TiC/Ti composite layer was obtained. The structural-phase state and microstructure were analyzed by X-Ray Diffraction (XRD) and Scanning Electron Microscopy (SEM). The micro-hardness and friction coefficient were measured by micro-hardmeter and tribometer, respectively.

2 EXPERIMENTAL PROCEDURE

Commercially pure titanium TA2 was used as the substrate material and the composition was given in Table 1. TA2 specimens with dimensions of $10 \times 20 \times 100$ mm were sand blasted and cleaned with acetone prior to laser treatment.

Experiment was performed using a 5 KW continuous wave CO_2 laser under argon atmosphere. Activated carbon powder was preplaced on the surface of TA2 with thickness about 1 mm. Laser process parameters were chosen as laser power of 1000~1200 W, defocused spot diameter of 4 mm and traverse speed of 6 mm/s and overlap width of 1.6 mm.

The alloyed specimens were polished and chemically etched in a solution of HF, HNO_3 and H_2O in volume ratio of 2:1:17. Microstructure and phase

Table 1. Chemical composition of the substrate (wt.%).

Fe	O	H	C	N	Si	Ti
0.1	0.12	0.001	0.01	0.02	< 0.04	Bal.

composition of the alloyed layer were characterized by S-520 type Scanning Electron Microscopy (SEM) and Rigaku Dmax-II A type X-Ray Diffraction (XRD). Micro-hardness measurement was taken by HVS-1000 type micro-hardmeter using 1000 g load and 15 s load time. Friction test was performed on a reciprocating ball-on-disk UMT-2MT tribometer under dry sliding condition with a load of 200 g and a sliding distance of 10 mm.

3 RESULTS AND DISCUSSION

3.1 *Microstructure of laser surface alloyed layer*

The X-ray diffraction pattern of the obtained alloyed layer with activated carbon powder in argon was given in Figure 1. TiC and α'-Ti peaks were found on the top surface and the depth of 0.2 mm below the top surface. The results of the X-ray diffraction analysis confirmed that TiC/Ti composite layers were fabricated. The fraction of TiC in TiC/Ti composite layers decreases with increasing depth. Moreover, the presence of a small quality of carbon was detected on the top surface. However, no carbon was detected at the depth of 0.2 mm.

Microstructure of the alloyed layer was shown in Figure 2. Figure 2a was low magnification SEM micrograph of the cross-section of the alloyed sample. It can be seen that the laser melted zone was about 0.7 mm in thickness. A few of pores were observed at the bottom of the alloyed layer. Figure 2b, c, d, e showed different TiC morphologies in different zones of the alloyed layer, respectively. The growth morphology of TiC is very dependent on the carbon content mixed with titanium and the cooling rate. Top surface of the alloyed layer exhibited a dense globular microstructure (Fig. 2b). From the top down of the alloyed layer, well-developed dendrite, cross-petal microstructure and cellular dendrite were observed, respectively (Fig. 2c, d, e). Figure 2e also showed the typical acicular martensite formed in the heat affected zone. This fine acicular martensite

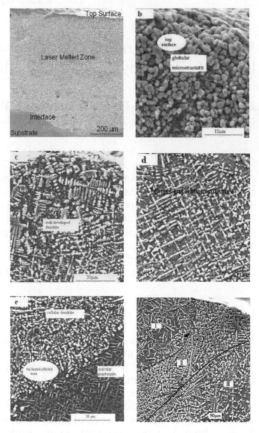

Figure 2. SEM micrographs of the alloyed layer a) low magnification of the alloyed layer, b) top surface of the alloyed layer, c) top of the alloyed layer, d) middle of the alloyed layer, e) bottom of the alloyed layer, f) overlapping zone of the alloyed layer multiple tracks.

product exhibited a hexagonal close-packed crystal structure and possessed relatively high hardness.

Figure 2f showed the typical TiC morphologies in the edge zone of the melted pool when multiple-run irradiation overlaps were applied. Acicular TiC was formed in zone I. Due to high carbon content and secondary thermal effect, dense and coarse dendrite and cellular dendrite were formed in zone II. Fine cross-petal microstructure was formed in zone III.

3.2 *Mechanical properties of laser surface alloyed layer*

Figure 3 showed the micro-hardness profile along the cross-section of the alloyed specimens. The mean micro-hardness was 425 Hv for the alloyed layer, and gradually decreased to 210 Hv for the substrate.

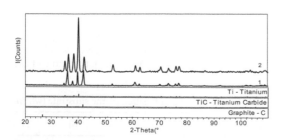

Figure 1. The XRD pattern of the laser alloyed layer; 1—the top surface, 2—the depth of 0.2 mm.

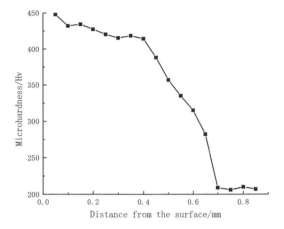

Figure 3. Micro-hardness distribution of the alloyed specimen.

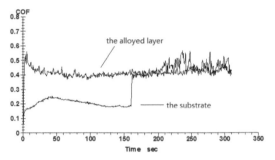

Figure 4. Variations of the friction coefficient of the alloyed layer and the substrate.

The variations of friction coefficients compared between the alloyed layer and the substrate were given in Figure 4. The average friction coefficients of the alloyed layer and the substrate were about 0.2 and 0.43, respectively. A lower friction coefficient of the alloyed layer was obtained due to the formation of hardening reinforcing phase TiC and the lubrication of a small quantity of residual carbon. Carbon could improve the tribological property as a solid lubricant. The mechanisms of friction and wear are clarified for this self-lubricating solid, and also its lubrication mechanism.

4 CONCLUSIONS

1. TiC/Ti composite layer about 0.7 mm in thickness was successfully fabricated by laser surface alloying with activated carbon powder. The alloyed layer was consisted of TiC and Ti (martensite).
2. The morphologies of TiC showed a globular microstructure, well-developed dendrite, cross-petal microstructure, cellular dendrite and acicular microstructure.
3. Micro-hardness of the laser surface alloyed layer was improved to 420 Hv as compared to 200 Hv of the substrate. The average friction coefficients of the alloyed layer and the substrate were about 0.2 and 0.43, respectively. A lower friction coefficient of the alloyed layer was obtained due to the formation of hardening reinforcing phase TiC and the lubrication of a small quantity of residual carbon.

REFERENCES

Adebiyi, D.I. & Popoola, A.P.I. 2015. Mitigation of abrasive wear damage of Ti-6 Al-4V by laser surface alloying, *Materials and Design,* 74: 67–75.
Filip, R. 2006. Alloying of surface layer of the Ti-6 Al-4V titanium alloy through the laser treatment, *Journal of Achievements in Materials and Manufacturing Engineering,* 15: 173–179.
Kulka, M. Makuch, N. Dziarski, P. Piasecki, A. & Mi-klaszewsk. A. 2014. Microstructure and properties of laser-borided composite layers formed on commercially pure titanium. *Optics & Laser Technology,* 56: 409–424.
Makuch, N. Kulka, M. Dziarski, P. & Przestacki. D. 2014. Laser surface alloying of commercially pure titanium with boron and carbon. *Optics and Lasers in Engineering,* 57: 64–81.
Tian, Y.S. Chen, C.Z. Wang, D.Y. & Lei, T.Q. 2005. Laser surface modification of titanium alloys—a review. *Surface Review and Letters,* 12: 123–130.
Wu, Y. Wang, A.H. Zhang, Z. Zheng, R.R. Xi, H.B. & Wang, Y.N. 2014. Laser alloying of Ti–Si compound coating on Ti–6 Al–4V alloy for the improvement of bioactivity, *Applied Surface Science,* 305: 16–23.

Specification of aeroelastic model design for further aerodynamic experiments

O.O. Egorychev & O.I. Poddaeva
MGSU, Moscow, Russia

ABSTRACT: Span cable-stayed bridges are increasingly being used in construction; however, such a suspension design has a number of features, such as pronounced exposure to wind effects, which can cause bridge destruction. It becomes necessary to conduct an experiment for the determination of wind loads on the projected bridge. One of the most difficult stages is the design and manufacture of the bridge aeroelastic model. This paper describes a method of the model designing of aeroelastic bridges, which allows playback modes and frequencies of oscillations of the mass distribution, moments of inertia and stiffness of structural elements in the model according to the theory of similarity.

1 INTRODUCTION

In today's world, span cable-stayed bridges are increasingly being used in the construction of roads and footpaths, and in urban and open areas. This architecture involves two tasks simultaneously: economic and, more often, stylistic. However, with the obvious merits in terms of style, this pendant design has a number of features; one of which is pronounced susceptibility to wind action. This susceptibility is so strong that it can cause bridge destruction and the application as a loss for the city's economy, and harmful to human health.

To overcome this problem, it is necessary to resort to the aerodynamic experimental methods for determining wind loads of the designed bridge. One of the most difficult stages of the experiment is the design and manufacture of the aeroelastic model.

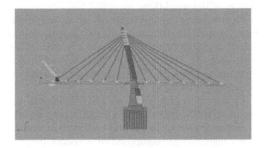

Figure 1. 3D model of a suspension bridge project.

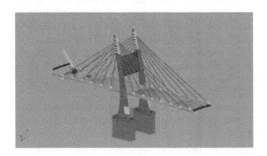

Figure 2. 3D model of a suspension bridge project.

2 FEATURES OF A CABLE-STAYED BRIDGE

A cable-stayed bridge with a span of 264 meters is under construction. A pylon consists of two supports by 126 meters in height interconnected by a bridge. The model is a 1/120 scale that represents a good compromise between the size of the model and size of the wind tunnel (4 m × 2.5 m). The model is made by aluminum to ensure a high accuracy of the geometric scaling, lightness and high rigidity, despite the small beam thickness.

The model design (pylon with a knee-fixed superstructure) was carried out on the basis of data on the size of the projected bridge forms and natural frequency distribution of mass, stiffness and moment of inertia by the structure.

Table 1 lists the scaling factors for the wind speed, time and oscillation frequency in accordance with the selected parameters of the wind tunnel (dimensions and wind speed).

It is assumed that taking into account the model scale and conditions of the construction site, the

Table 1. The scale factors for the wind speed, time and frequency of oscillation.

Name	Values
Geometric scale	1/120
Scale of the flow rate	1/6
Time scale	1/20
Scale of the oscillation frequency	20/1

environment will not have a significant impact on the construction aerodynamics.

Shapes and oscillation frequency, distribution of masses, moments of inertia and stiffness of structural elements in the pattern are reproduced in accordance with the similarity theory.

During the frame span design, deformation stiffness around the vertical axis was disregarded since it is so high that the corresponding waveform emerging from the frequency band is caused by the action of the wind flow.

Reproducing rigidity element is an aluminum plate treated in the form of the corresponding actual design geometry. To enable communication with shrouds, the aluminum plate has additional protruding elements. To limit the active damping element, it is made in one piece.

Elements reproducing weight structures are mounted to the frame using a single screw, in order to eliminate their influence on the deformation span. These elements are composed of a special polymer powder. They are used for reproducing the super-structure's geometry, mass and inertia moment in the model's scale. In order to reproduce most accurately, the shape and span of the oscillation frequency on elements are fastened to reproduce the model with weight fairings of 150 mm wide.

The analysis of the raw data and calculations is similar to those made for the span, which shows that the optimal material for the manufacture of the pylon model is also aluminum.

Ten mass elements are constructed and arranged on each pylon support. They are attached so as to reproduce the actual geometry of the pylon with respect to the total mass and mass distribution adjustment.

On the top, poles are provided for mounting the shrouds.

Mass elements are fixed to the frame using a single screw. To provide a given level of damping between the individual mass elements, gaps are provided.

The shrouds are modeled on the scale model using inextensible cables and springs at the ends of the cables in accordance with the characteristics of real shrouds.

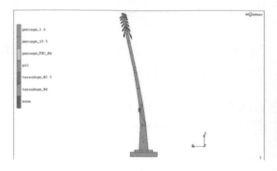

Figure 3. Design of the pylon frame.

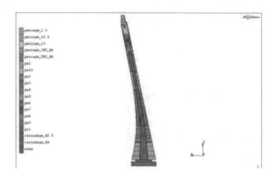

Figure 4. Design of mass pylon elements.

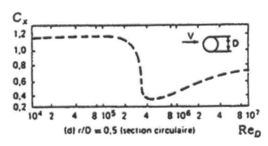

Figure 5. Effect of the Reynolds number on the drag coefficient of the cylinder.

In accordance with the shroud resistance at low wind speeds, the flow diameter must be adjusted in the cable model.

The shrouds in real designs have diameters ranging from 225 to 315 mm. Wind speed at the shrouds, according to a 50-year period, ranges between 35 and 45 m/s depending on the height. The Reynolds number is in the range of $5,2.10^5 < Re < 9,4.10^5$, which exceeds the critical value ($Rec \approx 2.10^5$). In this case, the drag coefficient is assumed to be 0.6 (Fig. 5).

Table 2. Characteristics of the selected springs.

Number of shroud	Real rigidity	Springs by brand SCHERDEL	Rigidity of spring of the model
1	4.0	TA1320	4
2	5.4	TA1820	5,43
3	5.5	TA1820	5,43
4	5.9	TA1950	6,14
5	6.1	TA1950	6,14
6	6.5	TA1560	6,37
7	6.1	TA1950	6,14
8	6.8	TA1180	6,95
9	6.0	TA1950	6,14
10	6.5	TA1560	6,37
11	6.9	TA1180	6,95
12	6.0	TA1950	6,14
13	5.5	TA1820	5,43

In accordance with a geometric model scale in the first approximation, the shroud diameter must be equal to the shroud diameter ranging from 1.9 mm to 2.6 mm, and the wind speed in the wind tunnel at the shroud should range from 5.8 to 7.5 m/s depending on the height. Thus, the Reynolds number for shrouds is in the range of $730 < Re < 1300$. At these values, the phenomenon of "Karman vortex street" is typical and there is a corresponding increase in the drag coefficient of 1.2.

To compensate for the fact that the coefficient of aerodynamic resistance is two times higher than the actual design model, the shroud diameter should be reduced to 2 times.

The average weight per unit length of the cable model is 7.6 g/m, and the linear density of the steel rope 6 is g/m^3.

Table 2 provides a choice of springs.

3 CONCLUSIONS

Thus, all the geometric and dynamic parameters of the model correspond to the parameters of the designed object model. This shows the feasibility of the proposed model during the aerodynamic experiments.

ACKNOWLEDGMENT

This work was financially supported by the Ministry of Education and Science of the Russian Federation (task no. 7.11.2014/K).

REFERENCES

Churin, P. & Poddaeva, O.I. 2014 Aerodynamic Testing Of Bridge Structures, *Applied Mechanics and Materials*. 477–478: 817–821.

Egorychev, O.O. Churin, P.S. & Poddaeva, O.I. 2015 Experimental study of aerodynamic loads on high-rise buildings, *Advanced Materials Research*. 1082: 250–253.

Egorychev, O.O. Dubinsky, S.I. & Fedosova, A.N. 2015 High-rise residential complex wind aerodynamics simulation, *Applied Mechanics and Materials*, 713–715: 1729–1732.

Gladkikh, V.A. Korolev, E.V. Poddaeva, O.I. & Smirnov, V.A. 2015. Sulfur-Extended High-Performance Green Paving Materials, *Advanced Materials Research*. 1079–1080: 58–61.

Poddaeva, O.I. Buslaeva, J.S. & Gribach, D.S. 2015. Physical model testing of wind effect on the high-rise, *Advanced Materials Research*. 1082: 246–249.

Shepovalova, O. Strebkov, D. & Dunichkin, I. 2012. *Energetically independent buildings of the resort-improving and educational-recreational complex in ecological settlement GENOM*, World Renewable Energy Forum, WREF 2012, Including World Renewable Energy Congress XII and Colorado Renew-able Energy Society (CRES) Annual Conference. Colorado. 3767–3772.

Advanced Materials and Structural Engineering – Hu (Ed.)
© *2016 Taylor & Francis Group, London, ISBN 978-1-138-02786-2*

An approach to view factor calculation for pool fires in the presence of wind

J.L. Zhao, H. Huang, M. Fu & B. Su
Department of Engineering Physics, Institute of Public Safety Research, Tsinghua University, Beijing, China

ABSTRACT: Thermal radiation from a pool fire is one of the main threats when fighting industrial fires. In the solid flame model, view factor is an important geometrical factor that represents received heat proportion. Due to the influence of the wind, the shape of the flame is considered to be a tilted cylinder. Using the Mudan's model, view factors in the downwind and upwind lines can be calculated. However, it is difficult to compute view factors at every point in the whole area to provide a spatial distribution. In the present paper, the tilted cylinder of flame is horizontally divided into a number of small cylinder sections with the same height, each of which is considered to be a vertical cylinder. The view factor between the tilted flame and the target is calculated by the weighted average of the view factors of all the small vertical cylinders. Using this method, the spatial distribution of the thermal radiation flux in windy condition can be easily calculated. The method is validated by Mudan's model in crosscut direction. In addition, the effects of wind speed, and wind direction on the heat flux distribution are discussed. The result shows that wind speed and wind direction have a significant influence on thermal radiation flux for a given target.

1 INTRODUCTION

Fires are the most frequent accidents in industrial installations, followed by explosions and gas clouds (Darbra et al. 2010). Many experiments have been conducted, and models are developed, to explore laws governing pool fires including the burn rate, the shape of flames and thermal radiation properties (Koseki et al. 2000). Many classical methods are referenced in SFPE (2008) and the 'Yellow book' to calculate the heat flux (DiNenno 2008, Book 1997). In the solid model, the view factor is a key for calculating the thermal radiation intensity. In the absence of wind, the shape of the flame is considered to be a vertical cylinder, and the view factor of storage tank areas is directly calculated using the formula given by Mudan (1984). For windy conditions, Mudan presents the formula to compute the thermal radiation flux for a target at ground level only for downwind and upwind lines (Mudan 1987). However, it is necessary to calculate the view factors for any location when fire-fighters are involved in firefighting.

In this study, a model is developed to calculate the view factor for a pool fire in the windy condition. The thermal radiation distribution can be obtained easily. Furthermore, the influences of wind speed and wind direction are studied.

2 METHOD

2.1 *The heat flux with no wind*

The solid flame model is applied to describe heat flux from hydrocarbon pool fires. In the solid flame model, the flame is considered to be a vertical cylinder (shown in Fig. 1), and the flame is assumed to radiate uniformly from the fire surface. The heat flux q is given by Mudan (1984) as:

$$q = E \times F \times \tau \qquad (1)$$

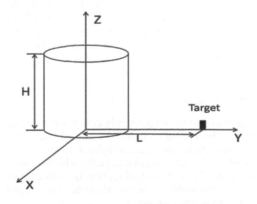

Figure 1. Flame shape under windless conditions.

where, E is the emissive power at the flame surface (W/m²), F is a geometrical view factor that defines the fractions of energy radiated by the fire that are intercepted by a receiving object; and τ is the atmospheric transmission ($\tau = 1$ in the study).

The emissive power is defined by Mudan (1984). This expression is widely used and accepted.

$$E = \frac{\eta \times \dot{m} \times \Delta h_c}{1 + 4L/D} \qquad (2)$$

where, η is a fraction of radiated energy from the fire; $\dot{m}$ is the burn rate of fuel (kg/(m²·s)) (Babrauskas 1983); L is the length of the flame (m), and D is the diameter of the pool fire (m). The parameter η was defined by Munoz and Planas (Munoz et al. 2007).

Using the new method, the study can get the distribution of thermal heat flux in accidents with the different wind speeds. From the result, it is concluded that the wind direction has a great influence on the distribution of heat flux. So firefighters can select an upwind area as the best rescue palace if do not consider other factors. The heat flux near the storage tank involved in accidents changes clearly with an increasing wind speed, particularly in the downwind direction.

In chemical industrial parks, people and facilities are protective targets, and they are all vertical, so the view factor in the horizontal direction is ignored. The view factors (F_v) in the vertical direction are given as follows (SFPE):

$$F_v = \frac{1}{\pi s} \tan^{-1}\left(\frac{h}{\sqrt{s^2 - 1}}\right) - \frac{1}{\pi s} \tan^{-1}\left(\sqrt{\frac{(s-1)}{(s+1)}}\right)$$
$$+ \frac{Ah}{\pi s \sqrt{A^2 - 1}} \tan^{-1}\sqrt{\frac{(A+1)(s-1)}{(A-1)(s+1)}} \qquad (3)$$

where, $S = L/R$, $h = H/R$, $A = (h^2 + S^2 + 1)/2S$, $B = (1 + S^2)/2S$; where, L is the distance between the center of the pool fire and the target (m). H is the height of flame (m). R is the radius of the pool fire (m).

2.2 The heat flux with wind

The flame shape under windy condition is considered to be a tilted cylinder, shown in Figure 2 (a). In this section, the tilted cylinder is crosscut into a number of smaller sections equally which is shown in Figure 2 (b). If each of section is small enough, so it is reasonable to assume that each micro cylinder is a vertical cylinder.

For each small part that is considered as a vertical cylinder, it is easy to calculate a view factor

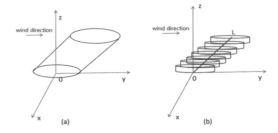

Figure 2. The shape of flame under the wind.

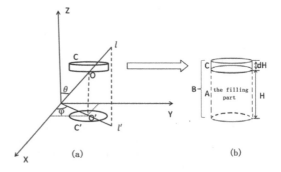

Figure 3. View factors calculation principle diagram.

between the target and the each small part using the model that is given by the above context.

In Figure 3(a), the micro cylinder C is part of the titled cylinder and point O is the center of the bottom surface of the micro cylinder C. Line l is the center axis of the tilted cylinder. Circle C', point O' and line l' are separately the horizontal projections at the XY plane. And the geometry relationship of the micro cylinders C is given as follows:

$$\begin{cases} (x - z\tan\theta\cos\phi)^2 + (y - z\tan\theta\sin\phi)^2 = R^2 \\ 0 < z \leq L\cos\theta \end{cases} \qquad (4)$$

where, φ is the angle between the central axis of the tilted cylinder at the horizontal plane projection and the x axis which is decided by the wind direction; θ is the tilted angle decided by the velocity of the wind; L is the length of the flame, (m); R is the radius of the circle C, (m); (x, y) represents a point on the circle C'.

To calculate the view factors, the micro vertical cylinder C is filled and extended to the horizontal XY plane, forming a new cylinder B (shown in Figure 3(b)). the thermal radiation flux for the vertical target caused by the extended vertical cylinders can be obtained from the windless model. According to energy conservation laws we can obtain:

$$E_{flux}S_cF_{C,da} = E_{flux}S_BF_{B,da} - E_{flux}S_AF_{A,da} \qquad (5)$$

where, E_{flux} is the surface radiation of the flame (kW/m²); $F_{B,da}$ is the view factors between the cylinder B and the vertical target at the ground; $F_{C,d}a$ is the view factor between the cylinder C and the target; $F_{A,da}$ is the view factor between the cylinder A and the target; S is the lateral area of the cylinders (m²).

$$E_{flux}S_{total}F_{Flame,da} = \sum_{i=1}^{n} E_{flux}S_i F_{i,da} \qquad (6)$$

where, $F_{i,da}$ is the view factor between the micro flame i and the vertical target; $F_{Flame,da}$ is the view factor between the actual flame and the target that is required in calculating the heat flux; S_i is the lateral area of the micro flame; S_{total} is the total lateral area of the flame. In the paper, each cylinder has the same height, so the $F_{Flame,da}$ is:

$$F_{Flame,da} = \frac{\sum_{i=1}^{n} E_{flux}S_i F_{i,da}}{E_{flux}S_{total}} = \frac{1}{n}\sum_{i=1}^{n} F_{i,da} \qquad (7)$$

The view factors between each micro cylinder and the fixed target are calculated using this method. For a fixed target, the view factor can be obtained easily by calculating the average of all $F_{i,da}$.

3 VALIDATION AND DISCUSS

3.1 Validation

In order to validate the developed method, we compare it to the Mudan's model (1987). The targets are positioned in the crosswind direction for convenience (i.e., perpendicular to the direction of tilt). The Wind blows toward the east. The thermal radiation flux in different wind speeds is shown in Figure 4.

From the Figure 4, it can conclude that the result from the developed method is very close to that obtained by the Mudan's method. Moreover, the results are not affected by the variation of the wind speed which illustrates the developed method is relatively stable. The linear correlation coefficients are all above 0.99 for the four cases.

It also shows that the greater the distance of the target from the pool fire, the less influence wind speed has, which is consistent with common sense.

3.2 Effect of wind speed on thermal radiation distribution

In this part, the influence of different wind speeds on the heat flux distribution is discussed.

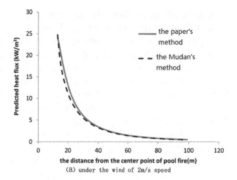

(a) under the wind of 2m/s speed

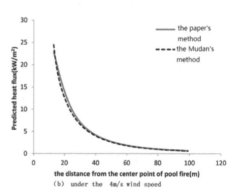

(b) under the 4m/s wind speed

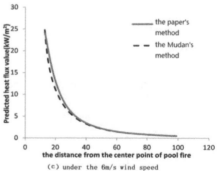

(c) under the 6m/s wind speed

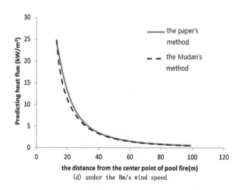

(d) under the 8m/s wind speed

Figure 4. The predicted heat flux comparison between the Mudan's wind-model and the presented method.

519

The results of heat flux distribution with wind speeds of 2 and 8 m/s are shown in Figures 5–6.

In Figures 5–6 the gray areas represent the locations that happen to pool fire, and the curves with different colors represent a different heat flux. We can see that wind has a great effect on thermal radiation. In general, when the distance is greater than 50 meters, heat flux values will decrease to 10 W/m². Under the windy conditions, the flame is closer to the ground and the closer regions absorb the most heat. The regions with flux levels over 20 kW/m² in the downwind direction at 4 m/s will become twice as large as those for wind speeds of 2 m/s. This fact needs to be considered when planning the distance between facilities.

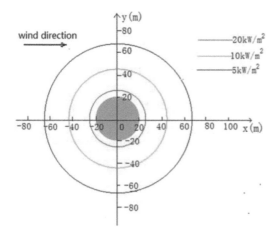

Figure 5. The thermal heat flux distribution under the wind speed of 2 m/s.

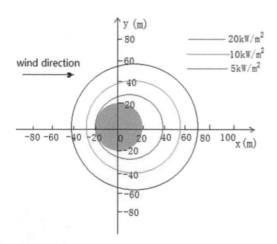

Figure 6. The thermal heat flux distribution under the wind speed of 8 m/s.

4 CONCLUSIONS

In this paper, we have developed a new method to calculate the radiation heat flux from a pool fire taking into account windy conditions. Under windy conditions, the shape of the flame becomes tilted. The tilted cylinder is crosscut into many small parts with the same height, each of which is considered to be a vertical cylinder. So the view factors at any point can be easily obtained by using the method. Meanwhile, the method has been validated by Mudan' model for some fixed direction. So using the new method, the study can get the distribution of thermal heat flux in accidents with the different wind speeds.

Using the new method, the study can get the distribution of thermal heat flux in accidents with the different wind speeds. From the result, it is concluded that the wind direction has a great influence on the distribution of heat flux. So firefighters can select an upwind area as the best rescue palace, if that does not consider other factors. The heat flux near the storage tank involved in accidents changes clearly with an increasing wind speed, particularly in the downwind direction.

ACKNOWLEDGEMENT

This work was supported by Tsinghua University Self Research Program No. 2012 THZ0124, and by the National "Twelfth Five-Year" Plan for Science & Technology Support under Grant No. 2012 BAK03B03.

REFERENCES

Babrauskas, V. 1983. Estimating large pool fire is burning rates. *Fire Technology*, 19(4): 251–261.
Book, T.Y. 1997. *Methods for the calculation of physical effects due to releases of hazardous materials (liquids and gases)*. CPR 14E, 3rd edn, TNO.
Darbra, R.M. Palacios, A. & Casal, J. 2010. Domino effect in chemical accidents: Main features and accident sequences. *Journal of hazardous materials*, 183(1): 565–573.
DiNenno, P.J. 2008. *SFPE handbook of fire protection engineering*. SFPE.
Koseki, H. Iwata, Y. & Natsume, Y. et al. 2000. Tomakomai large scale crude oil fire experiments. *Fire technology*, 36(1): 24–38.
Mudan, K.S. 1984. Thermal Radiation Hazards from Hydrocarbon Pool Fires. *Progress Energy Combustion Science*, 10: 59–80.
Mudan, K.S. 1987. Geometric view factors for thermal radiation hazard assessment. *Fire Safety Journal*, 12(2): 89–96.
Munoz, M. Planas, E. & Ferrero, F. et al. 2007. Predicting the emissive power of hydrocarbon pool fires. *Journal of hazardous materials*, 144(3): 725–729.

Advanced Materials and Structural Engineering – Hu (Ed.)
© 2016 Taylor & Francis Group, London, ISBN 978-1-138-02786-2

How construction industry safety climate affect migrant workers' safety performance: A structural equation model-based analysis

Q. Liu
School of Civil Engineering and Mechanics, Huazhong University of Science and Technology, Wuhan, China

Y.H. Wang
School of Civil Engineering and Mechanics, Huazhong University of Science and Technology, Wuhan, China
Henan Radio and Television University, Zhengzhou, China

L.M. Zhang & C.Y. Du
School of Civil Engineering and Mechanics, Huazhong University of Science and Technology, Wuhan, China

H.Y. Chen
Institute of International Education, Wuhan University of Technology, Wuhan, China

ABSTRACT: Taking migrant workers of a large construction enterprise in Wuhan as the main body in the survey, this paper sets up one-year safety climate training period and uses the same questionnaire to perform an empirical investigation twice. As a result, it builds a structural equation model about how construction industry safety climate influence on migrant workers' safety performance. Comparing the two models, it was shown that the effect of "safety climate" on "safety performance" increases substantially over time. Through strengthening the construction of safety climate, construction enterprises can improve migrant workers' compliance consciousness with safety regulations from the psychological cognition and the safety behavior, thus improving the safety performance of migrant workers.

1 INTRODUCTION

With the rapid development of urbanization in China, the number of migrant workers in the construction industry has been increasing year by year, and now it has reached 35.88 million. With a high risk factor, frequent casualties in the construction industry, how to improve the construction enterprises' safety management ability fundamentally and ensure the life security of migrant workers is worth pondering.

In 1994, Cooper et al. proposed that "safety climate is workers' common perception and beliefs to workplace safety." In 2004, Sivla et al. proposed that "safety climate, which can be observed and measured, is sharing ideas of the organization's values standard practices and norms." Chinese scholars Zhou Quan, Fang Dongping (2009) through empirical research and assumptions reasoned that there is a significant direct interaction between the safety climate and the safety accident rate. Nael et al. (2000) proposed that there is a close relationship between safety climate and safety behavior. Therefore, lots of research on the relationship between safety climate and safety performance have given some positive conclusions. Although many studies have conducted a qualitative analysis on how safety climate affects personal safety behavior, the support of quantitative research and specific data are lacking. On the other hand, many good scholars (Zhang & Zhang 2007) hold the opinion that the interaction process and specific relationship between safety climate and safety performance are not very clear.

The existing current research has weaknesses such as lack of quantitative research analysis and the unclear interaction mechanism involved. This paper presents the method of structural equation model analysis about how construction industry safety climate influence on migrant workers' safety performance.

2 RESEARCH METHODS AND STEPS

In view of the aforementioned factors, creating a culture of safety climate often takes time. Thus, taking migrant workers of a same construction enterprise as the main body in the survey, we set up one-year intervals and use the same questionnaire to perform an empirical investigation twice, which means the same model has been imported data twice. Specifically, the structural equation model analysis about how construction industry safety climate influence on migrant workers' safety performance presented in this paper includes the following four steps:

1. Determine the research objective: migrant workers, chosen from a large construction enterprise in Wuhan, whose project is under construction and the construction period of the project is over two years now, in order to ensure the stability of the basic characteristics of the research object themselves and the consistency and continuity created in the enterprise safety climate;
2. Questionnaire: in October 2013 and October 2014, 500 questionnaires were distributed to the construction sites for empirical research. Both hand-out surveys and mail were used for data collection. At the first time, we received 403 valid questionnaires back, reaching an effective rate of 80.6%, and at the second time, we received 421 valid questionnaires with a rate of 84.2%;
3. Principal component analysis: utilizing principal component analysis to extract the principal component of the "Organizational safety climate" sample, according to the characteristic value and contribution rate of each factor to get the influence factor.
4. Structural equation model: putting forward the model assumptions, building the structural equation model, importing questionnaire data and using AMOS21.0 to carry out the structural equation model analysis.

3 ASSESSMENT INDEX SYSTEM

3.1 *Index of organizational safety climate*

In this paper, a comprehensive consideration of 34 commonly used safety climate assessment indices (Lu et al. 2008) is taken, by KMO and Bartlett's sphere of inspection, we get KMO = 0.920, $P < 0.000$, the test results show that these indicators are not independent of each other, so a factor analysis is preferred. Then, a principal component analysis is used to extract 8 key factors. The following 8 dimensionalities are concluded as observable variables of organizational safety climate:

1. 1st Principal Component A1: the management department's commitments and concerns for safety
2. 2nd Principal Component A2: safety funding, staff working and living environment
3. 3rd Principal Component A3: safety education, training and supervision mechanism

4. 4th Principal Component A4: construction site labor protection measures and civil construction
5. 5th Principal Component A5: co-workers' influence and staff safety involved in motivation
6. 6th Principal Component A6: risk identification and emergency rescue
7. 7th Principal Component A7: sound safety regulations and rules, clear reward and punishment mechanism
8. 8th Principal Component A8: job satisfaction and communication effectiveness.

3.2 *Index of others*

3.2.1 *Individual safety climate*
In the safety climate assessment index of migrant workers themselves, referring to the research paper of Lan Rongxiang and other scholars (Lan 2004), 5 indicators are selected as the observable variables of the migrant workers themselves safety climate eventually, including B1-age and B2-whether alcoholism.

Table 1. The comprehensive assessment index system.

Organizational safety climate
1. A1: Management's concerns and commitments to safety
2. A2: Secure funding, staff working and living environment
3. A3: Safety education, training and supervision mechanism
4. A4: On-site labor protection measures and civil construction
5. A5: Satisfaction and communication effectiveness
6. A6: Risk identification and emergency rescue
7. A7: Sound safety regulations and clear incentive mechanisms
8. A8: Satisfaction and communication effectiveness
9. B1 Age

Individual
10. B2 Whether alcoholism

Safety
11. B3 Family responsibility

Climate
12. B4 Employment time
13. B5 Level of education

Compliance with
14. C1 Safety behavior sampling

Safety regulations
15. C2 Safety behavior self-reporting

Safety
16. D1 Standard rate of safety assessments

Performance
17. D2 Accomplishment rate of safety goals

3.2.2 Compliance with safety regulations

"Compliance with safety regulations" is not only an important prerequisite for the prevention of safety accidents, but also the key behavior status after migrant workers perceive safety climate. That is why it is considered as the intermediate variable between "safety climate" and "safety performance" in this paper. The observable values of "compliance with safety regulations" are C1-safety behavior sampling and C2-safety behavior self-reporting.

3.2.3 Safety performance

In view of the construction enterprises making their own safety management assessment solutions and safety goals, this paper sets the observable variable of the latent variable "safety performance" as D1—standard rate of safety assessments and D2—accomplishment rate of safety goals.

In summary, the comprehensive assessment index system constructed is listed in Table 1.

4 STRUCTURAL EQUATION MODELING

4.1 Proposed hypothesis

Hypothesis H_1: migrant workers' individual safety climate and organizational safety climate for compliance with safety regulations show a positive correlation. Hypothesis H_2: compliance with safety regulations for safety performance shows a positive correlation. Hypothesis H_3: migrant workers' individual safety climate and organizational safety climate are inter-related.

4.2 Model introduction

There are four potential variables in the model, and each potential is given a certain number of observed variables and error variables. We import the first survey questionnaire data and standardize path coefficients. Similarly, we import the second survey questionnaire data, thus obtaining the structural equation model shown in Figure 1 and Figure 2. The specific analysis of model outputs will be explained in detail in the next section.

4.3 Testing of the model's goodness of fit

With regard to the assessment indicators of the model's goodness-of-fit, the SEM model should pass through the following three types of assessment indices, namely absolute fit indices, fit incremental indices and parsimonious fit indices. Testing the goodness of fit of the model needs to meet requirements that the chi-square value is less than 0.5, the goodness of fit index is greater than 0.90, and the minimalist adaptation index is greater than 0.50. The goodness of fit index of the two models is fairly good.

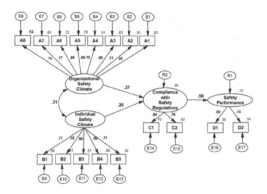

Figure 1. SEM based on the first survey data.

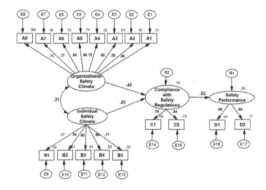

Figure 2. SEM based on the second survey data.

5 ANALYSIS OF MODEL RESULTS

5.1 Correlation analysis

The model path coefficient shows that the path coefficients for "Organizational safety climate", "individual safety climate" and "compliance with safety regulations" are 0.36 and 0.26 (0.45 and 0.25 for the second time), showing a positive correlation. The path coefficient for "compliance with safety regulations" and "safety performance" is 0.56 (0.53 for the second time), showing a positive correlation. The synergy factor for "migrant workers' individual safety climate" and "organizational safety climate" is 0.31, so that it verifies the above three assumptions (Table 2).

In the model, each observed variable and latent variables about safety climate show a positive correlation, and the influence on weight sorts results is presented in Table 3.

5.2 Causal effect analysis

Based on the comparative analysis of the structural equation modeling results obtained by the

Table 2. Validation of the model assumptions.

Variables path	1st	2nd	Hypothesis
Organizational climate on safety → compliance with safety regulations	0.36	0.45	Hypothesis H1
Individual safety climate → compliance with safety regulations	0.26	0.25	Hypothesis H1
Compliance with safety regulations → safety performance	0.56	0.53	Hypothesis H2
Organizational safety climate ↔ individual safety climate	0.31	0.31	Hypothesis H3

Table 3. Influence of the observed variables on weight sorts.

No.	Organizational safety climate	Weight
1	A5 Workers influence and staff safety involved in motivation	0.86
2	A6 Risk identification and emergency rescue	0.85
3	A3 Safety education, training and supervision mechanism	0.8
4	A7 Sound safety regulations and clear incentive mechanisms	0.77
5	A8 Satisfaction and communication effectiveness	0.73
6	A4 On-site labor protection measures and civil construction	0.72
7	A1 Management's concerns and commitments to safety	0.46
8	A2 Secure funding, staff working and living environment	0.32

No.	Individual safety climate	Weight
1	B3 Family responsibility	0.9
2	B4 Employment time	0.9
3	B2 Whether alcoholism	0.77
4	B1 Age	0.77
5	B5 Level of education	0.73

two surveys, we can obtain a similar trend and accordingly optimize the model. Changes in the structural equation model main path coefficients are presented in Table 4.

From the above analysis about how safety climate influence on safety performance over time, workers' perceptions and beliefs to workplace safety will gradually form, which can effectively promote migrant workers in their daily work to consciously abide by safety regulations, reduce unsafe behavior, and reduce the rate of accidents. Therefore, by strengthening the building of safety

Table 4. Comparative analysis of the path coefficients.

Path category	Variable relationship	1st	2nd	Changes
Direct effect	Organizational safety climate → compliance with safety regulation	0.36	0.45	Go up 25%
	Compliance with safety regulations → safety performance	0.56	0.53	Go down 5%
Indirect effect	Organizational safety climate → safety performance	0.177	0.234	Go up 32%
Total effect	Organizational safety climate → safety performance	0.204	0.234	Go up 15%

climate, construction enterprises can effectively improve the overall safety performance.

6 CONCLUSIONS

The proposed models in the two investigations both show that each observable variable has a positive correlation with "safety climate", and also verify the assumptions made before modeling. B3-Family responsibility and B4-Employment time show a significant positive correlation. Among the 8 factors of organizational safety climate after extraction by principal component analysis, A3-Safety education, training and supervision mechanism, A5-Co-workers' influence and staff safety involved in motivation and A6-Risk identification and emergency rescue have a significant positive correlation with organizational safety climate.

Comparing the two resulting models, we discover that the direct and indirect effects of "safety climate" on "safety performance" increase substantially over time after one year, and the final total effect on safety performance is also increasing. Empirical studies have shown that over time, the workers' perceptions and beliefs to the workplace safety has evolved, so the enterprise can improve the safety performance of migrant workers by enhancing the safety climate building.

REFERENCES

Cooper, M.D. & Phillips, R.A. 1994. Validation of a safety climate measure. *Occupational Psychology Conference of the British Psychological Society.* 3(5):104–116.

Flin, R. Mearns, K. & O'Connor, P. et al. 2000. Measuring safety climate: identifying the common features. *Safety science,* 34(1): 177–192.

Lan, R.X. 2004. *Influence of safety climate on safety behavior and development of safety climate survey tool*. Tsinghua University. Master's thesis.

Lu, B. Chen, P. & Fu, G. 2008. Research and design for the safety climate's questionnaire. *Journal of Safety Science and Technology*, 4(1): 47–50.

Neal, A. Griffin, M.A. & Hart, P.M. 2000. The impact of organizational safety climate, *climate and individual behavior. Safety Science*, 34(1): 99–109.

Silva, S. Lima, M.L. & Baptista, C. 2004. OSCI: an organisational and safety climate inventory. *Safety science*, 42(3): 205–220.

Wu, M.L. 2009. *Structural equation model: operation and application of AMOS*. Chongqing University Press, 2009.

Zhang, J. & Zhang, L. 2007. Study on the influence of enterprise safety climate on enterprise safety behavior. *Journal of Safety Science and Technology*, 3(1): 106–110.

Zhou, Q. & Fang, D.P. 2009. Mechanism of impact of safety climate on safety behavior in construction: an empirical study. *China Civil Engineering Journal*, 42(11): 129–132.

Advanced Materials and Structural Engineering – Hu (Ed.)
© *2016 Taylor & Francis Group, London, ISBN 978-1-138-02786-2*

A study on a turbine blade riveting device design by using finite element analysis and experiments

G.J. Kang & B.Y. Moon

Department of Naval Architecture Engineering, Kunsan National University, Kunsan, Korea

ABSTRACT: The purpose of this study is to analyze the effective process factors of the shape of a riveting device for riveting of the turbine blade rivet during the assembly between the turbine blade and the cover by using finite element analysis and experiments. The riveting system composed of an air hammer, which implied impulse force to riveting objectives and supporting structures were designed under the examination of frequency analysis to escape the resonance. The deformation of the structure was investigated by using finite element analysis. Also, the optimum set of parameters driving the effective factors in plastic forming can be determined to produce a sound riveting shape. The forming force was obtained during the forging process. From the results of the analysis, shape design of the device for turbine blade riveting was obtained by using finite element analysis.

1 INTRODUCTION

Nuclear energy steam turbine is a device that makes turning force and mechanical energy because it uses heat energy (Rolf, 2009). It needs high-quality material for its stress and high-speed accuracy of the device. The types of nuclear energy turbine depend on economic conditions, and have shown to have the difference between the steam conditions and the exhaust annulus area. At this time, the steam conditions and admissions, number, including conditions such as the extraction position 1,500 or 1,800 RPM, are to define the output limits on individual conditions. Because the atomic generation turbine operates under the stream of high temperature and high pressure, thermal stress analysis and thermal expansion should be accurately known. Thus, a seal device for minimizing steam leakage with a high-temperature and corrosion-resistant material is required. In addition, a high-precision unit operating with high efficiency is required. Also, the optimum gap should be designed corresponding to the temperature of the main area, pressure, and vibration monitoring for preventing vibration. After verification assembly plant for the production of pulp, over-speed test and balancing test are carried out. The two tests are essential processes for turbine operation that has a deep relationship with vibration reduction of the turbine and turbine life. During the turbine assembly process, as shown in Figure 1, the riveting assembly work of the turbine blade and the cover of the turbine proceed to manual operation by the air hammer. Thus, workers are exposed to the risk of musculoskeletal system damage. Some researchers have investigated the dynamic response of metallic plates under impulsive loading (Cui, 2012). Other researchers have studied riveting force and riveted joint life numerically by using the finite element method (Li, 2004). Also, Hossein investigated the effect of the aforementioned riveting parameters on the quality of a formed rivet using finite element simulation (Hossein, 2008). Kraoglu et al. performed stress analysis of a truck chassis with riveted joints by using FEM (Karaoglu, 2002). However, their studies have focused on theoretical research.

Therefore, mechanized operations that can replace manual operations are required. The objective of this study is to develop dedicated riveting machines for turbine assembly, which can be used to assemble the turbine generator of nuclear power, and to increase the efficiency of the turbine assembly technology.

Figure 1.   The turbine blade and the cover assembly.

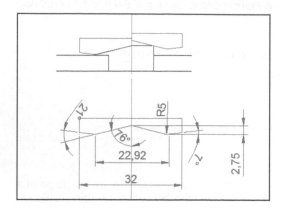

Figure 2. The shape of the rivet and the design of the pin.

The scope of the present study is to show that the installation is feasible as in automated machines and operation using an air hammer action.

2 MANUAL RIVETING ANALYSIS

2.1 Air hammer specifications

This study shows the feasibility of the automated device for riveting instead of the manual one using the air hammer. The air hammer specifications used in this study are as follows: piston diameter 28.5 mm; stroke 26 mm; frequency 3000 BPM; and reaction of air hammer 5 horsepower (5 ps).

2.2 Riveting shape

Figure 2 shows the shape of the rivet and the design of the pin for the air hammer.

2.3 Material properties

The hardness test of the pin, which is used in a conventional air hammer, was carried out. The hardness resulted in 57HRC of a Rockwell hardness measured according to ASTM A370-14. This value corresponds to the tensile strength value of 2180 N/mm².

3 MANUAL RIVETING ANALYSIS

3.1 Air hammer specifications

This study shows the feasibility of the automated device for riveting instead of the manual one using the air hammer. The air hammer specifications used in this study are as follows: piston diameter 28.5 mm; stroke 26 mm; frequency 3000 BPM; and reaction of air hammer 5 horsepower (5 ps).

3.2 Riveting shape

Figure 2 shows the shape of the rivet and the design of the pin for the air hammer.

3.3 Material properties

The hardness test of the pin, which is used in a conventional air hammer, was carried out. The hardness resulted in 57HRC of a Rockwell hardness measured according to ASTM A370-14. This value corresponds to the tensile strength value of 2180 N/mm².

4 DESIGN OF THE RIVETING SYSTEM

4.1 Natural frequency analyses

The resonance characteristic frequency analysis of the air hammer was carried out by its vibration. The plate and the air hammer are assumed to be of a cylindrical and a rectangular parallelepiped shape, respectively. Young's modulus of the plate and the air hammer is E = 2.1 × 10¹¹ N/m^2. An external force is assumed to concentrated forces in a material point in the air hammer. The power of the air hammer is 5 ps determined by the strength of the air compressor. The stiffness of the combined system of plates and the air hammer is assumed to be parallel, as shown in Figure 3. The mass, stiffness and natural frequency of the plate and the air hammer are listed in Table 1.

The total natural frequency is given by

$$\omega_d = \sqrt{\frac{K_{eq}}{M_{eq}}} \qquad (1)$$

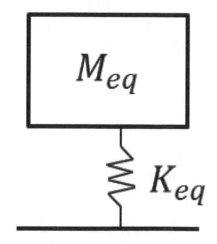

Figure 3. Frequency analysis model.

Table 1. Mass, stiffness and natural frequency of the plate and the air hammer.

	Plate	Air hammer
Mass, M (kg)	80	16.5
Stiffness, K (N/m)	5.70E+10	3.90E+08
Natural frequency, ω (rad/s)	26867	4923

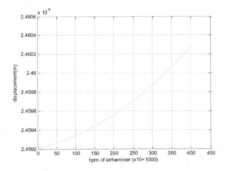

Figure 4. Displacement as bpm of the air hammer.

where K_{eq} is the stiffness of the plate and the air hammer; and M_{eq} is the mass of the plate and the air hammer. The external frequency is given by

$$\omega_d = 2\pi \times \text{bpm}. \qquad (2)$$

When the external force is excited, displacement is given by (Bishop, 1960)

$$X = \frac{F}{\sqrt{\left(K_{eq} - \omega_d^2 M_{eq}\right)^2}} \qquad (3)$$

where F is the reaction of the air hammer given as follows:

$$F = \frac{5 \text{ ps}}{\text{stroke velocity}} = \frac{3728.5 \text{ N} \cdot \dfrac{m}{s}}{\dfrac{2.6 \text{ m}}{s}} = 1.43 \text{ kN}. \qquad (4)$$

Figure 4 shows that the displacement of the riveting system varies from 2.4592×10^{-8} (mm) to 2.4603×10^{-8} (mm) in the range of the air hammer bmp as 3000 to 5000, which is in the range of 10^{-10} (mm). Therefore, the resonance of this system does not occur.

5 FINITE ELEMENT ANALYSIS

5.1 Displacement analysis

In order to determine the displacement of the riveting system according to the external force, finite

element analysis was performed. The analysis model has conditions such as mass 386.9 kg, area 4565810 mm², volume 49297000 mm³, average element size of mesh 0.1, and minimum element size 0.2. Static analysis was carried out under a concentrated load of 10,000 N. The stress analysis results of the riveting system are follows: Von misses stress 33.08 MPa, displacement 0.0067 mm, and contact stress 128.488 MPa. Figure 5 shows the displacement analysis results, suggesting that the maximum displacement is 0.000012 (m), which means the displacement is very small. As shown in Figure 6, the maximum displacements of the x axis and the z axis are 0.0008991 mm and 0.00309 mm, respectively.

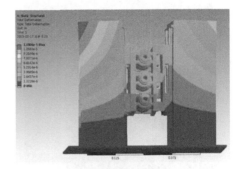

Figure 5. Displacement analysis of the riveting system.

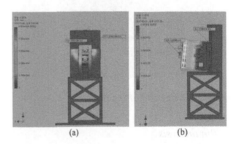

(a)   (b)

Figure 6. Maximum displacement: (a) x axis 0.0008991 (mm) and (b) z axis 0.00309 (mm).

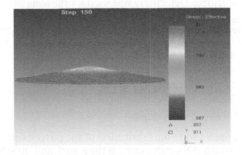

Figure 7. Effective stress distributions of the rivet at the final stage of riveting.

529

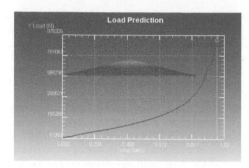

Figure 8. Load prediction of riveting.

Figure 9. Fabrication of the pin.

5.2 *Pin analysis*

To determine the stress distribution and load prediction of the riveting pin of the riveting device according to the external force, finite element analysis was performed. Deform3D was used for the simulation. Figure 7 shows the effective stress distributions of the rivet at the final stage of riveting. Figure 8 shows the load prediction of riveting. The ultimate load is 930,000 (N), which corresponds to 93Tonf.

6 EXPERIMENTAL RESULTS

6.1 *Fabrication of pin*

Figure 9 shows the fabrication of the pin consisting of riveting forming part, setting part and attachment part. The test result shown in Figure 10 indicates riveting shape with simulation.

Figure 10. Test results: (a) before riveting and (b) after riveting.

7 CONCLUSIONS

In this study, the riveting device for the riveting of the turbine blade rivet during the assembly between the turbine blade and the cover was designed by using finite element analysis and experiments. The resonance characteristic frequency analysis of the air hammer was carried out for its vibration, which showed that resonance may not occur. Also, in order to determine the displacement of the riveting system according to the external force, finite element analysis was performed and showed that the displacement is 0.000012 m, which means the displacement is very small. Riveting pin was designed and fabricated with finite element analysis. The test results of riveting showed a good agreement with the simulation results. This study showed that the installation is feasible as in automated machines and operation using an air hammer action.

ACKNOWLEDGMENT

This work (Grant No. C0184633) was supported by the Business for Cooperative R&D Between Industry, Academy, and Research Institute funded Korea Small and Medium Business Administration in 2014.

REFERENCES

Bert, C.W. & Jang, S.K. 1988. Two new approximate methods for analyzing free vibration of structural components, *AIAA*, 26(5): 612–618.
Blades, E.L. & Newman, J.C. 2007. *Aeroelastic Effects of Spinning Missiles,* AIAA 2243.
Wang, Y.L. 2001. *Differential Quadrature Method Theory and Application,* Nanjing University of Aeronautics and Astronautics.
Yang, S.Q. 2010. *Smart ammunition engineering,* Beijing National Defence Industry Press.
Zhao, B.B. Liu, R.Z. & Guo, R. 2014. Aerodynamic characteristics of the twist fin vehicle, *Journal of National University of Defense Technology.* 03: 19–24.

Advanced Materials and Structural Engineering – Hu (Ed.)
© *2016 Taylor & Francis Group, London, ISBN 978-1-138-02786-2*

Single-machine scheduling with past-sequence-dependent setup times and sum-of-logarithm-processing-times-based learning considerations

X.G. Zhang & Q.L. Xie
School of Mathematical Science, Chongqing Normal University, Chongqing, P.R. China

ABSTRACT: This paper studies single-machine scheduling problems with past-sequence-dependent setup times and general learning effect. The actual setup times are proportional to the length of the already processed jobs. The actual processing time of a job is a function of the sum of the logarithm processing times of the jobs already processed and its position. We show that the makespan minimization problem and the total completion time minimization problem can be solved in polynomial time, respectively. We also show that the problems to minimize the total weighted completion time and the maximum lateness are polynomial solvable under certain conditions.

1 INTRODUCTION

In many realistic situations, the efficiency of the production facility (e.g., a machine or a worker) improves continuously with time. As a result, the processing time of a given product is shorter if it is scheduled later (Pinedo 2008, Cheng et al. 2013). Biskup (1999) claimed that the repeated processing of similar tasks improves worker skills because workers are able to perform setup, deal with machine operations or handle raw materials and components at a faster pace. This phenomenon is known as the learning effect in the literature. He showed that the single-machine problems to minimize total deviations of job completion times from a common due date and to minimize the sum of job completion times were polynomial solvable. Kuolamas and Kyparisis (2007) presented that the actual processing time of a given job drops to zero precipitously when the normal job processing times are large in the sum of the processing-time-based learning model proposed. Due to the fact that the actual processing time of a given job drops to zero precipitously when the normal job processing times are large, Cheng et al. (2009) proposed a learning model where the actual job processing time is a function of the sum of the logarithm of the processing times of the jobs already processed. Wang et al. (2010) considered the processing time of a job is defined by an exponent function of the sum of the logarithm of the processing times of the jobs already processed. Wu et al. (2011) presented a truncation learning model where the actual job processing time is a function which depends not only on the processing times already processed but also on a control parameter. Wang (2008) studied

the processing time of a job is defined by a function of the total normal processing time of the already processed jobs, and the setup times are proportional to the length of the already processed jobs, i.e., the setup times are past-sequence-dependent (p-s-d).

Motivated by past-sequence-dependent setup times or sum-of-logarithm-processing-times learning effect, we consider single-machine scheduling problems with past-sequence-dependent setup times and a general learning effect in this paper. The setup time is proportional to the length of the already processed jobs, i.e., the actual setup time is past-sequence-dependent. The actual processing time of a job is a function of the sum of the logarithm of the processing times of the jobs already processed and its position. We will show that the makespan minimization problem and the total completion time minimization problem can be solved in polynomial time, respectively. It will be further showed that problems to minimize the total weighted completion time and the maximum lateness are polynomial solvable under certain conditions.

2 PROBLEM DESCRIPTION

There are n independent and non-preemptive jobs in a single machine. The job J_j has a normal processing time p_j and $\ln p_j \geq 1$, a weight w_j and a due date d_j. Let $p_{[k]}$ and $p_{[k]}^A$ be the normal processing time and the actual processing time of a job if it is scheduled in the kth position in a sequence, respectively. The actual processing time of job J_j scheduled in the rth position is defined as following:

$$p_{jr}^A = p_j \left(1 + \sum_{i=1}^{r-1} \ln p_{[i]}\right)^a g(r), \; r, \; j = 1, 2, ..., n, \quad (1)$$

where, $a < 0$ is the learning index, $g : [1, +\infty) \to (0,1]$ is a non-increasing function and $g(1) = 1$. Assume that the p-s-d setup time of job $J_{[r]}$ scheduled in position r is given by:

$$S_{[1]} = 0 \text{ and } S_{[r]} = b \sum_{i=1}^{r-1} p_{[i]}^A, \; r = 1, 2, ..., n, \quad (2)$$

where, $b \geq 0$ is a constant. For convenience, we denote by S_{psd} the p-s-d setup given by (2).

Let $C_{\max} = \max\{C_j \mid j = 1, 2, ..., n\}$, ΣC_j, $\Sigma w_j C_j$ and $L_{\max} = \max\{C_j - d_j \mid j = 1, 2, ..., n\}$ represent the make span, the total completion time, the total weighted completion time and the maximum lateness, respectively. In the remaining part of the paper, all the problems considered will be denoted using the three-filed notation scheme $\alpha|\beta|\gamma$ (Graham et al. 1979).

2.1 *Empennage unsteady dynamics*

In this section, we show that some single-machine scheduling problems remain polynomial solvable under the proposed model. We first present two lemmas as follows, they are useful for the following theorems (the methods of the lemmas proof take the first and the second derivatives of the functions with respect to the variable, we omit their proofs).

Lemma 1. Let $(\lambda - 1) + (1 + C_0 \ln \lambda + C_0 x)^a \theta - \lambda(1 + C_0 x)^a \theta \geq 0$, for $\lambda \geq 1$, $x \geq 1$, $1 < c_0 < 1$, $a \leq 0$ and $0 < \theta \leq 1$.

Lemma 2. Let $\lambda - 1 + \lambda_2(1 + \lambda t)^a \theta - \lambda_1 \lambda(1 + t)^a \theta \geq 0$, for $\lambda \geq 1$, $0 \leq \lambda_1 \leq \lambda_2 \leq 1$, $t \geq 0$, $0 < \theta \leq 1$ and $a \leq 0$.

Now, we will give two theorems to solve the makespan problem and total completion time problem optimally.

Theorem 1. For the problem $1 \mid p_{jr}^A = p_j \left(1 + \sum_{i=1}^{r-1} \ln p_{[i]}\right)^a g(r), S_{psd}, \mid C_{max}$, an optimal schedule can be obtained by sequencing the jobs in non-decreasing order of p_j (the SPT rule).

Proof: Assume that $\pi = \{S_1, J_j, J_k, S_2\}$ and $\pi' = \{S_1, J_k, J_j, S_2\}$ are two job sequence, where S_1 and S_2 denote partial sequences (note that S_1 and S_2 may be empty). Furthermore, we assume that there are $r - 1$ scheduled jobs in S_1. Then, the completion times of job J_j and J_k under π are

$$C_j(\pi) = \sum_{i=1}^{r-1}[b(r-i)+1]p_{[i]}^A + p_j \left(1 + \sum_{i=1}^{r-1} \ln p_{[i]}\right)^a g(r),$$

$$C_k(\pi) = \sum_{i=1}^{r-1}[b(r+1-i)+1]p_{[i]}^A + p_j \left(1 + \sum_{i=1}^{r-1} \ln p_{[i]}\right)^a g(r)$$

$$+ bp_j \left(1 + \sum_{i=1}^{r-1} \ln p_{[i]}\right)^a g(r)$$

$$+ p_k \left(1 + \sum_{i=1}^{r-1} \ln p_{[i]} + \ln p_j\right)^a g(r+1).$$

Similarly, the completion time of jobs J_j and J_k under π' are

$$C_k(\pi') = \sum_{i=1}^{r-1}[b(r-i)+1]p_{[i]}^A + p_k \left(1 + \sum_{i=1}^{r-1} \ln p_{[i]}\right)^a g(r),$$

$$C_j(\pi') = \sum_{i=1}^{r-1}[b(r+1-i)+1]p_{[i]}^A + p_k \left(1 + \sum_{i=1}^{r-1} \ln p_{[i]}\right)^a g(r)$$

$$+ bp_k \left(1 + \sum_{i=1}^{r-1} \ln p_{[i]}\right)^a g(r)$$

$$+ p_j \left(1 + \sum_{i=1}^{r-1} \ln p_{[i]} + \ln p_k\right)^a g(r+1).$$

Furthermore, we can obtain

$$C_j(\pi') - C_k(\pi) = p_j \left(1 + \sum_{i=1}^{r-1} \ln p_{[i]}\right)^a g(r) \left\{ \left(\frac{p_k}{p_j} - 1\right) \right.$$

$$+ \frac{\left(1 + \sum_{i=1}^{r-1} \ln p_{[i]} + \ln p_k\right)^a}{\left(1 + \sum_{i=1}^{r-1} \ln p_{[i]}\right)^a} \frac{g(r+1)}{g(r)}$$

$$\left. - \frac{p_k}{p_j} \frac{\left(1 + \sum_{i=1}^{r-1} \ln p_{[i]} + \ln p_j\right)^a}{\left(1 + \sum_{i=1}^{r-1} \ln p_{[i]}\right)^a} \frac{g(r+1)}{g(r)} \right\}$$

$$= p_j \left(1 + \sum_{i=1}^{r-1} \ln p_{[i]}\right)^a g(r)[(\lambda - 1)$$

$$+ (1 + c_0 \ln \lambda + c_0 x)^a \theta - \lambda(1 + c_0 x)^a \theta].$$

Let $\lambda = p_k / p_j$, $x = \ln p_j$, $c_0 = 1 / \left(1 + \sum_{i=1}^{r-1} \ln p_{[i]}\right)$ and $\theta = g(r+1) / g(r)$. Based on $p_j \leq p_k$, $g(r) \geq g(r+1)$, $\ln p_j \geq 1$, then $\lambda \geq 1$, $x \geq 1$, $0 < c_0 < 1$, $0 < \theta \leq 1$. From Lemma 1, then we have $C_j(\pi') - C_k(\pi) \geq 0$.

Theorem 2. For the problem $1 \mid p_{jr}^A = p_j \left(1 + \sum_{i=1}^{r-1} \ln p_{[i]}\right)^a g(r), S_{psd}, \mid \Sigma C_j$, an optimal schedule can be obtained by sequencing the jobs in non-decreasing order of p_j.

Proof: Here, we still use the same notations as in the proof of Theorem 1. In order to show that π dominates π', it suffices to show that

1. $C_k(\pi) \leq C_j(\pi')$,
2. $C_j(\pi) + C_k(\pi) \leq C_k(\pi') + C_j(\pi')$.

The proof of part (1) is given in Theorem 1. From $p_j \leq p_k$, we have $C_j(\pi) \leq C_k(\pi')$. Hence $C_j(\pi) + C_k(\pi) \leq C_k(\pi') + C_j(\pi')$.

Next, we will show that minimizing the total weighted completion time and minimizing maximum lateness can be solved in polynomial time under some special conditions.

Theorem 3. For the problem $1 \mid p_{jr}^A = p_j \left(1 + \sum_{i=1}^{r-1} \ln p_{[i]}\right)^a$ $g(r), S_{psd} \mid \sum w_j C_j$, if the jobs have agreeable weights, i.e., $p_j \leq p_k$, implies $w_j \geq w_k$ for all the jobs J_j and J_k, an optimal schedule can be obtained by sequencing the jobs in non-decreasing order of p_j / w_j (the *WSPT*).

Proof: Here, we still use the same notations as in the proof of Theorem 1. In order to show that π dominates π', it suffices to show that

$$w_j C_j(\pi) + w_k C_k(\pi) \leq w_k C_k(\pi') + w_j C_j(\pi').$$

From Theorem 1, we have

$$w_k C_k(\pi') + w_j C_j(\pi') - w_j C_j(\pi) - w_k C_k(\pi)$$

$$= b(w_j - w_k)\sum_{i=1}^{r-1} p_{[i]}^A + b(w_j p_k - w_k p_j)$$

$$\times \left(1 + \sum_{i=1}^{r-1} \ln p_{[i]}\right)^a g(r) + (w_j - w_k)$$

$$\times \left(1 + \sum_{i=1}^{r-1} \ln p_{[i]}\right)^a p_j g(r) \left\{\left(\frac{p_k}{p_j} - 1\right)\right.$$

$$+ \frac{w_j}{w_j + w_k} \frac{\left(1 + \sum_{i=1}^{r-1} \ln p_{[i]} + \ln p_k\right)^a}{\left(1 + \sum_{i=1}^{r-1} \ln p_{[i]}\right)^a} \frac{g(r+1)}{g(r)}$$

$$\left. - \frac{w_k}{w_j + w_k} \frac{p_k}{p_j} \frac{\left(1 + \sum_{i=1}^{r-1} \ln p_{[i]} + \ln p_j\right)^a}{\left(1 + \sum_{i=1}^{r-1} \ln p_{[i]}\right)^a} \frac{g(r+1)}{g(r)}\right\}$$

$$= b(w_j - w_k)\sum_{i=1}^{r-1} p_{[i]}^A + b(w_j p_k - w_k p_j)$$

$$\times \left(1 + \sum_{i=1}^{r-1} \ln p_{[i]}\right)^a g(r) + (w_j + w_k)\left(1 + \sum_{i=1}^{r-1} \ln p_{[i]}\right)^a$$

$$\times p_j g(r)[(\lambda - 1) + \lambda_2(1 + \lambda t)^a - \lambda_1 \lambda (1+t)^a \theta].$$

Let $\lambda_1 = w_k / w_j + w_k$, $\lambda_2 = w_j / w_j + w_k$, $t = p / 1 + \sum_{i=1}^{r-1} \ln p_{[i]}$, $\lambda = p_k / p_j$ and $\theta = g(r+1)/g(r)$.

Since $p_j \leq p_k \Rightarrow w_j \geq w_k$, we have $b(w_j p_k - w_k p_j)$ $\left(1 + \sum_{i=1}^{r-1} \ln p_{[i]}\right)^a g(r) \geq 0$ and $\lambda \geq 1, 0 \leq \lambda_1 \leq \lambda_2 \leq 1, t \geq 0$, $0 < \theta \leq 1$. From Lemma 2, we have

$$w_j C_j(\pi) + w_k C_k(\pi) \leq w_k C_k(\pi') + w_j C_j(\pi').$$

Theorem 4. For the problem $1 \mid p_{jr}^A = p_j \left(1 + \sum_{i=1}^{r-1} \ln p_{[i]}\right)^a$ $g(r), S_{psd} \mid L_{\max}$, if the jobs have agreeable conditions, i.e., $p_j \leq p_k$, implies $d_j \leq d_k$ for all the jobs J_j and J_k, an optimal schedule can be obtained by sequencing the jobs in non-decreasing order of d_j (the EDD).

Proof: Still use same notations mentioned above. We use the job interchanging technique to prove the theorem. From Theorem 1, the lateness of jobs J_j and J_k under π are

$$L_j(\pi) = \sum_{i=1}^{r-1} [b(r-i)+1]p_{[i]}^A$$

$$+ p_j \left(1 + \sum_{i=1}^{r-1} \ln p_{[i]}\right)^a g(r) - d_j,$$

$$L_k(\pi) = \sum_{i=1}^{r-1} [b(r+1-i)+1]p_{[i]}^A$$

$$+ (b+1)p_j \left(1 + \sum_{i=1}^{r-1} \ln p_{[i]}\right)^a g(r)$$

$$+ p_k(1 + \sum_{i=1}^{r-1} \ln p_{[i]} + \ln p_j)^a g(r+1) - d_k.$$

The lateness of jobs J_j and J_k under π' are

$$L_j(\pi') = \sum_{i=1}^{r-1} [b(r-i)+1]p_{[i]}^A$$

$$+ p_k \left(1 + \sum_{i=1}^{r-1} \ln p_{[i]}\right)^a g(r) - d_k,$$

$$L_k(\pi') = \sum_{i=1}^{r-1} [b(r+1-i)+1]p_{[i]}^A$$

$$+ (b+1)p_k \left(1 + \sum_{i=1}^{r-1} \ln p_{[i]}\right)^a g(r)$$

$$+ p_j(1 + \sum_{i=1}^{r-1} \ln p_{[i]} + \ln p_k)^a g(r+1) - d_j.$$

Based on $d_j \leq d_k \Rightarrow p_j \leq p_k$ and Theorem 1, we can obtain that $L_k(\pi) \leq L_j(\pi')$ and $L_j(\pi) \leq L_j(\pi')$. Therefore,

$$\max\{L_j(\pi'), L_k(\pi')\} \geq \max\{L_j(\pi), L_k(\pi)\}.$$

3 CONCLUSIONS

In this paper we consider single-machine scheduling problems with past-sequence-dependent setup times and general learning effect. The actual setup times are proportional to the length of the already processed jobs. The actual processing time of a job is a function of the sum of the logarithm of the processing times of the jobs already processed and its position. For the makespan minimization problem and the total completion time minimization problem can be solved by the SPT rule, respectively. In addition, we showed that minimizing the total weighted completion time problem can be obtained in polynomial time if jobs have agreeable weights, and minimizing the maximum lateness problem is polynomial solvable under agreeable due dates.

ACKNOWLEDGEMENT

We would like to thank the editor and reviewers for the time and effort they put in reviewing the manuscript. This work was supported by National Natural Science Foundation of China (11401065), the China Postdoctoral Science Foundation funded project (2013M540698, 2014T70854), The Chongqing Municipal Science and Technology Commission of Natural Science Fund Projects (cstc2014jcyjA00003) and the Key Foundation of Chongqing Normal University (2011XLZ05).

REFERENCES

Biskup, D. 1999. Single-machine scheduling with learning considerations. *European Journal of Operational Research*, 115: 173–178.

Cheng, T.C.E. Kuo, W.H. & Yang, D.L. 2013. Scheduling with a position-weighted learning effect based on sum-of-logarithm-processing-times and job position. *Information Sciences*, 221: 490–500.

Cheng, T.C.E. Lai, P.J. Wu, C.C. & Lee, W.C. 2009. Single-machine scheduling with sum-of-logarithm—processing-times based learning considerations. *Information Sciences*, 179: 3127–3135.

Graham, R.L. Lawer, E.L. Lenstra, J.K. & Rinnooy, Kan A.h.G. 1979. Optimization and approximation in deterministic sequencing and scheduling: A survey. *Annals of Discrete Mathematics*, 5: 287–326.

Koulamas, C. & Kyparisis, G.J. 2007. Single-machine and two-machine flow-shop scheduling with general learning functions. *European Journal of Operational Research*. 178(2): 402–407.

Pinedo, M. 2008. *Scheduling: Theory, Algorithms, and Systems. Prentice Hall*, Upper Saddle River, New Jersey,

Wang, J.B. 2008. Single-machine scheduling with past-sequence-dependent setup times and time-dependent learning effect. *Computers and Industrial Engineering*, 55: 584–591.

Wang, J.B. Sun, L.H. & Sun, L.Y. 2010. Single machine scheduling with exponential sum-of-logarithm-processing-times based learning effect. *Applied Mathematical Modelling*, 34: 2817–2819.

Wu, C.C. Yin, Y.Q. & Cheng, S.R. 2011. Some single machine scheduling problems with a truncation learning effect. *Computers and Industrial Engineering*, 60: 790–795.

Advanced Materials and Structural Engineering – Hu (Ed.)
© 2016 Taylor & Francis Group, London, ISBN 978-1-138-02786-2

A strategy for assembly using active compliant for industrial robots with F/T sensor

T. Zhang
School of Mechanical and Automotive Engineering, South China University of Technology, Guangzhou, Guangdong Province, China

B. Wang
School of Electronic Information Engineering, Tianjin University, Tianjin City, China

J.J. Lin
Guangzhou Machinery Scientific Research Inst. Co. Ltd., China

ABSTRACT: The paper gives an application to finish assembly task in an industrial robot with a six dimension Force and Torque (F/T) sensor. A new control mode is proposed for the robotic assembly which uses the tool coordination as the reference coordination. Considering zero point of the sensor would change by the gesture itself because it is influent by the gravity, a gravity compensation algorithm is researched to solve the problem. Finally, experiment result shows the theory is correct and the responded experimental results will be given.

1 INTRODUCTION

In the modern manufacturing industrial, automation equipment is growing fast. Since tasks are becoming more and more complex, the industrial robot functions are facing more challenges.

Assembly task is an important automation task part for a robot. In addition, assembly technology is the main research direction in this area (Feng 2007). Yuan F. Zheng (1989) established three models of dynamic and kinematic behavior of two coordinating robots for the whole process of assembly. Heping Chen et al. (2008) used vision and impedance control to complete the assembly of the automobile wheels. Cho et al. (2012) also used vision and impedance control to achieve connector assembly on an industrial robot that a camera and a force sensor are installed at the end of the robot.

In industrial robot assembly processing, we usually use passive compliant method, which use some compliant mechanical structures to absorb the impact force when the robot endpoint contact the environment, and the compliant mechanical can modify the tiny angle error in the process of the robot assembly task. But this method is mainly depended on the compliant structures, not by the active compliant of robot position modifying control. Actually, it could cause jamming and wedge-caulking easily (Wang 1999).

In this paper, a practical strategy for active compliance assembly based on force control is proposed.

The experiment was setup by a 6-DOF industrial robot, whose endpoint installed a six dimension F/T sensor. To meet different tasks of the holes to be assembled, the compensation algorithm of the gravity for the F/T sensor is given, as the zero point of the F/T sensor would change with the posture of the sense. Then, because the control system of the robot used the base coordination as the reference, it would make problems in describing the assembly algorithm. To simplify the process description, a method which used tool coordination as the reference is proposed. To overcome the relative position error between pin and hole assembly, an active compliant strategy is proposed.

The paper is arranged as follows. Section 2 would discuss the method of changing the referent coordination of the robot. Then, a strategy of the pin-in-hole assembly is discussed in Section 3. The compensation algorithm of the F/T sensor is discussed in Section 4. The experimental results would be given in Section 5. Finally, the conclusion and future works are shown in Section 6.

2 THE INVERSE JACOBIAN OF THE ROBOT

The advantages of changing the reference coordination to the tool coordination are described above. The robot's Jacobian matrix could be given by the parameters of the robot itself.

As the Jacobian of the endpoint of the industrial robot is a matrix consists of a fix expression and the joints' variables, to change the reference coordination of the endpoint of the robot, it should solve the problem as how to make relations of two different Jacobian matrixes between the endpoint coordination and the tool coordination.

Assume the D-H Matrix from the tool coordination to the endpoint coordination is

$$_T^6 T = \begin{bmatrix} R & p \\ 0 & 1 \end{bmatrix}$$

Here,

$$R = \begin{bmatrix} n_x & o_x & a_x \\ n_y & o_y & a_y \\ n_z & o_z & a_z \end{bmatrix}, p = \begin{bmatrix} p_x \\ p_y \\ p_z \end{bmatrix}$$

So we generate a skew symmetric matrix $S(p)$ as

$$S(p) = \begin{bmatrix} 0 & -p_z & p_y \\ p_z & 0 & -p_x \\ -p_y & p_x & 0 \end{bmatrix}$$

Then, the Jacobian of the tool coordination could be described as

$$J_T = \begin{bmatrix} R^T & -R^T S(p) \\ 0 & R^T \end{bmatrix} J \tag{1}$$

where, J is the Jacobian of the endpoint coordination.

The Jacobian of the tool coordination must be updated in every sample time, and calculate its inversion matrix J_T^{-1}.

3 PIN-IN-HOLE ASSEMBLY

Conditions: the material is high strength steel, whose elastic modulus E is 2.1e+11 Pa, μ is 0.3 and ρ is 7900 kg/m³, and the rotation velocity Ω is 0–200 r/s.

3.1 Method validation

The research aimed at the most traditional pin-in-hole model. The process of the assembly contains 3 steps: chamfer stage, one point contact stage, and two points contact stage, as shown in Figure 1 (Peng & Jin 1995, Huang et al. 1996, Wang 1999, Fei & Zhao 2001, Peng & Jin 1994). The assembly model can be seen in Figure 2. From Figure 2, if we used base coordination of

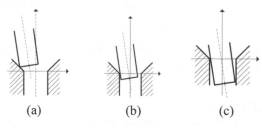

(a)   (b)   (c)

Figure 1.   Three phase of assembly.

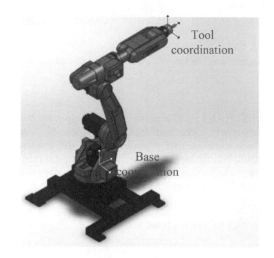

Figure 2.   Assembly model.

the robot to describe the process of the assembly, it would be problems in describing different gestures of holes. Once we used tool coordination of the robot, we can easily express the algorithm of the process of assembly as follows, assumed $F_m = [F_x \ F_y \ F_z \ M_x \ M_y \ M_z]^T$ as the measurement of the F/T sensor, $F_0 = [0 \ 0 \ F_{zr} \ 0 \ 0 \ 0]^T$ represents the desire assembly status.

Chamfer stage: This stage is shown in Figure 1 (a), we do not adjust the angle error at this stage because the pin is out of the hole, and we must make sure the pin can go into the hole theoretically. As we used the 6 dimension F/T sensor, and the endpoint of the pin is far from the surface of the F/T sensor. The control of the robot would be

$$\dot{x} = \begin{bmatrix} v \\ \omega \end{bmatrix} = \begin{bmatrix} k_f M_y \\ -k_f M_x \\ k_{fz}(F_z - F_r) \\ 0 \\ 0 \\ 0 \end{bmatrix} \tag{2}$$

where, $\dot{x} = [v \ \omega]^T = [v_x \ v_y \ v_z \ \omega_x \ \omega_y \ \omega_z]^T$ represents the velocity of the endpoint of the robot, k_f and k_{fz} are the coefficients of the impedance control in the responded axes.

One point contact stage: In the stage of Figure 1 (b). If the pin could go into the hole in this time, it had better use the same control method as the chamfer stage, as the error of the angle is tiny enough to make it go deeper.

Two points contact stage: We should consider the angle adjustment at this stage which is shown in Figure 1 (c), because if the angle errors do not get fixed, the process could not advance any more. At this time, the force diagram of the endpoint could be seen in Figure 3. The Figure shows that the direction of the torques M that the sensor measures, is negative of the fix direction ω in the responded axes. At this time, the contact point between the pin and the hole would become closer and closer to the surface of the F/T sensor, so the control method in this stage would be

$$\dot{x} = \begin{bmatrix} v \\ \omega \end{bmatrix} = \begin{bmatrix} k_f F_x \\ k_f F_y \\ k_{fz}(F_z - F_r) \\ -k_m M_x \\ -k_m M_y \\ 0 \end{bmatrix} \quad (3)$$

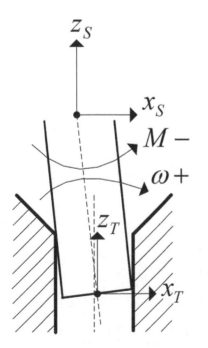

Figure 3. The torque diagram in the two points contacts stage.

The equations (1) and (2) satisfy any gestures of the holes because the axes z of the pin always point to the direction of the hole. According to the control method above, we can change the velocity of the operation space to the robot's joint space (Craig 2004),

$$\dot{q} = J^{-1}(q)\dot{x} \quad (4)$$

where, $\dot{q} = [\dot{q}_1 \ \dot{q}_2 \ \dot{q}_3 \ \dot{q}_4 \ \dot{q}_5 \ \dot{q}_6]^T$ refers the velocity of the joint space, which is a 1×6 matrix, represents the 6 joints' velocity respectively. $J^{-1}(q)$ is the inverse Jacobian matrix of the robot at the present time.

4 ALGORITHM IN THE ASSEMBLY

When the F/T sensor moves along with the endpoint of the robot, if the gesture of the endpoint is changed, the zero point of the sensor would also change deal with the gravity. Not only the gravity of the sensor itself, but also the tool which installed on the sensor would influent the zero point of the sensor. At this situation, the zero point of the sensor should be compensated. Considering the main reason that influent the change of the zero point is the gravity, a gravity compensation algorithm of the sensor is needed.

Assuming the present gesture of the endpoint of the robot is

$$T = \begin{bmatrix} {}^S x_T & {}^S y_T & {}^S z_T \end{bmatrix} = \begin{bmatrix} r_{11} & r_{12} & r_{13} \\ r_{21} & r_{22} & r_{23} \\ r_{31} & r_{32} & r_{33} \end{bmatrix}$$

The compensation formula of the F/T sensor is

$$\begin{cases} F_{xc} = r_{31} k_{xz} G_f + F_{x0} \\ F_{yc} = r_{32} k_{yz} G_f + F_{y0} \\ F_{zc} = r_{33} G_f + F_{z0} \\ M_{xc} = M_{x0} \\ M_{yc} = M_{y0} \\ M_{zc} = M_{z0} \end{cases} \quad (5)$$

where, $F_{xc}, F_{yc}, F_{zc}, M_{xc}, M_{yc}, M_{zc}$ represents the zero point of the F/T sensor in the initial place, G_f is the sum gravity of the sensor and the tools, $F_{xc}, F_{yc}, F_{zc}, M_{xc}, M_{yc}, M_{zc}$ refer to output value of the zero point of the sensor after calibration.

When the gesture of the robot is changed, we calculate the present gesture of the endpoint of the robot firstly. And then, use equation (5) to update the zero point of the F/T sensor. According to the present sensor zero point, we can know the present force that the sensor measured from the present sensor value.

Using the algorithm above, we can realize different gestures of assembly continuously without influence by the environment.

5 EXPERIMENTS

5.1 Experimental setup

We use Type RB-08 6-DOF industrial robot manufactured by GSK Inc. and its standard control system as experimental setup. The preparation of the experiment is shown in Figure 4. The hole is fixed on a table. The pin is installed on the surface of the F/T sensor, which attached on the endpoint of the robot. Add the algorithm that used tool coordination to the control system instead of the original base coordination.

At the beginning of the experiment, as the gesture of the hole is known, move the tool coordination to the place where made its 3 axes parallel to the task coordination and the pin is on the top of the hole. Then switch the control mode to the mode which is discussed in the Section 2, and use gravity compensation of the Section 3. Implement the assembly algorithm automatically and start the experiment.

5.2 Experimental results

The experiment shows that the pin could run into the hole safely. The tool coordination movement

in the assembly is shown in Figure 5, we can find that before the pin go into the hole, it contacts with the chamfer area and move along from the chamfer to the hole. When the chamfer area is passed, the pin undergoes a short period of one point contact and starts the two points contact stage.

The force and torque in the process of assembly is shown in Figure 6 and Figure 7. From the two figures, we can separate the process in 3 stages which is discussed above, and they are marked in the figures. Also, we can see that active complaint assembly could make the forces of the pin become

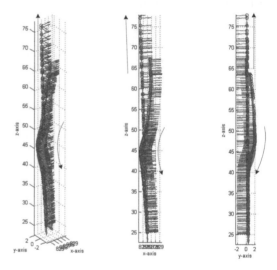

Figure 5. The movement of the endpoint of the robot in the process of assembly.

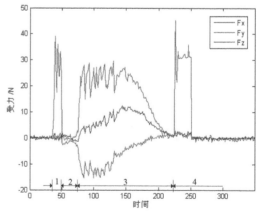

(1. Chamfer stage, 2. One point contact stage, 3. Two points contact stage, 4. Plug out)

Figure 6. Force diagram in the assembly.

Figure 4. Experimental setup.

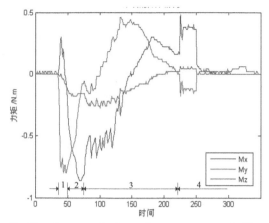

(1. Chamfer stage, 2. One point contact stage, 3. Two points contact stage, 4. Plug out)

Figure 7.   Torque diagram in the assembly.

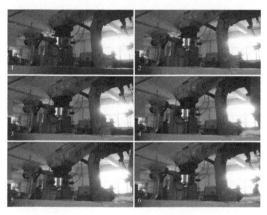

(1. Initial stage, 2. Chamfer stage, 3. One point contact stage, 4. Two points contact stage, 5. Assembly process complete, 6. Plug out)

Figure 8.   Process of the assembly.

zero except axis z in the final status. Figure 8 shows the procedure in the experiment.

6   CONCLUSION AND FUTURE WORKS

In this paper, a strategy of assembly based on passive complaint is proposed. The algorithm has been tested and proved on a RB-08, 6-DOF industrial robot and received a satisfied result. But the result can be improved if the control system adds an A/D convert model to translate the analog signals from the F/T sensor instead of the presented UART communication model. In this way, the sampling time would be decreased, and the respond time of the force would be decreased, which means decreasing the assembly time.

ACKNOWLEDGEMENT

This work was supported by the 863 Key Program (NO.2009 AA043901-3), Science and Technology Planning Project of Guangdong Province, China (No. 2012B090600028) and Industry, University and Research Project of Guangdong Province (No. 2012B010900076). Strategic and Emerging Industrial project of Guangdong Province (No. 2011 A091101001), Science and Technology Planning Project of Guangzhou City, 2014Y200014.

REFERENCES

Chen, H. Zhang, G. & Wang, J. et al. 2008. Flexible assembly automation using industrial robots, Technologies for Practical Robot Applications, 2008. TePRA. *IEEE International Conference on. IEEE,* 46–51.

Cho, H.C. Kim, Y.L. & Kim, B.S. et al. 2012. A strategy for connector assembly using impedance control for industrial robots, Control, Automation and Systems (ICCAS), *12th International Conference on. IEEE,* 1433–1435.

Craig, J.J. 2004. *Introduction to robotics: mechanics and control. Prentice Hall; 3 Edition.*

Fei, Y.Q. & Zhao, X.F. 2001. State analysis in Robot Automatic Assembly, *Journal of Mechanical Science and Technology.* 20(6): 881–883.

Feng, W.P.W.C.Z. 2007. The Research Progress on Robot Peg-in-Hole Assembly Control Technology, *Sci. Mosaic,* 3: 6–10.

Huang, X. Wang, M. & Hu, J. 1996. Geometric and mechanical analysis of unchamfer peg in hole assembly process, *Robot,* 18(2): 65–71.

Peng, S.X. & Jin, Z. 1994. Mechanism analysis of robotic compliance assembly, *Robot,* 16(1): 1–7.

Peng, S.X. & Jin, Z. 1995. Geometrical and force analysis for robot compliance assembly, *Journal of Mechanical Engineering.* 31(6): 53–60.

Wang, L. 1999. Force research of jam and wedge-caulking during compliance assembly, *Coal mine design,* 5: 32–35.

Wang, L. 1999. Mehanical analyses of Peg-Hole Compliant assembly, *J. Harbing University (Sci. Technol),* 4(1): 42–47.

Zheng, Y.F. 1989. Kinematics and dynamics of two industrial robots in assembly, Robotics and Automation, 1989. *Proc. 1989 IEEE International Conference on. IEEE,* 3: 1360–1365.

Advanced Materials and Structural Engineering – Hu (Ed.)
© 2016 Taylor & Francis Group, London, ISBN 978-1-138-02786-2

Longitudinal contact analysis of gear pair with alignment error

J. Xiao, X.C. Gui & Q. Sun
*Key Laboratory of Mechanism and Equipment Design of Ministry of Education, Tianjin University,
Tianjin, China*

Y.H. Sun
*Key Laboratory of Mechanism and Equipment Design of Ministry of Education, Tianjin University,
Tianjin, China*
*Engineering Research Center of Light-Duty Power Machine of Ministry of Education, Tianjin University,
Tianjin, China*

ABSTRACT: Alignment error of gear pair changes its meshing condition. A gear pair without machining error that transmits no load is used for analysis. While the two involute tooth profiles of gear pair just mesh on the end face of one side, there is a gap between the two involute tooth profiles on the end face of the other side possibly because of its alignment error. Taking the gap length along the meshing line as the influence evaluation index of alignment error on the evenness of longitudinal contact, an analysis of alignment error is conducted. The relationship between alignment error and contact line deviation is obtained. The influence degree of alignment error on longitudinal contact of a gear pair is concluded. Also, contact analysis of gear pair with alignment error is carried out with finite element simulation.

1 INTRODUCTION

Alignment error of gear pair is unavoidable in a real working condition. It generates bad effects on the meshing condition of gear pair.

Tang et al. (2012) conducted a contact trajectory analysis of gear pair, taking tooth profile modification and alignment error into consideration. Zhang et al. (2014) conducted a contact analysis of helical gear pair with a finite element software. Tong et al. (2014) analyzed the dynamic simulation of gear pair, considering its alignment error and machining error. Wang & Wu (1990) performed research on sensitiveness and showed that alignment had a point-contact tooth surface. Cao et al. (2014) analyzed a contact trajectory analysis of bevel gear pair with tooth profile modification. Zhou & Zhang (2007), Xu & Zhang (2012), and Filiz & Eyercioglu (1995) conducted research on how to build the 3D model of gear pair, and conducted a finite element contact analysis.

A gear pair without machining error that transmits no load is used for analysis. While the two involute tooth profiles of gear pair mesh on the end face of one side, there is a gap between the two involute tooth profiles on the other side possibly because of its alignment error. Taking the gap length along the meshing line as the influence evaluation index of alignment error on the evenness of longitudinal contact, an analysis of alignment error is conducted. The relationship between alignment error and contact line deviation is obtained, and the influence degree of alignment error on longitudinal contact of gear pair is gained. Also, contact analysis of gear pair with alignment error is carried out with finite element simulation.

2 TYPE OF ALIGNMENT ERROR

Alignment error of gear pair includes center-to-center distance error and parallel alignment error. Parallel alignment error includes alignment error in the axis plane and alignment error in the plane vertical to the axis plane. Alignment error in the axis plane is measured in the public plane of the ideal axis of gear pair. This public plane is defined by the axis of one gear and the center of one side of another gear. The deviation is the limit of the error. $\Delta F_{\Sigma\delta}$ refers to the alignment error in the axis plane, and $\Delta F_{\Sigma\beta}$ refers to the alignment error in the plane vertical to the axis plane, as shown in Figure 1.

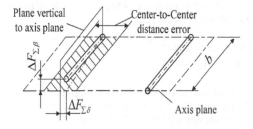

Figure 1. Alignment error.

Each parallel deviation is expressed as the value related to the effective gear width b.

3 EFFECT OF ALIGNMENT ERROR ON LONGITUDINAL CONTACT

3.1 *Definition of contact line deviation*

A gear pair without machining error that transmits no load is used for analysis. While the two involute tooth profiles of gear pair mesh on the end face of one side, there is a gap between the two involute tooth profiles on the end face of the other side possibly because of its alignment error. The gap length of the involute tooth profiles of two gears along the meshing line on the end face of this side is named the contact line deviation, referred to as Δl.

3.2 *Relationship between the center-to-center distance and the contact line deviation*

Center-to-center distance error is the gap between the actual distance and the ideal distance of the two center lines of gear pair, referred to as Δa. Based on the separability of involute gear pair, the two involute tooth profiles of gear pair with center-to-center distance error mesh on the end face of one side, while the two involute tooth profiles of gear pair mesh on the end face of the other side, that is, the value of the contact line deviation caused by the center-to-center distance error is zero.

3.3 *Relationship between alignment error in the axis plane, alignment error in the plane vertical to the axis plane and contact line deviation*

To determine the effect that the alignment error in the axis plane and that in the plane vertical to the axis plane has on the longitudinal contact of gear pair, the deviation is decomposed into the meshing plane and the plane vertical to the meshing plane, referred to as Δn and Δc, respectively. α refers to the pitch circle pressure angle of gear pair.

Based on the space relationship, after the decomposition of alignment error in the axis plane, the components in the meshing plane and the plane vertical to the meshing plane can be calculated by the following equation:

$$\begin{cases} \Delta n_\delta = \Delta F_{\Sigma\delta}\sin\alpha \\ \Delta c_\delta = \Delta F_{\Sigma\delta}\cos\alpha \end{cases} \tag{1}$$

After the decomposition of alignment error in the plane vertical to the axis plane, the components in the meshing plane and the plane vertical to the

meshing plane can be calculated by the following equation:

$$\begin{cases} \Delta n_\beta = \Delta F_{\Sigma\beta}\cos\alpha \\ \Delta c_\beta = \Delta F_{\Sigma\beta}\sin\alpha \end{cases} \tag{2}$$

While the two involute tooth profiles of gear pair without machining error that transmits no load just touch on one side, there is a gap between the two involute tooth profiles on the other side possibly because of its parallel alignment error. Assuming that gear pair only has parallel alignment error in the meshing plane, the relative position of the two involute tooth profiles is shown in Figure 2.

In Figure 2, r_{b1}, r_{b2}—the base circle radius of the pinion and the large gear;

N_1N_2—the theoretical meshing line of gear pair without alignment deviation. N_1, N_2 are extreme points;

$N_1'N_2$—the theoretical meshing line of gear pair with alignment deviation, N_1' is the extreme point;

A—the meshing point of gear pair without alignment deviation; and

A'—the crossover point of the involute tooth profile of the pinion and the new meshing line of gear pair with alignment deviation.

The contact line deviation equals to the length of $A'A$. So, the contact line deviation equals to the

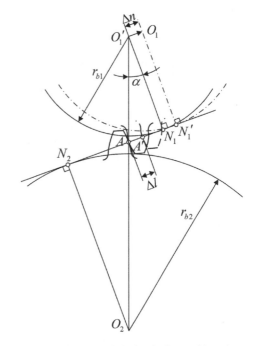

Figure 2. Alignment deviation in the meshing plane.

parallel alignment deviation if the alignment deviation only exists in the meshing plane.

Assuming that the gear pair only has parallel alignment error in the plane vertical to the meshing plane, the relative position of the two involute tooth profiles is shown in Figure 3.

In Figure 3, α'—the pressure angle of the pitch circle of gear pair with parallel alignment deviation;

θ—the intersection angle of the two meshing lines of gear pair with and without parallel alignment deviation;

$\overline{N_1'N_2'}$—the theoretical meshing line of gear pair with alignment deviation. N_1', N_2' are the extreme points;

C_1—the crossover point of tooth profile of the pinion and the new meshing line of gear pair with alignment deviation;

C_2—the crossover point of tooth profile of the large gear and the new meshing line of gear pair with alignment deviation.

The contact line deviation equals to the length of C_1C_2. So, the contact line deviation equals to the parallel alignment deviation, if the alignment deviation is in meshing plane. Based on the geometry relationship, the length of C_1C_2 satisfies the following equation:

$$
\begin{aligned}
\overline{C_1C_2} &= \overline{N_1'N_2'} - \overline{N_1'C_1} - \overline{N_2'C_2} \\
&= \overline{N_1'N_2'} - \left(\overline{N_1'C_1} - \overline{N_1C_1} + \overline{N_1C_1}\right) \\
&\quad - \left(\overline{N_2'C_2} - \overline{N_2C_2} + \overline{N_2C_2}\right) \\
&= \overline{N_1'N_2'} - \left(\overline{N_1'C_1} - \overline{N_1C_1}\right) \\
&\quad - \left(\overline{N_2'C_2} - \overline{N_2C_2}\right) - \overline{N_1N_2}
\end{aligned}
\tag{3}
$$

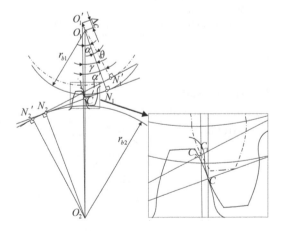

Figure 3. Alignment deviation in plane vertical to meshing plane.

In this equation, $\overline{N_1'N_2'} = (r_{b1} + r_{b2})\tan\alpha'$, $\overline{N_1'C_1} - \overline{N_1C_1} = \theta r_{b1}$, $\overline{N_2'C_2} - \overline{N_2C_2} = \theta r_{b2}$, $\overline{N_1N_2} = (r_{b1} + r_{b2})\tan\alpha$. Substituting these equations into Equation 3, we can obtain the length of C_1C_2, as follows:

$$
\overline{C_1C_2} = f(\Delta c) = (r_{b1} + r_{b2})(\tan\alpha' - \theta - \tan\alpha)
\tag{4}
$$

The center-to-center distance of the gear pair on the side that the two gear do not mesh satisfies the equation $a' = \{[(r_{b1} + r_{b2})\tan\alpha]^2 + (r_{b1} + r_{b2} + \Delta c)^2\}^{1/2}$. Thus, the pressure angle of pitch circle of gear pair with parallel alignment deviation can be calculated by the following equation:

$$
\alpha' = \arccos\left(\frac{r_{b1} + r_{b2}}{\sqrt{\left[(r_{b1} + r_{b2})\tan\alpha\right]^2 + (r_{b1} + r_{b2} + \Delta c)^2}}\right)
\tag{5}
$$

According to the geometrical relationship, we have $\gamma = \arctan\left[(r_{b1} + r_{b2})\tan\alpha/(r_{b1} + r_{b2} + \Delta c)\right]$ and $\theta = <N_1'ON_1 = \gamma - \alpha'$.

The radius of the base circle of the pinion and the large gear can be calculated by the following equations: $r_{b1} = z_1 m\cos\alpha$, $r_{b2} = z_2 m\cos\alpha$. Here, z_1 and z_2 represent the teeth number of the pinion and the large gear, respectively, and m represents the module of gear pair.

Finally, Δc can be expressed by C_1C_2 and gear parameters, including z_1, z_2, m and α.

So, the contact line deviation caused by alignment error in the axis plane can be calculated by the following equation:

$$
\Delta l_\delta = \Delta n_\delta + f\left(\Delta c_\delta\right) = \Delta F_{\Sigma\delta}\sin\alpha + f\left(\Delta F_{\Sigma\delta}\cos\alpha\right)
\tag{6}
$$

The contact line deviation caused by alignment deviation in the plane vertical to the axis plane can be calculated by the following equation:

$$
\Delta l_\beta = \Delta n_\beta + f\left(\Delta c_\beta\right) = \Delta F_{\Sigma\beta}\cos\alpha + f\left(\Delta F_{\Sigma\beta}\sin\alpha\right)
\tag{7}
$$

We assume that $z_1 = 1\text{~}200$, $z_2 = 1\text{~}200$, $m = 1.5$ mm, $\Delta c = 0 \sim \Delta c_{max}$, and $\alpha = 20°$. Δc_{max} equals to the length of the theoretical meshing line, that is, the length of $\overline{N_1N_2}$. MATLAB software was used to write the program. Finally, the formula $\overline{C_1C_2} - 0.01\Delta c < 0$ is obtained, that is, $f(\Delta c_\delta) < 0.01\Delta c$. The rightness of this in equation is independent of the given value of the module.

Thus, the contact line deviation caused by the alignment error in the meshing plane could be neglected, that is to say, the equation $\Delta l = \overline{C_1C_2} =$

543

$f(\Delta c) = 0$ is right. Based on this equation, the following equations are established:

$$\Delta l_\delta = \Delta n_\delta = \Delta F_{\Sigma \delta} \sin \alpha \qquad (8)$$

$$\Delta l_\beta = \Delta n_\beta = \Delta F_{\Sigma \beta} \cos \alpha \qquad (9)$$

If the alignment deviation meets the equation $\Delta F_{\Sigma \delta} = \Delta F_{\Sigma \beta}$ and pressure angle is 20°, we have the equation $\Delta l_\beta / \Delta l_\delta = \Delta n_\beta / \Delta n_\delta = \cot \alpha = 2.747$. When the alignment error in the plane vertical to the axis plane equals to the alignment error in the axis plane, the contact line deviation caused by alignment error in the plane vertical to the axis plane is 2~3 times larger than that caused by alignment error in the axis plane.

Based on the above analysis, it can be concluded that the center distance error has no effect on the longitudinal contact of gear pair, the parallel alignment error changes the even longitudinal contact of gear pair, and the alignment error in the plane vertical to the axis plane has a greater effect on the evenness of the longitudinal contact than the alignment error in the axis plane.

4 FINITE ELEMENT SIMULATION EXAMPLE

In the real working condition, all types of alignment error exist at the same time. When we explore the effect that one kind of alignment error has on the contact of gear pair, the other two kinds of alignment error are kept non-zero and constant. Eight alignment deviation combinations are listed in Table 1. Alignment deviation combination is designed to explore the contact status of gear pair without alignment error. It is designed to compare with the contact status of gear pair with alignment error. Alignment deviation combinations b, c, d are designed to explore the effect that alignment error in the axis plane has on the longitudinal

contact. Alignment deviation combinations e, c, f are designed to explore the effect that alignment error in the plane vertical to the axis plane has on the longitudinal contact. Alignment deviation combinations g, c, h are designed to explore the effect that the center-to-center distance error has on the longitudinal contact.

The contact analysis of gear pair with each alignment deviation combination listed in Table 1 is conducted using finite element software.

4.1 Finite element model

The parameters of gear pair for simulation are listed in Table 2. To make the analysis simple, we assume that it is an involute spur gear pair.

The gear body is simplified as a hollow cylinder. The contact ratio of the gear pair is 1.8. The gear pair is meshed at the pitch point for analysis. At this time, there is only one tooth pair in mesh. To save computation time, each gear has three teeth. The tooth in the mesh is the middle one of the three teeth.

Radial and axial degrees of freedom of all nodes of the inner hole cylinder surface of driving gear are constrained, and torque T is applied on this surface. All degrees of freedom of the inner hole cylinder surface of passive gear are constrained, that is, the surface is fixed, as shown in Figure 4.

Table 2. Values of parameters of gear pair.

Parameter	Value
Teeth number of pinion z_1	17
Teeth number of large gear z_2	30
Module m [mm]	1.5
Pressure angle α [°]	20
Width of the pinion b_1 [mm]	24
Width of the large gear b_2 [mm]	22
Transmitting torque T [N mm]	10000

Table 1. Alignment deviation combinations for simulation.

Alignment deviation combination	$\Delta F_{\Sigma \delta}$ [mm]	$\Delta F_{\Sigma \beta}$ [mm]	Δa [mm]
a	0	0	0
b	0	0.01	0.01
c	0.01	0.01	0.01
d	0.02	0.01	0.01
e	0.01	0	0.01
f	0.01	0.02	0.01
g	0.01	0.01	0
h	0.01	0.00	0.02

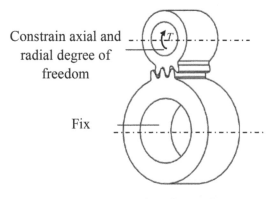

Constrain axial and radial degree of freedom

Fix

Figure 4. Loads and constraints of gear pair.

The gear pair is meshed, as shown in Figure 5. The grids of the tooth surface in contact and the region around the ideal contact line are refined. The size and number of grids are controlled to ensure the meshing quality. The total number of grids generated is 375657. The total number of nodes generated is 531150.

Both gears are made of the same material, which is 20CrMnTi. Its modulus of elasticity is 206GPa, Poisson's ratio is 0.2, and frictional coefficient is 0.2.

4.2 Finite element model

When the alignment deviation combination of the gear pair is a ($\Delta F_{\Sigma\delta} = 0$, $\Delta F_{\Sigma\beta} = 0$, $\Delta a = 0$) and c ($\Delta F_{\Sigma\delta} = 0.01$ mm, $\Delta F_{\Sigma\beta} = 0.01$ mm, $\Delta a = 0.01$ mm), the tooth surface contact stress distribution of gear pair are shown in Figure 6.

It can be seen that the longitudinal contact is even, when the gear pair has no alignment error. When the gear pair has alignment error, the teeth contact near the end face of one side and the maximum contact stress value increases. The maximum contact stress value reflects the evenness of the

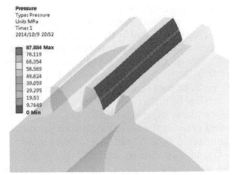

(a) Alignment deviation combination a

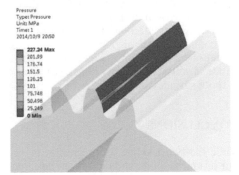

(b) Alignment deviation combination c

Figure 6. Tooth surface contact stress distribution of gear pair with alignment deviation.

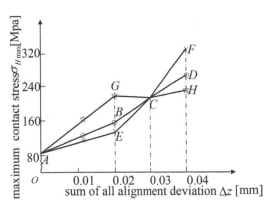

Figure 7. Maximum tooth surface contact stress of gear pair.

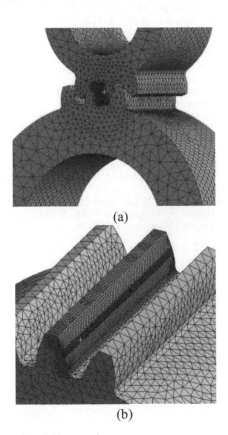

(a)

(b)

Figure 5. Grid generation.

longitudinal contact of gear pair. The maximum tooth surface contact stress of gear pair with alignment error listed in Table 2 is shown in Figure 7.

In Figure 7, the horizontal ordinate adopts the sum of three kinds of alignment deviation. The broken line connected by G, C and H shows that the center-to-center distance error has a little effect on the maximum tooth contact stress. The change in contact stress is caused by the change in the meshing position of the two involute profiles of the two gears. Comparing the broken line connected by B, C and D with the broken line connected by E, C and F, it can be seen that the alignment error in the plane vertical to the axis plane has a greater effect on the evenness of longitudinal contact than the alignment error in the axis plane. From the broken lines connected by B, C and D and that connected by E, C and F, it can be seen that the larger the parallel alignment deviation, the more serious the unevenness of the longitudinal contact becomes.

The results of finite element analysis reveal that the alignment error in the plane vertical to the axis plane has a greater effect on the evenness of longitudinal contact than the alignment error in the axis plane, causing a much more serious unevenness of longitudinal contact and generating a larger maximum tooth surface contact stress.

5 CONCLUSIONS

1. While the two involute tooth profiles of gear pair just mesh on the end face of one side, there is a gap between the two involute tooth profiles on the end face of the other side possibly because of its alignment error. The gap length along the meshing line is taken as the influence evaluation index of alignment error on the evenness of longitudinal contact, providing a thinking way for the analysis of alignment error.
2. Center-to-center distance error has no effect on the evenness of longitudinal contact of gear pair. The alignment error in the plane vertical to the axis plane and the alignment error in the axis plane cause the unevenness of longitudinal contact. The larger the alignment error is, the larger the maximum tooth surface contact stress value becomes.
3. The alignment error in the plane vertical to the axis plane has a greater effect on the evenness

of longitudinal contact than the alignment error in the axis plane, causing a much more serious unevenness of longitudinal contact and generating a larger maximum tooth surface contact stress value.

ACKNOWLEDGMENT

The authors gratefully acknowledge the financial support from the National Natural Science Foundation (program number 51175369).

REFERENCES

Cao, X.M. Jia, J. & Zhan, Y. 2014. Sensitiveness analysis of alignment error on tooth surface of bevel gear with tooth profile modification, *Journal of Mechanical Transmission,* 38(4): 40–43.

Filiz, I.H. & Eyercioglu, O. 1995. Evaluation of gear tooth stresses by finite element method, *Journal of engineering for industry,* 117(2): 232–239.

Tang, J.Y. Chen, X.M. & Luo, C.W. 2012. Contact analysis of cylinder gear considering tooth profile modification and alignment error, *Journal of Central South University,* 43(5): 1703–1709.

Tong, C. Sun, Z.L. & Ma, X.Y. 2014. Dynamic contact simulation of gear considering alignment error and machining error, *Journal of Northeastern University,* 34(7): 996–1000.

Wang, X.C. & Wu, X.T. 1990. Sensitiveness analysis of alignment error on contact characteristics of point-contact tooth surface, *Journal of Xi'an Jiao Tong University,* 24(6): 46–58.

Xu, S. & Zhang, Y. 2012. The Finite Element Modeling and Analysis of involute spur gear, *Advanced Materials Research,* 516–517: 673–677.

Zhang, F.J. Zhou, X.L. & Wang, J. 2014. Effect of alignment error on contact region and tooth bending stress of helical gear, *Journal of Mechanical Transmission,* 38(8): 6–10.

Zhou, H. & Zhang, Y. 2007. Parameterization design and realization of involute gear based on UG, *Machinery Design and Manufacture,* 2: 78–79.

Advanced Materials and Structural Engineering – Hu (Ed.)
© 2016 Taylor & Francis Group, London, ISBN 978-1-138-02786-2

The study on mechanical properties of Carboxymethyl Chitosan hybrid PVA hydrogels

J.P. Luo
College of Science, Hunan University of Technology, Zhuzhou, Hunan, China

H.Q. Hu
Institute of Machinery, Hunan University of Technology, Zhuzhou, Hunan, China

X.H. Zeng
Institute of Packaging and Materials Engineering, Hunan University of Technology, Zhuzhou, Hunan, China

ABSTRACT: A novel linear Carboxymethyl Chitosan (CMCS) hybrid organic nanocomposite hydrogels were synthesized by introducing water-soluble CMCS into the PVA hydrogels. The mechanical properties are obviously dependent on the composition of gels. With the increase of CMCs, the tensile strength and modulus increases significantly while the elongation break increases slightly. The gels exhibit higher compression strength than that of pure PVA hydrogels. This phenomenon is attributed to the effective entanglement of linear macromolecular chains around the polymer skeleton which enhances the interaction between PVA chains.

1 INTRODUCTION

In recent years, there has been a great deal of interest in the hydrogels due to their similarity with the human tissue, which is potential in the application of biotechnology and medicine fields (Lim et al. 2010, Sivakumaran et al. 2011, Demirel et al. 2009). For a better simulation of tissue, one of the main criteria governing the design of materials is the matching of mechanical properties of the graft with those of aim to prevent compliance related problems.

Polyvinyl Alcohol (PVA) has attracted much attention as a potential biomaterial for artificial blood vessels because it can be fabricated as hydrogel and can match the mechanical properties of the vascular system basically. For example, Wan and his co-workers (Wang et al. 2002) have optimized the tensile properties of PVA hydrogels to match those of the porcine aortic root. Chu and Rutt (Wan 2002) have applied PVA gel structures as vessel phantoms due to the similarity of mechanical properties between the PVA gels and porcine aortas. Millon et al. (2007) proved that the mechanical properties of porcine aorta can be matched by an anisotropic PVA conduit. The studies have established PVA hydrogels as good candidates of simulating the vessel.

However, due to the good biocompatibility of PVA hydrogels, to develop a new system with better mechanical properties in order to simulate other tissue in human being is meaningful.

Chitosan is a kind of biodegradable and biocompatible polymer, which has been investigated as a novel biomaterial. However, the application of chitosan was limited for its insolubility at neutral or high pH region. To improve the solubility of chitosan, a series of hydrophilic groups have been introduced into its skeleton. Carboxymethyl Chitosan (CMCS), is a water-soluble derivative of chitosan by introducing $-CH_2COOH$ groups onto $-OH$ groups along chitosan molecular chain. For its unique chemical, physical, and biological properties, especially its excellent biocompatibility, the CMCS has been widely used in the biological and medical fields such as drug release, tissue engineering and so on. Moreover, it has demonstrated good pH and ion sensitivity in aqueous solution due to its abundant $-COOH$ and $-NH_2$ groups. Hence, a lot of researchers focused on using CMCS to prepare responsive hydrogels applied in drug release. For example, CMCS/alginate hydrogels, CMCS/poly (acrylic acid-co-acrylamide) and CMCS/PNIPA interpenetrating hydrogels were developed and investigated deeply (Zhang et al. 2013, Kin et al. 2012).

In this research, the different content of CMCS was added into the PVA hydrogels crosslinked prepared by repeated freeze-thaw process to prepare semi-IPN hydrogels, the effect of CMCS content on the mechanical properties of hybrid gels were researched systematically.

2 EXPERIMENT

2.1 Materials

PVA provided by TCI Co., Japan, Carboxymethyl chitosan (CMCS) with 1.34 degree of substitution as determined by potentiometric titration was prepared by reacting chitosan (Mw = 300,000 Da, supplied by Aokang Co. China) with chloroacetic acid according to the literature method. Sodium phosphate monobasic dehydrate, sodium phosphate dibasic dodecahydrate, sodium hydroxide, hydrochloric acid for preparing buffer solution were all provided by Aladdin Co., China, and used without further purification. Deionized water was double distilled for using in all synthesis and analysis experiments.

2.2 Sample preparation

The preparation of pure PVA hydrogels: PVA/water solution was prepared by adding PVA into the water and being stirred at high speed for 30 min. The different concentration of PVA solutions was prepared and transferred to a columnar glass vessel ϕ50* 100 mm and sealed. The solution was put into the refrigeration cycle machine to polymerization. The freeze-thaw cycle was carried out from −20~20°C, the heating and cooling speed was controlled at 0.2°C/min, and the temperature −20 and 20°C was kept for 1 hour. The freeze-thaw cycle was carried out for 5 times. In this experiment, the concentration of PVA solutions was fixed 5%, 10%, 15%, 20%, 25%. The gels were named Pm gels. The m is the concentration of PVA solution. The preparation of CMCS hybrid PVA hydrogels is similar to that of pure, the distinction is that the CMCS was added into the PVA solution when the transparent PVA solution was prepared, the mass of added CMCS was 3%, 5%, 8%, 10%, 12%, 15% of water respectively. The resulting gels were named as Cn-Pm gels, the n and m correspond to the concentration of CMCS and PVA in water.

2.3 Characterization

Tensile mechanical measurements were performed on Cn-Pm gels of the same size ϕ5.5 mm *80 mm length using a UTM6000 universal mechanical tester (Suns Inc). The sample length between the jaws was 35 mm, and the crosshead speed was 100 mm/min, the initial cross section 23.75 mm² was used to calculate the tensile strength and the tensile modulus, the tensile strain was taken as the length change relative to the initial length of the sample. Compression tests were carried out by using samples of the same size (ϕ12 mm*10 mm), on the UTM6000 universal mechanical tester (Suns Inc).

The compression properties of all gels were obtained under the following conditions: compression speed, 10 mm/min, compression distance, 9 mm (90% strain).

3 RESULTS

3.1 The effect of CMCS content on the mechanical properties on gels

For a clear expression of the influence of CMCS content on the gels, the concentration of PVA at water solution was fixed at 10%. The change of tensile strength and modulus with the increased CMCS was shown in Figure 1. As shown in the

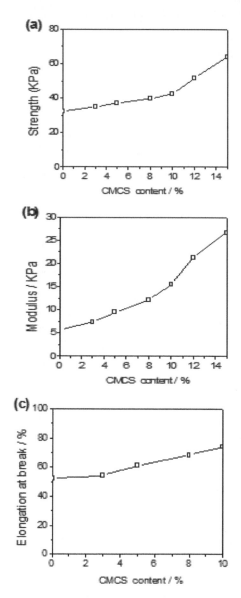

Figure 1. The tensile properties of Cn-P10 gels. (a) tensile strength; (b) tensile modulus; (c) tensile elongation at break.

figure, the strength and modulus show a significant enhancement, which is dependent on the content of CMCS. When n is higher than 10, the uptrend of strength and modulus intensifies drastically. It is may attributed to that the CMCS macromolecular chains in gels act as the role of linking bridge which loads the tensile stress. This may be attributed to the effective entanglement of CMCS around the PVA skeleton. Meantime, the elongation at break increases slightly, which may be ascribed to that the entanglement of CMCS around PVA network plays a crosslink effect. However, it could be concluded that the addition of CMCS is conductive to the improvement of the gels' mechanical properties.

On the other hand, as shown in Figure 2, the compression strength still shows obvious enhancement than CMCS free gels, and the strength is directly proportional to the CMCS content. When the concentration of CMCS is 10% of water, the compression strength is double of that of pure PVA hydrogels, revealing good pressure resistance. It could be explained by that the entanglement of CMCS marcromolecular chains around polymer network strengths the interaction of polymer chains and disperses the stress.

3.2 The effect of PVA concentration on the mechanical properties of gels

The concentration of PVA solution was controlled in polymerization and affected the mechanical properties directly. Figure 3 shows the tensile properties of gels with different concentration of PVA. In this research, the concentration of CMCS was fixed at 10% of water. As shown in Figure 3, with the increase of PVA, the strength and modulus of gels increases gradually, but the elongation at break of gels decreases drastically. This phenomenon is

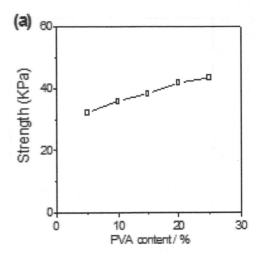

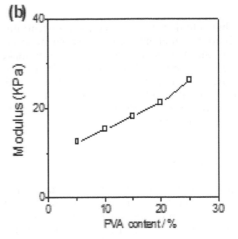

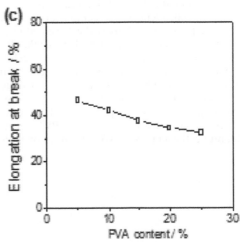

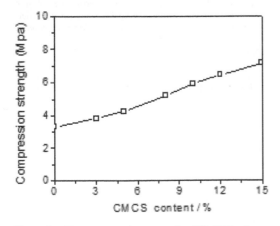

Figure 2.   The compression strength of Cn-P10 gels.

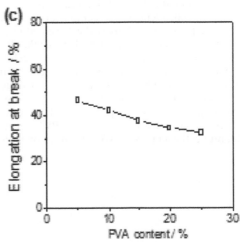

Figure 3.   The tensile properties of C10-Pm gels. (a) tensile strength; (b) tensile modulus; (c) tensile elongation at break.

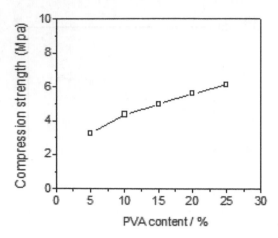

Figure 4. The compression strength of C10-Pm gels.

attributed to that the increased PVA concentration leads to a tighter interaction between molecular chains, through the higher interaction force makes the gels become hard to tensile, while it causes the stress concentration at the same time. On the other hand, it could be found from Figure 4 that the compression strength was effect slightly by the different concentration of PVA.

4   CONCLUSION

A novel linear Carboxymethyl Chitosan (CMCS) hybrid organic nanocomposite hydrogels were synthesized by introducing water-soluble CMCS into the PVA hydrogels. From the research of hybrid gels, it was found that the mechanical properties are obviously dependent on the composition of gels. With the increase of CMCs, the tensile strength and modulus increases significantly while the elongation break increases slightly. The gels exhibit higher compression strength than that of pure PVA hydrogels. This phenomenon is attributed to the effective entanglement of linear macromolecular chains around the polymer skeleton which enhances the interaction between PVA chains. Meantime, the increased PVA concentration is also conducive to the strength of gels, but the elongation at breaks of gels decreases.

ACKNOWLEDGEMENT

We acknowledge the financial supports of the National Natural Science Foundation (21104017), Hunan Province Natural Science Foundation (2015JJ4021).

REFERENCES

Demirel, G. Rzaev, Z. Patir, S. & Pişkin, E. 2009. Poly (N-isopropylacrylamide) Layers on Silicon Wafers as Smart DNA-Sensor Platforms, Journal of Nanoscience and Nanotechnology. 9: 1865–1871.
Kin, J.H. Kim, Y.K. Arash, M.T. Hong, S.H. Lee, J.H. Kang, B.N. Bang, Y.B. Cho, C.S. Yu, D.Y. Jiang, H.L. & Cho, M.H. 2012. Galactosylation of Chitosan-Graft-Spermine as a Gene Carrier for Hepatocyte Targeting In Vitro and In Vivo, Journal of Nanoscience and Nanotechnology. 12: 5178–5184.
Lim, S. Yang, D. & Lee, S.W. 2010. Fabrication of Magnetite-Hydrogel Nanocomposites with Clustered Magnetite Cores and Poly (N-isopropylacrylamide-co-acrylic acid) Shells for Drug Delivery Application, Journal of Nanoscience and Nanotechnology. 10: 7295–7299.
Millon, L.E. Nich, M.P. Hutter, J.L. & Wan. W. 2007. SANS Characterization of an Anisotropic Poly (vinyl alcohol) Hydrogel with Vascular Applications, Macromolecules, 40: 3655–3662.
Sivakumaran, D. Maitland, D. & Hoare, T. 2011. Injectable Microgel-Hydrogel Composites for Prolonged Small Molecule Drug Delivery, Biomacromolecules, 12: 4112–4120.
Wan, W.K. 2002. Optimizing the tensile properties of polyvinyl alcohol hydrogel for the construction of a bioprosthetic heart valve stent, Journal of Biomedical Materials Research. 63: 854–861.
Wang, H. Sun, X.Z. & Seib, P. 2002. You have full-text access to this content Mechanical properties of poly (lactic acid) and wheat starch blends with methylenediphenyl diisocyanate, Journal of Applied Polymer Science. 84: 1257–1262.
Zhang, Z.H. Abbad, S. Pan, R.R. Waddad, A.Y. Hou, L.L. Lv, H.X. & Zhou, J.P. 2013. N-Octyl-N-Arginine Chitosan Micelles as an Oral Delivery System of Insulin, Journal of Biomedical Nanotechnology. 9: 601–609.

Advanced Materials and Structural Engineering – Hu (Ed.)
© *2016 Taylor & Francis Group, London, ISBN 978-1-138-02786-2*

The study on mechanical properties of semi-IPN hydrogels consisting of crosslinked PNIPA network and linear Carboxymethyl Chitosan

J.P. Luo
College of Science, Hunan University of Technology, Zhuzhou, Hunan, China

H.Q. Hu
Institute of Machinery, Hunan University of Technology, Zhuzhou, Hunan, China

X.H. Zeng
Institute of Packaging and Materials Engineering, Hunan University of Technology, Zhuzhou, Hunan, China

ABSTRACT: The linear Carboxymethyl Chitosan (CMCs) was added into the PNIPA hydrogels crosslinked by BIS in situ radical polymerization to prepare a novel semi-IPN hydrogels. The effect of the content and BIS on the mechanical properties of prepared gels (named as C-PN gels) was researched. The results showed that the content of CMCs and BIS makes a non-ignorable and similar effect on the mechanical properties of gels. With the increase of CMCs and BIS content, the tensile strength, modulus and compression strength increases significantly while the elongation at break shows unobvious change. The phenomenon is attributed to the tighter network structure in gels.

1 INTRODUCTION

Hydrogels are hydrophilic three-dimensional polymer networks capable of absorbing a large volume of water or other biological fluid. Stimuli-sensitive hydrogels have the capability to change their swelling behavior, permeability or mechanical strength in response to external stimuli, such as small changes in pH, ionic strength, temperature and electromagnetic radiation (Peppas et al. 2006, Perale et al. 2011, Hou et al. 2008). Because of these useful properties, hydrogels have numerous applications, and they are particularly used in the medical and pharmaceutical fields.

Poly (N-isopropyl acrylamide) (PNIPAAm) hydrogel has been extensively studied as an intelligent polymeric matrix. The reversible phase transition of PNIPPAm hydrogel can be induced by a small external temperature change about its lower critical solution temperature (LCST, 33°C) in aqueous media (Takigawal et al. 2000, Zhang & Chu 2005, Kaneko et al. 2005). When the external temperature is below the LCST, the hydrogel hydrates and absorbs plenty of water, but it dehydrates quickly at temperatures above its LCST. Because of this unique property, significant attention has been focused on its application in the biotechnology and bioengineering fields. However, the poor mechanical properties and response rate caused by chemically crosslink limits its application significantly. Various modification methods were used to improve the properties of PNIPA hydrogels. Among all of the methods, adding other polymer chains into the gels to prepare semi-interpenetrating network gels is an effective method to improve the mechanical properties and increase the responsiveness.

Chitosan is kind of biodegradable and biocompatible polymer, which has been investigated as a novel biomaterial. However, the application of chitosan was limited for its insolubility at neutral or high pH region. To improve the solubility of chitosan, a series of hydrophilic groups have been introduced into its skeleton. Carboxymethyl Chitosan (CMCs), is a water-soluble derivative of chitosan by introducing $-CH_2COOH$ groups onto $-OH$ groups along chitosan molecular chain. For its unique chemical, physical, and biological properties, especially its excellent biocompatibility, the CMCs has been widely used in the biological and medical fields such as drug release, tissue engineering and so on. Moreover, it has demonstrated good pH and ion sensitivity in aqueous solution due to its abundant $-COOH$ and $-NH_2$ groups. Hence, a lot of researchers focused on using CMCs to prepare responsive hydrogels applied in drug release. For example, CMCS/alginate hydrogels, CMCs/poly (acrylic acid-co-acrylamide) and CMCs/PNIPA interpenetrating hydrogels were developed and investigated deeply.

In this research, the different content of CMCs was added into the PNIPA hydrogels crosslinked by BIS in situ radical polymerization to prepare semi-IPN hydrogels, the effect of CMCs content on the mechanical properties of gels were researched systematically.

2 EXPERIMENT

2.1 Materials

N-isopropylacrylamide (NIPA) provided by TCI Co. Japan, was purified by recrystallization from a toluene/n-hexane mixture (2/1 w/w) and dried under vacuum at 40°C. Crosslinker N, N'-methylene-bis-acrylamide (BIS) was provided by Aladdin Co, China. Carboxymethyl Chitosan (CMCs) with 1.34 degree of substitution as determined by potentiometric titration was prepared by reacting chitosan (Mw = 300,000 Da, supplied by Aokang Co. China) with chloroacetic acid according to the literature methodSodium phosphate monobasic dehydrate, sodium phosphate dibasic dodecahydrate, sodium hydroxide, hydrochloric acid for preparing buffer solution were all provided by Aladdin Co., China, and used without further purification. Deionized water was double distilled for using in all synthesis and analysis experiments.

2.2 Sample preparation

CMCs hybrid PNIPA hydrogels (C-PN gels) were synthesized by free-radical polymerization by using the method described previously. Briefly, monomer (NIPA), cross-linker BIS and deionized water were stirred at ice water temperature for at least 1 h to get a transparent solution. Then, the CMCs were added into the solution slowly with high speed continuous stirring. Next, the solution was ultrasonically oscillated for 20 min to be homogeneous and transparent. Finally, a solution with an initiator and a catalyst was added to the suspensions with stirring at ice water temperature for 5 min, pure nitrogen was bubbled, passing through the solution in all stirring processes. After the pretreatment, the solution was transferred to a columnar glass vessel φ5.5 * 120 mm and sealed. Free-radical polymerization was allowed to proceed in an ice water bath for 20 h. The prepared samples were immersed in deionized water at room temperature for at least 48 h, and the water was changed several hours to wash the unreacted monomer. The mass ratio of water/polymer ratio was fixed at 10/1 (w/w). The resulting hydrogels were named as Cn-PNm gels related to the gels, m and n correspond to the mole percent of BIS and mass percent of CMCs against NIPA monomer in gels.

2.3 Characterization

Tensile mechanical measurements were performed on C-PN gels of the same size φ5.5 mm * 80 mm length using a UTM6000 universal mechanical tester (Suns Inc). The sample length between the jaws was 35 mm, and the crosshead speed was 100 mm/min, the initial cross section 23.75 mm^2

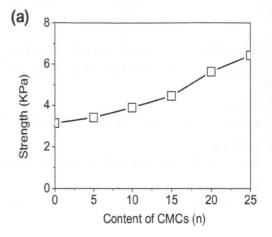

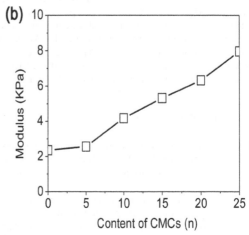

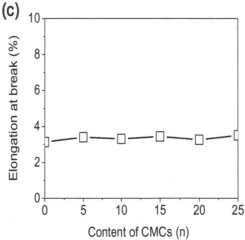

Figure 1. The tensile properties of Cn-PN4 gels. (a) tensile strength; (b) tensile modulus; (c) tensile elongation at break.

was used to calculate the tensile strength and the tensile modulus, the tensile strain was taken as the length change relative to the initial length of the sample. Compression tests were carried out by using samples of the same size (ϕ12 mm * 10 mm), on the UTM6000 universal mechanical tester (Suns Inc). The compression properties of all gels were obtained under the following conditions: compression speed, 10 mm/min, compression distance, 9 mm (90% strain).

3 RESULTS

3.1 The effect of CMCs content on the mechanical properties of gels

There is no doubt that the addition of linear molecular chains should affect the mechanical properties of gels. The tensile properties of Cn-PNm gels polymerized with different content of CMCs were shown in Figure 1. For a clear expression, the content of cross linker BIS was fixed at 4% of monomer (m = 4) As shown in Figure 1, the tensile strength and modulus shows an obvious increase with the increase of CMCs content. However, the elongation at break of gels is still low, which is easy to break in tensile. The enhanced strength of gels may be attributed to the effective entanglement of gels' skeleton and linear molecular, which shares the stress homogeneously, while, after the tensile, the entanglement loosen and the network break in C-C link, which is easier than other chemical bonds as the intrinsic of chemically cross linked gels.

Meantime, from the Figure 2, it could be found that the compression strength is improved drastically with the increase of CMCs. The enhanced strength is also attributed to the entanglement of CMCs and PNIPA network.

3.2 The effect of BIS content on the mechanical properties of gels

Except the content of CMCs, the crosslink density of gels is also very important to the mechanical properties of such semi-IPN gels. Figure 3 shows the effect of crosslinker BIS content on the

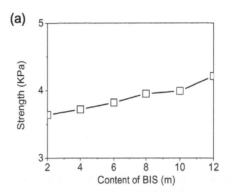

(a)

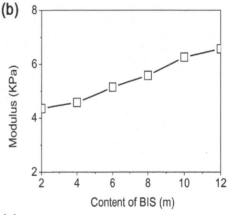

(b)

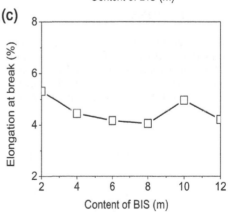

(c)

Figure 3. The tensile properties of C20-PNm gels. (a) tensile strength; (b) tensile modulus; (c) tensile elongation at break.

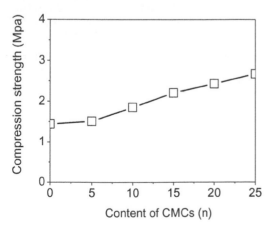

Figure 2. The compression strength of Cn-PN4 gels.

553

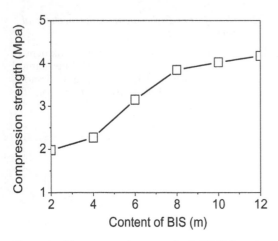

Figure 4. The compression strength of C20-PNm gels.

on the mechanical properties of gels. Comparing to the gels without CMCs, the resulting C-PN gels showed similar tensile and compression behavior. Moreover, with the increase of CMCs content, the tensile strength and modulus increases significantly due to the effective entanglement of linear macromolecule at the polymer network. Nevertheless, the elongation at break is still low. Meantime, the crosslink density is also affected the gels, with the increase of BIS, the tensile and compressive strength improved obviously.

ACKNOWLEDGEMENT

We acknowledge the financial supports of the National Natural Science Foundation (21104017), Hunan Province Natural Science Foundation (2015 JJ4021).

mechanical properties of gels, the content of CMCs is fixed at 20% mass ratio (n = 20) of NIPA. As shown in Figure 3, the tensile strength and modulus increases significantly with the increase of BIS, while the elongation at break of gels decreases. This could be reasonably explained that when the crosslink density increase, the network structure become tighter, which is helpful to the entanglement of CMCs molecular chains around the gels' network. However, the increased crosslink density makes the stress become more concentrated which is not conducive to the elongation at break in tensile obviously. Moreover, the compression strength of gels shows in Figure 4 exhibits an increasing trend with the increase of BIS content. The tighter network structure disperses the compression stress significantly.

4 CONCLUSIONS

In this paper, the different content of linear Carboxymethyl Chitosan (CMCs) was added into the PNIPA hydrogels crosslinked by BIS in situ radical polymerization to prepare semi-IPN hydrogels, the content of CMCs makes a non-ignorable effect

REFERENCES

Hou, Y.P. Matthews, A.R. Smitherman, A.M. Bulick, A.S. Hahn, M.S. Hou, H.J. Han, A. & Grunlan, M.A. 2008. Thermoresponsive nanocomposite hydrogels with cell-releasing behavior. *Biomaterial.* 29: 3175–3184.

Kaneko, T. Asoh, T.A. & Akashi, M. 2005. Ultrarapid molecular release from poly (N-isopropylacrylamide) hydrogels perforated using silica nanoparticle networks. *Macromolecular Chemistry and Physics*, 206: 566–574.

Peppas, N.A. Hilt, J.Z. Khademhosseini, A. & Langer, R. 2006. Hydrogels in biology and medicine: from molecular principles to bionanotechnology. *Advanced Materials.* 18: 1345–1360.

Perale, G. Rossi, F. Santoro, M. Marchetti, P. Mele, A. Castiglione, F. Raffa, E. & Masi, M. 2011. Drug release from hydrogel: a new understanding of transport phenomena. *Journal of Biomedical Nanotechnology.* 7: 476–481.

Takigawal, T. Araki, H. Takahashi, K. & Masuda, T. 2000. Effects of mechanical stress on the volume phase transition of poly (N-isopropylacrylamide) based polymer gels. *The Journal of Chemical Physics.* 113: 7640–7645.

Zhang, X.Z. & Chu, C.C. 2005. Fabrication and characterization of microgel-impregnated, thermosensitive PNIPAAm hydrogels. *Polymer.* 46: 9664–9673.

Advanced Materials and Structural Engineering – Hu (Ed.)
© *2016 Taylor & Francis Group, London, ISBN 978-1-138-02786-2*

The study of the transmission mechanism of landslide-generated waves and the research of the landslide surge effect on dam

S.L. Wu, H. Peng, J.L. Zhang, M. Li, J. Deng, G.D. Ma & Q.Q. Jiang
College of Hydraulic and Environmental Engineering, Three Gorges University, Yichang, Hubei, China

ABSTRACT: As the global environment has become increasingly severe, extreme natural disasters will cause a downturn of the slope with high speed in reservoir storage so that it will lead to surges. As a result of the propagation and evolution of surge in the river, which are regarded as a high-speed flow of water and a shock wave, it will threaten the safety of the dam structure. Therefore, the research of the security problems of dam caused by the landslide-generated waves is of great significance. In allusion to the problems that most of the time, dynamic water loads were obtained by the empirical formulas and but were not based on the exploration of the actual effect of dynamic water on the dam from the evolutionary mechanism of wave propagation in the existing specifications, first, an efficient domain decomposition algorithm to construct a new calculation model of water wave is proposed. Second, combined with the experiments of the numerical simulation and the physical model, the generation and propagation mechanism of the landslide surges in similar semi-infinite long channel bends under conditions of the high water level of reservoir is studied in depth and the accuracy of the calculation model of landslide surge is verified. Finally, the transmission mechanism of landslide-generated waves and the research of the landslide surge effect on dam from a qualitative and quantitative point of view are discussed.

1 INTRODUCTION

The hydropower project is not only a large project connecting with the people's livelihood, but also a construction project that involves an extremely large cost. Once hydropower project is wrecked, it would bring not only an enormous loss of hydropower projects to the people, but also an unpredictable disaster to the people, thereby threatening their personal safety. For instance, it led to more than 3,000 deaths and destroyed several towns and one city because of the Italy Vajont arch dam accident, which was due to surge triggered by the upstream geological disasters (Wang & Yin 2003). Therefore, under the condition of extreme precipitation, it is a significant theoretical and practical importance to study the propagation evolution of landslide surge at a high water level in the river, the damage destruction of landslide surge on the downstream of the dam for the rational development of hydropower resources and the comprehensive management of the reservoir area. Many scholars have studied seismic coupled dynamic problems of the dike—the base of a dam-reservoir. In calculation, we usually simulate seismic loads based on the seismic acceleration and seismic intensity (Hong & Cong 2000). Many scholars consider the water as an additional quality, ignoring the compressibility of water. In fact, the internal friction of water and water reflection in the riverbed not only

have a certain velocity in the natural river due to air resistance but also spread energy by the superficial wave and the internal shock wave (pressure energy). The propagation speed of shock wave in the water is far more than the flow rate of the water on the basis of earthquake acceleration, so the shock wave and the kinetic energy of water cannot be ignored. In addition, at present, relevant specifications list some computing method of the dynamic water pressure, but do not list the specific computing method against the problem of hydrodynamic loads and shock wave, which are caused by the surge. So, we must thoroughly study the approach of the landslide surge propagation and its function on the dam.

After summarizing the advantages and disadvantages of previous methods against landslide-generated waves, this article established a new calculation model of wave dynamics by using a regional Splitting Operator algorithm of alternating iterative solution of Navier-Stokes equations (hereafter referred to as N-S equations) and Euler equations (Noda 1970). On this basis, ignoring the impact of different media inside the landslide mass and regarding the landslide mass as a fluid, which consists of a three-phase unsteady flow of water and air, to establish a new calculation model of landslide-generated waves can not only reflect the interactions of the landslide mass and water, but also consider the different characteristics of the landslide mass through the introduction of the concentration and composition of the equation. By studying the interaction of the surge and river and the upstream face dam, and exploring the propagation process and the evolution mechanism of landslide surge wave propagation in a similar semi-infinite bending river (freedom of reservoir river, downstream of the dam block), we have understood the transmission mechanism of landslide-generated waves through tests and mathematical methods, and proved the distribution of dynamic pressures of river flowing water on the existing dam water and the rule of the energy conversion of its contact surface wave.

2 ESTABLISHMENT OF N-S WAVE EQUATION MODEL

Combined with the data of the reservoir level about the channel river of the dam, first, the project uses the Euler equation to simulate the whole flow field and proposes a class of efficient algorithms in the local area of the landslide body contacting with the water for the feature of actual river whose shape is complex, and then obtains the flow field in the whole domain solution by alternately running Euler equations and N-S equations. While quoting the governing equation of concentration field and boundary conditions on the surface of the dam body, tracing of the free surface of the liquid uses concentration methods. In addition, the tracing of the free surface of relatively fluid uses the VOF method (Kamphis & Bowering 1971).

Under the condition that the precision is assured, the free surface of the liquid uses concentration methods to make programming easier.

The specific programs are as follows:

A solution of N-S equations and Euler equations

The decomposition form of the dimensionless operator of N-S equations is given by

$$\varphi_t + F_{x^k}^k(\varphi, x, t) = \frac{1}{R_e} G_{x^k}^k(\varphi, x, t) \qquad (1)$$

Ignoring the equation on the right-hand side, the Euler equation can be obtained as follows:

$$\varphi_t + F_{x^k}^k(\varphi, x, t) = 0 \qquad (2)$$

Quoting dynamic decomposition algorithm of operator, complete answers to N-S equations can be decomposed into the following form:

$$\varphi(x,t) = \lim_{n \to \infty} \left| e^{\frac{1}{n}F_{x^k}^k} e^{\frac{1}{nR_e}G_{x^k}^k} \right|^n \varphi_0(x,t) \qquad (3)$$

The division of domain decomposition is based on the geometric shapes of the object and computational requirements, so it can use the grid of the structure on the complex shape. Each sub-domain (or considering the problem of multiple objects, such as landslide, water, contact surfaces of gas) set a fitted grid, the numerical solution will alternately go on within each grid and then transmit information through the boundaries of the domain of the adjacent grid. As mentioned above, the overall flow field will have a sub-area solution to use the Euler equations, and its computational grid used is called the Euler grid (Quecedo & Paster 2002). According to the above criteria, it will mesh refinement to solve the N-S equations in the dominated region of viscous effects, and the refined net can be called the N-S grid. The computational sub-domain of each piece of N-S equations will be included in the computational domain of the corresponding Euler equation, and the N-S grid of the encryption will exchange information with crude (Euler) grid, which only contains its sub-domain.

According to Equation 3, when solving the Euler equations into N-S equations, they regard the numerical solution of points on the Euler mesh around the N-S grid as their boundary conditions at a given coupled time steps. In the case of boundary conditions that remain constant, it will push the solution of N-S equations to advance to the same time of the numerical solution of the Euler equation, which is a rounding of the N-S grid. According to the volume average, it will push the solutions of fine mesh in N-S equations to pass to the points of the corresponding coarse grid and then give calibration solutions of the Euler field (Jiang et al. 2005). So, it can use them to solve the initial field of Euler equations and push the Euler computing equation to advance to the end of the next coupled time step. Finally, in order to establish the way of alternative iteration solutions of Euler equations and N-S equations, it will obtain the local N-S equation.

3 THE RESEARCH OF THE CALCULATION MODEL OF LANDSLIDE SURGE

Based on the solution of the wave model of N-S equations, it can obtain the regularities of distribution of the concentration field to calculate convection equations of the concentration field after obtaining the velocity field. Then, based on the conservation of the mass coefficient and the smooth coefficient of the free surface in order to achieve the evolution and propagation characteristics of the N-S equation of water waves over time, it will adjust the concentration field and rebuild the density and viscosity coefficient of N-S equations according to the concentration and the initial air and water density, and viscosity. The free surface of the fluid of tracking is dependent on the size of the density value (Song & Xing 2009), which is given by the following equation:

$$\frac{\partial \Phi}{\partial t} + v\frac{\partial \Phi}{\partial x} = \frac{a - b\Phi}{1 + \Phi} + k\frac{\partial^2 \Phi}{\partial x^2} \qquad (4)$$

In the N-S equation, this project intends to regard as a concentration equation to track the free surface. This formula represents the concentration value of computing element, the flow rate of the direction, the direction of the main channel in the river, physical constants (prepared according to the corresponding literature), the diffusion coefficient, and the time (unit of time is generally hr according to the distance of the wave transmission and computing scale in real time to determine the unit of time).

The project regards landslide as a kind of non-fixed-length stream of three phases, which consists of fluids, gases, and water phase. Based on Equation 3, the governing equation of concentration field of landslide can be introduced as follows:

$$\frac{\partial \Phi_i}{\partial t} + v\frac{\partial \Phi_i}{\partial x} = \frac{a - b\Phi_i}{1 + \Phi_i} + k\frac{\partial^2 \Phi_i}{\partial x^2} \qquad (5)$$

In addition, a three (landslide, water and gases) component equation is given by

$$\sum_{i=1}^{3} \Phi_i = 1 \qquad (6)$$

Landslide shape and physical parameters (such as density and viscosity) can be obtained according to the site survey of actual landslide and physical model tests (see Fig. 1). Density and viscosity, diffusion coefficient of the air and the gas can be determined based on the relevant information (Heinrich 1992). Its calculation process is similar to the calculation process of the wave model of N-S equations. The tracking of the free surface of fluids is based on the change in the concentration field of water bodies and gases.

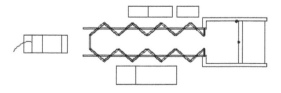

Figure 1. A plan view of the model of landslide surge.

4 THE VERIFICATION OF THE MODEL OF THE LANDSLIDE SURGE

In order to confirm the reliability of the results and obtain the change in swell through the curve and the reflection law of swell, interior tank modeling is designed for the model test of landslide surge. This model is a combination of the two boxes. In order to change the physical characteristics (thickness, shape features) and the speed of landslide, one method is to set landslide and ramps whose angles range from 30° to 90°, another method is to set up a river channel and a dam model in accordance with a certain valley shape and proportions (Watts & Grilli 1999). Wave propagation is recorded, which is due largely to a process that different slope masses are thrown into the water at different speeds through cameras on trail such as surge waveform, wave height and passed time while taking advantage of pressure sensors that are laid in the dam modeling to measure the amount of pressure on dams shaken by wave under the condition of different water levels. The test device could not only simulate different surge processes and the biggest surge height, which are due largely to a process that different slope angles and shapes of landslide are thrown into the water, but also measure different transmission rules of swell in a straight river channel or a bend of river (Raney & Butler 1976). Based on the propagation characteristics of swell and the variations of pressure obtained by sensors along with time and combining with the energy conservation law, the transmission mechanism of landslide-generated waves is inquired and the mechanism of energy conversion is explored.

5 CONCLUSIONS

Based on the solution of the wave model of N-S equations, the calculation model of landslide-generated waves is entirely feasible, combining indoor physical model tests and field parameters measured. It has formed certain results based on the propagation law of landslide-generated waves and the random nature of the shock wave on dam as follows:

1. When the slider just enters the water, its form swells. Then, the surge advancing to the surrounding will rapidly decay and the decline in waves becomes gradually smaller with the increase in its propagation distance. The decline in wave height is larger within about 1500 meters. Until 2.8 s, the free surface does not change significantly. After the production of the first wave, many inferior waves will be produced, but the height of the first wave is the

maximum. The results and the measured data are in good agreement. Therefore, during the prevention of the surge disaster, although the first wave will cause the greatest harm on both sides of the building, harms caused by inferior waves cannot be ignored.

2. The wave height is the maximum in the process of surge propagation in front of the slider when the wave height on both sides decreases gradually as the angle increases. The energy decay of surge propagation is slow. The ratio of generation to the decay rate of swell is in the range between 2 and 2.5. This result is of great significance for the generation and transmission range of the swell forecast. Once landslides occur, huge swells will spread along the coast and the rates of their decay will be very slow. It will cause a serious threat and loss to the residents on both sides and ships on the channel.

3. The sensitivity of the size of the maximum surge height factors affected in the following order: depth, the speed of landslide, the width of landslide, the length of landslide, and the thickness of landslide by the orthogonal experimental model analysis of variance. Therefore, in the swell forecast, the impact of these factors should be considered as the focus of the test results.

The dam construction, impoundment and operation is a complex and long-lasting process. In this process, the dam will be affected by many factors. The strength of different dam material, stiffness and resistance will decay and vary damage accumulation, and then will eventually lead to the reduction in its performance and life, or even to damage (Cooley & Moin 1976). The concern is the cracking of concrete in tension. Because tensile and compressive properties of concrete vary widely, subsequent research for the multi-parameter model of concrete materials will be the research emphasis on the project for the dam.

ACKNOWLEDGMENTS

This research was supported by the NSFC (National Natural Science Foundation of China, Grant No. 51379108) and the National Key Basic Research Development Program (973 Program) Sub-project (Grant No. 2013CB036401-2).

REFERENCES

Cooley, R.L. & Moin, S.A. 1976. Finite Element Solution of Saint-Venant Equations, *Journal of the Hydraulics Division*, 102(6): 759–775.

Harbitz, C.B. Pederson, G. & Gjevik, B. 1993. Numerical Simulations of Large Water Waves due to Landslide, *Journal of Hydraulic Engineering.* 119(12): 1325–1342.

Heinrich, P. 1992. Nonlinear Water Waves Generated by Submarine and Aerial Landslides, *Journal of Waterway, Port, Coastal and Ocean Engineering,* 118(3): 249–266.

Hong, W. & Cong, H.E.I. 2000. Numerical simulation and application of wave landslide in reservoir, *Journal of Water Resources of North China,* 1: 24–27.

Jiang, Z. Hanand, J. & Cheng, Z. 2011. Numerical simulations of water waves due to landslides, *Dams and Reservoirs under Changing Challenges,* Print ISBN: 978-0-415-68267-1.

Kamphis, J.W. & Bowering, R.J. 1971. *Impulse waves generated by landslides,* Proceedings Coastal Engineering Conference ASCE. *Reston.* 575–588.

Noda, E. 1970. Water waves generated by landslides, *J. of Waterways, Harbors, and Coastal Engineering Division,* 96(4): 835–855.

Quecedo, M. & Paster, M. 2002. A reappraisal of Taylor-Galerkin algorithm for dry-wetting areas in shallow water computations, *International Journal for Numerical Methods in Fluids,* 38(6): 515–531.

Raney, D.C. & Butler, H.L. 1976. Landslide Generated Water Wave Model, *Journal of the Hydraulics Division.* 102(9): 1269–1282.

Song, X.Y. & Xing, A.G. 2009. Two-dimensional numerical simulation of landslide waves by Flunt, *Hydrogeology and Engineering Geology.* 3: 90–94.

Wang, Y. & Yin, K.L. 2003. Analysis of movement process of landslide in reservoir and calculation of its initial surge height, *Earth Science Journal of China University of Geosciences.* 28(5): 579–582.

Watts, P. & Grilli, S.T. 1999. Modeling of waves generated by a moving submerged body. Application to underwater landslides, *Engineering Analysis with Boundary Elements.* 23 (8): 645–656.

Advanced Materials and Structural Engineering – Hu (Ed.)
© 2016 Taylor & Francis Group, London, ISBN 978-1-138-02786-2

Mechanical behaviors of three-dimensionally free-form titanium mesh plates for bone graft applications

M. Watanabe & J. He
Kogakuin University, Shinjyuku-ku, Tokyo, Japan

S. Suzuki
New X-national Technology K.K., Bunkyo-ku, Tokyo, Japan

ABSTRACT: Present metal artificial bones for bone grafts have problems such as too heavy and excessive elastic modulus compared with natural bones. In this study, three-dimensionally (3D) free-form titanium mesh plates for bone graft applications were introduced to overcome such problems. Fundamental mesh shapes and patterns were designed with different base shapes and design parameters using three-dimensional CAD tools with higher flexibility and strength points of view. Based on the designed mesh shape and patterns, sample specimens of titanium mesh plates with different base shapes and design variables were manufactured by laser processing. Tensile properties of the sample titanium mesh plates such as volume density and tensile elastic modulus were experimentally and analytically evaluated. The experimental results showed that such titanium mesh plates had much higher flexibility and their mechanical properties could be controlled to be similar to those of the natural bones. More details on the mechanical properties of titanium mesh plates including compression, bending, torsion and durability will be carried out in future study.

1 INTRODUCTION

Tailor-made tricalcium phosphate bone implants fabricated using a 3D ink-jet printer and used in clinical trial, shown in Figure 1, have mechanical properties such as very poor stiffness and fracture strength, and have been applied only for the non-load implant cases (Tessier et al. 2005, Eppley et al.

2005, Igawa et al. 2010). Titanium plate implants can be used as the reinforcements of such tailor-made tricalcium phosphate bone implants for many implant cases. On the other hand, the present titanium plates for implant applications have problems such as too heavy, mismatch-elasticity and excessive-strength compared with natural bones (Satoh et al. 2011, Wakui et al. 2012, Seki et al. 1994). In this study, improved titanium mesh plates with higher 3-dimensional flexibility (Abiko et al. 2011) were designed to solve such kind of problems.

Fundamental mesh shapes and patterns were designed under different base shapes (triangle, quadrangle and hexagon) and design parameters using three-dimensional CAD tools with higher flexibility and strength points of view. Based on the designed mesh shapes with different base shapes (triangle, quadrangle and hexagon) and mesh line widths, sample specimens of titanium mesh plates were manufactured through the laser beam machining for experimental and analytical studies.

Tensile experiments on sample titanium mesh plates were carried out to evaluate the effects of design parameters on the mechanical properties such as tensile elasticity and volume density. On the other hand, analytical approaches to the mechanical properties of titanium mesh plates were also

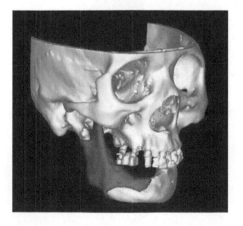

Figure 1. Bone implants fabricated from a 3D ink-jet printer.

carried out using the ANSYS finite element analysis code. Comparisons between the experimental and analytical results were made to validate the analytical approach method.

2 FUNDAMENTAL MESH SHAPE DESIGNS FOR 3D FREE-FORM TITANIUM MESH PLATES

To solve the too heavy, mismatch-elasticity and excessive-strength problems compared with natural bones in the present metal plate implants, mechanical properties of improved plate implants need to be similar to those of the natural bones. Then, the fundamental mesh shapes are considered under the following structural design conceptions:

1. Single fundamental mesh shape construction for simplification of manufacturing processing and cost-down purpose;
2. Higher three-dimensional flexibilities including expansion/contraction, bending and torsion for possibility of handily shape changes during surgery;
3. Easy-controllable mechanical properties such as elastic modulus and bending stiffness for approachability to natural bone's mechanical properties;
4. Desired uniform mesh line width and non-angle smooth shape to avoid the stress concentrations and lead to higher strengths and longer operating life; and
5. Ensured optional spaces for screw fixing.

Figure 2 shows three basic mesh shapes with basic five design parameters introduced in this study based on the above-mentioned design conceptions.

It is possible to create different varieties of mesh shapes from the five basic design parameters, as shown in Figure 3. Other dimensions, shown in Figure 4, can be deduced from the following equations with the five basic design parameters.

$$R_2 = R_1 + D \tag{1}$$

$$\theta_1 = \frac{90(n-2)}{n} \tag{2}$$

$$\theta_2 = \frac{180}{n} - \theta' \tag{3}$$

$$R_3 = \left(R_2 + \frac{L}{2}\right)\sec\theta_1 - R_2 \tag{4}$$

$$R_3 = \left(R_2 + \frac{L}{2}\right)\sec\theta_1 - R_2 \tag{5}$$

$$R_4 = (R_1 + R_2)\cos\theta_3 + (R_2 + R_3)\cos\theta_2 + R_2 \tag{6}$$

Basic mesh shapes, mentioned above, are designed from the regular triangle, quadrangle

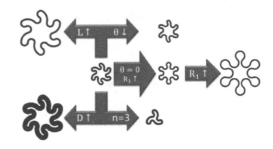

Figure 3. Different varieties of mesh shapes.

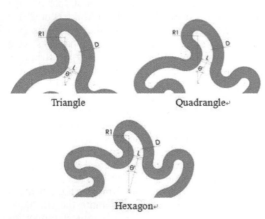

Figure 2. Mesh shapes with five basic design parameters: n: triangle = 3, quadrangle = 4 and hexagon = 6; D: mesh line width; R1: mesh space radius; L: minimum space length; θ: bending angle.

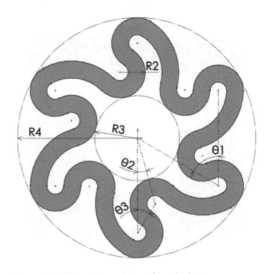

Figure 4. Other dimensions of mesh shapes.

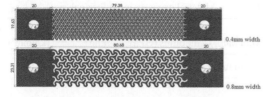

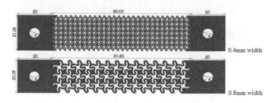

Figure 5. Meshed plate model from a triangle shape.

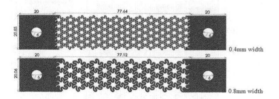

Figure 6. Meshed plate model from a quadrangle shape.

Figure 7. 3-dimensional meshed plate model from a hexagon shape.

and hexagon shapes; they have 120°, 90° and 60° axial symmetry with respect to the plate plane. Then, 3-deimensional meshed plate models using the designed mesh shapes with two different mesh line widths (0.4 mm and 0.8 mm) are obtained, and the CAD models for sample titanium mesh plates are shown in Figure 5 to Figure 7. From these models, we can see that the meshed plates have uniformed mesh line widths and smooth shapes according to the requirements of the design conceptions.

3 EXPERIMENTAL EVALUATIONS ON MECHANICAL PROPERTIES OF SAMPLE TITANIUM MESH PLATES

Based on the mesh shapes shown in Figure 5 to Figure 7, sample titanium mesh plate specimens with 0.6 mm plate thickness are fabricated for experimental evaluations by laser cutting processing. Six kinds of titanium mesh plate specimens are shown in Figure 8 to Figure 10. Volume densities of these sample titanium mesh plates are then obtained by dividing the specimen's total weight with the corresponding total unmeshed plate volume and taking the area ratio of the meshed part

to total specimens, and the results are shown in Figure 11. The comparison results of the volume densities, shown in Figure 11, indicate that the volume densities of sample titanium mesh plates are reduced to about 30% of same titanium plate and can be controlled to be similar to those of the natural bone ranging from 0.5 g/cm^3 to 1.1 g/cm^3.

Tensile properties of sample titanium mesh plates such as elastic stiffness and elastic modulus are evaluated through tensile tests. Figure 12(a) shows a typical measured load-displacement diagram of sample titanium mesh plates obtained from the triangle shape specimen's experiment, and Figure 12(b) shows the approached tensile elastic modulus of sample titanium mesh plates obtained by using the unmeshed cross-sectional areas. From these experimental results, we can see that titanium mesh plates show different tensile elasticity performances and reduce to the value by about 0.35% to 1.05% of the titanium plates (105.0 GPa). On the other hand, the elastic modulus of titanium mesh plates can be controlled to be similar to the natural cortical bones (2.0~30.0 GPa) by changing

Figure 8. Titanium mesh plates of triangle shape (0.4 mm width and 0.8 mm width).

Figure 9. Titanium mesh plates of quadrangle shape (0.4 mm width and 0.8 mm width).

Figure 10. Titanium mesh plates of hexagon shape (0.4 mm width and 0.8 mm width).

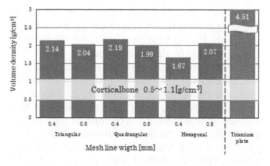

Figure 11. Volume densities of titanium mesh plates.

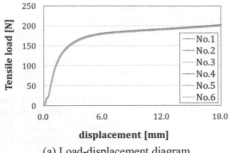

(a) Load-displacement diagram

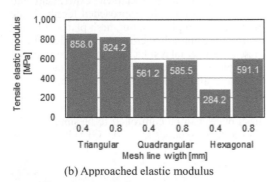

(b) Approached elastic modulus

Figure 12. Typical tensile experimental results (triangle 0.4 mm width specimen).

Table 1. Material property input for analysis.

Material	Elastic modulus (GPa)	Poisson's ratio	Density (kg/mm³)
Titanium	105.0	0.3	4.514

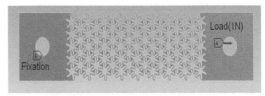

Figure 13. 3D analytical model on titanium mesh plate tensile tests.

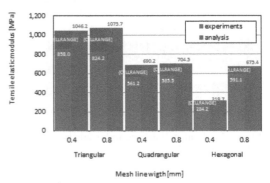

Figure 14. Comparisons on the tensile elastic modulus of titanium mesh plates.

different design variables such as mesh line width, plate thickness and base shapes.

4 ANALYTICAL APPROACHES ON TENSILE PROPERTIES OF SAMPLE TITANIUM MESH PLATES

Analytical approaches on the tensile experiments of the sample titanium mesh plates are carried out using Solid works 3D CAD software and ANSYS finite element analysis code. 3D meshed plate models with the same shapes and sizes of sample titanium mesh plate specimens on experimental evaluations are used, and material properties of titanium plate listed in Table 1 are used for analytical inputs. Figure 13 shows the loading and boundary conditions of the analysis model.

Comparisons with the experimental and analytical results on the tensile properties of sample titanium mesh plates are shown in Figure 14. From these results, it can be seen that more than 10% deviations between experimental and analytical tensile elastic modulus of titanium mesh plates exist, indicating that improvements of such analytical approach method were needed in further study.

5 CONCLUSIONS

In this study, experimental and analytical assessments of tensile properties on sample titanium mesh plates with higher 3-dimensional flexibility were executed, from which the following conclusions are obtained:

1. Different varieties of mesh shapes with fundamental design parameters can be obtained having the volume densities similar to the natural bone's densities.
2. Experimental results for the tensile properties of sample titanium mesh plates show the higher flexibility of titanium mesh plates compared with unmeshed titanium plates. Mechanical properties of such titanium mesh plates can be controlled to be similar to those of the natural cortical bones.
3. Analytical approach methods for tensile property evaluations of titanium mesh plates are introduced, and the comparisons between the

experimental and analytical results indicate that improvement of such analytical approach method will be done in future study.

REFERENCES

Abiko, R. Zama, K. Sinoda, T. Suzuki, S. & He, J. 2011. *A Primary Study on the Titanium mesh plate Structures for Reinforcement of Tricalcium Phosphate Bone Implants,* International Conference on Advanced Technology in Experimental Mechanics 2011 (ATEM'11, 2011), Kobe, Japan.

Eppley, B.L. Pietrzak, W.S. & Blanton, M.W. 2005. Allograft and alloplastic bone substitutes: a review of science and technology for the craniomaxillofacial surgeon, *Journal of Craniofacial Surgery,* 16(6): 981–989.

Igawa, K. & Chung U.I. et al. 2010. *Chin augmentation with inkjet-printed custom-made tricalcium phosphate implant,* Selected papers of the Congress of the European Association of Cranio-Maxillo-Facial Surgery, MEDIMOND/M914 C0611, 135–139.

Satoh, K. Murata, J. Tomioka, M. Seino, Y. Yokota, M. Mizuki, H. & Miura, H. 2011. Postsurgical stability following sagittal split ramus osteotomy with biodegradable Poly-L-lactide bone mini plate fixation, *Dental journal of Iwate Medical University,* 36(1): 46–52.

Seki, Y. Bessho, K. Sugatani, T. Kageyama, T. Inui, M. & Tagawa, T. 1994. Clinicopathological study on titanium miniplates, *Japanese Journal of Oral and Maxillofacial Surgery.* 40: 892–896.

Stryker Japan K.K., Meshed Plate 2014. (Approved documents of medical instruments), URL: http://www.info.pmda.go.jp/downfiles/md/pdf/730093_16200BZY01088000_A_LB_06.pdf, access on 16th Jan.

Tessier, P. Kawamoto, H. Matthews, D. Posnick, J. Raulo, Y. Tulasne, J.F. & Wolfe, S.A. 2005. Autogenous bone grafts and bone substitutes—tools and techniques: I. A 20,000-case experience in maxillofacial and craniofacial surgery, *Plastic and Reconstructive Surgery.* 116: 92S–94S.

Wakui, D. Nagashima, G. Ueda, T. Takada, T. Tanaka, Y. & Hashimoto, T. 2012. Three Cases of Scalp Rupture and Cosmetic Deformity caused by a Fixation Plate after Craniotomy, *Japanese Journal of Neurosurgery.* 21(2): 138–142.

Advanced Materials and Structural Engineering – Hu (Ed.)
© 2016 Taylor & Francis Group, London, ISBN 978-1-138-02786-2

Research on the basic mechanical properties of an axial pre-compressed piezo-electric bimorph actuator

K.M. Hu & L.H. Wen
School of Astronautics, Northwestern Polytechnical University, Xi'an, China

ABSTRACT: In order to obtain the static and dynamic characteristic behaviors of Post-Buckled Pre-compressed (PBP) piezoelectric bimorph actuators, simplified theoretical models were established and the PBP was simulated by finite element analysis. Furthermore, an experimental apparatus and platform were set up, based on which the static, dynamic and torque-measuring experiments were carried out to validate the theoretical predictions. It was shown that the experimental results and theoretical predictions were excellently correlated. Over 10° free peak-to-peak end rotation could be produced by the PBP actuator, which was more than three times of the available end rotation of the original bimorph under the same voltage, thus the design space of the actuator was increased significantly. The 1st natural frequency of PBP actuator reached up to 178 Hz, which was much higher than the bandwidth of conventional electromechanical sub-scale servo actuators. This work can provide the theoretical basis and test methods for the development of a PBP bimorph actuator prototype.

1 INTRODUCTION

Due to its advantages of the high bandwidth and low power consumption, the piezoelectric material has been predicted as the best choice for manufacturing aircraft drivers, with the premise of overcoming the defect of small strains (Wlezien et al. 1998). Recently, various methods to increase the displacement produced by piezo-drivers have been proposed, such as mechanical accumulation method (Loverich 2004), flexure hinge method (Yu et al. 2009), and leverage method (Zhang et al. 2009). However, none of these methods increases the electromechanical coupling coefficient.

With this background, Lesieutre G and Davis C proposed the LNPS (Low Net Passive Stiffness) technique, which can increase the electromechanical coupling coefficient by applying the axial compression force on the bimorph. When the axial compression force reached the buckling critical load, theoretically, the electromechanical coupling coefficient of the bimorph could reach 1, while that of a single piezoelectric patch was only 0.34 (Lesieutre & Davis 1997). This pre-compression method can increase the end rotation and output torque of bimorph actuators simultaneously. As a result, the design space of the bimorph, which is the available output torque and end rotation, was increased greatly compared with the conventional bimorph.

Based on this finding, a new class of Post-Bucked Pre-compressed (PBP) bimorph actuators was proposed by Barrett R in 2005 (Barrett et al. 2005).

Then, they were used successfully as the actuators of micro rotor aircraft, small UAV, miniature missiles and guided bullet (Vos & Barrett 2010, Barrett 2008). The end rotation of PBP bimorphs increased more than three times compared with the traditional bimorphs without affecting the output torque. Besides, they have the merits of low power consumption, high bandwidth control, less part count and small added mass; therefore, they are very suitable to be the servo actuator of the micro air vehicle.

In this paper, the static and dynamic characteristic behaviors of PBP actuators were analyzed theoretically and experimentally. This work can provide a theoretical basis and experimental methods for the development of a PBP bimorph actuator prototype.

2 ANALYTICAL MODELING

2.1 Analytical static model

A schematic representation of a piezoelectric bimorph actuator element is presented in Figure 1. It consists of two piezoelectric sheets that are adhesively bonded to each side of a substrate. The sizes along the 1st, 2nd, and 3rd directions are called as the width, length, and thickness of the element, respectively.

In the case of very thin adhesive layers, the deformation of transmission loss between an adhesive and the piezoelectric layer can be ignored. So, the

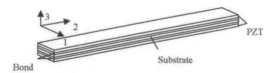

Figure 1. Structure of a piezoelectric bimorph.

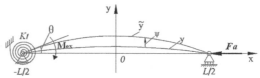

Figure 2. The analytical static model for PBP.

bimorph can be considered as a three-layer composite beam. Its bending stiffness EI can be evaluated by

$$EI = E_c t_c b \left(\frac{2}{3} t_c^2 + t_b t_c + \frac{t_b^2}{2} \right) + \frac{E_b b t_b^3}{12} \quad (1)$$

where b is the width of the bimorph; E_c is the elastic stiffness along the 1st direction of the piezoelectric sheet; E_b is the elastic stiffness of the substrate; t_c is the thickness of the piezoelectric sheet; and t_b is the thickness of the substrate.

The moment, which is produced by the piezoelectric effect with respect to the 1st axis, can be calculated by

$$M_\Lambda = E_c b \Lambda \int_{t_b/2}^{t_b/2+t_c} z dz - E_c b \Lambda \int_{-(t_b/2+t_c)}^{-t_b/2} z dz$$
$$= E_c b \Lambda \left[(t_b/2 + t_c)^2 - t_b^2/4 \right] \quad (2)$$

where $\Lambda = d_{31} \times E_3$ is the strain in the 1st direction, which is introduced by the electric-field strength E_3 in the 3rd direction.

The curvature in the 1st direction induced by the piezoelectric drive moment can be calculated by

$$\kappa = \frac{M_\Lambda}{EI} \quad (3)$$

The simplified model is shown in Figure 2. The loads applied to the beam consists of the axial force F_a and the external torque M_{ex} produced by a torsion spring with its torsion stiffness being K_t. y is the initial shape, which is caused by the piezoelectric effect. The ultimate deformed shape $\tilde{y}$ can be evaluated by

$$\tilde{y} = \kappa \frac{EI}{F_a} \left[\cos\left(\sqrt{\frac{F_a}{EI}} x \right) \Big/ \cos\left(\sqrt{\frac{F_a}{EI}} \frac{L}{2} \right) - 1 \right]$$
$$+ \frac{M_{ex}}{2 F_a} \left[\cos\left(\sqrt{\frac{F_a}{EI}} x \right) \Big/ \cos\left(\sqrt{\frac{F_a}{EI}} \frac{L}{2} \right) \right.$$
$$\left. - \sin\left(\sqrt{\frac{F_a}{EI}} x \right) \Big/ \sin\left(\sqrt{\frac{F_a}{EI}} \frac{L}{2} \right) \right] + \frac{M_{ex}}{F_a} \left(\frac{x}{L} - \frac{1}{2} \right)$$

$$(4)$$

The end rotation can be computed by

$$\theta \approx \frac{\kappa \sqrt{\frac{EI}{F_a}} \tan\left(\sqrt{\frac{F_a}{EI}} \frac{L}{2} \right)}{\left[1 + \frac{K_t}{F_a} \left[\frac{1}{2} \sqrt{\frac{F_a}{EI}} \left(\tan\left(\sqrt{\frac{F_a}{EI}} \frac{L}{2} \right)^{-1} \right. \right. \right.}{\left. \left. \left. - \tan\left(\sqrt{\frac{F_a}{EI}} \frac{L}{2} \right) \right) - \frac{1}{L} \right] \right]} \quad (5)$$

when M_{ex} and K_t are 0, $\tilde{y}$ and θ is the bimorph's end rotation under the free state.

Then, according to the variation of torsion stiffness, the relationship between the end rotation and the output torque can be obtained as follows:

$$M_{ex} = \theta(K_t) \cdot K_t \quad (6)$$

2.2 Analytical dynamic model

The 1st natural frequency of the analytical model for PBP can be calculated by (Vos & Barrett 2010, Schravendijk et al. 2009)

$$f = \frac{1}{2\pi} \sqrt{\int_0^L \frac{\left[EI(W''(x))^2 - F(W'(x))^2 \right] dx}{\left(\rho b t \int_0^L W(x)^2 dx \right)}} \quad (7)$$

where the shape function is $W(x) = C \times \sin(\pi/Lx)$ and the average density is $\rho = (2\rho_c t_c + \rho_b t_b)/(2t_c + t_b)$.

3 FEM MODEL OF THE PBP ACTUATOR

In order to verify the analytical formulations (5) and (7), the PBP's FEM model was developed using ANSYS® software. The solid 5 coupling field element was used for these two layers of piezoelectric sheets, and solid 45 was used for the substrate layer. The FEM model is shown in Figure 3. The width of the model is 10 mm; the length is 50 mm. The total thickness is 0.6 mm, with the thickness of each layer being 0.2 mm.

The properties of the materials in the FEM model are listed in Table 1.

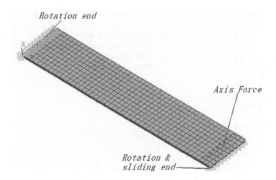

Figure 3. FEM model for the PBP actuator.

Table 1. Properties of the piezoelectric bimorph materials.

	Actuator sheet	Substrate
Material type	Pzt	Carbon fiber
Relative dielectric constant $\varepsilon^T_{33}/\varepsilon_0$	4400	
Piezoelectric charge constant d_{31} (pC/N)	−500	
Stiffness E (GPa)	40	10
Density ρ (kg/m³)	7600	1300

4 EXPERIMENTAL SETUP

A test bed was built to check the static and dynamic performance of the PBP actuator, as shown in Figure 4. The PBP actuator was pinned between an output shaft and a sliding shaft, allowing axial loads to be transferred to the PBP element. The axial force was regulated by changing the weight.

To measure the end rotation of the output shaft, a metal slice was fixed on the output shaft as a reflector. The laser displacement meter (Keyence® LK-G80) was mounted to measure the displacement Δ of the reflector in the x direction. Since the peak-to-peak end rotation of the output shaft is within 10°, it can be approximately computed by

$$\theta = arc\tan(\Delta/R) \qquad (8)$$

where R is the distance between the reflector and the axis of the output shaft.

To obtain the design space of the PBP actuator, a lever rod was fixed on the output shaft, with the other end tied to a tension spring, as shown in Figure 4. The output torque can be calculated by the stiffness of the spring, the deflection of the lever rod and the length of the lever.

The data acquisition system is shown in Figure 5. The system is composed of an industrial control

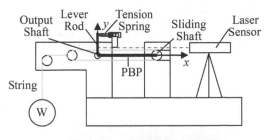

Figure 4. PBP actuator test bed.

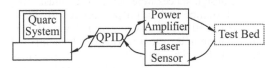

Figure 5. Schematic layout of data acquisition.

computer with Quanser®quarc/Simulink hardware-in-the-loop simulation system, a QPID data acquisition card, a power amplifier XMT®XE501-A600. A1, a laser sensor and the test bed.

5 RESULTS

5.1 Static results

The QDA60-10-0.6 bimorph actuators produced by Sinocera Piezotronics Inc were used as the specimens in our experiments. The size of the specimens is 60 mm × 10 mm × 0.6 mm. The length of the piezoelectric layer is 50 mm.

The end rotations of the actuator with various axial forces were measured with an applied sinusoidal voltage of ±90 V, 0.05 Hz. The results are shown in Figure 6. The free peak-to-peak end rotation can reach up to 10.13°, with the axial forces being 20.3 N, which is three times bigger than that produced by the bimorph without axial forces. The analytical and FEM results are in good conformity; the maximum relative error is 1%. This indicates that the analytical model can reflect the static characteristic of the actual PBP actuator accurately.

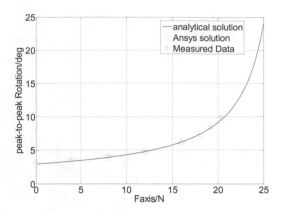

Figure 6. Numerical and experimental results of statics.

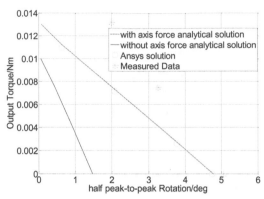

Figure 7. Numerical and experimental results of the design space.

The experimental results are slightly larger than the numerical results. It may be caused by the inaccurate material parameters in the numerical model and the axial forces error between the numerical model and the experiment.

The design spaces were achieved by illustrating the end rotation and the output torque with different torsion stiffnesses, as shown in Figure 7. The end rotation and output torque can be computed by the analytical model or the FEM analysis. Then, they are measured in the experiments, with the tension spring being one or two in parallel.

The theoretical predictions computed by Equation (5), FEM results and the experimental results are illustrated in Figure 7. It can be seen that the FEM results and the theoretical predictions of the output torque and the end rotation agree with each other very well, with the maximum misfit being about 2%. The block torque of the PBP actuator is 1.3 N·cm, slightly higher than the block torque of the original bimorph, which is 1 N·cm. The maximum end rotation reaches up to 5.1°, which is magnified 3 times. As a result, the design space of the bimorph actuator is increased more than three times. The divergence of the measured output torque and the numerical results are relatively high. This may be caused by relatively low values of the measured force and the nonlinear behavior of the tension spring under very low forces. The manufacturing errors of the spring and the experimental fixture may also be one of the reasons.

5.2 Dynamic results

Exposed to different axial loads, the linear sweep frequency tests of the PBP actuator were carried out, with the amplitude of the supplied sinusoidal voltage being 30 V and the sweep frequencies ranging from 1 Hz to 500 Hz. The harmonic responses

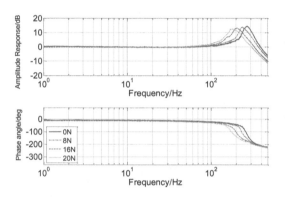

Figure 8. Normalized amplitude response and phase plot for various axial loads.

of the displacement of the mid-point of the PBP actuator are illustrated in Figure 8.

It is shown that the resonance frequency decreases with increasing applied axial force. This is expected since the equivalent bending stiffness will decrease when an axial force is applied, as shown in Equation (7). The magnification factor decreases, and the damping ratio increases with increasing applied axial force. From the phase-frequency curve, it can be observed that after the resonance frequency phase lag reaches up to −220°, it may be caused by the constant time delays in the measurement loop (Schravendijk et al. 2009).

The 1st natural frequencies of the PBP actuators under various axial forces are illustrated in Figure 9. It can be seen that the theoretical predictions computed by Equation (7) and the FEM results are in good correlation, with the maximum error being about 1.5%. When the axial force is less than 10 N, the experimental results are lower than the numerical results. While when the axial force

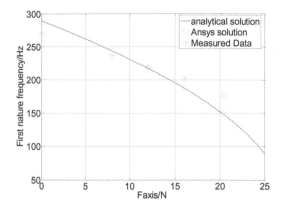

Figure 9. 1st natural frequencies of PBP actuators under various axial forces.

is larger than 10 N, the experimental results are higher than the numerical results, with the maximum error being 18%. This may be caused by the inaccuracy of material parameters in a numerical model. Anyway, in general, the experimental results conform well to the trend of the numerical results. In the case with the axial force being 20.3 N, although the 1st natural frequency of the PBP actuator drops to 178 Hz, it is still much higher than that of conventional sub-scale servo actuators due to the high control bandwidth of the PBP piezoelectric actuator.

6 CONCLUSIONS

The static and dynamic behaviors of PBP actuators are studied by theoretical analysis, FEM simulation and experiments. Based on the numerical and experimental results, the following conclusions can be drawn:

1. The numerical and experimental results show that under an axial force, the PBP actuator free peak-to-peak end rotation can reach up to 10.1° and the block output torque can increase slightly. Therefore, the design space of PBP actuator can increase more than three times of the original piezoelectric bimorph. The 1st natural frequency can reach up to 178 Hz, which is much higher than the bandwidth of conventional sub-scale servo actuators.

2. The experimental results coincide well with the numerical results, indicating that the test system established in this paper can be used as a test platform for the development of a PBP bimorph actuator prototype.

REFERENCES

Barrett, R. McMurtry, R. & Vos R. et al. 2005. *Post-Buckled Precompressed (PBP) Elements: A New Class of light Control Actuators Enhancing High Speed Autonomous VTOL MAVS*, In: PROC. of SPIE, San Diego, USA, 5762: 111–122.

Barrett, R. Vos, R. & DeBreuker, R. 2008. Post-Buckled Precompressed (PBP) Subsonic Micro-Flight Control Actuators and Surfaces, *Smart Materials and Structures*. 17: 055011.

Lesieutre, G. & Davis, C. 1997. Can a Coupling Coefficient of a Piezoelectric Device Be Higher Than Those of Its Active Material? *Journal of Intelligent Material Systems and Structures*. 8(10): 859–867.

Loverich, J.J. 2004. *Development of a New High Specific Power Piezoelectric Actuator*, PhD Dissertation, Pennsylvania State University.

Schravendijk, M. Groen, M. & Vos, R. et al. 2009. *Closed-Loop Control for High Bandwidth, High Curvature Post-Buckled Precompressed Actuators*, In: PROC. of AIAA, Palm Springs, USA, 2113.

Vos, R. & Barrett, R. 2010. Post-Buckled Precompressed Techniques In Adaptive Aerostructures: An Overview, *Journal of Mechanical Design*. 132(3): 031004-1-031004-11.

Wlezien, R.W. & Horner, G.C. et al. 1998. *The aircraft morphing program*, In: PROC. of AIAA, Long Beach, USA, paper, 1898–1927.

Yu, Z.Y. Yao, X.X. & Dai, R.Z. et al. 2009. Design of Micro-Displacement Amplifier of Piezoelectric Servo, *Acta Armamentarii*, 30(12): 1653–1657.

Zhang, X.L. Liang, D.K. & Lun, J.Y. et al. 2009. Design of Multiple Piezoelectric Bimorph Parallel Actuator, *Journal of Nanjing University of Aeronautics and Astronautics*, 41(3): 339–343.

Advanced Materials and Structural Engineering – Hu (Ed.)
© 2016 Taylor & Francis Group, London, ISBN 978-1-138-02786-2

Automatic control system for strengthening and polishing equipment

J.H. Tao, Z.F. Huang, Z.H. Guan, J.H. Zhang, Y.W. Bao & C.P. Jin
School Mechanical and Electrical Engineering, Guangzhou University, Guangzhou, China

ABSTRACT: This paper discusses the influences of strengthening and polishing methods on the workpiece surface quality, and presents the current existing problems of strengthening and polishing equipment, such as low machining precision, low degree of automation and low processing efficiency. Then, the design of a suitable control system for strengthening and polishing methods is proposed to solve these issues. This control system achieves the automatic processing by controlling four process parameters that affect the strengthening and polishing quality (moving speed, spray distance, injection time, and injection pressure).

1 INTRODUCTION

Strengthening and polishing technology is a precision machining technology based on the composite processing method, which can enhance the metal materials' characteristics, including anti-fatigue, anti-corrosion and anti-fraying. It is also a new machining method combining reinforcing elastic products processing and micro-abrasive cutting, which we call it "strengthening of grinding" (Liu 2011). The working principle is shown in Figure 1. The nozzle produces a three-phase hybrid jet, which consists of high strength steel balls, grinding fluid and abrasive powder by high pressure air injection system, to machine the workpiece surface. The model of strengthening and polishing is derived from the shot peening technology. On this basis, it combines the polishing technology and jet technology to form a new kind of processing technology. In this technology, the surface quality of workpiece is affected by main process parameters, such as injection pressure, injection velocity, and spray distance (HUST 2009). Based on these parameters, this article aims to design a

strengthening and polishing control system, which can be able to overcome the shortcomings of traditional shot equipment such as unsatisfied accuracy and low degree of intelligence.

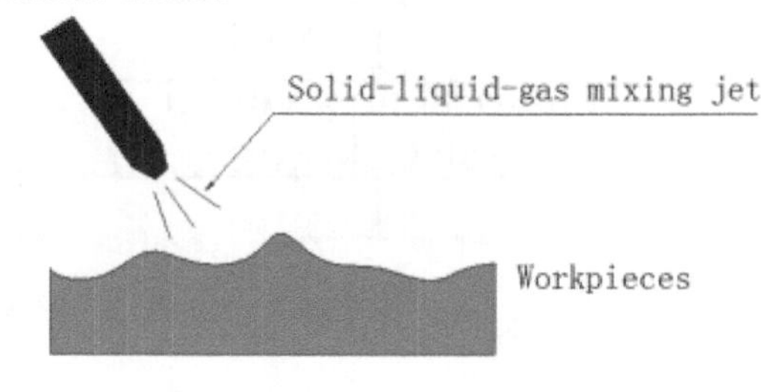

2 THE WORKING PRINCIPLE OF THE STRENGTHENING AND POLISHING CONTROL SYSTEM

In this paper, strengthening and polishing equipment is developed by the School Mechanical and Electrical Engineering, Guangzhou University, and its control system adopts Taiwan HUST company's HUST H6 general controller as the control core, which builds into the PLC module in the movement control cards. The equipment not only has the motion control card's characteristics of high-speed interpolation and rapid response, but also has the characteristics of flexibility and diversity of PLC in dealing with non-linkage (HUST 2009). The machining process is shown in Figure 2. The X, Y, and Z axes are the straight axis, which controls the nozzle's movement of the left and right, front and back, up and down, thereby regulating the control of the strengthening and polishing processing's spray distance; the A axis is the axis of rotation, which controls the jet angle in the process of machining; and the high pressure pump (not painted) compresses the air. By the air control valve, it jets the strengthening and polishing three-phase mixture injection at a high speed. In addition, the air control valve can control the injection pressure and injection speed. At the same time, in the control system, it uses the industrial PC computer as the upper computer. According to the need for processing, it generates the NC program automatically, and communicates with the controller to

Figure 1. Strengthening and polishing physical model.

Shower nozzle

Solid-liquid-gas mixing jet

Workpieces

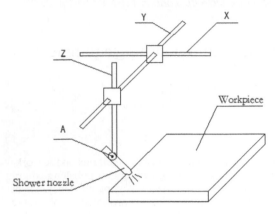

Figure 2. Strengthening and polishing mechanism sketch.

put the NC program into the control system, and then realizes the automation of processing. It is important to note that since the controller comes with the operation panel, the PC computer does not need to communicate with the controller in real time, and it can also communicate with more than one controller at the same time, thus reducing the cost of equipment of the factory.

3 THE OVERALL DESIGN OF THE STRENGTHENING AND POLISHING CONTROL SYSTEM

3.1 Functional requirements

1. *The system operation interface.* In the process of system running, there will always be a man-machine interface and the corresponding text prompts. Various operating modes can be switched by a simple button, thus it is very convenient to the operator of the non-professional. For example, if we want to pause in the process of workpiece machining, we just need to press the F key.
2. *The automatic generation of the processing path.* In the environment VC++ integrated development, it can design software that generates the path automatically according to the requirements of a system. The main functions of the software include providing a data interface with CAD drawing, generating the path automatically, and communicating with the control interface module. Then, achieving the function to generate NC programs automatically based on the existing CAD drawing and reduce labor costs.
3. *The alarm module.* It includes the stroke limit alarm, the software stroke limit alarm, the servo alarm, the fault alarm and the parameter setting error alarm.
4. *The other automatic control function.* It provides many functions, such as manual of cutter, the hand wheel testing, back to the mechanical origin, back to the knife point, automatic process, single step, I/O monitoring, processing and statistical, help, and other functions.

3.2 The flow chart

The strengthening and polishing machine's work process is shown in Figure 3. After the startup of

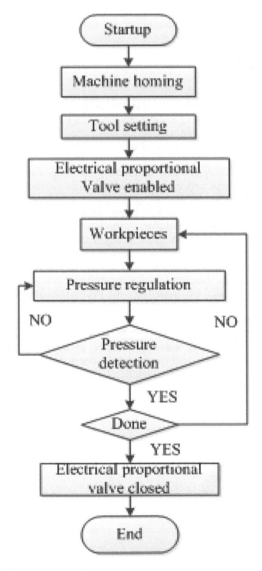

Figure 3. Workflow model.

the strengthening and polishing controller, first, it prompted a homing action. After the homing action, we can enter the parameters for the corresponding interface parameter settings. After setting the parameters, if we have previously been on the knife action, the knife returns back to the site; if we have never carried out on the knife action, the tool will need to be set manually. In order to ensure the safety during processing, the system designed from the beginning to add an electrical proportional valve enables function, adding electrical proportional valve enables the button in the operation panel; only in the case of the proportional valve enables can we carry out the injection process. During the machining process, it provided a pressure regulating feedback loop, detecting and real-time adjusting the processing of the injection pressure. After the process is complete, it enables closing of the electrical proportional valve.

4 THE ELECTRICAL CONTROL SYSTEM SOFTWARE DESIGN

The system uses the Taiwan HUST company's HUST H6 universal controller as a master key; it builds a PLC module in motion control cards. The controller has not only a high-speed interpolation motion control card and rapid response characteristics, but also the diversity and flexibility of PLC in handling the linkage (Tao & Fei 2013). By controlling the three-axis X, Y, Z-axis movement on the platform, AD/DA provide a voltage output, thereby controlling the output pressure of the electrical proportional valve to control the injection pressure. The operating system in the controller is responsible for the input data (e.g. machine parameters), the interpolation operation and position control, speed control and fault diagnosis, and other functions (Diao 2006).

The overall design of the hardware for the controller's electrical control system is shown in Figure 4. The HUST controller as the master core provides all major communication interfaces for external devices, such as communication interface RS232, through which programming the control system software to the controller, and it is also

the main communication interface of the upper computer and the MPG external interface. The axial interface allows a controller to control the servo axis. The I/O interface links the external I/O devices for input and output signals.

Strengthening and polishing method puts the strengthening and polishing mixture liquid to machine the workpiece surface by high pressure injection. This affects the surface quality of the main parameters such as the moving speed, spray distance, the injection time and injection pressure (HUST 2009). Moving speed, ejection distance and ejection time can be achieved by the motion control card through controlling the X, Y, Z movement, and injection pressure as an important parameter is not well controlled by previous strengthening and polishing equipment. For this reason, this system adopts the electric proportional valve to control the injection pressure. Electric proportional valve pressure output can be controlled stepless by the size of the voltage or current, and the HUST controller's AD/DA port provides a stepless voltage output. Therefore, the injection pressure can be controlled very precisely. To sum up, the important process parameters of the workpiece surface quality affected in the strengthening and polishing process can be directly controlled by the controller to achieve process automation and intelligent.

5 THE CONTROLLER AND THE UPPER COMPUTER COMMUNICATION

5.1 The electrical control system module

In this control system, the electrical control system software modules are shown in Figure 5. The electrical control system software is mainly divided into three parts: (1) the man-machine interface is composed of a different screen, which is written by editing software Screen-Editor. (2) CNC part

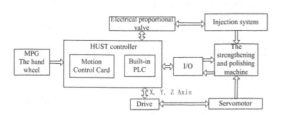

Figure 4. Overall design of the electrical control system hardware.

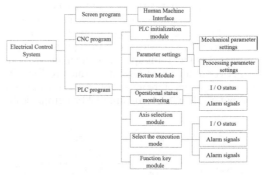

Figure 5. Block diagram of the electrical control system software.

program, this program is to realize the function of the processing path of the workpiece. In order to achieve the automatic generation of the path, this section has the PC communications import, which will be described in the upper computer module. (3) PLC functionality of the program modules, this part is written by HUST PLC-Editor, which mainly realizes the function that the controller controls the strengthening and polishing machine, including PLC initialization module, parameter setting module, the screen module, the operating status monitoring module, axis selection module, select the module and perform the F function key modules. Between the screening program, CNC and PLC programs interrelated to each other variables, and they can provide service for the HUST controller together.

5.2 *The upper computer module*

In this system platform, the upper computer is an automation module communicating with the controller, and its main functions are to automatically generate CNC programs, send CNC programs to control and real-time monitor the operating state of CNC. At the same time, according to the communication requirements, the HUST controller provides RS232 and USB data interface. It is easy to realize communications of the upper computer and controller.

When the upper computer communicates with the CNC system via the serial port, in general, the NC system is set to the host. The upper computer sends the communication request command to the CNC system. After receiving the information, the CNC system executes the corresponding action and returns the data needed by the upper computer (Diao 2006). In this control system, the HUST controller is the host and the PC computer is the upper computer. This section is divided into two parts: generation of NC programs and data interface. The process is shown in Figure 6.

From Figure 6, we can see that the data interface of the upper computer can be divided into two interfaces. One is the CAD graphics interface, which is to import CAD drawings into the upper computer systems. Another is the upper computer and the controller communication interface, which realizes the communications of the upper computer and the lower computer. The CNC program will be automatically imported into the controller. The generation of the CNC program is calculated from the CAD drawings, which is imported. In this system, we can use the VC ++ integrated development environment to achieve the development of the upper computer. By writing two interface programs, CAD drawings and control systems can be linked; at the same time, by writing a graphics program, the automatic generation of CNC programs can be achieved.

6 CONCLUSIONS

In this paper, the design of the strengthening and polishing automatic control system uses the HUST H6 universal controller as the control core. It has advantages as follows. (1) Highly intelligent. As long as the operators have working drawings, it could automatically generate toolpaths. (2) Simple operation. The operator can realize different operating modes simply by a few keystrokes to meet the processing needs. (3) Safe and reliable. In the control system, it is equipped with an alarm system and parameter setting error alarm system. (4) This control system achieves the automatic control of the pressure, overcomes the defects of manually pressure adjustment, which exists in the previous strengthening and polishing system, improves the accuracy of the strengthening and polishing process, and facilitates the promotion of this processing method.

ACKNOWLEDGMENT

This paper was supported by the Guangzhou Science and Technology Program (Grant No. 2013J4300009), the 2013 National University Students Innovation and Entrepreneurship Training Program of Local Universities (Grant No. 201311078022), and the Guangzhou Key Laboratory of High-performance Metal Materials Reinforced Grinding Machining ([2013]163-19).

REFERENCES

Diao, B. 2006. *Development of automatic material cutting system and milling machine system based on PC's management and CNC's control*, Nanjing: Southeast University.

HUST H6, 2009. *The NC wiring manuals*, HUST CNC Industry Corp.

Liu, C.J. 2011. *Finite Element Analysis and Equipment Development of Bearing Strengthening and Polishing Processing*, Guangzhou: Guangzhou University.

Tao, J.H. & Fei, X.J. 2013. Design of NC Tenoning Machine's Control System Based on HUST H2 controller, *Woodworking Mach.* 1: 13–16.

Tao, J.H. Bao, Y.W. & Liu, H. 2013. *Effect of Jet Pressure on Surface Roughness of Workpiece in Strengthening and polishing Process, Bearing*, 11: 30–33.

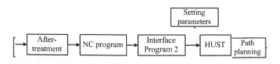

Figure 6.   The design module of the upper computer.

Advanced Materials and Structural Engineering – Hu (Ed.)
© *2016 Taylor & Francis Group, London, ISBN 978-1-138-02786-2*

Research on the dynamic simulation of armored vehicle's steering control based on man-machine interface matching

W.P. Liu, B.H. Fu & J.F. Nie
Department of Mechanical Engineering, Academy of Armored Forces Engineering, Beijing, China

ABSTRACT: According to data in GB and GJB, a steering control dynamic model of an armored vehicle driver was built by employing the Newton-Euler methodology. The driver's steering control was dynamically simulated on the simulating platform of Simulink. The factors influencing man-machine matching, such as human dimension and cabin layout, were analyzed. It was concluded that horizontal position parameters of the seat and the periscope should be given priority followed by the position optimization of the joystick when matching armored vehicle operation space. After considering both of the control comfort and accessibility, individuals with low height are more likely to be selected as drivers. Also, the control force should be decreased as far as possible. The method and result provide a reference for analyzing the cabin layout of armored vehicles and the driver's control comfort level.

1 INTRODUCTION

The safety and efficiency of vehicle's steering is closely related to the driver's control comfort level. Inversed kinematics was employed (Zhang & An 1993) to study the steering control comfort of armored vehicle by analyzing the law of each main link angle and driving posture during steering control. This method could determine the structural state of each limb; however, it only reflected the geometry compatibility between drivers and man-machine interface matching in cabin without revealing and fully reflecting the influence and reasons for driving posture on driver performance. Thus, this paper studied the driving and controlling problems of armored vehicles dynamically. A driver's control dynamic model was built by employing the Newton-Euler methodology. Driver's steering control was dynamically simulated and analyzed on the simulating platform of Simulink.

2 DYNAMIC MODEL

As is shown in Figure 1, the kinematic chain of the upper limb can be divided into 3 parts, namely the upper arm, forearm and joystick, which are connected by the shoulder, elbow and hand. After analyzing the kinds of typical armored vehicle cabin layout, the results indicate that steering joystick is arranged in the same plane with the driver's shoulder and elbow, and the scope of the wrist joint is small. Thus, the kinematic chain can be projected to the sagittal plane for analysis. Each limb of the upper limb should be regarded as a rigid body and

the Degree of Freedom (DOF) of the wrist joint and fingers should be ignored.

According to the human dimensions of armored vehicle crew (GJB 1835-93 1993), human dimensions of Chinese adults (GB/T 17245-2004 2004) and field test data (Zhang & An 1993), parameters of a rigid bar in the built model are defined. By analyzing the Newton dynamics and Euler equations of the kinematic chain of the upper limb, the sparse matrix equation set with 17 dimensions can be determined. The equation set is simplified and adjusted for an easy calculation to reduce the dimension of the matrix. The kinematic chain of the upper limb is divided into two parts: active member and follower. The position of the shoulder joint is defined by periscope. The upper arm

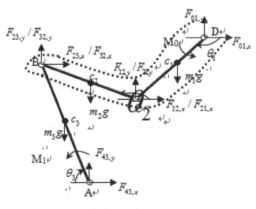

Figure 1. Force diagram of the upper limb multi-rigid-body model.

is regarded as the active member, while the forearm and joystick are regarded as the follower. The active member equation and follower equation are built separately. The kinematical equation and dynamic equation are divided. Therefore, the large matrix is divided into four matrices. Equation 1 is the kinematical equation of an active member, Equation 2 the kinematical equation of the follower, and Equation 3 is the dynamic equation of an active member, and Equation 4 is the dynamic equation of the follower:

$$\begin{bmatrix} a_{Cx} \\ a_{Cy} \end{bmatrix} = \begin{bmatrix} r_1\alpha_1\cos(\theta_1+\pi/2)+r_1\omega_1^2\cos(\theta_1+\pi) \\ r_1\alpha_1\sin(\theta_1+\pi/2)+r_1\omega_1^2\sin(\theta_1+\pi) \end{bmatrix} \quad (1)$$

$$\begin{bmatrix} r_2\cos(\theta_2+\pi/2)-r_3\cos(\theta_3+\pi/2) \\ r_2\sin(\theta_2+\pi/2)-r_3\sin(\theta_3+\pi/2) \end{bmatrix}\begin{bmatrix} \alpha_2 \\ \alpha_3 \end{bmatrix}$$

$$= \begin{bmatrix} -a_{Cx}-r_2\omega_2^2\cos(\theta_2+\pi)+r_3\omega_3^2\cos(\theta_3+\pi) \\ -a_{Cy}-r_2\omega_2^2\sin(\theta_2+\pi)+r_3\omega_3^2\sin(\theta_3+\pi) \end{bmatrix} \quad (2)$$

$$\begin{bmatrix} F_{01,x} \\ F_{01,y} \\ M_0 \end{bmatrix}$$

$$= \begin{bmatrix} m_1r_{c1}\alpha_1\cos(\theta_1+\pi/2)+m_1r_{c1}\omega_1^2\cos(\theta_1+\pi)+F_{12,x} \\ m_1r_{c1}\alpha_1\sin(\theta_1+\pi/2)+m_1r_{c1}\omega_1^2\sin(\theta_1+\pi)+F_{12,y}+m_1g \\ I_1\alpha_1-F_{12,x}r_1\sin\theta_1+F_{12,y}r_1\cos\theta_1+m_1gr_{c1}\cos\theta_1 \end{bmatrix}$$

$$(3)$$

$$\begin{bmatrix} 1 & 0 & -1 & 0 & 0 & 0 \\ 0 & 1 & 0 & -1 & 0 & 0 \\ -r_{c2}\cos\theta_2 & r_{c2}\sin\theta_2 & -(r_2-r_{c2})\cos\theta_2 & (r_2-r_{c2})\sin\theta_2 & 0 & 0 \\ 0 & 0 & 1 & 0 & 1 & 0 \\ 0 & 0 & 0 & 1 & 0 & 1 \\ 0 & 0 & r_3\sin\theta_3 & r_3\cos\theta_3 & 0 & 0 \end{bmatrix}\begin{bmatrix} F_{12,x} \\ F_{12,y} \\ F_{23,x} \\ F_{23,y} \\ F_{43,x} \\ F_{43,y} \end{bmatrix}$$

$$= \begin{bmatrix} m_2r_1\alpha_1\cos(\theta_1+\pi/2)+m_2r_1\omega_1^2\cos(\theta_1+\pi)+m_2r_{c2}\alpha_2\cos(\theta_2+\pi/2)+m_2r_{c2}\omega_2^2\cos(\theta_2+\pi) \\ m_2r_1\alpha_1\sin(\theta_1+\pi/2)+m_2r_1\omega_1^2\sin(\theta_1+\pi)+m_2r_{c2}\alpha_2\sin(\theta_2+\pi/2)+m_2r_{c2}\omega_2^2\sin(\theta_2+\pi)+m_2g \\ I_2\alpha_2 \\ m_3r_{c3}\alpha_3\cos(\theta_3+\pi/2)+m_3r_{c3}\omega_3^2\cos(\theta_3+\pi) \\ m_3r_{c3}\alpha_3\sin(\theta_3+\pi/2)+m_3r_{c3}\omega_3^2\sin(\theta_3+\pi)+m_3g \\ I_3\alpha_3+M_1 \end{bmatrix} \quad (4)$$

3 STEERING CONTROL SIMULATION

According to the above matrix, qudonggan.m, congdonggan.m, qudongli.m and congdongli.m functions can be written to conduct dynamic solving. Figure 2 shows the simulation model of the driver's upper limb bionic mechanism in Simulink.

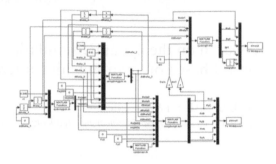

Figure 2. Dynamic simulated model.

3.1 Initial conditions of joints

According to the initial position of the kinematic chain of the upper limb end-effectors, initial angle of joints can be solved through inverse kinematics. Combined with the obtained joint angle, the initial angular acceleration is solved through the velocity equation of the system.

The velocity equation of the system is given by

$$T^0 = T_1^0(\theta_1)T_2^1(\theta_2)T_3^2(\theta_3) \quad (5)$$

$$T^0 = \begin{bmatrix} n_x & o_x & a_x & p_x \\ n_y & o_y & a_y & p_y \\ n_z & o_z & a_z & p_z \\ 0 & 0 & 0 & 1 \end{bmatrix} \quad (6)$$

where T is the transfer matrix and θ is the joint angle. According to the position of joystick, the kinematic chain of the upper limb end-effectors pose, i.e. the pose vectors $\vec{n},\vec{o},\vec{a},\vec{p}$, can be determined. Equation 6 is a set of nonlinear transcendental equations. Each joint angle is obtained through inverse kinematics with the methodology of Newton-Raphson.

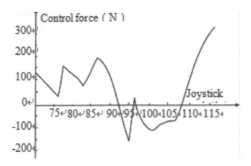

Figure 3. Control force curve of joystick of the armored vehicle.

System close-loop vector equation is given by

$$X = f(\theta) \tag{7}$$

The system velocity equation is obtained by taking the derivative of the above equation with respect to time:

$$X' = J(\theta) \cdot \theta' \tag{8}$$

where $X' = [v_x, v_y, v_z]^T$ is the end's velocity to the base coordinate; $\theta' = (\theta_1', \theta_2', \theta_3')^T$ is the angular velocity of each joint angular velocity; $J(\theta)$ is the Jacobian matrix to the base coordinate; X' is determined by operation, while $J(\theta)$ is determined by the member size and θ. Thus, θ' is obtained from Equation 8.

3.2 End-effectors moment of resistance

Figure 3 shows the relation curve between the control force and the joystick angle. Combined with the joystick length, caozongli.m function is written under the condition of maximum control force (joystick angle about 87°and control force about 200 N). As shown in Figure 2, a moment of resistance is obtained to provide input parameters for the function qudongli.m.

4 ANALYSIS OF THE SIMULATION RESULTS

Man-machine matching mainly consists of two aspects, i.e. "machine fits man" and "man fits machine". Therefore, this paper studies two aspects separately.

4.1 Influence of human dimensions

The operation posture of the armored vehicle driver's operation posture is constrained by many points. The actual static posture is at a compulsive state of fore raking, which makes the stress distribution of the lumbar vertebra uneven and ligament stretched or squeezed. Under this circumstance, discomfort of wrist will be quite strong, although there is no operation conducted.

Because of the difference in human dimensions, different percentile drivers' operating postures and their workload are different. Figure 4 shows different drivers' moment of force in the posture of static fore raking state.

Figure 4 shows that the wrist moment of force is larger, while the shoulder and elbow moment of force is smaller at the state of static fore raking. The wrist moment of force varies with height significantly. The larger the height is, the larger the wrist moment of force is. The shoulder and elbow moment of force varies with height non-significantly.

Figure 5 shows the relationship between the maximum moment of force of the main upper limb joints and the driver's height while pulling back joysticks. The results indicate that the wrist, shoulder and elbow moment of force increases as the height increases with the external force. However, the increment rate differs non-significantly. The driver's joint moment of force is influenced by the control force significantly as shown by comparing the above two figures.

Two conclusions can be drawn from the above results. First, individuals with low height are more likely to be selected as drivers after considering the static posture rationality and control accessibility. Second, the control force should be further decreased.

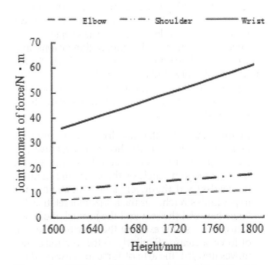

Figure 4. Different height of drivers' static fore raking influencing the joint moment of force.

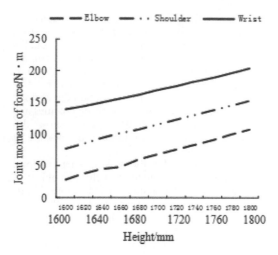

Figure 5. Different height of drivers' control force influencing the joint moment of force.

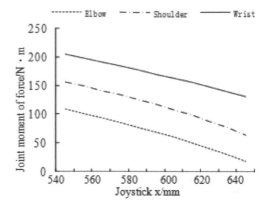

Figure 6. Influence of joystick horizontal position on the joint moment of force.

4.2 Influence of cabin layout parameters

The influence of human dimensions on the driver's workload comfort indicates the importance of armored vehicle driver selection. Besides, the layout position of the internal control device influences human workload comfort significantly under the condition of meeting demands of performance design.

Cabin layout devices involved with the kinematic chain of the upper limb mainly consist of seat, joystick and periscope. A coordinate is set up with the origin at the seat center, in which the horizontal positive direction is the body front direction and the vertical positive direction is the vertical upward direction. The 50th percentile individuals are selected as drivers and the relative distance between the seat and other two cabin layout devices is selected as the variable. Combined with human dimensions and armored vehicle cabin layout (Xie et al. 2008), the driver's body workload is analyzed.

Figures 6 to 9 show how the joint workload varies with the horizontal and vertical positions of joystick and periscope.

1. Figure 6 indicates that the driver's joint moment of force decreases as the horizontal relative distance between the joystick and the seat increases. Because the joint angle of the kinematic chain of the head is fixed, the driver's kinematic chain of the upper limb stretches as the horizontal relative distance between the seat and the joystick increases. Thus, it is easy for exerting the force and the arm of force decreases. Similarly, as the rotation center moves upward, the arm of force increases and the driver's joint moment of force increases as well, which is shown in Figure 7.

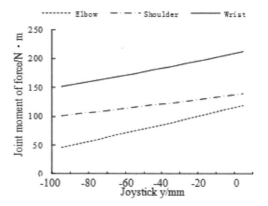

Figure 7. Influence of joystick vertical position on the joint moment of force.

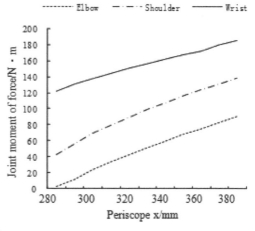

Figure 8. Influence of periscope horizontal position on the joint moment of force.

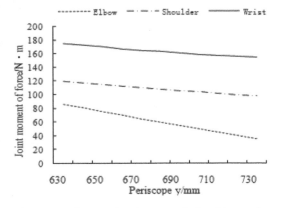

Figure 9. Influence of periscope vertical position on the joint moment of force.

2. Figure 8 indicates that the driver's joint moment of force increases as the horizontal relative distance between the seat and the periscope increases. Because the horizontal relative distance between the seat and the periscope increases, the body fore raking angle increases. Figure 9 indicates that the driver's joint moment of force decreases as the vertical relative distance between the seat and the periscope.

The armored vehicle cabin height is constrained by protection and transportation, so the vertical position range of the periscope and the seat is limited. Therefore, in order to decrease the driver's joint moment of force, parameters including the horizontal distance between the periscope and the seat, the horizontal and vertical downward distance between the joystick and the seat should be paid more attention when considering cabin layout matching. After meeting the control accessibility, the horizontal distance between the periscope and the seat should be decreased as far as possible. Also, the horizontal distance and the vertical downward distance between the joystick and the seat should be increased as far as possible.

5 CONCLUSIONS

Base on the Newton-Euler equation, this paper presented a matching method of cabin operation space by the driver's control dynamic simulation. The following conclusions can be obtained from the analysis of examples:

1. When matching armored vehicle operation space, horizontal position parameters of the seat and the periscope should be given priority followed by the position optimization of the joystick;
2. After considering both the control comfort and accessibility, individuals with low height are more likely to be selected as drivers. Also, the control force should be decreased as far as possible.

REFERENCES

GB/T 17245-2004, 2004. Inertial Parameters of Adult Human Body.
GJB 1835-93, 1993. Man-machine-environment System for Armored Vehicles-Requirements of Overall Design.
Qu, X.Q. 2007. *Dynamic Simulation of Planar Linkage Based on Matlab/Simulink*, Harbin Institute of Technology Press.
Wang, R. & Zhuang, D.M. 2007. Simulation and Analysis of Pilots' Manipulation Based on SimMechanics, *Journal of Beijing University of Aeronautics and Astronautics.* 33: 765–768.
Xie, C.L. Liu, W.P. & Du, L. 2008. Research on Man-machine Matching of Tank Driving Cabin Based on Static Comfort Analysis, *Journal of Academy of Armored Force Engineering.* 22: 69–72.
Zhang, S.Y. & An, G. 1993. Analysis of Tank Man-Machine Adaptability in Steering Control, *Journal of Academy of Armored Force Engineering.* 7: 6–13.

Advanced Materials and Structural Engineering – Hu (Ed.)
© *2016 Taylor & Francis Group, London, ISBN 978-1-138-02786-2*

Design and development of an expert system for rural biogas engineering

M. Zhou & Z.Y. Zou

School of Water Resources and Hydropower, Sichuan Agricultural University, Ya'an, Sichuan Province, China

ABSTRACT: An expert system for rural biogas engineering under the Web environment is designed and structured by applying the basic principles of the expert system and summarizing the experience of sustainable and stable operation of rural biogas engineering and biogas comprehensive utilization. This system comprises 6 major modules, namely biogas pretreatment, installation and use of biogas equipment, biogas engineering fault diagnosis, biogas comprehensive utilization, biogas manure comprehensive utilization, and knowledge base management. The overall structure, knowledge base and inference engine of this system is designed, and system operation examples are given. This system is published on the Web via ASP with a view to realizing the functions of remote fault diagnosis substituting senior biogas engineering experts. The test result shows that the deviation of temperature is controlled within ±0.9°C, that of pH is controlled within ±0.3, that of oxidation-reduction potential is controlled within ±30 mV, that of gas yield rate is controlled within ±8.5 L/m³, and that of methane concentration is controlled within ±4.5%. This system is easily expandable and applicable to biogas engineering at various scales.

1 INTRODUCTION

The rule biogas engineering is an important component of the modern eco-agricultural pattern (Song et al. 2014), which carries out innocuous disposal and recycling to livestock and poultry excrement as well as planting and breeding wastes by integrating renewable energy technologies and efficient eco-agricultural technologies. Its form and content involve multiple aspects of agricultural production and life, and it has achieved significant environmental, ecological and economical benefits (Chen et al. 2014).

Ordinary biogas users easily cause a variety of problems because they newly contact with the biogas and do not quite understand its detailed usage, comprehensive utilization and scope of application such that it not only wastes resources but also has influence on users' enthusiasm for using the biogas and their view to the biogas. Meanwhile, the faults in the stable operation of rural biogas engineering involve multiple aspects such as equipment, fermentation and utilization, so it is very difficult to rapidly and accurately determine the causes and solutions of a fault (Ling et al. 2005). A continuous and stable operation of biogas engineering can be ensured only under the guidance of experts with profound basic theories and abundant practical experience. Therefore, an expert system for rural biogas engineering is

developed in this study by applying the principles of the expert system with a view to substituting senior biogas experts to carry out default diagnosis, information consultation, decision supporting and technical guidance to operation, and management of rural biogas engineering.

2 OVERALL STRUCTURE OF THE SYSTEM

In order to realize favorable stability and response speed when multiple users consult simultaneously from different places at any time, the expert system for rural biogas engineering adopts the B/S three-layer technical architecture (Wu & Wen 2006). The features of this system are as follows: IE is used as the client software, core modules of the expert system such as consultation inference and interpretation are installed on the server, and the expert knowledge consultation service is provided to any user in any place at any time via LAN or WAN; authorized experts are also permitted to maintain the knowledge base via LAN or WAN from anywhere at any time. Management, storage and maintenance of the system's knowledge base are realized with the SQL server 2005 database. Access to the knowledge base and design of the inference engine is realized with the ASP technology. The structure of the system is shown in Figure 1.

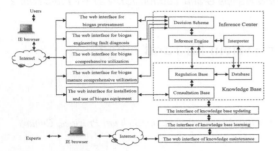

Figure 1. Basic structure of the expert system of rural biogas engineering.

3 FUNCTIONAL MODULES

3.1 *The module of biogas pretreatment*

As a result of middle and small biogas engineering, there is a problem of raw material pretreatment. In particular, when livestock and poultry excrement, breeding wastes and crop stalks are taken as raw materials for fermentation, it easily occurs with problems such as large granule and easy to float that seriously impact the biogas fermentation rate and gas yield or even make the whole biogas engineering stagnant. The basic knowledge of this module is provided to users in written message to make them have the basic understanding of biogas fermentation and also facilitate the operation of other modules.

3.2 *The module of installation and use of biogas equipment*

This module has the function of consultation. With this module, users can rapidly and accurately find all information of each kind of biogas equipment including structure, working principle, installation method, usage, and attention.

3.3 *The module of biogas engineering fault diagnosis*

This module has the function of diagnosis. It carries out inference diagnosis according to phenomena of faults of the biogas engineering's three major systems of gas production, gas transportation and gas use (Shi et al. 2013) to ascertain the causes of fault and give troubleshooting methods.

3.4 *The module of biogas comprehensive utilization*

With this module, users can search all information about biogas comprehensive utilization including basic principles, keys of technologies, attentions,

advantages and benefits. It can also search according to customized conditions such as the method for fertilizing greenhouse vegetables with CO_2 and attention. This module can assist the user in solving all kinds of problems occurring in the process of biogas comprehensive utilization and give correct methods and countermeasures.

3.5 *The module of biogas manure comprehensive utilization*

This module can instruct the user on carrying out scientific and reasonable comprehensive utilization of biogas manure. Users can consult key technologies and attention of biogas manure utilization, and corresponding pictures, animations and videos can provide more information to the users. The methods of knowledge browsing and searching are similar to those of the biogas comprehensive utilization module.

3.6 *The module of knowledge base management*

The module of knowledge base management comprises two parts, of which the first part is managing the knowledge base of the system, providing functions of knowledge base updating (addition, modification and deletion) and maintenance, and providing functions of searching and browsing to the user; the second part is knowledge learning, experts utilize this function to check the rules of the knowledge base and update the rule base with rules that they consider to be correct.

4 REALIZATION OF SYSTEM FUNCTIONS

The expert system normally consists of 6 components, namely knowledge base, rule base, human-machine interface, inference engine, knowledge acquisition mechanism and interpretation mechanism, the core of which is knowledge representation, rule base and inference engine design (Broner et al. 1990, Lin et al. 2007).

4.1 *Knowledge representation and rule base*

This system acquires knowledge in the mode of mutually cooperative talks between knowledge engineers and biogas experts, and it also sorts, summarizes and acquires knowledge from written documents. Knowledge classification is used according to biogas engineering operation. Through compiling and sorting, the collected information of all aspects related to rural biogas engineering is consolidated into knowledge points with unified rules of encoding. This paper adopts the knowledge representation method of production rule and represents them as IF P1, P2 ...... Pn, THEN Q1, Q2 ...... Qn (CF). The rule base is to be created in the knowledge representation method of the production rule. The rules are stored in the rule base, and each rule consists of antecedent and seccedent. In the rule base, the premise and conclusion forms a production rule and a lot of rules constitute a set of production rule.

4.2 Design of inference engine

The inference engine is actually a program for realizing fault inference, and it can realize fault inference easily. In view of the system, it also has to give reasons for the user to carry out fault diagnosis, the mode of integrated forward and backward inference is taken. In order to explain to the user, an interpretation mechanism is considered in the design of the inference engine. From the rule antecedent database, we can infer the rules and conclusion related to a condition through the RULE-NAME field with FACT-ID as the main keyword, and the corresponding rules are true only when such fact is true, while we can infer the premises of a conclusion from the rule seccedent database through the RULE-NAME field with FACT-ID as the main keyword. Both databases are related to the dictionary database, and bring out the best in each other (Lin et al. 2007).

4.3 Examples of system operation process

4.3.1 Fault diagnosis of biogas engineering
Now, it supposes that the fault of "low fire" is occurred in the operation of rural biogas engineering and, this system is utilized to carry out fault analysis. The user enters the interface of "biogas engineering operation fault diagnosis", and then the system prompts the user to input phenomenon reliability. Upon carrying out inference, the system will find out all causes of fault phenomenon. After the common user selects fault location in this module, the system will provide optional searching conditions and the user can combine any of them at will. Thereby, the computer acquires the information about fault signs. The fault phenomena that the user selects "low fire" are as below:

The gas is normally produced.
Combustion flame is small.
The flame is reddish yellow.

Upon selection, the system will infer and display all causes in the list box as follows:

Cause 1: the fire holes of the stove are blocked;
Cause 2: the fermentation liquid in the pit is over acidic resulting in low methane content;
Cause 3: the air inflow of the stove's air door is irrational;

If the user wonders why the system presents the above questions, he can select one of them in the cause list and requests the system to answer. When the first cause is selected, the system will explain.

Low fire may be caused by blocking of the stove's fire holes, while the selection of low fire is helpful for ascertaining whether this cause exists.

If the user wonders the solutions for each cause, he can select one of them in the cause list and requests the system to answer. When the second cause is selected, the system will answer.

Add the appropriate amount of plant ash, whitewash or cow dung; take out a part of the old material and replenish new material.

If the user wonders the reliability of each cause, he can select one of them in the cause list, and the system will give corresponding reliability for estimating the probability of such cause. Until now, the inferring process of the system is ended.

4.3.2 Consultation of biogas manure comprehensive utilization
When the user enters the main interface of the "expert system for rural biogas engineering", the system will first ask him to select the category of consultation, i.e. "biogas pretreatment", "installation and use of biogas equipment", "biogas engineering fault diagnosis", "biogas comprehensive utilization engineering" and "biogas manure comprehensive utilization engineering". If the user selects "biogas manure comprehensive utilization engineering", the system will display a hierarchical structure and ask the user to select "biogas manure for ecological planting" or "biogas manure for ecological breeding". If "biogas manure for planting" is selected, the system will display "using as fertilizer, making nutrient soil, planting edible fungus". If "planting edible fungus" is selected, the system will display "mushroom, oyster mushroom, ganoderma". The system will display key technologies and main points of operation according to the practical planting variety selected. The system also gives corresponding pictures or videos for providing more information to the user.

5 RESULTS AND ANALYSIS

Expert system is run on the PC, and the environmental parameters of biogas fermentation are set as follows: temperature 35°C, pH 7.0, oxidation-reduction potential −360 mV, gas yield rate per unit volume of digester 40.0 L/m^3, methane concentration 50%.

Table 1. Test data of temperature and pH, oxidation-reduction potential, gas yield rate, and methane concentration.

Time h	Temperature °C	Deviation	pH	Deviation	Potential mV	Deviation
0	35.2	0.2	7.1	0.1	−390	−30
2	34.8	−0.2	7.3	0.3	−385	−25
4	34.9	−0.1	6.8.2	−0.2	−386	−26
6	35.3	0.3	7.3	0.3	−381	−21
8	35.9	0.9	6.7	−0.3	−388	−28
10	34.7	−0.3	6.9	−0.1	−361	−1
12	35.5	0.5	7.0	0	−362	−2
14	35.1	0.1	7.2	0.2	−342	18
16	34.6	−0.4	6.8	−0.2	−344	16
18	34.8	−0.2	6.9	−0.1	−350	10
20	35.6	0.6	7.1	0.1	−359	1
22	35.1	0.1	7.2	0.2	−352	8

Time h	Gas yield rate L/m^3	Deviation	Methane %	Deviation
0	34.5	−5.5	45.5	−4.5
2	45.3	5.3	51.6	1.6
4	48.5	8.5	53.4	3.4
6	43.4	3.4	54.3	4.3
8	45.5	5.5	48.3	−1.7
10	38.6	−1.4	49.5	−0.5
12	37.6	−2.4	51.4	1.4
14	43.5	3.5	46.3	−3.7
16	40.2	0.2	48.7	−1.3
18	33.5	−6.5	49.8	−0.2
20	34.5	−5.5	51.5	1.5
22	45.3	5.3	50.6	0.6

To some extent, the oxidization-reduction potential, temperature, pH value, gas yield and gas component in the methane production process reflect the operation condition of the methane production process. Instruments for measuring and comparing oxidization-reduction potential, temperature, pH, gas yield and gas component respective employ AZ8651 oxidization-reduction potentiometer, AZ8891 thermometer with waterproof probe, CT2016 A biogas digester pH detector and MHZ92 gas component analyzer. Table 1 presents the data continuously controlled by expert and standard instruments for 24 h and the time interval of acquisition is 2 h. It can see from the table that the design of the expert system is reasonable, and the practical requirements of the application are satisfied.

6 CONCLUSIONS

With features of good interaction, simple operation and easy to use, the system is developed with ASP technology and database technology and suitable for vast rural biogas users and front-line technicians to use; the knowledge representation method of the production rule improves the capacity of knowledge representation and strengthens the inference function. While applying this expert system, it will also expand the knowledge base and strengthen the practicability of the system.

This system integrates biogas engineering experts' professional knowledge, scientific research achievements and practical experience and IT technology to build a comprehensive intelligent system. In the process of rural biogas comprehensive utilization, this system can not only provide engineering technical instruction, management and consultation services for users, but also be used as a multimedia teaching material for popularizing and teaching the knowledge of biogas comprehensive utilization. By setting up a platform that connects the rural biogas engineering expert system and vast users of rural biogas engineering through S&T information networks at the village level, it will make the biogas engineering expert system play the roles of "rural biogas guide" and "rural biogas tutor".

REFERENCES

Broner, I. King, J.P. & Nevo, A. 1990. Structured induction for agricultural expert systems knowledge acquisition, *Journal of Computers and Electronics in Agriculture.* 5: 87–99.

Chen, Y. Hu, W. Feng, Y. & Sweeney, S. 2014. Status and prospects of rural biogas development in China, *Journal of Renewable and Sustainable Energy Reviews.* 39: 679–685.

Lin, J. Cai, G.L. & Lu, J. 2007. Design and development of beer-spoilage microorganism control expert system, *Journal of Transactions of the Chinese Society of Agricultural Engineering.* 23(5): 173–176.

Ling, Q. Lanying, W. & Peng, Y. 2005. Study on expert system for the fault diagnosis of rural biogas project, *Journal of Transactions of the Chinese Society of Agricultural Machinery.* 36(2): 112–114.

Shi, J.G. Shang, X.Y. & Zhang, R.X. 2013. Expert system for fault diagnosis of biogas project based on BP neural network, *Journal of Hebei University: Natural Science Edition.* 33: 102–106.

Song, Z. Zhang, C. Yang, G. Feng, Y. Ren, G. & Han, X. 2014. Comparison of biogas development from households and medium and large-scale biogas plants in rural China, *Journal of Renewable and Sustainable Energy Reviews.* 33: 204–213.

Wu, B.G. & Wen, L.B. 2006. Expert consulting system for the diagnosis, prevention and control of important forest diseases and insect pests, *Journal of Beijing Forestry University.* 28: 113–118.

Research of an intelligent dynamic reactive power compensation device based on a single chip microcomputer

G.S. Zhang & M.R. Zhou
Electrical and Information Engineering College, Anhui University of Science and Technology, Huainan, Anhui, China

Y.N. Zhu
Economics and Management College, Anhui University of Science and Technology, Huainan, Anhui, China

ABSTRACT: According to the reactive power in power grid such as frequent fluctuations, network loss and low utilization rate of power grid, this paper designs an intelligent dynamic reactive power compensation device based on AT89C51 as the main controller. It introduces the dynamic reactive power compensation principle. In addition, it designs the compensation device based on a single chip microcomputer system structure and power factor compensation circuit. In the actual test, the circuit operates stability. The power factor has a substantial increase and the system controls precision.

1 INTRODUCTION

With the rapid increase in people's living standard, the power consumption is growing every year. Industrial, mining enterprises, agricultural production and the family's requirements of the amount of electricity and power quality are constantly higher. At the same time, the use of inductive load is also increasing. Many emotional loads absorb a lot of reactive power grid, reducing the grid power factor and increasing the energy loss (Shi 2014). Therefore, this paper designs a kind of reactive power compensation device called the "Research of intelligent dynamic reactive power compensation device based on single chip microcomputer". Based on the reactive power compensation equipment installed in the power grid to compensate the reactive power of inductive load required, its purpose is to reduce the electrical loss, ensure stable power grid and improve power supply reliability.

2 THE INTRODUCTION OF REACTIVE POWER COMPENSATION DEVICER

2.1 Synchronous condenser

As an early reactive power compensation, the synchronous condenser's structure is basically the same as that of the synchronous motor. Because of its complex operational structure and outdated technology, although it is still in use, it is no longer developed.

2.2 Fixed compensation capacitor

In the power grid system, the reactive power that the fixed compensation capacitors require contrasts with the reactive power of motors and other inductive loads required.

Reactive power supplied by the capacitor is very sensitive to voltage changes. Its ability to regulate reactive power is relatively poor.

However, due to its advantages such as more flexible and more easy to repair, the fixed capacitor compensation is still in use in the power system in China.

2.3 Dynamic reactive power compensation device

With the rapid development of power electronic technology, FACTS has been widely used in the electric power system. At present, the FACTS technology is mainly used in SVC. Compared with the synchronous condenser, SVC is a completely stationary equipment. However, its process of compensation is dynamic, that is to say, it can adopt the automatic tracking compensation technique according to the demand of reactive power and the changes in voltage.

2.4 Other

Because of some defects of SVC, people developed a Static Var Generator (SVG). SVG uses the new power electronic devices and detection technology. It can not only provide the lagging reactive power, but also provide the advance reactive power.

However, due to the huge investment, the large-scale application of SVG still has a long way to go.

With the rapid development of the power electronic inverter technology, passive power filter, active power filter, unity power factor converter and other new forms of compensation of the reactive power compensation control field have become a research hotspot.

In summary, SVC is the largest amount of control in reality. Although SVC has its own defects in control, because of its moderate cost and high technology, there is still much room for improvement. So, its development potential cannot be overlooked. The dynamic reactive power compensation controller is the core component of SVC. This paper mainly designs a kind of intelligent dynamic reactive power compensation device based on a single chip microcomputer.

3 THE PRINCIPLE OF DYNAMIC REACTIVE POWER COMPENSATION

In the process of power grid operation, power supply provides the right amount of reactive power to inductive load, which reduces the efficiency to a great extent, thereby causing much loss to the power supply circuit power. In order to improve the utilization rate of the power grid, this article through the installation of reactive power by the power grid method compensates the reactive power compensation device. The device can improve the power factor of the power grid, which can reduce the power loss. In addition, the device can also improve the power factor of the grid, which can reduce the power loss of transmission lines and transformers for conveying reactive power. This is the reactive power compensation. In the power grid, the output power can be divided into three types: active power, reactive power and apparent power (Qi et al. 2014).

1. Active power is also known as the average power, and the electrical resistance of the actual consumption is called the active power. The calculation formula is given by

$$P = \frac{1}{T}\int pdt = \frac{1}{T}\int uidt \qquad (1)$$

This equation represents the power average in a cycle, where p is the instantaneous power, which can be calculated by the following formula:

$$p = ui \qquad (2)$$

2. Reactive power: in a circuit containing inductance or capacitance, the power supply to recharge, they first convert electrical energy into

magnetic field or electric field energy stored up, to stay after the charge, inductor or capacitor stored energy release, it will throughout the cycle, energy loss. In the whole cycle, energy is not lost, but it is stored in power, and the inductance or capacitance at different times in different forms of energy. We call this amplitude the value of no loss of energy exchange the reactive power (Lao & Dong 2014).

When voltage waveform and current waveforms are the standard sine waves and the load is linear with the pure resistance load, the expressions of instantaneous values of voltage and current are given by

$$u = \sqrt{2}U \sin \omega_1 t \qquad (3)$$

$$i = \sqrt{2}I \sin(\omega_1 t - \Phi) \qquad (4)$$

where U—voltage RMS
I—current RMS
Φ—the current lag phase angle.

If in accordance with the division of work and no work, the current i can be divided into the following two parts:

$$i_p = \sqrt{2}I \cos \Phi \sin \omega_1 t \qquad (5)$$

$$i_q = \sqrt{2}I \sin \Phi \cos \omega_1 t \qquad (6)$$

where the current i_p and the voltage u are in the same phase. Also, the current i_q is the reactive current, which has a 90° phase angle difference with the voltage u.

By type (3) and type (4), we can roll out the expression of the instantaneous power p, which is given in type (7) as follows:

$$p = ui = 2UI \sin \omega_1 t \sin(\omega_1 t - \Phi)$$
$$= UI \cos \Phi (1 - \cos 2\omega_1 t) - UI \sin \Phi \sin 2\omega_1 t \qquad (7)$$

where $p_1 = UI \cos \Phi (1 | \cos 2)$ is the non-sinusoidal periodic power and the irreversible component. Also, $p_2 = -UI \sin \Phi \sin 2\omega_1 t$ is the sinusoidal periodic power, in which each of the positive time and negative time occupies half of one cycle. Therefore, we can roll out the following type (8):

$$\int_0^T p_2 dt = \int_0^T UI \sin \Phi \sin 2\omega_1 t dt = 0 \qquad (8)$$

According to type (8), we can know that the sine wave power in a cycle of integral is zero and a reversible component.

3. Apparent power: it is equal to the network port in the product of RMS voltage and current RMS, or it can be defined as the ratio of

active power and power factor, as represented in type (9) as follows:

$$S = UI \text{ Or } S = \frac{P}{\cos \Phi} \qquad (9)$$

where P—active power
cos Φ—power factor.

4 THE IMPLEMENTATION OF THE REACTIVE POWER COMPENSATION DEVICE

The intelligent dynamic reactive power compensation device is essential to provide safety of the power grid system. It can ensure the quality of the power and the safe operation of the power. Its principle is that the capacitive load and the inductive load are connected in parallel in the same circuit. The power provided by the power supply is changed between these two kinds of load of electric power. This method can be achieved by compensating for the inductive reactive power load required.

This device consists of main controller, A/D conversion, a detection circuit, power factor measurement system, the zero passage trigger module, man-machine interface circuit, synchronous switching devices and discharge protection system and other modules. The system overall structure block diagram is shown in Figure 1.

1. Master controller: this device is given priority with the AT89C51 controller, is used to control the whole circuit, and to measure the load voltage, current, reactive power and power factor, then the system analysis, logic judgment and real-time control, finally select the best compensation mode, and through the command control

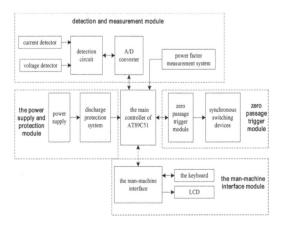

Figure 1. Unit overall structure block diagram.

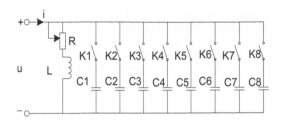

Figure 2. Power factor compensation circuit.

zero passage trigger module, used to judge synchronous switch conduction time, and thus can realize rapid and accurate of reactive power compensation.

2. Detection circuit: it mainly consists of current voltage detector, detector and multi-channel switch detects the current and voltage, and then converted to A digital signal by A/D, conveying to the main controller, and finally realize the reactive power compensation.

3. Power factor measurement system: electrical signal through the signal shaping, synchronous cycle, phase measurement after calculation, such as logic analysis into the single chip microcomputer, and it is concluded that the power factor of the circuit under test. This design simplifies the circuit structure to enhance the detection accuracy and quickness. The power factor compensation circuit is shown in Figure 2.

Synchronous switching device is used to control the opening and closing of the capacitor, and can be implemented without inrush current input and arc breaking. Synchronous switching devices than compound switch device canceled the thyristor components, and this not only simplified the device structure and reduced the production cost, but also reduced the failure rate, enhancing the reliability and accuracy of this switch.

5 THE CONTROL METHOD

Currently, the low-voltage dynamic reactive power compensation controller made in China generally uses univariate control (including three control modes such as variable voltage, power factor variable and reactive power variable), complex variable control (including two control modes such as the complex of power factor and voltage, and the complex of the amount of voltage and reactive power), al control (including three control modes such as fuzzy control, genetic algorithms and expert systems). Depending on the external circuit and control accuracy requirements, it uses different control strategies. At present, although the al control's procedures are complicated and its research costs are extremely higher, it is becoming a research direction of modern society to improve the low-voltage dynamic reactive power compensation controller. This passage utilizes the power factor variable. This univariate control has the properties of simple structure, precise control and low cost.

6 CONCLUSIONS

This article adopts AT89C51 as the main controller. Peripheral circuits include A/D conversion, detection circuit, power factor measurement system, the zero passage trigger module, man-machine interface circuit, synchronous switching devices and discharge protection system and other modules. The main controller and peripheral circuit can work well together.

In addition, due to the accurate control performance of AT89C51, the system control is more accurate and the response is faster than before. The use of the synchronous switch makes the switching surge smaller. The use of LCD makes the data readout more humanely. With a variety of performance, it realizes the dynamic tracking compensation in the low-voltage distribution network reactive power. It also achieves tracking and detection of grid operation state in real time.

This device is designed for power grid in the problem of the power loss and low utilization rate of power grid. The test results show that the power factor can well adapt to the complex circuit in the power grid, so as to ensure the stable operation of the plant and AT89C51 provide for high control performance. Detection module, power factor measurement system and zero passage trigger module performed well in the test process, with accurate measurement data and good control performance. This design is suitable for use in power grid and promotion.

REFERENCES

Jia, B. 2013. *Distribution network dynamic reactive power compensation device research*, Huazhong University of Science and Technology.

Lao, S.L. & Dong, H.J. 2014. Low-voltage dynamic reactive power compensation controller control strategy research, *Journal of Electronic Technology. Software Engineering* 14: 140.

Qi, Q.R. Zhang, C.P. & Yu, H.Y. 2014. The optimization thinking for smart grid reactive power compensation configuration, *Journal of Smart grid*, 01: 1–6.

Shi, Z.H. 2014. Dynamic reactive power compensation device and its application study, *Science Technology Enterprise*, 01: 262–264.

Advanced Materials and Structural Engineering – Hu (Ed.)
© 2016 Taylor & Francis Group, London, ISBN 978-1-138-02786-2

Research on the magnetic vibration control method of the electric hammer

H. Zhang

School of Naval Architecture and Ocean Engineering, Zhejiang Ocean University, Zhoushan, Zhejiang, China
Key Laboratory of Offshore Engineering Technology, Zhejiang Province, China

ABSTRACT: To minimize the vibration produced by the hammer for operators and equipment damages, a new method of the electric hammer magnetic damping is presented. First, the existing shortcoming of hammer vibration reduction methods is analyzed. Second, the magnetic field of permanent magnets is used as the elastic medium based on the principle of a magnetic levitation shock absorber. Finally, the magnetic damping mechanism is designed according to the two magnets relatively arranged in the same magnetic poles. The overall design of the electric hammer with the magnetic damping is proposed. The structural design sketch of the hammer handle is drawn in this paper. The experimental results show that the proposed method is effective.

1 INTRODUCTION

The electric hammer is a common electric tool. It has some advantages, such as small volume, lightweight, convenient to carry, simple operation, and reliable security. Due to produced HF vibrations, the arm of the operator easily fatigues. The muscle bone disease is also caused in the long-term use. The difficulty in the operation of high-power electric hammer is more obvious. Therefore, in order to reduce the injury caused by the electric hammer vibration on the human body and decrease the difficulty of operation hammer, the green design requirements (Liu et al. 2012, Du 2006, Yuan et al. 2012) about vibration, noise and safety need to be considered.

2 EXISTING METHODS OF VIBRATION REDUCTION IN THE HAMMER

The electric hammer consists of a motor-driven system, crank, gear-driven systems, cylinder components, front and rear handle, chassis, switches, cables and other components. When it is used, fatigue and injury can be caused due to the excessive vibration. The bit is prematurely worn and internal parts are damaged and prematurely discarded. Therefore, the related topic of the hammer vibration reduction has attracted researchers. According to the different parts of the vibration reduction, two methods are used to reduce electric hammer vibration: one is to reduce the vibration between the shell and handles, another is to reduce the vibration between the shell and the impact

mechanism. The concrete structures are described in detail below.

2.1 Vibration reduction between the shell and the handle

Elastic parts are used for vibration reduction, as shown in Figure 1. A spring is loaded on a connector of the handle and the shell of the hand-held tool with a vibration damper handle (Maikesina 2002). The handle is two legs shape extending roughly

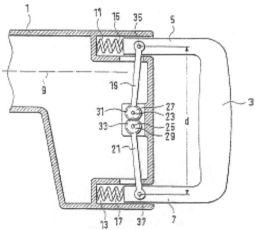

1-shell 3-handle 5, 7-leg 9- longitudinal axis 11, 13-slot 15, 17-spring 19, 21-lever 23, 25-joint 27, 29-circular 31, 33-gear 35, 37-joint

Figure 1.   Vibration damper handle.

parallel to the longitudinal axis of the body. It is elastically connected to the shell. This structure is simple and can reduce the vibration from the chassis to a large extent.

Another method for vibration reduction uses only spring in the shock absorbers rotary hammer (William 2004), as shown in Figure 2. The underside of the handle is connected to the body of the hammer. The handle pivot is formed. The upper end of the handle is connected to the rotary hammer main body by the elastic connecting mechanism. The upper pivot of the handle is formed. When the hammer works, the vibration is absorbed by small displacement changes in the handle under the influence of damping spring. Thus, the vibration is reduced.

2.2 *Vibration reduction between the shell and the shock*

The impact drill (Saito et al. 2005) is shown in Figure 3. A spring is added to the impact structure. It can reduce the vibration with the impact structure.

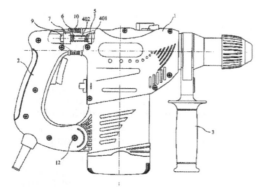

1-main 2-handle 3-auxiliary handle 4-spring 401-bezel 402-guide rod 5-screw 6-compression spring 7-slide 8-screws 9-screw 10-vibration reduction kits 11-ball bearings 12-screws

Figure 2. Shock absorbers rotary hammer (William 2004).

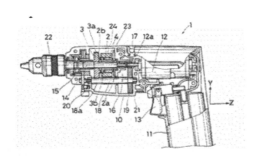

Figure 3. Spring damping of the impact structure (Saito et al. 2005).

By means of case analysis, the ordinary springs or elastic materials are mainly used to reduce the vibration in the current hammer handles. Vibration reduction is proposed in the form of a vibration reduction mechanism. The damping performance of the isolator will weaken with increasing time and frequency, and damping reliability will be reduced. Therefore, a method is proposed for reducing the vibration of an electric hammer handle by magnetic damping in this paper. The principle of permanent magnets, such as poles repel, is used to design a novel magnetic damping mechanism to reduce the vibration.

3 DESIGN OF THE MAGNETIC DAMPING MECHANISM

Because the non-contact magnetic levitation shock absorber technology can support an object, it has minimal friction and abrasion, low vibration, low noise and high-speed advantages (Wang et al. 2013, Cheng et al. 2013). In addition, special features such as the magnet itself enable it to be used in a particular environment. Taking into account the manufacturing economy, practicality and reliability, the magnetic damping mechanism is proposed based on the principle of a magnetic levitation shock absorber.

3.1 *Basic principles of the magnetic damping mechanism*

The mechanism consists of two cylindrical permanent magnets; one is a matrix; another is a floating body. The magnetic field of permanent magnets is used for the elastic medium. The same magnetic poles of two magnets are relatively arranged, in which the repulsive force between two magnetic poles is varied with the distance. In order to achieve

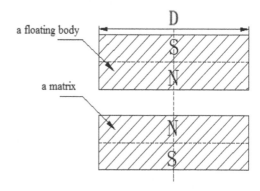

Figure 4. Basic principles of the magnetic damping mechanism.

the vibration function, repulsive magnetic springs are formed by at least two permanent magnets. The stored magnetic energy in the magnetic springs is the constant by determining the relative motion of a floating body in permanent magnets.

3.2 Design of the hammer vibration reduction mechanism

As shown in Figure 5, two permanent magnets are on the motor housing. They repel the magnetic poles of the permanent magnet in the handle, and make the handle always near the balance point O. When the hammer vibrates, the magnetic force of the permanent magnet makes the handle near the balance point. The damping effect can be achieved. In addition, the number of permanent magnets and inter-electrode distance can be adjusted according to the specific results of its work. If the damping effect is not obvious, the number of permanent magnets can be increased or the distance between the poles is decreased, so as to adjust the balancing force.

3.3 Structural design of hammer handles

The electric hammer handle directly contacts with the hand. The design and architecture of the handle has a great influence on the operator. The design factors such as shape, size and the bending angle can directly affect the users' safety and health. Handle structure needs to meet the user's requirements, and must conform to the structure, size and visual characteristics of the hand.

3.3.1 Design requirements

The overall structural design of the handle should comply with the design principles of man-machine engineering (Zhang 2008):

1. *The shape of the handle.* More gripping force is needed. In order to reduce hand pressure per unit area as far as possible, direct contact between the handle and hand area should be increased.

2. *The size of the handle.* Handle length should be in the range of 10~13 cm. The diameter of the handle for griping should be 3~4.5 cm. The width of the handle should be 4~6 cm.
3. The angle between the handle and the axis of the power tool should be at about 100 degrees.
4. The handle should provide sufficient gripping space for the hand. It should be consistent with the structure, size and visual characteristics of the hand.
5. The handle needs to make the wrist as straight as possible when the user holds the tool, and avoid bigger palm pressure.
6. The tool should be supported by a balancing mechanism. The quality 3 kg tools should be installed on the balancer and additional device should be installed in the tool's centre of gravity.

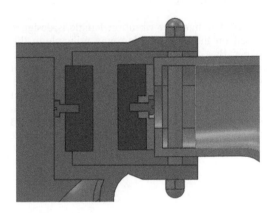

Figure 6. Vibration-organization of the handle.

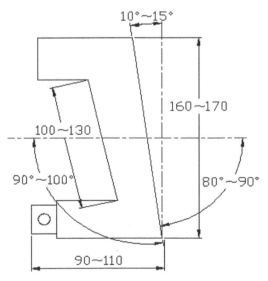

Figure 7. The design sketches of the handle.

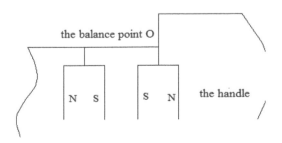

Figure 5. Vibration reduction mechanism in the handle.

3.3.2 *Total length and grip of the bounds*

Based on the above analysis of ergonomic design principles, the handle is designed with a rectangular shape to increase effective non-slip and stability. The design sketch of the handle is shown in Figure 7.

The vibration reduction mechanism is connected to the tail of the electric hammer drill body. It can not only reduce the electric hammer vibrations due to high-speed impact and operator fatigue, but also solve the problem of elastic elements decreases with increasing using time and frequency. Meanwhile, the mechanism can reduce the probability of premature failure of internal components caused by the vibration. To a certain extent, it reduces the noise caused by the vibration.

4 CONCLUSIONS

An electric hammer vibration handle is designed based on the magnetic damping principle in this paper. Based on this design, user fatigue and premature failure of the electric hammer and other issues caused by vibration are effectively solved. Next, leakage flux and magnetic springs and other effects of the external environment need to be considered. The magnetic damping mechanism needs to be analyzed through mechanical characteristic experiments.

REFERENCES

Cheng, Q.H. Li, Q.M. & Xu, X. 2013. Planar structures and decentralized control of magnetic suspension platform, *Machinery Design & Manufacture.* 29(2): 22–25.

Du, W. 2006. The analyses of the key contents of green design, *Machinery Design & Manufacture.* 2: 173–174.

Liu, Z.F. Gao, Y. Hu, D. & Zhang, J.D. 2012. Green innovation design method based on Triz and CBR principles, *Chinese Journal of Mechanical Engineering.* 23(9): 1105–1111, 1116.

Maikesina, G. 2002. *Hand-held tool with vibration damping shank machine,* China Patent, CN02801975.X.

Saito, T. Ohtsu, S. Watanabe H. & Toukairin, J. 2005. *Impact drill,* U.S. Patent 7,073,605.

Wang, W.Z. Jing F. & Zhou, H.B. 2013. Electromagnets in magnetic movement research on nonlinear effects of hysteretic characteristics, *Chinese J. Eng. Des.* 20(3): 212–217.

William, R. 2004. *Shock absorbers rotary hammer,* China Patent, 200420019835.3.

Yuan, S.S. Zhong, P.S. Liu, M. & Zhang, D.D. 2012. Research on key technologies of green manufacturing for mechanical product, *Machinery Design & Manufacture.* 4: 256–258.

Zhang, H.S. 2008. Study on ergonomic design of hand electric drill, *Packaging engineering.* 29: 102–115.

Advanced Materials and Structural Engineering – Hu (Ed.)
© 2016 Taylor & Francis Group, London, ISBN 978-1-138-02786-2

Flutter analysis of the rotating missile's variable cross-section empennage by the Differential Quadrature Method

B.B. Zhao
Nanjing University of Science and Technology, Nanjing, China
73917 Troops of Chinese People's Liberation Army, Nanjing, China

R.Z. Liu, R. Guo & L. Liu
Nanjing University of Science and Technology, Nanjing, China

X.C. Xu
The Ordnance Engineering College, Shijiazhuang, China

ABSTRACT: This study focuses on the tremor characteristics of cylinder-open fins under high rotational velocity conditions. To obtain the aerodynamic frequency, the empennage surface aerodynamic equations are established based on the piston theory. Aeroelastic partial differential equations of equivalent variable cross-section rotational beam are established based on the elastic theory, and partial differential equations are transformed to constant differential equations by using the Differential Quadrature Method to obtain the natural tremor frequency of empennage, which is consistent with the result obtained by the finite element method. Further analysis of the factors of natural frequency indicates that increasing the empennage section's reduction rate, rotational velocity, and thickness, or decreasing the empennage length can increase the basic frequency; however, the influence of common materials on empennage is limited, not obvious.

1 INTRODUCTION

In recent years, the smart ammunition that combines the advantages of traditional ammunition and missile has become a highlighted research hotspot. The suitable rotational velocity can not only reduce dispersion resulting from asymmetry factors, but also create conditions for stable scanning (Yang 2010). However, under relatively high velocity conditions, the gyroscopic effects, Coriolis force and many other complex nonlinear inertial loads on the empennage, combining variable cross-section and variable rotational velocity, will cause intense tremor and affect aeroelastic properties (Blades & Newman 2007), which are the main factors responsible for weakening the empennage strength and stability. Compared with the research conducted under low velocity conditions, the research on empennage tremor under complicated conditions has been a new issue. Zhao et al. (2014) focused their interest on the flexibility of improving rolling characteristics by designing different innovative empennage structures, and concluded that the twisted structure could increase the rolling velocity effectively. Bert & Jang (1988) developed the Differential Quadrature Method with a simple theory, which was easy to be applied on

computers, and could solve high-order partial differential equations efficiently, so this method was widely used in vibration or thermodynamics. This method has the potential for analyzing empennage aeroelastic effects.

This study illustrates the cycle changing factors on empennage surface aerodynamic characteristics, by analyzing unsteady aerodynamic on empennage, and solving the established vibration equations of the variable cross-section rotational empennage. These equations are solved based on the Differential Quadrature Method, and the rule of affecting (influence on) empennage tremor is developed, which is also consistent with the results obtained by the finite element method. These conclusions can be applied to design empennage geometrical structures.

2 MODELS AND DYNAMIC EQUATIONS

Four coordinate systems, namely velocity coordinate system (V), projectile axis coordinate system (A), second projectile axis coordinate system (O), and projectile body coordinate system (O'), are established on the empennage. Among them, system O can be obtained by translating system A with distance H, and the difference between system O′ and O is a rotational phase angle γ. Then, the empennage model can be simplified to a variable cross-section projecting beam, as shown in Figure 1, and the empennage deflection can be considered as the deformation in the z' direction.

2.1 Empennage unsteady dynamics

For the calculation of supersonic unsteady dynamics, the piston theory is used widely. Equation (1) illustrates the one-order piston theory:

$$p/p_\infty = 1 + kv_z'/c_\infty \qquad (1)$$

where p is the local pressure; p_∞ is the inflow pressure; k is the specific heat ratio; c_∞ is the sound velocity of inflow; and v_z' is the normal velocity of empennage surface flow.

The solution key is a normal velocity of surface flow, which can be obtained by translating the coordinate system. The velocity projection of mass center under the projectile axis coordinate system (A) is given by

$$v_A = \mathbf{A}_{AV}v = (a_{11}v, a_{21}v, a_{31}v)^T \qquad (2)$$

where $\mathbf{A}_{AV}$ is the transformation matrix and v is the velocity under the velocity coordinate system (V).

For translating v to that under the projectile body coordinate system (O′), which rotates with a body, the absolute velocity resulting from the convective velocity is necessary to obtain. Equation (3) is based on the free rigid body motion theory:

$$v_{O'} = \mathbf{B}_{O'O}(v_e + v_A) = (v_{O'x}, v_{O'y}, v_{O'z})^T \qquad (3)$$

where $\mathbf{B}_{O'O}$ is the transformation matrix; v_e is the convective velocity of the coordinate system B; and $v_e = (0, -\omega_z H, \omega_y H)^T$, where ω_z and ω_y are the rotational velocity in the z and y directions, respectively, during the rigid body motion. The empennage normal velocity developed from the free projectile motion is given by

$$v_{o'z} = b_{32}(a_{21}v - w_z H) + b_{33}(a_{31}v + w_y H) \qquad (4)$$

where $a_{21} = -\sin\delta_1$, $a_{31} = -\sin\delta_2\cos\delta_1$, $b_{32} = -\sin\Omega t$, and $b_{33} = \cos\Omega t$.

Considering the empennage rotation, the inflow velocity of the surface micro-unit is given by

$$v_z = v \cdot \varepsilon + v\partial b/\partial y - v_{o'z} - \dot{w} - \Omega y \qquad (5)$$

where ε is the empennage installation bias angle; b is the width of the empennage chord, and the second term is the inflow change (relevant to the width of the empennage chord, and if the width of the empennage chord is constant, then the value will be 0); and w is the empennage deflection.

Based on Equation (1), the empennage surface aerodynamic loads can be written as follows:

$$f = (p - p_\infty)b = \rho_\infty c_\infty v_z b \qquad (6)$$

In the empennage design, the one-order natural frequency must avoid the frequency of cycle load change, in order to avoid resonance.

2.2 Aeroelastic equations

In the plane y′Oz′, the empennage will bear the effects from the surface aerodynamic loads f and the centrifugal force F. F results from rotation, and the centrifugal force of any micro-unit at position y can be written as follows:

$$F(y) = \int_y^{l_w} \rho(y)\Omega^2 y\, dy \qquad (7)$$

where Ω is the projectile rotation velocity. The micro-unit mass and moment of inertia of the cross-section in position y are, respectively,

$$\rho(y) = \rho_0\left(1 - c\frac{y}{l_w}\right),\ I(y) = I_0\left(1 - c\frac{y}{l_w}\right)^3,$$
$$c = (c_a - c_e)/c_a \qquad (8)$$

where ρ_0 is the linear density of the empennage root; I_0 is the inertial moment of unit length of the

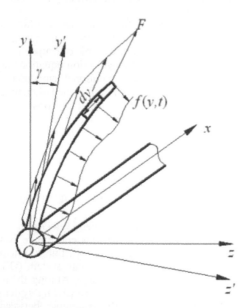

Figure 1. Coordinate systems and load distribution.

empennage root; l_w is the empennage span; c is the section reduction ratio (the reduction amplitude of the section); c_a is the thickness of the empennage root; and c_e is the thickness of the empennage tip.

Based on the variational method, Lagrange equations and Newton laws, the empennage surface aeroelastic equations are established. Based on the force analysis on micro-unit dy, as shown in Figure 2, simplifying tension as centrifugal force by ignoring the two-order driblet, and use of Lagrange equations, the forced vibration equation can be written as follows:

$$\frac{\partial^2}{\partial y^2}\left(EI\frac{\partial^2 w}{\partial y^2}\right) - \frac{\partial}{\partial y}\left(T\frac{\partial w}{\partial y}\right) + \rho\frac{d^2 w}{dt^2} = f + m\Omega^2 w$$

(9)

Equation (9) is a four-order differential equation, which needs to be simplified.

2.3 DQM for equation discretization

Using the dimensionless method on aeroelastic equations, and introducing dimensionless quantity, the dimensionless tremor equation of empennage can be written as follows:

$$C_0 w'''' + C_1 w''' + C_2 w'' + C_3 w' + C_4 w$$
$$+ C_5 \ddot{w} + C_6 \dot{w} = f_0$$

(10)

where

$$C_0 = EI_0(1 - cY)^3/l_w^4 \quad C_1 = -cEI_0(1 - cY)^2/l_w^4$$

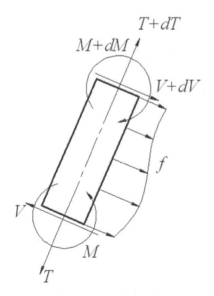

$$T+dT$$
$$M+dM$$
$$V+dV$$
$$f$$
$$V$$
$$M$$
$$T$$

Figure 2. Force analysis on micro-unit figure.

$$C_2 = \frac{6c^2 EI_0}{l_w^4}(1 - cY) - \rho_0\Omega^2\left(\frac{1}{2} - \frac{c}{3} - \frac{Y^2}{2} + \frac{cY^3}{3}\right)$$

$$C_3 = \rho_0\Omega^2(Y - cY^2) \quad C_4 = -\rho_0(1 - cY)\Omega^2$$

$$C_5 = \rho_0(1 - cY) \quad C_6 = \frac{P_\infty vb}{Ma}\Omega^2$$

The Differential Quadrature Method (DQM) (Wang 2001) is an efficient numerical method for solving problems with boundary initials, by discrete handling of differential equations and boundary conditions. Choosing dispersion m, discrete point Y_i can be determined as follows:

$$Y_i = \frac{1}{2}\left[\frac{1 + \cos(m - i)\pi}{m - 1}\right] \quad i = 1, 2, \ldots, m$$

(11)

After discrete handling, the aeroelastic equation (regarding Y_i) can be written as follows:

$$C_0(Y_i)A_{i,j}^{(4)} + C_1(Y_i)A_{i,j}^{(3)} + C_2(Y_i)A_{i,j}^{(2)} + C_3(Y_i)A_{i,j}^{(1)}$$
$$+ C_4(Y_i)A_{i,j}^{(0)} + C_5(Y_i)\ddot{w}_i + C_6(Y_i)\dot{w}_i = f_0(Y_i)$$

(12)

where $A_{i,j}^n$ is the interpolation weight of the n-order derivative of $w(Y)$ in the DQM.

Considering the empennage as a cantilever beam, with one end fixed and the other one vibrating freely, the boundary conditions can be written as follows:

$$w(0) = 0 \quad w(1) = 0 \quad \frac{\partial w(0)}{\partial Y} = 0 \quad \sum_{j=1}^{m} A_{1,j}^{(1)} w(j) = 0$$

$$\frac{\partial^2 w(l_w)}{\partial Y^2} = 0 \quad \sum_{j=1}^{m} A_{m,j}^{(2)} w(j) = 0 \quad \frac{\partial^3 w(l_w)}{\partial Y^3} = 0$$

$$\sum_{j=1}^{m} A_{m,j}^{(3)} w(j) = 0$$

(13)

where i, j = 1 ... m. Substituting Equation (11) into (10), and considering free vibration under no disturbance, the new equation just regarding w can be written as follows:

$$M\ddot{w} + C\dot{w} + Kw = 0$$

(14)

where M, C, K are the mass matrix, aerodynamic damping matrix and stiffness matrix, respectively. Also, the algebraic eigenvalue equation can be obtained as follows:

$$(\omega^2 M + \omega C + K)w = Dw = 0$$

(15)

597

where $w(y,t) = w(y)\exp(\omega t)$. The imaginary part of the characteristic root of the matrix D is the natural vibration frequency.

3 ALGORITHM VALIDATION AND RESULT ANALYSIS

Conditions: the material is a high strength steel, whose elastic modulus E is 2.1e+11Pa, μ is 0.3 and ρ is 7900 kg/m³, and the rotational velocity Ω is 0–200 r/s.

3.1 *Method validation*

For examining equations and programming solving methods, ANSYS is used to perform a modal analysis.

For simulating the effects on the natural frequency from the centrifugal force, the basic settings are as follows: in the global cylindrical coordinate system in ANSYS, the Z axis is the rotation axis, the Y axis is the rotation angle, and the X axis is the radial coordinate; the boundary conditions include setting the constraints on the circumferential and axial directions for the first-floor grid of the empennage root, and setting the constraints on rotation axial displacements for the whole grid; the rotational velocity is set as a special velocity along the rotation axis. Before the modal analysis, it is necessary to perform a static solution based on previous rotational velocity, for obtaining the effects on the stiffness matrix from the centrifugal force.

To perform algorithm validation, the thickness of the empennage root is set as 7 mm, and that of a tip is set as 2 mm for calculating the natural frequencies, respectively, under a rotational velocity of 0 and 100 r/s. In the DQM, the number of discrete points is set as 12, and in the FEM, the grid node number along the span is set as 30. The first 3 order terms are chosen for the analysis, and the result comparison is summarized in Table 1. It can be concluded that the natural frequency will increase with increasing rotation velocity; the 1-order natural frequency

is relatively low, playing a main role in resonance; the DQM coincides with the FEM for calculating the natural frequency.

Figure 3 shows the relationships between the tremor modal response frequency and the section reduction rate under different empennage root thicknesses. According to Figure 3a, with increasing section reduction rate, all basic frequencies of empennage will increase and the speed will increase quickly; the larger the thickness is, the higher the basic frequency is; there is also a little effect of the change rate of the section reduction rate on the basic frequency. The explanations of these conclusions are as follows. The stiffness matrix is proportional to the cube of thickness, but the mass matrix shows a linear relationship with thickness; the main factor for the basic frequency is the weight of root mass to the whole mass, and if the weight increases, the basic frequency will be higher, so the higher the section reduction rate is, the higher the basic frequency is. From Figure 3b, it can be known that in contrast to the basic frequency, 2-order and 3-order frequencies will decrease with increasing section reduction rate; the basic frequency is far lower than others, so the other two frequencies

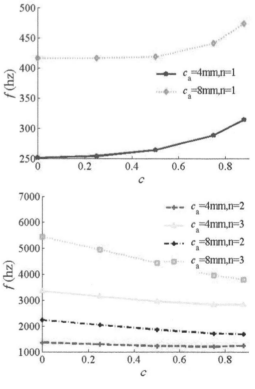

Figure 3. (a) The change curves of the basic frequency. (b) The change curves of 2- and 3-order frequencies.

Table 1. Comparison results between the DQM and the FEM.

Natural frequency	Ω 0 r/s DQM	FEM	Ω 100 r/s DQM	FEM
1-order	365	378	391	397
2-order	1485	1502	1507	1525
3-order	3462	3484	3493	3507

can be ignored when considering the empennage tremor. The explanation is that the average flexibility of the beam is the main factor for high-order frequencies.

3.1.1 Effects of the rotation velocity

As shown in Figure 4, the basic frequency will increase and increase quickly with increasing rotational velocity; the thicker the root is, the smaller the increasing extent of the basic frequency is; under the same root thickness, the larger the section reduction rate is, the more slowly the basic frequency increases. The explanation is that the rigid effect is obvious under the centrifugal force resulting from rotation, so the basic frequency increases; for a relatively thinner empennage, the rigid effect will be more obvious due to large flexibility; for the empennage with a relatively higher section reduction rate, the rigid effect will be less obvious, because the centrifugal force will focus on the root.

3.1.2 Other factors

Figure 5 shows the relationships between the basic frequency and different materials, thickness, and lengths of a span. M1 is the high strength steel, M2 is the high strength aluminum, and c_0 is the thickness. The figure shows a little effect on the basic frequency from two common materials; the thickness has obvious effects on the basic frequency and shows approximately a linear relationship. This is because the main factor of different materials for the basic frequency is the rate of the elastic modulus and density, and the values of two common materials are similar. From Figure 5b, we can see that the basic frequency decreases dramatically with increasing length of span, and this decrease will be to a very small extent when the

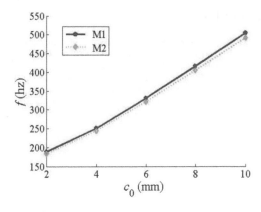

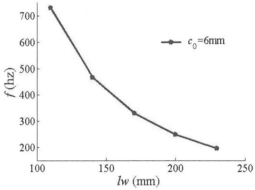

Figure 5. (a) The effect of different materials and thicknesses. (b) The effect of different lengths of a span.

span is long enough, because the basic frequency is inversely proportional to the square of the length of a span.

4 CONCLUSIONS

1. Base on the free vibration theory of variable cross-section rotation empennage, the forced vibration equations are obtained, and tremor properties must avoid the aerodynamic frequency.
2. Through the comparison of the analysis result of vibration properties based on the Differential Quadrature Method (DQM) and that obtained by the FEM, the accuracy of DQM is proved.
3. Many factors can affect the basic frequency. Increasing section reduction rate of empennage, rotation velocity and thickness (it has the most significant effect, and a 2 mm increment will bring a 60 hz increment on the basic frequency), or decreasing length of empennage will increase the basic frequency effectively.

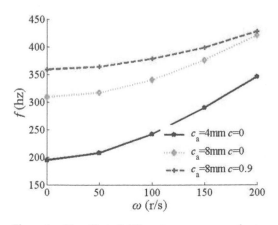

Figure 4. The effect of different empennages and rotation velocities.

REFERENCES

Bert, C.W. & Jang, S.K. 1988. Two new approximate methods for analyzing free vibration of structural components, *AIAA*, 26(5): 612–618.

Blades, E.L. & Newman, J.C. 2007. Aeroelastic Effects of Spinning Missiles, *AIAA*, 2243.

Wang, Y.L. 2001. *Differential Quadrature Method Theory and Application,* Nanjing University of Aeronautics and Astronautics.

Yang, S.Q. 2010. *Smart ammunition engineering*, Beijing National Defence Industry Press.

Zhao, B.B. Liu, R.Z. & Guo, R. 2014. Aerodynamic characteristics of the twist fin vehicle, *National University of Defense Technology,* 03: 19–24.

Advanced Materials and Structural Engineering – Hu (Ed.)
© 2016 Taylor & Francis Group, London, ISBN 978-1-138-02786-2

Impact and screw-holding strength of four bamboo-based panels

J.B. Li
College of Wood Science and Technology, Nanjing Forestry University, Nanjing, China

M.J. Guan
Bamboo Engineering and Research Center, Nanjing, China

ABSTRACT: Bamboo-based panels are mainly used as structural composites. Impact and screw-holding strength are important mechanical properties. This paper selected four bamboo-based panels including Bamboo-Wood Composite Panel (BWC), Laminated Bamboo Lumber (LBL), Carbonized Bamboo Scrimber (CBS) and Natural Bamboo Scrimber (NBS) to test their impact and screw-holding strength. The results showed that the average impact of bamboo-based panels were 18.97 KJ/m^2 (BWC), 58.34 KJ/m^2 (LBL), 119.87 KJ/m^2 (CBS), and 130.62 KJ/m^2 (NBS). The screw-holding strength value was 1.74 KN, 2.78 KN, 5.98 KN and, 4.84 KN, respectively. Pure bamboo panels were better than the bamboo-wood composite panel in impact and screw-holding strength. Fracture appearance of BWC showed a smooth cross section with interface debonding and it failed to brittle fracture. LBL and NBS showed the irregular cross damage and failed in short fiber pulling-out. Fracture pattern of CBS showed a lacerated cross section and failed in ductile fracture with long fiber pulling-out.

1 INTRODUCTION

Recently, bamboo-based products have been developed fast and they are widely used in house building industry and furniture manufacture field. The product can be used as bamboo flooring, container flooring and bamboo structural beams etc. (Sinha et al. 2014, Zhou & Bian 2014, Wei et al. 2013). The force of these panels in usage is comparatively complex. A. Sinha etc. tested flexural, shear and compressive properties of bamboo glulam beams (Sinha et al. 2014). However, structural panels are susceptible to impact loading as invisible damage occurs in the composites (Chen et al. 2014), meanwhile impact and screw-holding strength are also important mechanical properties to evaluate the quality of structure composites.

Impact properties of bamboo bundle laminated veneer lumber by preprocessing densification technology were tested by F.M. Chen et al. (2014). Z.X. Yu etc. explored low-velocity impact properties and damage mechanisms of bamboo scrimber and bamboo-wood scrimber with different densities (Yu et al. 2012). X.F. Mo etc. observed fracture pattern and measured toughness of Moso Bamboo (Mo et al. 2010). Y.X. Zhu etc. studied the impact performance of bamboo strengthened laminated veneer lumber of poplar (Zhu et al. 2005). Ultimate screw withdrawal loads of furniture-grade south pine (Pinus spp.) and sweetgum (Liquidambar L.) plywood were evaluated for five moisture content by Q.L. Wu (1999). G.Y. Li etc. measured the nail holding power used in the decoration of bamboo integrated timber and analyzed the influence of bolt category, diameter of guiding bore and screwing depth (Li & Wang 2013). However, research on the impact of bamboo-based panels and screw-holding strength of bamboo scrimber are comparatively rare.

Hence, this paper selected four bamboo-based panels to test the impact and screw-holding strength to supply fundamental parameters for their application as structure composites.

2 EXPERIMENTAL PROCEDURE

2.1 Materials and specimens

Bamboo-based panels used in this experiment were produced by Hangzhou ZhengTian Industrial Co., Ltd. (ZT INDUSTRY). The panels were Bamboo-Wood Composite panel (BWC), Laminated Bamboo Lumber (LBL), Carbonized Bamboo Scrimber (CBS) and Natural Bamboo Scrimber (NBS). BWC consisted of three-layer veneers including face, core and back veneer. Materials used for the panel were sliced thin bamboo strips, lumbers and poplar veneers separately. LBL was made of slightly carbonized bamboo strips with "Bamboo Yellow" side inward laying-up, exactly "Yellow-Yellow". CBS was made of extruded bamboo bundles after carbonization. NBS was made from extruded bamboo bundles directly.

Appearance of panels was showed in Figure 1. And Figure 2 displayed the structure schematic.

Density and Moisture Content (MC) of panels were listed in Table 1.

2.2 Methods and procedure

According to GB/T 1940 (2009), the specimens were sawn from the products with the dimension of 300 mm × 20 mm × h for impact test and 10 samples were repeated. Impact can be calculated by Equation 1 as below:

$$A = \frac{1000Q}{bh}. \tag{1}$$

where, A represented the impact of the specimen, while Q was the absorption energy of impact fracture, b was the width of specimens, and h was the thickness.

(a) Front side

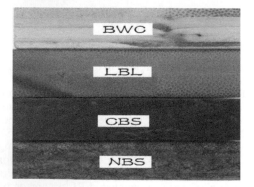

(b) Cross side

Figure 1. Appearance of bamboo-based panels.

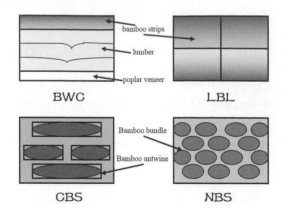

Figure 2. Structure schematic of bamboo-base panels.

Table 1. Density and MC of bamboo-based panels.

Item	BWC	LBL	CBS	NBS
Densities [g/cm^3]	0.5–0.6	0.5–0.8	1.15–1.2	1.1–1.15
Moisture content [%]	8–12	8–12	6–9	6–9

According to GB/T 17657 (2013), the dimension of screw-holding strength specimens were 75 mm × 50 mm × h and 10 samples were repeated. Before tests, the specimens were conditioned of 20 °C, 65% RH until the weight became constant.

3 RESULTS AND DISCUSSION

3.1 Impact

Figure 3 showed the average impact of four bamboo-based panels and the values were 18.97 KJ/m^2 (BWC), 58.34 KJ/m^2 (LBL), 119.87 KJ/m^2 (CBS) and 130.62 KJ/m^2 (NBS), respectively. Accordingly, pure bamboo panels were better than bamboo-wood composite panel in the impact. BWC had the smallest density as well as the smallest impact. Bamboo scrimber had the larger density and the higher impact. Therefore, a density of bamboo-based panels had a positive influence on the impact.

The impact of LBL was 3.08 times higher than that of BWC, but less than 50% of CBS' and NBS'. Compared to the impact of NBS, that of CBS decreased by 8.23%. However, CBS and NBS had approximate density. Thus, it suggested that the bamboo scrimber became brittle and impact decreased after the carbonization of bamboo bundles.

Figure 4 illustrated impact fracture appearance of specimens. In BWC, delamination occurred along weak boundary layer (bond line) and it failed

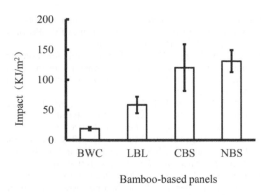

Figure 3. Impact of bamboo-based panels.

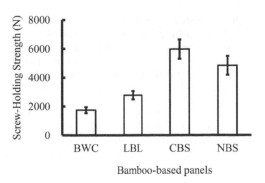

Figure 5. Screw-holding strength of bamboo-based panels.

(a) Front side

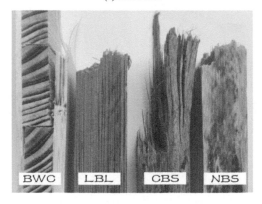

(b) Cross side

Figure 4. Impact fracture of bamboo-base panels.

mainly on face veneer and back veneer. While core lumbers were not obviously destroyed and bamboo strips in face veneer displayed a lacerated failure. The damage and fracture characteristics of BWC were interface debonding and fiber breakage.

Thus, BWC failed in brittle fracture. Some cracks in substrate happened to LBL along the bond line. Nevertheless, LBL showed slightly irregular ductile fracture morphology and mainly failed in short fiber pulling-out and bamboo fiber breakage.

Fracture pattern of CBS demonstrated a rough cross section with serious ductile damage such as tensile laceration and delamination. As shown in Figure 4, the failure mode of CBS was long fiber pulling-out and fiber breakage. The fracture appearance of NBS was similar to that of LBL. Hence, NBS mainly failed as ductile fracture in fiber breakage on an irregular cross section with short fiber pulling-out. Above all, BWC failed in brittle fracture, while LBL and NBS failed mainly in ductile fracture and CBS failed in ductile destruction.

3.2 Screw-holding strength

Figure 5 showed the average screw-holding strength of bamboo-based panels and the value were 1.74 KN (BWC), 2.78 KN (LBL), 5.98 KN (CBS) and 4.84 KN (NBS), respectively. Mean screw-holding strength of LBL exceeded 59.77% than that of BWC. However, the value of NBS and CBS were 2.15 and 1.74 times correspondingly higher than that of LBL. Therefore, pure bamboo panels had the larger screw-holding strength than bamboo-wood composite panel.

BWC had the smallest density as well as the smallest screw-holding strength. Bamboo scrimber had the larger density and the higher screw-holding strength. Therefore, density of bamboo-based panels had a positive influence on the screw-holding strength.

4 CONCLUSIONS

This paper selected four bamboo-based panels to test the impact and screw-holding strength. The results were shown as following:

1. Pure bamboo panels had the higher impact and screw-holding strength than bamboo-wood composite panel. NBS and CBS had the maximum impact and screw-holding strength separately, and BWC had the minimum impact and screw-holding strength. The density of panels had a positive effect on the impact and screw-holding strength.
2. BWC failed in brittle fracture. LBL and NBS failed mainly in a ductile fracture with short fiber pulling-out. And CBS failed ductile fracture with long fiber pulling-out.

ACKNOWLEDGEMENT

This research was sponsored by a project funded by the Priority Academic Program Development of Jiangsu higher education institutions (PAPD) and Natural Science Foundation of Jiangsu province, China (BK2011822).

REFERENCES

Chen, F.M. Deng, J.C. & Cheng, J.C. et al. 2014. Impact properties of bamboo bundle laminated veneer lumber by preprocessing densification technology, *Journal of Wood Science*. 60(6): 421–427.

GB/T 17657, 2013. *General Administration of Quality Supervision,* Inspection and Quarantine of the People's Republic of China, Beijing, China.

GB/T 1940, 2009. *General Administration of Quality Supervision,* Inspection and Quarantine of the People's Republic of China, Beijing, China.

Li, G.Y. & Wang, W.G. 2013. Study on nail holding power of bamboo integrated timber for decoration, *International Conference on Mechanical Structures and Smart panels.* 487: 324–327.

Mo, X.F. Guan, M.J. & Zhu, Y.X. et al. 2010. Fracture Pattern and Toughness of Maso Bamboo, *China Forestry Science and Technology.* 24: 45–47.

Sinha, A. Way, D. & Mlasko, D. 2014. Structural Performance of Glued Laminated Bamboo Beams, *Journal of Structural Engineering.* 140: 04013021.

Wei, J. Zeng, D. & Guan, M.J. 2013. Research on the bending properties of the bamboo-wood container flooring, *3rd International Conference on Machinery, panels Science and Engineering Applications.* 744: 366–369.

Wu, Q.L. 1999. Screw-holding capacity of two furniture-grade plywoods, *Forest Products Society,* 49: 56–59.

Yu, Z.X. Jiang, Z.H. & Wang, G. et al. 2012. Impact Resistance Properties of Bamboo Scrimber, *Journal of Northeast Forestry University.* 40: 46–48.

Zhou, A.P. & Bian, Y.L. 2014. Experimental Study on the Flexural Performance of Parallel Strand Bamboo Beams, *The Scientific World Journal,* 2014: 284929.

Zhu, Y.X. Guan, M.J. & Zhang, X.D. 2005. Studies on impact performance of bamboo strengthened laminated veneer lumber of poplar, *Journal of Nanjing Forestry University (Natural Sciences and Edition).* 29: 99–102.

Advanced Materials and Structural Engineering – Hu (Ed.)
© *2016 Taylor & Francis Group, London, ISBN 978-1-138-02786-2*

The effect of carbonized treatment on the wettability of poplar veneer and the shear strength of plywood

M.G. Xue
College of Wood Science and Technology, Nanjing Forestry University, Nanjing, China

M.J. Guan
Bamboo Engineering and Research Center, Nanjing, China

ABSTRACT: Being an environment friendly method of wood modification, carbonized treatment has already been widely used for poplar lumber, not poplar veneer. The effects of carbonized treatment on the wettability of poplar veneer and the shear strength of plywood were studied in this paper. The temperature of carbonized treatment was set as 160, 180 and 200°C. The processing time was 1, 1.5 and 2 h, respectively. Distilled water, methylene iodide, ethylene glycol and Melamine Urea Formaldehyde resin (MUF) were used to test the contact angle of the poplar veneer surface before and after the carbonized treatment, respectively. The shear strength of plywood with carbonized veneer and the control group were tested correspondingly. The results showed that the value of the contact angle of the four reagents was, in turn, ethylene glycol < methylene iodide < distilled water < MUF. The average equilibrium contact angle of MUF was 78.26°. The average shear strength of plywood was 1.18 MPa from carbonized veneer and 1.37 MPa from the control group, respectively. Compared with the control group, the shear strength of plywood with carbonized veneer was decreased by 13.63% in total.

1 INTRODUCTION

Poplar is an important and fast-growing species in China, which is widely distributed in the south and north of country, as the main raw material of low-density plywood (Zhu & Guan 2008). The basic density of wood is 0.35 g/cm³ with a light color and an unstable dimension (Liu 2013, Wu & Yu 2008). Because of its poor quality, poplar wood is commonly used only in the manufacture of low-grade wood products. If the wood could be improved by post-harvest treatment, it would be more useful. Furthermore, these problems also affect the quality and price directly, in order to make full use of wood, it is necessary to improve its dimension stability and enhance its properties. Mattos, B.D carried out a test by applying a two-step treatment with methacryloyl chloride. The wood samples were submerged in a solution of 10% methacryloyl chloride in dichloromethane for 15 h and then exposed to three temperature treatments. Dimensional stability and water repellence were improved by 40–50% and 60–75%, respectively (Mattos, B.D & Lourencon, T.V 2015). Deng, Yuhe studied the interface of modification by the treatment of silvergrass fibers with sodium hydroxide (NaOH) and Polymeric

Methylene Diphenyldiisocyanate (PMDI). The mechanical properties of the treated composites were compared well with those of untreated composites. A marked improvement of 49.0% in tensile strength and 47.35% in flexural strength for 40% SV fibers-reinforced high-density PE composites was observed (Xue & Deng 2015). As mentioned above, chemical modification is the main method to modify poplar wood dimension stability and mechanical properties. However, it is detrimental to the environment and decreases the bonding properties of wood and veneers. Therefore, in many countries such as France, Japan and Canada, a close attention has been paid to carbonized treatment on fast-growing lumber to enhance their dimension stability gradually (Yang et al. 2011). Carbonized treatment can improve the dimensional stability of poplar, prolong service life, prevent insect corruption, and change surface properties (Xu et al. 2011). Currently, the carbonized treatment of wood is widely applied to preserve lumber, but there are only a few studies available on veneers.

At present, poplar veneers are peeled from timber directly without hot water boiling treatment, producing stress in the veneer as well as leading to poor dimensional stability of veneers and plywood in China. So, it is necessary to improve the dimension stability of veneers to enhance that of plywood. Since chemically modified treatment is not good to the bonding of veneer, carbonized treatment on veneers is considered in this research.

This paper aims to study the effect of carbonized treatment on the wettability of poplar veneers and the shear strength of plywood, in order to understand the potential applications of carbonized treatment in enhancing plywood quality.

2 EXPERIMENTAL SPECIMENS AND PROCEDURE

2.1 Materials and specimens

Poplar (Populussimoniicarr) veneer was obtained from Guannan country of Jiangsu Province, China. The veneers were dried to 12% before carbonized treatment. The specimens were sawn into a dimension of $400 \times 400 \times 2.0$ mm.

2.2 Carbonized treatment

Using the super heating steam as the media, the carbonized temperature was set as 160, 180 and 200 °C. The carbonized time was 1, 1.5 and 2 h, respectively.

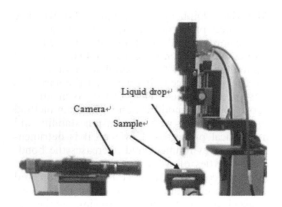

Figure 1. Contact angle test.

Before the contact angle was tested, the specimens were placed in the service environment for several hours, considering the change in moisture and temperature.

2.3 Plywood preparation

The 5-ply carbonized veneer and a control group were laminated to make plywood with a glue spread of 240 g/m² of the MUF with a solid content of 52.6% under the condition of hot pressing at 100 °C, for 10 min, at 1.0 MPa. The plywood was conditioned under 20 °C, 65% to equilibrium moisture content.

3 PERFORMANCE TEST

3.1 Wettability

Contact angle was tested by the surface-wetting method to reflect the change in wettability. The JCA2000A contact angle test (see Fig. 1) with a microscope and computer was adopted.

Experimental reagents were as follows: distilled water, ethylene glycol, methylene iodide and self-made Melamine Urea Formaldehyde resin (MUF) with a solid content of 52.6% and a viscosity of 89.24 mm²/s. In order to reduce the influence of the surface, each group had 3 specimens and each specimen was subjected to 3–5 tests, and the average value was obtained. Table 1 showed the basic information of four reagents under the condition of 20 °C, 65% for determination of the contact angle. Table 1 lists the purity, the source of the reagents, the volume of drop and the drop velocity.

3.2 Shear strength

According to GB/T 17657-1999, the shear strength of the specimens were tested on the universal mechanical machine (GB/T17657 1999). Figure 2 shows the size and shape of the shear strength of the specimens. These materials were sawn into specimens with the size according to the regulations of GB/T 9846.7 (2004).

Table 1. Test reagents information for the contact angle test.

Reagent	Purity [%]	Source	Volume [μL]	Drop velocity [μL/s]
Distilled water	–	Laboratory made	1.5	2
Ethylene glycol	99.0	The Company of Beijing Chemical Reagent	1.5	2
Methylene iodide	99.0	Company of Sigma	1.0	1.0
Melamine urea formaldehyde resin	–	Laboratory made	5	5

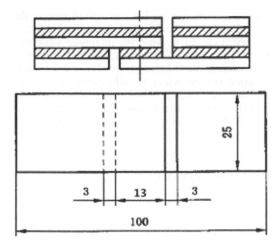

Figure 2. The shape and size of the shear strength specimens [mm].

4 RESULTS AND DISCUSSION

4.1 *Analysis of wettability*

Contact angle is one of the most important indices of wettability. The larger the angle is, the worse the wettability is. The contact angle of veneer before and after carbonized treatment is given in Table 2. When using distilled water, ethylene glycol and methylene iodide as reagents, the average contact angle of carbonized veneers was 70.25°, 28.01° and 31.49°, respectively. The contact angle of water was slightly larger than the other reagents. When the period of carbonized treatment was 1.5 and 2 hours, the contact angle became larger as the temperature increased.

Figure 3 shows the equilibrium contact angle of MUF on the veneer surface before and after carbonized treatment. The average equilibrium contact angle of MUF was higher than that of the control group. The equilibrium contact angle of carbonized veneer increased with increasing time at the same temperature. When the carbonized condition was 200 °C and 2 h, the average equilibrium contact angle reached a maximum value of 86.75°. Therefore, carbonized treatment decreased the wettability of MUF on the veneer surface to some extent.

4.2 *Analysis of shear strength*

The shear strength of plywood with the carbonized veneer and control group is listed in Table 3. The shear strength of plywood with carbonized veneer was lower than that of the control group in whole. At carbonized temperature of 160 °C, the

Table 2. The contact angle of veneers before and after carbonized treatment.

T [°C]	t/h	Distilled water [°]	Ethylene glycol [°]	Methylene iodide [°]
160	1	58.92	31.68	35.83
	1.5	49.34	32.34	32.34
	2	64.68	22.54	28.07
180	1	49.25	25.51	32.22
	1.5	54.13	31.94	26.99
	2	92.33	34.55	27.99
200	1	84.41	23.90	32.55
	1.5	76.38	28.79	27.84
	2	93.93	27.15	30.28
C	C	79.08	21.65	40.82

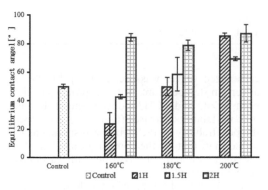

Figure 3. The equilibrium contact angle of MUF resin of veneers before and after carbonized treatment.

Table 3. The shear strength before and after carbonized treatment [MPa].

	160°C	180°C	200°C	C
1.0 h	0.92	1.35	1.22	–
1.5 h	1.12	1.27	1.17	–
2.0 h	1.21	1.11	1.28	–
C	–	–	–	1.37

shear strength of plywood with carbonized veneer increased with increasing time. However, at 180 °C and 200 °C, there was no obvious regulation on the shear strength of plywood with carbonized veneer. Compared with the control group (1.37 MPa), the maximum value of the shear strength of plywood with carbonized veneer was 1.35 MPa under the condition of 180 °C and 1 h, which decreased by 3.28%. While the minimum value was 0.92 MPa under the condition of 160 °C and 1 h, decreased by 36.68%. On the whole, compared with the control group, the shear strength of plywood with

Table 4. The Analysis of Variance for the shear strength of plywood with carbonized veneer [carbonized temperature].

	SS	df	MS	F	P	$F_{0.1}$ crit
Carbonized temperature	0.046	2	0.023	1.71	0.258	3.463
Error	0.080	6	0.013			
Total	0.126	8				

Table 5. The Analysis of Variance for the shear strength of plywood with carbonized veneer [carbonized time].

	SS	df	MS	F	P	$F_{0.1}$ crit
Carbonized time	0.002	2	0.001	0.050	0.951	3.463
Error	0.124	6	0.021			
Total	0.126	8				

carbonized veneer was decreased by 13.63%. At a certain temperature, with a prolonged time, both −C=O and −CH3 reduced, and at the same time wettability decreased (Qin 2014). As a result, the penetrability of the carbonized specimens was worse than that of the control group. Therefore, the shear strength of plywood with carbonized veneer decreased.

The Analysis of Variance for the shear strength of plywood with carbonized veneer is given in Table 4 and Table 5. From Table 4, it can be seen that $F_{0.1} >$ F, P-value > 0.1. From Table 5, it can be found that $F_{0.1} >$ F, P-value > 0.1. The shear strength of plywood with carbonized veneer was not significantly affected by the carbonized temperature and time.

5 CONCLUSIONS

Among the four reagents, the value of the contact angle was as follows: ethylene glycol < methylene iodide < distilled water < MUF. When using distilled water, ethylene glycol and methylene iodide as reagents, the average contact angle of carbonized veneers was 70.25°, 28.01° and 31.4°, respectively. For the reagent MUF, the average equilibrium contact angle was 78.26°.

The shear strength of all plywood groups with carbonized veneer was smaller than that of the control group. Among the carbonized groups, the maximum value was 1.32 MPa under the condition of 180 °C and 1 h, while the minimum value was 0.92 MPa under the condition of 160 °C and 1 h. The average shear strength of plywood with carbonized veneers was 1.18 MPa, while that of the control group was 1.37 MPa. In conclusion, compared with the control group, the shear strength of plywood with carbonized veneer was decreased by 13.63%.

ACKNOWLEDGMENT

This research was sponsored by a project funded by the Priority Academic Program Development of Jiangsu Higher Education Institutions (PAPD) and Innovation Project of Science and Technology Enterprise–Science and Technology Plan of Northern Jiangsu (BC2013425).

REFERENCES

GB/T 9846.7, 2004. *General Administration of Quality Supervision,* Inspection and Quarantine of the People's Republic of China, Beijing, China.

GB/T17657, 1999. *General Administration of Quality Supervision,* Inspection and Quarantine of the People's Republic of China, Beijing, China.

Gu, L.B. Tu, D.Y. & Yu, X.L. 2007. Characteristic and Application of Thermowood. *China Wood-Based Panels,* 2007(5): 30–32, 37.

Liu, J.L. 2013. *Study on Decay resistance of modified poplar.* Beijing Forestry University.

Mattos, B.D. & Lourencon, V. 2015. *Chemical modification of fast-growing eucalyptus wood,* Wood science and technology.

Qin, Y.Z. 2014. *Study on wood and adhesive surface/ interface wettability characterization and influencing factors.* Beijing Forestry University.

Wu, S. & Yu, Z.M. 2008. Review on Current Situation and Developing Trend of Wood Carbonization Technology. *China Wood-Based Panels,* 2008(5): 3–6.

Xu, X.W. Tang. Z.J. & Cui, Y. 2011. The Study on Color and Dynamic Mechanical Properties of carbonized treated Poplar wood. *Journal of Nanjing Forestry University (Natural Sciences and Edition),* 35: 65.

Yang, L. Li, H. & Yang, Z.J. 2011. Study on Carbonized and Physical and Mechanical Properties. *Hubei Forestry Science and Technology,* 2011(5): 32–35, 60.

Zhu, E.C. & Guan, M.J. 2008. Experiment on the Heat Treatment of Poplar Veneer, *China Forestry Science and technology,* 22: 79.

Advanced Materials and Structural Engineering – Hu (Ed.)
© *2016 Taylor & Francis Group, London, ISBN 978-1-138-02786-2*

Nonlinear undamped vibration frequency of rectangular pretension orthotropic membrane under impact force

Y.H. Zhang
Chongqing Jianzhu College, Chongqing, P.R. China

C.J. Liu
College of Environment and Civil Engineering, Chengdu University of Technology, Chengdu, P.R. China

Z.L. Zheng
College of Civil Engineering, Chongqing University, Chongqing, China

ABSTRACT: The undamped nonlinear vibration of pretension rectangular orthotropic membrane structure under impact force is investigated in this paper. Firstly, the governing equations of motion were obtained based on Von Kármán's large deflection theory. Then the governing equations were solved by the Bubnov-Galerkin method and L-P (Lindstedt-Poincaré) perturbation method, and the asymptotic analytical solution of vibration frequency was obtained. Finally, the frequency results were compared and analyzed through a example. The results obtained herein provide some computational basis for the vibration control and dynamic design of orthotropic membrane structures.

1 INTRODUCTION

The orthotropic membrane structure is widely applied in building structures, instruments and meters, electronic engineering and space and aeronautics fields, etc. Because of its lightweight and small stiffness, it is very easy to vibrate under external force (such as winds, rainstorm, hail, etc.), thus results in structural failure. Meanwhile, the amplitude of the membrane is much larger than its thickness, so the geometric nonlinearity caused by large amplitude must be concerned in its vibration process. At present, there are fewer reports about the nonlinear vibration problem of membrane structures. Wei and Shui (2008) analyzed the free vibration of a practical cable-membrane structure by using the finite element program, and the wind-induced vibration response of the structure is theoretically analyzed by using a self-developed program. Shin et al. (2005) investigated the geometric nonlinear dynamic characteristics of the out-of-plane vibration of an axially moving membrane by using the Hamilton principle and the Galerkin method. Sunny et al. (2012) studied the nonlinear vibration problem of a prestressed membrane by a domian decomposition. Soares and Gonalves et al. (2012, 2009, 2014) presented a detailed analysis of the nonlinear vibration of stretched annular, circular and rectangular membrane by using the Galerkin method and nonlinear finite element method. Wetherhold and Padliya (2014) presented a method for inferring the initial

tensions from measured vibration frequencies and demonstrated the sensitivity of the tensions with respect to imprecision in the measured frequencies. Zheng and Liu et al. (2014, 2009, 2010) studied the nonlinear free vibration of orthotropic rectangular and axisymmetric polar orthotropic circular membranes by applying the Galerkin method and perturbation method. However, they did not consider external excitations.

In this paper, we considered the external impact force acting on the membrane and obtained the governing equations of motion by Von Kármán's large deflection theory. The governing equations are solved by the Bubnov-Galerkin method and L-P perturbation method and the asymptotic analytical solution was obtained. The research results provide a computational basis for the vibration control and dynamic design of building membrane structures.

2 GOVERNING EQUATIONS AND BOUNDARY CONDITIONS

Assume that the rectangular membrane is fixed on its four edges and the membrane material is orthotropic. Its two orthogonal directions are x and y, respectively. a and b are the length of x and y, respectively. N_{0x} and N_{0y} are the initial tension in x and y, respectively. Assume that the impact loading is a pellet that can be considered as a particle. The initial velocity of the pellet is v_0. The mass of the pellet

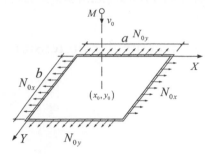

Figure 1. The studied model.

is M. The impact contact point is (x_0, y_0) on the membrane. The studied model is shown in Figure 1.

According to the Von Karman's large deflection theory and D'Alembert's principle (Zheng et al. 2009), the vibration partial differential motion equation and consistency equations are:

$$\begin{cases} \rho\dfrac{\partial^2 w}{\partial t^2} - (N_x + N_{0x})\dfrac{\partial^2 w}{\partial x^2} - (N_y + N_{0y})\dfrac{\partial^2 w}{\partial y^2} = p(x,y,t) \\ \dfrac{1}{E_1 h}\dfrac{\partial^2 N_x}{\partial y^2} - \dfrac{\mu_2}{E_2 h}\dfrac{\partial^2 N_y}{\partial y^2} - \dfrac{\mu_1}{E_1 h}\dfrac{\partial^2 N_x}{\partial x^2} + \dfrac{1}{E_2 h}\dfrac{\partial^2 N_y}{\partial x^2} \\ \quad -\dfrac{1}{Gh}\dfrac{\partial^2 N_{xy}}{\partial x \partial y} = \left(\dfrac{\partial^2 w}{\partial x \partial y}\right)^2 - \dfrac{\partial^2 w}{\partial x^2}\dfrac{\partial^2 w}{\partial y^2} \end{cases}$$

(1)

where, ρ denotes aerial density of membrane; N_x and N_y denote additional tension in x and y direction, respectively; N_{0x} and N_{0y} denote initial tension in x and y direction, respectively; N_{xy} denotes shear force; $w = w(x, y, t)$ denotes deflection; $p(x,y,t)$ denotes impact loading; h denotes membrane's thickness. E_1 and E_2 denote Young's modulus in x and y direction, respectively. G denotes shearing modulus. μ_1 and μ_2 denote Poisson's ratio in x and y direction, respectively. $p(x,y,t)$ denotes the impact loading.

If there is only the pellet impact loading on the membrane, the impact loading can be expressed as follows:

$$p(x,y,t) = F(t)\delta(x - x_0)\delta(y - y_0)$$

(2)

where, $F(t)$ denotes the impact force on the membrane; δ denotes the Dirac function.

According to the theorem of momentum, we obtain

$$F(t) = -M\dfrac{\partial w^2(x_0, y_0, t)}{\partial t^2}$$

(3)

The membrane is a plane before the impact contact. Therefore, the initial condition of the membrane is

$$w(x_0, y_0, t)\big|_{t=0} = 0,\ \dfrac{\partial w(x_0, y_0, t)}{\partial t}\bigg|_{t=0} = v_0',$$

(4)

where, v_0' denotes the velocity of membrane at the time of $t = 0$.

The corresponding displacement and stress boundary conditions of the governing equations can be expressed as follows:

$$\begin{cases} w(0,y,t) = 0,\ \dfrac{\partial^2 w}{\partial x^2}(0,y,t) = 0 \\ w(a,y,t) = 0,\ \dfrac{\partial^2 w}{\partial x^2}(a,y,t) = 0 \\ w(x,0,t) = 0,\ \dfrac{\partial^2 w}{\partial y^2}(x,0,t) = 0 \\ w(x,b,t) = 0,\ \dfrac{\partial^2 w}{\partial y^2}(x,b,t) = 0 \end{cases}$$

(5)

$$\begin{cases} \dfrac{\partial^2 \varphi}{\partial x^2}(0,y,t) = 0 \\ \dfrac{\partial^2 \varphi}{\partial x^2}(a,y,t) = 0 \end{cases} \begin{cases} \dfrac{\partial^2 \varphi}{\partial y^2}(x,0,t) = 0 \\ \dfrac{\partial^2 \varphi}{\partial y^2}(x,b,t) = 0 \end{cases}$$

(6)

3 ANALYTIC SOLUTION OF VIBRATION FREQUENCY

The effect of shearing stress is very small when the membrane is vibrating. In order to simplify the computation, we can take $N_{xy} = 0$. Meanwhile, introduce the stress function into Equation (1), then Equation (1) can be simplified as follows:

$$\rho\dfrac{\partial^2 w}{\partial t^2} + c\dfrac{\partial w}{\partial t} - h\left(\dfrac{\partial^2 \varphi}{\partial y^2} + \sigma_{0x}\right)\dfrac{\partial^2 w}{\partial x^2}$$
$$- h\left(\dfrac{\partial^2 \varphi}{\partial x^2} + \sigma_{0y}\right)\dfrac{\partial^2 w}{\partial y^2} = p(x,y,t)$$

(7)

$$\dfrac{1}{E_1}\dfrac{\partial^4 \varphi}{\partial y^4} + \dfrac{1}{E_2}\dfrac{\partial^4 \varphi}{\partial x^4} = \left(\dfrac{\partial^2 w}{\partial x \partial y}\right)^2 - \dfrac{\partial^2 w}{\partial x^2}\dfrac{\partial^2 w}{\partial y^2}$$

(8)

where, φ denotes stress function $\varphi(x,y,t)$, σ_{0x} and σ_{0y} denote initial tensile stress in x and y direction, respectively.

According to the vibration theory, assume that the functions that satisfy the boundary conditions (5) and (6) are

$$w(x,y,t) = \sum_{m=1}^{\infty}\sum_{n=1}^{\infty} T_{mn}(t) \cdot W_{mn}(x,y)$$

(9)

$$\varphi(x,y,t)=\sum_{m=1}^{\infty}\sum_{n=1}^{\infty}T_{mn}^2(t)\cdot\phi_{mn}(x,y) \tag{10}$$

where, $W_{mn}(x,y)$ is the mode shape function, and $\phi_{mn}(x,y)$ and $T_{mn}(t)$ are the unknown functions.

The mode shape function of rectangular membrane is

$$W_{mn}(x,y)=W(x,y)=W=\sin\frac{m\pi x}{a}\sin\frac{n\pi y}{b} \tag{11}$$

where, m and n are integer, which denote the sine half-wave number in x and y, respectively. Equation (11) satisfies the displacement boundary condition (5) automatically. We can take one term of Equations (9) and (10) for computation, i.e.

$$w(x,y,t)=T(t)\cdot W(x,y) \tag{12}$$

$$\varphi(x,y,t)=T^2(t)\cdot\phi(x,y) \tag{13}$$

and then superpose the final results. In this way, it will not affect the final computation results.

By substituting Equations (11), (12) and (13) into Equation (8), one obtains

$$\frac{1}{E_1}\frac{\partial^4\phi}{\partial y^4}+\frac{1}{E_2}\frac{\partial^4\phi}{\partial x^4}=\frac{m^2n^2\pi^4}{2a^2b^2}\left(\cos\frac{2m\pi x}{a}+\cos\frac{2n\pi y}{b}\right) \tag{14}$$

According to the calculus theory, solve Equation (14) and obtains

$$\phi(x,y)=\frac{E_2n^2a^2}{32m^2b^2}\cos\frac{2m\pi x}{a}+\frac{E_1m^2b^2}{32n^2a^2}\cos\frac{2n\pi y}{b}$$
$$+\frac{\pi^2E_2n^2}{16b^2}x^2+\frac{\pi^2E_1m^2}{16a^2}y^2 \tag{15}$$

The substitution of Equations (9), (10) and (15) into Equation (7), and according to the Bubnov-Galerkin method, furnishes

$$\iint_S\left[\rho W^2\frac{d^2T(t)}{dt^2}-h\left(\sigma_{0x}\frac{\partial^2W}{\partial x^2}W+\sigma_{0y}\frac{\partial^2W}{\partial y^2}W\right)T(t)\right.$$
$$\left.-h\left(\frac{\partial^2\phi}{\partial y^2}\frac{\partial^2W}{\partial x^2}W+\frac{\partial^2\phi}{\partial x^2}\frac{\partial^2W}{\partial y^2}W\right)T^3(t)\right]dxdy$$
$$=\int_0^a\int_0^b[F(t)\delta(x-x_0)(y-y_0)W(x,y)] \tag{16}$$

The substitution of all the aforementioned expressions into Equation (16) results in

$$\frac{d^2T(t)}{dt^2}+\frac{m^2\pi^2b^2N_{0x}+n^2\pi^2a^2N_{0y}}{\rho a^2b^2+4abM\sin^2\frac{m\pi x_0}{a}\cdot\sin^2\frac{n\pi y_0}{b}}T(t)$$
$$+\frac{6m^2n^2\pi^4h\beta+6m^2n^2\pi^4h\alpha}{\rho a^2b^2+4abM\sin^2\frac{m\pi x_0}{a}\cdot\sin^2\frac{n\pi y_0}{b}}T^3(t)=0 \tag{17}$$

It is clear that Equation (17) is a nonlinear differential equation with respect to $T(t)$. It is very difficult to obtain the accurate analytical solution, or unable to obtain. Therefore, we apply the L-P perturbation method to obtain its approximate analytic solution.

Because $(h^2/ab)\ll 1$ and it is a dimensionless parameter, we can take $\varepsilon=h^2/ab$ as the perturbation parameter. Then Equation (17) can be simplified as

$$\frac{d^2T(t)}{dt^2}+\omega_0^2\left[T(t)+\varepsilon\alpha_3T^3(t)\right]=0 \tag{18}$$

where, $\omega_0^2=\dfrac{m^2\pi^2b^2N_{0x}+n^2\pi^2a^2N_{0y}}{\rho a^2b^2+4abM\sin^2\frac{m\pi x_0}{a}\cdot\sin^2\frac{n\pi y_0}{b}}$,

$$\alpha_3=\frac{6m^2n^2\pi^2(\alpha+\beta)}{h\left(m^2\frac{b}{a}N_{0x}+n^2\frac{a}{b}N_{0y}\right)}.$$

By introducing a new variable $\tau=\omega t$, we obtain

$$\frac{d^2T(t)}{dt^2}=\omega^2\frac{d^2T(t)}{d\tau^2} \tag{19}$$

Spread ω and $T(t)$ as a power series with respect to ε:

$$\omega=\omega_0+\varepsilon\omega_1+\varepsilon^2\omega_2+0\left(\varepsilon^3\right) \tag{20}$$

$$T(t)=T_0(\tau)+\varepsilon T_1(\tau)+\varepsilon^2T_2(\tau)+0(\varepsilon^3) \tag{21}$$

The substitution of Equation (19), (20), (21) into Equation (18) yields

$$(\omega_0+\varepsilon\omega_1+\varepsilon^2\omega_2+\cdots)^2\left(\frac{d^2T_0(\tau)}{d\tau^2}+\varepsilon\frac{d^2T_1(\tau)}{d\tau^2}\right.$$
$$\left.+\varepsilon^2\frac{d^2T_1(\tau)}{d\tau^2}+\cdots\right)+\omega_0^2\left(T_0(\tau)+\varepsilon T_1(\tau)+\varepsilon^2T_2(\tau)\right.$$
$$\left.+\cdots\right)=-\omega_0^2\varepsilon\alpha_3\left(T_0(\tau)+\varepsilon T_1(\tau)+\varepsilon^2T_2(\tau)+\cdots\right)^3 \tag{22}$$

Take the first power of ε as an approximation. Spread Equation (22), and compare the coefficients of each power of ε yields

$$\varepsilon^0 : \omega_0^2 \frac{d^2 T_0(\tau)}{d\tau} + \omega_0^2 T_0(\tau) = 0 \qquad (23)$$

$$\varepsilon^1 : \omega_0^2 \frac{d^2 T_1(\tau)}{d\tau^2} + 2\omega_0\omega_1 \frac{d^2 T_0(\tau)}{d\tau^2} + \omega_0^2 T_1(\tau)$$

$$= -\omega_0^2 \alpha_3 T_0(\tau)^3 \qquad (24)$$

where, $T_0(\tau)$, $T_1(\tau)$ satisfy the periodic condition $T_i(\tau + 2\pi) = T_i(\tau)$, $i = 1,2 \dots$.

The membrane has no initial displacement when $t = 0$, so let the initial condition is

$$T_i(0) = 0, i = 1,2 \dots \qquad (25)$$

By solving Equation (23) by the initial condition of (25), one obtains

$$T_0(\tau) = A \cdot \sin \tau \qquad (26)$$

where, A is a constant.

The substitution of Equation (26) into Equation (24) yields

$$\frac{d^2 T_1(\tau)}{d\tau^2} + T_1(\tau) = \frac{A}{4\omega_0}(8\omega_1 - 3A^2\alpha_3\omega_0)\sin \tau$$

$$+ \frac{1}{4}A^3\alpha_3 \sin 3\tau \qquad (27)$$

Equation (27) is a linear forced vibration equation. A homogeneous general solution of Equation (27) is

$$g = B_1 \cos \tau + B_2 \sin \tau$$

where, B_1 and B_2 are two constants.

The inhomogeneous term of Equation (27) is

$$f = \frac{A}{4\omega_0}(8\omega_1 - 3A^2\alpha_3\omega_0)\sin \tau + \frac{1}{4}A^3\alpha_3 \sin 3\tau$$

According to the perturbation theory, in order to make the general solution of Equation (27) not contain secular terms, we must orthogonalize g and f, i.e.

$$\langle g,f \rangle = \frac{1}{2\pi}\int_0^{2\pi} g \cdot f d\tau = 0 \qquad (28)$$

The substitution of g and f into Equation (28) yields

$$\omega_1 = \frac{3\omega_0\alpha_3 A^2}{8}$$

By applying the initial condition of (25) to Equation (24), one obtains

$$T_1(\tau) = B\sin \tau - \frac{\alpha_3 A^3}{32}\sin 3\tau \qquad (29)$$

where, B is also a constant.

The substitution of Equation (26) and Equation (29) into Equation (21) yields

$$T(\tau) = T_0(\tau) + \varepsilon T_1(\tau) + 0(\varepsilon^2)$$

$$= (A + \varepsilon B)\sin \tau - \frac{\varepsilon\alpha_3 A^3}{32}\sin 3\tau \qquad (30)$$

By substituting $\tau = \omega t = (\omega_0 + \varepsilon\omega_1)t$ into Equation (30), one obtains

$$T(t) = (A + \varepsilon B)\sin \omega t - \frac{\varepsilon\alpha_3 A^3}{32}\sin 3\omega t \qquad (31)$$

where, $\omega = \omega_0(1 + (3\varepsilon\alpha_3 A^2/8))$.

Now, we determine the constant A according to the initial conditions. The impact actuation duration for the pellet impacting the membrane is very short. So the system formed by the pellet and membrane can be considered as a conservative system. According to the principle of conservation of momentum, we can obtain the following expression

$$Mv_0 = Mv_0' + \iint_s \rho v_0' W(x,y)ds \qquad (32)$$

The substitution of Equation (32) into Equation (33) yields

$$v_0' = \frac{Mv_0}{M + \frac{4\rho ab}{\pi^2}} \qquad (33)$$

The pellet and membrane have the same initial velocity v_0' at the time of $t = 0$. We can therefore obtain the following initial condition

$$\frac{\partial w(x_0,y_0,t)}{\partial t}\bigg|_{t=0} = \sin\frac{m\pi x_0}{a} \cdot \sin\frac{n\pi y_0}{b} \cdot \frac{dT(t)}{dt}\bigg|_{t=0} = v_0'$$

i.e.

$$\frac{dT(t)}{dt}\bigg|_{t=0} = \frac{v_0'}{\sin\frac{m\pi x_0}{a} \cdot \sin\frac{n\pi y_0}{b}} \qquad (34)$$

612

Table 1. Values of frequency (rad/s) under different initial velocities of pellet.

Order	Initial velocities of the pellet v_0 (m/s)					
	125	100	75	50	25	$v_0 \to 0$
First	1435.95	1400.47	1370.40	1347.26	1332.60	1327.57
Second	2344.56	2278.80	2222.35	2178.43	2150.33	2140.63
Third	3992.31	3870.70	3765.52	3682.98	3629.80	3611.37

The substitution of Equation (26) into Equation (34) yields

$$A = \frac{v_0'}{\omega \sin \dfrac{m\pi x_0}{a} \cdot \sin \dfrac{n\pi y_0}{b}} \quad (35)$$

By substituting Equation (35) into $\omega_0(1 + 3\varepsilon\alpha_3 A^2/8) = \omega$, and solving it, one obtains

$$\omega = \frac{\omega_0}{3} + \frac{2 \cdot 2^{1/3} \cdot C^2 \cdot \omega_0^2}{3\left(16C^6\omega_0^3 + 81C^4 v_0'^2 \omega_0 \alpha_3 \varepsilon + 9\sqrt{32C^{10}v_0'^2 \omega_0^4 \alpha_3 \varepsilon + 81C^8 v_0'^4 \omega_0^2 \alpha_3^2 \varepsilon^2}\right)^{1/3}} + \frac{\left(16C^6\omega_0^3 + 81C^4 v_0'^2 \omega_0 \alpha_3 \varepsilon + 9\sqrt{32C^{10}v_0'^2 \omega_0^4 \alpha_3 \varepsilon + 81C^8 v_0'^4 \omega_0^2 \alpha_3^2 \varepsilon^2}\right)^{1/3}}{6 \cdot 2^{1/3} \cdot C^2} \quad (36)$$

where, $C = \sin m\pi x_0/a \cdot \sin n\pi y_0/b$. Formula (36) is the nonlinear undamped vibration frequency of rectangular pretension orthotropic membrane under an impact force.

4 COMPUTATIONAL EXAMPLES AND DISCUSSIONS

Consider a commonly used membrane material where the Young's modulus in x and y are $E_1 = 1.4 \times 10^6$ kN/m^2 and $E_2 = 0.9 \times 10^6$ kN/m^2, respectively. The aerial density of membranes is $\rho = 1.7$ kg/m^2. The example membrane has a thickness $h = 1.0$ mm, a length $a = 0.4$ m and a width $b = 0.2$ m. The mass of the pellet is taken as $M = 10^{-2}$ kg. The pretension is $N_{0x} = N_{0y} = 10$ kN/m.

The frequencies of the first three orders under different initial velocities of the pellet are computed according to Equation (36). The results are presented in Table 1.

We can obtain some conclusions from the data in Table 1.

1. The frequency of the membrane increases with respect to increasing initial velocity of the pellet. The frequency increases with respect to increasing vibration order.

2. For each vibration mode, the frequency value is minimum when $v_0 \to 0$, and the frequency value equals the small amplitude (i.e. linear) free vibration frequency.

5 CONCLUSIONS

The governing equations of the undamped nonlinear vibration of pretension rectangular orthotropic membrane under impact force were obtained based on Von Kármán's large deflection theory. The governing equations were solved by Bubnov-Galerkin method and L-P (Lindstedt-Poincaré) perturbation method and the asymptotic analytical solution of vibration frequency was obtained.

The computational example proved that by this approach we can compute the nonlinear undamped vibration frequency of each order by using Formulas (36). The frequency of the membrane increases with respect to increasing initial velocity of the impact pellet and vibration order. This reflects the geometric nonlinearity of the vibration of membranes. The frequency value is minimum when $v_0 \to 0$, and the frequency value equals the small amplitude (i.e. linear) free vibration frequency.

The results obtained herein provide a simple and convenient approach to calculate the frequency of large amplitude nonlinear undamped vibration of rectangular orthotropic membranes under a impact force, and provide a computational basis for the vibration control and dynamic design of building or other practical membrane structures or components.

ACKNOWLEDGEMENT

This work is supported by the National Natural Science Foundation of China (No. 51178485), the personnel development project for young and middle-aged core teachers of Chengdu University of Technology and the Chongqing Municipal Construction Committee (Construction and Scientific 2014, No. 0-11-5).

REFERENCES

Goncalves, P.B. Soares, R.M. & Pamplona, D. 2009. Nonlinear vibrations of a radially stretched circular hyperelastic membrane, *Journal of Sound and Vibration*. 327(1–2): 231–248.

Liu, C.J. Zheng, Z.L. & He, X.T. et al. 2010. L-P perturbation solution of nonlinear free vibration of a prestressed orthotropic membrane in large amplitude, *Mathematical Problems in Engineering*. Article ID 561364, 17 pages.

Shin, C. Chung, J.T. & Kim, W. 2005. Dynamic characteristics of the out-of-plane vibration for an axially moving membrane, *Journal of Sound and Vibration*. 286(4–5): 1091–1031.

Soares, R.M. & Gonalves, P.B. 2012. Nonlinear vibrations and instabilities of a stretched hyperelastic annular membrane, *International Journal of Solids and Structures*, 49(3–4): 514–526.

Soares, R.M. & Goncalves, P.B. 2014. Large-amplitude nonlinear vibrations of a Mooney-Rivlin rectangular membrane, *Journal of Sound and Vibration*. 333(13): 2920–2935.

Sunny, M.R. Kapania, R.K. & Sultan, C. 2012. Solution of nonlinear vibration problem of a prestressed membrane by a domian decomposition, *AIAA Journal*, 50(8): 1796–1800.

Wei, D.M. & Shui, Y. 2008. Nonlinear wind-induced vibration response of cable-membrane structures, *Journal of South China University of Technology (Natural Science)*, 36(12): 1–6.

Wetherhold, R. & Padliya, P.S. 2014. Design aspects of nonlinear vibration analysis of rectangular orthotropic membranes, *Journal of Vibration and Acoustics, Transactions of the ASME*. 136 (3): Article ID 034506, 15 pages.

Zheng, Z.L. Guo, J.J. & Song, W.J. et al. 2014. Nonlinear free vibration analysis of axisymmetric polar orthotropic circular membranes under the fixed boundary condition, *Mathematical Problems in Engineering*, Article ID 651356, 8 pages.

Zheng, Z.L. Liu, C.J. & He, X.T. et al. 2009. Free vibration analysis of rectangular orthotropic membranes in large deflection, *Mathematical Problems in Engineering*. Article ID 634362, 9 pages.

Advanced Materials and Structural Engineering – Hu (Ed.)
© 2016 Taylor & Francis Group, London, ISBN 978-1-138-02786-2

Response to acoustic pressure of microstructured optical fibers: A comparison study

A. Abdallah, C.Z. Zhang & Z. Zhong
College of Information and Communication Engineering, Harbin Engineering University, Harbin City, Heilongjiang Province, P.R. China

ABSTRACT: Recently, Photonic Crystal Fibers (PCFs) have attracted many researchers because of their unique properties, and design flexibility that can't be realized by conventional fibers. One of the fruitful areas of research is the optical fiber hydrophones. In this paper, the Finite Element Solver (FES), COMSOL multi-physics, is used to study and compare the response to acoustic pressure of a Hollow-Core Photonic Bandgap Fiber (HC-PBF), a Solid-Core Photonic Crystal Fiber (SC-PCF), and a conventional Single-Mode Fiber (SMF) for different acoustic pressures in the frequency range from 10 kHz to 50 kHz. Simulation results of the investigated optical fibers show that the Normalized Responsivity (NR) of the HC-1550, LMA-5, and SMF are −344 dB, −367.5 dB, and −366 dB, respectively. The proposed results indicate the significance of the HC-PBFs in the future hydrophone systems.

1 INTRODUCTION

Optical fiber hydrophone is an important area of research. For many years, researchers have shown the feasibility of using the conventional SMF hydrophone as an alternative to the conventional Sound Navigation and Ranging (SONAR) technology (Cranch 2003). However, the conventional SMF is made of glass material that has high Young's modulus, and its pressure-induced change of the effective refractive index of the fundamental mode (n_{eff}) has opposite sign with respect to the length change, hence both reduce its NR (Pang & Jin 2009). These limitations motivated the researchers to search for alternatives to increase NR. One possible solution is to test the micro-structured optical fibers that can be classified into two categories; Solid-Core PCF (SC-PCF), and HC-PBF. It was shown experimentally that SC-PCF has the same response to acoustic pressure as that of SMF (Leguillona et al. 2011). However, SC-PCFs offer attractive advantages over SMFs such as the design flexibility, endlessly single-mode operation, and they are almost insensitive to bending allow the fabrication of hydrophones with larger number of arrays, smaller size, and lower cost (Nielsen et al. 2004). It was reported that Large Mode Area (LMA-5) PCF is almost bend insensitive up to bend diameter of 1 cm, that can be used to realize smaller hydrophone size (Nielsen et al. 2004, NKT Photonics website). The aforementioned advantages motivated us to study the feasibility of using LMA-5 in the optical hydrophone systems. It

was reported that HC-PBFs have many advantages over SMF that contribute to better responsivity to measurands and eligibility for many applications (Pang & Jin 2009, Kim et al. 2006). In this paper, the finite element solver COMSOL multi-physics, is used to study the response to acoustic pressure (*p*) of HC-1550 as HC-PBF, Large-Mode Area (LMA-5) as SC-PCF, and the conventional SMF. The proposed simulation results are in good agreement with the theoretical and experimentally measured results proposed in (Pang & Jin 2009, Leguillona et al. 2011).

2 MATHEMATICAL MODEL

The cross-section of a PCF is shown in Figure 1. The PCF is modeled as four circular regions; an air/silica core, a micro-structured inner cladding consists of array of cylindrical air holes, a solid silica outer cladding, and an acrylate coating. The acoustic pressure (*p*) is governed by the wave equation and is given by (COMSOL Multi-physics 2013);

$$\frac{1}{\rho_o c^2}\frac{\partial^2 p}{\partial t^2} + \nabla.\left(-\frac{1}{\rho_o}(\nabla p - q)\right) = Q \qquad (1)$$

where, *t* is the time, ρ_o is the density of the fluid, *q* and *Q* are the acoustic dipole and monopole source, respectively.

The wave equation can be solved in the frequency domain to expand the acoustic signal

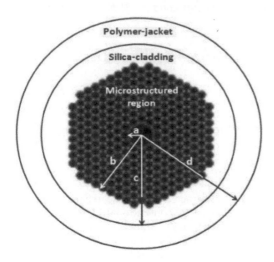

Figure 1. General cross-section of a PCF with an air/solid core, honeycomb air-silica inner cladding, a solid silica outer cladding, and a polymer coating, where (a-d) represent the radius of each layer.

into harmonic components by its Fourier series. A harmonic solution has the form;

$$p(x,y) = p(x)e^{i\omega t} \qquad (2)$$

where, ω is the angular frequency, and the actual physical value of the acoustic pressure is the real part of Equation (2), consequently, the time-dependent wave equation reduces to the Helmholtz equation given by;

$$\nabla \cdot \left(-\frac{1}{\rho_o} (\nabla p - q) \right) - \frac{\omega^2}{\rho_o c^2} p = Q \qquad (3)$$

In the homogenous case where the two source terms q and Q are zero, the solution to the Helmholtz equation is the plane wave given by;

$$p(x,y) = p_o(x)e^{i(\omega t - k.x)} \qquad (4)$$

where, p_o is the amplitude of the wave and k is the wave number. Considering the investigated fiber as the sensing arm of the interferometric hydrophone, the acoustic pressure primarily affect the fiber length (L), and n_{eff} of the phase (φ) of the travelling light through the fiber, where the phase is given by;

$$\varphi = \frac{2\pi}{\lambda} n_{eff} L \qquad (5)$$

where, λ is the wavelength of the propagating light. To calculate n_{eff}, the FES COMSOL multi-physics

are used to solve the vectorial electric field wave equation as an eigenvalue problem to calculate the propagation constant (β), and n_{eff}. NR, is a figure of merit which is independent of the wavelength and optical fiber dimensions, is commonly used to compare between different hydrophone designs. It normalizes the sensitivity of the hydrophone ($S = d\varphi/dp$) by the total optical phase shift, and is given by:

$$NR = \frac{d\varphi}{\varphi(dp)} = \frac{1}{L}\frac{dL}{dp} + \frac{1}{n_{eff}}\frac{dn_{eff}}{dp} = \frac{\varepsilon_z^2}{dp} + \frac{1}{n_{eff}}\frac{dn_{eff}}{dp} \qquad (6)$$

where, $\varepsilon_z^2 = dL/L$ is the axial strain of the micro-structured region of the HC-PBF, and the super-script (2) is used to denote the second region of the HC-PBF which is the micro-structured area.

The elasto-optic effect determines the index variation due to applied acoustic pressure. The general linear stress-optical relation is used to calculate the change of n_{eff} due to the applied acoustic pressure and is given by (COMSOL Multi-physics 2013, Szpulak et al 2004):

$$n_x = n_o - B_1 S_x - B_2(S_y + S_z)$$
$$n_y = n_o - B_1 S_y - B_2(S_x + S_z) \qquad (7)$$
$$n_z = n_o - B_1 S_z - B_2(S_x + S_y)$$

where, n_o is the refractive index of the undeformed material, B_1 and B_2 are the stress-optic coefficients and S_x, S_y, and S_z are the principal components of the induced stresses in the three directions.

The NR of the conventional SMF is calculated from (Cranch 2003);

$$\frac{d\varphi}{\varphi dP} = \frac{1}{dP}\left[\varepsilon_z - \frac{n^2}{2}[(p_{11} + p_{12})\varepsilon_r + p_{12}\varepsilon_z] \right] \qquad (8)$$

where, ε_z and ε_r are the axial and radial components of the strain.

3 SIMULATION RESULTS AND ANALYSIS

In this section, simulation results are introduced. The material's physical parameters of the investigated optical fibers were reported in (Pang & Jin 2009, Nielsen et al. 2004, NKT Photoncs website). The response of the investigated fibers to acoustic pressure is studied by coupling between the Acoustic Solid-Interaction (ASI), and the optics (OM) modules in the FES. This allows easy transfer of the required data between the two modules, and provides accurate calculations. After setting up the model geometry, and importing the required equations and parameters to the FES, the ASI is used to apply acoustic pressures of different amplitudes and frequencies that cause optical fiber's structural deformation and hence the induced stresses, strains of the investigated fibers are calculated. The ASI exchange data with the OM by coupling between them, then by performing mode analysis, n_{eff} corresponding to the deformed structure is calculated. This allows calculating the phase change and NR for the investigated fibers given by Equations (5), (6), and (8), respectively. The micro-structured regions of the PCFs are modeled as anisotropic materials while the silica outer cladding and the acrylate regions are modeled as isotropic materials (Pang et al 2010). The effect of the coating material and thickness on NR was proposed by many researchers and recently in (Yang et al. 2013). To avoid inaccurate comparison between the three investigated fibers because of different coating materials and thickness, the same coating material and thickness for all fibers are used.

Calculating the radial displacements of the investigated fibers can be used to study the flexibility of the fiber material for different acoustic pressures. Figure 2 and Figure 3 show the calculated radial displacement as a function of the optical fiber radius (r) for HC-1550, LMA-5, and SMF, for applied acoustic pressure of 5 Pa and acoustic frequency (f) of 10 kHz. In Figure 3, the radial displacement data of the investigated fibers is fitted by shape-preserving interpolant in MATLAB.

It can be seen from Figure 3, that the honeycomb region of the HC-1550 (5 < r < 35 μm) has peaks of displacement values in regions corresponding to the deformed silica between the air holes while the solid-silica region (35 < r < 60 μm) has very

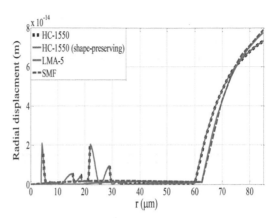

Figure 3. Radial displacement as a function of fiber radius for (a) HC-1550, (b) LMA-5, and (c) SMF, for P = 5 Pa, f = 10 kHz.

small values due to the high Young's modulus of the solid-silica material that limits its deformation. From Figure 3, it is shown that despite the presence of the air-holes in the microstructure region of the LMA-5, its radial displacement is approximately the same as that of the SMF. This can be attributed to the presence of the solid-core and the low air-filling ratio of the micro-structured area that limit its ability to deform. For SMF, it is seen that in the silica region (0 < r < 62.5 μm), the displacement is generally very small. This can be attributed to the high Young's modulus of the silica material of the core and cladding that limits the flexibility of the material. Generally, for the three optical fibers the region with the highest radial displacement is the polymer coating region (r > 60 μm) in which the acrylate material has low Young's modulus allows it to deform easily. As a conclusion, for the same applied acoustic pressure and frequency the radial displacement of the HC-1550 is higher than the LMA-5 and SMF. This indicates that generally, HC-PBFs are more compressive and consequently, have a higher sensitivity to acoustic pressure than that of SC-PCF and SMF (Pang & Jin 2009, Lagakos et al. 1990).

The FES enables us to calculate the NR of the investigated optical fibers as shown in Figure 4. It can be seen that the calculated average NR for applied acoustic pressure of 1 Pa in the frequency range from 10 kHz to 50 kHz for the HC-1550, LMA-5, and SMF are −344 dB, −367.5 dB, and −366 dB, respectively. HC-1550 has the highest NR due to its air-core and the high air-filling ratio in the honeycomb area. This reduces the amount of silica in the honeycomb area, reduces the index term, in Equation (6), of the NR and consequently, increases the NR.

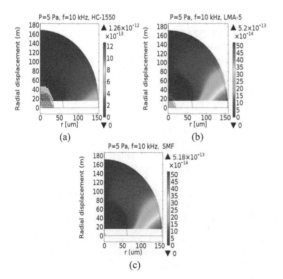

Figure 2. Radial displacement profile for (a) HC-1550, (b) LMA-5, and (c) SMF, for p = 5 Pa, f = 10 kHz.

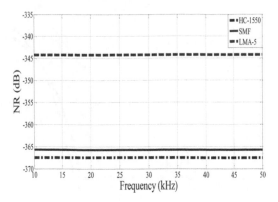

Figure 4. NR as a function of acoustic frequency for the HC-1550, LMA-5, and SMF for acoustic pressure of 1 Pa.

In (Pang & Jin 2009), the experimental results showed that the *NR* of HC-1550 is about 15 dB higher than that of the conventional SMF, while the proposed simulation results show a difference of about 22 dB. We attribute this 7 dB difference to the use of the real structure in (Pang & Jin 2009) that may contain imperfections that can reduce the *NR* of the HC-1550 compared to the ideal structure that we used in the simulation. Simulation results show that the *NR* of the LMA-5 is 1.5 dB lower than that of SMF, while it was expected that LMA-5 has better *NR* because of the presence of the micro-structured area that has smaller equivalent Young's modulus than that of the SMF. This can be attributed to the following reasons; (1) the large amount of silica in the micro-structured area because of the solid-core, and the low air-filling ratio (44%) that limits the ability of LMA-5 to deform, (2) as the light propagates in solid-core, the index term is significant and this reduces the overall *NR* of the LMA-5. In addition to the sensitivity to acoustic pressure, size of the sensor is also an important factor. Smaller size optical hydrophone requires an optical fiber that can be bent to very small diameters with minimum bending loss. It is expected that LMA-5 is suitable for this purpose as it was reported that it is almost insensitive to bending which is an advantage of LMA-5 compared to the conventional SMF (Nielsen et al. 2004).

4 CONCLUSION

As a conclusion, the FES COMSOL multi-physics are used to study and compare the sensitivity to acoustic pressure of HC-1550, LMA-5, and SMF. The proposed simulation results showed that for the same applied acoustic pressure and frequency the sensitivity to acoustic pressure of the HC-1550 is higher than that of LMA-5 and SMF by 23.5 dB and 22 dB, respectively. This is as a result of its air core and the high air-filling ratio in the honeycomb area allows the induced axial strain to increase and consequently, the NR. It was shown that despite the presence of the air-holes in the microstructure region of the LMA-5, its sensitivity to acoustic pressure is lower than that of SMF by 1.5 dB. However, LMA-5 has other advantages make them preferable as it allows the fabrication of smaller hydrophone size.

ACKNOWLEDGEMENT

This work is supported by National Natural Science Foundation of China (61102004), and Fundamental Research Funds for the Central Universities.

REFERENCES

COMSOL Multiphysics, 2013. Wave optics module user's guide.
Cranch, G.A. 2003. Large-scale remotely interrogated arrays of fiber-optic interferometric sensors for underwater acoustic applications, *IEEE Sensors Journal*, 3(1): 19–30.
Kim, H.K. Digonnet, M.J.F. & Kino, G.S. 2006. Air-Core Photonic-Bandgap Fiber-Optic Gyroscope, *Journal of Light Wave Technology*, 24(8): 3169–3174.
Lagakos, T.R.H.N. Ehrenfeuchter, T.R.H.N. Bucaro, J.A. Dandridge, A. 1990. Planar Flexible Fiber-optic Acoustic Sensors, *Journal of Lightwave Technology*, 8(9): 1298–1303.
Leguillona, Y. et al, 2011. *Phase sensitivity to axial strain of microstrustured optical silica fibers*, Proceedings of the 21st International Conference on Optical Fiber Sensors, Ottawa, Canada.
Nielsen, J.R.F.M.D. Mortensen, N.A. & Bjarklev, A. 2004. Bandwidth comparison of photonic crystal fibers and conventional single-mode fibers, *Optical Society of America*, 12(3): 430–435.
NKT Photonics website, http://www.nktphotonics.com/hollowcorefibers.
Pang, M. & Jin, W. 2009. Detection of acoustic pressure with hollow-core photonic bandgap fiber, *Optics Express*, 17(13): 11088–11097.
Pang, M. Xuan, H.F. Ju, H.F. & Jin, W. 2010. Influence of strain and pressure to the effective refractive index of the fundamental mode of hollow-core photonic bandgap fibers, *Optics Express*, 18(13): 14041–14055.
Szpulak, M. Martynkien, T. & Urbanczyk, W. 2004. Effects of hydrostatic pressure on phase and group modal birefringence in microstructured holey fibers, *Applied Optics*, 43(24): 4739–4744.
Yang, F. Jin, W. Ho, H.L. Wang, F. Ma, L. & Hu, Y. 2013. Enhancement of acoustic sensitivity of hollow-core photonic bandgap fibers, *Optics Express*, 21(13): 15514–15521.

Advanced Materials and Structural Engineering – Hu (Ed.)
© 2016 Taylor & Francis Group, London, ISBN 978-1-138-02786-2

Study on heat transfer and recuperator effectiveness for microturbine application: Channel of rectangular with semicircular ends

S. Rilrada, M. Thanate & N. Udomkiat

Department of Mechanical and Aerospace Engineering, King Mongkut's University of Technology North Bangkok, Bangkok, Thailand

ABSTRACT: This research paper presents the study on heat transfer and recuperator effectiveness of Primary Surface Recuperator (PSR) for microturbine application. The flow channel of study is like a rectangular channel with semicircular ends. The exhausted gas and the compressed air are in the counter-flow directions. The PSR is designed based on the microturbine recuperator requirements and the 30 kW microturbine-operating parameters. The exact solutions for temperature distribution, obtained from the thermodynamic modeling with the help of the analytical method and Laplace Transform, for both fluids are solved and are verified by computational fluid dynamic program package. The appropriate size of PSR is calculated through the thermal design process and built-in function of the MATLAB program. The temperature distributions of the exhausted gas and the compressed air along the length of PSR are visualized. The overall heat transfer coefficient, the convective heat transfer coefficient, the Nusselt number and the Reynolds number, the Number of Transfer Unit (NTU), the recuperator effectiveness, and pressure drop for distinct corrugated foils profiles are also investigated.

1 INTRODUCTION

A recuperator is a direct transfer kind of heat exchanger that is extensively used in microturbine systems. The thermal efficiency of the microturbine generator is about 20% or less if no recuperator was applied and increases to about 30% of the system with recuperator operating at 87% effectiveness (McDonald 2000). The recuperated microturbine generator system and the air standard Brayton cycle for the recuperated microturbine generator are shown in Figure 1 and Figure 2, respectively. As shown in the figures, the recuperator transfers heat from the turbine exhaust to preheat compressed high-pressure air before going into the combustion chamber. The utilization of the recuperator operating at high effectiveness leads to higher in the thermal efficiency of the microturbine. Moreover,

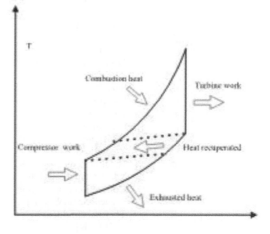

Figure 2. The air standard Brayton cycle for the recuperated microturbine system.

the recuperator also helps reduce the specific fuel consumption rate of the system.

Since the recuperator costs about 25–30% of the total power plant cost, therefore the recuperator must have high performance with minimum cost. The contemporary recuperator design for the microturbine system use prime surface geometry of hot fluid and cold fluid sides with no brazing, stacking, and welded at the side edges to form

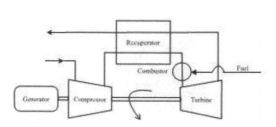

Figure 1. The recuperated microturbine generator system.

airflow passage. This configuration prevents the leaks and mixing of the fluids (McDonald 2003). The primary surface recuperator is a kind of recuperator that the surface geometry is 100% effective and can be built to meet all the desired performance requirements for low cost microturbine recuperators that are high exchange effectiveness, >90%, with low-pressure drop, <5%, counter-flow arrangement and uniform flow distribution, compact and lightweight, continuous/automated fabrication process, and matrix envelope flexibility.

In 2003, Muley and Sunden summarized the performance requirements for microturbine recuperator. They described the primary surface recuperator with counter flow arrangements. The plates have corrugation in the heat transfer region and are stacked and welded at the edges to form the flow passages. The exhausted gas and the compressed air are in the alternate flow passage. The design and analysis of a primary surface recuperator for microturbine are presented by applying the multi-objective optimization method (Liu 2008). The numerical study of cross-flow printed circuit heat exchange for advanced small modular reactors is also studied (Yoon 2014).

The paper developed the general methods for thermal design and cost estimation of printed circuit heat exchanger. The research paper developed the general methods for thermal design and cost estimation of printed circuit heat exchanger.

In this research paper, the study on heat transfer and recuperator effectiveness of the primary surface recuperator with distinct corrugated profiles are presented. The flow channel of study is as a rectangular channel with semicircular ends. The primary surface recuperator is designed based on the aforementioned requirements and the 30 kilowatts microturbine boundary condition.

2 PRIMARY SURFACE RECUPERATOR GEOMETRY

The domain of the study is the matrix contains the heat transfer surface area. The honeycomb core of primary surface recuperator is made up of 304 stainless steel corrugated foils. The transfer of enthalpy occurs by conduction through the corrugated foils. A schematic of the counter-flow primary surface

Table 1. Nomenclature.

a	Radius of the semicircular ends [m]	T	Temperature [°C]
A_o	Flow channel surface area [m²]	t	Plate thickness [m]
A_t	Overall heat transfer surface area [m²]	U_o	Overall heat transfer coefficient [W/m²K]
c_p	Specific heat capacity [kJ/kg K]		
D_h	Hydraulic diameter [m]	V	Fluid velocity [m/s]
f	Resistance coefficient		
h	Height of the corrugated foils [m]	*Greek and Symbols*	
h_j	Convection heat transfer coefficient [W/m²K]	ε	Recuperator effectiveness
K	Thermal conductivity [W/m K]	σ	The ratio of the flow sectional area to the total front face area
K	Pressure drop coefficient		
L	Length of the recuperator [m]		
ṁ	Mass flow rate [kg/s]	*Subscript*	
N	Number of flow channel	1,h	Exhausted gas (primary fluid)
NTU	Number of transfer unit ($= U_o A_t / \dot{m} c_p$)	2,c	Compressed air (secondary fluid)
Nu	Nusselt number (= hD/k)	x	X-direction
P	Pressure	y	Y-direction
Pr	Prandtl number ($= C_p \mu / k$)		
Re	Reynolds number ($= \rho VD/\mu$)		

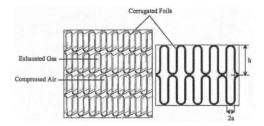

Figure 3. The cross section of the primary surface recuperator core.

recuperator is shown in Figure 3. The dimensions of the recuperator are L_X, L_Y, and L_Z.

3 THERMODYNAMIC MODELING AND ANALYTICAL METHOD

For a single-pass counter flow and both unmixed fluids, the energy balance for the flow channel can be written using the first law of thermodynamics with the assumption of uniform distribution of the total heat transfer surface area along the flow channel. With the help of dimensionless parameter

$$\theta_j = \frac{T_j - T_{1,in}}{T_{2,in} - T_{1,in}}, \quad \xi = \frac{x}{L} \quad \text{for } j = 1, 2 \tag{1}$$

The energy balance for the flow channel in dimensionless form are shown as

$$\frac{d\theta_1}{d\xi} + NTU_1(\theta_1 - \theta_2) = 0, \tag{2}$$

$$\frac{d\theta_2}{d\xi} - NTU_1 R_1(\theta_1 - \theta_2) = 0 \tag{3}$$

Applied the boundary conditions at the inlets ($\xi = 0$ and $\xi = L$), the general form of solutions can be obtained using the Laplace transforms method as shown.

$$\theta_1(\xi) = \frac{1 - e^{-\xi NTU_1(1-R_1)}}{1 - R_1 e^{-NTU_1(1-R_1)}} \tag{4}$$

$$\theta_2(\xi) = \frac{1 - R_1 e^{-\xi NTU_1(1-R_1)}}{1 - R_1 e^{-NTU_1(1-R_1)}} \tag{5}$$

where, $NTU_1 = \dfrac{UA}{(\dot{m}C_p)_1}$, $NTU_2 = \dfrac{UA}{(\dot{m}C_p)_2}$,

$$R_1 = \frac{NTU_2}{NTU_1}. \tag{6}$$

The exact solutions for temperature distribution of both fluids are obtained as

$$T_1(x) = T_{1,in} + \left(T_{2,in} - T_{1,in}\right)\left(\frac{1 - e^{\frac{-NTU_1 x(1-R_1)}{L}}}{1 - R_1 e^{-NTU_1(1-R_1)}}\right) \tag{7}$$

$$T_2(x) = T_{1,in} + \left(T_{2,in} - T_{1,in}\right)\left(\frac{1 - R_1 e^{\frac{-NTU_1 x(1-R_1)}{L}}}{1 - R_1 e^{-NTU_1(1-R_1)}}\right) \tag{8}$$

4 THERMAL DESIGN AND THERMAL DESIGN PROCESS

From the assumed geometry information, the number of compressed air flow channels (N_c) and exhausted gas flow channels (N_h) in both x and y directions are determined as shown. The heat transfer surface area (A_o) and the hydraulic diameter (D_h) of the flow channel are also developed as

$$N_{c,x} = \frac{L_x}{2a + t}, \tag{9}$$

$$N_{c,y} = \frac{L_y}{2h + 2t}, \tag{10}$$

$$N_{h,x} = N_{c,x} - 1, \tag{11}$$

$$N_{h,y} = N_{c,y} - 1, \tag{12}$$

$$A_o = 2L_z(2h + a(\pi - 2)), \tag{13}$$

$$D_h = \frac{2(4ah + a^2(\pi - 4))}{2h + a}. \tag{14}$$

The mass flow rate of the fluids is given based on the 30 kW-microturbine operating conditions. The velocity, the Reynolds number and the Prandtl number of the exhausted gas and the compressed air are calculated as follows

$$V_j = \frac{\dot{m}}{\rho_j A N_{j,x} N_{j,y}} \tag{15}$$

$$Re_j = \frac{\rho_j V_j D_h}{\mu_j} \tag{16}$$

$$Pr_j = \frac{c_{p,j} \mu_j}{k_j} \quad \text{for } j = 1, 2 \tag{17}$$

The heat transfer coefficient is calculated from the empirical heat transfer correlation. The heat transfer correlation applied in this paper is based on (Liu & Cheng 2008, Xin et al. 1995). The heat transfer coefficient for both fluids and also the overall heat transfer coefficient of counter-flow primary surface recuperator are determined as,

$$Nu = 0.0031 Re^{1.18} Pr^{0.4} \left(\frac{h}{a}\right)^{0.19} \tag{18}$$

where, $Re < 1000, 1 \le \dfrac{h}{a} \le 9$

$$h_j = \frac{Nu_j k_j}{D_h} \quad \text{for } j = 1, 2 \qquad (19)$$

$$U_o = \frac{1}{\left(\dfrac{1}{h_1} + \dfrac{t}{k} + \dfrac{1}{h_2}\right)} \qquad (20)$$

The heat exchanger effectiveness, ε, for a direct transfer kind of heat exchanger can be calculated as,

$$\varepsilon = \frac{1 - e^{-NTU(1-C^*)}}{1 - C^* e^{-NTU(1-C^*)}} \qquad (21)$$

The flow passages of the exhausted gas and the compressed air are divided into two zones that are the main flow zone and the diversion zone. The main flow zone is the primary zone that the transfer of enthalpy occurs. The corrugated foils are contained in this zone. Whereas the diversion zone is the area that distributes or collects the fluids when the fluids enter or exit the main flow zone. The inlet and outlet pressure drop of the diversion zone can be calculated by (Yoon 2014).

$$\Delta p_{in} = \frac{\rho V^2}{2}\left[\left(1 - \sigma^2\right) + K_{in}\right] \qquad (22)$$

$$\Delta p_{exit} = \frac{\rho V^2}{2}\left[\left(1 - \sigma^2\right) - K_{exit}\right] \qquad (23)$$

Since the local pressure loss in the diversion zone is extremely small when compare with the pressure loss in the main flow zone, therefore the pressure drop in this zone can be ignored. The total pressure drop of the exhausted gas and the compressed air in the main flow zone can be determined as,

$$\Delta p_j = 56 \rho_j v_j V_j \frac{L}{D_h^2} \quad \text{for } j = 1, 2 \qquad (24)$$

5 RESULTS AND DISCUSSIONS

The analytical solutions for temperature distributions along the PSR length, as shown in Equation 7–Equation 8, are verified against the solution from the computational fluid dynamic program package as shown in Figure 4.

The mass flow rate of the working fluid is equal to 0.21 kg/s. The temperature at the inlet of the primary surface recuperator in the exhausted gas side and the compressed air side are equal to 800 K and 460 K, respectively. The pressure condition at the inlet of the primary surface recuperator in the exhausted gas side and the compressed air side are equal to a bar and 2.5 bars, respectively.

To meet heat requirements of the 30 kW micro-turbine conditions, the size of PSR is calculated through the thermal design process (Liu & Cheng

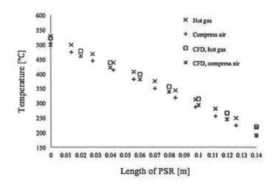

Figure 4. The analytical solutions for temperature distributions along the PSR length (Equation 7-Equation 8) are verified against the solution from computational fluid dynamic program package.

2008) and built-in function of the MATLAB program as shown in Figure 5.

The effect of distinct corrugated foil profiles on the Reynolds number, the Nusselt number, exhaust gas and compressed air velocity, the Number of Transfer Unit (NTU) and recuperator effectiveness (ε) are shown in Table 2. As shown in the table, the effectiveness and the NTU increase as the ratio of the height of the corrugated foils to the radius of the semicircular ends increase. The influence of the ratio h/a on PSR effectiveness is shown in the Figure 6.

Figure 7 shows the influence of the ratio h/a on the temperature at the PSR exit.

It is found that both the radius of the semicircular ends and the height of the corrugated foils have effects on the convection heat transfer coefficients, pressure drop on both fluid sides and also the recuperator effectiveness. The influences of the radius of the semicircular ends and height of the corrugated foils on convection heat transfer coefficient, pressure drop, and PSR effectiveness are presented in Figure 8 to Figure 13, respectively.

Figure 8 to Figure 10 shows the variation of convection coefficient, pressure drop and PSR effectiveness for different radius of the semicircular ends when the height of the corrugated foil is equal to 0.0045 m.

As shown in the figures, as the radius of the semicircular ends increases, the convection coefficient, pressure drop, and also the recuperator effectiveness decrease.

Figure 11 to Figure 13 shows the variation of convection coefficient, pressure drop and PSR effectiveness for different height of the corrugated foils when the radius of the semicircular end is equal to 0.0005 m. It can be seen that, as the height of the corrugated foils increase, the convection coefficients and also the recuperator effectiveness increase.

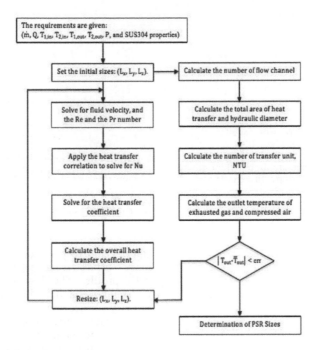

Figure 5.   PSR Thermal design process.

Table 2.   The effect of distinct corrugated foil profiles on Re, Nu, V, NTU and ε.

	Exhausted gas			Compressed air				
h/a	Re[h]	Nu[h]	V[h]	Re[c]	Nu[c]	V[c]	NTU	ε
9.0	119.54	6.88	3.89	119.39	6.87	1.45	12.53	0.938
8.5	119.46	6.50	3.91	119.61	6.50	1.46	11.88	0.924
8.0	119.39	6.11	3.94	119.87	6.13	1.47	11.24	0.920
7.5	119.35	5.73	3.97	120.16	5.77	1.48	10.61	0.916
7.0	119.31	5.34	4.01	120.49	5.40	1.49	9.97	0.911
6.5	119.26	4.96	4.05	120.86	5.03	1.50	9.33	0.905
6.0	119.21	4.58	4.10	121.29	4.66	1.52	8.70	0.899
5.5	119.18	4.19	4.16	121.79	4.29	1.54	8.06	0.892

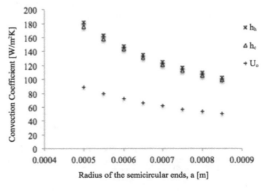

Figure 7.   The influences of the ratio h/a on the temperature at the PSR exit.

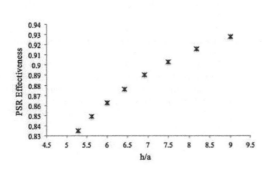

Figure 6.   The influences of the ratio h/a on PSR effectiveness.

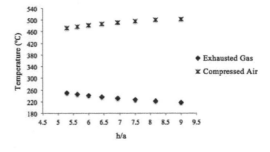

Figure 8.   Variation of convection coefficient [W/m² K] for different radius of the semicircular ends [m] (h = 0.0045 m).

623

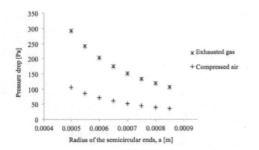

Figure 9. Variation of pressure drop [Pa] for different radius of the semicircular ends [m] (h = 0.0045 m).

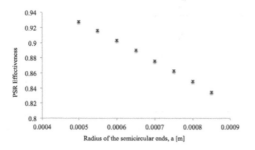

Figure 10. Variation of PSR effectiveness for different radius of the semicircular ends [m] (h = 0.0045 m).

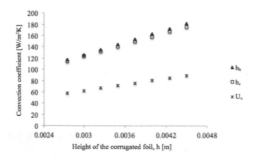

Figure 11. Variation of convection coefficient [W/m² K] for different height of the corrugated foils [m] (a = 0.0005 m).

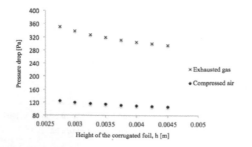

Figure 12. Variation of pressure drop [Pa] for different height of the corrugated foils [m] (a = 0.0005 m).

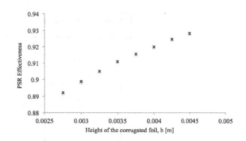

Figure 13. Variation of PSR effectiveness for different height of the corrugated foils [m] (a = 0.0005 m).

6 CONCLUSIONS

This research paper presents the study of heat transfer and recuperator effectiveness of the primary surface recuperator with distinct corrugated profiles. The flow channel of study is as a rectangular channel with semicircular ends. The primary surface recuperator is designed based on the aforementioned requirements and 30 kilowatts microturbine boundary condition. With the help of thermodynamic modeling, Laplace transform and MATLAB built-in function, the analytical and numerical solutions are inspected. The investigated data are valuable to the design of primary surface recuperator for microturbine application.

REFERENCES

Liu, Z.G. & Cheng, H. 2008. Multi-objective optimization design analysis of primary surface recuperator for microturbines, *Applied Thermal Engineering,* 28: 601–610.

McDonald, C.F. 2000. Low-cost compact primary surface recuperator concept for microturbines, *Applied Thermal Engineering,* 20: 471–497.

McDonald, C.F. 2003. Recuperator considerations for future higher efficiency microturbines, *Applied Thermal Engineering,* 23(12): 1463–1487.

Muley, A. & Sunden, B. 2003. *Advanced in recuperator technology for gas turbine systems,* ASME, IMECE 2003-43294.

Xin, M.D. Zhang, P.J. & Yang, J. 1995. Convective heat transfer of air in micro-rectangular channels, *Journal of Engineering Thermophysics,* 16(1): 86–90.

Yoon, S.J. Sabharwall, P. & Kim, E.S. 2014. Numerical study on crossflow printed circuit heat exchanger for advanced small modular reactors, *International Journal of Heat and Mass Transfer,* 70: 250–263.

Advanced Materials and Structural Engineering – Hu (Ed.)
© 2016 Taylor & Francis Group, London, ISBN 978-1-138-02786-2

Role of the hydration force in the swelling pressure of montmorillonite

Z.Q. Huang & X.C. Huang

School of Resources and Environment, North China University of Water Resources and Electric Power, Zhengzhou, China

ABSTRACT: Although the theoretical results calculated by the Gouy-Chapman model are close to the experimental results of long-range swelling pressures, this model could not simulate the adsorption of the cations on the clay layer surface. But the adsorption of the cations could be simulated by the Gouy-Chapman-Stern-Grahame (GCSG) model, and this model could also explain this experimental phenomenon that the zeta potentials of the montmorillonite layer surfaces are almost independent of the pH values. So, in this paper, the long-range swelling model based on the GCSG model is established to simulate the long-range swelling pressures. But the swelling pressures calculated by this model are rather lower than the observed data. Because it is maybe due to the hydration forces, which play a dominant role in the long-range swelling.

1 INTRODUCTION

Determination of the swelling pressure of expansive soils is required in many situations concerned with stability problems of foundations, retaining walls and slope stability of embankments. Recently two decades expansive soils such as bentonite have been used as a mixture backfill material, for example as backfill material for nuclear waste disposal systems (Pusch 1982, Bucher & Muller-Vonmoos 1989). The swelling behavior of expansive soils is due to the swelling mineral: montmorillonite. Because of the isomorphous substitution in the crystal lattice, in general, the montmorillonite crystal layers carry negative charges on the surfaces of the platelets. So cations are tightly adsorbed on the surface to balance the negative charges. In the presence of water, these cations tend to hydrate, thereby forcing the clay layers apart in a series of discrete steps, which cause the initial swelling of montmorillonite (Olphen 1963).

It is generally believed that the superimposed layers of montmorillonites exhibit two stages of expansion or swelling: crystalline swelling and osmotic swelling or diffuse double layer swelling (Olphen 1963, Norrish 1954). To avoid any inferences about the mechanisms involved, Low et al. (1995) used the term short-range swelling for the first stage of swelling and the term long-range swelling for the second stage of swelling. In this paper, the mechanism of the long-range swelling of montmorillonites is investigated systematically.

2 EXAMINATION OF THE DIFFUSE DOUBLE LAYER THEORY

Stern (Olphen 1963) has altered the classic diffuse double layer model for a solid wall by dividing the liquid charge into two parts. One part is thought of as a layer of ions adsorbed to the wall and is represented by a surface charge concentrated in a plane at a small distance from the surface charge on the wall. This distance is assumed to be of the order of magnitude of the radius of the adsorbed cations. The second part is then taken to be a diffuse space charge, as described in the Gouy-Chapman theory. The first part is called Stern layer. Then Grahame (Olphen 1963) gave a more precise model, Gouy-Chapman-Stern-Grahame model, in which the Stern layer is divided into two layers by Inner Helmholtz Plane (IHP) and Stern plane is also regarded as the Outer Helmholtz Plane (OHP).

Many experimental data (Nishimura et al. 2002, Avena & Pauli 1998, Miller & Low 1990, Chan et al. 1984) indicate that the zeta potential is almost independent on the pH values and the electrolyte concentrations. The zeta potential is commonly believed equal to the potential at the Stern plane or OHP plane (Nishimura et al. 2002). Nishimura et al. (2002) and Avena & Pauli 1998) respectively derived models based on the GCSG model to simulate their measured data. The calculated results with their model indicated that the majority of cations located in the Stern layer. This is similar to the computer simulation results obtained by de Carvalho & Skipper (2001). So the exists of the

Stern layer should be considered when derive the swelling model for the second swelling stage.

The model proposed by Nishimura (2002) are shown as follows:

$$K_h = \frac{[S^-][H^+]\exp(z_i)}{[SH]} \qquad (1)$$

$$K_N = \frac{[S^-][N_a^+]\exp(z_i)}{[SNa]} \qquad (2)$$

$$\sigma_i = e([SH]+[SNa]) \qquad (3)$$

$$\sigma_d = e([S^-]) \qquad (4)$$

$$\sigma_d = \sqrt{8n_0\varepsilon_r\varepsilon_0 kT}\sinh\left(\frac{z_d}{2}\right) \qquad (5)$$

$$\sigma = \sigma_i + \sigma_d \qquad (6)$$

$$\sigma_d = C(\phi_i - \phi_d) \qquad (7)$$

where, the K_h and K_N are the surface dissociation constant of H^+ and Na^+, $[S^-]$ is the surface density of negative surface sites, $1/m^2$, $[Na^+]$ and $[H^+]$ are the bulk concentration of Na^+ and H^+, respectively, $[SNa]$ and $[SH]$ denote the surface density of the bounded Na^+ and H^+, $1/m^2$, C is the capacitance of the OHP, $C^2N^{-1}m^{-3}$, σ, σ_i, σ_d are the charge densities, ϕ_i and ϕ_d are potentials at the plane of surface, IHP and OHP.

The zeta potential calculated by the model of Nishimura and the data measured by Miller & Low (1990) is shown in Table 1. The least square method was used to optimize parameters, and the value of K_h, K_N, and C for Miller's data are 0.0064, 0.335, 0.007, respectively.

In those experiments (Nishimura et al. 2002, Avena & Pauli 1998, Miller & Low 1990, Chan et al. 1984), the measured zeta potentials are for the single clay layer. So the Equation (5) should be replaced by Equation (8) in order to involve the effects of the interaction between parallel layers.

Table 1. Zeta potentials measured by Miller & Low (1990) and calculated by the model of Nishimura in a 10^{-4} M NaCl solution.

pH	Measured data (mV)	Calculated data (mV)
5.71	58.0	58.3
6.31	60.5	59.9
6.70	61.7	60.3
7.40	60.4	60.6
7.48	57.9	60.6
7.80	60.4	60.6

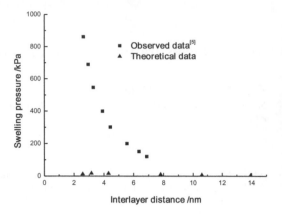

Figure 1. Theoretical and observed data of swelling pressure-interlayer distance for Upton montmorillonite at 10^{-4} M.

$$\sigma_d = \sqrt{2n_0\varepsilon_r\varepsilon_0 kT}\left(2\cosh(z_d)-2\cosh(u)\right) \qquad (8)$$

Then the long-range swelling model based on the GCSG is derived, which includes the Equation (1)–(4) and (6)–(8). The montmorillonites used by Miller & Low (1990) and Zhang et al. (1995) both are the Upton montmorillonite. Hence, the parameters determined from the Miller's data could be used to fit the measured results of Zhang. However, the calculated results by the long-range swelling model are very lower than the experimental data, shown in Figure 1. Similar deviation also found by Low (1987). It may be concluded that the diffuse double layer swelling are not the main effects even after the crystalline swelling. The hydration forces may cause the interaction of the clay layers at the second stage, which stage could be called hydration swelling.

3 THE HYDRATION SWELLING

The appearance of hydration forces has been ascribed to different mechanisms: interaction between hydrated ions in the diffuse layer (Marcelja 1997), presence of an adsorbed layer of hydrated cations at the interface (Pashley 1981), polarizations effects of surface dipoles or charges on the nearest water molecules (Schiby & Ruckenstein 1983), formation and breakage of hydrogen bonds between water molecules and surface groups (Attard & Batchelor 1988).

From the calculated results (Avena & Pauli 1998) and the computer simulation (Carvalho & Skipper 2001), there are little hydrated cations in the diffuse layer, so the possibility of hydration forces caused by the interaction between hydrated ions in the diffuse layer is very small.

626

The thicker of the Stern layer of a single layer is not exceeding 1 nm, so the range of the hydration forces caused by the adsorbed layer of hydrated cations may be not larger than 2 nm. So the second hypothesis may be useful to the crystalline swelling, but not the hydration swelling.

Ruckenstein and co-workers (Schiby & Ruckenstein 1983) think that the repulsive forces have their origin in the polarization induced by these dipoles on the neighboring water molecules propagates towards successive water layers. But their calculated results decay very quickly for any reasonable value of the bulk defect concentration (Attard & Batchelor 1988).

Marclja and Radic (1976) presented a phenomenological model that is considered the first theoretical model for hydration forces. According to this model, the origin of the hydration forces is associated with a larger order of the water molecules next to the surfaces than in the bulk. When the interlayer distance decreases, the ordering of the oriented water molecules is disturbed, so the free energy increases, hence the repulsive forces are larger. In that model, the system is described in terms of an order parameter that depends on the distance to the surface. The behavior of the order parameter follows from minimization of the free energy density and leads to the following expression of the repulsive pressure p between interfaces at a distance h:

$$p = 4p_0 \exp(-h/\xi) \qquad (9)$$

The value of p_0 (the pressure when the surfaces are at close contact) is determined by the extent to which the surface orders the water and, therefore, depends on the properties of the surface. The decay constant ξ is determined by the extent to which the ordering is propagated through water and therefore, according to the MR theory, is a property of water only. Kornyshev (1983) demonstrated that the decay constant is dependent on the nature of the surface. Forsman et al. (1997) made a conclusion that the range and amplitude of repulsive solvation force strongly depend on the nature of the surface-solvent interaction or by the range of the surface potentials. According to Besseling (1997), if the main effect of a surface is to influence the orientational distribution in adjoining water layers, the interaction between two such surfaces is repulsive, and this repulsive force is associated with the disruption of hydrogen bonding in the hydration layer. Hence in our view, the hydration force may be attributed to the formation and breakage of hydrogen bonds between water-water and between surface-water and the hydrogen bonds are affected by the surface potential. A detailed research of the mechanism of the hydration swelling will be continued in the future.

4 CONCLUSIONS

In this work, the long-range swelling mechanism of the montmorillonite is studied systematically. In the stage of long-range swelling, the majority of cations are adsorbed on the montmorillonite clay layers. Although this phenomenon could not be described by the Gouy-Chapman model, it is could be simulated by the Gouy-Chapman-Stern-Grahame (GCSG) model. So a long-range swelling model is established based on the GCSG model. But the swelling pressures calculated by this model are rather lower than the observed data. The reason for it is maybe due to the hydration forces.

ACKNOWLEDGEMENT

This work was financially supported by the Plan for Scientific Innovation Talent of Henan Province and the Key Scientific and Technological Project of Henan Province.

REFERENCES

Attard, P. & Batchelor, M.T. 1988. A mechanism for the hydration force demonstrated in a model system. *Chemical physics letters,* 149(2): 206–211.

Avena, M. & De Pauli, C.P. 1998. Proton adsorption and electrokinetics of an Argentinean Montmorillonite. *Journal of colloid and interface science,* 202: 195–204.

Besseling, N.A.M. 1997. Theory of hydration forces between surfaces. *Langmir,* 13: 2113–2122.

Bucher, F. & Muller-Vonmoos, M. 1989. Bentonite as a containment barrier for the disposal of highly radioactive wastes. *Applied clay science,* 4: 157–177.

Carvalho, R.J.F.L. & Skipper, N.T. 2001. Atomistic computer simulation of the clay–fluid interface in colloidal laponite. *Journal of Chemical Physics,* 114: 3727–3733.

Chan, D.Y.C. Pashley, R.M. & Quirk, J.P. 1984. Surface potentials derived from co-ion exclusion measurements on homonic Montmorillonite and Illite. *Clays and clay minerals,* 32(2): 131–138.

Forsman, J. Woodward, C.E. & Jonsson, B. 1997. The origins of hydration forces: Monte Carlo simulations and density functional theory. *Langmuir,* 13: 5459–5464.

Kornyshev, A.A. 1983. Non-local dielectric response of a polar solvent and Debye screening in ionic solution. *Journal of the Chemical Society, Faraday Transactions 2,* 79: 651–661.

Low, P.F. Structural component of the swelling pressure of clays. *Langmuir,* 3: 18–25.

Marcelja, S. & Radic, N. 1976. Repulsion of interfaces due to boundary water. *Chemical physics letters,* 42(1): 129–130.

Marcelja, S. 1997. Effects of ion hydration in double layer interaction. *Colloids and Surfaces A: Physicochemical and Engineering Aspects,* 129–130: 321–326.

Miller, S.E. & Low, P.F. 1990. Characterization of the electrical double layer of Montmorillonite. *Lanmuir*, 6: 572–578.

Nishimura, S. Kodama, M. Yao, K. Imai, Y. & Tateyama, H. 2002. Direct surface force measurement for synthetic smectites using the atomic force microscope. *Langmuir*, 18(12): 4681–4688.

Nishimura, S. Yao, K. & Kodama, M. et al. 2002. Electrokinetic study of synthetic smectites by flat plate streaming potential technique. *Langmuir*, 18: 188–193.

Norrish, K. 1954. The swelling of montmorillonite. *Discussions of the Faraday society,* 18: 120–134.

Olphen, V. 1963. *An introduction to clay colloid chemistry.* Wiley Interscience, New York.

Pashley, R.M. 1981. DLVO and hydration forces between mica surfaces in Li+, Na+, K+ and Cs+ electrolyte solutions: A correlation of double-layer and hydration forces with surface cation exchange properties. *Journal of colloid and interface science,* 83(2): 531–546.

Pusch, R. 1982. Mineral-water interactions and their influence on the physical behavior of highly compacted Na bentonite, *Canadian Geotechnical Journal,* 19: 381–387.

Rukenstein, E. & Manciu, M. 2002. The coupling between the hydration and double layer interactions. *Langmuir,* 18: 7584–7593.

Schiby, D. & Ruckenstein, E. 1983. The role of the polarization layers in hydration forces. *Chemical physics letters,* 95(4–5): 435–438.

Zhang, F.S. Low, P.F. & Roth, C.B. 1995. Effects of monovalent, exchangeable cations and electrolytes on the relation between swelling pressure and interlayer distance in montmorillonite. *Journal of colloid and interface science,* 173(1): 34–41.

Advanced Materials and Structural Engineering – Hu (Ed.)
© 2016 Taylor & Francis Group, London, ISBN 978-1-138-02786-2

A study on the experimental teaching reform of digital electronic technology offered by schools run in the model of Chinese-foreign cooperation

Y.L. Li & J. Gao
Department of Electrical Engineering and Information Technology, Shandong University of Science and Technology, Jinan, China

ABSTRACT: This paper aims to overcome the shortcomings of the traditional experiment teaching method of digital electronic technology and meet the demand of Chinese-foreign cooperation in running schools. The project-based mode is merged into its experiment teaching links such as experiment content, teaching methods and evaluation criterion. The new experiment teaching methods can be used to help the students stimulate study enthusiasm and initiative, establish the concept of engineering project, reach the experimental evaluation criteria set by foreign universities for this course, and meet the market demand for engineering and technical personnel.

1 INTRODUCTION

With the diversification of higher education development, Chinese-foreign cooperation in running schools has become a core of teaching management mode in many colleges and universities. Nowadays, the number of domestic colleges and universities that carry out the project of Chinese-Foreign cooperative education has gradually increased, and the professional fields are covered more widely, while the requirements of Chinese-Foreign cooperation in running schools on the domestic education have also intensified. However, because there is a huge difference in school-running ideas between domestic and foreign universities, the comprehensive evaluation system is also different. Therefore, the traditional domestic teaching model faces severe challenges. In order to make the students adapt to foreign study life faster on the base of domestic study, at the same time, to meet the market demand for engineering and technical personnel, the reform and development of engineering experimental teaching is imperative (Wang et al. 2011, Yuan 2011, Liu 2006).

Digital electronic technology is a compulsory professional basic course for power, automation and computer majors, which plays a pivotal role. However, at present, the domestic digital electronic technology experimental teaching has some drawbacks as follows. First, the teaching is dominated by verified experiments, ignoring the experimental design and the result analysis. Second, the content of each experiment is old, dull and isolated. In addition, students with different

perceptibilities must complete the same experimental requirements. For the above reasons, the students are unable to arouse enthusiasm for autonomous learning. Therefore, the digital electronic technology experimental teaching reform is imperative for Chinese-Foreign cooperation in running schools. In addition, we put forward the concrete measures of the digital electronic technique experimental reform of Chinese-foreign cooperation in running schools, in which project-based experiments are merged and become a reliable adhesive between theoretical knowledge and engineering project. This new experimental teaching mode can make students master a lifetime of self-learning and growing skills, reach the required evaluation standard of foreign universities, shorten the students' periods of adaptation to an abroad teaching concept and model, and meet the market demand for engineering and technical personnel.

2 CONCRETE MEASURES OF THE DIGITAL ELECTRONIC TECHNIQUE EXPERIMENTAL REFORM OF CHINESE-FOREIGN COOPERATION IN THE RUNNING MODE

2.1 Guiding ideology

For the digital electronic technology course of Chinese-foreign cooperation in the running mode, we fully improve the fusion of engineering education ideas and its experimental teaching, that is, according to the teaching program and the level of students, project-based experiment contents are refined from the practical application, and the experiment is managed and carried out as a project. Specifically, we establish the one-to-one correspondence of the whole experiment process and engineering chain parts. Namely, the parts of the engineering chain include task analysis, scheme demonstration, simulation, manufacture and adjustment, and finalized production. Accordingly, the process of the experiment consists of task analysis, scheme demonstration, simulation, debugging and summary report. Here, we discuss the experimental teaching reform of digital electronic technology and focus on its content selection, teaching methods, apparatus and assessment methods.

2.2 Experimental content selection

The experimental content selection is the key to merge the project-based mode into the digital electronic technology experimental teaching, which should follow several principles. First, the basic content of the teaching program should be covered as much as possible and key points of this course should be highlighted. Second, the experimental content should have modular design features that can reflect the whole course system structure. Finally, the experimental content should have a strong innovation that enables students to have a strong interest in learning.

According to the above principles, we first analyze the theory knowledge architecture of the digital electronic technology (Yan 2006), as shown within the dashed frame of Figure 1, and we choose a pre-reply device used in intelligence competition as the whole project-based experimental contents.

Then, the experimental contents are divided into some experimental units with different functions, which embody the different knowledge points of this course. As shown in Figure 2, the pre-reply device used in intelligence competition can be divided into the following experimental units: decoding display, code latch, timing circuit, answer keys, second pulse output, alarm circuit and control circuit. In addition, the corresponding relationship between theoretical knowledge points and experimental units is shown in Figure 1. So, the pre-reply device used in intelligence competition covers the main contents of the teaching program of digital electronic technology, and highlights the key points of this course. At the same time, the isolated and scattered knowledge points are tightly bound by one project-based experiment. In addition, due to its wide application and strong interest, the enthusiasm of study and innovation of all students are stimulated.

2.3 Experimental apparatus

In all experiments, whether whole or sub, the software Multisim 10 is used to simulate, and the electronics integrated experimental platform DZX-3 (Chen et al. 2008) is used to debug the hardware. Altogether, the experimental equipment consists of a one computer, Multisim 10 and one hardware experimental platform.

From the above-mentioned simulation tool, Multisim 10 is an electronic schematic capture and simulation program, which is widely used in the electronic schematic design and electronic circuit simulation, released by National Instruments, and is effective and cheap. In our experiments, Multisim 10 is used to help the students verify the theory knowledge learned, and the feasibility of their designed circuits can also be used to optimize the circuit design. By means of this practical application process, the students can understand that

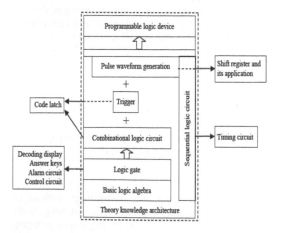

Figure 1. Theory knowledge architecture and the corresponding project-based experiment contents.

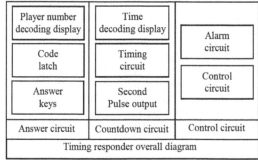

Figure 2. The division of the pre-reply device used in intelligence competition as a project.

the simulation analysis is an important auxiliary means of the engineering project design stage. In addition, it is able to greatly improve the reliability of hardware debugging and save the engineering cost.

The above-mentioned electronics integrated experimental platform DZX-3 is designed and produced by Zhejiang Tianhuang Technology Industrial Co. Ltd, consisting of an experimental control panel and an experiment table.

In addition, the experimental control panel includes two pieces of single side copper clad printed circuit board, power source, instruments and meters, which facilitates the students to test and measure. In our experiments, whether whole or sub, DZX-3 can be used to make a connection of the hardware.

Therefore, Multisim 10 and DZX-3 can help the students fully master the design and debugging skills required by the project-based experiment, and are an appropriate choice in this experimental teaching reform.

2.4 Experimental teaching method reform

The traditional Chinese experimental teaching methods of digital electronic technology and experiment are mostly explained in the order of aim, principle, steps and content. On the other hand, the experimental preview link is ignored; furthermore, the ability to organize the experimental summary report autonomously is not paid enough attention. This is considered to be the main problems existing in traditional experimental teaching methods, a new experimental teaching mode. The experiment is managed just in the project proposed. Specifically, teachers should guide students through the experiment according to the engineering project management model as follows: (1) the teacher should explain experimental functions, whether whole or sub; (2) according to one experiment, the students are required to discuss the experimental task and feasible scheme in groups; after that, the teacher comments on the schemes and determines the final one; and (3) the simulation analysis, experimental debugging and summary report should also be completed by the students under the guidance of the teachers. Figure 3 shows the flow chart of the project-based experimental teaching mode and the tasks corresponding with each link.

After finishing each sub-project-based experiment, the students are required to connect the corresponding simulation circuit into a whole one and ensure it achieves the functions of the pre-reply device used in intelligence competition. On this basis, the whole project-based experimental hardware is completed and debugged by connecting all the sub-project-based experimental hardware.

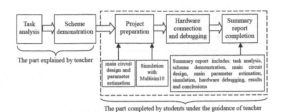

Figure 3. Flow chart of the project-based experimental teaching mode.

Finally, the students are required to complete the experimental summary report that includes task analysis, scheme demonstration, main circuit design, main parameter estimation, simulation, hardware debugging, results and conclusions. In addition, in order to exercise the students' ability of English expression and organization, laboratory reports are required to be completed in English. Through these experimental teaching methods, the students' abilities can be developed, such as creative thinking, communication skills, and engineering concept.

2.5 Experimental evaluation criterion

There is a great difference between project-based experiments and traditional experiments. Therefore, the traditional evaluation criterion, i.e. experiment scores = attendance scores + experiment achievement scores + experiment report scores, is no longer suitable for the reformed experimental teaching system. The reformed experimental evaluation should be set according to the completion of sub-project-based experiments and whole project-based experiment.

The sub-project-based experiments and whole project-based experiments account for the proportion of the experimental scores, as shown in Table 1. Specially, the sub-project-based experiments' scores are judged by the following aspects: experimental preparation, hardware circuit debugging and summary report. Also, the whole project-based experiments' scores are judged by four aspects as follows: hand drawings of the whole circuit, Multisim 10 simulation results, hardware circuit connection and summary report. It is worth to emphasize that the students are asked to complete the summary report in English because the reform of this experiment teaching is developed under the background of Chinese-foreign cooperation in running schools.

The establishment of the new experiment evaluation system of digital electronic technology is useful for the students to reach the experimental evaluation criteria set by foreign universities for this course, to adapt to study the following courses

Table 1. Allocation proportion of experimental scores.

Sub-project					
Code latch	Decoding display	Timing circuit	Second pulse output	Answer keys, control circuit and alarm circuit	Whole project
10%	10%	10%	10%	20%	40%

abroad as soon as possible, and to meet the market demand for engineering and technical personnel.

2.6 Effect of the experimental teaching reform

Digital electronic technology is a compulsory professional basic course of power, automation and computer majors, which has a pivotal role. Practice shows that project-based experimental teaching can help the students improve the studying interests, establish the concept of engineering project, and lay a good basis for the electronic design competition events and graduation design completion. So far, in the experiment reform, there are more than 60 students who won prizes in an electronic design contest; at the same time, the students pursuing further studies at the University of Tasmania have obtained approval. Moreover, some of them have been admitted to the degree of Bachelor of Engineering with Honors by University of Tasmania. Besides, this experimental reform has been approved by the 'School Stars Program' of the Shandong University of Science and Technology, and the name of the subject is called as 'Electronics Experimental Teaching Reform for Chinese-foreign cooperation in running mode', which prompts us to further explore and practise the digital electronic technology experimental teaching.

3 CONCLUSIONS

We have merged the project-based mode into the experimental teaching of digital electronic technology, which overcomes the shortcomings of the traditional experiment teaching such as single mode, poor association and weak creativity, and helps the students establish the concept of engineering project, stimulate study enthusiasm and initiative, reach the experimental evaluation criteria set by foreign universities for this course as soon as possible and meet the market demand for engineering and technical personnel. The good results achieved in this experimental reform can be seen from the students' performance. In addition, there are more

than 60 students who won prizes in an electronic design contest; at the same time, students pursuing further studies at the University of Tasmania have been accepted. Moreover, some of them have been admitted to the degree of Bachelor of Engineering with Honors by the University of Tasmania. Therefore, because the Chinese-foreign cooperation in running schools started late, we still lack practical experience of the foreign experimental teaching system. Therefore, the reform and exploration of the experimental teaching system should be further intensified.

ACKNOWLEDGMENT

We thank the following organizations for their financial support, namely the innovative experimental area construction project of personnel training mode for colleges and universities in Shandong Province in 2012, the innovative experimental area construction project of personnel training mode for Shandong University of Science and Technology in 2013, and the School Stars Program construction project of Shandong University of Science and Technology in 2013.

REFERENCES

Chen, Q. Li, G.F. & Lv Q.S. 2008. The trouble analysis and repairing of DC Power for DZX-3 electronics experimental equipment, *Instrumentation Technology*. 8: 57–59.

Liu, Y.P. & Chen, H.S. 2006. Discussion on the reform of the experimental teaching for digital electronic technology, *Research and Exploration in Laboratory*. 8: 981–983.

Wang, H. Liu, J.Y. & Chen, An. et al. 2011. Research on training innovative talents based on project leading electrical and electronic experiment teaching method, *Experiment Science and Technology*. 4: 75–78.

Yan, S. 2006. *Fundamentals of Digital Electronic Technology*, Higher education press, Beijing.

Yuan, P.T. 2011. Reform and explore of electrician experiment classification, *Coal Technology*. 12: 254–255.

Advanced Materials and Structural Engineering – Hu (Ed.)
© *2016 Taylor & Francis Group, London, ISBN 978-1-138-02786-2*

Methodology of the analysis of kinetic energy dispersion

J. Rajczyk, M. Rajczyk & J. Kalinowski
Faculty of Civil Engineering, Czestochowa University of Technology, Czestochowa, Poland

ABSTRACT: This paper presents the passenger safety mechanism to stop the vehicle on impact with an obstacle based on kinetic energy conversion and dissipation. The considerations took into account three types of braking systems: a) vehicle-mounted spring element, b) car-element elastic-free mass of the receiver, and c) vehicle-mounted spring element-rotating extractor. All of these types of systems can be securely installed as the solutions of the first and third types of moving vehicles. For each of the simulations, a model was made based on the principles of numerical simulation of continuous systems. Equations and diagrams are presented for the selected parameters specified in the braking time. The application of a progressive movement ratio at the rotating receiver for the kinetic energy through a worm gear with a high pitch is proposed. It is possible to use a variable pitch gear, so one can eliminate the spring element of the system using a big jump start and refrain from using self-locking transmission at a small jump.

1 INTRODUCTION

Research on traffic accidents has shown that the main cause is not the excessive speed but the poor quality and condition of the road system, which is not adapted to the amount of vehicles traveling on the roads. Polish roads are characterized by poor markings, with plenty of sharp turns, deep ditches located near them, culverts and other dangerous obstacles.

The occurrence of collisions on the following factors contributes to the obstacles:

- heavy traffic causes rapid fatigue and often impatient drivers, the presence of a large number of maneuvers with a small margin of error,
- driving at night or reduced visibility causes a delay in the driver noticing them, thus leaving less time to react to avoid obstacles,
- poor condition of the road with little traction, adverse weather conditions such as rain, snow, ice or sleet reduce braking capability and directional changes. The coefficient of friction of the tires depends on the parameters of the surface's texture (Rajczyk 2004).

Too many vehicles on the roads in conjunction with variable weather conditions are conducive to carrying out maneuvers borders on security, which increases the likelihood of slip and a collision with an obstacle (Ryszard 2009).

On the roads are particularly dangerous places where large amounts of recorded incidents of uncontrolled collisions occur outside the confines of the road. Such sites use barriers to energy-intensive hedges, which in case of contact with the vehicle, depending on the angle between the velocity vector of the vehicle to be safely directed on a correct path or to stop the vehicle. Interestingly, the phenomenon of kinetic energy conversion and dissipation is among others in Poland, designer and researcher Lucjan Łągiewka.

2 THEORETICAL BASICS

Vehicle that is in motion has a kinetic energy E_k associated with the motion of its entire mass at a certain speed, as well as additional energy resulting from the rotation of its wheels E_{ko}. The vehicle may also have a potential energy E_p if the stopping point of the vehicle is below the point of beginning to analyze inhibition. Stopping the vehicle requires a deterioration of the three types of energy.

The total energy of the system E consists of the potential energy E_p, traversing the kinetic energy E_k and kinetic energy rotation E_{ko}, which is described by the following formula:

$$E = E_k + E_{ko} + E_p \tag{1}$$

where
E_k—kinetic energy of progressive movement;
E_{ko}—rotational kinetic energy; and
E_p—potential energy.

The kinetic energy of progressive movement is given by the formula:

$$E_k = \frac{m_s v^2}{2} \tag{2}$$

where
m_s—vehicle mass and
v—speed of the vehicle.

The kinetic energy of rotation E_{ko} is defined by the formula:

$$E_{ko} = \frac{1}{2} I \omega^2 \qquad (3)$$

where
I—mass moment of inertia and
ω—rotational speed.

Potential energy E_p is determined by the formula:

$$E_p = m_s g h \qquad (4)$$

This energy is converted during braking into heat energy through friction on the brake disks, increasing the temperature of the disc brake and brake shoe, or the wheels lock and raise the temperature of the tire surface on which the tire is running. The amount of energy can be absorbed in this way depending on the force and the coefficient of friction with the ground material. The rest of the energy is used to do the work of permanent energy-intensive deformation on vehicle barriers, causing damage or destruction. Theoretically, it is possible to build a small rigidly attached or elastic deformable barrier and deformable or resilient little vehicle where the folded-speed collision with no permanent damage will occur and the transfer of power will occur almost elastically. The collision of these regimes would consist in the vehicle hitting the barrier at the stop a short distance dependent regimes deformability, and then the deflection of the vehicle from the barrier with the same velocity in a different direction, and hence an approximately equal amount of energy. Through high stiffness of the two bodies hitting, the vehicle stops were to be on a short way from the high acceleration of the vehicle, which would then be deflected exposing themselves to further impact. The energy of the vehicle after deflection would be reduced by the amount of energy absorbed by the brakes or other friction by braking and accelerating the vehicle. A small friction coefficient would be, for example, on an icy surface. The energy loss due to the very low coefficient of friction between the tire and the road surface is negligible. A very short braking distance is harmful because it causes the body to move due to elevated negative acceleration.

The stopping distance S with speed V_p from constant acceleration a is described by the formula:

$$S = \frac{V_p^2}{2a} \qquad (5)$$

Substituting into the above formula for acceleration a, we get the maximum allowable acceleration/delay, with the result being the minimum safe stopping distance on the road at the speed V_p.

The movement of any body of mass m uniformly accelerated motion causes the action on the body of inertial forces proportional to the value of acceleration.

This relationship is described by the formula:

$$F = m\,a \qquad (6)$$

where
F—force;
m—mass of the body; and
a—acceleration with which the body moves.

Rapid acceleration of the human body, as to the absolute value of an action on each of the major organs of the human body forces proportional to acceleration, causes internal injuries associated with the loss of tissue cohesion and can lead to death. Beyond the acceleration limit, humans will not be able to survive. The limit that is usually accepted is in the amount of ten times the acceleration of gravity (10 g).

The design of modern vehicles, in order to reduce the acceleration acting on the human body, has seat belts located and immobilized in vehicles, so that at the time of an impact with an obstacle, a so-called controlled crumple zone is provided. The crumple zone, in the event of a hard impact, visually reduces its length, which increases the braking motion path of the human body in a nondeformable protected cage. Increasing the actual stopping distance reduces the speed of the human body and thus reduces the value of mass forces. A well-designed crumple zone is permanently deformed when the traffic delays and associated forces take values close to the limit does not allow them to increase. Mass forces do work on the road longer, reducing their maximum values.

3 BRAKING WITH THE ELASTIC ELEMENT RIGIDLY ATTACHED

In the present case, the car's movement is restrained by the force generated by the spring as a result of the reduction caused by pushing on it with the return of the mass of the car opposite to the original motion of the car. According to the first principle of dynamics, the impact force on the car is equal to the force of the impact on the attachment, but these forces are opposed to returns (Fig. 1).

We assume that the used elastic element has a homogeneous stiffness in terms of shortening equal to the length needed to brake the car.

Figure 1. Diagram to stop a car with the elastic element attached.

The force applied to the car during braking on the elastic element is directly proportional to the stiffness of the spring element and the movement of the car, which is equal to the shortening of the elastic element.

The force with which the spring element acts on the vehicle is determined by the diffraction patterns, which is described in the following formula:

$$F = R_s * x \qquad (7)$$

where
R_s—element elasticity in N/m and
x—reduction of the resilient element in m.

The force at the start of the car contact spring element is set to a minimum of 0 and a maximum value of the car when it is stopped at a maximum displacement equal to the road on which the car is stopped. Since mass at a constant speed is proportional to the force, the absolute value of acceleration is a value from 0 at the time of contact with the spring element to a maximum value when the car stops. Car braking is done with a variable uniform negative acceleration.

The elasticity of the elastic element should be chosen for a given weight of the vehicle, and the maximum speed is less than or equal to the maximum allowable acceleration of 10 g.

Acceleration is described in the formula $\alpha = F/m$, where the force applied depends on the distance traveled $F = R_s$ and S can be written as $\alpha = (R_S/m)\,S$.

The equation that takes into account the differential relationship $\alpha = d^2S/dt^2$ can be represented in the form of differential equations:

$$\frac{d^2S}{dt^2} = \frac{R_s}{m}S \qquad (8)$$

$$\frac{d^2S}{dt^2} + \frac{R_s}{m}S = 0$$

By substituting $\beta = R_s/m$, we obtain the differential equation in the form of

$$\frac{d^2S}{dt^2} + \beta S = 0 \qquad (9)$$

The solution to this equation is in the form of a function, which is given by

$$S(t) = K_1 \sin\left(t\sqrt{\beta}\right) + K_1 \cos\left(t\sqrt{\beta}\right) \qquad (10)$$

Given the initial conditions $S(0) = 0$ and $S'(0) = V_{pocz}$, the solution is obtained in the form of

$$S(t) = V_{pocz}\frac{\sin\left(t\sqrt{\beta}\right)}{\sqrt{\beta}} \qquad (11)$$

Taking into account $\gamma = \sqrt{\beta} = \sqrt{R_s/m}$, we obtain a simpler form of the substitution patterns as follows:

$$S(t) = V_{pocz}\frac{\sin\left(t\sqrt{\gamma}\right)}{\sqrt{\gamma}} \qquad (12)$$

speed $V(t) = S'(t)$

$$V(t) = V_{pocz}\cos(t\gamma) \qquad (13)$$

acceleration $a(t) = V'(t)$

$$a(t) = V_{pocz}\gamma\sin(t\gamma) \qquad (14)$$

Stopping the car will come at a time t_k, when $V(t) = 0$

$$V_{pocz}\cos(t_k\gamma) = 0$$

$$\cos(t_k\gamma) = 0$$

$$\cos(t_k\gamma) = \cos\left(\frac{\pi}{2}\right)$$

$$t_k\gamma = \frac{\pi}{2}$$

$$t_k = \frac{\pi}{2\gamma} \qquad (15)$$

The maximum acceleration takes the value of

$$a_{max} = a(t_k) = a\left(\frac{\pi}{2\gamma}\right) = V_{pocz}\gamma$$

Taking into account the maximum allowable delay value a_{dop} of the elasticity of the spring R_s, the vehicle mass m should be

$$R_s \frac{a_{dop}^2}{V_{pocz}^2}m \qquad (16)$$

The limit of elasticity allows the car to stop on the shortest path without exceeding the limit a_{dop}.

The stopping distance S_z is described by the equation

$$S_z = S(t_k) = \frac{V_{pocz}}{\gamma} \qquad (17)$$

The minimum possible stopping distance with the selected optimum elasticity of the rail will be

$$S_z = \frac{V_{pocz}^2}{a_{dop}} \qquad (18)$$

3.1 Example analysis

Assuming that a car with a mass of $m_s = 900\ kg$ is braking from an initial speed of $V_{pocz} = 50\ km/h$. To calculate the elastic element's parameters, the car stops on the shortest path with an impassable delay value equal to 10 g is as follows:

$V_{pocz} = 50\ km/h = 13,89\ m/s^2$,
$a_{dop} = 10\ g = 98,1\ m/s^2$, $m_s = 900\ kg$

The shortest possible stopping distance S_z will be

$$S_z = \frac{V_{pocz}^2}{a_{dop}} = \frac{\left(13,89\ \frac{m}{s}\right)^2}{98,1\ \frac{m}{s^2}} = 1,967\ m$$

For this case, the spring must have the elasticity R_s as follows:

$$R_s = \frac{a_{dop}^2}{V_{pocz}^2} m_s = \frac{\left(98,1\ \frac{m}{s^2}\right)^2}{\left(13,89\ \frac{m}{s}\right)^2} 900\ kg = 44893\ \frac{N}{m}$$

The stopping time t_k will be

$$\gamma = \sqrt{\frac{R_s}{m}} = \sqrt{\frac{44893\ \frac{N}{m}}{900\ kg}} = 7,063\ \frac{1}{s}$$

$$t_k = \frac{\pi}{2\gamma} = \frac{\pi}{27,063 s^1} = 0,222\ s$$

Average delay $a_{śr}$ will be

$$a_{śr} = \frac{\Delta V}{\Delta t} = \frac{13,89}{0,222\ s} = 62,567\ \frac{m}{s^2} = 6,37\ g$$

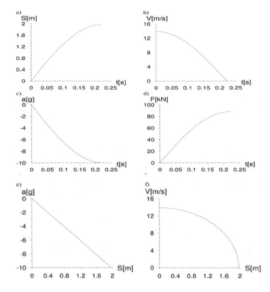

Figure 2. Graphs showing the movement parameters, for example the vehicle braking and door protector elastic, a) the distance dependency of time, b) the dependence of time c) the dependence of the delay time, d) the dependence of the braking force from time, e) the dependence of the delay on the distance traveled, and f) the dependence of the distance traveled.

The car's braking characteristics on an obstacle in the way of a mounted spring element are shown graphically in Figures 2 a–f.

4 ANALYSIS OF BRAKING WITH THE USE OF A SPRING ELEMENT WITH MOVEABLE MASS

The braking system diagram is shown in Figure 3. The system consists of a car with a mass m_s moving at the start of deceleration at an initial velocity V_{pocz}, which strikes the spring element with the elasticity R_s and on the other side of the receiver weighing m_o, which at the initial moment is at rest.

As a result of the contact of the moving car, the car spring element acts on the resilient force, which causes the reduction of the resilient element that generates a spring force in the element F_s resisting its compression.

Force, as a resilient force, acting on the system is given by

$$F_s = R_s s \qquad (19)$$

where
R_s—spring elasticity in N/m and
s—compression of the spring in m.

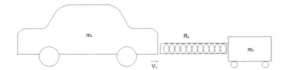

Figure 3. Diagram of the braking system with the use of the elastic element and free weight.

Reducing the counterforce spring element on the one hand acts on the vehicle, causing it to reduce the speed of the other value of the same acts on the mass and causing the receiver to increase its speed. The braking process is terminated when the vehicle speed reaches zero or the force in the elastic element reaches zero, that is, the resilient member to reach the original length.

The equations describing the behavior of the system during t take the form of

$$F_s(t) = R_s s(t) \qquad (20)$$

$$V_s(t + dt) = V_s(t) + \frac{F_s}{m_s} dt \qquad (21)$$

$$V_o(t + dt) = V_o(t) + \frac{F_s}{m_o} dt \qquad (22)$$

$$S_s(t + dt) = S_s(t) + \frac{V_s(t + dt) + V_s(t)}{2} dt \qquad (23)$$

$$S_o(t + dt) = S_o(t) + \frac{V_o(t + dt) + V_o(t)}{2} dt \qquad (24)$$

$$s(t) = S_s(t) \; S_o(t) \qquad (25)$$

with the initial conditions of

$$t = 0 \; S_s(0) = 0, S_o(0) = 0, V_s(0) = V_{pocz}, V_o(0) = 0$$

and the final condition of F_s 0 for $t > 0$.
where
R_s—value of the elastic modulus;
F_s—power of the elastic element;
V_s—speed of the car;
V_{pocz}—initial speed of the car;
S_s—path (location) of the vehicle relative to its initial position when $t = 0$;
m_s—mass of the car;
V_o—speed of the switching capacity;
S_o—path (location) of the receiver relative to its initial position at the time $t = 0$; and
m_o—mass of the energy receiver.

4.1   *The equations of motion were solved numerically*

In the analyzed system, assuming that there will be a collision of the car's elastic mass m_s moving at an initial speed of V_{ps} with the stationary receiver ($V_{po} = 0$) with a mass of m_o, the car speed of V_{ks} and the receiver V_{ko} after the collision are described in the following equations:

$$V_{ks} = V_{ps} \frac{m_s \; m_o}{m_s + m_o} \qquad (26)$$

$$V_{ko} = V_{ps} \frac{2m_s}{m_s + m_o} \qquad (27)$$

The characteristic behaviors of the system can be distinguished as follows:

$m_s > m_o$—there will be no complete braking of the vehicle $V_{ks} > 0 \; V_{ko} > 0$;
$m_s = m_o$—there will complete braking of the vehicle $V_{ks} = 0 \; V_{ko} > V_{ps}$ (speed exchange); and
$m_s < m_o$—the car will be reflected away from the receiver $V_{ks} < 0 \; V_{ko} > 0$.

When the weight of the car is equal to the mass of the receiver $m_s = m_o$, a speed exchange occurs, i.e., the overall deceleration of the car is $V_{ks} = 0$, and the speed of the receiver reaches the speed of the car before the collision.

After substituting into the formula, the ratio of weight to the weight of the car receiver will be $m_s/m_o = c$, and the dependence takes the form of

$$V_{ks} = V_{ps} \frac{1 \; c}{1 + c} = V_{ps} d_1 \qquad (28)$$

$$V_{ko} = V_{ps} \frac{2}{1 + c} = V_{ps} d_2 \qquad (29)$$

Elasticity of the resilient element should be such that it should not exceed the maximum acceleration of 10 g. The unit size is proportional to the stopping, which leads to the minimization. For the selection of the optimum value of the maximum acceleration of elasticity, the maximum established value should periodically reach up to 10 g.

$$d_1 = \frac{1 \; c}{1 + c} \qquad (30)$$

$$d_2 = \frac{2}{1 + c} \qquad (31)$$

The value of d_1 is the factor determining the ratio of the initial speed of the car to the speed of the car at the end of the final interaction of the system. The factor d_2 determines the final speed of the receiver in the same way.

Graphs of d_1 and d_2, depending on the mass ratio of the receiver to the weight of the car, are shown in Figure 4.

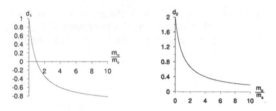

Figure 4. Graphs of d_1 and d_2, depending on the mass ratio of the receiver to the weight of the vehicle.

Figure 5. Diagram of the spring brake gear rotating screw extractor.

The value of d_1 multiplied by the speed of the car can start the car's residual value at the end of the impact. The d_2 value multiplied by the speed of the car gives the initial value of the receiver at the end of the final impact.

Figure 6. Braking system designed by the spring rotating screw extractor with gear to minimize the size of the device.

5 ANALYSIS OF THE BRAKING SYSTEM WITH ENERGY ABSORBER SPRING SPINNING

The system consists of a moving car from the start, which transmits its energy through the spring element to the guide starting the gear to rotate at the mass absorber, as shown in Figures 5 and 6.

Under the inspired work of the EPAR project (energy accumulating and dissipating converter), the receiver structures' proposed solution of the kinetic energy of a moving vehicle consisting of a system is shown in Figures 5 and 6.

The EPAR system performs in the case when a braking vehicle hits an obstacle depending on the change from the kinetic energy of the car traversing the kinetic energy of the rotation energy absorber mounted in a vehicle or with the obstacle. The energy is transmitted through the car's spring element, which is temporarily stored followed by a linear guide with circular toothed gears with a high ratio for a rotating mass, which is the kinetic energy receiver. The schematic diagram of the device is shown in Figure 7a and b. Since the time of the absorber reaching maximum energy, the energy is gradually dissipated by friction. High ratios are used to allow the absorber to achieve very high speeds and high-energy gathering device, a low-mass theory, which allows one to mount it in a vehicle without losing its functionality. The absorber is able to accumulate a large part of the kinetic energy of the vehicle, which automatically will not be used to damage him or obstacles. In addition, our solutions show that the forces acting on the body mass of the vehicle are braked by the EPAR system and are less than the forces due to the actual vehicle deceleration. However, the mechanism of this phenomenon is not clear.

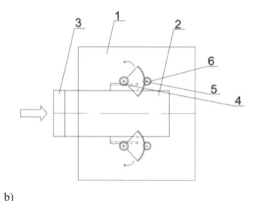

b)

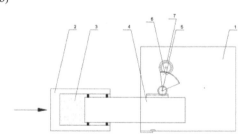

Figure 7. Conceptual diagrams of the EPAR system a) with a dual-battery kinetic energy (Gumula & Lagiewka 2006) and b) with a single-battery kinetic energy (Gumula & Lagiewka 2007).

The EPAR is also used as a dynamic brake where the braking power is transferred from the vehicle's onto the gears and then to the battery/kinetic energy absorber. At the time of acceleration, energy from the battery of the kinetic energy is transferred back to the wheels supporting the acceleration of the vehicle. In this principle, the KERS system that is used in formula 1 car works.

We assume that the transmission has the characteristics that the linear shift of the linear rack, with a length of f, followed by the full weight of the receiver rotation $2\pi rad$.

Apart from friction in the gearbox, work done by the force F moving through the bolt S will correspond to the work done by the momentum generated by the angle of rotation of the gear weight of the receiver α.

$$FS = M\alpha \qquad (32)$$

Substituting in the formula for S jump linear rack f for the angle α, we obtain the relationship defining the full torque acting on the axis of the receiver as follows:

$$M = \frac{Ff}{2\pi} \qquad (33)$$

According to the second principle of dynamics of the rotation, if the body is not mounted on a rotating axis in space, the moment of inertia is equal to the axis I, acting external forces, which cause the body to M resultant torque, as a result of which the body will rotate the angular acceleration of ε directly proportional to the torque acting M, and inversely proportional to the moment of inertia of the body I:

$$\varepsilon = \frac{M}{I} \qquad (34)$$

The present value of the angular acceleration gear is given by the formula:

$$\varepsilon = \frac{Ff}{2\pi I} \qquad (35)$$

The geometric properties of the ratio between the gears show a linear speed gears xxx linear and angular velocity of the receiver:

$$\frac{V_z}{\omega} = \frac{f}{2\pi} \qquad (36)$$

$$V_z = \frac{f\omega}{2\pi} \qquad (37)$$

In the beginning of the braking process, a linear rack and receiver are at rest. After some time, Δt linear rack reaches a speed of progressive V_z and receiver angular velocity ω.

Dividing both sides of the equation by Δt, we obtain a linear relationship between the acceleration of a_z and the angular ε as follows:

$$\frac{V_z}{\Delta t} = \frac{f}{2\pi}\frac{\omega}{\Delta t} \qquad (38)$$

$$a_z = \frac{f}{2\pi}\varepsilon \qquad (39)$$

Substituting for the angular acceleration of the dependence relationship $\varepsilon = Ff/2\pi I$, we obtain the linear acceleration linear rack depending on the force acting on it, the geometric properties of the transmission and mass moment of inertia of the receiver as follows:

$$a_z = F\frac{f^2}{4\pi^2 I} \qquad (40)$$

Instead, if the rotating canister is mounted to the receiver as a replacement power line replacement m_z weight to speed up the absorber according to the second principle of dynamics under the influence of the force F, it would amount to

$$a_z = F\frac{1}{m_z} \qquad (41)$$

Comparing the right sides of the weight of the previous model, the linear absorber replacement m_z absorber corresponding turnover can be determined as follows:

$$m_z = \frac{4\pi^2 I}{f^2} \qquad (42)$$

The inhibitory effect of the system is equivalent to the car-element elastic-receiver energy for identical elements of rigidity and weight m_z calculated using the above formula.

Comparable systems have different forms of kinetic energy storage, namely progressive movement and rotation.

The derived formulas also apply to the worm gear, including the condition that the symbol f in the formula corresponds to the pitch screw.

6 EXAMPLE

For a car with a mass $m_s = 900\ kg$ moving with the initial velocity of 50 km/h, to determine the optimal parameters of braking energy absorber with a rotary container, we assume that the linear shift gears with 0.04 m corresponding to a full rotation of the canister.

The car's energy should be completely transferred to the rotational energy of the receiver's mass moment of inertia that should correspond to a weight equal to the weight of a replacement vehicle previously designated by the following formula:

$$m_z = \frac{4\pi^2 I}{f^2}$$

The resulting moment of inertia of the absorber pattern is defined by the formula:

$$I = \frac{m_s f^2}{4\pi^2}$$

It was assumed that the canister will have the shape of a cylinder with a height h equal to 0.2 m and will be made of a material having a density of $\gamma = 2500 \ kg/m^3$.

Mass moment of inertia I for a cylinder of radius R is defined by the formula:

$$I = \frac{1}{2}mR^2 = \frac{1}{2}\pi R^4 h\gamma$$

Comparing the right sides of the two previous formulas, we obtain the formula for the radius of the canister as follows:

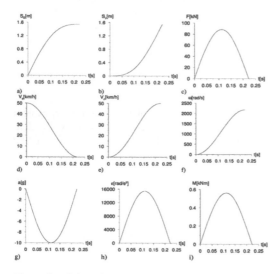

a) route traveled by the car, b) linear shift gears, c) braking force acting on the car, d) speed of the car, e) linear speed gears, f) speed switching capacity, g) vehicle deceleration, h) angular acceleration of the rotary power of the receiver, and i) moment acting on the power of the receiver axis.

Figure 8. Selected parameters specified in the braking time for the rotary system and the energy absorber a) route traveled by the car, b) linear shift gears, c) braking force acting on the car, d) speed of the car, e) linear speed gears, f) speed switching capacity, g) vehicle deceleration, h) angular acceleration of the rotary power of the receiver, and i) moment acting on the power of the receiver axis.

$$R = 4\sqrt{\frac{m_s f^2}{2\pi^3 h\gamma}} = 4\sqrt{\frac{900kg \ 0.04 \ m}{2 \ \pi^3 0.2 \ m \ 2500 \ \dfrac{kg}{m^3}}} = 0.08255 \ m$$

The absorber mass is given by $m_o = \pi R^2 h\gamma = \pi(0.08255 \ m)^2 \ 0.2 \ m \ 2500 \ kg/m^3 = 10.704 \ kg$ which is equivalent to the system of the linear absorber mass m = 900 kg, enabling the total transmission power.

The energy to the receiver was selected as the appropriate elastic element rigidity so that the maximum value of the delay does not exceed 10 g, which amounted to $R_s = 89.8596 \ kN/m$.

For the selected data, a numerical simulation was performed in a system, and selected results are shown in Figure 8. The analysis found a 10 g maximum acceleration and a complete transfer of kinetic energy of the car to the spinning energy absorber.

7 CONCLUSIONS

The proposed considerations were carried out based on the well-known laws of physics in the laws of motion and conservation of momentum and energy, neglecting friction. The relationships were derived through the analysis and computer simulation. For each of the simulations, a model was made based on the principles of numerical simulation of continuous systems (Ramowski & Bartkiewicz 1998).

The considerations took into account three types of braking systems: a) vehicle-mounted spring element, b) car-element elastic-free mass of the receiver, and c) vehicle-mounted spring element-rotating extractor. All of these types of systems can be securely installed as the solutions of the first and third types of moving vehicles. All solutions are sensitive to the mass of the vehicle braking. Due to the limitation of the size of the required space for the operation of the device, it is necessary to select the stiffness of spring elements and accordingly the mass or the mass moment of inertia based on the weight of the concerned inhibited vehicle. If the weight of the vehicle will be less inhibited than that assumed in the calculation of the vehicle, the deceleration will be greater than the limit value. With a greater mass, the assumed delay will be less, which will extend the braking distance, so one will need to use lever-limiting movement of the vehicle. Since the mass of moving vehicles on the roads has a spread from a few hundred pounds to several tons, it seems that on the basis of these solutions cannot be universally built to safely protect against the impact. These devices will work best if they are adapted to the weight of the vehicle moderated, which would compel them to consider installing the optimized version of the weight of the vehicle in the vehicle. Solution in each case, and solutions b and c for vehicles that weigh less than assumed in the selection of parameters would make the stopping vehicle to spin it in the opposite direction. This phenomenon can be eliminated by the use of additional components attached to the gear ratchet rod secured to allow movement in only one direction. After reaching the maximum displacement, gear energy is stored in the system as the potential energy would not be returned.

Systems b and c are similar in their inhibitory activity.

System c, by analogous operation with system b, can be adjusted for the replacement weight for system b, such that the braking effect of system c would be identical to that of system b.

For systems b and c, one can choose the appropriate mass and mass moment of inertia, such that there would be a complete transfer of energy from an elastic collision of equal weights. In system b, the receiver mass must be equal to the mass of the vehicle.

The application of a progressive movement ratio at the rotating receiver for the kinetic energy through a worm gear with a high pitch is proposed. It is possible to use a variable pitch gear, so one can eliminate the spring element of the system using a big jump start and refrain from using self-locking transmission at a small jump. The choice of a variable pitch screw in length will allow a more even distribution of braking force, which will shorten the stopping distance not exceeding the acceptable acceleration.

REFERENCES

Gumuła, S. & Łągiewka, L. 2006. Conceptual design of vehicles protection against the impacts of collisions using the energy transfer method. *Journal of KONES Powetrain and Transport*, 13(1): 269–277.

Gumuła, S. & Łągiewka, L. 2007. A method of impact and intertia force reduction during collision between physical objects. Results of experimental investigations, *Journal of Technical Physics*, 48(1): 13–27.

Rajczyk, J. 2004. *The scientific basis for the selection of geometric structure and kinematics of machines for surface treatment of concrete working tools*, University of Częstochowa Publishing, Częstochowa, Poland.

Rarnowski, W. & Bartkiewicz, S. 1998. *Mathematical modeling and computer simulation of continuous dynamic processes*, College University of Koszalin Publishing, Koszalin, Poland.

Ryszard, K. 2009. *Diagnosis of transport safety in Poland*. Warsaw: Communication and Connectivity Publishing, Poland.

Wrzesiński, T. 1978. *Inhibition of motor vehicles*, Communications and Connectivity Publishing, Warsaw. Poland.

Advanced Materials and Structural Engineering – Hu (Ed.)
© 2016 Taylor & Francis Group, London, ISBN 978-1-138-02786-2

Portable system for measuring torsional vibration in turbo-generators

J.Z. Liu
State Key Laboratory of Alternate Electric Power System with Renewable Energy Sources, North China Electric Power University, Beijing, China

T. Zhao
State Key Laboratory of Alternate Electric Power System with Renewable Energy Sources, North China Electric Power University, Beijing, China
North China Electric Power Research Institute, Xicheng District, Beijing, China

ABSTRACT: A portable system for measuring the torsional vibration of a turbo-generator was developed and analyzed. The speed signal–demodulating circuit of the system was designed on the basis of field programmable gate array technology, and a platform for torsional vibration analysis was established using virtual instrument technology. The torsional vibration components were directly extracted from the demodulated speed signals by a zero-phase filter. The torsional vibration characteristics were analyzed and the results were displayed in LabVIEW. Laboratory and site test results showed that the system met the engineering measurement requirements.

1 INTRODUCTION

In a high-power turbo-generator, a mechanical or electrical disturbance could cause a large current (voltage) impulse to occur. This would then cause a sharp increase in electromagnetic torque between the stator and rotor, which in turn would lead to a torsional vibration of the turbo-generator. In a power grid, subsynchronous resonance caused by Fixed Series Compensation (FSC) could also cause a torsional vibration of the shafting, which would seriously impact the safety of the power grid and turbo-generator. Recently in China, the requirements for large-capacity long-distance transmission at independent power plants have increased. In particular, for some newly built coal and electricity bases located far from load centers, long-distance electric transmission occurs mostly in the FSC mode. Consequently, torsional vibration of turbo-generators is becoming an increasingly serious problem (Subsynchronous Resonance Working Group of the System Dynamic Performance Subcommittee, Reader's guide to subsynchronous resonance 1992, Thomas 2012, Xie et al. 2011, Gu et al. 2001, Bao et al. 1998, Xiang et al. 2000).

In the early 1970s, countries abroad began comprehensive studies on torsional vibration of unit-grid coupling shafts. Some developed countries (i.e., the United States and Germany) have obtained practical results on issues such as the mechanics of torsional vibration responses of unit-grid coupling shafts and shafting systems to a disturbance

of the electric power system. In China, it was in the late 1980s that this issue attracted attention. Some results were obtained by theoretical analysis, simulations, actual machine testing, and on-line monitoring-approaches that are increasingly promoted by departments of design, manufacture, and operation.

Currently, many torsional vibrometers are in service overseas. These include, for example, the 2521, TVSC, and TVMS developed by the United States; high-precision systems DK-I and DK-II developed by Tsinghua University; and the TSA2.0 on-line monitoring system developed by North China Electric Power University. However, these highly specialized instruments are expensive and their maintenance costs are also very high (Xu & Zhang 2004, Zhang et al. 1996, Zhang et al. 2009, Yang & Hu 2008).

In this paper, we use Field Programmable Gate Array (FPGA) technology to develop equipment that demodulates high-frequency rotational speed signals, obtains torsional vibration components, reduces later computation, improves calculating efficiency, and reduces the cost and scale of the system. A torsional vibration measurement platform based on virtual instrument technology has advantages over traditional measurement instruments in terms of its simple data transmission, convenience of live debugging, and lower cost. This platform can also be used by electric power research institutes as a portable torsional vibration measurement system in site tests. By comparing the results obtained with the laboratory standard signal generator test and site test we find that the measurement system meets engineering measurement requirements.

2 ROTATIONAL SPEED SIGNAL DEMODULATING INSTRUMENT

Measurement of the angular displacement of torsional vibration is mostly based on a non-contact gear measuring principle, which allows precise separation of torsional vibration signals from rotational speed signals. When the speed measuring gear is rotating, the vibration sensor detects an impulse signal when each gear tooth passes the sensor. When the shaft is rotating stably, the output of the sensor is a symmetrical impulse wave. During torsional vibration of the shaft, the impulse wave is modulated to rarefaction-dense wave. Then, by demodulating the signal, the torsional vibration signal is conveniently obtained (Wang & Hu 2008).

The turbine rotational speed signal is a high-frequency signal. Its frequency f is proportional to the gear rotational speed n and number of teeth z. Current gears for measuring turbine speed has either 60 or 134 teeth. Hence, the fundamental frequency we measured directly was 3000 or 6700 Hz. The rotational speed was demodulating device demodulates the high-frequency signal; hence, the demodulated signal contains a slowly varying rotational speed component and a torsional vibration component (Fig. 1).

This device demodulates double-channel signals and uses FPGA technology to complete the calculations and detection. It is developed as a-design circuit of an Application-Specific Integrated Circuit (ASIC). The FPGA, which has an operating frequency of 200 MHz, is a parallel detection and calculation unit. Hence, the units in which the modulating signals of the two channels are demodulated and transformed are separated. This ensures fast, high-precision detection and transformation of each channel in the equipment. A schematic diagram of the hardware design is shown in Figure 2.

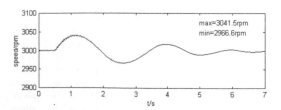

Figure 1. Demodulated speed signal.

Figure 2. Schematic diagram of the hardware design.

The input rotational speed signal may have a center frequency of 3000 or 6700 Hz, which is selected e depending on the numbers of measure speed gears, and a magnitude between −50 and 50 V. The analog signal converter unit filters, dampens, and shapes the input modulating signals. The optoelectronic isolator prevents the input signal to the center circuit from being disturbed. The FPGA detects and calculates the input signal, and transforms it to 12-bit code, which the Digital-to-Analog (D/A) unit requires. The D/A unit then transforms the 12-bit code from the FPGA unit to an analog signal with an amplitude range of 0 to 5 V. The amplitude transform unit then transforms this signal to one with amplitude range of −5 to 5 V, and low-pass filters the step wave from the D/A unit. The final output of the device is a demodulated rotational speed signal.

3 TORSIONAL VIBRATION ANALYZING PLATFORM

The torsional vibration analyzing platform performs data collection and analysis. It collects the rotational speed signal, modulates it, and analyzes the output signal. Subsequently, information on the torsional vibration, including the vibration component of the speed signal, frequency spectrum of the vibration component, and dynamic response properties of the amplitude and torsional angle of the vibration component, are obtained in each mode. The results can be saved in excel format.

In this paper, the PXI system (National Instruments Corp.) was chosen as the measurement system to implement the torsional vibration analyzing platform, and the data acquisition and analysis programs were designed with LabVIEW. PXI is a PC-based open platform whose advantage lies in its modularity and the characteristic of easy extension. LabVIEW is a kind of graphical programming language and specially suitable for the development of testing, analyzing and control programs (Chang et al. 2012).

3.1 Hardware system

As noted above, the main function of the torsional vibration platform is to collect and manage rotational speed data and thereby acquire information on torsional vibration. The basic configuration of the hardware system is shown in Figure 3.

The aforementioned PXI system, assembled with a PXI-8106 embedded controller and two PXI-4224 cards, served as the data acquisition system. The data analysis platform was displayed on a laptop computer. The signal to demodulate the rotational speed signal was injected into the data acquisition system through a cable. The data acquisition system and data analysis platform were linked by an Ethernet connection.

3.2 Software system

The software framework was developed in Lab-VIEW Real-time Module. Subsequently, sub-VI and shared variables were corrected and added to the mainframe and remote target system to complete development of the software.

In this paper, the PXI system was configured as the remote target system of the mainframe (Fig. 4). The shared variables are rotational speed, other signals, state of a target machine, commands, parameters, and tabs that are displayed by shared variables.

3.3 Algorithm implementation

The collected data on rotational speed comprise a slowly varying signal, three damped sinusoidal signals, and white noise. The trend signal is a slowly varying signal and is extracted by appropriate low-pass filtering. In this paper, a zero-phase-shift filter in LabVIEW was applied. The low-pass filtering was realized according to the following steps: first pass the time sequence through the filter and obtain the first-order filter sequence; next, invert

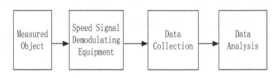

Figure 3. Hardware system structure.

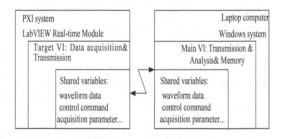

Figure 4. Software system structure.

the timing sequence and pass it through the filter to obtain the second-order filter sequence; finally, invert the second-order filter sequence to obtain the final filter output. We refer to this procedure as the FRR (Forward Reverse Reverse) method.

By using a low-pass zero-phase digital filter, we can remove high-frequency noise and obtain the original signal denoised. The trend signal is then obtained with a second-order low-pass zero-phase digital filter whose frequency is 5 Hz. Finally, by removing the trend signal from the denoised signal, the total torsional vibration component is obtained.

We obtained the frequency of vibration modes by analyzing the total spectrum and denoted them as mode 1, mode 2, and mode 3. We then extracted the vibration component of each mode with four-step band-pass zero-phase filtering. The center frequencies of the filters were equal to those of modes 1, 2, and 3, and their bandwidth was 3 Hz.

The amplitude of the modal component of the torsional vibration can be obtained by the Hilbert transformation. For a vibration modal component $m(t)$, its Hilbert transformation is given by Equation 1. The transformation can be achieved by the Fast Hilbert Transformation Function in LabVIEW.

$$\tilde{m}(t) = \frac{1}{\pi} \int_{-\infty}^{+\infty} \frac{m(\tau)}{t - \tau} d\tau \qquad (1)$$

The corresponding amplitude of the torsional vibration component is:

$$a(t) = \sqrt{m^2(t) + \tilde{m}^2(t)} \qquad (2)$$

The torsional angle in each mode can be obtained directly by integrating the modal component of the torsional vibration.

4 ANALYSIS OF AN EXAMPLE

4.1 Laboratory test

To test the reliability of the torsional vibration system, we used an arbitrary function generator (Tektronix AFG3022B, Tektronix Inc.) to generate a frequency modulated signal as the rotational speed signal and torsional vibration signal. The instantaneous frequency of the signal is:

$$f_i(t) = f_c + \Delta f \cos(2\pi f_s t) \qquad (3)$$

where, f_c is carrier frequency, the rotational frequency when the shafting is steady and free from torsional vibration; f_s is the modulated signal frequency, the frequency of torsional vibration;

Table 1. Test results of vibration parameters with standard signals.

Test	Test 1		Test 2		Test 3		Test 4		Test 5	
Parameter	Freq [Hz]	Amp [rad/s]	Freq [Hz]	Amp [rad/s]	Freq [Hz]	Amp [rad/s]	Freq [Hz]	Amp [rad/s]	Freq [Hz]	Amp [rad/s]
Set value	10	0.1046	10	0.0837	10	0.0418	20	0.0418	30	0.0418
Test value	10	0.1036	10	0.0827	10	0.0414	20	0.0414	30	0.0414
Relative error	0	0.95%	0	1.19%	0	0.96%	0	0.96%	0	0.96%

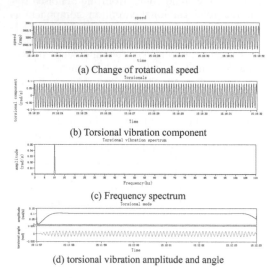

(a) Change of rotational speed

(b) Torsional vibration component

(c) Frequency spectrum

(d) torsional vibration amplitude and angle

Figure 5. Torsional vibration analysis results of standard signals.

and Δf is the frequency deviation, which is the amplitude of the torsional vibration.

When a torsional vibration occurs, the amplitude of the vibration is the torsional angular speed, whose unit is rad/s, which can be transformed by Equation 4, 5.

$$f = nz/60 \qquad (4)$$

$$\omega = 2\pi n/60 = 2\pi f/60 \qquad (5)$$

where, n is the rotational speed, whose unit is rpm (revolutions per minute), and z is the number of teeth on the gear (defined here as 60).

In our test, the frequency of the signal generator output f_c is 3000 Hz; f_s is 10, 20, or 30 Hz; and Δf is 1, 0.8, or 0.4 Hz. The torsional vibration amplitudes are transformed by Equation (5), giving an angular speed of 0.1064, 0.0837, or 0.0418 rad/s. The set value and test value are compared in Table 1. The relative error between set value and test value was approximately 1%. The results of the signal with f_s = 10 Hz and Δf = 1 Hz (Fig. 5) are

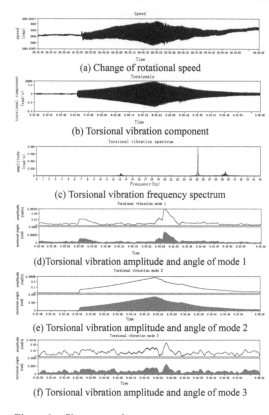

(a) Change of rotational speed

(b) Torsional vibration component

(c) Torsional vibration frequency spectrum

(d)Torsional vibration amplitude and angle of mode 1

(e) Torsional vibration amplitude and angle of mode 2

(f) Torsional vibration amplitude and angle of mode 3

Figure 6. Site test results.

consistent with the originally set value. The relative error between the set value and test value was 0 for the frequency (freq) and 0.95% for the amplitude (amp) of the torsional vibration (Table 1).

4.2 Site test

This system has been frequently used for torsional vibration site tests at power plants. Below are the analysis results for a power plant blocking filter for in-service and out-of-service test. Figure 6(a), (b), and (c) display the demodulated rotational speed signal, torsional vibration component (by de-noising the speed signal and removing the rotational trend signal), and frequency spectrum of the torsional vibration component, respectively. Three main vibration modes were identified at 13.19, 25.12, and 29.51 Hz; they were extracted and correspondingly denoted as mode 1, mode 2, and mode 3. Figure 6(d), (e), and (f) show the corresponding changes in vibration amplitude and torsional angle for each of the modes.

In this test, all three modes were motivated; however, the natural frequency of mode 2 was dominant because its amplitude was an order of magnitude higher than that of mode 1 or mode 3. The maximum torsional angle was 0.025% rad for mode 1, and 0.016% rad for mode 3. The fall time of mode 2, which was measured by the torsional vibration measurement system, was 40 s.

The results of the laboratory test and site test demonstrate that the torsional vibration measurement equipment meet the engineering measurement requirements.

5 CONCLUSIONS

In this paper, a rotational speed demodulating device was developed with FPGA technology, and a torsional vibration analysis system with virtual instrument technology. The Hilbert transformation and FRR (Forward Reverse Reverse) method are adopted in the signal processing. The system effectively captured the torsional vibration in laboratory and site tests. The test results also satisfied engineering measurement requirements.

REFERENCES

Bao, W. Wang, X.T. Yu, D.R. & Yang, K. 1998. The research summary of turbine-generator shart torsional oscillation. *Turbine Technology,* 40(4): 194–203.

Chang, G. Zhao, T. & Luo, L.Q. et al. 2012. Virtual instrumentation based measuring system of turbogenerator torsional vibration. *Electric power Automation Equipment,* 32(5): 1–5.

Gu, Y.J. He, C.B. & Yang, K. 2001. A monitoring and analyzing system for torsional vibrations of turbogenerator shafts. *Chinese Journal of Scientific Instrument,* 22(Z3): 269–270.

Subsynchronous Resonance Working Group of the System Dynamic Performance Subcommittee, Reader's guide to subsynchronous resonance, 1992. *IEEE Transactions on Power System,* 7(1): 150–157.

Thomas, W.R. 2012. *Turbine-generator shaft torsional vibrations resulting from transmission line transients.* Power and Energy Society General Meeting, San Diego, USA, 1–4.

Wang, X.T. Yu, D.R. & Bao, W. et al, 2000. A New Measuring Method for Subsynchronous Torsional Oscillations. *Automation of Electric Power Systems,* 24(17): 1–3.

Xiang, L. Tang, G.J. & Du, Y.Z. et al, 2000. Development of torsional vibration measuring system for turbine generating set. *Journal of Vibration, Measurement and Diagnosis,* 20(2): 134–138.

Xie, X.R. Guo, X.J. & Han, Y.D. 2011. Mitigation of multimodal SSR using SEDC in the Shangdu series-compensated power system. *IEEE Transactions on Power System,* 26(1): 384–391.

Xu, H.Z. & Zhang, Y.C. 2004. A new torsion vibration monitoring system for rotary power machine, *Journal of Vibration Engineering,* 17(S): 1137–1139.

Yang, J.G. & Hu, X.G. 2008. Development of torsional vibration measurement instruments based on LabVIEW. *China Measurement & Testing Technology,* 34(6): 28–32.

Zhang, Y. Jiang, Z.K. & Zhu, W. 1996. DK-IIA type torsional vibration monitoring system, *Journal of Tsinghua University (Sci & Tech),* 36(7): 19–23.

Zhang, Y.C. Li, C.X. & Shi, L.M. et al, 2009. Design and Implement of DK-III Shafting Torsional Vibration Measuring Instrument, *Process Automation Instrumentation,* 30(2): 70–72.

Advanced Materials and Structural Engineering – Hu (Ed.)
© 2016 Taylor & Francis Group, London, ISBN 978-1-138-02786-2

Design of a yaw damper for the aircraft model

M.X. Shen
School of Automatic Control and Mechanical Engineering, Kunming University, China

X.L. Li
Yuanmouxian Fire Services, China

X.Y. Ruan
Kunming Shipbuilding Design and Research Institute, China

ABSTRACT: The development of an automatic control system has played an important role in the growth of civil and military aviation. In order to increase the rate of a damper and improve the characteristic of a Dutch roll, it is crucial to design a damper with good performance. Dampers provide unique engineered solutions to the areas of design critical to the effectiveness and service life of the damper. In this paper, we design a damper using the state-model method, which is relatively simple to work with an MIMO system. Finally, we carry out the performance analysis before and after adjusting the system.

1 INTRODUCTION

Aircraft design relies heavily on the automatic control system that monitors and controls many subsystems. In order to meet the requirements of flying, generally, the vertical and horizontal movements of the aircraft need dampers that can work continuously. A damper is a device that deadens, restrains, or depresses. It is an important component that is used in many aircrafts (usually jets and turboprops) to damp (reduce) the rolling and yawing oscillations due to the Dutch roll mode. A damp involves yaw rate sensors and a processor that provides a signal to an actuator connected to the rudder. The use of the damper helps to provide a better ride for passengers and in some aircraft, it is a required piece of equipment to ensure that the aircraft stability remains within the certification values (http://en.wikipedia.org/wiki/Damper).

There are two kinds of damper, namely the pitch damper and the yaw damper. The pitch damper is responsible for the vertical movements of the aircraft, while the yaw damper is responsible for the horizontal movements of the aircraft. Most dampers are designed to restrict or stop the airflow. Due to the importance of dampers, more attention is now being focused by both academic and industrial groups for improving their performance (Du et al. 2005, Wang et al. 2008). Therefore, in recent years, flow-control, balancing, economizer, backdraft, face and bypass, and splitter dampers (Yu et al. 2009, Choi & Kim 2000) have been studied by many researchers and scientists.

Performance and design of construction are two key criteria for a high-quality damper. Performance can be measured by testing to ensure that the damper meets published leakage rates, and a good structural design and construction method can allow the damper provide a long-term good performance at specified pressures and a long service life.

In this paper, we study and design dampers to achieve the good performance of the airflow balance. We use the classical Matlab toolbox of a control system to design the dampers. The rest of this paper is organized as follows. Section 2 describes the mathematic model. Section 3 presents the analysis of system performance and simulation results. Section 4 concludes the paper.

2 MODEL DESCRIPTION

In the state of cruising, the state-space model of lateral motion is described as follows (Zhang 2009):

$$
\begin{bmatrix} \dot{x}_1(t) \\ \dot{x}_2(t) \\ \dot{x}_3(t) \\ \dot{x}_4(t) \end{bmatrix} = \begin{bmatrix} a_{11} & a_{12} & a_{13} & a_{14} \\ a_{21} & a_{22} & a_{23} & a_{24} \\ a_{31} & a_{32} & a_{33} & a_{34} \\ a_{41} & a_{42} & a_{43} & a_{44} \end{bmatrix} \begin{bmatrix} x_1(t) \\ x_2(t) \\ x_3(t) \\ x_4(t) \end{bmatrix}
$$
$$
+ \begin{bmatrix} b_{11} & b_{12} \\ b_{12} & b_{22} \\ b_{13} & b_{32} \\ b_{14} & b_{42} \end{bmatrix} \begin{bmatrix} u_1(t) \\ u_2(t) \end{bmatrix}
$$

$$\begin{bmatrix} y_1(t) \\ y_2(t) \end{bmatrix} = \begin{bmatrix} c_{11} & c_{21} & c_{31} & c_{41} \\ c_{21} & c_{22} & c_{23} & c_{24} \end{bmatrix} \begin{bmatrix} x_1(t) \\ x_2(t) \\ x_3(t) \\ x_4(t) \end{bmatrix}$$

$$(1)$$

where the state vectors are defined as follows: $x_1(t)$ is the side slip angle measured in rad; $x_2(t)$ is the rate of the yaw angle measured in rad/s; $x_3(t)$ is the angular velocity of rollover measured in rad/s; and $x_4(t)$ is the bank angle measured in rad. The input vectors are defined as follows: $u_1(t)$ is the draft angle of the rudder measured in rad and $u_2(t)$ is the draft angle of the aileron. The output vectors are defined as follows: $y_1(t)$ is the rate of the yaw angle measured in rad/s and $y_2(t)$ is the bank angle measured in rad.

The parameters of the model can be varied according to the flying speed and height of the flight.

The analysis proceeds as follows (Singh, 2014):

- Inputting the parameters of the state-space model;
- Defining state variables, input variables and output variables; and
- Setting up the state-space model.

3 PERFORMANCE ANALYSIS AND RESULTS

In order to achieve good performance, it is critical to adjust the system and perform performance analysis.

3.1 *Performance analysis before adjusting*

We analyzed the performance according to the state-space model described above. The steps are as follows:

- Calculating the open-loop characteristic values and
- Calculating the unit pulse response of the system.

From Figure 1, it can be seen that this model has a pair of conjugate poles, near the $j\omega$-axis, which corresponds to the Dutch roll state of the aircraft and with the minimum damper. From Figure 2, it can be seen that there is fluctuation during the transition process, the aircraft has a very low damper and the response time is long.

We take the angle of yaw as the input controller to obtain the frequency response, and the open-loop model is marginally stable and its characteristics are shown in Figure 3. Figure 4 shows the bode diagram.

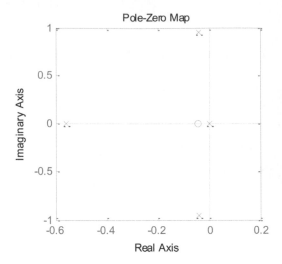

Figure 1.   Zeros and poles.

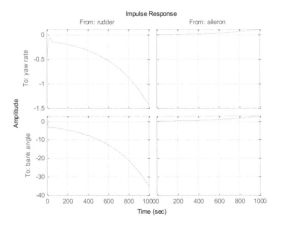

Figure 2.   Curves of the unit impulse response.

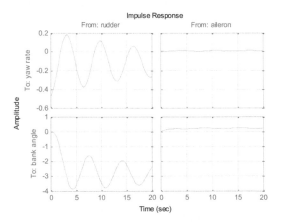

Figure 3.   Open-loop feedback during 20 sec.

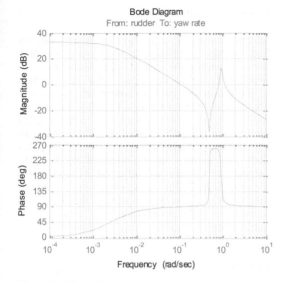

Figure 4.   Bode diagram.

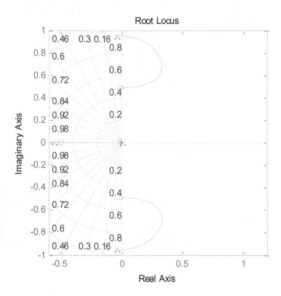

Figure 5.   Negative feedback.

3.2   *Adjusting the system*

In order to obtain a larger damper rate (more than 0.3), when $\omega_n < 1$ rad/s, we use the Matlab toolbox to simulate the root locus mapping to determine the ideal gain value. We then set the SISO (Single Input Single Output) closed-loop feedback system.

From Figure 5, it can be seen that the system is unstable due to the fact of adopting a negative feedback, so we select a positive feedback. From

Figure 6 and Figure 7, according to the gain and the rate of the damper, it is obvious that the positive feedback has better performance than the negative feedback.

Compared with the open-loop feedback, from Figure 8, it is obvious that the speed of the closed-loop feedback is much quicker and has the smaller amplitude of swing. We obtain Figure 8 by connecting output to input to set up the MIMO (Multiple Input Multiple Output) closed-loop feedback system. From Figure 9, it is obvious that the response rate of the yaw angle has a better value; however, the system do not deflect continuously, as the normal aircraft with the change in the

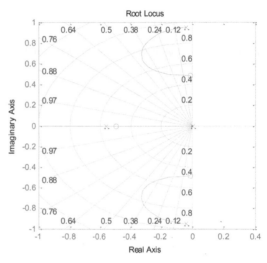

Figure 6.   Positive diagram $\xi \geq 0.3$.

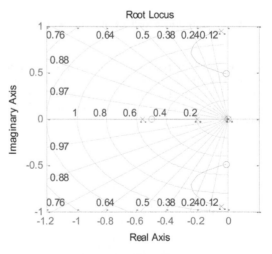

Figure 7.   Positive diagram $\xi \geq 0.5$.

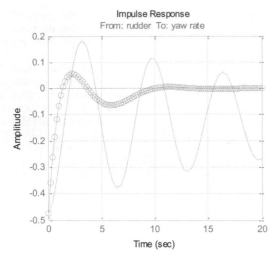

Figure 8. Closed-loop feedback during 20 sec (SISO).

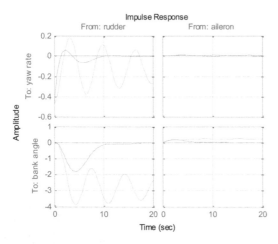

Figure 9. Closed-loop feedback during 20 sec (MIMO).

aileron angle, it presents a stable spiral rotation state. The spiral rotation state is a typically very slow movement state; in this state, the aircraft can deflect and rotate with no constant input of the aileron.

4 CONCLUSIONS

Automatic control system in the aircraft aids the flight crew in navigation, flight management, and augmenting the stability characteristic of the airplane. In order to improve the performance of the automatic control system of the aircraft, it is critical to design a good-quality damper. In this paper, we design the aircraft damper using the state-space method. This model has one pair of complex, conjugate lightly damped poles. They correspond to the "Dutch roll mode". The damper parameters are adjusted by comparing the open-loop feedback and closed-loop feedback circumstances. Finally, we provide the discussion of the results: the closed-loop responses are quite stable compared with the open-loop responses.

REFERENCES

Choi, S.B. & Kim, W.K. 2000. Vibration control of a semi-active suspension featuring electrorheological fluid dampers. *Journal of Sound and Vibration*, 234: 537–546.
Du, H. Sze, K.Y. & Lam, J. 2005. Semi-active H∞ control of vehicle suspension with magneto-rheological dampers. *Journal of Sound and Vibration*, 283: 981–996.
http://en.wikipedia.org/wiki/Damper.
Singh, S. & Rama Murthy, T.V. 2014, Design of an Optimal Yaw Damper for 747 Jet Aircraft Model. *Emerging Research in Electronics, Computer Science and Technology Lecture Notes in Electrical Engineering*, 248: 801–810.
Wang, J. Dong, C. Shen, Y. & Wei, J. 2008. Robust modelling and control of vehicle active suspension with MR damper. *Vehicle System Dynamics*, 46: 509–520.
Yu, M. Dong, X.M. Choi, S.B. & Liao, C.R. 2009. Human simulated intelligent control of vehicle suspension system with MR dampers. *Journal of Sound and Vibration*, 319: 753–767.
Zhang, D.F. 2009. *MATLAB/Simulink Modeling and Simulation*, Book, Electronic Industry Press, 349–357.

Advanced Materials and Structural Engineering – Hu (Ed.)
© *2016 Taylor & Francis Group, London, ISBN 978-1-138-02786-2*

Design of mechanical pre-stressing system for strengthening reinforced concrete members with pre-stressed NSM FRP bars

S.S. Abdulhameed
College of Civil and Transportation Engineering, Hohai University, Nanjing City, Jiangsu Province, China
College of Engineering, Al-Mustansiriya University, Baghdad, Iraq

E. Wu & B. Ji
College of Civil and Transportation Engineering, Hohai University, Nanjing City, Jiangsu Province, China

ABSTRACT: Strengthening of concrete structures by pre-stressing Fiber Reinforced Polymer (FRP) reinforcements by using Near Surface Mounted (NSM) technique proved to be efficient and practical because it combines the benefits associated with pre-stressing the FRP reinforcements with the NSM technique. A new mechanical system was developed to pre-stress FRP bar to strengthen reinforced concrete member by using NSM technique. The main features of the system are: (1) the pre-stressing forces are applied manually without using hydraulic jacks, (2) the pre-stressing level of the FRP bars can be controlled by using an electronic load cell, (3) the system is light weight and can be installed with no need to use heavy lifting equipment (4) the FRP bars are extended to the ends of the member, and (5) releasing of the pre-stressing FRP bar is accomplished under a slow strain rate. A Finite-Element Analysis (FEA) was used to verify the design of the mechanical device. Results showed that the mechanical pre-stressing system is sufficient to pre-stress and anchorage the FRP bar up to 1861.2 MPa.

1 INTRODUCTION

Strengthening of concrete members by pre-stressing Fiber Reinforced Polymer (FRP) reinforcements by using Near Surface Mounted (NSM) technique proved to be efficient and practical because it combines the benefits associated with pre-stressing the FRP reinforcements with the NSM technique.

The NSM technique consists of placing the FRP bars or plates into grooves precut into the concrete cover in the tension surface of the concrete member and bonding the FRP by using epoxy adhesive. This technique (NSM) was investigated by some researchers, among others Taljsten et al. 2003, Nordin & Taljsten 2006, Hajihashemi et al. 2011, Choi et al. 2011, Oudah & El-Hacha 2012d. Test results showed that, pre-stressing FRP reinforcements improve serviceability, reduce cracks width and delay onset of cracks, increase yielding of internal steel reinforcement at a higher proportion of the ultimate loads (El-Hacha et al. 2001).

Two methods were developed to pre-stress FRP bars or plates, namely external reaction frame method, and direct tension method.

The first method (external reaction frame method) was adopted by several researchers (Nordin et al. 2001, Casadei et al. 2006, Wu et al.

2007, Jung et al. 2007, Ines & Joaquim 2010). In this method, the FRP reinforcements are embedded inside grooves filled with epoxy adhesive. The ends of the FRP reinforcement are fixed to an external reaction frame and stressed to the desired pre-stressing level by using a hydraulic jack. The pre-stressing system was left until the epoxy adhesive was completely cured. The pre-stressed FRP reinforcement was released after curing the epoxy adhesive.

In the second method, the FRP bars or plates are tensioned against the strengthened beam itself. This method was adopted by several researchers (Badawi 2007, El-Hacha & Gaafar 2011, Kim et al 2014). The system consists of two steel anchors which are used to grip the ends of the FRP bars or plates. One of these anchors is fixed to the surface of the concrete beam and the other anchor is fixed to the hydraulic jack that is mounted on the surface of the beam at the other end. The pre-stressing force is applied using a hydraulic jack mounted on the beam.

Some shortcomings are associated with the above described pre-stressing systems, that is most of these pre-stressing systems are more suitable to use in the laboratory than in the field, they need power-operated hydraulic jacks; and the release method was carried out under high strain rates.

Recently Abdulhameed et al. (Abdulhameed et al. 2013) developed a mechanical pre-stressing system. This system is used to pre-stress FRP sheets to strengthen concrete members by using surface bonding technique as shown in Figure 1. They observed that using mechanical pre-stressing system is practical and efficient to strength reinforced concrete members.

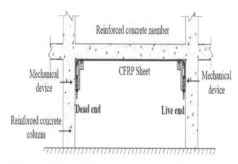

(a) Proposed method of assembling mechanical pre-stressing system in site application

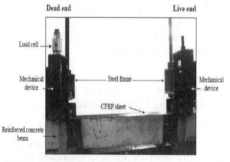

(b) Assemblage of mechanical pre-stressing system in the laboratory

Figure 1. Mechanical pre-stressing system of Abdulhameed et al., 2013.

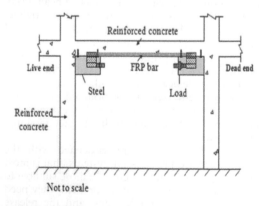

Figure 2. Assemblage of mechanical pre-stressing system.

In the current study, a new mechanical pre-stressing system was developed to pre-stress FRP bar. This system consists of two mechanical devices are fixed to the tension surface of reinforced concrete beam. One of these devices is used as a dead end, while the other is used as a live end as shown in Figure 2. The main features of this system are: (1) it can be used to strengthen reinforced concrete members (beams or slabs) of various lengths, (2) the FRP bars can be extended to the ends of the concrete member, (3) the pre-stressing transfer is achieved under slow strain rate, (4) the light-weight system can be fixed to concrete members with no need to heavy lifting equipment and (5) the pre-stressing forces can be controlled by using an electronic load cell.

A Finite-Element Analysis (FEA) was used to verify the design of the mechanical device.

2 DESCRIPTION OF THE MECHANICAL PRE-STRESSING SYSTEM

A new mechanical system was developed to pre-stress FRP bar to strengthen reinforced concrete member by using NSM technique as shown in Figure 2.

The system consists of two mechanical devices that are fixed at both ends of the member (Fig. 2). Each device consists of steel anchorage with a longitudinal slot and rounded steel rods. A summary of the main components and dimensions of the mechanical device are presented in Table 1 and Figures 3 and 4.

Table 1. Main components and dimensions of the pre-stressing mechanical device.

Component name	Component [ID]	Dimensions [mm]
Hole	A	Ø52
Slot	B	500 × 50
Threaded rod	C	Ø30
U-shape steel rod	D	Ø30 (outer diameter), Ø12 (inner diameter)
Hollow steel rod	E	Ø60 (outer diameter), Ø50 (inner diameter), 120 (length)
Steel plate	F	120 × 65 × 15
Solid steel rod	G	Ø36 × 90
Ball bearing	H	Ø36
Solid threaded rod	I	Ø50 × 600
Steel nut	J	Ø50
Plane bearing	K	Ø52
Load cell	L	–
Hollow steel rod	M	Ø20 (outer diameter), Ø10 (inner diameter)

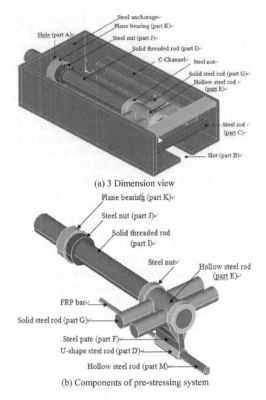

(a) 3 Dimension view

(b) Components of pre-stressing system

Figure 3. Main components of the mechanical pre-stressing system.

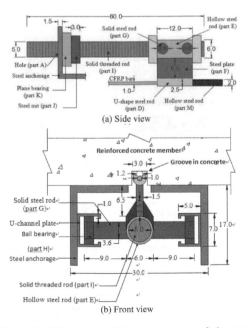

(a) Side view

(b) Front view

Figure 4. Dimensions of the components of the mechanical device.

3 PROCEDURE OF PRE-STRESSING FRP BAR

Strengthening of reinforced concrete member consists of preparation of the concrete surface by sawing up a groove in the tension surface of the concrete in the longitudinal direction. The dust should be removed from the groove by an air blower and cleaned by acetone.

Each end of the FRP bar is impregnated with epoxy adhesive using a brush and put a hollow steel rod (part M) and let it to fully cured (Fig. 5a).

The groove of the concrete member should be filled with epoxy adhesive, and then put the FRP bar inside it (Fig. 5b).

The solid steel rods (part G) are placed inside C-channels of the steel anchorage (Fig. 6) (ensure that the FRP bar is placed inside a U—shape steel rod (part D)).

The two steel anchorages are fixed to the concrete surface by using anchorage bolts. All the remaining components are installed successively:

1. Pass the solid steel rod (part I) through the hole (part A), plane bearing (part K), and the

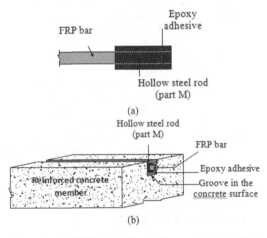

Figure 5. (a) FRP bar inside a hollow steel rod (part M); (b) Placing of the FRP bar inside a concrete's groove.

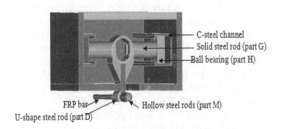

Figure 6. Solid steel rods (part G) inside C-channels.

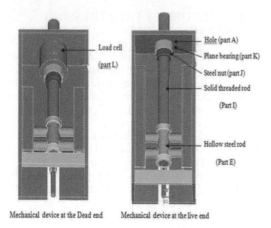

Figure 7. Mechanical devices at dead and live ends.

hollow steel rod (part E) and fix them with steel nuts (note that a load cell (part L) is assembled between the plane bearing and the anchorage wall at the dead end) (Fig. 7).
2. Fix steel rods (part C) to the steel anchorage by using steel nuts.
3. Tighten the steel nut (part J) to move the U-shape steel rod (part D) horizontally that push the steel rod (part M) and pre-stress the FRP bar.
4. Pre-stressing force in the FRP bar is monitored by a load cell that is assembled at the dead end of pre-stressing system.
5. After the desired pre-stressing level is achieved, additional epoxy adhesive paste should be added to the groove of the concrete surface. Excess epoxy should be removed using a spatula.
6. After curing the epoxy adhesive, transfer of the pre-stressing was carried out by slowly releasing of the steel nuts (part J) from both dead and live ends.
7. Remove the steel anchorage with all components.
8. Additional epoxy adhesive should be used to cover the steel rod (part M).

4 FINITE ELEMENT ANALYSIS (FEA)

Finite Element Analysis (FEA) simulation using the commercial software ANSYS was employed to verify the design of the device. An eight-node solid element, Solid 45 was used to simulate the steel plates and steel rods. The element is defined by eight nodes having three degrees of freedom at each node translations in the nodal X, Y, and Z directions (ANSYS manual version 8.0).

A horizontal force was applied to the U-shape steel rod (part D) until the desired pre-stressing

level was achieved (60% of the ultimate tensile stress (f_{fu}) of the Carbon Fiber Reinforced Polymer Bar (CFRP)) as shown in Figure 8. This value is upper than the permissible limit (55% f_{fu}) according to the recommendation of ACI. 440.2R-08 (ACI. 2002.) (creep-rupture of the CFRP reinforcement under sustained stresses was avoided by setting the pre-stress levels below 55% f_{fu}).

The effect of nonlinearities was included in the analysis using bilinear isotropic hardening (yielding of steel equal to 450 MPa) to ensure that yielding of steel will not occur in the device when the pre-stressing force is applied. The finite element mesh of the device model is shown in Figure 8.

Table 2 shows the main dimensions and properties of a CFRP bar that was used in the analysis (assume that the maximum length of the strengthened beam is 25 meter based on Yu et al. (2008).

Figure 9 shows Von-Mises stresses that were generated in the mechanical device due to the pre-stressing force. The maximum stress that generated in the mechanical device is 335.14 MPa and the maximum stress that generated at the point of contact between the C-channel of the steel anchorage and the steel rods (part G) is 90.43 MPa. Both values are lower than the yielding stress (450 MPa).

Figure 10 shows the deformed mesh in the X and Y directions. It can be seen that the maximum displacement in the X-direction is 0.15 mm for the short side of the steel anchorage; this value is lower than the displacement at the yielding load (0.22 mm).

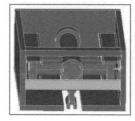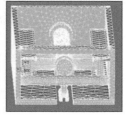

Figure 8. FEM discretization of pre-stressing mechanical device.

Table 2. Material properties of CFRP bar.

Material	Tensile strength [MPa]	Ultimate strain [%]	Elastic modulus [GPa]
CFRP bar*	2068	1.7	124

*A_f = 65 mm^2, Ø9.1 mm, L_f = 25000 mm.
where, A_f is the cross sectional area of the FRP bar, Ø is the bar diameter, and L_f is the CFRP bar length.

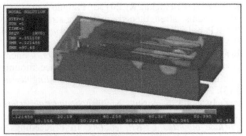

(a) Steel anchorage

(b) Steel rods

Figure 9. Von-Mises stress in the pre-stressing system.

(a) Displacement in X-direction

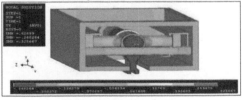

(b) Displacement in Y-direction

Figure 10. Variation of displacements in the X and Y directions.

5 CONCLUSIONS

In this research program, a new practical mechanical pre-stressing system was developed to pre-stress the FRP bar to strengthen reinforced concrete member by using NSM technique. The main conclusions of this analysis are that,

1. The system is practical and easy to use because the pre-stressing force can be applied manually without using hydraulic jacks, the level of pre-stressing force can be controlled by using an electronic load cell that was assembled on the threaded rod (part I), and the system is light weight and can be installed with no need to use heavy lifting equipment.
2. Based on the numerical results, the mechanical pre-stressing system is sufficient to pre-stress and anchorage the FRP bar up to 1861.2 MPa (pre-stressing level in the FRP bar $(\%f_{fu}) \leq 1861.2$ MPa).
3. Using a plane bearing (part K) can effectively reduce the frictional force between the steel nut (part J) and the steel anchorage.

REFERENCES

Abdulhameed, S. Wu, E. & Ji, B. 2013. Mechanical pre-stressing system for strengthening reinforced concrete members with pre-stressed carbon-fiber-reinforced polymer sheets, *Journal of Performance of Constructed Facilities*, 10.1061/(ASCE)CF.1943-5509.0000478, 04014081.

American Concrete Institute (ACI), 2002. *Guide for the design and construction of externally bonded FRP systems for strengthening concrete structures*, ACI 440.2R-08, Farmington Hills, MI.

ANSYS version 8.0 Houston [Computer software]. Swanson Analysis Systems, Texas.

Badawi, M. 2007. *Monotonic and fatigue flexural behavior of RC beams strengthened with pre-stressed NSM CFRP rods*, PhD Thesis, University of Waterloo, Waterloo, Ontario, Canada, 256.

Casadei, P. Galati, N. Boschetto, N. Tan, K.Y. Nanni, A. & Galecki, G. 2006. *Strengthening of impacted pre-stressed concrete bridge I-girder using pre-stressed near surface mounted CFRP Bar*. Federation International du Beton (FIB). In Proceedings of the 2nd International congress, Naples, Italy, 10: 10.

Choi, H.T. West, J.S. & Soudki, K.A. 2011. Effect of partial unbonding on pre-stressed near-surface-mounted CFRP-strengthened concrete T-beams, *Journal of Composites for Construction*, 15(1): 93–102.

El-Hacha, R. & Gaafar, M. 2011. Flexural strengthening of reinforced concrete beams using pre-stressed near—surface mounted CFRP bars, *PCI Journal*, 56(4): 134–151.

El-Hacha, R. Wight, R.G. & Green, M.F. 2001. Pre-stressed fibre-reinforced polymer laminates for strengthening structures, *Progress in Structural Engineering and Materials Journal*, 3: 111–121.

Hajihashemi, A. Mostofinejad, A. & Azhari, A. 2011. Investigation of RC beams strengthened with pre-stressed NSM CFRP laminates, *Journal of Composites for Construction ASCE*, 15(6): 887–895.

Inês, G.C. & Joaquim, A.O.B. 2010. *Flexural strengthening of RC beams with pre-stressed FRP laminates applied according to NSM techniques*, Engineering Week.

Jung, W. Park, J. & Park, Y. 2007. *A study on the flexural behaviour of reinforced concrete beams strengthened with NSM pre-stressed CFRP reinforcement*, In Proceeding of the 8th International symposium on Fiber Reinforced Polymer Reinforcement for Reinforced Concrete Structures (FRPRCS-8), University of Patras, Patras, Greece.

Kim, Y.J. Kang, J.Y. & Park, J.S. 2014. *Performance and Reliability of Bridge Girders Upgraded With Post-Tensioned Near-Surface-Mounted Composite Strips.* Probabilistic Safety Assessment and Management (PSAM 12), Honolulu, HI, USA.

Nordin, H. & Taljsten, B. 2006. Concrete beams strengthened with pre-stressed NSM CFRP, *Journal of Composites for Construction,* 10(1): 60–68.

Nordin, H. Taljsten, B. & Carolin, A. 2001. *Concrete beams strengthened with pre-stressed Near Surface Mounted Reinforcement (NSMR)*, In Proceeding of 1st International Conference on FRP Composites in Civil Engineering (CICE2001), Hong Kong, China. 2: 1067–1075.

Oudah, F. & El-Hacha, R. 2012d. Fatigue behavior of RC beams strengthened with pre-stressed NSM CFRP rods. *Journal of Composite Structures,* 94(4): 1333–1342.

Täljsten, B. Carolin, A. & Nordin, H. 2003. Concrete structures strengthened with near surface mounted reinforcement, *Advances in Structural Engineering,* 6(3): 201–213.

Wu, Z. Iwashita, K. & Sun, X. 2007. *Structural performance of RC beams strengthened with pre-stressed near-surface mounted CFRP tendons,* In Proceedings of ACI special Publication SP-245: Case Histories and Use of FRP for Pre-stressing Applications, Denver, Colorado, USA, 245-10: 165–178.

Yu, P. Silva, P.F. & Nanni, A. 2008. Description of a mechanical device for pre-stressing of carbon fiber reinforced polymer sheets, *Part I. ACI Structural Journal,* 105(1): 3–10.

Advanced Materials and Structural Engineering – Hu (Ed.)
© 2016 Taylor & Francis Group, London, ISBN 978-1-138-02786-2

A new shearer monitoring system based on ZigBee and wireless sensor network technology

W.H. Li & L.H. Zhou
School of Electrical and Information Engineering, Anhui University of Science and Technology, Huainan, China

ABSTRACT: In view of the present difficulties in mine, such as the inconvenience of shearer wiring, the long distance of data transmission, and the inefficiency in real-time monitoring, this paper proposes a new design scheme. The scheme takes RF Modules of low-dissipation CC2530 as the core; meanwhile it takes the CC2591 as the remote monitoring system of the wireless sensor radio frequency's front end. The system monitors the shearer's status according to the wireless data collection point, and then delivers the data to the terminal and deals with it. Meanwhile, it delivers the processed data to the monitor, taking advantage of the means that combine the wireless communication with the RS485 bus. Such a measure makes it possible to take remote monitoring of the shearer by wireless technology, which has an extraordinary significance.

1 INTRODUCTION

Nowadays, shearer underground mostly uses cable monitoring, which has the disadvantages of complicated routing, aging lines and non-extension. All defects have a negative impact on the data collection and transmission. This paper proposes a newly designed shearer monitoring system that is based on the ZigBee and wireless sensor network technology. ZigBee has become one of the most popular communication modes in recent years, which uses a bilateral communication mode to improve its efficiency, while the wireless sensor network monitoring technology is also a popular research direction. The primary functions of ZigBee and wireless sensor networks are to monitor the implementation of the shearer's real-time operating status, making use of wireless data collection points on the front and back of the shearer to gather the operating status of the shearer and the condition of the surroundings, and then deliver the information to each routing node by ZigBee.

The system monitors the shearer's status by using a wireless sensor, and then delivers the gathered data to the ground by RS485 bus, so that the controller on the ground can grasp the real-time operating status of the shearer as well as the gas concentration of the coal face, automatic alarms as soon as it passes the warning line, which improves the safety of the mine dramatically.

2 STRUCTURE AND OPERATION PRINCIPLE OF THE SYSTEM

The structure of the shearer monitoring system is shown in Figure 1.

The three kinds of nodes are included in the system such as wireless sensor nodes, routing nodes and coordinator nodes. Wireless sensor nodes refer to the wireless sensor on the shearer and especially on the back of the shearer. A large-capacity storage node is installed in front of the shearer, which is used to store the important parameters collected. Operating parameters (including shearer temperature, pressure, and flow) are delivered to the storage nodes in real time, while the shearer is

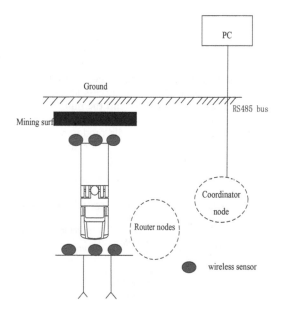

Figure 1. The structure of the shearer monitoring system.

operating and transmits to the routing nodes by the ZigBee network after pretreatment. Router node is responsible for data forwarding path to choose; at the same time, it can be connected to wireless sensors. After data transmission, the routing nodes can be set to hibernation by software, and waked up when the task is ordered. Coordinator node is responsible for establishing the general maintenance of the entire wireless network; a network can have only one coordinator node. The node has high-speed data processing capabilities, which can process and analyze the data uploaded by router nodes quickly, and then convert them into specific parameters and upload them to the PC on the ground by the RS485 bus. The operating parameters of the shearer will be displayed on the PC after the related processes (Levis & Gay 2009).

3 HARDWARE DESIGN OF THE SYSTEM

3.1 Terminal node design

Peripheral circuit design of the terminal node is not very complicated. The core module is CC2530, which is regarded as the network sensor notes. The main peripheral hardware device consists of a variety of sensors (including temperature sensors, pressure sensors, and flow sensors) and a communication device interface, alarm device interface and CC2591 auxiliary module. The structure is shown in Figure 2.

CC2530 is a wireless RF microcontroller. In addition to the basic functions of a general microcontroller such as 8051, there are other additional functions such as the support of the IEEE 802.15.4 protocol. Three kinds of storage can be used to access the memory bus by this MCU, which includes SFR (special function registers), DATA (data storage) and CORE/XDATA (code/external data memory). The mode of single-cycle access is used by CC2530 to access special function registers, data memory and the main SRAM. The advantage of using a single-cycle mode is that when the CC2530 is in the idle mode, in order to restore the MCU to the active mode, an interrupt signal needs to be sent (Ruan et al. 2008).

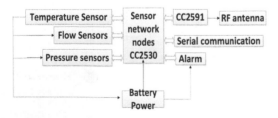

Figure 2. The structure of the network sensor nodes.

The function of data acquisition is realized by sensor network nodes controlled by the 8-channel 12-bit A/D port control sensor of CC2530, while the data transmission is completed by the CC2530 radio frequency module. The power is low when the CC2530 transmits the data. The sensitivity of the CC2530 is fixed, i.e., cannot be changed arbitrarily, which limits its communication distance. Considering the complex terrain of a coal mine tunnel, in order to increase the transmission power of a network of sensor nodes, a front-end RF chip CC2591 is added to improve the power transmission significantly. This design can ensure the transmission distance of sensor network node signals to achieve almost 1 km without attenuation.

Ordinary battery-powered models are selected to supply power for sensor nodes in the network, which means that two AA batteries are used as the power supply module. However, the voltage converter chip TPS60210 is also needed to convert 5 V DC power to a DC voltage of 3.3 V. When the sensor nodes in the network are in the working lifetime, that is, they are processing the data receiving and forwarding, the working current will achieve about 29 mA. In other times, the network of sensor nodes is basically dormant. The operating current is generally microamps that is almost negligible, which means the energy consumption of the network sensor node will be very low.

The hardware configuration of the router node is relatively similar to the terminal nodes that the relevant sensor needs to be removed in the specific design process.

3.2 Coordinator node hardware design

As an important part of the whole wireless network, the coordinator node needs to handle a large number of the passed data. Considering that the ordinary battery-powered models cannot reach the standard conditions of the normal work operations, electricity is selected as the power in its work. The role of the coordinator node can be simply summarized as a node that receives the data and then stores the data in the RAM of the node controller. Finally, the data is processed, analyzed, transferred to a control room on the ground through the RS485 bus. Considering the large amount and variety of data processing of the node, a processor with high speed and adequate external interface is needed. Under the comprehensive consideration, the core LPC2378 of NXP's ARM7TDMI-S is chosen as the core processor of the coordinator node.

3.3 PC hardware design

PC consists of two parts: 1) the processing platform of the information passed; 2) RS485 interface.

The information processing platform is the part of store, analyzes the process and displays the information transmitted. On the other hand, the role of the host computer is the connected Inoue coordinator node and processed information specifically.

4 SOFTWARE DESIGN

4.1 *Network node design*

The network node in a particular system includes a terminal node, a router node, and a coordinator node. Software design is based on the APL layer (application layer) of the ZigBee wireless network protocol and the stack of the terminal, and then loaded on to it to achieve. The three kinds of a node have a similar software architecture, except some differences in the specific subroutine design and related configuration information.

The flowchart of the network nodes is shown in Figure 3, Figure 4, and Figure 5.

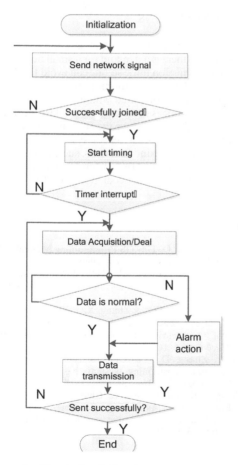

Figure 3.   Terminal node flowchart.

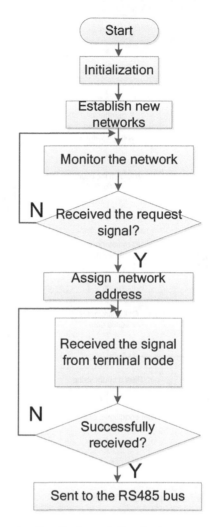

Figure 4.   Coordinator node flowchart.

Figure 3 shows a flowchart of the terminal node, which works as follows. CC2530 is initialized after power while power supply switch of sensors is connected when it turns on. Then, the ZigBee network underground is required to connect to an existing wireless network and join the network. If it connects successfully, the next stage of timed stages is accessed; otherwise it sends the signals continuously until it achieves. If the timer expires, the sensor installed in the front and rear of the shearer begins to collect data and preprocess it. If the collected data is normal, it sends it to the next node, otherwise it sounds an alarm, which shows that the shearer's working condition is deviated from the normal track (Yu et al. 2009).

Figure 4 shows a flow chart of the coordinator node, which works as follows. The modules are first

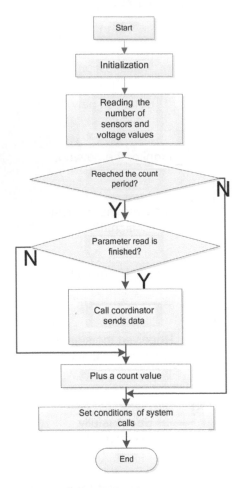

Figure 5. Routing node flowchart.

Figure 5 shows a flow chart of the routing node. Its role is to collect the data processing by the terminal node, and then transmit them to the coordinator node.

4.2 PC software

The PC software uses the VC ++ language. Thus, the following functions can be performed: real-time display, data query, remote upload and security.

Data query data flow are as follows: 1) to enter the underground shearer home monitoring system; 2) select the user login mode (including ordinary user login and administrator login) to enter the authenticity of the information, which will switch to functional interface automatically; 3) function interface includes the current operating parameters of the shearer and historical queries and query parameters set shearer and other options; any one of them can be operated as long as one clicks on the appropriate operating tips.

5 CONCLUSIONS

Since the design uses a wireless communication technology, there is a demand for a small number of signal transmission lines, so one can avoid the drawbacks that lines wired monitor brings more jumbled furnished. Compared with a wired network, a wireless network is more convenient, less hardware required, helpful to reduce the cost and much easier to maintain. A wireless sensor system installed on the shearer can expand functions, such as increasing environmental monitoring functions and achieving the multidimensional of coal mine monitor.

initialized in the CC2530 module and LPC2378 microprocessor after power up. Then, they start to build the new network node based on the current ZigBee protocol stack and its real-time monitor. After the formation of the new network, if a request joins the signal emitted by the terminal node and is received by the coordinator node, it assigns addresses for it. When the terminal node obtains a new network address assigned to the coordinator node, it suggests that the terminal node described in the normal group network state begins to transmit the collected data. If the collected data can be transferred successfully to the coordinator node, which illustrates that the coordinator can connect the power to receive information, it then puts the received data into processing and analysis; the data will determine the final classification after forwarding to the RS485 bus (Yu et al. 2009, Zhang et al. 2001).

REFERENCES

Levis, P. & Gay, D. Tiny OS Programming Manual. UK: Cambridge University Press, 2009.
Pottie, G.J. & Kaiser, W.J. 2000. Wireless integrated network sensors, Communications of the ACM. 43(05): 51–58.
Ruan, D.X. Tang, D.F. & Zhang, X.G. 2008. Application of ZigBee wireless sensor network technology in coal mine environmental monitoring. Coal Mine Machinery, 6: 163–164.
Yu, Y. Zhang, D.W. & Cui, J.J. 2009. Underground personnel positioning system based on ZigBee technology, Coal Mine Machinery, 12: 19–21.
Zhang, S.K. Zhang, W.J. & Chang, X. et al. 2001. Building and Assembling Reusable Components based on Software Architecture, Journal of Software, 12(09): 1351–1359.

Advanced Materials and Structural Engineering – Hu (Ed.)
© *2016 Taylor & Francis Group, London, ISBN 978-1-138-02786-2*

Fast spiral scanning based on Switched Capacitor Charge Pump

J. Zhang, L.S. Zhang & Z.H. Feng
Department of Precision Machinery and Precision Instrumentation, University of Science and Technology of China, Hefei, Anhui, China

ABSTRACT: The Switched Capacitor Charge Pump (SCCP) is a newly developed charge control method for the piezoelectric positioner used in Atomic Force Microscope (AFM). When the scanning of the AFM is performed in the traditional raster pattern, there is usually a trade-off between the low-frequency leakage and the positioner vibration. In this paper, a fast spiral-scan based on the SCCP method is proposed. Fast and vibration-free scanning can be achieved by using this SCCP-based spiral-scan strategy.

1 INTRODUCTION

Atomic Force Microscope (AFM) was invented by Binnig in 1986 (Binnig 1986). Since its emergence, it has become an essential tool in the areas such as micro- and nano-scale analyzing (Burnham & Colton 1989, Ducker et al. 1991), nanofabrication (Avouris et al. 1997, Tseng et al. 2008), and molecular investigation (Ando et al. 2001, Ando et al. 2008). As a key component used in AFM, the three-dimension positioner is usually constructed by piezoelectric actuators (PZAs) because of their advantages on solution and dynamics (Moheimani 2008). However, the performance of the piezoelectric actuator is largely limited by the hysteresis and nonlinearity between the output displacement and the applied voltage. A number of techniques have been proposed to solve these problems. They can be found in references (Clayton et al. 2009, Fleming 2012, Comstock 1981, Newcomb & Flinn 1982) for details. Specially, Huang et al. (2010) invented a Switched Capacitor Charge Pump (SCCP) method to linearize the displacement of the PZAs. In their study, a hysteresis of less than 2% was achieved over a frequency range from 0.01 Hz to 20 Hz. In spite of the improvements in hysteresis and linearity, there is still a crucial problem associated with the SCCP method, which is the low-frequency leakage.

When the SCCP method is utilized in an AFM, the scanning is normally performed in a raster pattern. The raster pattern is realized by applying a high frequency triangular waveform to the quick axis (normally x-axis) and a very low frequency triangular waveform to the slow axis (normally y-axis) of the positioner. For the charge control methods, one of the main drawbacks of the low-frequency driving of the slow axis is the charge leakage, which results in undesired displacement drift. In a raster scanning, the only way to increase the driving frequency of the slow axis is to increase the driving frequency of the fast axis. However, this would inevitably excite the mechanical resonance of the positioner since the triangular waveform contains all odd harmonics of the fundamental frequency whose amplitudes attenuates as $1/n^2$. Consequently, the mechanical resonance causes the positioner to vibrate and trace a distorted triangular displacement along the x-axis which can significantly distort the generated image. So this is a dilemma: if we want to reduce the low-frequency charge leakage, we have to increase the driving frequency of the fast axis and then face the undesired vibration of the positioner; if we want the positioner not to vibrate, then we have to tolerate the low-frequency leakage of the y-axis. Therefore there is a strong need for a new scanning strategy to achieve fast scanning and avoiding vibration simultaneously. For this purpose, many solutions were proposed. The unique spiral-scan (Mahmood & Moheimani 2009) scheme proposed by Mahmood satisfied the demand quite well. In this work, a fast spiral-scan based on switched capacitor charge pump was proposed, in which fast scanning without vibration was achieved.

2 THE SPIRAL SCAN

A spiral-scan scheme was applied in this work for fast scanning in the AFM. The spiral-scan trajectory is illustrated in Figure 1. The pattern is known as the Archimedean spiral. A property of this curve is that its pitch P, which is the distance between two consecutive intersections of the spiral curve with any line passing through the origin, is constant. This property is quite important for scanning purposes as it ensures that the sample

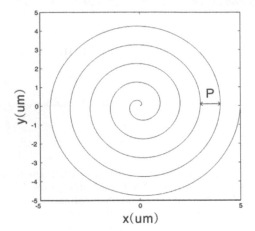

Figure 1. The spiral scanning.

surface is scanned uniformly. In spiral scanning, both axes follow sinusoidal signals of identical frequencies resulting in a smooth trajectory. This avoids the transient behavior that may occur in sinusoidal scans as the probe moves from one line to the next. Furthermore this scanning strategy does not require special hardware, and can be implemented in a standard AFM with minor software modification.

The equation that generates the spiral of pitch P and traced at an angular velocity of ω can be derived from a differential equation as

$$\frac{dr}{dt} = \frac{P\omega}{2\pi} \tag{1}$$

where, r is the instantaneous radius at time t. Equation 1 is solved for r by integrating both sides to obtain

$$r = \frac{P}{2\pi}\omega \cdot t \tag{2}$$

for $r = 0$ at $t = 0$. In this work, the proposed spiral-scan trajectory was realized by two orthogonal piezoelectric stack actuators. In order to move the scanner in a spiral trajectory, Equation 2 is transformed into Cartesian coordinates. The transformed equations are

$$x_s = r\cos\theta = \frac{P}{2\pi}\omega \cdot t \cdot \cos(\omega \cdot t) \tag{3}$$

and

$$y_s = r\sin\theta = \frac{P}{2\pi}\omega \cdot t \cdot \sin(\omega \cdot t) \tag{4}$$

where, x_s and y_s are the output displacements of the actuators in the x and y axes respectively and $\theta = \omega \cdot t$ is the angle. It can be inferred from Equation 3 and 4 that, to move the scanner in a spiral trajectory, one only needs to apply linearly-varying amplitude single frequency cosine and sine signals to the x- and y-axes of the scanner respectively.

3 THE SWITCHED CAPACITOR CHARGE PUMP

The switched capacitor charge pump method was firstly proposed by Huang et al. (2010). The circuit is shown in Figure 2(a). When the switch K is connected to the voltage source V_i, the capacitor C_1 is charged. When the switch K is shifted and connected the invert-input of the Operational Amplifier (OPA), a voltage difference appears between the positive and negative inputs of the OPA. To compensate for the voltage difference, the OPA will output a current pulse to balance the charges on the capacitor C_1, and charge the piezoelectric actuator C_{act} at the same time. So it likes the charges on the capacitor C_1 are "pumped" into the piezoelectric actuator. The actuator will then produce a small displacement during the charging process.

So when the switch K functions one time, an amount of charge

$$\Delta Q = V_i C_1 \tag{5}$$

is transferred into the piezoelectric actuator. If the switch K functions f_{CK} times per second, the piezoelectric actuator is charged with an amount of charge

$$Q = \Delta Q \cdot f_{CK} = f_{CK} \cdot V_i \cdot C_1 \tag{6}$$

in one second. So the equivalent charging current is

$$I_{eq} = Q/t = f_{CK} \cdot V_i \cdot C_1 \tag{7}$$

It should be noted that the charges are intermittently pumped into the piezoelectric actuator. However, when f_{CK} is largely higher than the highest frequency component of V_i, the charging

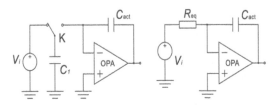

Figure 2. (a) Traditional SCCP circuit (b) Equivalent integrator.

664

current can be seen as continuous. Therefore the switched capacitor could be replaced by an equivalent resistor. The equivalent resistance is

$$R_{eq} = \frac{V_i - 0}{I_{eq}} = \frac{1}{f_{CK}C_1} \tag{8}$$

The switched capacitor charge pump is then transformed into an integrator as shown in Figure 2(b).

Traditionally, the charging voltage source in a SCCP circuit is set either constant positive or constant negative to achieve linearized charges on the piezoelectric actuator. By regarding the SCCP circuit as an integrator, as long as f_{CK} is largely higher than the highest frequency component of V_i, one only needs to apply a linearly-varying amplitude sine voltage as V_i to get a sinusoidal displacement with linearly-varying amplitude. The linearly-varying amplitude sine voltage can be easily realized by multiplying a single frequency sine signal by a low frequency triangular signal. The process is shown in Figure 3 and it can be easily realized in a LabVIEW program.

When the SCCP circuit is represented by the integrator shown in Figure 2(b), the charges on the piezoelectric actuator could be expressed as

$$Q_{act} = -\int_0^t \frac{V_i(t)}{R_{eq}}dt = -\frac{1}{R_{eq}}\int_0^t V_i(t)dt \tag{9}$$

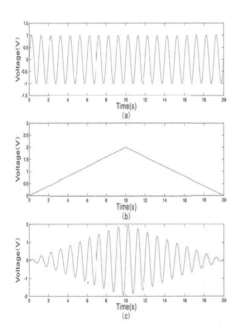

Figure 3. (a) Single frequency sine signal (b) Low frequency triangular signal (c) Varying amplitude single frequency sine signal.

in which $V_i(t)$ is the charging voltage source. If $V_i(t)$ is just the linearly-varying amplitude sine signal shown in Equation 3, Q_{act} could be calculated as

$$\begin{aligned}
Q_{act} &= -\frac{P\omega}{2\pi R_{eq}}\int_0^t t \cdot \sin(\omega t)dt \\
&= -\frac{P\omega}{2\pi R_{eq}} \cdot \left(\frac{\sin(\omega t)}{\omega^2} - \frac{t\cos(\omega t)}{\omega}\right)\Big|_0^t \\
&= \frac{P}{2\pi R_{eq}}\left(t\cos\omega t - \frac{\sin\omega t}{\omega}\right)
\end{aligned} \tag{10}$$

It is easy to see that Q_{act} includes two sinusoidal parts: the cos ωt part and the sin ωt part. The amplitude ratio of the two sinusoidal parts is $1/\omega t$. It is clear that with the scanning going the amplitude of the sin ωt part will be much smaller than that of the cos ωt part. Therefore it is reasonable to neglect the sin ωt part, and the charges on the piezoelectric actuator could be expressed as

$$Q_{act} = \frac{P}{2\pi R_{eq}} \cdot t \cdot \cos\omega t \tag{11}$$

It is obvious that Q_{act} is a linearly-varying amplitude sinusoidal signal. It results in a linearly-varying amplitude sinusoidal displacement of the piezoelectric actuator. If the two signals shown in Equation 3 and 4 are applied to two SCCP circuits, a spiral scanning can be achieved.

4 EXPERIMENT SYSTEM

In the former parts of this paper, we have proven that a SCCP-based spiral-scan can be achieved by applying two linearly-varying amplitude sinusoidal signals to the SCCP circuits. In this section, the proposed method is investigated by experiments.

A homemade three-dimension positioner, which is indicated in Figure 4(b), is employed to evaluate

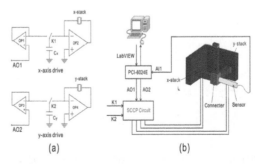

Figure 4. (a) Driving circuits for the x- and y-axes (b) Experiment setup.

the performance. The positioner consists of three orthogonal piezoelectric actuators, and the actuators for the x- and y-directions (x-stack and y-stack) are controlled by two SCCP circuits. The control signals for the switches K1 and K2 come from a signal generator (Rigol 1022, Rigol Corp, China). The input charging voltage sources for the SCCP circuits are provided by a data acquisition card (PCI-6024E, National Instruments, USA). AO1 and AO2 represent two analog output ports of the data acquisition card. The output displacements of the x-stack and y-stack are measured by a displacement sensor (SMT9700, KAMAN, USA). AI1 represents the analog input port of the data acquisition card. The communication between the computer and the data acquisition card is realized by a self-written Lab-VIEW program. Additionally, the amplifiers OP1 and OP3 are of the OP07 type, and OP2 and OP4 are of the OPA445 type. All the switches (K1 and K2) are realized by the MAX327 (Maxim Corp, USA). The capacitance of the charging capacitors (C_x and C_y) are 10 nF. The three piezoelectric stack actuators are of the AE0203D08 type (NEC Corp, Japan), which produce a displacement of 9 μ m under a DC voltage of 150 V.

The linearly-varying amplitude sinusoidal input signal for the x-direction is shown in Figure 5(a). It is realized by multiplying a sinusoidal signal with the frequency of 1 Hz by a triangular signal with the frequency of 25 mHz. The input signal for the y-direction is produced just like that of the x-direction. The only difference is that the input signal for the y-direction has a phase of 90°. The phase

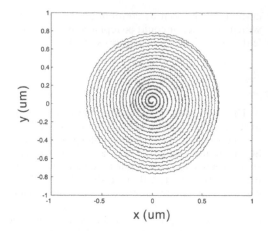

Figure 6.   Spiral scanning trajectory.

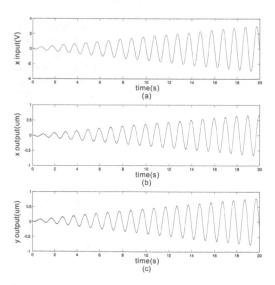

Figure 5.   (a) Input signal of x-direction (b) Output displacement of x-direction (c) Output displacement of y-direction.

difference between the x- and y-directions is the key to realizing spiral scanning. It is easy to see that the period the sinusoidal signal is 1 second and the half period of the low frequency triangular signal is 20 seconds. It means that the output displacements in the x- and y-directions increase from zero to the maximum value in 20 seconds and the trajectory runs for twenty circles at the same time. The output displacements of the x- and y-directions are measured and shown in Figure 5(b) and (c), respectively. Using the output displacement of x-direction as the x-axis data and the output displacement of y-direction as the y-axis data, the corresponding scanning trajectory is presented in Figure 6. It is obvious that the scanning trajectory basically follows the spiral scanning pattern, and the distance between two orbits maintains quite well.

For the SCCP-based spiral scanning, the maximum scanning range depends on both the maximum amplitude of the input signal and the gain of the SCCP circuit. And the gain of the SCCP circuit is calculated to be $1/2\pi \cdot C_i/C_{act} \cdot f_{CK}/f_i$, in which C_i is the capacitance of the charging capacitor, C_{act} is the capacitance of the piezoelectric actuator, f_{CK} is the clock frequency of the switch and f_i is the frequency of the charging voltage source. On the other hand, the scanning circles in an image is determined by the ratio of the half period of the triangular signal and the period of the sinusoidal signal, which is $f_x/2f_{tri}$. The parameters can be easily modified to achieve spiral scanning with different scanning range and resolution.

5   CONCLUSIONS

In this paper, a SCCP-based spiral-scan is proposed and investigated. The experimental results show that the SCCP-based spiral-scan is well realized. This method can be implemented in an AFM with a little sofware modificaiton in the near future.

REFERENCES

Ando, T. Kodera, T. Takai, E. Maruyama, D. Saito, K. & Toda, A. 2001. A high-speed atomic force microscope for studying biological macromolecules, *Proceedings of the National Academy of Sciences*, 98: 12468–12472.

Ando, T. Uchihashi, T. Kodera, N. Yamamoto, D. Miyagi, A. & Taniguchi, M. et al. 2008. High-speed AFM and nano-visualization of biomolecular processes, *Pflügers Archiv European Journal of Physiology*, 456: 211–225.

Avouris, P. Hertel, T. & Martel, R. 1997. Atomic force microscope tip-induced local oxidation of silicon: kinetics, mechanism, and nanofabrication, *Applied Physics Letters*, 71: 285–287.

Binnig, G. Quate, C.F. & Gerber, C. 1986. Atomic force microscope, *Physical review letters*, 56: 930–933.

Burnham, N.A. & Colton, R.J. 1989. Measuring the nanomechanical properties and surface forces of materials using an atomic force microscope, *Journal of Vacuum Science & Technology A*, 7: 2906–2913.

Clayton, G.M. Tien, S. Leang, K.K. Zou, Q. & Devasia, S. 2009. A review of feedforward control approaches in nanopositioning for high-speed SPM, *Journal of Dynamic Systems, Measurement, and Control*, 131: 061101.

Comstock, R.H. 1981. *Charge control of piezoelectric actuators to reduce hysteresis effects*, U.S. Patents 4263527.

Ducker, W.A. Senden, T.J. & Pashley, R.M. 1991. Direct measurement of colloidal forces using an atomic force microscope, *Nature*, 353: 239–241.

Fleming, A.J. 2012. A Review of Nanometer Resolution Position Sensors: Operation and Performance, *Sensors and Actuators A: Physical*, 190(1): 106–126.

Huang, L. Ma, Y.T. Feng, Y.T. & Kong, F.R. 2010. Switched capacitor charge pump reduces hysteresis of piezoelectric actuators over a large frequency range, *Review of Scientific Instruments*, 81(9): 094701.

Mahmood, I. & Moheimani, S.R. 2009. Fast spiral-scan atomic force microscopy, *Nanotechnology*, 20: 365503.

Moheimani, S.O.R. 2008. Invited Review Article: Accurate and fast nanopositioning with piezoelectric tube scanners: Emerging trends and future challenges, *Review of Scientific Instruments*, 79: 071101.

Newcomb, C. & Flinn, I. Improving the linearity of piezoelectric ceramic actuators, *Electronics Letters*, 18(11): 442–444.

Tseng, A.A. Jou, S. Notargiacomo, A. & Chen, T. 2008. Recent developments in tip-based nanofabrication and its roadmap, *Journal of nanoscience and nanotechnology*, 8: 2167–2186.

Advanced Materials and Structural Engineering – Hu (Ed.)
© 2016 Taylor & Francis Group, London, ISBN 978-1-138-02786-2

Experimental investigation on the ultimate strength of partly welded tubular K-joints

R. Cheng
School of Civil Engineering, Chongqing University, Chongqing, China
Key Laboratory of New Technology for Construction of Cities in Mountain Area (Chongqing University),
Ministry of Education, Chongqing, China

W. Chen, Y. Chen & B. Zhou
School of Civil Engineering, Chongqing University, Chongqing, China

ABSTRACT: An experimental investigation on the ultimate strength and behavior of the three types of steel tubular K-joints, namely partly welded specimens, fully welded specimens and strengthen specimens, was conducted in this paper. The failure modes of these tubular K-joints were investigated in the test and the ultimate strengths of the joints were compared. From the test, it was observed that two types of failure modes occurred. One failure mode was the excessive deflection of the chord member. The other failure mode just for the strengthen specimens was the tube concave at the stiffener end, which was determined by the widely accepted assumption that the ultimate load was reached when the deflection exceeded two percent of the chord member's diameter. For the ultimate strength of joints, it was shown that the strength for the fully welded and partly welded specimens was nearly the same, while that for the strengthen specimens is lower due to stress concentration on the tube at the end of the stiffener.

1 INTRODUCTION

Due to the appealing architectural appearance and rapid erection, the circular hollow section is commonly used in offshore structures, tower and long-span structures. Over the last forty years, CIDECT and other researchers have initiated many research programs in terms of the behavior of tubular structures (Packer et al. 2009, Dutta et al. 1998, Packer & Henderson 1997, Wardenier 1991). In the practical applications, the sections are profiled and welded to each other at the joints. The overlapped K joint is one of the commonly used types in tubular structures. A study on overlapped K joints was conducted (Serrano-López et al. 2013). The code for design of steel structures (CN-GB 2003) provides calculation methods for circular hollow section joints in China. The ASTM Standard Specification has provided useful advice for test investigation (US-ASTM 2014).

In general, these standards provide design rules for joints that are welded well. Most studies focus on the behavior of well-welded K joints. However, K joints are not welded well (partly welded) in some practical cases, due to the difficulty of construction. In that case, the behavior could be different. This paper presents an experimental investigation on the behavior of partly welded

K joints. A total of nine full-size specimens were tested. Three types of specimens were examined as follows: joints which were not welded well (partly welded), well-welded (fully welded) and not welded well but strengthened (strengthened). On the basis of the tests, the bearing capacity was assessed.

2 TEST PROGRAM

2.1 Scope

The strength of circular hollow section K-joints depends on the joint details. Three different details, as shown in Figure 1, were fabricated for the K-joint specimens in the tests: Partly Welded (PW), Fully Welded (FW) and Strengthened (STRSN). For PW specimens, circular hollow section members were welded only along the exterior intersection curves,

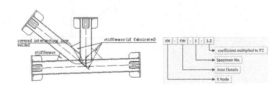

Figure 1. Specimen and naming.

welding not performed on the covered junction. For FW specimens, members were welded along all intersection curves. STRSN specimens were designed based on the PW specimen. In addition, some stiffeners were added to the conjunction area. This was aimed to investigate the behavior of the stiffener strengthened joints. Test loads were axial that were applied both to the brace and the chord members.

2.2 Specimens and detection scheme

A total of 9 full-scale K-joint specimens were tested: 3 PW specimens, 3 FW specimens and 3 STRSN specimens. Each specimen consists of a horizontal chord member, a vertical bracing member and a diagonal bracing member. The specimen material is Q235 steel (CN-GB 2003).

Figure 2 (a) shows the geometrical details. The diameter of the chord is 273 mm, while the thickness is 14 mm. The diameter of the bracing member is 245 mm, while the thickness is 10 mm.

The mechanical model is shown in Figure 2 (b). The diagonal bracing member was pinned to a hinged support, while a left end of the chord member was simply supported. The vertical bracing member and the right end of the chord were axial loaded. At least 3 or more displacements were measured (v1, v2, v3).

Rosette strain gauges were arranged along the superficial intersecting curves, the end points of the stiffeners, intersection points of the superficial intersecting curves, and crown and saddle points of the tubes. The direction of rosette strain gauges was parallel with superficial intersecting curves. Eight strain gauges were arranged at the mid-point of each branch. This was aimed to measure the normal force of each member. The position of the rosette strain gauges is shown in Figure 3.

As shown in Figure 3, three or more diameters were used in this test to capture the displacements in a plane. Three of them were located at the end of the vertical bracing member and on both ends of the chord member in order to capture the vertical displacement. The other two were located at the

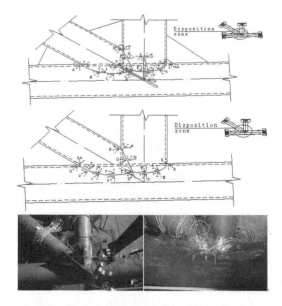

Figure 3.   Strain gauges arrangement.

mid-point of the chord member, which was used to capture the local concave (or convex). All the strain and displacement data were recorded by a computer data acquisition system.

2.3 Test set-up

Figure 4 shows the schematic sketches of the test arrangement of tubular K-joints. The diagonal bracing member of the specimen was simply supported. The ends of the chord member were supported by steel blocks anchored to the frame. The plane vertical movement was restrained. A point load was applied to both the vertical bracing member and the chord member.

2.4 Loading protocols

The ratio of load on vertical bracing members to load on chord members, i.e. P1/P2, was equal to 1.048, according to the design information. Besides, in the experimental investigation, P2 was set intentionally to the other values, so as to determine the influence of the ratio on the ultimate strength of the joint. The coefficient multiplied to P2 was shown in the name of the specimen.

Two hydraulic cylinders were simultaneously loaded during both the preloading and the test. The estimated value of the failure load P_y was obtained from the previous prediction by the numerical simulation analysis. Each specimen was preloaded at the level of $0.3P_y$ to eliminate the installation gap.

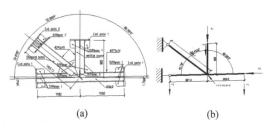

| (a) | (b) |

Figure 2.   Specimen details (a) and specimen mechanical model (b).

Figure 4.   Test set-up overview.

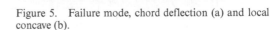

Figure 5.   Failure mode, chord deflection (a) and local concave (b).

Table 1.   Test results.

| | Specimen name | Failure load (kN) | |
		P 1	P 2
Partly welded	KN-PW-1-1.0	2200	2099
	KN-PW-2-1.2	2100	2405
	KN-PW-3-1.4	2000	2672
Fully welded	KN-FW-1-1.0	2160	2061
	KN-FW-2-1.4	2000	2672
	KN-FW-3-1.2	2040	2336
Strengthen	KN-STRSN-1-1.0	1920	1822
	KN-STRSN-2-0.5	2340	1116
	KN-STRSN-3-1.4	2020	2698

| | Failure displacement (10^{-2} mm) | | |
	Diameter 1	Diameter 2	Diameter 3
Partly welded	1458	5778	1165
	1424	1071	3394
	819	6263	143
Fully welded	864	1260	3987
	1075	957	4439
	1543	1277	5771
Strengthen	852	3845	1198
	45	1452	3811
	1326	1050	6529

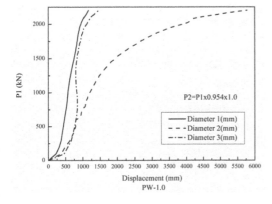

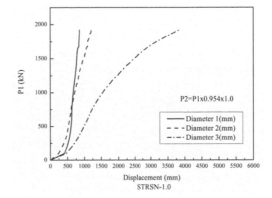

Figure 6.   (Continued).

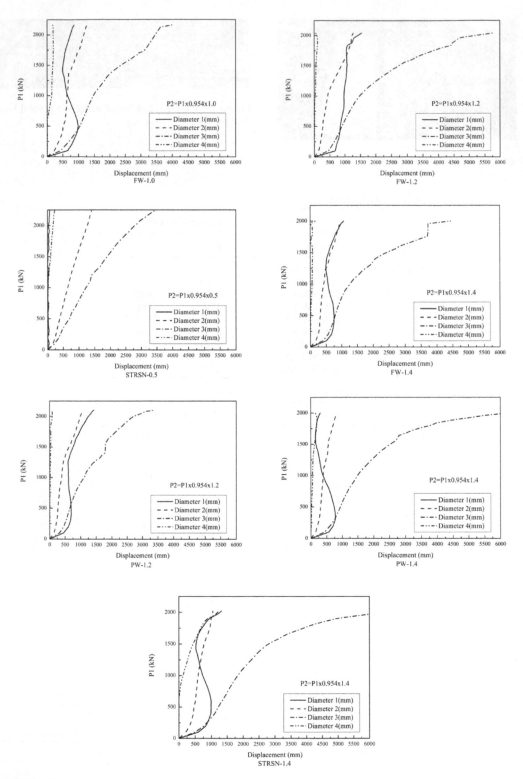

Figure 6.   Load vs. displacement curve.

2.5 Results and discussion

The failure mode in the test was the excessive deflection of the chord member. When the deflection exceeded the tolerable levels (2% of the diameter of the chord members), another failure mode just for the strengthen specimens was the tube concave at the stiffener end, as shown in Figure 5. The tests were ended. The failure loads and displacements are given in Table 1.

The ultimate load of fully welded specimens was higher than that of partly welded specimens that take the second place, but they were at about the same level, while the capacity of strengthen specimens (STRSN) is the lowest. For the failure mode of STRSN, the significant local convex or concave of the chord member should be the most important problem besides the excessive deflection. The stiffness of stiffeners is much greater than the chord member, and it provides the restrain against bending deformation of the tube, which leads to stress concentration on the chord member near the end of the stiffeners. Accordingly, the concave around the stiffeners on the chord member was observed.

The vertical brace member was in compression. In that case, the weld line did not transmit the tensile force. This could be the reason that the ultimate load of fully welded specimens and partly welded specimens was at the same level.

The ultimate states of the specimens were based on the deformation failure criteria in the experiment. In fact, the load could continue increasing for all the tests. The partly welded effect did not weaken the strength of the specimen. Both the partly welded and the fully welded specimen loading capacity approached the limitation of the loading equipment (2000 kN).

The slope of load vs. displacement curve at the beginning was larger because the installation of equipment existed some space. As the loads approached near to the failure level, the curves became steepened due to the increasing regional elastic plastic deformation.

3 CONCLUSION

North American, European, Australia and China provide their own specification for the hollow section structures. The CIDECT organization initiated many research programs in terms of the behavior of tubular structures. All these data provide the behavior of the fully welded tubular structure. In addition, the partly welded overlapped K-joint was not much investigated. This paper reported a static experimental investigation on the behavior of partly welded, fully welded and strengthens K-joints. From the test results, the failure load of the partly welded specimens and fully welded specimens was at the same level and that of the strengthen specimens is lower. For the strengthen specimens, the stiffness of stiffeners caused the stress concentration at the surface of the tubes. This lead to the local convex and affected the bearing capacity of the K joint.

ACKNOWLEDGMENT

This study was supported by the following projects: Project supported by the National Natural Science Foundation of China (Grant No. 51208535), Project supported by the Natural Science Foundation of Chongqing (Grant No. cstc2012 jjA30004), and Project No. CDJRC11200002 and No.CDJZR12200014 supported by the Fundamental Research Funds for the Central Universities. The authors gratefully acknowledge their support.

REFERENCES

CN-GB 2003. *Code for design of steel structures, in GB 50017–2003*. China Planning Press: China. 326.

Dutta, D. Wardenier, J. Yeomans, N. Sakae, K. Bucak, Ö. & Packer, J.A. 1998. *Design guide 7: for fabrication, assembly and erection of hollow section structures*, TÜV-Verlag.

Packer, J.A. & Henderson, J.E. 1997. *Hollow structural section connections and trusses, a design guide*, Canadian Institute of Steel Construction.

Packer, J.A. Wardenier, J. Zhao, X.L. Van der Vegte, G.J. & Kurabane, Y. 2009. *Design guide 2: for Circular Hollow Section (CHS) joints under predominantly static loading, 2nd ed.*, LSS Verlag.

Serrano-López, M.A. et al. 2013. Static behavior of compressed braces in RHS K-joints of hot-dip galvanized trusses. *Journal of Constructional Steel Research*, 89(0): 307–316.

US-ASTM 2014. Standard Practice for Verification of Testing Frame and Specimen Alignment Under Tensile and Compressive Axial Force Application, in ASTM E1012-2014. 18.

Wardenier, J. 1991. *Design guide for Circular Hollow Section (CHS) joints under predominantly static loading*. TUV Rheinland.

Advanced Materials and Structural Engineering – Hu (Ed.)
© 2016 Taylor & Francis Group, London, ISBN 978-1-138-02786-2

The influence of cooling intensity on structure of blow molded products

P. Brdlík & M. Borůkva
Department of Engineering Technology, Technical University of Liberec, Liberec, Czech Republic

ABSTRACT: The production of hollow products using blow molding technology is a non-stationary an isothermal process where thermal energy is removed via an external mold cooling system (drilled channels with circulating water) and gas internal cooling system. The external mold cooling system is more efficient than the internal gas system. The difference in cooling between internal and external product part boundaries could give rise to different thermo-kinetic crystallization conditions in the polymers and consequently to non-uniform structure distribution throughout product thickness. Such difference in conditions could cause increase in internal stress values or even product deformation. The main aim of this paper therefore focuses on investigating the impact of the internal cooling system intensity on the structural behavior of blow molded products.

1 INTRODUCTION

Product quality is a key issue in the blow molding processing. There are not many papers published throughout the literature on this topic even though companies put enormous emphasis on the quality of parts produced. Amongst the most interesting works are studies by D.M. Kalyona et al. (1983, 1991), and by S.B. Tan and P.R. Hornsby (2008). These researchers focused their efforts on investigating the influence of cooling intensity on structure development, deformation and optical properties. The results achieved indicate that an existence of pronounced differences in the thermal history of polymers has a considerable influence on product properties. The part of a polymer product with its outer surface close to the mold cavity is influenced by intensive thermal energy removal by the external water cooling system and cools down very fast. On the other hand, the inner surface of the part is cooled down significantly more slowly. It is because of low heat transfer by the blow molding medium—air, which under normal conditions has a very low heat transfer coefficient (5 W/m^2·°C) (Rosat et al. 2004). These obvious cooling rate differences could give rise to differences in thermo-kinetic solidification conditions and thus considerably influence structure development (Kalyon & Jeong 1991, Tan et al. 2008). In line with the studies cited above, the significant differences in relaxation time before wall solidification of both cooled surfaces (inner and outside) could cause differences in the distribution of orientation, birefringence and internal stress

(Kalyon & Jeong 1991, Tan et al. 2008). Therefore it is enormously important to get as uniform thermal removal by both cooling systems as possible. Consequently the application of progressive internal cooling systems could cause not only an increase in blow molding process productivity but also ensure more uniform cooling conditions on both sides of the product. This application could finally have a positive impact on product quality, as shown in Figure 1. The aim of this work therefore focuses on the influence of differences in the cooling abilities of the internal and external cooling systems, and its impact on the structural behavior of polymers, which in turn has considerable impact not only on the mechanical and optical properties, but also on the shape stability of products.

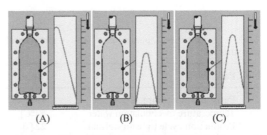

(A)   (B)   (C)

Figure 1. Temperature distribution in a product wall with different cooling alternatives (A) Common blow molding internal cooling process (air); (B) application of progressive internal cooling; (B1) increase in productivity by means of progressive internal cooling (FASTI 2012).

2 EXPERIMENTAL

The carbon dioxide (CO_2) internal cooling system was chosen to determine the influence on the intensity of thermal energy removal and resulting structure. It is because of the high cooling potential of CO_2 (Jorg 2006). Three technological setups were used to investigate the intensity of thermal energy removal (Table 1). The first was the application of a common cooling process to achieve maximum production without producing rejects (demolding temperature lower than 100°C, Fig. 2). To achieve this goal, the temperature of circulation water in the cooling channels was as low as possible (6°C). Further lowering of temperature would lead to condensation of moisture on the mold surface and consequently to producing parts with surface defects. For this blowing process, the standard industrial pressurized air was used under standard conditions (pressure 0,6 MPa and temperature 20°C). The high intensity external mold cooling system and the low efficiency of the internal cooling system would ensure getting expected results. An internal liquefied carbon dioxide injection cooling system was applied for the second experimental measurement. This cooling system markedly increased the cooling ability of the extrusion blow molding process on the inside part of products. The last experimental measurement consisted of applying internal CO_2 cooling while simultaneously increasing productivity. To explore the differences in cooling effects at the internal and external interface, a 0.3 liter bottle with square cross-section was made. 60 μm thick test specimens were taken at a 100 μm distance from both surfaces of the bottom part of the bottle (Fig. 3A). At this location, both the thickness and the cooling effect of the injected liquefied carbon dioxide were the highest. Therefore the influence of different cooling rates on the structure of polymers would be more apparent than in locations with lower thickness.

Table 1. Process parameter of experimental measurement.

No.	Parameter	Value
1	Parison temperature	190 [°C]
	Temperature of circulation water	6 [°C]
	Production cycle time for normal blowing process (maximal production)	22 [s]
2	Internal cooling time and quantity of injected CO_2	8 [s] (17,6 [g])
3	Productivity increase reached by application of CO_2 internal cooling	43 [%]
	Production cycle time reduction	12,5 [s]

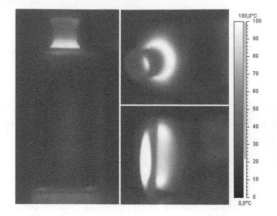

Figure 2. Thermo-vision images of common blow molding production at maximal process productivity.

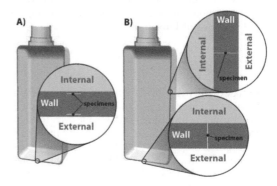

Figure 3. The locations where tested specimens were taken.

The most widely used polymer materials—polyolefins (Rosat et al. 2004) were chosen for testing. Specifically, these were linear high-density semi-crystalline polyethylene (PE—Liten BB 29) and homo-Polymer Polypropylene (PP—Mosten EH 01). Verification of the theoretical presumption introduced, on the influence of cooling rate on the structure of polymer products, was made using Differential Scanning Calorimetry (DSC—Mettler Toledo DSC 1 Star) analysis, where the degree of crystallinity is proportional to the melting enthalpy of the tested polymer and the melting enthalpy of the fully crystalline polymer. The value of the melting enthalpy can sufficiently explain the structure changes of the selected polymers (Stuart 2002).

Apart from analyzing the influence on the intensity of thermal energy removal on the structure of wall thickness distribution, the differences in cooling effect between the thickest and the thinnest wall were also explored. The thickest part of products is located at the bottom of the product due to sagging of extruded parison. The thinnest section is in the most formed part of parison. The locations of the test specimens taken are shown in Figure 3B.

3 RESULTS AND DISCUSSION

The measured endothermic curves together with the values of melting enthalpy evaluated are shown in Figures 4 up to 6 and Table 2. The results achieved on PE confirm the above introduced theory on the influence of differences in the cooling intensity of both cooling systems (internal and external) on the resulting structure. The lowest values of melting enthalpy were detected at the polymer/blow mold interface. This phenomenon is attributed to the high cooling effect of the external cooling systems. The lower cooling effect of pressurized industrial air caused increasing values of melting enthalpy and therefore higher degree of crystallinity at polymer/blow mold interface. Application of CO_2 internal cooling (8 seconds injection of CO_2) to the conventional extrusion blow molding process resulted in a marked decrease of melting enthalpy values in locations close to the internal product surface depending on the progressive cooling

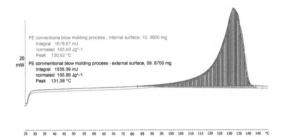

Figure 4. Endothermic curves of tested PE specimens taken from locations close to inner and outer product surface—conventional blow molding process.

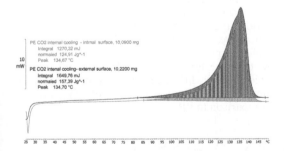

Figure 5. Endothermic curves of tested PE specimens taken close to inner and outer product surface—progressive application of CO_2 internal cooling.

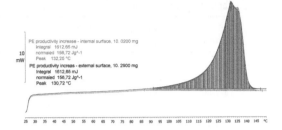

Figure 6. Endothermic curves of tested PE specimens taken from locations close to inner and outer product surface—43% increased productivity by means of CO_2 internal cooling.

effect ($-78°C$). Nevertheless, after an increase in productivity by 43%, the melting enthalpy results show very similar values due to more uniform temperature removal from both sides of the product. The results of thermal analysis in PP did not confirm the above theoretical presumptions and the values of the melting enthalpy are similar. The most likely reason lies in the speed of spherolite growth, which is generally thousand times lower in PP than the growing speed of spherolith in the case of PE (Cowie & Arrigh 2007). Due to the influence of the lower speed of crystallization, no crystallinity changes could be demonstrated directly due to the influence of cooling rate.

The endothermic curves do not show only melting and crystallization enthalpy, but there is also information about macromolecule orientation and magnitude of residual crystallization stress (Stuart 2002). A material with a narrower and higher DSC curve indicates higher values of these parameters. From the presented results (Figs. 4 up to 6) it is therefore obvious that applying CO_2 internal cooling to a conventional blow molding process resulted in a decrease in the degree of crystallinity and an increase of macromolecule orientation and residual crystallization stress due to the effect of intensive thermal energy removal. In the case of the blow molding extrusion process with higher productivity (increased by 43%), endothermic curves of the tested specimens taken from inside and outside locations in the product show a similar distribution caused by more uniform cooling. It means that the parameter of orientation and residual crystalline stress through the wall are uniform.

The second experimental measurement was made to explore cooling differences in the thickest and the thinnest locations of a product. In the conventional blow molding technological process with PE, the highest values of melting enthalpy and the highest crystallinity degree were detected in the bottom part of the product (the thickest section). This location with temperature extremes due to greater

Table 2. Evaluated parameters of melting enthalpy through product thickness.

Material	Surface location	Conventional blow molding	Application of CO_2 internal cooling	Increased productivity (CO_2)
PE	Outer	155,56 $[J \cdot g^{-1}]$	157,39 $[J \cdot g^{-1}]$	155,71 $[J \cdot g^{-1}]$
	Inner	163,93 $[J \cdot g^{-1}]$	124,91 $[J \cdot g^{-1}]$	156,72 $[J \cdot g^{-1}]$
PP	Outer	85,15 $[J \cdot g^{-1}]$	85,27 $[J \cdot g^{-1}]$	89,23 $[J \cdot g^{-1}]$
	Inner	87,68 $[J \cdot g^{-1}]$	86,68 $[J \cdot g^{-1}]$	87,12 $[J \cdot g^{-1}]$

Table 3. Evaluated parameters of melting enthalpy in locations with different thickness.

Material	Wall location	Conventional blow molding	Application of CO_2 internal cooling	Increased productivity (CO_2)
PE	Bootom	166,62 $[J \cdot g^{-1}]$	150,09 $[J \cdot g^{-1}]$	159,29 $[J \cdot g^{-1}]$
	Middle	162,22 $[J \cdot g^{-1}]$	157,06 $[J \cdot g^{-1}]$	157,75 $[J \cdot g^{-1}]$
PP	Bootom	93,87 $[J \cdot g^{-1}]$	94,93 $[J \cdot g^{-1}]$	91,58 $[J \cdot g^{-1}]$
	Middle	94,53 $[J \cdot g^{-1}]$	94,17 $[J \cdot g^{-1}]$	92,57 $[J \cdot g^{-1}]$

product wall thickness and lower cooling efficiency generated more suitable thermo-kinetic condition for spherolite growth than in thinner sections. On the other hand the product section with the smallest wall thickness, which is located at the center of the product wall reveals lower values of melting enthalpy (due to fast cooling) when compared to thicker and slower cooled locations. The application of progressive internal CO_2 cooling system to the conventional extrusion blow molding process (same as the previous experimental measurement) resulted in a decrease in melting enthalpy due to greater, more intensive cooling. This phenomenon was especially obvious in the bottom part of the product where the cooling effect of injected liquefied medium is the highest. After an increase in productivity the parameter of melting enthalpy reached similar values as in the conventional blow molding process. Also, this measurement was not confirmed beyond the results presented due to the lower crystallization speed of PP.

4 CONCLUSIONS

The experimental measurements on PE products confirmed the theoretical presumption of the influence of cooling differences in the blow molding process. The higher cooling efficiency of the external mold cooling system resulted in achieving lower degree of crystallinity in locations close to the outside product surface. On the other side of the product, the lower cooling efficiency of the conventional blow molding medium (air) resulted in, with regard to a slower cooling rate, better conditions for the growing of spherolith and consequently

in achieving a higher degree of crystallinity. The application of CO_2 internal cooling ensured a progressive improvement of cooling efficiency in internal parts of the products. The very fast rate of thermal energy removal at locations in the product had a significant effect on decreasing the degree of crystallinity, and on increased orientation of macromolecules and residual crystallization stress. But when the cooling potential of carbon dioxide was used to improve productivity, then uniform thermal removal from both sides of the product ensured uniform structure distribution through the wall thickness, a lower degree of orientation of macromolecules and residual crystalline stress. It means that applying internal cooling could have a positive effect on product quality rather than being negative, as it is stated in many publications (Rosat et al. 2004, FASTI 2012). From the experimental measurements it is also obvious that thickness distribution has considerable impact on the efficiency of the cooling system and location of temperature extremes, which could cause a further increase structure differences, differences in orientation, and could create residual crystalline stress. Therefore, for the blow molding process, uniform wall thickness is of utmost importance. The importance of the differences in cooling rates is considerable especially in the case of PE which has several times the faster speed of growing crystallites in comparison to PP.

ACKNOWLEDGEMENT

This paper was prepared thanks to financial support from the Student Grant Contest project

21005 (SGS 2015), from the TUL part within the framework of specific university research support.

REFERENCES

Cowie, J.M.G. & V. Arrighi. 2007. *Polymers: Chemistry and Physics of Modern Materials*. Broken Sound Parkway NW: Taylor & Francis Group.

FASTI. 1991. Internal Cooling for the Blow Molding Industry. Working paper. http://www.fasti.com/product3.php.

Jorg, C.H. 2006. Carboxyl Dioxide Cooling Method May Take the Waiting Out of Plastic Parts, *Journal of Automotive Engineering*, 31(7): 40–41.

Kalyon, D.M. & Jeong, S.Y. 1991. Microstructure Development in Blow Molded Amorphous Engineering Plastic. *Journal of Plastic, Rubber and Composites Processing and Applications*, 15(2): 95–101.

Kalyon, D.M. & Kamal, M.R. 1983. Heat Transfer and Microstructure Extrusion Blow Molding. *Journal of Polymer Engineering and Science*, 23(9): 503–509.

Rosat, D.V. Rosat, A.V. & Dimathiam, D.P. 2004. *Blow Moulding Handbook*. Munich: Hanser Gardner Publications.

Stuart, B.H. 2002. *Polymer Analysis*. Chichester England: John & Wiley Sons, LTD.

Tan, S.B. Hornsby, P.R. McAFee, M. Kearns, M.P. McCourt, M. & Hanna, P.R. 2008. *An Overview of Internal Cooling in Rotational Moulding*, ANTEC 2008, Milwaukee, USA.

Advanced Materials and Structural Engineering – Hu (Ed.)
© 2016 Taylor & Francis Group, London, ISBN 978-1-138-02786-2

Research on the monitoring measurement of using a shallow tunnel to expand to a large section multi-arch tunnel *in situ*

Y.D. Zhou & Y.L. Jia
Institute of highway, Chang'an University, Xi'an, China
Guangxi Transportation Research Institute, Nanning, China

Y.X. Xia & F. Ye
Institute of highway, Chang'an University, Xi'an, China

ABSTRACT: Research on the stability of the surrounding rock and supporting structure is carried out during the construction of an *in situ* expansion to a large section shallow-buried tunnel using the existing shallow tunnel as a middle guide drift. The weak parks, the stress characteristics and the influence of the supporting structure during the construction are analyzed through the on-site measurement of ground settlement, deformation of the surrounding rock, pressure of the surrounding rock and the steel stress in mid-partition. The results indicate the feasibility of existing construction technology and retaining parameter adopted for the *in situ expansion* to a large section tunnel using the existing shallow-buried tunnel as a middle guide drift. The results also indicate that the deformation of the surrounding rock is basically stable and controllable for the good supporting effect. The conclusion of this research will provide evidence for guiding construction and improving design, and will be a valuable experience and reference for design, construction, measurement and further research on tunnel engineering under similar conditions.

1 INTRODUCTION

With the rapid growth of traffic capacity, the existing roads have been unable to meet it day by day. As a throat of the road, the tunnel becomes a great challenge and it is imperative to expend or rebuilt a new tunnel on the basis of the original tunnel, such as the Taylor Williams Tunnel (Yu 2006), the Nazzano Tunnel (2000), the Chongqing Yuzhou Tunnel (Zhu et al. 2012), the Damaoshan Tunnel (Lin et al. 2013), the Pengtougou Tunnel (Gao 2012), and the Chongqing Eling Tunnel (Gao et al. 2010). These projects are mostly expansions to one tunnel or adding new neighborhood tunnels. The main styles of expansion to one tunnel include single-side expansion, double-side expansion and peripheral expansion. The main styles of adding new neighborhood tunnels include adding at the left side, adding at the right side and adding at the middle of the existing two tunnels.

Nowadays, most of the tunnel expansions have the characteristics such as neighboring closely to the existing tunnel, large span, tunnel groups and keeping open traffic during construction. The construction of expansion will change the existing tunnel structure performance, and have a negative effect on it such as descending of bearing capacity, even destruction and invasion of construction clearance owing to large deformation. At the same time, given the disturbance during the existing tunnel excavation, the surrounding rock will be disturbed once again during the expansion. The strength of the surrounding rock will be reduced and the integrality and self-stability will be weakened. The expansion is facing significant challenges in the disturbed surrounding rock (Zhang et al. 2010).

However, there is little experience and reference for the limited similar projects of *in situ* expansion to a large section tunnel using the existing tunnel as a middle guide drift at present. Given the disturbance of the surrounding rock during the existing tunnel excavation and the unsymmetrical stress for the unsymmetrical construction of multi-arch tunnels, it complicates the stress mechanism and the interaction to a large extent (Shen & Gao 2006). It is very difficult to search a correct physical mechanical model to reflect the rock performance (Xia et al. 2007). Thus, it becomes very necessary to take the monitoring measurement of such projects. In addition, it becomes very necessary to understand the rock deformation and the supporting structure's stress distribution timely after the monitoring measurement data have been analyzed. Furthermore, it is required to master the development tendency, guide to the design and construction further and make sure the safety construction.

Using the existing tunnel as a center drift, monitoring measurement about the changes in the surrounding rock and the linings after excavation was applied. After the comprehensive analysis and induction of the monitoring measurement data, it will help to feedback the reasonability of supporting parameters during design and construction, and then guide the construction and improve the design. It will also provide plenty of experience to the design and construction of the similar projects in the future.

2 ENGINEERING SITUATIONS

Having only a single double-lane hole with two driving directions, the Jingya Tunnel was built in 1993 lying in Yangshuo County. It is a second-class highway tunnel with a full width of 9 m and a length of 120 m. With the booming tourism market and the construction and development of the county town, the Jingya Tunnel is becoming a threat owing to the increasing traffic volume day by day.

If a neighborhood tunnel will be built on each other side of the existing tunnel, it will cause much removal for the project that is lying in the city. Based on the aerial photos, topographic maps and practical investigation data, it is very difficult to carry out for the levy land and removal. For the mountains towering stretches continuously, it is very difficult to design an optimum highway route to avoid the residential buildings if building a new road near the existing tunnel. It has little benefit to solve the traffic jams near the tunnel and in the tunnel because the highway route will be very long and will deviate far from the existing tunnel. Thus, it is imperative to expand the existing tunnel *in situ*.

Considering the related factors such as geomorphology, topography, geology and the existing tunnel's construction data, many plans conforming to the project with 4 motor lines, a non-motor line, and a sidewalk on both right and left sides were put forward after a thorough study of the Jingya Tunnel expansion plans combined with the current research of tunnel expansion both at home and abroad. An optimum plan was shown by comparing and analyzing those plans.

After the comparison analysis of the above expansion plans, it can be concluded that the 6-lines tunnel will be a large span shallow buried tunnel with complex stress. Without any related study findings and related projects, it has a very great risk to design and construct for the technology that is not yet mature. Mass blasting will be used at the cut and cover plan and the road cutting plan for the hard rocks. It will have a great influence on the safety of near buildings. It will also cause

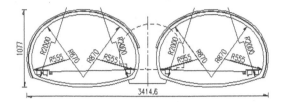

Figure 1. The general structure design of the tunnel.

over high slopes and collapse easily. Then, the whole environments will be severely destroyed and it will be very difficult to recover the vegetation. Lastly, it will also have a harmful influence on the tourism scenic area. Using the existing tunnel as a center drift, the multi-arch tunnel plan can well adjust measures to local conditions. It can reduce the risk effectively not only with the design but also at the construction stage by constructing the mid-partition just after the excavation of foundation and then expand on both sides. It can also protect the whole environment effectively. Based on practical engineering, the multi-arch tunnel plan is reasonable to be recommended as priority.

Having been built, the tunnel begins at K0+502 and ends at K0+627 with a length of 125 m. It will be a large span tunnel with an excavated width of 34.346 m and a net width of 14.803 m per hole. It will be a flat tunnel with a span ratio of less than 0.7, while including the invert and ratio of 0.54, while without the invert. Lastly, it will be a very shallow tunnel for a maximum buried depth of about 27 m. The tunnel's lining type is shown in Figure 1. Having been built, the tunnel will include 4 motor lines, a non-motor line and a sidewalk on both right and left sides. The tunnel will be a large span, large span ratio and shallow buried one. It will be a tunnel with complex stress as well as for the surrounding rock that was disturbed during the existing tunnel excavation. According to the geologic survey, the rock belongs to the II classification, and the tunnel will be buried mainly in carbonate limestone rocks.

3 EXCAVATION METHOD

The excavation is a dynamic and irreversible evolution process. It relates to the safety of construction and operation period closely. Owing to the advantages such as simple equipment, little disturbance, less process, convenient construction, high speeds and being fit for large machines, the bench cut method is widely used in good surrounding rock tunnels. It belongs to the II rock classification according to the tunnel geologic survey results. According to a good rock classification, it can be

Figure 2. Whole section excavation of the lower bench.

Figure 3. Step excavation of the lower bench.

excavated well by using the bench cut method with the upper bench of about 7.3 m. At the entrance region of the first excavated hole, the lower bench was exploded step by step on the left and the right parts. At the other sections, it was exploded at whole just, as shown in Figures 2 and 3. The excavate direction is from the entrance to the exit at the right hole. After the right hole has been exploded, it was excavated at a reverse direction with a way from the exit to the entrance at the left one.

4 MONITORING MEASUREMENT

Nowadays, it is a very important method to feedback the dynamics characteristics of the structure and the surrounding rock, optimize supporting parameters and make sure the safety of excavation through the amount of information about monitoring measurement (Li et al. 2013). Aiming at the geologic and topographic conditions of the Jingya Tunnel, the rock classification, the bury depth, the structure size and the excavation method the monitoring measurement about the stability of surrounding rock, the stress characteristics of structure and the safety of excavation method were applied during the excavation. It includes ground settlement, crown settlement, convergence, rock pressure, structure stress and mid-partition stress. The settlement of components is shown in Figure 4.

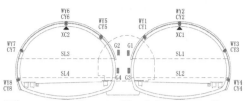

XC—crown settlement, SL—deformation of surrounding rock, WY—pressure of surrounding rock, CY—lining stress, G—steel stress)

Figure 4. Arrangement profile of monitoring points in a tunnel.

5 ANALYSIS OF MONITORING MEASUREMENT RESULTS

Up to now, the right hole of the tunnel has been exploded and the upper bench of the left hole has been exploded at the section of K0+585. The monitoring measurement results of the right hole are discussed below.

5.1 Analysis of monitoring measurement results of ground settlement

The measuring point for monitoring the ground settlement of the entrance of the hole is fixed on the ground surface. Based on the displacement of the ground settlement, the influence of excavation on the ground surface and the stability of the surrounding rocks and supporting structures can be obtained to provide information about the construction safety. Taking the section of K0+520 for example, the time-history curves are shown in Figures 5 and 6.

The construction log shows that, at the period of 28 days, the upper bench of the right hole was exploded thoroughly. The lower bench of the right hole was excavated at the K0+560 section and the lower bench of left hole was excavated at the K0+520 section. According to the measurement results, the ground settlement is nearly finished after 30 days. This indicates that due to the short length and shallow bury depth of the tunnel, the behavior of the ground surface is consistently influenced in the whole process of excavation. For the shallow buried tunnels, therefore, strengthening the monitor of ground settlement during the excavation process is very important, especially for the places, which are near houses and culture relics, and some constructing techniques, such as strictly controlled explosion, low excavation rate, early supporting and timely closed loop should be applied.

5.2 Analysis of monitoring measurement results of crown settlement and convergence

The monitoring measurements of crown settlement and convergence are very essential for

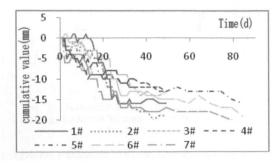

Figure 5. Time-history curves of ground settlement at the K0+520 section of the right tunnel.

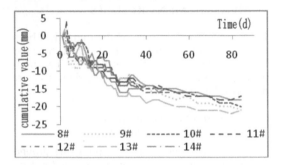

Figure 6. Time-history curves of ground settlement at the K0+520 section of the left tunnel.

the construction of tunnels. Based on the two measurement results, the deformation states of cross sections can be easily reflected, so that the stability of the surrounding rocks, the reasonability of initial supporting and construction method, and the proper casting time of the second liner can be judged reasonably. During the construction process of the Jingya Tunnel, after the explosion of the upper benches and the transportation of the slag, the layout and the measurement of initial values of monitoring points about crown settlement and convergence were rightly implemented. However, some monitoring points were damaged in the process of constructing, and it was failed to monitor the whole constructing process of the Jingya Tunnel. The time-history curves of monitoring results are shown in Figures 7–13.

From the monitoring measurement results, the crown settlements and the convergences of the upper and lower benches of all the monitoring cross sections are very small. The maximum value of the crown settlement is 15 mm. The maximum values of the upper bench and the lower bench are 3.08 mm and 1.16 mm, respectively. The main reason for this condition is that the layout and the measurement of monitoring points about crown settlement and convergence were conducted after the explosion of the

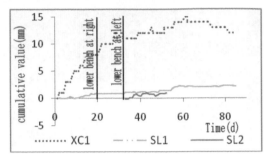

Figure 7. Time-history curves of crown settlement and convergence monitoring at the K0+520 section of the right tunnel.

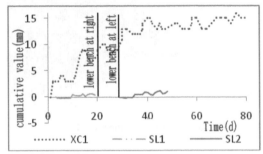

Figure 8. Time-history curves of crown settlement and convergence monitoring at the K0+530 section of the right tunnel.

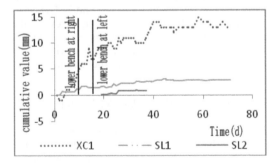

Figure 9. Time-history curves of crown settlement and convergence monitoring at the K0+552 section of the right tunnel.

upper benches and the transportation of the slag, and some deformation of the surrounding rocks was already partly finished. Furthermore, the surrounding rock belongs to the II classification and is relatively integrated, so that the settlement and deformation are relatively small.

In addition, the value of the crown settlement is bigger than the values of convergences of the

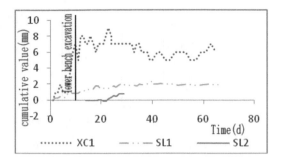

Figure 10. Time-history curves of crown settlement and convergence monitoring at the K0+571 section of the right tunnel.

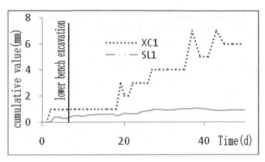

Figure 13. Time-history curves of crown settlement and convergence monitoring at the K0+614 section of the right tunnel.

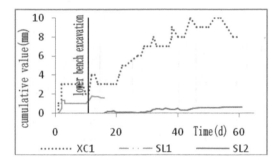

Figure 11. Time-history curves of crown settlement and convergence monitoring at the K0+593 section.

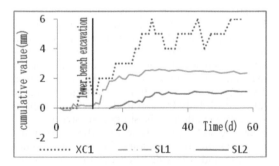

Figure 12. Time-history curves of crown settlement and convergence monitoring at the K0+605 section of the right tunnel.

upper benches and the lower benches. There are mainly two reasons for this condition. Due to the shallow buried depth, large span and large first blasting width and height with the bench cut method, the initial supporting system did not completely contact with the surrounding rocks and failed to support the surrounding rocks timely after the explosion. Therefore, the rock mass in the upper part of the hole generated the downward displacement due to self-gravity. Also, the surrounding rocks had already been disturbed by the excavation of the original existing tunnel. This expanding construction regards the center line of the existing tunnel as the axis to reconstruct the tunnel on both sides, and the expanding height is smaller than the width. Consequently, the impact of disturbance on height is more obvious than that on width, so that the values of crown settlement are bigger than those of convergences.

In the tunnel entrance section and middle section, the lower bench was divided into a right part and left part to be excavated. When the right part was excavated first, the crown settlement was obvious. Also, after the excavation of the left part, although it fluctuated, it became stable after 40 days. The main reason for this condition is that the excavation height of the upper bench is relatively large (about 7.3 m), which accounts for 68% of the total excavation height (10.77 m). Moreover, the surrounding rock belongs to the II classification and the covering layer is relatively thin, so the rock mass has good integrity and stability. The excavation of the lower bench has a little influence on the stability of the surrounding rock because its height is relatively small, and most of the settlements of the monitoring cross sections were almost finished after the excavation of the lower bench, as shown in Figures 7–10.

In the tunnel exit section, the covering layer is quaternary planting soil and is thicker than that in the entrance section and middle section. Therefore, the stability of the surrounding rock is poor. Also, in this section, the lower bench was excavated at a time rather than was divided into the right and left parts to excavate, so the excavation face is relatively large. The right and left sides are excavated at the same time. Therefore, the secondary explosion for excavating the lower bench had a significant influence on a crown settlement and upper bench convergence, especially on a crown settlement, as shown in Figures 11–13.

Figure 12 shows that the convergence of the upper bench has sudden changes before and after the excavation of the lower bench, which indicates that the vicinity of the boundary between the upper bench and the lower bench is a weak part of tunnel excavation, and attention should be paid to it in the construction process. In some cases, feet-lock bolts should be applied to enhance the stability of the surrounding rocks and initial supporting to prevent excessive deformation, which have a bad influence on the safety of the structure.

5.3 Analysis of monitoring measurement results of the surrounding rock

In order to learn the magnitude of surrounding rock loading and the stress state of the initial supporting and to analyze the safety of the supporting structure, some pressure cells were arranged in arch crown, haunch and springing areas. The time-history curves of monitoring measurement results are shown in Figures 14–17.

Monitoring measurement results show that the monitoring results of pressure cells in each part of the cross section increased greatly after the heavy rain, and the increasing range was from 0.01 MPa

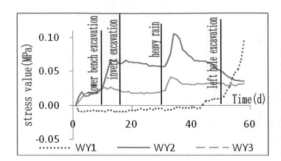

Figure 16. Time-history curves of rock pressure at the K0+587 section of the right tunnel.

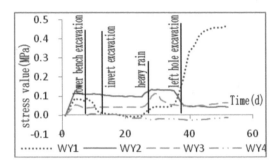

Figure 17. Time-history curves of rock pressure at the K0+607 section of the right tunnel.

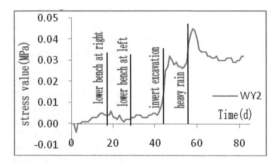

Figure 14. Time-history curves of rock pressure at the K0+521 section of the right tunnel.

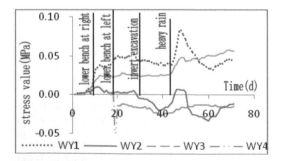

Figure 15. Time-history curves of rock pressure at the K0+552 section of the right tunnel.

to 0.03 MPa, which approximately accounts for 30% of the maximum pressure. Meanwhile, the increasing values of the arch crown and the right haunch in the K0+607 cross section are maximum, which accounts for 85.7% and 56.6% of the maximum pressure, respectively. The main reason resulting in this condition is that the shotcrete failed to fill fully the gap in the back of the pressure cell. In addition, the supporting structure also bears the water pressure, except the surrounding rock pressure, because some fissure water cannot flow out and is hoarded between the surrounding rock and the shotcrete due to the sealing effect of the shotcrete on the surrounding rock. After the rain has stopped for about a week, the pressure decreases gradually and the curve becomes flat, which is induced by the discharge of fissure water and the decrease in water pressure. At this moment, the supporting structure mainly supports the surrounding rock. Therefore, the monitoring measurement value of pressure presents a decline.

Figures 14 and 17 show that the monitoring measurement value of the surrounding rock pressure has some changes after the excavation of the lower bench in the tunnel portal section, but the change is very slight. The main reason for this is that the left and right parts of the lower bench in the portal section are excavated by steps, which reduces the free face of the upper bench and makes the change in the surrounding rock pressure very small.

Figure 16 shows that, in the construction process of the middle section of the tunnel, the full face excavation of the lower bench at a time make the upper bench become free face and leads to an abrupt increase in the surrounding rock pressure. Therefore, the support of feet-lock bolts should be strengthened to prevent collapses and even roof caving after the excavation of the upper bench, or the lower bench can be divided into the right and left parts to be excavated by steps and to build the supporting structures by steps to avoid the arch springing of both sides becoming free.

Figures 16 and 17 show that the excavation of the left hole, which is excavated latter, has a significant influence on the left arch haunch of the right hole and induces an abrupt increase in the pressure. However, it has slight impacts on the arch crown, right arch haunch and right arch springing of the right hole. Consequently, the weak part of the former excavated the right hole is mainly in the vicinity of the left arch haunch. There are three main reasons accounting for this case: ① the tunnel span is relatively large and the bench cut method makes the blasting free face also large, which results in an increase of the surrounding rock load; ② due to the excavation of the existing tunnel and its disturbance on the surrounding rock, the integrity of the surrounding rock is relatively poor, which leads to a decline in its bearing capacity; the width of the expanding construction is very large, and the arch haunches are so far from the existing tunnel that they encounter a small influence on the previous construction of the existing tunnel; ③ the existing tunnel was built in the early 1990s and the traditional mining method was mainly applied. The construction technology and equipment at that time were very backward and a lot of cavities (range of their heights 0.3–2 m) were generated between the lining and the surrounding rock, especially in the areas of the arch crown and haunch. These cavities make the load of the surrounding rock not to be transferred to the supporting structures but to be supported by themselves. After the explosion of the latter constructed hole, the free face increases dramatically and the first excavated hole bears more loads, which leads to an obvious increase in the pressure of the left arch haunch of the right hole. Therefore, in the construction process of the double-arch tunnel, some measurements, such as grouting slurry and adding bolts in the gap between the middle pilot tunnel crown and the initial lining and adding feet-lock bolts in the first excavated hole, should be implemented to strengthen the supporting system of the

area near the middle pilot tunnel so as to ensure the safety of the construction.

5.4 Analysis of monitoring measurement results about the stress of the intermediate supporting wall

For the double-arch tunnel, the middle supporting wall plays an essential role in the supporting system and is in complicated stress state. In the whole construction process of the tunnel, the force of the middle supporting wall adjusts dynamically to the tunnel structure accompanied by the bend, pull, pressure, torsion and shear stress state, and it is the part of stress concentration. The mechanical behaviors of the middle supporting wall have a direct impact on the whole double-arch tunnel stability. Therefore, stress meters should be installed in the middle supporting wall to monitor the behaviors of reinforcements and to guide the design and construction, in case of cracking damage, so as to ensure the structure safety. Apart from some of the monitoring cells are damaged, monitoring results are shown in Figures 18–20.

The monitoring measurement results present that, in the initial installment period of monitoring

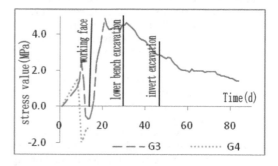

Figure 18. Time-history curves of steel stress at the K0+573 section.

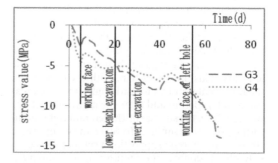

Figure 19. Time-history curves of steel stress at the K0+602 section.

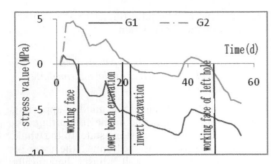

Figure 20. Time-history curves of steel stress at the K0+618 section.

cells, the monitoring result fluctuates with compressive stress and tensile stress, which is mainly induced by the concrete's self-weight, hydration heat, shrink and creep, after the intermediate supporting wall was casted. After that, regardless of whether the middle supporting wall is in compressive or tensile state, before and after the excavation of the working face, the monitoring stress result of the middle supporting wall always appears to have a huge fluctuation, for example, the peak value of the K0+573 working face appears while the K0+550 working face is excavated, and the peak value of the K0+618 working face appears while the K0+605 working face is excavated. All of this indicates that the middle supporting wall would be significantly affected by the excavation of the working face, especially the area in front of the working face 10~20 m. As a result, during the construction, much attention should be paid to observe and monitor the behaviors of the middle supporting wall in case of the wall's buckling failure.

Figure 19 shows that the monitoring cells are in a compressive situation and the compressive stress increases gradually. Stress in the middle supporting wall has an obvious change before and after every construction procedure, especially when the left or right working face passes it. Consequently, the excavation of the working face should be a weak explosion, short drilling footages and slight disturbances. Before the lower bench is exploded, the stress of the left part is smaller than that of the right part; then, after the lower bench is exploded, the increase of the stress in the right part is bigger than that of the left part. This illustrates that, with the excavation, the force center moves towards right (the first excavated hole). Before and after the excavation of the working face, the stresses of the left and right monitoring points are similar. However, the compressive stress of the middle supporting wall increases consistently. Therefore, the monitoring measurement of crown settlement and convergence should be enhanced in case of the excessive deformation of

the surrounding rock and its damaging impact on the safety of the tunnel structure.

Figure 20 shows that the monitoring cells in the right part of the monitoring cross sections are in compressive states, and the stress increases gradually. The monitoring cells in the left part of monitoring cross sections are in the tensile state before the excavation of invert, and then it becomes zero in the excavating process, and finally it is in compressive state after the excavation. These changes illustrate that the fulcrum of the primary lining of the middle supporting wall is already in the compressive state. However, due to the fact that the disturbance of the surrounding rock is very slight, the rock bolt on the top of the middle supporting wall plays a role in suspension, and the stress meters at the upper left of the fulcrum of the primary lining are in tension. After the excavation of the first excavated hole, the surrounding rock is disturbed and the compressive force on the middle supporting wall increases gradually under the effect of the upper load.

6 CONCLUSIONS

Until now, the monitoring measurement result of the Jingya Tunnel has achieved some preliminary achievements, which has provided a great help for the design and construction. However, due to the fact that the reconstruction of the tunnel has not been completed, some monitoring measurements of cross sections have not yet been carried out. The comprehensive analysis of all of the monitoring data about the right and left holes will achieve fruitful results, and provide useful references for subsequent similar engineering projects.

1. The monitoring measurement results of ground settlement, crown settlement, convergence and stress state of the middle supporting wall show that the bench cut method adopted by the Jingya Tunnel is applicable and its supporting parameters are feasible. In addition, they can control the deformation of the surrounding rock very well and ensure the stability of the surrounding rock and supporting structures.
2. For the shallow buried short tunnel, the ground settlement is always influenced by the excavation. The crown settlement is more obvious than the convergences of the upper bench and lower bench. The vicinities of the crown and the boundary between the upper bench and the lower bench are the weak parts of tunnel excavation, and attention should be paid to it in the construction process. In some cases, feet-lock bolts should be applied to enhance the stability of the surrounding rocks and initial supporting

to prevent excessive deformation, which have a bad influence on the safety of the structure. If the lower bench is excavated at a time rather than is divided into the right and left parts, this would result into the right and left sides of the upper bench becoming free at the same time and cause an abrupt release of surrounding rock stress and a drastic increase of stress in the supporting structure. Therefore, separated excavation of the right and left parts of the lower bench by steps should be adopted and the support in the arch springing area should be strengthened.

3. The supporting structures are not only under the surrounding rock loading, but also under water loading, especially in the areas of tunnel crown and haunch. Therefore, the fissure water should be drained by blind or half Ω drainage pipe into the longitudinal drain pipes in time in order to reduce the stress of the supporting structures.

4. Affected by the asynchronous construction of the left and right holes, the area around the middle pilot tunnel is relatively fragile. Especially in the condition of regarding the existing tunnel as the middle pilot tunnel to expand it to a double-arch tunnel, due to the disturbance of the excavation of the existing tunnel on the surrounding rock, the integrity and mechanical behaviors of the surrounding rock become relatively poor, which have a bad influence on the construction of the latter excavated hole, so the monitoring measurement and support should be strengthened in the construction process.

5. The excavation of the working face and construction procedure have an important influence on the behaviors of the middle supporting wall, making the stress center adjust constantly and each part of the middle supporting wall shows different stress states. Consequently, the stress state of the middle supporting wall for double-arch tunnel middle wall stress is extremely complicated and much attention and importance should be given to it.

REFERENCES

Gao, G. Liu, Y.X. & Zhou, J.Z. et al. 2010. Expansion form of underground space, *Modern Tunnelling Technology*, 46(6): 1–10.

Gao, Q. 2012. *Study on the construction technology and structural stress state during the reconstruction and extension process of loess highway tunnel*, Chang'an University.

Li, Y.Y. Zhang, Z.G. Xie, Y.L. & Liu, B.J. 2013. Study of site monitoring of vertical full displacement of vault surrounding rock of tunnels and its change law, *Rock and Soil Mechanics*, 06: 1703–1708+1715.

Lin, C.M. Zhang, Z.C. & Zheng, Q. et al. 2013. Study on the soft surrounding rocks stability and supporting parameters of two-to-fourlane tunnel of small clear-distance by CD method. *China Civil Engineering Journal*, 07: 124–132.

Nazzano Tunnel widening, 2000. *Tunnels and Tunnelling International*, 13.

Shen, Y.S. & Gao, B. 2006. Site monitoring and analytical research on mechanical characteristics under construction partial press in double-arch tunnel, *Rock and Soil Mechanics*, 27(1): 1733–1739.

Xia, C.C. Gong, J.W. & Tang, Y. 2007. Study on site monitoring of large-section highway tunnels with small clear spacing, *Chinese Journal of Rock Mechanics and Engineering*, 26(1): 44–50.

Yu, M. 2006. Redevelopment of Urban Transportation in The City Center—Boston's Central Artery/Tunnel Project. *Foreign urban planning*, 21(2): 87–91.

Zhang, G.H. Chen, L.B. & Qian, S.X. 2010. On-site supervision measure and analysis of Damaoshan tunnels with large section and small clear-distance, *Rock and Soil Mechanics*, 02: 489–496.

Zhu, G.Q. Lin, Z. & Zhu, Y.C. et al. 2012. Research on influences of in-situ tunnel extension project on adjacent buildings, *Rock and Soil Mechanics*, S2: 257–262.

Advanced Materials and Structural Engineering – Hu (Ed.)
© *2016 Taylor & Francis Group, London, ISBN 978-1-138-02786-2*

Research on product development gene oriented to process reuse

Z.H. Wang, S.R. Tong & J. Li
The School of Management, Northwestern Polytechnical University, China

ABSTRACT: In order to realize the reuse of the product development process, product development gene is analyzed and researched based on the theory of gene engineering. By analogy to the biological gene, the definition and characteristic of product development gene is described in detail. The genetic information driving the product development activity unit is extracted by decomposing gradually the product development process. Then, the model of product development gene is built. By means of the frame and production representation, the product development gene is represented formally. Finally, the operation process of PDG in the product development process reuse is put forward.

1 INTRODUCTION

The heredity and variation of the gene makes the similar but not identical hereditary feature between filial and parental generation. Through the study of genetic engineering, different biological genes can be processed and combined according to different requirements. Thus, new species are created by changing the biological hereditary feature. Similarly for the heredity and variation of gene, it is important to inherit and restructure original product development information in a new product development process. Product development gene can be extracted from the product development process. Then, it is modified and recombined with the gene engineering method to form the new product development process. The new product development method not only inherits original product development process characteristics, but also improves the performance according to new requirements and constraints. Therefore, it is meaningful to optimize and reuse the product development process.

To solve the problem of inheritance and innovation of product design, some scholars have already researched engineering design based on the correlation theory of genetic engineering in recent years. Professor Chen, K.Z. presented that a product has genetic information (Chen & Feng 2004, Chen et al. 2005), and the concept, feature and structure of virtual product gene was proposed. Professor Gero, J.S. and his research team applied the heredity theory to the field of engineering design (Gero 1998, Gero et al. 1996, Gero & Kazakov 2001). The concept of design gene was presented and was proved to be useful in product design. Autogenetic design theory was first propounded by professor Vajna et al. (2002, 2005). Meanwhile, he researched

the feature of the theory by comparing the product design process with the organism evolution process.

However, these studies mainly focused on the characteristics and types of product genes. The genes of similar products were extracted. Then, high-quality products were designed by the inheritance and reconstruction of product genes. So far, there has been little research on the product development process from the viewpoint of biological evolution.

In this paper, the product development process is regarded as a kind of special product. The information that guides the product development process is used as product development gene. The model and representation of product development gene are analyzed. This research provides support for the optimization and reuse of the product development process.

2 THE CONCEPT OF PRODUCT DEVELOPMENT GENE

Gene is a nucleotide sequence of DNA molecules that have genetic effects (Berg 1994). The gene contains specific genetic information, and is the minimum functional unit of genetic material (Higgins et al. 1998). By analogy with biological gene, the definition of product development gene is put forward by combining with the product development process characteristics. Product development gene is a set of genetic information that guides the product development process. It is a "blueprint" or "recipe" of a certain product development process, which stipulates the basic characteristics of the product development process and automatic generation mechanism. In the constraint of external condition, a new product development process can be generated by using product development genes. The product development gene has the following characteristics:

1. Product development gene is the basic functional unit of controlling product development characters. The gene is expressed by guiding the product development process, and then controls the characteristic and function of product development activities.
2. Product development gene is essentially a kind of information. It is a kind of special instruction related to the product development process. The product development process needs to be guided by information related to the product development process, such as product development approach, design strategy, personnel organization and resource allocation pattern.
3. Product development gene has the characteristic of heredity and variation. New product development often reuses the similar product development process. The original product development process information is passed to the new product development process, which is similar to the biological genetic characteristic. Meanwhile, the new product development needs to be modified and innovated according to the new requirements, which is similar to the biological gene variation characteristics.
4. Product development gene is represented by some structures or codes, which is similar to the biological gene represented by four bases of A, T, C, G. The representation of product development gene creates conditions for storage, retrieval and reuse of the gene.

3 THE ACQUISITION OF PRODUCT DEVELOPMENT GENE

The development of biology experienced the process of organism, organ, cell, protein, DNA, and gene. The process is gradually analyzed from the shallow to deep, from macroscopic to microscopic. Similarly, product development gene can be obtained by the analysis of the existing product development process (Veryzer 1998, Schilling & Charles 1998). Product development process is gradually decomposed into the product development activity unit similar to a cell. The genetic information that drives the product development activity unit is extracted from the unit. Then, the product development gene is obtained. As shown in Figure 1, the product development process is first divided into several stages such as product planning, concept design, structure design, detail design, process design, testing, refinement and product ramp-up phases (Ulrich 2003). Second, concept design is taken as an example to be decomposed into many tasks

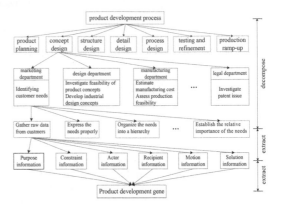

Figure 1. The acquisition process of the product development gene.

of a marketing department, design department, manufacturing department, and legal department. Third, according to customer requirements coming from marketing department, the task is decomposed into several activity units, such as gathering raw data, expressing the needs properly, organizing the demands into a hierarchy, and establishing the relative importance of the requirements. Then, the activity unit of gathering raw data from customers is used as an example, and the genetic information that contains purpose, constraint, actor, recipient, motion and solution is extracted. Finally, the information is aggregated as the product development gene that drives the activity unit of gathering raw data from customers.

4 THE MODEL OF PRODUCT DEVELOPMENT GENE

Product development activity unit can be described as "For certain purposes, actors carry out a series of motion to recipients under certain constraints and achieve certain solutions." Accordingly, product development activity unit includes some genetic information, such as purpose information, constraint information, actor information, motion information, recipient information, and solution information. The model of product development gene is constructed, as shown in Figure 2, which takes motion information as the core.

Product Development Gene (PDG) is described as follows:

$$PDG = \{P, C, A, M, R, S\} \qquad (1)$$

where P is the purpose information; C is the constraint information; A is the actor information; M is the motion information; R is the recipient information; and S is the solution information.

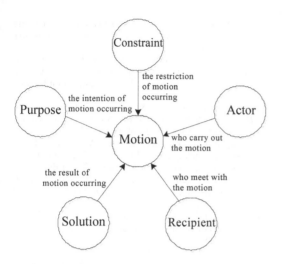

Figure 2. The acquisition process of the product development gene.

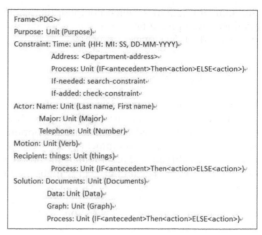

Figure 3. Coordinate systems and load distribution.

5 THE REPRESENTATION OF PRODUCT DEVELOPMENT GENE

Product development gene is represented based on the model of product development gene. The method lays the foundation for establishing the product development database and realizing the reuse of the product development process. The representation of product development gene is a formal and symbolized process of it. The most important factor is how to choose a proper form to represent the gene. Product development gene includes many forms of information, such as structure and process. Structural information is usually expressed by the frame representation (Minsky 1974, Rubenstein-Montano et al. 2001). However, production representation (Davis et al. 1977, Mylopoulos & Hector 1983) is used to describe process information. In this paper, the frame representation is proposed and then combined with production representation to represent the product development gene. As shown in Figure 3, the frame representation is treated as the main body and the production rule is embedded in the frame representation. The process information is described by using the production representation in the constraint slot, the recipient slot and the process side of solution slot.

6 THE OPERATION PROCESS OF PDG IN THE PRODUCT DEVELOPMENT PROCESS REUSE

Referring to the biological genetic regularity, product development process could be regarded as the replication, variation and expression process of product development gene. The process is depicted

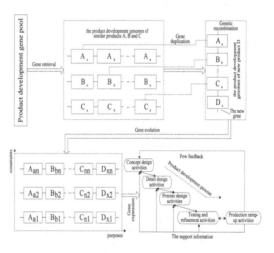

Figure 4. The operation process of the product development gene.

in Figure 4. The product development genomes of similar products A, B and C are first searched from the product development gene pool, then the new product development process inherits selectively the product development genes A_a, B_b and C_n of similar products, and the new gene D_x that does not exist in the gene pool is added according to the new product development demands. These genes are recombined to form the new product development genome. Second, according to the development purposes and constraints of new products, the new product development genes are mutated and evolved. The genes evolve from state 1 to state n and then the final new product development genome is obtained. Finally, product development genes are expressed as a series of product

693

development activities, which compose the whole product development process and realize the reuse of the product development process.

7 CONCLUSIONS

Product development gene is extracted from the product development process based on the biogenetics theory and method. The reuse and optimization of the product development process can be realized by modifying and recombining the product development gene. The model, representation and operation process of product development gene are studied in this paper. The research lays the foundation for the heredity and variation of product development gene. Furthermore it is the premise of the product development process reuse. However, the study of this paper is in the preliminary stage. How to realize product development gene variation and expression will be further studied.

REFERENCES

Berg, K. 1994. *Gene-environment interaction: variability gene concept. In Genetic Factors in Coronary Heart Disease*, Springer Netherlands, 373–383.

Chen, K.Z. & Feng, X.A. 2004. Virtual Genes of Manufacturing Products and Their Reforms for Product Innovative Design. Proceedings of the Institution of Mechanical Engineers, part C, *Journal of Mechanical Engineering Science*, 218(C5): 557–574.

Chen, K.Z. Feng, X.A. & Chen, X.C. 2005. Reverse Deduction of Virtual Chromosomes of Manufactured Products for Their Gene-Engineering-Based Innovation Design. *Computer-Aided Design*, 37(11): 1191–1203.

Davis, R. Bruce, B. & Edward, S. 1977. Production rules as a representation for a knowledge-based consultation program. *Artificial intelligence* 8(1): 15–45.

Gero, J.S. & Kazakov, V.A 2001. Genetic Engineering Approach to Genetic Algorithms. *Evolutionary Computation*, 9(1): 71–92.

Gero, J.S. 1998. *Adaptive Systems in Designing: New Analogies from Genetics and Developmental Biology.* Adaptive Computing in Design and Manufacture. Springer-Verlag, 3–12.

Gero, J.S. Kazakov, V.A. & Schnier, T. 1996. *Evolving Design Genes as well as Design Solutions.* Proceedings of the Third ASCE International Workshop on Computing in Civil Engineering. New York, N.Y., USA: American Society of Civil Engineers Press, 84–90.

Higgins, V.J. Huogen L., Ti X. Angie G. & Eduardo B. 1998. The gene-for-gene concept and beyond: interactions and signals. *Canadian Journal of Plant Pathology*, 20(2): 150–157.

Minsky, M. 1974. *A framework for representing knowledge.* Massachusetts Institute of Technology A.I. Laboratory.

Mylopoulos, J. & Hector. L. 1983. *An overview of knowledge representation, GWAI-83.* Springer Berlin Heidelberg, 143–157.

Rubenstein-Montano, B. et al. 2001. A systems thinking framework for knowledge management. *Decision support systems* 31(1): 5–16.

Schilling, M.A. & Hill, C.W.L. 1998. Managing the new product development process: Strategic imperatives. *The Academy of Management Executive*, 12(3): 67–81.

Ulrich, K.T. 2003. *Product design and development.* Tata McGraw-Hill Education.

Vajna, S. Clement, S. & Jordan, A. 2002. *Autogenetic design theory: an approach to optimise both the design process and the product.* ASME.

Vajna, S. Clement, S. & Jordan, A. et al. 2005. The Autogenetic Design Theory: An Evolutionary View of the Design Process. *Journal of Engineering Design*, 16(4): 423–440.

Veryzer, R.W. Discontinuous innovation and the new product development process. *Journal of product innovation management* 15(4): 304–321.

Computer aided for engineering application

Advanced Materials and Structural Engineering – Hu (Ed.)
© 2016 Taylor & Francis Group, London, ISBN 978-1-138-02786-2

Operation of functional parameters of video-verification integrated into I&HAS

J. Hart & V. Nidlova
Czech University of Life Sciences Prague, Prague, Czech Republic

ABSTRACT: The issue of integrating the video-verification systems to the security alarm and emergency systems is relatively new, but all the more important. To date, alarm verification has been conducted either in person or via a specifically-separated camera system. This alternative was functional, but it did not meet the comfort expectations of newly implemented systems. Nowadays, it is important that the video-verification systems directly integrated into the security alarm systems meet the basic conditions for their operation. This means that they should meet the basic functional requirements. Therefore, a comparison of security alarm and emergency systems with video-verification functions was carried out, and they were evaluated based on these results.

1 INTRODUCTION

Intrusion and Hold-up Alarm Systems (I&HAS) mainly serve as systems for protecting buildings against the illegal conduct of third parties and can also be used as monitoring and control systems. Thus, they are mainly a tool for achieving a state of security. They function in substantive areas (physical protection of values, life and health) and psychological areas (provide a sense of peace, security and a certain sense of confidence).

Therefore, it is important to avoid false alarms, which may even discourage end users from further use of I&HAS. Yet false alarms may occur despite all of the efforts of installers. The advent of modern technology allows for many options for how to verify a given alarm. Among others, these include security agencies with centralized protection panels (CSP), remote access to the camera system (CCTV), and I&HAS with photo-verification or video-verification. So the question remains just which option is best for this purpose.

2 MATERIALS AND METHODS

There are several ways to verify an I&HAS alarm, but not all of the selected methods are acceptable for everyone (Capel 1999, Cumming 1994). Most important in choosing this method is what the user intends to rely on, how much they intend to invest and whether this method is to be automated. An automatic verification system means a system without the need for personal or mediated inspections of a building. It is possible to say that the verification methods can be divided as follows:

- Human resources,
  - user,
  - security agency,
  - policy,
  - others,
- CCTV,
- I&HAS,
  - video-verification,
  - foto-verification,
  - audio-verification.

I&HAS manufacturers are currently focusing on the development of an intelligent verification alarm using audio-verification, photo-verification (Fig. 1) and video-verification. These types of verifications simplify alarm verification, and they can also automatically check them (Capel 1999, http://www.riscogroup.com, Petruzzellis 1994).

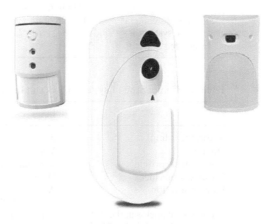

Figure 1. Photo-verification detectors.

Historically, audio-verification is the oldest of the aforementioned types of verification in I&HAS. It works on the simple principle of in-ear monitoring (almost like a normal phone) that is enabled via a Unified Telephone Network (UTN), or via GSM communication. Sometimes a speaker is added to it so that it is possible to communicate with the intruder.

Photo-verification allows for an image or several pictures of the disturbed area to be recorded,

which increases the chance of alarm confirmation. Most of these images are then sent to the user, or to a predetermined location.

Video-verification is similar to photo-verification. Unlike photo-verification, however, it sends video recordings instead of images, which can detect more than a few photographs can.

In order to compare video-verification, I&HAS were selected which are commonly available. The compared systems do not require a classic CCTV installation—all they need are cameras connected to a local network, or installation of special cameras directly from the manufacturer. These are the systems (Fig. 2) of manufacturers:

- RSI VIDEO TECHNOLOGIES
  - *Videofield*
- RISCO
  - *LightSYS 2*
- Ksenia Security
  - *Lares*
- Siemens
  - *Intrunet SPC*

Variables were selected for video-verification that a video-verification system should fulfill in order to be efficient and safe.[3,4] The first selected variable was an alarm recording. It is important that the system has an alarm recording so that

Figure 2. Compared systems.

Table 1. Comparison of systems Videofield and LightSYS 2.

	Videofield	LightSYS 2
Alarm recording	Yes	Yes
Stream	Only immediately after the alarm	Yes
Length of recording	10s	According to configuration
Placement of video	Central Security Panel (CSP)	SD card in camera, stream, cloud, CSP
Video resolution	*Outdoor black and white* 320 × 240 *Indoor colour and black and white* Up to 640 × 480	*Indoor, colour HD* 1,3 Mpx
Number of cameras in the system	Max. 25 *of all zones and devices in the system (including cameras)*	Unlimited
Integrated detector	Yes	No

Table 2. Comparison of systems Lares and Intrunet SPC.

	Lares	Intrunet SPC
Alarm recording	No	Yes
Stream	Yes	Yes
Length of recording	Without recording	According to alarm
Placement of video	Only stream	Stream, in the control panel, CSP
Video resolution	According to the type of used CCTV cameras	According to the type of used CCTV cameras
Number of cameras in the system	Unlimited	4
Integrated detector	No	No

it can be ascertained what triggered the alarm. Another variable is stream access. This access allows a video to be watched live, which means a direct transfer from cameras in real time. Another variable is the length of the recording, which is essential for further verification options of the movements of the intruder during an alarm. A very important variable is the location of the recording made by the system and video resolution. It is important that there is good access to the recording, that it is protected, and that the video provides sufficient resolution in order to identify potential offenders. The number of cameras that can be worked with is also essential. Last but not least, if the camera is integrated with the detector. Tables 1, 2 show the parameters of the compared systems (Powell & Shim 2012).

3 RESULTS AND DISCUSSION

Of the specified comparison criteria, and using a scoring method, a surveillance graph (Fig. 3) was created in the multi-criteria analysis of options. In this analysis, the emphasis was placed primarily on the variables that affect the alarm authentication function such as a storage location of the recording, alarm recording, video resolution and stream access.

It is clear from this graph that systems Videofield, Lares and Intrunet SPC are relatively on the same level, at least in terms of the results from the scoring method in the multi-criteria analysis of options. LightSYS 2 appears to be a very interesting system that allows for a functional method for verification of alarms using video-verification.

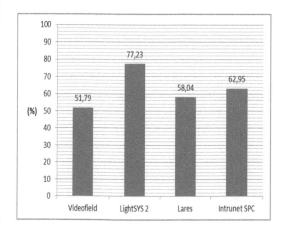

Figure 3. Percentage evaluation of video-verification for selected systems.

Like all control panels with alarm video-verification, LightSYS 2 has a specific method of connecting IP cameras to the system. It supports the so-called VUpoint, which is an alarm verification solution using Live Video. Via cloud, this platform allows us to provide streamed video from IP cameras. This real-time video can be run both during the alarm and upon authorized request. Another advantage of this system is that it has no limit on the number of IP cameras that can be installed in the system, which increases the possibility of better alarm verification. As the recording is saved on RISCO Cloud, the user can use a smartphone application to access it (Staff & Honey 1999).

4 CONCLUSIONS

Intrusion and hold-up alarm systems are primarily tools for achieving a state of security. We may perceive security in the wrong way. Security can be divided into several basic types: physical security and psychological security. Physical security means the real protection of property and health. No less important is psychological security, which is to induce a feeling of peace of mind. If false alarms occur at any frequency, this will usually cause psychological harm to the user; in fact, based on such an experience, the user may even stop using the system altogether. Therefore, when designing the system is important to consider that there should be a way to verify the alarm.

Although the use human resources have thus far been the oft-used method for verifying alarms, it is very risky. This means that the user is visited by the police or security agency. Whether the alarm is real or false, a personal visit always requires a lot of time. If it is real, then there is a risk of a confrontation with the intruder. Without prior preparation, the situation is especially risky for the user, as the user is not trained for similar situations and the user's life may be in danger. If the alarm is false, then it does not endanger the user or other persons arriving on the scene, but it can make life unpleasant, and a large financial burden may also arise for the user for paid visits by a security agency.

Video-verification integrated directly into the I&HAS system allows for quick and easy verification of the alarm via a video recording or Live Camera, and may thereby reduce the percentage of PCO trips to check on false alarms. At the same time, it can also improve the operational efficiency of the system and alleviate the psychological burden on the user when a false alarm occurs.

Although all of the systems are tested at a high level, the results of the scoring method in the multi-criteria analysis of options shows that according to its parameters, the LightSYS 2 system is the best system. It largely supports all of the evaluation criteria for alarm verification via video-verification.

The development of the method for verifying alarms via video-verification in alarm and security systems is in its infancy. This verification method is very efficient and effective. Of course, it is important that manufacturers do not give up on this type of verification and do not stop improving it and developing its other possibilities.

REFERENCES

Capel, V. 1999. *Security Systems & Intruder Alarms,* Elsevier Science, 301.

Cumming, N. 1994. *Security: A Guide to Security System Design and Equipment Selection and Installation,* Elsevier Science, 338.

http://www.riscogroup.com (on-line 28.1.2015).

Petruzzellis, T. 1993. *Alarm Sensor and Security,* McGraw-Hill Professional Publishing, 256.

Powell, S. & Shim, J.P. 2012. *Wireless Technology: Applications, Management, and Security,* Springer-Verlag New York, LLC, 276.

Staff, H. & Honey, G. 1999. *Electronic Security Systems Pocket Book,* Elsevier Science, 226.

Advanced Materials and Structural Engineering – Hu (Ed.)
© *2016 Taylor & Francis Group, London, ISBN 978-1-138-02786-2*

Model for selecting project members to minimize project uncertainties

C.C. Wei
Department of Industrial Management, Chung Hua University, Hsinchu, Taiwan

H.J. Tsai & Y.F. Lin
Ph.D. Program of Technology Management, Chung Hua University, Hsinchu, Taiwan

C.S. Wei
Department of Industrial Management, Lunghwa University of Science and Technology, Taiwan

ABSTRACT: Project risks come mainly from the future uncertainties, and new product and service development projects possessing the most uncertainty and therefore the highest risk. However, if the project team has gained experiences from previous similar projects, the risk of the project drastically declines. In other words, an inverse correlation exists between member experiences and project risk. Therefore, the guiding principle would be to assign an experienced member to execute familiar work for that member. This study aims to establish a mathematical model that uses member experience to minimize project uncertainty. The model assigns members with different experiences to the most suitable tasks to minimize project risk, and thus maximize project success rate under constraints of project cost and communication complexity. An experimental case is used to demonstrate the applicability of the proposed model, and results indicated that, under cost and communication complexity constraints, project uncertainty can truly be minimized by assigning the adequate member to the suitable task.

1 INTRODUCTION

Corporations continuously create and promote new products to widen the gap between their competitors and themselves. However, newer products create two unique problems: uncertainty in itself and less familiarity with members. A need exists to delegate team members efficiently for successful project management. But member experience and payment positively correlate. The number of affiliates directly affects the communication complexity. Therefore, corporations should pursue a more feasible project management mechanism instead. This study hopes to develop a quantitative mathematical model that takes into account the project cost and member communication complexity. It also assigns members with different experiences to the most suitable tasks. Integration of this model minimizes project uncertainty and elevates project success chances.

2 LITERATURE REVIEW

2.1 Project management

In general, project management is a kind of project management methodology (Yeh et al. 2012,

Wei et al. 2002). A good practical theory needs an effective plan. Strengthening of cost and time schedule must happen. One must also clearly realize the importance of member selection, delegation, and communication (Kerzner 2013). So, successful completion of high risk tasks relies on the project manager's skills. This person needs to utilize project management methods, assemble members of various professions, delegate them to appropriate tasks, and distribute tasks accordingly (Howard 2010, Katz & Allen 1985, Pinto & Slevin 1988).

2.2 Risk

Risk can be defined differently based on different tasks. It can be used to explain any possible event or action that will affect an organization. It can also be used to explain a result that is lower than expected (McNeil et al. 2010). Establishment of a new project inevitably holds some risk (Dey & Ogunlana 2004). Risk comes from various complex and unpredictable problems. Corporations can often not define risk clearly upon taking a project and signing a hedge futures contract (Motiar & Kumaraswamy 2005). Effective risk management decreases risk. The internal team efficacy factor controls at an easier level. A need for better member delegation exists to place the appropriate member in the most

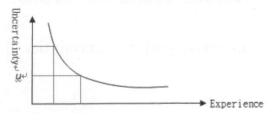

Figure 1. The Relationship between member experience and uncertainty.

suitable job. Successful placement improves communications and increases team efficacy. In short, the risk management methodology allows the correct adjustment methods to lower the chances of risk (Fan et al. 2008). Project executives are humans. Human experience highly correlates with project uncertainties. A need exists for a quantitative model. The model delegates project member experience and task distribution to lower project risks.

2.3 Uncertainty

Progression of any new project includes several uncertainties. Previous studies indicate a project's success or failure lies in the human factor. Technology makes a smaller impact (Lechler 1998, Belout & Gauvreau 2004). Member experience and task responsibility correlate strongly. Task details are similar to prior task experience, which induces lower uncertainty, reducing the risk, and increasing the chances of success. The relationship between member experience and uncertainties is shown as Figure 1. Scholars Mathiassen, L. and Pedersen, K. (2008) have divided uncertainties into fifteen categories. Man-made issues account for a large portion of uncertainties, which are represented by project-solving skills and development abilities. These areas equate strongly with member experience. Members with similar experiences to their current tasks predictably have better problem-solving skills and development abilities. This study aims to provide an effective mathematical model to solve this particular issue.

3 MODEL FORMULATION

If a new project's task is j ($j = 1 \dots n$), company members that can be delegated as i ($i = 1 \dots m$). The number of communication channels positively corresponds with the number of people squared. This assumes that project communication complexity will be positively correlated to project member number squared, with each project task having at least one member executing it. If the total budget of the project is C, communication

complexity thresh-hold is M, to prevent communication problems stemming from the excess of team members. Project manager expects to find the most appropriate member-task delegation to minimize project total uncertainty value U. The number of communication channels positively corresponds with the number of people squared. This assumes that project communication complexity will be positively correlated to project member number squared, with each project task having at least one member executing it. The whole problem could be expressed as a mathematical model:

target-oriented minimize project uncertainty
restricted cost $\leqq$ total budget
communication complexity $\leqq M$

which is,

$$\text{Min } U = \sum_{j=1}^{n}\sum_{i=1}^{m} x_{ij}u_{ij} \tag{1}$$

$$\text{S.T. } \sum_{j=1}^{n}\sum_{i=1}^{m} x_{ij}c_{ij} \leq C$$

$$\left(\sum_{j=1}^{n}\sum_{i=1}^{m} x_{ij}\right)^2 \leq M$$

$$x_{11} + x_{21} + x_{31} + \dots + x_{m1} \geq 1$$
$$x_{12} + x_{22} + x_{32} + \dots + x_{m2} \geq 1$$
$$x_{13} + x_{23} + x_{33} + \dots + x_{m3} \geq 1$$
$$\dots$$
$$x_{1n} + x_{2n} + x_{3n} + \dots + x_{mn} \geq 1$$
$$x_{ij} = 0, 1$$

Amongst which u_{ij} are uncertainties from personnel executing the task, c_{ij} represent the cost of executing the task.

4 CASE IMPLEMENTATION

To further elucidate the efficiency of the model, a hypothetical case will be used to describe the mathematical model application. If there are five members of different experiences being assigned to a four activity project, general budget of fifty, communication complexity value restricted to sixteen, with $C = 50$, $M = 16$. The uncertainty value of Member 1 executing four activities is $u_{11} = 2$; $u_{12} = 5$; $u_{13} = 1$; $u_{14} = 3$, expected cost is $c_{11} = 1$; $c_{12} = 5$; $c_{13} = 2$; $c_{14} = 3$. The uncertainty value of Member 2 executing four activities is $u_{21} = 4$; $u_{22} = 2$; $u_{23} = 1$; $u_{24} = 3$, expected cost is $c_{21} = 5$; $c_{22} = 3$; $c_{23} = 4$; $c_{24} = 2$. The uncertainty value of Member 3 executing four activities is $u_{31} = 7$; $u_{32} = 4$; $u_{33} = 2$; $u_{34} = 3$, expected cost is $c_{31} = 4$; $c_{32} = 2$; $c_{33} = 1$; $c_{34} = 2$. The uncertainty value of Member 4 executing four activities is $u_{41} = 2$; $u_{42} = 5$; $u_{43} = 3$; $u_{44} = 6$, expected cost is $c_{41} = 6$; $c_{42} = 3$; $c_{43} = 7$; $c_{44} = 4$. The uncertainty value of Member 5 executing four activities is $u_{51} = 3$; $u_{52} = 5$; $u_{53} = 8$; $u_{54} = 4$, expected cost is $c_{51} = 3$; $c_{52} = 2$; $c_{53} = 5$; $c_{54} = 4$. Input above assumption value into Equation (1), resulting in:

Min $U = (2x_{11}+5x_{12}+ x_{13} +3x_{14}) + (4x_{21}+ 2x_{22}$
$+ x_{23} + 3x_{24}) + (7x_{31}+ 4x_{32}+ 2x_{33}+ 3x_{34})$
$+ (2x_{41}+ 5x_{42}+ 3x_{43}+ 6x_{44})$
$+ (3x_{51}+ 5x_{52}+ 8x_{53}+ 4x_{54})$

S.T. $(x_{11}+ 5x_{12}+ 2x_{13}+ 3x_{14}) + (5x_{21}+ 3x_{22}+ 4x_{23}+$
$2x_{24}) + (4x_{31}+ 2x_{32}+ x_{33}+ 2x_{34}) + (6x_{41}+ 3x_{42}+$
$7x_{43}+ 4x_{44}) + (3x_{51}+ 2x_{52}+ 5x_{53}+ 4x_{54}) \le 50$

$((x_{11}+ x_{12}+ x_{13}+ x_{14}) + (x_{21}+ x_{22}+ x_{23}+ x_{24}) +$
$(x_{31}+ x_{32}+ x_{33}+ x_{34}) + (x_{41}+ x_{42}+ x_{43}+ x_{44})$
$+ (x_{51}+ x_{52}+ x_{53}+ x_{54}))^2 \le 16$

$x_{11} + x_{21} + x_{31} + x_{41} + x_{51} \ge 1$

$x_{12} + x_{22} + x_{32} + x_{42} + x_{52} \ge 1$

$x_{13} + x_{23} + x_{33} + x_{43} + x_{53} \ge 1$

$x_{14} + x_{24} + x_{34} + x_{44} + x_{54} \ge 1$

$x_{ij} = 0, 1$ (2)

Amongst them, u_{ij} are members i *executing tasks j* uncertainty, c_{ij} represent the cost of executing the task Equation (2) uses Lingo to get smallest uncertainty value as $U = 8$. Members delegating position are shown in Table 1 inner box, and delegating results matches the needs of the experiment. The main point of this study is the uncertainty value under which individual members were not restricted to only one task.

Table 1. Delegating position statement.

	Activity 1					Activity 2					Activity 3					Activity 4				
C										50										
M										16										
c_{ij}	c_{11}	c_{21}	c_{31}	c_{41}	c_{51}	c_{12}	c_{22}	c_{32}	c_{42}	c_{52}	c_{13}	c_{23}	c_{33}	c_{43}	c_{53}	c_{14}	c_{24}	c_{34}	c_{44}	c_{54}
	[1]	5	4	6	3	5	[3]	2	3	2	[2]	4	1	7	5	3	2	[2]	4	4
u_{ij}	u_{11}	u_{21}	u_{31}	u_{41}	u_{51}	u_{12}	u_{22}	u_{32}	u_{42}	u_{52}	u_{13}	u_{23}	u_{33}	u_{43}	u_{53}	u_{14}	u_{24}	u_{34}	u_{44}	u_{54}
	[2]	4	7	2	3	5	[2]	4	5	5	[1]	1	2	3	8	3	3	[3]	6	4
Selected person	Member 1					Member 2					Member 1					Member 3				
c_{ij} (Cost need)	1					3					2					2				
u_{ij} (Uncertainty)	2					2					1					3				
Communication complexity	1					1					1					1				
C_p										[8]										
M_p										[9]										
U										[8]										

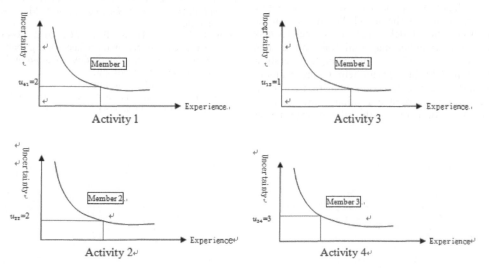

Figure 2. Distribution status.

703

Therefore, member 1 is assigned to Activity 1 and Activity 3.

Results of above assumption case indicate that using the model to delegate members holds several benefits. Task uncertainty effectively minimizes. The actual cost goes below the predicted cost. Each project task has at least one member executing it. It also improves communication.

5 CONCLUSIONS

Projects include various tasks, requiring organizations to delegate members of various experiences for the simultaneous implementation of tasks. One of the vital steps in project management is to assign suitable personnel to conduct the tasks. But, corporations often only consider whether a member is well-experienced or not during the selection process. However, work experience does not only imply seniority, but the capability of successfully completing the task being assigned. In other words, experience in this study refers to the ability to reduce the task uncertainty. Proper delegation mitigates uncertainty, lowering uncertainty to enable smoother control of project risk. Controlled project risk increases project's success rate. Also, member communications is another one important issue when executing tasks. Communications complexity should not interfere with project progression tolerance value so much as to delay it from successful completion. Therefore, this study combines experience, uncertainty, communication complexity and cost to form a mathematical model, which can minimize project uncertainty by assigning the most suitable personnel to be responsible for the most familiar task under budget constraint and within communication complexity threshold. An example case is used to demonstrate the applicability of the model. When overloading is allowed, the model assigns personnel to task that certain member may undertake more than one task. The results of the example case show that the model proposed can help the project manager, under the cost and communication complexity restrictions, assemble a suitable project team from a pool of people with different experiences to minimize project uncertainty, and therefore, increase project's success rate.

REFERENCES

Belout, A. & Gauvreau, C. 2004. Factors influencing project success: the impact of human resource management, *International journal of project management*, 22(1): 1–11.

Dey, P.K. & Ogunlana, S.O. 2004. Selection and application of risk management tools and techniques for build-operate-transfer projects, *Industrial Management & Data Systems*, 104(4): 334–346.

Fan, M. Lin, N.P. & Sheu. C.W. 2008. Choosing a project risk-handling strategy: An analytical model, *International Journal of Production Economics*, 112(2): 700–713.

Howard, W.R. Agile project management: creating innovative products, *Kybernetes*, 39(1): 155–155.

Katz, R. & Allen, T.J. 1985. Project performance and the locus of influence in the R&D matrix, *Academy of Management Journal* 28(1): 67–87.

Kerzner, H.R. 2013. *Project management: a systems approach to planning, scheduling, and controlling*. John Wiley & Sons.

Lechler, T. 1998. When it comes to project management, it's the people that matter: an empirical analysis of project management in Germany, *IRNOP III-The nature and role of projects in the next*, 20: 205–215.

Mathiassen, L. & Pedersen, K. 2008. Managing uncertainty in organic development projects, *Communications of the Association for Information Systems*, 23(1): 27.

McNeil, A.J. Frey, R. & Embrechts, P. 2010. *Quantitative risk management: concepts, techniques, and tools*, Princeton university press.

Motiar, R.M. & Kumaraswamy, M.M. 2005. Assembling integrated project teams for joint risk management, *Construction Management and economics*, 23(4): 365–375.

Pinto, J.K., & Slevin, D.P. 1988. *Project success: definitions and measurement techniques*, Project Management Institute.

Wei, C.C. Liu, P.H. & Tsai, Y.C. 2002. Resource-constrained project management using enhanced theory of constraint, *International Journal of Project Management*, 20(7): 561–567.

Yeh, J.Y. et al. 2012. The impact of team personality balance on project performance, *African Journal of Business Management*, 6(4): 1674–1684.

Advanced Materials and Structural Engineering – Hu (Ed.)
© *2016 Taylor & Francis Group, London, ISBN 978-1-138-02786-2*

Real-time simulation research on an electric drive system

Y.Y. Zhang, X.J. Ma, C.G. Liu & Z.L. Liao
Academy of Armored Force Engineering, Beijing, China

ABSTRACT: In this paper, the electric drive armored vehicle simulation system is built based on the distributed simulation technology. The simulation system is divided into three parts, namely the weak real-time part, the strong real-time part and the strict real-time part, according to the difference in the simulation step size. In the weak real-time part, driver operation and vehicle dynamic real-time simulation was connected by the CAN bus; In the strong real-time part, engine, generator and battery were modeled and simulated by two sets of RT-LAB software, and connected by the FlexRay bus. In the strict real-time part, the motor model and the inverter model were simulated in FPGA and DSP hardware, and connected by the parallel data bus.

1 INTRODUCTION

In future, the armored vehicle will be driven by an electric drive system, which is also the crucial basic of an all-electric battle vehicle. Compared with the traditional armored vehicle driven by the mechanic drive system, the electric driven system has an ideal drive, a better mechanical and a better steering performance. The transmission parts are connected by the wires flexibly. In addition, there is no constant ratio value between the motor and the driver wheel's speed; meanwhile, the motor is easier to control. It is also easier to realize the energy management and the control automation.

Previously, the simulations were restricted to the vehicle and its whole performance. Especially, the modeling of a motor and driving system are too simple to reflect the performance of an electric system. Moreover, the simulating tasks concentrate too much, which leads to the long simulation step size and low simulation accuracy. This paper adopted specific software to simulate a specific task based on distributed real-time simulation to realize the joint simulation between the vehicle dynamics and the motor vibration and electronic device.

2 ANALYSIS OF THE ELECTRIC DRIVE SIMULATION SYSTEM

2.1 *Layered modeling analysis*

The electric drive system is the driving system that can influence the vehicle's whole performance directly. If we study the electric drive system purely, it would not reflect the efficient performance of the system. Because of it, the vehicle model is usually established based on the real vehicle.

The devices inside the electric drive system have a large number of different areas such as electrical engineering, controlling, machinery and electro-magnetism. There is real-time simulation time constant of different components in different area steps over multiple levels. For example, the efficient simulation period of the motor and its controlling system is dozens of microseconds. However, the time in dynamic simulation is dozens of milliseconds. Hence, the simulating parts set a specific period according to its feature, and established a specific mode in different distributions and at different levels when establishing the real-time simulation system of an armored vehicle. Then, they exchange the data through different kinds of bus lines. According to the different simulation periods, the real-time simulation system can be divided into three parts, namely the weak real-time part, the strong real-time part and the strict real-time part, as shown in Figure 1.

2.2 *Real-time analysis of the simulation system*

In the distributed real-time simulation, each simulating task is divided into multi-computers to

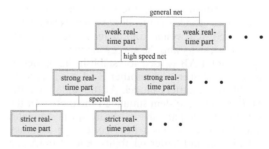

Figure 1.   Levels of the simulation system.

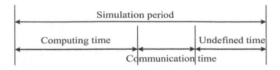

Figure 2. Time model of the real-time simulation.

simulate according to the independent feature. The real-time of the whole simulation is related to all the computers in the simulating group. The system must send the information from one computer to another to share the information and ensure the simulating security. The time model of the real-time simulation system is shown in Figure 2. The simulating step size is assumed to be T_p, which is given by

$$T_p = T_s + T_t + T_l \tag{1}$$

Here, T_s is the simulation time; T_t is the communication time; and T_l is the unsure time.

3 ESTABLISHMENT THE REAL-TIME SIMULATION SYSTEM

3.1 *Construction*

This real-time simulation system is divided into three parts to simulate in different software according to the weak real-time part, the strong real-time part and the strict real-time part. In the weak real-time part, the CAN bus line communication is adopted. In the strong real-time part, the FlexRay bus line communication is chosen. In the strict real-time part, the parallel data bus line communication is used.

In the weak real-time part, the driver's controlling system consists of the driver boilerplate and the signal collecting system. The system collects the signal of the accelerator pedal, brake pedal and steer. The real-time simulation of the vehicle's dynamic used Vortex to simulate. In this software, there is a full model of mechanical vehicle, but no electric drive system model. Hence, we have a secondary development in which we package the bottom function to form an electric drive model based on torque transmission. In the strong real-time part, the whole vehicle's controlling system is simulated in the dSPACE system and includes the drive controlling strategy and the energy management strategy. A RT-LAB is used to establish the model of motor-generator, and another RT-LAB is used to establish the battery model. At the same time, the DSP hardware system simulated the motor. In the strict real-time part, the hardware sprite simulates and calculates. It has a fast speed, high accuracy calculation and better reliability. Each FPGA can simulate two IGBT models in real-time simulation. The whole inverter is simulated with three FPGA.

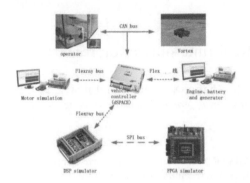

Figure 3. Structure of the electric drive simulation system.

3.2 *The operating principle of the system*

The basic operating principle of the real-time simulation system of the electric drive armored vehicle is that the driver's action signal is sent to the dSPACE simulation system through the CAN bus line after the AD exchange. In dSPACE, the energy management controlling strategy calculates the signal from the driver to deliver the controlling signal to the battery and rotate speed. This signal is sent to two RT-LAB computers through the FlexRay bus line. Meanwhile, the driving control strategy running in dSPACE will calculate the torque of the vehicle's driving motor. Then, the torque value is sent to the DSP motor simulation system through the FlexRay bus line. The DSP motor simulation system will calculate the PWM controlling signal of an inverter according to the given torque and the motor's speed signal. Then, the PWM controlling signal is sent to the FPGA simulation system through the parallel address bus line. After the simulation of an inverter simulated by the FPGA simulation system, the pulse voltage signal of the inverter's output is calculated. At the same time, the DSP motor simulation system calculates the pulse voltage to get the value of the input voltage of a motor and the real output torque. The torque is sent to the Vortex dynamic simulation system through dSPACE and added to the wheel as a directly driving torque.

4 SIMULATION AND ANALYSIS

4.1 *The real-time test of the simulation system*

The information on each simulation computer is shown as follows

The Vortex simulation computer: RAM 16 G, 12 cores processor, CPU 3.2 GHz.

The RT-LAB simulation computer: RAM 4 G, 6 cores processor, CPU 3.46 GHz.

The dSPACE simulation computer: DS1006 processor, CPU 2.2 GHz.

DSP: TMS320F2812, frequency of clock is 150 M.

FPGA: XC3S500E, the maximum frequency of clock is 305 M.

After the test, all the sections' simulation calculations time are given in Table 1.

In the three parts of communicating network, the data between FPGA and DSP connect with the stored device through the data address bus line. Hence, the maximum communicating delay is the period of DSP and FPGA. Its value is 6.7 ns and 3.3 ns. The FlexRay bus line in this paper has a 10 M baud rate and 40 M crystal oscillator. FlexRay used the time-division to transmit the data based on the recycle communication period. It can ensure the data arrived in time. In the protocol of FlexRay, the maximum value of cPropagationDelayMax is 2500 ns, which means that the maximum delay between the sections in the FlexRay system will not above 2500 ns when the system is working. In this paper, the CAN bus line network has four sections. Of these sections, three will feedback the data once they receive. The network delay is given in Table 2. The simulation platform has 14 messages and the maximum delay of the CAN bus line is 10.97 ms.

Substituting these values into Equation (1), it can be calculated that the simulation period T_p is 15 ms, and the minimum value of the unsure time T_{l_min} is 0.74 ms in the weak real-time part. The simulation period T_p is 40 μs, and the minimum value of the unsure time T_{l_min} is 12.97 μs in the strong real-time part. The simulation period T_p is 40 μs, and the minimum value of the unsure time T_{l_min} is 17.78 ns in the strict real-time part. The unsure time exists in every part of simulation, so that the real time of the whole system can be ensured.

4.2 Simulation results

The simulating object is the in-wheel armored vehicle driven by a motor, which is transformed from a certain kind of vehicle (8×8). There is an independent in-wheel motor inside each wheel, so that the mechanical connection between the wheels can be canceled. All the wheels can drive independently.

The time of simulation is set to be 25 s. The driver steps on the accelerator pedal at 1.3 s, as shown in Figure 4, to make the vehicle in a linear acceleration. The speed is shown in Figure 5. The acceleration is shown in Figure 6. The accelerating time from 0 km/h to 32 km/h is 4.7 s. The maximum value of the acceleration is 6.1 m/s2. The real vehicle's data is 5.5 s and 5.8 m/s2. The simulated data is almost equal to the real data. It also shows that the vehicle has a better performance in acceleration after its transformation because of the motor's faster response.

The motor simulation is taken as an example. In Figure 7 and Figure 8, it is the output torque curve and the rotate speed curve of the in-wheel motor. In Figure 9, it is the ideal output feature curve. At

Table 1. Simulation time of the sub-system.

Node	Simulation	T_p	$T_{s\,max}$	$T_{s\,min}$	$T_{s\,average}$
1	Dynamics simulation	15 ms	3.29 ms	1.97 ms	2.63 ms
2	Engine simulation	40 μs	2.02 μs	1.76 μs	1.83 μs
3	Generator simulation	40 μs	12.15 μs	10.36 μs	11.88 μs
4	Motor simulation	40 μs	24.53 μs	19.64 μs	22.37 μs
5	FPGA simulation	40 ns	15.52 ns	2.85 ns	8.58 ns

Table 2. Time delay of the CAN net.

Time delay	Section number		
	1	2	3
Number of messages			
5	2.29 ms	2.36 ms	2.23 ms
10	5.52 ms	5.68 ms	5.41 ms
15	9.75 ms	10.97 ms	9.63 ms
20	16.94 ms	16.63 ms	15.98 ms
25	24.36 ms	24.81 ms	24.03 ms

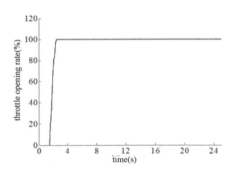

Figure 4. Throttle opening rate.

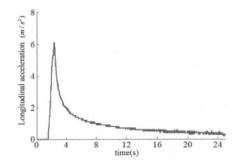

Figure 5. Longitudinal acceleration.

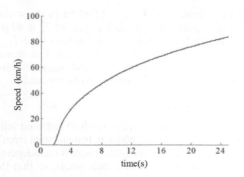

Figure 6.   Speed of the vehicle.

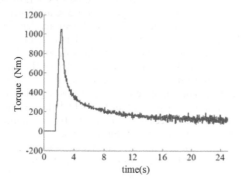

Figure 7.   Output torque of the third motor.

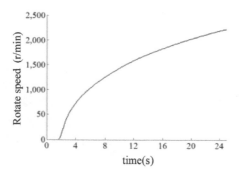

Figure 8.   Rotate speed of the third motor.

the beginning of deceleration, the improvement of the motor's torque equals to the accelerator pedal's improvement. It shows that the in-wheel motor has a fast response. In addition, this response fits in the real condition. When the accelerator pedal's aperture reaches 100%, the output torque can reach the peak value at 1100 Nm very fast. With the increasing motor's speed, the motor's output torque begins to decrease. These two variation trends fit in the motor's output feature. In Figure 7, it shows that the motor's output torque has a wave at 60 Nm.

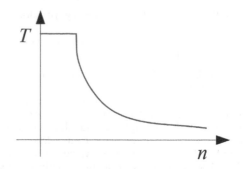

Figure 9.   Output characteristic of a normal motor.

5   CONCLUSIONS

In this paper, an armored vehicle electric driving real-time simulation system is divided into three parts, namely the weak real-time part, the strong real-time part and the strict real-time part, through the distributed real-time simulation environment, according to the difference in the simulation precision. The Can bus, FlexRay bus and parallel data bus are used in the data communication of the overall system. The simulation results for some electric drive vehicles show that this real-time simulation system runs well. It can move on to the combined simulation in terms of the vehicle performance, electric drive system performance and parts performance, and its real time meets the system simulation needs.

REFERENCES

Huang, Q. Huang, Y. Zhang, F.J. & Chen, Y.C. 2008. Co-simulation of Hybrid Power Plant of Electric Drive System for Armored Vehicles. *ACTA ARMAMEN-TARII*, 29(1): 11–14.
Li, B. Zhang, C.N. & Li, J.Q. 2009. Torque Control Strategies Based on RecurDyn and Simulink for Electric Drive Tracked Vehicle, *Transactions of the Chinese Society for Agricultural Machinery*, 40(7): 1–5.
Lu, L.J. Sun, F.C. & Zhai, L. 2006. Steering Performance Simulation for Electric Drive Tracke Vehicle Based on MATLAB SIMULINK, *Acta ArmamentarII*, 27(1): 69–74.
Wang, S.S. Zhang, Y.N. Yan, N.M. Zhao, Y.H. & Zhang, L. 2009. Simulation on Performance of Special Movement of Electric Transmission Tracked Vehicle Based on Virtual Prototyping, *Acta ArmamentarII*, 30(11): 1418–1422.
Zhang, C.N. Wu, J.B. Zou, Y. & Li, J.Q. 2009. Hardware-in-the-Loop Simulation on Control Strategy in Hybrid Electric Transmission Tracked Vehicle, *Transactions of Beijing Institute of Technology*, 29(9): 790–794.
Zou, Y. Sun, F.C. & Zhang, C.N. 2007. Electric Tracked Vehicle Real-time Simulation of Dual-motor Driving Control with Driver-global Controller In-loop, *Chinese Journal Of Mechanical Engineering*, 43(3): 193–197.

Advanced Materials and Structural Engineering – Hu (Ed.)
© 2016 Taylor & Francis Group, London, ISBN 978-1-138-02786-2

A research on the optimization of an anti-whiplash injury seat system

S.W. Zhou, S. Han, C. Sun, Y.R. Shi & C. Zhang
College of Mechanical Engineering and Automation, Northeastern University, Shenyang, China

ABSTRACT: The whiplash injury, which is hard to recover, is a neck injury resulting from the acceleration of automobile collision. The article analyzes the damage and the incidence of the whiplash injury, and conducts an in-depth study on the principle of injury and evaluation criterion. Based on the evaluation criterion, this paper analyzes the structure of the anti-whiplash injury, puts forward the optimizing strategy, and summarizes the basic method of the anti-whiplash injury and related theory, verifying the accuracy of the optimization strategy. The result of the study indicates that the optimization strategy can reduce the incidence of the whiplash injury, which increases the safety and stability of the automobile.

1 INTRODUCTION

The whiplash injury is a special spine and spinal cord injury, which is caused by physical violent acceleration or deceleration movement while the head movement is not synchronized, causing spinal cord injury by continuous and excessive flexion of the cervical spine (Luo et al. 2014). Rear-end collision is the main factor causing cervical spine injury. In the rear-end process, the seat drives torso moved forward with the impacting vehicle. While the head and neck are relatively free and under the force of inertia, the head moves backward, and the head and neck horizontal displacement results in the whiplash injury. In a variety of traffic accidents, neck injury is as high as 70%. While in the rear-end collision, the probability of crew neck injury is higher. Approximately 90% is caused by low speed collision ($\Delta V \leq 25$ km/h) (Li & Qian 1989). Currently, the incidence of the rear-end collision is increasing year by year. The traffic accident data analysis in 2006 showed that in the total number of traffic accidents, the frontal crash accounted for 18.5%, side crash accounted for 27%, and the frequency of the rear-end collision was just lower than the side collision, whose ratio was 25.5%. At the same time, in highway accidents, the frequency of the rear-end accident was much higher than the other types of collision, whose ratio was 44.86% (Shi 2006). The property loss caused by rear-end accidents in 2008 was 250 million yuan, accounting for 25.12% of the total accident loss, which was higher than the frontal crash (21.44%) and side crash (24.39%).

This article is based on the evaluation criteria of whiplash injury and proposes the optimization strategy, analyzing and comparing the anti-whiplash injury measures, and then having a further cognition about the accuracy of optimization.

2 THE EVALUATION CRITERIA AND OPTIMIZATION STRATEGY OF WHIPLASH INJURY

"C-NCAP Management Code (2012)" adds the flog test to the original project, which models the original car's seat belt restraint system and makes a dummy to sit on the seat that is fixedly mounted on a mobile trolley, which is launched at the 16.65 km/h acceleration waveform specific to the simulated rear-end collision. The extent of damage to the rear-end collision is detected on the crew neck (Li 2014). A modified version of the 2012 makes it difficult to obtain the five-star evaluation and the difficulty is increased by about 20%, which reflects the protection of the vehicle when the accidents occurred more fully and exactly. Compared with the 2012 edition, the 2015 version of the rules will pay more attention to four fields including active safety collision shape, measuring results, and safety assessment.

Based on the considerations of the safety of the seat, the structure of the seat includes the backrest, the headrest, the seat cushion, and the seat assembly fixing member connected to the body (Yao & Sun 2002). For mitigating flog injury damage, improving the headrests is the most direct way, and the back and headrest integrated design is the most effective way (such as Volvo's WHIPS whiplash protection system). Backrest includes flexible energy-absorbing backrest and rigid energy-absorbing backrest. Flexible energy-absorbing backrest has a significant effect to mitigate flog injury in the event of low-intensity tail hit, but has bad effects on the energy-absorbing capacity of occupant restraint for high-strength flexible collision and causes damage to the passenger. Headrest includes inductive headrest and the pre-responsive headrest. Inductive headrest means that in rear-end collisions, the passengers' weight on the seat makes headrest move forward and upward automatically (such as Volvo's WHIPS whiplash protection system). Pre-responsive headrest is that the car speed sensor and the acceleration sensor, which is installed in the rear of the car detecting a rear-end collision, sends move forward and upward instructions (e.g. Benz NECK-PRO crash pre-response headrest Technology).

3 PERFORMANCE COMPARISON BETWEEN THE EXISTING ANTI-WHIPLASH INJURY MECHANISMS

With the further research on the safety of the automobile, different automobile safety designs have been proposed. According to the different principles, it can be divided into the headrest protection and the other protections. According to the different power of the devices, it can be divided into the purely mechanical structure and the mechatronics structure. According to the different response modes, it can be divided into traditional passive protection and proactive protection. A description of the several existing safety seats is provided below.

3.1 The active headrest protection

According to the existing research, when the automobile is subjected to the rear collision, the most important measure to prevent the occurrence of the whiplash injury is to reduce the distance between the headrest and the head. The principle of the active safety headrest is that the movable headrest portion quickly moves forward and upward in the specific mechanical structure to catch the passenger's head to prevent its strong whiplash, when the vehicle detected the collision is impending or the collision has occurred, it not only ensures the comfort of the occupants, but also further protects the cervical of the occupants.

According to the research of the automotive safety, there are two kinds of active headrest: one is purely mechanical mechanism; the other is a mechatronic mechanism (Jin 2011).

Pure mechanical headrests depend on physical inertia drive through car crash. By triggering spring extending or special linkage pushing headrest forward, holding passengers heads, preventing backward whiplash, it does not need specific sensors. Its energy source of the structure is from physical inertia. Nowadays, there mainly exists two mechanical headrest as follows (Lin et al. 2011).

An active headrest in the patent (Fig. 1) (Changsha Lizhong automotive design company 2013)

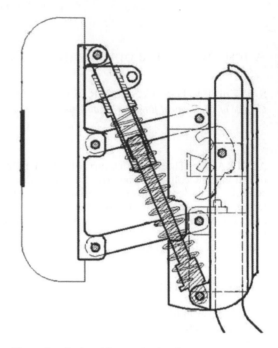

Figure 1. Spring-driven active headrest.

depends on the body shock trigger mechanical structure, striking the limited spring structure in the headrest, popping headrest, decreasing the distance between the headrest and the head, and reducing the effect of backward whiplash. Though the equipment's movement is reversible, it cannot renew the automobile without an outside force.

Another situation is shown in Figure 2, when the rear-end collision happens, the body will move backward due to inertia. With the linkage between the back and the headrest and pushing the headrest forward and decreasing the damage, the time of equipment reaction is longer than the former. Also, whiplash may occur before the headrest works. What's more, the result will change when the shock inertia changes.

Another active headrest is equipped with the mechatronics structure, while the active headrests combine the passive protection with the active protection, through the detection of road conditions. It determines whether to activate Protection Agency. It can collect the relative velocity and acceleration between cars by speed and acceleration sensor. In addition, a particular computer control system makes a rapid judgment whether there will be a rear-end according to the real-time collection of traffic information. At the same time, it will convey the information to perform the module, which can start the occupant protection device to protect the safety of the crew, according to the information

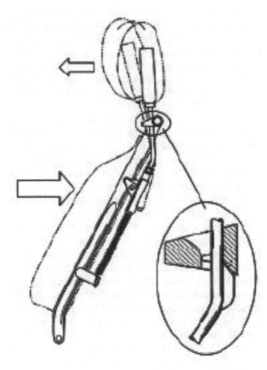

Figure 2. Active headrest of the spatial linkage mechanism.

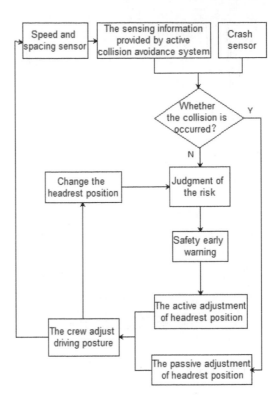

Figure 3. Flow chart of the active protection system.

passed by the judgment module. The flow chart of the system's response is shown in Figure 3.

Retractable active headrest mechanism should meet the following requirements: ① Stretching body should have a certain direction guidance; ② in the rear-end collisions, it can withstand loads applied thereto by the occupant's head; ③ its movement is reversible. According to the principle, there are two schemes about active headrest designs.

There is an active-headrest mechanism used for protecting the neck from the rear collision in the patent (Tsinghua University 2009) (Fig. 4). Its movable part is popped-up by the specific cross-connecting rod mechanism derived by the motor. According to the detects and judgments of the Automotive Safety Active-protection System to traffic and environment, the equipment will unfold the connecting rod mechanism just before the accident. In order to decrease the distance between the head and the headrest and avoid the whiplash of the neck, the connecting rod mechanism will be closed up after the situation. The merit of this equipment is that the more the connecting rod mechanism unfolds, the faster the headrest moves during the stronger load capacity it has. However, the headrest can only move forward and it can adjust the suitable pop-up angle to prop up the

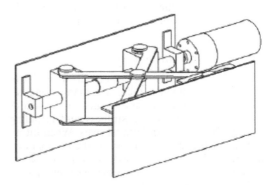

Figure 4. Motor drive active headrest.

head accurately, and, as a result, avoid the whiplash injury.

The second anti-whiplash head restraints hurt structure (Fig. 5) includes motorized faders, adjustable slide and pop pillow. There is a telescoping control unit in the headrest, which can response fast before rear-end. Besides, its drive expansion protection pillow extends a certain distance rapidly (the process will be stopped immediately if it contacts the head). So, it reduces the distance between

711

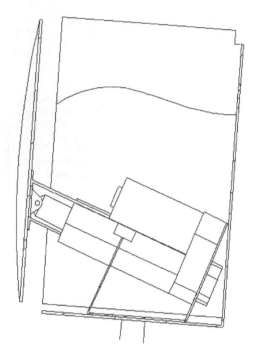

Figure 5. Electric active headrest.

the headrest while it reduces the head thrown back whip effect. Compared with the previous device, it can freely adjust the angle of the headrest pop. In a word, the protection is more accurate.

3.2 Other protective measures

In addition to the above-described initiative headrest adjustments, there are many measures to prevent whiplash injury, such as measures by changing the recliner stiffness and installing energy-absorbing translational seats. When hit by the car in the tail, the inertia of the occupant or a part of the car can provide force for the seats, which makes the preventing whiplash injury mechanism of car seat to take timely action to protect the neck of the occupant from injury. The following describes the protective principle of the translational energy-absorbing seat.

Translational energy-absorbing seat means that during the event of a collision at a low speed, the car seat can be moved along the fixed rail to a certain distance. Because of overcoming the resistance, it can absorb some of the energy. The core design is the design of the energy-absorbing mechanism (Zhang 2013). The main consideration factor of the design energy-absorbing mechanism is to make the energy-absorbing element absorb equal energy decreased best. Based on the design

requirements of translational energy-absorbing seat, it can be ensured that the energy absorption device can apply to the most current automobile. It only adds the translational energy-absorbing member on the rail part while the other parts of the seat do not change. The following briefly describe two common devices used to absorb energy.

The first is energy absorption by pipe crack (Fig. 6). First, round steel pipe in the axial direction is fixed on the guide rail above relative body position fixing, and rigid conical pipe fixed on the guide rail can move relative to the body. Thus, when the automobile rear-end collision occurs, with the backward of translational seat, the deformation of pipe is bracing crack and can provide a good platform force. In the design process, we need to make guide groove at one end processing failure. Thus, the circular pipe tensile stress failure is roughly along the axial direction of the pipe (Luo & Zhou 2010), crack dehiscence along a predetermined direction in order to ensure the structure safety and stability. The scheme can be in light of different vehicle parameters, such as vehicle load, the size of the space and car cockpit speed. Through the simulation experiments of finite element analysis, a reasonable set of round and ultimate tensile strength of deformation of thin steel pipe diameter, thickness, factors as well as the size and rigidity of tapered tube between the friction coefficients, in order to change the tear force size, thus changing the translational displacement of the vehicle.

The second is the energy absorption bending steel plate. Energy absorption bending steel plate works when the tail hit occurs, and the moving seat pulls the sheet steel around a fixed cylinder bending tensile force. So as to obtain the stable platform force, its structure is shown in Figure 6. A fixed circular tube and a fixed limit block limit the movement of the steel sheet, traction cylinder and the translational rail connected. The advantage of the structure is from the platform force, it can be increased rapidly on the steel plate bending, after stability, steel sheet moving distance relative to the seat buffering process. Its moving distance can be ignored. So, it can be considered that the

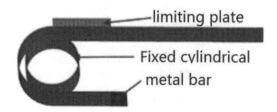

Figure 6. Steel plate bending energy-absorbing device.

platform force remains constant throughout the whole process.

4 CONCLUSIONS

With the development of science technology and the improvement of life, we have a higher request for the safety of the car. The safety of the car chair has attracted attention because our passengers' safety directly depends on the safety of the car chair. By reducing the distance from the head and headrest and increasing the rebound damping of the backs, the car safety chair can realize the function of the anti-whiplash injury. In the real life situation, active protection and passive protection can be combined. It can reduce the rate of the whiplash injury. Therefore, it is important for the development or production of the automobile.

REFERENCES

Changsha Lizhong automotive design company. 2013. *An active safety headrest device*. China. 201310041564.5.

Jin, J.X. 2011. *A research and improvement of automobile head restraint in whiplash*. Jilin University: 27–40.

Li, F. 2014. A Study on C—NCAP. *China Consumer News* 2014.1.10(B01).

Li, M.Q. & Qian, G. 1989. *The Identification of Damage Grade Vehicle Rear Impact*. BeiJing: China Communications Press.

Lin, Z. Du, H.L. & Zhou, Q. 2011. Research on Structural Design of Headrest in Rear End Collision. *Automobile Technology*: 4: 1–4.

Luo, J.L. Wang, A.L. Cai, J.C. Hu, S.Y. Gong, Z.Y. & Ye, H.Y. 2014. Clinical Observation of Whiplash Injury. *Chinese Journal of Traditional Medical Science and Technology* 2014(02).

Luo, M. & Zhou, Q. 2010. A vehicle seat design concept for reducing whiplash injury risk in low-speed rear impact. *Int. J. Crashworthiness*, 2010.

Shi, H.P. 2006. The traffic accident statistics report of 2006.

Tsinghua University. 2009. *Research on Structural Design of Active Headrest in Rear End Collision*; China. 200710175261.7.

Yao, W.M. & Sun, D.D. 2002. Safety performance in seat design is introduced. *Automobile Technology*. 2002(08): 5–8.

Zhang, X.W. 2013. *Analysis of Energy-Absorbing Sliding Seat Properties for Neck Injury Protection in Rear Impact*. Tsinghua University.

Advanced Materials and Structural Engineering – Hu (Ed.)
© 2016 Taylor & Francis Group, London, ISBN 978-1-138-02786-2

Calculation of a thick-walled inhomogeneous cylinder of a nonlinear-elastic material

V.I. Andreev & L.S. Polyakova
Moscow State University of Civil Engineering, Moscow, Russia

ABSTRACT: This paper considers the problem of the nonlinear theory of elasticity for an inhomogeneous thick-walled cylinder. The problem is solved in axisymmetric statement. In general, all the parameters of the nonlinear dependence between the intensities of the stresses and strains are functions of the radius. As an example, this problem solves the problem of stress distribution in the soil array with a cylindrical cavity.

1 INTRODUCTION

The development of methods for solving problems for physically nonlinear bodies is the accounting of inhomogeneity of their mechanical characteristics. If the parameters of a non-linear diagram $\sigma_i = f(\varepsilon_i)$ are continuously changing along the coordinate functions, the problems for such bodies should be classified as nonlinear inhomogeneous. The physical basis of touching upon such problems is the dependence of the mechanical properties of real materials on various factors. Along with the modulus of elasticity and Poisson's ratio, there may be variable ultimate strength σ_u and ultimate strain ε_u. In this case, for example, the non-linear relationship between stresses and strains is given by (Lukash 1978)

$$\sigma_i = f(\varepsilon_i) = E\varepsilon_i - A\varepsilon_i^\alpha \qquad (1)$$

We can express the parameter A, α through σ_u and ε_u that allows us to set A and α depending on the coordinates.

2 STATEMENT OF THE PROBLEM

Let us consider an axisymmetric problem of the equilibrium of a thick-walled cylinder, the behavior of the material at each point described by the Equation (1), and all three parameters of the diagram are arbitrary functions of the radius. Let the hollow cylinder, the inner and outer radii of which are denoted by a and b, be loaded with a uniform internal and external pressures p_a and p_b. We assume that the initial ratio of transverse strain v; in other words, we use the hypothesis about incompressibility of the material. We also assume

that the moment of the destruction corresponds to an extreme point on the diagram $\sigma_i - \varepsilon_i$ (Fig. 1). This allows using the condition:

When $\varepsilon_i = \varepsilon_u \to d\sigma_i/d\varepsilon_i = 0$, we can obtain the following equation:

$$\alpha = \frac{E\varepsilon_u}{E\varepsilon_u - \sigma_u}; \; A = \frac{E\varepsilon_u - \sigma_u}{\varepsilon_u^\alpha}. \qquad (2)$$

Furthermore, for α, it possible to introduce another definition as follows:

$$\alpha = 1/(1 - E_{sec,\,u}/E) \qquad (3)$$

where $E_{sec,\,u}/E = \sigma_u/\varepsilon_u$ is the secant modulus at the destruction point (Fig. 1).

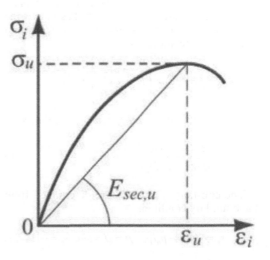

Figure 1. Determination of the parameters of the diagram.

3 SOLUTION FOR THE HOMOGENEOUS MATERIAL

Given that $\varepsilon_z = 0$ and $\sigma_z = v(\sigma_r + \sigma_\theta) = 0.5\,(\sigma_r + \sigma_\theta)$ using the expression for the intensities of the stress and strain in cylindrical coordinates, we get

$$\sigma_i = \frac{\sqrt{3}}{2}(\sigma_r - \sigma_\theta); \qquad \varepsilon_i = \frac{2}{3}\sqrt{\varepsilon_r^2 - \varepsilon_r\varepsilon_\theta + \varepsilon_\theta^2} \qquad (4)$$

The assumption of an incompressible material, i.e. the lack of volume deformations ($\varepsilon_r + \varepsilon_\theta + \varepsilon_z = 0$), we get $\varepsilon_r = -\varepsilon_\theta$, which allows us to integrate the condition of compatibility of strains as follows:

$$(\varepsilon_r = -\varepsilon_\theta)$$
$$d\varepsilon_\theta / dr = (\varepsilon_r - \varepsilon_\theta)/r = -2\varepsilon_\theta/r,$$

and

$$\varepsilon_\theta = B/r^2; \qquad \varepsilon_r = -B/r^2,$$

where B is the constant of integration.

From the equilibrium equation:

$$\frac{d\sigma_r}{dr} + \frac{\sigma_r - \sigma_\theta}{r} = 0$$

Using (4) and (1), we obtain

$$\frac{d\sigma_r}{dr} = -\frac{2\chi}{\sqrt{3}\,r}\left(\frac{2E\cdot|B|}{\sqrt{3}r^2} - \frac{A}{r^{2\alpha}}\cdot\left(\frac{|2B|}{\sqrt{3}}\right)\right) \qquad (5)$$

Here, to maintain the symmetry diagram $\sigma_i - \varepsilon_i$ in tension and compression, we enter the absolute values and parameter χ = sign ε_i. Passing to the dimensionless quantities $\rho = r/a$, $s_r = \sigma_r/p_b$ and introducing the notations:

$$K = Ap_b^{\alpha-1}/E^\alpha; C_1 = \left(E\cdot\left|\frac{2B}{\sqrt{3}}\right|\right)\Big/\left(p_b a^2\right)$$

The solution of (5) can be written as follows:

$$s_r = C_2 + \frac{\chi C_1}{\sqrt{3}\rho^2} - \frac{\chi K C_1^\alpha}{\alpha\sqrt{3}\rho^{2\alpha}} \qquad (6)$$

The constants C_1 and C_2 are determined from the boundary conditions:

$$\rho = 1, \ s_r = -\omega = -p_a/p_b; \ \rho = \beta = b/a, \ s_r = -1. \quad (7)$$

Substituting (6) into (7), we obtain a system of two nonlinear equations of degree α relatively

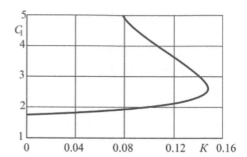

Figure 2. The dependence of the constant C_1 on the parameter K.

constants C_1 and C_2, which is generally solved numerically.

We then consider a particular case: the problem of the stress concentration near the cylindrical hole in an infinite array subjected to hydrostatic pressure $p_b = p$. Assuming $\beta = \infty$, $\omega = 0$, $\chi = +1$ [*], we find $C_2 = -1$ and for C_1 we obtain the equation:

$$\frac{K}{\sqrt{3}\alpha}C_1^\alpha - \frac{1}{\sqrt{3}}C_1 + 1 = 0 \qquad (8)$$

Thus, a constant C_1 is dependent on α, i.e. on the ratio $E_{sec,u}/E$ and the parameter K, which in turn α Equation (8) gives several solutions from which depends on the external pressure p and the tensile strength of the material σ_u. For some values, the one that converges to the solution for a linear-elastic material should be chosen.

Note that the transition to the linear problem is if we put $K = 0$, which corresponds to $A = 0$ or $E_{sec,u} = E$. In accordance with (3), $\alpha \to \infty$.

However, the transition $K \to 0$ can be viewed differently, considering that $E_{sec,u}/E = const$ ($\alpha = const$), and $\sigma_u \to \infty$ that for small values, ε closer the non-linear diagram to linear. Illustrating the above, in Figure 2 shows two branches of positive dependence of C_1 on K with $\alpha = 3$, which corresponds to the $E_{sec,u}/E = 2/3$. It is clear that the lower branch—$C_1^{(1)}$ is right, as at $K \to 0$ it provides to the solution corresponding the linear problem. This is easily verified by calculating the stresses σ_θ on the holes contour. Their dimensionless values are calculated according to the formula:

$$s_\theta = \frac{\sigma_\theta}{p} = C_2 - \frac{C_1}{\sqrt{3}\rho^2} + \frac{KC_1^\alpha(2\alpha-1)}{\alpha\sqrt{3}\rho^{2\alpha}}, \qquad (9)$$

For the considered case, we have

---

[*] In the compression array movements along a radius $u < 0$ from which it follows that $\varepsilon_\theta = u/r < 0$, and consequently $\varepsilon_i > 0$.

716

$$s_\theta(\rho=1) = -1 - \frac{C_1}{\sqrt{3}} + \frac{5KC_1^3}{3\sqrt{3}}.$$

It is easy to verify that if $K \to 0$ and $C_1 = \sqrt{3}$, the value $s_\theta = -2$, which corresponds to the known result for a linear-elastic problem (Andreev & Malashkin 1983). Figure 3 shows the dependence of C_1 on the ratio $E_{sec,u}/E$ for three load levels, determined by the ratio of p/σ_u. Figure 4 shows the diagrams of the dimensionless stresses s_θ calculated using Formula (9) with $p/\sigma_u = 0.5$ and for different values of the ratio $E_{sec,u}/E$. The dotted line shows the solution

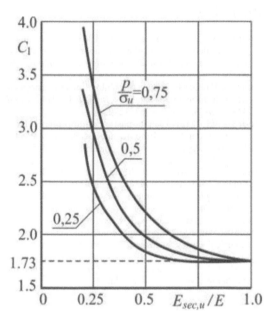

Figure 3. The dependence of the constant C_1 on the parameters of the diagram $\sigma_i - \varepsilon_i$.

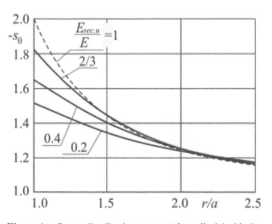

Figure 4. Stress distribution σ_θ near the cylindrical hole.

for a linear-elastic material. Note that with the increase in non-linearity (reduction ratio $E_{sec,u}/E$), there will be a decrease in stresses when compared with the elastic problem, which becomes more significant. A similar result is obtained by solving elastic-plastic problems (Andreev 2013a, b).

4 SOLUTION FOR THE INHOMOGENEOUS MATERIAL

The transition to a problem in which the non-linearity is taken into account along with inhomogeneity of the material is related to the replacement of the constants E, A and α involved in (1) for functions $E(r)$, $A(r)$ and $\alpha(r)$. Taking into account (2), the latter two dependence is due to the changes in physical characteristics $\sigma_u(r)$ and $\varepsilon_u(r)$. For the function $E(r)$, we will use the dependence form (Andreev 2013a), which can be written as follows:

$$E(r) = E_0[1 + (k_E - 1)(a/r)^{m_E}] \qquad (10)$$

where notations k_E and m_E emphasize that they relate to the function of the inhomogeneity $E(r)$.

Considering the problem of the stress concentration near the cylindrical hole, the dependence of the type (10), due to its local character, can also be used to describe the functions $\sigma_u(r)$ and $\varepsilon_u(r)$ with appropriate constants:

$$\sigma_u(r) = \sigma_{u,0}[1 + (k_\sigma - 1)(a/r)^{m_\sigma}];$$
$$\varepsilon_u(r) = \varepsilon_{u,0}[1 + (k_\varepsilon - 1)(a/r)^{m_\varepsilon}].$$

If we substitute these relations in (2), the resulting functions will be so complex that we can obtain the solution only numerically. Next, we consider some special cases when it is possible to obtain an analytical solution.

If, for example, $\varepsilon_u = \varepsilon_{u0} = const$, $k_E = k_\sigma$ and $m_E = m_\sigma$, then

$$A(r) = \frac{E_0\varepsilon_u - \sigma_{u,0}}{\varepsilon_{u,0}^\alpha}\left[1 + (k_E - 1)(a/r)^{m_E}\right]$$

and Equation (5) can be easily integrated.

Also, we can find a solution if $\alpha = const$ and $A(r) = A_0\left[1 + (k_A - 1)(a/r)^{m_A}\right]$. This case does not correspond to a specific dependencies $\sigma_u(r)$ and $\varepsilon_u(r)$, but may be obtained by the curve fit $A(r)$. The solution in dimensionless stresses is then given by

$$s_\theta = C_2 - \frac{C_1}{\sqrt{3}\rho^2} + \frac{2C_1(1-k_E)(m_E+1)}{\sqrt{3}(m_E+2)\rho^{m_E+2}}$$
$$- \frac{(1-2\alpha)KC_1^\alpha}{\sqrt{3}\alpha\rho^{3\alpha}} + \frac{2(2\alpha-1+m_E)K(1-k_E)C_1^\alpha}{\sqrt{3}(2\alpha+m_E)\rho^{2\alpha+m_E}},$$

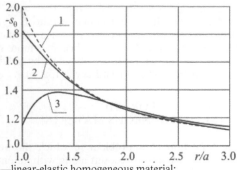

1—linear-elastic homogeneous material;
2—nonlinear elastic homogeneous material; and
3—nonlinear elastic inhomogeneous material.

Figure 5. Stress distribution in the array with a cylindrical hole.

where $K = A_0 p_b^{\alpha-1}/E_0^\alpha$. Constants C_1 and C_2 as usual are determined from boundary conditions (7). Figure 5 shows a diagram $s_\theta = \sigma_\theta/p$ (lower curve) for the case $\alpha = 3$; $k_E = k_A = 0.5$; $m_E = m_A = 2$ and boundary conditions: $p_a = 0$; $p_b(b \to \infty) = p = 0.5\sigma_{u,0}$.

For comparison, the same figure shows the diagrams for linear-elastic homogeneous and inhomogeneous materials, as well as for the nonlinear elastic homogeneous material. It may be noted that in this example, taking into account the nonlinearity and the inhomogeneity leads qualitatively to the same effect, reducing the stresses near the contour of the cavity. For other values of constants of inhomogeneity, for example, in the case of $K_E > 1$, stresses on cavities circuit may increase.

5 CONCLUSIONS

The resulting solution can be used to calculate cylindrical pressure vessels operating under a high temperature gradient and for determining stresses in the design and construction of subterranean wells. In the latter case, given that the Poisson's ratio of the soil can be close to 0.5, the neglect of the asymmetry of earth pressure has no effect on maximum stress. The results obtained are consistent with those reported previously (Andreev 2013c, Andreev et al. 2014).

REFERENCES

Andreev, V.I. & Malashkin, Yu. N. 1983. Calculation of thick-walled pipe from the nonlinear elastic material. *Building mechanics and calculation of structures.* 6: 70–72.

Andreev, V.I. 2013a. About the Unloading in Elastoplastic In-homogeneous Bodies. *Applied Mechanics and Materials.* 353–356: 1267–1270.

Andreev, V.I. 2013b. Elastic-plastic Equilibrium of a Hollow Cylinder from Inhomogeneous Perfectly Plastic Material. *Applied Mechanics and Materials.* 405–408: 3182–3185.

Andreev, V.I. 2013c. Equilibrium of a Thick-walled Sphere of Inhomogeneous Nonlinear-elastic Material. *Applied Mechanics and Materials,* 423–426: 1670–1674.

Andreev, V.I., Avershyev A.S. & Jemioło S. 2014. Elastic-plastic state of inhomogeneous soil array with a spherical cavity. *Advanced Materials Research.* 842: 462–465.

Lukash, P.A. 1978. *Fundamentals of nonlinear structural mechanics.* Moscow: Stroyizdat.

Advanced Materials and Structural Engineering – Hu (Ed.)
© *2016 Taylor & Francis Group, London, ISBN 978-1-138-02786-2*

Nonlinear modeling of the kinetics of thermal stresses in polymer rods

V.I. Andreev & R.A. Turusov
Moscow State University of Civil Engineering, Moscow, Russia

ABSTRACT: Articles (Turusov & Andreev 2014) have presented the results of experimental studies of thermal stresses in the polymer rods in their quasi-static uniform heating or cooling. The experiments revealed the following effects: thermal stress quasi-irreversibility at cyclical changing of the temperature, the growth stresses during isothermal stress relaxation, and extremum on the curves of isothermal relaxation. In this paper, to study the kinetics of thermal stresses in polymers, analysis of the impact of various factors on the relaxation processes and modeling of the relaxation behavior, we use a physically nonlinear constitutive equation in differential form.

1 INTRODUCTION

In solving problems, the following assumptions are introduced:

In the rod performed uniaxial stress state, $\sigma_x \neq 0$;

After the occurrence of thermal stresses, plane cross-sections remain plane;

Increase in the total deformation of the rod $d\varepsilon_x$ consists of the growth of elastic deformation $de_x = d\sigma_x/E$, induced rubbery (viscoelastic) deformation $d\varepsilon_x^*$ and temperature deformation $d\varepsilon_{xT}$:

$$d\varepsilon_x = de_x + d\varepsilon_x^* + d\varepsilon_{xT} \qquad (1)$$

Volumetric deformations have a little effect on the rate of the relaxation process.

Because of these assumptions for these problems, a generalized nonlinear Maxwell's equation for highly elastic deformation takes the form (Gurevich 1974):

$$\frac{\partial \varepsilon_{xs}^*}{\partial t} = \frac{\sigma_x - E_{\infty s}\varepsilon_{xs}^*}{\eta_{0s}^*} \exp\left(\frac{\left|\sigma_x - E_{\infty s}\varepsilon_{xs}^*\right|}{m_s^*}\right) \qquad (2)$$

where ε_{xs}^* $(s = 1,2)$ is the components of nonlinear viscoelastic deformation, corresponding to the two members of the spectrum relaxation times; η_{0s}^* are the coefficients of the initial relaxation viscosity; $E_{\infty s}$ is the high elasticity modules; and m_s is the speed modules.

Dependence of the relaxation viscosity on the stresses determines the non-linearity of the problem.

For the increment, the total strain in accordance with (8.1) we have

$$d\varepsilon_x = \frac{d\sigma_x}{E} + \sum_1^2 d\varepsilon_{xs}^* + \alpha dT, (T = T(x,t)) \qquad (3)$$

In general, $T = T(x,t)$, the equilibrium equation and the Cauchy formula are in the form:

$$\partial\sigma_x/\partial x; \quad \varepsilon_x = \partial u/\partial x \qquad (4)$$

Thus, for unknown $\sigma_x, \varepsilon_x, \varepsilon_x^*, u$ obtained the complete system of Equations (2)–(4), which describes the stress-strain state of the polymer rod exposed to temperature effects.

Assuming that the polymer rod before the experience is located in the unstressed state, we can write the following initial conditions:

$$t = 0: \quad \sigma_x = 0; \quad \varepsilon_{xs}^* = 0 \, (s = 1, 2); \quad T = T_0 \qquad (5)$$

Boundary conditions—absence of displacements at the ends of the fixed rod are in the form:

$$t \geq 0 \, \left.u\right|_{x=0} = \left.u\right|_{x=l}. \qquad (6)$$

2 THE HOMOGENEOUS CHANGE IN TEMPERATURE OF THE ROD

Here, we consider the change in temperature of the rod in time according to the formula: $T = T(t) = T_0 + kt$. Because of the nonlinearity of Equation (2) for the integration of Equations (2)–(4) with the initial conditions (5) and condition (6) is used step method of integration when $t_i = t_{i-1} + \Delta t$. As a result, we obtain a system of algebraic equations:

$$\Delta T = k\Delta t; \qquad \Delta\sigma_i = -E_i\left(\Delta\varepsilon_i^* + \alpha_i\Delta T\right);$$

$$\Delta\varepsilon_i^* = \sum_{s=1}^{n}\Delta\varepsilon_{si}^*; \qquad \Delta\varepsilon_{si}^* = \left(\frac{d\varepsilon_s^*}{dt}\right)_{i-1}\Delta t; \qquad (7)$$

$$\varepsilon_i^* = \varepsilon_{i-1}^* + \Delta\varepsilon_i^*; \qquad \sigma_i = \sigma_{i-1} + \Delta\sigma_i;$$

$$\left(\frac{d\varepsilon_s^*}{dt}\right)_i = \frac{\sigma_i - E_{\infty s,i}\varepsilon_{si}^*}{\eta_{0s,i}^*}\exp\left|\frac{\sigma_i - E_{\infty s,i}\varepsilon_{si}^*}{m_{si}^*}\right|.$$

The length of the rod was taken as $l = 50$ mm, and the time step is varied depending on the speed of the process, $\Delta x = 1$ mm.

Note that in the studied polymers, all parameters (elastic, viscoelastic, and thermophysical) are functions of the temperature.

In this paper (Babich 1966), results on how the mechanical properties of hard reticulated epoxy polymer in EDT-10 depend on the temperature are obtained. Here, the relevant formulas are as follows:

$$E = 4000\exp\left(-\exp\frac{T-339}{36.7}\right)\text{MPa};$$

$$\alpha = \begin{cases} 10^{-6}\left(0.46T - 58\right)\dfrac{1}{K} & \text{for } T \le 350K; \\ 10^{-6}\left\{102.3 + 110\exp\left[-\exp(6.3 - 0{,}04T)\right]\right\}\dfrac{1}{T} \\ \qquad \text{for } T \ge 350K; \end{cases}$$

$$E_{\infty 1} = \begin{cases} \left(2.4\cdot 10^6\,\dfrac{1}{T} - 6120\right)\text{MPa} & \text{for } T \le 370K; \\ \left(2.23T - 640\right)\text{MPa} & \text{for } T \ge 370K; \end{cases}$$

$$E_{\infty 2} = 0.1E_{\infty 1};$$

$$\eta_{01}^* = 36000\exp(9500/T - 20)\,\text{MPa}\cdot\text{s};$$

$$\eta_{02}^* = 36000\exp(35400/T - 90)\,\text{MPa}\cdot\text{s};$$

$$m_1^* = m_2^* = \left(-0.0155T + 7.73\right)\text{MPa};$$

$$\nu = 0.37 = const. \qquad (8)$$

Here, constants are obtained empirically. For a relatively short period of time and stresses enough account of "senior" member of the spectrum rubbery deformation ($i = 1$). This proved to be true with respect to the experiments with thermal stresses and isothermal relaxation.

All cross-sections of the rod at each time point are under the same conditions of temperature, and from conditions (6), we obtain an algebraic expression:

$$\varepsilon_x = \frac{\sigma_x}{E} + \varepsilon_{x.1}^* + \varepsilon_{x.2}^* + \alpha\Delta T = 0. \qquad (9)$$

The calculations were made for the cyclic temperature conditions. At each half cycle, the temperature was varied with time according to a linear law:

from T_0 to T_{fin} ($T_0 = 38°C$ $T_{fin} = 110°C$).

$$T_0 \nearrow T_{fin}: \quad T = T_0 + kt;$$
$$T_{fin} \searrow T_0: \quad T = T_k + k(t - \tau) \quad \text{and so on}$$

Results of the solution for the two rates of change in temperature $k = 4$ deg/min and $k = 0, 4$ deg/min when the whole process begins with heating from T_0 to T_{fin} be shown in the graphs (Fig. 1 a,b) in the coordinates temperature-stress.

As can be seen from Figure 1, the characteristic of the theoretical curves is similar to that of the experimental curves obtained in a previous report (Turusov & Andreev 2014).

The effect of the cooling rate on the value of thermal stresses in the polymer rod manifested at low temperatures, although the rate of relaxation processes are close to zero.

Figure 2 for comparison are theoretical curves of thermal stresses in a clamped rod by heating it from $T_0 = 38°C$) to $T_{fin} = 110°C$ as considering the relaxation processes in the polymer, so providing the calculations on the theory of elasticity.

It is clear that the calculation on the theory of elasticity cannot provide even a rough idea about the course of the thermal stresses in the polymer, since in this case, at cyclic temperature change the stresses will describe the same curve or a straight

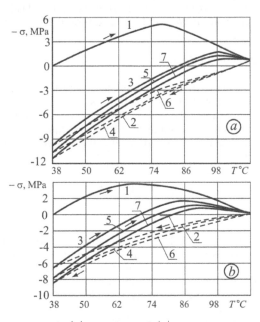

a) $-\left|k\right| = 4\,\text{deg/min}$;   *b -* $\left|k\right| = 0.4\,\text{deg/min}$.
The numbers on the curves indicate the number of half cycles.

Figure 1. Theoretical dependence of thermal stresses on the rod at cyclic temperature change.

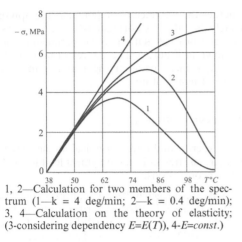

1, 2—Calculation for two members of the spectrum (1—k = 4 deg/min; 2—k = 0.4 deg/min); 3, 4—Calculation on the theory of elasticity; (3-considering dependency $E=E(T)$), 4-$E=const.$)

Figure 2. Comparison of the theoretical curves of thermal stresses.

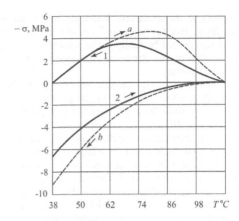

Figure 3. The temperature dependences of the stresses on the rod at cyclic temperature change: 1.2—Calculation with two members of the spectrum; a, b—calculation by the linearized theory.

line irrespective of the direction and rate of change of temperature.

To evaluate the role of the exponential factor in nonlinear Equations (2), we compared the results of calculations according to (2) with two members of the spectrum, with the results of calculations by the linearized theory, when in (2) exponent was formally replaced changed per unit.

In this case, we also take into account the dependence of the constants on temperature. It is known that the linearized theory qualitatively reflects quite well for most processes. However, quantitative differences from the experimental results are very significant and especially at high loads and for long-term experiments. There is no exception and thermal stresses (Figure 3).

Therefore, the linearized theory gives the same qualitative results as the non-linear. However, as the stress and temperature become a noticeable difference in the rates of relaxation processes, measured at the two theories. The same result is obtained when the reduction is in several ranges the rate of heating «k».

Figure 4 as an example shows for different heating rates a comparison of calculations on the nonlinear theory based on two components of the spectrum (curve 1) with the results of calculations with one member of the spectrum (curves 2 and 3). This gives an idea of what kind of constituents at what stage and at what speeds plays a dominant role in this process. It appears that the "senior" component predominates over most of the heating step to a temperature somewhat lower than T_g, yet the polymer is not softened.

The influence of the "younger" component ($i = 2$) significantly increases with increasing temperature, when the relaxation time $T_{0,2}^*$ decreases or equiva-

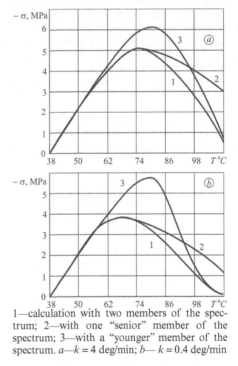

1—calculation with two members of the spectrum; 2—with one "senior" member of the spectrum; 3—with a "younger" member of the spectrum. a—$k = 4$ deg/min; b—$k = 0.4$ deg/min

Figure 4. Comparison of the theoretical curves of the temperature dependence of compressive stresses by heating the polymer rod.

lently the coefficient of the initial relaxation viscosity $\eta_{0,2}^*$ is close to $\eta_{0,1}^*$. Moreover, its effect occurs at a lower temperature and reduces the rate of heating.

A similar conclusion can be reached, if we estimate the value of accumulated highly elastic strains $\varepsilon_{x,1}^*$ and $\varepsilon_{x,2}^*$ in the calculation of the two members of the spectrum. When cooling is almost completely frozen "younger" component rubbery deformation that basically is the reason for changing the sign of the stress on the second half-cycle, i.e. upon cooling.

However, which of the components prevails in the study process depends on the polymer, i.e. on the ratio of the relaxation constants, in particular $\eta_{0,1}^*$ and $\eta_{0,2}^*$. If these values are close, the uniaxial stretching experiments determine the relaxation constants (Babich 1966), in which it is difficult to distinguish a particular component. As shown in experimental studies and theoretical calculations given below, to determine the thermal stresses, arising at heating rods from EDT-10 and PMMA are in the glassy state it is possible with sufficient accuracy in a wide range of temperatures restricts the "older" part.

A direct comparison of theoretical and experimental values of thermal stresses was performed on samples of two series of EDB-10 and PMMA. A preliminary theoretical analysis similar to that given above and the experimental values of the constants up to T_g for these parameters of polymers gave a reason to restrict the calculations to one "senior" member of the spectrum of relaxation times. This corresponded to the maximum values of high elasticity modulus $E_{\infty,s}$ and minimum value of the initial viscosity $\eta_{0,s}^*$.

From independent experiments, physical constants were found to be dependent on the temperature in the range of $30°C$ up to T_g, to study whether the parameters of polymers are approximately described by Formulas (8). The comparison of theoretical calculations and experimental results is shown in Figure 5. The figures show a satisfactory agreement between them at a rather wide range of temperatures.

3 CONCLUSIONS

The experimental and theoretical studies devoted to thermal stresses in the polymers lead to the following conclusions:

1. A nonlinear generalized equation of the Maxwell (Gurevich 1974) describes quite well the behavior of both linear (PMMA) and reticulated (EDT-10) polymers. This was also confirmed in previous work (Rabinovich 1970, Andreev 1968, Andreev & Malashkin 1983).
2. In the different modes of the temperature field, calculation is sufficient to consider the "older" part of the spectrum of relaxation times.
3. Using the dependence of the mechanical constants of polymers on temperature (Babich 1966) can be obtained diagrams of stress-temperature, which is in good agreement with the experimental data over a wide range of temperature.

ACKNOWLEDGMENT

This work was supported by the Ministry of Education and Science of Russia under grant no. 7.2122.2014/K.

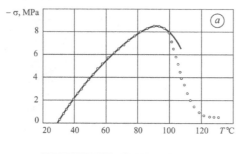

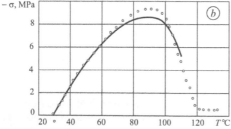

Figure 5. The temperature dependence of the stresses on the rod when heated (k = 4 deg/min): a—EDT-10; b—PMMA; —— - theoretical curve, o—experiment.

REFERENCES

Andreev, V.I. & Malashkin, Yu.N. 1983. Calculation of thick-walled pipes from the nonlinear elastic material. *Building mechanics and calculation of structures.* 6: 70–72.

Andreev, V.I. & Turusov, R.A. 2014. The thermal strength of adhesion bond. *Applied Mechanics and Materials.* 670–671: 153–157.

Andreev, V.I. 1968. On the stability of polymer rods at creep. *Mechanics of polymers,* 1: 22–28.

Babich, V.F. 1966. *Investigation of the effect of temperature on the mechanical properties of polymers.* Thesis of the candidate of technical sciences. Moscow.

Gurevich, G.I. 1974. *Deformable medium and propagation of seismic waves.* Moscow: Nauka.

Kuperman, A.M. & Turusov, R.A. 2012. Study of relaxation properties of reinforced plastics for half-disks tension of ring specimens. *Mechanics of Composite Material.* 48(3) 37–42.

Rabinovich, A.L. 1970. *Introduction to the mechanics of rein-forced polymers.* Moscow: Nauka.

Advanced Materials and Structural Engineering – Hu (Ed.)
© 2016 Taylor & Francis Group, London, ISBN 978-1-138-02786-2

The research of TD-Music location algorithm based on a virtual planar array model of space distributed nodes

P. Zhao, H.F. Yao, J. Liu, W.Z. Shi, W.H. Wang, K. Zhang, Y.H. Ma & B. Li
School of Electronics and Information, Northwestern Polytechnical University, Xi'an, Shaanxi, P.R. China

ABSTRACT: In this paper, an improved Transform Domain (TD), MUSIC location algorithm, is proposed for a 2D Direction-of-Arrival (DOA) estimation based on 3D distributed nodes with cooperative communication and synchronization. Most of the existing DOA algorithms are difficult to match with 3D distributed arrays due to the characters of array nodes, so the 3D distributed nodes mapping onto the 2D planar array through the virtual plane and orthogonal projection method are applied to form a virtual plane array, and then the TD-MUSIC algorithm is used through angle domain transformation to satisfy the 2D irregular plane array. The simulation results show that the algorithm model is successfully carried out with a better location performance of three source signals, and the angle mean square error is less than 0.6.

1 INTRODUCTION

Although the MUSIC algorithm has a high location precision, it is only applicable to the narrowband incoherent signal source to the equidistant of the uniform linear array structure model; this largely limits the range of the application of the MUSIC algorithm. Researchers have proposed many methods for this problem.

For a non-uniform linear array, the literature (Zhao et al. 2009, Dogan & Menden 1995, Rao et al. 2003), respectively, puts forward the virtual uniform linear array, the array interpolation method and the method of flow separation to make the MUSIC algorithm can be used with a non-uniform linear array, but the array interpolation method and the method of flow separation require that the virtual uniform array aperture's size is not less than 2 times the size of the non-uniform array aperture, otherwise it will result in poor precision.

When the number of signal source is greater than the number of array element, the MUSIC algorithm cannot distinguish the arrival angle of the signal effectively. Shan (2014) put forward MUSIC-like array extension method based on fourth-order cumulate to make array signal covariance matrix be full rank, so as to meet the requirements of the MUSIC algorithm. The limitations of the MUSIC spatial spectrum estimation are that it cannot handle the coherent signal, the spatial smoothing technique in the literature (2011) can make Eigen values matrix of the incidence coherent signal be full rank, thereby expanding the application range of the MUSIC spatial spectrum estimation algorithm.

For a wideband incoherent signal, the literature (Si et al. 2012, Hong et al. 2012) through the weighted processing to the signal energy distribution and improving signal structure proposed the ISM algorithm based on the maximum power frequency point. It selects the frequency point of sub-band of representing the highest power to construct the spectrum density array, then makes the matrix to be applied to narrow-band direction finding algorithm for direction finding.

Although MUSIC has a high precision for spectrum estimation, most of the researches are aimed at DOA estimation of planar array. So, in this paper, we put forward space distributed nodes spectrum estimation algorithm study based on the incoherent narrowband signal source. The main innovation point lies in making the distributed nodes map onto a virtual plane, and then use the TD-MUSIC algorithm to achieve the 2D spatial spectrum estimation.

2 VIRTUAL ARRAY MODEL

Consider that there are N narrow band source signals with a known common center frequency Ω_0 impinging from the directions $\vartheta_i = (\theta_i, \phi_i)$, $i = 1$, $2 \dots$ N on a space arbitrary array with M nodes, as shown in Figure 1, where θ_i is the DOA of the ith source relative to the X axis and ϕ_i relative to the Y axis.

By using the method of orthogonal projection, we map the 3D space distributed node onto the 2D irregular planar array, and then use the

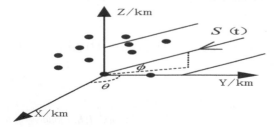

Figure 1. Space distributed array model.

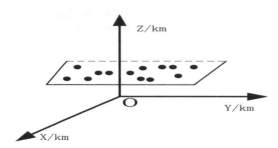

Figure 2. Planar array model.

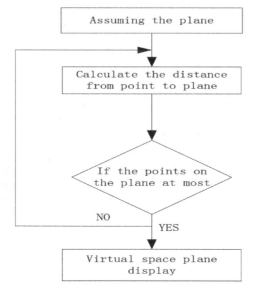

Figure 3. The flow chart of building the virtual space plane.

TD-MUSIC algorithm to estimate DOA. The model is shown in Figure 2.

Suppose that there is a space distributed array with M nodes, which can be described as $P_i(x_i, y_i, z_i)$, $i = 1, 2 \dots M$. We can build a space plane O: $Ax + By + Cy + D = 0$ by choosing three

points from M points and then can obtain C_M^3 planes in total. For the remaining M-3 array elements, the distance between the planes in which any one pickets to use the Formula (1) to compute:

$$d_i = \frac{|Ax_i + By_i + Cz_i|}{\sqrt{A^2 + B^2 + C^2}} \tag{1}$$

where N_i $(i = 1, 2 \dots C_M^3)$ represents the number of points on the plane. If $d_i = 0$ is true, then the value is $N_i + 1$. We select the largest N_i N_i and record as $N_{max} = MAX(N_i), (i = 1, 2 \dots C_M^3)$ to determine the virtual plane that contains the largest spatial points. The process can be shown in Figure 3.

When the virtual space plane is established, projecting all the points that are out of the plane to the virtual plane with the orthogonal projection method, so that all array elements is in one plane. Suppose that the centered of M points as an original point O (0, 0, 0) to construct a 3D coordinate system XYZ.

3 THE MUSIC LOCATION ALGORITHM

It is assumed that the array is linear and composed of M nodes and that N incoherent plane waves impinge on it from the direction $\vartheta_i = (\theta_i, \phi_i)$, $i = 1, 2 \dots N$. The incident plane waves are assumed to be narrowband with a center frequency Ω_0 under these assumptions, the output from the ith node is found to be

$$X_m(t) = \sum_{i=1}^{N} a_m(\theta_i, \phi_i) s_i(t) + n_m(t) \tag{2}$$

where $s_i(.)$ is a scalar complex waveform referred to as the ith signal; $a_m(\theta_i, \phi_i)$ is the complex gain of the ith node to the ith signal; and $n_m(t)$ is the gauss white noise. Stacking the M functions given by (2) into a vector and taking the first node as the reference node, we can write

$$x(t) = As(t) + n(t) \tag{3}$$

where $s(t)$ is the N × 1 vector

$$s(t) = [s_1(t), s_2(t), \dots s_N(t)]^T \tag{4}$$

and $n(t)$ is the M × 1 vector

$$n(t) = [n_1(t), n_2(t), \dots n_N(t)]^T \tag{5}$$

and A is the M × N matrix

$$A = [a(\theta_1, \phi_1), a(\theta_2, \phi_2), \dots a(\theta_N, \phi_N)] \tag{6}$$

and $a(\theta_i, \phi_i)$ is the M × N matrix

$$a(\phi_i,\theta_i) \overset{\Delta}{=} [\exp(-i\beta_{i,1}), \exp(-i\beta_{i,2}),$$
$$\exp(-i\beta_{i,3}), ... \exp(-i\beta_{i,M})] \qquad (7)$$

Suppose that λ is the signal wavelength and ith node coordinate is (x_m, y_m), and then $\beta_{i,m}$ can be written as follows:

$$\beta_{i,m} = \frac{2\pi}{\lambda}(x_m \cos\theta_i \sin\phi_i + y_m \sin\theta_i) \qquad (8)$$

The auto covariance matrix R of $x(t)$ can be obtained from Equation (9) as follows:

$$R = E\{xx^H\} = AE\{ss^H\}A^H + E\{nn^H\}$$
$$= AR_{SS}A^H + \delta_n^2 I \qquad (9)$$

For Eigen value decomposition for R, we can obtain $R = U\Lambda U^H$, and Diagonal matrix Λ can be $\Lambda = diag(\lambda_1, \lambda_2, ...\lambda_M)$, $\lambda_1 \geq ...... \geq \lambda_{M-N+1} = ... = \lambda_M = \delta^2$, where Eigenvectors is $e_i i = 1, 2 ... M$ and Eigenvectors matrix $U = [SG]$, where $S = [e_1, e_2, ... e_N]$, $G = [e_{M-N+1}, ... e_M]$ are the signal subspace and noise subspace, respectively. In the array processing, we can prove $S \perp G$ and $A = S$, so the following Equation (10) can be obtained:

$$a^H(\theta,\phi)G = 0_{M\times 1} \qquad (10)$$

Hence, the MUSIC space spectral function can be written as follows:

$$P_{MUSIC}(\theta,\varphi) = \frac{1}{a^H(\theta,\varphi)GG^H a(\theta,\varphi)} \qquad (11)$$

4 THE TD-MUSIC ALGORITHM

Compared with the MUSIC algorithm, TD-MUSIC using the noise subspace and conjugate noise subspace instead of noise subspace. Because $\beta_{i,m}$ function is nonlinear based on (θ,ϕ), it results in a lot of calculation. So, transform domain is proposed for DOA estimation. With respect to $\beta_{i,m}$, the transform expression can be obtained as follows:

$$\begin{cases} p = \cos\theta \sin\phi \\ q = \sin\theta \end{cases} \qquad (12)$$

From Equation (8) we get

$$\beta_{i,m} = \frac{2\pi}{\lambda}(x_m p_i + y_m q_i) \qquad (13)$$

As can be seen from the above equation, $\beta_{i,m}$ function is linear based on (p, q). For a given (θ,ϕ) has a unique transform domain (p, q). The relationship between them is expressed as:

$$\begin{bmatrix} \theta_1 = \sin^{-1}(q) \\ \phi_1 = \sin^{-1}\left(\frac{p}{\sqrt{1-q^2}}\right) \end{bmatrix} \begin{bmatrix} \theta_2 = \sin^{-1}(q) \\ \phi_2 = \pi - \sin^{-1}\left(\frac{p}{\sqrt{1-q^2}}\right) \end{bmatrix}$$
$$(14)$$

Therefore, there are two sets of values corresponding to a given (θ,ϕ); the real direction can be estimated from Equation (10). From Equation (12), we can rewritten Equation (10) as follows:

$$a(p,q) = a*(-p,-q) \qquad (15)$$

For Equation (10), replace $a(\theta,\phi)$ with $a(p,q)$ and take sides conjugate:

$$[a^*(p.q)]^H G^* = a^H(-p,-q)G^* = 0_{M\times 1} \qquad (16)$$

Comparing Equation (16) and Equation (10), we can get a truth that the transform domain steering vector and conjugate noise subspace orthogonal. If we use the noise subspace and conjugate noise subspace instead of noise subspace to construct TD-MUSIC spectrum function, the intersection will have dual orthogonal. Thus, TD-MUSIC estimation will produce extreme values in real and mirror position, simultaneously. Besides, we can just search for half of the range of (p, q) domain.

How to get intersection G_{inter} of G and G^* is the key step to construct TD-MUSIC. We can define $Q = G^T G^*$ and then get the SVD decomposition of Q to be

$$Q = C\sum D^H \qquad (17)$$

Thus, we can obtain $G_{inter} = GC(1, 1: M-2L) = W$, and the TD-MUSIC spectrum function can be written as follows:

$$P_{TD-MUSIC-SVD}(p,q) = \frac{1}{a^H(p,q)WW^H a(p,q)} \qquad (18)$$

4.1 The step of the TD-MUSIC algorithm

1. According to a given SNR, construct TD-MUSIC spectrum based on Equation (18).
2. Search in the half range of TD-MUSIC spectral; obtain the information of the (p,q) domain or the mirror (p_i,q_i) domain, for i = 1, 2.. M', M' < M.
3. Replace $a(\theta_i,\phi_i)$ of Equation (10) with $a(p_i,q_i)$ and then substitute (p_i,q_i) and $(-p_i,-q_i)$ to be

725

an extreme test, the real direction is on condition that $\|a^H(p,q)G\|^2 \approx 0$.

4. Substitute (p_i, q_i) into Equation (16) to compute probable direction (θ_{i1}, ϕ_{i1}) (θ_{i1}, ϕ_{i1}) and $(-\theta_{i1}, -\phi_{i1})$.

5. The extreme value test in a small area of between (θ_{i1}, ϕ_{i1}) (θ_{i1}, ϕ_{i1}) and $(-\theta_{i1}, -\phi_{i1})$.

5 SIMULATION AND ANALYSIS

An accurate and reliable synchronization scheme enables the irregular planar array to be used in applications that require tight synchronization for a precise event and dada correlation. So, this requires that all nodes in the network be synchronized with respect to each other or a common central node. Time synchronization ensures that all nodes in the network maintain the same time or are aware of their time difference with respect to a reference point. Therefore, the simulation is on the condition of the node synchronization, and the flow chart of simulation is shown in Figure 4.

5.1 Simulation of virtual plane

Suppose that there are 8 nodes and the coordinate can be expressed as ( ), (4,3,6), (2,5,5), (8,3,4), (7,9,5), (6,4,4), (7,5,3), (9,5,3), (4,6,3), the coordinate unit is 10^2 m. We choose 3 nodes arbitrarily from M nodes

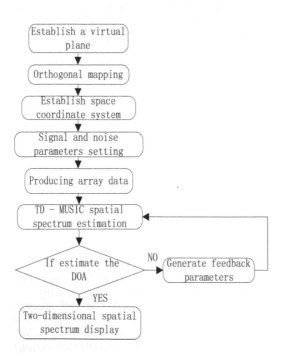

Figure 4. The flow chart of simulation.

to establish a plane and then compute the number of sensors on the plane. We repeat this process until we find the maximum number of points on the plane. The plane is shown in Figure 5.

As we can see from Figure 5, we get a virtual plane and its equation can be written as $Z = -0.5X - Y + 11$, and the largest number of nodes on the plane is 5.

5.2 Simulation of RSME

One array is the rectangle plane array and is composed of 8 nodes, and the other array is the irregular planar array and is composed 8 nodes.

Each node of the rectangle planar array can be expressed as

(3,2,2), (4,1,2), (2,4,2), (6,7,2), (7,9,2), (9,5,2), (5,3,2), (5,6,2)

Each node of the irregular planar array can be expressed as

(4,3,6), (2,5,5), (8,3,4), (7,9,5), (6,4,4), (7,5,3), (9,5,3), (4,6,3)

Each node of the projection coordinates can be obtained as

(4,3,6), (2,5,5), (8,3,4), (5.55,6.1,2.1), (6,4,4), (6.9,4.8,2.8), (8.65, 4.3, 2.3), (4, 6, 3)

Assuming that three sources located at, (20°,40°), (50°,30°), (60°,50°) both in the beam width area impinge on the array, the snapshot number is 512. Then, we analyze the RSME of rectangle planar array and irregular planar array with 100 Monte Carlo experiments. The Angle Mean Square Error can be expressed as

$$RSME = \sqrt{(\hat{\theta} - \theta)^2 + (\hat{\phi} - \phi)^2}$$

where (θ, ϕ) represent the signal azimuth and elevation of the true value, and (θ', ϕ') is the experimentally measured value. The relation curves of RSME of SNR are shown in Figure 6.

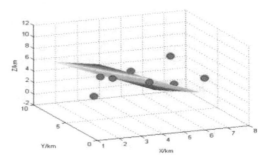

Figure 5. Space virtual plane.

726

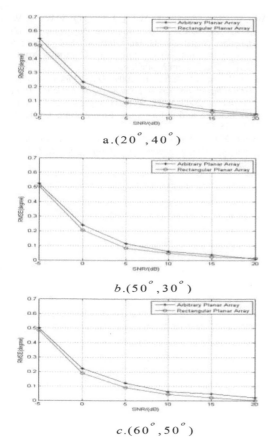

a.$(20^{o}, 40^{o})$

b.$(50^{o}, 30^{o})$

c.$(60^{o}, 50^{o})$

Figure 6.   Performance of RSME.

Estimation error is compared against SNR, with the numbers of samples being set at 100. Both arrays show an improvement in RSME as SNR is increased. As illustrated in Figure 6, it is clear that when the SNR is greater than 20 dB, the RSME performance is equal.

5.3   *Simulation of spatial-spectrum estimation*

We consider three sources located at (20°, 40°), (50°, 30°), (60°, 50°), respectively. The SNRs are all 20dB, the transform domain (p, q) is (0.6, 0.34), (0.3, 0.42), (0.65, 0.5) contrary to the (θ, ϕ) domain. The 2D DOA estimation based on TD-MUSIC with the arbitrary planar array is shown in Figure 7, and the 2D DOA estimation based on MUSIC with the rectangle planar array is shown in Figure 8.

Comparing Figure 7 with Figure 8, it is clearly noted that the TD-MUSIC algorithm for the irregular planar array can achieve a high-precision 2D DOA estimation and verify the correctness of the theoretical analysis.

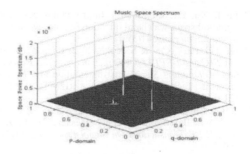

Figure 7.   Performance of the irregular planar array.

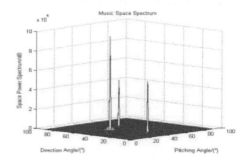

Figure 8.   Performance of the rectangle planar array.

6   CONCLUSIONS

From the analyses of the above simulation results, we can see that by using the TD-MUSIC algorithm based on virtual planar arrays with space distributed nodes, through virtual plane and orthogonal projection method can be achieved to satisfy requirement as the 2D irregular planar array DOA estimation. The simulation results also show that 2D the irregular planar array has a good estimation accuracy compared with the rectangle planar array. So, the configuration of the 2D location algorithm based on virtual planar arrays can be used to realize 2D location for space distributed nodes.

ACKNOWLEDGMENT

This paper was supported by the National Natural Science Foundation of China (No. 61271279 and No. 61201157).

REFERENCES

Dogan, M. & Mendel, J. 1995. Applications of Cumulates to Array Processing. Part II: Non-Gaussian Noise Suppression, *IEEE Trans on Signal Processing*, 43(7): 1663–1676.

Gong, B. Xu, Y.T. & Li, J. 2011. Improved non-coherent signal subspace wideband direction finding algorithm, *Radio Engineering,* 41(3): 11–13.

Hong, J.G. Park, C.S. & Seo, B.S. 2012. *Comparison of MUSIC and ESPRIT for Direction of Arrival Estimation of Jamming Signal.* Instrumentation and Measurement Technology Conference (I2MTC), 2012 IEEE International.

Rao, B.D. Engan, K. & Cotter, S.F. 2003. Subset Selection in Noise Based on Diversity Measure Minimization, *IEEE Transactions On Signal Processing,* 51(3): 760–770.

Si, W.J. Lan, X.Y. & Zou, Y. 2012. Novel high-resolution DOA estimation using subspace projection method, *The Journal of China Universities of Posts and Telecommunications.* 19(4): 110–116.

Wei, Y. Zhao, P. Li, L.F. Wu, L.J. Ji, Q.Q. & Wang, W. 2010. *An Acoustic Array Imaging Algorithm for 3D Display Based on MIMO And MUSIC Location Fusion.* International Conference on Computer Application and System Modeling.

Zhao, P. Shi, H.S. & Shi, K. 2009. *The 3D location algorithm based on Smart antenna with MUSIC DOA estimates,* 2009 WRI World Congress on Computer Science and Information Engineering, CSIE, 4: 750–753.

Zheng, X.J. & Wang, Q. 2014. *MUSIC spectral estimation of coherent signals spatial smoothing technique,* New Technology.

Advanced Materials and Structural Engineering – Hu (Ed.)
© *2016 Taylor & Francis Group, London, ISBN 978-1-138-02786-2*

Effect of the grid discrete error on the symmetry of topological optimization results

J. Fan, Z.Y. Yin & J.J. Wang
School of Energy and Power Engineering, Beihang University, China

Y. Xiao & C. Chen
Aviation Research Institute, China

ABSTRACT: The Solid Isotropic Material with Penalization (SIMP) approach is adopted to derive the mutative symmetric formula of topology optimization results caused by the grid discrete error and penalization parameter. The mechanism for the emergence of asymmetric results in topological optimization is analyzed using the relative perturbation analysis of the matrix eigenvalue. Therefore, the asymmetry coefficient is proposed to measure the asymmetric degree of topology optimization solution with the symmetric structure. Finite element models of symmetric spherical shell with different penalization parameters and shell thicknesses are presented. In these models, the effect of the grid discrete error on the asymmetry of topological optimization solution with the axisymmetric structure is studied. It is demonstrated that the grid discrete error and penalization parameter are significant factors in the symmetric effect of the topological optimization results. The results also point out that the consequence symmetry can be properly described by the asymmetry coefficient.

1 INTRODUCTION

There are many symmetrical and axisymmetric components in the aeroengine, such as disks, axis and fairing. When optimizing the structural topology of these components, engineers hope to get symmetrical optimization results according to the design index. However, some examples show that there are a few reasons that cause the asymmetrical results in the topological optimization process of the symmetrical structure. Watada et al. (2011) obtained asymmetrical results when optimizing the spherical shell structural topology. Kosaka et al. (1999) gained the asymmetrical topology of the trunk lid whose initial model was symmetrical. Kolanek et al. (2003) failed to get symmetrical results when optimizing the plate with symmetrical loads and supports.

Currently, asymmetrical analyses of symmetrical structural topology optimization are a hotspot of the topological optimization theory. Because the intermediate relative density is suppressed in the symmetrical structural topology optimization based on the SIMP method, asymmetrical results may be caused by different penalization parameters (Watada et al. 2011, Kosaka et al. 1999).

In theoretical research of topological optimization, truss optimization is important in studying the internal mechanism of the asymmetrical phenomenon. In the truss topological optimization, the reasons of asymmetrical results are the non-convex constraint and objective function (Cheng & Liu 2011), the dispersion of the cross-sectional area of truss (Stolpe 2010), the dispersion level of the design variable and the tankage of selectable variable set (Richardson et al. 2013). Of course, to ensure the symmetry of topological optimization results, the condition that the constraint and the objective function must be strictly convex could change to a semi-convex condition.

For the symmetrical continuum structure, topological optimization proceeded on the basis of the finite element method. In the process, few documents indicated that the grid discrete error had an important impact on the symmetry of the optimization results. These import errors may be magnified so that they cause asymmetrical results after entering the optimization program. Therefore, it is necessary to study the influence of the grid discrete error on the topological optimization results.

Based on the SIMP method, the influence formula of the grid discrete error and penalization parameter on the symmetry of topology optimization results was derived. The relative perturbation analysis of the matrix eigenvalue was used to explain the emergence mechanism of asymmetric topological optimization results. Considering the grid scale and the distance between load points and asymmetrical elements, the asymmetry coefficient is proposed to measure the asymmetric degree of the topology optimization solution with the symmetric structure. Finally, examples of the symmetric spherical shell with different penalization parameters and shell thicknesses are presented to analyze the effect of the grid discrete error on the asymmetry of topology optimization solution with the axisymmetric structure.

2 MATHEMATICAL MODEL BASED ON THE SIMP METHOD

The SIMP method is a density-based method, which can be broadly classified into the most widely used methodologies for structural topology optimization (Rong 2007, Zuo et al. 2004). Design variables are the relative density of elements. The method operates on a fixed domain of finite elements to identify whether each element should consist of a solid material or void.

In the SIMP approach, the relationship between the density variable and the material property is given by the power law. The relative density ρ_i is attached to each finite element "i" of the model, which can be obtained through the formula $\rho_i = \rho_e/\rho_0$, where the effective density is ρ_e and the base material property is ρ_0. By varying ρ_i between 0 and 1, the density interpolates between the void and the solid. The effective Young's modulus of the solid material $E_i(\rho_i)$ and the stiffness matrix $K_i(\rho_i)$ is related to the base material property E_0 and

$$E_i(\rho_i) = \rho_i^P E_0, K_i(\rho_i) = \rho_i^P K_i^0 \tag{1}$$

Here, p is the penalization parameter used to suppress the formation of intermediate densities in an optimal solution. The global stiffness matrix K can be assembled by the element stiffness matrix $K_i \in R^{n \times n}$:

$$K = \sum_{i=1}^{N} \left[\varepsilon + (1-\varepsilon)\rho_i^P \right] K_i \tag{2}$$

Here, ε represents a tiny positive value used to reduce the instability of numerical calculation in the optimization algorithm.

Equation (3) shows the common mathematical model of topological optimization. The objective is to minimize the compliance $W(\rho) = P^T U(\rho)$ and the volume is constrained:

$$\begin{cases} \rho = (\rho_1, \rho_2, \ldots \rho_N) \\ \min W(\rho) = F^T U(\rho) & (a) \\ s.t. \sum_{i=1}^{N} V_i^0 \cdot \left[\varepsilon + (1-\varepsilon)\rho_i \right] \le 0.5 V^0 & (b) \\ 0 \le \rho_i \le 1, (i = 1, \ldots, N) & (c) \end{cases} \tag{3}$$

Here, V_i^0 is the ith element volume; U is the displacement matrix of the nodes; and F is the concentrated load on the nodes.

3 HYPOTHESIS ABOUT THE GRID DISCRETE ERROR

A continuous solution domain was discrete to finite elements. The grid discrete error occurred when the field function of the global field was approximated by the element trial function (Zeng 2004). When a thin-walled structure was discrete by the shell element, it caused the grid discrete error during true values and calculated results of the element stiffness matrix that does not consider the nonlinear influence of the surface radius, the twist and the curvature deformation (Liu et al. 2010). Because the SIMP approach is based on the results of finite element analysis, the grid discrete error would have an important influence on the topological optimization results as an imported error.

To describe the influence of the grid discrete error, a hypothesis is proposed. The stiffness matrix and the displacement matrix change to a little extent because of the grid discrete error, which becomes a small perturbation source.

Considering the small perturbation, U_{ap} and K_{ap} described the displacement matrix and the stiffness matrix of the approximate solution, which means the influence of the grid discrete error also was associated with it:

$$K_{ap}U_{ap} = F \tag{4}$$

For an exact solution, the stiffness equation is described by the displacement matrix U and the stiffness matrix K:

$$KU = F \tag{5}$$

Let us suppose that C and D are constant matrices with least-norm, and there is $(I+C)^{-1} = I + O$, where I is the identity matrix. Because of the grid discrete error, the perturbation matrix can be represented by the matrix product CU and DK. The change between the approximate solution and the exact solution can be written as follows:

$$\begin{cases} K_{ap} = (I+C)K \\ U_{ap} = (I+D)U \end{cases} \tag{6}$$

Here, we assumed that matrices K, K_{ap} have the same rank. Equation (4) was adapted to the following equation:

$$(I+C)K(I+D)U = F \tag{7}$$

730

4 THE INFLUENCE OF PENALIZATION PARAMETERS ON THE ASYMMETRY OF TOPOLOGICAL OPTIMIZATION RESULTS CONSIDERING THE GRID DISCRETE ERROR

To analyze the influence of penalization parameters on the asymmetry of topological optimization results, the Lagrange equation of mathematical model (3) was calculated by taking the partial derivatives of ρ_i. The KKT condition can be written as follows:

$$G_i(\rho) \begin{cases} = 0 & for \ 0 < \rho_i < 1 \\ \leq 0 & for \ \rho_i = 1 \\ \geq 0 & for \ \rho_i = 0 \end{cases} \tag{8}$$

Here,

$$G_i(\rho) = \lambda V_i^0 (1-\varepsilon) - (1-\varepsilon) p \rho_i^{P-1} \\ \cdot U^T (I+D)(I+C) K_i (I+D) U \tag{9}$$

Here, λ is the Lagrange multipliers. Assuming that $I = \{i | 0 < \rho_i < 1\}$, where m is the number of set I, we can obtain $G_i(\rho) = 0$ and $m = |I|$. When $i \notin I, \rho_i' = 0$.

Taking the derivation of Equation (7) with respect to the penalization parameter P, the result is substituted into Equation (3) to obtain:

$$-(I+C)K(I+D)U'$$
$$-(I+C)\sum_{i=1}^{N}(1-\varepsilon) p \rho_i^{p-1} \rho_i' K_i (I+D) U$$
$$= (I+C)\sum_{i=1}^{N}(1-\varepsilon) \rho_i^p \ln \rho_i K_i (I+D) U \tag{10}$$

For $i \in I$, taking the derivation of Equation (3) with respect to the penalization parameter P, the results were multiplied by 0.5:

$$\frac{1}{2}\sum_{i=1}^{N}(1-\varepsilon) A_i h_i \rho_i' = 0 \tag{11}$$

Taking the derivation of Equation (9) with respect to the penalization parameter P and setting $H_i = U^T (I+D)(I+C) K_i (I+D) U$, we can obtain the following equation:

$$-\frac{1}{2}(1-\varepsilon) p \rho_i^{p-1} U^T (I+D)(I+C) K_i (I+D) U'$$
$$-\frac{1}{2}(1-\varepsilon) p(p-1) \rho_i^{p-2} \rho_i' H_i + \frac{1}{2}(1-\varepsilon) V_i^0 \lambda'$$
$$= \frac{1}{2}(1-\varepsilon) p \rho_i^{p-1} \ln \rho_i H_i + \frac{1}{2}(1-\varepsilon) \rho_i^{p-1} H_i \tag{12}$$

After reordering the elements in the collection $I, I = \{1, 2, \dots m\}$. The corresponding relative density has the order $\rho_0 = (\rho_1, \rho_2, \dots \rho_m)$. Equations (10), (11), (12) can be expressed in the matrix as follows:

$$\begin{pmatrix} -(I+C)K & B^{12} & 0 \\ B^{12T} & B^{22} & B^{23} \\ 0 & B^{23T} & 0 \end{pmatrix} \begin{pmatrix} (I+D)U' \\ \rho_0' \\ \lambda' \end{pmatrix} = \begin{pmatrix} b^1 \\ b^2 \\ 0 \end{pmatrix} \tag{13}$$

Here, $B^{12} \in R^{n \times m}$, $B^{22} \in R^{m \times m}$, and $B^{23} \in R^m$. Setting $K_{ji} \in R^n$, the i-th column in the stiffness matrix of the i-th element, there are:

$$B_{ij}^{12} = -(1-\varepsilon) p \rho_i^{p-1} K_{ji}^T (I+C)(I+D) U \tag{a}$$

$$B_{ii}^{22} = -\frac{1}{2}(1-\varepsilon) p(p-1) \rho_i^{p-2} H_i$$

$$B_{ij}^{22} = 0 \quad for \ i \neq j \tag{b}$$

$$B_i^{23} = \frac{1}{2}(1-\varepsilon) V_i^0 \tag{c}$$

$$b^1 = (1-\varepsilon)\sum_{i=1}^{N} \rho_i^p \ln \rho_i (I+C) K_i (I+D) U \tag{d}$$

$$b_i^2 = \frac{1}{2}(1-\varepsilon)(1 + p \ln \rho_i) \rho_i^{p-1} H_i \tag{e}$$

$$\tag{14}$$

Furthermore, according to the first line of Equation (13), it can be calculated as follows:

$$(I+D)U' = -K^{-1}(I+C)^{-1} b^1 \\ + K^{-1}(I+C)^{-1} B^{12} \rho_0' \tag{15}$$

Substituting Equation (15) into the second and third lines of Equation (13), we can get:

$$\begin{pmatrix} B^{22*} & B^{23} \\ B^{23T} & 0 \end{pmatrix} \begin{pmatrix} \rho_0' \\ \lambda' \end{pmatrix} = \begin{pmatrix} b^{2*} \\ 0 \end{pmatrix} \tag{16}$$

$$b^{2*} = b^2 + B^{12T} K^{-1} (I+C)^{-1} b^1 \tag{a}$$
$$B^{22*} = B^{22} + B^{12T} K^{-1} (I+C)^{-1} B^{12} \tag{b} \tag{17}$$

Making the assumption that the approximate solution B^{22*} corresponds to the exact solution B_e^{22*}, the symmetry of topological optimization results is closely related to the eigenvalue of B_e^{22*}. To contrast the eigenvalues of B_e^{22*} with those of B^{22*}, the equation can be adapted to as follows:

$$B^{22*} = B_e^{22*} + \delta B_e^{22*} \tag{18}$$

Here,

$$\delta B_e^{22*} = Q + V \qquad (19)$$

$$Q_{ii} = -\frac{1}{2}(1-\varepsilon)p(p-1)\rho_i^{p-2}$$

$$\cdot \begin{bmatrix} U^T K_i DU + U^T(D+C+DC)K_i U \\ + U^T(D+C+DC)K_i DU \end{bmatrix}$$

$$Q_{ij} = 0 \quad for \quad i \neq j$$

$$V_{ij} = \left[(1-\varepsilon)p\rho_i^{p-1} \right]^2$$

$$\cdot \begin{bmatrix} B_{ij}^{12T} K^{-1} O B_{ij}^{12} \\ + U^T(C+D+DC)K_{ji} K^{-1} K_{ji}^T \\ (C+D+CD)U \\ + U^T K_{ji} K^{-1} K_{ji}^T (I+C+D+CD)U \\ + U^T(C+D+DC)K_{ji} K^{-1} K_{ji}^T U \end{bmatrix} \qquad (20)$$

While B_e^{22*} is a symmetric matrix, δB_e^{22*} becomes the small perturbation. Assuming that the perturbation matrix cannot change the rank of the original symmetric matrix, according to the relative perturbation analysis of the matrix eigenvalue, there is the nonsingular matrix E in the content of eigenvalues decomposition, which restrict Equation (18) to:

$$B^{22*} = B_e^{22*} + \delta B_e^{22*} = E^T B_e^{22*} E \qquad (21)$$

Setting η_i and η_i' to represent the i-th eigenvalues of the symmetric matrix B_e^{22*} and B^{22*}, the variation in these eigenvalues is given by (Zhang 2004):

$$\left| \eta_i' - \eta_i \right| \leq |\eta_i| \left\| E^T E - I \right\|_F \qquad (22)$$

The equation shows that the variation range of the eigenvalue is determined by the deviation with the nonsingular matrix E deviating from an orthogonal matrix.

The symmetry of topological optimization results with the symmetrical structure is closely associated with η_i. While P = 1.78 and 2.94, $\eta_i \approx 0$, which makes the results with the maximum asymmetry. Considering the grid discrete error, the matrix B_e^{22*} is made into the matrix B^{22*} and the eigenvalue η_i is also made into η_i', which means a change occurs in the symmetry of topological optimization results. The reason for the change is that the small perturbation matrix C, D of the element stiffness and the displacement matrix is formed by the promotion of the grid discrete error. This change influences the eigenvalue of symmetric matrices, and finally the symmetry of the optimization results. The magnitude of the change is determined by C, D.

5 NUMERICAL EXAMPLES

In this section, the finite element model of the axisymmetric spherical shell with the aperture was used to study the internal mechanism of symmetry in topological optimization results. As Figure 1(a) and 1(b) shows, the dimension of the geometric model is $D = 40.0$ m, $\theta = \pi/3$, $\alpha = \pi/12$. The physical parameters are as follows: initial elastic modulus $E_0 = 2.10 \times 10^8 \, KN/m^2$ and Poisson's ratio 0.3. Degenerated solid shell elements were used to discrete the shell with axial 10 elements and circumferential 20 elements. Vertical concentrated loads $F = 1KN$ were imposed on the nodes of the outer ring, while the nodes of the inner ring were constrained. The optimization procedure is the symmetric problem because of the symmetric boundary, loads, supports and grid.

The grid discrete error of the surface shell was influenced by the surface radius and shear deformation. Shear deformation had a close relationship with the shell thickness (Liu et al. 2010). Therefore, different thicknesses were utilized to generate and represent different grid discrete errors. To analyze the influence of the grid discrete and the penalization parameter on the symmetry of topological optimization results, the change trends in results symmetry were calculated, while the thickness increased from 0.1 to 0.51 by a step growth of 0.01 and the penalization parameter increased from 1 to 3 by a step growth of 0.001.

5.1 The measure of asymmetry

To a certain extent, structural topological optimization was seen as a search for the best load path. The closer it gets to the symmetrical load, the more hope to have a symmetrical optimization result. Therefore, the distance was represented by the weighted coefficient. The weighted coefficients were bigger when the corresponding asymmetric elements had a closer distance with the symmetric

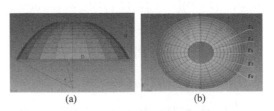

(a)   (b)

Figure 1. Spherical shell finite element model.

load. As shown in Figure 1(b), the asymmetric element in different rings had different weighted coefficients r_k. From the inner ring to the outer ring, the order is 1, 2, ... k, with the mark $r_1, r_2 \ldots r_k$. Because the loads were applied at the nodes of the outer ring, the elements in the outer ring were signed as $r_k = k$. The asymmetric coefficient was defined as Equation (23). The larger the asymmetric coefficient s, the greater the influence of the asymmetry:

$$s = \sum_{k=1}^{10} s_k \cdot r_k \Big/ \sum_{k=1}^{10} N_k \cdot r_k \qquad (23)$$

Here, N_k is the element number in the k-th ring and s_k is the asymmetric element number in the k-th ring.

5.2 Optimization results with shell thickness t = 0.50

Statics analysis results of the model with the shell thickness $t = 0.50$ are shown in Figure 2. It can be seen from the diagram that the displacement nephogram and the stress nephogram are axisymmetric. The axisymmetric problem obtained the axisymmetric results, which show that there is no human error in the model.

Increasing the penalization parameter p by a step growth of $\Delta p = 0.001$, this paper calculated the changing trend of the asymmetry coefficient s with p in the optimization problem (3), as shown in Figure 3. To be sure, in the light of the optimization problem of the spherical shell, other optimization algorithms, such as the Method of Feasible Direction (MFD), had the similar optimization results, the iterative steps with the SQP method. So, in this article, the SQP method is applied. Compared with the literature (Watada 2011), the calculation results show the following:

1. When $p = 1.717$ and 2.92, the asymmetrical coefficient was $s = 0.006$ and 0.032. The corresponding eigenvalue of the matrix B_e^{22*} was $\eta_i \approx 0$, which was similar to the calculation results in

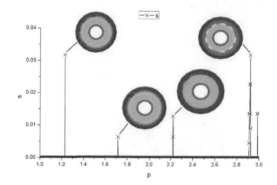

Figure 3. Variation tendencies of the asymmetrical coefficient S with P values when $t = 0.50$.

the literature. This shows that the calculation method and precision of the paper are correct.

2. The step length $\Delta p = 0.001$ obtained in this article was different from that reported in Watada (2011). More asymmetric penalization parameters are found, such as $p = 1.229$, 2.221 and 1.229 with the corresponding asymmetric coefficients $s = 0.031$, 0.013 and 0.013. It indicates that a smaller step length should be adopted than that used in the literature (Watada 2011) to get more comprehensive asymmetric points.

5.3 Optimization results with different shell thicknesses and penalization parameters

While the thickness increased from 0.1 to 0.51 by a step growth of 0.01 and the penalization parameter increased from 1 to 3 by a step growth of 0.001, the changing trend in the asymmetric coefficient is showed in Figure 4. As can be seen from the results:

1. The objective and constraint functions are convex functions in the symmetric structural topological optimization example. This should lead to achieving symmetric results (Guo et al. 2013) and, in fact, choosing most of p to obtain the symmetric results. However, under the influence of the grid discrete error, when choosing different t values of the finite element model, there are asymmetric topology optimization results with the particular p value such as that reported in the literature (Watada 2011).

2. The asymmetry phenomenon appears in succession with the continuous change in the value p. For example, choosing the p value in the entire segment $p \in [2.909, 2.923]$ with the shell thickness $t = 0.5$, asymmetric topology optimization results can be obtained. Numbers of asymmetric points Na were compared with that of the asymmetric segments with different shell

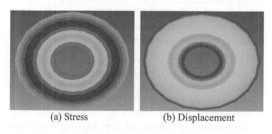

| (a) Stress | (b) Displacement |

Figure 2. Results of the statics analysis.

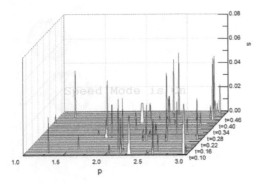

Figure 4. Variation tendency of the asymmetrical coefficient S with P values and shell thicknesses.

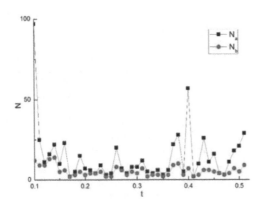

Figure 5. Comparison diagram of the numbers of asymmetric points and segments with t values.

thickness values $p \in [2.909, 2.923]$, as shown in Figure 5. It can be seen from the diagram that usually $N_a \geq N_b$. This proves the attitude.

In addition, when $N_a = N_b$, the numbers of asymmetric points were least. This means there are least asymmetry phenomenon and grid discrete error in the corresponding optimization results. In Figure 5, the minimum point occurred at $t = 0.17$ with $N_a = N_b = 2$ At this time, the grid discrete error has a minimum value.

3. It can be seen from Figure 5 that the number of asymmetric points increased dramatically to $N_a = 97$ when the shell thickness was $t = 0.1$. Because of the influence of the grid discrete error, the degenerated solid shell element or the flat shell element is used to divide the grid (Liu et al. 2010) according to the ratio of shell thickness to diameter curvature. This phenomenon shows that it is not proper to divide the grid into degenerated solid shell elements when the shell thickness is $t = 0.10$.

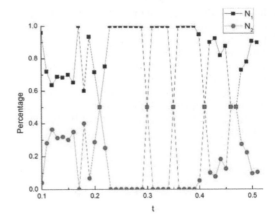

Figure 6. Percentages of asymmetric points in different intervals with t values.

4. Figure 4 shows that most of the asymmetric phenomenon appeared in the interval $p \in [2,3]$. N_1 was set to be the ratio of the asymmetric point number in the interval $p \in [2,3]$ to that in the interval $p \in [1,3]$, while N_2 was the corresponding ratio in the interval $p \in [1,2]$ to that. It can be seen from Figure 6 that usually $N_1 \geq N_2$. The phenomenon indicates that the influence of the grid discrete error on the symmetry of topology optimization results has a gradual increase with the increase of the p value.

6 CONCLUSIONS

To analyze the influence of the grid discrete error on the symmetry of topology optimization results, the relative perturbation analysis of the matrix eigenvalue was used to explain the emergence mechanism of asymmetric topological optimization results. Considering the grid scale and the distance between load points and asymmetrical elements, the asymmetry coefficient is proposed to measure the asymmetric degree of topology optimization solution with the symmetric structure. Finally, examples of the symmetric spherical shell with different penalization parameters and shell thicknesses are presented to analyze the effects of the grid discrete error on the asymmetry of topology optimization solution with the axisymmetric structure. The results show that:

1. Grid discrete error has an important influence on the symmetry of topology optimization results. According to Equation (22), the perturbation matrix C and D caused by the error changed the eigenvalue of the matrix B_e^{22*}. Lastly, the corresponding symmetry of topology optimization results also changed (Watada 2011).

2. The asymmetry phenomenon appears in succession with the continuous change in the p value when choosing the same t value.
3. Because of the influence of the grid discrete error, the degenerated solid shell element or the flat shell element is used to divide the grid (Liu 2010) according to the ratio of shell thickness to diameter curvature. An unreasonable shell element will produce a larger grid discrete error, which increases the asymmetric phenomenon in the optimization results.
4. The influence of the grid discrete error on the symmetry of topology optimization results has a gradual increase with the increase of the p value.

REFERENCES

Cheng, G.D. & Liu, X.F. 2011. Discussion on Symmetry of Optimum Topology Design, *Structural and Multidisciplinary Optimization*, 44: 713–717.
Guo, X. Du, Z.L. & Cheng, G.D. et al. 2013. Symmetry Properties in Structural Optimization: Some Extensions, *Structural and Multidisciplinary Optimization*, 47: 783–794.
Kolanek, K. & Lewinski, T. 2003. Circular and Annual Two-phase Plates of Minimal Compliance, *Computer Assisted Mechanics and Engineering Sciences*, 10177–10199.
Kosaka, I. & Swan, C.C. 1999. A Symmetry Reduction Method for Continuum Structural Topology Optimization, *Computers and Structures*, 70: 47–61.
Liu, J.T. Du, P.A. & Huang, M.J. 2010. Discrete Error Analysis and Correction Method of Finite Element for Curved Thin-walled Structures, 22(10): 2281–2286.
Richardson, J.N. Adriaenssens, S. & Bouillard, P. et al. 2013. Symmetry and Asymmetry of Solutions in Discrete Variable Structural Optimization, *Structural and Multidisciplinary Optimization*, 47: 631–643.
Rong, J.H. Tang, G.J. & Luo, Y.Y. et al. 2007. A Research on the Numerical Topology Optimization Technology of Large Three-dimensional Continuum Structures Considering Displacement Requirements, *Engineering Mechanics*, 24(3): 20–27.
Rozvany, G.I.N. 2011. On Symmetry and Non-uniqueness in Exact Topology Optimization, *Structural and Multidisciplinary Optimization*, 43: 297–317.
Stolpe, M. 2010. On Some Fundamental Properties of Structural Topology Optimization Problems, *Structural and Multidisciplinary Optimization*, 41: 661–670.
Watada, R. Ohsaki, M. & Kanno, Y. 2011. Non-uniqueness and Symmetry of Optimal Topology of a Shell for Minimum Compliance, *Structural and Multidisciplinary Optimization*, 43: 459–471.
Zeng, P. 2004. *Finite Element Analysis and Applications*, Beijing: Tsinghua University Press, 5: 226–227.
Zhang, X.D. *Matrix Analysis and Applications*, Beijing: Tsinghua University Press, 9: 408–409.
Zuo, K.T. Chen, L.P. & Zhong, Y.F. et al. 2004. New Theory and Algorithm Research about Topology Optimization based on Artificial Material Density, *Chinese Journal of Mechanical Engineering*, 40(12): 31–37.

Advanced Materials and Structural Engineering – Hu (Ed.)
© 2016 Taylor & Francis Group, London, ISBN 978-1-138-02786-2

Design of 16-QAM system based on MATLAB

J.H. Huang
School of Information Technology, Kunming University, China

ABSTRACT: QAM is a promising technique for wireless communication with very high data throughput. Due to the fact that this modulation technique can greatly enhance the spectral efficiency, it is widely used for modulating data signals onto a carrier used for radio communications. In this paper, we design 16-QAM modulation communication system based on Matlab. We present system block diagram and the process of system realization. The compared results of simulation and theory prove the validity of the 16-QAM system.

1 INTRODUCTION

In this day and age, due to the increasing demand of spectrum resources, it is urgent to develop spectrally efficient modulation schemes. Quadrature Amplitude Modulation (QAM) scheme plays a crucial role in saving spectrum resource. QAM techniques are used to transmit-bit symbols via a 2 signal point constellation, distributed on a complex plane. It is easy to attain bandwidth efficiency of 1.5 or 3.0 bits/s/Hz with 4-QAM or 16-QAM, respectively. Due to larger euclidean distances and larger state numbers, it can achieve advantages, such as higher data rates, higher reliability of communication, Variable-Rate signaling in the reverse link, and so forth (Garg et al. 2009).

Quadrature Amplitude Modulation (QAM) (Rossing & Tarokh 2001) is both an analog and a digital modulation scheme. It conveys two analog message signals, or two digital bit streams, by changing (modulating) the amplitudes of two carrier waves, using the Amplitude-Shift Keying (ASK) digital modulation scheme or Amplitude Modulation (AM) analog modulation scheme. The modulated waves are summed, and the final waveform is a combination of both Phase-Shift Keying (PSK) and Amplitude-Shift Keying (ASK), or (in the analog case) of Phase Modulation (PM) and amplitude modulation.

In this paper, we design a digital communication system based on 16-QAM scheme. Organization of the paper is as follows: in the next section is a brief presentation of 16-QAM and design of the system. The process of system's realization is in section III. Simulation results of BER and SER are given in section IV. At last, in section V we conclude our work.

2 DESIGN OF SYSTEM

In this section, we give a system block diagram of the designed system. In this design, we generate binary code as signal resources first, and then we adopt 16-QAM modulation and interpolate the modulated signal. After passing through the Low-pass filter, the noise can be filtered and the peak shape and peak height remained unchanged.

2.1 Empennage unsteady dynamics

On calculation for supersonic unsteady dynamics, the piston theory is used widely. Equation 1 illustrates the one-order piston theory:

$$p/p_\infty = 1 + kv'_z/C_\infty \qquad (1)$$

where, p is the local pressure, p_∞ is the inflow pressure, k is the specific heat ratio, c_∞ is the sound velocity of inflow, and v_z' is the normal velocity of empennage surface flow.

AWGN is used to simulate the practical channel. In AWGN, the signal is added to Gaussian white noise, which will be filtered a lot after it passing

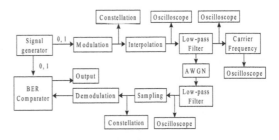

Figure 1. System block diagram.

Low-pass filter in receiver. They from sampling and demodulation in receiver, the final output signal are compared with the original signal to analysis the performance of the system.

The Matlab script performs as following:

- Generation of random binary sequence with equal 0 and 1 distribution;
- Assigning group of 4 bits to each 16-QAM constellation symbol;
- Interpolation to simulate the wide-band spectrum characteristic of AWGN;
- Filtering and addition of White Gaussian Noise;
- Filtering and sampling;
- Demodulation and de-mapping of 16-QAM symbols;
- Calculation of BER and SER.

3 THE REALIZATION OF SYSTEM

We generate a random binary signal using MATLAB. Figure 2 shows the waveform of random signal with uniform distribution. A constellation labeling map is an assignment of a bit pattern to each symbol in a signal-set constellation. Figure 3 shows the 16-QAM symbol mapping. The 4 bits in each constellation point can be considered as two bits each on independent 4-PAM modulation on I-axis and Q-axis respectively.

16-QAM is a generic modulation technique where the information is encoded into both the amplitude and phase of the sinusoidal carrier. It combines both ASK and PSK modulation techniques. QAM modulation technique is a two dimensional modulation technique and it requires two orthonormal basis functions (Garg et al. 2009):

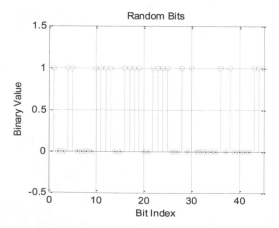

Figure 2. Waveform of random signal.

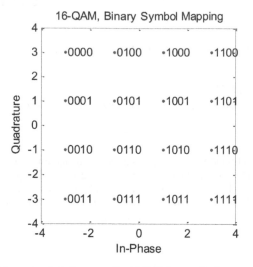

Figure 3. Labeling map for 16-QAM constellation.

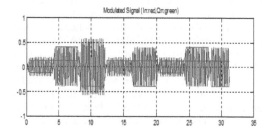

Figure 4. Waveform of modulated signal.

$$\Phi_I(t) = \sqrt{\frac{2}{T_s}} \cos(2\pi f_c t) \quad 0 \leq t \leq T_s$$

$$\Phi_Q(t) = \sqrt{\frac{2}{T_s}} \sin(2\pi f_c t) \quad 0 \leq t \leq T_s \qquad (2)$$

The 16-QAM modulated signal is represented as

$$S_i(t) = V_{I,i}\sqrt{\frac{2}{T_s}} \cos(2\pi f_c t) + V_{Q,i}\sqrt{\frac{2}{T_s}} \sin(2\pi f_c t)$$

$$0 \leq t \leq T \ i = 1, 2, \dots, M \qquad (3)$$

where, $V_{I,i}$ and $V_{Q,i}$ are the amplitudes of the quadrature carriers amplitude modulated by the information. In Figure 4 we simulate the waveform of modulated signal.

The advantage of moving to the higher order formats is that there are more points in the constellation and therefore it is possible to transmit more bits per symbol. Figure 5 is the spectrum of modulated signal and Figure 6 is binary symbol mapping for 16-QAM constellation.

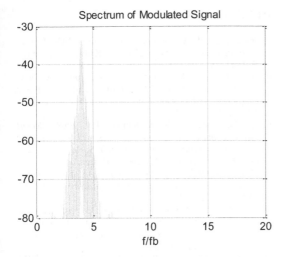

Figure 5.   Spectrum of modulated signal.

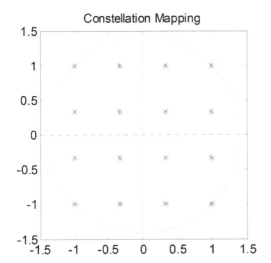

Figure 6.   Constellation mapping of 16-QAM.

4   BER AND SER OF 16-QAM

In this section, we compare the simulation results with the theory values to test the performance of our system.

Assuming that the additive noise n follows the Gaussian probability distribution function (John et al. 2004),

$$p(x) = \frac{1}{\sqrt{2\pi\sigma^2}} \exp\left\{\frac{-(x-\mu)^2}{2\sigma^2}\right\}$$

with $\mu = 0$ and $\sigma^2 = N_0/2$ (4)

Assuming that all the symbols are equally likely, the total probability of symbol error is (May et al. 1998),

$$P_{16QAM} \approx \frac{4}{16} \cdot 2erfc\left(\sqrt{\frac{E_s}{10N_0}}\right) + \frac{4}{16} erfc\left(\sqrt{\frac{E_s}{10N_0}}\right)$$
$$+ \frac{8}{16} \cdot \frac{3}{2} erfc\left(\sqrt{\frac{E_s}{10N_0}}\right) \approx \frac{3}{2} erfc\left(\sqrt{\frac{E_s}{10N_0}}\right)$$

(5)

From the (5), we can calculate the symbol error. $P_{s,16QAM} = (3/2)erfc\left(\sqrt{E_s/10N_0}\right)$, and then we can get the bit error rate for 16-QAM in Additive White Gaussian channel:

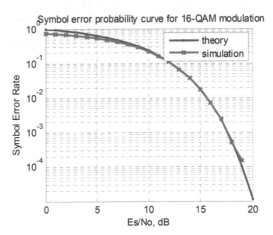

Figure 7.   SER for 16-QAM modulation.

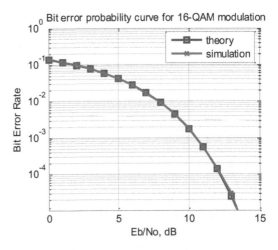

Figure 8.   BER for 16-QAM modulation.

$$P_{b,16QAM} = \frac{3}{2k} erfc\left(\sqrt{\frac{kE_b}{10N_0}}\right) \qquad (6)$$

5 CONCLUSIONS

The QAM scheme, which is a combination of in-phase and quadrature components, is a useful modulation technique in modern wireless communication systems especially in an OFDM system for achieving high data rate transmission. In this paper, we design and simulation study of the 16-QAM modulation communication system based on Matlab. A modified phase estimator was developed. It is found that with an optimum receiver at SNR below 13.5 dB, RR gives better performance than other. For large N, square 16-QAM has a smaller symbol error rate than star 16-QAM at moderate to high SNR. The approach described for calculating the SER and using it to determine optimum receiver performance can be used to design QAM systems with good signal constellations.

REFERENCES

Barry, J.R. Lee, E.A. & Messerschmitt, D.G. 2004. *Digital Communication: Third Edition*, Kluwer Academic Publishers.

Chong, C.V. & Venkataramani, R. 2003. A New Construction of 16-QAM Golay Complementary Sequences, IEEE Transactions on Information Theory, 49(11): 2953–2959.

Garg, G. Helleseth, T. & Kumar, P.V. 2009. *Recent advances in lowcorrelation sequences, New Directions in Wireless Communications Research, Springer*, 63–92.

Guo, D.H. & Hsu, C.Y. 2003. *Minimization of the PARP of OFDM by linear systematic block coding, Asia-Pacific Conference Communication*. APCC, Malaysia, 129–133.

May, T. Rohling, H. & Engels, V. 1998. Performance analysis of Viterbi decoding for 64-DAPSK and 64-QAM modulated OFDM signals, *IEEE Transactions on Communications*, 46: 182–190.

Robing, C. & Tarokh, V. 2001. A construction of OFDM 16-QAM Sequences Having Low Peak Powers, *IEEE Transactions on Information Theory*, 47(5): 2091–2094.

Advanced Materials and Structural Engineering – Hu (Ed.)
© 2016 Taylor & Francis Group, London, ISBN 978-1-138-02786-2

Wavelet analysis of the inter-annual change of typhoons striking Guangdong Province

Q.Y. Zhang, S.B. Zhong & Q.Y. Huang
Department of Engineering Physics, Institute of Public Safety Research, Tsinghua University, Beijing, China

ABSTRACT: Typhoons are one of the most destructive types of natural disasters striking coastal areas in the world. Guangdong Province is one such areas, which is the regional center with a high concentration of wealth in China, and suffers great loss from typhoons annually. Using the continuous wavelet time series, this paper analyzes the interannual change of typhoons striking Guangdong Province during 1949–2013. According to the continuous wavelet spectrum contour map and the global wavelet spectrum, the cycle periods of 23 years, 13 years and 4 years in the large scale, meso scale and small scale, respectively, can be obviously revealed. Based on the results from the wavelet variance test ($\alpha = 0.0$), the main cycle periods of 5-year, 8-year, 12-year and 23-year time scales are observed. Among them, the maximum peak corresponds to the 12 years, indicating the strongest shock of 12 years is the first period, which provides a useful clue for the risk evaluation of typhoons striking Guangdong.

1 INTRODUCTION

Typhoons are one of the most destructive types of natural disasters that most intensively strike the coastal areas in the word. Statistics shows that typhoons cause the highest insurance losses among all meteorological disasters (Gu 2012, Hsu & Li 2010). It has been found that the landfall of typhoons usually occurs in the coastal regions of southeast China. Guangdong Province is one such area, which is the regional center with a high concentration of wealth in China, and suffers great loss from typhoons annually. For example, Typhoon Usagi (No. 1319) that landed on Shanwei, Guangdong Province in 2013 caused an estimated 17.76 billion RMB in economic losses and killed at least 29 people. The property damage, economic losses, and causalities caused by typhoons have been increasing with population growth in Guangdong Province. Such alarming examples underscore the need for continuing scientific effort to study the characteristics of typhoon formation time and spatial migration. Since global climate change in recent decades has been disturbing the dynamic balance of the meteorological system, which in part appears as changes in the temporal characteristics of typhoons as well as variations in the frequency and intensity of typhoons in space, its impact on economy and environment should be seriously considered (Gu 2012, Ozger et al. 2010). Recognizing the temporal characteristics of typhoons can improve the accuracy of risk assessment, and enhance the capability of mitigating the disasters.

Recently, many kinds of research have been carried out on detecting and analyzing the temporal characteristics of the meteorological disasters. Based on the method of the objective weather chart analysis, Ren et al. created a new objective identification technique for persistent regional extreme events, which can effectively identify the properties of their spatial range and duration (Sang 2013). As to the typhoon disasters, according to the typhoon data of 1959–1994, Chan et al. studied the interannual variability and the long-term trend of the typhoon activities that occur in the northwestern Pacific, pointing out that during the early 1960s and the mid-70 s, the landfall frequency had a downward trend and then an upward one, but since 1994, the frequency showed a decreasing trend. The result has been widely accepted by relevant scholars (Torrence & Compo1998). Zhou et al. studied the typhoon cyclone activities and evaluated the cause of the typhoons from 1949 to 1999 in the northwestern Pacific (Ozger et al. 2010).

The wavelet analysis is one of the time-scale (i.e. time-frequency) signal analysis techniques. It has the characteristics of multi-resolution, and the ability to denote the local signal characteristics of both time and frequency domains. The wavelet analysis is the time-frequency localization analysis, which has the fixed window size, but with a variable shape for the width and the bandwidth (Torrence & Webster 1999, Vickery et al. 2009). In recent years, there has been an increasing interest in the use of wavelet analysis in a wide range of fields in science and engineering (Torrence & Webster 1999, Vickery et al. 2009).

The wavelet transform is a feature transformation, which preserves both time and frequency information. Wavelet transform is especially useful for signals that are non-stationary, have short-lived transient components, have features at different scales, or have singularities (Xiao et al. 2011). Considering the dynamic characteristics and non-uniform distribution of typhoon landing data and the need for regions in typhoon loss preventing management, we propose a framework to explore the temporal characteristics of typhoon landing data. For the change of typhoon landing times, there is no uniform period, but the local characteristics of multi-time scales of the time-frequency domain. Besides, even the Fourier analysis added window cannot analyze such characteristic. The wavelet analysis in the time-frequency domain has good localization characteristics, and can also detect the mutation point to analyze the multi-scale trend. The wavelet analysis is a good tool for detecting the change of the typhoon landing times (Yi & Shu 2012). Wavelet analysis has been applied in various areas, especially in climate and hydrological studies. Torrence and Compo introduced a practical step-by-step guide to wavelet analysis with examples taken from large-scale climate indices (Torrence & Webster 1999, Vickery et al. 2009). Antil and Coulibaly proposed a wavelet-based approach to describe the local inter-annual variability in steam flow, and to identify plausible climatic teleconnections that could explain these local variations (Xiao et al. 2011).

In this study, the temporal characteristics of typhoons striking Guangdong Province are explored, using time-series analysis techniques (e.g. trend, wavelet, and continuous wavelet analysis). Although several studies have analyzed the characteristics of typhoons, they are focused on the statistical methods (Ozger et al. 2010). By using the Morlet wavelet analysis as the basis function for one-dimensional continuous wavelet transform for the time-frequency localization analysis of the typhoon landing time series of Guangdong Province in 1949–2013, the characteristics of different time scales are revealed.

2 DATA AND METHODS

2.1 Data

In this paper, the needed data include statistics data of typhoons, meteorological data and the geographical data. The specific sources of the data are explained as follows:

1. Statistics data of typhoons: they collected from the Tropical Cyclone Yearbook of the Shanghai Typhoon Institute and the Chinese typhoon network, including the typhoon landing locations and the loss caused.
2. Meteorological data: they are acquired from the Best Track Dataset of the China Meteorological Administration (CMA) (during 1949–2013), containing the position and intensity information of typhoons in 6-hour intervals. Relevant properties include typhoon names and numbers (tropical depression was abbreviated as TD in the records), the position of typhoon centers (longitude and latitude), central pressures, and maximum sustained 10-min surface wind speeds.
3. Geographical data: they are provided by the surveying and mapping bureau, including city-level administrative boundary, the county-level administrative boundary data, and the vector graph of 1:104, and the attribute data of Guangdong Province.

2.2 The flow of data preprocessing and analysis

The main pre-processing of data was conducted in Excel of Microsoft office, and the calculation was made on the Matlab software platform. The one-dimensional continuous wavelet analysis has many varieties of models. This paper uses the Morlet transform analysis to draw maps and different frequency wavelet images.

In order to reduce the boundary effect, the symmetric method employed to extend the data on both ends is as follows:

Set the original data sequence:

$$f(1), f(2), ..., f(n)$$

Stretch forward n points:

$$f(-i) = f(i+1), \ i = 0, 1, ..., n-1$$

Extend backward n points:

$$f(i+n) = f(n+1-i), \ i = 1, ..., n$$

Standardized time-series data of typhoon landing numbers, and the Morlet wavelet analysis for one-dimensional continuous wavelet transform were used to get the corresponding wavelet coefficients, and the wavelet coefficients of the original data sequence were retained after the wavelet transform. When calculating the wavelet coefficients, the 2n mode was used in the extending time sequence, and the minimum scale parameter of the landing times is 2 years, and the maximum was set to 32 years. With the square interpolation of the wavelet coefficients, the contour map of the squared wavelet coefficients modulus was obtained, which is the global wavelet spectrum.

In order to determine the main cycle period of the time sequence, wavelet variance ($\alpha = 0.05$) was calculated and the wavelet variance map was drawn. Wavelet variance reflects the volatility of the energy distribution with the time scales, and the relative strength of different scales can be determined. The scale corresponding to the peak value was called the main time scale of the sequence, which is the main cycle period. Wavelet variance is an integration of the square modulus of the wavelet transform coefficients in the time domain. The data analysis and graphing were done in MATLAB7.0, and the formula is as follows:

$$Var(\alpha) = \int_{-\infty}^{\infty} |C(\alpha, \tau)|^2 \, d \tag{1}$$

2.3 Morlet wavelet analysis

The Morlet wavelet is a periodic function enveloped by a Gaussian function (Yi & Shu 2012). Therefore, the Morlet wavelet transform has been widely used to identify the periodic oscillations of the real-life signal (Özger, Mishr, and Singh, 2010).

In this study, we use a wavelet analysis function called Morlet to analyze typhoon disaster data. The Morlet wavelet analysis function is defined as:

$$g(t) = e^{\frac{-t^2}{2}} e^{i\alpha t} \tag{2}$$

Morlet wavelet is not the orthogonal function, and it can be used for the continuous wavelet transformation and the discrete wavelet transform (Sang 2013). In this study, the continuous wavelet transformation is used to analyze a discrete typhoon amount of time series.

The wavelet coefficients on the top and bottom of the data are subject to 'edge effects', because only half of the Morlet wavelet lies inside the data set. For long wavelengths, the edge effect can stretch across the whole time series according to the method mentioned above. Thus, the boundary of edge effects on the wavelet coefficients forms a wavelength-dependent curve for 'weak edge affected' areas known as the 'cone of influence'.

3 RESULTS AND DISCUSSION

3.1 Statistics characteristics of annual frequency

According to the Statistics Yearbook, in 1949–2013, the total number of tropical cyclones landing on Guangdong Province is 243, an average of 3.47, a maximum of 8, and a minimum of 1. On the whole, there is no long-term underlying trend through time, i.e. the landing number of

typhoon on Guangdong Province has not shown any marked shift over the last six decades (Fig. 1). What is clearly evident, however, is that the active period is in the early 1950s, the early mid-60 s, the mid 80 s and the early 90 s. Since the mid-50 s, and the mid—and late 80 s, the frequency of the typhoons landing on Guangdong Province is obviously less.

3.2 Statistics characteristics of monthly frequency

Based on the monthly frequency analysis of tropical cyclones during 1949–2013, the main influenced months are shown in Figure 2. As Figure 3 shows, the tropical cyclones frequently land on Guangdong Province in July and August. However, the frequency goes down sharply in May and November. During 1949–2013, the earliest landing on Guangdong is typhoon Neoguri, which migrated from Hainan to Guangdong. There is an obvious seasonal variation of typhoon land off, and 74% of the typhoon landing time is between

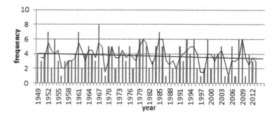

Figure 1.   Annual frequency analysis of tropical cyclone.

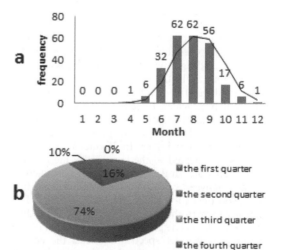

Figure 2.   Monthly frequency analysis of typhoons (a) and quarterly frequency analysis of typhoons (b).

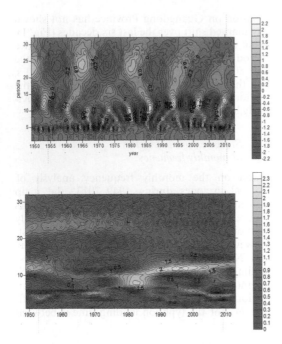

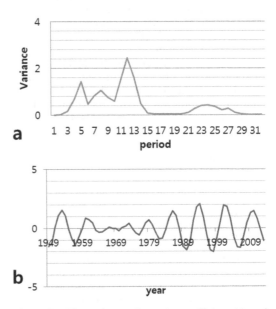

As shown in Figure 4, during the evolution of the landing number of typhoons, the value of 12-year is the most obvious periodic oscillations.

Wavelet variance ($\alpha = 0.05$) is drawn in Figure 4. Wavelet variance reflects the volatility of the energy distribution with the time scales, and the relative strength of different scales can be determined. The scale corresponding to the peak value is called the main time scale of the sequence, which is the main period. Wavelet variance is the integration of the square modulus of the wavelet transform coefficients in the time domain.

As Figure 4(a) shows, the wavelet variance of typhoon landing numbers has four relatively obvious peak values, which corresponds to 5-year, 8-year, 12-year and 23-year time scales. Among them, the maximum peak corresponds to the 12 years, indicating the strongest shock 12-year being the first period of the landing numbers; 5-year corresponds to the second peak, being the second period; the 8-year and 23-year correspond to the third and the fourth peaks, respectively, being the third and fourth periods. This shows that the fluctuation of the above four periods controls the typhoon landing numbers within the whole time domain. Based on the results of the wavelet variance test, the wavelet coefficients map of the landing numbers evolution of the first main period was drawn, as shown in Figure 4(b). From the trend of the main period, the characteristics of average period variation in different time scales are shown. According to Figure 4, in 12-year, there is an average period

Figure 3. Continuous wavelet spectrum contour map (a) and global wavelet spectrum of the continuous wavelet (b).

July and September. There is no tropical cyclone landing in the first quarter.

3.3 Continuous wavelet analysis

The key of the multi-period wavelet analysis is the absolute value of coefficients and the wavelet variance. The continuous wavelet contour map and the global wavelet spectrum of the continuous wavelet spectrum are shown in Figure 3. The contour map reflects the period variation of the typhoon landing times in different time scales and its distribution in the time domain, which can be used to judge the trend of the landing times in different time scales. Effective periodic events are seen as the light regions on the continuous wavelet transform image and the peaks of the global wavelet spectrum image. In Figure 3, the periodicities are located in the large scale, meso scale, and small scale. In the large scale, the periodicities located between 23-year scales level are seen during 1949–1981 for each continuous spectrum. The meso-scale periodic (13-year) events occurred during 1989–2013. The global wavelet spectrum of the continuous wavelet spectrum reflects the energy density distribution in the time domain. Besides, the greater coefficient indicates the stronger periodic of the corresponding period or the scale.

Figure 4. The variance of wavelet coefficient (a) and characteristics of 12-year time scales (b).

of 7-year of the landing numbers, approximately experiencing 9 distinctive qualities.

4 CONCLUSIONS

This study applied a time-series analysis tool, wavelet analysis, to evaluate the potential influence of long-term typhoon landing on Guangdong Province using the 65-year data of 1949–2013. The results show that the wavelet analysis is well suited to the time-series application of typhoon landing times and can easily find out the mutative trend of the landing times.

According to the result of the continuous wavelet spectrum contour map and the global wavelet spectrum of the continuous wavelet, the cycle periods of 23 years, 13 years and 4 years in the large scale, meso scale and small scale, respectively, can be obviously revealed. Based on the results of the wavelet variance test ($\alpha = 0.05$), the main periods of 5-year, 8-year, 12-year and 23-year time scales were shown. Among them, the maximum peak corresponds to the 12 years, indicating that the strongest shock 12-year was the first period of the landing numbers.

Typhoons are among the most deadly and destructive natural disasters in terms of loss of human life and economic destruction. Knowledge of such analysis can help the government reduce the loss caused by typhoon disasters. At the same time, it is possible to determine which areas and what time in Guangdong Province are more susceptible to the typhoon disasters. Our conclusion will be a useful tool for policy and decision makers (e.g. insurance industry, regional planners) to evaluate the vulnerability of coastal communities in order to emphasize the need for sound integrated management against typhoons in the regions.

ACKNOWLEDGMENT

The authors would like to thank the support of the National Natural Science Foundation of China (Study on Pre-qualification Theory and Method for Influences of Disastrous Meteorological Events, Grant No. 91224004), the youth talent plan program of Beijing City College (Study on Semantic Information Retrieval of Decision Analysis of Emergency Management for Typical Disastrous Meteorological Events, Grant No. YETP0117), and the National Natural Science Foundation of China (Key Scientific Problems and Integrated Research Platform for Scenario Response Based National Emergency Platform System, Grant No. 91024032).

REFERENCES

Gu, Y. 2012. Statistical Analysis on Tropical Cyclone of Northwestern Pacific from 1949 to 2010 Years, *Meteorology Journal of Inner Mongolia*, 1: 3–7.

Hsu, K. & Li, S. 2010. Clustering spatial–temporal precipitation data using wavelet transform and self-organizing map neural network. *Advances in Water Resources*, 33(2): 190–200.

Özger, M. Mishra, A.K. & Singh, V.P. 2010. Scaling characteristics of precipitation data in conjunction with wavelet analysis. *Journal of Hydrology*, 395(3–4): 279–288.

Sang, Y. 2013. A review on the applications of wavelet transform in hydrology time series analysis. *Atmospheric Research*, 122: 8–15.

Torrence, C. & Compo, G.P. 1998. A practical guide to wavelet analysis, *Bulletin of the American Meteorological society*, 79(1): 61–78.

Torrence, C. & Webster, P.J. 1999. Interdecadal changes in the ENSO monsoon system. *Journal of Climate*, 12: 2679–2690.

Vickery, P.J. Masters, F.J. & Powell, M.D. et al. 2009. Hurricane hazard modeling: The past, present, and future. *Journal of Wind Engineering and Industrial Aerodynamics*, 97(7–8): 392–405.

Xiao, Y.F. Duan, Z.D. & Xiao, Y.Q. et al. 2011. Typhoon wind hazard analysis for southeast China coastal regions. *Structural Safety*, 33(4): 286–295.

Yi, H. & Shu, H. 2012. The improvement of the Morlet wavelet for multi-period analysis of climate data. *Comptes Rendus Geoscience*, 344(10): 483–497.

Zhou, J.H. Shi, P.J. & Chen, X.W. 2002. Spatio-temporal variability of tropical cyclone activities in the western North Pacific from 1949 to 1999. *Journal of Natural Disasters*, 03: 44–49.

Zuki, Z.M. & Lupo, A.R. 2008. Interannual variability of tropical cyclone activity in the southern South China Sea, *Journal of Geophysical Research*, 113(D6).

Advanced Materials and Structural Engineering – Hu (Ed.)
© 2016 Taylor & Francis Group, London, ISBN 978-1-138-02786-2

Detection of radome defects with a new Holographic Subsurface Imaging Radar system

S.Z. Xu, M. Lu, C.L. Huang & Y. Su
School of Electronic Science and Engineering, National University of Defense Technology, Changsha, China

ABSTRACT: A novel non-destructive detection approach that can be used to detect the radome defects by using the microwave technology is introduced in this paper. The small defects in radome can cause serious threats to radars. Therefore, it is necessary to detect regularly the defects in radome. There are some methods to perform this task, such as inspection and ultrasound technology, but both of them are either fallibility or demanding. This paper introduces a new Holographic Subsurface Imaging Radar (HSIR) that has the advantages of small size, long battery life, fast scanning speed, and high resolution, which can be used to reveal the small internal defects in radome. Some experiments were performed to certify the applicability of the radome defects detection, and the results were displayed in the form of radar images in which defects in the radome provided a good contrast.

1 INTRODUCTION

As the basic telecommunication components, radars are widely used in the airplane, missile, and navigation system. Considering the fact that radars must be located outside the metal airframe and concealed from the public view, a structural, weatherproof enclosure called the radome is introduced to cover and protect the antenna.

Radome can protect the internal antennas from the influences and the disturbances of bad weather conditions, so as to ensure the proper working of radars in all weather and bad working conditions (Crone et al. 1981). The introduction of radome helps to prolong the life expectancy of the antennas, simplify the construction, and reduce the weight of the radar system.

There are two main demanding technical specifications for radome materials to work effectively and safely (Crone et al. 1981). First, radome must be constructed of material that minimally attenuates the electromagnetic signal transmitted or received by the antennas. Second, radome must be of great wind resistance and high heat resistance, especially when the airplane flies with a super high speed in the air. Figure 1 shows the sample of the radome composite materials.

However, the defects are inevitable in almost every stage of the radome's life, and present different forms in corresponding periods. Defects such as delaminations, debondings, and foreign debris may appear in the production process. Though these kinds of defects can result in heavy damages to the radome, they can be detected by X-ray before the radome is deployed and subsequently

the resulting damages can be avoided. However, the qualified products may also have some flaws when they are exposed to the external impact loadings and the poor working conditions. Some flaws such as cracks and watering always come from the rain erosion, hailstone shock, thunder damage, bird striking and misoperation of the users (Hanssen et al. 2006, Benjarmin 1985). These defects may modify the mechanical and electromagnetic properties of radomes. Watering should be paid special attention because water can absorb electromagnetic waves and attenuate the amplitude of the wave, and then reduce the performance of the whole radar system. Besides, especially when the aircraft moves with a high speed, cracks or debondings can put the aircraft in danger because the radome may be crunched by huge air pressure and the antennas would be destroyed. Therefore, regular radome

Figure 1. Sample of the radome composite material.

detection is very important. There exist some methods to perform the detection, such as X-ray, optical time domain reflectometry, fiber Bragg grating sensors, ultrasonic, shearography and stimulated infrared thermography (Bocherens et al. 2000); however, it is too time-consuming and labor-demanding to dismantle the radome for detection using these methods.

Therefore, two main methods are developed to execute online detection. A simple but common online method to detect these defects is inspection, which requires the inspectors to watch carefully on the surface of the radome to find the abnormality on the local surface such as cracks and hollows.

However, the inspection would fail when it comes to internal delaminations and cracks. On this occasion, another method called the tap test is applied (David & John 1999). A coin or some small metal objects could be used to knock the local surface around the position of possible defects. Inspectors need to distinguish the alteration in the reverberant sound. Theoretically, the frequency of the reverberant sound would change if there is any setback in the composite material such as cracks, delaminations or watering. However, this method faces an obvious weakness that it relies heavily on the experience of the inspector and is not always reliable.

Given the fact that not all the flaws can be avoided, some relative specific appliance is mostly necessary (Vinay et al. 2000). A detection tool called "woodpecker" was designed and produced in Japan. When the probe knock the radome, the amplitude and the frequency of the reflect sound could be recorded by the computer connected with the probe to examine the internal defects. This method is much more sensible and convincible, and could avoid the most disturbances of the environment.

However, the above methods are not distinct nor do they exactly locate and image the flaws. A novel nondestructive detection approach to probe the invisible defects inside the radome by using a microwave is introduced in this paper. The property that electromagnetic wave would be reflected when it comes through the heterogeneous objects is utilized to record the small internal defects of the radome, and the image of the radome could be displayed directly by using digital image processing techniques. Besides, the inspectors need no extra experiences but just read the images and locate the flaws on the screen. The new Holographic Subsurface Imaging Radar (HSIR) is used to perform this approach.

2 HOLOGRAPHIC SUBSURFACE IMAGING RADAR

The HSIR was designed and produced by the National University of Defense Technology.

Compared with other commercial products available in the market, the HSIR has the advantages of small size, long battery life, fast scanning speed, high resolution, and capability of non-metal detection.

The radar system consists of antennas, actuating device, Li-ion battery, digital processor and mini-computer. Figure 2 shows a set of the system. Thanks to good ergonomic design, the size and the weight of the system is well controlled, for which more specifically, the total weight is 5.7 kg, so it is very portable and can be operated by a single person. Unlike the conventional impulse radar using the wide band, the HSIR uses a narrow band continuous wave with low radiation intensity. Owing to advanced autofocus technologies, the HSIR can automatically search and image in the right depth of the targets.

Another outstanding feature is that the HSIR employs a circular scanning approach, by which a 16*16 cm² area could be scanned quickly with only one set of transmit and receive antennas. The match of the received data with the antenna position is guaranteed by a sophisticated synchronously scanning mechanic. Therefore, the device is much affordable than other counterparts with more array antennas. The original data received by the antenna can be displayed immediately on the screen intergraded in the device or personal computer transferred by wireless technology. After finishing one scanning period, the data are analyzed and calculated by relative software on the computer and then the results are displayed on the screen. The entire process takes only about 20 seconds.

Furthermore, this system consumes less power. Under typical operating conditions, the working time can be around 2 hours when powered by a 1800 mAh Li-ion battery, and there is no need of other electrical source, so it is very convenient to perform the task outdoors.

The device can be widely used due to the ability to detect metal as well as non-metal objects such

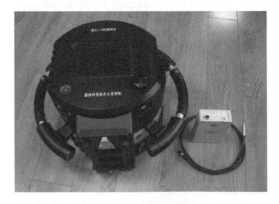

Figure 2. One set of the HSIR system.

as plastic or ceramic. It can be used in medical detection, archaeological probe, non-destructive evaluation, and unexplored ordnance probe. In this paper, we focus on the detection of invisible internal defects of the radome.

3 EXPERIMENTS AND RESULTS

In order to test the performance of the proposed approach, some experiments were conducted by using this new holographic subsurface imaging radar to visualize similar defects on the radome. The experimental materials were made by Acrylonitrile-Butadine-Styrene (ABS), with the thickness of 1 cm and 3 cm, respectively. The dielectric permittivity of this material is about 3, which is low and close to the radome composite material. To copy the defects in the radome, we made some similar flaws on the ABS board, such as cracks, holes, and watering.

For the first set of experiments, a series of small circular holes with a depth of 2 mm and a radius of 3 mm were drilled on the back side of the 3 cm-thick board. The distances of centers of these holes are 1 cm, 2 cm, and 3 cm, respectively, and the space of the closest hole brims was only about 4 mm. The board was put on the HSIR device with the front side on which the holes were invisible. Figure 3 shows a photograph of the experimental setup.

From the result image shown in Figure 4, the separate holes can be clearly distinguished with a good contrast. Because the closest holes can be seen separately, it certified the great resolution competence. As expected from the image, this device has a great potential to detect smaller holes with a radius less than 3 mm.

For the demonstration of the capability of watering detection, the word "S" was written on the back side of the board with a thin slice of water. For better vision by photos, the experimental fresh water was mixed with a little red ink water. Figure 5 shows the defect sample and the imaging result. A distinct

Figure 3.   Experimental scene.

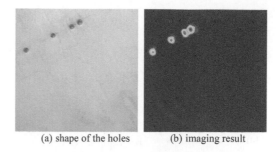

(a) shape of the holes   (b) imaging result

Figure 4.   Small holes imaging using the HSIR.

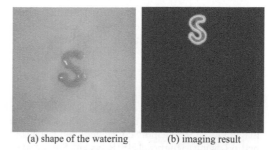

(a) shape of the watering   (b) imaging result

Figure 5.   Watering imaging using the HSIR.

Figure 6.   The experimental sample of cracks.

word "S" can be seen from the result image, which is identical with the word written with water.

In order to test the capability of this system on the small cracks detection, in the last experiment, some small cracks were made on the 1 cm board. Because the radome is a sandwich construction and the outermost paint covered the inside cracks, it is highly essential to use some sensitive appliances to detect invisible cracks under the paint. Some cracks were made by hand on the board and to conceal the cracks, we put another ABS board of 3 cm thick between the device and the defective board. Figure 6 illustrates the sample of the

 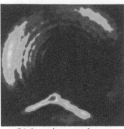

(a) shape of the cracks   (b) imaging result

Figure 7. Cracks imaging using the HSIR.

man-made cracks, and Figure 7 shows the detection result. From the result, the cracks can be clearly distinguished with a good contrast.

4 CONCLUSION

The paper has proposed a method for the radome defects detection by using the holographic subsurface imaging radar. It is a more convenient and visualized method owing to the novel HSIR system. Some experiments are performed and the results demonstrate the applicability of the new HISR system to detect the typical defects in the radome with clear and distinct images. However, all the experiments performed on the assumption that the detected surface is planar

while the top of the radome is always curved. The further study will focus on this problem. In addition, the pattern recognition and image processing could be used to automatically locate the defects from the images and warn the inspector. That is to say, the future defects detection can be fully automated without people's interpretation.

REFERENCES

Benjarmin, R. 1985. Problems of radome design for modern airborne radar, *Microwave Journal*. 28(1): 145–153.
Bocherens, E. et al. 2000. Damage detection in a radome sandwich material with embedded fiber optic sensors, *Smart Materials Structural*. 9: 310–315. UK.
Crone, G.A.E. Rudge, A.W. & Taylor, G.N. 1981. Design and Performance of airborne radomes—a review. Communications, *Radar and Signal Processing, IEE Proceedings*, 128: 451–464.
David, K. & John, J. 1999. *Imaging of flaws in composite honeycomb aircraft structures using instrumented tap test,* Nondestructive Evaluation of Aging Materials and Composites III Meeting: 236–245. Newport Beach.
Hanssen, A.G. Girard, Y. & Olovsson, L.A. 2006. Numerical model for bird strike of aluminium foam-based sandwich panels, *International Journal of Impact Engineering*, 32: 1127–1144.
Vinay, D. Gohar, M. & David, K. 2000. *Finite element modeling and advanced imaging by instrumented tap testing.* The 2000 ASME International Mechanical Engineering Congress and Exposition: 393–397. USA: Orlando.

Advanced Materials and Structural Engineering – Hu (Ed.)
© 2016 Taylor & Francis Group, London, ISBN 978-1-138-02786-2

The database design for the remote monitoring of an injection molding plant

O. Paudel & C.S. Gao

Nanjing University of Aeronautics and Astronautics, Nanjing, China

ABSTRACT: The Database Management System for Remote Monitoring in Molding Injection plant can provide different aspects of data throughout the whole process of injection with both machine variables and process variables of injection. First, a broader view of the database management system for Remote Monitoring in Molding Injection is provided in this paper. Second, the database design method and the design of this research project are described in detail. In this paper, the relational data model is mainly used. Third, the common techniques to develop the industrial database application, especially the ODBC API and ADO, are described, and in this paper we choose the ADO to develop database application for this remote injection molding. Last but not the least, application software is developed using VC++ combined with MS SQL Server 2005, and the main interfaces are provided in this paper for the readers.

1 INTRODUCTION

Information system management application is hyped up in these days due to increasing computing facilities and industrial communication hardware and software applications. Even injection molding is growing due to excessive plastic consumption. However, injection molding are categorized as follows: dynamics process with multi-variable, multi-phase, and cycle to cycle control (Agrawal et al. 1987). Therefore, information management based on the database system is one of the aspects of industries to accomplish their goal and requirements.

Database management usually refers to how data is structured, organized, stored, classified, protected, detected and accessed by querying, and this functionality of the database is a core of data processing (Date 1977). Currently, the process control and manufacturing process of injection molding is automated and intact with different intelligence, highly advanced and sophisticated technologies within the architecture of injection molding plants and factories and even industrial process control and monitoring (Han & Kim 2011). These monitored data and all the information generated from the plant and its environment by database application can be significant for the diagnosis of output qualities of products. In this paper, we will progress through different topics to follow systematic development of the Database Management System for Remote Monitoring in Molding Injection.

2 CONCEPTUAL DESIGN ARCHITECTURE

The initial phase of the database design is a conceptual design, and this phase uses the entity-relationship diagram (E-R diagram) to describe its model (Chen 1976). Conceptual design architecture is generally based on the user's requirements. This phase is used to construct the data model of a real situation of injection molding and its real world view or problems, and it is independent of physical consideration. In the process of design database, the designers need to analyze the system, and its relationships between all schema and data storage.

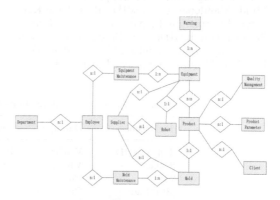

Figure 1. General ER diagram of the remote monitoring of injection molding.

The main concept behind the conceptual design is to outline three parts: attribute, entity and relationship. With constructing these three-part model data redundancy, the ER diagram of a conceptual model proposed in this paper could be achieved, as shown in Figure 1.

3 LOGICAL STRUCTURE DESIGN OF THE DATABASE

The logical structure of the database is the abstract design. Logical design involves arranging data into a series of logical relationship called attributes and entities. Basically, we can define the logical as collecting data (industrial), identifying primary key, foreign key and data items.

The logical design structures basically change the ER diagram into a logical structure, which is in line with the data model, while the data model is functionally constructed with the database management system. For the database design for the remote monitoring of the injection plant, we used MS-SQL server 2005 to develop for a data model. The relational models of remote monitoring injection molding are mentioned below with their physical structure.

System Users Information (UserName(Key), Password, TypeID);
The User type (TypeID(Key), Name, Note)
Employee (EmployeeID(Key), Name, Sex, Birthdate, ContactPhone);
Department (DepartmentID(Key), DepartmentName, Note);
Equipment(EquipmentID(Key), EquipmentName, Tonnage, ManufactureDate, Robot ID, SupplierID);
Equipment Maintenance (EquipmentID, Begin Date(Key), ServiceInformation, EmployeeID, ElapsedHour);
Mold(MoldID(Key), MoldName, FigureNo, WareHousing, StorageSite, SupplierId);
Mold Maintenance (MoldID, BeginTime(Key), ServiceInformation, EmployeeId, ElaspseHour);
Product (ProductID(Key), ProductName, Amount, Dailyoutput, NumberinCharge, BeginDate, EndDate, MoldId, EquipmentID, ClientID);

Parameter of Product (ProductID, Occurrence Time(Key), MaterialTemperature, MaterialHumidity, WaterTemperature, WaterPressure, Waterflow, OilTankTemperature, MoldTemperature, CoolingTime, CycleTime, Qualified);
Client Information (ClientID(Key), CompanyName, ContactName, ContactPhone, Address, Province, city);
Quality Problem (ProductId, Occurrence Time(Key), Description, Reason, TemporarySolution, Fundamentalsolution, Picture1, Picture2, Picture3);
Robot (RobotID(Key), RobotName, SupplierId, Note);
Supplier (SupplierID(Key), FactoryName, ContactName, Address, Province, City);
Warning (EquipmentID, OccurrenceTime(Key), WarningInformation, Endtime).

4 PHYSICAL DESIGN STRUCTURE OF THE DATABASE

After the logical design is completed, the first step is to create an initial physical data model by transforming the logical model into a physical implementation based on an understanding of the DBMS. In our case, we have chosen MS-SQL 2005 to be used for deployment (Zhang 2001). Therefore, a complete transformation includes following work entities transformation in tables, attributes transformation into columns and transforming domains into data types and constraints. Some of the modules of the database are described below with their attributes and objects.

5 IMPLEMENTATION OF THE PROTOTYPE

The database User interface is usually stated as a layer between the database application and the database. In this paper, it is stated as database application generated by the designer and MS-SQL server 2005 with database functionality,

Table 1. The employee information table.

Field name	Data type	Length	Remarks	NULL
Employee ID	Varchar	10	Primary key	No
Name	Varchar	20	Employee Name	No
Sex	Char	6	–	No
BirthDate	Date Time	–	–	No
DepartmentID	Int	–	Foreign key	No
Contactf Phone	Varchar	–	–	Yes

Table 2. The equipment maintenance table.

Field name	Data type	Length	Remarks	Null
EquipmentID	Varchar	10	Foreign key	No
Begin Time	Date Time	–	Primary key	No
Service Information	Varchar	50	–	No
EmployeeID	Varchar	10	Foreign key	No
Elapsed Hour	Real	–	–	No

Table 3. The product quality problem table.

Field name	Data type	Length	Remarks	Null
ProductID	Varchar	20	Foreign key	No
OccurrenceTime	DateTime	–	Primary key	No
Description	Varchar	50	–	No
Reason	Varchar	50	–	No
Temporary solution	Varchar	50	–	No
Fundamental solution	Varchar	50	–	No
Picture 1	Image	–	–	Yes
Picture 2	Image	–	–	Yes
Picture 3	Image	–	–	Yes

Table 4. The parameter of the product table.

Field name	Data type	Length	Remarks	Null
ProductID	Varchar	20	Foreign key	No
Occurrence Time	DateTime	–	Primary key	No
Water Temperature	Real	–	Temp of cooling water	No
Water Pressure	Real	–	Pressure of cooling water	No
Water Flow	Real	–	Flow of cooling water	No
Oil Tank Temp.	Real	–	Temperature of oil	No
Mold Temperature	Real	–	Temperature of mold	No
Cooling time	Real	–	Cooling time of mold	No
Cycle Time	Real	–	Cycle time of mold	No
Qualified	Bool	–	–	No

respectively. User Interface for the remote injection molding database consists of three major software components that are included in this system: DDRIM Database, data access and manipulation module, and user interface. In this paper, Microsoft Visual C++6.0 is used to develop the Microsoft Windows-based user interface (Li et al. 2008).

Database access technology abstracts the communication process between the DBMS and database application, and provides the access interface in order to simplify the access process. The common database access technologies used in VC++ are the ODBC (Open Database Connectivity API), MFC ODBC (Microsoft Foundation Classes ODBC), DAO (Object Link Embedding Database), and

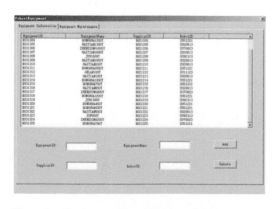

Figure 2. Equipment management module.

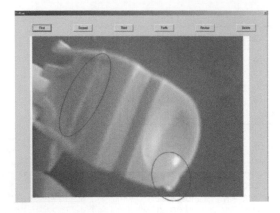

Figure 3. Product quality management module.

ADO (Active Data Object). Here, we choose the ADO.

The software is divided into five modules as follows: Authority Management, Equality Management, Mold Management, Product Management, and Client & Supplier Management (Li et al. 2010). The equipment management module and the product quality management module are shown in Figure 2 and Figure 3.

6 CONCLUSIONS

Obviously, this software has a user-friendly interface. Users can easily add, modify, delete and retrieve the data in the database, and analyze the data to reveal the entire process of the Molding Injection. By viewing different modules of the DDRIM, the users are able to get the information about all the molding injection.

Although the DDRIM helps a lot for the molding injection processing, the challenges that the molding injection faces still calls for management automation software to classify, store and sort the information in a systematic manner. Furthermore, analysis of the information stored in the database can be used for various decision-making proposes.

In conclusion, the DDRIM can be used to handle some rough task of molding injection in the plant. However, the DDRIM in the future must be more intelligent such as creating more user-friendly report and even analysis support for decision-making and printing support, support various kinds of multi-media information and feasible to integrate with other organizational software such as MES (Manufacturing Executive System) and ERP (Effective Resource Planning).

REFERENCES

Agrawal, A.R. Pandelids, I.O. & Pecht, M. 1987. Injection-Molding Process Control Review, *Polymer Engineering and Science*, 27(18): 1345–1357.

Date, C.J. 1977. *An Introduction to Database System*, Addison Wesley, Reading, Massachusetts.

Han, S.H. & Kim, Y.K. 2011. *An architecture of Real-time, Historical Database System for Industrial Process Control and Monitoring*, Proceeds International Conference on Computers, Networks, System and Industrial Engineering. 31–34.

Li, M.M. Wu, J. & Zhou, X.M. 2008. Visual C++6.0, The Navigation of the Database System Development Examples, *People's Post and Telecommunication Press*, 203–214.

Li, X.Z. Zhang, X.H. & Li, X.S. 2010. SQL server 2000 The Development of Management and Application system, *People's Post and Telecommunication Press*, 79–101.

Peter. C. 1976. The Entity—relationship Model—Toward a Unified View of Data, *ACM Transactions on Database System*, 1(1): 9–36.

Zhang, Q. 2001. Object-oriented Database system in Manufacturing: selection and application, *Industrial Management and Data system*, 101(3): 97–105.

Advanced Materials and Structural Engineering – Hu (Ed.)
© *2016 Taylor & Francis Group, London, ISBN 978-1-138-02786-2*

Hybrid method for prediction of centrifugal pump noise and optimization

G.Q. Liu & T. Zhang
School of Naval Architecture and Ocean Engineering, Huazhong University of Science and Technology, Wuhan, Hubei, China

W. Zhao
China Ship Development and Design Center, Wuhan, Hubei, China

Y.O. Zhang
Department of Mechanical and Aerospace Engineering, Jacobs School of Engineering, University of California, San Diego, California, USA

ABSTRACT: Flow-induced noise of the centrifugal pump is investigated with Large-Eddy Simulation (LES) and Lighthill's acoustic analogy. The results of the flow field are computed with LES, and treated as sound source data with Lighthill's acoustic analogy to predict the flow-induced noise. The flow field and the noise spectra of the centrifugal pump at different flow rates are computed and analyzed in details. It is concluded that the rotating stall and the flow separation result in broadband noise, while the interaction between the jet-wake from the impeller and the tongue causes tonal noise. The optimization of the centrifugal pump for good noise performance is conducted with the Orthogonal Optimization Method (OOM). With the $L_9(3^3)$ orthogonal table, nine numerical calculation schemes are established and computed. The range analysis is employed to analyze the most important factor for the flow noise in the pump. Finally, an optimized pump is achieved, whose sound pressure level at blade-passing frequency declines by a large margin (about 12 dB) compared with the original one.

1 INTRODUCTION

The flow field in the centrifugal turbomachinery has been studied by experiments and theories for several decades (Choi et al. 2006, Kearney et al. 2009, Bachert et al. 2010). In recent years, numerical simulations of unsteady flow field in the centrifugal pump have attracted much attention. Pump noise sources can be categorized into the following classification: hydraulic (such as flow turbulence, cavitation, blade passing interaction and rotating stall (Choi et al. 2003)), mechanical and electrical noise from motor. This paper is focused on the numerical study of the hydraulic noise in the centrifugal pump with Computational Fluid Dynamics (CFD) and computational aeroacoustics (CAA).

As is known, accurate prediction of the instantaneous turbulent flow field in a centrifugal pump is very important for the prediction of the hydraulic noise. Large-Eddy Simulation (LES) is currently applied in a wide variety of engineering applications. LES has the potential to predict unsteady flow fields that cover large-scale separation much more accurately than traditional Reynolds

Averaged Navier Stokes (RANS) approaches. LES is more feasible for a complex geometry such as a pump than Direct Numerical Simulation (DNS) because of its lower numerical costs. In a conclusion, LES has been widely used to compute turbulence and operates on the Navier-Stokes (N-S) equations to reduce the range of length scales and the computational cost.

The approaches used to compute acoustic problems can be categorized into direct computation and indirect, or hybrid computation (Cai et al. 2010, Ji & Wang 2010). The former for solving the huge area costs great amount of calculations, so that it is generally used to solve the 2D noise problems at a low Reynolds number. Seo et al. (2008) presented a DNS procedure for the cavitating flow noise and wrote the compressible N-S equations for the two-phase. Sandberg et al. (2009) used DNS to compute directly both the near-field hydrodynamics and the far-field sound generated by flow past NACA airfoils. Ikeda et al. (2012) numerically simulated the aero-acoustic sound generated from the flow around two NACA four-digit airfoils by using high-order finite-difference schemes to discretize compressible N-S equations. The latter uses CFD to calculate the main noise source to get the noise level of the far field with the integral or acoustic analogy strategy. One of the most fundamental assumptions of the indirect simulation is that it ignores the inter-actions between the acoustic source field and the flow field, so this method is more suitable for the low Mach number situations. The advantages of this method are to estimate the characteristics of the heterogeneous fluid and take the average flow effects on the sound field into account. Kato et al. (2007) simulated sound generated from flows with a low Mach number based on LES, Dynamic Smagorinsky Model (DSM) and Lighthill's acoustic analogy. Moon et al. (2010) predicted subsonic noise using a hybrid method, which was composed of LES and Linearized Perturbed Compressible Equations (LPCE). Wei et al. (2012) proposed CFD, Finite Element Method (FEM) and Refined Integral Algorithm (RIA) to predict the propeller excited underwater noise of the submarine hull structure.

With respect to the centrifugal pump noise, Chu et al. (1995) and Dong (1994) indicated that flow-induced noise in a centrifugal pump was mainly generated by blade-tongue interaction. Neise (1976) summarized the research work done by various experimenters of noise reduction in centrifugal fans during the last fifteen years. The results showed most of the works were aimed at reducing the blade passage sound. He (Neise & Koopmann 1980) also investigated experimentally a method which was applied to an acoustic resonator for reduction of the aerodynamic noise generated by turbomachinery. With appropriate tuning of the resonator, reductions in the blade passing frequency tones of up to 29 dB were observed. Liu et al. (2007) used the LES approach, acoustic analogy and vortex sound theory to study the noise of a centrifugal fan and the results of the sound fields were in good agreement with the experiments.

Thus the purpose of this study is to characterize the noise with numerical simulations and eventually to shed some light on the noise generating mechanism. The prediction of centrifugal pump noise is based on a LES/Lighthill's acoustic analogy hybrid method. The simulations are performed by solving the 3D-LES equations for the centrifugal pump with the commercial code CFX and the flow-induced noise is calculated by solving the variational formulation of Lighthill's acoustic analogy with ACTRAN, which is based on the finite and infinite methods. Firstly, the flow field and flow noise in the original pump at different flow rates are simulated and analyzed in details. Secondly, the optimization against the pump flow-induced noise is conducted with the Orthogonal Optimization Method (OOM) and the range analysis and

then the best configuration is achieved. Finally, the comparison between the original pump and the best optimized one is made in details. In comparison with the previous work by the authors (Dong 1994, Neise 1976, Qi et al. 2009), the flow-induced noise in the centrifugal pump decreases by a large margin of about 12 dB when the geometry of the impeller and the volute is modified.

2 NUMERICAL METHODOLOGY

2.1 Large-eddy simulation

LES is a popular technique for simulating turbulent flows and is used to compute the unsteady flow in this paper. The LES equations are proposed, which are derived from the N-S equations, by removing the smallest scales through a filtering operation. The filtering process effectively filters out eddies and explodes the flow variables into large scale and small scale parts. This process filters out the small eddies with size smaller than the filter width. In this way, LES can resolve large scales of the flow field with good fidelity and model the small scales of the solution.

The filtering equation is defined as:

$$G(x,x') = \begin{cases} \dfrac{1}{V}, & x' \in V \\ 0, & x' \notin V \end{cases} \tag{1}$$

The continuity equation is

$$\frac{\partial \rho}{\partial t} + \frac{\partial}{\partial x_i}(\rho \overline{u_i}) = 0 \tag{2}$$

The N-S equation is

$$\frac{\partial u_i}{\partial t} + \frac{\partial u_i u_j}{\partial x_j} = \nu \frac{\partial^2 u_i}{\partial x_j^2} - \frac{1}{\rho} \frac{\partial p}{\partial x_i} \tag{3}$$

Then, the filtered N-S equation can be written as:

$$\frac{\partial}{\partial t}(\rho \overline{u_i}) + \frac{\partial}{\partial x_j}(\rho \overline{u_i u_j}) = \frac{\partial}{\partial x_j}\left(\mu \frac{\partial \sigma_{ij}}{\partial x_j}\right) - \frac{\partial \overline{p}}{\partial x_i} - \frac{\partial \tau_{ij}}{\partial x_j} \tag{4}$$

where, τ_{ij} is the sub-grid Reynolds stress, and ρ_{ij} is the stress tensor caused by the viscous of molecular.

2.2 Lighthill's acoustic analogy theory

The method for predicting noise using Lighthill's equation (Lighthill 1952, 1954) is usually referred to as a hybrid method since noise generation and propagation are treated separately. Through the variational formulation of Lighthill's acoustic analogy, the acoustic solution is obtained at all discrete points inside the numerical domain. Furthermore, the interaction between the solid surfaces and the aerodynamic noise are taken into account by the acoustic solver without requiring explicit surface source terms as in the integral formulation. The variational formulation of Lighthill's acoustic analogy was first derived by Oberai et al. (2000). The variational equation in the ACTRAN manual is as follows:

$$-\int_\Omega \frac{\omega^2}{\rho_0 c^2}\psi\delta\psi d\Omega - \int_\Omega \frac{1}{\rho_0}\frac{\partial\psi}{\partial x_i}\frac{\partial\delta\psi}{\partial x_i}d\Omega$$
$$= \int_\Omega \frac{i}{\rho_0\omega}\frac{\partial\delta\psi}{\partial x_i}\frac{\partial T_{ij}}{\partial x_j}d\Omega - \int_\Gamma \frac{1}{\rho_0}F(\overline{\rho v}_i n_i)d\Gamma \quad (5)$$

where, $\delta\Psi$ is a test function, and Ω is the part of the computational domain (non-moving and non-deforming). Γ is the boundary surface where the contribution of the rotating CFD domain in the turbomachinery simulation can be taken into account.

3 NUMERICAL MODEL AND FLOW ANALYSIS

3.1 Numerical model

The centrifugal pump studied in this work is a three-dimensional low-specific-speed centrifugal pump (n_s = 78). The main parameters of the centrifugal pump and water at 20°C is summarized in Table 1.

The numerical domain is composed of several components: suction duct, impeller, volute, diffuser and outlet duct. The inlet and outlet ducts are extended to minimize boundary conditions effects. The front and rear axial gap between the impeller

Table 1. Main parameters of the centrifugal pump and water.

Impeller inlet diameter, m	D_1	0.075
Hub diameter, m	d_h	0.04
Impeller outlet diameter, m	D_2	0.188
Impeller outlet width, m	b_2	0.012
Number of impeller blades,	Z	6
Total head, m	H	10
Flow rate, m³/h	Q	25
Rotational speed, r min⁻¹	n	1450
Density, kg/m³	ρ	997
Viscosity, kg/(m·s)	v	8.899×10^{-4}
Sound velocity, m/s	c	1500
Specific heat capacity, J/(kg·K)	c_p	4220

and the volute casing sidewalls is not taken into account. And all the pipes involved are regarded as rigid. Therefore the noise induced by the vibration of the pipes is not considered. In addition, the pipes are assumed sound-insulating.

Some views of the numerical model of the centrifugal pump are shown in Figure 1. The mesh is structured with hexahedral cells in the whole model. In the volute, a mesh-refined zone is defined near the tongue and at the edge of the impeller blades.

The boundary conditions are a constant total pressure at the inlet section and a constant mass flow rate at the outlet section. The grids of the impeller and the volute are connected by means of a frozen-rotor interface for steady state calculations. The relative position of the impeller and the volute remains unchanged in the calculations. For the unsteady simulations, the grids are connected by using a rotor/stator interface, so the relative position would be changed according to the angular velocity of the impeller in the calculations. That is the so-called "sliding grid technique". The standard k–ε turbulence model is used for the steady simulation while the LES turbulence model is used for the unsteady simulation.

The simulations are performed for three flow rates corresponding to about 60%, 100%, and 140% of the nominal flow rate. Prior to the unsteady flow simulations, steady simulations are performed, and the steady numerical results are used as the initial condition for the unsteady calculation. For the bulk of the simulations, 30 time steps are considered per blade-passing period (180 time steps per impeller revolution), with a time step of 5.75×10^{-5} s. For each time step, the convergence criterion is that the scaled residuals are less than 10^{-5}, with at least 10 iterations per time step. The unsteady simulations are considered complete when the flow reaches a clear periodic regime, with a minimum of 10 full impeller revolutions.

The resulting grid size used in the simulation is determined after a mesh dependence analysis

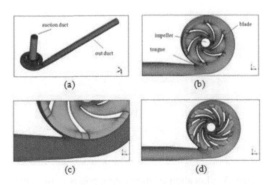

Figure 1. Meshing of the centrifugal pump.

on the total head (Fig. 2) carried out in the stationary regime. Observing the differences in the total head among the three cases, and considering the computational cost, the computational mesh (1,420,000 cells) is adopted to conduct the following simulations.

3.2 Flow analysis

In order to understand the pressure fluctuation characteristics of the internal flow of the centrifugal pump, the time-varying static pressures are recorded at the monitoring points located at various positions (every 10° at the mid-span) around the volute, especially near the interface of the impeller-volute (Fig. 3). After the computation of the unsteady flow is steady, the pressure data at 2000 steps is output per 2° of the pump rotation. To compared the pressure fluctuations at different flow rate, the pressure coefficient C_p is defined as follow,

$$C_p = p/(0.5\rho u^2) \qquad (6)$$

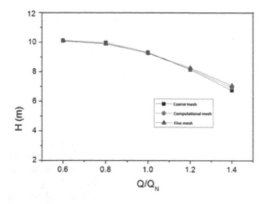

Figure 2. Influence of mesh on the total head for the original pump.

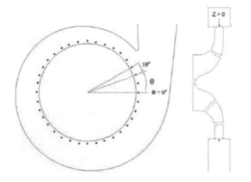

Figure 3. Location of monitoring points in the volute of the pump.

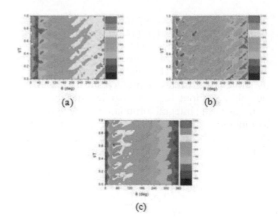

Figure 4. Time-spatial behavior of pressure fluctuations as function of angular position at impeller-volute interface for different rates, at mid-span impeller outlet: (a) at 60% flow rate; (b) at nominal flow rate; (c) at 140% flow rate.

where, p is the value of static pressure and u is the circumferential speed of the impeller outlet.

Figure 4 shows the pressure coefficient of the monitoring point of the outlet of centrifugal pump impeller as contours. The x-coordinate represents θ that is the angular position around the impeller and the y-coordinate corresponds to the dimensionless time t/T. The time history of the non-dimensional static pressure indicates that at any angular position the pressure fluctuations display periodic behavior related to blade passage. The largest amplitudes are located near the tongue (behind it) as a consequence of volute shape.

4 ACOUSTIC PREDICTION

The CAA sources from the CFD results interpolated between the CFD mesh and the acoustic mesh can be achieved with the interface named CFD in ACTRAN. Here, the results at 2000 steps are saved in the CFD and all are treated as sound sources in ACTRAN. The acoustic model is presented in Figure 5. The numerical domain is composed of two modules, the volume source domain and the sound propagation domain. The surface source is chosen on the interface between the impeller and volute. The free transmission of the noise is specified at the outlet of the volute.

In order to avoid the spurious noise, a spatial filter of cosine type is defined at the truncated surface as shown in Figure 6. The values are the weighting coefficients of the sound sources. It is used to eliminate the spurious noise caused at the truncated surface. Four lengths of the spatial filter

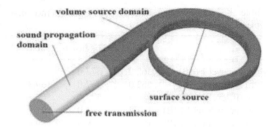

Figure 5. Acoustic computational domain.

Figure 6. Spatial filter used to damp source terms; red = 1: no damping is applied; blue = 0: sources are reduced to zero.

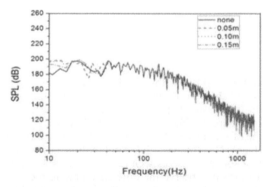

Figure 7. SPL spectra at the pump outlet under various lengths of spatial filter.

are adopted to predict the noise respectively, and the simulation results shows the length has little effect on the results, as seen in Figure 7. At last, 0.1 m is adopted as the length of the spatial filter.

5 OPTIMIZATION

5.1 The orthogonal optimization method

OOM is an effective method that can pick out the best factor combinations with fewer experiments and little simulation time. Orthogonal table $L_n(m^k)$

is introduced to arranges experiments as a tool, where n is the number of rows, that is the number of experiments; k is the number of columns, namely the number of factors; m is the number of factor levels. In OOM, all variables that can affect product's performance are called factors. The values that factors are assigned are levels. When all of the factors are assigned a set of particular values or levels, an experiment such as numerical simulation can be performed, and output would be recorded for statistical analysis, so that we can find out the optimal parameters combinations. OOM as a partial factor design method can be applied in multifactor experiments.

In this paper, the centrifugal pump parameters are orthogonal factors for the noise, which is mainly influenced by the radius of the tongue, the length of the splitter blade and the offset angle of the splitter blades et al. So the radius of the tongue, the length of the splitter blade and the offset angle of the splitter blades as the typical factors are picked out to be factors of orthogonal experiments in this study, 3 levels of each factor are given respectively. 3 factor and 3 levels orthogonal experiment is conducted, the factors and levels of orthogonal experiment are presented in the $L_9(3^3)$ orthogonal table. With the $L_9(3^3)$ orthogonal table, nine numerical calculation scheme—s are established and computed.

5.2 The results of the optimization

Low specific speed centrifugal pumps have been widely used due to its small flow rate and high head. In order to improve the performance of low specific speed centrifugal pumps, researchers have conducted a series of optimization research, and developed a number of low specific speed centrifugal pumps that have good hydraulic performance and low noise. According to many previous studies, increasing the gap between the tongue and the impeller up to about 20 percent of the impeller radius can reduce the internal flow noise in the centrifugal pumps to a certain extent. In this section, OOM is used to optimize against flow noise in the centrifugal pump. The details of the optimization factors are given in Figure 8 and Table 2, where

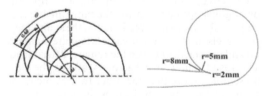

Figure 8. Optimization factors.

Table 2. Three levels of the optimization factors.

Factor	Level	Value	Factor descriptions
A	1	2 mm	The radius of the
	2	5 mm	tongue
	3	8 mm	
B	1	$0.5L$	The length of the
	2	$0.4L$	splitter blade
	3	$0.6L$	
C	1	0.5θ	The offset angle of
	2	0.4θ	the splitter blade
	3	0.6θ	

Table 3. Nine numerical calculation schemes and sound pressure levels at f_{BP}.

Schemes	Factor A	Factor B	Factor C	SPL (dB)
1	Level 1	Level 1	Level 1	191.7
2	Level 1	Level 2	Level 2	186.3
3	Level 1	Level 3	Level 3	184.2
4	Level 2	Level 1	Level 2	184.0
5	Level 2	Level 2	Level 3	187.4
6	Level 2	Level 3	Level 1	190.6
7	Level 3	Level 1	Level 3	197.2
8	Level 3	Level 2	Level 1	193.4
9	Level 3	Level 3	Level 2	198.9
Original pump	–	–	–	194.7

Table 4. The range analysis for the orthogonal optimization.

	Factor A	Factor B	Factor C
Sum of level 1 (dB)	562.2	573.0	575.7
Sum of level 2 (dB)	562.0	567.0	569.3
Sum of level 3 (dB)	589.5	573.7	568.8
Average of level 1 (dB)	187.4	191.0	191.9
Average of level 2 (dB)	187.3	189.0	189.8
Average of level 3 (dB)	196.5	191.2	189.6
Range (dB)	9.2	2.2	2.3

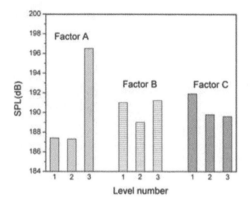

Figure 9. The difference among the levels for each factor.

numerical calculation schemes are established as the L_9 (3^3) orthogonal table.

In this paper, in order to save computing time and storage space, CFD results at only 1000 steps are saved as sound sources for the flow noises of the nine numerical schemes and the original pump consistently, and all the results are obtained at the nominal flow rate. The sound pressure levels at blade-passing frequency of the nine numerical schemes and the original pump are given in Table 3. It shows that most schemes reduce the noise to some extent, when compared with the original pump except schemes 7 and 9, and especially schemes 3 and 4, in which the noise declines about 10 dB.

In order to get further analysis of the factors for internal flow noise in centrifugal pumps, the orthogonal optimization and the range analysis are employed, as seen in Table 4, that is, the sound pressure levels of the levels of each factor are summed up and averaged, and then the range is determined. The results show that Factor A affects the flow noise much more than the other two. Namely, the radius of the tongue is the most important factor, followed by the offset angle of the splitter blade,

and the last one is the length of the splitter blade. To get the lowest noise, theoretically the radius of the tongue should be 5 mm (Level 2), and the length of the splitter blade should be $0.4L$ (Level 2), and the offset angle of the splitter blade should be 0.6θ (Level 3). However, this combination is not the best for the noise as seen in scheme 5 in Table 3. To directly observe the difference among the levels for each factor, Figure 9 came forward. For Factor A, there is little difference between Level 1 and Level 2, neither is between Level 2 and Level 3 for Factor C, as seen in Figure 9. The best pump configuration for producing the lowest flow noise is plotted in Figure 9 and Table 3, of which the radius of the tongue is 5 mm (Level 2), the length of the splitter blade is $0.4L$ (Level 2), and the offset angle of the splitter blade is 0.4θ (Level 2).

As a representative investigation, the best combination is computed at high resolution to compare with the original pump, where CFD results at 2000 steps are saved as sound sources. The comparison is made in details as below.

The pressure signals are analyzed and processed using Fast Fourier Transform (FFT) with a Hanning Window. The pressure fluctuation

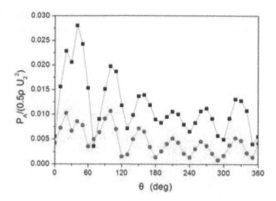

Figure 10. Numerical distribution of the unsteady pressure (amplitude at blade-passing frequency). Square: numerical data for the original pump; circles: numerical data for the optimized pump.

Table 5. Sound pressure levels of the original/optimized pump in dB.

Frequency (Hz)	Original pump	Optimized pump
145	193.6	181.2

at the blade-passing frequency is presented in Figure 10, for the original and optimized pumps at 36 positions distributed around the volute (Fig. 3). From the Figure the amplitude of the pressure fluctuation in the optimized pump is lower than that in the original one. There are six peaks in both the original pump and optimized one. The results show clearly that the splitter blades decrease the pressure amplitude levels. Additionally, there is a peak of the pressure fluctuations at the vicinity of the tongue for the original pump.

The sound pressure levels at blade-passing frequency of the original/optimized pumps are given in Table 5. Compared with the original pump, the optimized pump makes the sound pressure level at blade-passing frequency decline by about 12.4 dB.

6 CONCLUSIONS

The main objectives of the present work are to estimate the magnitude of the flow-induced noise in the centrifugal pump and obtain the optimal configuration of the pump. In the present work, flow-induced noise in centrifugal pumps is calculated by the hybrid method of combining LES and Lightlill's acoustic analogy theory. The process of simulation consists of two steps: firstly the LES is used for turbulent flow computation, then the results

of the flow field, such as the pressure fluctuations, is taken as sound source, and Lightlill's acoustic analogy theory is employed to simulation the flow-induced noise. The optimization of centrifugal pump is conducted with OOM which makes use of the L_9 (3^3) orthogonal table as a tool. OOM and the range analysis are employed to determine the optimal values of the length of the splitter blade, the offset angle of the splitter blade and the radius of the tongue.

The simulation results show that the flow field in the centrifugal pump is strongly unsteady and maybe has periodic behavior related to the blade passage. It is confirmed that the largest pressure fluctuation occurs at the vicinity of the tongue due to the interaction between the impeller and the volute. The splitter blades lower the non-uniform outflux from the impeller, which results in reduction in the amplitude of pressure fluctuation.

However, in the LES simulation, a poor wall resolution tends to over predict considerably wall pressure fluctuations downstream from the impeller. As a result, the tonal noise of the sound pressure level spectra is not obvious. But through analysis of the flow-induced noise at different flow rates, it can be seen that the rotating stall and the flow separation result in broadband noise, while the tonal noise is attributed to the interaction between the jet-wake from the impeller and the tongue.

By employing the orthogonal optimization and the range analysis, it is found that the pump configuration for producing the lowest flow noise should have the radius of the tongue at 5 mm, the length of the splitter blade at $0.4L$, and the offset angle of the splitter blade at 0.4θ; and the sound pressure level of the optimized pump at the blade-passing frequency declines by a large margin (about 12 dB) when compared with the original pump.

ACKNOWLEDGEMENT

This material is based upon work funded by Zhejiang Provincial Natural Science Foundation of China under Grant No. LQ12E09002; Project (51308497) supported by National Natural Science Foundation of China.

REFERENCES

Bachert, R. Stoffel, B. & Dular, M. 2010. Unsteady cavitation at the tongue of the volute of a centrifugal pump. *Journal of Fluids Engineering*, 132(6): 061301.

Cai, J.C. Qi, D.T. Lu, F.A. & Wen, X.F. 2010. Study of the Tonal Casing Noise of a Centrifugal Fan at the Blade Passing Frequency. Part I. Aeroacoustics. *Journal of Low Frequency Noise, Vibration and Active Control*, 29(4): 253–266.

Choi, J.S. McLaughlin, D.K. & Thompson, D.E. 2003. Experiments on the unsteady flow field and noise generation in a centrifugal pump impeller. *Journal of Sound and Vibration*, 263(3): 493–514.

Choi, Y.D. Kurokawa, J. & Matsui, J. 2006. Performance and internal flow characteristics of a very low specific speed centrifugal pump. *Journal of Fluids Engineering*, 128(2): 341–349.

Chu, S. Dong, R. & Katz, J. 1995. Relationship between unsteady flow, pressure fluctuations, and noise in a centrifugal pump. *Journal of Fluids Engineering*, 117(1): 24–29.

Dong, R.R. 1994. *The effect of volute geometry on the flow structure, pressure fluctuation and noise in a centrifugal pump*. Baltimore: Johns Hopkins University.

Ikeda, T. Atobe, T. & Takagi, S. 2012. Direct simulations of trailing-edge noise generation from two-dimensional airfoils at low Reynolds numbers. *Journal of Sound and Vibration*, 331(3): 556–574.

Ji, M. & Wang, M. 2010. *Aeroacoustics of turbulent boundary-layer flow over small steps*. American Institute of Aeronautics and Astronautics Pap, (2010–6).

Kato, C. Yamade, Y. Wang, H. Guo, Y. Miyazawa, M. Takaishi, T. & Takano, Y. 2007. Numerical prediction of sound generated from flows with a low Mach number. *Computers and Fluids*, 36(1): 53–68.

Kearney, D. Grimes, R. & Punch, J. 2009. An experimental investigation of the flow fields within geometrically similar miniature-scale centrifugal pumps. *Journal of Fluids Engineering*, 131(10): 101101.

Lighthill, M.J. 1952. On sound generated aerodynamically. I. General theory. *Proceedings of the Royal Society of London. Series A. Mathematical and Physical Sciences*, 211(1107): 564–587.

Lighthill, M.J. 1954. On sound generated aerodynamically. II. Turbulence as a source of sound. *Proceedings of the Royal Society of London A: Mathematical, Physical and Engineering Sciences*, 222(1148): 1–32.

Liu, Q. Qi, D. & Tang, H. 2007. Computation of aerodynamic noise of centrifugal fan using large eddy simulation approach, acoustic analogy, and vortex sound theory. *Proceedings of the Institution of Mechanical Engineers, Part C: Journal of Mechanical Engineering Science*, 221(11): 1321–1332.

Moon, Y.J. Seo, J.H. Bae, Y.M. Roger, M. & Becker, S. 2010. A hybrid prediction method for low-subsonic turbulent flow noise. *Computers and Fluids*, 39(7): 1125–1135.

Neise, W. & Koopmann, G.H. 1980. Reduction of centrifugal fan noise by use of resonators. *Journal of Sound and Vibration*, 73(2), 297–308.

Neise, W. 1976. Noise reduction in centrifugal fans: a literature survey. *Journal of Sound and Vibration*, 45(3): 375–403.

Oberai, A.A. Roknaldin, F. & Hughes, T.J. 2000. Computational procedures for determining structural-acoustic response due to hydrodynamic sources. *Computer Methods in Applied Mechanics and Engineering*, 190(3): 345–361.

Qi, D.T. Mao, Y.J. Liu, X.L. & Yuan, M.J. 2009. Experimental study on the noise reduction of an industrial forward-curved blades centrifugal fan. *Applied Acoustics*, 70(8): 1041–1050.

Sandberg, R.D. Jones, L.E. Sandham, N.D. & Joseph, P.F. 2009. Direct numerical simulations of tonal noise generated by laminar flow past airfoils. *Journal of Sound and Vibration*, 320(4): 838–858.

Seo, J.H. Moon, Y.J. & Shin, B.R. 2008. Prediction of cavitating flow noise by direct numerical simulation. *Journal of Computational Physics*, 227(13): 6511–6531.

Wei, Y.S. Wang, Y.S. Chang, S.P. & Jian, F. 2012. Numerical prediction of propeller excited acoustic response of submarine structure based on CFD, FEM and BEM. *Journal of Hydrodynamics, Ser. B*, 24(2): 207–216.

Advanced Materials and Structural Engineering – Hu (Ed.)
© 2016 Taylor & Francis Group, London, ISBN 978-1-138-02786-2

A new recognition algorithm of the lunar mare area based on the DEM contrast

X.L. Tian

Lunar and Planetary Science Laboratory/Space Science Institute, Macau University of Science and Technology, Macau, China
Faculty of Information Technology, Macau University of Science and Technology, Macau, China

A.A. Xu

Lunar and Planetary Science Laboratory/Space Science Institute, Macau University of Science and Technology, Macau, China

T.L. Xie, H.K. Jiang & J.L. Wang

Faculty of Information Technology, Macau University of Science and Technology, Macau, China

ABSTRACT: With the help of the local area contrast, the new algorithm proposed in this paper proved its advantage by identifying the lunar mare area of the DEM images. The accuracy of the new algorithm was tested in three different areas, and the results were found to be satisfactory. The corresponding kappa coefficients of the new algorithm with only the DEM data are better than those of a previous algorithm with the DEM and CCD data.

1 INTRODUCTION

Lunar mare area is a perfect area for landing and exploding, because the large and flat area provides safety to both rovers and astronauts. Not only the member of Apollo 11—Neil Armstrong—finished the first manned landing in the Sea of Tranquility (mare area), but also China's Jade Rabbit of Chang E-3 finished the first unmanned landing in Mare Imbrium. It is not hard to tell the importance of selecting a mare area in science exploration from these two examples. Almost all the dark and relatively featureless mare areas are located at the near side of the Moon, covering about 31% of it. However, only 2% of its far side is covered by mare areas.

A new algorithm for the recognition of the lunar mare area based on the DEM contrast is proposed in this paper, which will be of much use. The number of mare areas in the Moon can be highly recognized by the new algorithm. Although, in previous results, the Kappa coefficient was already 0.78 (with both DEM and CCD data) or 0.82 (with CCD data), both are not bad for identifying the lunar surface. However, a better Kappa coefficient can be offered by the new algorithm proposed in this paper, and the best result can reach up to 0.8601, calculated by the DEM after clustering with the DEM contrast.

2 A NEW RECOGNITION ALGORITHM FOR THE LUNAR MARE AREA

This section can be divided into three parts: the flowchart of the new algorithm, details about three features and the clustering.

2.1 The flowchart of the new algorithm

In order to show the whole processing of the new algorithm, the flowchart is shown in Figure 1.

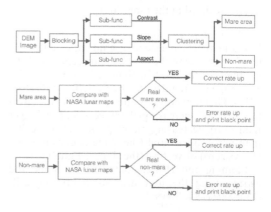

Figure 1. The flow chart of the new algorithm proposed.

2.2 Three features

In this part, three features of DEM data used for clustering in this paper will be discussed. One of them is the local contrast, which is the first time to be applied to the DEM data for recognizing the lunar mare area.

2.2.1 Feature 1—local contrast

The difference in color that makes an object's representation in a digital image different from others is called the contrast. It is a normal parameter in the digital image processing. Normally, it is a useful way to adjust the value of contrast to show more details in a digital image. However, this is the first contrast that is considered as a feature in a new algorithm to recognize the mare area on the surface of the Moon. In order to improve the quality of recognizing the result, the contrast is used in the algorithm to fix some problems caused by the image processing. There are many different methods for calculating the value of contrast due to different goals in their own fields, such as Weber contrast (Bex 2013) and Michelson contrast. The method applied in this paper is based on the pixel intensities called the Root mean square (Frazor & Geisler 2006) contrast, which does not depend on the spatial frequency content or the spatial distribution of contrast in the image. The RMS contrast is defined as the standard deviation of the pixel intensities and the equation is given by

$$C = \sqrt{\frac{1}{MN} \sum_{i=0}^{N=1} \sum_{j=0}^{M=1} (I_{ij} - \Gamma)^2} \qquad (1)$$

where intensities are the i-th, j-th element of the two-dimensional image of size M by N. The r is the average intensity of all pixel values in the image. The image I was assumed to have its pixel intensities normalized in the range [0, 1].

2.2.2 Feature 2—slope

In mathematics, the slope or gradient of a line is a number that describes both the direction and the steepness of the line (Clapham & Nicholson 2013). The slope is often denoted by the letter "m". (Eric 2013). There are only four situations about the direction of a line: increasing (the slope of a line is bigger than zero), decreasing (the slope of a line is smaller than zero), horizontal (the slope is constant and always zero) or vertical (the slope is undefined).

Fleming and Hoffer computed the slope and aspect using the four nearest-neighboring cells (Liu 2004) as follows:

$$f_x = (P7 - P1 + P8 - P2 + P9 - P3)/6 * r \qquad (2)$$

$$f_y = (P3 - P1 + P6 - P4 + P9 - P7)/6 * r \qquad (3)$$

P7	P8	P9
P4	P5	P6
P1	P2	P3

Figure 2. The window for calculating a slope.

In Formulas (2) and (3), "r" is the resolution of DEM; "f_x" is the slope of the north-south direction; and "f_y" is the slope of the east-west direction. Then, the slope of a point will be

$$S = \arctan((f_x)^2 + (f_y)^2)^{(1/2)} \qquad (4)$$

By following Formulas (2), (3) and (4) from the first calculation window to the last one, all the slopes of the whole DEM data can be calculated.

2.2.3 Feature 3—aspect

In geography, aspect is used to describe the direction that a slope faces. For example, a slope on the western edge of a big mountain toward a small plain is called as having a western aspect. The local temperature could be affected strongly by the aspect, The different angles of the Sun in the sky will cause different irradiation ranges and energy transmissions. For example, in the northern hemisphere of the Earth, the southern side can get more solar radiation easily than the northern side of a given surface, because the slope of the southern side is tilted toward the Sun. The scientist can utilize this special information, and obtain the high-accuracy temperature of the Moon by analyzing the aspect distribution.

The aspect A can be calculated by the formula $Z = (f_x, f_y)$, where f_x is the rate of change in latitude in the north-south direction and f_y is the rate of change in the east-west direction. (Liu 2004):

$$A = 270 + \arctan(f_y/f_x) - 90 * f_x/|f_x| \qquad (5)$$

where the "f_x" and the "f_y" have the same meaning in Equations (2) and (3). By iterating Formulas (2), (3) and (5) again and again, all the aspect of the whole DEM data can be calculated.

2.3 Clustering

2.3.1 Clustering distance

The weighted Euclidean distance model (independently by Horan, Bloxom, and Carroll and Chang) is considered as one of the most popular multidimensional scaling methods (Leeuw 1978). The formula of the Euclidean distance model is given as follows (Dong 2005):

$$d(i,j) = \sqrt{\left|X_{i1} - X_{j1}\right|^2 + \left|X_{i2} - X_{j2}\right|^2 + \cdots + \left|X_{ip} - X_{jp}\right|^2}$$

$$(6)$$

where $i, j = (1, 2, 3 \ldots p)$ are two objects data in the p dimension. Then, a weight is added for each variable according to the importance. The formula of the weighted Euclidean distance model is given by

$$dw(i,j)$$
$$= \sqrt{W_1\left|X_{i1} - X_{j1}\right|^2 + W_2\left|X_{i2} - X_{j2}\right|^2 + \cdots + W_p\left|X_{ip} - X_{jp}\right|^2}$$

$$(7)$$

2.3.2 Ward's method

Ward's method (Murtagh & Legendre 2011) is used as the clustering rule in this paper. There are different kinds of Ward's methods, and Ward's minimum variance criterion was used in this paper.

In each step, every pair of clusters will be merged. To accomplish this, a pair of clusters needs to be found, which will lead to a minimum increase. The formula of Ward's method is given by

$$W_{ij} = W(\{X_i\},\{X_j\}) = \left\|X_i - X_j\right\|^2 \qquad (8)$$

where W_{ij} is defined as the squared Euclidean distance between two object data.

3 TESTING RESULT

There are three test areas used in this paper, which are selected manually in order to compare with the previous algorithm (Jiang 2015). The test DEM data are 100 meters per pixel, computed from 69,000 WAC stereo models with digital photogrammetric techniques.

Then, there are 300 (20*15) points selected for each image to test how many areas can be recognized correctly by the new algorithm. Due to the size of the calculation window for clustering and other calculation of feature is 50*50, the test points are selected as 20*15 and evenly distributed in every testing area to avoid errors caused by the manual selection and other subjective reasons. However, the test points 20*15 cannot be guaranteed as a perfect condition for this kind

Table 1. The details of the testing area.

Name	Longitude	Latitude	Figure
W4	3.957° E ~ 18.138° E	18.121° N ~ 28.011° N	3
H010	0° W ~ 18° W	0° S ~ 14° S	4
SI	26° W ~ 39° W	40° N ~ 50° N	5

of area or the testing due to the limited experiment time.

More experiment for selecting the most suitable value of the testing windows will be conducted in the future. In order to show the error directly and compare with the original image, some black points are labeled in Figures 3, 4 and 5.

Figure 3. Testing results of the 'W4' area.

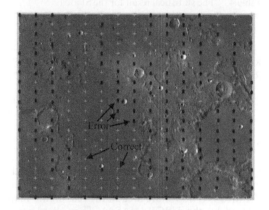

Figure 4. Testing results of the 'H010' area.

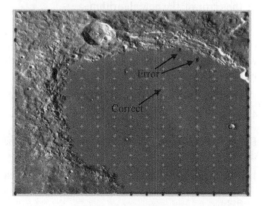

Figure 5. Testing results of the 'SI' area, where '+' indicates correct recognition, '●' indicates error in recognition.

Table 2. The statistical result for the W4 area.

	Mare (real)	Non-mare (real)	Sum	Rate
Result (mare)	187	4	191	97.91%
Result (non-mare)	15	94	109	86.24%
Correct rate	92.57%	95.92%	300	NULL

Table 3. The statistical result for the H010 area.

	Mare (real)	Non-mare (real)	Sum	Rate
Result (mare)	99	139	238	41.60%
Result (non-mare)	10	52	62	83.87%
Correct rate	90.83%	27.23%	300	NULL

Table 4. The statistical result for the SI area.

	Mare (real)	Non-mare (real)	Sum	Rate
Result (mare)	130	4	134	97.01%
Result (non-mare)	25	141	166	84.94%
Correct rate	83.87%	97.24%	300	NULL

For example, there is a testing point located in a mare area considered by the new algorithm according to the calculation of the clustering. However, the area is a non-mare area in the real digital map offered by the USGS (NASA). Then, that point will be marked as a black point in the area because of the error of the new algorithm.

Tables 2, 3 and 4 present the statistical results.

Among these three tables, Tables 2 and 4 provide a good result for recognizing the mare area, while Table 3 warns that this algorithm still needs to be improved in the future.

4 CONCLUSIONS

From the testing results, some conclusions can be drawn:

1. The new algorithm can recognize the mare area correctly if there is only mare and non-mare area.
2. The new algorithm finds it hard to differentiate between the mare area and the plain area or

some area with low height when the algorithm is required to classify more than these two classifications.
3. Errors in recognition can be found easily on the edge between the mare area and the non-mare area, no matter in what kind of classification.

Evaluating the result and making sure the high accuracy is not an accident. The value of Cohen's kappa can be calculated. It shows that the value for the 'W4' area is 0.8601 and for the 'SI' area is 0.8074. In the same or very similar areas, the above two results are both greater than that of Zhou (2011) (0.78), and Jiang (2015) (0.802). Based on the conclusion and the value of Cohen's kappa, the algorithm proposed in this paper has been proved that it can recognize more lunar mare areas than other algorithms can. This algorithm can be used for improving the precision in drawing the digital map for the Moon and other relative areas in the Moon.

However, the algorithm proposed in this paper still has some problems that need to be improved in the near future. The first one is how to improve the precision of recognizing the mare area on the edge. The second one is how to differentiate between a mare area and a plain area. The last one is how to recognize other kinds of surface in the Moon.

Due to the limited time, this paper only uses 50*50 pixel windows as a unit to calculate the feature value, which can be a possible problem leading to some errors in achieving a high precision for each feature. More experiments will be conducted in the future. Thus, this new algorithm can offer a better performance in recognizing the lunar surface.

ACKNOWLEDGMENT

This project was supported by the Science and Technology Development Fund, Macao SAR, China (No. 048/20 12/A2).

REFERENCES

Clapham, C. & Nicholson, J. 2013. *Oxford Concise Dictionary of Mathematics*, Gradient. Addison-Wesley.
Denis, G.P. & Peter, B. 2013. Measuring contrast sensitivity, *Vision Research*, 90: 10–14.
Dong, X. & Wei, Z.J. 2005. A clustering method of Euclid distance with weights. *Journal of information engineering university*, 6(1).
Fionn, M. & Pierre, L. 2011. Ward's Hierarchical Clustering Method: Clustering Criterion and Agglomerative Algorithm. http://arxiv.org/pdf/1111.6285.pdf
Jan, D.L. 1978. A new computational method to fit the weighted Euclidean distance model. *Psychometrika*, 43(4): 479–490.

Jiang, H.K. Tian, X.L. & Xu, A.A. 2015. A new segmentation algorithm for lunar surface terrain based on CCD image. *Research in Astronomy and Astrophysics 2015.*

Lalit, K. 2013. Effect of rounding off elevation values on the calculation of aspect and slope from a gridded digital elevation model. *Journal of Spatial Science.* 58(1): 91–100.

Liu, X.J. Wang, Y.F. Cao, Z.D. Li, J.F. & Tang, G.A. 2004. A study of spatial error distribution of slope and aspect derived from Grid DEM. *Journal of Bulletin of Surveying and Mapping*, (12): 11–13: 37.

Robert, A.F. & Wilson, S.G. 2006. Local luminance and contrast in natural images. *Vision Research*, 46: 1585–1598.

Spudis, P.D. 2007. Moon. World Book Online Reference Center, NASA. 2004.

The Lunar and Planetary Institute is a research institute that provides support services to NASA. http://www.lpi.usra.ed-u/resources/mapcatalog/usgs/

Weisstein, E.W. 2013. Slope. MathWorld—A Wolfram Web Resource.

Zhou, Z.P. Cheng, W.M. & Zhou, C.H. et al. 2011. Characteristic analysis of the lunar surface and automatically extracting of the lunar morphology based on CE-1 (in Chinese). *Chinese Science Bulletin (Chinese Ver)*, 56: 18–26.

Advanced Materials and Structural Engineering – Hu (Ed.)
© 2016 Taylor & Francis Group, London, ISBN 978-1-138-02786-2

The finite element analysis on a roll system of a 20-high Sendzimir mill based on ABAQUS

M.M. Zhang, X.B. Liu & F.Q. Feng
School of Aeronautic Manufacturing Engineering, Nanchang Hangkong University, Nanchang, China

ABSTRACT: According to the characteristics of the copper strip's rolling process, analysis of the elastic contact problem among the rolls and the difficulties in establishing the finite element model is performed. The space position, length and the diameter of the roll system are determined. The elastic deformation model of a roll system of a 20-high Sendzimir mill is established. The finite element analysis of a roll system of the 20-high Sendzimir mill is performed by using the finite element software ABAQUS. The results show that each roll in the roll system of the mill satisfies the strength requirements.

1 INTRODUCTION

As one of the most important species of the copper processing material, copper strip is widely used in aviation, electronics, energy and other fields. With the rapid development of the modern industry, the demand for the copper strip has been increasing in various fields, and the quality requirements of high performance, high surface quality and high precision for the copper strip have become more serious. The common 4- high or 6-high mill can no longer meet the demands for desired products, which only can be produced by a multi-roll mill (Gong 2011, Han et al. 2012).

Roll mill, the main bearing component, will directly affect the precision of the rolled copper strip. Under the comprehensive action of rolling force and rolling moment, roll will cause bending, shear, roll elastic flattening and other physical deformation. Therefore, it is very important to make the finite element analysis for the elastic deformation on the roll system. Traditionally, the analytic method and the influence function method are the main methods of analyzing the elastic deformation, but the finite element method was used in this paper. The finite element model of a roll system of a 20-high Sendzimir mill was established by using the large general finite element software ABAQUS, and the elastic deformation of rolls was analyzed by using the three-dimensional finite element method.

2 ANALYSIS OF DIFFICULTIES IN ESTABLISHING THE FINITE ELEMENT MODEL OF A ROLL SYSTEM OF A 20-HIGH SENDZIMIR MILL

2.1 *Empennage unsteady dynamics*

For solving the deformation problem of a roll system of the 20-high Sendzimir mill, it is difficult to analyze the contact pressure in the contact area between the two contact rolls. This is related to many factors, such as the size and the contact pressure of the contact area and roll shape. So, the iterative method, which is frequently used in engineering practice, was selected to solve the three-dimensional elastic problem (Chen 2014). The corresponding constraint equations, namely the penalty function equation and the Lagrange equation, were established by the different constraint conditions. The evidence shows that these calculations are usually done by iteration. Due to the increase in the roll number and its contact area, the computational complexity of the three-dimensional elastic problem will increase. Therefore, the key to solve it by finite element analyses is to solve the convergence problem of nonlinear inequality constraints in the three-dimensional elastic problem.

The difficulties in establishing the finite element model are as follows: (1) because the roll shapes of the 20-high Sendzimir mill contain the crown and cone angle, the position of the contact point is difficult to determine, and the model has a plurality of elastic contact, which is irregular face—face contact. (2) the difficulty in establishing the initial contact conditions: intermediate roll and work roll is rotated around their axis, resulting in clearance, which appear a Y-axis rigid body displacement within the rolling force; easy to cause the failure of solving the problem.

2.2 Determine the spatial location of the roll system

2.2.1 Determine the length of the roller in the roll system

According to the size of the maximum width B of the strip, which can be rolled by the 20-high Sendzimir mill, the size of the work roll length L of the rolling mill can be determined as follows:

$$L = B + a \qquad (1)$$

Because the tension is relatively large when the strip is rolled by the 20-high Sendzimir mill, its value is relatively small. For a large wide strip mill with B > 300 mm, a = 75 mm~125 mm, the maximum value is 200 mm. According to the specification of the mill, the width of the rolled strip ranges from 225 mm to 450 mm, by Formula (1): L = 450 + 75 = 525 mm.

The eight backup rolls in a roll system of the 20-high Sendzimir mill typing ZR33C-18 are composed of four backing bearings. So, the size of the backup roll length is equal to integer times of the width of the four backing bearings and the length of the saddle. To consult the relevant information, the structure value of the backup roll whose outer diameter is 160 mm can be obtained, that is, the design width value of the saddle is 40 mm and the width is 106 mm. Therefore, the value of the backup roll length in the 20-high Sendzimir mill is 544 mm.

2.2.2 The choice of the diameter of the roll system

According to the CTOYH formula, the minimum thickness of the rolled piece is determined by the maximum diameter of the work roll, and should satisfy the following formula:

$$D_{max} \leq \frac{0.28E}{\mu(k - \delta_{cp})} h_{min} \qquad (2)$$

where μ = coefficient of friction between the plate and strip and the surface of the work roll; k = plastic deformation resistance of the strip, equal to 1.15 δ_T; δ_T = yield limit; δ_{cp} = average of before and after tension; δ_0 = after tension; δ_1 = forward tension; and E = roll material elastic modulus. The size of the nip angle that can allow the mill to start

rolling also determines the size of the minimum diameter of the work roll, and should satisfy the following formula:

$$D_{min} \geq \frac{\Delta h}{1 - \cos \alpha} \qquad (3)$$

where Δh = absolute reduction and α = contact angle of the work roll and the rolled piece.

The size of the work roll diameter is determined as follows.

Consideration of two requirements: the minimum bending condition of the work roll under the action of an external force, and the work roll diameter in the extent permitted. In order to increase the service life of the roll and improve the precision and the surface quality of the rolled strip, we must choose the appropriate roll shape to match the work roll diameter.

According to the mill specification data in a copper factory, the minimum value of the work roll diameter is 23.77 mm; the maximum is 40.26 mm. The minimum export thickness of the strip rolled by the 20-high Sendzimir mill is 0.05 mm, so value of the work roll diameter is 28.5 mm. Because the ratio of the higher roll diameter to the lower roll diameter ranges from 1.5 to 2.1, thus

$$1.5 D_1 < D_2 < 2.1 D_1 \qquad (4)$$

The minimum value of the first intermediate roll diameter is 49.83 mm; the maximum is 58.42 mm. Therefore, the first intermediate roll diameter is set as 57 mm. The minimum value of the second intermediate roll diameter is 88.49 mm; the maximum is 95.70 mm. Therefore, the second intermediate roll diameter is set as 92 mm. The diameter of the backing bearings is 160 mm, so the diameter of the backup a roll is 160 mm.

Finally, all kinds of roll diameter can be set as follows: the value of the work roll $D_1 = 28.5$ mm; the value of the first intermediate roll $D_2 = 57$ mm; the value of the second intermediate roll $D_3 = 92$ mm; and the value of the backup roll $D_4 = 160$ mm.

3 THE FINITE ELEMENT MODEL OF A ROLL SYSTEM OF THE 20-HIGH SENDZIMIR MILL

According to the geometric parameters and material properties of the 20-high Sendzimir mill, the three-dimensional finite element analysis model was established by using the large general finite element analysis software ABAQUS. Because of the symmetry from up to down, or from left to right, the finite element analysis of a quarter of the roll system model is simply performed. The model is shown in Figure 1.

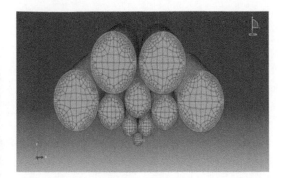

Figure 1. Rollers' finite element mode of the Sendzimir mill.

3.1 The material properties of the roll system

The material of the roll system is the same, so the function parameter of the roll material in the mill is as follows: elastic modulus = 2.156×10^5 MPa, Poisson's ratio = 0.3, and density: = 7.8×10^{-9} T/mm^3. The material of the roll is 9Gr2Mo. According to the hardness of the roll surface, the allowable contact stress of the roll is $[\sigma]$ = 2400 MPa and the allowable shear stress is $[\iota]$ = 730 MPa.

3.2 The mesh generation of the roll system

In the process of the model analysis, there will be a larger deformation region with the stress concentration, which will be meshed densely, otherwise sparsely. Therefore, the unit division is coarse inside the roll, and the more close to the roll surface, the smaller the unit division. The unit can be used to analyze the three dimensional problem with wedge units and hexahedron units. Thus, C3D8R is selected as the entity unit in the unit library. The mesh of the roll system is shown in Figure 1, containing 43764 nodes and 35896 units.

3.3 The boundary conditions and load

Because the contact among the rolls belongs to soft—soft contact, 6 contact pairs are defined in this paper. In contact properties, the normal behavior is set as "hard" contact; the tangential behavior is set as Kulun friction; and the friction coefficient is set to 0.2. In each contact pair, we specify the surface of the big diameter roller as the primary contact surface, and specify the surface of the small diameter roller as the assistant contact surface. In order to ensure that the contact pair can contact fully, a very small amount of interference is set in the small slip of every contact pair.

In the xoy symmetric plane, the symmetry constraint U_z is imposed on the midpoint of the roll, $U_z = 0$; the complete constraint is imposed on the position at which the backing bearing of the backup a roll is acted by the saddle force. In the yoz symmetric plane, the non-holonomic constraint is imposed in the $U_x = 0$ direction of the second intermediate roll (half roll) and work roller (half roll). The rollers' boundary condition of the 20-high Sendzimir mills is shown in Figure 2. Rolling force is exerted on the position of the equivalent node; rolling force is set up by the conic distribution in the copper belt width ranges, and each node in the contact area of the work roll and rolled piece is loaded by the concentrated force. The distribution form of the rolling force is changed by adjusting the value of A_P, the rolling force distribution coefficient (Yang et al. 2008, Yang, et al. 2006, Zhang et al. 2008).

According to one pass of the mill, the width of the rolled copper strip is 420 mm; the import thickness is 0.700 mm; the export thickness is 0.455 mm; the rolling direction is from right to left; the rolling force is 353.0 KN; the length of the contact arc

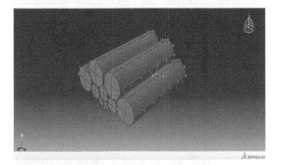

Figure 2. Rollers' boundary condition of the 20-high Sendzimir mill.

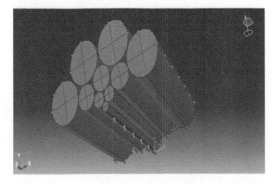

Figure 3. Rollers' rolling force of the 20-high Sendzimir mill.

can be calculated as 3.61 mm; and the load rolling pressure is 97.9 mm. The load rolling pressure is shown in Figure 3.

4 THE FINITE ELEMENT ANALYSIS OF A ROLL SYSTEM OF THE 20-HIGH SENDZIMIR MILL

Because the multi-supported beam structure is used in the 20-high Sendzimir mill, the backup roll will bear all the rolling force. Because all the material of the roll system is 9Gr2Mo, the geometric shape of the deformation occurs in the transition fillet of the backup roll whose diameter is discontinuous, which results in the occurrence of stress concentration easily in the rolling process, and the roll fracture and fatigue damage are most likely to occur in this area. From the deformed stress distribution of a roll system of the 20-high Sendzimir mill, as shown in Figure 4, the stress in the deformed area is below 550 MPa, except in the contact area of the rolls in the 6 contact pairs. So, the maximum stress deformation is 550 MPa, which is smaller than the allowable stress $[\tau] = 730$ MPa, and then the rupture and destruction will not occur.

In the rolling process, the pressure in the contact area of the roller in the 20-high Sendzimir mill is very large under the rolling force, which will produce a large contact stress. If this contact stress exceeds the allowable stress of the roll material, the contact fatigue failure will quickly occur in that area; even the roll surface falls off. Thus, we must check the strength in the contact area between the rolls. Although there are nearly 24 roll contact areas in a roll system of the 20-high Sendzimir mill, we just need to check the strength at which there is the largest contact stress because of the same material used in the roll. From the

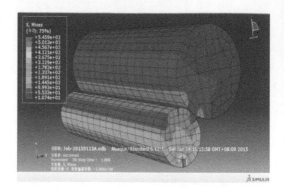

Figure 5. The contact stress distribution of the work roll and the first intermediate roll.

finite element analysis results shown in Figure 4, the deformation occurring in the contact area between the work roll and the first intermediate rolls is most serious, so the maximum contact stress occurs in this contact area. The contact stress distribution occurring in the contact area between the work roll and the first intermediate roll is shown in Figure 5.

From Figure 5, it can be seen that the maximum equivalent stress value in the contact area between the work roll and the first intermediate roll is $\sigma_{max} = 545.9$ MPa, which is less than the allowable contact stress $[\sigma]$. Although the value of the stress is relatively large, the roll can bear a relatively large contact stress because of the deformation of the material in the contact area under the compression stressed condition in the trip direction.

In the contact area, in order to avoid the fatigue failure, the maximum shear stress of each roll should be checked. The main shear stress should satisfy the following conditions:

$$\tau_{max} \leq [\tau] \tag{5}$$

The tangential stress should satisfy the following conditions:

$$\tau_{zy(max)} \leq [\tau] \tag{6}$$

According to the computational formula of the main shear stress, the value of the main shear stress can be obtained as $\tau_{max} = 0.304\sigma_{max} = 165.95$ MPa, which are significantly less than the allowable shear stress. Similarly, the value of the tangential stress can be obtained as $\tau_{zy(max)} = 0.256\sigma_{max} = 139.75$, which are also significantly less than the allowable shear stress $[\tau] = 730$ MPa. From the above

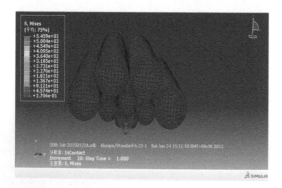

Figure 4. Rollers' deformation stress distribution diagram of the 20-high Sendzimir mill.

finite element analysis, each roll in the roll system of the mill satisfies the strength requirements.

5 CONCLUSIONS

According to the characteristics of the copper strip's rolling process, the space position, the length and the diameter of the roll system were determined. The elastic deformation model of a roll system of the 20-high Sendzimir mill was established. The finite element analysis of a roll system of the 20-high Sendzimir mill was performed by using the finite element software ABAQUS. The results show that the most serious deformation occurs in the contact area between the work roll and the first intermediate roll. The strength in the contact area was checked at the same time, and each roll in the roll system of the mill was found to satisfy the strength requirements.

ACKNOWLEDGMENT

This study was mainly supported by the International Science & Technology Cooperation Program of Jiangxi Province (20122BDH80023).

REFERENCES

Chen, J. 2014. *Kinematics analysis and calculation on a roll system of 20-High sendzimir mill*, Proceedings of the 24th Academic Conference on the committee of the mechanical Design and transmission, Hubei Mechanical engineering society and Wuhan mechanical Design and transmission society.

Gong, S.P. 2011. The status and development of copper strip products. *ShangHai Nonferrous metals*, 32(3):140–143.

Han, C. Yu, J.H. & Chen, J.S. 2012. Overview of domestic processing technology and equipment for copper strip and sheet. *Copper Engineering*, 02:12–16.

Yang, G.H. Cao, J.G. & Jie, Z. et al. 2008. The study on the backup roll contour of the Smart Crown tandem cold rolling mill. *Journal of Science and Technology Beijing*, 15(3):357–359.

Yang, G.H. Cao, J.G. & Zhang, J. et al. 2006. Optimization of application of Smart Crown on tandem cold rolling mill. *Iron and Steel*, 41(9):56–59.

Zhang, Q.D. Sun, X.M. & Bai, J. 2008. Analysis of rolls' elastic deformation on CVC 6-h mill by FEM. *China Mechanical Engineering*, 18(7):789–792.

Advanced Materials and Structural Engineering – Hu (Ed.)
© 2016 Taylor & Francis Group, London, ISBN 978-1-138-02786-2

Study on life cycle cost for the wheel re-profiling strategy

C.Y. Ren & C. Liang
China Academic of Railway Science, China

M. Zheng
Nanchang Railway Administration, China

ABSTRACT: In order to find a series of optimal strategies for the re-profiling of train wheels, the change in the wheels profile including the rim size and the tread diameter during the whole life cycle is investigated. Based on the repair database for the China Railways CRH1 model trains from Fuzhou south inspection and repair depot, the wear ratio of the train wheels is counted. Finally, the relation between the costs of re-profiling and the profile of wheel tread is studied. A new simulation method is proposed to analyze the costs of re-profiling throughout the whole lifecycle of the train wheels. This method can automatically judge whether the wheel profile needs to be repaired because of wears.

1 INTRODUCTION

The wheel-rail interaction is the main characteristic of locomotive; and the profile of the wheel tread (including rim size and tread diameter) plays a basic role in the wheel-rail interaction. The dominant failure mode of the train wheels is the wear occurring on the interface of wheel and rail. The wear results in the decrease of the tread diameter and flange thickness, which always influences the reliability, security, and riding comfort of the trains. Accordingly, the profile of the wheel tread should be repaired, when the trains are in the mileage between repairs. Therefore, it is important to develop a strategy to improve the operational quality of the CRH train at a low operational cost.

Braghin et al. set a mathematical model to predict railway wheel profile evolution due to wear, and found that the optimal mileage between repairs is 200,000 miles for the ETR500 and ORES1002 model express trains (Braghin et al. 2006). Pascual et al. reported the best time to repair the wheels on the trains manufactured by Spanish Talgo (Pascual et al. 2004). Company is the time when the flange thickness decreases to about 27.5 mm, and it should be resized to about 30.5 mm. Dong et al. investigated the re-profiling of the CRH3C model trains, and designed 18 kinds of the profile of the wheel tread to improve the reliability, security, and riding comfort of the trains (Dong et al. 2013). Wang et al. proposed a mathematical model to evaluate the remaining life of the subway in Guangzhou, and put forward some reasonable suggestions for the re-profiling of the wheels on the subway (Wang et al. 2011). Tao et al. analyzed the wear condition of the

wheel tread at various mileages of the train, and revealed that the wheels should be repaired when the mileage is bout 33×10^4 Km (Tao et al. 2013). Yuan et al. demonstrated that the re-profiling is a suitable and reasonable method for the CRH train (Yuan et al. 2006). Additionally, Zhao et al. and Xu et al. proposed some strategies of re-profiling based on the analysis of the wear ratio of train wheels (Zhao et al. 2014, Xu et al. 2010). Unfortunately, the strategies of the re-profiling have not been reported in detail, and the relations between the costs of re-profiling and the profile of wheel tread are still far from being well understood.

In this study, China Railways CRH1 model trains are taken as an example. The wear condition of the wheels profile including the rim size and the tread diameter is investigated. According to the relevant regulations, the mileage between repairs is about 300,000 Km; and the tread diameter and flange thickness are measured twice a day. Based on this database from Fuzhou south inspection and repair depot, the wear ratio of the train wheels is counted. In order to find a series of optimal strategies for the re-profiling of train wheels, a new simulation method is proposed to analyze the costs of re-profiling.

2 MODEL ESTABLISHMENT

Figures 1 and 2 show the variation of tread diameter and flange thickness along with the mileage of the trains, respectively. It can be seen that both the tread diameter and flange thickness decrease with the increase in the mileage of the trains.

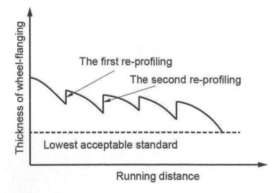

Figure 1. Flange thickness versus the mileage of the train.

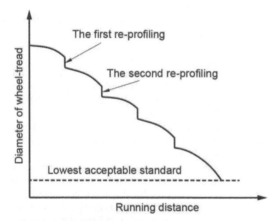

Figure 2. Tread diameter versus the mileage of the train.

The re-profiling is carried out on the train wheels many times throughout the whole lifecycle of train wheels, when the decrement of tread diameter and flange thickness reaches a critical value. Additionally, the diameter decreases; meanwhile, the flange thickness increases after re-profiling.

For ease of description, the decrement of the tread diameter is defined as ΔD. Similarly, the increment of flange thickness is described as ΔB. Then, the coefficient of re-profiling can be calculated as follows:

$$k = \frac{\Delta D}{\Delta B} \qquad (1)$$

where k is the coefficient of re-profiling.

The coefficient of re-profiling k keeps rising with the increase in the flange thickness (Liu et al. 2007). The coefficient of re-profiling k is investigated; and the optimal value of k is about 4.2, which is

qualified by the national standards (Yuan et al. 2006). Nevertheless, the coefficient of re-profiling varies at a wide range during the process of practical operation.

Fuzhou south inspection and repair depot measure the profile of the wheel twice a day. Additionally, complete systematical measurements are carried out when the mileage of the trains reaches 300,000 Km. Part of the results of the measurements accompanied with the wear ratio is presented in Tables 1 and 2. The wear ratio is calculated according to Equations 2 and 3:

$$v_B = \frac{B_i - B_{i+1}}{l_{i+1} - l_i} \times 10000 \qquad (2)$$

$$v_D = \frac{D_i - D_{i+1}}{l_{i+1} - l_i} \times 10000 \qquad (3)$$

where v_B and v_D are the wear ratio of the flange thickness and the tread diameter of the train wheels. B_{i+1} and D_{i+1} are used to describe the flange thickness and tread diameter before the $(i+1)$th times re-profiling. Similarly, the flange thickness and tread diameter before the ith times re-profiling are described as B_i and D_i, respectively. The l_{i+1} and l_i are the mileage of the trains before the $(i+1)$th and ith times re-profiling.

Table 1. Flange thickness and its wear ratio versus the mileage of the trains.

Mileage (10^4 Km)	Flange thickness (mm)	Wear ratio (mm/10^4 Km)
0–80	32–29.34	0.03
80–110	30.88–30.76	0.03
110–140	31.1–30.8	0.04
140–170	31.08–30.88	0.01
170–200	30.9–30.68	0.01

Table 2. Tread diameter and its wear ratio versus the mileage of the trains.

Mileage (10^4 Km)	Tread diameter (mm)	Wear ratio (mm/10^4 Km)
0–80	915–906.3	0.29
80–110	901.56–896.78	0.15
110–140	891.64–887.4	0.61
140–170	880.58–875.14	0.17
170–200	868.92–865.85	0.21

*The data listed in Tables 1 and 2 are selected from CRH1 model trains in Fuzhou south inspection and repair depot. The first time re-profiling is carried out as the mileage reaches 800,000 miles. Then, the re-profiling is carried out in every mileage of 300,000 miles.

It can be seen from Table 1 that the wear ratio of the flange thickness and tread diameter rises in advance and then falls down. The flange thickness is modified as 30 mm. This modification is rational and flexible with enough margin (30 mm to 31 mm). Additionally, the wear ratio is different when the flange thickness is the same, which results from the various conditions in the travel line of the trains. As shown in Table 2, the tread diameter varies during the process of running of the trains. The maximum wear ratio of the tread diameter is about 0.61 mm/104 Km.

3 COSTS OF THE RE-PROFILING

During the process of inspection and repairing of the train wheels, the strategies involving whether or when the re-profiling should be carried out by the train wheels is very important. It is directly related to the total costs of the re-profiling throughout the whole lifecycle of train wheels. Pursuant to maintenance instruction of the CRH train, the maximum and minimum thresholds of the flange thickness for re-profiling are 26 mm and 32 mm, respectively; and the maximum and minimum thresholds of the tread diameter for re-profiling are 835 mm and 915 mm, respectively. For security and stability reasons, the range of the flange thickness and tread diameter is revised in this study. The range of the flange thickness from 26 mm to 30 mm is adopted. The range of the tread diameter from 850 to 910 mm is set.

The cost of the re-profiling throughout the whole lifecycle of train wheels is the sum of the costs for every re-profiling, which depends on the co-action of the parameters of re-profiling including tread diameter, flange thickness and their thresholds. The total cost of the re-profiling throughout the whole lifecycle can be counted as follows:

$$C_A = \frac{\sum_{i=1}^{n} C_{Xi} + C_R}{L} \tag{4}$$

$$C_{Xi} = \frac{\Delta D_i}{\frac{D_0 - D_L}{2} - \Delta D \times T_i} \times C_L \times N \tag{5}$$

where C_A is the total cost of the re-profiling throughout the whole lifecycle; C_{Xi} is the cost of the ith times re-profiling; C_R is the labor cost of every time re-profiling; D_0 is identified as the initial value of the tread diameter; and D_L is the threshold value of the tread diameter; ΔD is used to show the decrement of the tread diameter; ΔD_i is on behalf of the increase of the tread diameter at every ith times re-profiling; T_i is the number of re-profiling

times; C_L is the cost of re-profiling for a pair of wheels; N is the number of the train wheels on the trains; and L is the running distance.

Matlab software is introduced into the simulation for counting the costs of re-profiling throughout the whole lifecycle of the train wheels. The structure of the main program, as well as the design thought, is shown in Figure 3. The initial conditions are L, B and D, which are set as 0, 32 and 910, respectively.

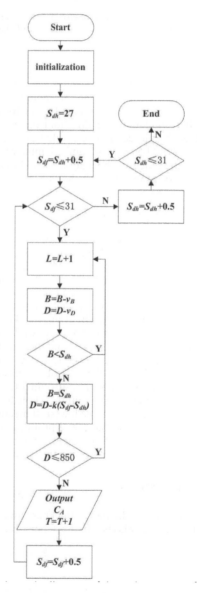

Figure 3. Schematic diagram of the main program for counting the costs of re-profiling throughout the whole lifecycle of the train wheels.

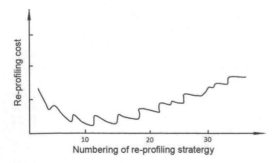

Figure 4. Result of simulation showing the total costs of re-profiling throughout the whole lifecycle of the train wheels.

S_{dh} is the threshold matrix of the tread diameter for re-profiling; and S_{dh} is the threshold matrix of the recovery value of re-profiling. During the process of the simulation, the threshold flange thickness must be picked from the value of 27, 27.5, 28, 28.5, 29, 29.5, 30, 30.5 and 31 mm, which is less than the practical application. In addition, the coefficient of re-profiling k is defined as 5.

4 RESULTS OF SIMULATION

The threshold matrix (S_{dh}, S_{df}) is input as 27 mm and 31 mm, which is divided into six parts of interval. The total number of kinds of the thresholds is about 36. The simulation result is shown in Figure 4. It can be seen that the total costs of re-profiling throughout the whole lifecycle of the train wheels depend on the threshold matrix. The total costs of re-profiling throughout the whole lifecycle of the train wheels decrease, and then increases with the increase in the threshold of the parameters of re-profiling. In the case of the threshold matrix (S_{dh}, S_{df}) is (27.5 mm, 30 mm), the total costs of re-profiling throughout the whole lifecycle of the train wheels tend to be lowest. It means that the optimal strategy is that the re-profiling should be carried out as the tread diameter reaches 27.5; meanwhile, it should be modified to 30 mm.

5 CONCLUSIONS

The wear ratio of the flange thickness and tread diameter of the wheels on the China Railways

CRH1 model trains has been worked out, based on the database from Fuzhou south inspection and repair depot.

A new simulation method has been proposed to analyze the costs of re-profiling throughout the whole lifecycle of the train wheels. By this means, 36 kinds of strategy have been tested, and an optimal strategy has been proposed. The optimal strategy is that the re-profiling should be carried out as the tread diameter reaches 27.5; meanwhile, it should be modified to 30 mm.

REFERENCES

Braghin, F. Lewis, R. & Dwyer-Joyce, R.S. et al. 2006. A mathematical model to predict railway wheel profile evolution due to wear. *Wear*, 261: 1253–1264.

Dong, X. Wang, Y. & Wang, L. 2013. Research on the reprofiling strategy for the wheel tread of high-speed EMU. *China Railway Science*. 01: 88–94.

Liu, H. 2007. Analysis of the economy in turning of wheelsets for qingzang railway passenger cars. *Rolling Stock*. 05: 31–32+36+46.

Pascual, F. & Marcos, J.A. 2004. *Wheel wear management on high-speed passenger rail: A common playground for design and maintenance engineering in the Talgo engineering cycle.* Proceedings of the 2004 ASME/IEEE Joint Rail Conference: 193–199.

Tao, G. Li, X. Deng, Y. Deng, T. Wen, Z. & Li, L. 2013. Wheel Wear Life Prediction Based on Lateral Motion Stability of Vehicle System. *Journal of Mechanical Engineering*. 10: 28–34.

Wang, L. Yuan, H. Na, W. Chen, X. & Li, Y. 2011. Optimization of the re-profiling strategy and remaining useful life prediction of wheels based on a data-driven wear model. *System Engineering—Theory & Practice*. 06: 1143–1152.

Xu, H. Yuan, H. Wang, L. Na, W. Xu, W. & Li, Y. 2010. Modeling of Metro Wheel Wear and Optimization of the Wheel Re-profiling Strategy Based on Gaussian Process. *Journal of Mechanical Engineering*. 24: 88–95.

Yuan, H. Xiao, S. & Wang, Y. 2006. Feasible Analysis of Wheel-set Re-profiling on the Statistics of Wheelset Wearing. *Urban Mass Transit*. 01: 43–45+49.

Zhao, W. Wang, L. Yuan, H. Chen, C. & Chen, X. 2014. Modeling of Metro Wheel Wear and Optimization of the Wheel Re-profiling Strategy Based on Markov Process. *Science Technology and Engineering*. 36: 116–119+132.

Advanced Materials and Structural Engineering – Hu (Ed.)
© 2016 Taylor & Francis Group, London, ISBN 978-1-138-02786-2

A condition warning method of primary equipment considering daily load regularity

J.S. Li, Y.C. Lu, C. Wei, F.B. Tao, P. Wu & M. Yu
Jiangsu Electric Power Company Research Institute, Nanjing, China

ABSTRACT: In order to detect the equipment condition changes timely, an algorithm considering daily load regularity is proposed, which has anti-noise capability, high sensitivity and self-learning ability. First, the parameters of its equivalent circuit are calculated using real-time electrical data at both ends and a parameter sequence can be get; Second, the sum of the parameter changes is calculated and compared with the threshold value to detect the equipment condition changes; Last, based on its day cycle operation law, parameter variation amplitude is quantified, which provides references to make maintenance measures. Take the detection of transformer winding short circuit fault as an example. Simulink simulations validate the effectiveness of this method.

1 INTRODUCTION

In 2009, the strategic goal of building the strong smart grid was put forward by State Grid Corporation of China. And many new requirements of the substation construction are put forward. Among them, the realization of online assessment of the primary equipment state is very important. In the substation, the SCADA/EMS system and the wide area measurement system based on the Phasor Measurement Unit (PMU) can provide rich data, which is helpful to analyze equipment status, forecast the equipment state changes online. Especially, the data provided by the PMU/WAMS system is conductive to analyze the state changing process, due to its unified time reference point (Xue et al. 2007, Xu et al. 2000, Huang & Quan 2001).

When a slight fault occurs inside the primary equipment, such as the transformer inter-turn short circuit fault, the equivalent circuit parameters change a little, which does not meet the protective conditions. If it operates continuously, the equipment may cause huge losses and large area blackout. The slight fault detection is helpful to warn equipment failure and ensure the safe operation of power grid. Nowadays, based on the SCADA and PMU data, many kinds of research have been done, which can be divided into two classes: (1) the equipment equivalent parameters identification and analysis from the point of probability, reducing the influence of environment, measurement and other factors (Wang et al. 2010, Bi et al. 2011, Wang et al. 2011, Bi et al. 2010, Chen et al. 2011, Cheng et al. 2006); (2) the equipment state changing process analysis based on SCADA

data, using the mean data and the variance data of the established indexes (Ni 2011). The above researches analyze the equipment state changes roughly and are difficult to detect the slight faults.

To improve the fault detection ability, a method based on probability thought considering daily load regularity is presented. In the method, the primary equipment is regarded as a port model and its equivalent circuit parameters can be identified by measured electrical data. And an integral algorithm is applied in the detection process, which can reduce the influence of errors. Through the analysis of the daily load regularity, the state changes are analyzed to assess the fault severity. This method has high sensitivity and can automatically adjust the threshold value. Take the transformer inter-turn short circuit detection as an example. the Simulink simulation results demonstrate the effectiveness of this method.

2 THE PARAMETERS IDENTIFICATION

As shown in Figure 1, the transformer, reactor, breaker and other primary equipment can be regarded as port models, and the voltage and current information can be measured. In Figure 1, $U_1(s)$, $I_1(s)$, $U_2(s)$, $I_2(s)$ are the frequency domain

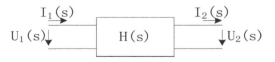

Figure 1.   Transfer function of primary equipment.

form. The equipment can be represented by the transfer function H(s).

In a two-winding single phase transformer, the equivalent circuit parameters changes caused by turn-to-turn faults are analyzed.

In Figure 1, the relation between the measured electrical information can be expressed as Formula (1).

$$\begin{bmatrix} U_1(s) \\ I_1(s) \end{bmatrix} = \begin{bmatrix} A & B \\ C & D \end{bmatrix} \begin{bmatrix} U_2(s) \\ I_2(s) \end{bmatrix} = H(s) \begin{bmatrix} U_2(s) \\ I_2(s) \end{bmatrix} \quad (1)$$

In a two-winding single phase transformer, the equivalent circuit parameters changes caused by turn-to-turn faults are analyzed.

2.1 The equivalent circuit in normal situation

Ignore the magnetizing branch, the equivalent circuit of the transformer is shown in Figure 2. The electrical information is imputed to one side. R_A and R_a are the leakage resistance of the two windings. X_A and X_a are the leakage reactance, representing the leakage magnetic effect.

2.2 The equivalent circuit in inter-turn short circuit grounding situation

When the secondary winding is grounded at some point, the transformer is changed to be a three-winding transformer, which is shown in Figure 3.

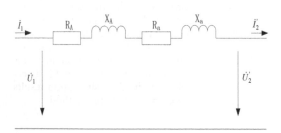

Figure 2. Equivalent circuit of a two-winding transformer.

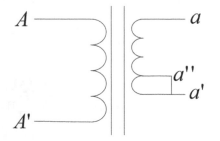

Figure 3. Inter-turn short circuit of secondary winding.

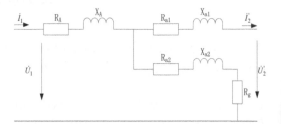

Figure 4. Equivalent circuit of transformer with inter-turn short circuit of secondary winding.

And the equivalent circuit of Figure 3 is shown in Figure 4. R_A, R_{a1} and R_{a2} are the leakage resistance. X_A, X_{a1} and X_{a2} are the leakage reactance. And R_g represents the arc resistance at the fault point.

2.3 Parameter identification

To identify the leakage resistance, a time window is chosen first and the parameters in the window are assumed to be the same. Then the fundamental frequency component of the measured electrical data is got by Fourier transform. The parameters can be calculated by Formula (2). Wherein, R_σ is the leakage resistance of the transformer, and the "*real*" represents the real part calculation.

$$R_\sigma = R_A + R_a = real\left(\frac{\dot{U}_1 - \dot{U}_2'}{\dot{I}_2''}\right) \quad (2)$$

When the turn-to-turn fault occurs, the parameters derived from Formula (2) are not the same as the value in Figure 2.

3 EQUIPMENT STATE CHANGES DETECTION

To detect the fault which doesn't meet the protection operation, an integral algorithm is proposed, which can remind the staff effectively and avoid black failure.

3.1 The integral algorithm

Influenced by environments, including voltage, load and other factors, the identified parameters show apparent randomness. The equipment condition is warned by the comparison of integral value of the parameter changes and the threshold value.

When the inter fault occurs, the equivalent circuit parameters change randomly around different values (Wang et al. 2010, Bi et al. 2011, Wang et al.

2011, Bi et al. 2010, Chen et al. 2011, Cheng et al. 2006). When the turn-to-turn faults occur, the leakage resistance and inductance will change. Take the detection of leakage resistance as an example, the detection process is shown as follows.

1. Set the proper length of the time window and the number in a time window to be *Num*.
2. Conduct the subtraction operation by Formula (3).

$$\Delta R_\sigma(i) = R_\sigma(i) - R_\sigma(i - Num) \qquad (3)$$

3. Conduct the criteria to judge the condition changes.

$$D_{R_\sigma} = \begin{cases} 0, & \sum\limits_{i=1}^{Num} \Delta R_\sigma(i) < \varepsilon_1 \\ 1, & \sum\limits_{i=1}^{Num} \Delta R_\sigma(i) > \varepsilon_1 \end{cases} \qquad (4)$$

where, ε_1 is the threshold value.

When $\sum\limits_{i=1}^{Num} \Delta R_\sigma(i) < \varepsilon_1$, $D_{R_\sigma} = 0$, which represents the transformer normal. Otherwise, the transformer is judged to be abnormal.

4. Move the time window at the interval of *Num*/2 points, and repeat the above steps to detect the condition changes.

$$\varepsilon_1 = k \bullet s_2 \qquad (5)$$

where in, k is the margin coefficient. And s_2 is the maximum absolute value of the sum data in normal situation.

3.2 Condition changes quantification

The condition changes quantification is very helpful to analyze the health of the primary equipment and determine the maintenance time. To reduce the influence of outer environment, a method considering daily load regularity is presented.

In this paper, the data is chosen by the correlation coefficient, as shown in Formula (6). In the formula, x and y are the data in different days. The superscript represents the mean data and the data i in the brackets represents the data at different time.

$$r = \frac{\sum\limits_{i=1}^{n}(x(i) - \bar{x})(y(i) - \bar{y})}{\sqrt{\sum\limits_{i=1}^{n}(x(i) - \bar{x})^2} \bullet \sqrt{\sum\limits_{i=1}^{n}(y(i) - \bar{y})^2}} \qquad (6)$$

where in, the x and y sequences contain n samples and they are similar when the coefficient r is close to 1.

Suppose that the data in the days from the first day to the *n*th day is represented by $\{x(1), x(2), \ldots, x(n-1), x(n)\}$. If the parameter changes in the *n*th day, obverse continuously and the data in the following k days can be got, which is expressed as $\{x(n+1), x(n+2), \ldots, x(n+k-1), x(n+k)\}$. The variation amplitude is calculated by the formula below.

$$\gamma = \frac{\dfrac{1}{n-1}\sum\limits_{i=1}^{n}x(i) - \dfrac{1}{k+1}\sum\limits_{i=n}^{n+k}x(i)}{\dfrac{1}{n-1}\sum\limits_{i=1}^{n-1}x(i)} \qquad (7)$$

4 EXPERIMENTAL VERIFICATION

4.1 The simulation model

In Figure 5, a model is established. The rated voltage of the transformer is 110 kV, the rated capacity is 40 MVA and the rated ratio is 110/10. The leakage resistance and inductance of the two windings are 0.605 Ω and 0.030812 H, and the excitation resistance and inductance are 60503 Ω and 192.58 H.

In the model, the voltage source is 110kV and the system impedance is 1+j0.314 Ω. The load is 20 MW. The sample frequency is 10 kHz and the secondary winding is grounded at some bottom position at 3 s.

In Figure 6, the leakage resistance of the transformer is shown. There are 300 samples and the noise in measured electrical data is set to be in normal distribution.

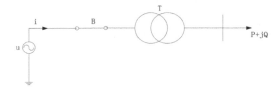

Figure 5. System model.

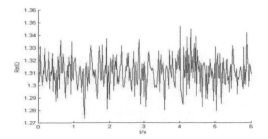

Figure 6. Curve of leakage current (mean 0, standard variance 3 Gauss white noise included in electrical data).

4.2 The parameter change detection

When the percentage of short circuit turns is 0.5%, the leakage resistance and the sum of changes are shown in Figure 7. The sum of leakage resistance changes in normal situation is between −0.1 and 0.1. The margin coefficient is set 1.4, and the threshold value is 0.14.

Similarly, when the percentage of short circuit turns is 0.8%, the leakage resistance and the sum of changes are shown in Figure 8. The sum of leakage resistance changes in normal situation is between −0.1 and 0.15. The threshold value is 0.21. The detection precision is 0.0042Ω.

From the above two figures, the integral value will be larger than the threshold value when the turn-to-turn faults occur. The arrival time is relative with the shorted turns. The parameter changes can be detected by different threshold considering the noise influence.

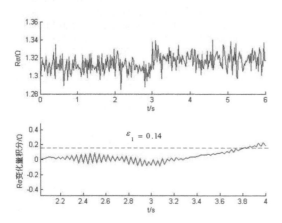

Figure 7. Integration of leakage current changes.

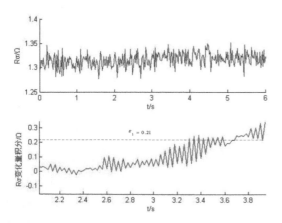

Figure 8. Leakage resistance variation and its wavelet coefficient curve.

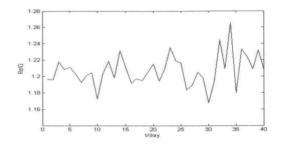

Figure 9. Parameter variation curve.

4.3 The parameter change quantification

Based on the daily load regularity, the leakage resistance at the same time in different days is simulated by Monte-Carlo simulation, considering the parameter randomness (Dong et al. 2003).

Suppose that the secondary winding is shorted at 8:00, and the leakage resistance in 30 similar days before the fault and 10 similar days after the fault is simulated. Before the fault, the mean data of the parameter in normal situation is 1.2 and the mean data of the parameter in fault situation increases 0.02. The parameter is shown in Figure 9.

According to the formula in the paper, the amplitude of the parameter changes is 1.71%, which is a little less than the setting value.

Based on the amplitude, the staff can assess the transformer condition and make proper maintenance measures.

5 CONCLUSIONS

To detect the condition changes of primary equipment and realize the condition warning, a method based on transverse detection and longitudinal quantification is proposed. In the method, the detection is accomplished by the integral algorithm and the quantification considers the daily operation regularity. The simulation results show that the minor change can be detected.

The method proposed in this paper can detect minor faults and has self-learning ability, which will reduce error times. The quantification can provide arguments for condition assessment.

REFERENCES

Bi, T.S. Ding, L. & Zhang, D.N. 2011. Transmission line parameters identification based on moving-window TLS, *Journal of Electric Power Science and Technology*, 26(2): 10–15.

Bi, T.S. Liu, H. & Wu, J.T. et al. 2010. On-line assessment on voltage consistency and frequency consistency of PMU measurement under steady state, *Automation of Electric Power Systems*, 34(21): 21–26.

Chen, J. Yan, W. & Lu, J.A. et al. 2011. A robust transformer parameter estimation method considering multi-period measurement random errors, *Automation of Electric Power Systems*, 35(2): 28–33.

Cheng, T. Yan, W. & Wen, Y. et al. 2006. Transformer parameter modifiability based transformer substation state estimation, *Journal of Chongqing University (Natural Science Edition)*, 29(3): 32–35.

Dong, Y.L. Gu, Y.J. & Yang, K. 2003. Criticality analysis on equipment in power plant based on monte carlo simulation, *Proceedings of the CSEE*, 23(8): 201–205.

Huang, J.H. Quan, L.S. 2001. Current status and development of condition-based maintenance of high-voltage electric power equipment in substation, *Automation of Electric Power System*, 25(16): 56–61.

Ni, L. 2011. *Study of substation operating performance evaluation based on SCADA*, Jinan: Shandong University.

Wang, M.H. Bao, J. & Xi, X. et al. Online estimation of transmission line parameters based on PMU measurements, *Automation of Electric Power Systems*, 34(1): 25–28.

Wang, M.H. Qi, X. & Niu, S.Q. et al. 2011. Online estimation of transformer parameters based on PMU measurements, *Automation of Electric Power Systems*, 35(13): 61–65.

Xu, J. Wang, J. & Gao, F. et al. 2000. A survey of condition based maintenance technology for electric power equipments, *Power System Technology*, 24(8): 48–52.

Xue, Y.S. Xu, W. & Dong, Z.Y. et al. 2007. A review of wide area measurement system and wide area control system, *Automation of Electric Power Systems*, 31(15): 1–6.

Advanced Materials and Structural Engineering – Hu (Ed.)
© *2016 Taylor & Francis Group, London, ISBN 978-1-138-02786-2*

Optimization design of automotive shroud part to reduce warpage in injection molding by using the finite element method

G.J. Kang & B.Y. Moon
Department of Naval Architecture Engineering, Kunsan National University, Kunsan, Korea

ABSTRACT: In this study, automobile shroud is considered as a plastic part example. To achieve the minimum warpage, the appropriate process condition parameters are determined. Considering the process parameters such as mold temperature, melt temperature, filling time, cooling time, packing pressure, and packing time, a series of mold analysis are performed. Finite element analyses are conducted for the combination of process parameters organized using the statistical three-level full factorial experimental design method. Orthogonal arrays of Taguchi, the Signal-to-Noise (S/N) and Analysis of Variance (ANOVA) are utilized to determine the optimum parameter levels, and to find out principal process parameters on warpage. ANOVA indicates that the mold temperature, melt temperature, filling time, cooling time, packing pressure, and packing time influence warpage by 5.6%, 8.1%, 20.5%, 32.4%, 6.1%, and 1.2%, respectively. The results show that the warpage is reduced. This indicates that the optimization methodology proposed in this study can be employed to improve other plastic parts to reduce the warpage.

1 INTRODUCTION

Plastic is widely used in the manufacturing process, and is a useful part in the structure. As one of the most significant evaluation indices of the quality of plastic parts, warpage of plastic during the injection molding process should be optimized. Warpage is an important factor affecting the product quality. Especially, automobile parts have been emphasized on the quality of the final product. The causes of warpage are attributed to the partial shrinkage of parts. The warpage problem can be improved by changing the geometry of parts, or by modifying the injection molding structure or adjusting the injection conditions.

There have been some studies focused on warpage reduction through optimization. An early study to reduce the warpage was proposed (Lee, 1995). They proposed an objective function for the optimization of warpage of injection-molded parts by deliberately varying each part wall thickness within the prescribed dimensional tolerance. Moreover, Lee and Kim optimized the gate location by a two-step search method in order to improve part quality in the areas of warpage, weld line, and percussive intensity (Lee, 1996). Although their research of optimization methods reduced the warpage effectively, the process was time-consuming because its numerical simulation analyses were conducted under numerous optimization iterations.

Recently, the Design of Experiment (DOE) analysis method has been widely applied for warpage reduction through process optimization (Wang, 2002; Dong, 2005, Oktem, 2007). Huang et al. studied the warpage of thin sell parts using commercial software C-MOLD for simulation (Huang, 2001).

However, various industries have employed the Taguchi method over the years to improve products or manufacturing processes. It is a powerful and effective method to solve challenging quality problems. Actually, the Design of Experiments (DOE) method has been used quite successfully in several industrial applications such as in optimizing manufacturing processes (Sofuoglu, 2006). From a previous literature review, the effective factors of warpage in the injection-molded products can be identified, including mold temperature, melt temperature, filling time, cooling time, packing pressure and packing time. Additionally, the Taguchi method was used to optimize the process design.

The aim of this research is to analyze the effective process factors of warpage in injection-molded items by the Taguchi techniques. Computer simulations using MoldFlow software of the injection molding process are carried out to obtain the warpage data. Then, the contribution percentage of each factor such as mold temperature, melt temperature, filling time, cooling time, packing pressure and packing time can be found. Also, the optimum set of parameters driving the effective factors in injection molding can be determined to produce a product with the minimum warpage.

2 METHODOLOGY OF THE TAGUCHI METHOD

2.1 Taguchi method

The Taguchi method has been successfully applied for its simple and robust technique for optimizing the process parameters (Leo, 1996). The experimental design proposed by the Taguchi method uses the orthogonal arrays to organize the parameters affecting the processes and varying levels. This allows testing only a limited collection of parameters combinations instead to check all possible combinations. The method also allows the determination of factors most influencing to the product quality with minimum experiments, which lead to less cost. The main objective of the Taguchi method is to produce a high-quality product at a low cost with respect to all noise factors. The Taguchi method uses a procedure that applies orthogonal arrays of statistically designed experiments to obtain the best results with a minimum number of experiments, thus reducing the time and cost of experimentation. The objective function for the matrix experiments is the Signal-to-Noise (S/N) ratio. Depending on the objective, there are three different mean square deviations for the signal–noise ratios that can be defined including nominal-the-better, larger-the-better, and smaller-the-better. The mean square deviation can be considered to be the average performance characteristic values for each experiment. The different signal–noise ratios, corresponding to n experiments, are presented below:

Nominal-the-better:

$$S/N = -10 \log\left(\frac{1}{nS}\sum_{i=1}^{n} y_i^2\right) = -10 \log\left(\frac{\bar{y}^2}{S^2}\right) \quad (1)$$

Larger-the-better:

$$S/N = -10 \log\left(\frac{1}{n}\sum_{i=1}^{n} \frac{1}{y_i^2}\right) \quad (2)$$

Smaller-the-better:

$$S/N = -10 \log\left(\frac{1}{n}\sum_{i=1}^{n} y_i^2\right) = -10 \log\left(\bar{y}^2\right) \quad (3)$$

where S is the standard deviation; y_i is the data obtained from the experiments; and n is the number of observations. The goal of the study presented in this paper is to optimize the warpage. Thus, the observed values of warpage are set to minimum.

2.2 Design of the experiment

The design of experiments is planned by using the Taguchi method, as it is considered to be a powerful

Table 1. Injection molding factors and their levels.

Column	Factors	Unit	Level 1	Level 2	Level 3
A	Mold temperature	°C	30	45	60
B	Melt temperature	°C	200	225	250
C	Filling time	Sec.	1.5	2	2.5
D	Cooling time	°C	25	35	45
E	Packing pressure	MPa	60	75	90
F	Packing time	sec.	2	4	6

tool when a process is affected by a number of parameters. In this study, Taguchi's orthogonal arrays were used, which consist of the ranges of plastic injection molding process parameters based on a three-level design of experiments. The ranges of four control factors in each level are given in Table 1. The values of parameters such as mold temperature and melt temperature were selected according to the recommended guidelines at the MoldFlow material library. Also, packing pressure, cooling time and packing time were selected based on industrial experiences.

2.3 Analysis of variance (ANOVA)

In order to determine the influence and relative importance of the factors, ANOVA was performed. ANOVA results are carried by separating the total variability into contributions by each of the design parameters and error. ANOVA demonstrates whether observed variations in the response are due to the alteration of level adjustments or experimental standard errors. ANOVA procedure results in the calculation of Sum of Squares (SS), Degree of Freedom (DOF), mean square (variance) and associated F-test of significance (F). The SS of factors is calculated as follows:

$$SS_A = \sum_{i=1}^{K_A}\left(\frac{A_i^2}{nA_i}\right) - \frac{T^2}{N} \quad (4)$$

3 FE SIMULATION OF AUTOMOBILE CLUSTER PART

3.1 3D FE Model

The steering column shroud utilized in this study is the automobile part. The shroud is held in place with retaining clips. It covers the steering column and electronic devices such as tilt control, cruise control and turn signal switch. The material of the shroud used in this study is Polypropylene (PP).

3.2 FE analysis

Finite element analysis of the automobile steering column shroud was carried out, and the parameters values during the analysis were set according to the Taguchi orthogonal arrays. The maximum warpage value of this part is 4.497 mm, occurring at the place far from the gates.

4 ANALYSIS OF EXPERIMENTAL DATA

4.1 Warpgae

Warpage for the each experiment set based on L27 orthogonal array of Taguchi is measured from finite element analysis, and the S/N ratio values for the warpage and volumetric shrinkage were calculated by using Equation (3).

4.2 Prediction of optimum performance

The control factor response for the warpage of each control factor to its individual level was calculated by averaging the S/N ratios of all experiments at each level for each factor. From the ANOVA results as shown in Figure 1, it is apparent that D is the most important parameter with a contribution value of 32.4% on warpage, which is followed by C with a contribution value of 20.5%, B with 8.1%, $A \times B$ with 7.0%, $B \times C$ with 6.1%, A with 5.6%, and F and e with 13.5%, respectively. The effect of the S/N ratio and warpage mean value of every factor is plotted in Figure 2. It can be seen from Figure 2b that the warpage increases with decreasing A from −59.7 to −62.7. As B decreases from −57.3 to −93.1, the warpage increases from 2.10 mm to 2.25 mm. By selecting the highest value of mean S/N ratio for each factor, the optimal level can be determined. On this basis, the optimum combination of levels in terms of the minimum warpage of the steering column shroud is A1B1C3D3E3F3, i.e. the mold temperature is 30 °C, the melt temperature is 200 °C, the filling time is 2.5 s, the cooling time is 45 s, the packing pressure is 90 MPa, and the packing time is 6 s.

4.3 Factor contributions

The contribution of each factor to the warpage of control factors can be determined by performing Analysis of Variance (ANOVA) based on Eq. (4). ANOVA results for the warpage are shown in Table 2. It can also be calculated from the ratio of the mean sum of squared deviations. The contribution P value shows the significance level and the percent (%) depicts the significance rate of parameters on the warpage in Table 2. Moreover, it can be seen that A (mold temperature), B (melt temperature), C (filling time), D (cooling time), E (packing pressure) and F (packing time) influence warpage by 6%, 8%, 21%, 32%, 6%, and 1%, respectively.

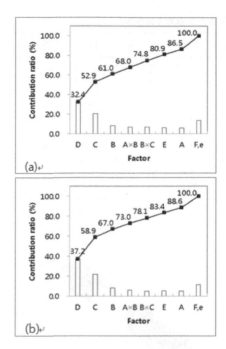

(a)

(b)

Figure 1. Contribution ratio of S/N ratio and warpage mean value: (a) S/N ratio, (b) warpage mean value.

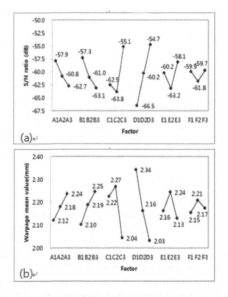

(a)

(b)

Figure 2. Effect of S/N ratio and warpage mean value: (a) S/N ratio, (b) warpage mean value.

4.4 Confirmation test

The confirmation test is very important in engineering analysis to validate the minimum warpage and volumetric shrinkage resulting from the optimization process. Consequently, confirmation

Table 2. ANOVA of S/N.

Factor	S	Φ	V	F0	F(0.05)	P(%)
A	1.33	2	0.67	2.287	4.1	0.06
B	1.93	2	0.96	3.312		0.08
C	4.86	2	2.43	8.353		0.21
D	7.68	2	3.84	13.19		0.32
E	1.45	2	0.72	2.491		0.06
F	0.30	2	0.15	0.508		0.01
A × B	1.65	4	0.41	1.418	3.48	0.07
B × C	1.61	4	0.40	1.383		0.07
e	2.91	10	0.29			0.12
T	23.71	30				

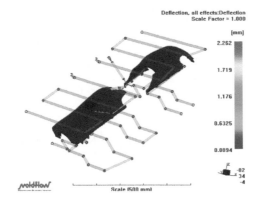

Figure 3. Optimization analysis result after optimization.

tests were carried out by utilizing the levels of optimal process parameters, such as A1B1C3D3E3F3 for minimum warpage using polymeric material PP, resulting from the optimization process. From the confirmation test, warpage is reduced from 4.497 mm before optimization to 2.262 mm after optimization by 2.235 mm, as shown in Figure 3.

4.5 Confirmation test

The confirmation test is very important in engineering analysis to validate the minimum warpage and volumetric shrinkage resulting from the optimization process. Consequently, confirmation tests were carried out by utilizing the levels of optimal process parameters, such as A1B1C3D3E3F3 for minimum warpage using polymeric material PP, resulted from optimization process. From the confirmation test warpage is reduced from 4.497 mm before optimization to 2.262 mm after optimization by 2.235 mm as shown in Figure 3.

5 CONCLUSIONS

Taguchi method was applied for optimization of warpage of automobile plastic part shroud.

From the research the following conclusions were made:

1. From the ANOVA results, mold temperature, melt temperature, filling time, cooling time, packing pressure, and packing time influence warpage by 5.6%, 8.1%, 20.5%, 32.4%, 6.1%, and 1.2%, respectively.
2. From the Signal-to-Noise (S/N) ratio using the smaller-the-better approach, it is clear that the optimal injection molding condition is A1B1C3D3E3F3, i.e. the mold temperature is 30 °C, the melt temperature is 200 °C, the filling time is 2.5 s, the cooling time is 45 s, the packing pressure is 90 MPa, and the packing time is 6 s.

ACKNOWLEDGMENT

This work (Grant No. C0184633) was supported by the Business for Cooperative R&D between Industry, Academy, and Research Institute-funded Korea Small and Medium Business Administration in 2014.

REFERENCES

Dong, B.B. Shen, C.Y. & Liu, C.T. 2005. The effect of injection process parameters on the shrinkage and warpage of PC/ABS'S part. *Polymeric Materials: Science and Engineering*. 21(4): 232–235.

Huang, M.C. & Tai, C.C. 2001. The effective factors in the warpage problem of an injection molded part with a thin shell feature. *Journal of Materials Processing Technology*, 110(1): 1–9.

Lee, B.H. & Kim, B.H. 1995. Optimization of part wall thicknesses to reduce warpage of injection-molded parts based on the modified complex method. *Polymer-Plastics Technology and Engineering*, 34(5): 793–811.

Lee, B.H. & Kim, B.H. 1996. Automated selection of gate location based on desired quality of injection molded part. *Polymer-Plastics Technology and Engineering*, 35(2): 253–269.

Leo, V. & Cuvelliez, C.H. 1996. The effect of the packing parameters, gate geometry, and mold elasticity on the final dimensions of a molded part. *Polymer Engineering & Science*, 36(15): 1961–1971.

Oktem, H. Erzurumlu, T. & Uzman, I. 2007. Application of Taguchi optimization technique in determining plastic injection molding process parameters for a thin-shell part. *Materials & Design*, 28(4): 1271–1278.

Sofuoglu, H.A. 2006. Technical note on the role of process parameters in predicting flow behavior of plasticine using design of experiment. *Journal of Materials Processing Technology*, 178(1–3): 148–153.

Wang, T.H. Young, W.B. & Wang, J. 2002. Process Design for Reducing the Warpage in Thin-walled Injection Molding. *International Polymer Processing*, 17(2): 146–152.

Advanced Materials and Structural Engineering – Hu (Ed.)
© 2016 Taylor & Francis Group, London, ISBN 978-1-138-02786-2

Parametric analysis of an airfoil aeroelastic system with hysteresis using precise integration method

C.C. Cui, J.K. Liu & Y.M. Chen
Department of Mechanics, Sun Yat-Sen University, Guangzhou, China

ABSTRACT: The aeroelastic system of an airfoil with a hysteresis nonlinearity based on precise integration method was investigated. As a piecewise linear system, the hysteresis model can be well analyzed by the precise integration method which has high computational efficiency and accuracy. The hysteresis constants and the initial conditions have important influence on the dynamic response of the system, such as the limit cycle oscillations, bifurcations and chaotic motion et al. Numerical examples showed that the solutions obtained by the precise integration method were in good agreement with the Runge-Kutta results and the exact results. The change of amplitude under different hysteresis constants and the initial conditions was illustrated.

1 INTRODUCTION

As one of the main problems of the aeroelastic system, airfoil flutter is a typical complex self-excited phenomenon (Dowell et al. 2004). The concentrate structural nonlinearities can be classified into three types, namely, cubic, freeplay and hysteresis. Because of the existence of these nonlinearity factors, the flutter system can deduce complex dynamic response, such as limit cycle oscillations and chaos. Therefore, the study of the dynamic behavior of the aeroelastic system is crucial.

Numerical methods for the nonlinear flutter analysis mainly include Runge-Kutta (RK) method and the finite difference method. Describing function technique, the equivalent linearization method and the harmonic balance method are the analytical or semi-analytical methods. Also other methods are developed, such as homotopy analysis method, perturbation-incremental method, the point transformation method, incremental harmonic balance method, et al.

Areoelastic system with cubic and free-play had been researched by many investigators. Chung used the perturbation-incremental method to investigate the bifurcation of a two-degree-of-freedom aeroelastic system with a freeplay (Chung et al. 2007). The Runge-Kutta method was used to study the influence of conditions on an airfoil with structural nonlinearities (Li et al. 2012). Liu used the point transformation method to analysis the nonlinear behavior of areoelastic system with freeplay (Liu et al. 2002). The limit cycle oscillation flutter could be accurately predicted by the Floquet multiplier for an aeroelastic airfoil with freeplay (Zhao & Zhang 2010). The precise integration

method initiated by Zhong (Zhong 2004) was proposed to analysis the aeroelastic response of an airfoil with freeplay (Cui et al. in press, Chen & Liu 2014).

Relatively few papers have been found to study the aeroelastic system with hysteresis. The point transformation method was developed to investigate the aeroelastic system (Liu et al. 2002). Liu also used the incremental harmonic balance method to analyze the bifurcation properties (Liu et al. 2012).

In this paper, we have investigated the influences of the hysteresis constants and the initial conditions on the dynamic response of the aeroelastic system with hysteresis by precise integration method.

2 MODELS AND DYNAMIC EQUATIONS

2.1 Aeroelastic model

Figure 1 is the model of a two-freedom-degree airfoil. The airfoil oscillates in pitch and plunge. The

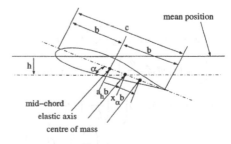

Figure 1. The sketch of an airfoil oscillating in pitch and plunge.

pitch angle about the elastic axis is denoted by α, positive with the nose up; the plunge deflection is denoted by h, positive in the downward direction. The elastic axis is located at a distance $a_h b$ from the mid-chord, while the mass center is located at a distance $x_a b$ from the elastic axis.

The coupled motions of the airfoil in incompressible unsteady flow can be written as follows

$$
c_0 \ddot{\xi} + c_1 \ddot{\alpha} + \left(c_2 + 2\zeta_\xi \frac{\bar{\omega}}{U^*} \right) \dot{\xi} + c_3 \dot{\alpha} + c_4 \xi + c_5 \alpha
$$

$$
+ c_6 w_1 + c_7 w_2 + c_8 w_3 + c_9 w_4 + \left(\frac{\bar{\omega}}{U^*} \right)^2 G(\xi) = f(t)
$$

$$
d_0 \ddot{\xi} + d_1 \ddot{\alpha} + d_2 \dot{\xi} + \left(d_3 + 2\zeta_\alpha \frac{1}{U^*} \right) \dot{\alpha} + d_4 \xi + d_5 \alpha
$$

$$
+ d_6 w_1 + d_7 w_2 + d_8 w_3 + d_9 w_4 + \left(\frac{1}{U^*} \right)^2 M(\alpha) = g(t)
$$

(1)

where, $\xi = h/b$ is the non-dimensional plunge displacement and τ is the non-dimensional time defined as $\tau = U^* t_1 / b$ (t_1 is the real time), the dot denotes the differentiation with respect to τ, U^* is a non-dimensional flow velocity defined as $U^* = U/b\omega_\alpha$ with U as the flow speed, and $\bar{\omega}$ is given by $\bar{\omega} = \omega_\xi / \omega_\alpha$. ω_ξ and ω_α are the natural frequencies of the uncoupled plunging and pitching modes respectively. ζ_α and ζ_ξ are the damping ratios. $G(\xi)$ and $M(\alpha)$ represent the nonlinear plunge and pitch stiffness terms, respectively. The coefficient $c_0, \dots, c_9$, $d_0, \dots, d_9$ are the related system parameters and the parameters $w_i's$ that depend on upon ξ and α are given by Liu & Dowell (Liu & Dowell 2004).

Introducing a vector $X = (x_1, x_2, \dots, x_8)^T$ with $x_1 = \alpha$, $x_2 = \dot{\alpha}$, $x_3 = \xi$, $x_4 = \dot{\xi}$, $x_5 = w_1$, $x_6 = w_2$, $x_7 = w_3$, $x_8 = w_4$, the coupled state space system given by Equation 1 can be rewritten as a set of eight first-order ordinary differential equations as Equation 2. The expressions for j, $a_{21}, \dots, a_{28}$, $a_{41}, \dots, a_{48}$ are also given by Liu & Dowell (Liu & Dowell 2004).

$$
\begin{cases}
\dot{x}_1 = x_2 \\
\dot{x}_2 = a_{21} x_1 + \left(a_{22} - 2 j c_0 \zeta_\alpha \frac{1}{U^*} \right) x_2 + a_{23} x_3 \\
\quad + \left(a_{24} + 2 j d_0 \zeta_\xi \frac{\bar{\omega}}{U^*} \right) x_4 + a_{25} x_5 + a_{26} x_6 \\
\quad + a_{27} x_7 + a_{28} x_8 + j \left(d_0 \left(\frac{\bar{\omega}}{U^*} \right)^2 x_3 \right. \\
\left. \qquad - c_0 \left(\frac{1}{U^*} \right)^2 M(x_1) \right)
\end{cases}
$$

$$
\begin{cases}
\dot{x}_3 = x_4 \\
\dot{x}_4 = a_{41} x_1 + \left(a_{42} + 2 j c_1 \zeta_\alpha \frac{1}{U^*} \right) x_2 + a_{43} x_3 \\
\quad + \left(a_{44} - 2 j d_1 \zeta_\xi \frac{\bar{\omega}}{U^*} \right) x_4 + a_{45} x_5 + a_{46} x_6 \\
\quad + a_{47} x_7 + a_{48} x_8 + j \left(c_1 \left(\frac{1}{U^*} \right)^2 M(x_1) \right. \\
\left. \qquad - d_1 \left(\frac{\bar{\omega}}{U^*} \right)^2 x_3 \right) \\
\dot{x}_5 = x_1 - \varepsilon_1 x_5 \\
\dot{x}_6 = x_1 - \varepsilon_2 x_6 \\
\dot{x}_7 = x_3 - \varepsilon_1 x_7 \\
\dot{x}_8 = x_3 - \varepsilon_2 x_8
\end{cases}
$$

(2)

The general sketch of the hysteresis is shown in Figure 2. The nonlinear functions are denoted by $M(x_1)$ given in

$$
M(x_1)
$$

$$
= \begin{cases}
x_1 - \alpha_f + M_0, & x_1 < \alpha_f, \dot{x}_1 > 0 \\
x_1 + \alpha_f - M_0, & x_1 > -\alpha_f, \dot{x}_1 < 0 \\
M_0 + M_f (x_1 - \alpha_f), & \alpha_f \le x_1 \le \alpha_f + \delta, \dot{x}_1 > 0 \\
M_f (x_1 + \alpha_f) - M_0, & -\alpha_f - \delta \le x_1 \le -\alpha_f, \dot{x}_1 < 0 \\
x_1 + M_0 - \alpha_f - \delta(1 - M_f), & x_1 > \alpha_f + \delta, \dot{x}_1 > 0 \\
x_1 - M_0 + \alpha_f + \delta(1 - M_f), & x_1 < -\alpha_f - \delta, \dot{x}_1 < 0
\end{cases}
$$

(3)

2.2 Method

The precise integration method was first proposed by Zhong (2004). This method has a great advantage to solve the homogenous linear equations.

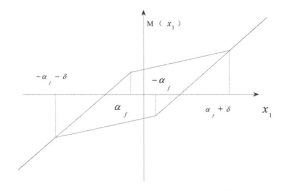

Figure 2. Sketch of a hysteresis stiffness in pitch.

The aeroelastic system (Equation 2) can be divided into six linear subsystems, as the hysteresis taken into account. Written in vector form as

$$\dot{X} = A_i X + B_i \qquad (4)$$

where, A_i, B_i is the coefficient matrixes for subsystem i (= 1, 2, 3, 4, 5, 6) respectively. Using the dimensional expansion method (Gu et al. 2001), Equation 4 will be transferred to a homogenous one.

$$\dot{H} = C_i H \qquad (5)$$

where, $H = [X, B]^T$,

$$C_i = \begin{bmatrix} A_i & I \\ 0 & 0 \end{bmatrix} \qquad (6)$$

For each sub-system, the analytical solution of Equation 5 can be expressed as

$$H = \exp(C_i \tau) H(0) \qquad (7)$$

$H(0)$ is the Initial Condition (IC). When the appropriate time step was selected, the precise integration method can solve the equation comparatively accurately. More details for the precise integration method were given in the article by Zhong (2004).

A predictor-correct algorithm was used to correct the solutions when the vibration state left from one sub-system to another. The switching point is so important that the correct is necessary for getting more accurate results.

3 NUMERICAL EXAMPLES

The system parameters are $\mu = 100$, $a_h = -0.5$, $x_\alpha = 0.25$, $r_\alpha = 0.5$, $\zeta_\xi = 0$, $\zeta_\alpha = 0$, $\bar{\omega} = 0.2$, the following examples show the differences in the change of hysteresis constants and initial conditions.

We can get the critical flutter speed as the system is linear, which is $U_0 = 6.2851$. The initial condition is chosen as $H(0) = [x_1(0), x_2(0), x_3(0), 0, 0, 0, 0, 0]$, with IC defined as $[x_1(0), x_2(0), x_3(0)]$ for convenience.

Figure 3 shows the time histories in pitch and plunge with $U/U_0 = 0.93$. The solutions obtained by the Precise Integration Method (PIM) are in excellent agreement with those obtained by Runge-Kutta (RK) method and the exact solutions. It is reliable for calculating the aeroelastic system with hysteresis by PIM.

Figure 4 shows the limit cycle oscillation amplitudes versus M_0. When M_0 is various from 0 to 1.5, the amplitude increases linearly with the changes of M_0. At $M_0 = 1.5$, the amplitude decreases

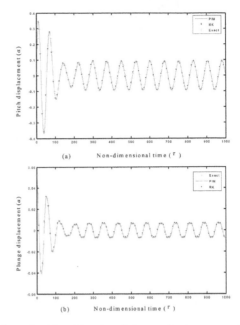

(a)　　Non-dimensional time (τ)

(b)　　Non-dimensional time (τ)

Figure 3. Time histories of the aeroelastic system with hysteresis at $U/U_0 = 0.9$, (a) pitch and (b) plunge.

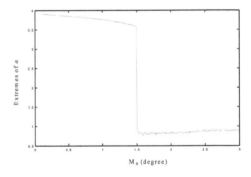

Figure 4. Limit cycle oscillation amplitudes versus M_0 by PIM. ($\delta = 1°$, $M_f = 0.5°$, $\alpha_f = M_0 - (\delta/2)(1 - M_f)$).

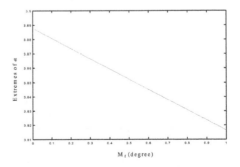

Figure 5. Limit cycle oscillation amplitudes versus M_f by PIM. ($M_0 = 0.5°$, $\delta = 1°$, $\alpha_f = M_0 - (\delta/2)(1 - M_f)$).

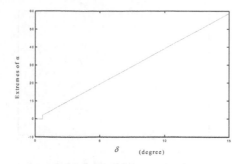

Figure 6. Limit cycle oscillation amplitudes versus δ by PIM. ($M_0 = 0.5°$, $M_f = 0.5°$, $\alpha_f = M_0 - (\delta/2)(1 - M_f)$).

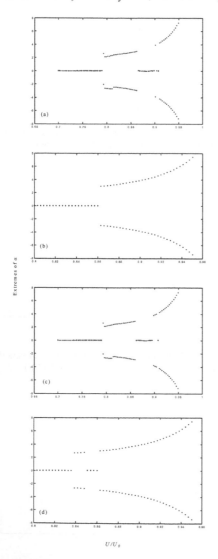

Figure 7. Bifurcations of areoelastic system in pitch with different initial conditions: (a) IC = [1, 3, 1], (b) IC = [1, 1, 1], (c) IC = [0.1, 3, 1], (d) IC = [0.1, 1, 1].

suddenly, and beyond 1.5, continue to increase linearly. Maybe when the M_0 is beyond 1.5, there are not LCOs. According to Figure 5 and Figure 6, we can also see the amplitude increases linearly with the change of M_f and δ.

Different initial conditions have important influence on the dynamic responses. Figure 7 shows the bifurcations of the areoelastic system in pitch. From Figure 7, when $U_0 < 0.78$, there are no LOCs no matter how the initial conditions change. However $U_0 > 0.93$, there are stable LCOs. When the initial displacement and the velocity are both small, the stable LCOs will appear early as shown in Figure 7 (b).

4 CONCLUSIONS

This paper presented a study on the areoelastic system with hysteresis nonlinearity by the precise integration method. Numerical examples show that, the results obtained by PIM are very accurate when compared with that obtained by RK and the exact solutions.

The influence of the hysteresis constants was studied by observing the limit cycle oscillation amplitude, which mainly followed the linear changes. The initial conditions had a great effect on the areoelastic system, and were investigated by using bifurcation diagrams in pitch.

The high accuracy and efficiency of the presented method make it effective to carry out the detailed parametric analysis, which implies this method could be applicable to more aeroelastic systems, especially those with non-smooth nonlinearities.

REFERENCES

Chen, Y.M. & Liu, J.K. 2014. Nonlinear aeroelastic analysis of an airfoil-store system with a freeplay by precise integration method. *Journal of fluids and structures*, 46: 149–164.

Chung, K. Chan, C. & Lee, B. 2007. Bifurcation analysis of a two-degree-of-freedom aeroelastic system with freeplay structural nonlinearity by a perturbation-incremental method. *Journal of sound and vibration*, 299: 520–539.

Cui, C.C. Liu, J.K. & Chen, Y.M. 2015. Simulating nonlinear aeroelastic responses of an airfoil with freeplay based on precise integration method. *Communications in Nonlinear Science and Numerical Simulation*, 22: 933–942.

Dowell, E.H. Clark, R. Cox, D. Curtiss, H. Edwards, J.W. & Peters, D.A. et al. 2005. *A modern course in aeroelasticity*. Springer Netherlands.

Gu, Y.X. Chen, B.S. Zhang, H.W. & Guan, Z.Q. 2001. Precise time-integration method with dimensional expanding for structural dynamic equations. *American Institute of Aeronautics and Astronautics*, 39: 2394–2399.

Li, D. Guo, S. & Xiang, J. 2012. Study of the conditions that cause chaotic motion in a two-dimensional airfoil with structural nonlinearities in subsonic flow. *Journal of fluids and structures*, 33: 109–126.

Liu, J.K. Chen, F.X. & Chen, Y.M. 2012. Bifurcation analysis of aeroealstic systems with hysteresis by incremental harmonic balance method. *Applied Mathematics and Computation*, 219: 2398–2411.

Liu, L. Wong, Y.S. & Lee, B.H.K. 2002. Non-linear aeroelastic analysis using the point transformation method, part 1: freeplay model. *Journal of sound and vibration*, 253: 447–469.

Liu, L. Wong, Y.S. & Lee, B.H.K. 2002. Non-linear aeroelastic analysis using the point transformation method, part 2: hysteresis model. *Journal of sound and vibration*, 253: 471–483.

Liu, L.P. & Dowell, E.H. 2004. The secondary bifurcation of an aeroelastic airfoil motion: effect of high harmonics. *Nonlinear Dynamics*, 37: 31–49.

Zhao, D.M. & Zhang, Q.C. 2010. Bifurcation and chaos analysis for areoelastic airfoil with freeplay structural nonlinearity in pitch. *Chinese Physics B*, 19: (030518)1–10.

Zhong, W.X. 2004. On precise integration method. *Journal of Computational and Applied Mathematics*, 163: 59–78.

Advanced Materials and Structural Engineering – Hu (Ed.)
© 2016 Taylor & Francis Group, London, ISBN 978-1-138-02786-2

A prediction method of dynamic cutting force in the milling process of S45C by flat-end mill cutter

N.T. Nguyen, M.S. Chen & S.C. Huang
Department of Mechanical Engineering, National Kaohsiung University of Applied Sciences, Taiwan, R.O.C.

Y.C. Kao
Advanced Institute of Manufacturing with Hi-Tech Innovations, National Chung Cheng University, Taiwan, R.O.C.

ABSTRACT: This study was performed to predict the dynamic cutting force in mill process by using HSS flat-end mill cutter and S45C workpiece. By mathematical modeling and experimental method, the dynamic cutting force model has been modeled and successfully verified by both simulation and experiment with very promising results. The investigated model can be applied in development of machine tool in industrial manufacturing and can also be extended to apply to other milling tool types and other workpiece materials.

1 INTRODUCTION

The cutting force models are very important components to predict the machining characteristics and develop the machining processes. In theories of metal-cutting mechanics, the cutting mechanics can be analyzed by orthogonal and oblique models (Altintas. 2012, Kao et al. 2015). The mathematics was used to model the cutting forces and other machining characteristics in machining processes. The procedure of cutting force modeling is generally realized by developing the experiential chip-force relationship based on the cutting force coefficients.

Several force models were used to predict the cutting force in milling process such as the exponential force model, the linear force model, and so on. In the exponential force model, the cutting force was modeled as an exponential function of the average chip thickness (Wan et al. 2010, Dang et al. 2010, & Wan et al. 2012). In the second force model, the cutting force was modeled as a linear function of chip thickness (linear-force model). This model is quite suitable to be applied to many types of milling tool such as the flat-end mill (Altintas. 2000, Budak. 2006, & Wang. 2014), ball-end mill (Narita. 2013) and the general-end mill (Gradišek. 2004).

By analysis of the effect of cutter's helix angle on the cutting force coefficient, the linear force model was used to predict the cutting force in the flat-end mill processes (Kao et al. 2015). It seem that the authors only investigated the static model force, that is to say, the dynamic cutting force model has not been investigated and verified.

In this study, the dynamic force model was investigated in milling processes. The main contributions of this study lie in three aspects: (1) The dynamic cutting force was modeled by mathematical model, (2) The dynamic structure of machine tool was determined by using the experimental method, and (3) the cutting force model was successfully verified by experiments.

2 MODELING THE DYNAMIC CUTTING FORCE

In the flat-end mill, the immersion is measured clockwise from the normal axis and the flutes are numbered counter-clockwise. Assuming that the bottom end of flute number one is designated as the reference immersion angle ϕ_1 and the bottom end point of the remaining flute number j is at an angle ϕ_j as shown in Figure 1. Then ϕ_j can be expressed as in Equation (1).

$$\phi_j = \phi_1 - (j-1)\phi_p, \quad j = 1 \sim N_f \tag{1}$$

where, ϕ_p is the cutter's pitch angle that can be expressed in Equation (2).

$$\phi_p = \frac{2\pi}{N_f} \tag{2}$$

When considering the cutter's helix angle β, the lag angle $\Psi_j(z)$ at axial depth of cut z, can be expressed in Equation (3), (Kao et al. 2015).

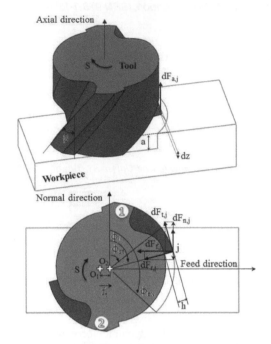

Figure 1. Geometry and processing of flat-end mill.

$$\Psi_j(z) = \frac{2\tan\beta}{D}z \qquad (3)$$

The flute number j, at an axial depth of cut z, the immersion angle is $\phi_j(z)$ that can be expressed by Equation (4), as shown in Figure 1.

$$\phi_j(z) = \phi_j - \Psi_j(z) = \phi_1 - (j-1)\phi_P - \frac{2\tan\beta}{D}z \qquad (4)$$

Assuming that the nose radius of the cutter is ze, for flute number j, at the time t and rotation angle ϕ_j, the tangential, radial, and axial forces acting on a differential flute element can be expressed as in Equation (5), (Kao et al. 2015).

$$\begin{cases} dF_{t,j}(\phi_j, z) = K_{te} * dz + K_{tc} * h_j(\phi_j(z)) * dz \\ dF_{r,j}(\phi_j, z) = K_{re} * dz + K_{rc} * h_j(\phi_j(z)) * dz \\ dF_{a,j}(\phi_j, z) = K_{ae} * dz + K_{ac} * h_j(\phi_j(z)) * dz \end{cases} \qquad (5)$$

The dynamic chip thickness $h_j(\phi_j(z))$ was calculated by Equation (6).

$$h_j(\phi_j(z)) = h_s(\phi_j(z)) + +h_d(\phi_j(z)) \qquad (6)$$

The static chip thickness was expressed as in Figure 1 and can be calculated by Equation (7).

$$h_s(\phi_j(z)) = f_t \sin(\phi_j(z)) \qquad (7)$$

The machine tool dynamic chip thickness was defined as in Figure 2 and can be calculated by Equation (8).

$$h_d(\phi_j(z)) = w_t(\phi_j) - w_{(t-\tau)}(\phi_j) \qquad (8)$$

where, $w_t(\phi_j)$ and $w_{(t-\tau)}(\phi_j)$ are the dynamic displacement at rotation angle $\phi_j(z)$ of current flute and previous flute, respectively, and can be calculated by Equation (9).

$$\begin{cases} w_t(\phi_j(z)) = x_t \sin(\phi_j(z)) + y_t \cos(\phi_j(z)) \\ w_{(t-\tau)}(\phi_j(z)) = x_{(t-\tau)} \sin(\phi_j(z)) + y_{(t-\tau)} \cos(\phi_j(z)) \end{cases} \qquad (9)$$

with x_t and y_t are vibration of machine tool at time t in x and y directions, respectively, and $x_{(t-\tau)}$ and $y_{(t-\tau)}$ are vibration of machine tool at time $t-\tau$ in x and y directions, respectively. The vibrations in x and y directions were calculated by Equation (10).

$$\begin{cases} m_x\ddot{x}(t) + c_x\dot{x}(t) + k_x x = F_x(t) \\ m_y\ddot{y}(t) + c_y\dot{y}(t) + k_y y = F_y(t) \end{cases} \qquad (10)$$

The dynamic cutting forces were simulated following the block diagram in Figure 3. The simulation procedure starts from static chip thickness. The cutting forces were calculated for the cutting processes based on the cutting force coefficients, the cutting conditions, and cutting force models. The machine tool vibrations were generated by the variations of cutting forces and the machine tool dynamic structure. By the effect of machine tool vibrations in x and y direction, the chip thickness changed as the dynamic chip thickness and the

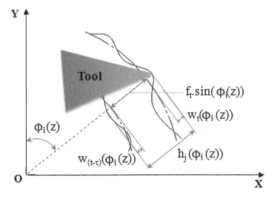

Figure 2. The dynamic chip thickness.

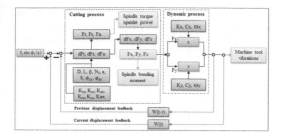

Figure 3. Block diagram of the integrated prediction procedures of dynamic cutting forces.

calculation process of cutting forces were repeated as a new loop. This calculation process was a closed-loop.

From Equation (5), the components of cutting forces can be calculated by Equation (11).

$$
\begin{cases}
dF_{t,j}\left(\phi_j, z\right) = \left[K_{tc}h_j\left(\phi_j(z)\right) + K_{te}\right] * dz \\
dF_{r,j}\left(\phi_j, z\right) = \left[K_{rc}h_j\left(\phi_j(z)\right) + K_{re}\right] * dz \\
dF_{a,j}\left(\phi_j, z\right) = \left[K_{ac}h_j\left(\phi_j(z)\right) + K_{ae}\right] * dz
\end{cases}
\tag{11}
$$

At each point in cutting edge, the cutting forces consist of three components, including radial force, tangential force, and axial force as expressed in Figure 1. So the elemental forces in feed, normal, and axial force were calculated by using the transformation as in Equation (12).

$$
\begin{cases}
dF_{f,j}\left(\phi_j, z\right) \\
dF_{n,j}\left(\phi_j, z\right) \\
dF_{a,j}\left(\phi_j, z\right)
\end{cases}
$$
$$
= \begin{bmatrix}
-\cos(\phi_j(z)) & -\sin(\phi_j(z)) & 0 \\
\sin(\phi_j(z)) & -\cos(\phi_j(z)) & 0 \\
0 & 0 & 1
\end{bmatrix}
\begin{cases}
dF_{t,j}\left(\phi_j, z\right) \\
dF_{r,j}\left(\phi_j, z\right) \\
dF_{a,j}\left(\phi_j, z\right)
\end{cases}
\tag{12}
$$

The differential cutting forces were integrated analytically along the in-cut portion of the flute number j. So, the total cutting forces in flute number j can be calculated by Equation (13).

$$
F_{q,j}\left(\phi_j\right) = \int_{z_1(\phi_j)}^{z_2(\phi_j)} dF_{q,j}\left(\phi_j, z\right), \quad q = f, n, a
\tag{13}
$$

The cutting forces exist only when the cutting tool is in the cutting zone, as expressed in Equation (14).

$$
\phi_{st} \le \phi_j \le \phi_{ex} + \Psi_a
\tag{14}
$$

where, ϕ_{st} and ϕ_{ex} are the entry and exit angles, respectively. Ψ_a is the maximum lag angle, it can be calculated by Equation (15).

$$
\Psi_a = \frac{2\tan\beta}{D}a
\tag{15}
$$

Considering the case that has more than one flute executing the cutting processes simultaneously, the total cutting forces on the feed, normal, and axial direction can be determined by Equation (16).

$$
\begin{cases}
F_f\left(\phi_j\right) = \sum_{j=1}^{N_f} F_{f,j}\left(\phi_j\right) \\
F_n\left(\phi_j\right) = \sum_{j=1}^{N_f} F_{n,j}\left(\phi_j\right) \\
F_a\left(\phi_j\right) = \sum_{j=1}^{N_f} F_{a,j}\left(\phi_j\right)
\end{cases}
\tag{16}
$$

In this study, feed direction coincidences to x direction, and normal direction coincidences to y direction. So, the cutting forces in x and y were calculated by Equation (17).

$$
\begin{cases}
F_x\left(\phi_j\right) = F_f\left(\phi_j\right) \\
F_y\left(\phi_j\right) = F_n\left(\phi_j\right) \\
F_z\left(\phi_j\right) = F_a\left(\phi_j\right)
\end{cases}
\tag{17}
$$

3 EXPERIMENTAL METHOD

The setup of the experiments in this paper includes workpiece and tool, machine tool dynamic measurement and cutting force measurement. The description of the setup is shown in Figure 3.

3.1 Workpiece, tool, and CNC machine

The cutter was chosen as follows. Cutter: a HSS flat-end mill with number of flute $N_f = 2$, helix angle $\beta = 30°$, rake angle $\alpha_r = 5°$, and the diameter was 10 mm.

The workpiece material was S45C. The compositions of S45C are listed in Table 1 and the The properties of the S45C were the following:

Table 1. Chemical compositions of S45C.

	Composite (%)					
	C	Mn	Si	P	S	Fe
Min	0.42	0.6	0.15	–	–	
Max	0.48	0.9	0.35	0.030	0.035	Balance

hardness 160–220 HB, Young's modulus = 190–210 GPa, Poisson's ratio = 0.27–0.30, tensile strength = 569 MPa. The experiments were performed at a Three-axis Vertical Milling Center (TMV-720A).

3.2 *Setup for determination of machine tool dynamic structure*

In order to determine the frequency response function and the dynamic structure of machine tool, an integrated measurement system that consisted of the acceleration sensor (ENDEVCO-25B-10668), hammer (KISTLER-9722 A2000), signal processing box (NI 9234), and CUTPRO™ software was used. The detail setting of the measurement experiment is illustrated in Figure 4.

3.3 *Setup and measurement of cutting forces*

A Dynamometer (Type: XYZ FORCE SENSOR, Model: 624-120-5 KN), signal filter and processing system, and a PC were used to measure cutting forces. The detail is illustrated in Figure 5.

a. Tool b. Acceleration sensor c. Hammer d. Signal processing box e. PC and CUTPROTM software

Figure 4. Setup of Frequency Response Function (FRF) measurement.

a. CNC machine b. Dynamometer
c. Processing System d. PC and Display System

Figure 5. Setup of cutting force measurement.

A series of cutting were performed to determine the cutting force coefficients of one pair of research cutter and workpiece. This process was detailed in reference (Kao et al. 2015). And then, the other cutting tests were performed to verify the dynamic cutting force model.

4 EXPERIMENTAL RESULTS AND DISCUSSIONS

4.1 *Determination of machine tool dynamic structure*

Determination of the machine tool dynamic structure is very important in simulation of dynamic cutting forces and other machining characteristics. In this study, the machine tool dynamic model is shown in Figure 6.

By using the measured results of the Frequency Response Function (FRF) of the machine tool dynamic system, the machine tool dynamic structure was analyzed by CUTPRO™ software. The machine tool dynamic structure was analyzed by the modals in x and y directions, as shown in Figure 7 and Figure 8. Finally, the parameters of the machine tool dynamic structure were determined and listed in Table 2.

4.2 *Verification of dynamic cutting force model*

The measured and simulated forces were compared for three cutting types (half-up, half-down, and slotting), in both normal and fed directions as presented in Figure 9 to Figure 11. The compared results showed that the predicted results of research model were close to the experimental

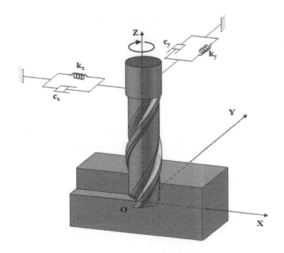

Figure 6. Machine tool dynamic structure modal.

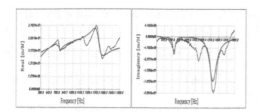
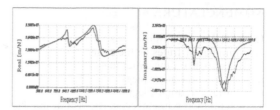

Figure 7. Analysis of dynamic structural modal in x direction.

Figure 8. Analysis of dynamic structural modal in y direction.

Table 2. Machine tool dynamic structure parameters.

Direction	Mode no	Natural frequency [Hz]	Damping ratio [%]	Modal stiffness [N/m]	Mass [kg]
X	1	842.9752	0.898	6.88E+8	24.5245
	2	1591.7397	5.067	6.073E+7	0.6072
Y	1	993.2973	2.165	2.4028E+8	6.1687
	2	148.9916	11.969	3.3983E+7	0.3919

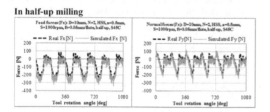
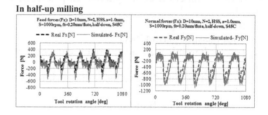

Figure 9. Comparison of measured and simulated cutting forces.

Figure 10. Comparison of measured and simulated cutting forces.

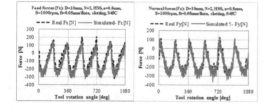

the friction, the deflection, the inconstancy of cutting depth, and so on.

5 CONCLUSIONS

The dynamic force model was successfully modeled by mathematical model to predict the cutting force in milling process.

By using the experimental method, the dynamic structure of machine tool was determined. The experimental results of the natural frequency, damping ratio, modal stiffness, and mass of dynamic system were used to predict the dynamic cutting force.

The dynamic cutting force model has been successfully verified by both simulation and experiment with very promising results.

This research results can be used during the machine tool development, in milling simulation, and milling operation optimization. This model could be extended to more complex type of milling tool for the ball-end mill, the bull-end mill, and so on.

Figure 11. Comparison of measured and simulated cutting forces.

results, and these research results show that the predicted results from research model agree satisfactorily with experimental results. Therefore, the cutting force models and cutting force coefficient values in this study could be used to predict the dynamic cutting forces and other machining characteristics in milling processes. These research results can be used in the development of machine tool in industrial manufacturing.

The reasons for the above differences were mostly originated from the noise, the temperature,

NOMENCLATURE

D = tool diameter [mm]
N_f = number of flutes on the cutter
β = helix angle on the cutter [deg]
ϕ_{st} = cutter entry angle [deg]
ϕ_{ex} = cutter exit angle [deg]
ϕ_j = instantaneous immersion angle of flute number j [deg]
a = maximum axial depth of cut [mm]
dz = differential axial depth of cut [mm]
Ψ_a = lag angle at the maximum axial depth of cut
$z = a$ [deg]
f_t = feed per flute [mm/flute]
K_{tc} = tangential shearing force coefficient [N/mm^2]
K_{rc} = radial shearing force coefficient [N/mm^2]
K_{ac} = axial shearing force coefficient [N/mm^2]
K_{te} = tangential edge force coefficient [N/mm]
K_{re} = radial edge force coefficient [N/mm]
K_{ae} = axial edge force coefficient [N/mm]
$h_j(\phi_j)$ = instantaneous chip thickness [mm]
$h_s(\phi_j(z))$ = static chip thickness [mm]
$h_d(\phi_j(z))$ = machine tool dynamic chip thickness [mm]
t = rotation time [Sec]
τ = tool passing period [Sec]
$w_t(\phi_j)$ = dynamic displacement of current flute [mm]
$w_{(t-\tau)}(\phi_j)$ = dynamic displacement of previous flute [mm]
x_t = machine tool vibration in x direction at time t [mm]
y_t = machine tool vibration in y direction at time t [mm]
$x_{(t-\tau)}$ = machine tool vibration in x direction at time $(t-\tau)$ [mm]
$y_{(t-\tau)}$ = machine tool vibration in y direction at time $(t-\tau)$ [mm]
m_x = mass model of machine tool dynamic structure in x direction [kg]
c_x = damping constant matrix of machine tool dynamic structure in x direction
ω_x = natural frequency matrix of machine tool dynamic structure in x direction [Hz]
k_x = stiffness matrix of machine tool dynamic structure in x direction [N/m]
m_y = mass matrix of machine tool dynamic structure in y direction [kg]
c_y = damping constant matrix of machine tool dynamic structure in y direction
ω_y = natural frequency matrix of machine tool dynamic structure in y direction [Hz]
k_y = stiffness matrix of machine tool dynamic structure in y direction [N/m]
$dF_{t,j}(\phi_j,z)$ = differential tangential cutting force [N]
$dF_{r,j}(\phi_j,z)$ = differential radial cutting force [N]
$dF_{a,j}(\phi_j,z)$ = differential axial cutting force [N]
$dF_{f,j}(\phi_j,z)$ = differential feed cutting force [N]
$dF_{n,j}(\phi_j,z)$ = differential normal cutting force [N]
$F_{t,j}(\phi_j)$ = total tangential cutting force in flute number j [N]
$F_{r,j}(\phi_j)$ = total radial cutting force in flute number j [N]
$F_{a,j}(\phi_j)$ = total axial cutting force in flute number j [N]
$F_{f,j}(\phi_j)$ = total feed cutting force in flute number j [N]
$F_{n,j}(\phi_j)$ = total normal cutting force in flute number j [N]
$F_f(\phi_j)$ = total feed cutting force of N_f flutes [N]
$F_n(\phi_j)$ = total normal cutting force of N_f flutes [N]
$F_a(\phi_j)$ = total axial cutting force of N_f flutes [N]
$F_x(\phi_j)$ = cutting force in x direction [N]
$F_y(\phi_j)$ = cutting force in y direction [N]
$F_z(\phi_j)$ = cutting force in z direction [N].

REFERENCES

Altintas, Y. 2000. Modeling Approaches and Software for Predicting the Performance of Milling Operations at MAL-UBC, *International Journal of Machining Science and Technology*, 4(3): 445–478.

Altintas, Y. 2012. *Manufacturing Automation: Metal cutting mechanics, machine tool vibrations, and CNC design*, Cambridge University Press, 2nd ed, ISBN 978-1-00148-0.

Budak, E. 2006. Analytical models for high performance milling. Part I: Cutting forces, structural deformations and tolerance integrity, *International Journal of Machine Tools & Manufacture*, 46(12): 1478–1488.

Dang, J.W. Zhang, W.H. Yang, Y. & Wan, M. 2010. Cutting force modeling for flat end milling including bottom edge cutting effect, *International Journal of Machine Tools & Manufacture*, 50(11): 986–997.

Gradišek, J. Kalveram, M. & Weinert, K. 2004. Mechanistic identification of specific force coefficients for a general end mill, *International Journal of Machine Tools & Manufacture*, 44(4): 401–414.

Kao, Y.C. Nguyen, N.T. Chen, M.S. & Su, S.T. 2015. A prediction method of cutting force coefficients with helix angle of flat-end cutter and its application in a virtual three-axis milling simulation system, *The International Journal of Advanced Manufacturing Technology*, 77(9–12): 1793–1809.

Narita, H. 2013. A Determination Method of Cutting Coefficients in Ball End Milling Forces Model, *International Journal of Automation Technology*, 7(1): 39–44.

Wan, M. Lu, M.S. Zhang, W.H. & Yang, Y. 2012. A new ternary-mechanism model for the prediction of cutting forces in flat end milling, *International Journal of Machine Tools & Manufacture*, 57: 34–45.

Wan, M. Zhang, W.H. Dang, J.W. & Yang, Y. 2010. A novel cutting force modeling method for cylindrical end mill, *Applied Mathematical Modelling*, 34(3): 823–836.

Wang, M. Gao, L. & Zheng, Y. 2014. An examination of the fundamental mechanics of cutting force coefficients, *International Journal of Machine Tools & Manufacture*, 78: 1–7.

Advanced Materials and Structural Engineering – Hu (Ed.)
© *2016 Taylor & Francis Group, London, ISBN 978-1-138-02786-2*

The development and philosophical thought for three generations of Artificial Neural Networks

Q. Liu, X.P. Yang, S.W. Han & X.S. Ma
School of Electronic Information and Electrical Engineering, Tianshui Normal University, Tianshui, China

ABSTRACT: Artificial Neural Network (ANN) is one of the major scientific and technological achieve-ments that is made by humans in the last century, and is an important milestone on the road to the human understanding nature and use of technology. The important features, the similarities and differences of the first generation to the third generation artificial neural networks are described with the development of ANN as a clue from the computing units of artificial neural network for the classification goal. Many natural dialectical thinking and methods of research and development in three generations of neural net-works are philosophically considered.

1 INTRODUCTION

Artificial Neural Networks (ANN) are inspired by the brain or natural neural network features, and the certain information processing tasks are accom-plished using a large number of simple process-ing units (neurons) through simulating the brain nervous system (Bian et al. 1999). ANN is an interdisciplinary research frontier disciplines on neurophysiology, microanatomy, mathematical sciences, psychology, electronics and information science, computer science, microelectronics and bioelectronics (Han 2007). ANN has been related to image processing, pattern recognition, speech processing, natural language understanding, and robotics nonlinear optimization in different areas, and has made exciting research results. In the mid-dle of the last century, it was found the different biological neuron's cell body, axons and dendrites and synapses of biological neural network in the biological neuroscientist and biological physicists and other scientists joint efforts. The artificial neu-ron model was abstracted and the limitations of digital electronic computer on the basis of the tradi-tional linear processing were made a breakthrough, and then the artificial neural network model was formed, which marked the beginning of the man-kind from the shackles of the linear theory (Wei & Zheng 2007).

In the artificial neural network model, the neurons are generally organized according to the hierarchical structure form; the processing unit in each layer is weighted way connected to other layer processing units. Neural network can achieve the nonlinear mapping, and adaptive self-learning and self-organization features. At the same time, the model also has a strong transformation, associative

storage, optimized combination, classification gen-eralization capabilities and fault tolerance.

2 THE DEVELOPMENT AND RESEARCH OF THE THREE GENERATIONS OF ANN

ANN is a very rapid development of interdiscipli-nary, and its origin can be traced back to the middle of the 20th century. The huge neural networks are composed with a large number of neurons in a cer-tain way, which can complete the complex informa-tion storage and processing, and the whole network system usually shows the characteristics of highly nonlinear and other superior. Simplified schematic mapping of the biological neurons is shown in Figure 1. ANN has many features of the parallel distributed processing, nonlinear mapping, strong adaptability, easy integration and hardware imple-mentation (Hu et al. 1993). Through the simulation,

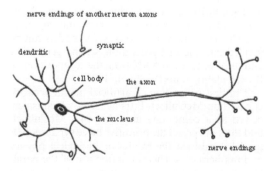

Figure 1.   Simplified schematic mapping of biological neurons.

biological neurons and formation of the artificial neuron have been found to have the following a few basic characteristics: (1) the neurons are interconnected according to certain rules; (2) connection strength between neurons determine the strength of the signal; (3) passed to the neuronal signal can produce an incentive effect, also inhibition; (4) connection strength between neurons can be varied through training; (5) the accumulation results of neuron to receive the signal and determine its state is in excitement, suppress, or stationary; (6) the threshold of each neuron compares with the accumulation results of the signal (Jiang 2003).

The development of neural networks in the scientific community is generally divided from the time course and roughly divided into the four periods. During the rise of the first period stage (40~60 s of the 20th century), some model and algorithm are proposed, which was the first time climax of the research of the artificial neural network. The second period is the low water period (60~80 s of the 20th century), which proved limitation that the perceptron's ability of the single-layer with theory because the perceptron wrote by M. Minsky and S. Papert, and result into a decade-long recession slow development of neural network. The third period is the recovery phase (1980s). During this period, a large number of pioneering works greatly to promote research and learn the algorithm of the neural network model, which further enhances the understanding of neural network system characteristics. The fourth period is the new phase (late 1980s to the present). On June 21, 1987, the First International Conference on neural network was held in San Diego, the International Neural Network Society (INNS) was established at this meeting, neural network research set off again within this marks a new world climax (Han 2007, Qing 2011, Zhang & Zhan 2009). In these different periods of development, a large number of the typical representative ANN model, such as BP network, Hopfield, Boltzmann machine, Radial Basis Function (RBF), Support Vector Machine (SVM) and Pulse Coupled Neural Network (PCNN), were produced, and widely used in the field of astronomy, geology, hydrology, engineering, physics and chemistry, and other areas of the science and engineering. However, these numerous ANN models can be divided into the first generation, second generation and third generation ANN from the composition of the calculating units (Wolfgang 1997).

The first generation of artificial neural network is based on McCulloch-Pitts neurons, which is perceived as a computing unit. In 1943, McCulloch and Pitts proposed the primitive basic synapses MP model combined the results of biological physics and mathematics. This is the first use of the parallel computing architecture to describe the collective artificial neural networks and network operating mode, which was proved to complete any finite logic operation. The first generation of ANN is characterized by the ability only to handle the digital input or output signals, and each Boolean function can be calculated by Multilayer Perceptron with a single hidden layer, the typical representatives Hopfield networks and Boltzmann machines.

The second generation of artificial neural network is the computing unit based on activation function, which can adjust the continuous output value set through the input signal weighted sum (or polynomial). The activation function is commonly used as a sigmoid function, sometimes this function can be used in a piecewise polynomial function. The second generation of ANN can deal with digital input and output signals, but has the very good processing capacity to an analog input and output signal. Another characteristic of this ANN is that the support gradient descent learning algorithm, which is typically the representatives of the BP network and Radial Basis Function (RBF).

Research on the neural network was concentrated on the first generation of ANN and the second generation of ANN before the mid-1990s. Although these two generations ANN vary, they only consider the cumulative network space in the design, without taking into account the cumulative effect of time. This kind of activity in the neurons of the network status is characterized by the average firing rate, so the first two generations of neural networks, also known as the average firing rate of ANN. Each neuron at a certain physiological level time has been continuously playing their role, and its activities with the average of the input and output signal to reflect, which is a first-order statistics of neurons release pulses. However, subtle structural features within that period a pulse are not fully demonstrated, a lot of useful information processing capacity is severely ignored.

With the development of biological neural science and related disciplines, more scholars are taking attention of the space time coding and the cumulative effect at the same time for studying the ANN's program. From the beginning of the last century, a new generation of artificial neural networks, the third generation of artificial neural networks, has began to be studied, which is also known as a pulse neural network in some kind of literature, the corresponding neurons are neurons pulse. As early as 1995, Hopfield (1995) and Sejnowski (1995) thought that the time code may be a new coding method, and is an important research direction in the future. On the basis of the second generation of ANN calculation unit, the third generation of ANN is given full consideration to the time delay, dynamic change information processing and other factors in dendrites and synapses for the real biological neurons, and allows it to work in the discrete time state. The third generation of ANN effectively uses the resources of time calculation that does not use in the ANN of the average firing rate. The third generation of ANN is closer to the biological treatment system, and the signal processing capacity is greatly enhanced, and this network of neurons ignition captures modulation and nonlinear coupling action such as accumulation of space and time coding information processing can be completed. At present, there are Eckhorn connection model, synchronous oscillation visual model and Pulse Coupled Neural Network (PCNN) model for the third generation of ANN (Qing et al. 2010).

It is worth mentioning in the late 1990s that the birth of the PCNN model is a typical representative of the third generation of artificial neural networks, which is the basis of the synchronous pulse distribution phenomena studied in cats, monkeys and other mammals in the visual cortex. PCNN has many series of features, such as variable threshold, capture and nonlinear modulation, dynamic impulses, and synchronous pulse compared with a traditional neural network, which can better simulate the biological visual neural system. Simplified pulse coupled the neural model, as shown in Figure 2. It is a good adaptive system that can be realized to the information from disorderly to orderly and to organizing the unstructured dynamic organizational process, and can complete different hierarchies of information processing. Meanwhile, there is analog signal processing ability in the receiving part and coupling modulation, and there is digital signal processing ability to pulse generator part that can produce pulse signal, which reflects the PCNN is a complex analog-digital hybrid processing system. In addition, the ignition PCNN neuron triggers its neighborhood similar neurons firing and forms a serial process. While a cluster of neurons will lead the process of issuing the synchronous pulse parallel processing, and constitutes a complex mix of series-parallel processing system. Therefore, this model has broad application prospects in artificial life, combinatorial optimization, image processing, automatic target recognition, and other fields.

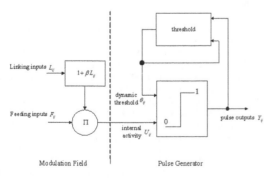

Figure 2.   Simplified pulse coupled neural model.

3   NATURAL DIALECTICAL THOUGHT FOR THREE GENERATIONS OF ANN

ANN is a highly integrated discipline. Its research is a serious challenge to the natural sciences, especially in the traditional artificial intelligence and computer science and philosophy. From the perspective of Dialectics of Nature to reflect on the development of three generations of ANN, we can draw the following conclusions.

3.1   The development of ANN reflects the trend of the development of science

The emergence of artificial neural network theory, and its development process of the rise, depression and climax embodies many other disciplines continue to cross-penetration and comprehensive development process. In the early years, biological neural science is the main research development of the brain function of physiological, psychological and neurons electrophysiological features. MP model has laid a new era of neuroscience research and makes a preliminary mathematics and neuroscience cross synthesis in 1943, and the Hebb rule is put forward, triggered a research upsurge in the '60 s neural network, and formed the first generation of neural network architecture. However, in the 1970s, Minsky and Papert's pessimistic conclusions and stability issues in neural networks are technically difficult to achieve, resulting in neural network research into the depression. Because ANN is a breakthrough in theory as well as with the biological technology, optical technology, the rapid development of VLSI technology in different fields, which provides the theoretical and technical support for the theory of neural networks, makes ANN once again become a hot international research, results in the second and third generation ANN system and forms a multidisciplinary integrated cross frontier discipline. Meanwhile, the rapid development of the ANN enrich the content of the scientific system.

3.2   The relationship between ANN and nonlinear science development

Nonlinear science is the study of nonlinear problems of an interdisciplinary. With the further development of society and science, linear methods encountered great obstacles in dealing with intelligent information processing such as image thinking and association memory complex problems. ANN from the early MP linear perceptron with some simple linear classification problem is the first generation of ANN model, and such models need to deal with the complex problems such as pattern recognition, which eventually encountered difficulties. With these problems on human understanding and improvement of the ANN theory, the non-linear calculation unit with the activation functions came into being and formed the second generation ANN, so that it is widely applied in various fields.

Later, with the development of many nonlinear sciences, the nonlinear characteristic of ANN and some kind of learning algorithm is more close to the characteristics of real biological neurons, and can handle many nonlinear problems, which fully proved the validity of the theory of ANN research methods. The current third generation ANN is a highly nonlinear adaptive large-scale information processing system, having the features of the human brain to some degree. Serial and parallel distributed processing, analog-digital hybrid processing and robust, adaptive and learning capabilities are brilliant achievements in recent years. It can be seen that the development of non-linear science promote continuous improvement and development of artificial neural networks, in turn, the generation and application of artificial neural networks for the formation of nonlinear science system also play a catalytic role.

3.3 The relationship between man and machine from ANN applications

Some commercial computer ANN and nervous system has been applied in defense, telecommunications, financial services, aviation and other fields. However, with the continuous development and improvement of the ANN, it also sparked anxiety and fear: a computer such as a neural computer will exceed the human brain it, highly developed intelligent neural robot in the near future could rule all mankind, and a series of problems. For example, in 1997, IBM's "Deep Blue" computer beat chess grandmaster Garry Kasparov, once again sparked concern about computers and the human brain problems.

The essence of the problem is the relationship between human and technology. First, fully demonstrating the wisdom of the human brain offers the possibility for an artificial machine to overcome the human brain. Therefore, ANN development and to explore the mysteries of the human brain is inextricably linked. Second, the relationship between man and machine is an asymptote, but cannot become contour in essence. As a result, the product of ANN theory, neural network computer or robot does not include all the brain thinking of subjective initiative, sociality and the human's subjective world. In addition, the research and development of three generations of ANN will also cause the following philosophical thinking: the interaction between ANN and philosophy of science, the research method of ANN will follow the process of practice, theory, practice and ANN theory is the result of intense competition and so was able to survival issue.

4 CONCLUSIONS

After half a century of human studies on ANN, the first generation to third generation of artificial neural network model was formed by perceptron cell, the cell activation function modulation and nonlinear coupling processing unit representation, respectively. Many philosophical questions are worth considering the trend of development of ANN, ANN perfect application and the relationship between man and machine. Although ANN future is unpredictable, there will make the artificial neural further research network to move forward, to make it better for the human, as long as the correct guidance of scientific research methodology, continue to introduce new expectations, new problems continue to find errors, correct errors, and continuously improve the theory into practice.

ACKNOWLEDGMENT

This work was supported by the National Natural Science Foundation of P.R. China (61461046), the Natural Science Foundation of Gansu Province (1010RJZE028), the Higher School in Gansu Province Basic Scientific Research Expenses (2050205) and the Tianshui youth scientific and technological support projects (TSK1201).

REFERENCES

Bian, Z. Zhang, X. et al. 1999. *Pattern recognition (Second Edition)*, Beijing: Tsinghua University press.

Ejnowski, T.J.S. 1995. Time for a new neural code, *Nature*, 376: 31–32.

Han, L. 2007. *The theory, design and application of artificial neural networks*. Beijing: Chem. Ind. Press.

Hopfield, J.J. 1995. Pattern recognition computation using action potential timing for stimulus representation, *Nature,* 376: 33–36.

Hu, S. Yu, S. & Dai, K. 1993. *Introduction of neural networks,* Changsha: National University of Defense Technology press.

Jiang, Z. 2003. *Introduction to artificial neural networks,* Beijing: higher education press.

Qing, L. 2011. *Research on Several Issues about Image Processing Based on Pulse Coupled Neural Networks.* Xian: Xidian University.

Qing, L. Luping, X. Yong, W. et al. 2010. A Novel Algorithm of Image Fusion based on Adaptive ULPCNN Time Matrix, *2010 WASE International Conference on Information Engineering.* 198–202.

Wei, H. & Zheng, X. 2007. Philosophical Reflecting on Imitation of Nervous Net in AI, *Studies in Dialectics of Nature,* 23(4): 22–25.

Wolfgang, M. 1997. Networks of spiking neurons: the third generation of a to neural network models, *Neural Networks*, 10(9): 1659–1671.

Zhang, J. & Zhan, Z. 2009. *Calculation of intelligent,* Beijing: Tsinghua University press.

Advanced Materials and Structural Engineering – Hu (Ed.)
© 2016 Taylor & Francis Group, London, ISBN 978-1-138-02786-2

Single Degree of Freedom (SDOF) and finite element analysis of steel column subjected to blast load

M.M. Abdallah
College of Civil and Transportation Engineering, Hohai University, Nanjing City, Jiangsu Province, China

B.H. Osman
College of Civil and Transportation Engineering, Hohai University, Nanjing City, Jiangsu Province, China
Department of Civil Engineering, Engineering College, Sinnar University, Sinnar, Sudan

ABSTRACT: Columns are the key load bearing members in any structure and the exterior columns usually exposed to transverse loads; in some cases as blast loads producing from terrorist attacks. A Single Degree of Freedom (SDOF) model is used to investigate the effect of steel column direction on the column's behavior during the blast event by studying the maximum midpoint displacement. ABAQUS finite element code was used to check the validity of SDOF results. Also, the effect of strain rate was introduced by determining the charge effects with different values of Dynamic Increase Factor (DIF).

1 INTRODUCTION

Determining the response of structures subjected to blast loads is a complicated process. In recent years, many papers have been published to study the behavior of structural elements under blast loading by comparing the results of; numerical finite element codes, experiments and the SDOF analysis. Nassr et al. (2011) conducted experimental work on steel beams under blast loading and discovered that the SDOF mid-span displacement results were nearly approached to the experiments results by increasing the effect of strain rate. The validity of the SDOF analysis in concrete columns under axial and blast loading was later confirmed by Serdar Astarlioglu et al. (2013) with the ABAQUS finite element code. In this particular study, two different steel columns were assumed in the analysis. SDOF method was used to investigate the effect of column direction on the behavior under blast loading. Also, the strain rate effect was included in three different cases, and the results compared with ABAQUS finite element code based on the maximum midpoint displacement. The charge weight used in this study was 100 Kg TNT at a 4.5 m standoff distance, and the blast loads were calculated for steel columns by using the UFC 3-340-02 (2008) technique method.

2 MODELS AND DYNAMIC EQUATIONS

Structural elements and systems such as walls, slabs, beams, columns, frames and shear wall can be represented by a single degree of freedom model (CCPS 1996). SDOF is a model that represents the dynamic.

Characteristic of a structure with a single mass and spring and the model component having mobility in only one axis Figure 1 shows a representation of the single degree of freedom system. John Biggs (1964) conducted the SDOF method to analyze the behavior of structural component's by using its properties to convert the member to an equivalent SDOF model, the findings of which were published in the Technical Manual UFC 3-340-02 (USDOD 2008), and were used to help design the Single-degree-of-freedom Blast Effects Design Spreadsheet (SBEDS) program designed by the United States Army Corps of Engineers-PD center. The SDOF model Equation of motion is represented by Equation 1.

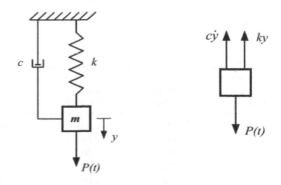

Figure 1. Single Degree of Freedom (SDOF) system.

$$K_{LM}M\ddot{y} + c\dot{y} + R(y) = P(t) \qquad (1)$$

where, $P(t), M, \ddot{y}, \dot{y}, R(y)$ are the beam loads, system mass, system acceleration, system velocity and beam resistance respectively. The load mass factor, K_{LM} is the ratio of mass a factor to load factor and is represented by K_M/K_L. The effect of damping c on the maximum deflection is extremely small (Baker et al. 1983), and can, therefore, be negated. The Equation of motion can be rewritten as Equation 2.

$$K_{LM}M\ddot{y} + R(y) = P(t) \qquad (2)$$

In blast load conditions, the load can be represented as a triangle load with a maximum force F_m and a time duration t_d as shown in Figure 2, and the blast impulse "I" can be approximated as the triangle area $(1/2\ F_m t_d)$. The elastic-plastic analysis of the beam resistance $R(y)$ is usually taken as (Ky) in the elastic stage, where K is the elastic stiffens, and R_m in the plastic stage, where $R_m = 8M_p/L$, M_p is the member's plastic moment capacity. The Equation for motion in the blast load condition can be expressed by Equation 3.

$$K_{LM}M\ddot{y} + R(y) = P(t) = F(t) = F_m\left(1 - \frac{t}{t_d}\right) \qquad (3)$$

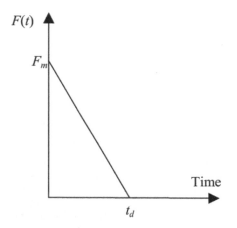

Figure 2.   Blast load as triangle load.

The SDOF factors which were introduced by Biggs for simply and fixed supported beams are shown in Table 1 and Table 2 respectively.

These factors can be expressed mathematically as:

$$K_M = \frac{\int_0^L m\varnothing^2(x)dx}{mL}, \text{ For distributed mass} \qquad (4)$$

$$K_R = K_L = \frac{\int_0^L p\varnothing^2(x)dx}{pL}, \text{ For distributed load} \qquad (5)$$

where, m, p, $\varnothing(x)$, and L are the member's mass per unit length, member's load per unit length, the assumed mode or displaced shape of the member and the member's length respectively.

Table 1.   Equivalent SDOF factors for simply supported beams (Biggs 1964, Yokoyama 2011).

Loading type	Strain range	Load factor K_L	Mass factor K_M	Maximum resistance R_M	Spring constant k
Uniform	Elastic	0.64	0.50	$8M_p/L$	$384EI/5L^3$
	Plastic	0.50	0.33	$8M_p/L$	0
Point	Elastic	1.00	0.49	$4M_p/L$	$48EI/L^3$
	Plastic	1.00	0.33	$4M_p/L$	0

Table 2.   Equivalent SDOF factors for beams with fixed ends (Biggs 1964, Yokoyama 2011).

Loading type	Strain range	Load factor K_L	Mass factor K_M	Maximum resistance R_m	Spring constant k
Uniform	Elastic	0.53	0.41	$12M_p/L$	$384EI/L^3$
	E-P	0.64	0.5	$8/L(M_{ps} + M_{pm})$	$384EI/5L^3$
	Plastic	0.50	0.33	$8/L(M_{ps} + M_{pm})$	0
Point	Elastic	1.00	0.37	$4/L(M_{ps} + M_{pm})$	$192EI/L^3$
	Plastic	1.00	0.33	$4/L(M_{ps} + M_{pm})$	0

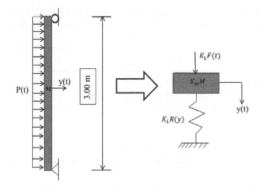

Figure 3. Equivalent SDOF system were a real structure is converted to its equivalent SDOF model.

3 SINGLE DEGREE OF FREEDOM (SDOF) OF STEEL COLUMN ANALYSIS

In SDOF analysis the real system, i.e. the steel column, is replaced by an equivalent spring mass system as shown in Figure 3. The SDOF analysis in this study was based on the solution of Equation 3. K_{LM} values for the elastic analysis and plastic range are 0.78 and 0.66 respectively. The resistance function of the steel column under blast loading $R(y)$ was taken as $Ky(t)$ in the elastic range and was taken as R_m in the plastic analysis, where $R_m = 8M_p/L$.

4 MATERIAL BEHAVIOR AT HIGH STRAIN RATE

To analyze the dynamic response of structures to blast loading, the effect of high strain rate should be taken into account, due to the increase of structural material strength under high rate of loading, i.e. blast loading, as compared to an identical material subjected to static loads. Blast loads typically produce very high strain rates in the range of $(10^2 - 10^4)S^{-1}$. Figure 4 shows the expected strain rates for different types of loading.

The Dynamic Increase Factor (DIF) is usually used to determine the strain rate effect which refers to the ratio of material dynamic strength to static strength. The DIF Equation as a function of the column axial strain rate $\dot{\varepsilon}$ is introduced in Equation 6 (Jones 1988).

$$DIF = 1 + \left(\frac{\dot{\varepsilon}}{D}\right)^{\frac{1}{q}} \qquad (6)$$

where, D and q are constants for the particular material. For steel $D = 40$ and $q = 5$.

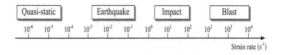

Figure 4. Strain rates associated with different types of loading (Ngo et al. 2007).

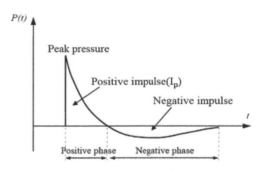

(a)

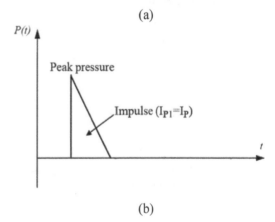

(b)

Figure 5. Simplification of blast load (a) real blast pressure history curve (b) simplified model (Dassault Systemes, 2012).

5 BLAST LOAD DETERMINATION

In this study, the simplified blast load as shown in Figure 5 is applied to the column's surface, when the column was in the major axis direction and in the minor axis direction respectively as shown in Figure 6. The blast load was determined for the columns based on UFC 3-340-02 (2008) and TM 5-1300. 100 kg TNT charge weight was used at a 4.5 standoff distance. The reflected pressure and time duration were 7280 KPa and 1.01 msec respectively and were determined by using appropriate Figures with scaling law Z. The SDOF spreadsheet created by the author is based on Biggs Equations and ABAQUS/Dynamic explicit finite

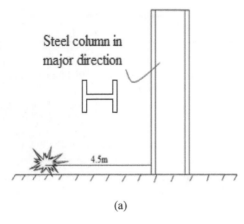

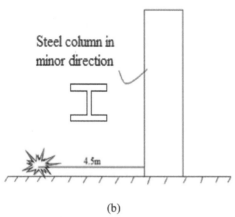

Figure 6. Steel column direction with charge weight source, (a) major axis (b) minor axis.

Table 3. Columns dimension details (The British Constructional Steelwork Association Ltd and The Steel Construction Institute 2007).

Column	D	B	t	T	r	d
203*203*46 (c_1)	203.2	203.6	7.2	11.0	10.2	160.8
254*254*73 (c_2)	254.1	254.6	8.6	14.2	12.7	200.3

Table 4. Steel material properties.

Density	7.86* 10^3 kg/m³	Plastic stress strain data			
Young modulus	2.1e11 Pa	Stress (Pa)	Strain	Stress (Pa)	Strain
Poisson ratio	0.3	0	0	400e6	0.06
Thermal conductivity	50 W/m-C	275e6	0.003	420e6	0.08
Specific heat	480 J/Kg-C	370e6	0.04	440e6	0.10

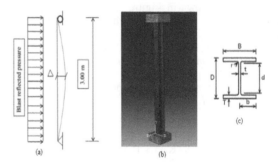

Figure 7. Column's arrangement (a) loading type (b) 3D view, and (c) column cross-section.

element code (2012) were used separately with the blast load data to determine the maximum deflection in the major and the minor direction of both columns.

6 NUMERICAL MODELING

Two different steel columns with I-shaped cross-sections have been numerically modeled, the details of each section and steel properties are illustrated in Table 3 and 4 respectively. The steel columns were numerically modeled as a solid element using ABAQUS finite element code. A rigid block was connected to each end of each steel column model to avoid the inauthentic distortion on the boundary (Ding et al. 2012). The boundary constraints are applied on the rigid block surfaces. Figure 7 illustrates the numerical steel column model, the

method of applying blast pressure, boundary constrains, the results being investigated (Δ), and the model meshing.

7 RESULTS AND DISCUSSIONS

The results obtained from SDOF analysis and ABAQUS finite element code are shown in Tables 5 and 6. Column one's (c1) behavior under blast loading with a DIF = 1 is shown in Figure 8. Figures 9 and 10 are shown the differences between ABAQUS and SDOF analysis for mid-point displacement of column one (c1) in the case DIF = 1.00 and column two (c2) in the case of DIF = 1.48.

Table 5. The maximum deflection of steel columns in major axis direction.

I_{xx} 100 kg TNT

Column	ABAQUS results in mm			SDOF results in mm		
	DIF = 1.00	DIF = 1.48	DIF = 1.76	DIF = 1.00	DIF = 1.48	DIF = 1.76
203*203*46 (c_1)	54.46	41.06	37.06	61.02	49.17	43.93
254*254*73 (c_2)	30.84	24.38	22.68	32.42	26.75	25.73

Table 6. The maximum deflection of steel columns in minor axis direction.

I_{yy} 100 kg TNT

Column	ABAQUS results in mm			SDOF results in mm		
	DIF = 1.00	DIF = 1.48	DIF = 1.76	DIF = 1.00	DIF = 1.48	DIF = 1.76
203*203*46 (c_1)	88.86	71.44	65.24	76.67	71.59	70.64
254*254*73 (c_2)	57.27	43.48	39.54	44.43	44.17	44.17

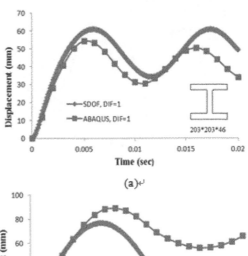

(a)

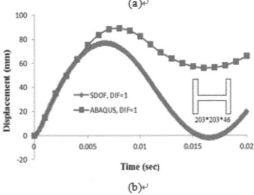

(b)

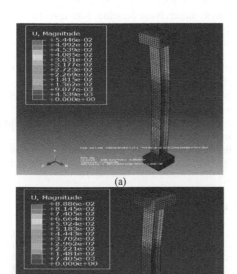

(a)

(b)

Figure 8. Column c1 behavior under blast load at (a) major axis I_xx (b) minor axis I_yy.

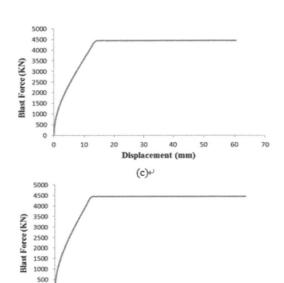

(c)

(d)

Figure 9. Column one behaviors under blast loading with DIF = 1.00, (a) Major axis direction. (b) Minor axis direction, (c) and (d) Blast load vs. displacement for each case.

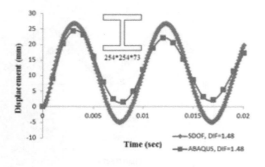

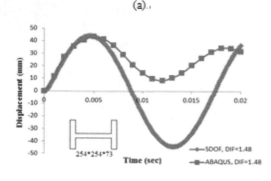

(a).

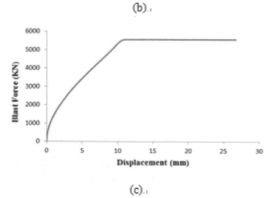

(b).

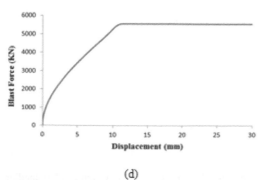

(c).

(d)

Figure 10. Column c2 behaviors under blast loading with DIF = 1.48, (a) Major axis direction. (b) Minor axis direction, (c) and (d) Blast load vs. displacement for each case.

Table 7. Differences between SDOF and ABAQUS results.

	I_{xx} differences			I_{yy} differences		
Column	DIF = 1.00	DIF = 1.48	DIF = 1.76	DIF = 1.00	DIF = 1.48	DIF = 1.76
203*203* 46(c_1)	6.96%	8.11%	6.87%	12.19%	0.15%	5.4%
254*254* 73(c_2)	1.58%	2.37%	3.05%	12.84%	0.69%	4.63%

Based on the column stiffness, the resistance of the column to blast loads in the major axis was superior to the resistance in the minor axis in all cases, and the results obtained from both methods, ABAQUS and SDOF, were nearly similar to each other with some minor differences, those of which are highlighted in Table 7. Additionally, the mid-point maximum displacement decreased with an increase in DIF, which is consistent with what actually occurs in the field due to the effect of dynamic loads.

8 CONCLUSION

The effect of steel column direction on the column behavior under blast loading was investigated by using the SDOF analysis method, and the obtained results compared to ABAQUS results to confirm its validity. The results which were obtained by different analysis methods for the maximum midpoint displacement of columns were approximately same with minor differences in some cases. The deflection values in the major direction were smaller than that of those in the minor direction due to column stiffness. Additionally the effect of strain rate was included in the study of three different cases, and the deflection decreased with an increase in DIF value.

REFERENCES

Astarlioglu, S. Krauthammer, T. Morency, D. & Tran, T.P. 2013. Behavior of reinforced concrete columns under combined effects of axial and blast-induced transverse loads, *Journal of Engineering Structure.* 55: 26–34.

Baker, W.E. Cox, P.A. Westin, P.S. Kulesz, J.J. & Strehlow, R.A. 1983. *Explosion Hazards and Evaluation,* Elsevier, New York.

Biggs, J.M. 1964. *Introduction to structural dynamics,* McGraw-Hill.

Center for Chemical Process Safety, 1996. *Guidelines for Evaluating Process Plant Buildings for External Explosions and Fires,* American Institute of Chemical Engineers.

Dassault Systemes, 2012. ABAQUS/CAE user manual.

Ding, Y. Wang, M. Li, Z.X. & Hao, H. 2012. Damage evaluation of the tubular steel column subjected to explosion and post-explosion fire condition, *Engineering Structures,* 55: 44–55.

Jones, N. 1988. *Structural impact, Cambridge*, New York: Cambridge University Press.

Nassr, A.A. Razaqpur, A.G. Tait, M.J. Campidelli, M. & Foo, S. 2011. Experimental Performance of Steel Beams under Blast Loading, *ASCE's J. Performance Construction Facilities,* 26(5): 600–619.

Ngo, T. Mendis, P. Gupta, A. & Ramsay, J. 2007. *Blast Loading and Blast Effects on Structures—An Overview*, EJSE Special Issue: Loading on Structures.

The British Constructional Steelwork Association Ltd and The Steel Construction Institute, 2007. *Hand book of structural steel work 4th edition,* Whitehall Court, London, SW1 A 2ES.

Unified Facilities Criteria, 2008. *UFC 3-340-02 Structures to Resist the Effects of Accidental Explosions*, U.S. Army Corps of Engineers, Naval Facilities Engineering Command, Air Force Civil Eng. Support Agency.

Yokoyama, T. 2011. *Verification and Expansion of Single-Degree-of-Freedom*, Transformation Factors for Beams Using a Multi-Degree-of Freedom Non-Linear Numerical Analysis Method, November.

Advanced Materials and Structural Engineering – Hu (Ed.)
© *2016 Taylor & Francis Group, London, ISBN 978-1-138-02786-2*

Improved Genetic Algorithm for solving large square jigsaw puzzles: An initial exploration

Y.D. Zhang, S.H. Wang & G.L. Ji
School of Computer Science and Technology, Nanjing Normal University, Nanjing, Jiangsu, China

S.W. Chen
State Key Laboratory of Millimeter Waves, Southeast University, Nanjing, Jiangsu, China
School of Communication and Information Engineering, University of Electronic Science and Technology of China, Chengdu, Sichuan, China

Q. Wang & C.M. Feng
Jiangsu Key Laboratory of 3D Printing Equipment and Manufacturing, Nanjing, Jiangsu, China

ABSTRACT: In this study, we propose a novel square jigsaw solver, which improves canonical Genetic Algorithm (GA). First, the region-growing technique is proposed to complete a candidate solution after crossover and mutation of two parent solutions. Next, we propose a new fitness function that combines both internal cost and external cost, so as to avoid shifted solution problems. Finally, we combine the fitness-scaling and elitism techniques to help improve solution correction rate. Experiments on seven jigsaw puzzle problems of pieces show the average values of the fraction of corrected neighbors of the proposed algorithm over ten runs are higher than GA. Therefore, the proposed jigsaw solver is accurate and rapid.

1 INTRODUCTION

A jigsaw puzzle requires the assembly of numerous small, often oddly shaped, interlocking and tessellating pieces. Each piece usually has a small part of a picture on it; when complete, a jigsaw puzzle produces a complete picture (Yao & Shao 2003). Solutions to the problem have been proved to be beneficial to the fields of image encryption and decryption (Kumar et al. 2010), collaborative learning (Pozzi 2010), protein (Liang & Dill 2001), packing (Lesh et al. 2004), and mobile application (Danado & Paterno).

In the past, scholars tend to use the shape feature (Chung et al. 1998), texture feature (Sagiroglu & Ercil 2006), and edge feature (Zhi & Ji 2009) to solve the jigsaw puzzle. Recently, the researchers focus more on regarding the problem to merely color-based solvers of square-piece jigsaw problems, i.e. given different non-overlapping square pieces of an image, the task is to reconstruct the original image with the help of merely chromatic information (Cho et al. 2010, Pomeranz et al. 2011, Tybon & Kerr 2009).

The core element for computationally solving a jigsaw puzzle is how to measure the correctness of a solution candidate. Shalomon et al. (2013) have proposed a measure (we called "internal cost" in this study), which sums up the pairwise dissimilarities over all neighboring pieces. However, the internal cost suffers from an obvious drawback of shifted solutions, due to its ignorance of edge information. In this study, we proposed a new measure that considers not only internal cost but also the external cost. On the other hand, the powerful computational capability of canonical Genetic Algorithm (GA) is harnessed as the search algorithm for piece placement, and is proven to be a powerful solver with state-of-the-art performance for large jigsaw puzzles (Cholomon et al. 2013, Kawechki & Niewierowicz 2014). However, GA was criticized for its slow speed and premature convergence.

To solve this, fitness-scaling and elitism mechanism are introduced. In all, we proposed a novel FItness-Scaling Elite Genetic Algorithm with Region Growing (FISEGARG) that takes advantages of both fitness-scaling mechanism and region-growing strategy.

2 PROBLEM DEFINITION

Given a puzzle image of ($M \times N$) pieces, we label each piece with a unique linear index. From another point of view, we vectorize the puzzle image by combining all columns from the matrix, each appended to the last. Therefore, it is unambiguous to represent an individual by a chromosome of a $M \times N$ matrix, where each element corresponds to the index of the piece.

The fitness function is evaluated for all individuals for the purpose of selection. The "dissimilarity" concept should be introduced beforehand. It relies on the premise that adjacent jigsaw pieces in the original image have a tendency towards sharing similar colors along their neighboring pieces. Therefore, the sum of color differences over all color channels should be minimal.

Assume pieces are presented in RGB space by a $H \times W \times 3$ matrix, where H is the height and W the width of a piece in the unit of pixels. $H = W$ for square-piece jigsaw problem. Given two pieces $\times$ and y, and a spatial relation between them $S \in \{l,r,u,d\}$, $D(x, y|S)$ represents the dissimilarity of piece $\times$ when placed to the left, right, up, or down side of piece y, respectively.

$$D(x,y\,|\,l) = \sqrt{\sum_{h=1}^{H}\sum_{c=1}^{3}\left(x(h,1,c)-y(h,W,c)\right)^2}$$

$$D(x,y\,|\,r) = \sqrt{\sum_{h=1}^{H}\sum_{c=1}^{3}\left(x(h,W,c)-y(h,1,c)\right)^2}$$

$$D(x,y\,|\,u) = \sqrt{\sum_{w=1}^{W}\sum_{c=1}^{3}\left(x(1,w,c)-y(H,w,c)\right)^2}$$ (1)

$$D(x,y\,|\,d) = \sqrt{\sum_{w=1}^{W}\sum_{c=1}^{3}\left(x(H,w,c)-y(1,w,c)\right)^2}$$

where, c represents one of the RGB color channels. The users need to pay particular attention that dissimilarity is not symmetric, so $D(x,y\,|\,S) \neq D(y,x\,|\,S)$. It is obvious the dissimilarity of the original image should be the least among all solutions.

The internal cost C_I is defined as the sum of pairwise dissimilarities over all neighboring pieces as

$$C_I = \sum_{m=1}^{M}\sum_{n=1}^{N-1}\left(D\big(J(m,n),J(m,n+1)\,|\,r\big)\right)$$
$$+ \sum_{m=1}^{M-1}\sum_{n=1}^{N}\left(D\big(J(m,n),J(m+1,n)\,|\,d\big)\right)$$ (2)

where, $J(m, n)$ represents a single puzzle piece at mth row and nth column in the unit of piece. However, C_I suffers from shifted solutions because of its ignorance of dissimilarity of edge. In this paper, we proposed the external cost C_E that measures the variance of the borders of the outermost pieces.

$$C_E = \sum_{n=1}^{N}\left(\text{var}\big(J(1,n)(1,:)\big)+\text{var}\big(J(M,n)(H,:|)\big)\right)$$
$$+ \sum_{m=1}^{M}\left(\text{var}\big(J(m,1)(:,1)\big)+\text{var}\big(J(m,N)(:,W)\big)\right)$$ (3)

where, $J(m, n)(h, w)$ equals to $x(h, w)$ that subjects to $x = J(m, n)$, and the colon represents the whole element along the dimension. Finally, we define the total cost C_T as the summation of internal cost and external cost, which is used as the fitness function. We guessed this kind of definition help to solve the problem.

3 IMPROVED GA

Traditionally, the crossover and mutation are independent with each other (Zhang et al. 2013), although both produce valid solutions (Fig. 1a). However, for the jigsaw problem, crossover and mutation will yield invalid candidates with duplicate and/or missing pieces, so we proposed a different improved scheme of GA, which uses a simplified crossover and mutation operators to initialize a seed (an incomplete solution), followed by a region-growing technique to get a complete solution (Fig. 1b).

Observing from Figure 1b, the position (m, n) of the initial seed of the child J_C is selected randomly by mutation operator.

$$(m,n) = \text{mutation}\,(M,N)$$
s.t. $1 \leq m \leq M,\ 1 \leq n \leq N$ (4)

Afterward, crossover operator is performed on the parent values at the location of seed.

$$J_C(m,n) = \text{Crossover}\big(J_{P1}(m,n),J_{P2}(m,n)\big)$$ (5)

where, J_{P1} and J_{P2} are two parents of jigsaw solutions. The region is then grown piece-by-piece

Figure 1. Generation of a new valid solution: (a) traditional GA and (b) the improved GA.

from these seed pieces to adjacent pieces of 4-connected neighborhood depending on some criterion.

$$J_C = \text{RegionGrow}(J_C(m,n)) \qquad (6)$$

The pseudocodes of the proposed region-growing method are depicted as follows:

Algorithm1 – Region Growing

Initialize a random seed, and generate a piece-list that contains all pieces
Repeat until the piece-list is empty
    Choose a piece at the boundary of known region randomly.
    Repeat until there is no empty piece in its 4 neighbors, the order of neighbors is randomly selected.
        If the neighbor is empty
        Then
            if both parents agree at the neighbor & the piece was not assigned
            then place the piece there
            otherwise choose the piece that was not assigned and has minimal total cost with current piece
        Otherwise go to next neighbor.

On the other side, the standard reproduction operator uses the fitness values to select the individuals of the next generation, which assigns a higher probability of selection to individuals with higher values, and a lower probability to those with lower values (Zhang et al. 2013, Okamoto et al. 2014, Chamorro-Posada et al. 2014). Two problems arise for this standard reproduction: (i) at the beginning of GA run, where there may be a local optimal individual with a very high fitness value. The search will have a bias toward it. (ii) nears at the end of GA run, when the population is converging, and there may be little separation among individuals.

To overcome those two undesirable phenomena, we employ fitness scaling technique so that the "selection pressure" remains the same throughout the run. Currently there exist bundles of fitness scaling methods (Zhang et al. 2013). The most common scaling techniques involve linear scaling, rank scaling, power scaling, top scaling, etc. Among those methods, power scaling finds a solution nearly the most quickly due to an improvement of diversity, but it is vulnerable to instability (Korsunsky & Constantinescu 2006). On the other hand, rank scaling shows superiority to other methods in terms of stability on different types of tests (Wang et al. 2010, Zhang et al. 2011). Zhang et al. proposed a power-rank fitness scaling strategy (Zhang et al. 2013) as:

$$f_i \leftarrow \frac{r_i^k}{\sum_{i=1}^{N} r_i^k}, \; r_i = \text{rank}\left(f_i \mid [f_1, f_2, \dots f_P]\right) \qquad (7)$$

where, f_i represents the fitness value of an i th individual, r_i its corresponding rank, P the number of population, and r^k represents r raised to the power of k. The strategy involves a three-step procedure. First, all individuals are sorted to obtain their corresponding ranks. Second, powers are computed for exponent k. Third, the scaled values are normalized by dividing the sum of the scaled values over the entire population. Finally, the original fitness values are replaced by the scaled values.

In addition, some of the individuals in the current population that have better fitness are chosen as elite. These elite individuals are passed to the next population directly (Behbahani & Dilva 2014, Chung et al. 2009). We dub the proposed algorithm as FItness-Scaled Elite Genetic Algorithm with Region-Growing (FISEGARG).

4 EXPERIMENTS AND RESULTS

All of the programs were in-house developed using Matlab2014a and were run on an IBM laptop with Intel Core i3 3GHz processor and 2GB RAM. The ratio of rows to columns of jigsaw pieces is fixed to 7:10. The performance of the proposed solver is verified by seven images, which are downloaded from the internet, with different sizes varying from 280 to 4480.

4.1 Region-growing result

Figure 2 illustrates the region-growing procedure. Figure 2 (a-b) shows the solutions of two parents. Figure 2c is the initial seed that is created by crossover and mutation operators on the two parents. Figure 2 (d-f) depict how the region is gradually grown until a complete child.

The region-growing technique employs both parents (Fig. 2a-b) as consultants. The operator starts with a single seed (Fig. 2c), which is the result of crossover and mutation of parents, and gradually combines other pieces at four-connected adjacent neighbors. From Figure 2d-e, we can observe the operator keeps adding new pieces from a bank of

(a) Parent 1   (b) Parent 2   (c) Initial Seed

(d) Step 1   (e) Step 2   (f) Final Complete Child

Figure 2. Illustration of region-growing.

available pieces, so it guarantees each piece appear only once in the final child (Fig. 2f). The boundary violation is altogether avoided, since the image size is prescribed in advance. The region-growing technique is simple. We can determine and revise the region-growing results by modifying the dissimilarity measure as depicted in Formula (1).

4.2 Algorithm performance comparison

We compared the proposed method with GA, which is proven to be a powerful solver with state-of-the-art performance for large jigsaw puzzles (Sholomon et al. 2013, Kawecki & Niewierowicz 2014). The size of population is assigned with a value of 500. The rate of the elite number to the whole population is assigned with a value of 5%. Crossover and mutation are performed for the generation of each seed.

The Fraction of Correct Neighbors (FCN) is employed to measure the performance of algorithms. The higher the value is, the better the solution is. We ran the algorithms 10 times on each image with different seeds each time. We recorded the results over those 10 runs. The average values of best, mean, worst, and Standard Deviation (SD) in Table 1 are calculated based on the score set of all individuals over 10 runs.

Results in Table 1 compare the proposed FISEGARG with GA in terms of average best, mean, worst, and SD. It is easily derived from the data that the proposed FISEGARG can obtain more accurate results than the GA. In addition, the SDs are quite small for all 10 runs, which indicates that the results of 10 runs are rather similar to each other, in spite of the randomness of initial seed generation.

5 CONCLUSIONS

In this study, we propose a novel accurate and rapid FISEGARG algorithm to solve square-piece jigsaw problems. The contributions of the paper lie in following four points. (1) We propose the region-growing technique to complete a candidate solution after crossover and mutation of two parent solutions. (2) We propose a new fitness function that combines both internal cost and external cost. (3) We propose the fitness-scaling technique and elitism to help improve the solution correction, and the average values of FCN of our solver are overall higher than GA. The future work will center in testing other global optimizations methods, such as particle swarm optimization (Zhang et al. 2010) and colony algorithm (Chang & Wu 2011) artificial bee colony, bacterial chemotaxis optimization, and firefly algorithm (Horng 2012).

ACKNOWLEDGMENT

This paper was supported by NSFC (610011024, 61273243, 51407095), Program of Natural Science Research on Jiangsu Higher Education Institutions (13KJB460011, 14KJB520021), Jiangsu Key Laboratory of 3D Printing Equipment and Manufacturing (BM2013006), Key Supporting Science and Technology Program (Industry) of Jiangsu Province (BE2012201, BE2014009-3, BE2013012-2), Special Funds for Scientific and Technological Achievement Transformation Project in Jiangsu Province (BA2013058), and Nanjing Normal University Research Foundation for Talented Scholars (2013119XGQ0061, 2014119XGQ0080).

Table 1. Average FCN comparison between GA and FISEGARG (averages are over 10 runs).

Size	No. of pieces	FCN (GA)			
		Best	Mean	Worst	SD
14 × 20	280	0.9692	0.9655	0.9616	0.0024
21 × 30	630	0.9702	0.9653	0.9603	0.0038
28 × 40	1120	0.967	0.9586	0.9491	0.0056
35 × 50	1750	0.9569	0.9514	0.9405	0.0049
42 × 60	2520	0.9361	0.9299	0.9208	0.0053
49 × 70	3430	0.9194	0.9122	0.9010	0.0054
56 × 80	4480	0.8997	0.8864	0.8713	0.0101

Size	No. of pieces	FCN (the proposed FISEGARG)			
		Best	Mean	Worst	SD
14 × 20	280	0.9742	0.969	0.9658	0.0028
21 × 30	630	0.9787	0.9672	0.9594	0.0062
28 × 40	1120	0.9722	0.9637	0.9552	0.0063
35 × 50	1750	0.9654	0.956	0.9473	0.0064
42 × 60	2520	0.9453	0.9374	0.9262	0.0067
49 × 70	3430	0.9300	0.9245	0.9117	0.0061
56 × 80	4480	0.9116	0.8987	0.8831	0.0095

REFERENCES

Behbahani, S. & Silva, C.W. 2014. Niching Genetic Scheme With Bond Graphs for Topology and Parameter Optimization of a Mechatronic System, Mechatronics, *IEEE/ASME Transactions on*, 19(1): 269–277.

Chamorro-Posada, P. Gomez-Alcala, R. & Fraile-Pelaez, F.J. 2014. Study of Optimal All-Pass Microring Resonator Delay Lines With a Genetic Algorithm, *Journal of Lightwave Technology*, 32(8): 1477–1481.

Cho, T.S. Avidan, S. & Freeman, W.T. 2010. *A probabilistic image jigsaw puzzle solver*, Computer Vision and Pattern Recognition (CVPR), 2010 IEEE Conference, 183–190.

Chung, J.W. Oh, S.M. & Choi, I.C. 2009. A hybrid genetic algorithm for train sequencing in the Korean railway, *Omega*, 37(3): 555–565.

Chung, M.G. Fleck, M.M. & Forsyth, D.A. 1998. *Jigsaw puzzle solver using shape and color*, in Signal Processing Proceedings, 1998. ICSP '98. 1998 Fourth International Conference, 2: 877–880.

Danado, J. & Paterno, F. 2014. Puzzle: A mobile application development environment using a jigsaw metaphor, *Journal of Visual Languages and Computing*. 25(4): 297–315.

Horng, M.H. 2012. Vector quantization using the firefly algorithm for image compression, *Expert Systems with Applications*. 39(1): 1078–1091.

Kawecki, L. & Niewierowicz, T. 2014. Hybrid Genetic Algorithm to Solve the Two Point Boundary Value Problem in the Optimal Control of Induction Motors. *Latin America Transactions, IEEE*, 12(2): 176–181.

Korsunsky, A.M. & Constantinescu, A. 2006. Work of indentation approach to the analysis of hardness and modulus of thin coatings, *Materials Science and Engineering: A*, 423(1–2): 28–35.

Kumar, M.R. et al. 2010. *Color image encryption and decryption based on jigsaw transform employed at the input plane of a double random phase encoding system*, in Ultra Modern Telecommunications and Control Systems and Workshops (ICUMT), 2010 International Congress, 860–862.

Lesh, N. et al. 2004. Exhaustive approaches to 2D rectangular perfect packings, *Information Processing Letters*, 90(1): 7–14.

Liang, J. & Dill, K.A. 2001. Are Proteins Well-Packed? *Biophysical Journal*, 81(2): 751–766.

Okamoto, Y. et al, 2014. Topology Optimization of Rotor Core Combined With Identification of Current Phase Angle in IPM Motor Using Multistep Genetic Algorithm, *Magnetics, IEEE Transactions on*, 50(2): 725–728.

Pomeranz, D. Shemesh, M. & Ben-Shahar, O. 2011. *A fully automated greedy square jigsaw puzzle solver*, in Computer Vision and Pattern Recognition (CVPR), 2011 IEEE Conference, 9–16.

Pozzi, F. 2010. Using Jigsaw and Case Study for supporting online collaborative learning, *Computers & Education*. 55(1): 67–75.

Sagiroglu, M.S. & Ercil, A. 2006. *A Texture Based Matching Approach for Automated Assembly of Puzzles*, in Pattern Recognition, ICPR 2006. 18th International Conference, 1036–1041.

Sholomon, D. David, O. & Netanyahu, N.S. 2013. *A Genetic Algorithm-Based Solver for Very Large Jigsaw Puzzles*, Comput. Vision Patt. Recog. 2013 IEEE Conference, 1767–1774.

Tybon, R. & Kerr, D. 2009. Automated solutions to incomplete jigsaw puzzles, *Artificial Intelligence Review*. 32(1–4): 77–99.

Wang, Y. Li, B. & Weise, T. 2010. Estimation of distribution and differential evolution cooperation for large scale economic load dispatch optimization of power systems, *Information Sciences*, 180(12): 2405–2420.

Yao, F.H. & Shao, G.F. 2003. A shape and image merging technique to solve jigsaw puzzles, *Pattern Recognition Letters*. 24(12): 1819–1835.

Zhang, Y. et al, 2010. Find multi-objective paths in stochastic networks via chaotic immune PSO, *Expert Systems with Applications*. 37(3): 1911–1919.

Zhang, Y. et al, 2013. Genetic Pattern Search and Its Application to Brain Image Classification, *Mathematical Problems in Engineering*. 2013: 880876.

Zhang, Y. Wu, L. & Wang, S. 2011. UCAV path planning based on FSCABC. *Information an international interdisciplinary journal*, 14(3): 687–692.

Zhang, Y. Wu, L. & Wang, S. 2013. UCAV Path Planning by Fitness-Scaling Adaptive Chaotic Particle Swarm Optimization, *Mathematical Problems in Engineering*. 2013: 705238.

Zhang, Y.D. & Wu, L.N. 2011. Bankruptcy Prediction by Genetic Ant Colony Algorithm, in New Trends and Applications of Computer-Aided Material and Engineering, *Trans Tech Publications Stafa-Zurich*. 459–463.

Zhang, Y.D. Wang, S.H. & Ji, G.L. 2013. A Rule-Based Model for Bankruptcy Prediction Based on an Improved Genetic Ant Colony Algorithm, *Mathematical Problems in Engineering*. 2013: 753251.

Zhi, L.Q.G. & Ji, Z. 2009. *Image Matching Algorithm Based on Edge Color Used in Automatic Computer Jigsaw Puzzle*, KAM '09. 2nd Int. Symp. 269–272.

Advanced Materials and Structural Engineering – Hu (Ed.)
© *2016 Taylor & Francis Group, London, ISBN 978-1-138-02786-2*

Structure design and knotting tests of a type of anthropopathic dual-finger knotter

W. Zhang & J. Yin
Key Laboratory of Modern Agricultural Equipment and Technology, Ministry of Education, Jiangsu University, Zhenjiang, China

ABSTRACT: Aiming at the problem of long-term dependence on imported knotter during the production of quadrate bale type of baller in China, a type of anthropopathic dual-finger knotter was developed to decrease the difficulty in the knotter production. The design object of the knotter was the simplification of space structure of the knotter stand and its actuating mechanisms. The spatial angles of five axle holes on knotter stand were optimized to be easy to manufacture as well as key parameters of compound driving gear plate. To test the performances of the developed knotters, empty knot and straw—bundling experiments were operated by straw forming and knotting test-bed. The tests showed that the knot-tied rate of forming normal φ shape of knots may reach 100 percent after each knotter finishes 500 empty knots. The field test showed that the knot-tied rate reaches 99.2 percent after the developed knotters bundled up 400 bundles of wheat straws and 600 bundles of rice straws. The knotter tests validate that the design of the knotter is successful and may provide some technical references for the knotter production.

1 INTRODUCTION

As a key assembly of field baler, D-knotter is an automatic device that can cord a knot rope to strap the stalks. It directly decides the density of bundles and ensures that the bundles do not loose during transportation. Knotter is a key device to measure the performance of field baler, and its performance directly affects the performance and competitiveness of the field baler. So far, Chinese enterprises depend on the import of expensive overseas knotter to produce a quadrate bale type of baller. Domestic manufacture of knotter has become a technical bottleneck problem that restricts the development of Chinese quadrate bale type of baller. Research on D-knotter has mainly focused on the analysis of the action principles (Yin et al. 2011, Wang et al. 2012, Li et al. 2011), structural analysis of the key component (Su et al. 2008), the design of the actuator of the knotter (Li et al. 2010, Li et al. 2012, Zhang et al. 2013) and the application in baler (Yang et al. 2011) in China. These studies paid more attention to the kinematic problem in the process of knotting than on the design method of knotter.

In this paper, the simplification of space structure of knotter stand and transmission parts was determined as the design object, and the spatial angle of five axis holes on knotter stand was optimized as well as key parameters of compound driving gear plate, biting hook and knotting hook.

These designs may decrease the difficulty in the domestic manufacture of knotter, and the straw-bundling test of the knotter prototype was carried out to validate the knotting rate of the designed knotter.

2 STRUCTURE COMPOSITION OF THE ANTHROPOPATHIC DUAL-FINGER KNOTTER

As shown in Figure 1, structure composition of the anthropopathic dual-finger knotter consists of multiple complex spatial mechanisms cooperating with each other accurately, including the knot-winding mechanism and coercive knot-tripping mechanism. The knot-winding mechanism is a kernel of the knotter, and made up of the knotting hook and its bevel gear transmission, the biting hook and cylindrical cam made on the knotter stand. The coercive knot-tripping mechanism is a spatial swinging cam mechanism, which consists of the grooved cam made on the compound gear plate and knife arm. All moving parts are installed on static knotter stand, except that the compound gear plate and their axle holes form a complex spatial angle relation between different planes. The complex spatial angle causes difficulty in precise manufacturing of the knotter stand and precise assembly of all moving parts.

1

3

4

5

6

(a) Front view of the knotter

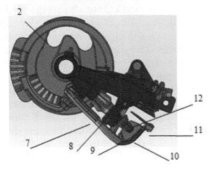

2

12

11

7

8

9

10

(b) Side view of the knotter

1. Knotter stand; 2. Compound gear plate; 3. Bevel gear of worm axle; 4. Rope-holding mechanism; 5. Helical gear; 6. Involute worm; 7. Knife arm; 8. Bevel gears of knotting hook axle; 9. Biting hook ; 10. Knotting hook ; 11. Knife of cutting rope; 12. Platen of biting hook.

Figure 1.   Structure composition of the anthropopathic dual-finger knotter.

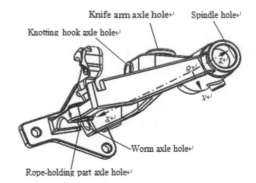

Knife arm axle hole

Spindle hole

Knotting hook axle hole

Worm axle hole

Rope-holding part axle hole

(a) Spatial structure of the knotter stand

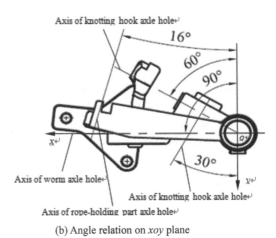

Axis of knotting hook axle hole

16°

60°

90°

30°

Axis of worm axle hole

Axis of knotting hook axle hole

Axis of rope-holding part axle hole

(b) Angle relation on *xoy* plane

Figure 2.   Spatial angle design of five axle holes on the knotter stand.

3   KEY STRUCTURE DESIGN AND PARAMETER DETERMINATION OF THE KNOTTER

3.1   *Spatial angle design of five shaft holes on the knotter stand*

Spatial structure of the knotter stand is shown in Figure 2. Spatial angle relation among five axle holes on the knotter stand is very important for the knotter to realize the knot-tied motions. The coordinate system *o-xyz* is built in Figure 2, and the axes of the worm axle hole and the spindle hole are chosen as the *x* axis direction and z axis direction. The sense of y is determined as a matter of fixed convention by the right-hand-thread rule. Spatial angle relation among five axle holes on the *xoy* plane is given in Figure 2, and the other angles related to the axis of the spindle hole on the *yoz* and *xoz* plane are 90 degrees. To make two pairs of spur bevel gear transmission mesh reliably, the axis of knotting hook axle hole intersects the

axis of worm axle hole at *o* point on the axis of spindle hole, and the angle between the two axles is 30 degrees. The helical gear is fixed slantwise on the axis of the rope-holding axle hole because the lead angle of the worm is determined as 16 degrees. So, the angle between the axis of worm axle hole and the axis of the spindle hole is also 16 degrees. The designed spatial angle relation among five axle holes on the knotter stand is beneficial to reduce the production difficulty of the knotter stand and guarantee assembly precision of the knotter.

3.2   *Design of the knot-winding mechanism and the knot-hooking mechanism*

The knot-winding mechanism is in charge of winding an annular rope loop on the knotting hook, and made up of outer bevel gear on the compound gear plate, bevel gear of knotting hook axle and the knotting hook. The knot-hooking mechanism is in charge of biting the rope and drawing out the rope end from the rope loop, and made up of the biting hook, the cylindrical cam on the knotter stand, late of biting hook and a compression spring. The biting hook is installed on the knotting hook by a fixed pin roll. It will open and close around the pin roll axle under the constraint of the cylindrical cam when the biting hook rotates with the rotation of the knotting hook. The compression spring may supply a pressure to a platen of a biting hook, and the platen of the biting hook makes the biting hook have a proper force of hooking the rope end of the rope loop.

The teeth number and module of the outer bevel on the compound gear plate and bevel gear of knotting hook axle are also 8 and 4, respectively. As shown in Figure 3, the knotting hook is designed as the shape of the olecranon, and the angle β' formed by the knotting hook axis and its inner surface may be determined, and the range is from 100 degrees to 105 degrees. The installation phase angle δ_4 between the knotting hook axis and the end line of the cylindrical cam on the knotter stand should be guaranteed to make roller of biting hook locate the lowest contour of the cylindrical cam on the knotter stand, and the range of δ_4 is from 120 degrees to 125 degrees. The contour of the cylindrical cam on the knotter stand is made up of two involutes with the limit of φ_1 and φ_1', two spirals of Archimedes with the range from φ_1 to φ_2 and from φ_2 to φ_2', as shown in Figure 3.

3.3 The coercive knot-tripping mechanism

The coercive knot-tripping mechanism is a spatial swinging cam mechanism, and made up of the grooved cam on the compound gear plate, knife arm and knife of cutting the rope. It is in charge of cutting off the rope clamped by the rope-holding mechanism and threading off the rope loop from the knotting hook, and finally forming a knot.

As shown in Figure 4, in order to thread off the rope loop from the knotting hook, the tripping jaw surface should keep tangency with the convex surface of the knotting hook. So, the convex surface of the knotting hook is a curved surface formed by a arc with the radius of r because the knife arm will swing to and fro around its axle, and the path of the tripping jaw is a arc. The width c of the tripping jaw should slightly be larger than the width of the knotting hook. The distance d between the knife arm axis and the bottom surface of the knife arm should ensure to avoid motion interference with the knotting hook. The installation phase angle δ_5 between the axis of the knotting hook and key groove direction is 138 degrees in this case. The installation angle γ of the knife of cutting the rope

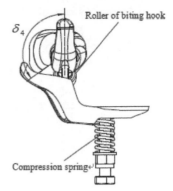

(a) Front view of the mechanism

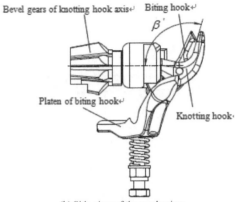

(b) Side view of the mechanism

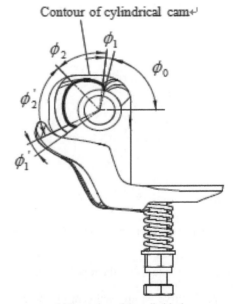

(c) Front view of the cylindrical cam

Figure 3. Key parameters of the knot-winding mechanism and the knot-hooking mechanism.

821

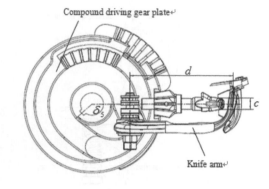

(a) Front view

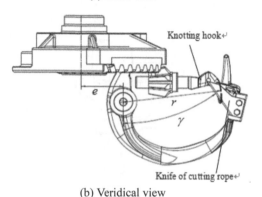

(b) Veridical view

Figure 4. Key parameters of the coercive knot-threading mechanism.

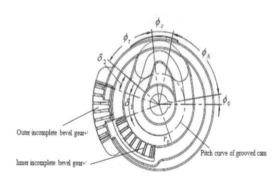

Figure 5. Phase design of compound driving gear plate.

is 36 degrees to ensure that the rope is cut off after the rope loop finishes.

The pitch curve of grooved cam on the compound gear plate depends on the swinging law of the knife arm and the distance e between the knife arm axis and the axis of the compound gear plate. As shown in Figure 5, the contour of the grooved

cam on the compound gear plate is made up of four curves, including two arcs with the limit of φ_s and φ_s' and two curves with the limit of φ_t and φ_h. The swinging law of the knife arm is adopted as sine acceleration.

The installation phase angle δ_1 between the inner incomplete bevel gear and the outer incomplete bevel gear on the compound gear plate is designed to 34 degrees. The installation phase angle δ_2 between the initial point of cam lift and outer incomplete bevel gear on the compound gear plate is designed to 20 degrees. The two phase angles will decide the motion sequence precision among the rope-holding mechanism, the knot-winding mechanism and the coercive knot-tripping mechanism.

4 KNOTTING AND STRAW-BUNDLING TESTS

Prototype assembly of the designed anthropopathic dual-finger knotter is shown in Figure 6. To test the performances of the developed knotter, knotting tests without bundling straws were operated by straw forming and knotting test-bed, as shown in Figure 7. The knot-tied rate of forming normal φ shape of knots is a test index, and it may be calculated by

Figure 6. Prototype assembly of the knotter.

Figure 7.   Knotting test without straw-bundling.

Figure 8.   Field test of bundling wheat straw.

$$S_h = \frac{n_d - n_s}{n_d} \times 100\% \tag{1}$$

where, S_h means the knot-tied rate; n_d means the number of overall tested bundles; and n_s means the number of failed knot-tied bundles.

The results of knotting without bundling straws on straw forming and knotting test-bed showed that the knot-tied rate of forming normal φ shape of knots may reach 100 percent after each knotter finishes 500 empty knots.

In order to further inspect the performance of the knotter, the two anthropopathic dual-finger knotters were installed on the 9YF-1000 type of pickup baler to do bunding-straw tests in the field. The polypropylene rope with a diameter in 2.5 mm was used in the baler and the driving shaft of the knotter was rotating at 90 revolutions per minute. Dimension of straw bundles was $320 \times 420 \times 600$ mm³. The Moisture content of wheat straws and rice straws were about 18% and 65%. Field bundling wheat straw and rice straw tests are shown in Figure 8 and Figure 9. The field tests showed that the knot-tied rate reaches 99.2 percent after the knotters bundled up 400 bundles of wheat straws and 600 bundles of rice straws. During the course of the tests, 50 bundles of straw were selected randomly to measure out the shape of the knot and stretching-length of the end of the knot. The results showed that the shape of the knot is nice and stretching-length of the end of the knot is between 25 mm and 29 mm. It indicated the position of the knife installed on the knife arm was controlled properly. The failure of the knot occurred during threading off the knot. The reason of the failure is that uneven straws fed into the compression room resulted in the vibration of the force imposed by the rope, smaller force of the rope leads to the failure of threading off the knot.

Figure 9.   Field test of bundling wheat straw.

The force imposed by the rope must be controlled properly to improve the rate of the knot-tied.

5   CONCLUSIONS

By optimizing spatial angle relation of five axle holes on the knotter stand and determining the installation phase angles among inner incomplete bevel gear, outer incomplete bevel gear and the initial point of cam lift on the compound gear plate, the production difficulty of the knotter stand may be reduced and assembly precision of the knotter may be improved. After knotting tests without bundling straws operated by straw forming and knotting test-bed, field bundling wheat straw and rice straw tests, the results showed that the knot-tied rate of forming normal φ shape of knots may reach above 99 percent. The knotter tests validate that the design of the knotter is successful and may provide some technical references for the knotter production.

ACKNOWLEDGMENT

This research was funded by the National Nature Science Foundation of China under Grant No. 51375215.

REFERENCES

Li, H. Li, H. & He, J. et al. 2010. Reconstruction and optimal design of driving dentate disc of D-bale knotter based on reverse engineering, *J. Trans. Chinese Soc. Agr. Eng*. 26: 96–102.

Li, H. Wang, Q. & He, J. et al. 2012. Experimental research on performance of different knotter driving pulleys, *J. Trans. Chinese Soc. Agr. Eng*. 28: 27–33.

Li, S. Yin, J. & Li, Y. 2011. Kinematic characteristic analysis of D-knotter and its ancillary mechanisms, *Journal of Machine Design and Research*. 27: 18–21.

Su, G. Shi, J. & Ge, J. 2008. Measurement of spatial angle of square knotter on the basis of reverse engineering, *J. Trans. Chinese Soc. Agr. Mach*. 39: 81–83.

Wang, L. Lv, H. & Wei, W. et al. 2012. Analytical conditions and visualized verification of knotter hook's rope-biting, *J. Trans. Chinese Soc. Agr. Mach*. 43: 96–100.

Yang, L. Liu, G. & Wang, Z. et al. 2011. Design and experiment of engine-driven constant frequency rectangular pickup baler, *J. Trans. Chinese Soc. Agr. Mach*. 42: 147–151.

Yin, J. Li, S. & Li, Y. 2011. Kinematic simulation and time series analysis of D-knotter and its ancillary mechanisms, *J. Trans. Chinese Soc. Agr. Mach*. 42: 103–107.

Zhang, S. Li, H. & Cao, Q. et al. 2013. Design of key transmission mechanism of double-α-knot knotter, *J. Trans. Chinese Soc. Agr. Mach*. 44: 74–79.

Advanced Materials and Structural Engineering – Hu (Ed.)
© *2016 Taylor & Francis Group, London, ISBN 978-1-138-02786-2*

Die forming of hollow pipe for wall thickness increasing and FEM analysis

H. Kamiyama, S. Nishida, R. Kurihara & M. Fujita
Gunma University, Ota City, Gunma, Japan

ABSTRACT: Automotive parts, especially drive-line parts such as drive shaft, are mostly lengthy parts. Therefore, the weight saving of these parts might be effective for the improvement of fuel efficiency. One of the weight saving methods is to make these parts from a hollow pipe. The hollow drive shaft is necessary to have lightweight properties and stiffness, so the center part of the shaft has a large diameter and thin wall and the end part has a small diameter and thick wall. There are some reports of products obtaining the above shape, for example, rotary swaging and friction welding. However, the production cost of these methods is high. Hence, the hollow pipe forming process was proposed to press to the narrow angle die. In this study, the mechanism of wall thickness increasing of steel pipe on the proposed process was investigated by experiment and analysis. The initial outer diameter, wall thickness and length of steel pipe are 39 mm, 7.6 mm and 160 mm, respectively. It was possible to achieve up to increase the wall thickness to about maximum 10%. At the constant region, the increasing was about 8%. The finite element method analysis results agree well to the experimental results.

1   INTRODUCTION

Currently, the prevention of global warming has become a global issue. The improvement of fuel efficiency has become an important issue in the automotive industry. The demand for weight saving for the improvement of fuel efficiency is increasing. Drive-line parts such as drive shaft are mostly lengthy parts. Therefore, the weight saving of these parts might be effective for the improvement of fuel efficiency. One of the weight saving methods is to make these parts from a hollow pipe. The hollow drive shaft is necessary to have lightweight properties and stiffness. The ideal shape of the drive shaft is described below:

Center part: large diameter and thin wall thickness for lightweight properties and stiffness.
End part: small diameter and thick wall thickness for strength and connecting other parts.

Typical production methods for obtaining the above shape are given below:

Friction welding: center part and end part are welded by friction welding. End parts are produced by machining or forging.
Rotary swaging: end part of steel pipe is formed by rotary swaging. The wall thickness of end part becomes thinner than a center part.
Hydroforming: pipe is expanded by inner hydraulic forming. It is difficult to apply the thick wall steel pipe. The wall thickness of center part decreases than the initial thickness.

However, there are some problems such as increasing in the number of processes and cost in these production methods. The other process (Kawabata et al. 2012, Kawabata et al. 2012) reported the wall thickness increasing on pipe shrink forming by finite element method analysis on the planetary roll (Kiuchi et al. 1989, Kotani et al. 2012) forming process.

In this study, the hollow pipe forming process was proposed to press to the narrow-angle die. The mechanism of wall thickness increasing in the steel pipe on the proposed process was investigated by experiment and analysis.

2   EXPERIMENT

Table 1 presents the experimental conditions. The initial pipe diameter (D) was 39 mm. The reduction ratio of pipe (γ) was changed. Figure 1 shows the schematic illustration of the proposed process. The taper angle was 20 degree. The feeding speed was 50 mm/sec.

Table 1.   Experimental condition.

Pipe diameter D [mm]	39.0
Pipe thickness [mm]	7.6
Pipe length [mm]	160.0
Taper angle [degree]	20
Feeding speed [mm/sec]	50
Reduction diameter [mm]	33.0, 33.8, 34.5, 35.5, 36.3

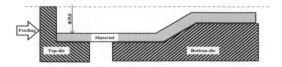

Figure 1. Schematic illustration of the molding.

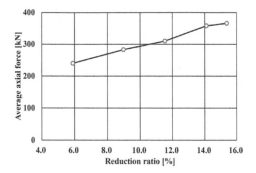

Figure 2. Relationship between pipe reduction ratio and average feeding load.

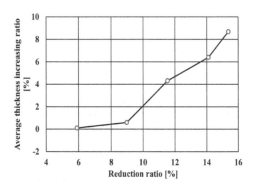

Figure 3. Relationship between pipe reduction ratio and average wall thickness increasing ratio.

Measurements of the pipe shape after forming were made by using three-dimensional measuring machines. The measuring probe of the machine was touched to the pipe surface along the longitudinal direction. The inner and outer surface shape was measured. The difference in the Z axis coordinate between the inner and the outer surface was the wall thickness of the pipe.

3 EXPERIMENTAL RESULT

Figure 2 shows the relationship between the pipe reduction ratio and the average feeding load (forming load). The average feeding load became large as the large pipe reduction ratio became large. It was revealed that the feeding load indicated a linearly increasing. Figure 3 shows the relationship

between the pipe reduction ratio and the average thickness ratio. The average wall thickness increasing ratio became large as the pipe reduction ratio became large. The maximum wall thickness increasing ratio obtained was about 9%.

4 FINITE ELEMENT ANALYSIS

Finite element analysis was conducted by using commercial software DEFORM-3D. Finite element method analysis results on die forming process were adapted to the experimental results. Figure 4 shows the analytical model. Table 2 presents the analytical conditions. These conditions were same to the experimental conditions. Flow stress and shear friction coefficient were obtained by the ring compression test. The ring compression test was generally used to measure the shear friction coefficient. The analytical model size was 1/8 model because of reducing the computation time.

5 ANALYTICAL RESULT

Figure 5 and Figure 6 show the comparing the experimental results with the analytical results.

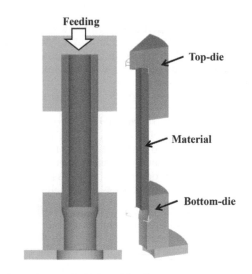

Figure 4. Analytical model (1/8).

Table 2. Analysis condition.

Analysis model size	1/8
Pipe diameter [mm]	39.0
Pipe thickness [mm]	7.6
Pipe length [mm]	160.0
Taper angle [degree]	20
Feeding speed [mm/sec]	50
Reduction diameter [mm]	33.0, 33.8, 34.5, 35.5, 36.3

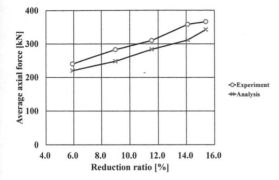

Figure 5.   Average axial force.

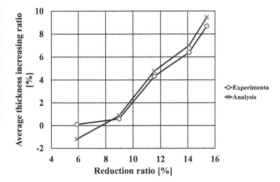

Figure 6.   Average thickness increasing ratio.

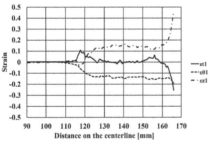

(a) Strain distribution of inner surface

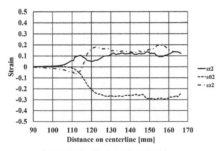

(b) Strain distribution of outer surface

Figure 7.   Strain distribution (D = 34.5 mm (γ = 11.5%)).

These figures indicated that both results conformed well. It was possible to analyze the proposed process such as a prediction of wall thickness increasing and a verification of mechanism of wall thickness increasing to make an adequate analytical model. So, the strain distribution of the pipe surface was analyzed.

Figure 7 shows the analytical strain distribution of the inner or outer surface of a pipe. D was 34.5 mm and γ was 11.5%. The additional character t indicates the wall thickness direction. The additional character θ indicates the circumferential direction. The additional character z indicates the longitudinal direction. Each strain obtained at a static state of Figure 7 was as follows: εt1, εt2, εθ1, εθ2, εz1, εz2 were 0, 0.12, 0.13, 0.26, 0.15 and 0.14, respectively. These results indicated that the pipe was extended to the longitudinal direction as the wall thickness increased. It is supposed that the reduction of ε_z might increase the ε_t (pipe wall thickness).

6   CONCLUSION

The hollow pipe forming process was proposed. The mechanism of wall thickness increasing of steel pipe was investigated by experiment and analysis. The results obtained were as follows:

1. The forming load increased when the pipe reduction ratio increased. The maximum wall thickness increasing ratio obtained was about 9% in this study.
2. It was possible to analyze the proposed process such as the prediction of wall thickness increasing and the verification of mechanism of wall thickness increasing to make an adequate analytical model.
3. Analytical results indicated that the pipe was extended to the longitudinal direction while the wall thickness increased.

REFERENCES

Kawabata, D. Kamiyama, H. & Nishida, S. 2012. Increase characteristics of wall thickness during pipe reduction process by using planetary rolls, *M & P* 20: 4.
Kawabata, D. Kamiyama, H. & Nishida, S. 2012. *Wall thickness in pipe reduction process using planetary roller*, The Japan Institute of Light Metals, 123 times Fall Conference, 393–394.
Kiuchi, M. & Shintani, K. 1989. Study on pipe Reducing by Planetary Roller Reducer 1st Report, *Production research*, 42(2): 14–17.
Kotani, Y. Kanai, S. & Watari, H. 2012. Characteristics of Wall Thickness Increase in Pipe Reduction Process, Using Planetary Rolls, *Advanced Material Research.* 430–432: 1241–1247.

Advanced Materials and Structural Engineering – Hu (Ed.)
© *2016 Taylor & Francis Group, London, ISBN 978-1-138-02786-2*

Equivalent strain analysis of piercing process in Diescher's mill using finite element method

L. Lu & E. Xu

College of Information Science and Technology, Bohai University, Jinzhou, Liaoning, China

ABSTRACT: In this paper, the simulation of the piercing process is performed by the three-dimensional finite-element method in Diescher's mill. The simulated results show that the strain rates are rapid changed, and the strains distribution is laminar. The maximum strain rate appears in the area where the moving work-piece contacts the rolls on the external surface, and it is in the area where the plug contacts the work-piece on internal surface. The maximum strains are both in the about 2/3 part of the plug on the external and internal surface of the work-piece.

1 INTRODUCTION

Pierce rolling is the first rolling performed in the production of the seamless pipes. There are three types of piercing: press roll piercing, barrel-type rotary piercing, and corn-type rotary piercing. The cross-sectional shape of the material has line symmetry in press roll piercing, while it has point symmetry in barrel-type rotary piercing and corn-type rotary piercing. The barrel-type rotary piercing is used widely in the world. Among the barrel-type rotary piercing, the best productivity and high quality of thick-walled tubes are obtained when two rolls rolling mills are used with two guiding discs of Diescher type (Pater et al. 2006).

Not only can it be used to proof the feasibility of the production process, but also to predict the micro-structure and beyond that the properties of the component (Sinczak et al. 1998). The piercing process in the production of the seamless steel tube is an important and complicated procedure. It is not only influenced by materials property, deformation temperature, rolling velocity, deformation amount and contact friction condition etc., but also a non-isothermal process with three-dimensional (3D) thermal dynamic coupling.

The Diescher tube piercing is very complex process of the material flow. The process relies on the cyclic mechanical loading of the material caused by the conical shape of the rolls and their rotation. Figure 1 shows the material is pulled in along a helical trajectory and a depressive mode causes a hole to form and develop in the billet.

2 NUMERICAL MODEL

The software SuperForm2005 (MSC 2005) is used as the basis for modeling the process.

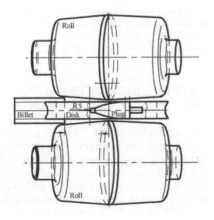

Figure 1. Schematics of the process and the tools (Yang et al. 2004).

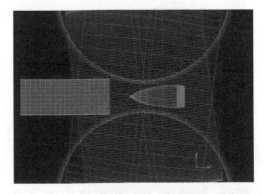

Figure 2. Position of the tools in the numerical model.

Five tools are included in the model: two rolls, two shoes and a plug, as depicted in Figure 2. The two rolls are rotating at 141 rotations/minute and their axes form a 7° angle. The diameter of the billet is 110 mm and it is assumed to be at 1255°C as it enters the roll bite. At the beginning of the deformation, the temperature of the rolls is taken as 100°C and the plug is 300°C. The constitutive law (Sellars 1966) for the steel grade of interest writes:

$$\sigma = m_0 \exp(m_1 T) \varepsilon^{m_2} \dot{\varepsilon}^{(m_3 T + m_4)} \qquad (1)$$

The material parameters are obtained from experimental tests:

$$\sigma = 1.1244 \exp(6243.9238/T) \cdot \varepsilon^{0.1571} \cdot \dot{\varepsilon}^{(0.0003T-0.1933)} \qquad (2)$$

The contact with the billet and the rolls is assumed to be close to sticking friction. Sliding modes are applied to model the contacts between the billet and the guides.

3 DISTRIBUTION OF STRAIN RATE ON WORK-PIECE EXTERNAL SURFACE

The FEM application allows for precise analysis of changes of strain during piercing in the skew rolls piercing mill. Given the kinematics of the rolls, the billet tends to twist. This is observed in practice. It makes the simulation very difficult since the position is very sensitive to the friction model for instance. Figure 3 presents the equivalent of plastic strain rate distributions on work-piece external surface for the stable piercing phase.

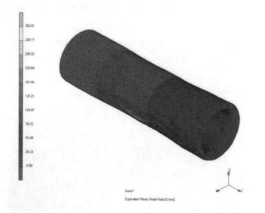

Figure 3. Strain rate distributions on the work-piece external surface.

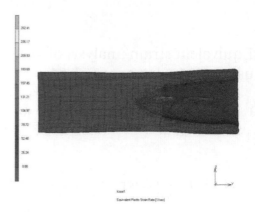

Figure 4. Strain rate distributions on the work-piece internal surface.

The analysis of data shows that the strain rate is very large in the piercing process and non-uniforms on the external surface of work-piece. Because the roll speed is very fast and the diameter ratio between the roll and the work-piece is very large in the piercing process. To enforce this stable motion of the billet in the numerical model, the discs are modified.

4 DISTRIBUTION OF STRAIN RATE ON THE INTERNAL SURFACE OF WORK-PIECE

The strain rate on the internal surface increases gradually with the plug piercing into the work-piece as shown in Figure 4. The maximum strain rate appears when the plug pierces into the work-piece completely. With holes cavity forming, the area where the plug contacts the internal surface of work-piece are very large. The viscosity appears on the surface layer of the area between the plug and the internal surface. So the strain rate is far high on the internal surface of work-piece.

5 DISTRIBUTION OF STRAIN ON WORK-PIECE EXTERNAL SURFACE

Due to the application of FEM, it is possible to precisely analyze the changes of work-piece shape present during piercing process. Figure 5 shows the equivalent strains evolution on the work-piece external surface during piercing processing. It shows the strains distribution in the rolled part of work-piece is layered. The strains increase gradually in the direction of the tube's axis with holes cavity forming. The maximum strain appears in

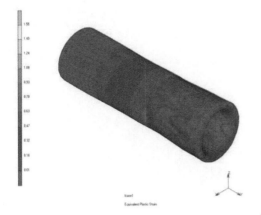

Figure 5. Strain distributions on the work-piece external surface.

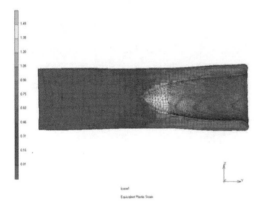

Figure 6. Strain distributions on the work-piece internal surface.

the 2/3 part of the plug, and then decreases in the direction of the tube's axis.

6 DISTRIBUTION OF STRAIN ON THE INTERNAL SURFACE OF WORK-PIECE

Figure 6 shows the distribution of equivalent strains on the internal surface of work-piece. On the basis of the Figure 6, it can be seen that the distribution of equivalent strains is very non-uniform and layered. The non-uniform distribution attributes to metal velocity, equivalent stress and temperature interaction. The maximum strain appears in the 2/3 part of plug, and then decrease in the direction of the tube's axis. Because with the hole cavity forming, the area where the plug contacts the internal surface of work-piece is very large. Not only is

the area inhomogeneous, but also the distribution of highest equivalent strain is stable.

7 CONCLUSIONS

In this paper, the FEM model of the piercing process was presented. Simulated results show that the strain rates and the strains are both non-uniform distribution on the external and internal surface of work-piece. The strain rates are rapidly changed on the internal and external surface of the work-piece. The maximum strain rate is in the area where the moving work-piece contacts the rolls on the external surface. The maximum strain rate is in the area where the plug contacts work-piece on the internal surface. Meanwhile simulated results show that the strains are layered in the rolled and the pierced parts of work-piece. The maximum strains are both in the about 2/3 part of plug on the external and internal surface of the work-piece. The maximum strain in the internal surface is higher than the one in the external surface.

The metal velocity, the evolution of stress and temperature and their distributions will be predicted via the FEM method. The results of prediction will be used as basic data for improving tool and designing, predicting, damaging and controlling the micro-structural evolution of processing tube piercing of the new steel 33Mn2V for oil well tubes.

ACKNOWLEDGEMENT

Research work was supported by the Liaoning Natural Science Foundation of China under Grant No. 2014020141. We are grateful to Wuxi Steel Tube Factory for financial support.

REFERENCES

MSC 2005. *Software Corporation*, MSC. SuperForm User's Guide Version 2005, South Coast Metro, USA.
Pater, Z. Kazanecki, J. & Bartnicki, J. 2006. Three dimensional thermo-mechanical simulation of the tube forming process in Diescher's mill, *Journal of Materials Processing Technology*. 177: 167–170.
Sellars, C.M. & McTegart, W.J. 1966. Tegart, On the mechanism of hot deformation, *Acta Metallurgica*. 14(9): 1136–1138.
Sinczak, J. Mahta, J. Glowacki, M. & Pietrzyk, M. 1998. Prediction of mechanical properties of heavy forgings, *Journal of Materials Processing Technology*. 80–81: 166–173.
Yang, J. Li, G. Wu, W. Sawamiphakdi, K. & Jin, D. 2004. Process modelling for rotary tube piercing application, *Journal of Materials Science & Technology*. 2: 137–148.

Advanced Materials and Structural Engineering – Hu (Ed.)
© 2016 Taylor & Francis Group, London, ISBN 978-1-138-02786-2

Logistics service quality evaluation model of B2C mode network shopping based on the AHP-BP neural network

C. Deng, P. Sun & R. Pan
School of Logistics Engineering, Wuhan University of Technology, Wuhan, China

ABSTRACT: Because of the defects of the current method of logistics service quality evaluation in the B2C mode network shopping platform, on the basis of the AHP evaluation method, a new method of logistics service quality evaluation based on the AHP-BP neural network is presented in this paper. Thus, logistics service quality evaluation index system of the B2C mode network shopping platform is established. First, the analytic hierarchy process method is used to determine the weights of the evaluation index system, according to each index weight calculation composite scores. Second, the MATLAB neural network toolbox is used to train the BP neural network model, with a combination of the AHP method to verify this evaluation system. The results show that the model for the logistics service quality evaluation is practical, thus the simulation results are right and effective.

1 INTRODUCTION

The research of service quality has been relatively perfect. Compared with the quality of service, research on logistics service quality under the background of the network shopping is slightly insufficient. As the executive of the activity of network shopping and terminal services, which have a direct contact with the customer, the quality of logistics service directly determines the customer acceptance of network shopping. Therefore, the research on logistics service quality evaluation, which is aimed at the B2C mode network shopping, is of great significance.

At present, two kinds of evaluation methods are commonly used; one is the Analytic Hierarchy Process (AHP), which contains many disadvantages described as follows:

1. The establishment of the index system is subjective;
2. It is apt to put different elements in the same level to compare them, thus affecting the accuracy of the results;
3. It is apt to overlook many dynamic factors (Lin et al. 1997).

Therefore, the veracity of evaluation results is limited by using this method. Another one is the fuzzy comprehensive evaluation method: a defect of this method is that the decision results are greatly influenced by subjective factors, thus the fittest degree (Mao 2008) is not high. Aiming at the above shortcomings, logistics service quality of the electronic shopping platform is evaluated by the AHP method and BP neural network with the combination of evaluation methods.

2 ESTABLISHMENT OF THE LOGISTICS SERVICE QUALITY EVALUATION INDEX SYSTEM

The process and characteristics of B2C mode e-commerce logistics service is based on the evaluation index of the SERVQUAL model (Franceschini & Rafele 2000), which can measure the logistics service quality of B2C mode e-commerce effectively, conducting importance analysis, giving weight for each indicator, and establishing logistics service quality index evaluation system of B2C mode e-commerce. It is given in Table 1.

3 ESTABLISHING THE MODEL OF AHP-BP NEURAL NETWORK MODE

BP neural network is a widely used type of neural network model currently (Franceschini & Rafele 2000), the essence is to adjust the weights of each layer to make it to memory learning sample set, the most basic network is a three-layer feed forward network, including the input layer, hidden layer and output layer, the number of neuron network in each layer is, as is shown in Figure 1.

Setting the input and output vectors of the network: $X = (x_1, x_2, ..., x_m)^T$ and $Y = (y_1, y_2, ..., y_n)^T$; in the situation, the number of neurons in each layer, also the weights and threshold between each layer

Table 1. Logistics service quality index evaluation system.

Objective	First class indicator	Secondary class indicator
B2C mode E-commerce logistics service quality	A1 Reliability	A11 accuracy of pickup location A12 accuracy of goods A13 integrity of goods
	A2 Timeliness	A21 response time of order A22 wait time of pickup A23 product return
	A3 Flexibility	A31 the variety of distribution mode A32 the variety of receiving mode A33 the flexibility of product return service
	A4 Empathy	A41 service attitude of the staff A42 error handling will A43 communication with customers
	A5 Informativity	A51 timeliness of logistics information A52 accuracy of logistics information A53 sufficiency of logistics information

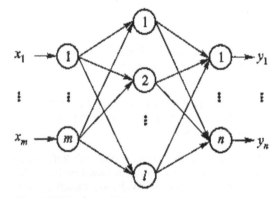

Figure 1. Topology of the three-layer BP neural network.

of the neurons, are definite, for a given input vector, the output value of the hidden layer and the output layer of the neurons are calculated, respectively, according to formulas (1) and (2) (Wang 2007):

$$O_j = f\left(\sum_{i=1}^{m} W_{ij}^{(1,2)} X_i - \theta_j\right) \quad j = 1, 2, \ldots, l \quad (1)$$

$$y_k = f\left(\sum_{j=1}^{l} W_{jk}^{(2,3)} O_j - \varphi_k\right) \quad k = 1, 2, \ldots, n \quad (2)$$

In the formula, $W_{ij}^{(1,2)}$, θ_j are, respectively, the connection weights of the input layer to hidden layer of the neurons, hidden layer threshold. $W_{jk}^{(2,3)}$, φ_k are, respectively, the connection weights of the hidden layer to input layer of the neurons,

output layer threshold. The basic calculation steps of the algorithm are to:

1. Determine the initial parameters: the number of neurons in each layer, the learning precision, the number of iterations, initial weights and thresholds, etc.
2. Provide the training sample, calculate the output value of the hidden layer according to formula (1), that is the input value of the output layer, calculate the output value of the output layer by formula (2).
3. Calculate the deviation between network output value and target value of the sample.
4. Adjust the weight and the threshold between the output layer and the hidden layer.
5. Calculate the deviation, readjust the weight and the threshold between the output layer and the hidden layer until the deviation meets the requirement.

4 SIMULATION AND VALIDATION OF THE MODEL USING MATLAB

We select the data and results of 15 business enterprises whose mode of logistics service quality are B2C mode. We then normalize the index as the basic data of the BP neural network training and testing, as is shown in Table 2.

We calculate the comprehensive score of each electronic commerce enterprise according to the index weight, and divide logistics service quality of the enterprises into five grades according to the comprehensive score by actual needs: very good quality (score is in the range of [0.9~1]), good quality (score is in the range of [0.8~0.9]), general quality (score is in the range of [0.7~0.8]), poor

Table 2. Table of the sample data.

	C1	C2	C3	C4	C5	C6	C7
A11	0.932	0.927	0.741	0.824	0.862	0.772	0.602
A12	0.911	0.943	0.810	0.855	0.874	0.752	0.638
A13	0.943	0.913	0.760	0.890	0.862	0.729	0.625
A21	0.876	0.894	0.771	0.878	0.855	0.767	0.643
A22	0.962	0.967	0.753	0.854	0.875	0.737	0.605
A23	0.947	0.923	0.777	0.885	0.833	0.753	0.668
A31	0.974	0.953	0.757	0.911	0.825	0.744	0.636
A32	0.951	0.917	0.834	0.923	0.887	0.757	0.659
A33	0.913	0.947	0.745	0.890	0.901	0.748	0.615
A41	0.935	0.912	0.731	0.912	0.821	0.706	0.653
A42	0.972	0.925	0.734	0.935	0.866	0.769	0.630
A43	0.925	0.925	0.767	0.880	0.801	0.717	0.657
A51	0.942	0.963	0.698	0.893	0.886	0.779	0.606
A52	0.915	0.953	0.740	0.901	0.800	0.767	0.659
A53	0.963	0.921	0.742	0.921	0.830	0.755	0.612
A61	0.925	0.884	0.733	0.843	0.834	0.733	0.631
A62	0.883	0.905	0.743	0.880	0.856	0.748	0.645
A63	0.973	0.913	0.734	0.854	0.812	0.777	0.639

	C8	C9	C10	C11	C12	C13	C14	C15
A11	0.603	0.768	0.525	0.756	0.824	0.607	0.935	0.615
A12	0.540	0.736	0.652	0.755	0.855	0.640	0.902	0.606
A13	0.563	0.740	0.612	0.762	0.844	0.661	0.938	0.636
A21	0.496	0.770	0.537	0.715	0.888	0.621	0.974	0.654
A22	0.608	0.742	0.505	0.756	0.838	0.641	0.929	0.647
A23	0.690	0.763	0.523	0.777	0.843	0.650	0.961	0.612
A31	0.587	0.705	0.445	0.719	0.875	0.617	0.905	0.663
A32	0.489	0.729	0.513	0.779	0.829	0.650	0.957	0.618
A33	0.590	0.724	0.403	0.751	0.874	0.608	0.934	0.636
A41	0.497	0.755	0.530	0.764	0.811	0.641	0.981	0.622
A42	0.432	0.779	0.410	0.761	0.840	0.633	0.931	0.632
A43	0.548	0.777	0.434	0.768	0.879	0.637	0.919	0.643
A51	0.526	0.727	0.549	0.742	0.852	0.656	0.923	0.646
A52	0.479	0.712	0.456	0.749	0.801	0.649	0.890	0.604
A53	0.546	0.733	0.509	0.751	0.811	0.645	0.921	0.662
A61	0.510	0.723	0.602	0.763	0.869	0.626	0.934	0.648
A62	0.529	0.746	0.456	0.700	0.848	0.603	0.931	0.628
A63	0.444	0.731	0.603	0.718	0.867	0.621	0.912	0.635

quality (score is in the range of [0.6~0.7]), and very poor quality (score is below the range of 0.6), as shown in Table 3.

According to the established quality evaluation model of B2C mode e-commerce logistics service, the model of three-layer BP neural network is established by using the MATLAB neural network toolbox. The number of neurons nodes of the input layer is 18 according to the secondary evaluation indicator of the index evaluation system. The number of neurons nodes of hidden layer is 14. The number of neurons nodes of output layer number is 1. Thus, we establish a model of BP neural network whose nodes number is 18, 14, 1

of each layer, respectively, which can be used for training and testing of the network. The schematic diagram of the model is shown in Figure 2.

For the logistics service quality evaluation index system of B2C mode e-commerce, the function tang is selected to activate the hidden layer, linear transformation function purely is selected to activate the output layer; the function trainer (Zhou & Kang 2005) is selected as the training function, and Bayesian normalization method is chosen to improve the generalization capability of the network to the training function. Select P = {C1 and C2, ..., C10} as training sample, P_test = {C11, C12, ..., C15} as a test sample from Table 2.

Table 3.   Table of comprehensive evaluation score.

C	C1	C2	C3	C4	C5
Comprehensive Score T0	0.9358	0.9304	0.7630	0.8804	0.8581
Grade	Very good quality	Very good quality	General quality	Good quality	Good quality
C	C6	C7	C8	C9	C10
Comprehensive Score T0	0.7474	0.6328	0.5564	0.7443	0.5312
Grade	General quality	Poor quality	Very poor quality	General quality	Very poor quality
C	C11	C12	C13	C14	C15
Comprehensive Score T0	0.7540	0.8465	0.6371	0.9370	0.6314
Grade	General quality	Good quality	Poor quality	Very good quality	Poor quality

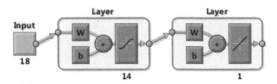

Figure 2.   Model of the BP neural network.

Parts of the MATLAB program codes are as follows:

```
for j = 6:14; %Build a circulate
net = newff(threshold,[j,1],{'tansig','purelin'},'train br');%Establish the neural network
net.trainParam.epochs = 500;%times of training
net.trainParam.goal = 1e-5;%precision of training
net = train(net,P,T);% training
y = sim(net,P_test);% output of the simulation
```

The default deviation is 0.00001, after 32 training steps, the deviation of the network reached the setting error, as is shown in Figure 3.

We put all the data in the data set, including training data, verification data and testing data, and then work out a linear regression analysis of the output of network and relevant expected output vector. The final results of the output of the linear regression are shown in Figure 4. The expected value of the output is relatively good, the value of R correspondingly almost reach 1, and matches the output of the network well.

Working out a network simulation for the untrained five testing values, the results are summarized in Table 4. The comparison between simulation results T1 and experts predicted results T0 is shown in Figure 5.

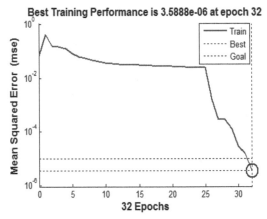

Figure 3.   Performance curve of network training deviation.

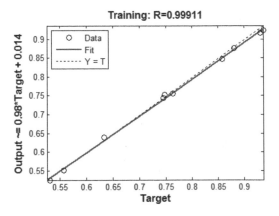

Figure 4.   Curve of linear regression.

Table 4. Result of the network simulation output.

	C11	C12	C13	C14	C15
T0	0.7540	0.8465	0.6371	0.9370	0.6314
T1	0.7531	0.8439	0.6382	0.9214	0.6373
Grade	General quality	Good quality	Poor quality	Very good quality	Poor quality

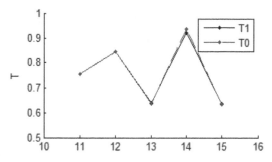

Figure 5. Fitting effects of the training sample.

The results of network simulation accords with the requirement, that is, the network model is suitable for the logistics service quality evaluation of B2C mode e-commerce.

5 CONCLUSIONS

Evaluation model of B2C mode network shopping service quality is established based on the combination of the AHP and BP neural network method. This calculation method of BP neural network, without apparent subjectivity and human factors, is nonlinear. Inputting the processed data into the model, then training and simulation it with the MATLAB software, we can get the evaluation results. We put all the data in the data set, including training data and test data, and then the network output and desired output vector of the linear regression analysis. High fitting degrees results, which avoid the simplicity of using the Analytic Hierarchy Process or Fuzzy Comprehensive Evaluation Method individually. This evaluation model has higher practicability for B2C online shopping logistics service quality evaluation.

REFERENCES

China's e-commerce market data monitoring report for 2012, 2013. The China electronic commerce research center [DB/OL]. http://wenku.baidu.com/View/403da524453610661ed9f42f.html.

Franceschini, F. & Rafele, C. 2000. Quality evaluation in logistic services, *International Journal of Agile Management Systems*, 2(1): 49–54.

Han, L. 2002. *Design and application of the artificial neural network theory*, Beijing, Chemical Industry Press.

Lin, Y. Gu, H. & Sheng, X. 1997. Fuzzy comprehensive evaluation misjudgment reasons discussed, *Appl. Syst. Eng. Theor. Method*, 6(2): 67–70.

Mao, Z. 2008. Based on the BP neural network of B2C e-commerce customer satisfaction evaluation model, *Science and technology in India*. 8(5): 49–52.

Wang, Z. 2007. *Logistics systems engineering*, Higher Education Press.

Zhou, K. & Kang, Y. 2005. *Neural network model and the designing of MATLAB simulation program*, Press of Tsinghua University, LTD.

Advanced Materials and Structural Engineering – Hu (Ed.)
© 2016 Taylor & Francis Group, London, ISBN 978-1-138-02786-2

Study and application of variation propagation of multi-station assembly processes with compliant autobody parts

A. Cui, S. Zhang, H. Zhang & Y. Luo
State Key Laboratory of Automotive Simulation and Control, Jilin University, Changchun, China

ABSTRACT: Most existing studies of the variation propagation are confined to a single station or multi-station assembly of rigid parts and a single station assembly of flexible parts. However, a large group of multi-station assembly processes consider compliant parts. It is important to take the impact of compliant parts deformation into account in the early design stage for improving the assembly quality and shortening the design cycle. A variation propagation model of multi-station flexible sheet metal parts assembly based on the deviation analysis of a single assembly station is applied in the paper. The model establishes a linear relationship between the input and output deviation state vectors. The influence of part variation, fixture variation and welding gun variation are considered in the model. The model is used for a practical application of multi-station assembly of body sheet metal parts. After a multi-station variation propagation model is established, the distribution of measuring point deviation in each station is obtained by using the Monte Carlo simulation. This method can significantly reduce the amount of computation and computing time compared with the direct finite element method.

1 INTRODUCTION

Many modern assembly systems are mostly of multi-station, and involve many complex manufacturing stations. Manufacturing errors in each station will be brought into and delivered in the system, and ultimately affect product quality. The deformation of compliant parts in the multi-station assembly process is taken into consideration in few studies in the early design stage (Wang & Ceglarek 2009). However, a large group of multi-station assembly processes consider compliant parts. For example, 37% of all assembly stations in autobody structure manufacturing assemble compliant parts (Shiu et al. 1997). It is important for improving the assembly quality and shortening the design cycle by taking the impact of compliant parts deformation into account in the early design stage.

At the single station level, Liu & Hu (1997) proposed a model to analyze the effect of deformation and spring back on assembly variation by applying linear mechanics and statistics. J. Grandjean et al. (2013) focused on surface defects of mechanical joint composed of two plane surfaces of two carter assemblies. P.K.S. Prakash et al. (2013) proposed a methodology to eliminate or reduce No-Fault-Found (NFF) Product failure in a field by adjusting manufacturing processes. Ding et al. (2000, 2002) developed a complete state space model for variation in the plane of rigidity for rigid components. Z. Yang et al. (2012) presented two assembly procedures of component stacks by

controlling variation propagation. Kang Xie et al. (2012) introduced a complete methodology for dimensional-related error compensation in compliant sheet metal assembly processes.

Comparatively, little research has been done in multi-station systems considering compliant, non-rigid parts. Chang & Gossard (1997) presented a graphic approach for multi-station assembly of compliant parts. Camelio & Hu (2003) developed a methodology to evaluate the dimensional variation propagation in a multi-station compliant assembly system based on linear mechanics and a state space representation.

In this work, a variation propagation model of multi-station flexible sheet metal parts assembly is accomplished based on the deviation analysis of a single assembly station (Camelio & Hu 2003). The model establishes a linear relationship between the input and output deviation state vectors. The influence of part variation, fixture variation and welding gun variation is considered in the model. The model is used for a practical application of multi-station assembly of body sheet metal parts.

2 VARIATION ANALYSIS OF COMPLIANT SHEET METAL WELDING PROCESS

2.1 *Sheet metal assembly process*

The single station variation propagation analysis of two parts is the base of compliant sheet metal assembly variation propagation modeling of multi-station. There are two kinds of sheet metal assembly processes: serial assembly and parallel assembly, as shown in Figure 1.

According to the locating, welding and releasing process of sheet metal parts, the whole process is modeled according to the following six sequential steps:

1. The parts are located by the 3-2-1 locating principle;
2. The additional locating points are added, and the deviation from the nominal position is smaller;
3. The parts are clamped to the nominal position and welded by welding guns;
4. The welding guns are released and the assembly is spring back;
5. The additional locating points are released and the assembly is spring back;

The 3-2-1 locating points are released and the assembly is spring back.

2.2 Variation modeling for sheet metal assemblies considering tooling deviation

Part deviation is only one of the variation contributors in the compliant parts assembly process. Also, it is necessary to consider the effect of tooling deviation over the assembly variation. Tooling deviation impact can be decomposed into two independent sources of variation: welding gun variation and fixture variation, including locators and clamps.

A position controlled welding gun is used to weld two parts at the gun/electrode position. It assumes that the welding gun can apply a sufficient force over the part to close the gap between the part deviation and its electrode position. As shown in Figure 2, part 1 has a deviation of v_1 and part 2 has a deviation of v_2. In addition, the welding gun has a deviation from the nominal v_g. The force required to close the gap in part 1 and 2 will be

$$F_1 = K_1(v_1 - v_g)$$
$$F_2 = K_2(v_2 - v_g) \tag{1}$$

where K_1 and K_2 are the stiffness of the two parts, respectively, which can be obtained by using the finite element methods.

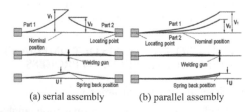

(a) serial assembly   (b) parallel assembly

Figure 1. The welding process of compliant sheet metal parts.

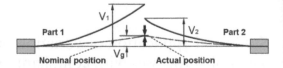

Figure 2. Welding gun deviation impact.

The assembly of part 1 and part 2 is assumed to be with a stiffness of K_w. The spring back force that is equal to the assembly force can be presented as $F = F_1 + F_2$, so the spring back can be presented as follows:

$$v_w = \frac{F}{K_w} = \frac{K_1}{K_w}v_1 + \frac{K_2}{K_w}v_2 - \frac{(K_1 + K_2)}{K_w}v_g \tag{2}$$

By using the sensitivity matrix definition, the assembly variation considering welding guns deviation can be presented as follows:

$$\{U\} = [S]\begin{Bmatrix} v_1 - v_g \\ v_2 - v_g \end{Bmatrix} = [S]\{V\} - [S]\{V_g\} \tag{3}$$

The variation caused by the (3-2-1) fixture can be obtained by using the method of influence coefficients. The sensitivity matrix of the welding points to the fixture variation is defined as $[S_{wf}]$, the sensitivity matrix of the measuring points to the fixture variation before welding is defined as $[S_{mf}]$, and the sensitivity matrix of the measuring points to the fixture variation after welding is defined as $[S_f]$. The variation caused by the (3-2-1) fixture can be presented as follows:

$$\{U_{(3-2-1)}\} = [S_f]\{V_{(3-2-1)}\}$$
$$= [S_{mf}]\{V_{(3-2-1)}\} + [S][S_{wf}]\{V_{(3-2-1)}\} \tag{4}$$

where $[S]$ is the sensitivity matrix without the tooling impact; $[S_{wf}]$ and $[S_{mf}]$ can be obtained by using the method of influence coefficients; and $[S_f] = [S_{mf}] + [S][S_{wf}]$.

The additional (N–3) locating points can be analyzed analogously as extra position-controlled welding guns. The clamps apply a force over the part. It produces a spring back when the force is released after the assembly. The additional locating points can be seen as an additional source of deviations. By using the method of influence coefficients, the variation caused by $V_{(N-3)}$ can be presented as follows:

$$\{U_{(N-3)}\} = [S_{f'}]\{V_{(N-3)}\}$$
$$= [S_{mf'}]\{V_{(N-3)}\} + [S][S_{wf'}]\{V_{(N-3)}\} \tag{5}$$

Synthetically, the assembly variation considering parts and tooling deviation can be presented as follows:

$$\{U\} = \{U_p\} + \{U_{(3-2-1)}\} + \{U_{(N-3)}\} - \{U_g\}$$

$$= [S]\{V\} + [S_f]\{V_{(3-2-1)}\}$$

$$+ [S_{f'}]\{V_{(N-3)}\} - [S]\{V_g\} \tag{6}$$

3 MODELING OF VARIATION PROPAGATION OF MULTI-STATION ASSEMBLY

The autobody assembly is a multi-level process, in which sheet metal parts formed to be subassembly or assembly, ultimately forming the body. During the assembly process, each part or subassembly is the input of the next station. While parts move from one station to another, dimensional variation of the parts and subassemblies propagates through the system. The station-to-station interactions cause an increase or sometimes decrease in the dimensional variation.

An assembly process can be viewed as a discrete time dynamic system of the station status, which can be expressed by independent time variables. State space model can be used to represent the deviation delivery of the parts assembly process. The deviation transfer process can be viewed as a discrete linear system, with each time variable corresponding to different stations; the discrete state space method can be used to model a multi-station assembly system. So, the dimension variation of the assembly parts can be presented as follows:

$$X(k) = A(k)X(k-1) + B(k)U(k) + w(k) \tag{7}$$

$$Y(k) = C(k)X(k) + w(k) \tag{8}$$

where Equation (7) is the state equation; Equation (8) is the output equation; $X(k)$ is the state vector; $A(k)$ is the state matrix; $B(k)$ is the input matrix; $U(k)$ is the input matrix; $Y(k)$ is the output vector; $C(k)$ is the output matrix; and $w(k)$ is the system noise.

3.1 *State vector X(K)*

Discrete system state vector contains related information of parts or sub-assemblies of each station. According to the complexity and precision of parts, a finite number of points are selected to analyze the deviation of the assembly process. The relevant points used to express the flexible parts assembly are as follows: welding points (W), locating points (L) and measuring points (M).

For an assembly system including m parts and N stations, the state vector of station k can be presented as follows:

$$X(k) = \begin{bmatrix} X^1(k) \\ X^2(k) \\ \vdots \\ X^m(k) \end{bmatrix} \quad k = 1, 2, \dots, N \tag{9}$$

3.2 *State matrix A(k) (without tooling variation)*

The assembly process at station k can be presented as: (1) the parts or subassemblies are relocated at station k by 3-2-1 fixture; (2) the additional (N-3) locating points are loaded with the parts or subassemblies, and the subassembly or assembly is formed by welding; (3) the welding guns and fixture are released, and the subassembly or assembly springs back.

All these processes can be represented by the form of matrices: the relocation effect is defined by the matrix M; the parts deformation before welding is defined by the matrix P; the spring back can be obtained by the matrix S. The state matrix can be presented as follows:

$$A(k) = f(M(k), P(k), S(k)) \tag{10}$$

$$A(k) = (S(k) - P(k) + I)(I + M(k)) \tag{11}$$

3.3 *Input matrix B(k) and output matrix U(k) (with tooling deviation)*

Using the state space method, defining the deviation of welding guns U_2, Equation (12) can be presented as follows:

$$X(k) = (S(k) - P(k) + I)(I + M(k))X(k-1)$$

$$- (S(k) - P(k))U_g + w(k) \tag{12}$$

Deviation of positioning fixture can be seen as rigid body translation and rotation, and can be obtained by kinematic analysis. The deviation of positioning fixture can directly affect the state space equation of the relocation process. Then, Equation (13) can be represented as follows:

$$X'(k-1) = X(k-1) + M(k)\left[X(k-1) - U_{(3-2-1)}(k)\right] \tag{13}$$

The (N-3) additional locating points can be viewed as additional positioning control guns, which can be loaded with the parts as force, released after assembly then the spring back is produced. Also, the (N-3) additional locating points can be viewed as additional deviation sources, using the method of influence coefficients. Then, Equation (14) can be presented as follows:

$$X(k) = (S(k) - P(k) + I)(I + M(k)) X(k-1)$$
$$- (S(k) - P(k) + I) M(k) U_{(3-2-1)}$$
$$- (S(k) - P(k))(U_{(N-3)} + U_g) + w(k) \quad (14)$$

So, the input matrix $B(k)$ and output matrix $U(k)$ can be, respectively, presented as follows:

$$B(k) = \left[-(S(k) - P(k) + I) M(k) - (S(k) - P(k)) \right]$$

$$U(k) = \begin{bmatrix} U_{(3-2-1)} \\ U_{(N-3)} + U_g \end{bmatrix}.$$

Finally, the state space model considering the parts deviation, fixture deviation and welding guns deviation can be presented as Equation (14).

Take the mean deviation and standard deviation of initial parts variation and tooling variation into account, by using the Monte Carlo simulation, the mean deviation and standard deviation of the assembly can be obtained.

4 APPLICATION WITH COMPLIANT AUTOMOTIVE BODY SHEET PARTS

An application of the variation propagation of multi-station assembly processes is conducted in this part. The welding of a front rail and an inner panel of a pillar of a type of SUV is taken as an example. In actual production, generally, the side frame is welded together first, and then the front rail is welded to the side frame. In order to simplify the model calculation, the impact of assembly order to the assembly variation is not considered in this paper. However, the application of the model is without loss of generality.

The assembly process is as follows: at station 1, the inner panel of front rail (A) and the outer panel of front rail (B) are welded to be a subassembly front rail (A+B), which is a parallel assembly; at station 2, the inner panel of A pillar (C) and the reinforcement panel of A pillar (D) are joined together to be a subassembly (C+D), which is a parallel assembly; at station 3, (A+B) and (C+D) are joined together to be the assembly (A+B+C+D), as shown in Figure 3.

The finite element analysis is conducted in the analysis software ABAQUS. The thickness of the inner panel of front rail is 1.0 mm, the thickness of the outer panel of front rail is 1.3 mm, the

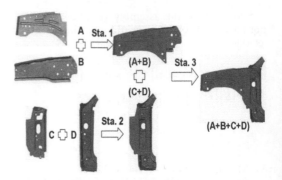

Figure 3.   Assembly process of 3 stations.

Table 1.   Initial state vector (welding points) mm.

Points no.	Part 1		Part 2		Part 3		Part 4	
	m	s	m	s	m	s	m	s
W 1	0.15	0	−0.2	0	0.25	0.25	0	0.25
W 2	0.2	0.5	−0.2	0	0.25	0	−0.3	0
W 3	0	0.25	0	0.5	0	0	0	0.15
W 4	0	0	0.25	0.25	0.25	0	0	0.25
W 5	0.25	0.5	0	0	0	0.25	0.2	0
W 6	−0.2	0.25	0.25	0	−0.2	0	0.2	0
W 7	0.15	0	0	0.2	0.25	0	0	0.2
W 8	0	0.25	0.25	0	0	0.2	−0.2	0
W 9	0.25	0.5	0	0	0.15	0	0	0
W 10	0	0	0	0	0	0	0	0
W 11	0	0	0	0	0	0	0	0
W 12	0	0	0	0	0	0	–	–
W 13	–	–	0	0	–	–	–	–

thickness of the inner panel of A pillar is 1.2 mm, and the thickness of the reinforcement panel of A pillar is 0.8 mm. The material of every part is mild steel with Young's modulus E = 210 Gpa, and Poisson's ratio $\mu = 0.3$. The model is built with shell elements, and the Multi-Point Constrain (MPC) is used to simulate the welding points join. The parameters of the state equation are obtained by using the method of influence coefficients. The variation output of each station can be obtained by the state equation. The initial state vectors including the variation of each welding point (W), locating point (L), measuring point (M) are given in Table 1 and Table 2.

Figure 4 shows the comparison of 30 measuring points of the parts with and without the tooling deviation. From the figure, we can see that the tooling deviation has a bigger impact on the measuring points 3, 8, 15, 20, and 23, while has a smaller impact on the measuring points 5, 7, 16, 19, and 28.

Table 2. Initial state vector (location points and measure points) mm.

Points no.	Part 1 m	Part 1 s	Part 2 m	Part 2 s	Part 3 m	Part 3 s	Part 4 m	Part 4 s
L 1	0.15	0.25	0.5	0.25	0.25	0.25	0.35	0.25
L 2	0	0.25	0.5	0.25	0.25	0.25	0.15	0.25
L 3	0.25	0.25	0	0.25	0.5	0.25	0.25	0.25
L 4	0.55	0.25	0	0.25	0.25	0.5	0	0.5
M 1	1.5	0.5	0.8	0.3	1	0.25	1.8	0.5
M 2	1.2	0.25	1.5	0.25	0.9	0.3	1.2	0.3
M 3	1.8	0.25	1.2	0.5	1.2	0.5	1	0.3
M 4	0.8	0.2	1.8	0.5	1.8	0.25	1.2	0.5
M 5	1	0.2	2	0.25	1.4	0.5	1.8	0.5
M 6	1.6	0.25	1.2	0.25	1.2	0.5	1.2	0.5
M 7	2	0.5	0.8	0.5	1.6	0.25	0.8	0.3
M 8	–	–	–	–	1	0.25	1.5	0.25

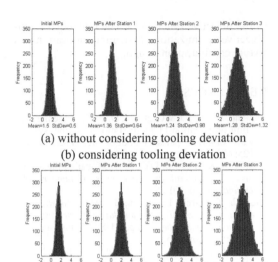

(a) without considering tooling deviation

(b) considering tooling deviation

Figure 5. Variation distribution of measuring point 1.

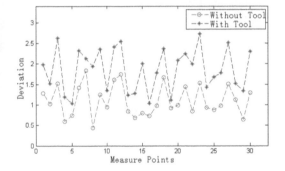

Figure 4. Deviation of measuring points at station 3.

After the variation propagation model is built, input the mean deviation and standard deviation of initial vector at station 1, by using the Monte Carlo simulation, the mean deviation and standard deviation of each station can be obtained. Figure 5 is the variation distribution of measuring point 1 at each station after 4000 times of Monte Carlo simulations, respectively, with and without the tooling deviation. If the simulation is taken by using the direct finite elements simulation, 4000 times of finite elements simulations will be needed. A large amount of calculation will be required and the calculation will be very complicated. Compared with the direct finite elements simulation, the application of variation propagation model of multi-station can greatly reduce the amount of computation and calculation time.

However, the inadequacy of this method is that when the fixture configuration changes, the state matrix and input matrix will change correspondingly, which should be obtained again by a new calculation. In addition, as the number of parts and stations increases, the dimension of the state vector, state matrix, input vector $A(k)$, and input matrix $B(k)$ will increase.

5 CONCLUSIONS

Based on the deviation analysis of a single assembly station, a variation propagation model of multi-station flexible sheet metal parts assembly is developed. The model establishes a linear relationship between the input and output deviation state vectors. The influence of part variation, fixture variation and welding gun variation is considered in the model. The model is used for a practical application of multi-station assembly of body sheet metal parts. Compared with the direct finite elements simulation, the application of variation propagation model of multi-station can greatly reduce the amount of computation and calculation time.

Using the variation propagation model of multi-station, some further expanded research can be conducted. ① Tolerance allocation optimization of multi-station assembly. By the covariance relationship of state equation, building the variation—tolerance model of sheet metal assembly, tolerance allocation optimization of multi-station assembly can be conducted with the appropriate tolerance—cost. ② Fixture locating is design and optimization. The state equation of the assembly process provides a linear quantitative relationship between the deviation of locating deviation and assembly deviation. Fixture design and optimization can be conducted by minimizing the assembly variation as the objective function. ③ The integrated multi-objective optimization of tolerance allocation and fixture configuration design.

REFERENCES

Camelio, L. Hu, S.J. & Ceglarek, D. 2003. Modeling variation propagation of multi-station assembly systems with compliant parts, *Journal of Mechanical Design.* 125: 673–681.

Chang, M. & Gossard, D.C. 1997. Modeling the assembly of compliant, no-ideal parts, *Computer-Aided Design.* 29(10): 701–708.

Ding, Y. Ceglarek, D. & Shi, J. 2000. *Modeling and diagnosis of multistage manufacturing process: part I—state space model,* Japan-USA Symposium on Flexible Automation.

Ding, Y. Ceglarek, D. & Shi, J. 2002. Design evaluation of multi-station assembly processes by using state space approach, *ASME Journal of Mechanical Design.* 124(3): 408–418.

Grandjeana, J. Ledoux, Y. Samper, S. & Favreliere. H. 2013. Form Errors Impact in a Rotating Plane Surface Assembly, *Procedia CIRP*, 10: 178–185.

Kang, X. Jaime, A. Camelio, L. & Eduardo, I. 2012. Part-by-part dimensional error compensation in compliant sheet metal assembly processes, *Journal of Manufacturing Systems.* 31(2): 152–161.

Liu, S.C. & Hu, S.J. 1997. Variation simulation for deformable sheet metal assemblies using finite element methods, *ASME Journal of Manufacturing Science and Engineering.* 119: 368–374.

Prakash, P.K.S. & Ceglarek, D. 2013. Multi-step Functional Process Adjustments to Reduce No-fault-found Product Failures in Service Caused by In-tolerance Faults, *CIRP*, 11: 38–43.

Shiu, B.W. Ceglarek, D. & Shi, J. 1997. Flexible beam-based modeling of sheet metal assembly for dimensional control, *T. NAMRI/SME*, 25: 49–54.

Wang, H. & Ceglarek. D. 2009. Variation propagation modeling and analysis at the preliminary design phase of multi-station assembly systems, *Assembly Automation.* 29(2): 154–166.

Yang, Z. Popov, A.A. & McWilliam, S. 2012. Variation propagation control in mechanical assembly of cylindrical components, *Journal of Manufacturing Systems.* 31(2): 162–176.

Advanced Materials and Structural Engineering – Hu (Ed.)
© *2016 Taylor & Francis Group, London, ISBN 978-1-138-02786-2*

Numerical simulation of the whole failure process of rock under various confining pressures

J. Jia, B.L. Xiao & C.R. Ke
Hubei University of Technology, Nanhu Lake, Hongshan District, Wuhan City, Hubei Province, China

ABSTRACT: Virtual internal bond model (abbreviated as VIB) that considers solid materials at the meso scale are constructed by discrete material particles, between materials particles are connected by a VIB. Through special cohesive force law and interactions, we deduce macroscopic constitutive material equations and failure criterion embedded in the constitutive equation. The Monte-Carlo method is adopted to describe the rock material's heterogeneity, and assumed elements are supposed to yield to the Weibull distribution. By the density or stiffness in evolution of VIB, the whole failure process of rock under various confining pressures has been conducted. According to numerical simulation results, an equation to describe rock intensity under various confining pressures is obtained to describe the relationship curve of peak stress and the confining pressures by the least squares method in this paper.

1 INTRODUCTION

Generally, underground rock is always in the state of triaxial compression before excavation disturbance; the rock's lateral pressures will change by different burial depth. As a rule, the deeper the burial depth is, the greater the lateral pressures are. Studying the failure process of rock under various lateral pressures and grasp its law is the important significance of evaluation, design and construction of geotechnical engineering. The research of fracturing process and the law of rock under various confining pressures has made a lot of achievements. Zhenyu Tao et al. tested the influence of mechanical properties of rocks under various confining pressures (Tao & Zhu 1996, Ge & Ren 2004, Li 2003); however, due to the limitation of samples and research funds, their studies were only in view of the limited sample tests and then analyzed the results. Chunan Tang et al. developed a set of rock failure and instability analysis system, and conducted a large number of studies on the rock failure process under confining pressures (Tang et al. 2003). However, the system is based on the theory of mechanics of continuous field, in the failure process of a rock. The original continuous field was destroyed, and became a discontinuous field. Therefore, research and development of the new theory of rock failure has both theoretical and practical significance.

Recently, Gao et al. proposed a theory of VIB model in simulating the solid material's failure (Gao & Klein 1998, Klein & Gao 1998). The model considers that continuous solid materials at the micro-scale are composed of discrete material particles, every particle is connected by a VIB, and macroscopic constitutive equations of material are based on the interaction between particles. The model is a multi-scale model, which embeds the failure criterion into the constitutive equations through the cohesive force law of material particles at the meso scale, achieving the numerical simulation of material failure without other additional strength criterions. Zhenna Zhang put forward a new method for the analysis of material failure behavior on the basis of VIB, namely the macroscopic response of material attributed to the evolution of the micro VIB, and developed into the multi-dimensional Virtual Internal Bond Model (VMIB) (Zhang & Ge 2005). However, according to the process of rock failure meso test analysis (Ge & Ren 2004), the fracture mode of the rock's failure process is the opening mode. The fractured zone is a macroscopic collection of multiple meso units after being fractured. The space lattice model is made of flexible deformation of VIB between material particles, which is more concise than VMIB. So, the VIB model is applied to simulate the process of rock failure, and successfully carried out the simulation of the rock failure process curve (Ke & Ge 2009, 2012). On the basis of the previous research, this paper attempts to study the whole process of failure simulation of rock under various confining pressures by VIB.

2 MODEL ESTABLISHMENT AND PARAMETER SELECTION

2.1 *Theory of virtual internal bond model*

VIB model is shown in Figure 1.

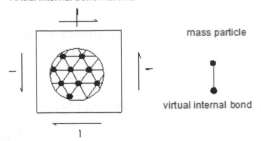

virtual internal bond material

mass particle

virtual internal bond

1

Figure 1. Virtual internal bond model.

In level one, which is the level of dividing the finite element mesh, VIB uses the continuum mechanics model. In the second level, solid materials are constructed by material particles at meso scale, and between particles through interactions of lattice structure consists of VIB. According to the results of related research (Ke & Ge 2009, 2012), the fourth-order elastic tensor for material unit is given by

$$C_{ijmn} = \int_0^{2\pi} \int_0^{\pi} k \xi_i \, \xi_j \, \xi_m \, \xi_n \, D(\theta,\varphi) \sin\theta d\theta d\varphi \qquad (1)$$

where k is the stiffness coefficient of VIB and $\xi_i, \xi_j, \xi_m, \xi_n$ are direction vectors of VIB. In the spherical coordinate system (r, θ, φ),

$$\xi = (\sin\theta\cos\varphi, \sin\theta\sin\varphi, \cos\theta)^T \qquad (2)$$

$D(\theta,\varphi)$ is the distribution density of VIB.

If the aggregation of the representative infinitesimal particles is material particle in mechanical properties at the macro scale, and equivalent to the corresponding continuum elements, the result is given by

$$k = \frac{15E}{8\pi(1+\mu)} \qquad (3)$$

where E denotes the material elastic modulus and μ is Poisson's ratio.

For the plane stress problem, it can be written as:

$$k = \frac{2E}{(1-\mu)\pi} \qquad (4)$$

For the plane strain problem, it can be written as:

$$k = \frac{2E}{(1-2\mu)(1+\mu)\pi} \qquad (5)$$

2.2 Description of the rock material model

As a kind of geological materials, one basic characteristic of rock is heterogeneity, and natural defects and pores existed inside rock. In order to describe the property of the rock, this paper adopts the way that combines the Monte-Carlo method with statistical description, to use random assignment for elements, and the elements are supposed to yield to the Weibull distribution. It is a level selection problem to choose the level of VIB or the level of elements. This paper chooses the elements level to use random assignment for the elastic tensor of the elements. The specific methods are as follows: elements are supposed to yield to the Weibull distribution. The first step produces N random numbers χi ($I = 1, 2, ..., N$) in accordance with the Weibull distribution, where N is the number of elements. Then, the N values of the random numbers are assigned to the meshed units. That is, in the element stiffness matrix, the elastic tensor of i unit is given by

$$C^i_{jmn} = x_i C_{jmn} \qquad (6)$$

3 MODEL VERIFICATION

In order to verify the validity of the model, the numerical simulation of the triaxial tests of the rock under various confining pressures is made.

The data of model are derived from the sample tests on the MTS815.03 electric servo rock mechanical test machine at the Institute of Rock and Soil Mechanics, Chinese Academy of Sciences. The constants of materials are listed in Table 1, and the dimension and boundary conditions of the rock sample are shown in Figure 2.

The model uses plane the 3-node constant strain unit, the total number of nodes is 496 and the total number of units are 900. The numerical simulation is assumed to the state of plane stress. Using the displacement control method in the process of the numerical simulation, each displacement load step is 0.03 mm. For the macroscopic mechanical behavior of the materials depending on the evolution of the micro VIB, this

Table 1. Rock parameters and the corresponding model parameters.

Rock parameters				Model parameters	
$E/G\,Pa$	μ	σ_b/MPa	$\varepsilon_b \times 10^3$	c_1	c_2
27.35	0.322	86.03	8.623	7.0	0.39

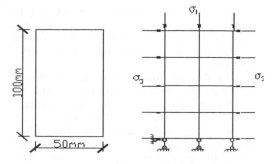

Figure 2. Rock sample and boundary conditions.

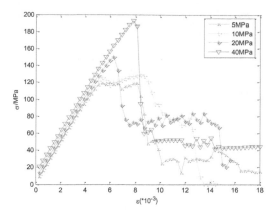

Figure 3.   Stress-strain curves under various confining pressures.

paper assumes that the density evolution equation of VIB is (Ke & Ge 2012):

$$k = k_0 \exp\left[-c_1 \frac{l}{\varepsilon_b \left(c_2 \frac{\sigma_2}{\sigma_b} \right)\left(1 + 2.0 \frac{\sigma_2}{\sigma_b} \right)} \right]^{4.0} \quad (7)$$

where k_0 is the initial stiffness of VIB, according to the Eqn (2) to Eqn (4); i is the deformation of VIB, and $l = \left| \xi_i \varepsilon_j^d \xi_j \right|$, ε_j^d is the deviatoric tensor of strain; ε_b is the uniaxial compressive strain intensity, determined by the compression tests of rock specimens; σ_2 is the principal stress; c_1 and c_2 are model parameters. In the above equation, they, respectively, determine the different macroscopic characteristics of the material.

3.1 Characteristics of the stress-strain relationship

The whole stress-strain relationships under different confining pressures are shown in Figure 3. We can find under the various confining pressures that the shapes of the whole axial stress-strain curves are similar to each other. They can be divided into three stages: elastic deformation, plastic deformation, and failure. Comparing with the whole process curve obtained from the rock physical experiment, they lack the first stage called the compression phase. This is mainly due to the fact that this model does not have closed pores to be pressured, but the basic characteristics of the other three stages coincide with the physical model.

3.2 Relationship between peak stress and confining pressure

It can be seen from Figure 3, the performance of the intensity characteristics under different confining pressures is as follows: with the increasing confining pressure, the compressive strength increases. By the tracing point method, we found approximately the linear relationship between peak stress and confining pressures, so we adopt the linear relationship between them. Through the least squares regression analysis, the relationship between σ_1 called the rock axial failure stress (strength) and σ_3 called the confining pressures is shown in Figure 4. Fitting function is given in the following form:

$$\sigma_1 = \sigma_c + k\sigma_3 \quad (8)$$

where σ_1 is the axial peak stress (MPa); σ_3 is the lateral stress (MPa); σ_c is the uniaxial compressive strength of the rock (MPa); and k is the rock intensity influence coefficient.

According to regression analysis, the expression can be written as:

$$\sigma_1 = 108.471 + 2.101\sigma_3 \quad (9)$$

3.3 Analysis of the intensity characteristics of the rock

The final failure mode of rock specimens under different confining pressures is shown in Figure 5. In general, under confining pressures, the failure of rock is shear failure characteristics, which is almost consistent with the conclusions of rock physical experiments. With the increase in confining pressures, shear failure is more obvious, and shear plane is more neat. However, the angles between the shear plane and the axial pressures show that the influence of the confining pressures' regularity is weak.

847

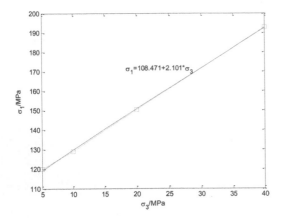

Figure 4. Axis peak stress at various confining pressures.

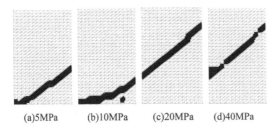

(a)5MPa   (b)10MPa   (c)20MPa   (d)40MPa

Figure 5. Failure of simulated rock specimens under various confining pressures.

4 CONCLUSIONS

1. This paper has adopted the way that combines the Monte-Carlo method with statistical description, and then has valued the element intensity of VIB, and elements were supposed to yield to the Weibull distribution. It can better describe the heterogeneity of rock material and natural defects and pores, so as to realize the numerical simulation of rock failure.
2. Under various confining pressures, the deformation of the rock is divided into three stages as follows: elastic deformation, plastic deformation, and destruction. With the increasing confining pressure, the peak intensity of the rock increases, and there is a linear relationship between the peak intensity of the rock and the confining pressures. The paper has obtained the relationship curve for the peak stress and the confining pressures by the least squares fitting method.
3. The final failure mode of rock under different confining pressures is of shear type. With the increasing confining pressure, shear failure is more obvious, and shear plane is more neat. However, the angles between the shear plane and the axial pressures show that the influence of confining pressures' regularity is weak. In a word, the model that is proposed in this paper is worth improving, and it is a way to research in the future.

REFERENCES

Gao, H.J. & Klein, P. 1998. Numerical simulation of crack growth in an isotropic solid with randomized internal cohesive bond, *Journal of the Mechanics and Physics of Solids*, 46(2): 187–218.
Ge, X.R. & Ren, J.X. 2004. *Experimental study of geotechnical macro mesoscopic damage mechanics*, Beijing: Science Press.
Ke, C.R. & Ge, X.R. 2009. Complete curve simulation of rock under uniaxial compression based on virtual internal bond model, *Rock and Soil Mechanics*. 30(5): 1509–1514.
Ke, C.R. & Ge, X.R. 2009. Numerical simulation of rock fracturing under uniaxial compression using virtual internal bond model, *J. Shanghai Jiaotong University (Sci.)*, 14(4): 423–428.
Ke, C.R. & Ge, X.R. 2012. Numerical simulation of heterogeneous rock dynamic fracture based on Virtual Internal Bond (VIB) model, *J. Shanghai Jiaotong University (Sci.)*, 46(1): 142–145.
Klein, P. & Gao, H. 1998. Crack nucleation and growth asst rain localization in a virtual—bond continuum, *Engineering Fracture Mechanics*. 61: 21–48.
Li, X.W. 2003. *The properties of rock mechanics*, Beijing: China Coal Industry Publishing House.
Tang, C.N. Wang, S.H. & Fu, Y.F. 2003. *Numerical experiments of rock failure process*, Beijing: Science Press.
Tao, Z. & Zhu, H.C. 1996. *Geological and physical fundamentals of rock mechanics*, Wuhan: China University of Geosciences Press.
Zhang, Z.N. & Ge, X.R. 2005. A new quasi continuum constitutive model for crack growth in an isotropic solid, *European Journal of Mechanics—A/Solids*. 24: 243–252.

Advanced Materials and Structural Engineering – Hu (Ed.)
© 2016 Taylor & Francis Group, London, ISBN 978-1-138-02786-2

Computation and analysis of the caisson wharf

Y. Shi, Q.C. Ning & X. Yang
Coastal and Offshore Engineering, College of Harbour, Hohai University, Nanjing, Jiangsu, China

R. Chen
The 28th Research Institute of China Electronics Technology Group Corporation, Nanjing, Jiangsu, China

K. Cui
Jiangsu Province Communications Planning and Design Institute Limited Company, Nanjing, Jiangsu, China

ABSTRACT: In the traditional design of caisson wharf, a simplified method is always adopted to calculate the internal force, which causes the deviation. This paper introduces an effective design method that uses the finite element modeling software to analyze the structural internal force of the caisson wharf. Supposing that the load is constant, the article utilized the ANSYS to calculate the stress and displacement situation of the caisson wharf at different height-width ratios. The comparative results show that the most reasonable height-width ratio in the project is 1:1~1:1.2. The proposed method in this paper can be used for reference of engineering design.

1 INTRODUCTION

Caisson wharf is a kind of gravity wharf, usually composed of outer wall, floor and partition wall (Li 1990). It is widely used for its good integrality and rapid construction speed.

Currently, in the design and construction code of gravity wharf, the floor and outer wall can be calculated by the following factors (JTS167-2-2009 2009): the outer wall is divided into the inner part and outer part in the location of 1.5 times of the inner partition wall distance. In the inner part, the calculation is made in the light of three edges fixed and one edge simply-supported slab. This method assumes that the constraint from the partition wall and floor to the outer wall is the same, but because the difference in the thickness of the partition wall and floor, the constraints are different, so this method leads to the deviation. In the outer part, the computation is made on the basis of two ends fixed continuous slab (Yao et al. 2009). The floor is computed as a slab whose four edges are fixed.

Besides, the simplified method gets the internal force of slab with the help of the static structural calculation handbook, uses the look-up table method and obtains the mid-span moment, support moment and maximum moment under the action of concentrated load, uniform load and triangular load, and then adds together. However, under the different forms of load, the extreme values are frequently not on the same point, so it adds together directly, which causes the deviation.

In order to design a more reasonable caisson wharf structure, the model is established by the finite element modeling software. Assuming that the load is fixed, this article compared the ANSYS with the simplified method in terms of the internal force of the important component, and calculated the reasonable ratio of height to width.

2 ENGINEERING EXAMPLE

We take a 100,000 DWT caisson wharf as an example, as shown in Figure 1 and Figure 2. There are 9 cells in a caisson. A single caisson is a structure

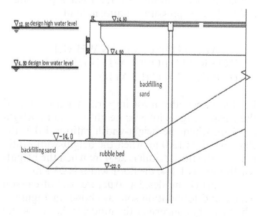

Figure 1. Cross-section of the caisson wharf.

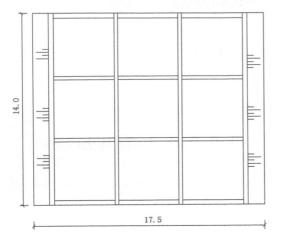

Figure 2. Sectional view of the caisson wharf.

part, the height of the caisson is 18.5 m; the length of the caisson is 14 m; the thickness of the partition wall is 0.5 m; the thickness of the floor is 1.0 m; and the length of the fore toe and hind toe is 1 m.

3 DESIGN LOADS AND LOAD COMBINATION

The mooring force acts on the breast wall in the form of horizontally uniform force; the handling machinery load acts on the breast wall in the form of vertically uniform force; the back filling pressure is based on the active earth pressure theory (Lu 2002); the filling pressure inside the cell is calculated according to the silo pressure; the residual head is 0.5 m; and the uniform load on the wharf surface is 20 kPa.

This paper obtains the contrasting results by the following load combination: dead weight + ship unloader load + mooring force + earth pressure + residual water pressure + uniform load.

4 ANALYSIS OF CAISSON WHARF'S INTERNAL FORCE WITH DIFFERENT HEIGHT-WIDTH RATIOS

In this paper, the most important parameter of the caisson are as follows: height is 18.5 m; length is 14 m; when the height-width ratio is 1:0.7, the floor width is 13.0 m; when the height-width ratio is 1:1.1, the floor width is 20.4 m; when the height-width ratio is 1:1.4, the floor width is 25.9 m.

This article modeled a structure part of caisson wharf and foundation soil, as shown in Figure 3. The primary concern of the model is the stress and displacement situation of the caisson structure with different height-width ratios.

4.1 Analysis of the displacement

The displacement in the X-scale of the caisson is given in Table 1.

The following table shows that with the increasing width of the caisson, the displacement will decrease. This is due to the fact that the increase in weight can increase the anti-sliding force. Meanwhile, the integral rigidity of the caisson in the X-scale increases, the height and the earth pressure are invariant, so the displacement of the caisson in the bottom and top decreases.

4.2 Stress analysis of the fore toe and front panel

The stress in the Z-scale of the fore toe and the front panel is given in Table 2 and Table 3.

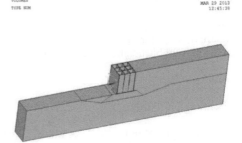

Figure 3. Finite-element model of the caisson wharf.

Table 1. The displacement of the caisson (mm).

Height-width ratio	Top	Bottom
1:0.7	0.32	0.11
1:1.1	0.18	0.05
1:1.4	0.13	0.02

Table 2. Stress of the fore toe (MPa).

Height-width ratio	Tensile stressMax	Compressive stressMax
1:0.7	0.81	1.13
1:1.1	0.67	0.76
1:1.4	0.53	0.69

Table 3. Stress of the front panel (MPa).

Height-width ratio	Tensile stressMax	Compressive stressMax
1:0.7	0.15	0.27
1:1.1	0.20	0.32
1:1.4	0.28	0.43

The above table shows that with the increase in the caisson's width, the stress of the fore toe will decrease. This is because the increase in the caisson's width causes the increase in the anti-overturning stability torque, and then the resultant force's eccentricity will decrease and the subgrade reaction becomes even, so the stress of the fore toe decreases.

The stress of the caisson's front panel increases as the caisson's width grows. This is because the silo pressure in the X-scale increases with the growth of the cell's size, which causes the increase in the stress of the front panel.

4.3 Stress analysis of the floor

The stress in the Y-scale of the floor is given in Table 4 and Table 5.

The above table shows that with the caisson's width increase, the stress in the Y-scale will decrease. This is because the subgrade reaction is distributed uniformly with the increased width of the floor.

With the increase in the caisson's width, the shear stress of the floor and wall joints increases gradually. This is because the silo pressure increases as the cell's size grow, which causes the shear stress to increase.

4.4 Stress analysis of the back panel

The stress in the Z-scale of the back panel is given in Table 6.

Table 4. Stress of the floor (MPa).

Height-width ratio	Tensile stressMax	Compressive stressMax
1:0.7	0.82	0.94
1:1.1	0.63	0.81
1:1.4	0.42	0.71

Table 5. Shear stress of the joints (MPa).

Height-width ratio	Shear stressMax
1:0.7	0.24
1:1.1	0.38
1:1.4	0.42

Table 6. Stress of the back panel (MPa).

Height-width ratio	Tensile stressMax	Compressive stressMax
1:0.7	0.93	1.21
1:1.1	0.76	0.93
1:1.4	0.45	0.73

Table 7. Comparison of the tensile stressMax between the two methods (MPa).

	Fore toe	Floor	Back panel
ANSYS	0.67	0.63	0.76
Simplified calculation	0.48	0.41	0.52

The above table shows that the back panel's stress in the Z-scale reduces along with the increase in the caisson's width. This is due to the bigger the cell's lateral dimension, the higher the silo pressure. While the back of filling pressure is constant, the resultant force declines with the increase in the caisson's width.

5 COMPARATIVE ANALYSIS OF THE CONVENTIONAL SIMPLIFIED CALCULATION

Take the caisson with the height-width ratio of 1:1.1 as an example. The comparison of the results between the conventional simplified calculation and the ANSYS is presented in Table 7.

1. Through the above comparison, the result from the ANSYS is slightly larger than the simplified calculation. In the latter method, the component junction is always being processed as consolidation, but as a matter of fact, the rigidity of floor and panel is different, especially the thickness of the floor is much bigger than other components. So, the accuracy of the results is being affected.
2. The result from ANSYS takes the caisson as a whole to calculate the stress. However, the simplified calculation based on the load of a single component depends on the static structural calculation handbook. This is another reason why the two results are different. We take the floor as an example. The external force is decomposed into a downward uniform load and an upward triangular load. Then, we calculate the internal force of the floor, respectively, and add two forces together. However, under the two kinds of load, the maximum moment inside the floor is a positive and a negative; moreover, they do not appear in the same point, so we add two forces together directly to make the maximum to decrease.

6 CONCLUSIONS

1. The above results indicate that under the same load, height and length situation, the lateral displacement, the normal stress of the fore toe,

851

the floor and the back panel decline with the increase in the caisson's width; while the normal stress of the front panel, the shear stress of the floor and wall joints increase as the width of the caisson grows.

2. In this paper, the ANSYS is used to analyze the caisson's structure in three different height-width ratios. The final analysis conclusion shows that under the same load, the most reasonable height-width ratio of the caisson in the project is 1:1~1:1.2.

3. This article shows that the simplified method will directly affect the accuracy of the calculation result. So the best partial coefficient should be adjusted appropriately in the engineering design.

REFERENCES

Design and Construction Code for Gravity Quay (JTS167-2-2009), 2009.

Li, Y.Y. 1990. Strength Calculation of Various Kinds of Caisson, *Port Engineering Technology*, 2: 34–39.

Lu, T.H. 2002. *Soil Mechanics (II)*. Nanjing: Hohai University Press, 198–200.

Yao, J. Gan, G. & Chen, S. et al, 2009. *Building Structure Static Calculation Practical Handbook*. China Archit. Build. Press, 151–178.

Advanced Materials and Structural Engineering – Hu (Ed.)
© *2016 Taylor & Francis Group, London, ISBN 978-1-138-02786-2*

Numerical simulation of particles' acting force on the blades of the rotary valve

X.S. Zhu & L.W. He

School of Mechanical and Power Engineering, East China University of Science and Technology, Shanghai, China

ABSTRACT: The rotary valve is one of the transportation equipment of particles. The stress and deformation of the blades of the rotary valve are analyzed in this paper. First, the discharging process of the rotary valve is simulated with the discrete element method to obtain the particles' acting force on the blades of the rotary valve. Then, the load data are imported into the ANSYS workbench to calculate the stress and deformation of the blades. The result can provide a reference for the rotary valve design.

1 INTRODUCTION

Rotary valve is a rotary machine for discharging, mixing, and quantitative delivery used in a pneumatic conveying system. Many scholars have conducted research on granular flow in the rolling container; for example, M.E. Sheehan (Sheehan et al. 2005, Britton et al. 2006) and Sergio M. Savaresi et al. (2001) researched the rolling granular desiccant flow with the signal plate. M. Silvina. Tomassone's (Silvona et al. 2006) experiment reflected the flow status of dispersion in rotary drying container. T. Jules et al. (1996) used the ideal mixture model and improved the Cholette-Cloutier model to simulate particles in axial rotary drum dryer transmission characteristics. Boateng & Barr's (1996) research reflected the granular flow state in the rotary cylinder without copy board case. However, studies of the particles' acting force on a rotary machine have been relatively less so far. Although the particles' force is considered in Gao's research (Gao 2012), in fact, he regarded the particles' gravity as the force on the rotary machine, which is only an approximate treatment. Based on the discrete element method, the particles' acting force on the blades of the rotary valve is obtained more accurately in this paper, which is necessary to study the structure of the rotary valve. The load data can be imported into the ANSYS workbench to calculate the stress and deformation of the blades. The result can provide a reference for the rotary valve design.

2 THE DISCRETE ELEMENT METHOD AND HERTZ-MINDLIN (NO SLIP) CONTACT MODEL

2.1 *The discrete Element Method (DEM)*

DEM (Cundall & Strack 1979) makes the following assumptions: 1) Particles are a rigid body; 2) contacts between particles occur in a small area, which is the contact point; 3) deformation of the particle system is the sum of the particles' contact deformation; and 4) contacts between particles are soft contact, which allows the overlap at the contact point. The discrete element method provides advanced 3D simulation for a discontinuous system with complex interactions.

2.2 *Contact model*

Particle contacts can be divided into soft contact and hard contact. Hard contact is only applicable to a sparse rapid granular flow. Soft contact allows particle collision persisting for some time, including collisions between multiple particles. The overlap is allowed at the contact point. The overlap at the contact point, physical properties of contact particles, relative impact velocity and contact information of the last time step are connected by two forces that are equal in size and opposite in direction, to calculate the acting force on particles. Then, the acceleration is calculated by Newton's second law before updating the velocity and displacement of particles:

$$\sum F_i = m \cdot g + F_n + F_t \qquad (1)$$

$$m_i \frac{dv_i}{dt} = \sum F_i \qquad (2)$$

2.3 Hertz-Mindlin contact model

The Hertz-Mindlin contact model (Wang et al. 2010) is applied in this paper. Normal force in the model is based on the Hertz contact theory, while the tangential force is based on the research of Mindlin-Deresiewicz, which makes the force calculation more accurately and efficiently. Specific equations are as follows.

Assuming that two spherical particles whose radii are, respectively, R_1 and R_2 are elastic contact, the normal overlap α is given by

$$\alpha = R_1 + R_2 - |r_1 - r_2| \qquad (3)$$

where r_1, r_2 are the two particle sphere center position vectors

$$a = \sqrt{\alpha R^*} \qquad (4)$$

where R^* is the equivalent particle radius, which can be calculated by the formula

$$\frac{1}{R^*} = \frac{1}{R_1} + \frac{1}{R_2} \qquad (5)$$

1. The normal force between particles F_n

$$F_n = \frac{4}{3} E^* (R^*)^{\frac{1}{2}} \alpha^{\frac{3}{2}} \qquad (6)$$

where E^* is the equivalent elastic modulus, determined by the elastic modulus and Poisson's ratio of particle 1 and particle 2.

2. Normal damping force F_n^d

$$F_n^d = -2\sqrt{\frac{5}{6}} \beta \sqrt{S_n m^*} v_n^{rel} \qquad (7)$$

where m^* is the equivalent mass, determined by m_1, m_2. v_n^{rel} is the relative speed of the normal component value. β and normal stiffness S_n is determined by the recovery coefficient and normal stiffness.

3. The tangential force between particles F_t

$$F_t = -S_t \delta \qquad (8)$$

where δ is the tangential overlap. S_t is the tangential stiffness, determined by the shear modulus between two particles.

4. The tangential damping force F_t^d

$$F_t^d = -2\sqrt{\frac{5}{6}} \beta \sqrt{S_t m^*} v_t^{rel} \qquad (9)$$

where v_t^{rel} is the tangential component value of the relative velocity.

2.4 Time step

Selecting an appropriate time step is the key to ensure the accuracy and efficiency of the simulation. When the particles contact, surfaces are influenced by variable stress, generating a polarized wave propagating along the surface of particles, which is called the Rayleigh wave. The experiment showed that 70% of the total energy consumption is consumed by Rayleigh wave when particles collide. So, the critical time step should be determined by the Rayleigh wave velocity spreading along a solid spherical particle surface. The time step of the system with different particles is given by

$$\Delta t = \pi \left[\frac{R}{0.163v + 0.877} \sqrt{\frac{\rho}{G}} \right]_{\min} \qquad (10)$$

The time step Δt determined can guarantee the calculation stability of the particle system. An appropriate time step should be chosen according to the particles' movement intensity to ensure the stability of the calculation; generally the time step is 10%~40% of Δt.

3 SIMULATION OF THE DISCHARGING PROCESS OF THE ROTARY VALVE

The discharging process of the rotary valve is simulated in EDEM. The meshes of the rotary valve are shown in Figure 1a and Figure 1b. The particles' acting force on the blades of the rotary valve can be extracted by post-processing function of EDEM, where the load file can be imported into the ANSYS workbench to calculate the equipment of stress and deformation of rotary blades.

A virtual panel is built above the entrance as the particle factory, which is used to generate particles. Besides, a circular baffle on both sides of the rotor is added to prevent the solids from overflowing along the Z axis. PE particles are adopted as the simulated media whose basic parameters are given in Table 1.

The time step is 1e⁻⁵ s, which is 25% of Δt. The rotor speed is 20 rpm. The rotor has twelve blades.

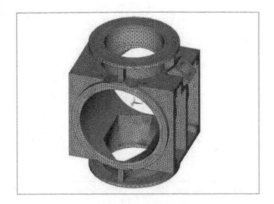

Figure 1a.   Mesh of the valve body.

Figure 1b.   Mesh of the rotor.

Table 1.   DEM model parameters.

Parameters	Value
Poisson's ratio of granular	0.203
Shear modulus of granular [pa]	8.31e8
Density of granular [kg/m³]	945
Poisson's ratio of steel	0.3
Shear modulus of steel [pa]	7e10
Density of steel [kg/m³]	7850
Restitution coefficient between granular	0.2
Coefficient of static friction between granular	0.5
Coefficient of rolling friction between granular	0.01
Restitution coefficient between granular and steel	0.5
Coefficient of static friction between granular and steel	0.5
Coefficient of rolling friction between granular and steel	0.01
Rotor blade radius [m]	0.16
Rotor speed [rpm]	20
Radius of spherical particles [mm]	3

The degree between two adjacent blades is 30°. Considering that particles' movements in each sector are the same as those during the process of discharging, the model can be simplified into one sector. The discharging model is shown in Figure 2.

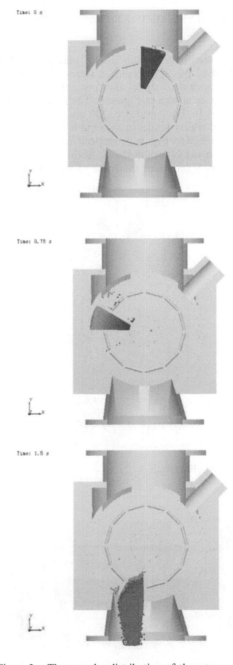

Figure 2.   The granular distribution of the rotary valve at 0 s, 0.75 s, and 1.5 s.

855

The particles are filled in one sector of the rotor, where particles are supported by two blades on the left and right. Then, the rotor rotates in the counterclockwise direction to simulate for 3 s (turning a circle). The total number of particles for calculating is 28469, whose total mass is 3 kg.

4 ANALYSIS OF PARTICLES' ACTING FORCE

Using the bin group function of selection module in EDEM post-processing, a series of small bins are imported to detect the force on the contact surface, as shown in Figure 3. The number of bin group is 16. Each grid size is 20 mm × 25 mm. The force of rotary blades in each bin is exported.

Figure 5 to Figure 8, respectively, express the force on the blades in the X, Y direction. Different curves represent the force of the lattice with

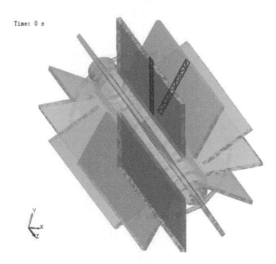

Figure 3.   Bin group.

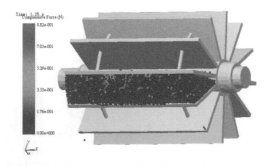

Figure 4.   Compressive force of particles at 1.25 s.

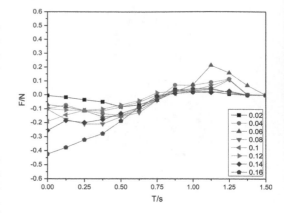

Figure 5.   The force on the left blade in the X direction.

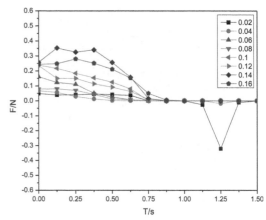

Figure 6.   The force on the right blade in the X direction.

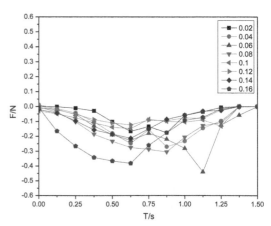

Figure 7.   The force on the left blade in the Y direction.

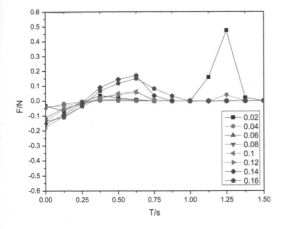

Figure 8. The force on the right blade in the Y direction.

different distances from the top to the root of the blade. The horizontal axis represents the distance from the top to the root of the blade. The vertical axis is the value of the total force. To describing conveniently, the initial position of the left blade is defined at 0°, while the right side is −30°.

Figure 5 shows the forces on the left blade in the X component. At the beginning of discharging, the forces in different locations are negative. The maximum occurs at the root of the blade. The forces decrease with time, reducing nearly to zero at 0.75 s. Then, the value increases gradually along the positive direction of the X axis, reaching the maximum at 1.25 s. Finally, the forces reduce to zero again at 1.5 s.

Figure 6 represents the forces on the right blade in the X component. The forces in different locations are positive at the start of discharging. The maximum occurs in the location of 0.14 m. The forces are close to zero at 1 s, which is delayed by 0.25 s when compared with the left blade because of 30° included angle. Then, the force reaches up to 0.35 N at 120° because of the force chain between particles, as shown in Figure 4.

Figure 7 indicates the forces on the left blade in the Y component. All forces on the blade are negative during the discharging process. At the beginning of discharging, the forces are near to zero. Then, they increase along the negative direction of the Y axis, running up to a maximum at 0.625 s. Theoretically, the maximum should occur at 0.75 s when the blade is in horizontal direction. Particle' compressive forces at 0.75 s are larger due to the internal surface of the valve body, which lead to the acting force being less. Then, the forces decrease, reaching the minimum at 1.5 s.

Figure 8 shows the forces on the right blade in the Y component. The forces are negative when the blade's location is −30° at 0 s. Then, they reduce to zero at 0.25 s in the position of 0°. After that, the forces increase along the positive direction of the Y axis, running up to a maximum at 0.625 s. Then, they reduce to zero in the position of 90°. First, the value increases before 1.25 s and finally reduces to zero.

5 SIMULATION OF ROTOR'S STRESS AND DEFORMATION

The load data from EDEM are loaded into the ANSYS workbench to calculate the rotor's stress and deformation. The load distribution of the rotor at 0 s in the ANSYS workbench is shown in Figure 9.

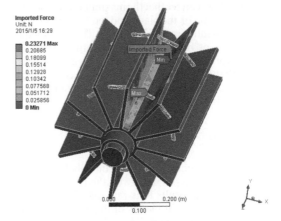

Figure 9. Load distribution.

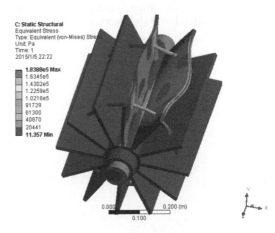

Figure 10. Stress nephogram of the blade.

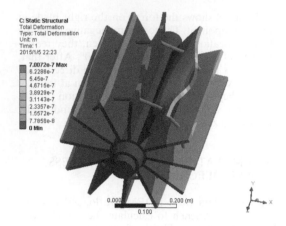

C: Static Structural
Total Deformation
Type: Total Deformation
Unit: m
Time: 1
2015/1/5 22:23

7.0072e-7 Max
6.2286e-7
5.45e-7
4.6715e-7
3.8929e-7
3.1143e-7
2.3357e-7
1.5572e-7
7.7858e-8
0 Min

0.000   0.200 (m)
      0.100

Figure 11.   Deformation nephogram of the blade.

From the calculated stress nephogram of Figure 10, we can see that the maximum stress occurs at the junction of the blade and reinforcement. The maximum value is 0.16345 MPa. The stress value at the root of blade is far less than the maximum. The deformation nephogram in Figure 11 illustrates that the maximum deformation is at the top of the blade in the middle. The maximum reaches 6.2206×10^{-7} m.

6   CONCLUSIONS

1. Particles' acting force on the blades of the rotary valve is obtained with the Hertz-Mindlin model. The results illustrate that the forces on the left and right blades in the X and Y direction are all different. The value of the left and right blade in the X direction decreases with time; however, the left one increases along the opposite direction after rotating 90°. The value in the Y direction mainly acts on the left blade,

adding up to a maximum when rotated to the position of 75°.
2. The load data from DEM are imported into the ANSYS workbench to calculate the rotor's stress and deformation. The maximum stress occurs at the junction of the blade and reinforcement, while the maximum deformation is at the top of the blade in the middle.

REFERENCES

Boateng, A.A, & Barr, P.V. 1996. Modeling of particle mixing and segregation in the transverse plane of a rotary kiln. *Chemical Engineering Science,* 51(17): 4167–4181.
Britton, P.F. Sheehan, M.E. & Schneider, P.A. 2006. A physical description of solid transport in flighted rotary dryers. *Powder Technology,* 165: 153–160.
Chaudhuri, B. Mehrotra, A. Muzzio, J. & Silvina, T.M. 2006. Cohesive effects in powder mixing in a tumbling blender. *Power Technology,* 165: 105–114.
Chaudhuri, B. Muzzio, J. & Silvina. T.M. 2006. Modeling of heat transfer in granular flow in rotary vessels. *Chemical Engineering Science,* 61: 6348–6360.
Chengyan, G. 2012. Static and dynamic analysis of rotary valve in thermal and mechanical loads, *Shanghai,* 44–51.
Cundall, P.A. & Strack, O.L. 1979. A discrete numerical model for granular assembles. *Geotechnique,* 29(1): 47–65.
Duchesne, C. Thibault, J. & Bazin, C. 1996. Modeling of the solids transportation within an industrial rotary dryer: a simple model. *Industrial & Engineering Chemistry Research,* 35: 2334–2341.
Savaresi, M. Bitmead, R. & Peirce, R. 2001. On modeling and control of a rotary sugar dryer. *Control Engineering Practice,* 9: 249–266.
Sheehan, M.E. Britton, P.F. & Schneider, P.A. 2005. A model for solids transport in flighted rotary dryers based on physical consideration. *Chemical Engineering Science,* 60: 4171–4182.
Wang, G.J. Hao, W.J. & Wang, W.X. 2010. *Discrete element method and its application in EDEM.* Northwestern Polytechnical University press, 16–18.

Advanced Materials and Structural Engineering – Hu (Ed.)
© *2016 Taylor & Francis Group, London, ISBN 978-1-138-02786-2*

Behavior of prestressed composite steel-concrete beam during assembly: Numerical modeling

M. Karmazínová & T. Vokatý
Faculty of Civil Engineering, Brno University of Technology, Brno, Czech Republic

ABSTRACT: This paper presents the basic results of the FEM numerical modeling of prestressed composite steel-concrete beam during the assembly stage. The aim of this paper is to analyze stresses and deformations of the particular type of the prestressed beam using the FEM numerical model, especially from the viewpoint of the influence of prestress affecting the steel beam in the phase of assembly. It is one of the possible ways how to carry out the prestress of composite steel-concrete beam, which means applying the prestress of steel part (steel beam only), before the casting of the concrete slab. The results of the calculation obtained from the numerical model created using the software of DLUBAL are compared with the usual engineering procedures of the calculation based on the rules given in the European Standards.

1 INTRODUCTION

Composite steel-concrete beams are of long term and commonly being used for bridge structures, for example approximately over the last 30 years for ceiling structures of buildings. Recently, there is the endeavor to apply high-strength steels and high-performance concretes in such composite structures, to increase their resistance. Therefore, the usage of high-strength steel in the composite beam can lead to an increase in their load-carrying capacity. On the other hand, the modulus of elasticity of steels is the same, regardless of the steel grade given according to the steel strength. The elasticity modulus of concrete is also increasing more slowly than its strength. For these reasons, the usage of high-strength materials does not always cause a significant increase in the flexural stiffness, nor does it cause a decrease in beam deflections.

The possible method, how to increase the resistance of mentioned composite beams, can also be the prestress of steel beam. In the mentioned case, the so-called stiff prestressing reinforcement is applied as one of the useable possibilities. The stiff reinforcement can be, for example, represented by a high-strength steel rod, located under the bottom flange (see Fig. 1). For the solution, the methods based on the principles of the structural mechanics, elasticity and plasticity can be used. However, for the detailed investigation of the stress distribution in the steel beam and concrete slab, the finite element method can be suitably applied. In this connection, the paper shows some results of the FEM solution compared with the analytical methods.

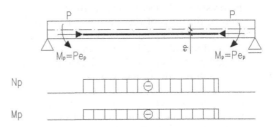

Figure 1. Forces in the beam given by the effect of prestress.

2 DESIGN OF PRESTRESSED COMPOSITE STEEL-CONCRETE BEAM

2.1 *Basic principles of prestressing*

The prestressing, as a general technology in structural engineering, is commonly used in bridge and structural engineering over the last 50 years. The technology of prestressing is very widespread in the area of concrete structures, but less in the case of steel structures. The essence and aim of the prestress is to obtain the reserve capacity, which should lead to increasing the load-carrying capacity.

In the process of the design of composite steel-concrete beams, the design conditions of the reliability must be satisfied for all load combinations and for all assembly stages, and within the both limit states, i.e. in the Ultimate Limit State (ULS), as well as in the Serviceability Limit State (SLS).

An efficient structural and static design of the beam, which satisfies the conditions of the ultimate limit state, does not always have to satisfy the conditions of the serviceability limit state also. This problem is usually caused by the excessive deflection of the beam (e.g. just in the assembly stage). The most simple and commonly used solution of this problem is to increase the dimensions of steel beam, which could, however, lead to less effective static design of the beam (smaller utilization of the cross-section) from the viewpoint of the ultimate limit state. This problem is more evident in the case of using high strength steel, because the modulus of elasticity is the same as for commonly used steel, e.g. S235. There are several approaches how to decrease the deflection of steel beam, e.g. the supporting of the beam in the assembling stage leading to the shortening of its span during concreting, or camber of the beam. Unfortunately, these steps do not always have to lead to the effective design of composite beam, but they can also cause the problems during assembly. Therefore, finally, these steps usually do not lead to the increase in the load-carrying capacity of the beam.

The prestress of steel beam, i.e. the prestress of only one of the parts of the composite beam before concreting, can be a possibility how to increase the flexural stiffness of steel beam during the assembly stage and how to increase the load-carrying capacity of the composite structure as a whole. One of the prestress methods, which can bring the contribution for the increase in load-carrying capacity, is based on the prestress of steel beam using high-strength steel external bars parallel with the bottom flange of the steel beam. Thus, the beam is prestressed before concreting, and needs no supports during assembly.

2.2 Design principles

The static and structural design of the prestressed composite steel-concrete beam depends on many factors. One of the advantages of prestress is the active resistance acting against the load. The effect of prestress on the beam has to create a reserve for some portion of the load, e.g. 100% of the dead load, so the prestressing force depends on the load value. The force in the prestressing rods has an eccentricity that leads to the negative bending moment and camber of the beam. The increase in the load-carrying capacity and flexural stiffness (decrease of the deflection) is thus basically caused by that effect. For the design of prestressed composite steel-concrete beams, the method of equivalent load, which is usually used for prestressed concrete structures, could be applied (Nie et al. 2011, Li et al. 1995). The principle of this method is in the substitution of prestressing effects by external force (see Fig. 1).

The value of bending moment caused by the prestressing has to eliminate some predetermined value of the load. The load value, which has to be reduced by the effect of prestress, is only one of the several boundary conditions, as well as the distance between the starting point and the endpoint of prestressing rods.

2.3 Actual model of prestressed composite steel-concrete beam

For the purpose of theoretical and future experimental studies, the prestressed composite steel-concrete beam, typical for the floor structures of usual buildings, is designed. Such prestressed composite beams could be potentially applied to the floor structures of larger spans with higher loading, e.g. in parking houses, administrative buildings, libraries.

As the default conditions, the values of the total load and span are chosen. The beam span depends on the test room possibilities; therefore, the span is chosen as L = 8000 mm. The permanent load (including the weight of the beam and floor) is g = 3.0 kNm^{-2} and the transient load is q = 5.0 kNm^{-2} (for parking houses, for example, according to the relevant standard). Depending on these conditions, steel-concrete beam is composed of steel beam of IPE 300 cross-section and steel grade of S 235, and the concrete slab of the effective width of b = 2 000 mm and the thickness of h = 120 mm with concrete class C 20/25. The value of the prestressing force is P = 200 kN.

The circular cross-section with the diameter of 30 mm with threads on ends is chosen for the

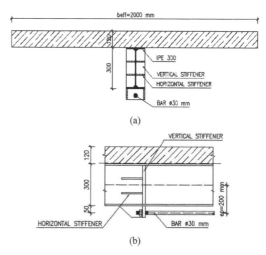

Figure 2. a) Cross-section of the actual model of prestressed composite steel-concrete beam; b) Detail of the anchorage of the prestressing rod to the steel beam.

prestressing rod. The material of the rod is steel of the maximum nominal tensile strength of 600 MPa. The rod is anchored at the distance of 900 mm from the supports of the beam on both ends (see Fig. 2). The anchorage of the rod is realized by the stiffening plates (horizontal and vertical). The rod is anchored by the contact between the vertical stiffener and the washer with nut (see Fig. 2b).

The prestressed composite steel-concrete beam is exposed to many load combinations in particular assembly stages. The main assembly stages are as follows: the prestress of steel beam using the rods before concreting, and further the load of prestressed steel beam, caused by wet concrete during concreting. In both cases, the steel beam is stressed by the combination of normal force (i.e. compressive force) and hogging bending moment. In these assembly phases, the steel beam has to satisfy the reliability conditions from the viewpoint of the strength and stability.

3 NUMERICAL MODEL

The stage of assembly is the critical phase for the design of the steel beam. The beam is exposed to the combination of normal force and bending moment, which can cause the compression buckling and lateral torsional buckling. Normal force and bending moment caused by prestress do not act on all of the length of the beam, but only on the distance between the anchorages of the prestressing rod (see Fig. 1). The assessment of the beam is made using the formulas mentioned in the standards (see e.g. (EN1993 2008, EN1994 2008)), but usually does not lead to the correct results. However, the numerical modeling by the FEM method could bring better information about the stresses depending on the prestress steel beam during the assembly stage. The other reason for the utilization of the numerical model is to give the base for the design of the suitable test specimens for experimental verification in the testing room.

The numerical model is created utilizing the software of RFEM created by the DLUBAL Company. Steel beam is modeled using 2D finite elements. Special attention was paid to the modeling of the stiffeners in the area of anchorage, where the loads from the prestressing rods are transferred.

The prestressing rods are also modeled using 2D elements. The prestressing rod is modeled using the 1D element—the beam. The ends of these elements are connected to the stiffeners through the hinges in the created node. This solution is designed as an easy approach how to join the 1D element to the 2D element. The disadvantage of this method can be singularity occurring in the output data. This problem can be partly solved by mesh refinement in critical areas. The mesh is created using triangular and square finite elements. The supports of the beams are modeled as for the simple beam without the rotation around the longitudinal axis. The bottom and upper flanges are supported in the direction of the x axis (see Fig. 3).

Also, the imperfections are included in the numerical model. All real imperfections are substituted by the equivalent geometrical imperfections, which included both global and local imperfections. The geometrical imperfection is substituted as the initial curving of the beam, according to EN 1993 2008. The initial eccentricity e_0 depends on the buckling curve type. In the case of steel profile of IPE 300 made of S235 steel, for the buckling about the z axis (weak), the buckling curve of b is assigned (EN 1993 2008). The initial eccentricity is taken as $e_0 = L/250$. The initial bow imperfection of the beam in the direction of the weak axis is introduced using an equivalent horizontal load given as $q_{z,im}p = (8\ N_{Ed}\ e_0)/L^2$.

The value of e_0 is calculated as $e_0 = 32$ mm. Then, the corresponding horizontal force is $q_{z,imp} = 0.8$ kNm^{-1}. The behavior of the material of steel beam and prestressing rod is considered as

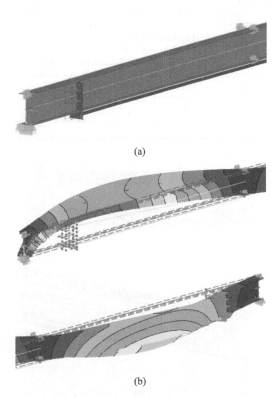

(a)

(b)

Figure 3. a) One half of modeled steel beam with prestressing rod during the assembly stage; b) Initial bow imperfection of beam in the direction of the weak axis.

linear elastic and isotropic. The prestressing force is input as the final prestress to the rod modeled as the 1D element. The corresponding horizontal force $q_{z,imp}$ is input as the uniform load on the line of the bottom flange modeled by the 2D element. The numerical calculation is made as nonlinear based on the large deformation analysis. For the solution of nonlinear algebraic equations, the Newton-Raphson method is chosen.

The results of the numerical calculation are listed in Tables 1 and 2. The distribution of the normal

Table 1. Comparison of normal stresses obtained from the FEM model and analytical calculation.

Normal stress	Calculation method	Upper flange	Web	Bottom flange
σ_x [MPa]	FEM numerical model	+14/+39	+29/−109	−67/−150
	Analytical model	+26	+26/−194	−194

Table 2. Beam deformations obtained from the FEM model.

Deformation [mm]	Horizontal direction		Vertical direction
	Longitudinal deformation u_x	Transverse deformation u_y	Vertical deflection u_z
	2.7	4.4	17.0

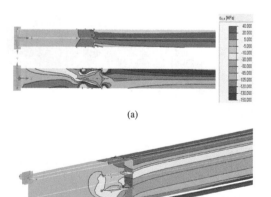

(a)

(b)

Figure 4. Normal stress σ_x during the assembly stage with prestress influence: a) upper, bottom flange, b) web.

Figure 5. Deflection (camber) of the beam caused by the effect of prestress during the assembly stage.

stresses on the beam is shown in Figure 4 (4a, 4b). The details of normal stresses in the anchorage area are shown in Figure 4b. The value of stress approaches to the singularity in several points, e.g. in the area of the prestressing rod anchorage and in the area of the support.

The deflections of the beam caused by the prestress effect during the assembly stage, which correspond with the stresses mentioned in Figure 4, are shown in Figure 5. The values of all deformations, i.e. not only vertical deflections, but also horizontal deformations (transversal and longitudinal), are listed in Table 2 (see "Summary" below).

4 ANALYTICAL MODEL

In parallel with numerical modeling, the analytical calculation according to EN1993 2008 is made. The bending moment causes the compression in the bottom flange of steel cross-section; thus, it is necessary to consider the lateral torsional buckling, and therefore the steel beam is assessed according to Equations (1) for the combination of the buckling with the bending moment including lateral torsional buckling (EN1993 2008). In addition, normal stresses on the upper and bottom edges of the cross-section are assessed using Equation (2):

$$\frac{N_{Ed}}{N_{b,Rd}} + k_{yy}\frac{M_{y,Ed}}{M_{y,b,Rd}} \leq 1.0 \quad (1)$$

and

$$\frac{N_{Ed}}{N_{b,Rd}} + k_{zy}\frac{M_{z,Ed}}{M_{z,Rd}} \leq 1.0,$$

$$\sigma_x = \frac{N_{Ed}}{\chi_z\,A} \pm \frac{M_{y,Ed}}{\chi_{LT}\,I_y}z \pm \frac{M_{z,Ed}}{I_z} \leq f_{yd}. \quad (2)$$

According to the above-mentioned formulas, normal force and bending moment caused by prestressing that act in a specific distance from the support of the beam may be neglected. The aim of the analytical calculation is to find the difference between the results obtained from the numerical

FEM modeling and those obtained from the analytical calculation, with neglecting the locations of acting prestressing force.

5 CONCLUSIONS

The values of normal stresses σ_x obtained by the numerical modeling and the analytical calculation are listed in Table 1. The comparison of the analytical and numerical results served as an answer to the question, whether the analytical calculation is apposite for the assessment of the prestressed composite steel-concrete beam during the assembly stage. Analytical calculation is also a function of the control framework for numerical calculation, yet before the experimental results will be obtained and correctly evaluated.

The accordance between stresses obtained from the numerical and analytical calculations in the upper flange (in tension) is very good. The stress values in bottom flange (in compression) obtained from the numerical calculation compared with the analytical calculations show larger differences. The higher stress values in the bottom flange obtained from the analytical calculation are probably caused by the higher degree of the reliability determined according to the standards. For the derivation of correct conclusions, the results of numerical FEM models should be verified, helping the loading tests. The conclusions mentioned here should be understood as partial results, which have the aim to study the behavior of prestressed steel beam during the assembly stage by the numerical calculation. The knowledge obtained from the numerical analysis will be used for the preparation of the test specimens and subsequent experimental verification.

In Table 2, the deformations corresponding to the normal stresses (see Table 1) calculated by the FEM model are only listed. From these deformations, especially the vertical deflections, it is evident that their values reach approximately the values of L/470, which is, for example, on the level of about one half of the usual limit for the deflections of ceiling structure.

ACKNOWLEDGMENT

This paper was elaborated within the solution of the university specific research project No. FAST-S-14-2544, funded by the Ministry of Education, Youth and Sports of the Czech Republic.

REFERENCES

EN 1990. 2004. *Basis of Structural Design*, CEN: Brussels.
EN 1991. 2004. *Loading Actions*, CEN: Brussels.
EN 1993, 2008. *Design of Steel Structures*, CEN: Brussels.
EN 1994, 2008. *Design of Steel and Concrete Composite Structures*, CEN: Brussels.
Li, W.L. Albrecht, P. & Saadatmanesh, H. 1995. Strengthening of composite steel-concrete bridges, *Journal of Structural Engineering, American Society for Concrete Engineering ASCE*, 121(12): 1842–1849. ISSN 0733-9445.
Nie, J.G. Tao, M.X. & Li, S.J. 2011. Analytical and numerical modelling of prestressed continuous steel-concrete composite beams, *Journal of Structural Engineering, American Society Concrete Engineering ASCE*, 137(2): 1405–1418. ISSN 0733-9445.
Vokatý, T. 2013. *Strengthening of steel and steel-concrete floor structures*, In Proceedings of the WTA Conference, Brno University of Technology—Faculty of Civil Engineering, 6, Brno (2013).

Advanced Materials and Structural Engineering – Hu (Ed.)
© 2016 Taylor & Francis Group, London, ISBN 978-1-138-02786-2

Study on EME denoising based on adaptive EEMD and Improved Wavelet Threshold

Z. Yang, J. Cai, G.F. Liu, X. Li & D.D. Ye
College of Electrical and Engineering Control, Liaoning Technical University, Huludao, China

ABSTRACT: In order to make up numerous shortcomings of traditional wavelets denoising of electromagnetic radiation signals, we propose to take the combination of adaptive EEMD and improved wavelet threshold algorithm as the denoising method. The new algorithm not only solved the mode mixing problem of EMD efficiently, but also overcame the shortcomings that threshold de-noising is selective to wavelet bases. IWT, EMD, EMD-Wavelet and EEMD-IWT are exerted to two kinds of MATLAB inner signals with adding noise to simulate denoising and carry out a case study using EME data collected from loaded coal and rock by full frequency-domain antenna. The results show that the denoising effect with EEMD-IWT is obviously increased.

1 INTRODUCTION

The electromagnetic emission prediction of technology is a kind of promising non-contact method that can effectively forecast the earthquake and coal and dynamic rock disasters. EME signal, a kind of electromagnetic wave, is inevitably interfered by noise in the experiment and field test. Therefore, how to get the real signal feature through effectively handling the electromagnetic signal is one of the key investigations. At present, the commonly used EME signal denoising methods are wavelet denoising, and fast CIA denoising (Nie et al. 2006, Yang et al. 2014, Yang et al. 2015).

The effective signal amplitude of actually measured electromagnetic radiation signals under field conditions is small, the ambient noise would submerge the true signal generally; in this situation, it is not ideal to only adopt the wavelet algorithm. Empirical Mode Decomposition (EMD) put forward by Huang et al. (1998) is a new method of data processing. It has the capability of linearization and stationary processing to nonlinear, nonstationary signal, which makes up for the shortage of wavelet bases selectivity of wavelet denoising. It can decompose a complex signal into a number of intrinsic mode functions arranged by high and low frequencies, can reconstruct the original signal effectively and losslessly. However there is also a drawback: if interval signals, pulse interference and noise exist in the signals, the IMF will generate the mode mixing problem (Meng 2013). Mode mixing would make EMD lose the capacity of decomposing non-stationary signals into stationary signals.

This article improves the existing electromagnetic radiation signal denoising method in view of its shortcomings, and puts forward a new algorithm that combines the Adaptive Ensemble Empirical Mode Decomposition (Adaptive EEMD) and the Improved Wavelet Threshold (IWT). The algorithm has the advantages of Adaptive EEMD and IWT, and is used into electromagnetic radiation signals collected on the spot so as to explore new ways of electromagnetic radiation signal denoising.

2 ADAPTIVE EEMD METHOD

The principle of EEDM is to add several times white noise, conduct Empirical Mode Decomposition (EMD) to the signal combined with measurement signal and noise. Figure 1 shows the flow diagram of the EEMD algorithm.

EMD adaptively decomposes nonlinear and non-stationary signals into a finite frequency band from a high frequency to low frequency, that is IMF. The frequency components and bandwidth of the frequency bands are not fixed values, but are determined by the local features of the signals (Kopsinis & McLaughlin 2009).

The size and set average number of auxiliary white noise in EEMD are the key parameters that decide the effect of eliminating the mode mixing problem and offsetting added auxiliary white noise. Generally, most of the literature choose $N = 100$, $\alpha = 0.2$, but the evaluations of N and α have different effects on different signals. In order to reduce interference errors from auxiliary white noise acting on the final EME signal to a greater degree and eliminate mode mixing more effectively, the Adaptive EEDM algorithm is designed to improve the accuracy of the signal decomposition.

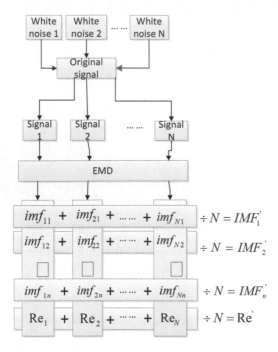

Figure 1. EEMD algorithm flow diagram.

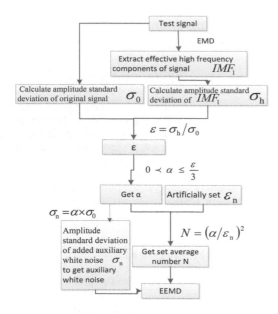

Figure 2. Adaptive EEMD algorithm flow diagram.

The algorithm can determine the corresponding N and α according to the characteristics of measured EME waves.

Figure 2 shows the flow diagram of the Adaptive EEMD algorithm. To obtain the amplitude standard deviation of high-frequency components σ_h, IMF1 decomposed from original noise by EMD is chosen as high-frequency components of the original signal approximately; the literature (Chen et al. 2011) proves that even mode mixing appears in IMF1, and the impacts on the calculation is negligible.

The literature (Chen et al. 2011) has contrasted EEMD with Adaptive EEMD on the multi-frequency signal and then concluded that the adaptive EEMD algorithm has good adaptability and is capable to decompose complex noisy signals containing surge components; meanwhile, the amount of the remaining auxiliary white noise is next to zero through decomposition, greatly improving the accuracy of signal decomposition consequently, so that the effect of signal denoising is improved.

3 IMPROVED WAVELET THRESHOLD ALGORITHM

Commonly used threshold processing methods are hard threshold and soft threshold algorithms (Zhang et al. 2014). In order to improve the shortcomings of the soft and hard threshold methods and keep its advantages, a new threshold function is constructed as follows:

$$
\hat{w}_{j,k} = \begin{cases} w_{j,k} - \lambda(1-e^{-N}) & w_{j,k} \geq \lambda \\ \beta w_{j,k}^{2} & 0 \leq w_{j,k} < \lambda \\ w_{j,k} + \lambda(1-e^{-N}) & w_{j,k} \leq -\lambda \\ -\beta w_{j,k}^{2} & -\lambda \leq w_{j,k} < 0 \end{cases} \quad (1)
$$

Here,

$w_{j,k}$—wavelet transform coefficients;
$\hat{w}_{j,k}$—processed wavelet transform coefficients;
λ—threshold value, $\lambda = \sigma\sqrt{2\ln N}$, σ is the mean square error of noise, N is the length of the signal; and
β—form factor, $\beta < 1/\lambda$, β is used to control the shape of threshold function in the area of $w_{j,k} < \lambda$, that is to control the degree of attenuation.

4 COMBINATION DENOISING ALGORITHM OF ADAPTIVE EEMD AND IWT

This article combines Adaptive EEDM and IWT, and the new algorithm has advantages of the two denoising algorithms. The steps for combining the denoising algorithm of adaptive EEMD and the improved wavelet threshold are as follows:

1. Decompose the original signal into several IMFs from high frequency to low frequency

IMF_1', IMF_2' ... IMF_k' ... IMF_n' and remainder Re' by the adaptive EEMD algorithm;

2. The energy of noise mainly focuses the on signal's high frequency, set the first k IMFs as dominant modes, conduct improved wavelet threshold denoising to IMF_1', IMF_2', ... IMF_k', so as to get IMF_1', IMF_2', ... IMF_k', set $k = n/2$ based on the result;

3. Reconstitute the signal with denoised dominant modes IMF_1'', IMF_2'', ... IMF_k'' and IMF_{k+1}'' ... IMF_n'', Re', so that denoised waves are obtained.

5 DENOISING SIMULATION EXPERIMENT OF NOISE-ADDED SIGNAL

In order to verify the effectiveness of the new denoising algorithm, we make use of MATLAB

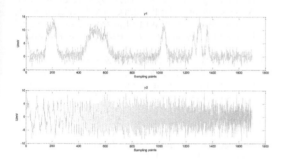

Figure 3.   Corrupted signals.

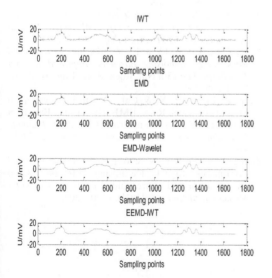

Figure 4.   Denoising oscillogram of noise-added bumps signal.

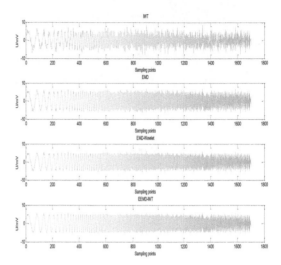

Figure 5.   Denoising oscillogram of noise-added Quad-chirp signal.

inner function to generate corrupted signals with different Signal Noise Ratios (SNR), which include y1 (Bumps) and y2 (Quadchirp), and the length is 1800, add noise with a certain SNR to them, and then, respectively, adopt IWT, EMD, EMD-Wavelet and EEMD-IWT to study denoising simulation. Corrupted signals are shown in Figure 3.

Figures 4 and 5 show the denoising oscillogram of adopting IWT, EMD, EMD-Wavelet and EEMD-IWT to noise-added signals y1 (Bumps) and y2 (Quadchirp). All the algorithms select the layer 3 and the wavelet db6. The comparison results indicate that wavelet denoising apply to signals with a higher SNR, and the EMD, EMD-Wavelet and EEMD-IWT under both high and low SNR have more stable denoising effect, EEMD-IWT has a perfect denoising effect under both high and low SNR.

6 EME DENOISING EXPERIMENT RESEARCH

In order to verify the EEMD-IWT denoising effect of EME denoising, we adopt full frequency-domain antenna to collect EME data generated from loaded coal and rock, and magnify received EME by the self-designed amplifying circuit, and take the magnified signals as the sample to verify the denoising effect. As is shown in Figure 6, respectively, IWT, EMD, EMD-Wavelet and EEMD-IWT were exerted to denoise the collected EME signals, and the effect of the four denoising methods were measured by SNR and MSE (Mean Square sampled Error). Table 1 presents the SNR and MSE of the noisy

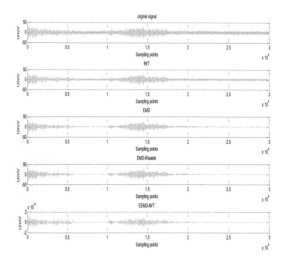

Figure 6.   Denoising comparison of the EME signal.

Table 1.   Comparison effects of different denoising methods.

Denoising method	SNR	MSE
IWT	5.6234	9.3423
EMD	10.2231	5.4326
EMD-Wavelet	13.4124	2.5856
EEMD-IWT	17.5641	0.5256

signals processed by four denoising algorithms. The SNR increased from 5.6234 by IWT to 17.5641 by EEMD-IWT. The MSE decreased from 9.3423 by IWT to 0.5256 by EEMD-IWT. Thus, the denoising effect is improved significantly.

7   CONCLUSIONS

1. Combine adaptive EEDM and IWT to denoise the EME signals. The new algorithm has not only efficiently solved the mode mixing problem of EMD, but also overcame the shortcomings that threshold denoising is selective to wavelet bases.
2. Exert IWT, EMD, EMD-Wavelet and EEMD-IWT, respectively, to the signals of Bumps, and Quadchirp to conduct denoising simulation

research. EEMD-IWT has a perfect denoising effect under both high and low SNR.
3. Exert IWT, EMD, EMD-Wavelet and EEMD-IWT, respectively, to denoise the collected EME signals. The results show that the denoising effect using EEMD-IWT is improved significantly.

ACKNOWLEDGMENT

This work was supported by the National Natural Science Foundation of China (51204087, 51274114) and the Funding Project of School Market Research Fund (SCDY2013023, 14-T-006).

REFERENCES

Chen, J. 1999. *Machine Vibration Monitor and Faults Diagnosis*, Shanghai: Shanghai Jiao Tong University Press.
Chen, L. Tang, G.S. Zi, Y.Y. Feng, Z.N. & Li, K. 2011. Application of Adaptive Ensemble Empirical Mode Decomposition Method to Electrocardiogram Signal Processing, *Journal of Data Acquisition & Processing*. 03: 361–366.
Huang, N.E. Wu, M.L. Long, S.R. & Shen, Z. et al. 1998. The empirical mode decomposition and the Hilbert spectrum for nonlinear and non-stationary time series analysis, *Proceedings of the Royal Society of London*. 454A: 903–993.
Kopsinis, Y. & McLaughlin, S. 2009. Development of EMD-Based Denoising Methods Inspired by Wavelet Thresholding, *IEEE Transactions on Signal Processing*. 04: 1351–1362.
Meng, F.L. 2013. *Research on Theories and Applications of Ensemble Empirical Mode Decomposition*, Jiangsu University Science Technology.
Nie, B.S. He, X.Q. He, J. & Zhai, S.R. 2006. Study on wavelet transform de-noising of EME signal, *Taiyuan Technology University Journal*. 05: 557–560.
Yang, Z. Li, X. Cai, J.Y. & Li, Y. 2014. *Research on Coal Rock EME De-noising Based on Improved Wavelet Transform*, The 13th Nat. Rock Mech. Eng. Academic Conf. 266–269.
Yang, Z. Li, Y. Li, X. & Cai, J.Y. 2015. Research on Method of Coal Rock EME De-noising Based on Improved Wavelet Transform, *Journal of Liaoning Technology University (Natural Science)*, 34(3): 410–413.
Zhang, P. Xie, Y. Sun, N. & Wei, Z. 2014. Wind Tunnel Continuous Signal De-noising Method Based on an Improved Threshold Function, *Computer Measurement Control*. 22(04): 1300–1302.

Advanced Materials and Structural Engineering – Hu (Ed.)
© *2016 Taylor & Francis Group, London, ISBN 978-1-138-02786-2*

Mathematical approaches to evaluating plastic hinge region of flexure-governed shear walls

K.H. Yang

Department of Plant Architectural Engineering, Kyonggi University, Suwon, Kyonggi-do, South Korea

ABSTRACT: This paper proposes a simple model for determining the potential plastic hinge length for seismic design of shear walls. From the idealized curvature distribution along the shear wall length, a fundamental equation of plastic hinge length was derived as a function of the yielding moment, maximum moment, and additional moment due to the diagonal shear crack. Using the generalized moment equations, the plastic hinge length of a shear wall was simply formulated as a function of the longitudinal tensile reinforcement index, vertical shear reinforcement index, and axial force index. The predicted plastic hinge lengths were in good agreement with test results, with the mean and standard deviation of the ratios of plastic hinge length between experiments and predictions as 1.02 and 0.10, respectively.

1 INTRODUCTION

Plastic hinges in a Reinforced Concrete (RC) shear wall generally develop in the vicinity of the critical region, where inelastic rotation occurs in addition to the elastic rotation at the ultimate state. Hence, the ductile behavior of shear walls significantly depends on the inelastic rotation capacity in the plastic hinge region (Park & Paulay 1933, Paulay & Priestley 1992, Kang & Park 2002). Considering this fact, most code provisions (ACI 318-11 2011, EN 1998-1 2004) critically deal with the potential plastic hinge length in seismic design of RC shear walls to determine the details of boundary element region and the amount of transverse reinforcement at the boundary element for a required ductility. However, the method used to determine the plastic hinge length in shear walls is still controversial because of the lack of information about the inelastic deformation capacity of the boundary element and reliable design models. The present study aimed to establish a reasonable and simple model for determining the potential plastic hinge length of shear walls.

2 SIMPLE MODEL OF PLASTIC HINGE LENGTH

2.1 Fundamental equation

At the ultimate state, the actual distribution of a curvature can be idealized as a bilinear shape with a kink point at the yielding moment, as shown in Figure 1 (a). Considering the additional curvature due to the diagonal shear crack, the bilinear

distribution of the curvature is transformed into a trilinear shape, as shown in Figure 1 (b). Ultimately, the total length of the plastic hinge region (l_p) is determined by the individual integration of inelastic rotation regions due to flexure and the diagonal shear crack. Hence, the l_p of a shear wall can be written as follows:

$$l_p = l_{p(flex)} + l_{p(shear)} \qquad (1)$$

where, $l_{p(flex)}$ is the length of inelastic rotation region under flexure and $l_{p(shear)}$ is the length of additional inelastic rotation region due to the diagonal shear crack. From the general moment–curvature relationship, $l_{p(flex)}$ can be expressed as follows:

$$l_{p(flex)} = h_w \left\{ 1 - \left(\frac{M_y}{M_n} \right) \right\} \qquad (2)$$

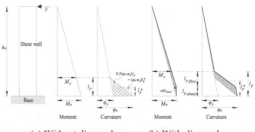

(a) Without diagonal shear cracks   (b) With diagonal shear cracks

Figure 1. Distribution of moment and curvature along shear wall length at the ultimate state.

where, M_y and M_n are the yielding and ultimate moment capacities, respectively, of the shear wall, and h_w is the wall height. As the additional curvature distribution due to the diagonal shear crack is idealized as a linear variation, $l_{p(shear)}$ is calculated from:

$$l_{p(shear)} = h_w \left(\frac{\Delta M_{shear}}{M_n} \right) \qquad (3)$$

where, ΔM_{shear} is the additional moment due to the diagonal shear crack. Substituting Equations (2) and (3) into Equation (1), the following equation is obtained:

$$l_p = h_w \left(1 - \frac{M_y}{M_n} + \frac{\Delta M_{shear}}{M_n} \right) \qquad (4)$$

The above equation indicates that l_p decreases with a decrease in the ratio of yielding and ultimate moment capacities of the shear wall, whereas it increases with an increase in the ratio of additional moment due to the diagonal shear crack and ultimate moment capacity. This implies that l_p depends on the ratio of neutral axis depths at yielding and ultimate states because the resultant forces of longitudinal tensile reinforcement remain constant after the yielding moment.

2.2 Yielding moment capacity (M_y) and ultimate moment capacity (M_n)

Calculations of a yielding moment and ultimate moment capacities are accompanied by extremely tedious and complex procedures owing to the confinement effect at the boundary element and differences in longitudinal reinforcement arrangements between the boundary element and the web. Hence, these capacities are calculated using Yang's simple model (Yang 2013) as follows:

$$M_y = 0.87 \lambda^{1.1} f'_c b_w d_w^2 \qquad (5)$$

$$M_n = 0.96 \lambda f'_c b_w d_w^2 \qquad (6)$$

$$\lambda = \omega_s + \omega_v^{1.3} + \omega_p^{1.4} \qquad (7)$$

where $\omega_s (= A_s f_y / f'_c b_w d_w)$ and $\omega_v (= A_v f_{yv} / f'_c b_w d_w)$ are the longitudinal tensile and vertical reinforcement indices, respectively, $\omega_p (= N_u / f'_c b_w d_w)$ is the axial load index, A_s, A'_s, and A_v are the areas of the longitudinal tensile and compressive reinforcing bars and vertical reinforcing bar, respectively, b_w is the wall thickness, d_w is the effective depth of the longitudinal tensile reinforcement, f'_c is the compressive concrete strength, and f_y and f_w are the yield strengths of longitudinal and vertical web reinforcements, respectively.

2.3 Additional moment due to diagonal shear crack (ΔM_{shear})

Based on the truss mechanism (Park & Paulay 1933), ΔM_{shear} is obtained from

$$\Delta M_{shear} = V_n \cdot e_v \qquad (8.a)$$

$$e_v = \left\{ \cot \alpha - 0.5 \left(\frac{V_n - V_c}{V_n} \right) (\cot \alpha + \cot \beta) \right\} jd \geq 0 \qquad (8.b)$$

where, V_c is the shear transfer capacity of concrete, V_n is the shear strength of the shear wall, which can be calculated using M_n/h_w for concentric lateral load at the free end, α and β are the angles of the diagonal crack and shear reinforcement, respectively, relative to the longitudinal axis of walls, and $jd (\approx 0.9(0.8 l_w))$ is the distance between resultant internal compressive and tensile forces. For the diagonal shear crack in a shear wall, the following assumptions (Oesterle et al. 1976, Oesterle et al. 1979, Shiu et al. 1981) can be made: shear transfer of vertical shear reinforcement along the diagonal shear crack is negligible and the inclination of the crack is close to 45°. Hence, for the lateral load applied to the free end of a shear wall, Equation (8) can be expressed as follows:

$$\Delta M_{shear} = 0.36(V_n + V_c) l_w \qquad (9)$$

where, l_w is wall length.

2.4 Simplification of additional moment for plastic hinge length

The plastic hinge length of a shear wall is calculated by substituting the values of M_y, M_n, and ΔM_{shear} into Equation (4). However, calculating ΔM_{shear} is accompanied by some tedious processes such as the determination of the shear transfer capacity of concrete and shear strength of the wall. To eliminate such processes and establish a simple equation for l_p, a parametric study was conducted to simplify the ΔM_{shear} equation. The equation of additional moment is formulated using these nondimensional parameters through Nonlinear Multiple Regression (NLMR) analysis. Each variable was coupled to the others and solved iteratively until a relatively high correlation coefficient (R^2) was obtained. The additional moment (ΔM_{shear}) is affected by l_w/h_w as well as nondimensional parameters (ω_s, ω_v, ω_p) and. From the NLMR analysis of these influencing parameters, ΔM_{shear} can be simplified as follows (Fig. 2):

$$\Delta M_{shear} = \left\{ 0.365\lambda^{0.85}\left(\frac{l_w}{h_w}\right)\right\} f'_c b_w d_w^{\,2} \qquad (10)$$

Finally, by substituting Equations (5), (6), and (10) into Equation (4), l_p is easily calculated from the following equations:

$$l_p = h_w\left(1 - 0.91\lambda^{0.1} + \xi_{shear}\right) \qquad (11.a)$$

$$\xi_{shear} = 0 \text{ for } V_n \le V_c \qquad (11.b)$$

$$\xi_{shear} = 0.38\lambda^{-0.15}\left(\frac{l_w}{h_w}\right) \text{ for } V_n > V_c \qquad (11.c)$$

where, ξ_{shear} is a factor for representing the ratio of $\Delta M_{shear}/M_n$.

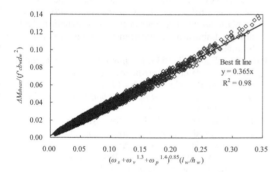

Figure 2. Regression analysis for simplifying the additional moment due to the diagonal shear crack.

2.5 Comparison of predicted and test results

The measured l_p data were compiled from available literature (Oesterle et al. 1976, Oesterle et al. 1979, Shiu et al. 1981, Adebar et al. 2007) and compared with those predicted using different equations, including the current model and previous empirical equations (Bohl & Adebar 2011, Paulay & Priestley 1993). The details of shear wall specimens and the ratios, $\gamma_{lp} = l_{p(Exp)}/l_{p(Pre)}$, of experiments and predictions are summarized in Table 1. Bohl and Adebar (2011) and Paulay and Priestley (1993) proposed l_p on the basis of the assumption that the areas of the equivalent plastic hinge region and inelastic region are equal. Furthermore, their equations do not consider the effect of the primary parameters on the inelastic rotation capacity of shear walls. Paulay and Priestley's equation does not consider the vertical reinforcement within the web and axial load in the influencing parameters in the regression analysis of beam data. Owing to these shortcomings, the previous equations are found to highly underestimate the experimentally obtained l_p values, showing that the values of γ_{lp} tend to decrease with an increase in ω_p. The means of the γ_{lp} values are 1.907 and 1.875 for Bohl and Adebar's equation and Paulay and Priestley's equation, respectively. The values of γ_{lp} determined from Bohl and Adebar's equation decrease as the longitudinal tensile reinforcement index and vertical shear reinforcement index increase. The predictions obtained from the model proposed in this paper are in better agreement with test results than what could be obtained using the other equations (Bohl & Adebar 2011, Paulay & Priestley 1993); the

Table 1. Comparison of measured and predicted plastic hinge lengths.

| Researcher | Specimen | Details of shear wall specimens | | | | $l_{p(Exp.)}/l_{p(Pre.)}$ | | |
		f'_c (MPa)	ω_s	ω_v	ω_p	Bohl and Adebar	Paulay and Priestley	This study
Oesterle et al.	R2	46.5	0.034	0.017	0	2.244	2.358	0.995
	B3	47.3	0.051	0.018	0	2.151	2.261	1.019
	B4	45.0	0.054	0.020	0	2.238	2.352	1.071
	B5	45.3	0.174	0.020	0	1.908	2.005	1.154
Oesterle et al.	B6	21.8	0.358	0.040	0.136	1.758	1.464	1.077
	B7	49.3	0.164	0.019	0.076	1.732	1.612	0.949
	B8	42.0	0.189	0.022	0.090	1.680	1.526	0.935
	B9	44.1	0.181	0.021	0.086	1.689	1.545	0.935
	B10	45.6	0.094	0.020	0.083	2.046	1.881	1.005
	F2	45.6	0.198	0.026	0.081	1.619	1.698	1.048
Shiu et al.	CI-1	23.3	0.153	0.034	0.010	2.125	2.204	1.203
Adebar et al.	1	49.0	0.023	0.018	0.150	1.692	1.594	0.833
Mean						1.907	1.875	1.019
Standard deviation						0.239	0.346	0.102
Coefficient of variation						0.126	0.184	0.100

mean, standard deviation, and coefficient of variation of γ_{lp} are 1.019, 0.102, and 0.100, respectively (Table 1). In addition, the variation in γ_{lp} is marginally dependent on ω_s, ω_v, and ω_p. Overall, the comparison of predictions with experimental results indicates that the effect of the primary parameters on the plastic hinge length of a shear wall is adequately accounted for the proposed model.

3 CONCLUSIONS

Using mathematical models to determine the plastic hinge length of a shear wall, the following conclusions were drawn:

1. The additional moment and curvature due to the diagonal shear crack were calculated using an analogous truss. The nonlinear distribution of curvature over the inelastic rotation region at the ultimate state was then idealized as a bilinear shape with a kink point at the yielding moment, whereas the bilinear shape was transformed into a trilinear shape when a diagonal shear crack occurred in the web of the shear wall.
2. Using the generalized moment equations, the plastic hinge length of the shear wall was simply formulated as a function of the longitudinal tensile reinforcement index of the boundary element, vertical shear reinforcement index of the web, and axial force index. The plastic hinge lengths predicted by the current simple model were in better agreement with test results than what could be obtained using the previous equations, with the mean, standard deviation, and coefficient of variation of the ratios of the plastic hinge length between experiments and predictions being 1.019, 0.102, and 0.100, respectively.

ACKNOWLEDGEMENT

This work was supported by a Kyonggi University Research Grant 2014.

REFERENCES

ACI Committee 318, 2011. *Building code requirements for structural concrete (ACI 318-11) and commentary*, Farmington Hills, MI, American Concrete Institute.
Adebar, P. Ibrahim, A.M.M. & Bryson, M. 2007. Test of a high-rise core wall: effective stiffness for seismic analysis, *ACI Structural Journal*, 104: 549–559.
Bohl, A. & Adebar, P. 2011. Plastic hinge lengths in high-rise concrete shear walls, *ACI Structural Journal*, 108: 148–157.
Kang, S.M. & Park, H.G. 2002. Ductility confinement of RC rectangular shear wall, *Journal of the Korea Concrete Institute*, 14: 530–539.
Oesterle, R.G. Aristizasbal-Ochoa, J.D. Fiorato, A.E. Russell H.G. & Corley, W.G. 1979. *Earthquake resistant structural walls—tests of isolated walls phase II*, Report to National Science Foundation, Construction Technology Laboratories, Portland Cement Association, Skokie, IL.
Oesterle, R.G. Fiorato, A.E. Johal, L.S. Carpenter, J.E. Russell H.G. & Corley, W.G. 1976. *Earthquake resistant structural walls—tests of isolated walls, Report to National Science Foundation*, Construction Technology Laboratories, Portland Cement Association, Skokie, IL.
Park, R. & Paulay, T. 1933. *Reinforced concrete structures*, New Jersey, Wiley Interscience Publication.
Paulay, T. & Priestley, M.J.N. 1992. *Seismic design of reinforced concrete and masonry buildings*, New York, John Wiley & Sons, Inc.
Paulay, T. & Priestley, M.J.N. 1993. Stability of ductile structural walls, *ACI Structural Journal*, 90: 385–392.
Shiu, K.N. Daniel, J.I. Aristizasbal-Ochoa, J.D. Fiorato, A.E. & Corley, W.G. 1981. *Earthquake resistant structural walls—tests of walls with and without openings*, Report to National Science Foundation, Construction Technology Laboratories, Portland Cement Association, Skokie, IL.
The European Standard, EN 1998–1:2004. *Eurocode 8: design of structures for earthquake resistance*, London, British Standards Institution.
Yang, K.H. 2013. *Development of performance-based design guideline for high-density concrete walls*, Technical Report (2nd year), Kyonggi University.

Advanced Materials and Structural Engineering – Hu (Ed.)
© 2016 Taylor & Francis Group, London, ISBN 978-1-138-02786-2

Design of an IOT-based intelligent control system for flower greenhouse

Z. Zhiyong & Z. Man
College of Mechanical and Electronic Engineering, Sichuan Agricultural University, Ya'an,
Sichuan Province, China

ABSTRACT: The parameters of flower greenhouse including atmospheric temperature and humidity, illumination intensity, carbon dioxide (CO_2) concentration and soil water content have direct influences on yield and quality of flowers. The monitoring system designed in this paper with the ZigBee network, embedded controller and 3G network as cores can realize intelligent adjustment of each parameter. With advantages of high precision and stability, the design of the sensor circuit mainly employs digital module sensors. In order to save energy, the sensor circuit is controlled by relay switch to work at the proper time. The gateway node is designed by employing a high-performance 32-digit embedded controller, and WinCE6.0 embedded OS is self-customized. Embedded SQlite database is realized on WinCE6.0 for effectively managing data. The closed loop control is realized by employing the fuzzy Neural Network Algorithm (NNA). The test result shows that the deviation of atmospheric temperature is controlled within ±0.6°C, that of atmospheric humidity is controlled within ±1.2%, that of illumination intensity is controlled within ±276 LUX, that of CO_2 concentration is controlled within ±23 PPM, and that of soil water content is controlled within ±0.9%. Thus all parameters fully meet the practical requirements of flower greenhouse.

1 INTRODUCTION

With constant improvement of people's life quality and increase of all kinds of ceremonies and rite, the market demand for flowers is increasing substantially, and currently it cannot be satisfied by the traditional modes of flower production (Wu et al. 2007, Wu et al. 2013). For the purpose of improving the production efficiency and quality of flower greenhouse, this paper designs a kind of intelligent monitoring system for flower greenhouses on the basis of IOT (internet of things).

As an emerging technology in the 21st century, great importance is now attached to the IOT technology worldwide. The IOT mainly consists of sensing layer, transport layer and application layer. The development of IOT technology will affect the development process of modern agriculture, and play an important role in the transformation of the whole agricultural production system (He et al. 2013).

2 OVERALL DESIGN OF THE SYSTEM

Main parameters that have an influence on the growth of flowers include atmospheric temperature and humidity, illumination intensity, CO_2 concentration and soil water content. In order to effectively detect and control each parameter, the intelligent monitoring of IOT, as shown in

Figure 1, is proposed (Han et al. 2009, Jennifer et al. 2008, Liu et al. 2012).

The system consists of sensor nodes, control nodes, gateway node, remote monitoring PC and cell phone terminal of a user, and each node is designed by taking CC2530 wireless SCM that follows the ZigBee protocol as core. With excellent sensitivity and anti-interference capacity and powerful GPIO interface, the CC2530 SCM is a kind of system-on-chip developed by American TI Corporation, which integrates an advanced RF transceiver with the industrial standard enhanced 8051CPU. The coordinators of the gateway node and the routing nodes are designed by employing

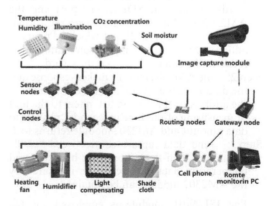

Figure 1. Structure of the intelligent monitoring system.

the minimum applied system circuit structure recommended by TI Corporation, and such circuit is simple and reliable. The sensor nodes are designed combining the sensor signal conditioning circuit. The control nodes are designed by combining relay driving circuit. The monitoring program of remote monitoring PC and that of the user's cell phone is developed with Java. The design of the CC2530 wireless SCM program is the secondary application and development on the basis of the TI Corporation's ZStack protocol stack. The sensor nodes and control nodes are powered by 12 V 5 AH storage batteries, and the routing nodes are powered by 3 V dry batteries; with the mobile power source as power, it can be used flexibly and the gateway node is powered by 12 V DC that is regulated by power adapter from a 220 V AC power source. In order to save energy, the power of the sensor circuit is controlled with relay switch, for which the power of the sensor circuit is switched on when the time is to capture and the power is cut off in the other time.

3 DESIGN OF SENSOR NODES

Employing a high-cost efficient AM2302 module, the atmospheric temperature and humidity sensor is a complex temperature and humidity sensor containing a calibrated digital signal output. This sensor applies a special digital module capture technology and humidity-temperature sensing technology to ensure ultra-high reliability and superior long-term stability of a product. The communication of the AM2302 module takes the mode of the SDI-12 bus protocol, i.e. bidirectional data transfer is realized on a piece of the data line. The CC2530 wireless SCM of sensor nodes reads humidity and temperature data by programming through its P1.0 port, and the data is obtained in 3 steps as follows:

Step 1: write in capture command control words through the SDI-12 bus port and the AM2302 module makes response, the format of control word is "aM!", of which 'a' refers to address, 'M' refers to capture command and '!' refers to the end mark of a control word.

Step 2: the AM2302 module makes response and the data format is "atttn", of which 'a' refers to the address of this sensor, 'ttt' refers to the unit of measuring of this time in seconds, 'n' refers to data amount and AM2302 module returns to 2.

Step 3: send data capture control word "aD0!" again, and then the atmospheric humidity and temperature data of greenhouse measured by the AM2302 module can be obtained.

The ISL29010 module is employed for the detection of illumination intensity. The ISL29010 module is an integrated photoelectric sensor with the I^2C interface. The CC2530 wireless SCM of the sensor node obtains illumination intensity data by writing control word into the I^2C interface of ISL29010 module through its P1.1 and P1.2 ports. The control words of this system's ISL29010 module are set as COMMAND $= 0 \times 88$, CONTROL $= 0 \times 0C$. In addition, the illumination intensity parameter of flower greenhouse is captured by reading the illumination intensity values at addresses of 0×04 and 0×05 of data register of the ISL29010 module.

The CO_2 concentration is measured with the AJD-VCO2 module, which is an integrated module with $0 \sim 5$ V DC output. The CC2530 wireless SCM of the sensor node captures the parameter of CO_2 concentration of flower greenhouse by connecting its P2.1 port with the analog voltage output port of the AJD-VCO$_2$ module, and the P2.1 port of CC2530 wireless SCM is the input port of its internal analog-digital conversion.

The FDS100 soil water content sensor module is employed. The output signals of the FDS100 module are current signals, so it is handled by transmitting-amplifying circuit of two levels, of which the first level uses precision transmitting-amplifying OP07 to convert current into voltage, and the second level uses high-gain transmitting-amplifying UA741 and DC to perform summation. The output voltage of UA741 is sent to the P2.2 port of CC2530 wireless SCM to carry out an analog-digital conversion, and then the parameter of soil water content of flower greenhouse is captured.

4 DESIGN OF CONTROL NODES

The control nodes aim to achieve mainly the control of heating fan, light-compensating lamp, humidifier and shade cloth. Four kinds of adjusting equipment are placed in different areas of flower greenhouse, so they are controlled with 4 pieces of CC2530 wireless SCMs. The P1.0 port of CC2530 wireless SCM drives 9013 triodes to control the relay switch, and control of the corresponding adjustment equipment is realized by the relay switch.

5 DESIGN OF GATEWAY NODE

As the bridge for the communication between the sensor nodes and control nodes and the remote monitoring PC, the gateway node is responsible for collecting sensor data and on-site management of the ZigBee network, so it is designed with the 32-digit S3C6410 embedded controller with strong data processing capacity as core. In order to improve the electromagnetic compatibility of circuit design and convenience of maintenance, the design thought of core-board plus baseboard is adopted for the network nodes. The core-board consists of

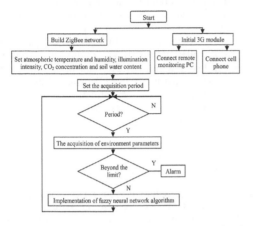

Figure 2. Flow chart of the intelligent monitoring system.

S3C6410 embedded controller, 256 GB SDRAM made by connecting two pieces of K4X1G163PC-FGC6 storage and 4 GB NAND Flash made by connecting two pieces of K9GAG08U0E-S storage. Besides, the baseboard consists of a power supply module, USB interface, JTAG interface, LED interface and URAT interface (Zou et al. 2014).

The UART interface of S3C6410 embedded controller is connected with the serial port of CC2530 wireless SCM for realizing communication among nodes of the ZigBee network. Meanwhile, the CC2530 wireless SCM on the gateway node serves as the coordinator of the ZigBee network.

The image capture module employs T6836 WIP, which captures images with 0.3 Mega pixel progressive CMOS sensor, and has the features of automatic white balance, automatic gain control, build-in controller and WiFi communication. By connecting the I^2C interface of S3C6410 embedded controller with the WiFi communication module, it will be able to capture the real-time image data of greenhouse.

In order to display intuitively and facilitate control, the display of AT070TN83V 7″ LCD touch screen is employed so as to realize real-time display and touch control.

Table 1. Test data of atmospheric temperature and humidity, illumination intensity, soil water content, and CO_2 concentration.

Time h	Temperature °C	Error	Humidity %	Error	Illumination LUX	Error
0	9.0	−0.5	68.5	0.5	8236	236
2	9.3	−0.2	67.3	−0.7	7842	−158
4	9.7	0.2	68.3	0.3	8090	90
6	10.1	0.6	68.1	0.1	7880	−120
8	9.2	−0.3	67.6	−0.4	8070	70
10	9.9	0.4	67.2	−0.8	7820	−180
12	9.7	0.2	69.2	1.2	8213	213
14	9.4	−0.1	69.1	1.1	7932	−68
16	9.8	0.3	67.5	−0.5	8099	99
18	9.1	−0.4	67.1	−0.9	7828	−172
20	9.3	−0.2	67.8	−0.2	8133	133
22	9.6	0.1	68.6	0.6	7724	−276

Time h	Soil %	Error	CO_2 PPM	Error
0	7.1	−0.4	415	15
2	8.1	0.6	406	6
4	6.7	−0.8	391	−9
6	7.2	−0.3	398	−2
8	8.2	0.7	412	12
10	7.7	0.2	407	7
12	6.6	−0.9	401	1
14	7.0	−0.5	392	−8
16	7.6	0.1	382	−18
18	7.8	0.3	423	23
20	6.9	−0.6	412	12
22	7.7	0.2	379	−21

WinCE6.0 embedded OS is self-customized with Platform in a design of the system program for gateway node. In order to manage data effectively, SQlite embedded database is transplanted toWinCE6.0 embedded OS. SQlite has the features of small size, open sources, low RAM requirement and convenient operation, two library files of SQlite.lib and SQlite.dll are added into application project of WinCE 6.0 and the operation of SQlite embedded database can be realized by calling API function of SQlite. Intelligent control of each parameter adopts fuzzy NNA to carry out closed loop control (Han & Junfei 2010). The flow chart of an intelligent monitoring system is shown in Figure 2.

6 TESTING OF THE SYSTEM

An intelligent monitoring system is set up as per the structure shown in Figure 1 for testing the system. On the gateway node, the environmental parameters of flower greenhouse are set as follows: atmospheric temperature 9.5°C, atmospheric humidity 68%, illumination intensity 8000 LUX, soil water content 7.5% and CO_2 concentration 400 PPM. The atmospheric temperature is adjusted with heating fan, the atmospheric humidity is adjusted with humidifier, the illumination intensity is adjusted with light-compensating lamp and shade cloth, the soil water content is adjusted by sprinkling irrigation, and the CO_2 concentration is adjusted by air inlet and outlet motors. Comparison of test data and data detected with standard instruments is presented in Table 1, and the data show that the deviation of atmospheric temperature is controlled within ±0.6°C, that of atmospheric humidity is controlled within ±1.2%, that of illumination intensity is controlled within ±276 LUX, that of soil water content is controlled within ±0.9%, and that of CO_2 concentration is controlled within ±23 PPM.

7 CONCLUSIONS

1. A kind of IOT-based intelligent monitoring system is designed for the flower greenhouse, the hardware design realizes sensor signal capture circuit for atmospheric temperature, atmospheric humidity, illumination intensity, soil water content, and CO_2 concentration and

gateway node; the design of the ZigBee network program realized secondary application and development on the basis of ZStack protocol stack of TI Corporation, WinCE6.0 embedded OS is self-customized on gateway node and SQLite embedded database is realized on WinCE6.0 for realizing effective data management.

2. Fuzzy intelligent NNA is realized on the gateway node. The closed loop control result shows that the deviation of atmospheric temperature is controlled within ±0.6°C, that of atmospheric humidity is within ±1.2%, that of illumination intensity is within ±276 LUX, that of soil water content is within ±0.9% and that of CO_2 concentration is within ±23 PPM, and the effect of control is favorable.

REFERENCES

Han, H.F. Du, K. & Sun, Z.F. 2009. Design and application of ZigBee based telemonitoring system for greenhouse environment data acquisition, *Transactions of the Chinese Society of Agricultural Engineering.* 25: 158–163.

Han, H.G. & Junfei, Q. 2010. A self-organizing fuzzy neural network based on a growing-and-pruning algorithm, *Fuzzy Systems, IEEE Transactions on.* 18: 1129–1143.

He, Y. Nie, P.C. & Liu, F. 2013. Advancement and Trend of Internet of Things in Agriculture and Sensing Instrument, *Transactions of the Chinese Society for Agricultural Machinery.* 44: 216–226.

Jennifer, Y. Biswanath, M. & Dipak, G. 2008. Wireless sensor network survey, *Journal of Computer Networks.* 52: 2292–2330.

Liu, D.H. Zhou, J.W. & Mo, L.F. 2012. Applications of Internet of Things in Food and Agri-food Areas, *Transactions of the Chinese Society for Agricultural Machinery.* 43: 146–152.

Wu, X.L. Zhao, H.Q. & Zhang, Y. 2007. Research on the effect of controlled environment on flowering timer regulator of protected cultivation of flowers, *Journal of Inner Mongoli Agriculture. University: Natural Science Edition.* 28: 302–305.

Xu, H.L. Zhang, H. Shen, Y. & Ren, S.G. 2013. Low Power Transmission and Fuzzy Control of Environment Parameters for Facilities Flower, *Transactions of the Chinese Society for Agricultural Machinery.* 44: 236–241.

Zou, Z.Y. Zhou, M. Xu, L.J. Kang, Z.L. & Liu, M.D. 2014. Design of Methane Fermentation environmental factor detection system based on ZigBee, *Applied Mechanics and Materials.* 539: 567–571.

Advanced Materials and Structural Engineering – Hu (Ed.)
© 2016 Taylor & Francis Group, London, ISBN 978-1-138-02786-2

Probabilistic loss estimates of non-ductile reinforced concrete frames

C.L. Ning
College of Civil Engineering, Tongji University, Shanghai, China

ABSTRACT: The loss ratio of non-ductile reinforced concrete frames has to be estimated in probabilistic manner because of the large number of epistemic and aleatory uncertainties. Based on the fragility analysis of three non-ductile reinforced concrete frames, which corresponds to the low-rise, middle-rise and high-rise building respectively, a probabilistic method is proposed by using the obtained fragility curves in peak ground acceleration. The damage factors are assumed to be a set of random variables by the normal distribution to generate the probabilistic loss ratio curves. After comparing with the loss ratio table provided by ATC-13, the confidence interval of the loss ratio can be computed with a predefined confidence level of 98% for non-ductile reinforced concrete frames.

1 INTRODUCTION

Stakeholders such as investors, city planners and building owners need to know the expected seismic loss of engineering structures under the potential seismic excitation. Currently, there are two alternative methods to estimate the seismic loss ratio of building structures, namely the expert opinion and engineering parameter (King & Liremidjian 1994, ATC 1985, Rojahn et al. 1997, Thiel & Zsutty 1987). The expert opinion relies on the expert judgment and past experience to develop a relationship of ground motion intensity and structural loss ratio but the engineering parameter quantifies the loss ratio based on the seismic behavior of each facility class. Recently, the trend of estimating the seismic loss ratio has been towards the use of engineering parameter studies because it is more rational to consider the parameters such as the structural type, ground motion and building behavior etc. The engineering parameter method is usually expressed as a probability that a specified damage level will be exceeded for a given ground motion intensity in the form of lognormal function. The quantitative measure of the earthquake intensity and structural seismic response are therefore adopted, taking the ability to integrate with a large number of epistemic and aleatory uncertainties in a statistically consistent manner.

2 FRAGILITY ESTIMATES

Generally, the conditional probability of being in or exceeding a particular damage state given the peak ground acceleration can be defined by the fragility function as follows (HAZUS 99 user's manual 2001):

$$F\left(IM; a_i, b_i\right) = \Phi\left[\frac{\ln(IM/a_i)}{b_i}\right] \qquad (1)$$

where, IM is the earthquake ground motion intensity, a_i is the median value of peak ground acceleration at which the building reaches the threshold of damage state i, b_i is the standard deviation of the natural logarithm of peak ground acceleration for the damage state i and Φ is the standard normal cumulative distribution function. It is observed that the equation can be described completely by the median value that corresponds to the threshold of each damage state and by the variability. Therefore, three non-ductile reinforced concrete frames with 2-storey, 6-storey and 12-storey are first designed, respectively representing the low-rise, middle-rise and high-rise building. The incremental dynamic analysis is then

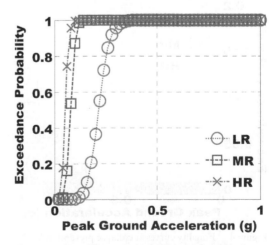

Figure 1.   Fragility curves at fully operational level.

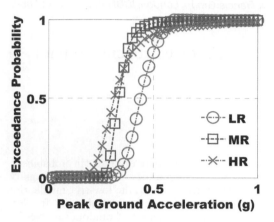

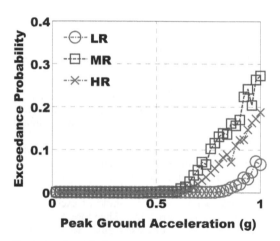

Figure 2. Fragility curves at immediate operational level.

Figure 3. Fragility curves at life safety level.

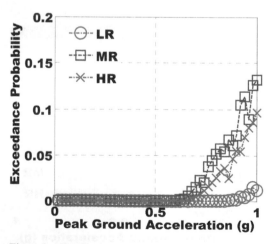

Figure 4. Fragility curves at collapse prevention level.

introduced to estimate the two statistical parameters by selecting the maximum inter-storey drift ratio as the structural damage index. Figures 1–4 illustrates the generated fragility curves of three non-ductile reinforced concrete frames at four damage states, namely the Fully Operational (FO) Level, Immediate Operational (IO) level, Life Safety (LS) level and Collapse Prevention (CP) level, where LR, MR and HR denotes the low-rise, middle-rise and high-rise building respectively. It is observed from the figures that the exceedance probability of middle-rise building is higher than that of high-rise building when the structural damage becomes severe.

3 DAMAGE FACTOR RANGE

The Applied Technology Council (ATC) conducted a survey on building structures by collecting 530 results from 31 strong motion recording stations after the Northridge, California, earthquake of January 17, 1994. Based on the collected data, the earthquake engineering experts categorized the building damage with the damage states defined by ATC-13 to develop a function of a percentage of replacement cost. Hereby, the function is named by the damage factors. The ranges of corresponding damage factors and seven ATC-13 damage states are presented in Table 1, providing a possibility to estimate the expected loss caused by earthquake ground motions for each facility class

Table 1. ATC-13 damage states and corresponding damage factor ranges.

Damage state	Damage factor range (%)	Central damage factor (%)
1-None	0	0.0
2-Slight	0–1	0.5
3-Light	1–10	5.0
4-Moderate	10–30	20
5-Heavy	30–60	45
6-Major	60–100	80
7-Destroyed	100	100

Table 2. Relationship between proposed damage categories and ATC-13 damage states.

Proposed damage categories	ATC-13 damage states	Damage factor range (%)
Fully operational level	1, 2, 3	0–10
Immediate operational level	4	10–30
Life safety level	5	30–60
Collapse prevention level	6, 7	60–100

in deterministic manner. Because the previous estimation procedure is a deterministic approach and uncertainties are not taken into account, which however plays an important role in defining the damage state and estimating the final damage factors. Therefore, the large number of epistemic and aleatory uncertainties would be considered in this paper to estimate the seismic loss of non-ductile reinforced concrete frames.

As such, the seven damage states associated with the damage factors are firstly mapped to the four damage categories used in fragility estimates. Table 2 presents the mapping of four damage categories and damage states of ATC-13 as well as the damage factors in ATC-13.

4 LOSS RATIO ESTIMATES

A probabilistic approach is then developed to account for the uncertainties, which are associated with the hazard definitions, structural capacities estimations, damage state descriptions and analytical models simplifications etc. For simplicity, they are represented by an overall uncertainty in this paper. Specifically, the damage factor of each damage category is assumed to be a random variable, where the normal distribution is adopted without loss of generality. Figure 5 shows the histogram of damage factors for each damage category. The probability distribution of each damage factor is bounded by the range of damage factors.

Following this, the probability of being in each damage category is computed as the difference of the conditional probabilities of the bounding fragility curves. The loss ratio of building structures is then computed by multiplying the discrete probabilities of each damage category with the corresponding damage factor at the given earthquake intensity IM as follows (Bai et al. 2009).

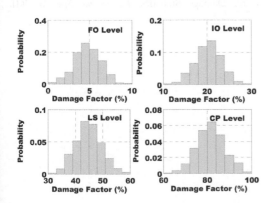

Figure 5. Probability distribution for each damage factor.

$$L = P_{CP\backslash IM} \times L_{CP} + (P_{LS\backslash IM} - P_{CP\backslash IM}) \times L_{LS} \\ + (P_{IO\backslash IM} - P_{LS\backslash IM}) \times L_{IO} + (P_{IO\backslash IM} - P_{FO\backslash IM}) \times L_{FO}$$

$$(2)$$

where, $P_{CP|IM}$, $P_{LS|IM}$, $P_{IO|IM}$ and $P_{FO|IM}$ is the discrete probability at the CP level, LS level, IO level and FO level for the given earthquake intensity respectively; L_{CP}, L_{LS}, L_{IO}, L_{FO} represents the damage factors at the corresponding damage category. After that, by multiplying the loss ratio at the given earthquake intensity with the floor area and repair cost per square meters for each building, the total repair and replacement cost can be easily obtained.

5 LOSS RATIO VALIDATION

A prediction interval for the loss ratio for non-ductile reinforced concrete frames can be further constructed by using the Monte Carlo simulation method with a predefined confidence level of 98%. The prediction bands of loss ratio for non-ductile reinforced concrete frames, respectively corresponding to the low-rise, middle-rise and high-rise buildings are illustrated in Figure 6 as a function of Peak Ground Acceleration (PGA), where the MLR as shown in Table 3 represents the mean loss ratio which is calculated from fragility estimates.

Given a comparison to the loss ratio estimated from the fragility estimates, the loss ratio table of

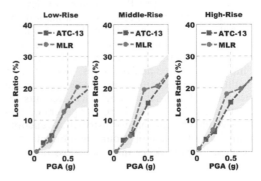

Figure 6. Expect loss ratio with confidence bands for reinforced concrete frames.

Table 3. Mean loss ratio of non-ductile reinforced concrete frames calculated from fragility curves.

PGA (g)	Low-rise	Middle-rise	High-rise
0.04	0.00	0.07	0.98
0.24	3.94	5.48	7.25
0.44	12.8	19.5	18.1
0.64	20.3	20.6	19.7
0.84	20.5	25.5	23.4

Table 4. Mean loss ratio of non-ductile reinforced concrete frames in ATC-13.

MMI	PGA Range (g)	Medium Value of PGA (g)	Low-rise	Middle-rise	High-rise
VI	0.092–0.18	0.136	2.80	3.60	3.78
VII	0.18–0.34	0.260	5.10	5.41	6.18
VIII	0.34–0.65	0.495	14.42	15.13	15.40
IX	0.65–1.24	0.945	22.00	27.96	26.77

non-ductile reinforced concrete frames in ATC-13 has been reviewed as shown in Table 4, where the relationship between the Modified Mercalli Intensity (MMI) and PGA range was developed by the United States Geological Survey (USGS). It is observed from the table that the mean loss ratio calculated from fragility curves agrees well with the statistical values of ATC-13. However, the developed method takes the ability to cover the loss ratio of ATC-13 by the prediction interval of loss ratio at the 98% confidence level, indicating the advantage of engineering parameter method in estimating the loss ratio caused by ground motions.

6 CONCLUSIONS

During the past decades, studies of earthquake induced structural losses have mainly focused on expert opinion. More recently, researchers have started to use the PEER framework to develop methods for building-specific loss estimation. As an application, a probabilistic method is proposed to estimate the loss ratio of non-ductile reinforced concrete frames. The proposed method is comprised of three analysis steps: generation of fragility curves, definition of damage factors and estimation of loss ratios. Each step accounts for the uncertainties involved. So the generated loss ratios would be a probability distribution function, defined by a function of ground motion intensity. As such, the confidence interval of loss ratio can

be predicted easily with a predefined confidence level from the probability distribution function of loss ratios, which provides an option for structural engineers to estimate the seismic loss ratio of non-ductile reinforced concrete frames in confidence manner.

REFERENCES

Applied Technology Council. 1985. *Earthquake damage evaluation data for California*. No. ATC-13, Applied Technology Council, Redwood City, Calif.

Bai, J.W. Hueste, M.B.D. & Gardoni, P. 2009. Probabilistic assessment of structural damage due to earthquakes for buildings in Mid-America, *Journal of structural engineering*, 135(10): 1155–1163.

HAZUS99 user's manual; service release 2. 2001. *Federal Emergency Management Agency (FEMA)*, Washington, D.C.

King, S.A. & Kiremidjian, A.S. 1994. *Regional Seismic Hazard and Risk Analysis Through Geographic Information Systems*. The John A. Blume Earthquake Engineering Center Report No. 111. Department of Civil Engineering, Stanford University. Stanford, California.

Rojahn, C. King, S.A. Scholl, R.E. Kiremidjian, A.S. Reaveley, L.D. & Wilson, R.R. 1997. Earthquake Damage and Los Estimation Methodology and Data for Salt Lake County, Utah (ATC-36). *Earthquake Spectra*. 13(4): 623–642.

Thiel, C. & Zsutty, T. 1987. Earthquake Characteristics and Damage Statistics. *Earthquake Spectra*. 3(4): 747–792.

Advanced Materials and Structural Engineering – Hu (Ed.)
© 2016 Taylor & Francis Group, London, ISBN 978-1-138-02786-2

A simplified swelling pressure models for expansive soils based on a nonlinear fitting function

C.W. Yan
Shaying River Valley Authority of Henan Province, Luohe, China

Z.Q. Huang & X.C. Huang
School of Resources and Environment, North China University of Water Resources and Electric Power, Zhengzhou, China

ABSTRACT: The swelling pressure model developed by former researchers based on the Gouy-Chapmann interacting diffuse double layer theory and osmotic pressure theory is effective in simulating swelling pressure of expansive soils. But it is difficult to combine equations of the model to calculate swelling pressure directly from void ratio because one of the equations is elliptic integral. So some researchers proposed the best-fit linear equation for the parameters u and Kd in the elliptic integral and then presented simplified swelling pressure model composed of the best-fit linear equation and other equations. However, when the range of Kd is big, the linear relationship does not fit u-log (Kd) well. Here the second order exponential function is applied to fit u-Kd relationship, and the results indicate that it fits u-Kd exactly without the influences of the range of Kd and other factors. At last, based on the u and Kd values obtained by numerical integration of the interacting double layer model and the second order exponential function for u-Kd, the feasibility to use the single double layer theory to establish swelling pressure model is discussed.

1 INTRODUCTION

The expandable mineral montmorillonite mainly contributes to the swelling of expansive soils. Several theories have been proposed for the description of ion distributions adjacent to charged surfaces in colloids. The Gouy-Chapmann theory of the diffuse double layer has received the greatest attention, and it has been applied to the behavior of clays with varying degrees of success (Mitchell 1976). Based on the Gouy-Chapmann interacting diffuse double layer theory and osmotic pressure theory, the equations developed by Bolt (Bolt 1956), van Olphen (1977), Sridharan & Jayadva (1982), to determine swelling pressure are as follows:

$$e_0 = G\rho_w Sd \tag{1}$$

$$\left(\frac{dy}{d\xi}\right)_{x=0} = \frac{B}{S}\frac{1}{\sqrt{2n_0\varepsilon kT}} \tag{2}$$

$$(2\cosh z - 2\cosh u)^{\frac{1}{2}} = -\left(\frac{dy}{d\xi}\right)_{x=0} \tag{3}$$

$$\int_z^u (2\cosh y - 2\cosh u)^{-1/2}\,dy = -Kd \tag{4}$$

$$p = 2n_0 kT(\cosh u - 1) \tag{5}$$

where, K is the double layer parameter given as:

$$K = \sqrt{\frac{2n_0 v^2 e^2}{\varepsilon kT}} \tag{6}$$

where, x is the distance from the clay surface, d is half the distance between parallel clay platelets, B is capacity of exchangeable cations (C/g), S is the specific surface area of soil (m²/g), ε is the dielectric constant of the pore fluid (is the product of permittivity of vacuum $\varepsilon_0 = 8.854 \times 10^{-12}$ C² J⁻¹ m⁻¹ multiplies relative dielectric constant $\varepsilon_r = 80.4$), T is the absolute temperature, v is the valance of cations, e is the elementary electric charge ($= 1.602 \times 10^{-19}$ C), n_0 is the molar concentration of ions in bulk fluid, p is the swelling pressure, e_0 is the void ratio, G is the specific gravity of soil solids, ρ_w is the density of water, u is the non-dimensional mid-plane potential, z is the non-dimensional potential at the clay surface, y is the non-dimensional potential at x, and ξ is the distance function ($= Kx$).

For any given pore fluid medium, determination of the swelling pressure using Equation (5) requires the non-dimensional mid-plane potential u. Since Equation (4) is an elliptic integral with the unknown integration domain, determination of u

is an indirect and time-consuming process, so it is difficult to combine above equations to calculate swelling pressure from e_0 directly. Hence a relationship between u and the non-dimensional distance function, Kd, must be established to determine u for any given values of Kd. At present, best-fit linear equations for the parameters u and Kd in the elliptic integral are proposed (Sridharam & Jayadva 1982, Sridharan 2002, Tripathy et al. 2004) and then simplified swelling pressure models composed of best-fit linear equations and other equations, i.e. Equation (1), (5) and (6), are presented. However, when the values of Kd are relative big, the linear relationship does not fit u and Kd well. Here the second order exponential function is applied to fit u-Kd relationship, and the results indicate that it fits u-Kd exactly without the influence of the values of Kd and other factors.

The equations for the single double layer model are as follows (Mitchell 1976, Olphen 1977, Komine & Ogata 1996):

$$y = 4 * \tanh^{-1}[\exp(-Kx) * \tanh(z/4)] \quad (7)$$

$$z = 2 * \sinh^{-1}\left[\frac{1}{2} * \left(\frac{dy}{d\xi}\right)_{x=0}\right] \quad (8)$$

where, all terms are previously defined. van Olphen (1977) has showed that when the interaction between two parallel layers is weak, that is, the value of Kd is big, the mid-plane potential can be taken as the sum of the double layer potentials at distance d based on the solutions for a single plate, that is,

$$u = 2y_d = 8 * \tanh^{-1}[\exp(-Kd) * \tanh(z/4)] \quad (9)$$

So the simplified model based on the single double layer model could be established with Equation (1), (5) and (6) (Komine & Ogata 1996, 2004). However, the feasibility of applying the simplified model is confused. Hence, It is significant to discuss, with the help of numerical values of u and Kd obtained from the interacting double layer model and the second order exponential best-fit function for u-Kd, how to use the swelling pressure model based on the single double layer model.

2 MODELS AND DYNAMIC EQUATIONS

Four coordinate systems, including velocity coordinate system (V), projectile axis coordinate system (A), second projectile axis coordinate system (O), projectile body coordinate system (O′), are established on an empennage. Among them, system O can be obtained by translating system A

with distance H, and the difference between system O′ and O is a rotation phase angle γ. Then the empennage model can be simplified to variable section projecting beam as shown in Figure 1, and the empennage deflection can be considered as the deformation in z′ direction.

3 THE SECOND ORDER EXPONENTIAL BEST-FIT FUNCTION FOR U-KD

To establish the relationship between u and Kd, Equations (2)–(6) are used for this purpose. For any pressure, u can be found from Equation (5). For known B, S, and u values, $(dy/d\xi)\,x = 0$ and then z can be found from Equation (2) and (3). From Equation (4), for known u and z values, Kd can be found by numerical integration. Then the best-fit equations for u-Kd could be got. For any void ratio, e_0, knowing K from Equation (6) and d from Equation (1), Kd can be found. Then the u value for the corresponding Kd value can be determined by established u-Kd relationship (Sridharan & Choudhury 2002, Tripathy et al. 2004). In this paper, the authors introduce another method, which is described later, based on the single double layer model to establish the best-fit equation for u-Kd.

From above description for determining the u-Kd relationship, we can conclude that $(dy/d\xi)_{x=0}$ is a dominant factor to determining the u-Kd relationship. Olphen (1977) has provided numerous values of u and Kd for various $(dy/d\xi)_{x=0}$. The u-Kd values calculated by them are showed in Figure 1.

In Figure 1, when Kd are smaller than 3, the relationship between u and Kd are linear, but when Kd are larger than 3, the deviations from linear curve occur. We can conclude that the linear equation does not fit u-log(Kd) well, especially when $Kd > 3$.

In Figure 2, u and Kd values are plotted in linear coordinate system and show exponential

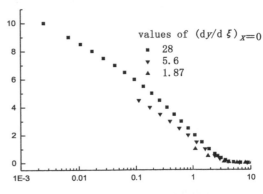

Figure 1. u-Kd in semi-logarithm coordinate system.

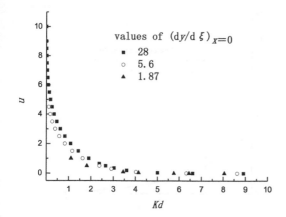

values of $(dy/d\xi)_{x=0}$

■ 28
○ 5.6
▲ 1.87

Figure 2. u-Kd in linear coordinate system (Olphen 1977).

Table 1. The best-fit results for u-Kd using the second order exponential function.

$(dy/d\xi)_{x=0}$	Second order exponential equations to fit u-Kd	R^2
1	$u = 3.940*\exp(-Kd/0.304)$ $+ 3.739*\exp(-Kd/0.007)$ $+ 0.384$	0.99284
5	$u = 4.874*\exp(-Kd/0.716)$ $+ 3.723*\exp(-Kd/0.016)$ $+ 0.295$	0.99537
10	$u = 5.607*\exp(-Kd/0.703)$ $+ 4.096*\exp(-Kd/0.012)$ $+ 0.377$	0.99171
50	$u = 6.797*\exp(-Kd/0.635)$ $+ 4.356*\exp(-Kd/0.013)$ $+ 0.517$	0.99329
100	$u = 6.807*\exp(-Kd/0.694)$ $+ 4.429*\exp(-Kd/0.024)$ $+ 0.431$	0.99549
28	$u = 3.699*\exp(-Kd/0.032)$ $+ 6.083*\exp(-Kd/0.896)$ $+ 0.112$	0.99737
5.6	$u = 4.175*\exp(-Kd/1.132)$ $+ 1.787*\exp(-Kd/0.124)$ $- 0.005$	0.99998
1.87	$u = 1.474*\exp(-Kd/1.025)$ $+ 1.474*\exp(-Kd/1.025)$ $- 0.0004$	0.99999
1.87	$u = 1.474*\exp(-Kd/1.025)$ $+ 1.474*\exp(-Kd/1.025)$ $- 0.0004$	0.99999

relationship. So the values of u and Kd are fitted by the second order exponential function, listed in Table 1. The results indicate that they fit u-Kd exactly without the influence of the values of Kd and other factors. So it is workable to replace

Equation (2), (3), (4) by the second order exponential function fitting for u-Kd to establish the simplified swelling pressure model with a high accuracy.

4 DISCUSSION ON THE SWELLING PRESSURE MODEL BASED ON THE SINGLE DOUBLE LAYER THEORY

From previous authors' analysis on the single double layer theory, it is known that only when Kd is big enough the theory could be used to establish the swelling pressure model. Although Komine (1996, 2004) has applied the single double layer to a swelling pressure model, the feasibility of the model has not been discussed. Immediately, we will make a discussion on it with the numerical solutions of u and Kd obtained from the interacting double layer model and the second order exponential function for u-Kd.

Some parameters of the bentonite used in Komine (1996) are given in Table 2. The valence in Table 2 is a mean of Na^+ and Ca^{2+}. The method building the second order exponential best-fit equation based on the single double layer theory is introduced as follows: $(dy/d\xi)_{x=0} = 21.7869$ and $z = 6.16682$ could be obtained from Equation (2) and (8). And then a series of values of z in interacting double layer model can be postulated by taking the values larger than the single double layer z. Known z of interacting double layer model, u and Kd can be found from Equation (3) and (4), and then be listed in Table 3. The values of u calculated by Equation (3) and the values of Kd calculated by the numerical integration of Equation (4) here are called true values of u and Kd respectively.

The best-fit equation fitted by the second order exponential function for true values of u and Kd in Table 3 can be derived as following equation:

$$u = 2.505*\exp(-Kd/1.204)$$
$$+2.505*\exp(-Kd/1.204) - 0.052 \qquad (10)$$

$R^2 = 0.99996$. Substitute true values of Kd into above equation, the best-fit values of u can then be obtained.

Table 2. Some parameters of bentonite used in Komine (1996).

S	388.8 m^2/g
B	70.64 C/g
v	1.5
T	295 K
ε	$80 \times 8.854 \times 10-12$ C$^2 \cdot$ J$^{-1} \cdot$ m^{-1}
n_0	0.02 mol/l

Table 3. True values of u and Kd calculated by interacting double layer model and u values calculated by the second order exponential function and the single double layer model.

$z^{(1)}$	$u^{(2)}$	$Kd^{(3)}$	$u^{(4)}$	$u^{(5)}$	$u^{(2)}/u^{(5)}$
6.1950	2.7447	0.7064	2.7348	3.8796	0.7075
6.1900	2.5728	0.7782	2.5735	3.5719	0.7203
6.1840	2.3187	0.8967	2.3274	3.1276	0.7414
6.1790	2.0429	1.0447	2.0522	2.6619	0.7674
6.1750	1.7478	1.2297	1.7525	2.1871	0.7992
6.1730	1.5571	1.3671	1.5579	1.8948	0.8218
6.1720	1.4447	1.4560	1.4433	1.7283	0.8359
6.1712	1.3435	1.5417	1.3406	1.5825	0.8490
6.1709	1.3023	1.5782	1.2990	1.5244	0.8543
6.1700	1.1655	1.7077	1.1613	1.3355	0.8727
6.1697	1.1144	1.7596	1.1101	1.2668	0.8797
6.1694	1.0598	1.8173	1.0557	1.1947	0.8871
6.1691	1.0012	1.8824	0.9974	1.1184	0.8953
6.1688	0.9377	1.9567	0.9346	1.0373	0.9040
6.1685	0.8683	2.0433	0.8661	0.9504	0.9135
6.1682	0.7911	2.1470	0.7903	0.8561	0.9242
6.1679	0.7038	2.2758	0.7049	0.7519	0.9359
6.1676	0.6016	2.4460	0.6051	0.6337	0.9492
6.1673	0.4749	2.6977	0.4811	0.4923	0.9647
6.1670	0.2935	3.1976	0.2999	0.2984	0.9837
6.1669	0.2502	3.3609	0.2552	0.2534	0.9875
6.1669	0.2455	3.3805	0.2503	0.2485	0.9879
6.1669	0.2254	3.4673	0.2292	0.2278	0.9895
6.1669	0.1974	3.6019	0.1995	0.1991	0.9917
6.1668	0.1573	3.8315	0.1558	0.1582	0.9943
6.1668	0.1235	4.0755	0.1176	0.1240	0.9963
6.1668	0.1025	4.2633	0.0931	0.1027	0.9973

(1) Values of z postulated for interacting double layer model
(2) True values of u
(3) True values of Kd
(4) Values of u calculated by Equation (10)
(5) Values of u calculated by Equation (9).

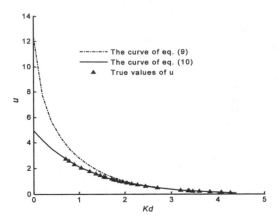

Figure 3. Various relationships of u-Kd obtained by various methods.

Values of u calculated by various methods are listed in Table 3, and observes from Table 3 indicate that:

1. The differences between best-fit values fitted by the second order exponential function and true values of u are negligible, so it is workable to apply the second order exponential function to fit u-Kd.
2. The curve of the single double layer model get closer to the true values of u with the increasing of Kd values, shown in Figure 3. If the ratio of true value u to single double layer model u needs to reach 0.8, the Kd should be 1.2297 at least. However when the model based on the single double layer theory built by Komine (1996) was using, the largest value of Kd is 0.9, according to which, the ratio of true value u to single double layer u is only 0.74.
3. The curves of Equation (9) and (10) are also plotted in Figure 3. There is a big gap between two curves when Kd is smaller than 1, and only when Kd is larger than 1, two curves are closer.

5 CONCLUSIONS

From the results presented in this paper it is could be concluded that:

1. Despite the values of $(dy/d\xi)_{x=0}$ and Kd, the second order exponential functions fit u-Kd well and the relevant coefficients R^2 are bigger than 0.99. The fitted results using the second order exponential functions are better than those using the linear functions. So the second order exponential function could be used to build the simplified swelling pressure model, which could derive the swelling pressure p directly from e_0.
2. Only when the values of Kd is relative large, the curve of the single double layer model are close to the true values of u, and then the single double layer could be used to established the swelling pressure model.

ACKNOWLEDGEMENT

This work was financially supported by the Plan for Scientific Innovation Talent of Henan Province and the Key Scientific and Technological Project of Henan Province.

884

REFERENCES

Bolt, G.H. 1956. Physico-chemical analysis of the compressibility of pure clays. *Geotechnique*, 6(2): 86–93.

Komine, H. & Ogata, N. 1996. Prediction for swelling characteristics of compacted bentonite, *Canadian Geotechnical Journal*, 33(1): 11–22.

Komine, H. & Ogata, N. 2004. Predicting swelling characteristics of bentonites, *Journal of Geotechnical and Geoenviromental Engineering*, 130(8): 818–829.

Mitchell, J.K. 1976. *Fundamentals of soil behavior.* New York: John Wiley & Sons.

Olphen, Van. 1977. *An introduction to clay colloid chemistry.* New York: Wiley.

Sridharan, A, & Jayadva, M.S. Double layer theory and compressibility of clays. *Geotechnique*, 32(2): 133–144.

Sridharan, A. & Choudhury, D. 2002. Swelling pressure of sodium montmorillonites. *Geotechnique*, 52(6): 459–462.

Tripathy, S. Sridharan, A. & Schanz, T. 2004. Swelling pressure of compacted bentonite from diffuse double layer theory. *Canadian Geotechnical Journal*, 41(3): 437–450.

Advanced Materials and Structural Engineering – Hu (Ed.)
© 2016 Taylor & Francis Group, London, ISBN 978-1-138-02786-2

The basic study of introduction for the certification system of infill in South Korea

E.K. Hwang
Research Fellow, Korea Institute of Civil Engineering and Building Technology, Korea

S.R. Park
Department of Architecture, Hanyang University, Seongdong-gu, Seoul, South Korea
Researcher, Korea Institute of Civil Engineering and Building Technology, South Korea

S.A. Kim
Research Fellow, Korea Institute of Civil Engineering and Building Technology, Korea

ABSTRACT: Long-life Housing is aimed to provide a solution to the need to flexibly respond to diversified demands for housing. In this system, the support is separated from infill. For the flexibility, the private sector satisfied each household's needs in terms of infill using. However, there are no clear criteria for infill in Korea. In addition, it is a narrow infill market. So, this study aims to identify the current infill level through the analysis of the existing infill studies, and to take advantage of this result as the basis for the introduction and activation of infill system.

1 INTRODUCTION

The proportion of apartments among all housing in Korea exponentially increased from 22.7% in 1990 to 46.8% in December 2012. However, the life of architecture is extremely short compared with advanced countries, due to poor handling of consumer's various demands and changes. There are many restrictions in supplying houses through the current method of reconstruction, due to the appearance of high-level apartments that exceed 25 floors, enhanced supply rate of housing, increased speed of popularization and decreased number of houses, and the use of limited resources and the treatment of architectural waste.

Thus, the Ministry of Land, Infrastructure and Transport established standards related to "long-life housing construction standards and certification policy, etc." pursuant to Article 21(6) of the Housing Law through a revision of the Housing Law in December 2013, and consequently, a pre-announcement of a legislation related to the "long-life housing construction and certification standards" that will oblige the obtainment of the long-life housing certification upon constructing joint housing complexes having more than 1,000 units that have been made in preparation for specific standards.

This announcement defines long-life housing as housing assessed and certified of the performance thereof by the director of a long-life housing performance rating certification organization, in terms of the durability, variability, and convenience of repairs. Further, it regulates that long-life housing should be planned and constructed to handle social changes, technological changes, changes in units, changes and diversification of family compositions by increasing physical and technical life in comparison to standard housing.

To achieve this, long-life housing is fundamentally constructed by separating the support and infill to secure technical life, which can handle the various demands of the residents, and physical life of the support.

Specifically, in order to increase technical life, it is very important to develop and apply infill, which can provide spatial diversity that can handle various lifestyles and life stages of the residents. Thus, a partial development of infill, which considers long-life housing, is being conducted in certain corporations. However, there is no progress since there are no markets to apply such infill, in addition to there being no specific standards for the performance of infill.

Accordingly, this study has the purpose of being used as basic data for the implementation and activation of the Certification System of Infill by apprehending the current standard of infill in Korea through an analysis of existing studies and policies related to infill.

Table 1. Assessment categories of long-life housing.

Category	Assessment categories	Related infill
Durability	– Thickness of iron sheath – Strength of design standards, unit quantity of cement, water to cement ratio, chloride content (concrete quality)	

Variability	– Support-structure method (essential): Existing bearing wall to column length ratio (%)
	– Infill-wall lining material and construction method (essential)
	  • Ratio of dry wall from the total internal wall of the interior of units
	  • Previous method of variability
	– Infill-pipes (optional): Pipes in the bathroom (toilet) of the applying floor
	– Support-story height (optional): Points added from 3,000 mm
	– Infill-space variability (optional): Double floor
	– Infill-water usage space variability (optional): Bathroom, kitchen
	– Infill-exterior wall variability, etc. (optional): Exterior wall
		– Dry wall lining and construction method
		– Piping and construction method
		– (Dry) Double floor and construction height
		– Bathroom, kitchen piping method
		– Exterior wall

Convenience of repairs	① Exclusive area
	– Convenience of repairs and inspection
	  • Security of independency of communal pipes and exclusive equipment spaces
	  • Easy planning of repairs and replacement of pipes and wires (essential)
	  • Underground laying of pipe support forbidden (optional)
	  • Dry floor heating (optional)
	– Horizontal separation plan of unit (optional)
	  • Space plan: Separation of front door, security of telecommunication panelboard for each unit upon separation of front door
	  • Equipment plan: Standpipes for the washing machine, etc.
		– Piping, wiring
		– Dry floor heating

Convenience of repairs	② Communal area
	– Convenience of repairs and inspection
	  • Placement plan for piping (shaft) in the communal area (essential)
	  • Inspection area of the communal piping space (essential)
	  • Placement of pipes within the piping area (shaft) (optional)
	  • Piping structure (optional)
	– Handling of changes in future demands and energy sources (optional)
	  • Security of 20% extra space in the main communal pipe shaft
	  • Installation of more than 1 additional preliminary shaft
		– Pipe structure

2 LONG-LIFE HOUSING CONSTRUCTION CERTIFICATION STANDARDS AND INFILL

The long-life housing certification standards of the recently announced "long-life housing construction certification standards" can be summarized as below. First, the assessment categories are largely categorized into durability, variability, and convenience of repairs. Among which, the sections related to infill include variability and convenience of repairs, and the types of related infill include furniture, panel-type, combined dry wall, double flooring, ceiling lining, stratiform pipes and wall surfaces, bathroom (toilet), kitchen, exterior wall,

and dry floor heating. Such variability and convenience of repairs not only apply infill but also are assessed on the construction methods between infill and infill, and infill and support. Herein, infill excludes the support of the housing, and refers to such that can be standardized and mass-produced due to its properties of allowing repairs and replacements based on the changes in user demands, in addition to having a short service life than the support.

Meanwhile, the scores based on the assessed categories are comprised of 35 points each for durability and variability, and 30 points each for convenience of repairs, totaling 100 points, and each assessment category is categorized into four ratings. The entire certification rating is also categorized into four ratings: outstanding, excellent, satisfactory, and standard. In addition, the corresponding test scores are 90 points, 80 points, 60 points, and 50 points, respectively.

3 ANALYSIS OF KOREAN INFILL-RELATED POLICIES

Currently in Korea, the Industrial Standard (KS) Certification system pursuant to the Industrial Standardization Act is the only infill-related policy, Industrial standardization is a "standard for industrial standardization" and refers to simplifying and unifying the following terms.

Type, form, dimensions, structure, device, quality, rating, substance, performance, function, durability related to the safety of mining and manufacturing products.

Working methods and safety conditions related to the methods of production, designing, planning, use, operation, and unit resources for mining and manufacturing products.

The type, form, dimensions, structure, performance, rating, and method related to the packaging of mining and manufacturing products.

The methods of testing, analysis, auditing, inspection, assessment, statistical review, and measurement and the terms, abbreviations, symbols, signs, standard numbers, and units related to mining and manufacturing products and technologies used in the mining and manufacturing industry.

Methods of planning and construction or safety conditions of architectural constructions of other structures.

Commercial transactions based on information systems and telecommunication media managing the procurement, planning, production, operation, repairs, and disposal of articles related to corporate activities.

Terms related to providing procedure, method, system, and evaluation method of services (excludes telecommunication services) related to industrial activities.

The scope of KS certification is categorized into 24 fields, including basic, mechanical, electronic and electric, metal, mining, and construction, and the construction field is indicated as "KS F".

In order to promote the use of KS products, the Korean industrial standards must be adhered to by national organizations, local governments, public organizations, and public groups in terms of procurement, production, management, and facility construction for products and services, and it is legally regulated to purchase KS products as a priority. Further, KS products are regulated as being wholly or partially exempt from inspection, assessment, testing, authorization, reporting, and approval pursuant to 14 legislations, including the Quality Management and Industrial Product Safety Management Act and the Electrical Product Safety Management Act. The KS F standard regulations particularly regulate the standards of the dimensions and performance of architectural parts and materials.

Meanwhile, the Housing Construction Promotion Act, which was abolished in February 1999,

Table 2. Main KS related to architectural building materials.

Standard KS F	Main architectural materials and infill
– F 1010: Category of performance for each part of an architectural structure	– F 2223: Residential combined sanitary unit
– F 1503: Regulations and standards for architectural module adjustment	– F 2224: Residential storage wall system
	– F 3109: Door set
– F 1505: Determining the basic allowance and dimensions for architectural building materials	– F 3117: Window set
	– F: 4722: Assembled concrete wall panel
– F 1510: Priority dimensions for module adjustment of architectural building materials	– F 4729: Assembled concrete roof panel
	– F 4726: Assembled concrete floor panel
– F 1513: Standard nominal module dimensions of wainscot stress	– F 4741: Combined steel panel
– F 1514: Standard nominal module dimensions of flooring stress	– F 4760: Double flooring
– F 1515: Arrangement of window module dimensions	– F 6308: Residential toilet unit
– F 1517: Standard module dimensions of building materials for shifting dividers	– F 6309: Residential bathroom unit
	– F 6310: Residential air-conditioning and heating unit
– F 1523: Arranging dimensions of housing kitchen equipment	– F: 6313: System kitchen
– F 1525: Planning standard of arranging architectural modules	
– F 2222: Standard nominal module dimensions of assembled bathrooms for housing	
– F 2290: Nominal module dimensions of piping units for housing	
– F 2291: Nominal module dimensions of air-conditioning and heating units for housing	

Table 3. Items of housing materials to register (attachment 2 regulated by housing construction standard).

Category	Items	Other
Plastic clay products	Standard bricks, standard blocks	– Attachment 6 of the Implementing Decree thereof (material dimensions and quality): standard bricks, cement bricks, empty cement blocks, concrete assembly parts, lightweight bubble concrete assembly parts
Cement processed products	Cement bricks, empty cement blocks, standard cement roof tiles	
Concrete products	Concrete assembling parts (parts for the main structure, such as the wall plate, floor plate, roof plate, column, bow, stairs, and other similar parts), lightweight bubble concrete assembly parts (wall plate, floor plate, roof plate)	– Attachment 6(2) of the Implementing Decree thereof self-quality management testing categories: • Standard Bricks, etc.: Dimensions test, compression stress test • Concrete Assembly Parts: Dimension test, compression stress test, internally distributed pressure test, transverse strength test, aggregate salt content test
Window products	Wooden window and frame, concrete window and frame, aluminum window and frame, steel window and frame, synthetic resin window and frame	
High-pressure products	High-pressure bricks, high-pressure blocks	– Attachment 7 of the Implementing Decree thereof (housing materials quality test standards): dimension quality, lot size, and sample number for 6 items, such as the common terms and standard bricks
Earthwork products	Earthwork bricks, earthwork blocks, earthwork roof tiles	
Pressurized cement plate product	Pressurized cement plate roof plate	

Table 4. Legislations and main regulations related to the excellent housing material certification system.

Related legislations	Main regulations

– Article 45(2) of the Housing Construction Promotion Act (excellent housing material certification, etc.)
– Article 62 of the Regulations related to Housing Construction Standards (excellent housing material certification, etc.)
– Rules related to Housing Construction Standards, etc.
  • Article 24 (applicable subjects of excellent housing materials)
  • Article 24 (standards of certification of excellent housing materials)
  • Article 25 (standards for applying for the excellent housing materials certification, etc.)
	– Applicable Subjects for Certification
	  • Assembled housing parts
	  • Housing windows and frames
	  • Housing materials developed as an industrial-based technological development project pursuant to the Technological Development Promotion Act, etc.
	  • Housing materials determined by the Minister of Construction and Transportation (materials that may contribute to industrialized construction method or new industrial developments)
	– Standards of Certification
	  • Must have excellent safety, durability, and functionality
	  • Must allow standardized production based on a mass-production system
	  • Must be easily constructible
	  • Must have a suitable price
	  • Must be suitable for the standards of performance and production of industrialized housing (restricted to assembled housing materials)

but currently a part of the Housing Act, operated the excellent housing material certification system and the housing material manufacturer registration system.

First, Article 41 (Registration of Main Structure Housing Material Manufacturers) of the Housing Construction Promotion Act regulates each housing material item wished to manufacture be registered with the Minister of Construction and Transportation, and the party constructing the housing to use housing materials manufactured by a registered housing material manufactured pursuant to this regulation. The applying housing materials herein are as shown in the table below. Further, Attachment 6 of the Implementing Decree thereof regulates the dimensions and quality of housing materials; Attachment 6(2) regulates the self-quality management testing items; and Attachment 7 regulates the housing material quality testing standards.

Table 5. Main materials for a long-term repair plan (Attachment 5 of implementing decree of the housing act).

Category	Sub-category	Remarks
Building exterior	Roof, exterior, external window, and others	
Building interior	Ceiling, interior walls, floor, interior windows, stairs, and others	
Electrical, fire-extinguishing, elevator, and intellectual home network equipment	Preliminary power (self-development) equipment, transformer equipment, wiring equipment within the housing, automated fire-sensing equipment, fire-extinguishing equipment, elevator and hoist, lightning protection equipment and exterior lights, communication and broadcasting equipment, boiler room and machinery room, security and crime prevention facility, intellectual home network equipment	
Water supply, sanitation, gas, and ventilation equipment	Water supply equipment, gas equipment, piping equipment, sanitation equipment, ventilation equipment	
Heating and hot water equipment	Heating equipment, and hot water equipment	
Additional exterior facilities and exterior welfare facilities		

The excellent housing material certification system, which was abolished together with the housing material manufacturer registration system, was operated pursuant to Article 45(2) of the Housing Construction Promotion Act, Article 62 of the Regulations related to Housing Construction Standards, and Article 24, 25 and 26 of the Implementing Decree thereof. Upon review of the specific details, the applying subjects for the excellent housing certification are restricted to materials that can contribute to assembled housing parts, housing windows and frames, and the industrialization method. Additionally, the standards of the certification are regulated based on excellent safety, durability, and functionality; the possibility of standardized production; constructability; and suitability of price. Unlike the housing material manufacturer registration system, however, no special incentives were granted upon obtaining the excellent housing materials certification.

Meanwhile, Article 47 of the Housing Act, which is a current legislation, regulates that a long-term repair plan must be established with respect to joint housing for the prolongation of life and maintenance of the performance of quality architectural structures. Further, Article 26 and Attachment 5 of the Implementing Decree thereof regulates that the standards of establishing a long-term repair plan be categorized into construction type, repair method, and interval of repairs for the exterior and interior of the building.

4 COMPARATIVE ANALYSIS OF THE INFILL POLICY

Currently, there are no infill policies that consider the properties of long-life housing.

Accordingly, there are restrictions of comparative analysis between infill policies. Thus, the restrictions on infill-related policies were analyzed with respect to long-life housing.

First, KS, which is operated based on the Industrial Standardization Act of the Ministry of Industry, Commerce, and Resources, is applied on the specification, performance, and experiment method of each product, and lacks standards related to the performance and assessment method for various connections when connecting the support and infill, and an infill and another infill, which are the characteristic features of long-life housing. However, KS regulates industrial standards of various housing infills, and thus, there is a need for a connection thereof upon preparing certification standards for infill types.

Meanwhile, the housing material manufacture registration system and the excellent housing material certification system are policies that induce reinforced price competitiveness through mass-production of factory-based housing infill and enhanced performance of housing materials and infill. However, these systems were abolished in February 1999 due to the restrictions of the housing infill market.

Lastly, the long-term repair plan operated by the currently Housing Act is considered a very important policy for the prolongation of life and security of performance of joint housing. However, there is a need for the preparation of standards with respect to exclusive areas because such standards are limited to communal areas of joint housing. Further, there is a need for the preparation of standards related to separation construction of the body and infill through the application of various infills that exceed the monolithic wet construction method focused on materials.

5 SEEKING OF A CERTIFICATION SYSTEM OF INFILL IMPLEMENTATION PLAN

For the activation of supply and early stabilization of long-life housing, there is a need to activate the application of various infills that can handle user demands, in addition to the longevity of the life of the support. However, not only the infill market in Korea is small, but also consumer awareness on infill is extremely poor. Accordingly, there is an urgent need for the implementation of the Certification System of Infill for the activation of quality infill production and market, and the following plans are required in order to implement the Certification System of Infill.

The current types of infill are primarily restricted to those used in order to secure the variability and convenience of repair of long-life housing, such as double flooring, double ceiling, and panel and storage type divided walls, which are technologically developed and applied on site. However, for the construction of long-life housing, various parts or materials composing of the housing must be applied as infill and be repaired and replaced according to the term of use for each infill. Accordingly, there is a need to apply various materials and parts composing of housing as infill.

For the construction of quality long-life joint housing through the application of infill, there is a need to apply a method of connection that enables easy repair and replacement, and security of performance of interfaces between a support and infill, and an infill and another infill, in addition to the performance of the infill itself. To achieve this, there is a need to prepare certification standards of each infill, and each infill must be evaluated to verify the performance thereof. Accordingly, there is a need for each individual infill certification standard to specifically regulate the basic standards of the material composition and dimensions of infill, as well as the performance standards of infill production, planning, construction, the infill itself, and the interface between each infill, and an assessment method that can assess each standard of performance.

Korea displays the technical and financial difficulty in constructing housing infill, in addition to having restrictions in mass-supplying and producing housing due to the generalization of the wet-unified construction method for the past few decades. Accordingly, there is a need to seek a plan for providing various incentives, such as financial and tax support for infill manufacturers and construction companies, and expansion of various bidding opportunities in order to stabilize infill production and construction.

6 CONCLUSIONS

This study has sought to find a plan to implement the Certification System of Infill through the analysis of existing infill-related studies and policies. First, the implementation of the Certification System of Infill has proposed the implementation of a system that is suitable for the properties of current long-life housing based on the housing material manufacturer registration system and the outstanding housing material certification system, which were abolished in 1999. Further, for the implementation of the Certification System of Infill, there is a need to prepare a variability of the types of infill, a construction of a standard and an assessment system of the Certification System of Infill, and a plan to provide incentive for infill manufacturers and construction companies. It is anticipated that a more affordable supply and activation of long-life housing will be achieved through the implementation of the Certification System of Infill.

ACKNOWLEDGMENT

This study was made possible by the financial support from part of results a major research project conducted by the Korea Ministry of Land, Infrastructure and Transport, Residential Environment Research Project in 2014. Project No.: 14RERP-B080965-01.

REFERENCES

Kim, S.A. et al. 2000. *Technological Development of 21st Century Standardization and Informatization for Enhanced Productivity of Construction*, Ministry of Construction and Transportation.

Kim, S.A. et al. 2007. *Development of a Joint Long-Life Housing Planning System, Technological Development of PLUS 50 Environmentally Symbiotic Architectural Technologies*, The Korea Institute of Civil Engineering and Building Technology.

Korea Land & Housing Corporation et al. 2010. *Joint Research Report on the Technological Development of Long-Life Housing having Durability and Variability, Ministry of Land, Transport and Maritime Affairs*, The Korea Institute of Civil Engineering and Building Technology.

Advanced Materials and Structural Engineering – Hu (Ed.)
© 2016 Taylor & Francis Group, London, ISBN 978-1-138-02786-2

Analysis of geometric nonlinear free vibration of pretensioned rectangular orthotropic membrane structure

Y.H. Zhang
Chongqing Jianzhu College, Chongqing, P.R. China

C.J. Liu
College of Environment and Civil Engineering, Chengdu University of Technology, Chengdu, P.R. China

Z.L. Zheng
College of Civil Engineering, Chongqing University, Chongqing, China

ABSTRACT: This paper presented the computational example for the homotopy perturbation solution of the geometrical nonlinear free vibration of pretensioned orthotropic membrane with four edges fixed. According to the approximate analytical solution of the vibration frequency and displacement function, the frequency, vibration mode and displacement and time curve of each feature point are analyzed, and compared them with other solutions of the geometrical nonlinear free vibration of pretensioned orthotropic membrane with four edges fixed. The analysis results proved that homotopy perturbation method is an effective, simple and high-precision method for solving the nonlinear free vibration problem of membrane structures. The results obtained herein provide some valuable computational basis for the vibration control and dynamic design for membrane structures or components.

1 INTRODUCTION

At present, there are some reports about the nonlinear vibration problem of membranes. Sunny et al. (2012) studied the nonlinear vibration problem of a prestressed membrane by adomian decomposition. Soares and Gonalves et al. (2012, 2014) presented a detailed analysis of the nonlinear vibration of stretched annular, circular and rectangular membrane by using the Galerkin method and nonlinear finite element method. Reutskiy (2009) adopted a new numerical method to study nonlinear vibration of arbitrarily shaped membranes. The method is based on mathematical modeling of the physical response of a system that was excited over a range of frequencies. The vibration problems of rectangular membranes placed in a vertical plane were solved and the exact solutions of the vibration frequencies were obtained by Wang (2011). The geometric nonlinear dynamic characteristics of the out-of-plane vibration of an axially moving membrane were studied by using the Hamilton principle and the Galerkin method by Shin et al. (2005). Wetherhold and Padliya (2014) presented a method for inferring the initial tensions from measured vibration frequencies and demonstrated the sensitivity of the tensions with respect to imprecision in the measured frequencies.

Liu et al. (2014) obtained the frequency solution of geometric nonlinear free vibration of rectangular flat pretensioned orthotropic membrane with fixed boundary by homotopy perturbation method. But they didn't compare the solution with other solutions. In this paper, we will give the computational example and compare the results of homotopy perturbation solution with the results of other solutions to verify the accuracy of the homotopy perturbation solution.

2 HOMOTOPY PERTURBATION SOLUTION OF FREQUENCY AND DISPLACEMENT

According to paper (Liu et al. 2014), the analytical homotopy perturbation solution of the vibration frequency and displacement function of the geometrical nonlinear free vibration of pretensioned orthotropic membrane with four edges fixed are as follows:

$$\omega = \sqrt{\lambda}\alpha = \sqrt{\frac{10\lambda + 7\varepsilon a_0^2 + \sqrt{64\lambda^2 + 104\lambda\varepsilon a_0^2 + 49\varepsilon^2 a_0^4}}{18}}$$

$$(1)$$

$$w(x,y,t) = \sum_{m=1}^{\infty} \sum_{n=1}^{\infty} \sin\frac{m\pi x}{a}$$

$$\sin\frac{n\pi y}{b} \left(\begin{array}{c} \dfrac{3\varepsilon a_0^3}{4\lambda(\alpha^2-1)}\cos\sqrt{\lambda}\alpha t \\ \\ + \dfrac{\varepsilon a_0^3}{4\lambda(9\alpha^2-1)}\cos 3\sqrt{\lambda}\alpha t \end{array} \right) \quad (2)$$

where,

$$\varepsilon = \frac{3h\pi^4}{16\rho}\left(\frac{E_1 m^4}{a^4} + \frac{E_2 n^4}{b^4}\right), \quad \lambda = \frac{h\pi^2}{\rho}\left(\frac{m^2}{a^2}\sigma_{0x} + \frac{n^2}{b^2}\sigma_{0y}\right)$$

$$\alpha = \sqrt{\frac{10\lambda + 7\varepsilon a_0^2 + \sqrt{64\lambda^2 + 104\lambda\varepsilon a_0^2 + 49\varepsilon^2 a_0^4}}{18\lambda}} ; \; m \text{ and}$$

n are natural numbers; σ_{0x} and σ_{0y} denote initial tensile stress in x and y direction, respectively; denotes aerial density of membrane; h denotes the thickness; E_1 and E_2 denote Young's modulus in x and y direction, respectively; a and b denote the length of x and y directions, respectively.

Assuming initial displacement of membrane is $w_0(x,y) = w_0$, According to the orthogonality of vibration mode, we obtain.

$$a_0 = \begin{cases} \dfrac{16w_0}{mn\pi^2} & (m,n = 1, 3, 5 \ldots) \\ 0 & (m,n = 2, 4, 6 \ldots) \end{cases} \quad (3)$$

The exact frequency solution is (Zheng et al. 2009)

$$\omega_{ex} = \frac{\sqrt{\lambda + \dfrac{\varepsilon}{2}a_0^2}}{\displaystyle\sum_{p=0}^{\infty}(-1)^p\left(\frac{(2p-1)!!}{(2p)!!}\right)^2\left(\frac{\varepsilon a_0^2}{2\lambda + \varepsilon a_0^2}\right)^p} \quad (4)$$

where, $p = 0, 1, 2, 3 \ldots$.

The frequency obtained by the L-P perturbation method is (Liu et al. 2010).

$$\omega_{pert} = \sqrt{\lambda} + \frac{3\varepsilon a_0^2}{\sqrt{\lambda}8} \quad (5)$$

3 COMPUTATIONAL EXAMPLES AND DISCUSSIONS

Take the membrane material commonly applied in a project as an example. The Young's modulus in x and y are $E_1 = 1.4 \times 10^6$ kN/m^2 and $E_2 = 0.9 \times 10^6$ kN/m^2, respectively; the aerial

density of membranes is $\rho = 1.7$ kg/m^2; the membrane's thickness is $h = 1.0$ mm, $a = 1$ m, $b = 1$ m, $\sigma_{0x} = \sigma_{0y} = 5.0 \times 10^3$ kN/m^2.

According to the initial displacement w_0, we can figure out a_0 by Equation (3). Then substitute a_0 into Equations (1), (4) and (5) to figure out the frequency of the first three orders. The results are shown in Table 1.

The comparison and analysis of Table 1 are as follows.

All of the frequency values calculated according to Equations (1), (4) and (5) increase with the increase of initial displacement. This reflects the geometric nonlinearity characteristic of the vibration of the membrane. When the initial displacement approaches zero, namely $w_0 \rightarrow 0$, the frequency values calculated according to Equations (1), (4) and (5) are the same and equal to the small amplitude vibration frequency values. The frequency values calculated according to Equation (1) are slightly larger than the corresponding ones calculated according to Equation (4). The relative differences between them become larger and larger with the increase of initial displacement, and the maximal relative difference is 3.33%. The frequency values calculated according to Equation (5) are larger than the corresponding ones calculated according to Equation (4). The relative differences between them increase dramatically with the increase of initial displacement, and the maximal relative difference is 78.94%. Obviously, the precision of Equation (1) is higher than that of Equation (5).

Substituting the material and geometric parameters in computational example and the frequency values calculated by Equation (1) (while $w_0 = 0.05$ m) into Equation (2), we obtain the displacement function of the vibration of the first three orders.

① The first order vibration mode

$$w(x,y,t) = (0.0789866\cos 427.8t \\ + 0.00207038\cos 1283.4t) \\ \sin\pi x \sin\pi y$$

② The second order vibration mode

$$w(x,y,t) = (0.026399\cos 858.84t \\ + 0.000620067\cos 2576.52t) \\ \sin\pi x \sin 3\pi y$$

③ The third order vibration mode

$$w(x,y,t) = (0.0263092\cos 858.84t \\ + 0.000709343\cos 2971.02t) \\ \sin 3\pi x \sin\pi y$$

Table 1. Frequency values (rad/s) under different initial displacement.

Order	Formula	Initial displacement w_0 (m)				
		0.10	0.09	0.08	0.07	0.06
1	(41)	749.32	682.31	616.22	551.38	488.31
	(42)	725.63	661.75	598.82	537.18	477.30
	(43)	1251.66	1059.62	887.80	736.20	604.80
2	(41)	1447.47	1323.23	1201.25	1082.34	967.63
	(42)	1405.22	1286.99	1171.08	1058.21	949.45
	(43)	2161.18	1852.93	1577.12	1333.76	1122.84
3	(41)	1752.38	1594.08	1437.77	1284.18	1134.44
	(42)	1695.88	1544.89	1395.98	1249.87	1107.64
	(43)	3034.62	2560.41	2136.12	1761.74	1437.28

Order	Formula	Initial displacement w_0 (m)				
		0.05	0.04	0.03	0.02	0.01
1	(41)	427.80	371.13	320.38	278.82	250.88
	(42)	419.93	366.21	317.97	278.09	250.82
	(43)	493.63	402.66	331.91	281.38	251.06
2	(41)	858.84	758.58	670.71	600.64	554.80
	(42)	846.33	751.16	667.32	599.71	554.73
	(43)	944.38	798.36	684.80	603.68	603.68
3	(41)	990.34	854.77	732.60	631.70	563.27
	(42)	970.97	842.48	726.44	629.80	563.10
	(43)	1162.74	938.11	763.40	638.61	563.74

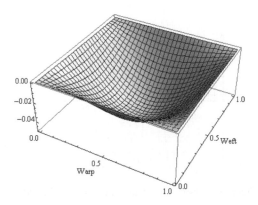

Figure 1. The first order vibration mode.

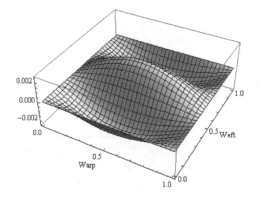

Figure 2. The second order vibration mode.

According to these displacement functions, we can draw the vibration mode figures of the first three orders while $t = 0.02$ s in Figures 1 to 3. Superposing the vibration mode of the first four orders, we can obtain the superposed vibration mode figure in Figure 4.

From the result of the vibration mode analysis, we can conclude that the amplitude decreases with orders, namely the contribution of high orders for total mode decreases gradually. In addition, using the displacement function (2) we can compute the vibration mode of each order and obtain the total superposed vibration mode conveniently. From Figures 1 to 4, we can see that the total mode is axsymmetric.

Substituting the material and geometric parameters in computational example and the frequency values calculated according to Equation (1) (while $w_0 = 0.05$ m) into Equation (2), we can compute the displacement time histories of the feature points on membrane surface and draw the displacement and time curves. The feature points are A

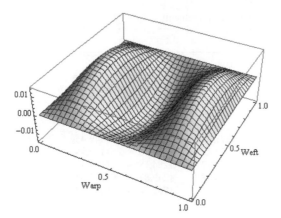

Figure 3.   The third order vibration mode.

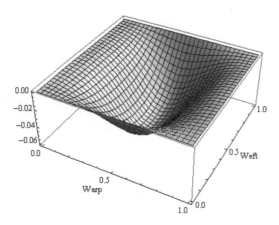

Figure 4.   The superposed vibration mode of the first three orders.

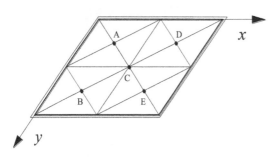

Figure 5.   Feature points on membrane surface.

($x = 0.25$ m, $y = 0.25$ m), B ($x = 0.25$ m, $y = 0.75$ m), C ($x = 0.5$ m, $y = 0.5$ m), D ($x = 0.75$ m, $y = 0.25$ m) and E ($x = 0.75$ m, $y = 0.75$ m), and they are shown in Figure 5. The first single-order displacement time histories are shown in Figures 6–10 and the

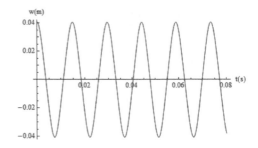

Figure 6.   Displacement and time curve of A point.

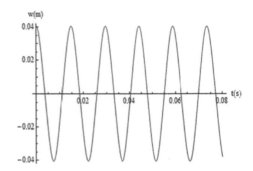

Figure 7.   Displacement and time curve of B point.

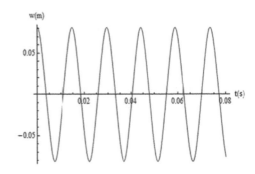

Figure 8.   Displacement and time curve of C point.

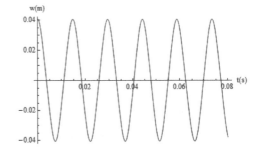

Figure 9.   Displacement and time curve of D point.

896

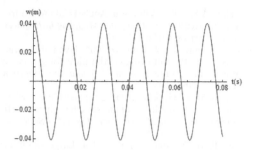

Figure 10.   Displacement and time curve of E point.

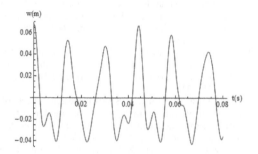

Figure 11.   Superposed displacement and time curve of A point.

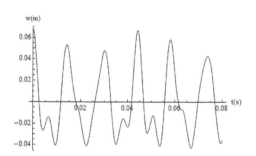

Figure 12.   Superposed displacement and time curve of B point.

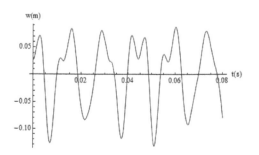

Figure 13.   Superposed displacement and time curve of C point.

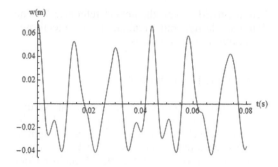

Figure 14.   Superposed displacement and time curve of D point.

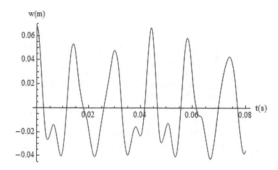

Figure 15.   Superposed displacement and time curve of E point.

superposed displacement time histories of the first three orders are shown in Figures 11–14.

From the analysis of displacement time histories, we can obtain the following conclusion.

We can obtain the single-order and superposed displacement time history of each feature point by using Equation (2) conveniently. The displacement time histories of point A, B, D and E are the same. This is because the displacements of point A, B, D and E are symmetrical. The displacement time history of C is different from the other four points, and the amplitude of C point is maximal. This is because C point is the center of the membrane, and the amplitude is the maximum spontaneously.

4   CONCLUSIONS

Through comparison, we find that Equation (1) is a simple and high-precision formula for the practical application. The precision of Equation (5) will decrease with the increase of initial displacement, and Equation (4) is too complex. Therefore, the practicability of Equation (1) is much better than that of Equations (4) and (5). Results obtained herein provide some theoretical reference for the dynamic design and manufacture of membrane structures and components.

ACKNOWLEDGEMENT

This work is supported by the National Natural Science Foundation of China (No. 51178485), the personnel development project for young and middle-aged core teachers of Chengdu University of Technology and the Chongqing Municipal Construction Committee (Construction and Scientific 2014, No. 0-11-5)

REFERENCES

Liu, C.J. Yang, X.Y. Zhao, H. & Zheng, Z.L. 2014. *Homotopy perturbation solution for strong geometrical nonlinear vibration of flat prestressed orthotropic membrane structure*, Shell Structures: Theory and Applications-Proceedings of the 10th SSTA 2013 Conference, 12: 313–316.

Liu, C.J. Zheng, Z.L. & He, X.T. et al. 2010. L-P perturbation solution of nonlinear free vibration of prestressed orthotropic membrane in large amplitude, *Mathematical Problems in Engineering*. 2010: Article ID 561364, 17.

Reutskiy, S.Y. 2009. Vibration Analysis of Arbitrarily Shaped Membranes, *Computer Modeling in Engineering & Sciences*. 51(2): 115–142.

Shin, C. Chung, J.T. & Kim, W. 2005. Dynamic characteristics of the out-of-plane vibration for an axially moving membrane, *Journal of Sound and Vibration*. 286(4–5): 1091–1031.

Soares, R.M. & Gonalves, P.B. 2012. Nonlinear vibrations and instabilities of a stretched hyperelastic annular membrane, *International Journal of Solids and Structures*, 49(3–4): 514–526.

Soares, R.M. & Goncalves, P.B. 2014. Large-amplitude nonlinear vibrations of a Mooney-Rivlin rectangular membrane, *Journal of Sound and Vibration*. 333(13): 2920–2935.

Sunny, M.R. Kapania, R.K. & Sultan, C. 2012. Solution of nonlinear vibration problem of a prestressed membrane by adomian decomposition, *AIAA Journal*, 50(8): 1796–1800.

Wang, C.Y. & Wang, C.M. 2011. Exact solutions for vibrating rectangular membranes placed in a vertical plane. *International Journal of Applied Mechanics*, 3(3): 625–631.

Wetherhold, R. & Padliya, P.S. 2014. Design aspects of nonlinear vibration analysis of rectangular orthotropic membranes, *Journal of Vibration and Acoustics, Transactions of the ASME*. 136(3): 15–19.

Zheng, Z.L. Liu, C.J. He, X.T. & Chen, S.L. 2009. Free vibration analysis of rectangular orthotropic membranes in large deflection, *Mathematical Problems in Engineering*, 2009: Article ID 634362, 9.

Advanced Materials and Structural Engineering – Hu (Ed.)
© 2016 Taylor & Francis Group, London, ISBN 978-1-138-02786-2

Virtual reconstruction: A brief analysis on the architecture developments of the digital technology era

Y.H. Zhang & Q. Yan
Chongqing Jianzhu College, China

ABSTRACT: Scientific and technological progresses have impacted heavily on traditional architectural design theories, processes, and methods in the information age. Nowadays, an increasing number of architects have begun to use digital technology to achieve their goals. Their designs move from the virtual world to reality, from traditional architectural forms to the reconstruction style. Architects can use digital technology to simulate the architecture generative process and final conditions by using software in the building design phase. This paper begins with discussed the features of digital technology and then explored the development trend of virtual building and reconstruction in order to provide a basic for the design theory of architecture, and help designers to create more era significance buildings by using digital methods and technologies.

1 INTRODUCTION

Digital technology is a new product in the information technology era which promotes the development of the architecture industry through its applications in the field of construction. It affects the architecture industry more deeply. Digital technology influences the process of design, and has risen from purely technology to design concept (Wang 2007). The implementation and construction process becomes more abundant with digital technology. It provides more possibilities for technical and operational for architects to achieve their goals. It also offers a more open vision to promote building development at the same time. Applications of Computer-Aided Design (CAD) in the architecture field began in Western universities in the 1960s.

Columbia University pioneered Paperless Studios in the 1990s, which was followed by many universities in developed countries, like Harvard University, the British Architectural Association (AA), and the Netherlands Architecture Berlage Institute. They also pioneered the digital lab and began to research a new direction in architecture and urban design for the Western countries (Ding 2013). Digital display is also a very effective way to discuss building space styles and effect and indoor and outdoor environments. American architect Frank Gary was the first to suggest the use of digital technology in architectural design. He had a sensitive and accurate view on the times. Gary had made the impossible aspects of architectural design come true. It was possible that he used the

digital technology to analyze architectural rationality, which came from Gary's digital technology in real architectural design and construction process (Xu & Zhang 2011).

2 THE CHARACTERISTICS OF DIGITAL TECHNOLOGY

We are entering into an information civilization, characterized by computer technology. This is also called the "digital era." Digital technology is based on computer data processing technology, information transmission technology, intelligent control, etc. The building process of digital technology includes computer technology design, calculation, demonstration, construction, monitoring, control, maintenance, and other applications related to the entire building technology field (Lu & Xiang 2005). The application of digital technology has provided buildings with automatic control abilities that change according to the external environment. It is dependent on innovative theories and methods in architectural design. The computer era develops not only a new style but also a new design method, which use the new technology to make the system logic diagram into reality. Architectural form would become less important under this new vision. We should explore the potential function and focus of the design process to be more intelligent and logical (Yuan 2012).

Patrick Schumacher brought up the idea of parametricism, or autonomous architecture (Zhang 2012) (Fig. 1, Fig. 2). The new digital technology

Figure 1. Patrick Schumacher's the Milky Way SOHO (Baidu).

Figure 2. Patrick Schumacher's the Milky Way SOHO (Baidu).

method was promoted and deepened through the control of the design process. "Digital" is a descriptive word for the process, not for the results. Using the associative information system based on the BIM (building information model) is more accurate due to parameter descriptions for system integration. These technologies could control the flow of design more effectively. It also provided more accurate information as well as digital construction. Gary architecture CATIA software making by building technology controls time and costs

more effective, its biggest advantage is that is provides a platform that allows communication with construction teams, and it is at (Yuan 2012).

3 ARCHITECTURE'S VIRTUAL REALITY BY DIGITAL TECHNOLOGY

3.1 Virtual reality of digital technology

The human space concept has changed with the rise of the network at the end of the 20th century. Virtual reality and cyberspace triggered the creation of digital technology. People used the virtual space instead of entity space forming by dots, lines, and surfaces. The concept of this new digital space has a deep influence on architectural design (Guo & Xiao 2007). Virtual reality is usually abbreviated as VR. It is used in computer graphics technology to simulate environments through a variety of sensing device that users input into the environment. Users can interact directly with the environment of natural technology (Yu & Huang 2000).

Digital technology can use computer technology and network technology to build realistic, virtual architectural spaces. This removes the limitations of architectural space on the physical form. The external style of architecture has trended towards the digital era's virtual and nonlinear characteristics under the impetus of digital technology. Designers can use input parameters to calculate and generate virtual 3D building models in the design process. This could result in immersive feelings and experiences. The parametric design of virtual reality is not only a design form but also a virtual concept from architects. They carry on the wonderful ideas been came true by creation and re-creation of the model examination (Yu 2001).

3.2 The establishment of the digital technology of virtual architecture

Virtual architecture has a visual sense of space that is referred to and established by computer and internet technology. The original digital virtual space couldn't establish any connection to traditional architectural space concepts. It especially did not have a proper vision of 3D space. It lacked of architectural character; therefore, virtual space cannot be called "virtual architecture." A lot of virtual planes have been replaced with spaciousness and interactive cyberspace through the improvement of virtual computer technology and interactive functions. The construction of digital virtual architecture had begun. Although virtual architecture only exists virtually, it has similarity spatial perceptions as real architecture, such as visual spatial dimensions, materials, texture, light, and shadow. Therefore, this virtual space can complete the partial function of physical architecture (Lu & Xiang 2005). Virtual architecture first started and was fashioned in three major areas, including digital media, the internet, and film and television works. They started with special effects production and the scene design in the electronic entertainment industry (Yu 2001). Architects can model the real world in every possible way and interact with their models directly by using virtual reality technology. Therefore, an architect can design a building directly through virtual building processes and finish his design (Yu & Huang 2000).

Figure 3. Heydar Aliyev Cultural Center (SJ33).

4 RECONSTRUCTION ON DIGITAL TECHNOLOGY ARCHITECTURE

4.1 Reconstruction on architectural style of digital technology

Digital technology brings vitality and development power. It also changes the designs of architecture and urban planning. Digital technology brought about the technical revolution for architecture design. It will enrich the ways and means of current architectural design to promote the diversification trend of architectural design development. The emergence of digital technology developed this architectural vision. Designers' unlimited ideas and architectural concepts can come true. The ideas greatly changed the architectural design process and the method of formation. At the same time, digital technology uses logical tools to help design the complex curved surfaces that can be seen in the figures. Parametric design can translate these complex issues to model. Parametric design provides the accurate control model platform for the designer and can move designs from the perceptual to the rational (Peng 2014). Take the Heydar Aliyev Cultural Center for example. It presents a fluid appearance (Fig. 3, Fig. 4) using geographical terrain and natural extension stacks, coils, and a separate function area. All functional areas and the entrance existed in different stack with a single and continuous surface of the building.

4.2 The abundant process of architectural design using digital technology

Architects often adopt the method of hand-drawn sketches to express the preliminary conceptions of their architectural designs. Then they use auto CAD software to further and more deeply express the building's exterior style

Figure 4. Heydar Aliyev Cultural Center (SJ33).

using modeling methods. The rendering company responds to the expression of the design. However, sketch ideas often far differ from the final architectural models; the owner's idea joins the program through the problem-solving process. Software also is the limit of architectural design. However, digital technology will be the way to solve these problems through the use of functional configurations, computer language input, and the parameterized model. Architects don't have to worry about the current form of the conditions. Architects can pay attention to the parameters of the relationship between the input and output based on the random computer results. People need to choose the optimal results; therefore such work efficiency can improve several times than ever before. Architects can see the whole process and efficiently create a variety of options in parametric design. The former methods focus on the design results and tend to be converted to processing design patterns (Peng 2014).

5 CONCLUSIONS

1. Architects have more methods, resources, and conditions for architectural design at their disposal in the current digital age. Architects can use digital tech-ology to help maximize architecture and widen our vision and trains of thought. Then they can make the design come true. Architects have to learn the method of architectural design.
2. Digital technology makes the generation process of construction more abundant. The development of technology promotes the deployment of society, which improves people's aesthetic consciousness. They do their best to create a building designer spare artistic flavor by using advanced technology.
3. Digital technology has become the new milestone in the history of the catalyst. Innovations not only include the architectural developing design concepts and processes but also historic reforming, which is closely related to the construction industry. The development trend of digital era buildings will be closely related to them.

REFERENCES

Ding, J.F. 2013. *Pragmatistic Coding, 1970s—Born Chinese Architects' Perspectives on Digital Design in Architectural Fields*, Time & Architecture.

Guo, B.F., & Xiao, D.W. 2007. Digital Architecture—A New Conception of Space, *Journal of Chongqing Jianzhu University*. 3: 17–21.

http://image.baidu.com/i?tn=baiduimage&ps=1&ct=20
1326592&lm=-1&cl=2&nc=1&ie=utf-8&word=%E9
%93%B6%E6%B2%B3SOHO%E8%AE%BE%E8%
AE%A1.

http://www.sj33.cn/architecture/slsj/wenhua/201311/
35667.html.

Lu, A.M. & Xiang, B.R. 2005. Ecological Building Following Digital Technology, *New Architecture*, 6.

Peng, X. 2014. Analysis of the Application of Digital Construction in Architectural Design, *HuaZhong Architecture*, 2: 18–21.

Wang, H. 2007. *Digital era, architecture, the architect—The influence of digital era to the Architecture and architects*, The fifth national conference for graduate students in architecture and planning.

Xu, M. & Zhang, J. 2011. Analysis of the Pritzker Architecture Prize Winning Works with the Characteristic Architectural Forms of the Digital Age, *New Architecture*, 6.

Yu, C.F. 2001. Buildings with no one to inhabit Architects without clients—Discussing on Virtual architect and Virtual architecture in practice area, *HuaZhong Architecture*, 6: 12–15.

Yu, H. & Huang, T. 2000. *Internet, Virtual Reality (VR) and Architecture Design*, Architecture and Culture.

Yuan, P.F. 2012. *Digital Fabrication Paradigm Shifting under the New Methodology*, Time & Architecture.

Zhang, S.J. 2012. *Formation and Materialisation Integral Design*, Time & Architecture.

Advanced Materials and Structural Engineering – Hu (Ed.)
© 2016 Taylor & Francis Group, London, ISBN 978-1-138-02786-2

Identifying resolution controlling parameters in the design of Coefficient of Thermal Expansion measurement tool

S.A. Hassan & H.M. Ali
Department of Mechanical Engineering, Institute of Space Technology, Islamabad, Pakistan

W. Aslam, M.A.A. Khan & M.H. Ajaib
Department of Materials Science and Engineering, Institute of Space Technology, Islamabad, Pakistan

H. Moin
Department of Aerospace and Aeronautical Engineering, Institute of Space Technology, Islamabad, Pakistan

ABSTRACT: The ever increasing demand for the high end reliable engineered products has put the materials metrology industry on much pressure, and has resulted in the development of various tools and practices to tailor and craft the end products with high precision and accuracies. The coefficient of thermal expansion is one such material property. In this study, mathematical models are indigenously developed and presented to predict the movement of the laser spots on the surface of the lever resting on two material columns as a function of its thickness and the mirror's leg height. These two parameters are shown to control intricately the overall tool's resolution. The laser passing through the mirror is altered when it gets tilted due to the change in a resting surface's dimension. This paper discusses the relative change in the position of the spots as a fraction of the dimensional change in material length.

1 INTRODUCTION

The material design and manufacturing industry is always intrigued by the idea of exploring advanced materials and associated processes, which may result in high technology products and overall yield. However, in recent times, more attention is given to improving existing systems and processes and making them more efficient and capable. Hence, mechanical systems are now designed, developed and run with more control with resulting products finely tuned, reliable and predictable and their useful life enhanced to meet customer demands satisfactorily. This can only be done when materials and components used in the systems or machines are designed and manufactured with near zero or apparently flawless perfection (http://www.pinkbike.com/news/Chris-Kings-vision-Products-2012.html). Dimensional accuracies are tuned and controlled up to submicron and potentially in the nanoscales (Sitti 2001). However to manufacture systems with such characteristics, it is necessary to develop and explore tools that can actually measure or determine such properties of the materials precisely. One such material property under varying temperature fields is the Coefficient of Thermal Expansion (CTE). It is the parameter that decides the strain produced in the material at elevated or reduced temperatures

(Zhu et al. 2002). This parameter needs to be accurately predicted for the components to function in the high demanding applications such as in aerospace, military and sensitive optical cum electrical systems.

Various methods researched for the determination of the CTE based on different principles include: Push rod quartz dilatometer, Michelson Interferometer Probing on 2 sides of sample, Michelson Interferometer Probing on 1 side of sample, Absolute Michelson Interferometer Probing on 1 side of sample, and Diverging beam interferometer probing on flat mirror on sample and the Capacitance cell (Olanrewaju et al. 2011, Progelhof et al. 1976, Christensen & Lo 1979). They are all high-resolution CTE measurement techniques. One such approach being manifested by the authors in the past is by using the optical comparator principle to help take measurements for the changes in the lengths of the columns and material blocks to calculate the linear CTE of the samples (Hassan et al. 2013).

In this study, the optical lever principle is investigated further, and a new mathematical formulation is generated to predict the movement related behavior of the laser spots incident on the optical lever when in a total horizontal/zero tilt position and when the material expands or contracts under the action of temperature change. The newly found analytical formulas can be further employed to predict the maximum resolution range that can be attained from this sort of CTE measurement tools, presented in the study. The spot dynamics is shown to be a function of Optical lever's Thickness T and its Leg height LH, which rests on top of the samples.

2   THE BASIC TOOL UNDER CONSIDERATION

The CTE measurement tool based on the principle of the optical lever is shown in Figure 1, where the optical lever or the double-sided glass slab is the fundamental part of the whole tool. This lever rests on the two columns of materials: one being the specimen whose CTE needs to be determined and the other one is the reference material Zerodur with known or relatively fixed CTE values. The diode laser is used as a profiling tool, as shown below, and would hit the lever at a fixed point when in a total horizontal position, and would come out of the lever after going through the mirror, from possibly two positions.

The mirror is specifically coated as such so as to acquire only two beams out of the mirror. Hence, two laser spots would be observed, as shown in the Figure 1. The separation distance between the two outgoing laser beams is the distance that can be mapped onto the piece of hardware capable of recording the movement of the laser spots. This then can be programmed, translated and fed into the computer to calculate the CTE of the material. The tilt of the mirror would be produced when one resting surface (i.e. specimen with unknown CTE) when subjected to temperature change, expands or contracts, hence, resulting in the disturbance of the lever from its initial total horizontal rest position.

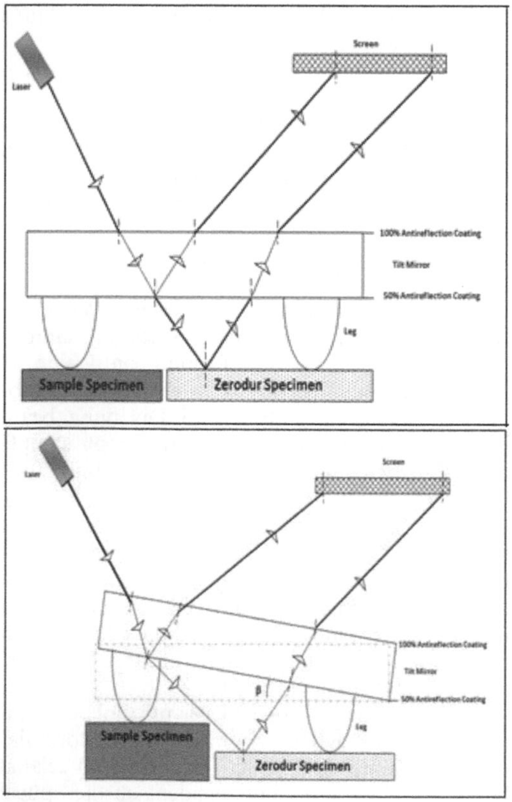

3   MATHEMATICAL MODEL TO PREDICT LASER SPOT MOVEMENT

There are two cases that need to be considered. One already described earlier, the total horizontal position of the lever and the other is the tilted case, as shown in Figure 2 and Figure 3, respectively.

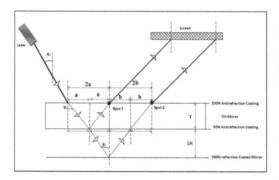

Figure 2.   Spot dynamics for the lever at the rest position.

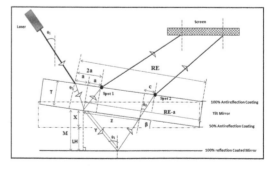

Figure 3.   Spot dynamics for the lever at the rest position.

Figure 1.   Two basic positions of the optical lever and the respective positions of the laser spots.

In both cases, the first laser beam after it has entered into the lever strikes the 50% polished internal lower side of the lever and after reflection gets out of the lever as spot 1. Then, 50% of the initial beam gets out of the lever when it strikes its previously mentioned lower side. This beam after hitting the totally polished face of the reference material reflects back to re-enter the lever as beam 2. Then, again traveling through the thickness of the lever and getting out at the top as spot 2.

3.1   Case 1

For the calculation of spot 1 distance from the initial laser strike point which will be a fixed point for each tilt position, $2a$ needs to be calculated as follows:

$$T \cdot \tan \theta_2 = a \qquad (1)$$

where θ_2 can be determined by using Snell's law, which is defined as the angle with which the laser enters the glass slab (taken with normal), and T is defined as the thickness of the lever:

$$\theta_2 = \sin^{-1}\left(\frac{n_1}{n_2} \cdot \sin \theta_1\right) \qquad (2)$$

where n_1 and n_2 are the refractive indices of the air and glass, respectively, and θ_1 is the angle by which the laser hits the glass slab.

Now by putting the value of θ_2 in Equation 1, it becomes

$$a = T \cdot \tan\left[\sin^{-1}\left(\frac{n_1}{n_2} \cdot \sin \theta_1\right)\right] \qquad (3)$$

where $2a$ is the total separation distance from the incident spot, so

$$2a = 2T \cdot \tan\left[\sin^{-1}\left(\frac{n_1}{n_2} \cdot \sin \theta_1\right)\right] \qquad (4)$$

For the calculation of spot 2 distances from the initial laser strike point, which will be a fixed point for each tilt position, $2a+2b$ is the required distance. As we have calculated $2a$, $2b$ needs to be calculated as follows:

$$LH \cdot \tan \theta_1 = b \qquad (5)$$

where LH is the leg height, which is placed on the 100% reflected coated mirror. While $2b$ is

$$2b = 2 \cdot (LH \cdot \tan \theta_1) \qquad (6)$$

So, the final equation for the total distance from the laser to spot 2, which is $2a+2b$, can be written as follows:

$$2 \cdot \left\{T \cdot \tan\left[\sin^{-1}\left(\frac{n_1}{n_2} \cdot \sin \theta_1\right)\right]\right\} + 2 \cdot (LH \cdot \tan \theta_1) \qquad (7)$$

3.2   Case 2

In the tilted position to find spot 1 distance, $2a$ needs to be calculated first as follows:

$$2T \cdot \tan \theta_2 = 2a \qquad (8)$$

While θ_2 can be determined by using Snell's law, which is defined as the angle with which the laser enters the glass slab (taken with normal):

$$\theta_2 = \sin^{-1}\left(\frac{n_1}{n_2} \cdot \sin(\theta_1 + \beta)\right) \qquad (9)$$

where β the angle by which the tilt mirror or lever goes to be tilted upon the expansion or contraction of the test specen with an unknown CTE:

$$|Spot\ 2| = a' + c + z \qquad (10)$$

where a' is the same distance as calculated above for the spot 1 distance, but with addition of the tilt in the total angle, while c and z are the unknown distances, as defined in Figure 3:

$$a' = T \cdot \tan\left[\sin^{-1}\left(\frac{n_1}{n_2} \cdot \sin(\theta_1 + \beta)\right)\right] \qquad (11)$$

While θ_2 can be expressed as follows:

$$\theta_2 = \sin^{-1}\left(\frac{n_1}{n_2} \cdot \sin(\theta_1 + \beta)\right) \qquad (12)$$

To Solve z, various unknown quantities need to be determined, which include X, Y and M:

$$(RE - a) \cdot Sin\beta = X \qquad (13)$$

where RE is the distance from the right edge to the point where the laser enters the tilt mirror:

$$Y = \frac{M}{Cos\theta_1} \qquad (14)$$

$$M = X + LH \qquad (15)$$

where X is the distance from the lower surface of the tilt mirror after tilting to the lower surface of the tilt mirror before tilting.

Putting the value of M in Equation (13), it becomes

$$Y = \frac{X + LH}{Cos\theta_1} \qquad (16)$$

Then, after putting the value of X from Equation (13) in Equation (16), it becomes

$$Y = \frac{(RE - a)\cdot Sin\beta + LH}{Cos\theta_1} \qquad (17)$$

By using the Law of Sines we obtain an equation having the desired unknown z function:

$$z = \frac{Y \cdot Sin2\theta_1}{Sin\{90 - (\theta_1 - \beta)\}} \qquad (18)$$

Putting the value of Y from Equation (17) in Equation (18), it becomes

$$z = \frac{\{(RE - a)\cdot Sin\beta + LH\}\cdot Sin2\theta_1}{Cos\theta_1 \cdot Sin\{90 - (\theta_1 - \beta)\}} \qquad (19)$$

Now, calculating distance c as follows:

$$T \cdot \tan\theta_3 = c \qquad (20)$$

where θ_3 can be determined by using Snell's law, which is defined as the angle with which the laser beam exits the glass mirror and, respectively, enters the air medium (taken with normal):

$$\theta_3 = \sin^{-1}\left(\frac{n_1}{n_2}\cdot\sin(\theta_1 - \beta)\right) \qquad (21)$$

Putting the value of θ_3 in Equation (20) it becomes

$$c = T \cdot \tan\left[\sin^{-1}\left(\frac{n_1}{n_2}\cdot\sin(\theta_1 - \beta)\right)\right] \qquad (22)$$

By putting the values of a, z and c in Equation (12) it becomes

$$|Spot\ 2| = \left[T\cdot\tan\left\{\sin^{-1}\left(\frac{n_1}{n_2}\cdot\sin\theta_1 + \beta)\right)\right\}\right]$$
$$+ \left[\frac{\{(RE - a)\cdot Sin\beta + LH\}\cdot Sin2\theta_1}{Cos\theta_1 \cdot Sin\{90 - (\theta_1 - \beta)\}}\right]$$
$$+ \left[T\cdot\tan\left\{\sin^{-1}\left(\frac{n_1}{n_2}\cdot\sin(\theta_1 - \beta)\right)\right\}\right]. \qquad (23)$$

4 CONCLUSIONS

The proposed mathematical model can precisely calculate the two spot distances, respectively, from the initial laser hit point, which will be kept constant and stationary in the whole experimental procedure. As the temperature in the holding assembly the chamber is raised or lowered, the specimen material having an unknown CTE will expand or contract depending on the heating condition. This produces deflection on the part of a lever or the mirror placed on top of it. To control the resolution of the whole measurement tool, say less than 1 micron, certain aspects of the assembly need to be profiled and set accordingly. The final formulae obtained in Equation (7) and Equation (23) for spot 2 distances from the initial laser hit point include terms for the mirror thickness T and the height of the legs LH, whereas in the spot 1 distance formulae, i.e. in Equation (4) and Equation (8), only T appears. This clearly suggests that spot 1 dynamics is a function of the mirror thickness, whereas spot 2 dynamics is a function of both mirror thickness and a height of mirror legs. Apart from controlling these two variables, one other variable that can be manipulated to obtain the desired resolution or capability to detect even the minutest of details in length change is the distance with which the screen or camera lens is placed for the laser dots to fall upon after they come out from top of the whole assembly, as shown in Figure 1. For the proposed case, the more the distance of the screen from the stationary mirror leg placed on the reference sample, the more closer the laser dots will be, as the beam is converging as can be seen from Figure 1. However, a compromise needs to be achieved so as to consider for the lowering of the laser spots intensity if and when the distance of the screen is increased substantially. Hence, collectively, all of these three independent variables can be manipulated to provide with the desired output resolution expected from the whole tool, and can provide with more intricate design control and benefits for the industry to achieve precise and accurate measurement assemblies. Apart from these factors, there are, however, limitations of the setup such as the requirement of a thermally stabilized laser source, control over the temperature fluctuations and vibration and disturbance free environment in which to place the assembly.

REFERENCES

Article available at http://www.pinkbike.com/news/Chris-Kings-vision-Products-2012.html.
Beck, J.V. Cole, K.D. Sheikh, A.H. & Litkouhi, B. 1992. *Heat Conduction Using Green's Function*, Hemisphere Publishing Corporation, London, 346–351.

Christensen, R.M. & Lo, K.H. 1979. Solutions for Effective Shear Properties in Three Phase Sphere and Cylinder Models. *Journal of the Mechanics and Physics of Solids*, 27: 315–330.

Hassan, S.A. Ahmed, H. & Israr, A. 2013. An Analytical Modeling for Effective Thermal Conductivity of Multi phase Transversely Isotropic Fiberous Composites using Generalized Self Consistent Method, *Applied Mechanics and Materials*, 249–250: 904–909.

Olanrewaju, P.O. Gbadeyan, J.A. Hayat, T. & Hendi, A.A. 2011. Effects of internal heat generation, Thermal radiation and buoyancy force on a boundary layer over a vertical plate with a convective surface boundary condition, *South African Journal of Science*, 107(9–10): 1–6.

Progelhof, R.C. Throne, J.L. & Ruetsch, R.R. 1976. Methods for Predicting the Thermal Conductivity of Composite Systems: A Review, *Polymer Engineering and Science*, 76(9): 615–619.

Sitti, M. 2001. *Survey of nanomanipulation systems*, Proc. of the IEEE Nanotechnology Conference, Maui, USA, 75–80.

Zhu, H. Li, W. Tseng, A.A. & Phelon, P. et al. 2002. Measurement of CTE at reduced temperature for stressed specimens, *Composite Materials: Testing, Design and Acceptance Criteria ASTM STP*, 1416: 212–220.

Advanced Materials and Structural Engineering – Hu (Ed.)
© *2016 Taylor & Francis Group, London, ISBN 978-1-138-02786-2*

Summary of intelligent power grid fault diagnosis methods

Z. Yu & C. Chen

College of Electrical Engineering and New Energy, China Three Gorges University, Yichang,
Hubei Province, China

ABSTRACT: Together with the present power grid fault diagnosis, fault diagnosis methods are important to improve the grid work efficiency and ensure failure alarm real-time and effectiveness. We summarized the rough set theory, cloud computing, distributed system, fault recording, temporal fuzzy theory, geographic information system and multi-agent system diagnosis methods. The principles, advantages and disadvantages of different diagnosis methods were researched.

1 INTRODUCTION

With the construction and development of smart grids in China, the system complexity has been increasing. When the electric power system failure occurs, the control center receives a huge amount of information. At the same time, interconnected power grid blackouts risk is increasing. The power grid operation security and stability has been an indispensable prerequisite for national economic development and social normal operation. In order to improve the grid work efficiency and ensure the normal operation and failure alarm real-time and effectiveness, an intelligent power system fault diagnosis analysis system must be established. Through analyzing a large amount of alarm filtering and logical reasoning, it can diagnose actual events in the power system. With the development of artificial intelligence, computer network and electronic technology, all kinds of fault diagnosis methods have been put forward constantly, based on the rough set theory, cloud computing, distributed system, fault recording, temporal fuzzy theory, geographic information system and multi-agent system diagnosis methods.

2 COMMON FAULT DIAGNOSIS METHODS INTRODUCTION

2.1 *Rough set*

As a theory to deal with uncertain, inaccurate and incomplete data in the mathematical model, the rough set is a mathematical tool to describe incompleteness and uncertainty. It can deal with all kinds of uncertain information effectively, discover the implicit knowledge, and reveal the potential regularity.

Its advantage involves the powerful data analysis ability. It can analyze and deal with imprecise, inconsistent, incomplete information effectively, and eliminate noise and understand without prior knowledge. It has a strong fault tolerant ability in the attribute reduction process. Its disadvantage involves the difficulty in establishing and dealing with the decision table for a large power grid.

Chao (2011) formed a grid alarm decision table by using the rough set theory. It has the decision table attribute reduction, value reduction and automatic extraction of grid intelligent alarm rules. It provided integrated alarm information to the operation staff timely and accurately by using the extracted rules and grid topology results for logical reasoning alarm information. Thus, it achieved the alarm initial intelligent information processing. Zhao (2005) proposed the attribute reduction algorithm and improvement value reduction algorithm, combined with the rough set theory and binary logic operations, by using the knowledge mining method with the rough set for fault diagnosis. Chao (2011) studied the decision tree method in data mining and the rough set method. On this basis, it analyzed power grid alarm information attributes, and improved alarm message denoising classification decision tree model and the noise information filtering rate. Then, it ensured rules completeness by using the rough set method for digging the grid alarm inference rules.

2.2 *Cloud computing*

Cloud computing is a computing resource amount paid access and an increase-use-delivery pattern based on an Internet-related service. It usually involved dynamical, scalable and virtualized resources provided by the Internet. Cloud is a metaphor of Internet or network. Computing resource pool (e.g. network, servers, storage, applications and services) can be obtained and configured quickly. It can reduce management inputs and interact with suppliers significantly. It can also integrate resources and services better in different locations and different network devices to realize the maximum range resources-data cooperation and sharing.

Its advantages are low cost, high reliability, huge data processing amount, flexible-extensible-storage capacity, fast calculation speed and high equipment utilization rate. Its disadvantages are immaturity in the architecture, virtual machine, data management, energy management, resource scheduling, programming model and cloud security technology.

Chen (2013) put forward ideas and methods to cloud computing architecture construction of a county electric power dispatching automation system. It realized unified scheduling work platform, secure mobile and collaborative scheduling and advanced applications with the desktop virtualization technology. Li (2013) presented the Mobile-agent power system information interaction mechanism as a cloud computing platform, and built the power system cloud computing collaboration model based on Mobile-agent. Wang et al. (2010) proposed a smart grid state monitoring cloud computing platform solution, according to the smart grid monitoring characteristics, and combined with the Hadoop cloud computing technology. It researched virtualization, distributed storage and parallel programming model in cloud computing to achieve smart grid information reliable storage and fast parallel processing.

2.3   Distributed system

Distributed system is a software system built on a network, with high cohesion and transparency. Cohesion refers to autonomy of each database distribution node. Besides, there is a local database management system. Transparency means that each database node distribution is transparent for user applications, without local or remote recognition. Grid distributed diagnosis framework consisted of four subsystems: communication, topology processing, comprehensive treatment and external data acquisition. The bottom is connected to a data grid. Each could communicate via a dedicated network in a power system. Grid distributed diagnosis emphasizes the comprehensive distribution of resources, function, task, control and management.

Its advantages are distribution, high reliability, scalability, economy, cohesion, transparency, high response speed, and ability of fault, fault check and load balancing. The disadvantage is the lack of contact line fault diagnosis rules.

Li et al. (2010), based on the foundation of distributed idea, proposed an association rules mining algorithm based on the rough set theory to extract the association rules effectively from the distributed information system. Wang (2013) presented

a distributed fault diagnosis program operation mode to provide a high-performance distributed computing strategy for a more complex fault diagnosis algorithm and ensure the diagnosis time to meet the needs of practical engineering. Hong (2009) presented a distributed power system fault diagnosis method based on the concept lattice to diagnose the faults effectively among the local power grid tie line, and provide the possible fault types in case of losing the core property.

2.4   Fault recording

Based on a microcomputer, the fault recording device can record electrical quantities and state change process information before and after the fault occurs, and reflect fault transient changes and relay protection actions. It has the data archiving and analysis ability. With the communication technology development, power dispatching can collect fault recorder information distributed in each station at any time, that is, the fault recorder network system.

Its advantages are fault recording accuracy and continuity, process and sequential logic, powerful analysis and many kinds of high-precision algorithms in the computation. Its drawback is the difficulty in data storage and transmission in the face of a large amount of data.

Zhang (2011) presented the phase power system fault diagnosis method based on recorded fault data, and developed the corresponding program. He also presented the practical and effective fault diagnosis method by using the phase characteristics results, namely the horizontal and vertical comparative methods. He (2001) presented virtual protection diagnostic thinking, and established the corresponding comprehensive diagnosis model to compensate for the diagnosis limitations effectively with switch protection information, and built a universal sharing diagnosis system by using the dynamic component object model COM and Windows link library DLL. Xu et al. (2011) designed the real-time and off-line substation fault analysis system based on the fusion of fault recorder and protection information, and developed the corresponding software. Based on the COMTRADE protocol, it solved the fault recorder information and protection information format conversion problems. The recorded information is preprocessed by using the analog low-pass filter and digital filter technology and Fourier transform and wavelet transform filtering compression methods.

2.5   Temporal fuzzy theory

In the alert information, in addition to the event logic, there is temporal logic. The time of the incident called the event. The sequence is called the order. The temporal logic sequence of event time and event sequence is called the timing constraints. Sequential logic alerts can be used to handle the protection or breaker misoperation and false alarm and other uncertain factors. At the same time, the time sequence of event information with millisecond resolution, based on the global positioning system, recognizes event signal state metabolic order and provides objective conditions for using the alarms temporal information.

Its advantage is that the complete process is helpful to the power system fault analysis. Thus, it ensures the diagnosis validity. Its disadvantage is that one has to ask for time accurate definition. So, it is not conducive to the practical application.

Zeng et al. (2014) presented a power system fault diagnosis method combined with temporal reasoning and fuzzy reasoning. It can deal with two kinds of uncertain factors. They are the reliability of protective relays and circuit breakers, and the correctness and completeness of the alarm signal receiving in the dispatching center. Guo (2010) presented an analytical model for power system fault diagnosis, concerning alarm information timing characteristics. It also proposed a fault diagnosis for a high-voltage transmission line algorithm, based on waveform matching, namely the harmony search algorithm. The high-voltage transmission line fault diagnosis is described as a hybrid optimization model, consisting of discrete and continuous optimization variables. Li (2013) improved the fault diagnosis model based on the temporal constraint network, and analyzed the alarm node timing constraint rationality, and came up with a method to measure warning information entropy. He also constructed the power system fault diagnosis model based on the alert timing constraints rationality. Guo et al. (2009) constructed the alarm processing analytical model, which made full use of timing information, and studied the events' uncertain problem. On this basis, Liu et al. (2014) built a fault diagnosis improved optimization model, concerned with alarm timing information, and set the failure time as a continuous change real variable, and protection and circuit breaker status as a real function. Thus, this can change over time to reflect the continuous state and cooperative relationship of a member, protection and circuit breaker during the failure occurrence development process.

2.6 Geographic information system

Geographic Information System (GIS) is a comprehensive discipline. The system can be an organic fusion of geospatial information and computer system. Through the unified model system establishment, it realizes the unified display, operation and analysis functions of spatial data and attributes data. It is a computer system with input, storage, query, analysis and display functions of geographic data.

Its advantages are rapid positioning in the face of sudden failure, improving the query efficiency with figure mutual check data, and data sharing. Its disadvantages are spatial data structure and data management problems, complex graphics and attribute data input, difficulty in computerizing the geographic information system, not perfect development model. Thus, it is not a distributed system uniform standard.

Jin (2013) designed the city domain network management system based on GIS. It also designed the visualization-imaging GIS network management system of power grid operation real-time parameters automatic acquisition and power facilities parameters unified management, with real-time information acquisition technology, GIS technology, component technology, database and network technology. Tang (2009) introduced a drawing method based on the MapInfo power system geographical wiring and principle diagram, and analysis method based on the grid topological structure. It also researched the realization method of the grid information management and network planning management system based on GIS. Hu (2005) researched the grid data visualization based on the GIS, with the geographic information system, digital elevation model and scientific visualization technology as the support, by means of geographical information system data structure based on geographic elements, spatial object management and analysis function, plus scientific visualization technique for efficient data science and explanation.

2.7 Multi-agent system

Multi-agent system is a system of intelligent and flexible response to changes in working conditions and demand around the process. It is composed of multiple agents through cooperation. The basic unit is the agent. The agent can interact with the environment. It is composed of three functional layers: management and organization layer, coordination layer and execution layer. It focuses on how to coordinate the multiple agent actions with logical or physical separation-different target to act cooperatively or solve problems. So, it can maximize the realization of respective and overall aims.

Its advantages are flexibility, scalability, adaptability, autonomy, persistence, mobility, rational communication ability, reasoning ability, coordination, planning, perception and simple expression. Its shortcoming is that the system is more complex. In addition, there is no planning to describe Agent behavior. Establishing and improving the

Table 1. Comparison of the power grid fault diagnosis methods.

fault diagnosis methods	advantages	disadvantages
rough set	powerful data analysis ability,analyze and deal with imprecise, inconsistent, incomplete and so on incomplete information effectively, eliminate noise and understand without prior knowledge, and strong fault tolerant ability in attribute reduction process	difficult to establish and deal with decision table for large power grid
cloud computing	low cost, high reliability, huge data processing amount, flexible-extensible- storage capacity, fast calculation speed and high equipment utilization rate	immaturity in the architecture, virtual machine, data management, energy management, resource scheduling, programming model and cloud security technology
distributed system	distribution, high reliability, scalability, economy, cohesion, transparency, high response speed, and ability of fault,fault check and load balancing	lack of contact line fault diagnosis rules
fault recording	fault recording accuracy and continuity, process and sequential logic, powerful analysis and many kinds of high-precision algorithms in the computation	difficult in data storage and transmission in the face of a large amount of data
temporal fuzzy theory	complete process,effective diagnosis	asking for time accurate definition, and difficulty to practice
geographic information system	rapid positioning in the face of sudden failure, improving the query efficiency with figure mutual check data, and data sharing	spatial data structure, data management problems, complex graphics and attribute data input, difficulty to computerize geographic information system, not perfect development model and no distributed system uniform standard
multi-agent system	flexibility, scalability, adaptability, autonomy, persistence, mobility, rational communication ability, reasoning ability, coordination, planning, perception and simple expression	more complex system, no planning to describe agent behavior, and hardware and software problem

distributed simulation system, and test platform based on MAS is still faced with the hardware and software problem.

Sun et al. (2005) presented a power grid dispatch intelligent decision model, by using the intelligent agent, online analysis and data mining technology, to solve the distribution, collaborative and interactive processing problem.

3 SORTING SUMMARY

Compared with the advantages and disadvantages of various types, power grid fault diagnosis methods are listed in Table 1.

4 POWER GRID FAULT DIAGNOSIS SYSTEM DEVELOPMENT TENDENCY

Throughout the power grid study, though the power system alarm fault diagnosis processing field has fruitful theoretical results and some methods have been into the practical phase, there are still some shortcomings. It has a further development space.

Most power grid fault diagnosis research fails to advance the grid fault diagnosis research progress in modeling, just based on solving a certain point in diagnosis and introducing new intelligent algorithm on existing models. So, some problems cannot be resolved fundamentally.

By increasing the power grid scale and protection configuration complexity stress, the inference model and knowledge base are established. Also, the fickle operation makes the grid topology change. So, the knowledge base often needs to be reconstructed. It is difficult to complete and maintain the knowledge base.

Diagnostic methods focus on the results' accuracy rather than interpretability. Considering the real-time computation efficiency, it needs to show more detailed breakdown, recreate the, and provide the fault diagnosis specific basis.

5 CONCLUSIONS

The smart grid will become the focus of future power grid development. The intelligent alarm system plays an important role in guaranteeing the safe power grid operation. This paper introduced the grid fault intelligent diagnosis methods simply and compared the advantages and disadvantages of various methods, to try to sort out the above methods, for reference to the related workers and even researchers. So, it can provide more effective power grid alarm, and improve the operation, information processing and management level of the scheduling system, to complete the power grid intelligent scheduling system.

ACKNOWLEDGMENT

This work was financially supported by the College of Electrical Engineering and New Energy Leading Academic Discipline Project (51177088) in China Three Gorges University.

REFERENCES

Chao, J. 2011. *Research on Power Grid Intelligent Alarm System Based on Data Mining Technology*, North China Electric Power University.

Chao, J. Liu, W.Y. Liu, Y.Z. & Zhao, L.B. 2011. Automatic extraction and application of network alarm rule based on the rough set, *Power System Protection and Control*, 39(8): 1–5.

Chen, M. 2013. *Study and Implementation on Key Technologies of Power Dispatching Automation System Based on Cloud Computing*, Guangxi University.

Guo, W.X. 2010. *Power System Alarm Processing and Fault Diagnosis: Analytic Models, Methods and Applications*, South China University of Technology.

Guo, W.X. Wen, F.S. Liao, Z,W, Wei, L.H. & Xin, J.B. 2009. Configuration and Principle Analysis of Microcomputer Protection for Classified High-voltage Controllable Shunt Reactor, *Automation of Electric Power Systems*, 33(21): 1–8.

He, Z.F. 2001. *Research on Power System Fault Diagnosis Method Based on FRD and Development of FRD Analysis Software*, Beijing, North China Electric Power University.

Hong, X.D. 2009. *Application of Concept Lattices in Distributed Power Networks Fault Diagnosis*, North China Electric Power University.

Hu, Z.W. 2005. *The Research of Grid Management System Data Visualization Methods Based on GIS*, Zhejiang University.

Jin, H.Q. 2013. *The Research and Design of Grid Management System Based on GIS*, Yangzhou University.

Li, J.L. 2013. *Synergistic Model of Cloud Computing for Power System Based on Mobile-agent*, North China Electric Power University.

Li, M. 2013. *Research on Fault Diagnosis Models for Substation Centralized Control Center of Power System*, South China University of Technology.

Li, R. Zhang, L.Y. Gu, X.P. & Li, H.M. 2010. Distributed Fault Diagnosis of Power Networks Applying the United Rules Mining Algorithm Based on Rough Set Theory, *Proceedings of the CSEE*, 30(4): 1–7.

Liu, S.H. & He, F. 2014. Optimization Model for Power Network Fault Diagnosis with Temporal Information of Alarm Messages, *Proceedings of the CSU-EPSA*, 26(1):1–5.

Sun, C.J. Zhang, J.W. & Zhao, X.Q. 2005. Multi-Agent intelligent decision support system model for grid dispatching, *East China Electric Power*, 33(8): 1–3.

Tang, B.Y. 2009. *GIS-based Planning and Management of Power Grid System*, Zhengzhou University.

Wang, D.W. Song, Y.Q. & Zhu, Y.L. 2010. Information Platform of Smart Grid Based on Cloud Computing, *Automation of Electric Power Systems*, 34(22): 1–6.

Wang, L. 2013. *Research on Power Grid Fault Diagnosis and Its Framework*, Shandong University.

Xu, Y. Zhang, X.M. Wu, Z.J. & He, R.M. 2011. Monitoring and control system based on the touch screen and PLC for shipboard power station, *Electric Power Automation Equipment*, 31 (1): 1–5.

Zeng, F. Zhang, Y. Liu, Y. Zhang, Y. Zhang, X.Y. Yuan, Y.B. & Wen, F.S. 2014. A temporal and fuzzy logic inference based method for power system fault diagnosis, *Journal of North China Electric Power University*, 41(1): 1–8.

Zhang, J.H. 2011. *Research on Power System Fault Diagnosis Method Based on Fault Recorder Data*, North China Electric Power University.

Zhao, D.M. 2005. *Research on Power Grid Fault Diagnosis Method Based on Multi-information Sources*, North China Electric Power University.

Advanced Materials and Structural Engineering – Hu (Ed.)
© *2016 Taylor & Francis Group, London, ISBN 978-1-138-02786-2*

A novel modified UWB model for multiple cluster

M.X. Shen
School of Automatic Control and Mechanical Engineering, Kunming University, China

X.L. Li
Yuanmouxian Fire Services, China

X.Y. Ruan
Kunming Shipbuilding Design and Research Institute, China

ABSTRACT: Ultra-Wideband (UWB) is a promising technology for signal transmission in automatic systems. This paper carries out a simulation study of the modified UWB propagation channel model for multiple cluster. We develop and test the modified UWB model using MATLAB, and compare our simulation results with the measurement data published in the literature. The result shows that the proposed model can achieve good performance.

1 INTRODUCTION

Channel modeling is the first and most important step to investigate signal transmit for various systems. The ultimate performance limits of automatic systems are determined by the signal channel it operates in. Meanwhile, Ultra-Wideband (UWB) as a fast emerging technology, which is based on ultra-short waveform, offers low cost, low power high data rate solutions and thus is a promising candidate for WUSB, is a radio technology pioneered by Robert A. Scholtz and others. It may be used at a very low energy level for short-range, high-bandwidth communications using a large portion of the radio spectrum and supports high-speed data rates up to 448 Mbps (Wikipedia).

By the USA radio regulation authority Federal Communications Commission (FCC), the UWB signal has a bandwidth greater than 500 MHz, which means that the overlapping with the other RF systems is evident (Oppermann 2004). The standard spectrum band of UWB is 3.1–10.6 GHz, which is super high compared with a narrow band. The main distinguishing features of the UWB propagation channel are its extremely multipath-rich profile and non-Rayleigh fading amplitude characteristics.

In this paper, we investigate the UWB channel model based on the Saleh-Valenzuela (S-V) Model, the Neyman-Scott (N-S) cluster model and the Δ-K model. Then, we study the UWB system performance in the presence of multi-band interference.

The main focus on the channel modeling is for UWB multiple cluster scenarios.

In Section 2, we briefly introduce the UWB propagation model. Three classical UWB channel models, namely the S-V model, N-S cluster model and Δ-K model, are studied in Section 3. Section 4 describes the simulation flow chart for the modified UWB model and presents the simulation results. Finally, the conclusion is provided in Section 5.

2 UWB PROPAGATION MODEL

Wireless propagation channels have been investigated extensively in the literature, particularly in the cellular communications context, and a large number of channel models are available in the literature.

Mobile radio channel models are usually studied in terms of deterministic models, empirical models. Deterministic models are based on a fixed geometry (e.g. buildings, streets) and used to analyze particular situations. They provide high accuracy, but require actual path profiles and time-consuming computations. Empirical as well as pure statistical models are based on the measurement. They linked to the environment and the parameters of the measurement campaign. Such models are characterized by simple input data and low computational effort, but are usually very poor at extrapolating outside of the measured parameter range and suffer from a classification problem involved in describing the environment.

Due to the lack of physical background, such models, however, only apply with good results in environments that are very close to the one they have been inferred from.

With the employment of a wide-band signal, a channel model that describes the radio propagation in an indoor medium can be described by three kinds of models:

- Tap-delay line Rayleigh fading model
- SalehValenzuela (S-V) model
- Δ-K model

In this paper, we study the Saleh-Valenzuela (S-V) Model, the Neyman-Scott (N-S) cluster model and the Δ-K model, and design our model based on them.

3 S-V MODEL, N-S CLUSTER MODEL AND Δ-K MODEL

In this section, we briefly introduce the Saleh-Valenzuela (S-V) Model, the Neyman-Scott (N-S) cluster model and the Δ-K model.

3.1 Saleh-Valenzuela (S-V) model

S-V model (Saleh & Valenzuela 1987) is a kind of indoor propagation channels model. It works with 1.5 GHz frequency band and the bandwidth is 200 MHz. It is a statistical model whose basic assumption is that Multi-Path Components (MPCs) arrive in clusters, rather than on each sampling time interval, such as the 802.11 model. The arrived clusters are formed by the multiple reflections from the objects in the vicinity of the receiver and the transmitter.

There are two Poisson models that are employed in the modeling of the arrival time. The first Poisson model is for the first path of each path cluster and the second Poisson model is for the paths (or rays) within each cluster. We define

T_l = the arrival time of the first path of the l-th cluster;

$\tau_{k,l}$ = the delay of the k-the path within the l-th cluster relative to the first path arrival time, T_l;

Λ = cluster arrival rate; and

λ = ray arrival rate, i.e., the arrival rate of the path within each cluster.

By definition, we have $\tau_{0l} = T_l$. The distributions of the cluster arrival time and the ray arrival time are given by

$$p(T_l \mid T_{l-1}) = \Lambda \exp\left[-\Lambda(T_l - T_{l-1}), l > 0\right]$$
$$p(\tau_{k,l} \mid \tau_{(k-1),l}) = \lambda \exp\left[-\lambda\left(\tau_{k,l} - \tau_{(k-1)l}\right)\right], k > 0$$
$$(1)$$

The magnitude of the k-th path within the l-th cluster is denoted by β_{kl}. It is Rayleigh distributed with a mean given by

$$\overline{\beta^2_{kl}} = \overline{\beta^2(0,0)} \exp(-T_l / \Gamma) \exp(-\tau_{kl} / \lambda), \qquad (2)$$

where $\overline{\beta^2(0,0)}$ is the average power of the first arrival of the first cluster.

The limitation of the S-V model is that it is only for indoor scenarios.

3.2 Neyman-Scott (N-S) cluster model

There are various natural phenomena exhibiting clustering behavior that occur in clusters, and modeled as such by Neyman and Scott (Cressie 1993).

- Invisible parent events are Poisson distributed with intensity λ (per unit area).
- Each parent independently produces a Poisson (μ) number of offspring.
- The positions of the offspring relative to their parents are independent and have an isotropic bivariate normal distribution with variance ρ^2 in both the x- and y-directions.

The detection function $g(x)$ is the probability of detecting an offspring at a distance x from the transect line. We will assume a normal detection function:

$$g(x) = g_0 \cdot \exp\left(-x^2/2\sigma^2\right) \qquad (3)$$

where $g_0 = g(0)$ is the detection probability at $x = 0$. The two parameters g_0 and σ are typically estimated from external data, and are assumed to be known in the present context. The effective strip width is given by

$$\int_{-\infty}^{\infty} g(x) = \sqrt{2\pi}\sigma g_0 = 2\omega \qquad (4)$$

where ω is the effective strip half-width.

Suppose that a line transect survey of infinite length is carried out along $x = 0$, and consider a parent located at $x = c$. Let T denote the detected offspring in that cluster. The expected number of detected offspring under the condition on c is given by

$$E(T \mid c) = g_0 \mu p_c$$

where $p_c = \sigma(\sigma^2 + \rho^2)^{-1/2} \exp\{-c^2 / 2(\sigma^2 + \rho^2)\}$. Further, the conditional distribution of T is a Poisson distribution, i.e., $T \mid c \sim Poisson(g_0 \mu p_c)$.

3.3 Δ-K model

The Δ-K model (Hayar & Vitetta) is a two state Markov model, which has two states: the first state S_1 and the second state S_2. In this model, the average arrival frequencies of channel echoes are $\lambda_0(t)$ and $K\lambda_0(t)$, respectively. The initial state of the process is assumed to be S_1. If at the instant t, a signal echo arrives, the process state becomes S_2, and if at the end of the interval $[t, t+\Delta]$, no new signal echo has appeared, then the process state changes again, returning to S_1.

The modified Poisson Δ-K model (Nikookar & Prasad 2009), used to describe the mobile channel arrival time, was first presented by Turin and later developed by Suzuki. This model takes into account the cluster character result from scatters.

4 THE MODIFIED UWB MODEL

In this section, we develop a modified UWB model based on the S-V model, the N-S cluster model and the Δ-K model.

4.1 Model description

According to the measured data, we adjust the parameters of the S-V model, N-S cluster model and Δ-K model, and take a simulation study in a small scale multipath environment on both LOS and NLOS scenarios. We then analyze the characteristics of different propagation channels based on the measurement and simulation results, and adjust the parameters again. Finally, we test the performance of the modified UWB model through measurement data. The flow chart of the modified UWB model for multiple clusters is shown in Figure 1.

4.2 Simulation results

UWB signals are produced by pulsed emissions, in order to have a better understanding of UWB signals and systems. We make a simulation study of the UWB pulse both in the time and frequency domains, as shown in Figure 2. From Figure 2, we can see that UWB signals have wide bandwidth signals, whose waveform is similar to the Gaussian function.

Next, a comparison performance based on the BER calculations for the modified UWB model and measured data is presented.

Figure 3(a) and Figure 3(b) show the UWB Rayleigh pulse both in the time and frequency domains; its shapes are similar to those shown in Figure 2 both in the time and frequency domains.

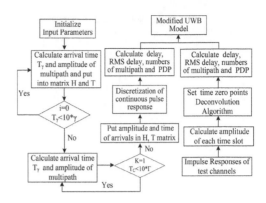

Figure 1.   Flow chart of the modified UWB model.

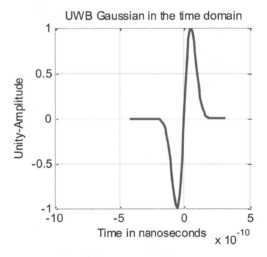

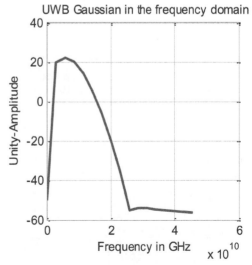

Figure 2   (a). UWB Gaussian pulse in the time domain.
(b). UWB Gaussian pulse in the frequency domain.

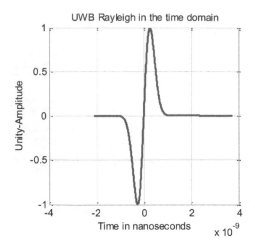

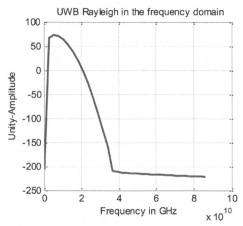

Figure 3 (a). UWB Rayleigh pulse in the time domain. (b). UWB Rayleigh pulse in the frequency domain.

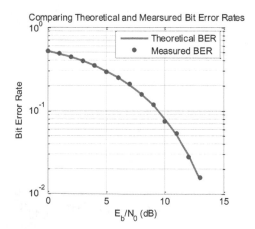

Figure 4. Comparing theoretical and measured bit error rates.

From Figure 4, we can see that the theoretical and measured data fit very well. It should be noted that this modified channel model is used without any assistance such as the channel equalizer or receiver diversity techniques or even channel coding.

5 CONCLUSIONS

In this paper, we characterized the UWB propagation using the designed UWB channel model. The characterization shows that UWB Gaussian and UWB Rayleigh have similar waves both in time domain and frequency domains. We also examined the question of UWB-based ranging within data centers, and showed a kind of ranging errors that one can expect in this environment. Finally, the performance of the modified UWB model in different environments is analyzed and compared based on MATLAB. The results show that the designed model is valid and accurate in UWB propagation.

REFERENCES

Cressie, N.A.C. 1993. Statistics for Spatial Data (Revised Edition). New York: Wiley.

Ghassemzadeh, S.S. et al. 2004. Measurement and Modeling of an UltraWide Bandwidth Indoor Channel, *IEEE Transactions on Communications*. 52(10): 1786–1796.

Hashemi, H. 1993. Impulse Response Modeling of Indoor Radio Propagation Channels, *IEEE Journal on Selected Areas in Communications*, 11(7): 967–978.

Hayar, A.M. & Vitetta, G.M. *Channel Models for Ultra-Wideband Communications: an Overview*, This work has been developed in the framework of the IST Network of Excellence "NEWCOM" (507325).

Karedal, J. & Wyne, S. et al. 2004. *Statistical analysis of the UWB channel in an industrial environment*, Vehicular Technology Conference, 1: 81–85.

Nikookar, H. & Prasad, R. 2009. *Introduction to Ultra Wideband for Wireless Communications*, Signals and Communication Technology, 52–61.

Oppermann, I. Hamalainen M. & Iinatti, J. 2004. *UWB Theory and Applications*, John Wiley & Sons.

Patwari, N. & Ash, J.N. et al. 2005. *Locating the nodes-cooperative localization in wireless sensor networks*, IEEE Xplore: Signal Processing Magazine. 54–69.

Patwari, N. & Hero, A.O. et al. 2003. *Relative location estimation in wireless sensor networks*, IEEE Xplore: Signal Processing, 2137–2148.

Saleh, A. & Valenzuela, R. 1987. A Statistical Model for Indoor Multipath Propagation, *IEEE Journal on Selected Areas in Communications*, SAC-5(2): 128–137.

Saleh, A.A.M & Valenzuela, R.A. 1987. A statistical model for indoor multipath propagation, *IEEE Journal on Selected Areas in Communications*, 5: 128–137.

USC Viterbi School of Engineering. Archived from the original 2012-03-21. http://en.wikipedia.org/wiki/Ultra-wideband.

Advanced Materials and Structural Engineering – Hu (Ed.)
© *2016 Taylor & Francis Group, London, ISBN 978-1-138-02786-2*

Contribution to quality research: A literature review of the SERVQUAL model from 1998 to 2013

Y.L. Wang

Graduate School of Management, National Taiwan University of Science and Technology, Taipei, Taiwan

T.Y. Luor & P. Luarn

Department of Business Administration, National Taiwan University of Science and Technology, Taipei, Taiwan

H.P. Lu

Department of Information Management, National Taiwan University of Science and Technology, Taipei, Taiwan

ABSTRACT: Although many past studies have focused on service quality research, specifically using the "SERVQUAL model," knowledge of the mechanics of this model is still limited. Motivated by the need to gauge the contribution of the SERVQUAL model, this study reviews 367 articles from the Social Sciences Citation Index (SSCI) and Sciences Citation Index (SCI) between 1998 and 2013 that are related to the model. We identify the key factors and conduct a survey to search for related articles in the ISI Web of Science (WOS) database. This study provides evidence that the SERVQUAL model is useful to academic researchers, and significantly contribute to service quality research. We have also discovered certain trends of research on the model from our results.

1 INTRODUCTION

SERVQUAL is the abbreviation for "Service Quality." The SERVQUAL scale was developed based on the ten requisites of quality service in the "Conceptual Model of Service Quality—PZB Method" (Parasuraman et al. 1985). In 1988, PZB conducted further research (Parasuraman et al. 1988), and categorized their findings into five determinants to be SERVQUAL scale: Tangibles, Reliability, Responsiveness, Assurance, and Empathy. It provides a complete scoring system for every industry, to assist management with credibility and efficiency, and to serve the purpose of service improvement.

This study differs from traditional literature reviews, which design SERVQUAL scales according to the PZB model and measure the clients' satisfaction of service quality. Our research aims to generalize and probe changes in the research on service quality and the areas of analysis over the past fifteen years to provide an overview to researchers interested in service quality (Luor 2008). When selecting references, most researchers value the impact factor as an influential criterion in addition to research methods and theories. We use the ISI Web of Science (WOS) database as a major research tool. This database collects all recent papers in the Sciences Citation Index (SCI) and Social Sciences Citation Index (SSCI). We analyze past papers by institutional researchers, countries, number of times cited, subject areas, and other aspects in preparation for future research.

2 RESEARCH METHODS

2.1 Scope of the study

Our study uses keywords to search the ISI WOS database. It includes 367 SERVQUAL model-related articles published in 61 SCI and SSCI journals. The result (Fig. 1) shows that there is an

Trend of number of articles and times cited (1998-2013)

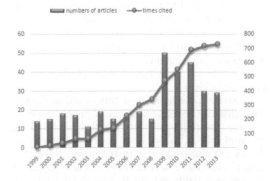

Figure 1.   Number of articles and times cited publications (1998–2013).

increasing focus on the area of service quality. In particular, the number of articles and times cited suddenly increased between 2009 and 2011. Since then, the number of articles on the follow-up discussion regarding this topic has dropped marginally, but the number of times cited has increased. Thus, it can be seen that service quality is still a significant research area among service-related articles.

2.2 *Assessment of research contribution*

We calculate the number of articles and times cited from the ISI WOS database to measure the "contribution" of authors and articles published. Times cited is a crucial index in measuring the contribution of the articles, while the number of articles reflects important trends in the relevant research, which helps generate the value for study and research (Cote et al. 1991). By generalizing and analyzing their emphasis, we can see (1) the productivity of the journal and the author, (2) the updates of publications as the reference for research expansibility, and (3) the impact on the field by the number of times the author has been cited.

3 RESULTS AND DISCUSSION

3.1 *Frequency of research on the SERVQUAL model by journal, country, institution, and researchers*

We analyze the research fields of the articles (see Table 1). The first five items of the top ten fields cover people's lifestyles, core topics of corporations, and life trends. Among these, the most discussed field is that of management. The information age also results in the rapid development of e-business and related technologies. It not only helps innovation in service values, but also modifies business models. Correspondingly, it is also influential in customer satisfaction. Whether or not this can be converted to profit for companies has been a topic of discussion in many recent papers. Even traditional industries participate in service quality initiatives, and no longer use production and yields to measure performance.

The articles investigated may overlap across multiple fields.

In general, the number of times cited in the top ten fields can indicate changes in people's overall attitude toward life and value systems. For instance, an emphasis on health leads to the popularity of the leisure and tourism industry, in which the customers also value service quality. Throughout all the research fields, we can also see the diversity of articles.

Table 1. The top ten categories according to the number of WOS articles (1998–2013).

Research field	Number of articles	Percentage of total articles
1. Management	156	42.5
2. Business	62	16.9
3. Information science; library science	34	9.3
4. Computer science; information systems	27	7.4
5. Hospitality; leisure; sport; tourism	26	7.1
6. Operations research management science	17	4.6
7. Computer science; artificial intelligence	13	3.5
8. Engineering; industrial	12	3.3
9. Health care science services	12	3.3
10. Economics	11	3.0

Note: In total, 75 categories are presented according to the number of WOS articles.

3.2 *The top ten journals, countries, and institutions*

A. The top ten journals are [journal name (number of articles, %)]: Total Quality Management Business Excellence (37, 10.1%), Service Industries Journal (18, 4.9%), African Journal of Business Management (14, 3.8%), Managing Service Quality (12, 3.2%), Total Quality Management (12, 3.2%), Tourism Management (11, 3.0%), International Journal of Service Industry Management (10, 2.7%), Journal of Business Research, (10, 2.7%), Expert Systems with Applications (8, 2.0%), and Quality Quantity (8, 2.0%). As shown above, there were 367 articles published in 61 journals, 130 (39%) of which are in the top ten journals. There is an emphasis on articles in the service categories in the practical application and research areas. This emphasis is constant over time, and articles have been published and discussed in several professional areas.

B. The top ten countries with the most published articles are [country (number of articles, %)]: USA (103, 28%), Taiwan (58, 15.8%), England (29, 7.9%), Turkey (24, 6.5%), People's Republic of China (22, 6.0%), Spain (19, 5.2%), Brazil (15, 4.1%), South Korea (13, 3.5%), Australia (11, 3.0%), and Greece (11, 3.0%). A total of 52 countries were involved in the publication of articles; 305 (82.3%) articles are from the top ten countries, while the remaining 17.7% are from the other 42 countries. It also shows that three

Table 2. The top ten frequently cited SERVQUAL articles (1998–2013).

Rank	Authors	Times cited	Article title	Journal
1	Devaraj, S., Fan, M., & Kohli, R. (2002)	269	Antecedents of B2C channel satisfaction and preference: Validating e-commerce metrics	Information Systems Research
2	Baker, D.A., & Crompton, J.L. (2000)	254	Quality, satisfaction and behavioral intentions	Annals of Tourism Research
3	Dabholkar, P.A., Shepherd, C.D., & Thorpe, D.I. (2000)	177	A comprehensive framework for service quality: An investigation of critical conceptual and measurement issues through a longitudinal study	Journal of Retailing
4	Mentzer, J.T., Flint, D.J., & Hult, G.T.M. (2001)	152	Logistics service quality as a segment-customized process	Journal of Marketing
5	Collier, J.E. & Bienstock, C.C. (2006)	114	Measuring service quality in e-retailing	Journal of Service Research
6	Brady, M.K., Cronin, J.J., & Brand, R.R. (2002)	113	Performance-only measurement of service quality: A replication and extension	Journal of Business Research
7	Jiang, J.J., Klein, G., & Carr, C.L. (2002)	94	Measuring information system service quality: SERVQUAL from the other side	MIS Quarterly
8	Barnes, S.J. & Vidgen, R. (2001)	86	An evaluation of cyber-bookshops: The WebQual method	International Journal of Electronic Commerce
9	Cao, M., Zhang, Q., & Seydel, J. (2005)	77	B2C e-commerce web site quality: An empirical examination	Industrial Management & Data Systems
10	Wakefield, K.L. & Blodgett, J.G. (1999)	76	Customer response to intangible and tangible service factors	Psychology and Marketing

Asian countries (Taiwan, China, and South Korea) have 93 (25.3%) articles in total, indicating the significant contribution and interest of Asia in this field.

C. We analyze the articles by active research institutions. There are 113 institutional contributions, with 73.8% from the top ten institutions. The result is as follows [institution (number of articles, %)]: Eastern Mediterranean University (6, 1.6%), Hong Kong Polytech University (6, 1.6%), National Chiao Tung University (6, 1.6%), Fatih University (5, 1.4%), Florida State University (5, 1.4%), National Taiwan University of Science and Technology (5, 1.4%), Penn State University (5, 1.4%), Texas A&M University (5, 1.4%), University of Granada (5, 1.4%), and Yonsei University (5, 1.4%).

3.3 The top ten most frequently cited articles

The number of times cited is a crucial index in measuring the contribution of the articles. In Table 2, we analyze the top ten most cited articles. The most cited article is "Antecedents of B2C channel satisfaction and preference: Validating e-commerce metrics" by Devaraj and Kohli (Devaraj et al. 2002), with 269 citations. It is considered a classic among articles on service quality, and shows by empirical research that client satisfaction requires reliability, responsiveness, and empathy. The result also shows that articles related to electronic information account for one-third of the top ten articles.

4 CONCLUSION

We identified key factors and surveyed related articles in the ISI WOS database. The results include:

1. SERVQUAL model attracted several researchers from Asia, Oceania, and Europe. The more the developing countries, the more the scholars for undertaking further research.
2. We propose a systematic approach to measure the contributions of service quality research. While USA leads, Asian countries are also very active, having increasing economic growth and government support.
3. Important trends were identified by studying the top ten most cited publications:
   – Internet popularity and e-commerce increase service quality research.

- B2B/B2C interactions demand quality service in logistics and operations.
- More innovation in quality service can increase product value and customer satisfaction.

Our future work includes identifying more impact factors to enable researchers to further understand and develop the SERVQUAL model.

REFERENCES

Baker, D.A. & Crompton, J.L. 2000. Quality, satisfaction and behavioral intentions. *Annals of Tourism Research*, 27(3): 785–804.

Barnes, S.J. & Vidgen, R. 2001. An evaluation of cyberbookshops: The Web Qual method. *International Journal of Electronic Commerce*, 6(1): 11–30.

Brady, M.K., Cronin J.J. & Brand, R.R. 2002. Performance-only measurement of service quality: A replication and extension. *Journal of Business Research*, 55(1): 17–31.

Cao, M. Zhang, Q. & Seydel, J. 2005. B2C e-commerce web site quality: An empirical examination. *Industrial Management & Data Systems*, 105(5): 645–661.

Collier, J.E. & Bienstock, C.C. 2006. Measuring service quality in e-retailing. *Journal of Service Research*, 8(3): 260–275.

Cote, J.A. Leong, S.M. & Cote, J. 1991. Assessing the influence of Journal of Consumer Research: A citation analysis. *Journal of Consumer Research*, 18(3): 402–410.

Dabholkar, P.A., Shepherd, C.D. & Thorpe, D.I. 2000. A comprehensive framework for service quality: An investigation of critical conceptual and measurement issues through a longitudinal study. *Journal of Retailing*. 76(2): 139–173.

Devaraj, S. Fan, M. & Kohli, R. 2002. Antecedents of B2C channel satisfaction and preference: Validating e-commerce metrics. *Information Systems Research*, 13(3): 316–333.

Jiang, J.J. Klein, G. & Carr, C.L. 2002. Measuring information system service quality: SERVQUAL from the other side. *MIS Quarterly*, 26(2): 145–166.

Luor, T.T. Johanson, R.E. Lu, H.P. & Wu, L.l. 2008. Trends and lacunae for future Computer Assisted Learning (CAL) research: An assessment of the literature in SSCI journals from 1998–2006. *Journal of the American Society for Information Science and Technology*, 59(8): 1313–1320.

Mentzer, J.T., Flint, D.J. & Hult, G.T.M. 2001. Logistics service quality as a segment-customized process. *Journal of Marketing*, 65(4): 82–104.

Parasuraman, A. Zeithaml, V.A. & Berry, L.L. 1985. A conceptual model of service quality and its implications for future research. *Journal of Marketing*, 49(4): 41–50.

Parasuraman, A. Zeithaml, V.A. & Berry, L.L. 1988. SERVQUAL: A multiple-item scale for measuring consumer perceptions of service quality. *Journal of Retailing*, 64(1): 1240.

Wakefield, K.L. & Blodgett, J.G. 1999. Customer response to intangible and tangible service factors. *Psychology & Marketing*, 16(1): 51–68.

Civil material and hydrology science application

Advanced Materials and Structural Engineering – Hu (Ed.)
© *2016 Taylor & Francis Group, London, ISBN 978-1-138-02786-2*

Waterproof curtain stability analysis of deep foundation pit near lakes or rivers

F.T. Lu, Y.J. Zheng, L.X. Li & S.L. Zhang
School of Civil Engineering, Shandong University, Jinan, Shandong, China

ABSTRACT: A lot of deep foundation pit projects near lakes or rivers have appeared. Waterproof curtain behavior has great influence on safety and progress of this kind of project. So it is necessary and urgent to study waterproof curtain action in this kind of foundation pit project. Based on the deep foundation pit engineering of south-to-north water diversion project in Baliwan pump station, this paper analyzed fluid-solid interaction of the deep foundation pit near lakes by ABAQUS. The role of waterproof curtain in foundation pit was studied, so did the influence of curtain elastic modulus and curtain effective thickness on curtain stress. The results showed that curtain stress increased with the increase of curtain modulus, and decreased with the increase of curtain thickness and the decreasing speed was slowing with the increase of curtain thickness. Considering above results, this project dewatering design was optimized.

1 INTRODUCTION

With the development of our national economy and the increase of social needs, more and more deep foundation pit projects near lakes or rivers, which have close hydraulic contact with nearest lakes or rivers, have appeared. The seepage field is very complicated, thus the stress field and the displacement field of this kind of foundation pit is more complicated than that of general foundation pit. In order to improve the seepage condition of the foundation pit near lakes or rivers and lower the drawdown outside the pit and reduce the influence of surroundings, waterproof curtain is often adopted in projects, which can dramatically reduce the water amount from foundation pit, as well as the labor-hour of draining equipment. But waterproof curtain can cause hydraulic head difference between the inside and outside of waterproof curtain. The head difference is increasing during the drainage engineering, leading to the increase of curtain stress and deformation, threatening the stability of waterproof curtain (Gong 2008).

Numerical modeling is a common method of analyzing this kind of subject (Ding et al. 2005, Sun et al. 2012). Based on the deep foundation pit engineering of south-to-north water diversion project in Baliwan pump station, this paper analyzed fluid-solid interaction of the deep foundation pit near lakes by ABAQUS, analyzing the influence of waterproof curtain on water level outside the pit and deeply studying the role of waterproof curtain in foundation pit and influencing factors of curtain stability, including the effect degree of each factor on curtain stress and deformation, and then providing theoretical basis for the optimal design of waterproof curtain.

2 CALCULATION MODEL

Under the action of drainage and excavation, fluid-solid coupling calculation of the foundation pit was did for analyzing the influence of waterproof curtain on foundation pit and surrounding environment, showing how waterproof curtain worked in foundation pit projects.

2.1 Geometric model

The waterproof curtain depth was 200 mm, the dewatering well depth was 30 m, well spacing was 20 m, and the well diameter was 400 mm. Considering that it's hard to work by computer if making meshing for the whole foundation pit, so picking 20 m wide of the pit at the center line for computational modeling. Simplifying circular well hole as square well hole by equivalent area method to improve the quality of meshing that was

$$B = \sqrt{\pi} \cdot R \approx 354 \,(\text{mm}) \tag{1}$$

where, B was the length of the side of the equivalent square, R was the diameter of well. The profile drawn of the geometric model was shown in Figure 1.

2.2 Model parameters

According to geological survey data, the soil constitutive model used Mohr-Coulomb model. The reconnaissance report didn't provide deformation modulus E_0. The soils were not ideal elastic body, and layered property, porosity and moisture content have appreciable impact on elastic property. In addition, laboratory tests had unavoidable sample disturbance, and could break the natural structure of soils, so that the tested compression modulus was smaller than the real modulus. Based on statistical materials, E_0 can be several times of $\beta \cdot E_S$ (Liang 2004). In general, more stiff the soil was, bigger the multiple was, and E_0 of soft soil is close to $\beta \cdot E_s$. Normally, $E_0 = \lambda \cdot E_S$, λ was 4 in this paper. The parameters of all soils and waterproof curtain were shown in Table 1 and 2, and soil layer distribution was shown in Figure 2.

2.3 Meshing

The model was meshed based on overall consideration of staged excavation, arrangement of dewatering wells, and waterproof curtain and soil distribution. The model used 8-nodes solid elements (C3D8P) with the degree of freedom of pore pressure. The global size of elements was 2 m. There were 30810 elements and 35442 nodes.

2.4 Boundary conditions

Right boundary was the boundary of fixed water head, height of water head was 41.29 m. Left boundary, front boundary, back boundary, upper boundary and lower boundary were impervious boundaries. The boundary of well was drainage boundary with fixed pore water pressure, which was zero. It is a Limiting degree of freedom of lower

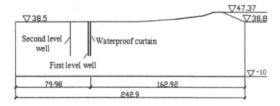

Figure 1. Model section drawn.

Table 1. Mechanical parameter of soils and curtain.

	Soil	Dry density (g/cm³)	Permeability coefficient Kx (m/s)	Permeability coefficient Ky (m/s)
1	Medium silty loam	1.44	5.68e-7	1.35e-7
2	Mud and muddy loam	1.10	4.88e-8	4.58e-6
3	Clay	1.31	1.58e-8	8.32e-9
4	Light silty loam	1.58	1.22e-7	4.04e-7
5	Heavy silty loam	1.55	1.38e-7	1.11e-7
6	Light silty loam	1.58	2.43e-7	2.82e-8
7	Medium fine sand	1.87	3.55e-5	3.55e-5
8	Heavy silty loam	1.19	7.62e-7	2.63e-7
	Waterproof curtain	1.6	1.0e-9	1.0e-9

Table 2. Physical parameter of soils and curtain.

	Soil	C (KPa)	Φ (°)	Compression modulus E_S (MPa)	Deformation modulus E_0 (MPa)	Poisson's ratio λ
1	Medium silty loam	20.8	20.6	6.95	27.8	0.28
2	Mud and muddy loam	19.7	2.2	2.51	10.04	0.30
3	Clay	24.8	24.7	5.96	23.84	0.28
4	Light silty loam	18.2	20.0	10.39	41.56	0.28
5	Heavy silty loam	28.1	9.1	6.07	24.28	0.28
6	Light silty loam	18.2	20.0	10.39	41.56	0.28
7	Medium fine sand	0.1	30.8	38.97	155.88	0.26
8	Heavy silty loam	48.6	10.2	4.61	18.4	0.28
	Waterproof curtain	1000	35	25	100	0.24

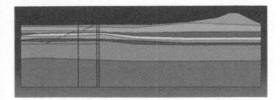

Figure 2. Soils profile.

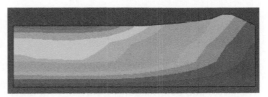

Figure 3. Displacement after 7 days dewatering (no curtain).

boundary in three directions and degree of freedom of the four side boundaries in normal direction.

3 COMPARATIVE ANALYSIS WITH CURTAIN AND WITHOUT CURTAIN

3.1 Excavation steps in finite element simulation

Considering the actual construction sequence of the deep foundation pit engineering in Baliwan pump station, the three-dimensional fluid-solid interaction of the excavation was analyzed by the following steps.

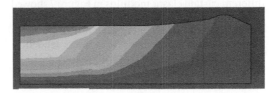

Figure 4. Displacement after 7 days dewatering (curtain).

1. Building the initial stress field by the Geostatic module in ABAQUS.
2. Dewatering wells started working and lasted for 7 days.
3. Starting the first step excavation till the excavation depth was about 5 m under the ground surface, the time was 15 days.
4. Starting the second step excavation till the excavation depth was about 9 m under the ground surface, the time was 15 days.
5. Starting the third step excavation till the excavation depth was about 15 m under the ground surface, the time was 15 days.

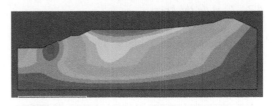

Figure 5. Displacement of excavating to the bottom of foundation pit (no curtain).

3.2 Calculation results and analysis

The results of finite element simulation showed in Figure 3–6. It can be seen that deformation of foundation pit was mainly formed in the dewatering stage. And during dewatering, waterproof curtain helped reduce the drawdown outside the curtain, thus reducing the ground settlement to a great extent and greatly migrate the impact on surroundings.

In the excavation stage, the waterproof curtain had unconspicuous influence on the pit deformation. The overall trend of deformation was top sink and bottom heave. Comparative analysis of the total displacement of foundation pit in two conditions showed that the curtain reduced top settlement, but the heave deformation at the bottom of pit was large due to the high water head outside curtain, and the deformation of the pit without curtain was relatively small, but the top

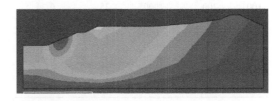

Figure 6. Displacement of excavating to the bottom of foundation (curtain).

settlement was larger and would adversely affect the surrounding buildings.

In conclusion, it's hard to enhance mechanical property of soils during the excavation stage with curtain, but it could clearly reduce the influence of dewatering inside foundation pit on the dam.

4 INFLUENCING FACTORS ANALYSIS OF CURTAIN STABILITY

The factors affecting the stress and deformation of waterproof curtain were complicated (Yang et al.

2004, Liu 2006). This paper analyzed two of them, the elastic modulus and the effective thickness of curtain.

4.1 Elastic modulus

The stress and deformation of curtain with different elastic modulus were analyzed according to the waterproof curtain parameters shown in Table 3 by referring to the value range of elastic modulus of curtain in the practical engineering. The stress of different curtain elastic modulus in Figure 7 showed that, the curtain stress increased due to the increase of elastic modulus. The increasing rate of the vertical positive stress S22 and Mises stress tended to be gentle and the horizontal positive stress S11 and shear stress S12 increased linearly with the increase of the elastic modulus. In other words, the bigger elastic modulus of curtain was, the more unfavorable the curtain can carry loads.

4.2 Curtain thickness

The paper calculated the model based on different curtain thickness in Table 4. The results showed

Table 3. Calculation cases of different modulus of elasticity.

Elastic modulus E/Mpa	Curtain thickness/ mm	Cohesion C/Kpa	Frictional angle φ/°	Poisson's ratio v
100	200	1000	35	0.24
200	200	1000	35	0.24
400	200	1000	35	0.24
700	200	1000	35	0.24
1000	200	1000	35	0.24
2000	200	1000	35	0.24
3000	200	1000	35	0.24

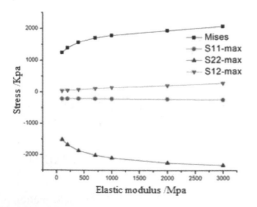

Figure 7. Stress of different modulus of elasticity cases.

Table 4. Calculation cases of different curtain thickness.

Curtain thickness/ mm	Elastic modulus E/Mpa	Cohesion C/Kpa	Frictional angle φ/°	Poisson's ratio v
100	100	1000	35	0.24
300	100	1000	35	0.24
500	100	1000	35	0.24
700	100	1000	35	0.24

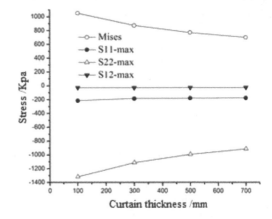

Figure 8. Stress of different curtain thickness cases.

that the resistance of curtain cross section increased and the curtain stress decreased because of the augment of curtain effective thickness. However, as depicted in the Figure 8, the decreasing tendency of curtain stress went slowly. Therefore, the larger the curtain thickness was, the more guaranteed the security was. Considering the economy of the curtain, the suggested thickness was 200~400 mm.

5 CONCLUSIONS

According to the analysis of the influence of the foundation pit dewatering and excavation on waterproof curtain and surroundings around foundation pit, several conclusions were acquired as follow.

1. Waterproof curtain can obviously reduce the drawdown outside the foundation pit and the ground settlement outside the foundation pit, and it can greatly decrease the influence range of dewatering, thus reducing the impact on the surrounding environment and ensuring the safety of the nearby dam.
2. The water head outside the foundation pit and heave deformation of the under-surface of the

foundation pit were much larger with water-proof curtain. While without curtain, the deformation of the under-surface of the foundation pit was relatively small but the settlement on the top of the foundation pit are much larger, which could adversely affect to the buildings around the foundation pit.

3. It was suitable that applying soft waterproof curtain rather than rigid waterproof curtain to the foundation pit projects near lakes or rivers. The excavation would cause larger curtain stress if its rigidity was too large, leading to higher requirements on construction quality. Suggested curtain elastic modulus was 100~300 MPa, and the compressive strength should not be less than 0.5 MPa.

4. The augment of effective thickness of curtain would increase the resistance of curtain cross section and decreased the internal stress of curtain, and the decreasing degree tended to be slow. Considering the actual engineering practice and the economy of the curtain, the suggested thickness was 200~400 mm.

ACKNOWLEDGEMENT

This project was supported by Promotive Research Fund for Excellent Young and Middle-aged Scientists of Shandong Province (Item No. BS2013SF024).

REFERENCES

Ding, X.Z. et al. 2005. Finite element analysis of environmental effects of waterproof wall on foundation pit. *Rock and Soil Mechanics,* 26:1147–1150.

Gong, X.N. 2008. *Foundation treatment manual (3rd Edit).* Beijing: China Building Industry Press.

Liang, F.Y. 2004. An approach for estimating deformation modulus of subsoil based on porous medium theory. *Rock and Soil Mechanics,* 25(7):1147–1150.

Liu, A.J. 2006. *The optimal design and engineering application of water sealing curtain in foundation pit.* Master Dissertation of North China University of Water Resources and Electric Power.

Sun, X.S. et al. 2012. Coupling numerical analysis of seepage and stress fields after excavation of slope. *Procedia Engineering,* 28(0): 336–340.

Yang, B. et al. 2004. Characters analysis and design optimization of super-deep anti-seepage purdah. *Geotechnical Engineering Technique,* 18(5):267–270.

Advanced Materials and Structural Engineering – Hu (Ed.)
© *2016 Taylor & Francis Group, London, ISBN 978-1-138-02786-2*

Explicit method of solving critical water depth and critical slope of the triangular channel section

H.Y. Gu
Shandong Provincial Research Institute of Water Resources, Jinan, China

Y.C. Han, T.Q. Peng, S.H. Wang & L. Fu
College of Resources and Environment, University of Jinan, Jinan, China

ABSTRACT: Triangular section is one important type of cross section of channel, and the calculation of critical water depth is important for flow state judgment and water surface profile computation. However, the explicit formula is not presented. This paper derives and presents an explicit formula to calculate the critical water depth of the triangular channel section directly. The results show that the critical water depth of the triangular channel section is the function of flow and side slop. The explicit formula makes the calculation easily. According to the definition of critical bottom slope, this paper also derives the explicit formula for the critical bottom slope.

1 INTRODUCTION

Critical water depth calculation is important for flow state judgment and water surface profile compute. However, the studies are only limited to rectangular, trapezoid, and circle. As one important type of cross section of channel, the studies of the triangular section are lacked. There is no direct explicit equation to calculate its critical water depth. Thus, it is essential to derive the direct explicit equation for the triangular section as the rectangular section.

Among the studies of critical water depth, Kanani (2008) presented an evolutionary algorithms based on a Genetic Algorithm (GA) for the calculation of critical depth in conduits. Zhengzhong Wang (1998) studied the critical depth of trapezoidal open channel. Prabhata K Swamee (1999, 2005, 2006) researched the exact equations for the critical depth of the trapezoidal section.

2 EXPLICIT EQUATION OF CRITICAL WATER DEPTH AND CRITICAL SLOPE DERIVATION

2.1 *Explicit equation of critical water depth*

The cross section of triangular channel is shown in Figure 1, and section parameters are given as follows:

$$A = mh_0^2 \tag{1}$$

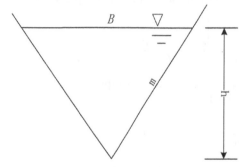

Figure 1. Triangular channel section of calculated critical flow.

$$B = 2mh_0 \tag{2}$$

$$P = 2\left(m^2 + 1\right)^{0.5} h_0 \tag{3}$$

where A = flow area (m²);
B = water surface width;
P = wetted length;
h = water depth (m); and
m = side slope.

The critical flow has these characters (Chow, 1959): the section energy is minimum value for a given discharge or the discharge is largest for a given energy. The section energy of channel is expressed as

$$E_s = h + 1/2\,\frac{\alpha v^2}{g} \qquad (4)$$

where E_s = section energy;
v = velocity of flow (m/s);
h = water depth (m);
α = kinetic energy correction factor; and
g = acceleration of gravity.

Substituting h, Q into Equation (5), E_s can be expressed as:

$$E_s = h + 1/2\,\frac{\alpha Q^2}{gA^2} \qquad (5)$$

where Q = flow discharge and
g = acceleration factor of gravity.

Putting Equation (1) into Equation (5), E_s will become as

$$E_s = h + 1/2\,\frac{\alpha Q^2}{gh^4 m^2} \qquad (6)$$

According to the maximum and minimum principle, the critical water depth is expressed as

$$\frac{dE_s}{dh} = \frac{d\left(h + 1/2\dfrac{\alpha Q^2}{gh^4 m^2}\right)}{dh} = 0 \qquad (7)$$

The differentiation for Equation (7) can be simplified as follows:

$$\frac{dE_s}{dh} = 1 - 2\,\frac{\alpha Q^2}{gh^5 m^2} = 0 \qquad (8)$$

Solving Equation (8), the explicit equation to calculate the critical water depth h_c is presented as follows:

$$h_c = \frac{\sqrt[5]{2}\sqrt[5]{\alpha Q^2 g^4 m^3}}{gm} \qquad (9a)$$

where h_c = critical water depth.

Equation (9a) shows that the critical water depth of the triangular channel section is the function of flow, Q and side slope m. There is no relationship between h_c and the bottom slope i.

Using the average velocity expressing the critical water depth, it can be written as

$$h_c = \frac{\sqrt[5]{2}\sqrt[5]{\alpha V_c^2 m^5 h_c^4 g^4}}{gm} \qquad (9b)$$

Sometimes, the discharge flow is needed to calculate when h_c is given. From Equation (8), we can obtain

$$Q = 1/2\,\frac{\sqrt{2}\sqrt{\alpha gh_c}\,h_c^2 m}{\alpha} \qquad (10a)$$

$$V_c = 1/2\,\frac{\sqrt{2}\sqrt{\alpha h_c g}mh_c}{\alpha(mh_c + b)} \qquad (10b)$$

Because $A = (mh_c + b)h_c = mh_c^2$, substituting Equation (10b) into Equation (10), the section energy of channel can be expressed as

$$E_s = h_c + 1/2\,\frac{\alpha Q^2}{gA^2}$$

$$= h_c + 1/2\,\frac{\alpha\left[1/2\dfrac{\sqrt{2}\sqrt{\alpha gh_c}\,h_c^2 m}{\alpha}\right]^2}{gA^2} = \frac{5}{4}h_c \quad (11)$$

It is known that the critical water depth and section energy of the rectangular section can be computed by an explicit formula (chow, 1959). From Equation (9) and Equation (11), it shows that the critical water depth and section energy of the triangular section can be calculated by an explicit formula. The comparisons are listed in Table 1. Obviously, the section energy of the rectangular section is larger than that of the triangular section under the same conditions.

2.2 Explicit equation of critical slope

When the normal depth is equal to the critical water depth, the bottom longitudinal slope is called the critical bottom slope. It is expressed as

$$Q = A_c C_c \sqrt{R_c i_c} \qquad (12)$$

where A_c = flow area when the water depth is equal to the critical water depth;
C_c = Chezy coefficient when the water depth is equal to the critical water depth;
R_c = hydraulic radius when the water depth is equal to the critical water depth; and
i_c = critical bottom slope.

Table 1. Comparison of critical water depth between the triangular section and the rectangular section.

Section type	h_c	E_s	Remarks
Triangular section	$\dfrac{\sqrt[5]{2}\sqrt[5]{\alpha Q^2 g^4 m^3}}{gm}$	$\dfrac{5}{4}h_c$	
Rectangular section	$\sqrt[3]{\dfrac{\alpha Q^2}{gb^2}}$	$\dfrac{1}{2}h_c$	

Note: b = bottom width of the rectangular section.

According to the Manning formula, C_c can be expressed as

$$C_c = \frac{1}{n} R_c^{1/6} \qquad (13)$$

where n = roughness of channel bed.

Substituting Equation (13) into Equation (12), we can obtain

$$Q = \frac{A_c^{5/3} \sqrt{i}}{n P_c^{2/3}} \qquad (14)$$

According to the definition of the critical bottom slope, the critical bottom slope can be computed by the following equations:

$$\begin{cases} Q = \dfrac{A_c^{5/3} \sqrt{i}}{n P_c^{2/3}} \\ 1 - 2 \dfrac{\alpha Q^2}{g h^5 m^2} = 0 \end{cases} \qquad (15)$$

From Equations (15), we can obtain

$$1/2 \frac{\left(m h_c^2\right)^{5/3} \sqrt{i_c} \sqrt[3]{2}}{n \left(\sqrt{m^2 + 1} h_c\right)^{2/3}}$$

$$= 1/2 \frac{\sqrt{2} \sqrt{\alpha g h_c} h_c^2 m}{\alpha} \qquad (16)$$

Solving Equation (16), the critical bottom slope can be obtained as follows:

$$i_c = \frac{g h_c \left(\sqrt{m^2 + 1}\, h_c\right)^{4/3} n^2 \sqrt[3]{2}}{\alpha \left(m h_c^2\right)^{4/3}} \qquad (17)$$

3 APPLICATIONS

3.1 Example 1

Let us consider a channel with a triangular cross section, $Q = 0.8$ (m³/s), side slope, $m = 1.5$; channel roughness $n = 0.014$. Now, we calculate the critical water depth.

The solving process: substituting all known conditions into Equation (9a), the critical water depth can be derived directly as follows:

$$h_c = \frac{\sqrt[5]{2} \sqrt[5]{\alpha Q^2 g^4 m^3}}{g m}$$

$$= 0.56593 \text{ (m)}.$$

3.2 Example 2

Let us consider a channel with a triangular cross section, $Q = 1.5$ (m³/s), side slope, $m = 2$. Now, we compute the critical water depth.

The solving process: substituting all known conditions into Equation (9a), the critical water depth can be obtained directly as follows:

$$h_c = \frac{\sqrt[5]{2} \sqrt[5]{\alpha Q^2 g^4 m^3}}{g m} = 0.6486124630 \text{ (m)}.$$

3.3 Example 3

Let us consider a channel with a triangular cross section, $h_c = 0.9$ (m), side slope, $m = 2$. Now, we compute the flow discharge Q.

The solving process: substituting all known conditions into Equation (10a), the flow discharge Q can be derived directly as follows:

$$Q = 1/2 \frac{\sqrt{2} \sqrt{\alpha g h_c}\ h_c^2 m}{\alpha} = 3.402 \text{ (m)}.$$

4 CONCLUSIONS

This paper aimed at the triangular channel section, and derived the explicit equation to calculate the critical water depth. The results show that the critical water depth of the triangular channel section is a function of Q and m, and is independent of the parameters i, n, m. The explicit equations will bring us great convenience and efficiency to compute the critical water depth of the triangular channel cross section.

ACKNOWLEDGMENT

This project was supported by the Development of Science and Technology Plan of Jinan City of China (201302052) and the "Five-twelfth" National Science and Technology Support Program of China (2015BAB07B02).

REFERENCES

Chow, V.T. 1959. *Open Channel Hydraulics*, McGraw-Hill: New York.
Kanani, A. Bakhtiari, A. & Borghei, S.M. et al. 2008. Evolutionary Algorithms for the Determination of Critical Depths In Conduits. *Journal of irrigation and drainage engineering*, 134(6), 847–852.
Liu, Q.G. 1999. Iteration Method for Calculation of Normal Depth In Trapezoidal Open Channel. *Water Resources & Hydropower of Northeast China*, 4: 26–27.

Swamee, P.K, Rathie, P.N, & Achour, B. et al. 2004. Exact Solutions for Normal Depth Problem. *Journal of Hydraulic Research*, 42(5): 541–547.

Swamee, P.K, Wu, S. & Katopodis, C. 1999. Formula for Calculating Critical Depth Of Trapezoidal Open Channel. *Hydrocarbon Engineering*, 125(7): 785–785.

Swamee, P.K. & Rathie, P.N. 2005. Exact Equations for Critical Depth in Trapezoidal Canal. *Journal of irrigation and drainage engineering*, 131(5): 474–476.

Wang, S.J. Hou, Y. Zhang, X.L. & Ding, J. 2002. Genetic Algorithm for Calculating Normal Water Depth of Horseshoe Cross Section, *Journal of Irrigation and Drainage*, 21(3): 43–45.

Wang, Z.Z. Song, S.B. & Wang, S.M. 1999. A Direct Formula Calculating Normal Depth for Open Trapezoidal Channel with Spherical Bed. *Journal of Yangtze River Scientific Research Institute*, 16(4): 31–34.

Wang, Z.Z. Xi, G.Z. Song, S.B. & Wang, Y.G. 1998. A Direct Calculation Formula for Normal Depth in Open Trapezoidal Channel. *Journal of Yangtze River Scientific Research Institute*, 15(6): 1–3.

Zhao, Y.F. Wang, Z.Z. & Zhang, K.D. 2007. Direct Calculation Method For Critical Depth of Open Trapezoidal Channel. *Journal of Shandong University of Technology*, 37(6): 101–105.

Zheng, Z.W. 1998. Formula for Calculating Critical Depth of Trapezoidal Open Channel. *Journal of Hydraulic Engineering ASCE*, 124(1): 90–92.

Advanced Materials and Structural Engineering – Hu (Ed.)
© *2016 Taylor & Francis Group, London, ISBN 978-1-138-02786-2*

Simulation and analysis of the liquid cargo replenishment of the ship

W. Deng, D.F. Han & W. Wang
Harbin Engineering University, Harbin, Helongjiang, China

ABSTRACT: The decision of liquid cargo replenishment based on the ship's stability, strength and the floating state is a multi-objective optimization problem. Thus, a mathematical model of multi-objective decision problem is built. The TOPSIS algorithm modified by permutation and combination is utilized to solve the supply plan. In addition, artificial selection and modification module are associated with the existing example to analyze the results, thus indicating the practicability of this system. Through calculating the actual supplying work, the most suitable schemes are found, which can meet the ship's stability constraint. The results demonstrated that the proposed method is good and suitable.

1 INTRODUCTION

More multi-function ships have began to execute different tasks for enhancing the military power, so replenishment on the sea for a large number of reserve consumption should be more accurate and quick (SOLAS 1997, Shields 1984, John 1990).

The research object of this article is a large replenishment ship. It can replenish cargo and fuel to other ships at any time. There are 8 fuel oil tanks, 6 diesel oil tanks, 2 fresh water tanks and 1 jet fuel tank on the ship (Fig. 1). In order to satisfy the demand for ships that need to be replenished and also satisfy the replenishment ship's stability and strength at the same time, fast replenishment schemes are worked out. The liquid cargo replenishment event can be considered to be a discrete system. In this article, liquid cargo replenishment schemes are analyzed. Simulation modeling of the discrete event is discussed and original replenishment schemes are determined. The TOPSIS approach ideal solution sorting algorithm is used to find the optimal solutions.

2 ANALYSIS AND SEARCHING FOR REPLENISHMENT SCHEMES

2.1 Analysis of the principles of replenishment schemes problem

Before searching the replenishment schemes, several principles should be brought forward including the symmetry principle. The symmetry principle needs to put 2 symmetrical tanks merged into 1 tank. Following up the principle, the contents of Figure 1 can be simplified into the contents of Figure 2.

After searching suitable schemes, the stability of the ship should be considered, which is discussed in Section 3.1. Every kind liquid cargo is defined a different series logo. A, B, C and D represent 4 symmetry fuel tanks; W, J represents the jet fuel and fresh water tank; 1, 2 and 3 represent 3 symmetry diesel oil tanks.

2.2 Searching the replenishment schemes with permutation and combination theory

Permutation and combination theory (Botter & Brinati 1992) is used as a preliminary search method to find possible schemes. Permutation and

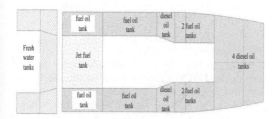

Figure 1. Arrangement plan of oil tanks.

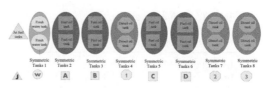

Figure 2. Oil grouping schematic.

combination formula is described in Equation (1) as follows:

$$A_n^m = n(n-1)(n-2) \ldots (n-m+1) \quad (1)$$

The value of n and m is given in Table 1.

In this condition, the residual amount of each symmetrical tank is known, and all possible schemes can be obtained. All initial schemes can be obtained through Equation (1).

If different kinds of oil are required to replenish at the same time, we also should use permutation and combination to find the replenishment schemes based on the above method. Let us assume that the ship that comes to be supplied needs both fuel oil and diesel oil together, and there are M schemes for supplying fuel oil and N schemes for diesel oil. So, M × N schemes are found.

2.3 Searching a reasonable solution with the adaptive step traversal algorithm

Adaptive step traversal algorithm is used for increasing the precision of the calculation. In the actual situation, we set the traversal step to be transformable. In the process of searching, the first big step length is set and then smaller step length for searching until the best solutions are found.

First, the amount of each oil tank is divided into 10 parts, and every part is set to be a step length for replenishment. There are 11 kinds of oil tank residual quantity state: 0; 1/10; 2/10; 1,0 represents oil empty and 1 represents oil is not needed to supply. At the same time, another kind of symmetrical tank state can be determined.

Let us assume that the amount of port and starboard oil tank are 1000t and 600t (Fig. 3). Each tank is divided into 10 parts. Then, all 11 kinds of the state are calculated. When a heeling angle is the smallest, this state can be used as the first state for a small step length binary search.

In a binary search, based on the best solution which is found in the first search, the step length is changed to 1t, then continue searching with the replenishment quantity is already determined in the first search. Searching will not stop until a better solution is found, so the solution is one of the best solutions that the step length is 1t.

Table 1. Value of n and m.

Kinds of supply cargo	n	m
Jet fuel oil	1	0;1
Fresh water	1	0;1
Diesel oil	3	0;1;2;3
Fuel oil	4	0;1;2;3;4

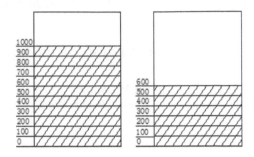

Figure 3. Separation of oil tanks.

3 OPTIMIZATION DECISIONS OF LIQUID CARGO REPLENISHMENT SCHEMES

3.1 Mathematic model of multi-objective decision problem

Liquid Cargo Supply's project can be considered to be a Multiple Attribute Decision-Making (MADM) problem. It is a kind of optimization controlling project.

Assuming that x_j represents the jth attribute, a_i represents the ith scheme, and $u_{ij} = f(a_i, x_j)$ represents the preference value of the i_{th} scheme, which is under the jth objective attribute, the mathematic model of MADM can be expressed as follows:

$$\begin{cases} DR[a_1(u), a_2(u), \ldots a_m(u)], \ u \in U \\ U = \left[u \mid u_{ij} = f(a_i, x_j) \right], \ i=1, 2, \ldots m; \ j=1, 2, \ldots n \end{cases}$$
$$(2)$$

Before calculating the MADM's mathematical model, the constraints in the process of supplying about the floating state and stability of the ship should be set. The constraints are as follows:

1. The absolute value of the heeling angle is less than 0.5°;
2. GM >0.75 m;
3. Trim angle is less than <0.4%L (ship length), and as big as better. In this paper, the value ranges from −0.9 to 0.

3.2 Searching solutions for multi-objective optimization problem

3.2.1 Determining a feasible scheme with the TOPSIS method

Finding out the best and worst scheme as the first step in the TOPSIS theory is called the ideal solution and negative ideal solution.

Making all various measures of each scheme compares with the measures of the ideal and negative ideal solution as the second step. If one scheme's calculation results, which are the closest to the ideal solution and the farthest to the negative ideal solution at the same time, this scheme is called the optimal decision for supplying. If not, it is referred to as the worst decision.

According to the constraints of supplying, a data structure of analysis for the replenishment schemes problem should be built (Dubrovsky et al. 2002). The TOPSIS theory is used to determine all possible schemes, which are called the Evaluation Unit (EU). We then determine the Evaluation Attribute (EA) together. The two elements constituting the data structure are given in Table 2.

3.2.2 The improvement of the TOPSIS decision algorithm

The TOPSIS algorithm is improved in this part based on the case that the final value is close to the target value. We changed the MOP (multi-objective problem) model to the SOP (single-objective problem) model (Xiong & Li 2003). The mathematical model is as follows:

$$Subject\ to \begin{cases} |a| < 0.5° \\ -0.9\ m\ \le t \le 0\ m \\ \overline{GM} > 0.75\ m \end{cases} \quad (3)$$

$$f_i(\overline{GM}) = \frac{\overline{GM}_{max} - \overline{GM}_{min}}{\overline{GM}_i}$$

$$f_i(t) = \frac{|t_i|}{\max\{|t_i|\} - \min\{|t_i|\}}$$

$$f_i(\alpha) = \frac{|\alpha_i|}{\max\{|\alpha_i|\} - \min\{|\alpha_i|\}}$$

where
$F(f)$—objective function;
f_i—ith possible scheme in the data structure;
$\overline{GM}$—initial stability height of the ship;
α—heeling angle of the replenishment ship;
t—trim angle of the replenishment ship;
w_1—weight value of the objective function's heeling angle;
w_2—weight value of the objective function's trim angle; and

Table 2. Data structure of decision analysis.

| EU | EEA | | |
	Heeling angle	$\overline{GM}$	Trim angle
No. 1	x_{11}	x_{12}	x_{14}
No. 2	x_{21}	x_{22}	x_{24}
...	...	...	...
No. N	x_{m1}	x_{m2}	x_{m4}

w_3—weight value of the objective function's initial stability height.

The judgment matrix of 3 sub-goals is as follows:

$$A = \begin{bmatrix} 1 & 2 & 2 \\ 1/2 & 1 & 1 \\ 1/2 & 1 & 1 \end{bmatrix} \quad (4)$$

Normalizing the above characteristic vector of matrix above is given by $w = (w_1, w_2, w_3) = (0.5, 0.25, 0.25)$.

So, Equation (3) can be expressed as follows:

$$\min_i\ F(f_i) = 0.5f_i(a) + 0.25f_i(t) + 0.25f_i(\overline{GM})$$

$$Subject\ to \begin{cases} |a| < 0.5° \\ -0.9\ m\ \le t \le 0\ m \\ \overline{GM} > 0.75\ m \end{cases} \quad (5)$$

Equation (5) is the improved mathematical model based on the objective theoretical value, which can solve the evaluation of all indices (Lu & Yen 2003). The improved multi-objective decision formula is as follows:

$$Object[i] = \frac{1}{Ya[i] + 1}$$
$$+ \frac{Max_{GM} - Min_{GM}}{Max_{GM} - Min_{GM} + |GM[i] - GM_{goal}|}$$
$$+ \frac{Max_t - Min_t}{Max_t - Min_t + |t[i] - t_{goal}|}$$

$$(6)$$

where
$Ya[i]$—represents the ith possible plan;
Max_{GM}—represents the max $\overline{GM}$ value of all schemes;
Min_{GM}—represents the min $\overline{GM}$ value of all schemes;
$GM[1]$—represents the ith plan of all schemes;
GM_{goal}—represents the target value of all schemes;
max_t—represents the max value of the trim angle;
min_t—represents the min value of the trim angle;
$t[i]$—represents the value of the ith plan; and
t_{goal}—represents the target value of the trim angle.

4 ANALYSIS AND SIMULATION OF LIQUID CARGO REPLENISHMENT SCHEMES EXAMPLE

4.1 Calculation example

Assuming a full load departure, the needed amount of replenishment is listed in Table 3.

Table 3. The replenishment quantity of the ship.

Kinds	Amount
Fuel oil	600 t
Diesel oil	400 t
Jet fuel	200 t
Fresh water	100 t

Table 4. Data of the full load departure.

Description	Value
Displacement	48000 (t)
Fore draft	10.516 (m)
Aft draft	11.023 (m)
Barycentric coordinate	(−2.3,0,12.1)
α	−0.18 (°)
t	−0.507 (m)
GM	1.553 (m)

Table 5. Analysis of optimization-selected schemes.

	Plan 1	Plan 2	Plan 3
Whether meet the replenishment demand	YES	YES	YES
Number of fuel oil tank for replenishment	5, 6	5, 6	1, 2
Number of diesel oil tank for replenishment	5, 6	3, 4	5, 6
$\overline{GM}$	1.575	1.616	1.602
Heel angle	−0.083	−0.095	−0.087
Trim angle	−0.754	−0.853	−0.622

According to the calculated stability of the replenishment ship, the solution is given in Table 4.

4.2 Analysis of the simulation calculation result

The multi-objective decision is calculated based on the ship's work condition to find the possible schemes that satisfy the constraints. Finally, the possible schemes are presented in Table 5.

We take Plan 1 as the real replenishment scheme, and the state curve of the float and stability are shown in Figures 4, 5 and 6.

From Figures 4, 5 and 6, in the process of supplying the range of the trim angle value is (−0.9, 0), and the heel angle value changed in the range from −0.5 to 0.5 and GM is more than 0.75 m. The status of the ship in the whole process is suitable.

After replenishment, a state of the ship has changed, and the updated state is given in Table 6.

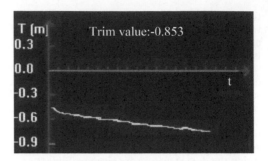

Figure 4. Trim curve.

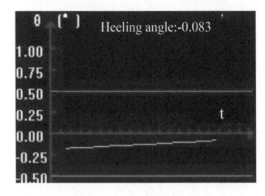

Figure 5. Heeling angle curve.

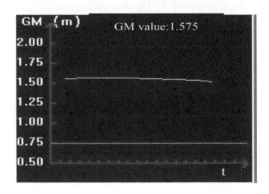

Figure 6. GM value curve.

Table 6. The ship's state after supplying liquid cargo.

Displacement (t)	Fore draft (m)	Aft draft (m)
45500	9.944	10.699

Heel angle (°)	Trim angle (m)	GM (m)
−0.083	−0.853	1.575

The calculation result shows that the picked scheme can suit the principles and constrains of the replenishment problem, and the values of the ship's stability are good and state curves changed in a reasonable range.

5 CONCLUSIONS

The calculation methods of floating state, stability and strength are summarized in this paper. In addition, according to real engineering research projects, the problem is turned into multiple optimization mathematical models. The adaptive step ergodic of permutation and combination is applied in terms of the liquid cargo tanks.

Based on the application of TOPSIS, the algorithm alternatives are solved in the same trend and the dimensionless standard matrix is established. Then, the relatively optimal replenishment solution can be obtained by checking the Euclidean distance between each scheme and ideal solution. Examples of a special replenishment quantity calculation are analyzed. The calculation result verifies that the proposed algorithm is good and suitable.

REFERENCES

Botter, R.C, & Brinati, M.A. 1992. Stowage Container Planning: A Model for Getting an Optimal Solution. Computer Applications in the Automation of Shipyard Operation and Ship Design, *IFIP Transactions B (Application in Tech.)* 1992: 217–229.

Dubrovsky, O, Levitin, G, & Penn, M.A. 2002. Genetic Algorithm with a Compact Solution Encoding for the Container Ship Stowage Problem. *Journal of Heuristics*, 8: 585–599.

John, J.D. 1990. Expert System Applications to Ocean Shipping-A Status Report. *Marine Technology*, 27(5): 265–284.

Lu, H.M. & Yen, G.G. 2003. Rank-density-based multiobjective genetic algorithm and benchmark test function study. *IEEE Trans on Evolutionary Computation*, 7(4): 56–71.

Shields, J.J. 1984. Containership stowage: a computer-aided preplanning system. *Marine Technology*, 21(4): 370–383.

SOLAS, 1997. *Chapter XII Additional Safety Measures for Bulk Carriers Amendments to the International Convention for the Safety of Life at Sea.* IMO, MSC68.

Xiong, S.W. & Li, F. 2003. *Parallel strength pareto multiobjective evolutionary algorithm*, The 2003 Congress on Evolutionary Computation, CEC' 2003, IEEE 2003:223–232.

Advanced Materials and Structural Engineering – Hu (Ed.)
© *2016 Taylor & Francis Group, London, ISBN 978-1-138-02786-2*

The study on swelling behavior of semi-IPN hydrogels consisting of crosslinked PNIPA network and linear Carboxymethyl Chitosan

Y. Chen, W.Y. Liu, G.S. Zeng & J.H. Yang

Key Laboratory of New Packaging Materials and Technology, Hunan University of Technology, Zhuzhou Hunan, China

ABSTRACT: The different content of Linear Carboxymethyl Chitosan (CMCs) was added into the PNIPA hydrogels crosslinked by BIS in situ radical polymerization to prepare semi-IPN hydrogels the content of CMCs makes a non-ignorable effect on the swelling behavior of gels. Comparing to the gels without CMCs, the resulting gels (named as C-PN gels) showed similar LCST and temperature response behavior, while the pH responsiveness appears. With the increase of added CMCs, the swelling ratio decreased considerably around the Isoelectric Point (IEP) of CMCs, while increased in both strong acidic and alkaline condition. Meantime, both deswelling and reswelling rate increases significantly.

1 INTRODUCTION

Hydrogels are hydrophilic three-dimensional polymer networks capable of absorbing a large volume of water or other biological fluid. Stimuli-sensitive hydrogels have the capability to change their swelling behavior, permeability or mechanical strength in response to external stimuli, such as small changes in pH, ionic strength, temperature and electromagnetic radiation (Sivakumaran et al. 2011, Xu et al. 2007, Zhang et al. 2004, Demirel et al. 2009). Because of these useful properties, hydrogels have numerous applications, and they are particularly used in the medical and pharmaceutical fields.

Poly (N-isopropylacrylamide) (PNIPAAm) hydrogel has been extensively studied as an intelligent polymeric matrix. The reversible phase transition of PNIPPAm hydrogel can be induced by a small external temperature change about its lower critical solution temperature (LCST, 33°C) in aqueous media (Hirokawa & Tanaka 1984, Pekcan & Kara 2000, Inomata et al. 1995, Grinberg et al. 2000). When the external temperature is below the LCST, the hydrogel hydrates and absorbs plenty of water, but it dehydrates quickly at temperatures above its LCST. Because of this unique property, significant attention has been focused on its application in the biotechnology and bioengineering fields. However, the poor mechanical properties and response rate caused by chemically crosslink limits its application significantly. Various modification methods were used to improve the properties of PNIPA hydrogels. Among all of the methods, adding other polymer chains into the gels to prepare semi-interpenetrating network gels is an effective method to improve the mechanical properties and increase the responsiveness.

Chitosan is kind of biodegradable and biocompatible polymer, which has been investigated as a novel biomaterial. However, the application of chitosan is limited for its insolubility at neutral or high pH region. To improve the solubility of chitosan, a series of hydrophilic groups have been introduced into its skeleton. Carboxymethyl Chitosan (CMCs), is a water-soluble derivative of chitosan by introducing $-CH_2COOH$ groups onto $-OH$ groups along chitosan molecular chain. For its unique chemical, physical, and biological properties, especially its excellent biocompatibility, the CMCs has been widely used in the biological and medical fields such as drug release, tissue engineering and so on. Moreover, it has been demonstrated that the CMCs shows good pH and ion sensitivity in aqueous solution with different pH value due to its abundant $-COOH$ and $-NH_2$ groups. Hence, a lot of researchers focused on using CMCs to prepare responsive hydrogels applied in drug release. For example, CMCS/alginate hydrogels, CMCs/ poly (acrylic acid-co-acrylamide) and CMCs/ PNIPA interpenetrating hydrogels were developed and investigated deeply (Zhang et al. 2013, Kim et al. 2012).

In this research, the different content of CMCs was added into the PNIPA hydrogels crosslinked by BIS in situ radical polymerization to prepare semi-IPN hydrogels, the effect of CMCs content on the swelling behavior of gels was researched systematically.

2 EXPERIMENT

2.1 *Materials*

N-isopropylacrylamide (NIPA) provided by TCI Co. Japan, was purified by recrystallization from a toluene/n-hexane mixtrure (2/1 w/w) and dried under vacuum at 40°C. Crosslinker N, N′-methylene-bis-acrylamide (BIS) was provided by Aladdin Co, China. Carboxymethyl chitosan (CMCs) with 1.34 degree of substitution as determined by potentiometric titration was prepared by reacting chitosan (Mw = 300,000 Da, supplied by Aokang Co. China) with chloroacetic acid according to the literature methodSodium phosphate monobasic dehydrate, sodium phosphate dibasic dodecahydrate, sodium hydroxide, hydrochloric acid for preparing buffer solution were all provided by Aladdin Co., China, and used without further purification. Deionized water was double distilled for using in all synthesis and analysis experiments.

2.2 *Sample preparation*

CMCs hybrid PNIPA hydrogels (C-PN gels) were synthesized by free-radical polymerization following the method described previously. Briefly, a monomer (NIPA), cross-linker BIS and deionized water were stirred at ice water temperature for at least 1 h to get a transparent solution. Then, the CMCs were added into the solution slowly with high speed continuous stirring. Next, the solution was ultrasonically oscillated for 20 min to be homogeneous and transparent. Finally, a solution with an initiator and a catalyst was added to the suspensions with stirring at ice water temperature for 5 min, pure nitrogen was bubbled, passing through the solution in all stirring processes. After the pretreatment, the solution was transferred to a columnar glass vessel ϕ 5.5* 120 mm and sealed. Free-radical polymerization was allowed to proceed in an ice water bath for 20 h. The prepared samples were immersed in deionized water at room temperature for at least 48 h, and the water was changed several hours to wash the unreacted monomer. The mass ratio of water/polymer ratio was fixed at 10/1 (w/w). The resulting hydrogels were named as Cn-PNm gels related to the gels, m and n correspond to the mole percent of BIS and mass percent of CMCs against NIPA monomer in gels.

2.3 *Characterization*

The swelling experiment was performed by immersing the dried gels in water and solution with different pH value at least for 48 h to reach the equilibrium at every particular temperature. The weight of gel was measured after wiping off the excess water from the surface with moistened filter paper. The swelling ratio was calculated from the following formula: Swelling Ratio (SR) = (Ws − W_d)/W_d. Where Ws is the weight of equilibrium swollen hydrogels and W_d is the dry weight of the hydrogels.

The deswelling kinetics of the hydrogels after a temperature jump from the equilibrated swollen state at 20°C to the hot water at 60°C were measured after wiping off the excess water from the surface with moistened filter paper. The water retention corresponding to deswelling ratio was calculated from the following formula: Water Retention (WR) = (Wt − Wd)/Ws. Where, Wt is the weight of swollen hydrogels at different time, Ws is the weight of equilibrate hydrogels.

3 RESULTS

3.1 *The effect of CMCs content on the swelling ratio of gels*

Figure 1 and Figure 2 exhibits the swelling ratio of Cn-PN4 gels in water at different temperature and in solution with different pH value at 20°C respectively. As shown in Figure 1, the swelling degrees of all gels decrease with the increase of temperature and exhibit a steep downtrend when the temperature reaches the special critical value which corresponds to their lowest Lower Critical Solution Temperature (LCST). Obviously, the LCST of gels is unchanged by the addition of CMCs basically. Moreover, the swelling ratio of gels increases with the increase of CMCs

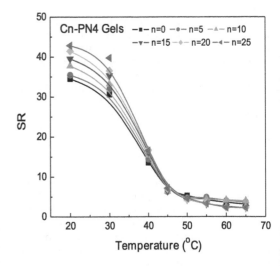

Figure 1. Swelling ratios of Cn-PN4 gels at different temperatures in water.

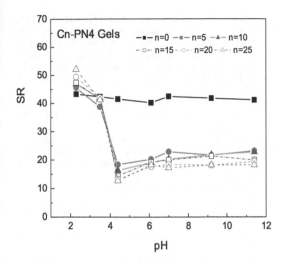

Figure 2. Equilibrium swelling ratios of Cn-PN4 gels at solution with different pH value.

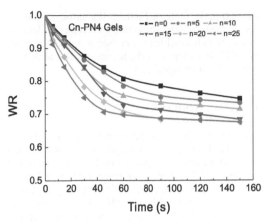

Figure 3. Deswelling kinetics of Cn-PN4 gels.

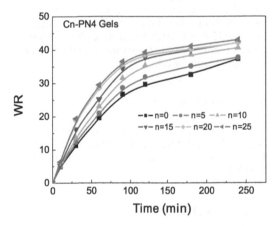

Figure 4. Swelling kinetics of Cn-PN4 gels.

this is ascribed to the increased hydrophily from CMCs.

Figure 2 shows the equilibrium swelling ratios of Cn-PN3 gels immersed in buffer solution as a function of pH value at 20°C. It could be observed that the gels show biggest swelling ratio in acid condition, then the swelling ratio decreases drastically and reached the lowest at the solution with pH = 4.4, after that, the swelling ratio increases and stays on a low platform in alkaline solution. This result is ascribed to the intrinsic character of CMCs. In acidic condition, NH^+ and $NH^+ (CH_3)_2$ appears caused by electrolysis of CMCs and interacts each other which improves the ability to swelling. With the increase of pH, the pH reaches the Isoelectric Point (IEP) of CMCs (about pH = 4–5), the equivalent NH^+ and COO^- in CMCs molecular chains interacts and combines, thereby resulting in the shrinkage of network. In alkaline solution, the swelling ratio increases slightly again with the increase of CMCs.

3.2 The effect of CMCs content on the response rate of gels

Figure 3 shows the shrinking kinetics of Cn-PN4 hydrogels after a temperature jump from an equilibrated swollen state at 20°C to ultrapure water 50°C. It can be clearly seen that the shrinking rate of Cn-PN3 gel increases significantly with the increased CMCs content and are much faster than that of C0-PN4 gel. For example, after being immersed in ultrapure water with a temperature of 50°C for 1 min, about 12 wt% of the water is freed from C0-PN4 gel. As a comparison, C25-PN3 gel

shrinks and lost water to the extent of over 40 wt%, which is almost three times than that of C0-PN4 gel. It is known that the diffusion rate of the freeze water through the porous network determines the deswelling rate of the hydrogel. After adding the CMCs, the linear molecular exists in the gels, and could active without obvious limitation due to the incorporation way. When the shrinking happens, the linear molecular in gels should produce the nano-sized channels for the diffusion of water.

Meantime, the swelling rate is also affected by the incorporation of CMCs. Figure 4 exhibits the plots of water uptake as a function of time for Cn-PN4 gels in swelling process, which reflects the reswelling rate of the gels. For all Cn- PN4 gels, the rate of water uptake increases with increasing CMCs and is faster than that of conventional CMCs-free gel. This could also be attributed to the two main factors. First, as the same reason for fast deswelling rate, the activity of linear molecular causes the channel for water. Second, the hydrophily of CMCs increases the incorporation ability of gels with water significantly.

4 CONCLUSIONS

In this research, the linear carboxymethyl chitosan was added into the PNIPA hydrogels in situ radical polymerization to prepare semi-IPN hydrogels, the content of CMCs makes a non-ignorable effect on the swelling and response behavior of gels. Comparing to the gels without CMCs, the swelling ratio of resulting C-PN gels increases with the increase of added CMCs, the gels showed similar LCST and temperature response behavior. Moreover, the pH responsiveness appears, after adding CMCs into the gels, the swelling ratio of gels decreases in high pH condition. The lowest swelling ratio appears near the IEP of CMCs. Meanwhile, the addition of CMCs is conducive to the improvement of deswelling and reswelling rate.

ACKNOWLEDGEMENT

We acknowledge the financial supports of the National Natural Science Foundation (21104017), Hunan Province Natural Science Foundation (2015JJ4021).

REFERENCES

Demirel, G. Rzaev, Z. Patir, S. & Pişkin, E. 2009. Poly (N-isopropylacrylamide) Layers on Silicon Wafers as Smart DNA-Sensor Platforms, *Journal of Nanoscience and Nanotechnology*. 9: 1865–1871.

Grinberg, V.Y. Dubovik, A.S. Kuznetsov, D.V. Grinberg, N.V. Grosberg, A.Y. & Tanaka, A.Y. 2000. Studies of the Thermal Volume Transition of Poly (N-isopropylacrylamide) Hydrogels by High-Sensitivity Differential Scanning Microcalorimetry. 2. Thermodynamic Functions, *Macromolecules*. 33: 8685–8692.

Hirokawa, Y. & Tanaka, T.J. 1984. Volume phase transition in a nonionic gel, *Chemical Physics*. 81: 6379–6380.

Inomata, H. Wada, N. Yagi, N. Goto, S. & Saito, S. 1995. Swelling behaviours of N-alkylacrylamide gels in water: effects of copolymerization and crosslinking density, *Polymer*. 36: 875–877.

Kin, J.H. Kim, Y.K. Arash, M.T. Hong, S.H. Lee, J.H. Kang, B.N. Bang, Y.B. Cho, C.S. Yu, D.Y. Jiang, H.L. & Cho, M.H. 2012. Galactosylation of Chitosan-Graft-Spermine as a Gene Carrier for Hepatocyte Targeting In Vitro and In Vivo, *Journal of Nanoscience and Nanotechnology*. 12: 5178–5184.

Pekcan, O. & Kara, O. 2000. Lattice heterogeneities at various crosslinker contents—a gel swelling study, *Polymer*. 41: 8735–8739.

Sivakumaran, D. Maitland, D. &. Hoare, T. 2011. Injectable Microgel-Hydrogel Composites for Prolonged Small Molecule Drug Delivery, *Biomacromolecules*, 12: 4112–4120.

Xu, X.D. Wei, H. Zhang, X.Z. Cheng, S.X. & Zhuo, R.X. 2007. You have full text access to this content Fabrication and characterization of a novel composite PNIPAAm hydrogel for controlled drug release, *Journal of Biomedical Materials Research Part A*, 81: 418–426.

Zhang, X.Z. Wu, D.Q. & Chu, C.C. 2004. Synthesis, characterization and controlled drug release of thermosensitive IPN–PNIPAAm hydrogels, *Biomaterials Science*. 25: 3793–3805.

Zhang, Z.H. Abbad, S. Pan, S. Waddad, A.Y. Hou, L.L. Lv, H.X. & Zhou, J.P. 2013. N-Octyl-N-Arginine Chitosan Micelles as an Oral Delivery System of Insulin, *Journal of Biomedical Nanotechnology*. 9: 601–609.

Advanced Materials and Structural Engineering – Hu (Ed.)
© 2016 Taylor & Francis Group, London, ISBN 978-1-138-02786-2

Analysis on installation accuracy system of fluid floating coupling

Z. Li & W.C. Wang
China Academy of Space Technology, Beijing, China

ABSTRACT: In order to determine the floating ability of fluid floating coupling which could compensate the installation error, the conventional analysis method which accumulates multi link maximum errors was put forward in the paper. And then an optimization design method which directly controls the installation accuracy of fluid floating coupling under the docking mechanism coordinate was given out. So that the error transfer link of the complex system reaches minimum. The installation accuracy system of fluid floating coupling was established eventually. According to this method, the floating ability target of a floating coupling was determined. The effectiveness of the method was verified by the ground inserting and separating test. This method could be used in the process of space station construction.

1 INTRODUCTION

Fluid floating couplings of manned spacecraft are installed on docking mechanism, and used for the connection and separation of fluid pipeline between transfer vehicle and space station. Fluid floating coupling is an important mechanism for the construction and operation of a space station. The locking connection error between the active and passive docking mechanisms, and the installation error between the fluid floating coupling and docking mechanism both could cause the lateral error and angle error between the active and passive fluid floating couplings. In order to compensate for these errors, and realize the connection of fluid pipelines, fluid floating coupling must be provided with floating function.

To set a reasonable floating ability target for fluid floating coupling is an important work. In this paper, the conventional analysis method which accumulates multi link maximum errors was put forward first, but the method contains some limitations. Then an optimization analysis method which directly controls the installation accuracy of fluid floating coupling under the docking mechanism coordinate was proposed. So that the error transfer link of the complex system reaches minimum. The installation accuracy system of fluid floating coupling was established eventually. The floating ability target of a real fluid floating coupling was determined through this way, and the ground inserting/separating tests used to verify the method were carried out. The result proved that this method is correct and effective.

2 FLOATING FUNCTION OF FLUID FLOATING COUPLING

Fluid floating coupling has an insertion structure, and could be divided into an active side and a passive side. The passive side adopts the sleeve structure which is simple and reliable. The active side is more complex relatively, usually consisted of the insertion tube, driving module and floating module 3 parts. Insertion tube is the pathway of the fluid pipe, and droved by the driving module to extend and insert into the passive sleeve. Driving module is responsible for driving the insertion tube and the floating module to extend or back, so as to realize the insertion and separation function. The floating module realizes the lateral and angle floating

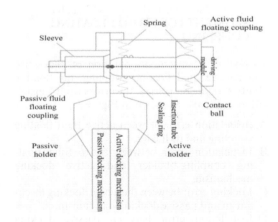

Figure 1. Principle schematic of fluid floating coupling.

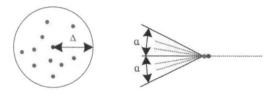

Figure 2. Lateral and angle floating ability of fluid floating coupling.

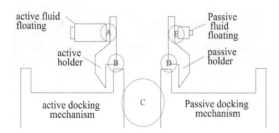

Figure 3. Installation segments of fluid floating coupling.

function of the insertion tube which could make up the installation error (Cardin 1991, Hamilton 1989, Farrell 1995, Ma et al., 2011). As shown in Figure 1.

Suppose the lateral floating ability of a fluid floating coupling is Δ, the angle floating ability is α. So that if the lateral error between active and passive fluid floating coupling less than or equal to Δ, and the angle caused by the active and passive fluid floating coupling less than or equal to α, the insertion tube can be inserted into the sleeve reliably. As shown in Figure 2.

3 ANALYSIS ON THE INSTALLATION SEGMENTS OF FLUID FLOATING COUPLING

To realize the insertion/separation of the active and passive fluid floating coupling, the floating ability should compensate the errors from 5 installation segments. As shown in Figure 3.

A. Installation error between active fluid floating coupling and its holder;
B. Installation error between active fluid floating coupling holder and active docking mechanism;
C. Locking error between the active docking mechanism and passive docking mechanism;
D. Installation error between passive docking mechanism and passive fluid floating coupling holder;

E. Installation error between passive fluid floating coupling and its holder.

4 ANALYSIS ON THE ERROR SOURCES

4.1 Installation error between the fluid floating coupling and its holder

The installation errors include the lateral error, longitudinal error and angle error.

In Figure 4 (a), the displacement error on X axis witch caused by fluid floating coupling reference point O' and its designed installation point O is defined as a lateral error, and the maximum is Δ_{1X}. The displacement error on Y axis is defined as longitudinal error, and the maximum is Δ_{1Y}.

In Figure 4 (b), the angle caused by the reference line of fluid floating coupling and the X axis is defined as angle error. And the maximum is α_1.

4.2 Installation error between docking mechanism and fluid floating coupling holder

The installation errors include the radial direction error, circumferential error and angle error.

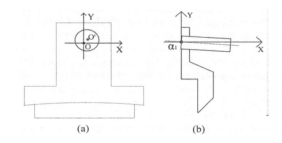

Figure 4. Installation error between the fluid floating coupling and its holder.

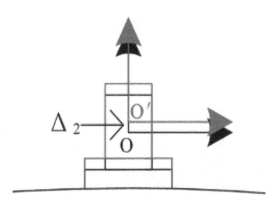

Figure 5. Radial direction error between holder and docking mechanism.

The displacement error in the direction along the radius of the docking mechanism which is caused by the holder reference point O' and its designed installation point O is defined as radial direction error, and the maximum is Δ_2. As shown in Figure 5.

The angle error along the circumferential direction of docking mechanism which is caused by the holder reference point O' and its designed installation point O is defined as circumferential error, and the maximum is θ. As shown in Figure 6.

The angle error caused by the holder reference line and the center line of docking mechanism is defined as angle error. And the maximum is α_2. As shown in Figure 7.

4.3 Locking error between the active docking mechanism and passive docking mechanism

Locking error between the active docking mechanism and passive docking mechanism includes lateral error, circumferential error and angle error.

The displacement error caused by the centers of two docking mechanisms after they locked is defined as lateral error. And the maximum is Δ_3. Thereby

$$\Delta_3 = \frac{c}{\cos 30°} + 2 \cdot d \cdot \sin 30°.$$

c is a minimum clearance of the docking mechanism's guide pin and hole, as shown in Figure 8;

d is the installation error in the direction along the radius of the docking mechanism when the guide pin and hole are installed on the docking mechanism. As shown in Figure 9.

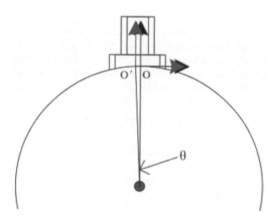

Figure 6. The circumferential error between holder and docking mechanism.

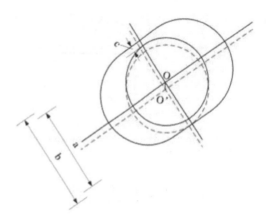

Figure 8. Clearance of the docking mechanism guide pin and hole.

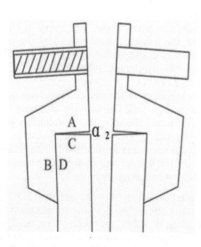

Figure 7. Angel error between holder and docking mechanism.

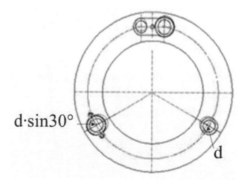

Figure 9. Lateral error due to the radial distribution error of guide pins and holes.

The angle error caused by the two Y axes of the active and passive docking mechanisms is defined as circumferential error. Suppose the maximum distribution angle error of a guide pin (guide hole) is δ, then maximum circumferential error is 2δ. As shown in Figure 10.

The angle error caused by the X axles of active and passive docking mechanisms coordinates is defined as angle error. And the maximum is α_3. As shown in Figure 11.

$$\text{And } \alpha_3 = 2 \cdot \arcsin\left(\frac{e}{2L}\right) + 2 \cdot \arctan\left(\frac{f}{L}\right) \qquad (1)$$

e is a minimum clearance of the active and passive docking mechanisms after the locked;

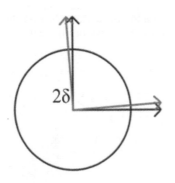

Figure 10. Circumferential error of active and passive docking mechanisms.

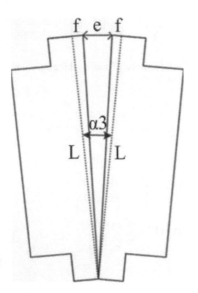

Figure 11. Angel error of active and passive docking mechanisms.

f is the planeness of the active and passive docking mechanisms;

L is the diameter of docking mechanism.

5 CONVENTIONAL ANALYSIS METHOD

The lateral floating ability of fluid floating coupling should be analyzed under the extreme condition that the lateral error caused by active and passive docking mechanisms and the circumferential error caused by holders and docking mechanisms and the lateral error caused by the holders and fluid floating couplings are in the same direction.

The angle floating ability of fluid floating coupling should be analyzed under the extreme condition that the angle error caused by active and passive docking mechanisms and the angle errors caused by holders and docking mechanisms and the angle errors caused by the holders and fluid floating couplings are in the same direction.

The conventional analysis method just accumulates multi link maximum errors, so the maximum lateral error and the maximum angle error of fluid floating coupling could be got. As shown in follows.

$$\Delta_{\max} = \left[\left(L \cdot \theta + \Delta_3 + L \cdot \delta + 2\Delta_{1X} \right)^2 + 4\left(\Delta_2 + \Delta_{1Y} \right)^2 \right]^{1/2}$$

$$\alpha_{\max} = 2\alpha_1 + 2\alpha_2 + 2\arcsin\left[e/(2L) \right] + 2\arctan(f/L)$$

So the lateral floating ability of fluid floating coupling should $\geq \Delta_{\max}$, and the angle floating ability should $\geq \alpha_{\max}$.

6 OPTIMIZATION ANALYSIS METHOD

The conventional analysis method has its limitation. Because the installation accuracy control measures are carried out on every installation segment, the measurement results of the installation accuracy could be far away from its real value (Moog 1990, Studenick 1990, Milton et al. 1993, Zhang et al. 2011).

If the installation accuracy of fluid floating coupling were controlled under the docking mechanism coordinate directly, the installation segment caused by the holder could be ignored. So the installation segments reduce from 5 to 3, and the error transfer link of the complex system reaches minimum. As shown in Figure 12.

A. Installation error between active fluid floating coupling and active docking mechanism;
B. Installation error between passive docking mechanism and passive fluid floating coupling;
C. Locking error between active docking mechanism and passive docking mechanism.

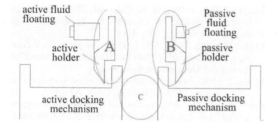

Figure 12. Installation segments of fluid floating coupling.

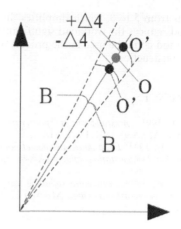

Figure 13. Installation errors of fluid floating coupling under the docking mechanism coordinate.

Under the docking mechanism coordinate, the displacement error in the direction along the radius of the docking mechanism which is caused by the fluid floating coupling reference point O′ and its designed installation point O is defined as radial error, and the maximum is Δ_4. It's certain that $\Delta_4 \leq \Delta_{1Y} + \Delta_2$.

The angle error along the circumferential direction of docking mechanism which is caused by the fluid floating coupling reference point O′ and its designed installation point O is defined as circumferential error, and the maximum is β. It's certain that $\beta \leq \theta + 2\Delta_{1X}/L$. As shown in Figure 13.

Suppose the angle error caused by the reference line of fluid floating coupling and the X axis of docking mechanism coordinate is α_4. It's certain that $\alpha_4 \leq \alpha_1 + \alpha_2$. As shown in Figure 14.

The lateral floating ability of the fluid floating coupling should be analyzed under the extreme-condition that the lateral error caused by active and passive docking mechanisms and the circumferential errors caused by fluid floating couplings and docking mechanisms are in the same direction.

The angle floating ability of the fluid floating coupling should be analyzed under the extreme-condition that angle error caused by active and passive docking mechanisms and the angle errors caused by fluid floating couplings and docking mechanisms are in the same direction.

Then the maximum lateral error and the maximum angle error of the active and passive fluid floating couplings could be got. As shown in follows.

$$\Delta_{max} = \left[\left(L \cdot \theta + \Delta_3 + L \cdot \beta \right)^2 + 4\Delta_4{}^2 \right]^{1/2}$$

$$\alpha_{max} = 2\alpha_4 + 2\arcsin\left[e/(2L) \right] + 2\arctan(f/L)$$

So the lateral floating ability of fluid floating coupling should $\geq \Delta_{max}$, and the angle floating ability should $\geq \alpha_{max}$.

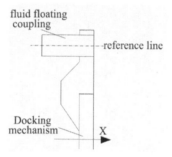

Figure 14. Angle error between fluid floating coupling and docking mechanism.

7 EXPERIMENTAL VERIFICATION

According to the optimization method, the floating ability of a fluid floating coupling was determined. The lateral floating ability is Δ and the angle floating ability is α. Fluid floating coupling was examined in the single level and system level insertion/separation test. The results show that the fluid floating coupling could insert and separate correctly and every index could meet the requirement.

8 CONCLUSIONS

In this paper, the conventional and optimization analysis method on determining the floating ability of a fluid floating coupling both were proposed. Compared with the conventional method, the optimization method reduces the installation

949

segments from 5 to 3, which simplifies the design work and reduces the workload significantly. This new method could be used in the process of space station construction.

REFERENCES

Cardin, J. 1991. *A standardized spacecraft resupply interface,* AIAA 91-1841, USA, AIAA.

Farrell, W. Jr. 1995. *Fluid quick disconnect coupling for international space station alpha,* AIAA-9-2353, USA, AIAA.

Hamilton, W. 1989. *Automatic refueling coupling for on-orbit spacecraft servicing,* AIAA89-2731, USA, AIAA.

Ma, H. Chen, J. Wei, Q. & Tang, M. 2011. Design of floating coupling for on-orbit resupply, *Journal of Rocket Propulsion,* 37(4): 45–49.

Milton, M.E. & Tyler, T.R. 1993. *Development and Testing of the Automated Fluid Interface System.* The 27th Aerospace Mechanisms Symposium.

Moog, I. 1990. *An automated fluid interface system for on orbit servicing.* AIAA-90-1938, USA, AIAA.

Studenick, R.M. 1990. *Automated fluid interface system (AFIS) for remote satellite refueling,* AIAA 90-1884, USA, AIAA.

Zhang, H. Shi, D. & Li, Z. 2010. Research of Disconnect Coupling Simulation Based on ADAMS, *Journal of Spacecraft Engineering,* 19(3): 43–48.

Advanced Materials and Structural Engineering – Hu (Ed.)
© 2016 Taylor & Francis Group, London, ISBN 978-1-138-02786-2

Study on the two-dimensional magnetic buoyancy force fields inside ferrofluids

X.Q. Jiang

Department of Physics, Naval University of Engineering, Wuhan, China

R.L. Jiang

Department of Electronic and Computer Engineering, University of Houston, TX, USA

ABSTRACT: The non-existence of the non-uniform magnetic field that only varies in one direction $\vec{H} = H_x(z)\vec{i}$ has been proved based on the electromagnetic theory. Real non-uniform magnetic fields are always distributed in three-dimensional spaces, which can be represented as $\vec{H}(x,y,z)$. The edge effect can be ignored for the non-uniform magnetic field generated by a long wedge-shaped magnetic yoke (in the y-direction), so that we can describe its magnetic field by a two-dimensional spatial distribution function $\vec{H} = H_x(x,z)\vec{i} + H_z(x,z)\vec{k}$. In this research, we build the dynamic equations for the motion of non-magnetic particles immersed in the magnetic buoyancy field generated by ferrofluids. The phenomena that non-magnetic particles tend to vertically stratified according to density and horizontally agglomerate are discussed based on the dynamic equations. It has been shown that the degree of vertical stratification of non-magnetic particles is unrelated to their volumes.

1 INTRODUCTION

Ferrofluids are novel functional material, which has many applications in different areas. Ferrofluids are colloidal liquids made of nanoscale ferromagnetic particles suspended in a carrier liquid (usually oil or water). Each tiny particle is coated with a surfactant to inhibit clumping. As magnetic particles are uniformly suspended in the carrier liquid, ferrofluids can be regarded as liquid with a uniform density. In the absence of any magnetic field, ferrofluids behave like normal fluids. In the presence of externally applied magnetic fields, ferrofluids show varieties of properties, rendering their wide used in different areas such as ore sorting (Zhang et al. 2013, Guo & Bao 2009, Xiong et al. 2007), medical instrument (Aviles et al. 2007, Yung et al. 2009), aerospace, and mechanical sealing (Kvitantsev et al. 2002, Naletova et al. 2003).

Particle sorting based on the magnetic buoyancy force has multiple advantages including high throughput, high accuracy, and eco-friendly. This property especially fits for sorting rare metal ores (Zhang et al. 2013). Zhu (2010) built a micro-fluidic device for continuous separation of non-magnetic particles inside ferrofluids based on the magnetic buoyancy force. Publications on ore sorting inside ferrofluids have investigated the suspended position of non-magnetic particles by simplifying the magnetic field generated by wedge-shaped magnets to a one-dimensional non-uniform magnetic field $\vec{H} = H_x(z)\vec{i}$, which only varies in the vertical direction (Zhang et al. 2013, Guo & Bao 2009, Xiong et al. 2007). The proposed one-dimensional simplification is not proper because it does not even satisfy the Maxwell Equations. In this paper, we are concerned about the two-dimensional magnetic buoyancy force exerted by ferrofluids to the non-magnetic particles among them in response to externally applied two-dimensional magnetic fields. This property can be applied to ore sorting and separation of biological cells. In Section 1, the one-dimension proximate model has been analyzed and its non-existence is proved according to the electro-magnetic theory. In Section 2, the two-dimensional simplify model of magnetic field caused by a wedge-shaped magnet is proposed. We found that the magnetic buoyancy force is proportional to the gradient of magnetic energy density. In Section 3, we investigate the motion of non-magnetic particles driven by the magnetic buoyancy force, and discuss the phenomenon of vertical stratification and horizontal agglomeration of non-magnetic particles inside ferrofluids.

2 PROXIMATE MODEL OF ONE-DIMENSIONAL NON-UNIFORM MAGNETIC FIELD

When investigating non-magnetic particles sorting based on the magnetic buoyancy force, researchers usually treat the non-uniform magnetic field between long wedge-shaped magnets as a one-dimensional non-uniform magnetic field $\bar{H} = H_x(z)\bar{i}$ for simplification, thus to study the vertical steady state positions of non-magnetic particles inside ferrofluids (Zhang et al. 2013, Guo & Bao 2009, Xiong et al. 2007). However, according to the Maxwell equations, it can be proved that the one-dimensional non-uniform magnetic field does not exist.

2.1 *Proximate physical image of the one-dimensional non-uniform magnetic field*

When two wedge-shaped long magnets are placed horizontally, assuming that the length of the magnets is far greater than the distance between them, the non-uniform magnetic field between them has a relatively high vertical gradient. Researchers usually treat the magnetic field between wedge-shaped magnets as a one-dimensional non-uniform magnetic field $\bar{H} = H_x(z)\bar{i}$ (Fig. 1), thus to get the expression of the one-dimensional magnetic buoyancy force (Zhang et al. 2013, Guo & Bao 2009, Xiong et al. 2007) as follows:

$$F_m = -V_p \mu_0 \chi_m H_x(z) \frac{dH_x(z)}{dz} \qquad (1)$$

where V_p is the volume of non-magnetic particle; μ_0 is the vacuum's magnetic permeability; and χ_m is the magnetic susceptibility of ferrofluids. The minus sign in Equation (1) indicates that the direction of the magnetic buoyancy force is opposite to the gradient of the magnetic field. As shown in Figure 1, the magnetic buoyancy is vertically upward. The motions and steady state positions of non-magnetic particles inside ferrofluids are determined by the magnetic buoyancy distribution.

However, this approximate model not only ignores the edge effect of the magnetic pole, but also simplifies the curved magnetic field lines between the wedge-shape magnets to be straight parallel lines, indicating that the resultant magnetic buoyancy force has only a vertical component. This one-dimensional approximation can only be applied to investigate the vertical motion and distribution of non-magnetic particles in ferrofluids. Furthermore, one can prove that the one-dimensional non-uniform magnetic field $\bar{H} = H_x(z)\bar{i}$ does not exist because it is in conflict with the Maxwell equations.

2.2 *Proof for the non-existence of the one-dimensional non-uniform magnetic field*

Any electro-magnetic field $\bar{E}, \bar{H}$ must satisfy the Maxwell equations. For the static magnetic field $\bar{H}(x, y, z)$ between the wedge-shaped magnets, we have

$$\oint_l \bar{H}(x, y, z) \cdot d\bar{l} = 0 \qquad (2)$$

Proof by contradiction: assuming that there is a one-dimensional non-uniform magnetic field $\bar{H} = H_x(z)\bar{i}$, and integrating along the path shown in Figure 2, we have

$$\oint_l \bar{H}(x, y, z) \cdot d\bar{l}$$
$$= \int_a^b \bar{H}(x, y, z) \cdot d\bar{l} + \int_b^c \bar{H}(x, y, z) \cdot d\bar{l}$$
$$+ \int_c^d \bar{H}(x, y, z) \cdot d\bar{l} + \int_d^e \bar{H}(x, y, z) \cdot d\bar{l}$$
$$\oint_l \bar{H}(x, y, z) \cdot d\bar{l} = H(z) \cdot L - H(z + \delta z) \cdot L \neq 0 \qquad (3)$$

This contradicts Equation (2). As a result, the one-dimensional non-uniform magnetic field does

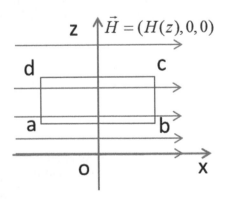

Figure 1. One-dimensional approximation of the non-uniform magnetic field between wedge-shaped magnets.

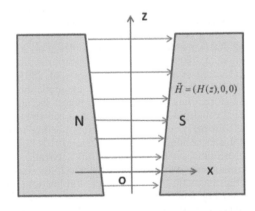

Figure 2. Integral loop inside a one-dimensional non-uniform magnetic field.

952

not exist. The one-dimensional approximation of the magnetic field between wedge-shaped magnets is not rigorous in theory. To comprehensively study the temporal movement and spatial distribution of non-magnetic particles in response to the magnetic buoyancy force, it is necessary to consider the two-dimensional approximation model of the non-uniform magnetic field between the poles.

2.3 Physical model of the two-dimensional magnetic buoyancy force

For the static magnetic field, $\nabla \times \vec{H} = 0$, with this equation, we can obtain the relationships between the partial derivatives of magnetic fields as follows:

$$\frac{\partial H_z}{\partial y} = \frac{\partial H_y}{\partial z}; \quad \frac{\partial H_x}{\partial z} = \frac{\partial H_z}{\partial x}; \quad \frac{\partial H_y}{\partial x} = \frac{\partial H_x}{\partial y} \tag{4}$$

Equation (4) indicates that the partial derivative of H_x with respect to z equals to the partial derivative of H_z with respect to x. That is to say, the one-dimensional non-uniform magnetic field $\vec{H} = H_x(z)\vec{i}$ does not exist. In another word, the non-uniform magnetic field must have at least two dimensions. As a result, the non-uniform magnetic field between the wedge-shaped magnets should be simplified to a two-dimensional field to satisfy the Maxwell equations.

2.4 Physical image of the two-dimensional non-uniform magnetic field

Assuming that the wedge-shaped magnets are long enough in the y-direction, so that we can ignore the effect of the ends of the magnetic, and simplify the three-dimensional magnetic field $\vec{H}(x, y, z)$ generated by the magnets to a two-dimensional magnetic field, which is given by

$$\vec{H} = H_x(x, z)\vec{i} + H_z(x, z)\vec{k} \tag{5}$$

The magnetic field lines between wedge-shaped magnets are not a straight line, but rather a set of curves symmetrical about the vertical centerline, as shown in Figure 3.

2.5 Relationship between the magnetic buoyancy force and the gradient of magnetic energy density

In the presence of the externally applied magnetic field, non-magnetic particles inside ferrofluids experience a magnet buoyancy force given by (Rosensweig 1985)

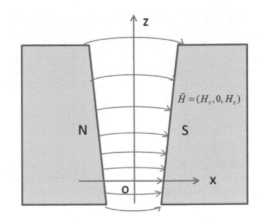

Figure 3. Two-dimensional approximation of the non-uniform magnetic field between wedge-shape magnets.

$$\vec{F}_m = -V_p \mu_0 (\vec{M} \cdot \nabla)\vec{H} \tag{6}$$

where V_p is the volume of the non-magnetic particle and $\vec{M} = \chi_m \vec{H}$ is the effective magnetization of ferrofluids. Applying to Equation (6), we obtain

$$\vec{F}_m = -V_p \mu_0 \chi_m (\vec{H} \cdot \nabla)\vec{H} \tag{7}$$

According to the relationship of the vector field: $\nabla(\vec{A} \cdot \vec{C}) = (\vec{C} \cdot \nabla)\vec{A} + (\vec{A} \cdot \nabla)\vec{C} + \vec{A} \times (\nabla \times \vec{C}) + \vec{C} \times (\nabla \times \vec{A})$, when $\vec{A} = \vec{C}$, we have

$$\nabla(\vec{A} \cdot \vec{A}) = 2(\vec{A} \cdot \nabla)\vec{A} + 2\vec{A} \times (\nabla \times \vec{A}) \tag{8}$$

For a static magnetic field, $\nabla \times \vec{H} = 0$, and denoting $\vec{A} = \vec{H}$ and applying it to Equation (8), we get

$$(\vec{H} \cdot \nabla)\vec{H} = \frac{1}{2}\nabla(H^2) \tag{9}$$

Combining with the equation of magnetic energy density, we get:

$$w_m = \frac{1}{2}\mu H^2 \tag{10}$$

where $\mu = \mu_0 \mu_r, \mu_r = 1 + \chi_m$, Equation (7) can be rewritten as

$$\vec{F}_m = -V_p \mu_0 \chi_m \nabla\left(\frac{1}{2}H^2\right) \tag{11}$$

$$\vec{F}_m = -V_p \frac{\chi_m}{\mu_r}\nabla w_m = -V_p \frac{\chi_m}{1 + \chi_m}\nabla w_m \tag{12}$$

Equation (12) illustrates that the magnetic buoyancy force exerted on non-magnetic particles is proportional to the volume of the particle and the inverse of the gradient of magnetic energy density. When $\chi_m \ll 1, \chi_m/(1+\chi_m) \approx \chi_m$, the magnetic buoyancy force is proportional to the magnetic susceptibility of ferrofluids χ_m. When $\chi_m \gg 1, \chi_m/(1+\chi_m) \approx 1$, the magnetic buoyancy force is independent of x_m.

2.6 Two-dimensional magnetic buoyancy force field

Combining Equations (5) and (7) to obtain the expressions of components of the two-dimensional non-uniform magnetic buoyancy force, we get

$$F_{mx} = -V_p\mu_0\chi_m\left(H_x\frac{\partial H_x}{\partial x} + H_z\frac{\partial H_x}{\partial z}\right) \quad (13a)$$

$$F_{my} = 0 \quad (13b)$$

$$F_{mz} = -V_p\mu_0\chi_m\left(H_x\frac{\partial H_z}{\partial x} + H_z\frac{\partial H_z}{\partial z}\right) \quad (13c)$$

According to the irrotational property of the static magnetic field, $\nabla \times \vec{H} = 0$, Equation (13a&c) can be rewritten as follows:

$$F_{mx} = -V_p\mu_0\chi_m\left(H_x\frac{\partial H_x}{\partial x} + H_z\frac{\partial H_z}{\partial x}\right) \quad (13a')$$

$$F_{mz} = -V_p\mu_0\chi_m\left(H_x\frac{\partial H_x}{\partial z} + H_z\frac{\partial H_z}{\partial z}\right) \quad (13c')$$

Equation (13a' & c') can also be derived from Equation (11). For the two-dimensional non-uniform magnetic field shown in Figure 3, the dominant part in Equation (13a' & c') is $-V_p\mu_0\chi_m H_x \partial H_x/\partial z$. The z-component of the magnetic buoyancy force will drive the non-magnetic particles to be distributed vertically according to their density. This is the theoretical foundation of ore sorting in a static non-uniform magnetic field. Although the x-component of the magnetic buoyancy force is relatively small, it still may effectively drive the non-magnetic particle moving toward the area with less magnetic energy density, thus altering the horizontal distribution of the non-magnetic particle. This phenomenon also might be applied to other areas such as cell sorting and manipulation.

3 KINETIC EQUATIONS OF NON-MAGNETIC PARTICLES INSIDE FERROFLUIDS

3.1 Kinetic equations of non-magnetic particles

The magnetic susceptibility of non-magnetic particles is negligible, i.e. $\chi_{mp} \approx 0$. As a result,

non-magnetic particles experience no magnetic force in the non-uniform magnetic field. The motion of non-magnetic particles inside ferrofluids is affected by the following four forces: gravity $V_p\rho_p\vec{g}$, buoyancy force $-V_p\rho_m\vec{g}$, magnetic buoyancy force $\vec{F}_m$ and fluidic drag force $\vec{f}_m = -6\pi\eta R\vec{v}_r$. Here, R is the radius of spherical non-magnetic particles; η is the viscosity of ferrofluids; v_r is the velocity of the particle relative to the fluid; ρ_p is the density of the non-magnetic particle; and ρ_m is the density of ferrofluids. In the presence of an externally applied magnetic field, the kinetic equations of non-magnetic particles inside ferrofluids can be expressed as follows:

$$m\frac{d\vec{v}}{dt} = \vec{F}_m + \vec{f}_m + m\vec{g} - V_p\rho_m\vec{g} \quad (14)$$

$$V_p\rho_p\frac{d\vec{v}}{dt} = \vec{F}_m + \vec{f}_m + V_p\rho_p\vec{g} - V_p\rho_m\vec{g} \quad (15)$$

Applying Equation (11), we obtain

$$\rho_p\frac{d\vec{v}}{dt} = -\mu_0\chi_m\nabla\left(\frac{1}{2}H^2\right) + \frac{\vec{f}_m}{V_p} + (\rho_p - \rho_m)\vec{g} \quad (16)$$

Assuming that the ferrofluids are static, that is $\vec{v}_r = \vec{v}$, then Equation (16) becomes

$$\rho_p\frac{d\vec{v}}{dt} + \frac{9\eta\vec{v}}{2R^2} = -\mu_0\chi_m\nabla\left(\frac{1}{2}H^2\right) + (\rho_p - \rho_m)\vec{g} \quad (17)$$

Equation (17) is the kinetic equation of spherical non-magnetic particles inside static ferrofluids in response to the non-uniform magnetic field.

3.2 Agglomeration of non-magnetic particles in response to the magnetic buoyancy force

As shown in Figure 3, we consider the ferrofluids in the two-dimensional non-uniform magnetic field generated by the wedge-shaped magnets, and assume that the non-magnetic particles are dilute, so that their movement will not bring the ferrofluids to motion. Then, Equation (17) can be applied to describe the motion of non-magnetic particles. Driven by the two-dimensional magnetic buoyancy force, non-magnetic particles will be vertically stratified according to their density, and will tend to horizontally agglomerate where the magnetic energy density is lower. Density-based particle separation can be achieved according to the proposed stratification and agglomerate phenomenon. The static position of non-magnetic particles inside ferrofluids in response to the non-uniform magnetic field is independent of the volume of the particle, but determined by the density of the particle and the magnetic energy gradient. This property allows for sorting different kinds of ore. The static position of non-magnetic particles can be determined by setting the velocity as well as the acceleration equal to zero in Equation (17):

$$0 = -\mu_0 \chi_m \nabla \left(\frac{1}{2} H^2 \right) + (\rho_p - \rho_m)\vec{g} \qquad (18)$$

The two-dimensional non-uniform magnetic field is shown in Figure 3.

$$F_{mz} \approx -V_p \mu_0 \chi_m H_x \frac{\partial H_x}{\partial z} \gg |F_{mx}| \qquad (19)$$

We define apparent density of the ferrofluids as follows:

$$\rho^* = \rho_m + \frac{\mu_0 \chi_m}{g} H_x \frac{\partial H_x}{\partial z} \qquad (20)$$

Then, the vertical static condition for a non-magnetic particle is given by

$$\rho_p = \rho^* = \rho_m + \frac{\mu_0 \chi_m}{g} H_x \frac{\partial H_x}{\partial z} \qquad (21)$$

As $F_{mz} \gg |F_{mx}|$, the stratify process will be much faster than agglomeration.

4 CONCLUSIONS

This research investigates the physical phenomena of non-magnetic particles inside ferrofluids in response to an externally applied magnetic field. On the basis of the above analysis, we draw the following conclusion:

1. One-dimensional non-uniform magnetic field $\vec{H} = H_x(z)\vec{i}$ does not exist because it violates the Maxwell equations.
2. The magnetic buoyancy force experienced by non-magnetic particles is proportional to the magnetic energy density gradient, in the opposite direction.
3. Non-magnetic particles are stratified inside ferrofluids due to the two-dimensional magnetic buoyancy force. Static positions of the particles are determined by Equation (21), where

the density of non-magnetic particles is set to be equal to the apparent density. The degree of vertical stratification of non-magnetic particles is unrelated to their volumes.
4. Non-magnetic particles tend to agglomerate to the area with a lower magnetic energy density.

In the future study, we will investigate the trajectory of non-magnetic particles inside ferrofluids using Equation (17), and simulate the evolution of the spatial distribution of particles along with time.

ACKNOWLEDGMENT

This work was financially supported by the National Natural Science Funds (10272115).

REFERENCES

Aviles, M.O. Chen, H. Ebner, A.D. Rosengart, A.J. Kaminski, M.D. & Ritter, J.A. 2007. In vitro study of ferromagnetic stents for implant assisted – magnetic drug targeting, *Journal of Magnetism and Magnetic Materials*. 311: 306–311.

Guo, T.L. & Bao, L.H. 2009. Analysis of the Magnetic Force of Non-Magnetic Mineral Particles in MHSS, *Metal Mine*, 394(4): 35–39.

Kvitantsev, A.S. Naletova, V.A. & Turkov, V.A. 2002. Levitation of Magnets and Paramagnetic Bodies in Vessels Filled with Magnetic Fluid, *Fluid Dynamics*. 37(3): 361–368.

Naletova, V.A. Kvitantsev, A.S. Turkov, V.A. 2003. Movement of a magnet and a paramagnetic body inside a vessel with a magnetic fluid, *Journal of Magnetism and Magnetic Materials*. 258–259: 439–442.

Rosensweig, R.E. 1985. *Ferrohydrodynamic,* Cambridge, New York: Cambridge University Press.

Xiong, L.G. Ping, P.Y. & Chen, X. 2007. Magnetic Force Analysis of Immersed Body in magnetic Fluids and Its Magnetic Levitation Characteristic, *J. Funct. Mater.* 38: 1234–1237.

Yung, C.W. Fiering, J. Mueller, A.J. & Ingber, D.E. 2009. Micromagnetic microfluidic blood cleansing device, *Lab Chip*, 9: 1171–1177.

Zhang, Y.Q. Wei, J.T. Wu, Z.G. & Wang, T.Y. 2013. Mechanical Model of Non-Magnetic Particles in Magneto Hydrostatic Static Separation, *Nonferrous Metal (mineral processing)*, 4: 49–51.

Zhu, T. Marrero, F. & Mao, L. 2010. Continuous separation of non-magnetic particles inside ferrofluids, *Microfluid Nanofluid*, 9: 1003–1009.

Advanced Materials and Structural Engineering – Hu (Ed.)
© *2016 Taylor & Francis Group, London, ISBN 978-1-138-02786-2*

Effect of drop size on the rheology of water-in-oil emulsion

W.Y. Wang & C.W. Liu

College of Petroleum Engineering, China University of Petroleum, Qingdao, China

ABSTRACT: The present study aims to increase the current understanding of the effect of drop size on the rheology of water-in-oil emulsions. Water-in-Oil (W/O) emulsions were prepared using two methods and the dependences of water content and rotor speed on the drop size were determined. It was found that emulsions with smaller drop size had higher viscosity and displayed a more remarkable shear-thinning effect. At low water content, drop size was independent of water content while viscosity and the shear-thinning effect of the emulsions increased with water content. With further increase of water content, an emulsion-gel network spanned the sample, drop size will increase and accompanied by the formation of a distorted emulsion-gel structure, which both can contribute to the anomalous rheological behavior of high water content emulsions.

1 INTRODUCTION

Emulsions are a popular subject of rheological studies due to their practical applications in food products, crude oil recovery, polymer blends, biological liquids, etc (Aomari et al. 1998, Chanamai & McClements, Masalova et al. 2003). It is well established that the rheology of emulsions are influenced by multiple factors such as volume fraction of dispersed phase, viscosity of continuous phase, shear rate, temperature, average drop size and, nature and concentration of emulsifying agents, etc (Dan & Jing 2006, Urdahl et al. 1997, Farah et al. 2005). Among these factors, the effect of drop size on emulsion rheology has received less attention despite its practical significance (Pal 1996, Malkin et al. 2004, Rodriguez & Kaler 1992).

Publications on the effect of drop size on emulsion rheology are limited to oil-in-water emulsions. Research on the effect of drop size on the water-in-oil emulsion is significantly underexplored. This study aims to develop the current understanding of the dependence of drop size on the rheology of water-in-oil emulsion. First, the relationship between emulsification parameters and the drop size distribution of W/O emulsion was investigated. Subsequently, the effect of drop size on the rheology of water-in-oil emulsions was discussed.

2 EXPERIMENTAL SYSTEM

2.1 Materials

The non-ionic surfactant Span 80 (Zibo HaiJie Chemical CO., LTD) was chosen as the emulsifier for the preparation of water-in-oil emulsions. The emulsifier concentration in the oil solutions was 5% by weight, which was sufficiently high to suppress drop coalescence under emulsification and stagnant conditions. The oil supplied by Shengli oilfield was a Newtonian fluid with a viscosity and density of 33.6 mPa·s and 0.85 g/ml at 35 °C. The dispersed phase was a mixture of distilled water and 0.2 wt.% Na_2CO_3.

2.2 Emulsion preparation

2.2.1 Method 1
The water-in-oil emulsions were prepared by mixing the pre-established amounts of the dispersed phase and the continuous phase in a high shear rotor-stator homogenizer (FJ200-S, Shanghai Specimen and Model Factory) at three different rotor speeds of 2500, 3500 and 4500 rpm. For each emulsion set (with same rotor speed), the dispersed phase volume fraction was increased gradually until phase inversion was observed and each sample was stirred for 10 minutes to achieve a stable state.

2.2.2 Method 2
Three emulsions were first prepared using Method 1 with a rotor speed of 2500, 3500, 4500 rpm. Emulsions with different water content were prepared by diluting the original high concentration emulsions with different amounts of 5 wt.% Span 80 oil phase solution. The dilute samples were stirred for three minutes to achieve uniform state upon gentle agitation.

2.3 Drop size analysis

The drop size of an emulsion was determined by taking photomicrographs using an optical microscope equipped with a camera (Chongqing Optec Instrument Co., Ltd). The samples were diluted with 5 wt.% Span 80 oil phase solution before taking photomicrographs. The images were analyzed using automated image analysis software (OPTPro).

2.4 Rheological measurements

All rheological measurements were carried out in a digital viscometer (DV-2+Pro, Brookfield, Middleboro, MA) with a circulating heating/cooling water bath. The viscometer was equipped with a concentric cylinder SSA setup and spindle S34 #. The temperature was controlled by the circulating water bath with an accuracy of ±0.2 °C. The emulsions flow curves were measured as the shear rate was increased in steps. At each shear rate, the samples were equilibrated for ten seconds before the measurement. Every measurement was repeated two times and insignificant viscosity change was found. The results present in this study are the average of the two measurements.

3 RESULTS AND DISCUSSION

3.1 Drop size distribution

Figure 1 shows the effect of water content on the Volume Drop Size Distribution (VDSD). It indicates that, when water content, $\varphi \leq 32.5\%$, the VDSD seems independent of the water content which agrees well with Boxall's research. When φ is increased from 32.5% to 37.3%, a distinct increase of drop size was observed and multiple drops were observed near the inversion point (~47.1%). A sharp increase in the drop size as well as changes in drop morphology near the inversion point was also reported by Pal (1993).

Figure 2 shows the VDSD for the three rotor speeds. All volume DSDs were monomodal and the size distribution shifts to smaller drop size with increasing rotor speed. Similar phenomena have been observed by other researchers. The reason for

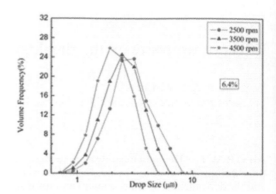

Figure 2. Effect of rotor speed on the VDSD of emulsion with water content of 6.4%.

this trend can be attributed to the increasing energy dissipation rate resulting from increasing rotor speed, leading to the breakup of larger drops.

3.2 Drop size distribution

Figure 3 (a) and (b) show the typical flow curves (shear stress vs. shear rate) of diluted and non-diluted emulsions, respectively. Figure 3 (a) shows that the flow behavior of diluted emulsions changed qualitatively around characteristic water content φ^*. For $\varphi < \varphi^*$, sample flow curves exhibited a constant increase in shear stress in the range of the testing shear rate which corresponded to Newtonian fluid. At water content above φ^*, emulsions showed typical non-Newtonian characteristics. When the shear rate was less than some critical value, given as γ^*, the emulsions behaved as Newtonian fluids and an increase in the dispersed phase volume fraction led to systematically higher values of the critical shear stress at γ^*. Beyond γ^*, flow curves initially exhibited qualitatively parallel straight-line segments and then followed a slight upward segment on the log-log plots. The slope of the curves decreased slightly with increasing water content and formed a fanning of these segments. Similar flow behaviors were also observed for non-diluted emulsions except when the water content was higher than 32.5%. According to Figure 3 (b), when the water content was higher than 32.5%, the flow curves of each emulsion exhibited a new peculiar feature. At low shear rate, the shear stress (including the critical shear stress at γ^*) decreased with the water content which resulted in a lower viscosity for the higher water content emulsions. At high shear rate, the flow curves intersect the lower-φ curves.

The viscosity vs. shear rate curves of emulsions with different drop sizes are shown in Figure 4. By combining the drop size data, it can be concluded that emulsions with smaller drop size have higher viscosity, especially at lower shear rate. The observed

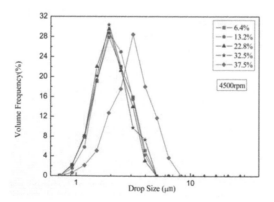

Figure 1. Effect of water content on volume drop size distribution.

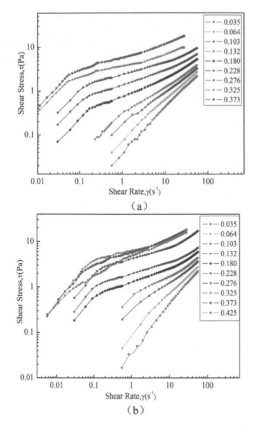

(a)

(b)

Figure 3. Shear stress vs. shear rate curves of (a) diluted and (b) non-diluted emulsions with different water content at rotor speed of 4500 rpm.

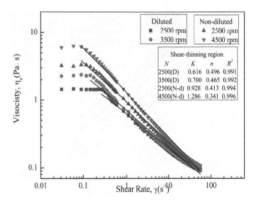

Figure 4. Viscosity vs. shear rate curves of emulsions with different drop sizes at water content of 0.180.

increase in viscosity upon reduction of drop size could be due to several possible reasons. First, with a decrease in drop size, the concentration of drops and the mean distance of separation between the drops increases and decreases respectively. Meanwhile, Brownian motion and attractive forces led to an increase in flocculation tendency and hydrodynamic interaction and consequently an increase in viscosity. Second, the thickness of the absorbed surfactant layer with respect to drop size becomes more important as the drop size is decreased, resulting in higher effective dispersed phase content and viscosity. As discussed in Figure 3, when the shear rate was above a critical shear rate, the emulsions showed shear-thinning tendency which could also be explained in terms of the flocculation of the drops. At lower shear rates, the local break-up rate of the original structure by shear deformation was comparable to the structural growth or recovery rate due to flocculation, so that the overall resistance to flow was constant. At higher shear rates, the rate at which the original structure was destroyed was greater than the

rate of structural growth, leading to a decrease in cluster size and effective dispersed phase volume fraction, which consequently decreased the viscosity. Furthermore, the decreasing effective dispersed phase volume fraction decreased the difference between the effectively dispersed phase volume fraction and the real dispersed phase volume fraction leading to the observed gradually closing viscosity curve with shear rate. Therefore, the shear-thinning effect is generally more pronounced in fine emulsions and the power law model was employed to give a quantitative description of the shear-thinning tendency. The higher consistency constant, K and smaller flow index, n, were observed in emulsions with smaller drop size and agreed well with the higher viscosity and remarkable shear-thinning effect of fine emulsions. Although the power law model can fit the shear-thinning region with high satisfaction, a certain degree of deviations from the power law model at high shear rates and higher experimental viscosity date were observed for all samples ($>\sim 20$ s^{-1}). This is because, at high shear rates, the cluster/ flocculation size became smaller and more compact with the change in the spatial structure of cluster (e.g. fractal dimension), leading to different dependence of viscosity on shear rate. These deviations corresponded to the slight upward segments observed in Figure 3 and can be regarded as a universal feature of shear-thinning emulsions.

Figure 5 shows the viscosity vs. shear rate for diluted and non-diluted emulsions with different water content. It also indicates that the shear-thinning region can be well described by the power law model. In both systems, when $\varphi \leq 32.5\%$, the consistency constant K and flow index n increase and decrease with water content, respectively, and result in the slight fanning of non-Newtonian flow segments in Figure 3. When $\varphi \geq 32.5\%$, different trends were observed for the diluted and non-diluted systems. When $\varphi \geq 32.5\%$ for the diluted system, both

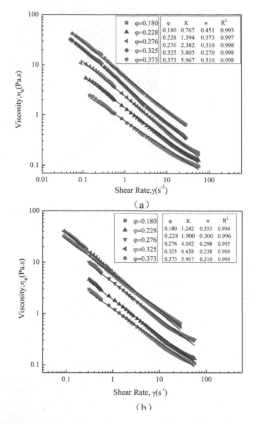

Figure 5. Viscosity vs. shear rate curves of (a) diluted and (b) non-diluted emulsions with different water content at rotor speed of 4500 rpm.

the consistency constant and flow index increased with φ. When φ ≥ 32.5% for the non-diluted system, the consistency constant decreased and the flow index increased with φ, corresponding to the intersection of flow curves observed in Figure 3.

4 CONCLUSIONS

Systemic experiments have been conducted to investigate the dependence of emulsification parameters on drop size and the resulting change of water-in-oil emulsions rheology. Based on the experimental results and analysis, the following conclusions can be drawn:

1. The emulsification process in the present study was performed in the turbulent viscous regime where the drop size strongly depends on the rotor speed. At low water content (<32.5%), the drop size is determined by the balance between the viscous stress and the drop capillary pressure and shows insignificant dependence on the water content; when the water content is higher

than the glass-transition volume fraction, an emulsion-gel network will span the sample, further increasing the water content and compelling some of the water to become trapped in the small-drop gel in the form of larger drops.

2. With an increase in water content, emulsions transform from a Newtonian fluid to a non-Newtonian fluid. For non-Newtonian emulsions, significant deviations from the power-law model at high and low shear rates were observed.

3. For a given concentration, emulsions with smaller drop size have higher viscosity. With the increase in shear rate, the effect of drop size on emulsion viscosity weakens, resulting in a more significant shear-thinning effect in fine emulsions. In the low water content range (<32.5%), emulsions increasing water content results in increasing viscosity and a more pronounced shear-thinning effect. When the water content is higher than 32.5%, the drop size increases along with an accompanied emulsion microstructure change and leads to an anomalous dependence of emulsion rheological behavior on water content.

REFERENCES

Aomari, N. Gaudu, R. Cabioc'h, F. & Omari, A. 1998. Rheology of water in crude oil emulsions. *Colloids and Surfaces A: Physicochemical and Engineering Aspects.* 139: 13–20.

Chanamai, R. & McClements, D.J. 2000. Dependence of creaming and rheology of monodisperse oil-in-water emulsions on drop size and concentration. *Colloids and Surfaces A: Physicochemical and Engineering Aspects.* 172: 79–86.

Dan, D. & Jing, G. 2006. Apparent viscosity prediction of non-Newtonian water-in-crude oil emulsions. *Journal of Petroleum Science and Engineering.* 53: 113–122.

Farah, M.A. Oliveira, R.C. Caldas, J.N. & Rajagopal, K. 2005. Viscosity of water-in-oil emulsions: Variation with temperature and water volume fraction. *Journal of Petroleum Science and Engineering.* 48: 169–184.

Malkin, A.Y. Masalova, I. Slatter, P. & Wilson, K. 2004. Effect of drop size on the rheological properties of highly-concentrated w/o emulsions. *Rheologica Acta.* 43: 584–591.

Masalova, I. Malkin, A.Y. Slatter, K. & Wilson, K. 2003. The rheological characterization and pipeline flow of high concentration water-in-oil emulsions. *Journal of Non-Newtonian Fluid Mechanics.* 112: 101–114.

Pal, R. 1993. Pipeline Flow of Unstable and Surfactant-Stabilized Emulsions. *AIChE Journal.* 39: 1754–1764.

Pal, R. 1996. Effect of Drop Size on the Rheology of Emulsions. *AIChE Journal.* 42: 3181–3190.

Rodriguez, B.E. & Kaler, E.W. 1992. Binary Mixtures of Monodisperse Latex Dispersions. 2. *Viscosity. Langmuir.* 8: 2382–2389.

Urdahl, O. Fredheim, O. & Loken, K.P. 1997. Viscosity measurements of water-in-crude-oil emulsions under flowing conditions: A theoretical and practical approach. *Colloids and Surfaces A: Physicochemical and Engineering Aspects.* 123: 623–634.

Advanced Materials and Structural Engineering – Hu (Ed.)
© 2016 Taylor & Francis Group, London, ISBN 978-1-138-02786-2

Optimization of viscosity model for water-in-oil emulsions

W.Y. Wang

College of Petroleum Engineering, China University of Petroleum, Qingdao, China

ABSTRACT: Systemic experiments have been conducted to investigate the rheology of Water-in-Oil (W/O) emulsions by using a concentric viscometer. The rheological behavior of W/O emulsions was discussed, and also the comparisons with theoretical viscosity models were presented. Results indicated that all the shear rate-independent models underestimate the emulsion viscosity due to flocculation of the drops. Two local remarkable increases of the emulsion viscosity with dispersed phase volume fractions correspond to the percolation and glass-transition, respectively. When compared with the Cross model, the Carreau model showed better agreement with the flow curves. Furthermore, the Mooney equation was proven to be the most reasonable model for predicting the high shear limit viscosity of the shear-thinning emulsion.

1 INTRODUCTION

Water-in-Oil (W/O) emulsions are encountered at nearly every step of the petroleum production and recovery operations, e.g. in reservoir pores, in the chokes at the wellheads, in phase separators, in flotation units, in crude oil transport facilities, and at various stages of the refining process (Pal 1997, Evdokimov et al. 2008). The rheology of emulsions is a very popular subject due to their practical applications and a good knowledge of it is necessary for oil field development and petroleum transportation (Dou & Gong 2006, Urdahl et al. 1997, Oliveira & Goncalves 2005). In the past, numerous viscosity models have been developed for the emulsions (Sjoblem 2006). This study aims at optimizing the viscosity models by the measured viscosity data of W/O emulsions.

2 EXPERIMENTAL SYSTEM

2.1 Materials

A crude oil collected from Shengli Oilfield is used to prepare all emulsions. This oil is a Newtonian fluid. The viscosity and density are measured to be $33.6\,mPa\cdot s$ and $0.85\,g/cm^3$, respectively, at temperature of 35 °C. The water phase used in this study is prepared by mixing distilled water with 0.2 wt% Na_2CO_3. The non-ionic surfactant Span 80 (Zibo HaiJie Chemical Co., LTD) is chosen as the emulsifier for the preparation of water-in-oil emulsions.

2.2 Experimental procedures

2.2.1 Emulsion preparation
The emulsions were prepared by two methods. The first method is presented as the water-in-oil emulsions were prepared by mixing the pre-established amounts of the dispersed phase and the continuous phase in a high shear rotor-stator homogenizer (FJ200-S, Shanghai Specimen and Model Factory) at three different rotor speeds of 2500 (set 1), 3500 (set 2), and 4500 (set 3) rpm. The second method is that, three emulsions were first prepared using method 1 with a rotor speed of 2500, 3500, 4500 rpm. Emulsions with different water contents were prepared by diluting the original high concentration emulsions with different amounts of 5 wt.% Span 80 oil phase solution.

2.2.2 Rheological measurements
All rheological measurements were carried out in a digital viscometer (DV-2+Pro, Brookfield, Middleboro, MA) with a circulating heating/cooling water bath. The viscometer was equipped with a concentric cylinder SSA setup and spindle S34 #. The temperature was controlled by the circulating water bath with an accuracy of ±0.2 °C. The emulsions flow curves were measured as the shear rate was increased in steps. At each shear rate, the samples were equilibrated for ten seconds before the measurement. Every measurement was repeated two times and insignificant viscosity change was found. The results present in this study are the average of the two measurements.

3 RESULTS AND DISCUSSION

3.1 Typical rheological curve

Figure 1 shows the typical flow curves (shear stress vs. shear rate) of diluted emulsions respectively. It indicates that the flow behavior of diluted emulsions changed qualitatively around a characteristic water

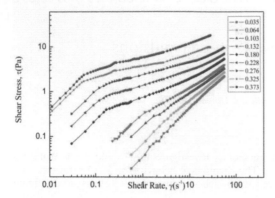

Figure 1. Typical rheological curves of water-in-oil emulsions with different water contents at a rotor speed of 4500 rpm.

content φ^*. For $\varphi < \varphi^*$, sample flow curves exhibited a constant increase in shear stress in the range of the testing shear rate which corresponded to Newtonian fluid. At water contents above φ^*, emulsions showed typical non-Newtonian characteristics. When the shear rate was less than some critical value, given as γ^*, the emulsions behaved as Newtonian fluids and an increase in the dispersed phase volume fraction led to systematically higher values of the critical shear stress at γ^*. Beyond γ^*, flow curves initially exhibited qualitatively parallel straight-line segments and then followed a slight upward segment on the log-log plots. The slope of the curves decreased slightly with increasing water contents and formed a fanning of these segments.

3.2 Comparison with the viscosity models

3.2.1 Shear rate-independent models

In 1906, Einstein proposed a thermodynamic viscosity model for the diluted suspensions. New models have been developed since for emulsions with a higher dispersed phase volume fraction.

Batchelor extended Einstein's model to involve interactions between the drops over a wider concentration range of the dispersed phase given as:

$$\eta_r = \left(1 + 2.5\varphi + c\varphi^2\right) \tag{1}$$

By considering the effect of the addition of one solute-molecule to an existing solution, Brinkman obtained a simple expression for the viscosity of an emulsion of finite concentration as follow:

$$\eta_r = \left(1 - \varphi\right)^{-2.5} \tag{2}$$

Krieger and Dougherty proposed an empirical model for the viscosity of emulsions with high

concentrations of the dispersed phase. The model introduced the maximum packing fraction, φ_{max}, as a crowding parameter to better consider the contributions of finite drop size:

$$\eta_r = \left(1 - \frac{\varphi}{\varphi_{max}}\right)^{-2.5\varphi_{max}} \tag{3}$$

Mooney also considered the crowding effect of the dispersed phase and suggested an exponential relationship to predict the flow behavior of the emulsion (Mooney 1951):

$$\eta_r = \exp\left(\frac{2.5\varphi}{1 - \varphi/\varphi_{max}}\right) \tag{4}$$

By using a free cell model, Mills derived an equation for a concentrated suspension of hard spheres with only hydrodynamic interactions; the maximum packing fraction φ_{max} was also introduced as:

$$\eta_r = \frac{1 - \varphi}{\left(1 - \varphi/\varphi_{max}\right)^2} \tag{5}$$

Based on the comparison of experimental data (Newtonian flow regime) with predictions made by the above viscosity models (see Fig. 2), all models underestimate the emulsion viscosity and the results strongly depend on the drop size, water content. When compared with the other models, the Mooney model gave the best prediction. The disparity between the viscosity models' predictions with experimental data can be attributed to the strong flocculation tendency of the drops. As is well known, the flocculation effect, resulting from

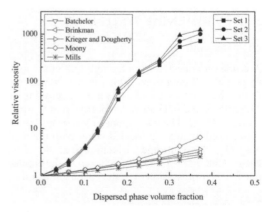

Figure 2. Relative viscosities of Newtonian flow regime vs. water content for three diluted sets. (For Krieger and Dougherty, Mooney and Mills models, $\varphi_{max} = 0.74$).

the micelle-induced depletion mechanism, will lead to the continuous phase being entrapped in the space of the cluster, resulting in a higher effective dispersed phase volume fraction. The viscosity models discussed above were developed for concentrated emulsions and focus on the hydrodynamic interaction of the dispersed phase. These models do not consider flocculated systems which leads to large deviations observed in this work. Jansen et al. (2001) scaled the viscosity of surfactant stabilized emulsions and also found that no satisfactory model was available for the lower shear rate viscosity of flocculent emulsions.

In order to quantify the flocculation effect, the rescale viscosity defined as the ratio of experimental values to prediction date with the Mooney model was presented in Figure 3. It indicated that both the non-diluted sets and diluted sets followed the same trend, marked as two local distinct increases in water content close to 0.15–0.20 and 0.30–0.35, respectively. The first notable increase at φ among 0.15–0.20 is attributed to the formation of a dynamic percolative cluster, while the second notable increase may be identified with the glass-transition volume fraction. In emulsions of non-flocculated drops and glass-transition phenomena are usually observed at volume fractions of the dispersed phase close to 0.58, while for the attractive drops system, local flocculation of the drops to form clusters arrest due to either caging or residual attraction. The clusters are viewed as compact (spherical or quasi-spherical) renormalized drops, whose effective volume fraction is significantly larger than the real dispersed phase volume fraction. In this case, the glass-transition threshold should be ascribed to the effective

volume fraction of the clusters. In the present experiment, the internal volume fraction of drops within the clusters, $\varphi_{int} = \varphi/\varphi_{eff}$ (φ_{eff} can be calculated from Mooney equation with the experimental viscosity data) is about equal to 0.56 at $\varphi = 0.30$, so the glass-transition phenomenon will be expected at $\varphi \approx 0.32$. At this water content, the cluster will become crowded and a 3D gel structure is formed. Similar rescaled viscosity anomalies, peaking at water contents below 0.2 and close to 0.4, were also found by other researchers, and they suggested that this phenomenon appeared to be a universal feature of native W/O petroleum emulsions.

3.2.2 Shear rate-dependent viscosity models

The concentrated emulsions in this study show significant deviations from the power law model at high and very low shear rates (Fig. 1). It is necessary to use a model to which takes account of the limiting values of viscosity η_0 (low shear rates) and η_∞ (high shear rates) to fit the experimental data.

One notable model in this class is the Carreau viscosity equation:

$$\frac{\eta_{app} - \eta_\infty}{\eta_0 - \eta_\infty} = \left(1 + (\gamma/\lambda)^2\right)^{(n-1)/2} \quad (6)$$

where, the parameter λ is the critical shear rate at which viscosity begins to decrease. The power law slope is $(n-1)$.

Similar in form to the Carreau relation, another four-parameter model was proposed by Cross (1965) which is given as:

$$\frac{\eta_{app} - \eta_\infty}{\eta_0 - \eta_\infty} = \frac{1}{1 + (\gamma/\lambda)^m} \quad (7)$$

Both models include η_0, λ and η_∞, though η_0 and λ, can be obtained from experimental observations. However, due to limitations (e.g. shear rate) associated with the viscometer, it is difficult to obtain η_∞. Jansen et al and Evdokimov et al recommend the Mooney model as the appropriate choice to predict the higher shear limit. In the following comparison, the Mooney model was employed to determine η_∞. The additional parameter in the model can then be determined reasonably well by using best visual fits, employing a spreadsheet-type program such as Microsoft Excel. The comparison results and the corresponding parameters were presented in Figure 4 and Table 1, respectively. From Figure 4, it can be concluded that the Carreau model has excellent agreement with the experimental data over the whole shear rate range. By contrast, the Cross model can give a good fit at a high shear rate but underestimates the emulsion viscosity at low shear rate significantly.

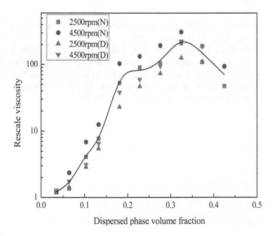

Figure 3. Experimental viscosities for non-diluted and diluted sets rescaled via dividing by the Mooney model. Trendlines provided to guide the eye.

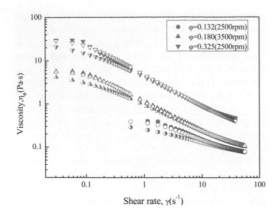

Figure 4. Comparison of experimental data (filled dot) with Carreau model (open dot) and Cross model (half-filled dot).

Table 1. Fitting parameters of Carreau model and Cross model presented in Figure 4.

Model	Sample	Fitting parameters				
		η_0	η_∞	λ	n	m
Carreau	$\varphi = 0.132$	0.396	0.050	1.670	0.280	–
	$\varphi = 0.180$	5.700	0.061	0.112	0.225	–
	$\varphi = 0.325$	30.00	0.143	0.089	0.219	–
Cross	$\varphi = 0.132$	0.396	0.050	1.670	–	0.714
	$\varphi = 0.180$	5.700	0.061	0.112	–	0.776
	$\varphi = 0.325$	30.00	0.143	0.089	–	0.779

Furthermore, the good fits of both models at high shear rates imply that it is reasonable to use the Mooney model to predict the high shear limit viscosity of shear-thinning emulsions.

4 CONCLUSIONS

Systemic experiments have been conducted to investigate the rheology of Water-in-Oil (W/O) emulsions by using a concentric viscometer. Due to the flocculation of drops, all five shear rate-independent viscosity models employed in this study underestimated the emulsion viscosity. The viscosity of the Newtonian flow regime rescaled via dividing by the Mooney model show two local distinct increases in water content close to 0.15–0.20 and 0.30–0.35, corresponding to the dynamic percolation and glass-transition, respectively. The Carreau model provides an excellent fit of the experimental data over the whole shear rate range, while the Cross model only gives a good fit to high shear rate flow behavior. The Mooney equation was proven to be the most reasonable model to predict the high shear limit viscosity of the shear-thinning emulsion.

REFERENCES

Cross, M.M. 1965. Rheology of non-Newtonian fluids: a new flow equation for pseudoplastic systems. *Journal of Colloid Science.* 20: 417–437.

Dou, D. & Gong, J. 2006. Apparent viscosity prediction of non-Newtonian water-in-crude oil emulsions. *Journal of Petroleum Science and Engineering.* 53: 113–122.

Evdokimov, I.N. Efimov, Y.O. & Losev, A.P. 2008. Morphological transformations of native petroleum emulsions. 1. Viscosity studies. *Langmuir.* 24: 7124–7131.

Jansen, K.M.B. Agterof, W.G.M. & Mellema, J. 2001. Viscosity of surfactant stabilized emulsions. *Journal of Rheology.* 45: 1359–1371.

Mooney, M. 1951. The viscosity of a concentrated suspension of spherical particles. *Journal of Colloid Science.* 6: 162–170.

Oliveira, R.C.G. & Goncalves, M.A.L. 2005. Emulsion Rheology-Theory vs. Field Observation. *OTC.* NO. 17386.

Pal, R. 1997. Flow behavior of emulsions containing small and large droplets. *Petroleum Society of CIM.* 97: 169.

Sjoblom, J. 2006. *Emulsions and emulsion stability, 2nd ed.;* CRC Press: New York.

Urdahl, O. Fredhein, A.O. & Loken, K.P. 1997. Viscosity measurements of water-in-crude-oil emulsions under flowing conditions: a theoretical and practical approach. *Colloids and Surfaces A: Physicochemical and Engineering Aspects.* 124: 623–634.

Advanced Materials and Structural Engineering – Hu (Ed.)
© *2016 Taylor & Francis Group, London, ISBN 978-1-138-02786-2*

Effect of sintering conditions on performances of LiFePO$_4$ cathode material obtained by liquid-phase method

X.M. Zu, P.F. Bai, J.J. Ma & X.Y. Wang
Department of Chemistry, School of Sciences, Tianjin University, Tianjin, China

W.R. Zhao
College of Environmental and Energy Engineering, Beijing University of Technology, Beijing, China

ABSTRACT: LiFePO$_4$ cathode materials were successfully synthesized by a liquid-phase method. The influences of sintering conditions on the structure, morphology and electrochemical performances of LiFePO$_4$ were investigated. The structure and morphology of samples were characterized using powder X-Ray Diffraction (XRD), Scanning Electron Microscopy (SEM). The electrochemical performances were conducted by galvanostatic charge/discharge test. The electrochemical results show that LiFePO$_4$ sintered at 600 °C for 2 h exhibits good rate and cycling performance. At a low rate of 0.1 C, the LiFePO$_4$ sample presents a high discharge capacity of 164.8 mAh/g which is near the theoretical capacity (170 mAh/g), and it still attains to 163 mAh/g after 30 cycles. It also exhibits an excellent rate capacity with high discharge capacities of 153.4 mAh/g, 140 mAh/g and 100.2 mAh/g at 0.2 C, 0.5 C and 1 C, respectively.

1 INTRODUCTION

Olivine-structure lithium iron phosphate (LiFePO$_4$) has been considered as the most promising candidate positive electrode material for Li-ion batteries materials. It has been widely used in portable electronic devices, electric vehicles and hybrid electric vehicles, owing to its high theoretical capacity of 170 mAh/g, abundance of raw materials, thermal stability and environmental benign (Wang et al. 2011). However, LiFePO$_4$ has two obvious obstacles: (1) low intrinsic electronic conductivity (~10^{-9} S/cm) and slow diffusion of Li-ions (~1.8 × 10–14 cm^2/s) (Hu et al. 2014) leading to fast capacity decay, especially under high current densities. (2) Low tap density is resulting in low volume specific capacity, which hinders the practical application of LiFePO$_4$ materials used in small and large applications. Currently, many investigations have been undertaken to improve the performances of LiFePO$_4$ by coating an electronically conductive phase (Konarova & Taniguchi 2010), doping supervalent metal ions (Li et al. 2009), developing a new synthesis method.

Many preparation methods have been used to fabricate LiFePO$_4$, such as solid-state reaction (Cheng et al. 2011), hydrothermal process (Miao et al. 2014), solve-thermal method (Murugan et al. 2008), sol-gel method (Hsu et al. 2004) and co-precipitation (Zhu et al. 2006). Although the traditional solid-state method has been industrialized, the grinding and long times sintering cycles usually lead to the formation of larger particles with lower electrochemical performances. In order to synthesize homogeneous nano-size LiFePO$_4$ and control its morphology, people have devoted to researching the liquid-phase method. W.F. Zhang et al. (Yang et al. 2011) prepared CoLiFePO$_4$/C composites using a hydrothermal route the first discharge capacity of sample developed from 115 mAh/g to 132 mAh/g at 0.1 C. Guo et al. (2009) and Honma et al. (2010) made LiFePO$_4$ nano-architectures using the solve-thermal method. But unfortunately, those materials did not show any excellent electrochemical performances. Qiu et al. (2007) synthesized olive-type LiFePO$_4$ via chemical reduction and lithiation of FePO$_4$, and discussed the influences of sintering temperatures on the physical and electrochemical properties of resulting LiFePO$_4$. But the cycle performances at different current densities were not investigated. Moreover, the morphology and the discharge capacities at different currents were unsatisfactory.

In this paper, a liquid-phase method was developed for high performance applications of LiFePO$_4$. As a feature of this work, we investigated the effect of sintering conditions on the properties of LiFePO$_4$, which has not been reported in the previous paper. The method would be attractive in the commercial application of LiFePO$_4$ cathode material.

2 EXPERIMENTAL

2.1 Preparation

All the chemical reagents were of analytically pure grade and used as received. $LiFePO_4$ were synthesized by a two-step solution phase method. In the first step, the as-prepared $FePO_4 \cdot xH_2O$ was prepared via chemically-induced precipitation. 0.1 mol $FeSO_4 \cdot 7H_2O$ and 0.1 mol $NH_4H_2PO_4$ were dissolved sequentially with vigorous stirring to form a solution. The proper amount of H_2O_2 solution (30 wt%) was added drop wise. Then 1:1(v/v) ammonia was gradually added into the solution to controlling the pH in the range of 2.0–2.5. After precipitating 20 min, the resulting $FePO_4 \cdot xH_2O$ was filtered, washed, and then oven-dried for 12 h at 80 °C. In the second step, stoichiometric amounts of as-synthesized $FePO_4 \cdot xH_2O$, CH_3COOLi, glucose and ascorbic acid were dispersed in absolute ethanol (63 mL). Then, the solution was stirred vigorously for 6 h at 60 °C in a round bottom flask, and stored at the room temperature for another 2 h. After that, the $LiFePO_4$ precipitate was filtered and washed with ethanol solution, and dried at 80 °C for 12 h. Finally, heat treatment under Ar atmosphere was carried out to develop the properties of $LiFePO_4$.

2.2 Characterization

The crystal structures and phases of the $LiFePO_4$ samples were characterized by X-ray powder diffraction (XRD, Rigaku D/MAX 2500 VPC, Japan) using Cu-Kα radiation with $\lambda = 1.5406$Å, operated at 40 kV and 20 mA, scanning range (2θ): 10–80°. Scanning electron microscopy (SEM, S-4800, Japan) was applied to examine the morphology of the samples. Thermogravimetry differential scanning calorimetry (TG-DSC) analysis (TA Instruments Corporation STA449F3, Germany) measurement was carried out from room temperature to 800 °C with a heating rate of 10 °C/min in N_2 atmosphere.

The electrochemical experiments were performed using CR2032 coin-type cells with Li metal as the counter electrode. Cathodes were fabricated by mixing the active materials, acetylene black and Polyvinylidene Fluoride (PVDF) with the mass ratio of 80:10:10. N-Methyl Pyrrolidinone (NMP) was the solvent to form slurry. The slurry was spread onto an aluminum foil, dried in vacuum at 120 °C for 12 h. The electrolyte was1 M $LiPF_6$ dissolved in Ethylene Carbonate (EC) and Dimethyl Carbonate (DMC) (1:1,v/v), this parador was Celgard 2400. The CR2032 cell was assembled in an argon-filled glove box. The charge/discharge testing was conducted in the range of 2.5–4.2 V using a Land test instrument (LAND, CT2001 A).

3 RESULTS AND DISCUSSIONS

3.1 TG-DSC analysis

To determine the suitable sintering temperatures, TG-DSC analysis was carried out in N_2 atmosphere. As shown in Figure 1, the TG curve presents three main steps of weight loss and DSC curve displays several corresponding endothermic peaks. The weight loss below 240 °C is related to the absorbed moisture, there is an obvious endothermic peak at this temperature of around 80 °C, which may be related to the evaporation of the adsorbed ethanol molecules. The main weight loss between 240 and 450 °C can be attributed to a complicated process including the decomposition of the reactants and reducing agents, the iron-related redox reaction and so on. The final small weight loss from 450 to 550 °C may be caused by the crystallization of phosphate, and corresponding endothermic peak appears in DSC curve. Above 550 °C, there is nearly no weight loss in TG curve, but the endothermic peak in DSC curve could be ascribed to the heat absorption of the crystal phase transition. Therefore, it is possible to calcine $LiFePO_4$ above 550 °C to obtain well-crystallized products. In this study, $LiFePO_4$ samples were post-treated by heating at 600, 700 and 800 °C.

3.2 Structure analysis

The phase and purity of the samples were determined by XRD, The XRD patterns of $LiFePO_4$ calcined for 2 h at 600, 700 and 800 °C are shown in Figure 2. On one hand, all the intense peaks in the spectrum clearly show a single phase formation of an olivine structure, which indicates that three samples are well indexed to an orthorhombic, Pnmb space group (JCPDS Card NO. 40–1499) without any impurity phase. On the other hand, with the increasing of temperature (from 600 to

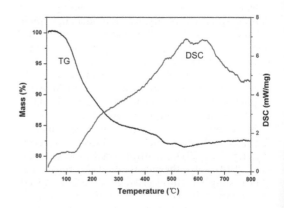

Figure 1. TG-DSC curves of $LiFePO_4$.

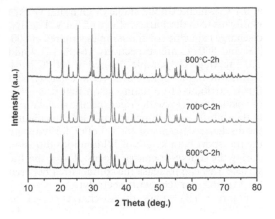

Figure 2.   XRD patterns of LiFePO$_4$.

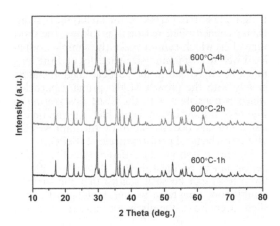

Figure 3.   XRD patterns of LiFePO$_4$.

Figure 4.   SEM images of LiFePO$_4$ sintered at (a) 600 °C, (b) 700 °C, (c) 800 °C for 2 h.

Figure 5.   SEM images of LiFePO$_4$ sintered at 600 °C for (d) 1 h, (e) 2 h, (f) 4 h.

800 °C), the diffraction peaks were sharper, and the crystallinity of LiFePO$_4$ became more perfect.

Figure 3 shows the XRD patterns of LiFePO$_4$ sintered at 600 °C for 1, 2 and 4 h. The diffraction peaks of all the samples match well with the standard LiFePO$_4$ spectrum. Compared to 1 h, the samples calcined for 2 h and 4 h possess sharper intense diffraction peaks and narrower half peak widths, this illustrates their higher order of crystallinity.

3.3   Morphology characterization

Figure 4 shows SEM images of LiFePO$_4$ sintered at different temperatures for 2 h. The particle sizes of LiFePO$_4$ samples increase as sintering temperature increases. At the temperature of 600 °C (a), quasi-spherical and small size particles with minimal agglomeration are obtained for LiFePO$_4$. Along with the temperature increasing (b: 700 °C, c: 800 °C), the particle sizes are remarkable different.

And there are a great number of agglomerate particles that are most likely grown on the account of small grains.

Figure 5 shows SEM images of LiFePO$_4$ sintered at 600 °C for the different time. The particles show variation in their dimensions with increasing sintering time. Figure 5(d) reveals that 1 h is too short for the grains to grow well, which leads to a broad size distribution. Figure 5(e) shows the image of the sample sintered for 2 h, and big particles can be rarely seen in the image. The particles are small and uniform without obvious aggregations, and have good surface topography. By increasing the time to 4 h (f), the particles agglomerated greatly with the growth of the primary particles which is comparable to the theoretically calculated growth morphology of LiFePO$_4$. Therefore the best feasible sintering condition is 600 °C for 2 h.

3.4 *Electrochemical characteristics*

Figure 6 (A) gives the initial discharge curves of LiFePO$_4$ sintered at different temperatures for 2 h

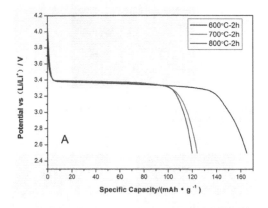

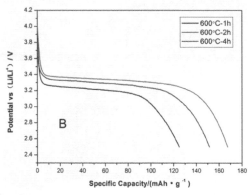

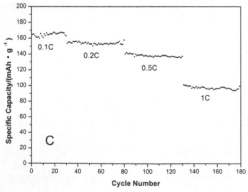

Figure 6.   Initial discharge curves of LiFePO$_4$ at 0.1 C (A, B) and cycling performance of LiFePO$_4$ (C).

at 0.1 C, which exhibits stable discharging voltage platforms with discharge voltage of 3.4 V. The first discharge capacities of the samples sintered at 600, 700 and 800 °C are respectively 164.8, 123.9 and 119.9 mAh/g. The sample sintered at 600 °C has a significant improvement in specific capacity, and this is attributed to a main reason that it consists of smaller particles with larger specific surface area in the sintering step. Smaller particles can enhance the discharge capacity of the LiFePO$_4$ for reducing the transport path length of lithium ions diffusion during the charge-discharge process. Large specific surface area can increase the contact area between the LiFePO$_4$ particles with electrolyte.

Figure 6 (B) shows the discharge curves of LiFePO$_4$ sintered at 600 °C for different time at the rate of 0.1 C. The first discharge capacities of the samples sintered for 1 h, 2 h and 4 h at 600 °C are 125, 167.4 and 151.5 mAh/g, respectively. Obviously, the sample sintered at 600 °C for 2 h has the best discharge capacity. The reason is as follows: the short time (1 h) which cannot make the sample crystallized well to form nano-size particles and the long time (4 h) may lead to the particles agglomerating greatly with the growth of the primary particles, which is consistent with the SEM observation in Figure 5. It indicates that suitable sintering time is also a primary decisive factor on the improvement of electrochemical performance of LiFePO$_4$.

Figure 6 (C) shows the different rates life cycling performances of the LiFePO$_4$ sintered at 600 °C for 2 h. It can be seen clearly that the samples exhibit good long term cycling stability with a very slight decrease of the discharge capacity. The first discharge capacities at 0.1, 0.2, 0.5 and 1 C are 164.8, 153.4, 140 and 100.2 mAh/g, respectively. At 0.1 C, the sample maintains a discharge capacity of 163 mAh/g after 30 cycles, with the capacity retention of 98.91%. The discharge capacities are as high as 152.3, 137 and 95.7 mAh/g after 50 cycles at 0.2 C, 0.5 C and 1 C and their capacity retention is 99.28%, 97.86% and 95.51%, respectively. In a word, the capacity retention of the battery fades very slightly. This retention of the LiFePO$_4$ possesses an excellent rate capability and good life cycle. These results illustrate that the sintering condition of 600 °C for 2 h can optimize the properties of the LiFePO$_4$ material, which facilitates the intercalation and deintercalation of lithium ions.

4   CONCLUSIONS

In summary, the quasi-spherical LiFePO$_4$ materials were successfully synthesized by a simple and economical liquid-phase method and the final products were obtained by the subsequent heat treatment. The patterns of XRD and SEM demonstrate well-crystallized LiFePO$_4$ materials and well-distributed nano-size particles. The heat-treated LiFePO$_4$ materials show its excellent electrochemical properties of specific discharge capacity and cycling stability. The LiFePO$_4$ material calcined at 600 °C for 2 h shows preferable specific discharge capacity of 164.8 mAh/g at 0.1 C and the voltage platform is 3.4 V. Their specific capacity retention is 99.28%, 97.86% and 95.51% respectively after 50 cycles at 0.2 C, 0.5 C and 1 C.

ACKNOWLEDGEMENT

This work was financially supported by the National Natural Science Foundation of China (No. 21276185).

REFERENCES

Cheng, F.Q. Wan, W. Tan, Z. Huang, Y.Y. Zhou, H.H. Chen, J.T. & Zhang, X.X. 2011. High power performance of nano-LiFePO$_4$/C cathode material synthesized via lauric acid-assisted solid-state reaction, *Electrochimica. Acta,* 56: 2999–3005.

Hsu, K.F. Tsay, S.Y. & Hwang, B.J. 2004. Synthesis and characterization of nano-sized LiFePO$_4$ cathode materials prepared by a citric acid-based sol-gel route, *Journal of Materials Chemistry,* 14: 2690–2695.

Hu, Y.K. Ren, J.X. Wei, Q.L. Guo, X.D. Tang, Y. Zhong, B.H. & Liu, H. 2014. Synthesis of rod-like LiFePO$_4$/C materials with different aspect ratios by polyol process, *Acta Physico-Chimica Sinica,* 30: 75–82.

Konarova, M. & Taniguchi, L. 2010. Synthesis of carbon-coated LiFePO$_4$ nanoparticles with high rate performance in lithium secondary batteries, *Journal of Power Sources,* 195: 3661–3667.

Li, L.J. Li, X.H. Wang, Z.X. Wu, L. Zheng, J.C. & Guo, H.J. 2009. Stable cycle-life properties of Ti-doped LiFePO$_4$ compounds synthesized by co-precipitation and normal temperature reduction method, *Journal of Physics and Chemistry of Solids,* 70: 238–242.

Miao, C. Bai, P.F. Jiang, Q.Q. Sun, S.Q. & Wang, X.Y. 2014. A novel synthesis and characterization of LiFePO$_4$ and LiFePO$_4$/C as a cathode material for lithium-ion battery, *Journal of Power Sources,* 246: 232–238.

Murugan, A.V. Muraliganth, T. & Manthiram, A. 2008. Rapid microwave-solvothermal synthesis of phospho-olivine nanorods and their coating with a mixed conducting polymer for lithium ion batteries, *Electrochemistry Communications,* 10: 903–906.

Qiu, Y.L. Wang, B.F. & Yang, L. 2007. Synthesis and electrochemical properties of LiFePO$_4$ cathode material, *Journal of Inorganic Materials,* 22: 79–83.

Rangappa, D. Sone, K. Kudo, T. & Honma, I. 2010. Directed growth of nanoarchitectured LiFePO$_4$ electrode by solvothermal synthesis and their cathode properties, *Journal of Power Sources,* 195: 6167–6171.

Wang, Y.G. He, P. & Zhou, H.S. 2011. Olivine LiFePO$_4$: Development and future, *Energy and Environmental Science,* 4: 805–817.

Yang, H. Wu, X.L. Cao, M.H. & Guo, Y.G. 2009. Solvothermal synthesis of LiFePO$_4$ hierarchically dumbbell-like microstructures by nanoplate self-assembly and their application as a cathode material in lithium-ion batteries, *The Journal of Physical Chemistry C,* 113: 3345–3351.

Yang, J.M. Bai, Y. Qing, C.B. & Zhang, W.F. 2011. Electrochemical performances of Co-doped LiFePO$_4$/C obtained by hydrothermal method, *Journal of Alloys and Compounds,* 509: 9010–9014.

Zhu, B.Q. Li, X.H. Wang, Z.X. & Guo, H.J. 2006. Novel synthesis of LiFePO$_4$ by aqueous precipitation and carbothermal reduction, *Materials Chemistry and Physics,* 98: 373–376.

Advanced Materials and Structural Engineering – Hu (Ed.)
© 2016 Taylor & Francis Group, London, ISBN 978-1-138-02786-2

Viscosity measurement of Newtonian fluids using an in-plane torsional piezoceramic discal resonator

G.J. Xiao
School of Instrument Science and Opto-Electronics Engineering, Hefei University of Technology, Hefei, Anhui, China
Department of Precision Machinery and Precision Instrumentation, University of Science and Technology
of China, Hefei, Anhui, China

C.L. Pan
School of Instrument Science and Opto-Electronics Engineering, Hefei University of Technology, Hefei, Anhui, China

Y.B. Liu & Z.H. Feng
Department of Precision Machinery and Precision Instrumentation, University of Science and Technology
of China, Hefei, Anhui, China

ABSTRACT: In this paper, a new torsional piezoelectric resonator is presented to measure the viscosity of Newtonian fluids. The key component of proposed resonator is a piezoceramic disk with spiral interdigitated electrodes, which can generate in-plane torsional vibration. Based on the impedance analysis, principle of the viscosity measurement is explained. Prototype viscometer is fabricated and tested in glycerol/water solutions. Experiment results show a good linear relationship between the dynamic resistance of resonator R_2 and square root of fluid density and viscosity product $(\rho\mu)^{0.5}$ with a R-square of 0.9979. The linearity between the shift of resonant frequency Δf and $(\rho\mu)^{0.5}$ is also acceptable with a R-square of 0.9796. The resonator can be used as a competitive probe for portable viscometer with simple and compact structure.

1 INTRODUCTION

Viscosity reflects the internal friction property of fluids, which is a critical material parameter for hydrodynamics. Conventional viscometers, such as capillary, falling ball, and revolving cup types (Lucena & Kaiser 2008), have been commercialized with the advantages of mature technologies and reliable performances. However, most of them are expensive, bulky, and usually operated in the laboratory conditions, which seriously hamper their applications for miniature portable systems.

Based on vibration coupling between solid elements and fluids, vibrating type viscometers can extract the information of fluid viscosity from the shift of resonant frequency or Q factor of the vibration mode. With the advantages of small size, quick response, and on-line operating capability, many types of resonators have been used as probes of vibrating type viscometers, include torsional vibrating crystals (Willmott & Tallon 2007), thickness-shear quartz crystals (Doy et al. 2010), magnetoelastic plates (Stoyanov & Grimes 2000), and piezoelectric cantilevers (Mather et al. 2012). Among them, only resonators with torsional vibration modes can provide pure shear couplings

with fluids, which is an essential factor to achieve a high relative accuracy of viscosity measurement.

We previously presented a torsional piezoelectric fiber for viscosity measurement of Newtonian liquids (Pan et al. 2012). In this paper, a new alternative resonator is proposed with improved data processing method for the calculation of fluid viscosity. Compared to the resonators with quartz crystal materials, the piezoceramic resonators have much higher electromechanical coupling coefficients, which are more suitable for the viscosity measurement of viscous fluids. In the following sections, principle of the viscosity measurement will be explained firstly. Then, fabrication and test of prototype resonator will be introduced. Experiment results of the prototype viscometer in standard glycerol-water solutions will be discussed. Finally, conclusions will be drawn.

2 PRINCIPLE OF VISCOSITY MEASUREMENT

As shown in Figure 1(a), the proposed viscometer is composed of an in-plane torsional piezoelectric discal resonator and a holder. The piezoceramic disk is attached with spiral interdigitated electrodes (SIDEs)

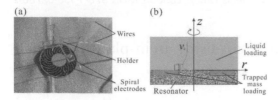

Figure 1. (a) Prototype viscometer, (b) schematic of operation.

on one surface, which has been validated to generate in-plane torsional vibration efficiently (Pan et al. 2014). The resonator shares the same SIDEs for polarizing piezoceramic material and exciting vibration mode. The in-plane torsional vibration comes from the combination of d_{31} and d_{33} effects, and the first torsional vibration mode is selected as the working mode of the resonator.

When the resonator is immersed and operated in liquid, shear wave will be generated in the adjacent liquid due to the in-plane torsional motion of the discal surfaces. As shown in Figure 1(b), the liquid entrained by the resonator interface acts as a liquid loading of the resonator. In Newtonian fluids, the shear wave velocity $v_\theta(z,t)$ of the liquid loading can be expressed as:

$$v_\theta(z,t) = v_{\theta 0} \exp\left[-\frac{(1+j)z}{\delta}\right] \exp\left(j\omega t\right) \quad (1)$$

where, $v_{\theta 0}$ is the particle velocity of the disk interface, δ is defined as the penetration depth of the shear wave, which is equal to $(2\mu/\omega\rho)^{1/2}$, ω is angular frequency of the vibration, ρ and μ are the liquid density and shear viscosity, respectively. The coupling liquid loading is actually a thin layer of liquid near the surface of resonator, for example, δ is about 60 μm in 20 °C pure glycerol solution with vibration frequency of 109 kHz.

As described in Equation 1, the liquid loading does not move synchronously with the resonator interface, but with both an amplitude decrement and a progressive phase delay along the distance from the interface. The shear mechanical impedance of the resonator per unit area of the interface due to liquid loading can be calculated as:

$$Z_L = \frac{T_{r\theta}}{v_\theta}\Big|_{z=0} = -\frac{\mu}{v_{\theta 0}}\frac{\partial v_\theta}{\partial z}\Big|_{z=0} = \sqrt{\frac{\omega\mu\rho}{2}}(1+j) \quad (2)$$

Due to the non-smooth surface of the resonator with SIDEs, some liquids might be trapped by the groove of SIDEs on the resonator surface, which will vibrates synchronously with the resonator. Theoretically, there is no internal friction energy consumption among these liquids and they can be treated as a mass loading. The mechanical

impedance per unit area due to the trapped mass loading is given by (Martin et al. 1994):

$$Z_M \cong j\omega\rho h \quad (3)$$

where, h is defined as effective thickness of trapped liquid. If h is comparable to or larger than δ, the massing loading will result in a significant additional shift of the resonant frequency.

Considering the disk thickness t is much smaller than radius r by two orders of magnitude, the effect of the disk side surface area is ignored. The total mechanical impedance of the resonator can be expressed as:

$$Z_S = \pi r^2 (2Z_L + Z_M) \quad (4)$$

where, r is the radius of the disk.

Electrical characteristics of a piezoelectric resonator near its vibration mode can be described by a simplified Butterworth-Vas Dyke equivalent circuit. The equivalent circuits for the in-plane torsional piezoceramic discal resonator under vacuum and liquid conditions are shown in Figure 2. The piezoelectric resonator itself is generally equivalent to a static capacitance C_0 and a parallel mechanical branch (series RLC portion with C_1, L_1, and R_1). C_1, L_1, and R_1 represent the inertial mass, elastic stiffness, and internal mechanical damping of the resonator, respectively. The liquid loading is equivalent to a resistance R_2 as additional mechanical damping and inductance L_2 as additional mass loading. The trapped liquid mass, without damping loss, is equivalent to another inductance L_3 (Martin et al. 1994). The relationship between Z_s and mechanical branch of equivalent circuits can be described as:

$$Z_S = \frac{1}{K}[R_2 + j\omega(L_2 + L_3)] \quad (5)$$

where, K is an electromechanical coupling constant for the in-plane torsional vibration. From Equations 2–5, a linear relationship between R_2 and $(\rho\mu)^{0.5}$ can be approximatively deduced as:

(a)   (b)

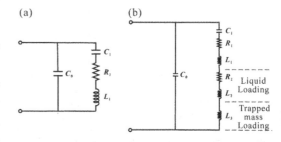

Figure 2. Equivalent circuits for the piezoelectric resonator under (a) vacuum and (b) liquid conditions.

$$R_2 = 2\pi^{1.5} r^2 K f_0^{0.5} \cdot \rho\mu^{0.5} \approx k' \rho\mu^{0.5} \qquad (6)$$

here, f_0 is the resonant frequency of the vibration in liquid. k' is approximate to a constant and can be calibrated by standard liquids.

In addition, leakage current of the resonator and conductivity of the liquid also influence the electrical characteristics of the resonator. They can be consider as an addition of parallel resistance R_0 in the equivalent circuit (Muramatsu et al. 1988). In general, the value of electrical resistance R_0 should be much large than the value of mechanical resistance R_1 and R_2 to achieve a good measurement accuracy of latter. And it means that the leakage current of the resonator must be limited and the resonator is only suitable for the viscosity measurement of nonconductive liquids. Then, the total electrical impedance of the resonator in liquid can be expressed as:

$$Z = \frac{1}{j\omega C_0} // [R_1 + R_2 + j\omega(L_1 + L_2 + L_3) + 1/(j\omega C_1)]$$

$$(7)$$

3 FABRICATION AND EXPERIMENTS

A prototype resonator for the viscosity measurement of liquids was fabricated, as shown in Figure 1(a). The piezoceramic disk was supplied by Kunshan Pant Piezoelectric Tech. Co. Ltd., China. It had a radius of 13.5 mm and a thickness of 0.2 mm. The piezoceramic material was PT-300. Silver film of 5 μm thickness was previously deposited on one surface of the disk and the SIDEs were fabricated by a laser etching process. Another surface of the disk was bare. Thin steel tubes were used as holders and fixed on the vibration nodes of the first in-plane torsional vibration mode with epoxy adhesive (DP460). Leading wires were threaded through the steel tubes and connected with the SIDEs. The disk was polarized in the thermostatic silicon oil at 90 °C, under an average electric field intensity of 2500 V·mm⁻¹, lasting for 30 min.

Figure 3 shows the schematic of experiment setup for testing the prototype viscometer. Several glycerol/water solutions with volume fraction from 50% to 100% were chosen as the standard viscosity liquids with distinct products of viscosity and density, as listed in Table 1. While the densities of these solutions change slightly, their viscosities change from 6 to 700 mPa·s. Water bath with temperature stability of 0.1 °C was used to maintain the temperature of tested solutions at 28 °C. Admittances Y (equal to $1/Z$) of the prototype resonator immersed in different tested solutions were measured by a precision impedance analyzer (GW8101, Good Will Instrument Co., Ltd., Taiwan) with a linear frequency

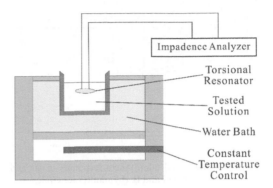

Figure 3. Experiment setup for testing prototype viscometer.

Table 1. Material properties of tested glycerol/water solutions (at 28 °C).

Glycerol conc, lv%	Viscosity (10⁻³ Pa·s)	Density (10³ kg·m⁻³)	(Dens·Visc)⁰·⁵ (Pa·s kg·m⁻³)⁰·⁵
(Dry air)	0.018	0.0012	0.005
50%	6.146	1.143	2.650
60%	11.24	1.168	3.623
70%	23.15	1.192	5.253
80%	55.94	1.215	8.244
90%	168.8	1.238	14.46
100%	702.4	1.259	29.74

sweep from 105 kHz to 113 kHz near the first in-plane torsional vibration mode of the resonator.

4 RESULTS AND DISCUSSIONS

The frequency responses of conductance G and susceptance B in air and tested solutions are shown in Figure 4(a) and their corresponding admittance circles are plotted in Figure 4(b). Fitting the experiment data with the admittance circle calculated from Equation 7 by the least squares method, the parameters of the equivalent circuit can be obtained (Muramatsu et al. 1988). The resonant frequent f_0 is located at the point where G reaches its maximum value. Under the air condition, the resonant frequency of prototype viscometer is 109.4 kHz and the corresponding Q factor is 94. The value of R_1 is calculated as 301 Ω. While the effect of air viscosity is negligible, the dynamic resistance R_1 can be regard as the characterization of resonator inherent mechanical damping. While the total resistance of R_1 and R_2 is calculated from the admittance circle in liquid, the added resistance R_2 can be calculated. Thus, the fluctuation of R_1 will significantly influence the accuracy of the viscosity measurement.

The results of dynamic resistance R_2 (liquid loading) as a function of $(\rho\mu)^{0.5}$ are shown in

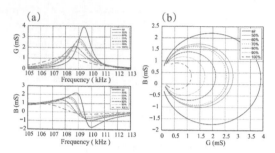

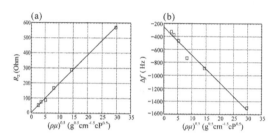

Figure 4. Measured admittances responses in air and tested solutions: (a) G and B, (b) admittance circle.

Figure 5. Dynamic resistance R_2 and shift of resonance frequency Δf as a function of $(\rho\mu)^{0.5}$.

Figure 5(a). Solid line is the linear fitted curve. It shows a good linear relationship between R_2 and $(\rho\mu)^{0.5}$ with a R-square of 0.9979. Figure 5(b) shows the linearity between resonant frequency shift Δf and $(\rho\mu)^{0.5}$ with a R-square of 0.9796. The liquid loading has two effects on admittance, reducing response peaks and changing resonant frequency. In another aspect, the mass loading also results in the shift of resonant frequency but no peak damping. The fitted line in Figure 5(a) passes through the zero point reveals that energy dissipation from the trapped mass loading is very little compared with the liquid loading. Since both the liquid loading and trapped mass loading have impact on the shift of resonant frequency, the fitted line in Figure 5(b) does not pass through the zero point. And the linearity between Δf and $(\rho\mu)^{0.5}$ declines compared with that of R_2 and $(\rho\mu)^{0.5}$, because the densities of tested solutions change slightly (Itoh & Ichihashi 2011). However, the linearity between Δf and $(\rho\mu)^{0.5}$ is also acceptable in this case.

5 CONCLUSIONS

In-plane torsional piezoceramic disk can be used as a resonator to measure the viscosity of Newtonian fluids. A prototype viscometer is fabricated and tested in glycerol/water solutions. The method

based on fitting of admittance circle is used to obtain parameters of equivalent circuit. Experiment results show a good linear relationship between dynamic resistance R_2 and $(\rho\mu)^{0.5}$ with a R-square of 0.9979. Due to the influence of mass loading, the linearity between Δf and $(\rho\mu)^{0.5}$ declines slightly with an acceptable R-square of 0.9796. The resonator can be used as a competitive probe for portable viscometer with simple and compact structure.

ACKNOWLEDGEMENT

The authors would like to acknowledge the financial support from Projects 51205338 supported by National Natural Science Foundation of China and Project 2014M561818 funded by China Postdoctoral Science Foundation.

REFERENCES

Doy, N. McHale, G. Newton, M.I. Hardacre, C. Ge, R. MacInnes, J.M. Kuvshinov, D. & Allen, R.W. 2010. Small volume laboratory on a chip measurements incorporating the quartz crystal microbalance to measure the viscosity-density product of room temperature ionic liquids, *Biomicrofluidics.* 4: 014107.

Itoh, A. & Ichihashi, M. 2011. Separate measurement of the density and viscosity of a liquid using a quartz crystal microbalance based on admittance analysis (QCM-A), *Measurement Science & Technology.* 22: 015402.

Lucena, S.E. & Kaiser, W. 2008. Stepping-motor-driven constant-shear-rate rotating viscometer, *IEEE Transactions on Instrumentation and Measurement.* 57: 1338–1343.

Martin, S.J. Frye, G.C. & Wessendorf, K.O. 1994. Sensing liquid properties with thickness-shear mode resonators, *Sensors and Actuators A-Physical.* 44: 209–218.

Mather, M.L. Rides, M. Allen, C.R.G. & Tomlins, P.E. 2012. Liquid viscoelasticity probed by a mesoscale piezoelectric bimorph cantilever, *Journal of Rheology.* 56: 99–112.

Muramatsu, H. Tamiya, E. & Karube, I. 1988. Computation of equivalent circuit parameters of quartz crystals in contact with liquids and study of liquid properties, *Analytical Chemistry.* 60: 2142–2146.

Pan, C.L. & Liao, W.H. 2012. *Torsional piezoelectric fiber for viscosity measurement of Newtonian fluids,* Proceedings of the 2012 international conference on Mechatronics and Automation, Chengdu, 635–640.

Pan, C.L. Xiao, G.J. Feng, Z.H. & W.H. Liao, 2014. Electromechanical characteristics of discal piezoelectric transducers with spiral interdigitated electrodes, *Smart Materials and Structures.* 23: 125029.

Stoyanov, P.G. & Grimes, C.A. 2000. A remote query magnetostrictive viscosity sensor, *Sensors and Actuators A-Physical* 80: 8–14.

Willmott, G.R. & Tallon, J.L. 2007. Measurement of Newtonian fluid slip using a torsional ultrasonic oscillator, *Physical Review E.* 76: 066306.

Advanced Materials and Structural Engineering – Hu (Ed.)
© 2016 Taylor & Francis Group, London, ISBN 978-1-138-02786-2

Research on the clamping device of the underwater wire rope cutter

Z. Hui
School of Naval Architecture and Ocean Engineering, Zhejiang Ocean University, Zhoushan, Zhejiang, China
Key Laboratory of Offshore Engineering Technology, Zhejiang Province, China

ABSTRACT: In order to improve the efficiency of underwater cutting, a new clamping device for the underwater wire rope cutter is proposed in this paper. First, the types and characteristics of the underwater wire rope cutter are introduced, and the advantages and disadvantages of the existing structure of the clamping device are analyzed. Then, based on the conditions of use, a new type of clamping devices is presented, designed and calculated. Finally, according to the man-machine engineering principle, the model of a clamping device for the underwater wire rope cutter is established. The designed model proves that the proposed method is feasible and effective.

1   INTRODUCTION

With the speed development of human development and the uses of the oceans, marine engineering, such as ocean oil exploitation, scientific investigation, involved in salvage, has increased. The underwater tasks that need to be completed is gradually increasing. Underwater wire rope cutter is a commonly used underwater cutting tool, which is mainly used for cutting steel tube, steel bar and wire rope on the ocean floor. However, due to work under harsh conditions, at high-pressure and high-temperature environments, the cutting efficiency is low. Ergonomics is a discipline that has been widely applied in practice (Ulrikkeholm & Hvam 2013, Ji & Bing et al. 2014), which takes into account not only the product features in product design and good maneuverability, but also a better environment applicability (Xu & Liu et al. 2014). Therefore, a method to resolve the cutting efficiency of the underwater wire rope cutter based on the man-machine engineering theory is produced in this paper.

2   DESIGN REQUIREMENTS FOR THE UNDERWATER WIRE ROPE CUTTER

2.1   *Analysis of the existing underwater cutting tools*

The underwater work is usually done by the underwater robot, which plays an increasingly important role in Ocean Engineering (Jiang 1997, Peng 2004). Underwater robots perform underwater work by carrying all kinds of general and special underwater tool. Underwater cutting tools are commonly used.

Hydraulic shearing machine: hydraulic cutter is a type of underwater cutting tool used to cut a variety of underwater cables and wire ropes. Its characteristic is to cut short, but it cannot be used for cutting sophisticated steel pipes and steel bars.

Abrasive cutting: grinding wheel type cutter is an underwater cutting tool in rotary motion. It can cut not only the cable, but also steel bars and steel tubes. Wheel cutter can cut metals and non-metals, such as concrete, wood, rubber, and plastic. However, cutting the wire is more difficult.

Therefore, a tool used for cutting steel wire need to be designed to reduce the clamping phenomena, as well as to improve the surface accuracy and rigidity of the cutting tools, and to improve the effectiveness of wire cutting.

2.2   *Structure of the underwater wire rope cutter*

The underwater wire rope cutter is commonly used, as shown in Figure 1. It consists of a swinging arm, a wrist, a base plate, hydraulic motors, cylinders, clamps, a blade and a fixed arm. One end of the swing arm is mounted on the retention arm, and the other end has a rotating saw blade. The swing

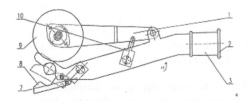

1-wing arm 2,3-wrist 4,5-base plate 6-hydraulic motor 7-cylinder 8-clamp 9-saw blade 10-feed cylinder 11-fixed swing arm

Figure 1.   Underwater wire rope cutter (Zhang et al. 2000).

arm and fixed arms are fitted with a feed cylinder clamp. The clamp with the clamping cylinder is hinged with a fixed arm. It is located on the blade. The other end of the fixed arm is connected to the wrist. The upper and lower substrate is located on either side of the fixed arm.

This structure has the following advantages:

1. The layout structure of clamps can be effectively cut without clamping.
2. Cut range: it can cut not only the cable, but also steel bars and steel tubes. It is suitable for all kinds of underwater cutting environment.
3. Using the underwater seal technology makes it light weight, small size, seawater resistant and saw blade service long life.

2.3 Design requirements of the wire rope cutter

To improve the cutting efficiency of the underwater wire rope, the design factors related to work and underwater environment need to be considered (Wang 2009):

1. *The pressure of the water.* Underwater tools must withstand the external pressure expected performance of water depth. When the water depth increases to 10 m, the pressure would increase by approximately 0.1 MPa.
2. *The viscosity of water.* The operation of water is more difficult than on land. The effect of water resistance must be considered while designing, and try to minimize this impact.
3. *The reliability of the clamping device.* When in high-speed cutting, the wire rope needs to be kept clamping in order to ensure safe and efficient cutting.

3 DESIGN OF THE UNDERWATER WIRE ROPE CUTTERS CLAMPING DEVICE

3.1 Existing clamping devices

The clamping device can be used in metallurgy, mining, oil field, port and forest of large material handling and palletizing jobs. It can also be used in laying pipe, drill pipe transit rules, stations, terminals, and other places, such as the item shipment of long and irregular-shaped objects. It is a widely used tool with high efficiency and very strong adaptability. The main types are discussed below.

3.1.1 Automatically closed clamping device
The automatically closed clamping device works through the bridge type crane. First, under the control of the lock closed device, the clip tight device placed put above the steel volume, and makes the mouth of a convex clamp to orient cone alignment steel volume center hole, as shown in Figure 2a. Then, the locked heart of the lock closed device

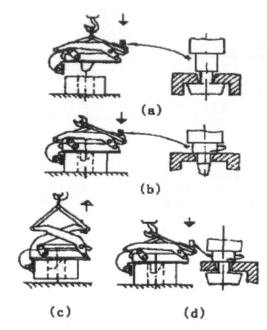

(a)

(b)

(c)   (d)

Figure 2. Automatic closing type clamping unit (Zhang et al. 2000).

is locked up over, as shown in Figure 2b. While the lock closed device is opened, eventually steel volume clip will tighten, as Figure 2c shows. When the steel volume put with volume handling to specific locations, the closed device is closed, as shown in Figure 2d. The automatically closed clip tight device completes a work cycle.

3.1.2 A clamping device for electrical conductivity and sheet metal
The clamping device is composed of a floor plate and a clip horizontal extended by the two ends. At the other end of the clips back to the floor form an end. The base plate and clips have a width at the bottom of the vertical. The width of the base plate is less than the width of the clip. This makes workers to facilitate the effectiveness of actions during construction, as shown in Figure 3.

3.2 A clamping device

The clip clamp structure consists of a base, a clip solid grip handle, and a clip solid guide block. A moving clip holding block lies on the clip solid guide block. The jaw end parts of the clip extended by the solid guide block has a clip business block positioning hole and a capacity reset hole. A positioning hole and the clamp slot are designed to a moving clip holding block. A capacity reset hole is located in the clamp slot. The moving clip holding of the moving clip holding block appears plane-like. The moving clip holding block and clamp slot sets

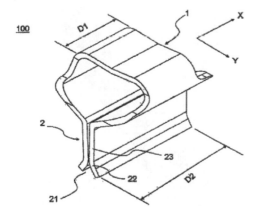

Figure 3. Clamping device for electrical conductivity and sheet metal (Lin 2011).

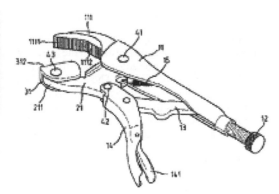

Figure 4. A clamping device (Wang 1998).

is set in jaw end parts of the clip solid guide block, as shown in Figure 4.

4 DESIGN OF THE CLAMPING DEVICE

4.1 Design parameters of the clamping device

According to the current cutter design parameters, the cutting speed is 50–750 mm/min, the cutting diameter is $\Phi20$–$\Phi600$ mm, the maximum cutting length is 1200 mm, and the cutting precision is less than or equal to 1.5 mm. According to the parameters, the structure of the clamp can be designed.

4.2 Structural design of the clamping device

Clamping device generally consists of four parts: (1) the original part takes a direct action, (2) the original part contacts with the working surface; (3) mechanicals change the acting force; (4) increase the role to the original clamping contact links. The design of a clamping device based

on the principle of hydraulic clamping devices is produced. The hydraulic clamping device is a clamping structure through liquids plastic or clamping force transmitted to each clamping by liquid. Liquid plastic is commonly used, it is passing links, and will transmit the force to the original part that withstands the force.

Figure 5a shows the principle diagram. The liquid plastic is in a closed shell. A piston 1 is in a shell hole and associated with the pressure plate 2. When screw 4 is rotated, the pressure is produced in liquid plastic by the piston 3 and acts round the shell hole. The plate 2 clamps down when piston 1 moves down.

Figure 5b shows many plunging clamping fixtures using hydraulic. It works in the same way as described above. In this design, the clamping of the piston displacement is not only down, but also upward. Its clamping force direction is up.

In the hydraulic clamping devices, the axial force acting on the passive piston P_{oo} is given by

$$P_{oo} = \frac{P_{\pi P} \times (D^2 - d^2)}{4} \qquad (1)$$

where
$P_{\pi P}$—units pressure on the end surface of the piston;
D—diameter of a passive piston; and
d—diameter of a passive piston rod.

The unit pressure P_C acted on the active piston surfaces can be calculated according to the following formula:

$$P_c = 4Q/\pi D_{HCX}^2 \qquad (2)$$

where,
Q—an axial force acted on the initiative piston by screw and
D_{HCX}—diameter of an active piston.

According to the formula of spiral clamping force (2), Q is substituted as follows:

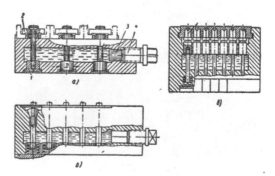

Figure 5. Diagram of a hydraulic clamping device.

$$P_c \approx \frac{10.5 P_H \cdot l}{d_{cp} \cdot D_{HCX}^2} \qquad (3)$$

where,

P_{H-}—tangential force on the pull handle (or lever);
l—distance from the axis of screws to the P_H; and
d_{cp}—average diameter of the screw threads.

Due to the friction between the piston and the casing wall, the friction between the piston rod and the liquid plastic and the friction of the entire interior parts, the pressure produced by the driving pistons in the liquid plastic is slightly reduced.

The unit pressure value $P_{\pi P}$ on the end surface of the passive piston can be decided by the following approximate formula (if and when the working surface of the body cavity is $\nabla 3$):

$$P_{\pi p} = \left| P_c - \frac{0.05L}{s} \right| \text{kg/cm}^2 \qquad (4)$$

where

L—is the distance between the active piston and the passive piston, in cm and
S—cross-sectional area storage of the liquid plastic hole, in cm^2.

The above values are substituted in Equation (1), and the axial clamping force of the passive piston is given by

$$P_{oc} \approx \frac{\pi}{4}(D^2 - d^2)\left(P_c - \frac{0.05L}{s} \right) \qquad (5)$$

According to the axial clamping force calculated, the structure of the gripper is designed. The design model of clamps is shown in Figures 6 and 7.

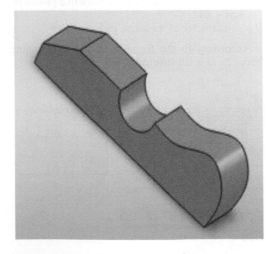

Figure 6.   Effect of the clamping part.

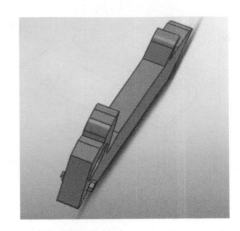

Figure 7.   Clamping effect.

5   CONCLUSIONS

The underwater wire rope cutter is a new kind of cutting tool. A clamping device used in the underwater wire rope cutter is presented. It is with excellent stability, security, and control. The actual working conditions and influences of various factors, such as the external environment, need to be considered. So, the next step is to analyze the operating characteristics and theoretical model of the underwater wire rope cutter based on the extensive experiments in order to obtain optimal design parameters.

REFERENCES

Ji, M.H. & Bing, Y.H. et al. 2014. Application of modern design method in the design and develop process of mechanical product. *Advanced Manufacturing and Information Engineering*, 602: 189–192.

Jiang, X.S. 1997. The direction for future Robotics, *Robotics Technology Applications*, 2: 2–5.

Lin, Z.X. 2011. *Clamp structure*, CN201120023048.6.

Peng, X.L. 2004. Underwater research present situation and development trend, *International Journal of Robotics Applications*. 2: 2–4.

Ulrikkeholm, J.B. & Hvam, L. 2013. *Including product features in the development of engineering design processes.* Proceedings of the 19th International Conference on Engineering Design (ICED13), 1: 1–10.

Wang, L.Q. Zhang, B. Wang, G. & Zhang, L. 2008. *An underwater wire saw cutting machine*, China, CN200810209831.

Wang, Z.J. 1998. *A clamp structure*, CN98203975.

Xu, H.Y. & Liu, R.Q. et al. 2014. *A machine learning-based design representation method for designing heterogeneous microstructures.* ASME 2014 International Design Engineering Technical Conferences and Computers and Information in Engineering Conference. Buffalo, New York, USA, 17–20.

Zhang, J. Sun, B. Zhang, A.Q. & Kang, S.Q. 2000. *Underwater wire rope cutters*, China, CN00123284.

Advanced Materials and Structural Engineering – Hu (Ed.)
© *2016 Taylor & Francis Group, London, ISBN 978-1-138-02786-2*

A new type system of pressure regulating and water conveyance based on small and middle hydropower station

S. Feng, Z.M. Su, Y.Z. Yan & F.H. HuangFu
Kunming University of Science and Technology, Kunming, Yunnan, China

ABSTRACT: For some pressure regulating and water conveyance system of small and middle hydropower station in complex terrain, this paper proposes a new system of pressure regulating and water conveyance. The traditional diversion mode of pressure tunnel between upstream buildings and the surge chamber is changed for the open channel with no pressure water diversion mode. The new type water diversion and power generation system surge chamber not only has the effect of pressure regulating, but also is to be the switch room from water diversion without pressure to water conveyance with pressure. In this paper, the general layout of the hydropower station water supply system is given out. Then the mathematical model of the whole transition process of water conveyance is provided. And the section size and elevation of new switch room is calculated by bernoulli equation, continuity equation, partial water head loss formula and darcy formula. The shape and size of water conveyance open-channel without pressure are drafted according to the water requirement of a downstream powerhouse. Last, the engineering example is calculated to prove the feasibility of the new pressure regulating and water conveyance system. It can avoid the too long excavation of pressure tunnel and optimize the general layout of water diversion and power generation system.

1 INTRODUCTION

At present, hydropower station consists of upstream water building, water diversion pressure tunnel, surge chamber, steel penstock, workshop and downstream river course (Ma & Wang 2007). The water diversion pressure tunnel is filled with water when there is water in it. So the top of the tunnel is under some pressure. There is not free water surface in pressure tunnel. The flow regime is good. However, the water pressure is high in the tunnel. It needs a hard rock to resist water pressure, or it is not economic. The pressure tunnel is applied when the Geological conditions are good. However, the environmental conditions and working mechanism of the hydraulic high pressure headrace tunnel is complicated. There are a lot of problems that are not solved well in the aspects of crack spacing and width, high pressure in the inland waters extravasation and water load, analysis and calculate method after tunnel lining cracking. At the same time, the tunnel belongs to underground engineering. The construction process of it is very complicated and it needs a lot of manpower and material resources (Lin & Wang 2008). Thus, it can be seen that the existing water diversion—regulating system in engineering does not meet the requirements for the long tunnel, difficult excavation or unable to excavation hydropower station. If the traditional way of pressure diversion tunnel water can be instead of unconfined water flow in an open channel, it will avoid the problem that pressure tunnel being long, difficult excavation and unable to excavate. However, the hydropower workshop needs the pressure head. The existing water diversion—regulating system in engineering does not meet the requirements for it. So it needs a new type of pressure regulating water conveyance system to solve the problem of hydropower station water diversion—regulating system transformed from non-pressure water delivery to pressure water delivery.

Compared with the traditional surge conveyance system, the new surge conveyance system has some different. The surge of traditional surge conveyance system is changed for the switch room of non-pressure water delivery to pressure water delivery during the whole design. The design of the switch room is on the basis of the previous mature theory. In the article, the effective elevation and section size of the switch room is calculated according to the hydraulics related theory. The engineering example is calculated. And the feasibility of the new type of pressure regulating water conveyance system is verified. The new type of pressure regulating water conveyance system for some small and medium-sized hydropower station has the very vital significance in the aspects of saving capital construction and optimizing the overall layout of a water power generation system.

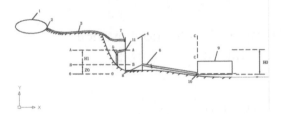

Figure 1. The overall layout char of the new accent presses and the water—carrige system.

2 NEW DESIGN MODEL OF THE PRESSURE REGULATING WATER SYSTEM

In the article, a new type of pressure regulating water conveyance system is given. The layout of the new type pressure regulating water conveyance system is set as shown in Figure 1. First the workshop 9 is established on the location of the downstream flat of the upstream water intake structures 1 (river, galaxy, lake, dams). The switch room 4 is never established the workshop 9. The switch room 4 should never be the workshop as far as possible. So the water-hammer pressure of penstock 6 and turbine can be reduced (Ma & Wang 2007). The weir 11 is set in the switch room. The drainage pipeline 5 is used to connect weir 11 and canal or spillway. The interface 8 is set at the bottom of switch room 4. The workshop 9 is connected to the water pressure steel pipe 6 by the interface 10. The non-pressure derivation 3 can be excavated along the edge of the mountain. So it can avoid the excavation of pressure runnel. After the excavation of the molding, the pulp stone paving is used on both sides of non-pressure 3. The top of non-pressure derivation 3 is connected to water-fetching building 1 by gate 2 that the building set itself, or by gate 2 that the weir of water-fetching building 1 side excavation set itself. The water inlet 7 of the switch room is set on the top of the switch room 4.

3 MATHEMATICS CALCULATION MODEL (WU 2007)

3.1 The determination of switch room section size and height

According to Figure 1, the height H_1 of the switch room is determined by the Bernoulli equations of section A-A and section C-C. The Bernoulli equations of section A-A and section C-C are as follows:

$$H_1 + z_0 + \frac{\alpha_1 v_s^2}{2g} = H_0 + \frac{\alpha_2 n v_c^2}{2g} + h_f + h_\xi \qquad (1)$$

Among it: section A-A is the section 11 of the weir. Section C-C is the vertical section of the water pressure steel pipe in a workshop. H_1 is the switch room effective height that it guarantees the normal work required a storage capacity of the hydropower plant. z_0 is the elevation difference between workshop and switch room. α_1 and α_2 are the kinetic energy correction coefficient. v_s is the section velocity of the switch room. g is an acceleration of gravity. H_0 is the effective water height of workshop. n is the number of water pressure steel pipe. v_c is the velocity of water pressure steel pipe. h_f is the Frictional head loss of water pressure steel pipe. h_ξ is the local head loss of interface 10.

If there are three tube parallel from section A-A to section C-C, energy loss is the same from section A-A to section C-C when unit weight of the liquid are in section A-A or section C-C. If h_{f1}, h_{f2} and h_{f3} are each tube frictional head loss respectively, thus:

$$h_{f1} = h_{f2} = h_{f3} = h_f \qquad (2)$$

Frictional head loss is calculated by Darcy formula:

$$h_f = \lambda \frac{l}{d} \frac{v_c^2}{2g} \qquad (3)$$

Local head loss is calculated by local head loss formula:

$$h_\xi = \xi n \frac{v_c^2}{2g} \qquad (4)$$

Among them: λ is the frictional head loss coefficient. n is the number of water pressure steel pipe. l is the length of single water pressure steel pipe. d is the diameter of the water pressure steel pipe. ζ is local head loss coefficient.

Cross section area of the switch room is as following. Cross section area is determined by the height H_1 of switch room, Bernoulli equation of section A-A and section C-C and continuous equation. The continuous equation is as follows:

$$A_s v_s = n A_1 v_c \qquad (5)$$

Among it: A_s is the section area of the switch room. v_s is the section velocity of switch room. n is the number of water pressure steel pipe. A_1 is the section area of water pressure steel pipe (single). v_c is the velocity of water pressure steel pipe.

3.2 The section size calculation of non-pressure water diversion open channel

The non-pressure water diversion open channel is designed to use the rectangle form. The size is calculated by plant water requirement, Bernoulli equation and Darcy formula. The cohesive paragraph between non-pressure water diversion open channel and top of switch room is in the form of diffusion that is as following Figure 2. So there is some effect that part of the effective energy is eliminated.

The calculation formula of hydropower installed capacity is as follows:

$$N = 9.82\, H_0 Q \ (kw) \tag{6}$$

Among it: N is the installed capacity. H_0 is the effective water head that hydropower station needs. Q is the water consumption that hydropower station needs.

The hydropower station cited flow Q can be calculated by hydrological data. The width and height of non-pressure water diversion open channel can be proposed preliminarily by the value of Q. After the section size of non-pressure water diversion proposed, it needs open channel uniform flow equation to calculate water flow Q_j. Then it is used to check whether it meets the surge flow Q. If $Q_j \geq Q$, it shows that the proposed flow meets the flow required. If $Q_j \leq Q$, it shows that the proposed flow cannot meet the flow required. Then it needs to propose the size. And check it until it meets the conditions $Q_j \geq Q$. The whole work is as Figure 3.

The flow formula to calculate open channel uniform flow is as follows:

$$Q = AC\sqrt{Ri} \tag{7}$$

Among it: A is the section area of rectangular open channel. C is the Xie Qi coefficient. R is the water conservancy radius of open channel. i is bottom slope.

Among it:

$$C = \frac{1}{n} R^{\frac{1}{6}} \tag{8}$$

$$A = bh \tag{9}$$

$$R = \frac{bh}{b + 2h} \tag{10}$$

Among it: b is the width of rectangular section. h is the height of rectangular section. n is the roughness coefficient of open channel.

4 THE INSTANCE CALCULATION

The installed capacity of a hydropower station is 30000 kw. The other parameters are as follows:

The cited flow $Q = 51\ m^3 \cdot s^{-1}$, $d = 2$ m, $v_c = 5.414\ m \cdot s^{-1}$, $\alpha_1 = \alpha_2 = 1$, $\lambda = 0.025$, $\xi = 0.5$, $H_0 = 60$ m, $Z_0 = 8$ m, $l = 80$ m (The length of single water pressure steel pipe), n = 3; The length of non-pressure diversion open channel are $L = 2000$ m, $S_1 = 15$ m, $g = 9.8\ m \cdot s^{-2}$. The bottom slope i is 1/800. The concrete tank bottom is used on open channel. The pulp ashlaring is built on both sides of the slope. The roughness coefficient n_0 is 0.015.

4.1 The determination of switch room section size and height

Three water pressure steel pipes are set on the bottom of switch room. According to $h_f = \lambda (l/d)(u_c^2/2g)$ and $h_f = \xi n(v_c^2/2g)$, it can be calculated that $h_f = 1.5$ m, $h_g = 2.24$ m. According to Bernoulli equation that $H_1 + z_0 + (a_1 v_s^2/2g) = H_0 + (a_2 n v_c^2/2g) + h_f + h_g$, it can be calculated that $H_1 = 60.23$ m. Among them, $a_1 v_s^2/2g$ is small value. So it can be ignored in the equation, so it can be taken 0. I have been ignored, so the calculation is slightly larger than the actual. So take the $H_1 = 60$ m.

$H_1 = 60$ is taken into $H_1 + z_0 + (a_1 v_s^2/2g) = H_0 + (a_2 n v_c^2/2g) + h_f + h_g$. Although $a_1 v_s^2/2g$ is small value, v_s cannot be ignored. Then it is calculated that $v_s = 2.123\ m \cdot s^{-1}$. Then we can get that $A_s = 24.02$ m^2 from $A_s v_s = n A_1 v_c$. The switch room is a square. So the side length of switch room 4 is 4.901 m. It is as 5 m.

4.2 The calculation of non-pressure water diversion open channel section size calculation

The shape of the open channel is rectangular section. The cited flow of the workshop is

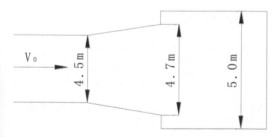

Figure 2. The entry floor plans of open channel flowing into conversion room.

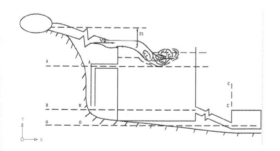

Figure 3. Working drawing of conversion room.

$Q = 51\ m^3 \cdot s^{-1}$. The width and height of open rectangular channel are preliminarily drafted as 4.5 m and 4 m by Q. Then check if the flow meets the requirements of hydropower plant.

$$Q = AC\sqrt{Ri}.$$

Among it, $A = bh = 4.5 \times 4 = 18\ m^2$,

$$R = \frac{bh}{b+2h} = \frac{4.5 \times 4}{4.5 + 2 \times 4} = 1.44\ m,$$

$$C = \frac{1}{n}R^{\frac{1}{6}} = \frac{1}{0.015} \times 1.44^{\frac{1}{6}} = 70.8439\ m^{\frac{1}{2}}/s$$

It can be get that

$$Q_j = AC\sqrt{Ri} = 18 \times 70.8439 \times \sqrt{1.44 \times \frac{1}{800}} = 54.$$

$Q_j \geq Q$. So the drafted size of open rectangular channel meets the requirements.

The new type of pressure regulating water conveyance system has been adopted in some small hydropower station in Yunnan. It saves amount of cost during the construction. It has been put into use and the running effect is good.

5 CONCLUSIONS

In the article, a new type of pressure regulating water conveyance system is set up for some small and medium-sized hydropower stations. It avoids many problems. The section size and height of the new switch room are calculated by Bernoulli equation, the continuity equation, Darcy formula and the local head loss formula. And the shape and size of non-pressure water delivery open channel are drafted by the downstream plant water requirement. The shape of the switch room is a square. The side length of the square is 5 m. The effective water head H_0 of hydropower station is 60 m. The shape of the open channel is rectangular. The width and height of the rectangular are 4.5 m and 4 m respectively.

The rated flow Q is 51 m³/s. The calculated flow Q_j is 54.10 m³/s, which is larger than rated flow Q. The new type of pressure regulating water conveyance system is feasible according to the instance calculation. The principle of the new system is simple. And there is good mature theoretical basis. The key structure of the new system is switch room 4. It not only avoid water diversion mode which is pressure tunnel in the upstream, but also greatly reduced the cost. The mode of non-pressure water delivery is more safety. And the switch room of the new system can play a role as surge well. It is a project of kill two birds with one stone. The weir is set on the top of the switch room so that the excess water can flow out free. This can be a very good guarantee requirement for an effective constant head of hydropower station. No dam water delivery can be used for upstream of small and medium-sized hydropower station. It does not need dam for the mode of no dam water delivery. It can greatly save investment.

ACKNOWLEDGEMENT

The National Natural Science Fund (51169009).

REFERENCES

Lin J.Y. & Wang, G.L. 2008. *Hydraulic Structure (fifth edition)*. Beijing: China Water Power Press, 382–443.

Ma, S.D. & Wang, R.Z. 2007. *Hydraulic station structure (second edition)*. Beijing: China Water Power Press, 144–148.

Wu, C.G. 2007. *Hydraulics (fourth edition)*. Beijing: Higher Education Press, 62–212.

Advanced Materials and Structural Engineering – Hu (Ed.)
© 2016 Taylor & Francis Group, London, ISBN 978-1-138-02786-2

A study on the characteristics and trends of domestic long-life housing research in South Korea

S.R. Park
Department of Architecture, Hanyang University, Seongdong-gu, Seoul, South Korea
Researcher, Korea Institute of Civil Engineering and Building Technology, South Korea

E.Y. Kim
Researcher, Korea Institute of Civil Engineering and Building Technology, South Korea

E.K. Hwang
Research Fellow, Korea Institute of Civil Engineering and Building Technology, South Korea

ABSTRACT: According to the data from the Korean Statistical Information Service announced in 2013, the housing supply ratio in South Korea has already exceeded 100%, and the improvement of people's living standards has led to an increase in the demands for pleasant residential environments and the maintenance of the existing houses. In this regard, the need to research on and supply long-life housing has been highlighted at a national level. To meet the demands, the Ministry of Land, Infrastructure, and Transport introduced the long-life housing construction certification standards in December 2014, and began to operate the long-life housing certification system. The sublevel regulations of the long-life housing system, however, have yet to be clearly established, and the people's awareness of the matter is still low. Also, institutional and economic obstacles persist. In light of such circumstances, this study sought to provide the basic data needed to expand the supply of domestic long-life housing based on the analysis of the domestic research trends and the structural elements for the commercialization of long-life housing.

1 INTRODUCTION

According to the data from the Korean Statistical Information Service announced in 2013, the domestic housing supply ratio has already exceeded 100%, and the improvement of the general quality of living has expanded the demand for pleasant residential environments, beyond the basic residential conditions. In this regard, the need to research on and supply long-life housing has been highlighted at a national level.

To meet the aforementioned demands for pleasant residential environments, the Ministry of Land, Infrastructure, and Transport began to implement the Housing Performance Grade Indication System by announcing the grades for key housing performances in January 2006. Subsequently, on December 30, 2011, the Nature-friendly Building Certification System and the Housing Performance Grade Indication System were unified, and the Green-Building Construction Support Act changed the name of the unified system to the Green-Building Certification System in February 2013.

Despite the recent re-separation of the Housing Performance Grade Indication System from the Nature-friendly Building Certification System, however, the sublevel regulations have yet to be clearly established, and the people's awareness of the matter is still low. Also, other institutional and economic obstacles persist.

In light of such circumstances, this study sought to analyze the domestic research trends and thereby provide the basic data needed to expand the supply of long-life housing for its commercialization in South Korea.

2 THEORETICAL CONSIDERATIONS OF LONG-LIFE HOUSING

Long-life housing refers to housing that lasts for a long time. Housing consists of various materials and components, and the deterioration of even one of these may affect the life of the entire housing. Therefore, housing must be designed in such a way as to avoid such issue. The materials and components (infill) that may affect the life of a housing include furniture/panel/mixed/dry walls, dual floors, ceiling materials, wall structures and floor piping, bathroom (toilet), kitchen, outer wall structures, and dry ondol (floor heating system). Here, infill refers to components other than the housing structure with shorter life spans than the structure, which can be repaired or replaced to accommodate the changes

Table 1. Fixed and variable factors of the long-life housing.

Segment		Content
Fixed	Support	Columns, beams, slabs, bearing walls inside the households, patrician walls between the households
	Common equipment	Water supply pipe, drainage pipe, gas pipe electric wiring, information wiring, elevator, etc.
	Public space	Public corridors, public stairs, etc.
	Others	Building facades
Variable	Interior parts	Ceilings, patrician walls, flooring materials, etc.
	Dedicated pipes	Water supply pipe, drainage pipe, gas pipe, electric wiring, information wiring, etc.
	Exclusive facilities	System kitchen, prefabricated bathroom
	Others	Floor plan (size, space, equipment, interior parts, uses)

in the user's requirements, and can be standardized and mass-produced. Therefore, long-life housing refers to housing to which a technology for maintaining the physical long life of the supports and other components that are less affected by social or functional changes has been applied, while allowing for changes in the rest of the parts (infill) pursuant to the changes that may occur over time and the quality required. Such housing is designed with a focus on ensuring that the value of the building will be preserved for more than 100 years.

Long-life housing has to have high durability and high capacity for change in accordance with the social or functional changes or the changes in the family or residents that may occur over a long time, or has to have components that share such characteristics. Without the capacity to accommodate functional durability, such housing cannot be used for a long time even if it is physically durable.

Therefore, long-life housing needs to have high comprehensive durability, which means having both physical durability and functional durability. At the same time, to ensure that the housing stock becomes the society's asset with high quality, the high quality of the housing stock should be secured as a social/national asset capable of accommodating the changes in social situations and in housing plans and production technologies that may occur over time and the social demands related to housing.

The factors that comprise long-life housing can be classified into two main categories: fixed factors (support) and variable factors (infill). The purpose of such a classification is to ensure variability or renewability to satisfy the residents' demands or the changes that may occur over time.

The fixed factors (support) are those construction materials comprising housing that are controlled in accordance with the public intent and are not directly affected by an individual's intention. These factors need to have long-term durability and comprise the very foundation of housing.

The variable factors (infill) are those construction materials comprising housing that can be used and controlled by an individual resident; a resident may freely design or easily renovate such factors. These factors need to be renewable and consist of non-pillar solid components.

3 RESEARCH METHOD

The Open Housing Theory, since it was conceived by Habraken, Holland, and SAR in the 1920s, has spread to various countries, resulting in the conduct of further research studies on it. Various research studies were conducted in the 2000s, along with the active construction of experimental housing. Therefore, in this paper, the research on the concept, method, or composing factors of long-life housing among the long-life housing research studies that were conducted in the last 10 years are analyzed, along with the existing literature related to such. The housing factors and characteristics discussed by previous research studies are also analyzed herein.

In this paper, the architecture research papers on long-life housing that had been published in journals from 2000 to 2014 are analyzed. The source journals mostly consist of journals published by academic architecture associations, including the Architectural Institute of Korea, the Korean Housing Association, and the Korea Institute of Interior Design.

The articles were sampled from the Research Information Search Service (RISS, http://www.riss.kr) provided by the Korea Education and Research Information Service (KERIS) by searching journal-published research articles related to long-life housing with the use of keywords associated with long-life housing ("long-life housing," "open housing," and "flexible housing"). The number of research articles collected from the search was 59. Of those articles, 29 were published by the Architectural Institute of Korea, 10 by the Korean Housing Association, 6 by the Korea Institute of Interior Design, and the remaining 14 by other associations.

4 ANALYSIS OF THE RESEARCH STATUS OF DOMESTIC LONG-LIFE HOUSING

The investigation of the yearly distribution of research studies on long-life housing revealed that few research studies were conducted in the early 2000s, followed by an increase in the number of research studies that began in 2003 (4 research studies) and peaked during the period from 2005 to 2008 (27 research studies). Such distribution seems to have been caused by the increase in the number of research studies on experimental housing after 2005. The number of research studies began to dwindle after 2011 (6 research studies), demonstrating the recent recess of research and development efforts on long-life housing.

The research articles on long-life housing were also analyzed in terms of their distribution in dif-

ferent areas. For this analysis, the articles were classified into six areas: building system method for long-life housing, spatial configuration, element and component, production and cost, and system and maintenance.

In terms of areas, the largest number of research studies (16) was on factors and components of long-life housing, followed by spatial composition (14) and construction methods (11). This analysis shows that as research and development efforts on long-life housing have unfolded in South Korea, research studies on the development of element technologies for long-life housing have been continuing owing to the studies conducted by the long-life Joint Housing Research Team and other institutes.

On the other hand, only eight, four, and six research studies were found on the production and cost of the commercialization of long-life housing, drawing, and maintenance, respectively. The research studies in these areas have been relatively lacking. In particular, no research on maintenance has been undertaken since 2011, despite its importance in terms of the maintenance of long-life housing performance and usability for the residents.

Likewise, research on systems for revitalizing long-life housing has been done only intermittently, with only four research studies done after 2008.

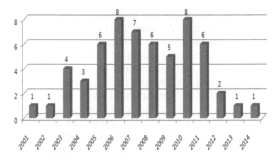

Figure 1. Domestic long-life housing research distribution in South Korea (by year).

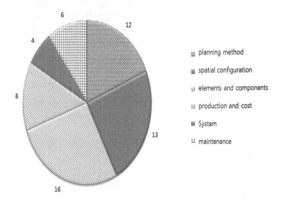

Figure 2. Long-life housing research distributions in South Korea (by area).

5 DOMESTIC LONG-LIFE HOUSING RESEARCH TRENDS

The research trends in long-life housing planning methods can be divided into three stages. Most of the research studies produced during the early stage were on the development of plane types and the systematization of the support and infill planning factors. During the middle stage, most research studies focused on the performance of each factor of long-life housing planning and the development and categorization of the systems. During the late stage, research studies on the introduction of nature-friendly planning methods and the improvement of equipment technologies were predominant.

The research studies on the spatial composition of long-life housing can be divided into two stages: early stage and latter stage. While the research studies during the early stage mostly consisted of research studies proposing plane types with variability, the main research studies in the latter stage analyzed the variability characteristics through the categorization of the variable plane types and the analysis of real-world cases.

As for the research trends on the factors and components of long-life housing, the relevant

Table 2. Domestic long-life housing research distribution in South Korea (by area).

Segment	Planning method	Spatial configuration	Elements/ components	Production/ cost	System	Maintenance	Total
2001	1						1
2002		1					1
2003	1		1			2	4
2004	1	1	1				3
2005	1	4				1	6
2006	2	3	2			1	8
2007		2	4	1			7
2008			2	2	2		6
2009	1		3	1			5
2010	1	1	2	2		2	8
2011	2	1	1	1	1		6
2012	1			1			2
2013					1		1
2014		1					1
Total	11	14	16	8	4	6	59

research studies touch various topics, including the spatial factors, component factors, and connection factors. The early stage saw many research studies aimed at categorizing the spatial/component factor technologies and drawing their characteristics. Into the middle stage, the number of research studies on the long-life housing factors and component connections (interface) began to increase. During the late stage, an increasing number of research studies derived the characteristics of the architectural factors (e.g. wall structure, furniture, wall) and studied the design factors.

The analysis of the research studies on the production cost of long-life housing revealed that most of the research studies were related to the frame construction methods and economic-feasibility analysis, while those on the production and construction method of the infill factors were lacking.

Lastly, as for the research trends on the systems and maintenance of long-life housing, the research studies on systems mainly focused on analyzing largely unorganized systems and performance criteria. These articles, however, did not propose related legal criteria in a comprehensive manner. As for the research studies on maintenance, research in 2005 studied the durability of the long-life housing components and the derived performance indicators, but follow-up studies of the 2005 research were lacking.

6 CONCLUSIONS

Below is a summary of the results of the analysis of the domestic research trends for long-life housing.

First, a total of 59 research articles related to long-life housing were collected using the selected keywords associated with long-life housing, such as "long-life housing," "open housing," and "flexible housing."

Second, the research studies on long-life housing have increased since 2003, and were most actively conducted from 2005 to 2008, focusing on the long-life housing planning techniques, spatial configuration, and component parts.

Third, the past long-life housing research studies focused on the separation of the fixed and variable parts, flexible design, planning methods, and individual elements, but the integrated review of the performance maintenance and commercialization in response to the life cycle divided into construction and maintenance was not sufficiently taken into account.

Fourth, there is a critical lack of research on the systems for the activation of long-life housing.

This paper has a limitation in that it analyzes long-life housing with a focus on the related keywords, and thus only describes the general trends of the existing published articles. For the future work, there is a need to include the detailed contents of the existing literature for a comparative review in the analysis of the research on long-life housing.

ACKNOWLEDGMENT

This study was made possible by the financial support from part of results a major research project conducted by the Korea Ministry of Land, Infrastructure and Transport, Residential

Environment Research Project in 2014. Project No.: 14RERP-B080965-01.

REFERENCES

Information on Architecture & Urban Research Institute (http://www.auric.or.kr).
Information on Ministry of Land, Infrastructure, and Transport (http://www.molit.go.kr).
Information on Research Information Sharing Service (http://www.riss.kr).

Kim, S.A. et al. 2003. A Study on the Design Method for long-life Housing: PLUS50 environmental symbiosis building construction technology development (2nd year), *Korea Institute of Civil Engineering and Building Technology*, 28–29.
Kim, Y.N. Kong, S.K. & Chu, B. 2010. A Study on Interior Design Planning of Long-Life Housing—Focus on Case Studies-, *Journal of the Korean Institute of Interior Design*, 19(3): 136–144.
Lee, B.R. Kim, S.A. & Hwang, E.K. 2008. A Study on the Systematic Improvement Plans for Facilitating long-life Housing, *Journal of the Architectural Institute of Korea Planning and Design*, 24(3).

Advanced Materials and Structural Engineering – Hu (Ed.)
© 2016 Taylor & Francis Group, London, ISBN 978-1-138-02786-2

Explicit normal water depth formulae and optimal economical section of the triangular channel

H.Y. Gu
Shandong Provincial Research Institute of Water Resources, Jinan, China

Y.C. Han, T.Q. Peng & Y. Gao
College of Resources and Environment, University of Jinan, Jinan, China

ABSTRACT: Normal water depth calculation is very important in an open channel design. Triangular channel section is one of the important types of channel section. However, no research or explicit formulae of the triangular channel section have been studied. This paper derived the explicit equations to calculate the normal water depth of the triangular channel cross section. The examples of using explicit equations to compute and verify the normal water depth are presented. The results show that the explicit equations can bring us great convenience and efficiency. Finally, the model to compute the optimal economical section for the triangular channel section is presented.

1 INTRODUCTION

Normal water depth calculation is very important in an open channel design. Over the past years, scholars have conducted a large number of studies on the normal water depth for the rectangular, trapezoidal, U-shaped (Zhang Xin-Yan 2013), and horseshoe (Wang Shun-Jiu 2002, Lu Hong-Xing 2001) sections, giving the explicit equations to compute the normal depth. This brings us great convenience and efficiency. Triangular channel section is one of the important types of channel section. However, no research or explicit formulae of the triangular channel section have been studied. This paper deduced the explicit equations of the normal water depth for the triangular channel cross section.

Liu Qing-Gu equations (1999) presented the iteration method to compute the normal depth of the trapezoidal open channel; Wang Shun-Jiu (2002) studied the genetic algorithm for calculating the normal water depth of the horseshoe cross section; Wang Zheng-Zhong (1998) researched a direct calculation formula for the normal depth of the open trapezoidal channel. Wen-Hui (2012) presented a further study on an explicit formula for the normal water depth of uniform flows in circular pipes.

2 EXPLICIT NORMAL WATER DEPTH FORMULA DERIVATION

The flow in the triangle section channel can be expressed using the Chézy and Manning formula:

$$Q = AC\sqrt{Ri} \qquad (1a)$$

where Q = flow discharge (m³/s);
A = flow area (m²);
R = hydraulic radius (m);
i = longitudinal bottom slope; and
C = Chézy coefficient,

which can be computed by the Manning formula:

$$C = \frac{1}{n_t} R^{1/6} \qquad (1b)$$

Substituting Equation (1b) into Equation (1a), we get

$$\frac{1}{n} \frac{A^{5/3} i^{1/2}}{P^{2/3}} \qquad (1c)$$

where
n = channel roughness and
P = wetted perimeter (m).

If the normal water depth is denoted by a variable, the uniform flow area A, water width B and the wetted length P can be expressed as

$$A = m h_0^{\,2} \qquad (2)$$

$$B = 2 m h_0 \qquad (3)$$

$$P = 2\left(m^2 + 1\right)^{0.5} h_0 \qquad (4)$$

where A = flow area (m²);

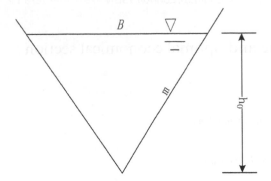

Figure 1. Triangular channel section.

h_0 = normal water depth (m); and
m = side slope.

Substituting Equation (2) and Equation (4) into Equation (1), the equation to solve the normal water depth is given as follows:

$$Q = 1/2 \frac{\left(mh_0^2\right)^{5/3} i^{0.5} \sqrt[3]{2}}{n\left(\left(m^2+1\right)^{0.5} h_0\right)^{2/3}} \qquad (5)$$

Solving the above formula, the explicit equation to calculate the normal water depth h will be deduced as follows:

$$h_0 = \sqrt[4]{2} \frac{\sqrt{i}}{Q^2 n^2 m^8 \sqrt{m^2+1.0}} \left(\frac{Q^3 n^3 \left(m^2+1\right) m^{11}}{i^{3/2}}\right)^{\frac{5}{12}}$$

$$\times \sqrt{Qmn\sqrt{i}\left(\frac{Q^3 n^3 \left(m^2+1\right) m^{11}}{i^{3/2}}\right)^{\frac{5}{12}}}$$

$$(6)$$

Simplifying Equation (6), below the equivalent formula to calculate the normal water depth h can be obtained as follows:

$$h_0 = \frac{\sqrt[4]{2}\sqrt{\dfrac{Q^{9/4} n^{9/4}}{\sqrt[8]{i}} m^{\frac{67}{12}}\left(m^2+1\right)^{\frac{5}{12}} m^{-\frac{41}{12}}}}{\sqrt[8]{i} Q^{3/4} n^{3/4}\left(m^2+1\right)^{1/12}} \qquad (7)$$

Equation (7) is the implicit equation to calculate the normal water depth h of the triangular section.

From Equation (5), the equation to compute the bottom slope i is given as follows:

$$i = 2\frac{Q^2 \sqrt[3]{2} n^2 \left(\left(m^2+1\right)^{0.5} h\right)^{4/3}}{\left(h^2 m\right)^{10/3}} \qquad (8)$$

3 APPLICATIONS OF THE EXPLICIT NORMAL WATER DEPTH FORMULA

3.1 Example 1

Let us consider a channel with a triangle cross section, $Q = 1.2$ (m³/s), side slope $m = 1.5$, channel bottom longitudinal slope $i = 1/15000$, and channel roughness $n = 0.012$. Now, we calculate the normal water depth.

The solving process: Substituting all the known conditions (Q. m. n. i) into Equation (7), the normal water depth can be obtained directly as follows:

$$h_0 = \frac{\sqrt[4]{2}\sqrt{\dfrac{Q^{9/4} n^{9/4}}{\sqrt[8]{i}} m^{\frac{67}{12}}\left(m^2+1\right)^{\frac{5}{12}} m^{-\frac{41}{12}}}}{\sqrt[8]{i} Q^{3/4} n^{3/4}\left(m^2+1\right)^{1/12}}$$

$$= 1.323088601 \text{ (m)}.$$

Using Equation (6), the same result can be obtained.

For testing and verifying the correctness, $h_0 = 1.323088601$ (m) and $i = 1/15000$, $m = 1.5$, and $n = 0.012$ are substituted into Equation (2) and Equation (4), thus we then obtain

$A = 2.625845170$ (m²)
$P = 4.770463794$ (m)

Then, A, P is substituted into Equation (1), and we get

$Q = 1.200000001$ (m³/s)

Through the reversed test and verification, it shows that the explicit equations (Equation (6) and Equation (7)) to calculate are right.

3.2 Example 2

Let us consider a triangle cross section channel, $Q = 0.3$ (m³/s), side slope $m = 1.0$, channel bottom longitudinal slope $i = 1/5000$, and channel roughness $n = 0.010$. Now, we calculate the normal water depth.

The solving process: Substituting all the known conditions (Q. m. n. i) into Equation (7), the normal water depth can be obtained directly as follows:

$$h_0 = \frac{\sqrt[4]{2}\sqrt{\dfrac{Q^{9/4} n^{9/4}}{\sqrt[8]{i}} m^{\frac{67}{12}}\left(m^2+1\right)^{\frac{5}{12}} m^{-\frac{41}{12}}}}{\sqrt[8]{i} Q^{3/4} n^{3/4}\left(m^2+1\right)^{1/12}}$$

$$= 0.7250427170 \text{ (m)}.$$

Using Equation (6), the same result can be obtained.

3.3 *Example 3*

Let us consider a triangle cross section channel, $Q = 1.2$ (m³/s), side slope $m = 2$, the water depth for uniform flow $= 1.8$, and channel roughness $n = 0.012$. Now, we calculate the channel bottom longitudinal slope i.

The solving process: Substituting all the known conditions into Equation (8), the channel bottom longitudinal slope i can be obtained directly as follows:

$$i = 2\frac{Q^2\sqrt[3]{2}n^2\left(\left(m^2+1\right)^{0.5}h\right)^{4/3}}{\left(h^2m\right)^{10/3}} = 0.000041217$$

4 THE OPTIMAL ECONOMICAL SECTION

The construction cost of the channel is composed by the earthwork excavation, lining and land acquisition costs. The total construction cost are expressed as follows:

$$C = C_A A + C_B B + C_X P \qquad (9)$$

where
C_A = earthwork excavation cost unit area and per length channel;
C_R = acquisition cost per length channel; and
C_X = lining cost unit area and per length channel.

Substituting Equation (2), Equation (3) and Equation (4) into Equation (9), we get

$$C = mh^2C_A + 2mhC_B + 2C_X\left(m^2+1\right)^{0.5}h \qquad (10)$$

The economical section is a section in which the total construction cost is the minimum under the design discharge. So, the optimal triangular cross section can be expressed as follows:

$$\text{Min } C = mh^2C_A + 2mhC_B + 2C_X\left(m^2+1\right)^{0.5}h \quad (11)$$

Subjecting to

$$Q = 1/2\frac{\left(mh_0^2\right)^{5/3}i^{0.5}\sqrt[3]{2}}{n\left(\left(m^2+1\right)^{0.5}h\right)^{2/3}} \qquad (12)$$

Let us consider a channel with the design discharge of 6 (m³/s). The roughness is 0.014, the longitudinal slope is 1/12000, and the side slope coefficient is 2. The earthwork excavation cost unit area and per length channel is 18 (CNY, China Yuan). The acquisition cost per length channel is 12 (CNY, China Yuan). The lining cost unit area and per length channel is 42 (CNY, China Yuan). Now, we design a channel using the most economic triangle section.

Solving the nonlinear optimization problems Equation (11) and Equation (12) using the algorithm, the result obtained is $h = 2.167$ m.

5 CONCLUSIONS

This paper deduced the explicit equations for the normal water depth of the triangular channel cross section. The results show that the normal water depth is connected only with the following parameters: side slope m, bottom slope i and discharge Q. The explicit equations will bring us great convenience and efficiency to compute the normal water depth of the triangular channel cross section. The model to compute the optimal economic section for the triangular channel section is obtained, which will help designer to design the economical section.

ACKNOWLEDGMENT

This project was supported by "the Science and Technology Development Plan Project of Jinan (201302052, 201303082), and the "Five-twelfth" National Science and Technology Support Program of China (2015BAB07B02).

REFERENCES

Aisenbrey, A.J.Jr. Hayes, R.B. Warren, H.J. Winsett, D.L. & Young, R.B. 1978. *Design of small canal structures. Bureau of Reclamation*, Denver, Colorado, U.S.

Bhattacharjya, R.K. 2006. Optimal Design of Open Channel Section Incorporating Critical Flow Condition. *Journal of Irrigation and Drainage Engineering*, 13(5): 513–518.

Chow, V.T. 1959. *Open channel hydraulics*. McGraw-Hill, New York.

Kananil, A. Bakhtiari, M. & Borghci, S.M. et al. 2008. Evolutionary Algorithms for the Determination of Critical Depths In Conduits. *Journal of Irrigation and Drainage Engineering*, 134(6): 847–852.

Liu, Q.G. 1999. Iteration Method For Calculation Of Normal Depth In Trapezoidal Open Channel, *Journal of Water Resources & Hydropower of Northeast China*, 4: 26–27.

Monadjemi, P. 1994. General Formulation of Best Hydraulic Channel Section. *Journal of Irrigation and Drainage Engineering*, 120(1): 27–35.

Morris, H.M. & Wiggert, J.M. 1972. *Applied hydraulics in engineering*. John Wiley, New York.

Prabhata, K.S. & Pushpa, N.R. 2005. Exact Equations for Critical Depth in Trapezoidal Canal. *Journal of Irrigation and Drainage Engineering*, 131(5): 474–476.

Wang, S.J. Zhang, H.Y. & Xi-Li, D.J. 2002. Genetic Algorithm for Calculating Normal Water Depth of Horseshoe Cross Section, *Journal of Irrigation and Drainage,* 21(3): 43–45.

Wang, Z.Z. Song, S.B. & Wang, S.M. 1999. A Direct Formula Calculating Normal Depth for Open Trapezoidal Channel with Spherical Bed, *Journal of Yangtze River Scientific Research Institute,* 16(4): 31–34.

Wang, Z.Z. Xi, G.Z. Song, S.B. & Wang, Y.G. 1998. A Direct Calculation Formula for Normal Depth in Open Trapezoidal Channel, *Journal of Yangtze River Scientific Research Institute*, 15(6): 1–3.

Wen, H. & Li, F.L. 2002. Further study on explicit formula for normal water depth of uniform flows in circular pipes, *Advances in Science and Technology of Water Resources,* 32(6): 15–17.

Zhang, X.Y. Lü, H.X. & Zhu, D.L. 2013. Direct calculation formula for normal depth of U-shaped channel, *Transactions of the Chinese Society of Agricultural Engineering,* 29(4): 115–119.

Zhao, Y.F. Wang, Z.Z. & Zhang, K.D. 2007. Direct, Calculations method for critical depth of open trapezoidal channel, *Journal of Shandong University of Technology,* 37(6): 101–105.

Zheng, Z.W. 1998. Formula for Calculating Critical Depth of Trapezoidal Open Channel. *Journal of Hydraulic Engineering ASCE,* 124(1): 90–92.

Author index